Lecture Notes in Computer Science 8294

Commenced Publication in 1973
Founding and Former Series Editors:
Gerhard Goos, Juris Hartmanis, and Jan van Leeuwen

Benoit Huet Chong-Wah Ngo Jinhui Tang
Zhi-Hua Zhou Alexander G. Hauptmann
Shuicheng Yan (Eds.)

Advances in Multimedia Information Processing – PCM 2013

14th Pacific-Rim Conference on Multimedia
Nanjing, China, December 13-16, 2013
Proceedings

Springer

Volume Editors

Benoit Huet
EURECOM, Sophia Antipolis, France
E-mail: benoit.huet@eurecom.fr

Jinhui Tang
Nanjing University of Science
and Technology, China
E-mail: tangjh1981@acm.org

Alexander G. Hauptmann
Carnegie Mellon University
Pittsburgh, PA, USA
E-mail: alex@cs.cmu.edu

Chong-Wah Ngo
City University of Hong Kong
E-mail: cwngo@cs.cityu.edu.hk

Zhi-Hua Zhou
Nanjing University, China
E-mail: zhouzh@nju.edu.cn

Shuicheng Yan
National University of Singapore
E-mail: eleyans@nus.edu.sg

ISSN 0302-9743 e-ISSN 1611-3349
ISBN 978-3-319-03730-1 e-ISBN 978-3-319-03731-8
DOI 10.1007/978-3-319-03731-8
Springer Cham Heidelberg New York Dordrecht London

Library of Congress Control Number: 2013953933

CR Subject Classification (1998): H.5.1, H.3-4, I.4, I.5, I.2.10, F.2.2

LNCS Sublibrary: SL 3 – Information Systems and Application,
incl. Internet/Web and HCI

Typesetting: Camera-ready by author, data conversion by Scientific Publishing Services, Chennai, India

Printed on acid-free paper

Springer is part of Springer Science+Business Media (www.springer.com)

Preface

Welcome to the proceedings of the Pacific-Rim Conference on Multimedia (PCM 2013), held December 13–16, 2013, in Nanjing, China. The Pacific-Rim Conference on Multimedia (PCM) is a leading international conference for researchers and industry practitioners to share and showcase their new ideas, original research results, and engineering development experiences from all multimedia-related areas.

PCM 2013 featured a comprehensive program including three keynote talks, four tutorials, six oral presentation sessions, a best paper candidates session, a poster session, and five special sessions. The 153 submissions included a large number of high-quality papers on multimedia content analysis, multimedia signal processing and communications, as well as multimedia applications and services. We thank our 135 Technical Program Committee members and reviewers who spent many hours reviewing papers and providing valuable feedbacks to the authors. On the basis of the three or four reviews per paper, the Program Chairs decided to accept only 30 as oral papers and 27 as poster papers. The acceptance rate for the oral presentation is therefore 19% accepting only the papers of the highest technical quality Including the poster presentation, the acceptance rate becomes 37%. Furthermore, one award for the Best Paper was chosen among the four best papers candidates.

The technical program is an important aspect but only provides its full impact if surrounded by challenging keynotes. We were pleased with and thankful to our keynote speakers, Ramesh Jain, Wen Gao, and Yong Rui, for having accepted to present their work at PCM 2013.

We are also heavily indebted to many individuals for their significant contribution. We thank the PCM Steering Committee for their invaluable input and guidance in crucial decisions. We wish to acknowledge and express our deepest appreciation to the Organizing chairs, Yong-Dong Zhang, Zhenmin Tang and Ting Rui, the panel chairs, Marcel Worring and Qi Tian, the special session chairs, Yong Man Ro, Shuqiang Jiang and Junsong Yuan, the tutorial chairs, Qingshan Liu and Tao Mei, the publicity chairs, Rainer Lienhart, Shin'ichi Satoh, Winston Hsu, Dacheng Tao, Jianfei Cai and Ke Lu, the demo chairs, Richang Hong, Yo-Sung Ho and Ke Lu, the finance chair, Yan Song, and the registration chair, Xinguang Xiang. Without their efforts and enthusiasm, PCM 2013 would not have been made a reality.

We gratefully thank NSF China for their generous support of PCM 2013, which made several key aspects of the conference possible.

Finally, we wish to thank all committee members, reviewers, session chairs, student volunteers, and supporters. Their contribution is much appreciated.

December 2013

Benoit Huet
Chong-Wah Ngo
Jinhui Tang
Zhi-Hua Zhou
Alexander G. Hauptmann
Shuicheng Yan

Organization

Technical Program Committee

Hezerul Abdul Karim	Multimedia University, Malaysia
Bingkun Bao	Institute of Automation, Chinese Academy of Sciences, China
Marco Bertini	Università degli Studi di Firenze, Italy
Liu Bin	Dalian University of Technology, China
Laszlo Boeszoermenyi	University of Klagenfurt, Austria
Miguel Carrasco	Universidad Diego Portales, Santiago de Chile, Chile
Xiujuan Chai	ICT, CAS
Yoong Choon Chang	Faculty of Engineering, Multimedia University, Malaysia
Kuan-Ta Chen	Academia Sinica, Taiwan
Xiangyu Chen	Institute for Infocomm Research, Singapore
Yao Chen	Google Inc.
Zhineng Chen	Institute of Automation, Chinese Academy of Sciences, China
Jian Cheng	Chinese Academy of Sciences
Wen-Huang Cheng	Academia Sinica, Taiwan
Gene Cheung	National Institute of Informatics, Japan
Chen-Kuo Chiang	National Tsing Hua University, China
Sunghyun Cho	Adobe Systems
Wei-Ta Chu	National Chung Cheng University, Chiayi, Taiwan
Michael Cree	University of Waikato, New Zealand
Liya Ding	Harvard Medical School, USA
Zhao Dong	Cornell University, USA
Gianfranco Doretto	West Virginia University, USA
How-Lung Eng	Institute for Infocomm Research, Singapore
Ralph Ewerth	Jena University of Applied Sciences, Germany
Jianping Fan	UNCC
Jiashi Feng	National University of Singapore
Haiyan Fu	Dalian University of Technology, China
Chiou-Shann Fuh	National Taiwan University, Taiwan
Shenghua Gao	ADSC of UIUC, USA
Yue Gao	National University of Singapore
Margrit Gelautz	Vienna University of Technology, Austria
Bo Geng	Peking University, China

Luis Herranz	ICT
Changbo Hu	University of Texas, Austin
Gang Hua	Stevens Institute of Technology, USA
Chun-Rong Huang	Institute of Networking and Multimedia, National Chung Hsing University, Taiwan
Gaofeng Huang	Google Inc.
Qingming Huang	Graduate University of the Chinese Academy of Sciences
Rui Huang	Rutgers University, USA
Benoit Huet	EURECOM, France
Tan Hung Khoon	Tunku Abdul Rahman University, Malaysia
Naoyuki Ichimura	National Institute of Advanced Industrial Science and Technology, Japan
Ichiro Ide	Nagoya University, Japan
Daisuke Iwai	Osaka University, Japan
Shen Jialie	Singapore Management University
Yu-Gang Jiang	Fudan University, China
Xin Jin	Tsinghua University, China
Alexis Joly	Inria, France
Li-Wei Kang	Institute of Information Science
Mario Koeppen	Kyushu Institute of Technology, Japan
Congyan Lang	Beijing Jiaotong University, China
Duy-Dinh Le	National Institute of Informatics, Japan
Tung-Ying Lee	National Tsing Hua University, China
Haojie Li	Dalian University of Technology, China
Jenn-Jier James Lien	National Cheng Kung University, Taiwan
Guo-Shiang Lin	Da-Yeh University, Taiwan
Damon Shing-Min Liu	National Chung Cheng University, Taiwan
Dong Liu	Columbia University, USA
Huiying Liu	Institute for Infocomm Research, Singapore
Jing Liu	NLPR of CASIA
Xueliang Liu	EURECOM, France
Yan Liu	The Hong Kong Polytechnic University
Hong Lu	Fudan University, China
Yijuan Lu	University of Texas at San Antonio, USA
Yasushi Makihara	Osaka University, Japan
Fabrice Meriaudeau	Université de Bourgogne - Le2i - UMR CNRS 5158, France
Vasileios Mezaris	Informatics and Telematics Institute/Centre for Research and Technology Hellas, Greece
Weiqing Min	
Philippe Mulhem	LIG-CNRS
Hajime Nagahara	Kyushu University, Japan
Atsushi Nakazawa	Osaka University, Japan

Chong-Wah Ngo	City University, Hong Kong
Naoko Nitta	Osaka University, Japan
Noel E. O'Connor	Dublin City University, Ireland
Wei Tsang Ooi	National University of Singapore
Vincent Oria	NJIT, USA
Junbiao Pang	BJUT, China
Yuxin Peng	Peking University, China
Dinh Phung	Deakin University, Australia
Christian Pieringer	PUC
Yanyun Qu	Xiamen University, China
Isaac Rudomin	Barcelona Supercomputing Center, Spain
Klaus Schöffmann	University of Klagenfurt, Austria
Jie Shao	The University of Melbourne, Australia
Donggyu Sim	Kwangwoon University, Korea
Cees Snoek	University of Amsterdam, The Netherlands
Mingli Song	Zhejiang University, China
Chao Sun	
Yongqing Sun	NTT Media Intelligence Laboratories
Yu-Wing Tai	KAIST, Korea
Masayuki Tanaka	Tokyo Institute of Technology, Japan
Jinhui Tang	Nanjing University of Science and Technology, China
Sheng Tang	Institute of Computing Technology, Chinese Academy of Sciences
Kong-Wah Wan	Institute for Infocomm Research, Singapore
Feng Wang	East China Normal University, China
Jinjun Wang	Xian Jiaotong University, China
Meng Wang	AKiiRA Media Systems Inc.
Yi Wang	Dalian University of Technology, China
Yu-Chiang Frank Wang	Academia Sinica, Taiwan
Xiao-Yong Wei	Sichuan University, China
Fei Wu	Zhejiang University, China
Xiao Wu	Southwest Jiaotong University, China
Changsheng Xu	Institute of Automation, Chinese Academy of Sciences
Dong Xu	Nanyang Techonological University, Singapore
Jizheng Xu	Microsoft
Keiji Yanai	University of Electro-Communications, Japan
Ming Yang	NEC Laboratories America
Xiaoshan Yang	CASIA
Yang Yang	The University of Queensland, Australia
Ting Yao	City University of Hong Kong
Qixiang Ye	University of Maryland, USA

Long Ying	CASIA
Lei Yu	
Zinan Zeng	Advanced Digital Sciences Center of UIUC
Zheng-Jun Zha	National University of Singapore
Guangtao Zhai	University of Erlangen Nuremberg, Germany
Tianzhu Zhang	University of Alabama at Birmingham
Wei Zhang	City University of Hong Kong
Wan-Lei Zhao	Inria, France
Yao Zhao	Beijing Jiaotong University, China
Yiliang Zhao	National University of Singapore
Bo Zheng	The University of Tokyo, Japan
Yan-Tao Zheng	Google Inc.
Yantao Zheng	Google
Xiangdong Zhou	Fudan University, China
Cai-Zhi Zhu	National Institute of Informatics, Japan
Shi-Ai Zhu	City University of Hong Kong
Roger Zimmermann	National University of Singapore

Additional Reviewers

Arzate, Christian
Breitbarth, Andreas
Chen, Heng-Chou
Chen, Xiangyu
Chen, Xiaogang
Chen, Yao
Cheng, De
Escudero, Marcos
Fang, Quan
García Martín, Alvaro
Guo, Long
Gupta, Sunil
Huang, Gaofeng
Huang, Xiangqi
Hudelist, Marco A.
Lai, Kuan-Ting
Li, Yingbo
Liang, Yudong
Lu, Min
Min, Weiqing
Motiian, Saeid
Nga, Do Hang
Ngo, Thanh Duc
Nguyen, Tien-Vu
Nguyen, Tu
Ning, Huazhong
Poullot, Sébastien
Primus, Manfred Jürgen
Rivalcoba, Jorge-Iván
San Miguel, Juan Carlos
Sanmiguel Avedillo, Juan Carlos
Sharlemin, Sajid
Siyahjani, Farzad
Sun, Jichao
Waltl, Markus
Wu, Fei
Yang, Xiaoshan
Yuan, Zhaoquan
Zhang, Shizhou
Zhang, Shun
Zhang, Yifan
Zhao, Xuran
Zhao, Yiliang
Zhu, Guangyu
Zhu, Shi-Ai

Abstracts of Keynote Talks

Smart Social Systems

Ramesh Jain

Dept of Computer Science
University of California
Irvine, 92697
United States
jain@ics.uci.edu

Abstract. Availability of enormous volumes of heterogeneous Cyber-Physical-Social (CPS) data streams may allow design and implementation of networks to connect various data sources to detect situations with little latency. In fact, in many cases it may even be possible to predict situations well in advance. This opens up new opportunities in designing smart social systems for specific tasks. Such systems may be very useful for many important problems at local as well as regional and even global level. We believe that such systems offer many novel challenges to researchers in multimedia, particularly in social and cross-modal media systems. We will present our ideas and early approach towards building smart social systems.

Domain-Oriented Video Coding

Wen Gao

Peking University
Beijing 100871, China
wgao@pku.edu.cn

Abstract. Image and video data is becoming a majority in the big data, a reasonable improvement of video coding efficiency may get a big cost saving in transmission and/or storage, that is why so many researchers work on the new coding technologies and standards. For example, a join video team between ISO and ITU-T works on HEVC/H.265 standard which achieves about 50% bits saving over H.264/MPEG-4 AVC with the comparable visual quality. However, the limitation on further performance improvement under the traditional hybrid coding framework is recognized, researchers are looking for new model/technology. Recently, a surveillance domain-oriented video coding standard has been published by IEEE Standards Association in June of 2013, known as IEEE Std 1857 standard. IEEE 1857 doubles the performance in the coding efficiency compared with AVC/H.264 and HEVC/H.265, by using a model based coding schema. Similar idea is happening internet video coding area. This talk will discuss the recent developments of domain-oriented video coding technology, specifically on background picture model based surveillance coding and cloud-based image coding, and the challenging problems and future directions.

Knowledge Mining and Semantic Search

Yong Riu

Microsoft Research Asia (MSRA)
yongrui@microsoft.com

Abstract. In this talk, we will discuss how to extract and mine knowledge from various data sources. We will talk about the traditional library, the first generation of web search engines, todays search engines, and tomorrows search engines. Specifically, we will discuss the difference and relationship between term-based search vs. entity-based search, document-centric view vs. query-centric view, and query understanding vs. query answering. We will use three examples to illustrate entity-based knowledge mining: in the academic domain, in the open text domain, and in the multimedia domain.

Table of Contents

A Cloud-Based Multi-functional e-Meeting System by Flash-Based Multimedia Technology for Higher Education on the WebELS System

Mohamed Osamnia[1,2], Arjulie John Berena[2], Sila Chunwijitra[1,2], Hitoshi Okada[1,2], and Haruki Ueno[2]

[1] The Graduate University for Advanced Studies
[2] National Institute of Informatics, Tokyo, Japan
{osamnia,berena,sila,ueno,okada}@nii.ac.jp

Abstract. With the recent advances of ICT technologies, teaching methodologies in higher education are changing from the traditional classroom-based methodology to the so-called Virtual Learning Environment (VLE). Implementing an effective virtual classroom system, requires understanding the characteristics and the activities in higher education, in terms of characterized self-learning, group meetings and discussion. A cloud-based multi-functional e-Meeting system is an essential feature to fulfil the former demands, hence allowing effective academic activities. In this paper, we discuss the design and implementation of a multi-functional e-Meeting system that based on cloud-computing built on the Web-based e-Learning System (WebELS) platform. Using Flash-based multimedia technology, it is possible to conduct on-line meetings and lectures through the Internet. Specifically, the proposed system offers important features, such as on-line presentations, on-line annotations, video conferencing, and chat massaging systems, which are the basic components for an effective virtual classroom methodology in higher education. Uploading various documents was presented for enabling a rich multimedia content. The design concerns non-IT users, and low bandwidth Internet users, such as in the developing countries.

Keywords: e-Meeting, Higher Education, Multimedia, Flash-based, WebELS.

1 Introduction

As a new trend, facilitating the interaction and communication between teachers and students, and to meet the social demands for a flexible educational system, has now backed up by integrating the informatization of society with the modern rapidly growing communication technology. Such as changing the teaching methodologies of higher education, from the traditional classroom-based methodology to the use of information and communications technologies.

Internet-based e-Learning and on-line video meeting systems are applicable technologies that are used to encourage the Virtual Learning Environment (VLEs) activity. Where various benefits from these technologies can boost up the

B. Huet et al. (Eds.): PCM 2013, LNCS 8294, pp. 1–12, 2013.

learning curve. Such systems can be easily achieved because of the advanced in inter-networking, multimedia and software technologies at lower cost and higher quality in global scale. They can help organizing a meeting comfortably via the Internet. This high-technology is used to link members to join from any place to meet in a virtual environment. It also changes the business communication, what can reduce operational administrative costs and can make stronger business competition [17] . To implement an effective virtual collaborative learning system, we should understand the characteristics and activities in higher education which are different from middle school or undergraduate program where course management and automated student assessment are typical features. In the graduate school, it is characterized as self-learning, group meetings and discussion. Some studies have analyzed the characteristics of higher education based on the information technology perspective [8,10]. To respond to the demands of higher education, a cloud-based multi-functional e-Meeting system is strongly requested rather than a typical Management System (LMS). The on-line meeting systems can help speed up the learning process and make faster expertise-based decisions. Furthermore, they can allow rich social collaboration between teachers and students around the world without travel costs or waste of time [13].

However, an e-Meeting system should include some primary functions and features such as on-line slide presentations, on-line annotations, web conferencing, chat massaging to be effective tools for higher education mechanisms [16]. Furthermore, the system must provide variety of document types a user can upload in order to create a rich multimedia contents for discussion. Since the cloud computing system is being a hot technology, it becomes the best solution for major e-Learning platforms [11] . The cloud computing can provide many benefits to all e-Lectures systems, such as reliable, flexible, cost-efficient, without the requirement of many hardware and software resource.

In this paper, we discuss the design and implementation of a multi-functional e-Meeting system based on cloud-computing, built on the Web-based e-Learning System (WebELS) platform, by Flash-based multimedia technology to enhance the on line meetings and lectures through the internet. Unlike the majority of e-Meeting systems that require IT skills to use, the design of the proposed e-Meeting system is addressed to non-IT users offering a simple and flexible interface. e-Meeting is made possible by combining the on-line slide presentation and video conferencing system to create a virtual classroom, where the meeting participants congregate via the internet. Hence, the proposed system offer important features such as, on-line presentation, on-line annotation, video conferencing, and chat massaging system, and use of various document types, which are the required features for a virtual collaborative learning system. We briefly discuss some of the present e-Meeting systems, and present practical usage of the system in its successful performance in domestic and international collaboration.

2 Review of Related Works

With regard to the rapid growth of ICT technologies and internet applications, various e-Meeting systems have become widely available. We describe some popular related technologies on on-line presentation, video conferencing and discuss the advantages and drawbacks of each technology in the view of e-Meeting solution for higher eduction.

Screen sharing is a famous technology for on line presentations, that allows the user to transmit the contents on his computer to one or more connected users. Some of this screen sharing systems allow remote control, which gives the authority to control keyboard and mouse. The advantages of using screen-sharing systems is that the presenter can share and display multiple applications on his computer with other users. However, the presenter should keep the privacy of his personal information and make sure that is not displayed to other participants. Another drawback of screen-sharing technique is that it works well in broadband, but not in the narrow-band. As it utilizes greater bandwidth for streaming of encoded pixel data of screen image, in case of a lot of pixels change at the same time, such us when scrolling a window or viewing full-screen video the bandwidth demands gets very high, some of the systems that use screen sharing are GoToMeeting [3], Real VNC [6] and WebEx [2].

Moreover, TV conference system such as the polycom [5] is also a popular system to be used for real-time distant learning in universities and companies. The limitations of those systems are that they require high bandwidth network, and the users have to find a place where the systems are equipped. Besides, using these systems needs proxy and firewall setting. There are also some open sources of web-based conference system which can be used for on-line meeting and lectures such as openmeeting [4] and BigBlueButton [1]. The drawback in this system is that some configurations set-up and privacy information are required before creating a meeting or a class-room, which is a big barrier for non-IT users [12].

Considering the above issues and requirements, we have developed a multi-functional e-Meeting system based on cloud computing. Using the Flash-based multimedia technology on the WebELS platform, we can provide on-line presentation and video conferencing in one single platform to be an (all in one) multimedia system.

3 Cloud Computing for e-Learning

The Internet and network technologies are now transforming to the next generation. The development of those technologies are fast delivering and rapidly evolving advanced utilization to support daily life usage. There are many modules, services and sections which depend on the functionalities and other factors used to sustain any internet activities. Cloud computing systems are being a popular technology at the present use to serve the server side services [11]. This technology allows much more efficient computing by centralizing data storage, processing and bandwidth. It also allows consumers and businesses to use applications without installation and access their personal files at any computer. Cloud computing technologies can be implemented in a wide variety of architectures, under different services and deployment models, and can coexist with other technologies and software design approaches. The proposed system, was created based on this technology, in order to guarantee the teaching and learning activities. Moreover, the quality and the running of the ecosystem can make the system reliable, flexible and cost-efficient for all type of users.

4 Concept and Background of WebELS Meeting

WebELS is designed to provide high quality e-Learning environment for higher education in science and technology using advanced information and communication technology [8]. The key concepts of the WebELS system are as follows:

- Lecture is based on slide presentation and on-line annotation.
- WebELS is a client server architecture.
- The system is suitable to work under low speed internet environment [14].
- The users can connect from any operating system to use WebELS such as Windows, MacOS and Linux.
- WebELS doesn't require any specific tool or software, all what a client need is a browser and internet access.
- Powerful authoring features can allow professors to edit their own contents on their personal computers and upload them to the WebELS server to be share with students.
- The system interface is easy to understand and manipulate even without any tutorial which makes it convenient for non-IT users.
- WebELS is an open source platform for academic usage.

The previous design of WebELS Meeting separated into video conferencing panel and on-line presentation panel, which are independent with each other. However, there were several difficulties in operating the system, especially for the usage in the international conference distribution, such as (1) confusing for the user to determine which content to download in order to join the video meeting, and (2) difficulty in relaying the changing of the conference presenter and presentation. Thus, we developed a new version of the e-Meeting system, where the on-line slide presentation and video conferencing system are combined into an "all-in-one" system to resolve the problems stated above.

5 Flash-Based e-Meeting System

The overview of the WebELS meeting system is shown in Fig. 1. According to this figure, there are three servers at the server side system: database server, contents server, and streaming server. Database and contents server are used for contents and user data management, while streaming server is used for real-time audio-video communication.

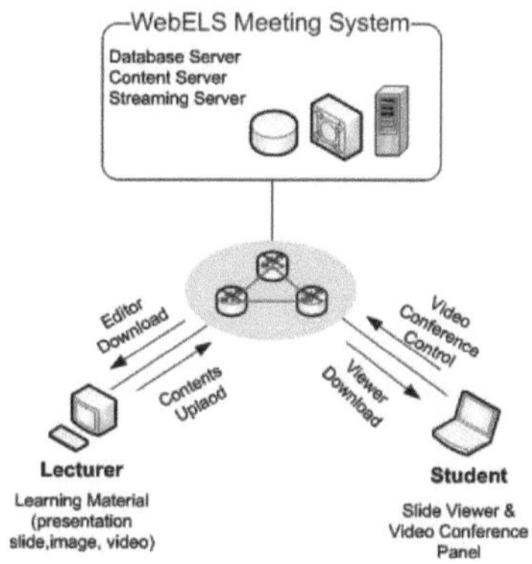

Fig. 1. System Overview of WebELS Meeting

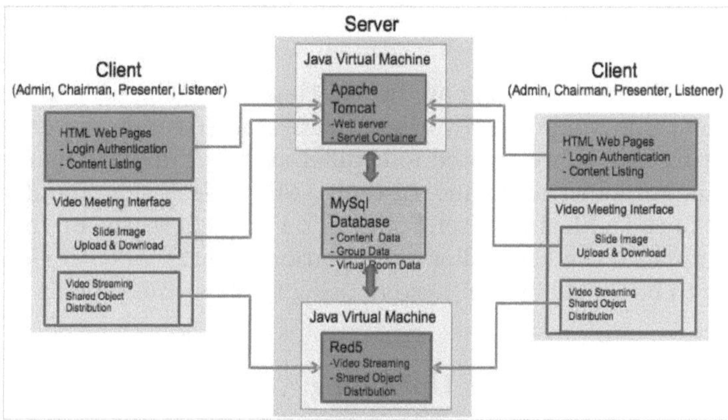

Fig. 2. Flash-based e-Meeting system Diagram

5.1 System Design and Implementation

The proposed system merges two different multimedia technologies (JAVA and Flash) in one technology (Flash), to provide a powerful tool for on-line lecture to support higher education. The goal behind that is to combine the whiteboard technology and the video conferencing technology. Fig. 2, shows the system diagram of the Flash-based e-Meeting system and demonstrates the different technologies we used on both client and server side such as: Apache tomcat, MySql and RED5 for server side. The client needs to open the browser and access to the WebELS Meeting server through the internet.Real Time Messaging Protocol(RTMP) Real Time Messaging Protocol Tunnelled(RTMPT) is used for both slide synchronization and audio-video communication. Since WebELS meeting is client server architecture, all the events in the meeting system that include the whiteboard for on-line presentations and on-line annotations, video

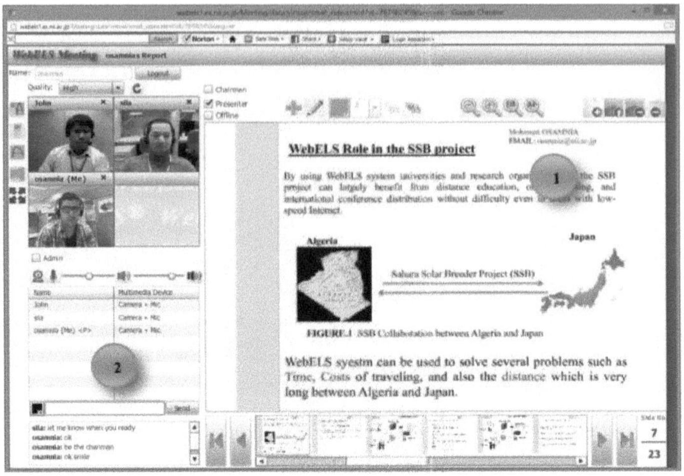

Fig. 3. Flash-based e-Meeting Panel Design

conferencing, and chat messaging are in real-time streaming using flash technology on the client side, supported by using the Red5 as streaming sever. According to the e-Meeting events, we created four different types of participants: Presenter, Chairman, Administrator, and Listener. More discussion about user roles are described in the e-Meeting User Policy 5.5. Fig. 3 shows the e-Meeting panel design that supports the (1) on-line presentation and (2) video conferencing simultaneously to be as "all in one" system using the flash multimedia technology.

5.2 Flash-Based Document Uploading

To create a virtual classroom, it is necessary to create a content by uploading various types of documents. Users just need to create their own presentation document using their personal computers, and upload the presentation document to the WebELS Meeting server. This process require only opening a browser and connect to the Internet, without the need of any special software. WebELS meeting content authoring is made by using the flash-based multimedia content editor in Fig. 4.

The first step is shown in Fig. 4 (a), the user have to fill in the title of the content and if needed insert a password to protect the content for privacy. Then, browse to upload the document that can be PDF, presentation (odp, sx, ppt, pptx), document (odt, sxw, doc, docx), or only image file (png, jpg, tiff, bmp). Since WebELS Meeting was designed to support low-speed internet, the user can select the output of conversion of the selected document. We developed three conversion outputs to support different bandwidth environments, such as PNG (low quality image, for internet speed of 100Kbps and below), JPG (medium quality image, for internet speed between 100Kbps and 1Mbps) and SVG (High quality vector graphics, for internet speed of 1Mbps and higher). At the server-side, the

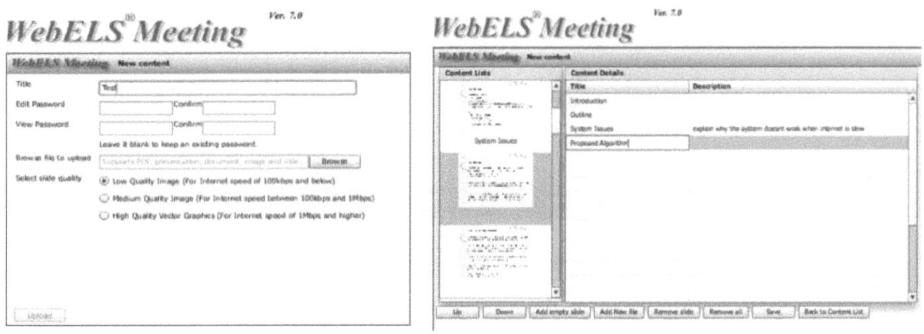

(a) (b)

Fig. 4. Flash-based Document Uploading (a) and (b)

program will choose the proper conversion tools (e.g, Ghostscript application for PDF, presentation, and document, PDF2SVG for SVG format, etc.). The conversion process is based on specific parameters of conversion according to the user output selection. Flexibility of XML-based storage allows the uploaded document materials to be delivered in many different ways.

Moreover, most e-Meeting systems prefer to use PDF format as the standard document for on-line presentation. In case of postgraduate student, in a specific field of studies, the above document types might be not enough. Thus, we proposed the video clip format, which is very important in case of engineering discussion. We developed a new and unique feature in the e-Meeting systems to make the system support on-line video clip presentations. It is possible to upload a variety of video clips format such as (mov, avi, wmv, mpg, ogg, mp4, flv). At the server side, all video clips will be converted to FLV format using FFMPEG. We also defined the conversion parameters to an acceptable rate to support the low-speed internet environments.

The second step is when the conversion process is done at the server side, another interface of the content authoring will be downloaded from the server to the user machine. Fig. 4 (b) shows, the interface containing the slide images and videos converted at the server side. The user can edit the slides, images by inserting a title or description for each slide. Furthermore, the user can change the order of slides up or down, and also can add empty slides, or remove slides which are important features to edit the multimedia content. In case a user would like to insert another document file (image, video, PDF, etc.) within the content creation process, it is possible to do it without going back to the first interface of authoring. This interface was enriched by "Add File" function that can enable browsing any file in the users computer. Thus, it will upload a document and execute the conversion process at the server side. The uploaded document will be downloaded to the user machine, in order to create the final

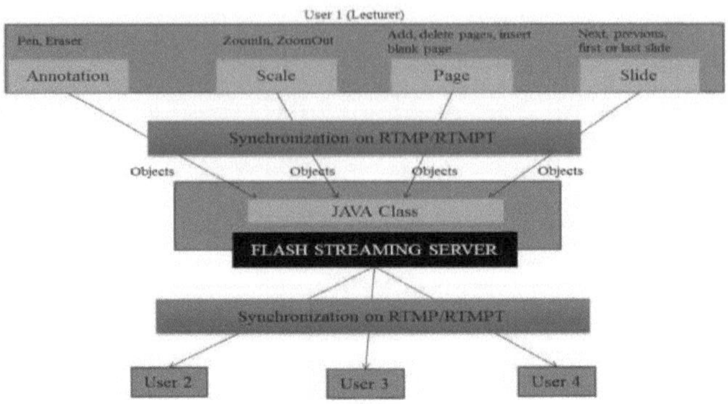

Fig. 5. The System Diagram of the Real-Time Distant Whiteboard Technology

multimedia content which contains all the types of uploaded documents. Once the content is created, a virtual room at the server side is automatically generated, where the participants simultaneously via the internet in the same place, convene and discuss the same content.

5.3 Whiteboard Technology and Features

Fig. 5, shows the system diagram of the WebELS Real-Time whiteboard technology. All the objects such as annotation, scale, page, and slide will be send to the server via RTMP or RTMPT protocol same as the video conference system described in section 5.4. Subsequently, the server will store all the information as files, where the objects will be sent to other clients for the synchronization among the whiteboards in the same virtual room [9].

The "Annotation" objects are pen colors, thickness and eraser information. The "Scale" objects are zoom-in, zoom-out and fit to the normal parameters. The "page" objects are the current page and the total number of existing page, a user can insert a new blank page to use it for drawing information on the whiteboard. Moreover, a user can select any page to delete or delete all pages. The "slide" objects are the presentation slides, where the presenter who has the control on the presentation can switch among slides. Adaptation in a User Interface (UI) can be served by adaptivity, adaptability or dynamicity in the aspects of interaction [7]. Therefore, Flash-based e-Meeting system on the WebELS platform was designed to be all-in-one system, by mean that users can select the proper option in the same panel by changing the interface layout from only video conferencing, to only whiteboard with audio discussion. Full-screen whiteboard presentation and wide-video meeting in case of a conference.

Besides the object features mentioned above on the whiteboard, the system offers uploading on-fly-way function for the users in the same virtual room. The function enables the presenter to add any type of document during online discussion which will be automatically merged with the original content.

Moreover, as mentioned above the whiteboard can be used for on-line video clip presentation. The unique feature allows the presenter to control the streaming of the video clip by a simple "play-pause" button commands and use all the object features such as: zooming, Cursor, annotation, and etc. without influencing the playback of the video clip, in order to enrich the discussion and the multimedia contents.

5.4 Video Conferencing

The aim of the proposed system is to provide a powerful on-line meeting platform to support higher education under low-speed internet. Therefore, we merged the whiteboard system with a video conferencing system. What provides audio-video communication among the users in a shared virtual room. Because Flash-based e-Meeting system is based on streaming, then the whiteboard data, the audio, and the video are all on streaming over the internet between flash player on the client side, and the streaming server which uses RTMP. The panel (2) in Fig. 3 demonstrates the use of the video conferencing, during on-line presentation to enable the face-to-face communication. The main features of the video conferencing system are summarized as the following:

- H. 264 is used for video and audio codec.
- Network Status Monitor is used for monitoring the status of network connection while the meeting session is ongoing. After the learner logged-in to the system, this function always checks the network connection between server and client nodes. In case of connection loss, the reconnection manager function will be triggered for handling the connection.
- Reconnection Manager is used to keep the network connection when there is a loss of connection. This function clears the old connection handler,and waits for the new connection status. Once connection signal appears again, the login process is automatically done by using the latest meeting information [17].
- Device Manager is used to manage the basic input devices, such as microphone and web camera (or video camera).
- Quality Controller is used for controlling the quality of streaming data from the input devices. Several parameters are used for configuring the conference streaming data such as video size, video compression rate, video frame rate, audio gain, audio sampling rate and audio bit-rate. Those qualities can be selected in manual or automatic mode [15].

5.5 e-Meeting User Policy

During a conference, each participant has a role such as being a Presenter, a Listener, or a Chairman. Therefore, we designed different roles for users in the same virtual room as follow:

- Presenter: By checking the presenter, the user will send a shared object signal to other users via the server, announcing that the presenter was selected.

A presenter has the right to control the slides presentation by switching slide images, drawing annotation, using cursor pointer, using zoom functions, and also uploading other document in-fly-way if necessary.
- Chairman: The role of chairman during the conference is very important to change the presentations and presenters order. We developed this function to allow the chairman user of the open lecture to choose and select any presentation from the presentation list. Hence the presentation file will be changed for the next presenter, while all the other users keep connecting in the same virtual room.
- Listener: This user role is to attend the e-Meeting with no control on the slides. However, a listener can participate in the discussion by activating his microphone and camera.
- Administrator: This role is very important who has the full control on the e-Meeting event. The system can have one administrator in each virtual room. The administrator can control the users devices, mute a specific user or all the users, set a presenter, or block users.

6 Some Practical Uses of the e-Meeting System and Discussion

The WebELS meeting has been tested and used in many organizations and universities such as National Institute of Informatics (NII), NECTEC, UNESCO Jakarta, and etc. WebELS Meeting is officially used for JST-JICA collaboration project on Sahara Solar Breeder (SSB) between Japan and Algeria. Furthermore, the Flash-based e-Meeting system was tested in an open lecture organized by

Fig. 6. The System Diagram of the Real-Time Distant Whiteboard Technology

UNESCO. Fig. 6 shows Prof Okada in Japan giving a lecture using the proposed system. The system shows the ability of broadcasting international conferences and lecturers, which works well for around 20 users in one virtual room without any connection problem. However, we found out that in case of large number of users more than 20, too many updates on user client can degrade the performance of the meeting system. Because each user will send data to the server to be broadcast to other users, so in case of 50 users. On the server should manage 50 data uploaded and in the same time broadcast it back, where the client will receive a big amount of data from the server to update his status. Hence, we limited the shared objects to some functions only to decrease the transfer of data between clients and the server. We enlarged the memory reserved for JAVA and RED5 to boost up server performance for video streaming and web serving. In order to make the server robust to receive big amount of data sent from clients, we enlarged the size of the receiving and sending buffers in the RED5 to respond to this demand.

7 Conclusion and Future Work

We presented the design and implementation of multi-functional e-Meeting system by Flash-based multimedia technology for a flexible and global higher education. WebELS meeting provides powerful features for on-line slide presentation with audio-video communication created for non-IT users using friendly interface. We have enriched the variety of document types that a user can upload including PDF, presentations, documents, videos and image files. The system is highly stable under low-speed Internet. Where open lecture conducted by UNESCO illustrated useful information for refining the system. Our future target is to serve around 100 individual connection for the on-line lecture using such system. As future work, we will develop an embedded automated authoring tool system within the e-Meeting system to record the on-line lecture events such as presenter video and on-line slides presentation with annotation, to be stored and used as e-Learning multimedia content.

Acknowledgements. The authors would like to express sincere thanks to all persons who supported the WebELS project of NII Japan, especially to Dr. Vuthichai Ampornarambeth, Prof. Nobuo Shimamoto, Dr. Zheng Ho and Dr. Pao Sriprasertsuk for contributions in designing and implementing the WebELS system. The project is funded by Science Research Foundation of Japan, The Telecommunications Advancement Foundation, Amada Foundation for Metal Work Technology, Japan Science and Technology Agency (JST) and The Graduate University of Advanced Study (SOKENDAI). We express sincere thanks to the E-Learning project of UNESCO Jakarta office and the Sahara Solar Breeder (SSB) project for collaborations using WebELS.

References

1. BigBlueButton, http://bigbluebutton.org/
2. Google Docs, http://docs.google.comm/
3. GoTOMeeting, http://www.realvnc.com/
4. Openmeeting, http://code.google.com/p/openmeetings/
5. Polycom, http://www.polycom.co.jp/
6. RealVNC, http://www.gotomeeting.com/
7. Abdul, A., Otman, B., KKhaled, H.: Adaptive users interfaces for intelligent e-learning: Issues and trends. In: Proceedings of The 4th International Conference on Electronic Business, pp. 925–934 (2004)
8. Ampornaramveth, V., Zhang, T., Hadiana, A., Shimamoto, N., Ueno, H.: A web-based e-learning platform for post-graduate education. In: Proceedings of the 5th IASTED International Conference on Web-Based Education, pp. 388–393. ACTA Press (2006)
9. Berena, A., Chunwijitra, S., Okada, H., Ueno, H.: Shared virtual presentation board for e-meeting in higher education on the webels platform. Human-Centric Computing and Information Sciences 3(1), 1–17 (2013)
10. Berena, A.J., Sriprasertsuk, P., He, Z., Ueno, H.: An open source integrated e-learning/e-communication platform for posytgraduate education and corporate cyber meeting. In: Proceedings of the 5th International Workshop on Ontology Matching, pp. 33–37 (2009)
11. Deepanshu, M., Ashish, P., Suneet, K., Arjun, A.: E-learning based on cloud computing. International Journal of Advanced Research in Computer Science and Software Engineering (2012)
12. Erol, B., Li, Y.: An overview of technologies for e-meeting and e-lecture. In: Proceedings of IEEE International Conference on Multimedia and Expo., pp. 1000–1005. IEEE (2005)
13. Ueno, H., He, Z., Yue, J.: WebELS: A content-centered E-learning platform for postgraduate education in engineering. In: Jacko, J.A. (ed.) HCI International 2009, Part IV. LNCS, vol. 5613, pp. 246–255. Springer, Heidelberg (2009)
14. Ueno, H., He, Z., Sriprasetsuk, P., Berena, A.J.: Webels: Content-centered general purpose e-learning platform f or higher education in science and technology for low speed internet. In: MacTeer, C.F. (ed.) NOVA Science Publishers, in Distance Education, pp. 33–66 (2010)
15. Osamnia, M., Berena, A., Sila, C., Hitoshi, O., Ueno, H.: Automatic adaptation of streaming data for webels meeting for low-speed internet. In: Proceedings of The IEICE Technecal Report, pp. 25–30 (2012)
16. Al Saiyd, N.A., Al Sayed, I.A.: Multimedia distance e-learning system for higher education students. In: Snasel, V., Platos, J., El-Qawasmeh, E. (eds.) ICDIPC 2011, Part I. CCIS, vol. 188, pp. 356–367. Springer, Heidelberg (2011)
17. Chunwijitra, S., Arjulie, B., Okada, H., Ueno, H.: Design of suitable meeting management model for webels meeting to meet the business situations. In: Proceedings of the 1st International Conference on Advanced Collaborative Networks, Systems and Applications, pp. 52–57 (2011)

Disparity-Constrained Resolution Adaptation for Stereoscopic Images

Yi Liu[1,2], Lifeng Sun[1], and Shiqiang Yang[1]

[1] Department of Computer Science, TsingHua University, Beijing, China
htcliuyi@163.com, {sunlf,yangshq}@tsinghua.edu.cn
[2] Beijing Aerospace Control Center, Beijing, China

Abstract. This paper addresses the problem of retargeting the stereoscopic images. We propose a disparity-constrained resolution adaptation method. In our method, we try to preserve the shapes of the prominent contents in the stereo images when they are displayed in the target resolution. Moreover, in order to improve the visual comfort, viewers can adjust the disparity of the stereo image pair linearly in our method. We first model the retargeting energy considering the images' significance map, content distortion and disparity constraints. Then we minimize the total energy and solve it as a nonlinear least square problem. The optimization is automatic without any post work, such as inpainting or occlusion handling. Finally, we demonstrate our method on the Middlebury stereo datasets and 3D animation screenshots. For stereo evaluation, the disparity map comparison results are provided.

Keywords: Stereoscopic image, resolution adaptation, image retargeting.

1 Introduction

Stereoscopic media have become very popular in our lives recently. More and more 3D photos are produced by cell phones or stereo cameras and shared in the cloud storage. The images will be downloaded and displayed in various devices. However, the resolution of the display device is more likely to mismatch that of the images. To resolve these problems, the routine solution in 2D field is to linearly stretch the images. But it often brings uncomfortable visual experience.

At first, in the 2D image process field, the uniform stretch to images cannot produce good enough results in many cases. Fixing the aspect ratio, the uniform stretch is the simplest and most efficient method to modify the resolution of images. But if the aspect ratio is changed, the uniform stretch method is not enough. At this time the combination of scaling and cropping may be work in some simple situations. But if the image contains some objects near the edges, by using this combined method, we cannot get the satisfied results either. Viewers can not accept losing of the important information.

Unlike the display of a single image, in general, stereoscopic images contain two different images for viewers. Viewers see one image with the left eye and see the

B. Huet et al. (Eds.): PCM 2013, LNCS 8294, pp. 13–24, 2013.

other with the right eye. The distance between a pair of correspondence point is called disparity, which is one of the important depth cues. With the positive disparity the perceived objects could be pushed behind the screen, while with the negative disparity, they will be placed between the viewer and the screen. The bigger the positive disparity is, the farther the point is perceived behind the screen. And the bigger negative disparity could bring the perceived points closer to the viewer[3]. So it seems that the bigger disparity is wonderful because it can produce the stronger stereoscopic experience. However, in fact, it often results in fatigue and disgust. Human eyes verge on a point, which gives the absolute depth to the viewer. In an area around this point, called Panum's area, the brain merges the two images to form a single image, thus perceiving relative depth [20]. According to the distance between the eyes, the distance from the screen, and some other parameters, different viewers need different disparities in order to place all of perceived points inside the Panum's fusional area. The routine solution in 2D field does not include the consideration of disparity and can not be used alone to resolve stereoscopic images retargeting problems.

In this paper, we present a disparity-constrained resolution adaptation method, in which we resize the stereoscopic image pair to the required resolution and meanwhile maintain the comfortable stereo visual effect.

We extract the important features from images, such as the gradient information, saliency information, and so on. Using these features we can build a significance map for every image. The contents of image in the significant areas should be deformed less in the retargeting process. And the contents in the less significant regions should be deformed more to meet the new resolution. We design a set of equations for the retargeting, which could avoid the distortion of the detail parts. And then we get the optimal solution using the energy minimization algorithm, which means the important information are remained as much as possible.

The more critical problem is how to make the retargeted image pair looked comfortable in the new resolution. Since the disparity is closely linked to the stereo visual effect, we hope to maintain the disparity of the stereoscopic image pair. In order to improve the efficiency of the algorithm, we extract the sparse correspondence points first. And then we filter the outlier point pairs using RANSAC algorithm. But the correspondence points always cluster in some areas. We evaluate the sparse points and redistribute them more uniformly. The well-distributed sparse point pairs could provide enough detailed disparity values nearly throughout of the whole images. We then design the constraints for the disparity values at the key points. The horizontal disparity is allowed to be adjusted linearly, and the vertical disparity is not allowed. Thus, constraining the disparity, the highly consistent retargeted image pair is produced, which provide more comfortable stereoscopic experience in the new resolution.

To summarize, in our method, we focus on two problems:

- To improve the warping result of images.
- To manipulate the retargeted disparity.

The main contribution in this paper is a retargeting technique for the stereoscopic image pairs. In our technique, we constrain the disparity distortion and adjust the disparity value using the well-distributed sparse features in the retargeting process. The sparse features can not only ensure the comfortable visual effect but also reduce the unnecessary computation. As the second contribution, this paper also provides a novel significance measurement method for the images. The proposed significance map can better prevent the important contents from deforming.

The rest of this paper is organized as follows. Section 2 summarizes related works. Section 3 presents our approach in detail. Section 4 discusses the experiment results and Section 5 briefly concludes the work presented in this paper.

2 Related Work

Many methods are used to resolve the problem how to adapt the images in various display devices with different resolutions. For a single image, two classes of methods are summarized in [2]: discrete approaches and continuous approaches. These methods are also applied to stereoscopic images. In this section, we focus on the stereoscopic image authoring and processing. Some methods are proposed in [4-6], [12], [14], [17], [18], [22], [23].

In [4], the proposed system is fully automatic. The images are cropped and exploited under 3D viewing constraints. But if there are many objects in the images, the method will lose some information. [5] and [17] focus on the manipulation of the disparities. [6] propose an approach for stereo images resizing. When the display aspect ratio is changed, they attempt to retarget the images and hold the disparity. However, a limited number of feature points could result in the deformation of some image contents. [12] and [22] are based on the seam carving method, which could lose much important information of image contents and change their shapes, especially when the resolution changes greatly. In [14], a new layer-based resizing method is proposed. The layers make retargeting more flexible and allow the greater resolution change. But it does not discuss how to deal with occlusion and its retargeting result relies on the image segmentation. Recently, warping-based methods have appeared to resolve the stereoscopic retargeting problems, such as [18], [23], etc. Inspired by these works, we propose a novel disparity-constrained method to resolve the resolution adaptation problem for the stereoscopic images in Section 3.

Our work belongs to the stereo image retargeting algorithm. By limiting the disparity adjustment and preventing the deformation of images, our method produces more comfortable stereo outputs, which accords to the stereoscopic principles in [8].

3 The Proposed Method

As mentioned in [2], the spatial retargeting methods, for the images, include two basic classes. We choose the continuous solution. However, the stereoscopic constraints to be met to produce flawless 3D contents are many, which are summarized in [4]. The disparity is one of the most important components. So, being different from previous works, we apply some novel disparity constraints in the retargeting process. We preserve the disparity carefully and obtain good experimental results.

The basic idea is first comput the significance map (S) for every image (I_L, I_R) (Section 3.1). We divide the images into grids with the rectangle meshes and compute the grid significance values (S_{gL}, S_{gR}) (Section 3.2). We then obtain the sparse and well-distributed correspondence points (p_L, p_R) in the image pair (Section 3.4). Finally, we measure the mesh distortion levels with the grid deformation energy (E_{GD}) (Section 3.2), edge distortion energy (E_{ED}) (Section 3.3) and disparity constrant energy (E_{HD}, E_{VD}) (Section 3.4). With the equation set, we get the satisfied retargeting result using the optimization algorithms. In fact, we need to find the most suitable locations of the mesh vertexes that cost the least energy. In the method, we don't have to inpaint and will not lose a great deal of information of the images either.

3.1 Significance Map

In the proposed method, we first obtain the significance map (A) of the images respectively. With the saliency algorithms, every pixel will be assigned a value which indicates its importance. In our experiments we used the algorithm proposed in [15]. Itti et al. apply various filters to extract color, intensity and orientation properties. They look for the regions that have more different properties than their surroundings. Secondly, the edges are always the important cues for the human's vision, while no edge information given in the saliency agorithm. So we compute the gradient map (G) of images respectively, which could be used to add edge information to the significance maps. We denote the basic significance map S_o as

$$S_o = A \times G \tag{1}$$

Line Maintenance. For human's vision, the lines in an image are always easy to be understood. If the lines in the image are distorted in the retargeting steps, we can also find them easily. However, not all the lines in the image need to be preserved. Some lines in the areas with lower significance values can be bent a little. So we substitute the line features into the pixel significance value and decide whether the line is important and should be prevent from distortion.

In our method, to detect the line segments in the image pair, we adopt the LSD algorithm proposed in [19]. And then we assign the higher significance values for the pixels in the line segments. In our experiments, we first obtain the significance map, and then double the significance value of the pixels on the lines. In Fig.1, in the top row: the left is the original image, and the right is produced by algorithm [24]. In the bottom row, the left is significance map of the original image, and the right is our result. We could find the contents deformed less and the lines maintained better in our result. For example, the chairs deformed less obviously in our result.

Fig. 1. Line-maintained retargeting

Fig. 2. Significance map

We define the final significance value of the pixel at the coordinate (x, y): $S(x, y)$, where, L_k is the k-th line segment, and n is set to 2 in our experiments.

$$S(x,y) = \begin{cases} n * S_{\delta}(x,y) & ,(x,y) \in L_k \\ S_{\delta}(x,y) & ,(x,y) \notin L_k \end{cases} \quad (2)$$

At last, we normalize the significance values within 0.05 and 1. An example of the significance map shown in Fig.2. The top row shows the original image (left), the saliency map (middle), and the gradient map (right). The bottom row shows the line segment map (left) and the significance map (right).

3.2 Grid Deformation Energy

In an image, the pixels with higher significance value are always the important information for our eyes, and we hope to preserve them carefully. We divide the images into N grids with the rectangle meshes. The edge length is defined as 20 pixels in the experiments. We obtain the average significance values of pixels in every grid, S_g, which indicates the importance of the grid. The more important grid should avoid the distortion as far as possible, and the less important grid should be deformed more to meet the retargeting need.

Being similar to the [24], we denote the deformation energy of the i-th grid, $E_g(i)$ as

$$E_g(i) = \sum_{k=1}^{4} \left\| e_k^1 - s_g e_k^0 \right\|^2, \quad i \in [1, M] \tag{3}$$

where e_k^0 is the k-th edge vector of an original grid, e_k^1 is the k-th edge vector of the corresponding deformed grid. The coefficient s_g indicates the deforming scale, see [24]. We compute s_g by

$$\frac{\partial E_g(i)}{\partial s_g} = 0 \tag{4}$$

In order to measure the grid deforming energy E_{GD} according to the grid significance , we define it as

$$E_{GD} = \sum_{i=1}^{N} E_{gL}(i) S_{gL}(i) + \sum_{i=1}^{N} E_{gR}(i) S_{gR}(i) \tag{5}$$

where E_{gL} is the deforming energy of the grids in the left image, E_{gR} is that in the right image. And, in the left image and the right image respectively, the significance value of grid is denoted as S_{gL} or S_{gR}.

3.3 Edge Distortion Energy

Although the seriously deformed grids are not the very important parts of images, they also may cause the artificial distortion because of the inclination of their edges. We hope to maintain the edges' original directions as far as possible at the same time to modify the edges of the grids. The jagged edges can distort the image contents.

As the grid deforming energy mentioned above, we define the scale coefficient s_e for every edge, which can change the length of the edge. In order to measure the angle between an original edge e_i^0 and its deformed version e_i^1, we calculate the distance between them, see [6] . The longer the distance is, the more seriously the edge is distorted. Thus we could define the edge distortion energy as

$$E_e(i) = \left\| s_e e_i^0 - e_i^1 \right\|^2, \quad i \in [1, M] \tag{6}$$

where M is the number of edges. We compute the value of s_e by taking the partial derivative of $E_e(i)$ with respect to it, which is similar to the s_g in Section 3.2. The sum of all edges' distortion energy in the image pair should be minimized. The edge distortion energy E_{ED} is defined as

$$E_{ED} = \sum_{i=1}^{M} E_{eL}(i) + \sum_{i=1}^{M} E_{eR}(i) \tag{7}$$

where E_{eL} is the sum of all edges' distortion energy of the left image, E_{eR} is the sum of all edges' distortion energy of the right image.

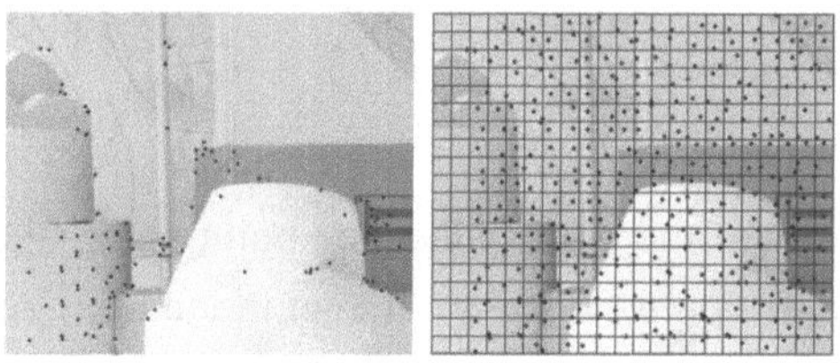

Fig. 3. Distribution of correspondence points

3.4 Disparity Constraint Energy

Disparity is one of the key parameter in stereoscopy. Assuming that the original image pair is rectified, the retargeted images should remain the good disparity values for the contents in the images. In the process of warping the images, we hope to constrain the disparity adjustment both in the horizontal direction and in the vertical direction.

We first detect SIFT features for two images using the algorithm in [9]. We verify all the point pairs and select the well-matched pairs using the fundamental matrix estimated according to RANSAC algorithm [16]. However, some point pairs often cluster together. They cannot provide more disparity information but reduce the efficiency of the optimization algorithm. We hope that the distribution of the point pairs is sparse and uniform approximately in both images. So we verify the significance values of the feature points in one of image and, in every mesh grid, select the most important one. In our experiment, we choose the left image. In the large textureless regions, where few feature points are found, we exploit down sampled dense correspondence information using the algorithm [7]. Similarly the most important point will be selected in every grid. As for the occluded regions or the margin, which exist in the left image but not in the right image, no correspondence point is assigned for the grids. Fig.3 shows the distribution of correspondence points in one of stereo image. The points are clustered in some regions in the left one. The points in the right one are well distributed.

We define the sparse correspondence points in the two images as $p_L(i)$ and $p_R(i)$, where i is the sequence number of point pairs. And the disparity between a pair of points is defined as $D(i)$, which is the x-coordinate difference of $p_L(i)$ and $p_R(i)$. After retargeting, the coordinates of the correspondence points are changed accordingly. The disparity between the new point pairs $p'_L(i)$ and $p'_R(i)$ is defined as $D(i)$.

Horizontal Disparity. In order to prevent the disorder of the depth information, the disparities of the contents in two images should change consistently. The disparity cues should not go against the other cues, such as occlusion cue, shadow and so on. We propose the energy equation to constrain the manipulation of disparity. The equation is defined as

$$E_{HD} = \frac{1}{K}\sum_{i=1}^{K}(s_d D(i) - D(i))^2 \tag{8}$$

where K is the number of the correspondence point pairs, and s_d is the disparity scale factor. In some situation we hope to retarget the image pair and meanwhile hold the disparity unchanged. Then we set s_d as 1. Otherwise, the disparity values will be modified linearly based on the original ones. Certainly we could realize the non-linear disparity manipulation if we assign different scale value for the different correspondence point pairs.

Vertical Disparity. Vertical disparities can be introduced as the corresponding feature points in the left and right image have different coordinates and are mapped to different locations although the same warping function. And the unconstrained warping could locate the deformed feature points, which result in the vertical disparities. However, the vertical disparity always causes 3D fatigue for the viewers [3]. By using of the energy equations, we hope to reduce the vertical disparities as far as possible. We define the equation as

$$E_{VD} = \sum_{i=1}^{K}(p'_L(i)[y] - p'_R(i)[y])^2 \tag{9}$$

where [y] represent the y component of the i-th point in the image. Minimizing this energy, the retargeting result will introduce few vertical disparity, which will not attract the viewer's attention.

3.5 Total Energy

Finally we minimize the sum of the energy mentioned above. The energy equation is

$$E = aE_{GD} + bE_{ED} + cE_{HD} + dE_{VD} \tag{10}$$

where a, b, c, d are the weights of the different energy items. In our experiments, we define that a=10, b=1, c=1, d=1. We do not change the coordinates of the vertexes at the four corners. The other vertexes on the boundaries are only allowed to move along the boundary line.

Fig. 4. Decreasing the size of the stereoscopic images

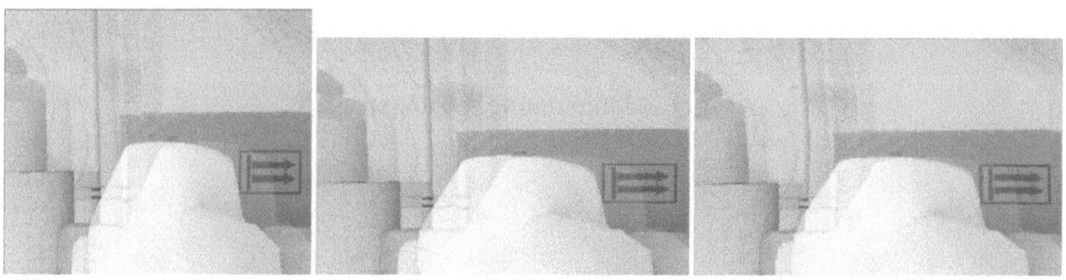

Fig. 5. Retargeting the images and preserving the disparity

4 Experimental Results

For our experiments, we collected binocular images from the Middlebury stereo datasets [25] and 3D animation. The experimental results are presented as red/cyan color anaglyph images. To observe the stereo effect after retargeting, an Acer GD245HQ 3D monitor and nVIDIA GeForce 3D VISION devices were used.

In Fig.4, we change the resolution of the input images from 400*229 to 280*206 (approximatively from 16:9 to 4:3) and try to maintain the original disparity. In the top row, the images, from left to right, include the input image pair, correspondence point map and the original red/cyan anaglyph. In the bottom row, the right one is the retargeted red/cyan anaglyph, and the four on the left are the retargeted left and right images and the warped mesh map respectively. Comparing two anaglyphs, we can observe intuitively that the disparity is preserved. And, by using the significance map, the shapes of prominent contents in the images are preserved as well. In Fig.5, we retarget the images' resolution from 4:3 to 16:9. From left to right, they are the original image, the linear stretch result and our result. We observed that the disparity increased linearly in the linear stretch image, while, in our result, by setting S_d as 1 in Eq. (8), the disparity values are preserved, especially in the significance area.

In Fig.6, we decrease the disparity and meanwhile widen the images. In the experiment, we set the S_d as 1.2. That is to say, the disparity of retargeted image pair would become almost 0.8 times smaller. The middle image in the figure shows the retargeting result of the input image on the left. The right image shows the vertical disparity difference at the correspondence points before and after the retargeting. We describe the new vertical disparity values in the original left image by using blue line at every correspondence point. Observing the experiment data, our result introduces few vertical disparities by using Eq. (9). If the original image pair is rectified, our retargeting method will constrain the vertical disparity to be close to zero.

Fig. 6. Manipulating the disparity

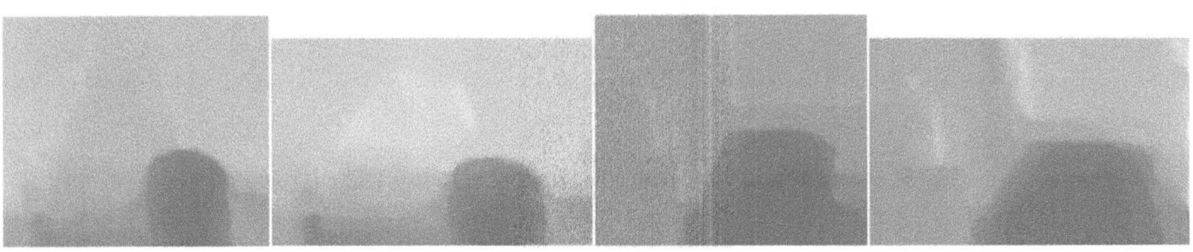

Fig. 7. Comparison of the depth maps

We show the experimental results by means of red/cyan anaglyphs in Fig.4-6. However, for the stereoscopic images, the depth maps are necessary to show their depth distributions, which are also strong demonstrations for the evaluation of stereoscopic effect. We provide the dense depth estimation by using the optical flow method [7]. Fig.7 shows the depth maps. The two images on the left are the depth maps of "Midd2" before and after retargeting, and the two on the right are that of "Lampshade2". Comparing the depth maps, we observe that the retargeted image pairs are geometrically consistent approximately, by using the well-distributed and sparse correspondence point pairs, which preserve the disparity information well. We do not find the disparity distortion and no vertical disparity is produced. Hence, the constrained disparity can provide more comfortable stereo visual experience.

5 Discussion

First, because of the characteristics of the warping-based retargeting algorithm, we have to make a trade off between getting the better result and cost more time. Segmenting the images with the smaller mesh grid can bring better retargeting result and preserve more detailed features of the original images. But the optimization process will costs much more time. Although the segmentation with the bigger mesh grid can accelerate the computation, we could not finely control the shape preservation of the small prominent contents. Hence, in our experiments, we set the edge length of grid to 20 pixels for the images and get good experimental results.

Secondly, the significance map is one of most important parts. The accurate significance values are very helpful to improve the optimization result. On the contrary, using a bad significance algorithm, maybe all of the optimization is in vain. So the quality of the retargeting result depends on the significance map in some way. Third, our method cannot produce good results when the original images contain too many structured objects, because there is no enough space is left to warp for the algorithm.

6 Conclusion

In this paper, we presented a disparity-constrained resolution adaptation method for stereoscopic images, which can retarget the images to a new resolution and preserve the depth cues at the same time. The proposed method considers image contents and depth coherence through the optimization steps. As a result of optimization, the prominent contents are deformed less, which prevents the image from distortion. And the change of disparity is constrained, which ensures the stereoscopic effect. To improve the stereoscopic visual experience, the disparity can be manipulated linearly, which provides users more freedom to custom the comfortable disparity range. We obtain the well-distributed correspondence points and use them to guide the disparity adjustment. The experimental results demonstrate the effectiveness of our method.

In future research we would improve the algorithm parallelism and decrease the computational complexity by refining the optimization method. Furthermore, since the current significance map is very dependent, we want to develop the more robust algorithms to generate the significance map.

References

1. Mansfield, A., Gehler, P., Van Gool, L., Rother, C.: Scene Carving: Scene Consistent Image Retargeting. In: Daniilidis, K., Maragos, P., Paragios, N. (eds.) ECCV 2010, Part I. LNCS, vol. 6311, pp. 143–156. Springer, Heidelberg (2010)
2. Shamir, A., Sorkine, O.: Visual Media Retargeting. ACM SIGGRAPH Asia Courses (2009)
3. Mendiburu, B.: 3D Movie Making: Stereoscopic Digital Cinema from Script to Screen. Focal Press (2009)

4. Chamaret, C., Boisson, G., Chevance, C.: Video Retargeting for Stereoscopic Content Under 3D Viewing. SPIE-IS&T Electronic Imaging (2012)
5. Wang, C., Sawchuk, A.A.: Disparity Manipulation for Stereo Images and Video. In: Proc. SPIE, Stereoscopic Displays and Applications XIX, vol. 6803, p. 68031E (2008)
6. Chang, C.H., Liang, C.K., Chuang, Y.Y.: Content-aware Display Adaptation and Interactive Editing for Stereoscopic Images. IEEE Transactions on Multimedia (2011)
7. Liu, C.: Beyond Pixels: Exploring New Representations and Applications for Motion Analysis. Massachusetts Institute of Technology (2009)
8. Liu, C.W., Huang, T.H., Chang, M.H., Lee, K.Y., Liang, C.K., Chuang, Y.Y.: 3D Cinematography Principles and Their Applications to Stereoscopic Media Processing. IEEE Transactions on Multimedia (2011)
9. Lowe, D.G.: Distinctive Image Features from Scale-invariant Keypoints. International Journal of Computer Vision (2004)
10. Banterle, F., Artusi, A., Aydin, T.O., Didyk, P., Eisemann, E., Gutierrez, D., Mantiuk, R., Myszkowski, K.: Multidimensional Image Retargeting. ACM SIGGRAPH Asia (2011)
11. Lin, H.S., Guan, S.H., Lee, C.T., Ouhyoung, M.: Stereoscopic 3D Experience Optimization Using Cropping and Warping. ACM SIGGRAPH Asia (2011)
12. Lin, H.Y., Chang, C.C., Hsieh, C.H.: Cooperative Resizing Technique for Stereo Image Pairs. In: International Conference on Software and Computer Applications (2012)
13. Kiess, J., Guthier, B., Kopf, S., Effelsberg, W.: SeamCrop for Image Retargeting. In: Proc. SPIE, Multimedia on Mobile Devices, vol. 8304 (2012)
14. Lee, K.Y., Chung, C.D., Chuang, Y.Y.: Scene Warping Layer-based Stereoscopic Image Resizing. In: IEEE Conference on Computer Vision and Pattern Recognition (2012)
15. Itti, L., Koch, C., Niebur, E.: A Model of Saliency-based Visual Attention for Rapid Scene Analysis. Pattern Analysis and Machine Intelligence (1998)
16. Fischler, M.A., Bolles, R.C.: Random Sample Consensus: a Paradigm for Model Fitting with Applications to Image Analysis and Automated Cartography. Communications of the ACM (1981)
17. Lang, M., Hornung, A., Wang, O., Poulakos, S., Gross, M.: Nonlinear Disparity Mapping for Stereoscopic 3D. ACM SIGGRAPH (2010)
18. Krähenbühl, P., Lang, M., Hornung, A.: A System for Retargeting of Streaming Video. ACM SIGGRAPH Asia (2011)
19. von Gioi, R.G., Jakubowicz, J., Morel, J.M., Randall, G.: LSD: a Line Segment Detector. Image Processing On Line (2012)
20. Koppal, S.J., Zitnick, C.L., Cohen, M.F., Kang, S.B., Ressler, B., Colburn, A.: A Viewer-centric Editor for 3D Movies. IEEE Computer Graphics and Applications (2011)
21. Sugimoto, S., Shimizu, S., Kimata, H., Kojima, A.: Multi-layered Image Retargeting. In: IEEE International Conference on Image Processing (2012)
22. Basha, T., Moses, Y., Avidan, S.: Stereo Seam Carving a Geometrically Consistent Approach. IEEE Transactions on Pattern Analysis and Machine Intelligence (2013)
23. Niu, Y., Feng, W.C., Liu, F.: Enabling Warping on Stereoscopic Images. ACM SIGGRAPH Asia (2012)
24. Wang, Y.S., Tai, C.L., Sorkine, O., Lee, T.Y.: Optimized Scale-and-stretch for Image Resizing. ACM SIGGRAPH Asia (2008)
25. Scharstein, D., Szeliski, R.: A Taxonomy and Evaluation of Dense Two-frame Stereo Correspondence Algorithms. International Journal of Computer Vision (2002)

Facial Point Detection with Occlusion Insensitive Visibility-Aware Part Model

Yuanqian Li, Yanfei Liu, and Xi Zhou

Chongqing Institute of Green and Intelligent Technology,
Chinese Academy of Sciences, 401122, Chongqing, China
{liyuanqian,liuyanfei,zhouxi}@cigit.ac.cn

Abstract. In this paper, we describe a method named visibility-aware part model for facial point detection in static images based on the pictorial structure model. A binary part visibility term is introduced to describe the occlusion state of each part, which can determine which facial points are occluded. The introduction of the term enhances the representation power of the model especially for the occlusions. The combining of the structure constrains and the powerful appearance model makes the model more robust and reduces the possibility of model crashing in some extent. Experimental results show that our proposed model can detect facial feature points accurately and robustly under occlusions.

1 Introduction

The facial feature points are the prominent landmarks around facial component such as eyebrows, eyes, nose, mouth and face contour. Facial feature point detection plays foundational roles in many face analysis tasks, such as face recognition, pose estimation and 3D face reconstruction, etc. In these tasks, Accurate and efficient facial feature point detection is desirable to make automated face analysis systems more effective.

Existing methods for facial feature point detection can be categorized into three categories: texture-based, shape-based and methods combining texture and shape.

Texture-based methods model the local texture around a given feature point, for example the pixel values in a small region around an eye corner. Typical texture-based methods include Eigenfaces [1], Elastic Bunch Graph Matching (EBGM) [2], and Local Feature Analysis (LFA) [3], etc. Shape-based methods regard all facial feature points as an holistic shape, which is learned from a set of labeled faces, and try to find the proper shape for any known face. Typical shape-based methods include Active Shape Models (ASM) [4], Active Appearance Models (AAM) [5] and 3D Morpharable Model [6]. The researchers also proposed methods combining texture and shape information. For example, [7] uses principal component analysis (PCA) on the grey level images combined with ASM, [8] extends the AAM with Constrained Local Models, and [9] applies a boosting algorithm to determine facial feature point candidates for each pixel in an input image and then uses a shape model as a filter to select the most probable position of five points.

B. Huet et al. (Eds.): PCM 2013, LNCS 8294, pp. 25–34, 2013.

Fig. 1. Illustration of typical occlusion conditions (images from the LFW database [10])

Overall, most of the existing methods are only feasible under conditions without occlusion, and few previous works have reported to be able to robustly handle large occlusions such as glasses, beards, and hair which partially cover the eyes or the mouth. As illustrated in Figure 1, occlusion is a common phenomenon in the real world. However, it is very challenging for facial feature point detection, and makes many face processing tasks based on feature point detection cumbersome, such as face recognition. On one hand, even the most sophisticated feature point detection model will crash under the condition of partial occlusions. Consequently the points without occlusion cannot be precisely detected, which seriously affects the robustness of the detection model. On the other hand, assuming the points without occlusion are detected accurately, location prediction of the occluded points is also a tough problem, although it is usually useful to many face processing tasks. Given these issues, accurate and efficient detection of facial feature points under partial occlusions remains challenging.

Pictorial structure model is an influential approach in object recognition, which decomposes the appearance of an object into local part templates, together with geometric constraints on pairs of parts [11]. Inspired by the pictorial structure model, we introduce a novel representation for modelling part-occluded faces, Visibility-aware Part Model (VPM), to address the occlusion problem. Unlike traditional models for object recognition using parts parameterized solely by location, we introduce a visibility state for each part to describe the occlusion state of the part, which can determine which of the facial points are occluded. Firstly, the introduction of the visibility state makes the appearance model more powerful to represent the occlusions, which allows our proposed model to predict the occluded points more reliable. Secondly, since our model is motivated by the pictorial structure, in our model the feature point is defined to be a spatial relation between two facial feature points rather than being a facial feature point itself. The model combining of the structural constrain and the appearance model of powerful representation ability will automatically produce corrected point configurations that preserve well-estimated points. Extensive experiments on facial point detection show that the proposed model can localize facial feature points accurately even under severe occlusions.

The paper is structured as follows. Section 2 proposes our visibility-aware part model. Section 3 presents the algorithm for learning model parameters and the method for fitting the learned model to test images. Section 4 discusses the experimental results. The paper concludes in Section 5.

2 Model

In the Pictorial Structure an object is first decomposed into parts and then the best part candidates are searched subject to some spatial constraints such that the likelihood of generating the concerned image is maximized. The pictorial structure model can usually be expressed as an N-node relational but undirected graph $G=(Pt,E)$, where $Pt=\{Pt_1,Pt_2,...,Pt_N\}$ corresponds to the N parts and the edges $E=\{(Pt_i,Pt_j),i\neq j\}$ specify which pairs of parts have consistent relations. Let θ to be a set of parameters that define an object model, I denote an image, and $L=(l_1,l_2,...,l_N)$ denote a configuration of the object, where each l_i specifies the location of each part Pt_i on the image plane. To infer the locations of the parts of an object from this model, we can search for the maximum a posterior $p(L|I,\theta)$, i.e., the probability that a face configuration is L given the model θ and an image I. Using Bayes rule, the posterior can be written as

$$P(L|I,\theta)\propto p(I|L,\theta)p(L|\theta) \tag{1}$$

where $p(I|L,\theta)$ is the generative model of the appearance and $p(L|\theta)$ measures the prior probability that a face appears at the location L. In general, the model parameter is denoted by $\theta=(u,c)$, where u expresses the appearance while c expresses structural constraints on edges. This provides the opportunity to predict the location of occluded points by reconstructing new model. Many objects, including faces and people can be represented by such multi-part models in its simplicity and generality, thus, the pictorial structure formulation is appealing in many object recognition fields [12-15].

Specially in our facial feature point detection task, we model every facial feature point as a part Pt_i. For the purpose of precise detection of feature points under occlusions, we introduce a binary visibility state term $s=\{s_1,s_2,\cdots,s_N,\}$, $s_i\in\{0,1\}$ to present whether the part is occluded or not, where $s_i=1$ denotes the part is visible while $s_i=0$ denotes the part is not visible or occluded. Thus, the posterior can be written as $P(L,s|I,\theta)$ in our model,

$$P(L,s|I,\theta)\propto p(I|L,s,\theta)p(L,s|\theta) \tag{2}$$

Our visibility-aware part model is parameterized by $\theta=(u,c)$, where $u\in\{u_1,u_2,\cdots,u_N\}$ are appearance parameters, and $c=\{c_{ij}\mid(Pt_i,Pt_j)\in E\}$ are connection parameters which denote the spatial relationships between parts. Obviously, the first term in Eq. (2) depends only on the appearance of the parts, while the second term depends only on the connection parameters. Assuming that the parts are statistically independent for appearance, we have

$$p(I\mid L,s,\theta)=p(I\mid L,s,u)=\prod_{i=1}^{N}p(I\mid l_i,s_i,u_i) \tag{3}$$

The appearance of each part can be modelled by unimodal Gaussian distribution $p(I\mid l_i,u_i)\propto\mathcal{N}\left(\alpha(l_i),\mu_i,\Sigma_i\right)$, where $u_i=(\mu_i,\Sigma_i)$ and $\alpha(l_i)$ is a high-dimensional feature vector of an image patch centered at the position l_i in [13], which collects all the responses of a set of filters of different scales at the point l_i. Then we have

$$\prod_{i=1}^{N}p(I\mid l_i,s_i,u_i)=\prod_{i=1}^{N}(s_i\cdot\mathcal{N}\left(\alpha(l_i),\mu_i,\Sigma_i\right)+(1-s_i)\cdot\mathcal{N}(\alpha(l_i),\mu_i^{'},\Sigma_i^{'})) \tag{4}$$

And in our model, $u_i=(\mu_i,\Sigma_i,\mu_i^{'},\Sigma_i^{'})$, (μ_i,Σ_i) is the appearance parameter of unoccluded point and $(\mu_i^{'},\Sigma_i^{'})$ denotes the appearance parameter of point under occluded state. The value of s_i determines which distribution is used as the detecting model. The appearance parameters $(\mu_i^{'},\Sigma_i^{'})$ can be learned from the information provided by occluded example images. Obviously, the introduction of s_i enhances the representation ability of the appearance model and makes the model more robust to occlusions.

With a similar independent assumption on the edge constraints between components, we have the structure model

$$p(L,s\mid\theta)=p(L,s\mid c)=\prod_{(Pt_i,Pt_j)\in E}p(l_i,l_j,s_i,s_j\mid c_{ij}) \tag{5}$$

Similarly, the spatial relationships between pairs of parts can also be modelled by Gaussian distribution $p(l_i,l_j\mid c_{ij})\propto\mathcal{N}\left(l_i-l_j,\mu_{ij},\Sigma_{ij}\right)$ [13], where $c_{ij}=\left(\mu_{ij},\Sigma_{ij}\right)$. The spatial relationships between pairs of parts will not change no matter if the points are occluded or not. Therefore, the structure model has nothing relationship with s_i and s_j, and can then be rewritten as

$$\begin{aligned}&\prod_{(Pt_i,Pt_j)\in E} p(l_i,l_j,s_i,s_j \mid c_{ij})\\&=\prod_{(Pt_i,Pt_j)\in E} p(l_i,l_j \mid c_{ij})=\prod_{(Pt_i,Pt_j)\in E}\mathcal{N}\left(l_i-l_j,\mu_{ij},\Sigma_{ij}\right)\end{aligned}\tag{6}$$

Plugging Eq. (3), Eq. (4), Eq. (5) and Eq. (6) into Eq. (2), we get the global objective function

$$\begin{aligned}&p(L,s \mid I,\theta)\propto\{\prod_{i=1}^{N}p(I \mid l_i,s_i,u_i)\prod_{(Pt_i,Pt_j)\in E}p(l_i,l_j,s_i,s_j \mid c_{ij})\}\\&=\prod_{i=1}^{N}(s_i\cdot\mathcal{N}\left(\alpha\left(l_i\right),\mu_i,\Sigma_i\right)+\left(1-s_i\right)\cdot\mathcal{N}(\alpha\left(l_i\right),\mu_i^{'},\Sigma_i^{'}))\cdot\\&\prod_{(Pt_i,Pt_j)\in E}\mathcal{N}\left(l_i-l_j,\mu_{ij},\Sigma_{ij}\right)\end{aligned}\tag{7}$$

Facial feature points are frequently occluded by beard, hair, glasses and other objects such as tennis ball. Since occlusions often do not happen at random, the locations of occluded points (parts) may have consistent appearance. We model occlusions by learning separate appearance parameters $(\mu_i^{'},\Sigma_i^{'})$ for occluded points.

3 Learning and Inference

3.1 Learning Model Parameters

Suppose we have a set of M example images $\{I^1,I^2,\cdots,I^M\}$ including images with various face-partial occlusions and also images without any occlusion on the face. The corresponding feature point locations $\{L^1,L^2,\cdots,L^M\}$ and visibility states $\{s^1,s^2,\cdots,s^M\}$ are also labeled for these training images. By definition, the maximum likelihood estimate of θ is the value θ^* that maximizes $p(L^1,\cdots,L^M,s^1,\cdots,s^M \mid I^1,\cdots I^M,\theta)$. Assuming each example is generated independently, it can be rewritten as

$$\begin{aligned}\theta^*&=\arg\max_{\theta}\prod_{k=1}^{M}p(L^k,s^k \mid I^k,\theta)\\&=\arg\max_{\theta}\prod_{k=1}^{M}p(I^k \mid L^k,s^k,\theta)\prod_{k=1}^{M}p(L^k,s^k \mid \theta)\end{aligned}\tag{8}$$

The first term in this equation depends only on the appearance of the parts, while the second term depends only on the connection parameters. As a consequence, we can use this framework as long as there is a maximum likelihood (ML) estimation procedure for learning the model parameters for a single part from training images.

For the appearance parameters u, we have

$$u^* = \arg\max_{u} \prod_{k=1}^{M} p(I^k \mid L^k, s^k, u) \tag{9}$$

From Eq. (3) and Eq. (4), we get

$$\begin{aligned} u_i^* &= \arg\max_{u_i} \prod_{k=1}^{M} p(I^k \mid l_i^k, s_i^k, u_i) \\ &= \arg\max_{\mu_i, \Sigma_i, \mu_i', \Sigma_i'} \prod_{k=1}^{M} (s_i^k \cdot \mathcal{N}\left(\alpha\left(l_i^k\right), \mu_i, \Sigma_i\right) + (1 - s_i^k) \cdot \mathcal{N}(\alpha\left(l_i^k\right), \mu_i', \Sigma_i')) \end{aligned} \tag{10}$$

This is exactly the ML estimation of the appearance parameters for part Pt_i, given independent examples $\{(I^1, l_i^1, s_i^1), (I^2, l_i^2, s_i^2), \cdots, (I^M, l_i^M, s_i^M)\}$. Similarly for the connection parameters c, we have

$$c^* = \arg\max_{c} \prod_{k=1}^{M} p(L^k, s^k \mid c) \tag{11}$$

From Eq. (5) and Eq. (6), we get

$$c_{ij}^* = \arg\max_{c_{ij}} p(l_i^k, l_j^k, s_i^k, s_j^k \mid c_{ij}) = \arg\max_{\mu_{ij}, \Sigma_{ij}} \prod_{k=1}^{M} \mathcal{N}\left(l_i^k - l_j^k, \mu_{ij}, \Sigma_{ij}\right) \tag{12}$$

This is the ML estimation for the joint distribution of l_i and l_j, given independent training examples $\{(I^1, l_i^1, s_i^1), (I^2, l_i^2, s_i^2), \cdots, (I^M, l_i^M, s_i^M)\}$. Learning our model involves picking labeled landmarks on a number of human face. Then using the ML estimation procedure, the appearance models for each part and spatial relationships between parts are automatically estimated from these training examples.

3.2 Inference

Inference corresponds to maximizing $p(L \mid I, \theta)$ from Eq. (6) given learned parameters $\theta = (u^*, c^*)$. Simply enumerate all locations and all values of s (0 or 1), and find the best configuration of parts.

$$\begin{aligned} L^* &= \arg\max_{L} p(L \mid I, \theta) \\ &= \arg\min_{L} \Big(\sum_{i=1}^{N} -\log p(I \mid l_i, s_i, u_i) + \sum_{(Pt_i, Pt_j) \in E} -\log p(l_i, l_j, s_i, s_j \mid c_{ij})\Big) \end{aligned} \tag{13}$$

Where $-\log p(I \mid l_i, s_i, u_i)$ is a match cost measuring how well part Pt_i matches the image data at location l_i, and $-\log p(l_i, l_j, s_i, s_j \mid c_{ij})$ is a deformation cost measuring how well the relative locations of the part Pt_i and Pt_j agree with the deformable model. In this work, we use a tree structure to construct the model. Therefore, the dynamic programming (DP) algorithm [13] can be adopted to find a location L with maximum posterior probability, which is used to detect facial feature points in novel images. We omit the message passing equations for lack of space.

4 Experiments

We firstly test how well the database copes with occlusions. Then, we perform a benchmark comparison of our proposed method with the existing state of the art.

The images are selected from LFW [10] and CIGIT (Chongqing Institute for Green and Intelligent Technology) databases to form a new database including images without occlusion and images with many occlusions, such as the mouth area occluded by beards and the eye areas occluded by glasses and hair, etc. Then this database is taken as the training set of our proposed algorithm. The CIGIT dataset is being collected by our team as a continual project, whose goal is to simulate the partially-controlled surveillance scenarios, including 91 poses, 6 facial movements, 4 kinds of occlusions, and 5 sets of combined indoor, controlled and outdoor, uncontrolled lighting. The LFW and CIGIT dataset not only have many occlusions, but also cover almost all the complex conditions in the real world. Thus, we select images from these two datasets to train our model.

To evaluate the precision of feature point detection, we adopt the most popular measure m_{e17} proposed by [8], which is a mean error over all internal points (17 points). And to compare our work with the current state of the art, we evaluated our method on the BioID database [16] which is one of the benchmark databases used by most facial point detection works. We compare our approach with the traditional pictorial structure model (PSM) [13], and the constrained local model (CLM) [8].

Figure 2 plots the cumulative error distribution curves of the compared methods, where the horizontal axis is the normalized Euclidean distance (m_{e17}), while the vertical axis is the cumulative localization score, showing the percentage of images that have been successfully processed corresponding to the detection error. As expected, our proposed method outperforms the traditional PSM and CLM.

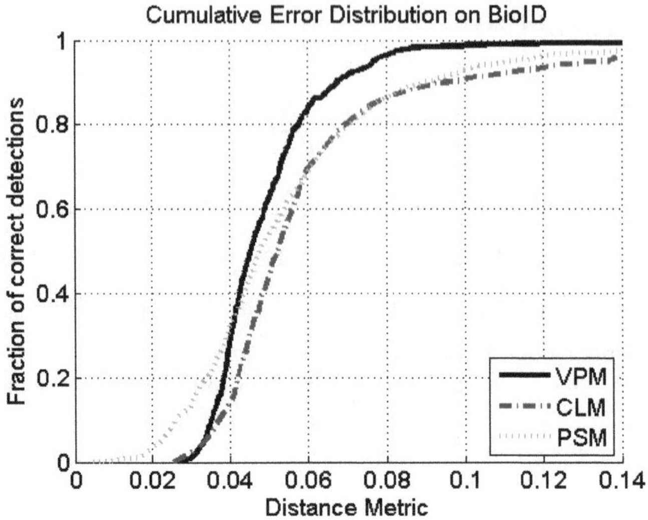

Fig. 2. Comparing the cumulative error distribution curves

Figure 3, Figure 4 and Figure 5 illustrate detection results on images with partially occluded faces. As shown in Figure 3, Figure 4, and Figure 5, the detection algorithm based on our model automatically handles partial occlusion in a robust way.

Fig. 3. Detection results on LFW database

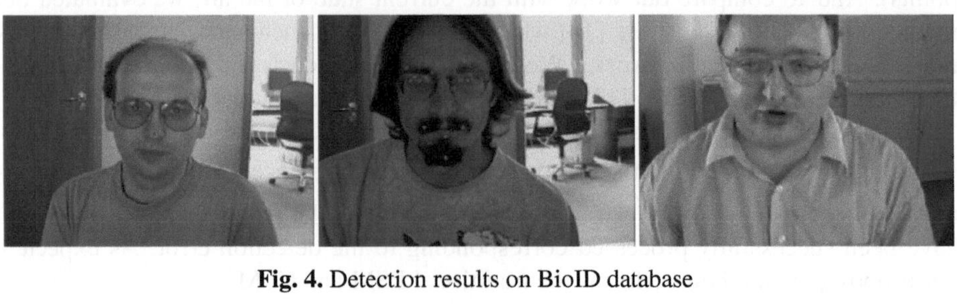

Fig. 4. Detection results on BioID database

Fig. 5. Detection results on Yamaha database

To make this clear, we tabulate the percentage of successful localization subject to $m_{e17} < 0.05$ and $m_{e17} < 0.1$ in Table 1. It can be observed that the successful detection rate is promoted from at most 61.79% to 89.5% at $m_{e17} < 0.05$, and our algorithm achieves the best correct localization rate of 98.15% at $m_{e17} < 0.1$. Obviously, our method provides substantial performance compared to other methods.

Table 1. Percentages of successful detection to $m_{e17} < 0.05$ and $m_{e17} < 0.1$, respectively

Method	$m_{e17} < 0.05$	$m_{e17} < 0.1$
Traditional PSM	54.36%	92.91%
CLM	43.33%	90.89%
EASM	61.79%	97.32%
MKL-SVM	45.26%	92.5%
Our method	89.5%	98.15%

5 Conclusion

Aiming at resolving the problem of occlusion, we presented a model for facial point detection in this paper, which we refer to as the visibility-aware part model. Instead of modelling an object using parts parameterized solely by location, we introduce a "visibility" for each part to describe the occlusion state of the part. The introduction of the "visibility" provides us the accurate detection of facial feature points under occlusion. We show that the proposed visibility-aware part model can obtain better facial feature point detection results under partial occlusion. We also show that our proposed method outperforms the traditional PSM and CLM method when applied to the BioID data sets.

Acknowledgement. This research was supported by two projects from committee on science and technology of Chongqing, with Grant No. cstc2011ggC40009 and cstc2012gg-sfgc0079, respectively.

References

[1] Turk, M., Pentland, A.: Eigenfaces for Recognition. Journal of Cognitive Neuroscience 3, 71–86 (1991)

[2] Wiskott, L., Fellous, J.M., Kuiger, N., von der Malsburg, C.: Face recognition by elastic bunch graph matching. IEEE Transactions on Pattern Analysis and Machine Intelligence 19, 775–779 (1997)

[3] Penev, P.S., Atick, J.J.: Local feature analysis: a general statistical theory for object representation. Network: Computation in Neural Systems 7, 477–500 (1996)

[4] Cootes, T.F., Taylor, C.J., Cooper, D.H., Graham, J.: Active Shape Models-Their Training and Application. Computer Vision and Image Understanding 61, 38–59 (1995)

[5] Cootes, T.F., Edwards, G.J., Taylor, C.J.: Active appearance models. IEEE Transactions on Pattern Analysis and Machine Intelligence 23, 681–685 (2001)

[6] Blanz, V., Vetter, T.: Face recognition based on fitting a 3D morphable model. IEEE Transactions on Pattern Analysis and Machine Intelligence 25, 1063–1074 (2003)

[7] Milborrow, S., Nicolls, F.: Locating facial features with an extended active shape model. In: Forsyth, D., Torr, P., Zisserman, A. (eds.) ECCV 2008, Part IV. LNCS, vol. 5305, pp. 504–513. Springer, Heidelberg (2008)

[8] Cristinacce, D., Cootes, T.: Feature detection and tracking with constrained local models. In: Proc. British Machine Vision Conference, pp. 929–938 (2006)

[9] Chen, L., Zhang, L., Zhang, H., Abdel-Mottaleb, M.: 3D shape constraint for facial feature localization using probabilistic-like output. In: Sixth IEEE International Conference on Automatic Face and Gesture Recognition, pp. 302–307 (2004)

[10] Huang, G.B., Mattar, M., Berg, T., Learned-Miller, E.: Labeled Faces in the Wild: A Database for Studying Face Recognition in Unconstrained Environments. In: Workshop on Faces in 'Real-Life'Images: Detection, Alignment, and Recognition (2008)

[11] Fischler, M.A., Elschlager, R.A.: The representation and matching of pictorial structures. IEEE Transactions on Computers 100, 67–92 (1973)

[12] Felzenszwalb, P.F., Girshick, R.B., McAllester, D., Ramanan, D.: Object detection with discriminatively trained part-based models. IEEE Transactions on Pattern Analysis and Machine Intelligence 32, 1627–1645 (2010)

[13] Felzenszwalb, P.F., Huttenlocher, D.P.: Pictorial structures for object recognition. International Journal of Computer Vision 61, 55–79 (2005)

[14] Yang, Y., Ramanan, D.: Articulated pose estimation with flexible mixtures-of-parts. In: IEEE Conference on Computer Vision and Pattern Recognition (CVPR), pp. 1385–1392 (2011)

[15] Felzenszwalb, P.F., Girshick, R.B., McAllester, D.: Cascade object detection with deformable part models. In: IEEE Conference on Computer Vision and Pattern Recognition (CVPR), pp. 2241–2248 (2010)

[16] Jesorsky, O., Kirchberg, K.J., Frischholz, R.W.: Robust face detection using the hausdorff distance. In: Audio-and Video-Based Biometric Person Authentication, pp. 90–95 (2001)

3-D Geometry Enhanced Superpixels for RGB-D Data

Jingyu Yang, Ziqiao Gan, Xiaolei Gui, Kun Li, and Chunping Hou

Department of Electronic Information Engineering, Tianjin University, China
{yjy,lik,hcp}@tju.edu.cn, {xiaoqiao086,gxl0616}@gmail.com

Abstract. This paper introduces a novel 3-D geometry enhanced superpixels for RGB-D data. First, we reconstruct the 3-D geometry of the scene by projecting the depth map into 3-D coordinates. Then, a distance metric for superpixel clustering is constructed using 3-D geometry and color information. Finally, pixels are iteratively clustered into superpixels using the proposed distance metric. The proposed method is able to distinguish objects in similar colors due to the introduced 3-D geometry. The oversegmentation results on RGB-D pairs in the Middlebury datasets demonstrate that our approach shows better performance than other three state-of-the-art superpixel methods. The proposed superpixels are also evaluated in the application of segmentation, and we achieve the best segmentation results compared with three state-of-the-art segmentation methods.

Keywords: RGB-D, Superpixels, Segmentation.

1 Introduction

In recent years, many depth sensing technologies, e.g., structured light and time of flight (ToF), are developed beyond laser scanning, and commodity real-time depth camera such as PMD Camcube and Microsoft Kinect are available. Such depth sensing devices are often mounted with color cameras in practical applications, creating color-depth data (referred to as RGB-D data hereafter). RGB-D data provide a more comprehensive description of the captured scene by providing 3D geometry information in addition to the texture information. Therefore, depth cameras have been successfully used in many fields such as object recognition [1], scene analysis [2,3], and 3-D reconstruction [4,5]. In many such applications, segmentation is required to separate interested regions or objects to facilitate the subsequent processing. There exist some segmentation methods of RGB-D data [6,7], but they heavily rely on the geometry of the indoor scenes, e.g., planar objects. In image segmentation, images are often first divided into small regions (called *superpixels*, a kind of over-segmentation) to reduce the complexity of segmentation. However, existing superpixel methods are designed for color or grey images, and thus they cannot fully exploit the strong structural correlation of RGB-D data as shown in examples in Fig. 1(a) and 1(b).

B. Huet et al. (Eds.): PCM 2013, LNCS 8294, pp. 35–46, 2013.

Superpixel algorithms can be mainly classified into two categories: graph-based and gradient-ascent-based. In graph-based algorithms, a cost function is defined on a graph and minimized to generate superpixels. Each pixel is treated as a node and its edge weight contributes to the cost function. Normalized cut [8] and superpixel lattice [9] are two representative graph-based methods. Superpixel lattice produces superpixels that are forced to conform to a grid (a regular superpixel lattice). Despite this added topological constraint, this algorithm is comparable to other graph-based superpixel methods in terms of speed and accuracy.

In gradient-ascent-based algorithms, a rough initial clustering (e.g. rectangular tiling) is first constructed and then refined recursively. Examples of this kind are Quick-shift [10] and SLIC [11]. Quick-shift is a mode-seeking segmentation scheme but cannot explicitly control the size or number of superpixels. SLIC achieves promising superpixel segmentation results by a K-means-like clustering method based on a distance metric. The number of superpixels can be flexibly controlled in SLIC. Previous superpixel methods are mainly designed for RGB data. With only RGB data, it is difficult to distinguish adjacent objects with similar colors, which generates semantically-incoherent superpixels as shown in Fig. 1 (a) and (b).

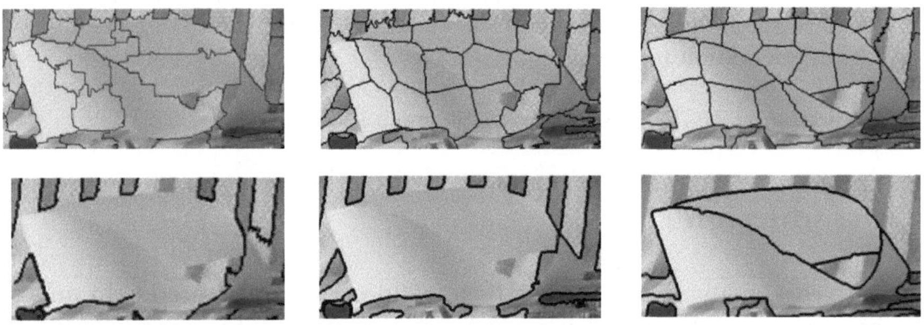

(a) **(b)** **(c)**

Fig. 1. Visual comparison of the superpixels (at the top) and their merging results (at the bottom). (a): Quick-shift [10] with max distance 10. (b): SLIC [11] using compactness 10. (c): our method with compactness 10.

This paper proposes a superpixel method for RGB-D data with 3-D geometry information, to address the difficulty in distinguishing adjacent objects in similar colors. First, we project the depth map into 3-D space to get the 3-D coordinates. Then, we calculate the weights for 2-D position and 3-D coordinates to construct an 8-D distance metric. Finally, we iteratively cluster pixels based on the proposed distance metric. As the example shown in Fig. 1 (c), the over-segmentation results on Middlebury Stereo Datasets [12] demonstrate that our approach can distinguish adjacent objects in similar colors and achieve a more semantically-coherent result. In addition, we apply our superpixels to segmentation and obtain more object-level-consistent results compared with other three segmentation algorithms.

The rest of this paper is organized as follows. Our method is introduced in Section 2. Section 3 evaluates the proposed method with both over-segmentation results and segmentation results. Finally, the paper is concluded in Section 4.

2 Our Method

In Sec. 2.1, we first give the motivation of our work by analyzing the correlations between color images and their associated depth maps. Then, we present the 3-D geometry reconstruction from depth map in Sec. 2.2, and the proposed distant metric in Sec. 2.3. Finally, the proposed algorithm is detailed in Sec. 2.4.

2.1 Color-Depth Correlations

In RGB-D data, a color image and the associated depth map capture different information of the same scene. However, they have strong structural correlations as shown in Fig. 2(a) and 2(b). Many edges and contours in color images and depth maps are consistent with each other, e.g., the region highlighted with a red rectangle in Fig. 2. However, as the region marked by the blue rectangle in Fig. 2, adjacent objects may have very similar colors, but are well distinguished by their position in the 3-D space. Therefore, superpixel methods use only color information, such as Turbopixels[13], Normalized cut [8], and SLIC [11], can hardly distinguish adjacent objects in similar colors. This motivates us to generate more semantic-consistent superpixels by incorporating 3-D geometry with color information based on the SLIC method.

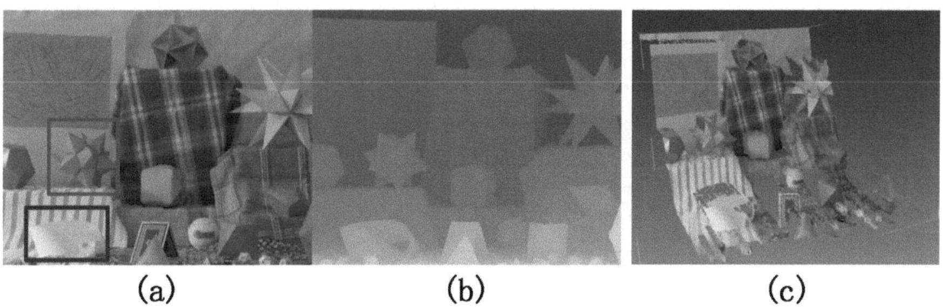

Fig. 2. A RGB-D dataset: (a) color image, (b) depth map, and (c) its 3-D reconstruction

2.2 3-D Geometry Reconstruction

Without loss of generality, assume that the world coordinate system coincides with the camera coordinate system. Given the internal parameters, the 3-D geometry of the captured scene can be reconstructed from the depth map. Denote by (x, y) the location of a pixel, and by d its depth. The 3-D coordinate of the pixel, denoted by (X, Y, Z), can be obtained as

$$\begin{bmatrix} X \\ Y \\ Z \\ 1 \end{bmatrix} = \begin{bmatrix} f_x & \gamma & 0 & x_0 \\ 0 & f_y & 0 & y_0 \\ 0 & 0 & 1 & 0 \\ 0 & 0 & 0 & 1 \end{bmatrix}^{-1} \begin{bmatrix} x \\ y \\ d \\ 1 \end{bmatrix} \tag{1}$$

where f_x and f_y are focal length of the x-axis and y-axis, respectively, γ is the skew ratio between the two axes, and (x_0, y_0) is the principle point.

2.3 Distance Metric

Our approach clusters pixels by both their color similarity and space relativity. Therefore, we perform segmentation in the eight-dimensional (8-D) [Lab-XYZ-xy] space, where [Lab] means the pixel color vector in CIELAB color space, and [XYZ] stands for the reconstructed 3-D position, and [xy] denotes the pixel coordinate in the image plane. As the eight parameters describe two different kinds of information, it is not possible to model the distance between different pixels simply using Euclidean distance in the 8-D space. We normalize the Lab, XYZ, and xy data into the range of [0, 1], respectively. Denote by D_s the distance between the ith and jth pixels. Then, the distance D_s is designed as the fusion of the distances in three space, i.e., color distance, 3-D space distance and 2-D distance:

$$D_{\mathrm{s}} = d_{\mathrm{Lab}} + \alpha d_{\mathrm{XYZ}} + \beta d_{\mathrm{xy}} \tag{2}$$

where d_{Lab}, d_{XYZ}, and d_{xy} are the three distances defined in Eq. (3), respectively, and α and β are the weights to trade off the importance of the distance in each space.

$$\begin{aligned} d_{\mathrm{Lab}} &= \sqrt{(L_i - L_j)^2 + (a_i - a_j)^2 + (b_i - b_j)^2} \\ d_{\mathrm{XYZ}} &= \sqrt{(X_i - X_j)^2 + (Y_i - Y_j)^2 + (Z_i - Z_j)^2} \\ d_{\mathrm{xy}} &= \sqrt{(x_i - x_j)^2 + (y_i - y_j)^2} \end{aligned} \tag{3}$$

The weighting parameter β in Eq. (2) is given by

$$\beta = \frac{m}{s} \tag{4}$$

where m is a compactness parameter depending on image complexity. $s = \sqrt{N/K}$ is the interval between two adjacent initial superpixels, where N is total number of pixels in the image and K is the output number of superpixels. For a given image, α is a fixed value and β depends on m and K.

In the designed distance metric, we incorporate the 2-D distance d_{xy} in the image plane besides the distance d_{XYZ} in 3-D space. The 3-D geometry is reconstructed from the depth information. Due to the imperfect depth sensing, the transitions between adjacent depth layers of different depths are not ideal jumps, but present some kinds of gradual changes. This would results in some intermediate 3-D points between the two distinct objects when the 2D points are mapped into the 3-D space using the depth map. We observe that the boundaries of segments tend to be jaggy due to the the presence of the noisy 3-D points. Therefore, we use the 2-D distance d_{xy} in the image plane to weaken the interference of noisy 3-D points.

2.4 Algorithm

We firstly calculate the 3-D coordinates and the weights, α and β, for our distance metric. The color image and depth maps are divided into non-overlapped rectangular regions as the initial superpixels. Then, the initial superpixels are iteratively refined by the K-means with the proposed distance metric until converges, similar to SLIC [11]. Finally, we enforce connectivity of the generated superpixels. The proposed algorithm is summarized in Algorithm 1.

Algorithm 1. Proposed Algorithm

1: Project depth map into 3-D space and calculate the 3-D coordinates (Eq. (1))
2: Choose the suitable weights α and β (Eq. (4))
3: Initialize K cluster centers
   $C_k = [L_k, a_k, b_k, X_k, Y_k, Z_k, x_k, y_k]^T$, by sampling pixels at regular grid steps S.
4: **repeat**
5:    **for** each cluster center C_k **do**
6:       Assign the best matching pixels from a $2S \times 2S$ square neighborhood around the cluster center according to the distance measure (Eq. (2) $\sim$ Eq. (3)).
7:    **end for**
8:    Update cluster centers
9: **until** convergence
10: Enforce connectivity.

3 Experimental Results

3.1 Evaluation of Our Superpixels

In this section, the proposed method is evaluated on a set of six Middlebury depth-texture pairs, *Art, Book, Moebius, Laundry, Baby,* and *Plastic*, as shown in Fig. 3, and is compared with three representative superpixels: Quick-shift [10], Turbopixels [13] and SLIC [11]. In our experiment, we set the compactness to

Fig. 3. Test images in our experiment. From left to right: the color image with segmentation ground-truth(top) and its depth map(bottom) for images: *Art*, *Book*, *Moebius*, *Laundry*, *Baby*, *Plastic*.

10 and the number of superpixels to 600 for both SLIC and our method, and the same number of superpixels for Turbopixels. As the Quick-shift algorithm can not control the output number directly, we set the parameter maximal distance at 10, which produces the best results according to our observation for the test dataset. For quantitative evaluation, the ground-truth segmentations of the dataset are obtained by manual segmentation.

Superpixel segments obtained by the four methods are shown in Fig. 4. The results show that our method outperforms other methods in regions which contain salient depth discontinuities. Take the *Art* in Fig. 4 for example, the barrels of art brushes (at the bottom-right part) have similar colors to the background. Around the barrels, the superpixels obtained by the three compared methods mix up with background pixels since they solely use color information. On the contrary, our method is able to generate coherent superpixels thanks to the incorporation of 3-D geometrical information. For regions with smooth depth, our method is comparable to other methods.

In addition to the subjective results in Fig. 4, we calculate the boundary precision and accuracy to evaluate the superpixel methods using the method in [11]. The precision is defined as the hit rate of boundaries, and the accuracy is calculated as the proportion of superpixels that lie in the groundtruth segments. The precision and accuracy of superpixel segments with respect to the number of superpixels are shown in Fig. 5. As K-means cannot control the number of superpixels, we compare our method with Turbopixels [13] and SLIC [11]. The reported quantitative results in Fig. 5 are consistent with the results in Fig. 4. Our method significantly outperforms other two methods in accuracy for depth-texture pairs containing rich depth edges such as *Art* and *Moebius*. For example, our method has a higher accuracy than SLIC by up to 18% for *Art*. For the precision, our method is significantly better than Turbopixels and competitive to SLIC. We observe that our method has a slightly lower precision than SLIC for *Book* and *Laundry*. This may be because the depth around the object boundaries

Fig. 4. Visual comparison of superpixels for image *Art*, *Book*, *Moebius*, *Laundry*, *Baby*, and *Plastic*. For each row, images from left to right are obtained by Quick-shift , Turbopixels, SLIC and our method.

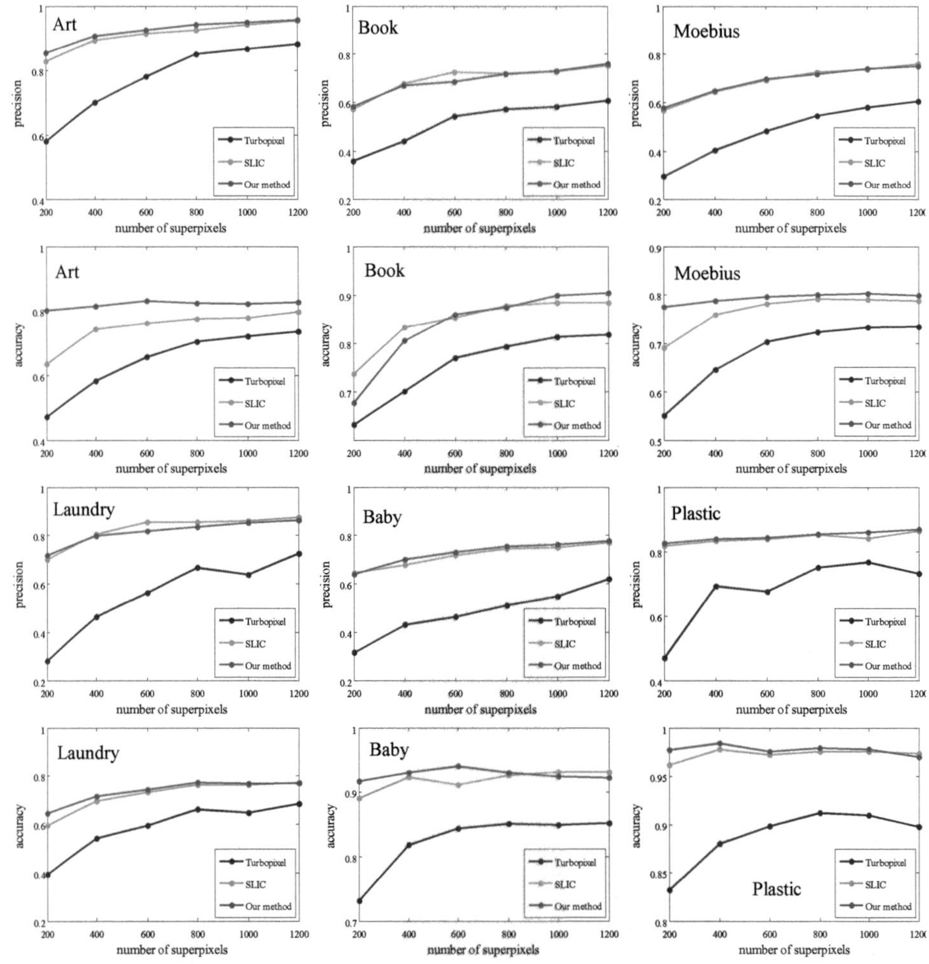

Fig. 5. Evaluation of superpixels for image *Art*, *Book*, *Moebius*, *Laundry*, *Baby*, *Plastic*. Row 1 and Row 3: the precision result. Row 2 and Row 4: the accuracy result.

are too flat to make a good distinction. Both the visual quality in Fig. 4 and the quantitative results in Fig. 5 show that the proposed method produces more sematically-coherent superpixels for RGB-D data, which is quite useful in other applications as shown in the following section.

3.2 Image Segmentation Results

Superpixels are useful in many applications such as detection [14], segmentation [15] and localization [16,17]. In this section, we evaluate the proposed superpixel method in the segmentation of RGB-D data. For segmentation, the obtained superpixels are merged into larger sematic regions in a bottom-up manner. Specifically, starting from the obtained superpixels, the similarity of each pair of adjacent segments are measured by both color intensity and 3-D distance. Then, adjacent regions with high similarity are merged into a larger segment. The merging procedure is repeated until convergence.

Table 1. Performance comparison of precision(Pre.) and recall(Rec.) with efficient graph-based method [18], multiscale spectral method [19], and contour detection and hierarchical segmentation method [20], for six test images from Middlebury Dataset [12]

	Art		*Book*		*Moebius*		*Laundry*		*Baby*		*Plastic*	
method	Pre.	Rec.	Pre.	Rec.	Pre.	Rec.	Pre.	Rec.	Pre.	Rec.	Pre.	Rec.
[18]	0.68	0.51	0.64	0.37	0.64	0.39	0.72	0.61	0.72	0.22	0.81	0.25
[19]	0.39	**0.60**	0.48	**0.43**	0.38	0.42	0.45	**0.65**	0.50	0.31	0.57	0.34
[20]	0.66	0.56	0.57	0.35	0.60	0.39	0.56	0.50	0.61	0.38	0.79	0.54
ours	**0.85**	0.55	**0.73**	0.36	**0.72**	**0.53**	**0.90**	0.63	**0.85**	**0.72**	**0.94**	**0.58**

In Fig. 6, our segmentation results are compared with those produced by the efficient graph-based method [18], multiscale spectral [19] and contour detection and hierarchical segmentation method [20]. As shown in Fig. 6, the efficient graph-based method tends to give oversegmented results; many segments provided by multiscale spectral covers different sematic objects; and the method in [20] generates undersegmented results. Our method generates more accurate segmentation results thanks to the use of 3-D geometrical information. However, the proposed method is limited in distinguishing adjacent objects in similar colors and depths since it is hard to find a suitable threshold for merging.

In Table 1, we further evaluate the segmentation results in Fig. 6 using the precision and boundary recall metrics. As shown in Table 1, using only a simple merging strategy, our method outperforms other three methods in precision by 15% on average. This suggests that the segmentations benefit a lot from the semantically-coherent superpixels. We observe that our method is slightly lower

Fig. 6. Segmentation results. For each column, from top to bottom: the original image, efficient graph-based result [18], multiscale spectral result [19], contour detection and hierarchical segmentation result [20] and our superpixels merging result.

than the multiscale spectral [19] in accuracy for *Art*, *Book* and *Laundry*. The reason may be that our merging method only considering the adjacent superpixels, remaining the bitty regions and redundant edges in segmentation result. We believe that a better merging method would further improve the segmentation results. In summary, the visualization in Fig. 6 and the quantitative results in Table 1 demonstrate that the proposed superpixel methods significantly boost the segmentation performance due to its capability to create semantically-coherent superpixels.

4 Conclusion

We present a novel 3-D geometry enhanced superpixel algorithm for RGB-D data. The depth map is first converted into 3D geometrical information. Then, superpixels are iteratively clustered according to a distance metric designed from the color information and 3-D geometry. By introducing the geometrical information, the proposed superpixel method overcomes the difficulty in distinguishing adjacent objects with similar colors. Both the quantitative results and visual comparisons demonstrate that the proposed method is able to generate more semantically-coherent superpixels than several state-of-the-art methods. The successful application to RGB-D segmentation verified the effectiveness of the proposed superpixels. The reported results are preliminary since the available depth-texture dataset are created by a 3-D laser scanner from small-scale scenes. The effectiveness of the proposed method would be even more prominent when applied to RGB-D data from real generic scenes, which is considered as our further work.

Acknowledgement. This work is supported by the General Program (No. 61072062) and Key Program (No. 60932007) of NSFC, the Doctoral Fund (20110032110029) and the NCET Program (No. NCET-11-0376) from the Ministry of Education of China and the Tianjin Research Program of Application Foundation and Advanced Technology (No. 12JCYBJC10300).

References

1. Lai, K., Bo, L., Ren, X., Fox, D.: Sparse distance learning for object recognition combining RGB and depth information. In: ICRA, pp. 4007–4013 (2011)
2. Herbst, E., Ren, X., Fox, D.: RGB-D object discovery via multi-scene analysis. In: Intelligent Robots and Systems, pp. 4850–4856 (2011)
3. Ren, X., Bo, L., Fox, D.: RGB-D scene labeling: Features and algorithms. In: CVPR, pp. 2759–2766 (2012)
4. Henry, P., Krainin, M., Herbst, E., Ren, X., Fox, D.: RGB-D mapping: Using depth cameras for dense 3D modeling of indoor environments. In: The 12th International Symposium on Experimental Robotics, vol. 20, pp. 22–25 (2010)

5. Izadi, S., Kim, D., Hilliges, O., Molyneaux, D., Newcombe, R., Kohli, P., Shotton, J., Hodges, S., Freeman, D., Davison, A., et al.: Kinectfusion: real-time 3D reconstruction and interaction using a moving depth camera. In: Proceedings of the 24th Annual ACM Symposium on User Interface Software and Technology, pp. 559–568 (2011)
6. Koppula, H.S., Anand, A., Joachims, T., Saxena, A.: Semantic labeling of 3D point clouds for indoor scenes. In: Proceedings of the Advances in Neural Information Processing Systems (2011)
7. Silberman, N., Hoiem, D., Kohli, P., Fergus, R.: Indoor segmentation and support inference from RGBD images. In: Fitzgibbon, A., Lazebnik, S., Perona, P., Sato, Y., Schmid, C. (eds.) ECCV 2012, Part V. LNCS, vol. 7576, pp. 746–760. Springer, Heidelberg (2012)
8. Shi, J., Malik, J.: Normalized cuts and image segmentation. IEEE Transactions on Pattern Analysis and Machine Intelligence 22(8), 888–905 (2000)
9. Moore, A.P., Prince, S., Warrell, J., Mohammed, U., Jones, G.: Superpixel lattices. In: CVPR, pp. 1–8 (2008)
10. Vedaldi, A., Soatto, S.: Quick shift and kernel methods for mode seeking. In: Forsyth, D., Torr, P., Zisserman, A. (eds.) ECCV 2008, Part IV. LNCS, vol. 5305, pp. 705–718. Springer, Heidelberg (2008)
11. Achanta, R., Shaji, A., Smith, K., Lucchi, A., Fua, P., Susstrunk, S.: SLIC superpixels compared to state-of-the-art superpixel methods. IEEE Transactions on Pattern Analysis and Machine Intelligence (2012)
12. Middlebury stereo vision page, http://vision.middlebury.edu/stereo/data/
13. Levinshtein, A., Stere, A., Kutulakos, K.N., Fleet, D.J., Dickinson, S.J., Siddiqi, K.: Turbopixels: Fast superpixels using geometric flows. IEEE Transactions on Pattern Analysis and Machine Intelligence 31(12), 2290–2297 (2009)
14. Gong, X., Liu, J.: Rock detection via superpixel graph cuts. In: 19th ICIP, pp. 2149–2152 (2012)
15. Li, Z., Wu, X.-M., Chang, S.-F.: Segmentation using superpixels: A bipartite graph partitioning approach. In: CVPR, pp. 789–796 (2012)
16. Fulkerson, B., Vedaldi, A., Soatto, S.: Class segmentation and object localization with superpixel neighborhoods. In: ICCV, pp. 670–677 (2009)
17. Xu, Y., Liu, J., Cheng, J., Yin, F., Tan, N.M., Wong, D.W.K., Cheng, C.Y., Tham, Y.C., Wong, T.Y.: Efficient optic cup localization based on superpixel classification for glaucoma diagnosis in digital fundus images. In: ICPR, pp. 49–52 (2012)
18. Felzenszwalb, P.F., Huttenlocher, D.P.: Efficient graph-based image segmentation. IJCV 59(2), 167–181 (2004)
19. Cour, T., Benezit, F., Shi, J.: Spectral segmentation with multiscale graph decomposition. In: CVPR, vol. 2, pp. 1124–1131 (2005)
20. Arbelaez, P., Maire, M., Fowlkes, C., Malik, J.: Contour detection and hierarchical image segmentation. IEEE Transactions on Pattern Analysis and Machine Intelligence 33(5), 898–916 (2011)

Graph-Based Label Propagation with Dissimilarity Regularization

Haixia Zheng and Horace H.S. Ip

Centre for Innovative Applications of Internet and Multimedia Technologies, Department of Computer Science, City University of Hong Kong, Hong Kong
Haixia.Zheng@my.cityu.edu.hk, cship@cityu.edu.hk

Abstract. Recent studies have shown promising performance of graph-based semi-supervised learning. But one of major limitations of most graph-based semi-supervised learning approaches is that they did not explore the label dissimilarity knowledge. In this paper, we presented a novel graph-based label propagation framework that effectively incorporates similarity and dissimilarity information into semi-supervised classification. The class mass normalization is utilized to make the label decision rule match class priors. The function induction algorithm is also proposed to predict the labels of test data. In particular, by solving quadratic optimization, our approach can give rise to closed-form solution for classification functions of unlabeled data and out-of-sample data. The proposed framework has been extensively evaluated over three widely used datasets for image classification task. The experimental results in terms of classification accuracy demonstrate that the proposed method can achieve significant performance improvements with respect to the state-of-the-arts.

Keywords: Label propagation, dissimilarity regularization, function induction, graph-based semi-supervised learning.

1 Introduction

One of the central issues in machine learning research communities is that only a few labeled data are available and large amounts of data remain unlabeled. It is very expensive and time consuming to label data. For example, obtaining a single labeled example for protein shape classification, which is one of the grand challenges in biological and computational science, requires months of expensive analysis by expert crystallographers. Therefore, how to efficiently combine abundant unlabeled data with limited labeled data is an important research field, which is the focus of semi-supervised learning techniques.

The semi-supervised learning problem has attracted an increasing amount of interest in the past decade, and many semi-supervised learning approaches have been proposed [1]. Most of semi-supervised learning algorithms focus on the transductive setting where labels of unlabeled data are estimated by learning a function defined over the point cloud data [2][3]. As graph has proved to be

B. Huet et al. (Eds.): PCM 2013, LNCS 8294, pp. 47–58, 2013.

successful in characterizing pairwise data relationship and manifold exploration, in recent years the development of the graph-based semi-supervised learning methods is a prominent achievement in the semi-supervised learning area. The graph-based semi-supervised learning approaches model the whole dataset as a graph in which the vertices represent the data and the edges represent the pairwise relationships. The basic idea behind the graph-based semi-supervised learning is label propagation on graph with the cluster consistency, which means two data points on the same structure (cluster or manifold) are likely to have the same class label [4][5].

Recent studies have shown promising performance of graph-based semi–supervised learning approaches, however, most graph-based approaches are limited to explore positive similarity, which can be interpreted as the confidence of assigning two different examples to the same class. In many situations, we may encounter dissimilarity, or negative similarity, which expresses the confidence of assigning two examples to different classes. For instance, consider the problem of predicting a persons political view (left, right) from his/her posting to online blogs. The fact that person B quotes person A and uses expletives near the quote is a strong indication that B disagree with A [6]. The dissimilarity pair (A, B) can be created to reflect our knowledge that A and B have different labels (political view). Another application of negative similarity is semi-supervised data clustering, in which one has side information of must link pairs and must not link pairs. One way to explore the side information is to associate every must link pair with a positive similarity, and every must not link pair with a negative similarity [7].

Although such dissimilarity knowledge has been studied in semi-supervised clustering, most existing graph-based semi-supervised classification approaches cannot handle dissimilarity knowledge easily, because the dissimilarity relations are non-transitive and therefore cannot be propagated directly. Moreover, dissimilarity could result in unbounded energy function, which makes most graph-based algorithms inapplicable. To address this problem, our paper proposed a new framework for graph-based label propagation, which can learn from similarity and dissimilarity information on labeled and unlabeled data. We employ the class mass normalization (CMN) to adjust the class distributions, so as to make the decision rule, which predicts the label according to classification function, incorporate the class prior knowledge. The function induction algorithm is also presented to predict the labels of the out-of-sample examples without solving again a linear system of dimension n (the number of unlabeled and labeled training examples), which can cost $O(n^3)$. In particular, we can generate analytical closed-form solutions for classification functions of unlabeled data and new test examples, respectively.

The remainder of this paper is organized as follows: in section 2 we first briefly review the basic framework of graph-based semi-supervised learning. In section 3, we introduce the framework of label propagation with dissimilarity regularization. Section 4 presented the function induction algorithm. Experimental results are shown in section 5. Section 6 draws the conclusion on this paper.

2 Basic Framework

Semi-supervised learning is designed to fully utilize large amount of unlabeled data together with the limited labeled data, such as spectral graph transducer [2], the harmonic approach [8], the Greens function approaches [9], the online approach [10], the Gaussian process [11], and so on. The key to semi-supervised learning is based on the prior assumption of consistency, also called cluster assumption, which states that: (i) nearby points are likely to have the same label; (ii) points on the same structure (typically referred as a cluster or a manifold) are likely to have the same label [4].

Given a dataset $X = \{x_1, \cdots, x_\ell, x_{\ell+1}, \cdots, x_n\} \subset R^m$ and a label set $\mathcal{L} = \{1, \cdots, c\}$, the first ℓ data $x_i (i \leq \ell)$ are labeled as $y_i \in \mathcal{L}$ and the remaining data $x_i (\ell + 1 \leq i \leq n)$ are unlabeled. A weighted undirected graph G=(V, E, W) can be constructed to characterize the pairwise similarities among data points, where V is the vertices set and each node v_i corresponds to a data point x_i; E is the set of edges; W is the set of weights measuring the strength of the pairwise similarity. To predict the label of the unlabeled data through graph-based semi-supervised learning, the cost function associated with classification function F can be defined as:

$$\min_F \frac{1}{2} \sum\nolimits_{i,j=1}^{n} W_{ij} \left\| F_i - F_j \right\|^2 + \gamma \sum\nolimits_{i=1}^{n} \left\| F_i - Y_i \right\|^2 \tag{1}$$

where, $\gamma > 0$ is the regularization parameter; Let $\mathcal{F}$ denote the set of $n \times c$ matrices with nonnegative entries. Matrix $F = [F_1^T, \cdots, F_n^T]^T \in \mathcal{F}$ corresponds to a classification on the dataset X by labeling each point x_i as the label $y_i = \arg\max_{j \leq c} F_{ij}$. The $n \times c$ matrix $Y \in \mathcal{F}$ is defined with $Y_{ij} = 1$ if x_i is labeled as $y_i = j$ and $Y_{ij} = 0$ otherwise. Clearly, Y is consistent with the initial labels according to decision rule.

In Eq.(1), the first term represents the smoothness constraint, also called energy function, which means that a good classification function should not vary too much between nearby points; the second term stands for the fitting constraint, which means a good classification function should not vary too much from the initial label assignment [4]. Therefore, the basic framework only encodes label similarity domain knowledge, and did not consider label dissimilarity knowledge.

3 The Framework of Label Propagation with Dissimilarity

In order to incorporate label dissimilarity domain knowledge into the framework of label propagation, the cost function must simultaneously meet the following two conditions: (i) minimize the energy function that measures the positive similarity; (ii) maximize the energy function which measures the negative similarity. Following the idea of Linear Discriminative Analysis (LDA) [12][13], the energy function in Eq.(1) can be transformed into:

$$\frac{\sum_{(i,j)\in S} P_{ij} \|F_i - F_j\|^2}{\sum_{(i,j)\in D} N_{ij} \|F_i - F_j\|^2} \tag{2}$$

where, $(i,j) \in S$ denotes data x_i and x_j have similarity domain knowledge; $P_{ij} > 0$ stands for their positive similarity weight between data x_i and x_j. Likewise, $(i,j) \in D$ denotes data x_i and x_j have dissimilarity domain knowledge; $N_{ij} > 0$ stands for their negative similarity weight between data x_i and x_j.

In order to minimize Eq.(2) with respect to classification function F, we can further transform Eq.(2) into the following corresponding Lagrangian formula [12]:

$$\sum_{(i,j)\in S} P_{ij} \|F_i - F_j\|^2 + \lambda(1 - \sum_{(i,j)\in D} N_{ij} \|F_i - F_j\|^2) \tag{3}$$

where, Lagrangian parameter $\lambda > 0$.

Therefore, the final cost function including similarity and dissimilarity domain knowledge can be defined as:

$$\begin{aligned} &\min_F \frac{1}{2}(\sum_{(i,j)\in S} P_{ij} \|F_i - F_j\|^2 - \lambda \sum_{(i,j)\in D} N_{ij} \|F_i - F_j\|^2) \\ &+ \gamma \sum_{i=1}^{n} \|F_i - Y_i\|^2 \\ &= \text{trace}(F^T(L_p - \lambda L_N)F) + \gamma \|F - Y\|^2 \end{aligned} \tag{4}$$

where, trace$(\cdot)$ denotes the trace of a matrix; L_P and L_N are graph Laplacian matrices for positive similarity matrix P and negative similarity matrix N, respectively.

The selected parameter λ should make the matrix $(L_P - \lambda L_N)$ positive semidefinite, so as to ensure there exists an optimal solution in Eq.(4). Through solving quadratic optimization problem of Eq.(4), we can obtain the closed-form solution for classification function F as:

$$F^* = \gamma(L_P - \lambda L_N + \gamma I)^{-1} Y \tag{5}$$

where, I is an identity matrix.

To predict labels of unlabeled data $x_i(\ell + 1 \leq i \leq n)$ according to the classification function F, the general decision rule is to label data x_i as $y_i = \arg\max_{j \leq c} F_{ij}$. This decision rule works well when the classes are well separated. However, in real datasets, classes are often not ideally separated, and this decision rule tends to produce severely unbalanced classification. The class priors can be utilized to improve the general decision rule [8]. We adopt the class mass normalization (CMN) to adjust the class distributions so as to match the priors. Lets assume the desirable proportion of class j is $q_j (0 < q_j < 1)$. Define the mass of class j to be $\sum_{\ell+1 \leq i \leq n} F_{ij}$. Class mass normalization scales these masses so that an unlabeled point $x_i(\ell + 1 \leq i \leq n)$ is assigned to the following label:

$$y_i = \arg\max_{j \leq c} q_j \frac{F_{ij}}{\sum_{\ell+1 \leq i \leq n} F_{ij}} \tag{6}$$

4 Function Induction Algorithm

As the test examples are included in the unlabeled dataset, the graph-based label propagation framework is usually utilized in a transductive manner. It cannot provide the obvious prediction for new examples. To obtain function induction without solving the linear system for each new test data, one approach is to parameterize F with a flexible form such as a neural network or a linear combination of non-linear bases [14]. Another is to use the induction formula, which will be described in detail.

In order to transform the above transductive algorithms into function induction algorithm, two adjustments are needed: (i) continue to use the cost function in Eq.(4), but include a new test example x; (ii) after the new test example x is added, we keep the values of F_{x_i} on training examples x_i fixed, as in ordinary function induction [15]. The second adjustment is motivated by the prohibitive cost of solving again the linear system, and the reasonable assumption that the value of the function over the unlabeled examples will not change much with the addition of a new point. This is clearly true asymptotically when $n \to \infty$. For the non-asymptotic case, we should expect transductive method to perform better than induction formula.

After adding the terms about a new test data x in Eq.(4) and keeping the values of F_{x_i} on the training data x_i fixed, we can get the following optimization problem with respect to F_x:

$$\begin{aligned} \min_{F_x} \frac{1}{2} & \sum\nolimits_{(x,x_j)\in S} P_{x,x_j} \left\| F_x - F_{x_j} \right\|^2 \\ & - \frac{\lambda}{2} \sum\nolimits_{(x,x_j)\in D} N_{x,x_j} \left\| F_x - F_{x_j} \right\|^2 + \gamma \left\| F_x \right\|^2 \end{aligned} \tag{7}$$

Let $Q(F_x)$ denote Eq.(7), and differentiate $Q(F_x)$ with respect to F_x, then let $\frac{\partial Q(F_x)}{\partial F_x} = 0$, we can obtain the optimal analytical solution:

$$F_x^* = \frac{\sum_{(x,x_j)\in S} P_{x,x_j} F_{x_j} - \lambda \sum_{(x,x_j)\in D} N_{x,x_j} F_{x_j}}{\sum_{(x,x_j)\in S} P_{x,x_j} - \lambda \sum_{(x,x_j)\in D} N_{x,x_j} + 2\gamma} \tag{8}$$

After obtaining the optimal classification function F_x^* for the test data x, similar to unlabeled data, we can compute its label y_x of the test data x by the following Eq.(9):

$$y_x = \arg\max_{j \leq c} q_j \frac{F_x^*(j)}{\sum_{\ell+1 \leq i \leq n} F_{ij} + F_x^*(j)} \tag{9}$$

Based on the previous analysis, the proposed algorithm for the label propagation with dissimilarity regularization is summarized in Algorithm 1.

Algorithm 1. Label propagation with dissimilarity regularization

Input: Dataset $X = \{x_1, \cdots, x_\ell, x_{\ell+1}, \cdots, x_n\} \subset R^m$. Label set $\mathcal{L} = \{1, \cdots, c\}$. The first ℓ data $x_i (i \leq \ell)$ are labeled as $y_i \in \mathcal{L}$ and the remaining data $x_i (\ell + 1 \leq i \leq n)$ are unlabeled. Test data x.

Output: Labels for the unlabeled data $x_i (\ell + 1 \leq i \leq n)$ and test data x

1. **Training phase**
   (i) Compute classification function F for the unlabeled data by Eq.(5).
   (ii) Calculate the labels of the unlabeled data through Eq.(6).
2. **Testing phase**
   For a new test data x, first compute its classification function F_x by Eq.(8), then get its label y_x via Eq.(9).

5 Experiments

In this section, the proposed label propagation with dissimilarity regularization method will be extensively evaluated on three widely used datasets: USPS handwritten digit dataset, UCI Isolet spoken letter dataset, and Caltech-101 dataset [16]. We first describe the experimental setup, and then empirically demonstrate the advantages of our proposed framework in several classification tasks.

5.1 Experimental Setup

In our implementations, we introduce dissimilarity edges between randomly sampled data points with different labels. Because the edges represent ground-truth dissimilarity, we disallow edges to touch labeled points, to prevent the true labels propagating throughout the unlabeled data. Note that the actual label values are not revealed—just the fact that the points should receive different label classifications. Simulating domain knowledge in this manner is common for cannot-link clustering and related work [17].

Similar to [3], we encode similarity using k-nearest-neighbor graphs with Gaussian weights. Specifically, the weight between kNN points x_i and x_j is $\exp(-\|x_i - x_j\|^2 / 2\sigma^2)$, while all other weights are zero. We then add the above dissimilarity edges, and assign them a relatively large weight. In the experiments, we set fitting constraint parameter $\gamma = 100$. The other parameters are set to be their respective optimal values according to the 5-fold cross validation.

The proposed approach is compared with the following state-of-the-arts: supervised SVM and RLS, transductive SVM (TSVM), semi-supervised LapSVM and LapRLS [18], locality preserving semi-supervised SVM (LPSSVM) [19]. The classification accuracy is employed to evaluate the performance of these approaches. Following the common benchmarking procedures, the final classification accuracy is obtained by averaging the results over 10 different trials. For each trial, again, the labeled and unlabeled data, dissimilarity edges, and test data are randomly chosen in the experiments.

5.2 USPS Handwritten Digit Recognition

The USPS dataset contains grayscale handwritten digit images scanned from envelopes by the U.S. Postal Service. The size of the images is 16×16, with pixel values in the range 0 to 2. Some sample images from each category are shown in Fig. 1. The original training set contains 7291 images, and the test set contains 2007 images. Following [18], during this set of experiments, the first 400 images for each digit in the original training set were taken to form our training set. The remaining images formed the test set. In our training set, 2 images for each class were randomly labeled and the rest were left unlabeled.

We first conduct experiment with different number of dissimilarity edges, where the weights of all the dissimilarity edges are fixed to be 1. The effect of the number of dissimilarity edges on the classification results is shown in Fig. 2. Compared with the baseline without dissimilarity edges, the classification accuracies of unlabeled and test data improve with the number of dissimilarity edges increasing.

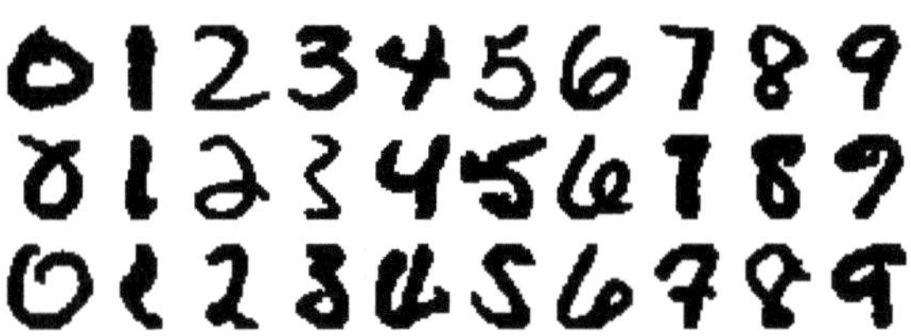

Fig. 1. Sample images per category from USPS dataset

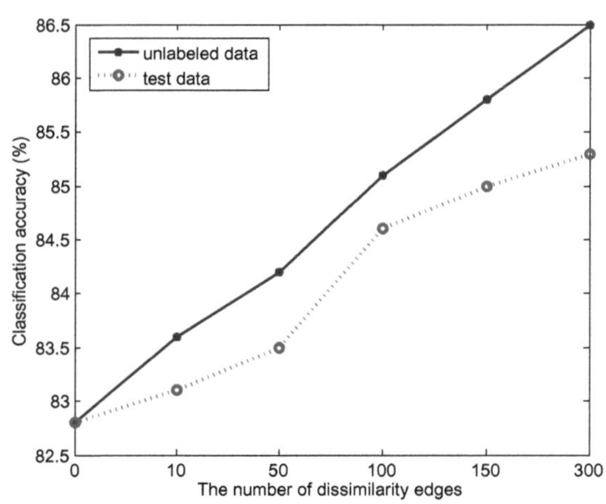

Fig. 2. Classification comparison with respect to different number of dissimilarity edges

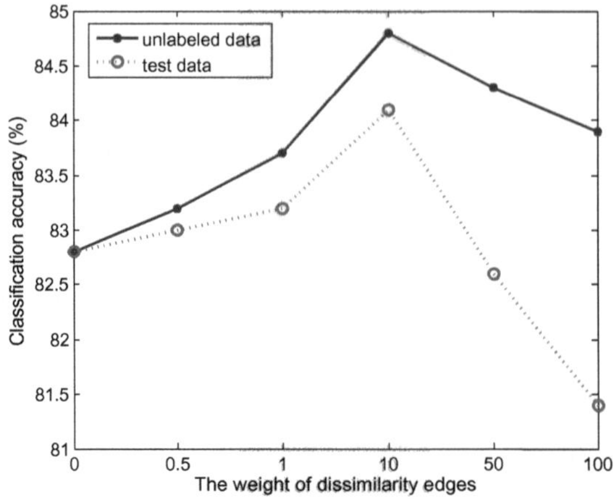

Fig. 3. Classification comparison with respect to different weights of dissimilarity edges

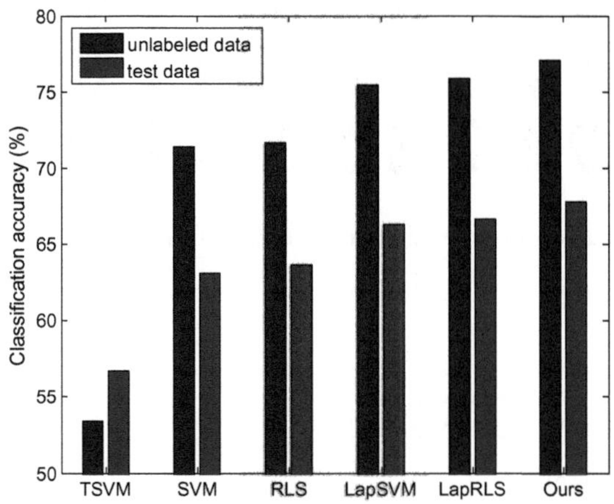

Fig. 4. Performance comparison of different methods on UCI Isolet dataset

Fig. 5. Example images from Caltech-101 dataset

Then we fix the number of dissimilarity edges at 30, and perform the experiment when varying the weight of dissimilarity edges. Fig. 3 presented the classification results with different dissimilarity weights. It can be observed that the maximal classification accuracy appears at a weight of approximately 10, when the weight is bigger than 10 the classification accuracy slightly decreases. The reason is that the larger weight on dissimilarity edges leads to overfitting problem—the learning process relies too heavily on the dissimilarity edges, and much useful similarity information is being ignored.

5.3 Spoken Letter Recognition

This experiment was performed on the UCI Isolet dataset (available from the UCI machine learning repository). This dataset contains utterances of 150 subjects who spoke the name of each letter of the English alphabet twice. The speakers are grouped into 5 sets of 30 speakers each, and are referred to as isolet1, isolet2, isolet3, isolet4, and isolet5. We chose to train on the first 30 speakers (isolet 1) which formed a training set of 1560 examples, and test on isolet 5 containing 1559 examples (1 utterance is missing in the database due to poor recording). In the training set, all 52 utterances of one speaker were labeled and the rest were left unlabeled [18]. We randomly constructed 500 dissimilarity edges, and assigned the weight of each dissimilarity edge to be 10. In Fig. 4, we reported the average classification accuracies of different methods over 26 classes for the unlabeled training data and the test data, respectively. It can be seen that our approach outperforms other existing state-of-the-arts.

5.4 Performance over Caltech-101 Dataset

The Caltech-101 dataset [16] contains 9144 images in 101 classes (including airplanes, butterflies, chairs, etc.) with significant variance in shape. Some randomly selected example images are shown in Fig. 5. The number of images per category varies from 31 to 800. Most images are in medium resolution, i.e. about 300×300 pixels.

The spatial pyramid bag-of-feature representation with local SIFT descriptors has been proven effective for classifying this dataset [20]. Similar to [21], a set of local SIFT features of 16×16 pixel patches are uniformly sampled from these images over a grid with spacing of 8 pixels. Then a visual codebook is constructed by clustering all SIFT features into 1024 codewords. Local features in each image block are mapped to the codewords to compute codeword histograms. The spatial pyramid representation is also used to preserve the spatial information. We utilize the top three layers and set the same weight for each layer. The maximum side (width or height) of each image is resized to 300 pixels with preserved aspect ratio.

Following the experiment setup for Caltech-101 in [19], for each image category, 5 images are randomly selected as labeled data and 25 images are randomly selected as unlabeled data for training. The remaining images are used as test data, and the maximum number of test images per category is limited to be 30.

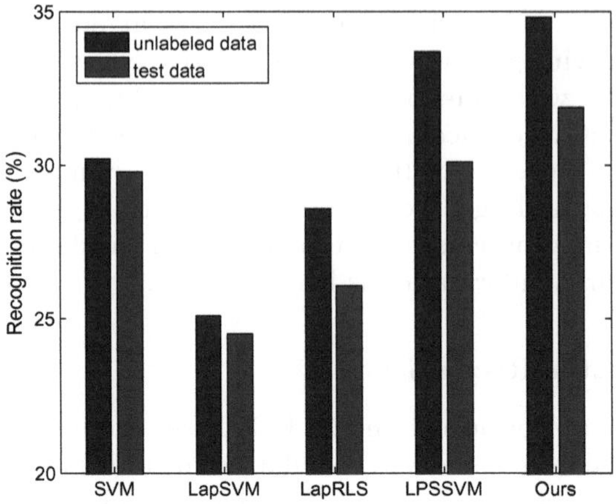

Fig. 6. Performance comparison of different methods on Caltech-101 dataset

We randomly constructed 5000 dissimilarity edges, and the weights of all the dissimilarity edges are set to be 10. Fig. 6 presented the average recognition rates of different algorithms over 102 classes (i.e. 101 classes and a background class) for the unlabeled training data and the test data, respectively. It can be observed that our approach leads to significant improvement of classification performance over other competing methods.

6 Conclusion and Future Work

In this paper, we presented a convex algorithm that can simultaneously encode similarity and dissimilarity information into label propagation framework. It is illustrated that when such dissimilarity domain knowledge is available, our algorithm can make good use of it and improve classification results. The promising experimental results evaluated in three well-known datasets clearly demonstrate the effectiveness of our proposed label propagation framework.

In the experiments we observed that the variances of the parameter λ are very large for different datasets, therefore, how to efficiently choose the optimal parameter λ is a direction for our future research. Furthermore, although the proposed framework is evaluated on image classification task in this paper, we can also apply it to recommendation system, such as collaborative filtering.

References

1. Zhu, X.: Semi-supervised learning literature survey. Technical Report 1530, Computer Sciences, University of Wisconsin-Madison (2005)
2. Joachims, T.: Transductive learning via spectral graph partitioning. In: Proceedings of the Twentieth International Conference on Machine Learning, pp. 290–297. AAAI Press (2003)

3. Sindhwani, V., Niyogi, P., Belkin, M.: Beyond the point cloud: from transductive to semi-supervised learning. In: Proceedings of the 22nd International Conference on Machine Learning, pp. 824–831. ACM, New York (2005)
4. Zhou, D., Bousquet, O., Lal, T.N., Weston, J., Schölkopf, B.: Learning with local and global consistency. In: Advances in Neural Information Processing Systems, vol. 16. MIT Press (2003)
5. Wang, F., Zhang, C.: Label propagation through linear neighborhoods. In: Proceedings of the 23rd International Conference on Machine Learning, pp. 985–992. ACM, New York (2006)
6. Mullen, T., Malouf, R.: A preliminary investigation into sentiment analysis of informal political discourse. In: AAAI 2006 Spring Symposium on Computational Approaches to Analysing Weblogs (2006)
7. Kulis, B., Basu, S., Dhillon, I., Mooney, R.: Semi-supervised graph clustering: a kernel approach. In: Proceedings of the 22nd International Conference on Machine Learning, pp. 457–464. ACM, New York (2005)
8. Zhu, X., Ghahramani, Z., Lafferty, J.: Semi-supervised learning using Gaussian fields and harmonic functions. In: Proceedings of the Twentieth International Conference on Machine Learning, pp. 912–919 (2003)
9. Zhou, D., Schölkopf, B., Hofmann, T.: Semi-supervised learning on directed graphs. In: Advances in Neural Information Processing Systems, vol. 17 (2004)
10. Herbster, M., Pontil, M., Wainer, L.: Online learning over graphs. In: Proceedings of the 22nd International Conference on Machine Learning, pp. 305–312. ACM, New York (2005)
11. Altun, Y., McAllester, D., Belkin, M.: Maximum margin semi–supervised learning for structured variables. In: Advances in Neural Information Processing Systems (2006)
12. Mika, S., Ratsch, G., Weston, J., Schölkopf, B., Mullers, K.R.: Fisher discriminant analysis with kernels, pp. 41–48 (1999)
13. Tong, W., Jin, R.: Semi-supervised learning by mixed label propagation. In: Proceedings of the 22nd National Conference on Artificial Intelligence, vol. 1, pp. 651–656. AAAI Press (2007)
14. Belkin, M., Niyogi, P.: Using Manifold Structure for Partially Labelled Classification. In: Advances in NIPS (2002)
15. Delalleau, O., Bengio, Y., Roux, N.L.: Nonparametric function induction in semi-supervised learning. In: Proc. Artificial Intelligence and Statistics (2005)
16. Fei-Fei, L., Fergus, R., Perona, P.: Learning generative visual models from few training examples: An incremental bayesian approach tested on 101 object categories. In: IEEE CVPR Workshop of Generative Model Based Vision (WGMBV) (2004)
17. Goldberg, A.B., Zhu, X., Wright, S.: Dissimilarity in graph-based semi-supervised classification. Journal of Machine Learning Research 2, 155–162 (2007)
18. Belkin, M., Niyogi, P., Sindhwani, V.: Manifold regularization: A geometric framework for learning from labeled and unlabeled examples. Journal of Machine Learning Research 7, 2399–2434 (2006)
19. Jiang, W., Chang, S.F., Jebara, T., Loui, A.C.: Semantic concept classification by joint semi-supervised learning of feature subspaces and support vector machines. In: Forsyth, D., Torr, P., Zisserman, A. (eds.) ECCV 2008, Part IV. LNCS, vol. 5305, pp. 270–283. Springer, Heidelberg (2008)

20. Lazebnik, S., Schmid, C., Ponce, J.: Beyond bags of features: Spatial pyramid matching for recognizing natural scene categories. In: Proceedings of the 2006 IEEE Computer Society Conference on Computer Vision and Pattern Recognition, vol. 2, pp. 2169–2178. IEEE Computer Society, Washington, DC (2006)
21. Yang, J., Yu, K., Gong, Y., Huang, T.S.: Linear spatial pyramid matching using sparse coding for image classification. In: Computer Vision and Pattern Recognition, pp. 1794–1801 (2009)

Localize and Segment Scene Text

Xiaoqian Liu[1] and Weiqiang Wang[1,2]

[1] School of Computer and Control Engineering,
University of Chinese Academy of Sciences, Beijing, China
[2] Key Lab of Intell. Info. Process.,
Institute of Computing Technology, CAS, Beijing China

Abstract. In this paper, we present a scene text extraction approach which can realize text localization and segmentation simultaneously. Two popular paradigms (machine learning method and rule-based method) are combined to achieve competitive performance. For a given image, a sliding window is used to detect scene text. The texture feature Local Binary Pattern is extracted to represent the content of each window, and an unbalanced SVM classifier is designed to identify candidate text regions. Then, candidate text windows are further verified using color contrast and binarized by an adaptive local thresholding computation to get candidate text connected components. Further, non-text ones among them are removed utilizing some empirical rules. Finally, text connected components are linked into text lines according to their spatial relationships and appearance similarities. The evaluation results on two challenging and standard datasets ICDAR 2003 and ICDAR 2011 demonstrate that the proposed approach can effectively detect and segment scene text with different sizes, fonts, colors and arrangement directions.

Keywords: LBP, SVM, rule-based method, text detection, text extraction.

1 Introduction

With the advent of the information age, more and more intelligent devices flood into the life of humans. Semantic understanding is an indispensable part of realizing various intelligent applications on the devices, so automatically detecting and recognizing text in images and videos as a key technique has become a hot research topic which attracts a lot of researchers' attention. In the last decade, text extraction technology has been exploited in many applications, such as intelligent traffic management system [1], postal letter sorting [2], helpful devices for blind and visually impaired people [16], and automatic caption extraction system [4]. Graphic text in images and videos has two forms, artificial text and scene text. The artificial text is manually overlaid on images or video frames and generally owns uniform color, size, position and arrangement direction, e.g., captions. Scene text refers to those captured by cameras as a part of natural scenes. Compared with artificial text, the colors, fonts, sizes and locations of scene text are diverse and they are also vulnerable to the influence of illumination variance

B. Huet et al. (Eds.): PCM 2013, LNCS 8294, pp. 59–68, 2013.

and shooting angles. So, figuring out an effective scene text extraction approach is still a challenging problem.

Most text extraction algorithms can be decomposed into two parts, text detection and text segmentation, which aim to localize text regions and identify text pixels respectively. Overall, the existing mainstream methods for text extraction can be broadly classified into two categories, the rule-based method and machine learning method. The rule-based methods are relatively simple and intuitive in system construction, and they generally utilize some distinct characteristics summarized from colors, edges or structures of characters to filter out non-text areas or pixels based on some empirical rules. For instance, Li *et al.* [5] first extract connected components by the Maximally Stable Extremal Region(MSER) algorithm and then filter out non-text components based on some geometric rules. Finally, some false positives are eliminated according to stroke width of text. Chen *et al.* [6] first exploit a multi-scale Laplacian of Gaussian (LoG) edge detector to obtain the edge set, and then combine both edge and color information to localize text regions. The rule-based methods usually involve many thresholds and too much dependence on the choice of thresholds is prone to reduce the robustness and objectivity of the system. To overcome this disadvantage, some text extraction approaches integrate the machine learning techniques, like SVM, Neural Networks, into their system. In [7], a cascade with four strong classifiers are trained by an AdaBoost machine learning algorithm to classify text candidate regions. Thillou *et al.* [8] exploit an unsupervised clustering algorithm to separate text pixels from background by combining color feature with spatial information. Lukas *et al.* [9] use the SVM to classify the extracted Maximally Stable Extremal Regions (MSERs) into character and non-character regions. In [10], text detection and recognition are realized based on a scalable unsupervised feature learning method.

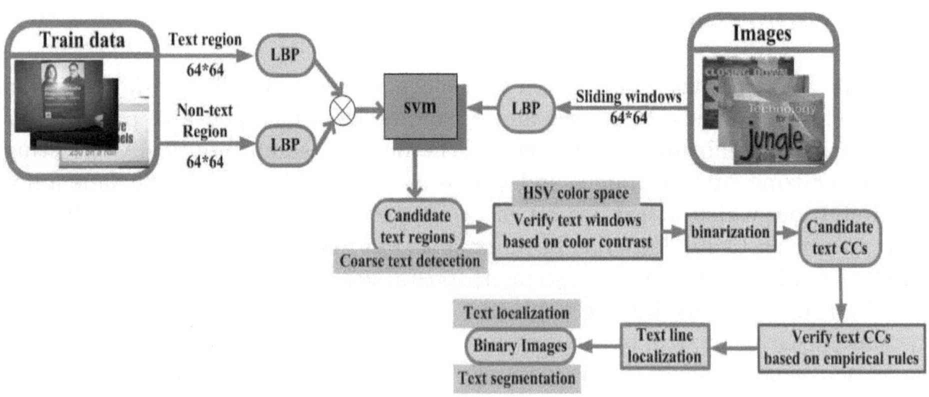

Fig. 1. The framework of the proposed approach

In this paper, we present a scene text extraction approach combining the advantages of the rule-based algorithm and the machine learning algorithm together. First, an unbalanced SVM classifier is designed to efficiently detect the existence of scene text at each position of sliding windows, so that candidate text regions can be coarsely localized. Next, the color contrast information is exploited to further verify and identify text windows, and the binarization computation based on adaptive local thresholds are conducted on them to obtain candidate text connected components. Then, non-text connected components are removed out based on some empirical rules. Finally, text lines are obtained by linking the text components with similar appearance and close spatial distance together.

The rest of this paper is organized as follows. Section 2 presents the proposed method. The experimental results are reported in Section 3. Section 4 concludes the paper.

2 Methodology

The framework of the proposed approach is shown in Figure 1. For a given image, a sliding window is used to detect scene text. At each window position, the Local Binary Pattern (LBP) feature is extracted and exploited by an SVM classifier to coarsely localize candidate text regions. Next, candidate text windows are further verified based on color contrast and binarized by an adaptive local thresholding computation to get candidate text connected components. Further, non-text ones among them are removed utilizing some empirical rules. Finally, text connected components are linked into text lines according to their spatial relationships and appearance similarities. The more details are described in the corresponding subsection respectively.

2.1 Localizing Candidate Text Regions Using SVM

For a given image, a m by n window slides over it from top to bottom and from left to right with a certain step, and at each window position an SVM classifier is exploited to judge whether scene text exists in it. In our system, the size of the sliding window is 64 by 64 pixels and the sliding step is 32 pixels. In each window, the texture feature of Local Binary Patterns (LBP) is extracted, which has been acknowledged to be effective and has a good invariance to lightness. The basic representation of LBP is utilized in our approach based on the consideration of computational efficiency. Concretely, for each pixel in the window, the grayscale values of its eight neighboring pixels are compared with its value. The neighboring pixel is marked as 1 if it is greater than the current pixel; otherwise, it is marked as 0. Thus, an 8 bit unsigned number is formed through concatenating a sequence of 1s and 0s corresponding to the eight neighboring pixels as the LBP code of the current pixel. Further, for the current window, the statistical histogram of these LBP codes is computed to describe its content as the feature representation $\mathbf{x}$.

To train the classification model, we randomly sample 10140 positive instances and 100000 negative instances from the training images of ICDAR2003. Figure 2 gives some examples of positive instances and negative instances in our training datasets. Each training instance has the size of 64 $\times$ 64 pixels and different instances are permitted to be partially overlapped. The SVM is chosen as the classification model, and the linear kernel is used due to its low computation complexity. Formally, the classification function $y = sign(\mathbf{w}^T\mathbf{x}+b)$. If $\mathbf{w}^T\mathbf{x}+b > 0$, the current window is classified as text window with $y = 1$; otherwise, it is a non-text window with $y = 0$. In our system, the SVM is exploited to quickly filter out most non-text regions. At the same time, we also expect that the obtained SVM classifier has a recall as high as possible, so as to guarantee that as many text windows as possible are remained. To realize this, after we have obtained the classifier $y = sign(\mathbf{w}^T\mathbf{x} + b)$, the bias b is adjusted. Apparently a lower bias b corresponds to a higher recall and a lower accuracy. Our experiments show that through the classifier, lots of non-text region can be removed. To avoid filtering out text sub-images in this step, we adjust the parameter of classifier to ensure that the recall is equal to 1. Exploiting the SVM classifier obtained, each window is classified, and all the windows which are classified as text windows are merged together to form candidate text regions. We assume that scene text consist of more than one word, so if the text sub-image is isolated, it will be taken as non-text one and removed according to the constraint of size.

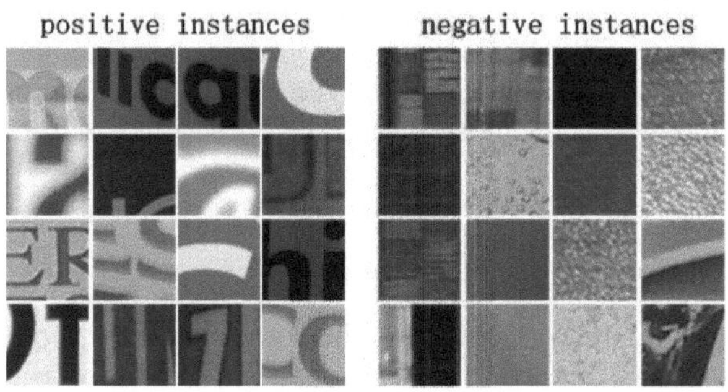

Fig. 2. The examples of positive instances (text windows) and negative instances (non-text windows)

2.2 Extracting Text Connected Components

The scene text on the billboards and signs usually owns strong color contrast with background so that they can be easily captured and recognized by human eyes. The color characteristics of scene text can be exploited to extract text connected components in the following computation. For those candidate text regions identified by the SVM classifier, the windows containing scene text are further verified, and the binarization computation is conducted on them to segment out the text connected component through an adaptive local thresholding computation. For an image I containing k candidate text regions, each text region $R_i, (i = 1, 2, \ldots, k)$ are formed by a group of candidate text windows R_i^j, where j denotes the jth windows in R_i. Here we should note that these text windows R_i^j are just those labeled by the SVM classifier. Then, the system calculates the mean $m_{i,j}$ and variance $\sigma_{i,j}^2$ of the hue values of pixels in R_j^i in the HSV color space. If the following condition

$$\sigma_{i,j} > \theta \tag{1}$$

is satisfied, the candidate text window R_i^j is binarized with threshold $m_{i,j}$, where threshold $\theta = 0.3$ in our system. Through the above computation, a set of candidate text connected components can be obtained.

Further, five rules are established by us and utilized to identify text connected components. They are summarized as follows.

- The text connected components can not border on the boundaries of images.
- The area N_c of text connected components, i.e., the number of pixels forming the connected component, should not be too small. In our system, $N_c \geq 10$.
- The height h_c and width w_c of the minimum enclosing rectangle (MER) of a text connected component should not be too small. In our system, both h_c and w_c should exceed 8 pixels.
- The aspect ratio $R_a = h_c/w_c$ of a text connected component should satisfy a certain ratio according to the structure of text strokes. In our system, $R_a \in [0.5, 2]$.
- If the occupation ratio $R_o = N_c/(h_c * w_c)$ is more than 0.8 or less than 0.1, the corresponding connected components is removed out as background.

The above rules are designed based on the geometric characteristics of characters, and the specific choice of thresholds is based on trial and error. Through these simple rules, non-text connected components can be effectively removed.

2.3 Text-Line Localization and Text Segmentation

Finally, these isolated characters segmented are linked into text lines according to their spatial relationships and appearance similarities. The spatial location of each text connected component c_i is represented by its centroid. We used the Euclidean distance d_{ij} to measure the proximity of two connected components c_i and c_j. Let w_i, w_j and h_i, h_j denote the width and height of MERs of c_i and c_j respectively. If the $d_{ij} \leq (w_i + w_j)$, c_i and c_j will be considered to be close enough. Appearance similarity requires the MERs of two linked connected components own similar shapes, i.e.,

$$|w_i - w_j| \leq \beta * \max\{w_i, w_j\} \tag{2}$$

$$|h_i - h_j| \leq \beta * \max\{h_i, h_j\} \tag{3}$$

where $\beta = 0.2$ in our system. Although the arrangement directions of text may be arbitrary, we assume that the characters in the same word or sentence should be approximately in a straight line. In our implementation, we combine the connected components whose centroids are approximately collinear. Only when the connected components satisfy all of the three constraints above, they are combined into a text line and marked by its minimum enclosing rectangle with the angle consistent to the arrangement direction.

3 Experiments

To validate the proposed approach, our evaluation experiments use two challenging datasets ICDAR 2003 [14] and ICDAR2011 [16], which are the standard public datasets in the competition of scene text localization. They contain 251 and 491 images for testing respectively.

3.1 Results of Localizing Candidate Text Regions

In this group of experiments, we evaluate whether the unbalanced SVM classifier can effectively filter out non-text windows by using the simple LBP feature. A window is labeled as a text window if it contains some pixels belonging to text regions in the ground-truth. Through adjusting the bias b in the classification function, two performance curves of precision vs. recall can be obtained corresponding to ICDAR 2003 and ICDAR 2011 datasets respectively, as shown in Fig. 3. The recall is the ratio of the number of text windows correctly classified to the total number of text windows in the ground-truth, and the precision is the ratio of the number of windows correctly classified to the total number of windows. It can be seen that when the recall approaches 100%, the precision is still above 55%, which means that most non-text windows are filtered out due to the computation of this step. Since the classification model is established based on the images of ICDAR 2003, the result on ICDAR 2011 is slightly lower. Whatever, it shows that the SVM classifier based on the LBP can effectively filter out most of non-text regions.

Table 1. Performance comparison of text localization on ICDAR 2003 dataset

Method	P	R	F
Our method	**0.72**	**0.69**	**0.705**
Epshtein*et al.* [11]	0.73	0.60	0.66
Lukas *et al.* [12]	0.72	0.62	0.67
Becker [13]	0.62	0.67	0.64
Ashida [14]	0.55	0.46	0.50

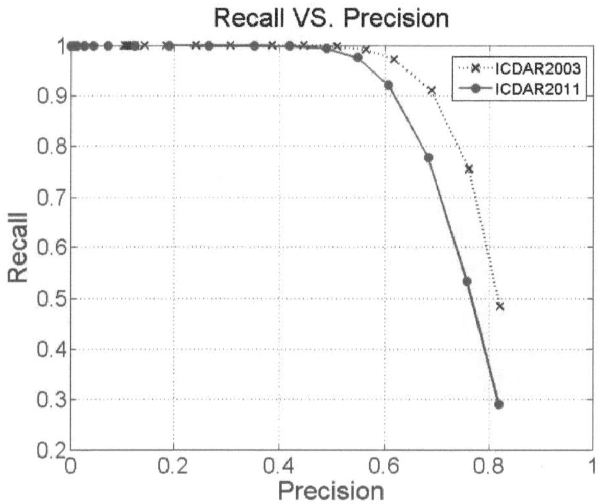

Fig. 3. The results of text sub-images classification. The blue dashed line and the solid red line describe the results on ICDAR 2003 and ICDAR 2011 respectively.

Table 2. Performance comparison of text localization methods on ICDAR 2011 dataset

Method	P	R	F
Kim* [16]	0.83	0.625	0.713
Our method	**0.70**	**0.68**	**0.69**
Lukas *et al.* [15]	0.647	0.731	0.687

Fig. 4. The results of text localization and extraction on ICDAR 2003 dataset

Fig. 5. The results of text localization and extraction on ICDAR 2011 dataset

3.2 Results of Localizing and Segmenting Scene Text

Three metrics are adopted to evaluate the proposed approach, i.e., precision P, recall R and F-measure F. P is the ratio of area of the correctly localized text regions N_r to area of the whole detected regions N_d by our system, where N_r can be calculated by taking an operation "And" between the area of text regions localized and the text regions as groundtruth. R is the ratio of area of N_r to area of the text regions in ground-truth. The area of a region is represented by the number of pixels in it. $F = \frac{2 \times P \times R}{P+R}$ is a composite indicator combining P and R. The localization results on ICDAR 2003 and ICDAR 2011 are shown in Table 1 and Table 2 respectively. For ICDAR 2003 dataset, the localization performance of our approach is the best according to R. Although the corresponding precision P is slightly lower compared with the best one, the F-measure is still the best compared with other state-of-the art methods, where [13] and [14] are the champions of ICDAR 2005 competition and ICDAR 2003 competition respectively. The performance of our approach is also competitive on the latest dataset ICDAR 2011 and they are comparable with the winner of ICDAR 2011 Robust Reading competition marked by asterisk in Table 2.

In addition to localizing scene text regions, text segmentation can be realized simultaneously in our approach. The examples of results of text localization and extraction on both datasets are shown in Fig. 4 and Fig. 5 respectively. The red rectangles enclosing the text regions own the same inclination angles with the arrangement direction of text lines.

4 Conclusion

In this paper, we present a scene text extraction approach which can realize text localization and segmentation simultaneously. Two popular paradigms (machine learning method and rule-based method) are combined to achieve competitive performance. The evaluation experiments on two challenging public datasets demonstrate the proposed approach can effectively handle with scene text with different colors, sizes, fonts and arrangement directions.

Acknowledgments. This work was supported by the National Natural Science Foundation of China under Grant No. 61232013, No. 61271434, No. 61175115.

References

1. Park, S.H., Kim, K.I., Jung, K., Kim, H.J.: Locating car license plates using neural networks. Electronics Letters 35(17), 1475–1477 (1999)
2. Liu, C.-L., Koga, M., Fujisawa, H.: Lexicon-driven segmentation and recognition of handwritten character strings for japanese address reading. IEEE Transactions on Pattern Analisis and Machine Intelligence 24(11), 1425–1437 (2012)
3. Ezaki, N., Bulacu, M., Schomaker, L.: Text detection from natural scene images: towards a system for visually impaired persons. In: Proc. of 17th International Conference on Pattern Recognition, Cambridge, England, UK, August 23-26 (2004)
4. Liu, Q., Jung, C., Kim, S., Moon, Y., Kim, J.: Stroke filter for text localization in video images. In: Proc. of IEEE International Conference on Image Processing, Atlanta, GA, USA, October 8-11, pp. 1473–1476 (2006)
5. Li, Y., Lu, H.: Scene text detection via stroke width. In: Proc. of 21st International Conference on Pattern Recognition, Tsukuba, Japan, November 11-15, pp. 681–684 (2012)
6. Chen, X., Yang, J., Zhang, J., Waibel, A.: Automatic detection and recognition of signs from natural scenes. IEEE Transactions on Image Processing 13(1), 87–99 (2004)
7. Chen, X., Yuille, A.L.: Detecting and reading text in natural scenes. In: Proc. of Computer Vision and Pattern Recognition, Washington, DC, USA, June 27-July 2, pp. 366–373 (2004)
8. Mancas-Thillou, C., Gosselin, B.: Spatial and color spaces combination for natural scene text extraction. In: Proc. of the 13th International Conference on Image Proceedings, Atlanta, GA, October 8-11, pp. 985–988 (2006)
9. Neumann, L., Matas, J.: A method for text localization and recognition in real-world images. In: Proc. of Asian Conference on Computer Vision, New Zealand, November 8-12, pp. 30–35 (2010)
10. Carpenter, B., Case, C., Satheesh, S., Suresh, B., Wang, T., Wu, D.J., Ng, A.Y.: Text detection and character recognition in scene images with unsupervised feature learning. In: Proc. of the 11th International Conference on Document Analysis and Recognition, Beijing, China, September 18-21, pp. 440–445 (2011)
11. Epshtein, B., Ofek, E., Wexler, Y.: Detecting text in natural scenes with stroke width transform. In: Proc. of the 23rd IEEE Conference on Computer Vision and Pattern Recognition, San Francisco, CA, June 13-18, pp. 2963–2970 (2010)
12. Neumann, L., Matas, J.: Text localization in real-world images using efficiently pruned exhaustive search. In: Proc. of the 11th International Conference on Document Analysis and Recognition, Beijing, China, September 18-21, pp. 687–691 (2011)
13. Lucas, S.M.: Icdar 2005 text locating competition results. In: Proc. of the 8th International Conference on Document Analysis and Recognition, Seoul, Korea, August 29-September 1, pp. 80–84 (2005)

14. Lucas, S.M., Panaretos, A., Sosa, L., Tang, A., Wong, S., Young, R.: ICDAR 2003 robust reading competitions. In: Proc. of the 7th International Conference on Document Analysis and Recognition, Edinburgh, Scotland, UK, August 3-6, 2003, pp. 682–687 (2003)
15. Neumann, L., Matas, J.: Real-time scene text localization and recognition. In: Proc. of the 25th IEEE Conference on Computer Vision and Pattern Recognition, Providence, Rhode Island, June 16-21, pp. 3538–3545 (2012)
16. Shahab, A., Shafait, F., Dengel, A.: ICDAR 2011 robust reading competition challenge 2: Reading text in scene images. In: Proc. of the 11th International Conference on Document Analysis and Recognition, pp. 1491–1496 (2011)

Multi-class Cosegmentation with Pairwise Active Learning

Anran Wang[1], Hongyuan Zhu[2], Jianfei Cai[2], and Jianxin Wu[2]

[1] BeingThere Centre, Institute for Media Innovation,
Nanyang Technological University, Singapore
[2] School of Computer Engineering,
Nanyang Technological University, Singapore

Abstract. Jointly segmenting common objects from multiple images remains a challenging problem. In this paper, we propose a multi-class cosegmentation method based on correlation clustering, which requires no prior knowledge of the number of clusters. Our method can handle large number of images because of the flexible graph structure and scalable clustering method. Moreover, we use active learning to intelligently recommend pairs of regions to users, in order to get pairwise *must-link* and *cannot-link* constraints. Then a novel dimensionality reduction method is proposed to produce an affinity matrix which reflects both the intrinsic structure of data and the constraints. Finally, correlation clustering is applied on the newly generated affinity matrix to acquire refined results. Experimental results show that our system can correct errors of initial segmentation and personalize segmentation result according to user preferences.

Keywords: Cosegmentation, active learning, pairwise constraints.

1 Introduction

Fully bottom-up unsupervised segmentation of single image is difficult due to its ill-posed nature. When given more images containing common objects, the saliency of these objects are boosted with the concurrence, thus the current trend is to analyze a set of related images simultaneously to exploit the enriched context in the images. One of the methods is cosegmentation, which jointly extracts two or more common objects from multiple images.

Since the seminal work of Rother et al. [1], many variants have been proposed [2,3], but most of them focus on the binary class problem. Recently, Kim et al. [4] proposed the first method aimed at handling multiple object classes and images. They modeled different object classes as heat sources in a heat diffusion system and maximized the overall system temperature to generate multi-class cosegmentation. Joulin et al. [5] proposed the discriminative clustering framework for image cosegmentation which can be extended to multi-class.

However, existing works on cosegmentation dealing with multi-class problems require a predefined number of clusters, which is sometimes difficult to estimate.

B. Huet et al. (Eds.): PCM 2013, LNCS 8294, pp. 69–78, 2013.

Furthermore, handling a large-scale image set remains challenging. Imposing strong pairwise foreground appearance constraint makes the algorithms unscalable to a large number of images and produces difficulty in optimizing energy functions. Moreover, with the bottom-up cues abstracted from images, the quality of the segmentation result is hardly satisfactory due to the fact that local cues cannot capture the global shape of the objects. Hence high-level knowledge is needed to guide the process of cosegmentation. Overall, much improvement is needed.

In order to tackle the above problems, we propose an efficient multi-class cosegmentation system with active pairwise metric learning. First, it is based on an efficient correlation clustering method which can automatically determine a suitable clusters number and scale well to a large number of images. Second, active learning is applied which can automatically request users for the true labels to learn high-level cues. Unlike the active learning method which has been used in previous cosegmentation tasks [7,8], in which they recommended uncertain regions to be annotated, our method recommends top ranked uncertain pairs of regions to handle a large number of classes. Based on the pairwise must-link and cannot-link constraints from users, we develop a novel dimensionality reduction for our system to propagate user preferences to all regions in correlation clustering. The experiments demonstrate the efficacy of our method in terms of speed and quality.

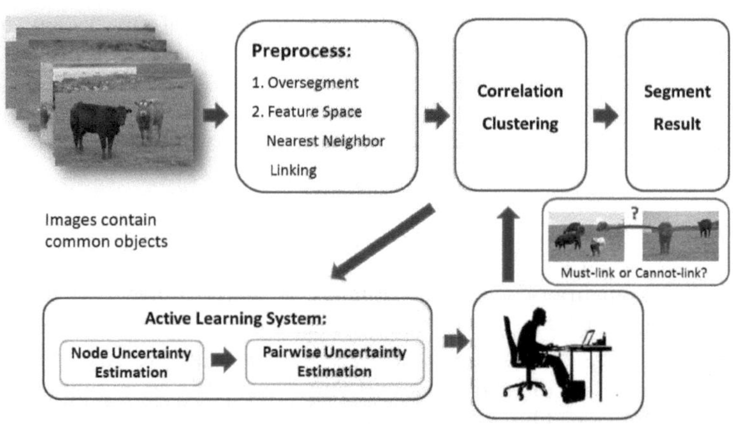

Fig. 1. System diagram of proposed cosegmentation method. **Best viewed in color.**

The rest of the paper is organized as follows. We describe the proposed system in Section 2, while the experimental results are shown in Section 3. Finally, we conclude the paper in Section 4.

2 Framework

The work flow of our method is shown in Fig. 1. The initial input to the system is a set of images containing common objects, and the output is the cosegmentation result, where the regions which are assigned the same color belong to the same class. The images were first over-segmented and we use generated superpixels as the atomic units in the system. Then, a superpixel adjacency graph is constructed across the image set whose weight reflects pairwise similarity. There are two types of edges in the graph: intra-image edges and inter-image edges. Next, graph-based correlation clustering is conducted to get the initial result. Later in the recommendation stage, the method first estimates superpixels uncertainty and recommend uncertain pairs of superpixels to users using the active learning theory to get pairwise constraints. The constraints are propagated in the feature space with extended metric learning techniques. Finally, segmentation is refined by correlation clustering with learned features.

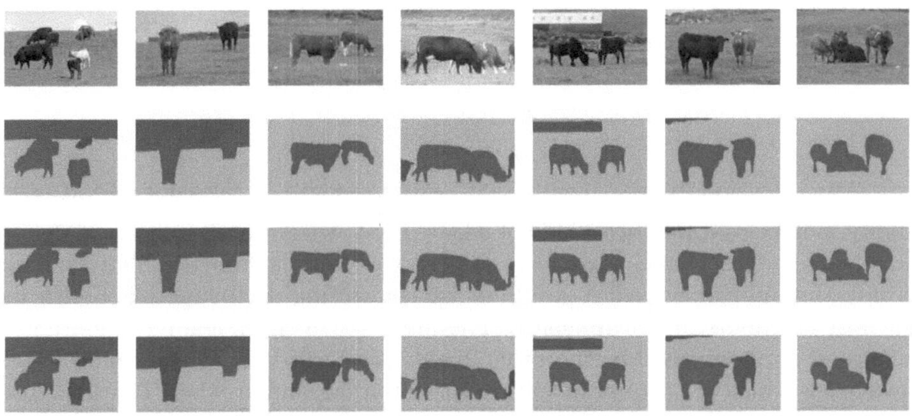

Fig. 2. The first row: input images. The second row: initial segmentation result. The third row and the fourth row are results after applying user constraints. According to active learning recommendation, we show pairs of regions to users like those in Fig. 3 and let the user determine whether the pairs belong to the same class (*must-link* constraints) or different classes (*cannot-link* constraints). The third row shows that if users think black cows and yellow cows as one cluster, our system can give results in which cows are in a cluster. The fourth row shows that if users regard black cows and yellow cows as different clusters, the result will assign yellow cows and black cows different labels. **Best viewed in color.**

2.1 Correlation Clustering

Correlation clustering seeks to maximize the intra-cluster agreement (attraction) and the inter-cluster disagreement (repulsion). This optimization algorithm can

Fig. 3. The recommended pairs of regions. Regions are colored by blue. **Best viewed in color.**

automatically recover a suitable number of clusters : K [10]. Given an affinity matrix $W \in \mathbb{R}^{n \times n}$, n is the number of superpixels. If $W_{ij} > 0$ we say that superpixels i and j attract each other and for those edges $W_{ij} < 0$, we say that i and j repel each other. Any K-way partition of n superpixels can be written as $U \in \{0,1\}^{n \times K}$ $s.t.$ $U_{ic} = 1$ if point i belongs to cluster c. An optimal partition U minimizes:

$$\begin{aligned} E(U) &= -\sum_{ij} W_{ij} \sum_{c} U_{ic} U_{jc} \\ s.t. \quad U_{ic} &\in \{0,1\}, \quad \sum_{c} U_{ic} = 1 \end{aligned} \tag{1}$$

Note that $\sum_{c} U_{ic} U_{jc}$ equals 1 if superpixels i and j belong to the same cluster. The affinity matrix contains two kinds of edges. First, within each image, we define intra-image edges between adjacent superpixels with a four-neighbor system; second, the inter-image edges are built by connecting each superpixel with its l nearest neighbors in feature space. The intra-image edges enforce smoothness among adjacent pixels. On the other hand, the inter-images edges encourage superpixels which are similar in feature space to take similar label. The structure of the graph is illustrated in Fig. 4. The weight of edges is defined as:

$$W(i,j) = log \frac{P(i,j)}{1 - P(i,j)} \tag{2}$$

$P(i,j)$ measures the similarity of superpixels i and j. The whole energy function can be efficiently solved by using the graph-based method proposed by Bagon and Galun [10].

We conduct correlation clustering of superpixels to get the segmentation result. To generate the superpixels, we use the implementation of Arbelaez et al. [9] to get superpixels which can respect the salient object contours in the image.

For Intra-image Edges: We use UCM (Ultrametric Contour Map) from the superpixel method [9]. Each pixel has a score of being part of edges $\Pr(m) \in [0,1]$. For two neighboring superpixels i and j:

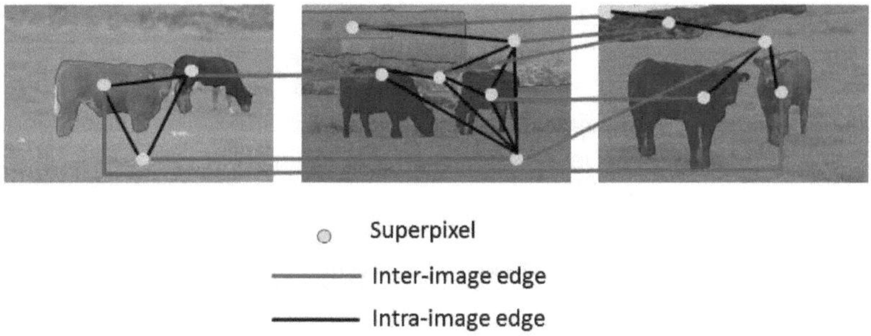

Fig. 4. Illustration of the structure of built graph. **Best viewed in color.**

$$P(i,j) = 1 - \frac{1}{Num} \sum_{m=1}^{Num} Pr(m) \tag{3}$$

Num is the number of pixels on the common boundary of superpixels i and j.

For Inter-image Edges: We model the edge weight by the difference between the features of superpixels i and j:

$$P(i,j) = \exp\{-Dis(i,j)\} \tag{4}$$

$Dis(m,n)$ is the Euclidean distance of two features. The inter-image edges are established by linking the l nearest neighbor in the feature space. Fig. 2 shows the initial segmentation result.

2.2 Active Learning

Bottom-up cues are sensitive to lighting variation and perspective change. Involving users in the system to acquire high-level cues is essential for improving segmentation results. While labeled data is relatively expensive to obtain, active learning techniques aim to query the user to label the most informative entries. In existing applications of active learning in cosegmentation work, they recommend regions based on uncertainty [7,8]. To minimize user effort, a recent study has shown that suggesting pairwise constraints can maximize performance [11]. Given initial result of cosegmentation, we first estimate the node labeling uncertainty of all superpixels. The system can automatically recommend top uncertain pairs of superpixels according to the estimated uncertainty of edges connecting to uncertain superpixels.

Node Uncertainty Estimation. Given current labeling L, a node is uncertain if it causes few energy changes in the energy function by switching its labels. To measure the confidence of whether the initial segmentation result of a certain superpixel i is correct, we extract the expected change of energy. We compute the increase in energy by flipping different labels to a superpixel. If the expected change is small, the system is uncertain of its label. The expected change of superpixel i can be expressed as

$$EC_i = \max_{v \in \{1,2,\cdots,K\}, v \neq u} |E_i(u,v) - E| \tag{5}$$

E is the energy of the initial correlation clustering of superpixels. $E_i(u,v)$ refers to the energy after changing the label of superpixel i from u to v. After initial segmentation, superpixel i is labeled to the u-th cluster. Then we change the label of superpixel i to v which belongs to the set of labels L. We choose the maximum change as the expected change of superpixel i.

In the example of Fig. 2, the yellow cow in picture 1 has similar color feature as that of grass, and it has similar SIFT feature as that of the black cow. So in correlation clustering, labeling it the same as the black cow or grass would have similar energy cost. This type of superpixels are what we want to find in node uncertainty estimation.

Edge Uncertainty Estimation. Having top N uncertain superpixels, we estimate the uncertainty of edges around these nodes. Each pair of superpixels will be scored with the expected changes of energy. We calculate the expected change of these two states: Link and Not-Link. If an edge has a small expected change, whether to label these two superpixels the same makes little difference on the clustering energy, so it requires user annotation. We recommend pairs by top uncertainty measured by expected change. The user can add as many constraints as desired before re-clustering.

2.3 Metric Learning with Pairwise Constraints

Users pairwise constraints give us intuitive information: When two superpixels are given the same label, we want the distance between them to be small. Similarly, if labeled differently, we tend to separate them from each other. We use the metric leaning technique to solve this problem. The goal of metric learning is to learn a distance metric that best fits a known set of property. Thus we can make the best use of the information on pairwise constraints and propagate it to the whole clustering system. Here we follow the semi-supervised dimensionality reduction of Zhang et al. [12] and make modifications to make it perform better in our cosegmentation system.

As described in [12], given a set of d dimensional feature vectors $X = [x_1, x_2, \cdots, x_n]$ of n superpixels, together with some pairwise must-link constraints (M) and cannot-link constraints (C), find a projective matrix Q which is d-by-d' such that the transformed low-dimensional representations $y_i = Q^T x_i$

(d' dimensional) can convey the information of M and C. We use y_i as the new features.

The object function [12] is defined as maximizing $J(Q)$

$$J(Q) = \frac{1}{2n^2}\sum_{i,j}\left(Q^T x_i - Q^T x_j\right)^2 + \frac{\alpha}{2n_C}\sum_{(x_i,x_j)\in C}\left(Q^T x_i - Q^T x_j\right)^2 - \frac{\beta}{2n_M}\sum_{(x_i,x_j)\in M}\left(Q^T x_i - Q^T x_j\right)^2 \tag{6}$$

$w.r.t \quad Q^T Q = 1$. Here n_C and n_M are the numbers of cannot-link and must-link constraints. n is the number of superpixels. The first term of Eq.6 expresses the average squared distance between all data samples in the transformed space. The motivation for considering unlabeled data is to use them to make the system stable when the constraints are few. Second term refers to the squared distance between features of cannot-link constraints. We want to maximize the distance between cannot-link constraints to fit user requirements. The same goes for must-link constraints as the third term shows. α and β are used to balance the contribution of both kinds of constraints.

In our framework, we make full use of the initial segmentation result. We make modification of the object function:

$$J(Q) = \frac{\alpha_1}{2n_D}\sum_{(x_i,x_j)\in D}\left(Q^T x_i - Q^T x_j\right)^2 - \frac{\beta_1}{2n_S}\sum_{(x_i,x_j)\in S}\left(Q^T x_i - Q^T x_j\right)^2 + \frac{\alpha}{2n_C}\sum_{(x_i,x_j)\in C}\left(Q^T x_i - Q^T x_j\right)^2 - \frac{\beta}{2n_M}\sum_{(x_i,x_j)\in M}\left(Q^T x_i - Q^T x_j\right)^2 \tag{7}$$

here $(x_i, x_j) \in D$ means that in the initial segmentation stage, x_i and x_j are in different clusters. $(x_i, x_j) \in S$ means in the initial segmentation stage, x_i and x_j are in the same cluster. n_D and n_S are the numbers of superpixel pairs labeled differently and same respectively. So we change the first term in Eq.6 to two terms in Eq.7 to preserve the structure of the initial segmentation result.

3 Experimental Result

In this section, we present our experiments with multiple object categories using the recently released iCoseg database, along with four MSRC classes.We evaluate the result both on initial segmentation and improved segmentation after active learning. We also present quantitative comparisons between our segmentation result with CoSand from Kim et al. [4] and discriminative clustering method from Joulin et al. [5].

Datasets. There are only foreground and background labels for images in iCoseg dataset. Furthermore, the MSRC dataset is not suitable for multi-label test, because its class does not correspond to a well-defined visual category. Hence we regard main object categories as the foreground and the rest the pixels as

Table 1. Accuracies of segmentation for image sets from iCoseg dataset and MSRC dataset. **Best results in bold.**

Dataset	Class (%)	Ours Before Active Learning	Ours After Active Learning	Kim et al. [4]	Joulin et al. [5]
iCoseg	Baseball Player	64.4	**66.1**	59.7	53.5
	Gymnastics 1	64.0	**70.1**	40.3	18.8
	Goose	76.7	77.1	**77.5**	27.4
	Kite Panda	58.0	**81.8**	50.4	57.8
	Monk	66.5	70.3	67.7	**76.9**
	Skating	68.7	**75.8**	34.2	47.2
	liberty statue	93.1	**96.6**	46.9	88.3
MSRC	Cow	60.3	**65.6**	60.8	42.6
	Tree	72.4	**77.4**	59.7	57.9
	Sheep	62.1	**66.2**	62.3	52.3
	Bike	45.7	**52.9**	36.2	34.2
	Average	66.5	**72.4**	54.2	50.6

the background. We use the standard intersection-over-union score of PASCAL challenges to measure segmentation accuracy.

$$\max_p \frac{1}{I} \sum_{i \in I} \frac{GT_i \cap R_i^p}{GT_i \cup R_i^p} \tag{8}$$

where GT_i is the ground truth and R_i^p the region associated with the p-th class in image i.

Baseline. The number of meaningful "objects" K must be set by hand. In practice, we run CoSand from Kim et al. [4] and the discriminative clustering method from Joulin et al. [5] by changing K from 2 to 5, and report the maximum intersection-over-union scores. We use the source codes provided by original authors.

User Study. To demonstrate the effectiveness of the active learning independently of the specific user input, we performed a study where we asked five participants to cosegment image sets in Table 1. Each user will be shown recommended pairs and make the choice of must-link or cannot-link. Empirically, we set the number of pairs to five, because it is enough to catch user preferences. Also, the metric learning method in our framework can perform in a stable manner when the constraints are few. The 4th column of Table 1 shows the accuracy of cosegmentation after active learning, averaged for all the user sessions.

Result. We user color and SIFT as the features and obtain best performance for 9 of the 11 classes, and achieve best average accuracy. We get significant accuracy improvement after applying user constraints. We can observe from Table 1 that the performance of our result after applying user constraints is better than the initial result, and can achieve up to 23.8% improvement. Fig.5 shows some examples of cosegmentation from five groups. In each set, we show input images and color-coded cosegmentation output. As the affinity matrix in correlation clustering is sparse, it takes an average of $27s$ to run correlation clustering on an image set containing 30 images.

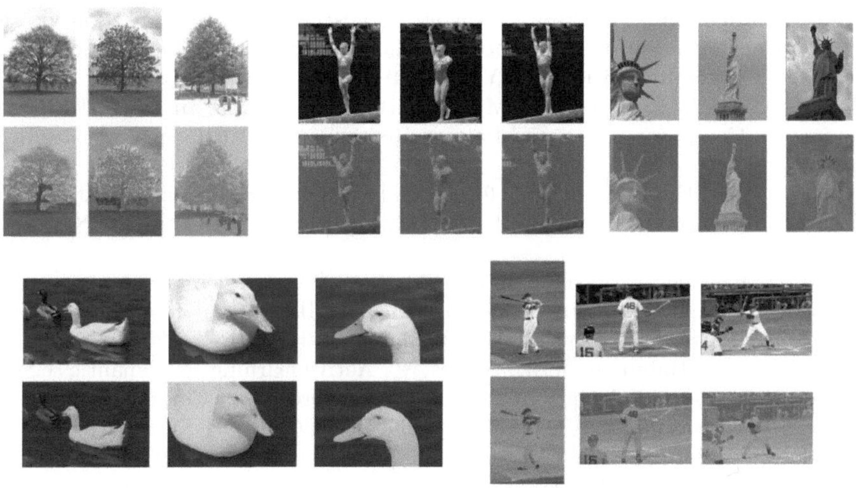

Fig. 5. Example results for the cosegmentation. The images taken from iCoseg and MSRC. We sampled three images per group. **Best viewed in color.**

4 Conclusion

In this paper, we have proposed a multi-class cosegmentation approach which requires no prior knowledge of K (number of classes). After initial segmentation, we derive user pairwise constraints from the active learning process. This enables our system to correct errors and personalize segmentation results according to user preferences. The experiments show that our cosegmentation method performs well and obtains significant improvement after applying user constraints.

Acknowledgment. This research, which is carried out at BeingThere Centre, is supported by the Singapore National Research Foundation under its International Research Centre @ Singapore Funding Initiative and administered by the IDM Programme Office.

References

1. Rother, C., Minka, T., Blake, A., Kolmogorov, V.: Cosegmentation of image pairs by histogram matching-incorporating a global constraint into mrfs. In: 2006 IEEE Computer Society Conference on Computer Vision and Pattern Recognition, pp. 993–1000. IEEE (2006)
2. Hochbaum, D.S., Singh, V.: An efficient algorithm for cosegmentation. In: 2009 IEEE 12th International Conference on Computer Vision, pp. 269–276. IEEE (2009)
3. Mukherjee, L., Singh, V., Dyer, C.R.: Half-integrality based algorithms for cosegmentation of images. In: IEEE Conference on Computer Vision and Pattern Recognition, CVPR 2009, pp. 2028–2035. IEEE (2009)
4. Kim, G., Xing, E.P., Fei-Fei, L., Kanade, T.: Distributed cosegmentation via submodular optimization on anisotropic diffusion. In: 2011 IEEE International Conference on Computer Vision, ICCV, pp. 169–176. IEEE (2011)
5. Joulin, A., Bach, F., Ponce, J.: Multi-class cosegmentation. In: 2012 IEEE Conference on Computer Vision and Pattern Recognition, CVPR, pp. 542–549. IEEE (2012)
6. Kim, G., Xing, E.P.: On multiple foreground cosegmentation. In: 2012 IEEE Conference on Computer Vision and Pattern Recognition, CVPR, pp. 837–844. IEEE (2012)
7. Batra, D., Kowdle, A., Parikh, D., Luo, J., Chen, T.: icoseg: Interactive cosegmentation with intelligent scribble guidance. In: 2010 IEEE Conference on Computer Vision and Pattern Recognition, CVPR, pp. 3169–3176. IEEE (2010)
8. Vezhnevets, A., Buhmann, J.M., Ferrari, V.: Active learning for semantic segmentation with expected change. In: 2012 IEEE Conference on Computer Vision and Pattern Recognition, CVPR, pp. 3162–3169. IEEE (2012)
9. Arbelaez, P., Maire, M., Fowlkes, C., Malik, J.: Contour detection and hierarchical image segmentation. IEEE Transactions on Pattern Analysis and Machine Intelligence 33, 898–916 (2011)
10. Bagon, S., Galun, M.: Large scale correlation clustering optimization. arXiv preprint arXiv:1112.2903 (2011)
11. Klein, D., Kamvar, S.D., Manning, C.D.: From instance-level constraints to space-level constraints: Making the most of prior knowledge in data clustering (2002)
12. Zhang, D., Zhou, Z.-H., Chen, S.: Semi-supervised dimensionality reduction. In: Proceedings of the 7th SIAM International Conference on Data Mining, pp. 629–634 (2007)

Cross-View Gait-Based Gender Classification by Transfer Learning

Zhenjun Yao[1], Zhaoxiang Zhang[1], Maodi Hu[2], and Yunhong Wang[1]

[1] Laboratory of Intelligent Recognition and Image Processing (IRIP), Beijing Key Laboratory of Digital Media
[2] School of Computer Science and Engineering, Beihang University, Beijing, China

Abstract. The gender of a person is easily recognized by his/her gait when training data and test data are from the same view. However, when it comes to cross-view gender classification, traditional methods can not deal with large view variation without enough labeled data in the target view. In this paper, we solve this problem by introducing a transfer learning based framework. Firstly, Gait Energy Image (GEI) of each gait sequence for all views is generated, and Principal Component Analysis (PCA) is carried out to obtain efficient gait representations. Subsequently, an inductive transfer learning approach, TrAdaBoost, is adopted to transfer knowledge from the source view to the target view, which significantly improves the performance of gait-based gender classification. Abundant experiments are conducted and experimental results demonstrate the superiority of the proposed method over traditional gait analysis methods.

Keywords: gait-based gender classification, cross-view, transfer learning.

1 Introduction

Gait has attracted more and more attentions due to the fact that it can identify people at a distance as well as non-contact. Gait-based gender classification is an important branch of gait recognition for its social applications. Many work [1] [2] have been done on gender classification. However, there are still many factors including clothing, variations of view, carrying conditions and injuries, which may affect the recognition performance [3]. Among all these factors, view changes has been considered as one of the biggest challenges, which leads to sharp accuracy decrease since training samples in the source view and test samples in the target view do not share the same distribution.

Researchers have tried to find view-independent methods to solve the cross-view gait recognition. Generally, they can be divided into three categories : 1) extracting view-invariant gait features. C. Benabdelkader et al. [4] made use of Self Similarity Plot (SSP) to enhance the robustness to view changes. However, SSP can only work in a limited range of viewing angles. Robert Bergevin et al. [5] normalized the body trajectories of arbitrary views to a standard plane

B. Huet et al. (Eds.): PCM 2013, LNCS 8294, pp. 79–87, 2013.

so that their similarities can be compared directly. However, this viewpoint invariant feature may suffer from body overlaps, which made it difficult to track body parts such as foot. 2) reconstructing gait information of arbitrary view using 3D calibration. Amit Kale et al. [6] exploited perspective projection and planar imaging geometry to synthesize side-view gait features from other arbitrary views. But it assumed that the person was far away from the camera and camera calibrations must be calculated as well. Furthermore, camera calibration limits the practicability of gait recognition in real applications. 3) utilizing view transformation model (VTM). VTM has been prevalent in recent years since it can deal with large view changes without any calibration. In [7], researchers used VTM which was based on High-Order Singular Value Decomposition (HOSVD) to transform gait features from the source view to the target view.

However, VTM-based method requires sufficient labeled samples in both source and target view for training. In real applications, it is difficult to obtain enough labeled target samples. It is a critical issue to precisely classify the target view data by making use of the source view data.

In this paper, we argue that in cross-view gender classification scenarios, although source view and target view instances are under different distributions, they do intrinsically share the same type of features and labels. This motivates us to develop a strong classifier for the target view with labeled source samples and insufficient labeled target samples. In this case, inductive transfer learning which assumes that certain parts of the source domain data can be reused for learning in the target domain by re-weighting is an appropriate choice. There are many inductive transfer learning methods such as [8] and [9]. In this paper, we propose a cross-view gender classification method based on the former one, TrAdaBoost [8], since it is more suitable to our framework. Experimental results validate our previous statements and demonstrate the advantages of the proposed scheme over traditional methods on the task of cross-view gait-based gender classification.

The remainder of this paper is organized as follows: The process of feature extraction is described briefly in Section 2. We introduce the mechanism of TrAdaBoost and how to exploit it for cross-view gender classification in Section 3. In Section 4, we demonstrate the experimental results. Finally, we draw our conclusions in Section 5.

2 Gait Representation

Assume that a multi-view gait database is available and silhouettes from all views have been extracted from original human walking sequences. Silhouettes processing [10] including size normalization and alignment is then applied on the extracted silhouettes. After estimating the gait cycle of each silhouette sequence by using method in [11], we can obtain the Gait Energy Image (GEI) [12] for cross-view gait analysis.

Given the preprocessed binary gait silhouette images $S_t(x, y)$ at time t in a sequence, the gray-level gait energy image is defined as :

$$G(x, y) = \frac{1}{N} \sum_{t=1}^{N} S_t(x, y) \tag{1}$$

where N is the number of frames in a complete gait cycle, t is the frame number in the sequence and x and y are values in the 2D image coordinate. GEIs of a female and a male from 0° to 180°are shown in Fig. 1. The reasons why we adopt GEI as gait representation are 1) it contains global information of the periodic gait sequences such as body shape, temporal and spatial changes of human body, which is discriminative for gender classification. 2) synthesized by multiple gait sequences in a period, it is insensitive to noise from individual image.

The actual feature of gait can be described as a 1D vector composed of gray values along all consecutive rows and dimensionality reduction is necessary to obtain efficient gait representations. We adopt the classical statistical approach, Principal Component Analysis (PCA), since it is an unsupervised one that can significantly reduce dimension of gait feature.

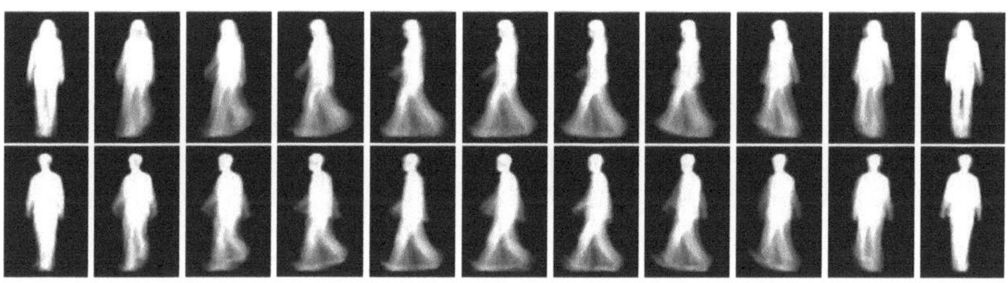

(a)0° (b)18° (c)36° (d)54° (e)72° (f)90° (g)108° (h)126° (i)144° (j)162° (k)180°

Fig. 1. GEIs of a female (top row) and a male (bottom row) for different views

3 Cross-View Gender Classification by Transfer Learning

To address the inductive transfer learning problem, Dai et al. [8] extend the AdaBoost algorithm to TrAaBoost. AdaBoost [13] is an iterative algorithm which aims to train a strong classifier containing certain selected weak classifiers. During the iteration, the weights of instances wrongly classified by the current weak classifier become higher so that more attentions are paid to them in the next round. However, like other traditional methods, AdaBoost cannot be applied to cross-view gait analysis directly because of different-distribution problem. TrAdaBoost tries to bridge the gap between source and target domain by making full use of limited target domain data and enough source domain data. For the target domain, TrAdaBoost processes the data just as AdaBoost does. But for the source domain, wrongly classified instances are getting lower weights in the next round since they are considered the most dissimilar to the target domain. After rounds of iteration, instances similar to the target domain are better utilized and boost the classify ability together with the target domain data.

Algorithm 1. Cross-view gender classification via TrAdaBoost

Input:
Training data with gender label from the source view T_s and target view T_t. The maximum number of iterations N and a base learning algorithm Learner.

1: Initialize the weight vector $w = (w^s, w^t)$, where $w^s = (w_1^s, ..., w_n^s)$ and $w^t = (w_1^t, ..., w_m^t)$, to the desired distribution. Set $\beta_s = 1/(1 + \sqrt{2 \ln n/N})$.

2: **for** $t \leftarrow 1$ to N **do**

3: Normalize the weight vector $w = (w^s, w^t)$ to 1.

4: Call Learner, providing it the combined training set T with the distribution w. Then, get a weak classifier $h_t : X \to Y$.

5: Calculate the error of h_t on T_t:

$$\epsilon_t = \frac{\sum_{j=1}^{m} w_j^t \left| h_t(x_j^t) - y(x_j^t) \right|}{\sum_{j=1}^{m} w_j^t} \tag{2}$$

6: Set $\beta_t = \epsilon_t/(1 - \epsilon_t)$, where $\epsilon_t < \frac{1}{2}$.

7: Update the weight vector:

$$\begin{aligned} w_i^s &\leftarrow w_i^s \beta_s^{\left| h_t(x_i^s) - y(x_i^s) \right|}, \quad 1 \leq i \leq n \\ w_j^t &\leftarrow w_j^t \beta_t^{-\left| h_t(x_j^t) - y(x_j^t) \right|}, \quad 1 \leq j \leq m \end{aligned} \tag{3}$$

8: **end for**

Output: Target view gender classifier function:

$$f_T(x) = \begin{cases} 1, & \sum_{t=\lceil N/2 \rceil}^{N} \ln(1/\beta_t) h_t(x) \geq \frac{1}{2} \sum_{t=\lceil N/2 \rceil}^{N} \ln(1/\beta_t) \\ 0, & \text{otherwise} \end{cases} \tag{4}$$

Here, we introduce how to exploit TrAdaBoost to achieve cross-view gait-based gender classification in detail. After extracting gait features as described in Section 2, let X_s be the set of source-view instances space and X_t be the set of target-view instances space. $Y = \{0, 1\}$ is the set of category labels where 0 and 1 represent female and male respectively. The training data set $T \subset \{X, Y\}$ can be divided into two parts, $T_s = \{(x_i^s, y(x_i^s))|_{i=1}^n\}$ and $T_t = \{(x_j^t, y(x_j^t))|_{j=1}^m\}$, where $X = X_s \cup X_t$, $x_i^s \in X_s$ and $x_j^t \in X_t$. $y(x)$ returns the label of the data instance x. The size of T_s and T_t are n and m respectively. We denote the final target view classifier function as f_T, detailed procedure of the mechanism is presented in Algorithm 1.

In the initial step, set the weights of the source-view instances and target-view instances to the desired distribution. During each round of iteration, the Learner is invoked to train a classifier h_t on the combined training data T. Next, we calculate the error of instances on target-view training data T_t by using equation (2). When obtaining the new error, update the weights of training instances according to equation (3). Note that both β_s and β_t are in the interval $(0, 1)$. In equation (3), we can see: 1) weights of the source-view instances become lower when they are wrongly classified as $\beta_s^{|h_t(x_i^s)-y(x_i^s)|} \in (0, 1)$. 2) when it comes to target view, weights of instances become higher when they are misclassified as $\beta_t^{-|h_t(x_i^t)-y(x_i^t)|} > 1$. After rounds of iterations, the last $\lceil N/2 \rceil$ classifiers are used to give the final prediction, see equation (4).

The theoretical analysis of TrAdaBoost is similar with AdaBoost, and the upper-bound of the generalization error on the target domain is:

$$\epsilon + O(\sqrt{\frac{Nd_{VC}}{m}}) \tag{5}$$

where N is the number of iterations, m is the size of target domain training data T_t, d_{VC} is the VC-dimension of hypothesis space and ϵ is the error on the target domain data from f_T.

4 Experimental Results

CASIA Gait Database B [14] is chosen to evaluate the performance of our proposed method as it is the largest publicly available multi-view gait dataset. The dataset contains 124 subjects, 32 females and 92 males. There are 6 normal walking sequences for each subject. All gait walking sequences are captured from 11 different views, namely 0°, 18°, 36°, 54°, 72°, 90°, 108°, 126°, 144°, 162°and 180°. Sample images of a female and a male from the database for all the 11 views are shown in Fig. 2.

(a)0° (b)18° (c)36° (d)54° (e)72° (f)90° (g)108° (h)126° (i)144° (j)162° (k)180°

Fig. 2. Sample images of a female (top row) and a male (bottom row) from CASIA gait database B for different views

In order to balance the gender ratio, we select 30 females and 30 males randomly. Therefore, we have got $60 \times 6 \times 11$ walking sequences totally. After silhouettes processing and cycle detection for each silhouette sequences, we extract efficient gait features based on the GEI just as Section 2 described. In each experiment, one source view and one target view are selected from all the 11 views. For the source view, we select 30 subjects, 15 females and 15 males, to be training data T_s. The rest 30 subjects are all in the target view. Limited instances of them are randomly chosen as training data T_t, the gender ratio of which is also balanced. The remaining instances from the target view are for test data S. During training process, SVM^{light}[1] with linear kernel is chosen as the base learner for TrAdaBoost.

SVM and Transductive SVM (TSVM) algorithms are chosen for comparison. When only T_t is used for SVM, we denote the situation as SVM_t. If both T_s and T_t are included for SVM, we denote it as SVM_{st}. Besides T_t, test data S without label is required for TSVM as it is a semi-supervised learning method. AdaBoost algorithm is excluded for its limitation in handling the cross-view classification.

In our experiments, the number of iterations is set to 100 and the ratio of T_t/T_s is set from 0.01 to 0.5. Some of the results are given in Table 1. The accuracy of is calculated based on the average of all 11×10 settings since we select one view from the 11 views as source view and one view from the other 10 views as target view in a single trial.

Table 1. Accuracy of different methods under different T_t/T_s ratios

T_t/T_s	SVM_t	SVM_{st}	TSVM	TrAdaBoost
0.01	0.600	0.720	0.689	**0.897**
0.05	0.609	0.754	0.801	**0.907**
0.08	0.614	0.795	0.846	**0.915**
0.10	0.635	0.809	0.852	**0.920**
0.30	0.723	0.875	0.883	**0.938**
0.40	0.894	0.934	0.924	**0.949**
0.50	0.918	0.940	0.931	**0.951**

From Table 1, we can see method based on TrAdaBoost always outperforms others, especially when T_t/T_s is low, about 20% better than TSVM as $T_t/T_s \approx 0.01$. In other words, the difference of distribution is obvious for cross-view gender classification, which makes it difficult for traditional methods to deal with without enough labeled target data. When the ratio is larger than 0.4, the performance of other methods can be comparative with TrAdaBoost. However, as the ratio decreases, our transfer learning based approach starts to demonstrate a distinctive advantage because the labeled target data is hardly sufficient to train a strong classifier by traditional methods.

[1] `http://svmlight.joachims.org/`

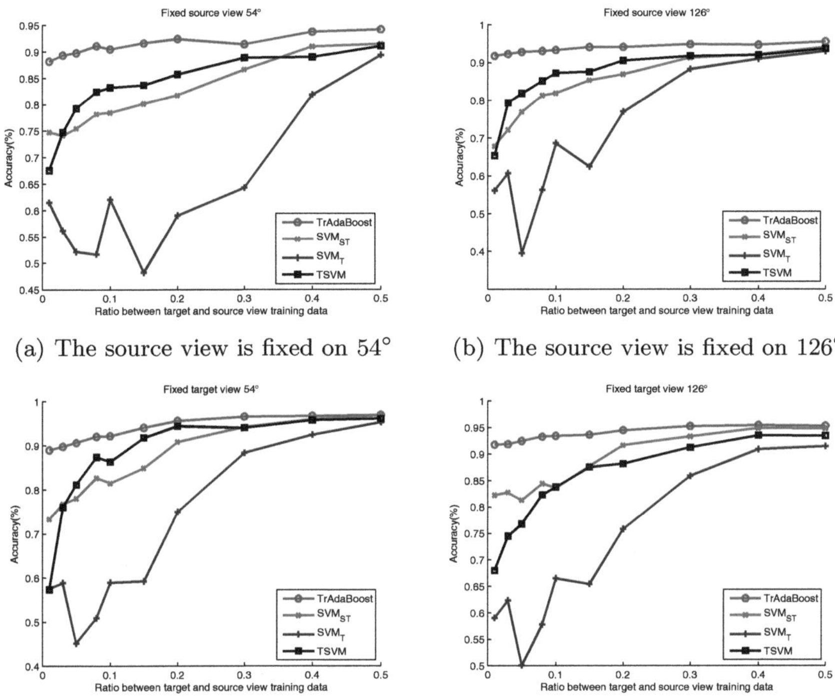

(a) The source view is fixed on 54°

(b) The source view is fixed on 126°

(c) The target view is fixed on 54°

(d) The target view is fixed on 126°

Fig. 3. Comparison of approaches on gender classification

We also conduct a series of experiments to evaluate the performance of different methods when one view, either source view or target view, is fixed. The result is the average of the 10 trials as one of the 10 other views is selected as the corresponding view. Since all the cross-view results demonstrate a similar trend. Without losing simplicity and generality we take 54 °and 126 °for demonstration as they are "medium" views that neither too horizontal nor vertical to the camera. The comparison results are illustrated in Fig. 3. It can be observed that the performance of the method based on transfer learning is always the best while the fluctuation of SVM_t's performance mainly dues to the uncertain classifier trained on the insufficient target-view training data. Meanwhile, the accuracy of TrAdaBoost can mostly achieve 90% even when the ratio T_t/T_s is 1% and rise steadily as T_t/T_s increases, which means that the performance of the proposed method depends little on the increasing number of target-view labeled data. In practical applications, this characteristic is of vital significance for that it can greatly free the system from manually labeling of massive target data which can be extremely time and resource consuming. Finally, there is no negative transfer case (transfer learning lowers the original performance) in our experiments. As a result, TrAdaBoost can be applied on gait-based gender classification safely.

Table 2. Accuracy for typical views with $T_t/T_s = 0.08$, in which the row (S) and column (T) denote the source view and target view, respectively

$S \setminus T$	0°	36°	90°	144°	180°	Average
0°	—	0.941	0.913	0.943	0.905	0.926
36°	0.926	—	0.974	0.944	0.873	**0.929**
90°	0.886	0.940	—	0.902	0.877	0.901
144°	0.928	0.966	0.942	—	0.924	**0.940**
180°	0.908	0.945	0.865	0.905	—	0.906
Average	0.912	**0.948**	0.924	**0.924**	0.895	—

The accuracy of TrAdaBoost method for some typical views are shown in Table 2 when T_t/T_s ratio is set to 0.08. It could be observed that 1) according to the average accuracy comparison, source views of 36° and 144° perform better than the others. The reason is probably that in the two groups, less appearance and dynamic information for human gait is lost than the other views. 2) for each source view group (each row in Table 2), experiments of 36° and 144° target views get the highest transfer accuracy as well. So we can conclude that, medium angles such as 36° and 144°, are better choices for both target and source views in cross-view gait-based gender classification.

5 Conclusions

In this paper, we introduce a framework based on inductive transfer learning to solve the problem of cross-view gait-based gender classification. By the embedding of the TrAdaBoost to our framework, we can effectively reuse the training sample of the source view to enhance the performance in the target view. Experimental results demonstrate that our method outperforms other traditional approaches such as SVM and TSVM. In the further study, we will focus on the influence of transferring knowledge from multiple source views to one target view, namely multi-view transfer learning for cross-view gait-based gender classification, instead of concerning only one source view.

Acknowledgement. This work is funded by the National Basic Research Program of China (No. 2010CB327902), the National Natural Science Foundation of China (NO. 61005016), the Open Projects program of National Laboratory of Pattern Recognition, and the Fundamental Research Funds for the Central Universities.

References

1. Yu, S., Tan, T., Huang, K., Jia, K., Wu, X.: A study on gait-based gender classification. IEEE T-IP 18(8), 1905–1910 (2009)
2. Hu, M., Wang, Y., Zhang, Z., Wang, Y.: Combining spatial and temporal information for gait based gender classification. In: ICPR, pp. 3679–3682 (2010)
3. Bouchrika, I., Goffredo, M., Carter, J.N., Nixon, M.S.: Covariate analysis for viewpoint independent gait recognition. In: Tistarelli, M., Nixon, M.S. (eds.) ICB 2009. LNCS, vol. 5558, pp. 990–999. Springer, Heidelberg (2009)
4. Ben Abdelkader, C., Cutler, R., Davis, L.: Gait recognition using image self-similarity. EURASIP Journal on Advances in Signal Processing (4), 572–585 (2004)
5. Jean, F., Bergevin, R., Branzan, A.: Trajectories normalization for viewpoint invariant gait recognition. In: ICPR, pp. 1–4 (2008)
6. Kale, A., Chowdhury, A., Chellappa, R.: Towards a view invariant gait recognition algorithm. In: AVSBS, pp. 143–150 (2003)
7. Hu, M., Wang, Y., Zhang, Z., Zhang, D.: Multi-view multi-stance gait identification. In: ICIP, pp. 541–544 (2011)
8. Dai, W., Yang, Q., Xue, G.R., Yu, Y.: Boosting for transfer learning. In: ICML, pp. 193–200 (2007)
9. Wu, P., Dietterich, T.G.: Improving svm accuracy by training on auxiliary data sources. In: ICML, pp. 871–878 (2004)
10. Sarkar, S., Phillips, P., Liu, Z., Vega, I., Grother, P., Bowyer, K.: The humanID gait challenge problem: data sets, performance, and analysis. IEEE T-PAMI 27(2), 162–177 (2005)
11. Liu, Z., Sarkar, S.: Improved gait recognition by gait dynamics normalization. IEEE T-PAMI 28(6), 863–876 (2006)
12. Han, J., Bhanu, B.: Individual recognition using gait energy image. IEEE T-PAMI 28(2), 316–322 (2006)
13. Freund, Y., Schapire, R.E.: A desicion-theoretic generalization of on-line learning and an application to boosting. In: Vitányi, P.M.B. (ed.) EuroCOLT 1995. LNCS, vol. 904, pp. 23–37. Springer, Heidelberg (1995)
14. Yu, S., Tan, D., Tan, T.: A framework for evaluating the effect of view angle, clothing and carrying condition on gait recognition. In: ICPR, pp. 441–444 (2006)

Hierarchical Block Size Decision for H.264/AVC Intra Prediction

Hao Zhang[1], Fengfeng Tan[1], and Zhan Ma[2]

[1] Central South University, School of Information Science and Engineering, Changsha, Hunan, 410083, China
[2] Samsung Research America Dallas, Samsung Electronics Richardson, TX 75082 USA

Abstract. In H.264/AVC intra coding, rate-distortion optimization (RDO) is employed to determine the best block size. It demands a huge amount of computing resource, especially for the High Profile which incorporates the additional 8×8 block wise prediction. Obviously, block size prediction with high accuracy could reduce the computational complexity significantly. For this purpose, a two-stage block size decision scheme for H.264/AVC intra prediction is proposed in this work. At first stage, the sum of absolute adjacent pixel error (SAAPE) is used to classify the macroblock into texture and homogeneous groups. Then, the SAAPE of sub-blocks are further applied to determine the best block size in the second stage. With such two-stage design, only one or two block sizes for each macroblock is selected for subsequent RDO process. Extensive experimental results have shown that proposed algorithm leads to over 45% reduction in encoding time on average with negligible degradation in rate-distortion performance.

Keywords: H.264/AVC, block size decision, intra prediction, block features.

1 Introduction

The H.264/AVC standard [1] improves the coding efficiency significantly in comparison to its predecessors, where its superior performance is achieved by introducing spatial-domain directional intra prediction, variable block-size motion estimation, long term reference frames, rate-distortion optimization (RDO), and etc. As for intra coding, multiple directional predictions at different block sizes (i.e., 16×16, 8×8, and 4×4[1]) is adopted to extensively exploit the spatial correlation. Normally, the optimal one at each block size is determined by the expensive RDO process. As there are 4 directional prediction modes for 16×16 block (noted as I16MB) and 9 for the 4×4 and 8×8 block (noted as I4MB and I8MB respectively), original RDO based intra prediction demands tremendous computational

[1] Only luma component is discussed here.

B. Huet et al. (Eds.): PCM 2013, LNCS 8294, pp. 88–97, 2013.

resources. It is unbearable in many scenarios, especially for the real-time video applications. Hence, fast intra prediction has been studied intensively for many years with few of them named here [2–5].

Among those published works, many of then focus on the best prediction mode decision at each block size, such as the most cited work [2] and most recent work [4][2,3]. They are able to speed up the encoder through eliminating some unlikely prediction modes according to a certain pre-defined criteria. Specifically, Pan *et. al* [2] propose a fast intra mode decision algorithm based on the edge detection where Sobel detector is used to calculate the directional amplitude and aggregated amplitude histogram is utilized to pick few modes for RDO calculation. Similarly, Tsai *et al.* [3] develop a intensity gradient filter in intra prediction which exploits the spatial direction and energy of the local texture. In [4], Zeng *et al.* use the sum of absolute Hadamard transform difference (SAHTD) to form a mode subset from the complete prediction modes as well as an early I16MB decision to avoid I4MB process.

These algorithms and others are not mentioned here examined all three intra block sizes to finally reach the optimal coding mode. Intuitively, rich texture region of a frame tends to be encoded with a small block size, while the homogeneous region are better with the large block size. Hence, it is straightforward to further reduce the computational complexity by eliminating the unlikely block sizes according to the local texture distribution. Towards this goal, a *variance*-based (σ^2) fast block size decision is proposed in [6]. However, the calculation of *variance* still requires significant computation because of the large amount of the multiplications involved.

To overcome the issues aforementioned, a new two-stage block size decision is developed in this paper according to the sum of absolute adjacent pixel error (SAAPE). Different thresholds are used to compare the SAAPE or its local variation (via sub-block SAAPE calculation), so as to eliminate the unnecessary process on unlikely block size. As we can see, SAAPE derivation is much easier than the σ^2 where multiplication is completed removed in SAAPE process. On the other hand, SAAPE can better reflect the local texture distribution so as to accurately reduce the unnecessary block size evaluation. This will be confirmed by the simulations shown later.

The rest of the paper is organised as follows. Section 2 briefly introduces the variance-based fast intra block size decision developed in [6]. Section 3 proposes a new two-stage block size decision scheme based on novel SAAPE idea. Experimental results will be presented in Section 4 and conclusions is given in Section 5.

[2] To the best of our knowledge, [4] probably also demonstrates the best performance in terms of the trade-off between coding efficiency loss and encoding time reduction.

[3] Even though [4] implements an early 16×16 block size decision, its main idea is using the Hadamard cost to select few prediction modes for later RDO.

2 A Brief Review of Variance Based Fast Intra Block Size Decision

For luma intra prediction, H.264/AVC introduces three block sizes, i.e. I16MB, I8MB and I4MB. There are 9 directional prediction modes for I8MB and I4MB and 4 modes for I16MB, respectively. A total number of $4 + 4\times9 + 16\times9 = 184$ RDO-based iterations are performed for a luma macroblock to choose the best intra prediction block size and prediction mode. As we can see, it demands the tremendous amount of computation. Obviously, an early skip of sub-optimal block size would lead to significant encoding complexity reduction.

A *variance*-based classification of texture complexity is proposed in [6] to skip one unlikely block size for each macroblock. Macroblock variance σ^2 is defined as:

$$\sigma^2 = \sum_{i=0}^{15}\sum_{j=0}^{15}[Y(i,j)]^2 - \frac{1}{256}\left[\sum_{i=0}^{15}\sum_{j=0}^{15}Y(i,j)\right]^2 \quad (1)$$

where $Y(i,j)$ is the luminance value of the pixel at (i,j)-th position in a macroblock. A macroblock can be categorized into either high or low complexity according the variance derivation via Eq. (1). For high complexity macroblock, I8MB and I4MB are considered as candidate block sizes while I16MB and I8MB for low complexity case on the other hand. As we can see, for each macroblock, one block size will be skipped without going through the RDO. Referring to the Eq. (1), we can see that over 256 multiplications and 512 additions are required to derive the σ^2 (for the most-straightforward implementation). Even after applying the algorithm, there still exists two candidate block sizes for each macroblock, which is still requiring large amount of computing power, in particular for the case that most of the macroblocks are classified as the "high complexity" category (i.e., I8MB and I4MB processing are involved).

Therefore, a two-stage low-cost block size decision scheme is proposed in this paper and will be detailed in the following section. Since computational complexity of chroma intra prediction is much lower than the luma part, and its block size is fixed at 8×8, only luma component is discussed in this paper.

3 Two-Stage Block Size Decision Scheme

Intuitively, rich texture regions of a frame tend to be encoded using a smaller block size while homogeneous regions prefer a larger block size. The error between adjacent pixels could be a good measurement of the texture complexity for a macroblock. Therefore, we define the the sum of absolute adjacent pixel error (SAAPE) as,

$$\text{SAAPE}_{n\times n} = \sum_{i=0}^{n}\sum_{j=0}^{n}\left[|Y(i,j) - Y(i+1,j)| + |Y(i,j) - Y(i,j+1)|\right] \quad (2)$$

where $Y(i, j)$ is the pixel value at (i, j)-th position and $n = 15$ representing the SAAPE of a macroblock. Please note that SAAPE could be derived for any block size with the specific input n. Compared with the σ^2 derivation in Eq. (1), it has much lower computational complexity. Meanwhile, we conduct the experiments using eight different sequences at QCIF and CIF resolutions. SAAPE is collected for each macroblock coded with different block sizes, and shown in Table 1 with the averaged value. As we can see, the SAAPE is quite distant for different encoded block size. A well-designed threshold could differentiate the optimal block size easily. However, for different video content, we can see the SAAPE range overlaps, which brings the difficulty to select the optimal block size with simple thresholds. Therefore, in the following sections, we will introduce the two-stage mechanism to select the optimal coding block so as to avoid the redundant computing.

Table 1. Averaged $\text{SAAPE}_{16\times16}$ for I16MB, I8MB and I4MB

Res.	Sequence	I16MB	I8MB	I4MB
QCIF	Akiyo	746	1353	3855
	Foreman	1781	2336	4501
	Mobile	378	4081	8812
	Salesman	1614	3178	4191
CIF	Coastguard	2688	4576	6212
	News	667	2022	4568
	Hallmonitor	1201	1873	5513
	Silent	1414	2647	3888
Ave.		1311	2758	5193

3.1 First Stage: Rich Texture or Homogeneous

We first introduce a threshold to differentiate whether current block belongs to the rich texture or homogeneous group. If it belongs to the rich texture group, I16MB mode is skipped while I4MB is skipped if it is with the homogeneous group. Let T_{12} be the threshold to differentiate the rich texture and homogeneous group. We collect the hit ratio by varying the T_{12}, where the hit ratio is the percentage of successful block size (either I16MB or I4MB)removal between using the T_{12} and brute force RDO method. As shown in Figure 1, $T_{12} = 2000$ is a good balance to achieve the decent hit ratio and complexity reduction. Moreover, we compare the hit ratio, BD-Rate increase and time reduction between proposed SAAPE and *variance*-based solution [6], as shown in Table 2 . Results demonstrate that our proposed low-complexity SAAPE outperforms the variance based solution with better block size differentiation capability.

In general, if $\text{SAAPE}_{16\times16}$ is low enough or high enough, it has a high probability that corresponding macroblock would choose I16MB or I4MB as the best block size. Based on this, another two threshold T_{11}, T_{13} is set to filter out a

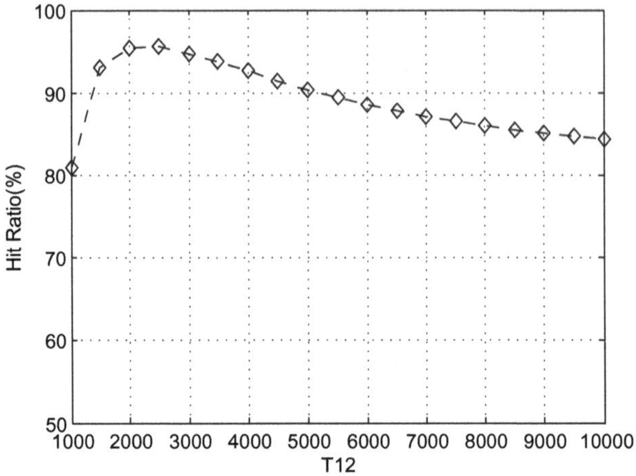

Fig. 1. The hit ratio of T_{12}-based block size removal versus RDO-based solution

portion of macroblocks whose best block size is determined as I16MB or I4MB directly. Since studying the accuracy of T_{11}-based and T_{13}-based decision would make them more dependable, one way is to examine the probability that selected block size is same as the RDO-based decision. As shown in Figure 2 and 3 , the probability could reach as high as 90%. Therefore, in this paper, T_{11} and T_{13} is set to 300 and 10000 empirically. Meanwhile, block size decision flow using $SAAPE_{16\times16}$ is shown in Figure 4 .

Table 2. Hit ratio, BD-Rate increase and time reduction between proposed SAAPE and variance-based classification

Res.	Sequence	σ^2 [6](%)			SAAPE(%)		
		HIT	**BD-Rate**	**ΔTime**	**HIT**	**BD-Rate**	**ΔTime**
QCIF	Akiyo	94.63	0.49	-37.43	95.34	0.30	-35.67
	Mobile	98.62	0.11	-13.50	99.52	0.07	-13.26
	News	89.72	0.71	-25.36	93.28	0.59	-24.86
CIF	Container	87.12	1.23	-34.85	89.57	1.09	-31.16
	Foreman	85.63	1.61	-25.33	92.79	1.20	-35.84
	Paris	90.91	0.64	-24.35	94.36	0.43	-21.09
720P	Vidyo1	95.14	0.96	-50.26	94.79	1.37	-52.85
	Vidyo3	86.20	2.05	-49.13	90.02	0.96	-47.97
	Vidyo4	88.95	2.19	-46.62	95.50	0.65	-51.78
Ave.		90.77	1.11	-34.09	93.91	0.74	-34.94[4]

[4] Compared with *variance*-based solution, SAAPE requires much less arithmetical operations. However, as RDO process occupies such a huge computation, it leads to slight improvement in the overall time reduction.

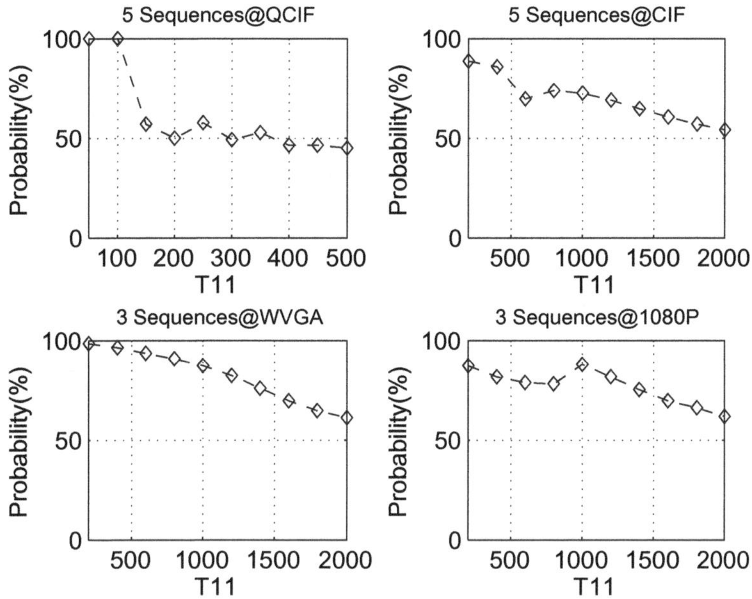

Fig. 2. The probability of T_{11}-based block size is same as the RDO-based decision

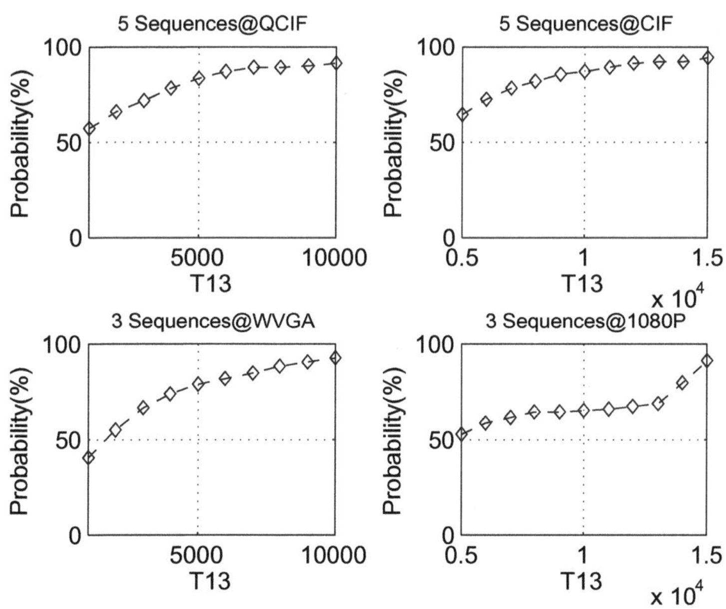

Fig. 3. The probability of T_{13}-based block size is same as the RDO-based decision

3.2 Second Stage: Refinement Using Local Texture Variation

In this stage, the $\text{SAAPE}_{8\times8}$ and $\text{SAAPE}_{4\times4}$ are further employed to determine macroblock's best block size without RDO process. Intuitively, if the SAAPE of the adjacent subblocks varies intensely in one macroblock, it probably has much more texture and tend to be encoded using smaller block size. According to this assumption and our further observations, for homogeneous group (i.e., I16MB and I8MB are candidate block sizes), macroblocks with low mean error of adjacent $\text{SAAPE}_{8\times8}$ (noted as $\text{ME}_{8\times8}$) are tend to be encoded as I16MB, while macroblocks with high $\text{ME}_{8\times8}$ prefer I8MB as the best block size. For texture group (i.e., I16MB and I8MB are candidate block sizes), the mean error of adjacent $\text{SAAPE}_{4\times4}$ (noted as $\text{ME}_{4\times4}$) has the similar behavior as the $\text{SAAPE}_{8\times8}$ in homogeneous group.

$\text{ME}_{8\times8}$ and $\text{ME}_{4\times4}$ are defined as follows,

$$\text{ME}_{8\times8} = \sum_{i=0}^{1}\sum_{j=0}^{1}[|S_{8\times8}(i,j) - S_{8\times8}(i+1,j)| + |S_{8\times8}(i,j) - S_{8\times8}(i,j+1)|] \quad (3)$$

$$\text{ME}_{4\times4} = \sum_{i=0}^{3}\sum_{j=0}^{3}[|S_{4\times4}(i,j) - S_{4\times4}(i+1,j)| + |S_{4\times4}(i,j) - S_{4\times4}(i,j+1)|] \quad (4)$$

where $S_{8\times8}(i,j)$ and $S_{4\times4}(i,j)$ is the SAAPE of subblocks, and (i,j) is the position relative to the macroblock. Please note that the calculation of $\text{SAAPE}_{8\times8}$ and $\text{SAAPE}_{4\times4}$ could be embedded when we are derive the $\text{SAAPE}_{16\times16}$ in the first stage, thus there is not additional computational overhead involved here.

Based on the statistical data provided in Table 3 , two thresholds are preset empirically for each group of macroblocks classified by the first stage. For homogeneous group, the macroblock with $\text{ME}_{8\times8}$ lower than T_{21} is determined as I16MB while higher than T_{22} is determined as I8MB. And for texture group, T_{23} and T_{24} is set to determine the best block size as the homogeneous group. T_{21},T_{22},T_{23} and T_{24} are set to 100, 1000, 1500 and 3000 respectively in this paper. The flow chart of second stage is shown in Figure 4 as well. The hit ratio of these thresholds based block size decision is provided combined with the first stage in the next section.

4 Experimental Results

The proposed scheme is implemented on the latest H.264 reference software [7]. Main encoder configurations are given in Table 4. The simulations run at a Linux machine with an AMD quad-core 1.9GHz CPU and 32GB RAM.

The hit ratio of proposed scheme, BD-Rate [8] and time saving (noted as ΔTime) against the RDO-based block size decision are tabulated in Table 5. The "Checked" represents the percentage of macroblocks whose one best block size is determined by the proposed solution, and $\Delta\text{Time} = (T_{\text{PROP}} - T_{\text{RDO}})/T_{\text{RDO}} \times$

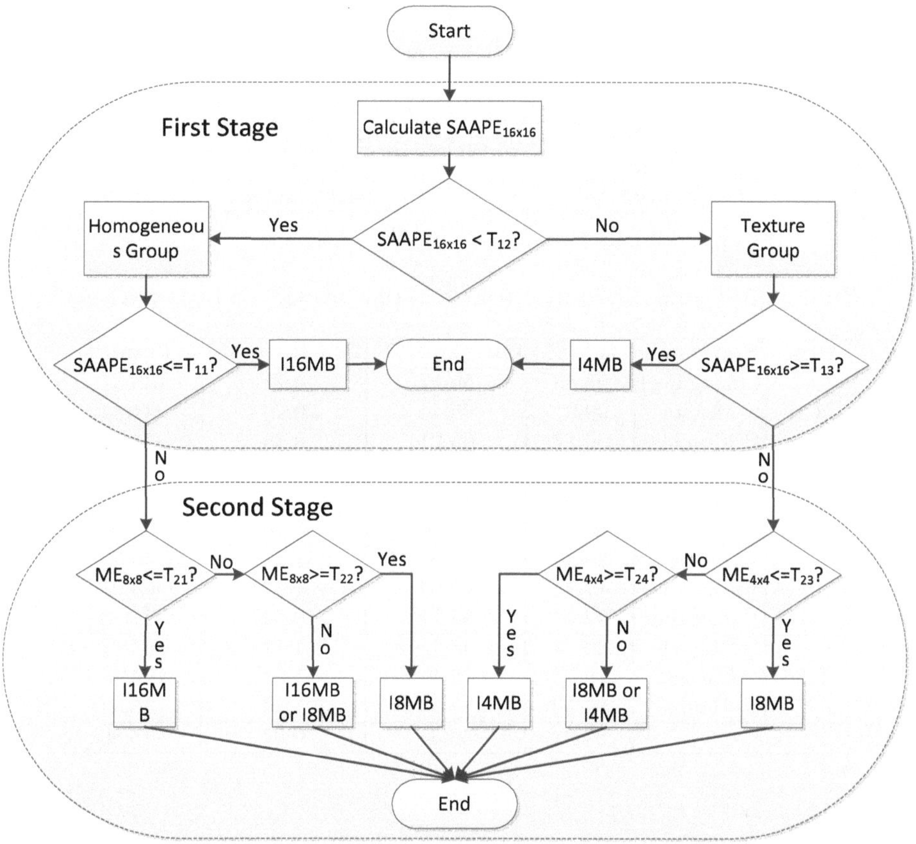

Fig. 4. Flowchart of hierarchical block size decision based on SAAPE

Table 3. Mean error of adjacent $SAAPE_{8\times8}$ and $SAAPE_{4\times4}$ for three macroblock types

Res.	Sequence	$ME_{8\times8}$			$ME_{4\times4}$		
		I16MB	I8MB	I4MB	I16MB	I8MB	I4MB
QCIF	Akiyo	315	745	2295	477	853	3472
	Mobile	225	2066	3460	268	2375	6019
	Salesman	947	1103	1793	970	1661	2799
CIF	News	206	1003	2544	324	1218	3603
	Hallmonitor	467	774	2737	595	1009	3896
	Paris	944	1060	3721	1118	1280	4970
Ave.		517	1125	2758	625	1399	4127

Table 4. Encoder Configuration

Parameter	Value	Description
JM	18.4	Latest version
IntraPeriod	1	IIIII sequence
QPISlice	28,32,36,40	4 Quantization parameters
ProfileIDC	100	High profile
SymbolMode	1	CABAC
RDOptimization	1	RDO enabled

Table 5. Hit ratio, BD-rate increase and time saving of proposed scheme

Res.	Sequence	HIT(%)	Checked(%)	BD-Rate(%)	ΔTime(%)
QCIF	Coastguard	92.43	59.17	0.71	-35.37
	Mobile	99.01	95.46	0.26	-31.84
	Bridgeclose	90.17	65.21	0.95	-34.73
CIF	Coastguard	84.52	58.54	1.18	-35.83
	Waterfall	91.21	52.15	0.46	-38.64
	Paris	88.58	70.10	0.80	-38.58
WVGA	Keiba	90.30	44.50	1.21	-40.33
	Flowervase	87.72	67.54	1.42	-68.25
	PartyScene	92.49	64.79	0.64	-31.75
HD	Vidyo4	82.18	46.74	1.77	-64.88
	Cactus	88.75	42.17	1.01	-54.14
	Tennis	90.49	34.54	1.24	-66.04
Ave.		89.82	58.41	0.97	-45.86

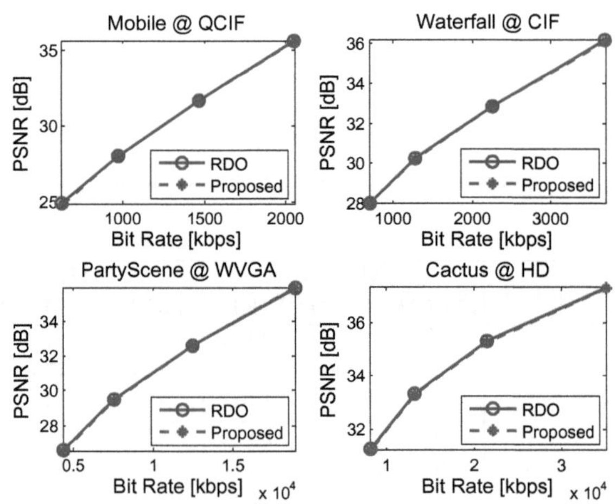

Fig. 5. Rate-distortion performance for different video sequences

100%. As we can see, the hit ratio of proposed scheme reach up to 90%, and about 60% of macroblocks are "Checked", and the other 40% of macroblocks only need to examine two block sizes. The encoding time saves 45.86% on average with 0.97% increase in BD-Rate. Figure 5 shows the rate-distortion curves for four sequences and both curves almost coincide with each other. All these results demonstrate that proposed scheme could reduce the computation complexity with negligible loss in performance. Combining it with the fast prediction mode decision would be our future work to further improve the encoding efficiency.

5 Conclusion

In this paper, a hierarchical fast block size decision is proposed to speed-up the encoding without noticeable coding efficiency loss. A two-stage scheme is proposed to eliminate sub-optimal block sizes with a high accuracy up to 90% on average. Experimental results demonstrate a significant encoding time (i.e., 45.86% on average) reduction with negligible performance loss (i.e., 0.97% in terms of BD-Rate). Since the latest High-Efficiency Video Coding (HEVC) implements the recursive Coding Tree Unit (CTU) from 64×64 to 8×8 rather macroblocks any more, it will be interesting to see the performance by applying the proposed algorithm for HEVC intra coding. This is deferred as our future work.

References

1. Joint Video Team, "Advanced video coding for generic audiovisual services," ITU-T Rec. H.264 & ISO/IEC 14496-10 AVC (March 2005)
2. Pan, F., Xiao Lin, L., Rahardja, S., Lim, K., Li, Z.G., Wu, D., Wu, S.: Fast Mode Decision Algorithm for Intraprediction in H.264/AVC Video Coding. IEEE Trans. Circ. and Sys. for Video Tech. 15(7), 813–822 (2005)
3. Tsai, A.-C., Paul, A., Wang, J.-C., Wang, J.-F.: Intensity Gradient Technique for Efficient Intra-Prediction in H.264/AVC. IEEE Trans. Circ. and Sys. for Video Tech. 18(5), 694–698 (2008)
4. Zeng, H., Ma, K.-K., Cai, C.: Hierarchical intra mode decision for h.264/avc. IEEE Transactions on Circuits and Systems for Video Technology 20(6), 907–912 (2010)
5. Wang, F., Fan, Y., Lan, Y., Liu, W.: Fast intra mode decision algorithm in H.264/AVC using characteristics of transformed coefficients. In: Visual Information Engineering (2008)
6. Huang, Y.-H., Ou, T.-S., Chen, H.H.: Fast decision of block size, prediction mode, and intra block for H.264 intra prediction. IEEE Transactions on Circuits and Systems for Video Technology 20(8), 1122–1132 (2010)
7. H.264/AVC JM (2013), `http://iphome.hhi.de/suehring/tml/`
8. Bjontegaard, G.: Calculation of average PSNR differences between RD-curves. In: 13th VCEG Meeting, Doc. M33, Austin TX (April 2001)

Scale Selection in SIFT Flow for Robust Dense Image Matching

Shimiao Li and Wei Xiong

Institute for Infocomm Research, Agency for Science,
Technology and Research (A*STAR), Singapore
{lis,wxiong}@i2r.a-star.edu.sg

Abstract. SIFT flow is a method to align an image to its neighbors in a large image collection consisting of a variety of scenes. Originally proposed for cross-scene alignment, it has good potential for robust dense matching between images of a scene taken under large viewing condition changes. However, this potential has not been fully exploited due to dense SIFT in SIFT flow is extracted under single pre-defined scale. In this paper, we explore this potential and propose new algorithms to select proper scales while applying SIFT flow for dense image matching. By studying the behavior of SIFT flow under scale changes, we propose two new concepts, namely, the optimal relative scale factor (ORSF) and the optimal matching scale (OMS) using gradient-enhanced normalized mutual information. ORSF and OMS define the most proper scales during SIFT flow matching. Suboptimal and optimal methods for estimating ORSF and OMS are proposed. It is shown that by applying ORSF and OMS, the accuracy of SIFT flow for dense image matching is greatly improved on images with significant scale changes.

Keywords: SIFT flow, dense image matching, scale selection.

1 Introduction

Image correspondence matching is a crucial problem with wide applications in computer vision. Finding correct correspondence points is important in computer vision tasks such as object recognition, image retrieval, robot navigation, 3D reconstruction etc. Generally, there are two approaches for finding correspondences. Firstly, the interest point approach, such as SIFT [1] etc, detects local feature points and represents them using robust feature descriptors which are invariant to certain image transformations. This kind of techniques is able to find correspondences under large viewing condition changes but only sparse correspondences can be obtained. Often many outliers occur due to the lack of spatial constraint. Secondly, the dense sampling approach, such as optical flow techniques [2], aims to find dense correspondences on densely sampled grid on images. Traditionally, only simple local image information such as pixel intensity or local patch intensity is used in this approach. Such information can be very sensitive to viewing condition changes which makes this kind of techniques unable to handle large image deformations. Neither of the above two approaches is

B. Huet et al. (Eds.): PCM 2013, LNCS 8294, pp. 98–109, 2013.

able to deal with dense correspondence matching under large image deformation caused by viewpoint changes, lighting condition changes and occlusions.

SIFT flow, proposed by Liu et al [3], is a hybrid approach in which SIFT feature descriptors are extracted on densely sampled grid. Global optimization method, which is often used in optical flow techniques, is applied to find dense correspondences. The use of robust feature descriptor in a densely sampling way, together with global optimization, brings SIFT flow the potential ability of finding dense correspondences under large viewing condition changes. However, this potential has not been fully exploited because SIFT flow was originally proposed for cross-scene alignment in large database, which aims to match similar structures in different scenes of the same category. The large database used by SIFT flow in [3] guarantees that the scale changes between a query image an its neighbors would be small thus a single pre-selected scale would be sufficient in most cases. However, in correspondence matching, we usually do not have such a large database and proper scale selection is necessary. In another aspect, the key difference between cross-scene alignment and correspondence matching on images of the same scene is that in the former problem, pixel-level accuracy matching is not defined and thus is not required. So in cross-scene alignment, SIFT flow can obtain reasonable alignment without carefully selecting the scale of SIFT. A pre-selected scale is used for extracting dense SIFT. While in the latter problem, correspondences are well defined. To obtain accurate matching, the scale of SIFT descriptor needs to be carefully selected in the two images especially when there is significant scale change between them. Even in cross-scene alignment, blindly selecting a single scale does not generate reasonable alignment when scale change is big.

However, choosing the proper scale for dense SIFT features in SIFT flow is not a well-defined or well-understood task. In sparse SIFT, scale is selected by finding the local extrema in the Laplacian of Gaussian (LOG) function computed over scales [1]. The repeatability of SIFT features computed in such way is largely guaranteed in spite of viewing condition changes. While in the SIFT flow case, most of points on the densely sampled grid do not lie on extrema of LOG and scale cannot be selected for them. This poses great difficulty in selecting the proper scales for feature extraction in SIFT flow.

In this paper, we propose new algorithms for selecting the proper scales while applying SIFT flow for dense image matching of images taken with different viewing conditions in a scene , especially with large scale change. We firstly explore the behavior of SIFT flow under different selected scales. Then based on the observation, we propose two new concepts, namely, *the optimal relative scale factor (ORSF)* and *the optimal matching scale (OMS)*. ORSF and OMS define the most proper scales during SIFT flow matching. Optimal and suboptimal methods for estimating ORSF and OMS are proposed. It is shown that by applying ORSF and OMS, the performance of SIFT flow for dense image matching is greatly improved on matching images with significant scale changes.

The rest of this paper is organized as follows. Section 2 reviews recent related works. Section 3 analyzes the behavior of SIFT flow under scale changes and proposes ORSF and OMS for proper scale selection. Section 4 proposes two new algorithms for scale selection during SIFT flow matching. Section 5 shows experimental results. Section 6 concludes the paper and discusses about future working directions.

2 Related Works

There are several recent works closely related to the proposed method in this paper. Hassner et al [4] studies the scale problem in SIFT flow and proposes an alternative feature descriptor, the scale-less SIFT (SLS), which does not change with scale,by computing the linear subspace of SIFTs under a range of scales. SLS seems to be effective in handling scale changes. However, it is very time-consuming to compute the linear subspace for all the pixels on the densely sampled grid. Tuytelaars [5] obtains dense feature points with scale by searching for the locally maximal value of LOG function within a bounded searching area. It is shown that repeatability of features is improved. But it is unclear if large scale change can be handled by the method. Xu et al [6] solves the scale change problem in optical flow by adding scale as an additional optimization variable in the objective function. This method has the potential of being adapted to SIFT flow, but only when the behavior of SIFT flow under scale changes is better understood. The generalized patch match algorithm [7] handles scale change during matching by randomly searching over scales. But due to the lack of spatial constraint, the method generates large numbers of matching outliers, which limits its usage in dense image matching. To our knowledge, so far in the literature, there is no work directly handles the scale selection problem in image matching using SIFT flow.

3 Optimal Scales for SIFT Flow Matching

3.1 SIFT Flow and Scale Changes

Fig. 1. The example image pair (the two men pair) (images are taken from [4]). Left: image 1; Middle: image 2; Right: original SIFT flow matching result (image 2 is wrapped to image 1 using the estimated SIFT flow).

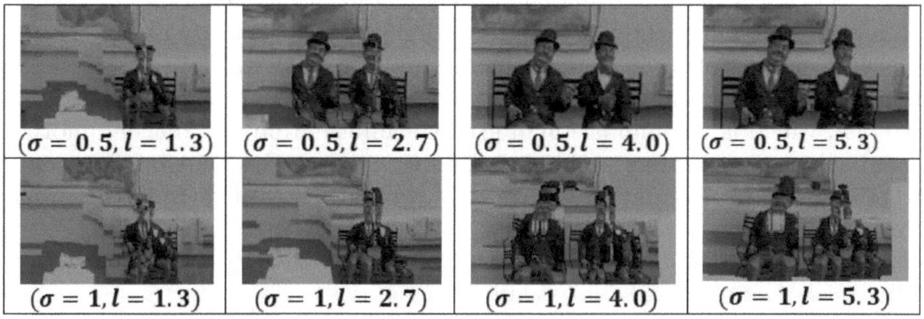

Fig. 2. SIFT flow matching results under different selected scales

Fig. 3. Matching result under optimal scales ($\sigma_0 = 0.6, l_0 = 4.0$)

Consider using SIFT flow to match two images of a scene which are of significant scale change, as the example in Fig.1(left) and Fig.1(middle). Original SIFT flow with default window size of dense SIFT cannot give reasonable matching result for this example due to the large scale change (Fig.1(right)). However, with proper scale selection, SIFT flow may be able to handle this scale change. To explore this potential of SIFT flow, let l_1 and l_2 be the scales on which dense SIFT features are extracted from the two images respectively (i.e. the actual descriptor window size is $12l_1$ and $12l_2$). Let $l = l_1$, $\sigma = \frac{l_2}{l_1}$. We call l *the matching scale (MS)* and σ *the relative scale factor (RSF)*. Fig.2 shows SIFT flow matching results by wrapping image 2 to image 1 using SIFT flow estimated under different selected scales (σ, l). It is shown that under some proper scales, SIFT flow is able to give reasonable dense matching result. Now our questions are: how to define the proper scales mathematically? How to find these proper scales in an efficient way?

3.2 Defining Optimal Scales Using Gradient-Enhanced Mutual Information

As it is shown in Fig.2, under some proper scales, SIFT flow gives visually reasonable matching results. One way to measure the matching accuracy is computing the mutual information [8] between image 1 and the wrapped image from image 2 to image 1. Matching on the proper scales should give higher mutual information than on non-proper scales. However, as pointed out in [8] [9], mutual information is not the optimal measure for matching thin structures. As a result, as we will show in Section 5, although matching under large scales generates larger mutual information value, the matching is actually not as accurate as under small scales at locations of high gradient magnitude. To deal with this problem, we define proper scales for matching using gradient-enhanced mutual information (GMI) [9], which is a combination of gradient information and mutual information, to enhance accurate matching on thin structures. Gradient information in GMI measures the alignment accuracy by comparing gradient magnitude and direction.

Specifically, we define *the optimal matching scale (OMS)* l_0 and *the optimal relative scale factor (ORSF)* σ_0 as follows:

$$(\sigma_0, l_0) = \arg\max_{(\sigma,l)} GNMI(\mathbf{I_1}, \mathbf{I_2'}(\sigma, l)) \tag{1}$$

where $\mathbf{I_2'}(\sigma, l)$ is the wrapped image from image 2 to image 1 using SIFT flow estimated on scales (σ, l). $GNMI(*, *)$ is the gradient-enhanced normalized mutual information [9] between two images. For simplicity, we write $GNMI(\mathbf{I_1}, \mathbf{I_2'}(\sigma, l))$ as $GNMI(\sigma, l)$ in the later discussions. In the above example, $GNMI(\sigma, l)$ has a maximum value $GNMI(0.6, 4.0)$ in a range of possible scales $(0.1 < \sigma < 1, 1.3 < l < 14)$. By our definition, $(\sigma_0 = 0.6, l_0 = 4.0)$ are the optimal scales for SIFT flow matching between the two images in this example case (Fig.3).

Ideally, ORSF σ_0 corresponds to the physical relative scale factor between the two images. Suppose we have an object of size L in the real world. In the first image, this object has size l_1. In the second image, size l_2. The physical relative scale factor between the two images is $\sigma_0 = \frac{l_2}{l_1}$. In sparse SIFT matching, σ_0 can be estimated by comparing the scales of matched SIFTs. However, when viewing condition change is large, sparsely matched SIFTs may contain many outliers, making the estimation of σ_0 unreliable.

OMS l_0 may not have such a clear physical meaning as ORSF σ_0. Let us try to explain it in this way. With too small l, fine scale dense SIFT features cannot characterise the common visual content of the two images. A metaphor of this phenomenon is one cannot compare the appearance of people by using microscope looking into the pores on the faces. On the other aspect, with too big l, although large scale dense SIFT features may still give reasonable matching result, the matching accuracy decreases due to by using large descriptor window, SIFTs can be of little changes in a large image region. This will cause blurring effect on edges in the wrapped image as we will see in Fig.8 in Section 5. Therefore, a proper matching scale is important for accurate matching and l_0 is the optimal value for it.

3.3 Observation on the GNMI Surface

However, finding the optimal scales (σ_0, l_0) is not easy. If we have M possible values for σ and N possible values for l, we need to evaluate $M \times N$ $GNMI(\sigma, l)$ values to find the optimal scales. Fortunately, by examining the surface $GNMI(\sigma, l)$ (Fig.4), we have the following observation: On large matching scale $l = l'$ where $l' > l_0$, the curve $GNMI(\sigma, l')$ achieves maximum value on $\sigma_{0'}$ which is not too far away from σ_0. In more formal terms, the above observation can be expressed as follows:

Observation 1. *Let* $\sigma_{0'} = \arg\max_\sigma GNMI(\sigma, l')$, *where* $l' > l_0$. *We have* $|\sigma_{0'} - \sigma_0| < \epsilon$, *where* ϵ *is a small number.*

Observation 1 tells that we can obtain a sub-optimal RSF $\sigma_{0'}$ by only evaluating the $GNMI(\sigma, l')$ curve on a large matching scale $l' > l_0$. Note that although Fig.4 is plotted based on the example in Fig.1, the above observation generally holds for SIFT flow matching under scale changes. We have observed it from all the image pairs used in our experiment. One possible explanation for Observation 1 is that on large scale, where big descriptor window is used, SIFT flow matching accuracy is more sensitive to RSF, due to a non-proper RSF will cause very different image content being represented in the descriptors of two images.

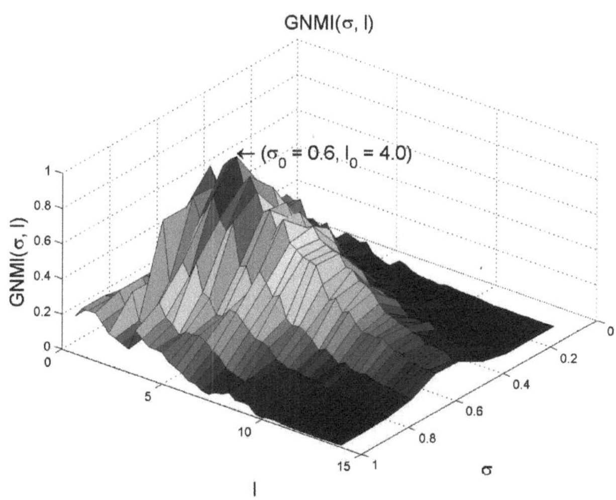

Fig. 4. GNMI surface: $GNMI(\sigma, l)$

4 The Scale Selection Algorithms

Based on the analysis in Section 3, we firstly propose the suboptimal scale selection algorithm for selecting the proper scale when SIFT flow is used for matching two images of significant scale change (Algorithm 1). This algorithm reduces the search space of proper scales from $M \times N$ to $M + N$. Suboptimal scales can be selected by the algorithm, which gives high gradient-enhanced mutual information between image 1 and the wrapped image from image 2 to image 1.

To find the optimal scales which gives maximum gradient-enhanced mutual information, we propose the optimal scale selection algorithm (Algorithm 2).

Input: two images to be matched
Output: selected suboptimal scales $(\sigma_{0'}, l_{0'})$
SIFT flow under the selected suboptimal scales
Initialization: $l \leftarrow l'$, where l' is a large scale;
for $\sigma \leftarrow \sigma_1$ *to* σ_2 **do**
  estimate SIFT flow using scales (σ, l');
  compute and record $GNMI(\sigma, l')$;
end
$\sigma_{0'} \leftarrow \arg\max_\sigma GNMI(\sigma, l')$;
for $l \leftarrow l_1$ *to* l_2 **do**
  estimate SIFT flow using scales $(\sigma_{0'}, l)$;
  compute and record $GNMI(\sigma_{0'}, l)$;
end
$l_{0'} \leftarrow \arg\max_l GNMI(\sigma_{0'}, l)$;

Algorithm 1. Suboptimal algorithm for scale selection in SIFT flow

Input: two images to be matched
Output: selected optimal scales (σ_0, l_0)
SIFT flow under the selected optimal scales
Initialization: $l \leftarrow l'$, where l' is a large scale;
$GNMImax \leftarrow 0$;
$\sigma_{0'} = \sigma_1$;
repeat
  for $\sigma \leftarrow \sigma_1 - \triangle\sigma$ *to* $\sigma_1 + \triangle\sigma$ **do**
    estimate SIFT flow using scales (σ, l');
    compute and record $GNMI(\sigma, l')$;
  end
  $\sigma_{0'} \leftarrow \arg\max_\sigma GNMI(\sigma, l')$;
  if $GNMI(\sigma_{0'}, l') > GNMImax$ **then**
    $GNMImax \leftarrow GNMI(\sigma_{0'}, l')$; $\triangle\sigma \leftarrow \triangle\sigma/2$; $\sigma_0 = \sigma_{0'}$; $\sigma_1 = \sigma_{0'}$;
    for $l \leftarrow l_1$ *to* l_2 **do**
      estimate SIFT flow using scales $(\sigma_{0'}, l)$;
      compute and record $GNMI(\sigma_{0'}, l)$;
    end
    $l_{0'} \leftarrow \arg\max_l GNMI(\sigma_{0'}, l)$;
    if $GNMI(\sigma_{0'}, l_{0'}) > GNMImax$ **then**
      $GNMImax \leftarrow GNMI(\sigma_{0'}, l_{0'})$; $l_0 \leftarrow l_{0'}$; $l' \leftarrow l_{0'}$;
    end
  end
until *$GNMImax$ does not change* **or** *$GNMImax > Threshold$*;

Algorithm 2. Optimal algorithm for scale selection in SIFT flow

This algorithm alternatively updates σ and l by searching for the maximum value on $GNMI(\sigma, l')$ and $GNMI(\sigma', l)$ curves iteratively. Experiments in Section 5 show that the algorithm converges in a few iterations. The search space of proper scales reduces from $M \times N$ to $kM + 2N(1 - \frac{1}{2^k})$, where k is the number of iterations. Suboptimal scale selection algorithm corresponds to the $k = 1$ case.

5 Experiments

We evaluate the proposed methods on 8 image pairs (Fig.5). Each pair of images is taken from a scene under very different viewing conditions, including large scale, lighting and viewpoint changes. We compare the dense matching results of the original SIFT flow method (with default window size), the proposed suboptimal algorithm and the proposed optimal algorithm. Our program selects over relative scale factor between 0.1 to 1 and matching scale between 0.7 to 10.0. We choose initial matching scale to be 8.0 for both suboptimal and optimal algorithms.

Fig.6 shows the suboptimal algorithm results of the first six pairs of images, comparing with result of the original SIFT flow. For these six pairs, the optimal algorithm obtains the same results with the suboptimal algorithms with longer computation time. It is shown that the suboptimal algorithm produces significantly improved dense matching results compared to the original SIFT flow method. Note that when large scale is selected, we cannot get the matching on near image boundary regions. This is a limitation of selecting large scale for matching. Fig.7 shows results of two pairs on which the optimal algorithm outperforms the suboptimal algorithm. It can be seen that optimal algorithm generates better matching on some detailed image regions. Table 1 shows estimated ORSF and OMS together with computation time (on computer with Intel 3.2GHZ CPU). The computation speed can be improved by using parallel computing on different relative scale factor and matching scale.

Table 1. Estimated ORSF, OMS and computation time of the proposed suboptimal and optimal algorithms

Image Pair	suboptimal $(\sigma_{0'}, l_{0'})$	optimal (σ_0, l_0, k)	image resolution	time used (suboptimal) (seconds)	time used (optimal) (seconds)
two men	(0.6, 4.0)	(0.6, 4.0, 2)	400×265	441.9	494.7
Notre Dame	(0.5, 5.3)	(0.5, 5.3, 2)	550×412	1117.0	1243.7
Summer Place	(0.7, 7.3)	(0.7, 7.3, 2)	382×256	388.4	429.8
Trevi fountain	(0.6, 9.7)	(0.6, 9.7, 2)	600×427	1198.7	1296.6
Chinese garden	(0.7, 9.0)	(0.7, 9.0, 2)	640×265	1369.3	1489.0
New Swan castle	(0.5, 10.0)	(0.5, 10.0, 2)	500×350	815.0	887.0
Fullerton hotel	(0.6, 10.0)	(0.7, 10.0, 3)	500×334	782.0	1478.2
Louvre	(0.6, 10.0)	(0.7, 10.0, 3)	400×269	451.9	867.3

Fig. 5. Testing image pairs used for the experiments. Left column: target image; Right column: source image. Source image is expected to be warpped to target image according to dense matching result. From top to bottom: the two men pair, the Notre Dame pair, the Summer Place pair, the Trevi fountain pair, the New Swan castle pair, the Chinese garden pair, the Fullerton hotel pair, the Louvre pair.

Fig. 6. Results comparison of the original SIFT flow (shown in the left column) and the proposed suboptimal algorithm (shown in the middle column). For visual comparison, original target images are shown in the right column. From top to bottom: the two men pair, the Notre Dame pair, the Summer Place pair, the Trevi fountain pair, the New Swan castle pair, the Chinese garden pair. The optimal algorithm produces the same results as the suboptimal algorithm on these pairs.

Fig. 7. Results comparison of the original SIFT flow (shown in 1st column), the proposed suboptimal algorithm (shown in 2nd column) and the proposed optimal algorithm (shown in 3rd column). For visual comparison, original target images are shown in 4th column. From top to bottom: the Fullerton hotel pair, the Louvre pair.

Fig. 8. Part of the absolute difference image between the target image and the wrapped image for the Trevi fountain pair (left: $l = 6.0$; right: $l = 10.0$), showing edge blurring effects when large matching scale ($l = 10$) is used

Fig. 8 shows the absolute difference image between the target image and the wrapped image for a small region in Trevi fountain image pair. When large matching scale l is used, we can see obvious blurring effects on the edges. This illustrates the importance of selecting the proper matching scale. The use of gradient-enhanced mutual information suppresses choosing too large scales and improves the accuracy of matching on the edges, whereas using mutual information alone cannot achieve this purpose.

6 Conclusion and Future Works

New algorithms for selecting the proper scales in the SIFT flow matching are proposed in this paper. It is shown that the proposed methods significantly improves the performance of the SIFT flow for dense matching of images under significant scale changes. Future work will be focused on improving the efficiency by parallel computing. Besides, the problem of local scale selection will be studied. Dense 3D reconstruction will be performed based on the matching result.

References

1. Lowe, D.G.: Distinctive image features from scale-invariant keypoints. International Journal of Computer Vision 60(2), 91–110 (2004)
2. Baker, S., Scharstein, D., Lewis, J., Roth, S., Black, M.J., Szeliski, R.: A database and evaluation methodology for optical flow. In: IEEE International Conference on Computer Vision (ICCV), pp. 1–8 (2007)
3. Liu, C., Yuen, J., Torralba, A., Sivic, J., Freeman, W.T.: SIFT flow: Dense correspondence across different scenes. In: Forsyth, D., Torr, P., Zisserman, A. (eds.) ECCV 2008, Part III. LNCS, vol. 5304, pp. 28–42. Springer, Heidelberg (2008)
4. Hassner, T., Mayzels, V., Zelnik-Manor, L.: On sifts and their scales. In: IEEE Conf. on Computer Vision and Pattern Recognition (CVPR), pp. 1522–1528 (2012)
5. Tuytelaars, T.: Dense interest points. In: IEEE Conf. on Computer Vision and Pattern Recognition (CVPR), pp. 2281–2288 (2010)
6. Xu, L., Dai, Z., Jia, J.: Scale invariant optical flow. In: Fitzgibbon, A., Lazebnik, S., Perona, P., Sato, Y., Schmid, C. (eds.) ECCV 2012, Part II. LNCS, vol. 7573, pp. 385–399. Springer, Heidelberg (2012)
7. Barnes, C., Shechtman, E., Goldman, D.B., Finkelstein, A.: The generalized patch-match correspondence algorithm. In: Daniilidis, K., Maragos, P., Paragios, N. (eds.) ECCV 2010, Part III. LNCS, vol. 6313, pp. 29–43. Springer, Heidelberg (2010)
8. Pluim, J.P.W., Maintz, J.B.A., Viergever, M.A.: Mutual information based registration of medical images: a survey. IEEE Transactions on Medical Imaging (22), 986–1004 (2003)
9. Pluim, J.P.W., Maintz, J.B.A., Viergever, M.A.: Image registration by maximization of combined mutual information and gradient information. IEEE Transcations on Medical Imaging (19), 809–814 (2000)

Adaptive Weighted Prediction for Scalable Video Coding Based on HEVC

Gang Wu, Wenpeng Ding, Yunhui Shi, and Baocai Yin

Beijing Key Laboratory of Multimedia and Intelligent Software Technology
College of Computer Science and Technology
Beijing University of Technology
100124 Beijing, China
wu136gang@163.com,
{wpding,syhzm,ybc}@bjut.edu.cn

Abstract. Scalable video coding provides an efficient way to serve video contents at different visual quality levels and different resolutions. In this paper, the spatial scalability and quality scalability are evaluated on a base of the new Scalable High-efficiency Video Coding (SHVC) extension of HEVC under development. To further improve the coding efficiency of SHVC, we present a set of coding schemes to exploiting inter-layer redundancies for efficient compression of the enhancement layer. The coding schemes include inter-layer frame prediction, combined weighted prediction and inter-layer adaptive filter, which aim at fully utilizing the reconstructed base pictures in the enhancement layer prediction. Experimental results show the effectiveness of our scheme compared with the simulcast case and other HEVC-based scheme. In average, these schemes can achieve 1.5% and 1.7% BD-rate gain for RA HEVC 2x spatial scalability case and HEVC SNR scalability case, respectively.

Keywords: scalable video coding, combined weighted prediction, inter-layer adaptive filter, inter-layer frame prediction, HEVC.

1 Introduction

With the rapid development of the internet and wireless network, video applications become more and more popular, such as video conference, multimedia messaging, wireless and wired video streaming and digital TV broadcasting. Most of these applications involve different devices that have different screen resolutions, compute capabilities, and demand for different network bandwidths. So how to meet these various needs efficiently. The traditional video coding schemes cannot meet these various demands effectively. Scalable video coding presents a highly attractive solution to these applications, since it enables a single bitstream to simultaneous serve various devices with different screen resolutions and qualities. Thus the term "scalability" in this paper refers to the removal of parts of the video bit stream in order to adapt it to the various needs or preferences of terminal users as well as to varying terminal capabilities. So the investigation on SVC [1], especially spatial or quality scalable video coding, is necessary and important.

B. Huet et al. (Eds.): PCM 2013, LNCS 8294, pp. 110–121, 2013.

Scalable video coding introduces a unique problem – how to explore the correlation between efficiently so as to achieve comparable coding performance. Additional inter-layer prediction mechanisms should be integrated in the coding of the enhancement layers to exploit the redundancy between base layer and enhancement layer and improve the coding efficiency.

In this paper, we focus on the High Efficiency Video Coding (HEVC) [2]-based spatial and quality scalable video coding solution and propose several coding schemes to improve the compression efficiency of the enhancement layer. Inter-layer frame prediction, combined weighed prediction and inter-layer adaptive filter. The main goals of these schemes are to enable the usage of as much lower layer information as possible for improving compression efficiency of the enhancement layer.

The rest of this paper is organized as follows. Section 2 provides a background on scalable video coding and introduces HEVC standard briefly. The overview and details of our scheme is specified in Section 3. To validate our scheme and evaluate compression efficiency, simulation results are given in Section 4. Finally, we draw the conclusion in Section 5.

2 Background

In this section, the background on Scalable Video Coding (SVC) and HEVC standard are briefly introduced. In addition, the most important coding tool in SHVC is also briefly introduced.

2.1 Scalable Video Coding

The scalable video coding has been an active research and standardization area for at least twenty years. The prior international video coding standards H.262 I MPEG-2 Video [3], H.263 [4], and MPEG-4 Visual [5] already include several tools by which the most important scalability modes can be supported. However, due to the significant loss of coding efficiency and large increase of decoder complexity, those SVC solutions have limited applications.

In the latest scalable video coding standard – the scalable extension of H.264/Advanced Video Coding (AVC) [6], employs a so-called single-loop (SL) motion compensation framework to better balance the coding efficiency and decoding complexity than prior standards. But the gap between SVC solutions and non-SVC ones in terms of coding efficiency is still wide. Recently, MPEG has been developing the next generation video coding standard—High Efficiency Video Coding (HEVC). New and state-of-the-art techniques and coding tools are being adopted to significantly improve coding performance compared with H.264/AVC. New techniques, such as quad-tree partition and sample adaptive offset are employed to further improve coding efficiency. Because scalable techniques largely depend on the underlying single layer compression, with the HEVC standard, new possibilities for scalable improvement should be investigate. Currently, MPEG is calling for proposals on scalable extension [6].

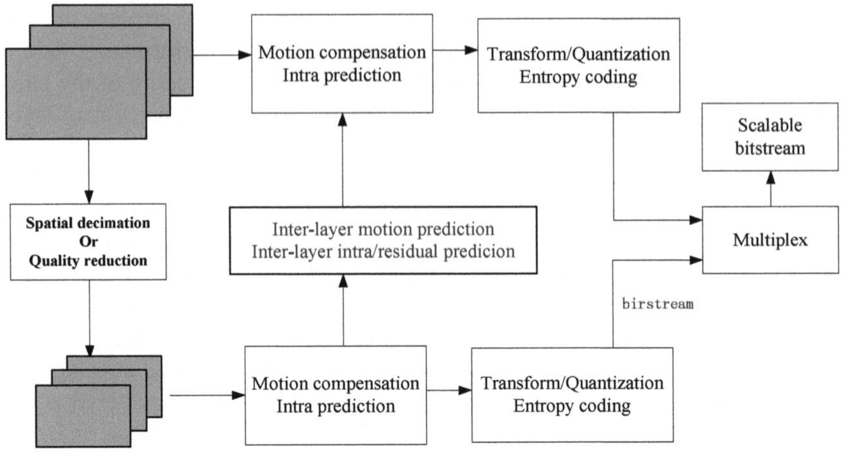

Fig. 1. Typical encoder structure for the scalable extension of H.264/AVC

Generally, there are three types of scalabilities: temporal, quality, and spatial. HEVC supports temporal scalability naturally due to the hierarchical prediction structure. The quality scalability can be treated as a special case of spatial scalability with the same resolution in different layers. In this paper, we focus on the HEVC-based spatial and quality scalable video coding solution. Fig. 1 shows a typical coder structure with 2 spatial layers. The lowest spatial layer employs a standard H.264/AVC encoder to generate the base layer bitstream. Higher spatial enhancement layers adopt additional inter-layer prediction mechanisms to exploit the redundancy between two successive layers to improve coding efficiency.

2.2 High Efficiency Video Coding Standard

High Efficiency Video Coding (HEVC) is the new video coding standard developed by a joint effort between ITU-T and ISO/IEC MPEG standardization organization, working together in a partnership known as the Joint Collaborative Team on Video Coding (JCT-VC). The goal of HEVC is have a bit rate reduction of 50% at the same subjective image quality compared to the H.264/MPEG-4 AVC High profile and computational complexity ranging from 1/2 to 3 times that of the High profile.

HEVC adopts a similar but more complicated coding framework than H.264/AVC. New technologies and coding tools are employed to make a significant coding performance improvement. One of the main characteristics of HEVC is the large-size and partitions adaptive coding blocks. Corresponding to the macro-block in H.264/AVC, the basic process unit in HEVC is called a coding unit (CU). As shown in Fig.2 (a). The largest CU (LCU) with size 64x64 can be divided into smaller CUs according to the quad-tree structure [7] [8] recursively. It results to a quad-tree segmentation of one LCU. Once a CU is not allowed to be split, it will be considered as a prediction unit (PU). A PU may have multiple partitions and multiple transform units (TU). Similar to CU, a TU may be split into four smaller TUs recursively, which leads to

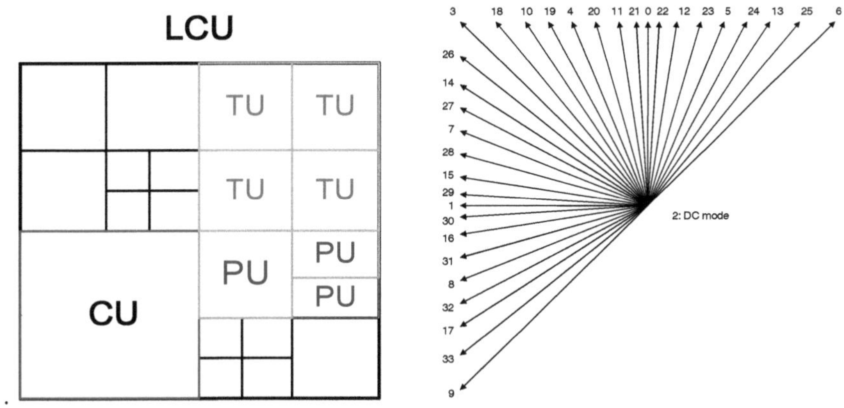

(a)Quad-tree of LCU (b) The 33 intra prediction directions

Fig. 2. HEVC new technologies

another quad-tree segmentation of the residual signal. The quadtree-based coding structure helps to achieve more flexible and accurate prediction signals and more compact residuals, especially for high-resolution video sequences.

HEVC employs more complicated intra prediction modes [9] and more flexible inter motion compensation than H.264/AVC to reduce the temporal and spatial redundancy. For example, HEVC design supports a total of 35 intra prediction modes. Intra prediction mode 0 refers to the planar intra prediction, mode 1 to DC prediction and mode 2 to 34 to angular prediction modes with different directionalities. Fig.2 (b) further illustrates the prediction directions associated with the angular modes. In addition, HEVC adopted a large number of extra tools. Such as Sample Adaptive Offset (SAO). Details on these new techniques can be found in [2] [10].

2.3 Inter-layer Texture Prediction

Inter-layer texture prediction [11] utilizes the collocated intra coded CUs in the base layer. When an enhancement-layer CU is coded in inter-layer intra prediction mode, the reconstructed signals from base layer intra-coded CUs are used as the prediction. inter-layer texture prediction is performed on CU level and partition size is 2Nx2N in PU. Thus, as shown in Fig.3 there are three CU modes, intra mode, inter mode and proposed inter-layer texture prediction mode, respectively. The best CU mode is determined by calculate RD cost among best intra mode, best inter mode and inter-layer texture prediction mode. A flag, named ILTP_flag is inserted after the skip flag to indicate whether this mode is selected.

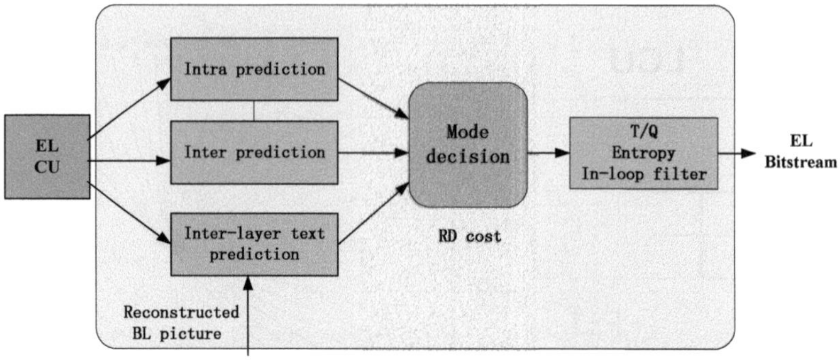

Fig. 3. CU encoding flow in enhancement layer

3 Proposed Scheme

The basic idea of the proposed scheme is to reduce inter layer redundancy by utilizing the reconstructed sample and motion information from base layer (BL). In this section, we will introduce the framework of scheme first. Then some detailed techniques based on HEVC are discussed in subsection 3.1 to 3.3.

Since the base layer picture and the enhancement layer picture have the similar textural information. So decoded pixel information from the lower layer can be used during the inter-layer prediction process of the enhancement layer.

3.1 Inter-layer Frame Prediction

In scalable video coding, the current base layer (BL) reconstructed picture may predict better than the enhancement layer (EL) reference picture for regions that newly appear or have light changes. If the BL quantization distortion is not large, the EL original samples should be close to the BL reconstructed samples.

To further reduce the inter-layer redundancy by utilizing reconstructed sample from base layer picture. We can insert the reconstructed base layer picture into the list of reference pictures of the collocated enhancement layer frame. Interlayer frame prediction takes advantage of the high textural similarity between BL and EL.

The framework of inter layer frame prediction is shown as in Fig .4. For the enhancement layer, the additional inter-layer frame prediction is realized by inserting the corresponding BL picture into the list of reference pictures. The corresponding BL picture is the base-layer picture with the same picture order count (POC) as the enhancement layer picture.

HEVC supports multiple references [2] and up to four references can be used in motion estimation. The reference index decision will affect the coding efficiency as well as the reference selection since the referred index will be coded and transmitted to the decoder. When inserting the BL picture into the reference list of enhancement layer, one problem should be solved is that which reference index should be given to the BL picture.

According to the basic principle of motion estimation, the higher correlation between the reference and the target frame, a higher possibility the reference is selected. We are observed that the most referred picture should be given the smallest reference index. So the BL picture should be inserted into the reference list with index zero. That is to say BL picture should replace the reference picture which POC closest to current encode frame's POC.

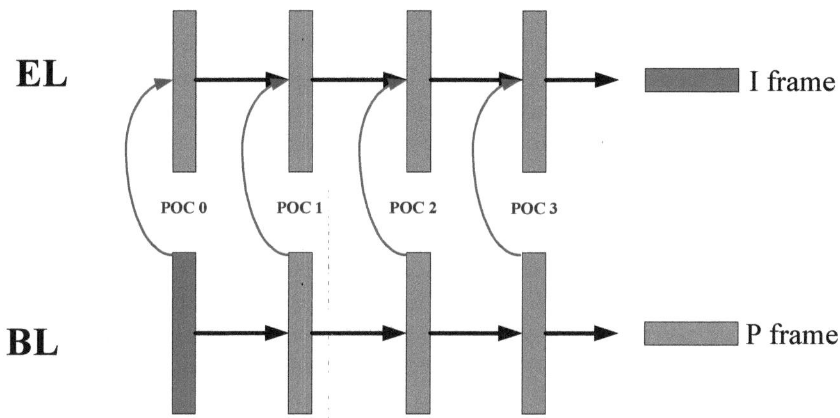

Fig. 4. Coding structure of inter layer frame prediction

Obviously, when the motion is large and rough, the base layer picture is more correlated to the enhancement layer picture than the enhancement layer reference list picture. Besides, the prediction from base layer may not be better than that from the enhancement layer references.

3.2 Combined Weighted Prediction

The major coding tool is inter-layer texture prediction. This method can bring large gain of BD-rate compared to simulcast case based on HM8.0 [12]. So in spatial scalable video coding, the reconstructed picture is one of the most important inter-layer predictions. More specifically, the reconstructed low-resolution picture is up-sampled to form a prediction for the same picture at a higher resolution level. However, the up-sampled base-layer picture often suffers low quality due to the resolution re-sampling error and the quantization error from base-layer encoding. Consequently, these errors degrade the interlayer prediction performance and reduce the overall coding efficiency of the enhancement layer.

Using up-sampling signal we cannot distinguish fine details of signal which correspond high band in frequency domain. From the other side we can expect that the boundary samples ofELwhich are used for Intra prediction provide necessary information about high frequencies.

In this paper, we propose anew intra prediction that is mixture of inter-layer texture prediction and enhancement layer (EL) intra prediction. Since, the up-sampled BL pictures usually have more low frequencies information while EL intra prediction has more high frequencies information. Proposed scheme is applied for luma component only in our implementation. For each Intra PU on enhancement layer we modify Intra prediction as follows Equation (1)

$$\mathrm{Pred}^{*}_{EL} = \alpha * \mathrm{Pred}_{EL} + \beta * \mathrm{Pred}_{BL} \quad (1)$$

As indicated in Equation (1), Pred_{EL} is enhancement layer intra prediction for current PU, Pred_{BL} is up-sampled collocated base layer block, α and β are weight value. In order to distinguish correspond low frequencies information and high frequencies information better. Wavelet transformation is employed in combined prediction. We also use least square method (LSM) [13] to obtain the best α and β. In this paper, considering the complexity of the 9/7 wavelet and the effect is similar with 5/3 wavelet, 5/3 wavelet are used, but the experimental data of 9/7 wavelet is still given in section 4.We treated separately for four blocks with different frequency bands in frequency domain. The proposed scheme is only applied to each intra-coded CU with a partition mode equal to 2Nx2N. Only coding block with size of 32 are consideration to avoid bitrate increase largely. Since weight values should be transmitted to the decoder, different weight values need to be calculated according to different frequency bands, as shown in Fig. 5. It should be note to only one level for our wavelet decomposition. Because experimental results demonstrate that the coding efficiency of EL decreased when two level wavelet decomposition into account.

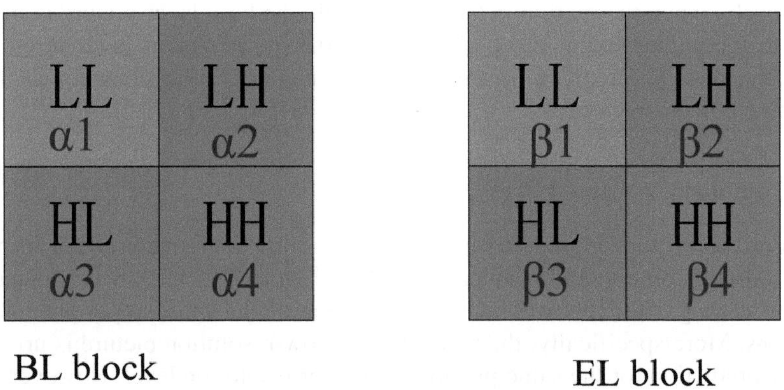

Fig. 5. Combined prediction scheme

3.3 Inter-layer Adaptive Filter

Inter-layer adaptive filter is introduced to further improve coding efficiency of the enhancement layer. In this contribution, inter-layer adaptive filtering on up-sampled base layer (BL) pictures is implemented by using Wiener filtering to reduce distortions between up-sampled base layer pictures and original enhancement layer pictures. The filter footprint is 9x7Cross+3x3Square, and only one filter is allowed per color component per slice.

The inter-layer adaptive filter is applied after the-up-sampling process of the BL picture and before inter-layer intra prediction. The purpose of the inter-layer adaptive filter is to improve sample predictors of inter-layer intra prediction by reducing the distortion between the up-sampled based-layer picture and the original picture in the enhancement layer (EL).

$$\begin{aligned} D5^{el} = &(A5+G5)*h0+(B5+F5)*h1+(C4+E6)*h2+ \\ &(C5+E5)*h3+(C6+E4)*h4+(D1+D9)*h5+(D2+D8)*h6 \\ &+(D3+D7)*h7+(D4+D6)*h8+D5*h9 \end{aligned} \tag{2}$$

In our proposed inter-layer adaptive filter, only single filter is allowed per color component per slice, and the selected filter footprint is 9x7Cross+3x3Square, we added CU-level on/off flags on top of the inter-layer adaptive filter to enable or disable it at CU level. As shown in Fig.6, where 1/2-symmetry is applied and the numbers denote filter coefficient indices. This filter is 2-D non-separable, represented by ten distinct filter coefficients $\{h_i\}$i=0, …, 9. For example, pixel “D5” is filtered with Equation (2).

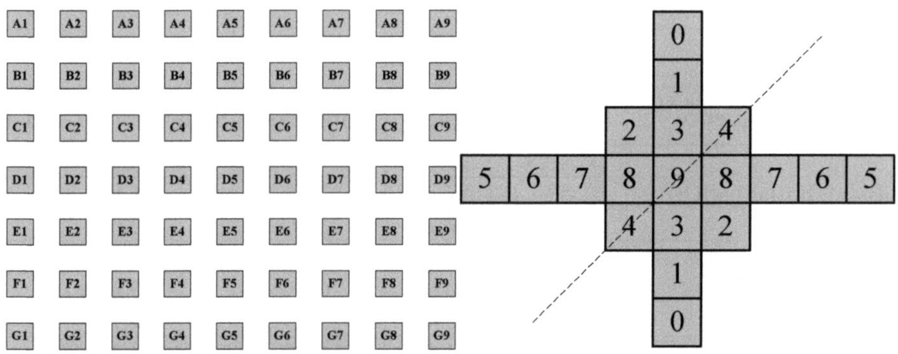

(a). Pixel of up-sampled BL picture (b). 2D-non-separable filter shape

Fig. 6. The proposed filter footprint is 9x7Cross+3x3Square with 1/2-symmetry

4 Performance Evaluation

We implement our proposed schemes in SMuC-0.1.1[14] and perform extensive experiments to evaluate the performance in comparison with simulcast, single-layer, and the SMuC-0.1.1.In the experiments, two spatial layers [one base layer (BL) and one enhancement layer (EL)] are encoded with a spatial radio of 1:2. The QPs used in two layers are denoted as BL−QP and EL−QP, respectively. For each video sequence, BL−QP is fixed whereas EL−QP is set to a series of different values so that a PSNR-rate curve of EL is achieved. The input videos for BL coding are generated from EL input using down-sampling tool provided in JSVM [15]. The 1-D separate filters provided by SMuC-0.1.1 are adopted for up-sampling in the tests. For the spatial and quality solutions, the BL is encoded by the HEVC encoder with high efficiency configuration

[2]. For the simulcast solution, two layers are both coded by HEVC without the inter-layer prediction.

For comparison, we also implement two additional wavelet based schemes: haar wavelet based and 9/7 wavelet based. These two schemes are labeled as "HAAR" and"97" in our experimental results, respectively. And our proposed schemes are integrated together, not separate. In the experiment, Class B test sequences are used for performance evaluation. They are:

(1) Kimono(1080P)
(2) ParkScene(1080P)
(3) Cactus(1080P)
(4) BQTerrace(1080P)
(5) BasketballDrive(1080P)

The coding conditions are:
(1) Random Access with GOP size 8 and MAIN configuration used for coding.
(2) The quantization parameter of enhancement layer is set to 22, 24, 26 and 28; the quantization parameter of enhancement layer is set to 30. The BD-rate [16] results are shown in Table 1 and Table 2.

Table 1. Spatial Scalability Case

Class	Sequence	Proposed	Single	Simulcast	Smuc0.1.1(EL+BL)
B	BasketballDrive	HAAR	17.8%	-21.2%	**-0.3%**
		53	17.3%	-22.1%	**-1.8%**
		97	17.1%	-22.3%	**-2.0%**
B	Cactus	HAAR	11.2%	-26.1%	**0.1%**
		53	10.6%	-26.8%	**-0.6%**
		97	10.4%	-27.0%	**-0.7%**
B	BQTerrace	HAAR	13.1%	-18.3%	**-0.2%**
		53	12.4%	-18.9%	**-1.0%**
		97	12.2%	-19.1%	**-1.1%**
B	Kimono	HAAR	16.2%	-36.1%	**-0.1%**
		53	15.7%	-37.8%	**-2.6%**
		97	15.0%	-38.2%	**-2.7%**
B	ParkScene	HAAR	17.2%	-25.2%	**0.1%**
		53	16.4%	-25.9%	**-1.4%**
		97	16.3%	-26.0%	**-1.4%**
Overall		HAAR			**0.0%**
		53			**-1.5%**
		97			**-1.6%**

Since the goal of HEVC is have a bit rate reduction of 50% at the same subjective image quality compared to the H.264/MPEG-4 AVC High profile. So, SMuC0.1.1 as the HEVC-based test model, the performance of HEVC-based scalability is better than the performance of H.264/AVC-based scalability firmly. In the last column of

Table 2. Quality Scalability Case

Class	Sequence	Proposed	Single	Simulcast	Smuc0.1.1(EL+BL)
B	BasketDrive	HAAR	17.9%	-28.1%	**-0.9%**
		53	17.2%	-30.9%	**-1.1%**
		97	16.4%	-31.3%	**-1.3%**
B	Cactus	HAAR	20.1%	-26.5%	**-0.5%**
		53	18.6%	-29.3%	**-1.0%**
		97	18.3%	-29.8%	**-1.0%**
B	BQTerrace	HAAR	18.5%	-19.1%	**-1.2%**
		53	17.8%	-21.9%	**-3.4%**
		97	17.5%	-22.2%	**-3.5%**
B	Kimono	HAAR	16.2%	-32.1%	**-0.5%**
		53	15.5%	-35.6%	**-2.2%**
		97	15.1%	-36.3%	**-2.3%**
B	ParkScene	HAAR	17.7%	-28.0%	**-0.3%**
		53	16.4%	-29.4%	**-1.0%**
		97	16.0%	-30.2%	**-1.1%**
Overall		HAAR			**-0.7%**
		53			**-1.7%**
		97			**-1.8%**

Table 2, a comparison with SHVC test model 0.1.1 is made. We need to pay more attention to the last column data. Overall, we observed that our present schemes are more suitable for quality scalability.

(a). ParkScene (b). Kimono

Fig. 7. Frequencydomain information

In addition, it should be noted that better coding efficiency is achieved for Kimono sequences which have fast motion. In addition, it should be noted that better coding efficiency is achieved for Kimono sequences which have lots of high frequency information.

In terms of inter layer adaptive filter scheme, we observed not all of the images are suitable for filtering. The characteristic of Wiener filter is better to save the image edges and high frequency details information. For images with less high frequency information, filtering may lead to image distortion. Here, frequency domain information of the Kimono sequence and ParkScene sequence are illustrated in Fig.7.

Compared with ParkScene, Kimono contains more high frequency information. And the experimental data also show that Kimono sequence have a higher coding efficiency than ParkScene sequence. Hence, the proposed schemes would be more effective in the region with fast motion and more high frequency information.

Finally, comparison was made between the proposed schemes and the JCTVC-M0091 named Combined Intra Prediction with low-pass filter [17]. BD-rate of the [17] scheme is illustrated in Table 3.

Table 3. BD-rate of Class B

	AI HEVC 2x		
	Y	U	V
Class B	-0.6%	-2.0%	-2.1%
Overall (Test vs. Ref)	-0.6%	-2.0%	-2.1%

In this table, on average, 0.6% BD-rate saving is realized for luma by using the [17] proposed method. Thus the performance difference between our proposed schemes and [17] technology is apparent.

5 Conclusion

In this paper we present several inter-layer prediction schemes to obtain efficient compression of the enhancement layer. The coding schemes are including adaptive inter-layer frame prediction, combined weighted prediction and inter-layer adaptive filter. The main purpose of these coding schemes are fully utilized the reconstructed base pictures in the enhancement layer prediction. Experimental results show the effectiveness of our scheme compared with the simulcast case and other HEVC-based scheme.

To further improve the coding performance of our scheme, we would like to put efforts on the following works in the future. First, the quadtree structure and the correlation between two layers should be further exploited to reduce the overhead bits of enhancement layer. Second, combined prediction should be further investigated to improve coding efficiency of the EL.

References

1. Schwarz, H., Marpe, D., Wiegand, T.: Overview of the Scalable VideoCoding Extension of.264/AVC Standard. IEEE Trans. on Circuitsand Systems For Video Technology 17(9), 1103–1120 (2007)
2. Bross, B., Han, W.-J., Sullivan, G.J., Ohm, J.-R., Wiegand, T.: High efficiency video coding (HEVC) text specification draft 8. In: ITU-T/ISO/IEC Joint Collaborative Team on Video Coding (JCT-VC) document JCTVC-J1003 (July 2012)
3. Generic Coding of Moving Pictures and Associated Audio Information – Part 2: Video, ITU-T Rec. H.262 and ISO/IEC 13818–2 (MPEG-2 Video), ITU-T and ISO/IEC JTC 1 (November 1994)

4. Video coding for Low Bit Rate Communication, ITU-T Rec. H.263, ITU-T, Version 1: November (1995), Version 2: January (1998), Version 3: November (2000)
5. Coding of audio-visual objects – Part 2: Visual, ISO/IEC 14492-2 (MPEG-4 Visual), ISO/IEC JTC 1, Version 1: April (1999), Version 2: February (2000), Version 3: May (2004)
6. Advanced Video Coding for Generic Audiovisual Services, ITU-T Rec.H.264 and ISO/IEC 14496-10 (MPEG-4 AVC), ITU-T and ISO/IECJTC 1, Version 8 (including SVC extension): Consented in (July 2007)
7. Han, W.-J., Min, J., Kim, I.-K., et al.: Improved Video Compression Efficiency Through Flexible Unit Representation and Corresponding Extension of Coding Tools. IEEE Trans. Circuits Syst. Video Technol. 20(12), 1709–1720 (2010)
8. Karczewicz, M., Chen, P., Joshi, R., et al.: A Hybrid Video Coder Basedon Extended Macroblock Sizes,Improved Interpolation, and FlexibleMotion Representation. IEEE Trans. Circuits Syst. Video Technol. 20(12), 1698–1708 (2010)
9. Lainema, J., Bossen, F., Han, W.-J., Min, J., Ugur, K.: Intra Coding of the HEVC Standard. IEEE Trans on Circuits and Systems For Video Technology 22(12), 1792–1801 (2012)
10. Sullivan, G.J., Ohm, J.R., Han, W.J., et al.: Overview of the High Efficiency Video Coding (HEVC) Standard. IEEE Trans. on Circuits and Systems For Video Technology 22(12), 1649–1668 (2012)
11. JCTVC-L1008, SHVC Working Draft 1, Joint Collaborative Team on Video Coding (JCT-VC) of ITU T SG16 WP3 and ISO/IEC JTC1 SC29/WG1111th Meeting: Geneva, October 10-19 (2012)
12. Available at, `http://hevc.kw.bbc.co.uk/trac/browser/tags/HM-8.0` HM-8.0, reference software
13. Kotropoulos, C., Pitas, I.: Adaptive LMS L-filters for noise suppression in images. IEEE Transactions on Image Processing 5(12), 1596–1609 (1996)
14. Available at, `https://hevc.hhi.fraunhofer.de/svn/svn_SHVCSoftware/`
15. JSVM 9, `http://ip.hhi.de/imagecom-G1/savce/downloads/`
16. Bjøntegaard, G.: Calculation of average PSNR differences between RD-curves. In: Document VCEG-M33 of ITU-T Video Coding Experts Group, VCEG (April 2001)
17. JCTVC-M0091, Combined Intra Prediction with low-pass filter, Joint Collaborative Team on Video Coding (JCT-VC) of ITU-T SG 16 WP 3 and ISO/IEC JTC 1/SC 29/WG 11 13th Meeting: Incheon, KR, April 18-26 (2013)

Saliency Detection Using DCT Coefficients and Superpixel-Based Segmentation

Chi-Yu Hsu and Jian-Jiun Ding

Graduate Institute of Communication Engineering, National Taiwan University, Taipei, Taiwan
r00942039@ntu.edu.tw, djj@cc.ee.ntu.edu.tw

Abstract. The salient region is the area of an image that attracts the attention of viewers. In this paper, a very effective saliency detection algorithm is proposed. Our algorithm is mainly based on two new techniques. First, the discrete cosine transform (DCT) is used for constructing the block-wise saliency map. Then, the superpixel-based segmentation is applied. Since DCT coefficients can reflect the color features of each block in the frequency domain and superpixels can well preserve object boundaries, with the two techniques, the performance of saliency detection can be significantly improved. The simulations performed on a database of 1000 images with human-marked ground truths show that our proposed method can extract the salient region very accurately and outperforms all of the existing saliency detection methods.

Keywords: saliency detection, saliency map, image segmentation, computer vision.

1 Introduction

Salient regions are parts of an image that a person pays more attention to. It can be used in a number of content-based image processing applications, such as adaptive image compression [1], similarity measurement, image retrieval [2], image segmentation [3][4], object recognition [5], graph cut [6], image resizing [7-9], etc. For example, when compressing an image adaptively, a large quantization step can be used in the non-salient region. When measuring the similarity of two images or two objects, more weights can be assigned to the salient region. In image resizing, saliency detection can be adopted to extract the low energy part of an image. In these applications, saliency map detection plays a very important role.

Based on the regional concept, we propose a framework that uses boundary scoring and the border measurement to construct the saliency map. Boundary scoring gives a region a higher saliency weight if it has a higher contrast in the boundary part. The border measurement counts the number of image border pixels in a region, and gives a higher saliency weight if the region has fewer image border pixels. Before applying them, the input image is segmented using our proposed superpixel-based segmentation method. We use different parameters to produce the segmentation results

B. Huet et al. (Eds.): PCM 2013, LNCS 8294, pp. 122–133, 2013.

with coarse and fine scales, for achieving the best performance in the two methods. In addition, a block-wise saliency map is generated using the DCT-based context aware saliency detection method. The flow chart of our framework is shown in Fig. 1.

This paper is organized as follows. In Section 2, the recent works of saliency detection are described. In Section 3, we present the three proposed techniques: (1) the DCT-based context aware saliency detection method, (2) the superpixel-based segmentation algorithm, and (3) precision-enhanced integration. In Section 4, several simulations are performed to compare the accuracies of the saliency map generated by our method and that of eleven state-of-the-art methods on the database provided by [11] that consists of 1000 images. The simulations show that our proposed approach outperforms all of the existing saliency detection methods.

2 Related Work

There are two types of computational models for saliency detection. The first one is the top-down saliency model, which uses high-level features based on the knowledge from neurosciences, biology, computer vision, etc. Which high-level feature is used is dependent on applications [10]. The second one is the bottom-up saliency model, which uses low-level stimulus such as intensity, color contrast, orientation, and motion. The bottom-up saliency model is more popular since it requires less computation time and is efficient in memory.

Most of the previous methods scale down the input image during saliency detection for the sake of efficient computation and generate spotlight saliency maps. However, the result has a lower resolution and does not well match object the boundaries. Spotlight saliency maps are useful for predicting eye fixation, but are not accurate enough for content-based applications, such as salient object segmentation and content-aware image retargeting. To overcome this problem, a pre-segmentation process is needed. In other words, saliency detection should be computed at the region-level instead of the pixel-level. In 2011, Cheng *et al.* [12] simply used a color contrast concept with graph-based segmentation [13] to compute saliency maps and achieved an excellent result with high precision and recall rates. Jiang *et al.* [14] adopted the similar concept and evaluated saliency scores in many scales to get more robust performance. In 2012, Perazzi *et al.* [15] segmented the input image into compact and homogeneous pieces and compute contrasts and saliency scores in a unified way using high-dimensional Gaussian filters. With their method, a very accurate pixel-wise saliency map can be obtained with linear complexity.

The context aware (CA) saliency detection method was proposed in [16]. Although its complexity is high since it computes the color differences between all blocks in three scales, it has very good performance especially on the boundaries of objects, since higher contrasts always occur in the region edges.

3 Proposed Methods

3.1 Block-Based Saliency Detection Using DCT Coefficient

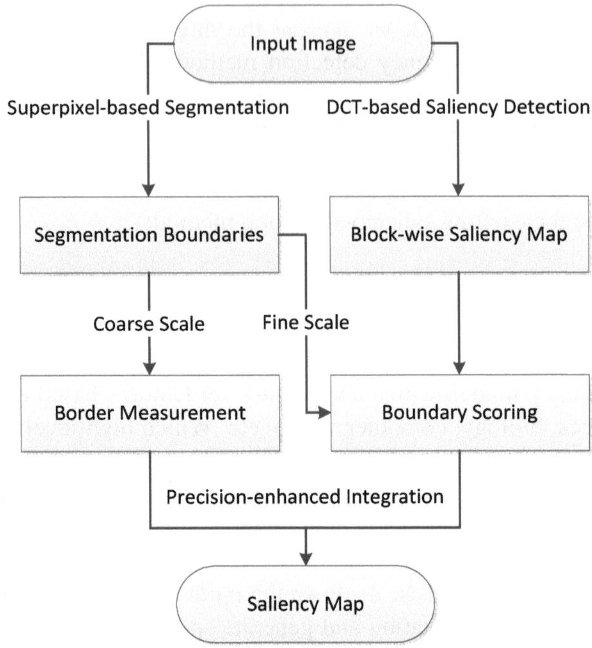

Fig. 1. Flowchart of the proposed saliency detection algorithm

The flowchart of the proposed saliency detection algorithm is shown in Fig. 1. As [16], our framework is also block-wise and uses the concept of context aware. However, the DCT and superpixels are applied to reduce the computation complexity and further improving the performance of saliency detection.

The image is first partitioned into 8 x 8 blocks. Then, the DCT is applied to each block. We adopt the DCT because it can well separate the low and the high frequency parts and the high frequency part. The high frequency part is always related to noise or tiny details and may worsen the performance of saliency detection.

We preserve only the DC term and the first five AC values for each color channel. The high frequency AC terms are ignored due to the considerations of computation efficiency and reducing the effect of noise. Then, the color features of each block are represented as a $(1+5)\times3 = 18$-tuple vector.

Then, the color distance between two blocks i and j is defined as the Euclidean distance in the 18-D space $d_{color}(i, j)$, and the spatial distance $d_{position}(i, j)$ is defined as the Euclidean distance between the centers of blocks i and j. If there are K blocks in an image, then the DCT-based saliency score of each block is defined as follows. The term $\exp[-d_{position}(i, j)]$ is applied because the block j should has a smaller effect on computing $S_{DCT}(i)$ if the distance between blocks i and j is larger.

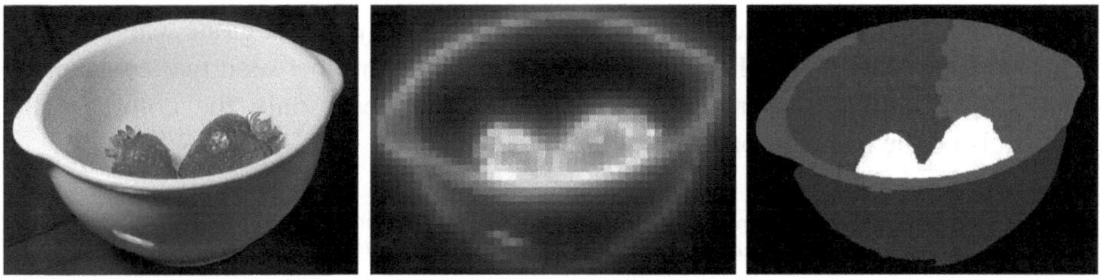

(a) (b) (c)

Fig. 2. (a) The input image. (b) The DCT-based saliency score of the input image. (c) The final saliency map constructed by the proposed algorithm (together with border measuring and boundary scoring).

$$S_{DCT}(i) = \frac{1}{K}\sum_{j=1}^{K}\left\{d_{color}(i,j)\times\exp\left[-d_{position}(i,j)\right]\right\} \quad (1)$$

In Fig. 2(b), the DCT-based saliency score (defined in (1)) of an image is shown. Larger intensity means a larger score. From Fig. 2(b), one can clearly see that the DCT-based saliency score is high in the two strawberries and the edges of the bowl. Therefore, DCT coefficients are indeed good features to conclude whether a region is a salient part of an image.

There is another advantage for using DCT coefficients in saliency detection. Since the DCT has a fast algorithm, the computation loading is much less than that of the original CA approach.

3.2 Superpixel Based Segmentation

In addition to DCT coefficients, we also adopt superpixels for saliency detection. A superpixel is a perceptually meaningful atomic region [17]. It is a combination of several pixels and the color and the intensity are consistent within a superpixel. An example of the superpixel can be seen from Fig. 3(b). Note that the edges of objects match the boundaries of several superpixels.

There are two advantages of using superpixels for saliency detection. First, since the number of superpixels is much less than that of pixels, the computation complexity can be reduced. Moreover, with superpixels, a better image segmentation result can be achieved, which is helpful for constructing a more accurate saliency map.

In the proposed algorithm, the input image into N SLIC superpixels using the method in [17]. Then, a superpixel graph is constructed to replace the rigid structure of the pixel grid. Then, the spectral clustering technique is used to cluster superpixels into C clusters or R non-split regions, as in Fig. 3(c). Finally, a region merging process, which is called boundary-focused region merging, is performed to merge smaller regions into L larger regions, as the example in Fig. 3(d).

For the spectral clustering process, we use the implementation of [18]. Each superpixel is represented as a 5-D vector in *Labxy* space. For the region merging process, we use a proposed boundary-focused region merging algorithm, which is to compute the difference between adjacent regions and merge the adjacent regions whose difference is smaller than a threshold T. The difference measure between two adjacent regions is not computed by mean colors of regions. Instead, only the colors of the superpixels on the adjacent boundaries are computed.

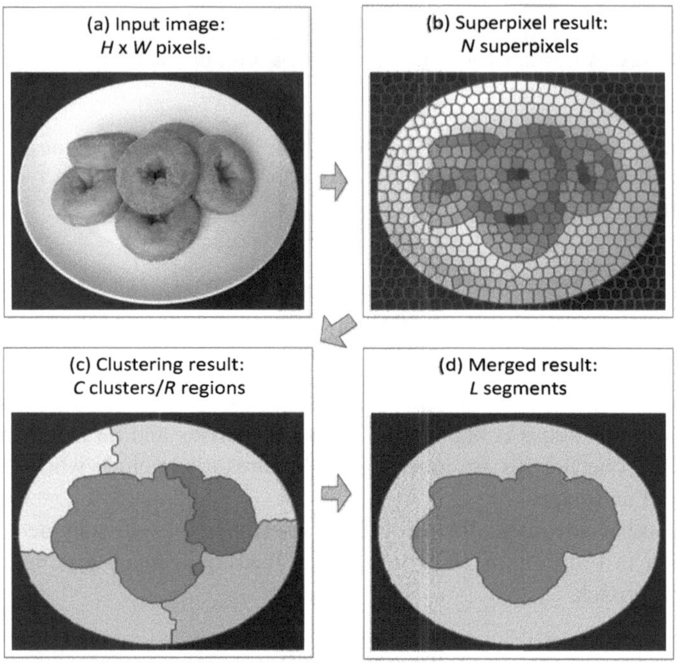

Fig. 3. Using superpixels to segment an image into several regions. (a) Original image. (b) The SLIC superpixels of the input image. (c) The segmentation result after spectral clustering in the 5-D space. (d) The segmentation result after boundary-focused region merging.

For two superpixels k and h belonging to different regions, their distance $E_{k,h}$ is defined as

$$E_{k,h} = \sqrt{(l_k - l_h)^2 + (a_k - a_h)^2 + (b_k - b_h)^2} \tag{2}$$

where (l_k, a_k, b_k) and (l_h, a_h, b_h) are the mean color values of superpixels k and h in the *Lab* color space. Then, the difference measure D between two adjacent regions R_i and R_j is defined as

$$D(R_i, R_j) = \frac{1}{|Adj(i, j)|} \sum_{k,h \in Adj(i,j)} E_{k,h} \tag{3}$$

where $Adj(i, j)$ is the set of distances between the superpixels on the adjacent boundaries of R_i and R_j, and $|Adj(i, j)|$ is the number of $Adj(i, j)$. It means that, when determining whether two regions should be merged, we consider only the color difference of the superpixels on the adjacent boundaries. After applying superpixel-based segmentation with different clustering numbers, we obtain a coarse scale segmentation result and a fine scale segmentation result. Then, the border measurement is applied to the coarse scale segmentation result and the boundary scoring is applied to the fine scale segmentation result, as in Fig. 1.

3.3 Border Measurement and Boundary Scoring

As most state-of-the-art saliency detection methods, we make an assumption that the regions in the center of an image are more important than those near the image border. Therefore, after performing image segmentation using superpixels, we calculate the value of *border*(*i*) for each region of the coarse scale segmentation result:

$$border(i) = \frac{B_1(i)}{2(H \times W)}$$

where $B_1(i)$ means that in region i there are $B_1(i)$ pixels on the image border. H and W are the height and the width of the image. Then, the *border measurement based saliency value* is defined as:

$$S_{BM}(i) = \exp(-\sqrt{2} \cdot border(i)) . \tag{4}$$

The exponential function is applied because the value of $S_{BM}(i)$ should be smaller if the boundary of a region highly overlaps with the image border.

Then, we calculate boundary scoring for each region of the fine scale segmentation result. First, the DCT-based saliency score in (1) is converted into the following form:

$$S_1(m,n) = S_{DCT}(j) \tag{5}$$

if the pixel (m, n) is in the jth 8×8 block of the image. Then, the boundary scoring of each region is defined as the average value of $S_1[m, n]$ on the region boundary. That is, for region i, the *boundary scoring based saliency value* is determined from

$$S_{BS}(i) = \frac{1}{|B_i|} \sum_{(m,n) \in B_i} S_1(m,n) \tag{6}$$

where B_i denotes the boundary of region i and $|B_i|$ is the number of pixels of B_i. It means that a region is more salient if it has higher DCT-based saliency scores on its boundary.

3.4 Precision-Enhanced Integration

From the processes in Sections 3.1, 3.2, and 3.3, two saliency values are calculated in our framework. One is the border measurement based saliency value, S_{BM} (defined in (4)). The other one is the boundary scoring based saliency value, S_{BS} (defined in (6)).

Then, we use a precision-enhanced integration method to define the final saliency map S from the two saliency values:

$$S = N\left[S_{BS} + S_{BS} \times S_{BM}\right] \tag{7}$$

where N is a normalizing factor used for making the value of S in the range of [0, 1]. Note that, in (7), S_{BS} (related to DCT coefficients) and S_{BM} (related to border measurement) are both used for determining the saliency score and S_{BS} plays a more important role than S_{BM}, since S_{BS} has higher precision when only one saliency value is used. We first choose S_{BS} as a basic map. Then, the product of S_{BS} and S_{BM} is used to enhance the intersection area of two saliency maps. Theoretically, applying the intersection can reduce the area of salient regions but increase accuracy, since the probability that the values of S_{BS} and S_{BM} for a non-salient part are both high is very low.

4 Simulation Results

4.1 Database

In our simulations, we used the publicly available database provided by Achanta *et al.* [11], which consists of 1000 images from the MSRA dataset together with the ground truth for each image. The ground truths are binary masks obtained by drawing the contour of the salient object manually. This database is widely used in saliency detection simulations, since the number of images is sufficiently large and well-defined human-marked ground truths are included.

4.2 Precision, Recall, and *F*-measure

The proposed algorithm is compared with 11 state-of-the-art saliency detection methods: Itti *et al* (IT) [19], fuzzy growing (MZ) [20], graph-based visual saliency (GBV) [21], spectral residual (SR) [22], Achanta *et al* (AC) [23], context-aware (CA) [16], frequency-tuned (FT) [11], Zhai *et al* (LC) [24], histogram contrast (HC), region contrast (RC) (both were proposed by Cheng *et al.* [12]), and the saliency filter (SF) [15]. We use precision and recall to measure the performance where

$$precision = \frac{TP}{N_1}, \qquad recall = \frac{TP}{N_2}, \tag{8}$$

N_1 is the number of pixels in the detected salient region, N_2 is the number of pixels in the salient region of the ground truth, and TP (true positive) is the number of pixels of the intersection of the detected salient region and the salient region in the ground truth. A higher precision rate means that fewer pixels in the non-salient part of the ground truth are misidentified to be in the salient region. A higher recall rate means that more pixels in the saliency part of the ground truth are correctly identified to be in the saliency part by the saliency detection algorithm.

A higher recall rate can be achieved using a lower threshold, but the precision is then reduced, and vice-versa. In order to generate a fair comparison result, we binarized the saliency maps at a threshold T_f where T_f was varied from 0 to 255.

Then, he precision-and-recall curve was generated by computing the precision rate and the recall rate for each T_f.

In Fig. 4, the precision-and-recall curves of the proposed algorithm and 11 existing saliency detection methods are shown. From Fig. 4, it can be seen clearly that our proposed algorithm outperforms other 11 methods. The result shows that the proposed algorithm indeed has a very good performance for detecting the salient regions of images.

Similarly to the works of [11][12][15], we also adopted a weighted harmonic mean measure, which is called the *F*-measure, to compare the performance. It computes the precision and recall rates based on a binarized saliency map and uses the image-dependent adaptive threshold T_a proposed by [11] for the thresholding procedure, which is defined as

$$T_a = \frac{2}{W \times H} \sum_{x=1}^{W} \sum_{y=1}^{H} S(x, y) \tag{9}$$

where *W* and *H* are the width and height of an image, respectively. This value is twice of the mean saliency of the image. The *F*-measure is defined as

$$F_\beta = \frac{(1+\beta^2) \cdot Precision \cdot Recall}{\beta^2 \cdot Precision + Recall}. \tag{10}$$

As in [11][12][15], we set $\beta^2 = 0.3$.

In Fig. 5, we show the precision rates, the recall rates, and the *F*-measures of the proposed algorithm and 11 existing algorithms. We also use the 1000 image database provided by Achanta *et al.* [11]. From Fig. 5, the proposed algorithm still outperforms all of the existing methods when the *F*-measure is compared.

4.3 MAE and MSE

In [15], another comparison method was introduced. It evaluates the mean absolute error (MAE) between the continuous saliency maps *S* (the values range from 0 to 1) and the binary ground truth *GT* (the values are either 0 or 1) and is defined as

$$MAE = \frac{1}{W \times H} \sum_{x=1}^{W} \sum_{y=1}^{H} \left| S(x, y) - GT(x, y) \right| \tag{11}$$

where *W* and *H* are the width and height of an image, respectively.

Moreover, to emphasize the larger error case, we also used another well-known measurement, the mean square error (MSE), to measure the performance

$$MSE = \frac{1}{W \times H} \sum_{x=1}^{W} \sum_{y=1}^{H} \left[S(x, y) - GT(x, y) \right]^2. \tag{12}$$

In Fig. 6, we show the MAEs and the MSEs of 11 existing saliency detection methods and the proposed algorithm (denoted by OURS). From Fig. 6, one can see that the MAE and the MSE of proposed algorithm are much lower than other methods. It proves that the proposed algorithm has very accurate saliency detection results.

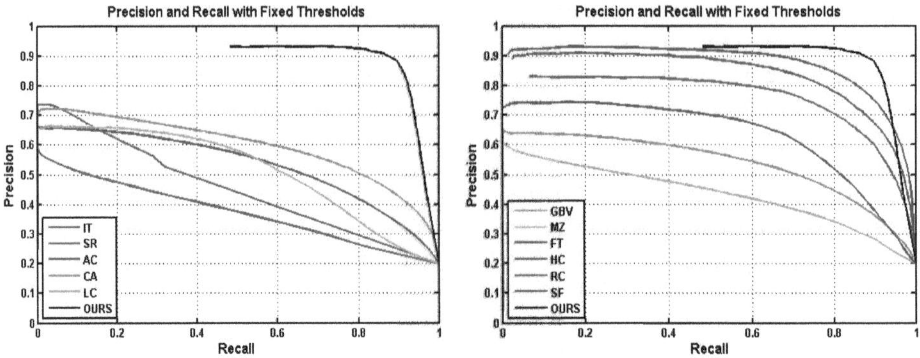

Fig. 4. Precision-and-recall curves of existing saliency detection methods and our proposed algorithm (OURS) when using the database provided by [11]. It can be seen that the proposed algorithm outperforms ALL of the existing methods for saliency detection.

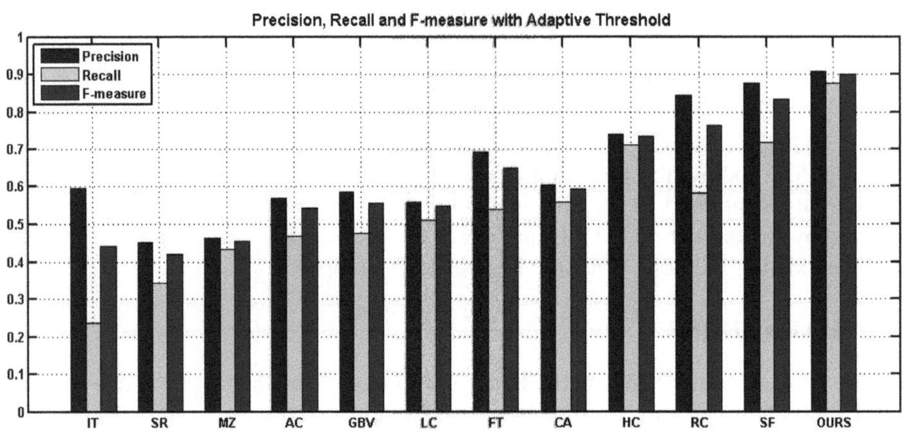

Fig. 5. Precisions, recalls, and F-measures for the existing saliency detection methods and the proposed algorithm (OURS) when using the image-dependent adaptive threshold T_a proposed by [11]

4.4 Analysis

The proposed saliency detection algorithm adopts two important concepts: color contrast and boundary information. With the DCT coefficient method proposed in Section 3.1, the color contrast can be measured in a more precise way. With the segmentation algorithm based on superpixels described in Section 3.2, the boundary information can be extracted in a more accurate way. Therefore, using the proposed techniques of DCT-based color contrast and superpixel-based segmentation, more accurate saliency detection results can be achieved.

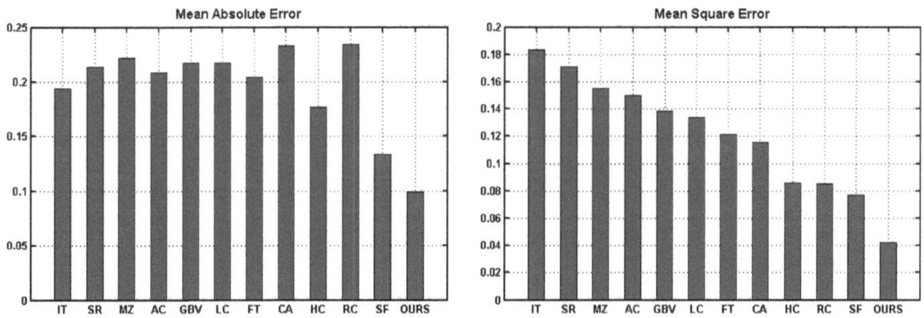

Fig. 6. Left: MAE of the existing saliency detection methods and our proposed algorithm (OURS). Right: MSE of the existing saliency detection methods and our proposed algorithm.

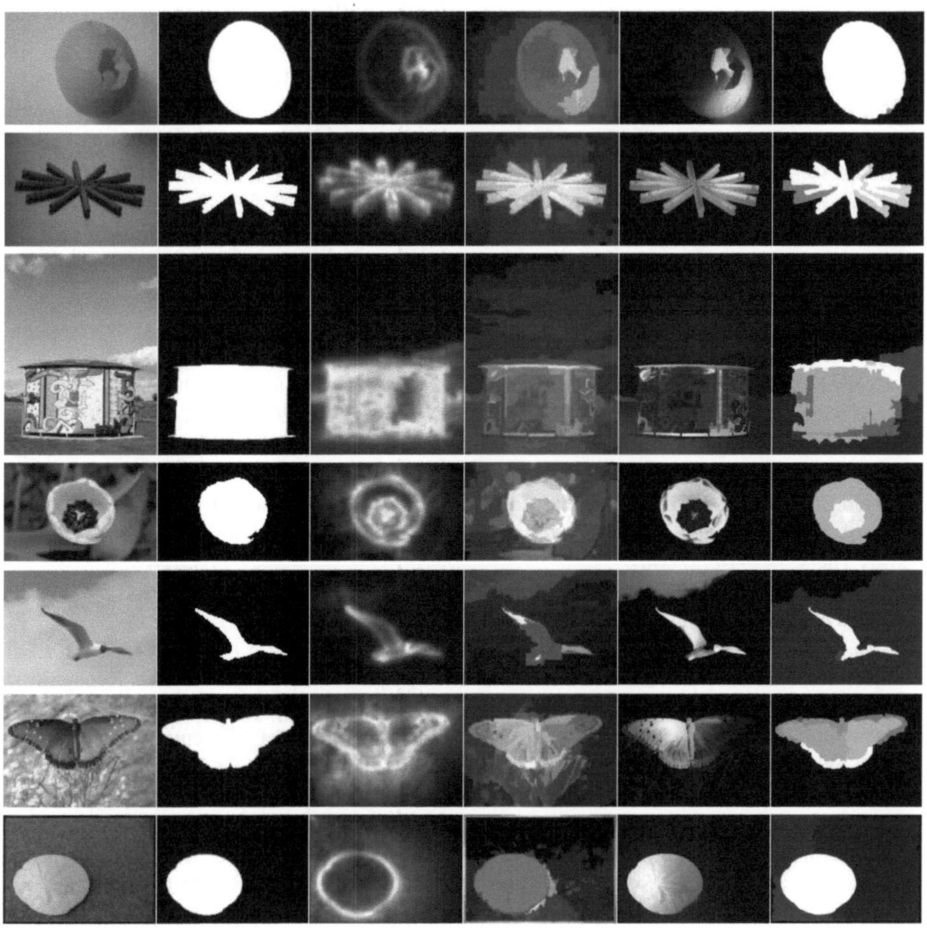

(a) (b) (c) (d) (e) (f)

Fig. 7. Saliency detection examples for visual comparison. (a) Original images, (b) Ground truths, (c) Goferman *et al.* [16], (d) Cheng *et al.* [12], (e)Perazzi *et al.* [15], and (f) our method.

4.5 Visual Comparison

Visual comparisons of the saliency map detection results are shown in Fig. 7. Here, the methods for comparison are the CA method [16], the RC approach [12] and the SF method [15]. It can be seen that our proposed method performs even better than these state-of-the-art saliency map detection methods.

5 Conclusion

In this paper, a very accurate saliency detection framework is proposed. We adopt two novel techniques, DCT-based color contrast and superpixel-based segmentation. A block-wise saliency map is first generated using the DCT-based color contrast. Then, by employing the superpixel-based segmentation with the border measurement and boundary scoring, we obtain two saliency values with full-resolution. Finally, the precision-enhanced integration method is applied to calculate the saliency score.

We evaluated our proposed approach on the largest publicly available data set with a well-defined ground truth and compared our scheme with 11 state-of-the-art saliency detection methods. Simulation results show that the proposed algorithm achieves the best performance for saliency map generation in terms of the precision rate, the recall rate, the F-measure, the MAE, and the MSE.

References

1. Xue, J., Li, C., Zheng, N.: Proto-object Based Rate Control for JPEG2000: An Approach to Content-based Scalability. IEEE Trans. Image Processing 20(4), 1177–1184 (2011)
2. Chen, T., Cheng, M.M., Tan, P., Shamir, A., Hu, S.M.: Sketch2photo: Internet Image Montage. ACM Transactions on Graphics 28(5), article number 124 (2009)
3. Han, J., Ngan, K.N., Li, M., Zhang, H.J.: Unsupervised Extraction of Visual Attention Objects in Color Images. IEEE Transactions on Circuits and Systems for Video Technology 16(1), 141–145 (2006)
4. Chen, O.C., Chen, C.C.: Automatically-determined Region of Interest in JPEG 2000. IEEE Transactions on Multimedia 9(7), 1333–1345 (2007)
5. Rutishauser, U., Walther, D., Koch, C., Perona, P.: Is Bottom-up Attention Useful for Object Recognition? In: IEEE Conference on Computer Vision and Pattern Recognition, vol. 2, pp. 37–44 (2004)
6. Boykov, Y., Kolmogorov, V.: An Experimental Comparison of Min-cut/ Max-flow Algorithms for Energy Minimization in Vision. IEEE Transactions on Pattern Analysis and Machine Intelligence 26(9), 1124–1137 (2004)
7. Avidan, S., Shamir, A.: Seam Carving for Content-aware Image Resizing. ACM Transactions on Graphics 26(3), article number 10 (2007)
8. Wang, Y.S., Tai, C.L., Sorkine, O., Lee, T.Y.: Optimized Scale-and-stretch for Image Resizing. ACM Transactions on Graphics 27(5), article number 118 (2008)
9. Wu, H., Wang, Y.S., Feng, K.C., Wong, T.T., Lee, T.Y., Heng, P.A.: Resizing by Symmetry-summarization. ACM Transactions on Graphics 29(6), article number 159 (2010)
10. Itti, L., Koch, C.: Computational Modeling of Visual Attention. Nature Reviews Neuroscience 2(3), 194–203 (2001)

11. Achanta, R., Hemami, S., Estrada, F., Susstrunk, S.: Frequency-tuned Salient Region Detection. In: IEEE Conference on Computer Vision and Pattern Recognition, pp. 1597–1604 (2009)
12. Cheng, M.M., Zhang, G.X., Mitra, N.J., Huang, X., Hu, S.M.: Global Contrast Based Salient Region Detection. In: IEEE Conference on Computer Vision and Pattern Recognition, pp. 409–416 (2011)
13. Felzenszwalb, P.F., Huttenlocher, D.P.: Efficient Graph-based Image Segmentation. International Journal of Computer Vision 59(2), 167–181 (2004)
14. Jiang, H., Wang, J., Yuan, Z., Liu, T., Zheng, N., Li, S.: Automatic Salient Object Segmentation Based on Context and Shape Prior. In: BMVC, vol. 3, p. 7 (2011)
15. Perazzi, F., Krahenbuhl, P., Pritch, Y., Hornung, A.: Saliency Filters: Contrast Based Filtering for Salient Region Detection. In: IEEE Conference on Computer Vision and Pattern Recognition, pp. 733–740 (2012)
16. Goferman, S., Zelnik-Manor, L., Tal, A.: Context-aware Saliency Detection. IEEE Transactions on Pattern Analysis and Machine Intelligence 34(10), 1915–1926 (2012)
17. Achanta, R., Shaji, A., Smith, K., Lucchi, A., Fua, P., Susstrunk, S.: SLIC Superpixels Compared to State-of-the-art Superpixel Methods. IEEE Transactions on Pattern Analysis and Machine Intelligence 34(11), 2274–2282 (2012)
18. Chen, W.Y., Song, Y., Bai, H., Lin, C.J., Chang, E.Y.: Parallel Spectral Clustering in Distributed Systems. IEEE Transactions on Pattern Analysis and Machine Intelligence 33(3), 568–586 (2011)
19. Itti, L., Koch, C., Niebur, E.: A Model of Saliency-based Visual Attention for Rapid Scene Analysis. IEEE Transactions on Pattern Analysis and Machine Intelligence 20(11), 1254–1259 (1998)
20. Ma, Y.F., Zhang, H.J.: Contrast-based Image Attention Analysis by Using Fuzzy Growing. In: Proceedings of the 11th ACM International Conference on Multimedia, pp. 374–381 (2003)
21. Harel, J., Koch, C., Perona, P.: Graph-based Visual Saliency. In: Advances in Neural Information Processing Systems, vol. 19, pp. 545–552 (2007)
22. Hou, X., Zhang, L.: Saliency Detection: A Spectral Residual Approach. In: IEEE Conference on Computer Vision and Pattern Recognition, pp. 1–8 (2007)
23. Achanta, R., Estrada, F., Wils, P., Süsstrunk, S.: Salient Region Detection and Segmentation. In: Gasteratos, A., Vincze, M., Tsotsos, J.K. (eds.) ICVS 2008. LNCS, vol. 5008, pp. 66–75. Springer, Heidelberg (2008)
24. Zhai, Y., Shah, M.: Visual Attention Detection in Video Sequences Using Spatiotemporal Cues. In: Proceedings of the 14th Annual ACM International Conference on Multimedia, pp. 815–824 (2006)

High Quality Image Deblurring Scheme Using the Pyramid Hyper-Laplacian L2 Norm Priors Algorithm

Yu Chen[1], Jian-Jiun Ding[1], Wei-Sheng Lai[1], Ying-Jou Chen[1], Chir-Weei Chang[2], and Chuan-Chung Chang[2]

[1] G. Inst. Communication Engineering, National Taiwan University, Taipei, Taiwan
{r00942049,r01942068,r01942046}@ntu.edu.tw, djj@cc.ee.ntu.edu.tw
[2] Electronics and Optoelectronics Research Laboratories, Industrial Technology Research Institute, Hsinchu, Taiwan
{ChirWeeiChang,ChuanChungChang}@itri.org.tw

Abstract. In this paper, a very effective image deblurring algorithm is proposed. It combines the techniques of the pyramid structure, the Hyper-Laplacian model, the hybrid norm priors for gradients, and the half-quadratic penalty method. In image deblurring, it is always a tradeoff between the goals of making the edge part sharp and reducing the ringing and the noise effects in the non-edge part. However, using the proposed algorithm, the two goals can be achieved at the same time. Many state-of-art image deblurring algorithms take the gradient of the reconstructed image into account to reduce the effect of noise. In the proposed algorithm, since the pyramid structure and the hybrid norm are applied, the image deblurring performance can be further improved. Simulations show that the proposed algorithm can successfully reconstruct the original image and outperforms the state-of-art methods for image deblurring.

Keywords: image deblurring, hyper-Laplacian model, L_2 norm priors, optimization, pyramid structure.

1 Introduction

Image deblurring is very important in digital cameras, the microscope system, smart phones, telescopes, and medical image processing. A typical blurred image can be modeled as follows [1][2]:

$$b = l \otimes k + m \tag{1}$$

where $\otimes$ is the convolution operator, l is the original image, k is the point spread function (PSF), m is the noise, and b is the blurred image. The goal of image deblurring is to reconstruct the original image l from the blurred image b.

The image reconstruction problem can be categorized into two types: blind deconvolution and non-blind deconvolution. If the PSF is unknown, the problem is called blind deconvolution. An example of blind deconvolution is the motion blurring

B. Huet et al. (Eds.): PCM 2013, LNCS 8294, pp. 134–145, 2013.

problem; if the PSF is known, the problem is called non-blind deconvolution, and it usually occurs in the coded aperture system. A comprehensive literature review can be found in [2].

Image deblurring can be used in many applications, such as microscopy images, consumer photography, and computational photography. Ideally, the process of image restoration can be seen as a decoding process to compensate for the blurring PSF. An inaccurate PSF will also amplify the artifacts and result in undesired image structures. Moreover, the unpleasant artifacts are produced even if the PSF is known. Since the PSF is often band-limited with a sharp frequency cut off, there will be zero values or near-zero values in its frequency response. Therefore, the magnitude will be amplified greatly in those frequencies causing a substantial increase in the signal and noise.

In this paper, we proposed an image reconstruction algorithm for medical microscopic image deblurring and point spread function estimation. By using the identification pattern which is similar to [3], we can say that the identification pattern provides measurable frequencies at all orientations. With this feature, we can estimate the microscopy system's point spread function (PSF) with the actual shooting image. We use the Tikhonov regularization to estimate the system PSF.

After we acquire the system PSF, we adopt the PSF to our proposed pyramid structure hyper-Laplacian L_2 norm priors image reconstruction algorithm to restore the blurred medical microscopic images. Since medical microscopic images are much more complicated than the natural images which taken by normal camera system, our algorithm are time-saving and can adjust the suitable parameters to restore sharp images.

This paper is organized as follows: In Section 2, related works are shown. Our proposed method is shown in Section 3, and the experiment results are presented in Section 4. The conclusion is shown in Section 5.

2 Related Work

In the literature, it has been shown that image deblurring can be categorized into two types: non-blind problems and blind problems. For the application of a known target image and PSF, non-blind deconvolution is used; On the contrary, blind deconvolution is used.

In [4], Chen et al. proposed a pyramid structure with the Wiener filter, used the tested image to generate a suitable filter to reconstruct the original images, and used pyramid-based adaptive restoration (PBAR) with noise suppression. Based on their research, we developed a new algorithm to expand the application to the microscopy systems.

Levin et al. [5] proposed an MAP algorithm with a high-order prior involving a second order derivative filter, using the specialty of the sparse property; they also proposed the coded aperture system to produce an arbitrary focused image.

However, using the sparse specialty is usually very time-consuming. In [6], Tanaka et al. proposed a high-usability deblurring filter, which enables users to choose the appropriate parameters easily.

Xiong et al. [7] presented a new method to reduce ringing artifacts without introducing any image models. They established the link between iterative back-projection and belief propagation and proposed an error-variance aware deblurring algorithm. However, the most significant problem with their method is that it is time-consuming deblurring process; the same problem also occurs in many reconstruction algorithms.

Even with a known kernel, non-blind deconvolution is still under-constrained. The ringing artifacts are inevitable because of high frequency loss in the blurred image. The noise due to the sensor also amplifies the ringing artifacts. For instance, the RL algorithm results in ringing artifacts because of the noise propagation with too much iteration.

In [8], Shen et al. proposed a spatially-varying deblurring method to remove the out of focus blur. There proposed method mainly contains three parts: blur map generation, image deblurring, and scale selection. Our proposed method is focused on the spatially invariant problem which is very different form [8], and they assumed the out-of-focus blur is Gaussian blur which is not suit for our case.

In [9], authors proposed a variable-splitting and penalty techniques to solve the nonlinear l_1 norm optimization problem. By using a new variable w and adopting the iterative method, the l_1 norm non-linear problem can be solved easily and efficiently.

In [10], Krishnan et al. proposed a fast deconvolution method by using look-up table (LUT) and the analytical solution to solve a much challenge Hyper-Laplacian problem. Without using the time-consuming traditional methods like CG or IRLS, by using the LUT and the analytical solution in some special cases, we can easily solve the complicated non-convex problem.

3 Our Proposed Method

In this section, we proposed an image reconstruction algorithm for medical microscopic image deblurring and point spread function (PSF) estimation. In the microscopy system, the object will be amplified by thousands or million times, therefore, in the microscopy system slightly rugged will cause dramatically blurring because the delicate design of the microscopy system.

According to [4], we build an extended depth of field (EDoF) system. By using the coded lens, we are able to make the shooting image become to a spatially-invariant blurred image, after restoring the coded image, we can acquire a sharp image result.

The rest of this section is outlined as follows. In Section 3.1, we introduce the residual deconvolution which proposed by [11] and the improved pyramid structure used in our algorithm structure. In Section 3.2, we perform a system PSF estimation using Tikhonov regularization, and our proposed method is presented in Section 3.3 and Section 3.4.

3.1 The Residual Deconvolution and Improved Pyramid Structure

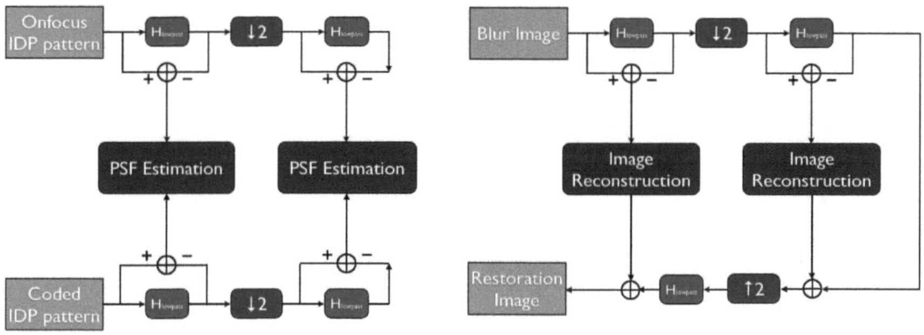

(a) (b)

Fig. 1. The system flowchart of (a) PSF estimation and (b) the deblurring algorithm

We adopted the pyramid structure for much more efficient manipulate the restoration process at each level. In the literature [12], the concept of the Gaussian pyramid is proposed to reduce the aliasing effect by convolving with the Gaussian function, and the highpass frequency response is reduced so the overlap yield from the down sampling processes can be diminished. In our method, we want to prevent the aliasing effect or reduce it; hence, the low-pass filter is used to accomplish the task. Applying the low-pass filter, we can obtain the non-overlapped frequency response when making use of the down sampling processes.

Different from the previous pyramid structure with three-layer decomposition, our improved pyramid structure has only two-layer structure. Owing to down sampling is the simplest deblurring method, when the image getting through down sampling 4x , the blur is so insignificant that we can ignore it. With this improved pyramid structure, the whole structure became much simpler and also save one extra low-pass filter. The system structure can be found in Fig. 1.

3.2 System PSF Estimation

By using the coded lens, we have the ability to extend the depth of field; however, adding the coded lens will cause spatially-invariant blur. In order to overcome this problem, first, we need to estimate the PSF of the coded system. On account of the manufacture problem, the coded lens will not exactly fit the simulation PSF (Fig. 3(a)). Therefore, how to estimate the accurate system PSF is the critical process to build the EDoF system successfully.

We have designed an identification pattern (IDP) which is similar to [3]; moreover, we can say that the IDP pattern provides measurable frequencies at all orientations. With this feature, we can estimate the microscopic system's PSF with the actual shooting image. The IDP pattern is shown in Fig. 2(a) and the coded (blur) IDP pattern is shown in Fig. 2(b). Reconsidering the blur image model as follows:

(a) (b)

Fig. 2. (a)The on focus IDP pattern. (b) The coded IDP pattern.

$$b_{\text{cidp},j} = k_j \otimes l_{\text{idp},j} + m \quad \text{where } j = 1,2 \tag{2}$$

where k_j is the system PSF, $\otimes$ is the convolution operator, m is the additive noise, $b_{\text{cidp},j}$ are the coded IDP patterns and $l_{\text{cidp},j}$ are the onfocus IDP patterns as follows:

$$b_{\text{cidp},1} = b_{\text{cidp}} \otimes H_{\text{lowpass}}$$
$$l_{\text{cidp},1} = l_{\text{cidp}} \otimes H_{\text{lowpass}}$$

and

$$\begin{aligned} b_{\text{cidp},2} &= (b_{\text{cidp}} \otimes H_{\text{lowpass}})_{\downarrow 2} - (b_{\text{cidp}} \otimes H_{\text{lowpass}})_{\downarrow 2} \otimes H_{\text{lowpass}} \\ l_{\text{cidp},2} &= (l_{\text{cidp}} \otimes H_{\text{lowpass}})_{\downarrow 2} - (l_{\text{cidp}} \otimes H_{\text{lowpass}})_{\downarrow 2} \otimes H_{\text{lowpass}} \end{aligned} \tag{3}$$

where H_{lowpass} is the low-pass filter, or we can express (3) in frequency domain

$$B_{\text{cidp},j} = K_j L_{\text{idp},j} + M \quad \text{where } j = 1,2\,. \tag{4}$$

Since we seek to estimate the system PSF $\hat{k}_j$ of the coded microscopic system, a naive estimate $\hat{k}_j$ is obtained using the operator inverse $L^{-1}_{\text{idp},j}$ as follow:

$$\hat{k}_j = IFFT(L^{-1}_{\text{idp},j} B_{\text{cidp},j} = K_j + L^{-1}_{\text{idp},j} M)\,. \tag{5}$$

However, it is very rarely stable enough. For example, the inverse is not defined in frequencies for which $L^{-1}_{\text{idp},j}$ = 0. Even in case $L^{-1}_{\text{idp},j}$ is not exact 0 but small, the inverse is very sensitive to noise. That is, when $L^{-1}_{\text{idp},j}$ = 0 or small the noise contribution is increased. Moreover, it is difficult to practically estimate the actual optical PSF of the imaging system due to the manufacturing inaccuracy. An inaccurate system PSF will also amplify the unpleasant artifacts and result in undesired image structures.

To overcome these difficulties, we use the Tikhonov regularization to estimate the system PSF. The objective function of this problem can be estimated in least square sense as

$$\hat{k}_j = \min_{k_j} \| b_{\text{cidp},j} - k_j \otimes l_{idp,j} \|_2^2 + \mu \| \partial_\phi k_j \|_2^2 \tag{6}$$

where μ is the weighting function and ∂_ϕ is the gradient operator, where $\phi \in \{h, v\}$ denoting two directions. The simplest way to solve (6) is in the frequency domain. The solution is expressed as

$$F(\hat{k}_j) = \frac{L^*_{\text{idp},j} \circ B_{\text{cidp},j}}{L^*_{\text{idp},j} \circ L_{\text{idp},j} + \mu F(\partial_\phi)^* \circ F(\partial_\phi)} \tag{7}$$

where $F(\bullet)$ is the FFT operator, $*$ denotes complex conjugate, $\circ$ denotes component-wise multiplication and the division is also component-wise. The kernel estimation procedure, which described above, should notice that the kernel pixel values are non-negative and the sum of kernel values should be one by energy conserving condition. The results of system PSF estimation are shown in Figs. 3(b) and (c).

3.3 Deblurring Method

Numerous deconvolution approaches exist. A simple filtering operation is very fast but typically has poor results. Most of the best-performing approaches solve globally for the corrected image, and use the marginal statistics acting as a prior to regularize the problem.

In [10], if the marginal distributions are assumed to be Gaussian, a closed form exists in the frequency domain and FFTs can be used to reconstruct the image quickly as mention before in PSF estimation section. However, the real-world images typically have marginal that are non-Gaussian. Studies of real-world image have shown the marginal distributions have heavier tails than a Laplacian distribution, which is called the Hyper-Laplacian distribution. However, since the objective function is no longer a quadratic function or the l_1-norm problem, meaning that the previous research methods are no longer applicable. Instead, standard optimization methods such as conjugate gradient (CG) must be used.

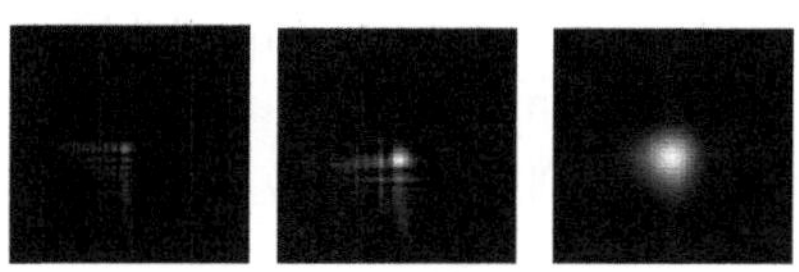

(a) (b) (c)

Fig. 3. (a) The simulation PSF. (b) The estimation system PSF $\hat{k}_1$ and (c) $\hat{k}_2$

In [5], the iteratively reweighed least squares (IRLS) have been used; however, these methods are very time-consuming which is not appropriate to use in our system.

Similar to [10], we proposed a hyper-Laplacian L_2 norm prior to solve such problem. In our proposed, we not only used the Hyper-Laplacian term but also add L_2-norm to help us on image deconvolution problem. Since some image are naturally have partly smooth regions, by adding the Gaussian term can improve the flexibility of deblurring algorithm. By using look-up table (LUT) and the analytical solution, the final result can be solved efficiently.

3.4 Hyper-Laplacian L_2 Norm Priors Image Reconstruction Algorithm

Since we already have estimated PSF, we rewriting (3) as

$$b_j = \hat{k}_j \otimes l_j + m \quad \text{where } j = 1,2 \tag{8}$$

where b_j and l_j are coded blur image and latent image, respectively, m is the additive white Gaussian noise. b_j is express as follows:

$$\begin{aligned} b_1 &= b - b \otimes H_{\text{lowpass}} \\ b_2 &= (b \otimes H_{\text{lowpass}})_{\downarrow 2} - (b \otimes H_{\text{lowpass}})_{\downarrow 2} \otimes H_{\text{lowpass}} \end{aligned}.$$

From a probabilistic perspective, we would like to find the maximum a-posteriori probability (MAP) for l_j given $\hat{k}_j$ and b_j:

$$l_j = \arg\max_{l_j} P(l_j \mid \hat{k}_j, b_j) \propto P(b_j \mid l_j, \hat{k}_j) P(l_j).$$

In our proposed, the first term being a Gaussian likelihood and second begin the hyper-Laplacian L_2 norm image prior. We can recover l_j using our proposed model:

$$\min_{l_j} \frac{\mu}{2} \| \hat{k}_j \otimes l_j - b_j \|_2^2 + \sum_{i=1}^{N} (\lambda_1 \| \partial_{\phi,i} l_j \|^{\alpha} + \lambda_2 \| \partial_{\phi,i} l_j \|_2^2) \tag{9}$$

where i is the pixel index with total pixel N, $\partial_{\phi,i} l_j$ denotes the discrete gradient of l_j at pixel i, μ is a parameter, λ_1 and λ_2 are weighting term control the strength of the regularization, and $0 \le \alpha < 1$.

Similar to [8], [9] and [10], we adopt the half-quadratic penalty method for our objective function (9). We introduce an auxiliary variable w_i at each pixel that allows us to move the $\partial_{\phi,i} l_j$ terms outside $\| \, \|^{\alpha}$. We now get a new cost function:

$$\min_{l_j, w_i} \frac{\mu}{2} \| \hat{k}_j \otimes l_j - b_j \|_2^2 + \lambda_1 \left(\frac{\beta}{2} \sum_{i=1}^{N} \| w_i - \partial_{\phi,i} l_j \|_2^2 + \| w_i \|^{\alpha} \right) + \lambda_2 \sum_{i=1}^{N} \| \partial_{\phi,i} l_j \|_2^2 \tag{10}$$

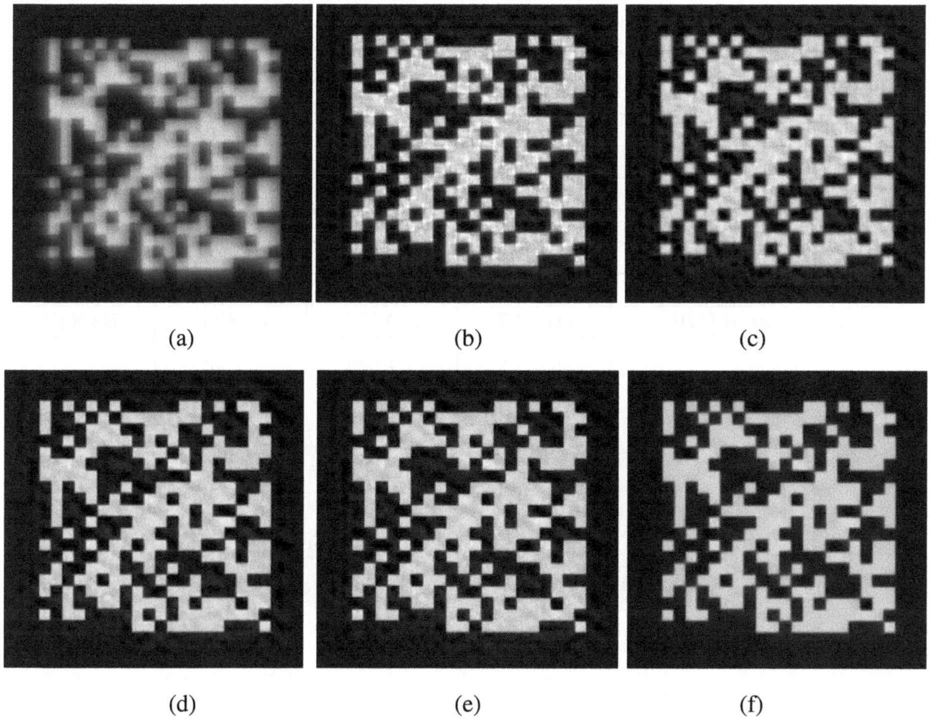

Fig. 4. (a) Blurred IDP pattern. (b) Reconstructed by classical RL. (c) Reconstructed by Shen et al. [8] (d) Reconstructed by Levin et al. [5] (e) Reconstructed by Krishnan et al. [10] (f) Reconstructed by our proposed.

where β is a penalty parameter that vary during the optimization. As β is large enough such that $w_i \rightarrow \partial_{\phi,i} l_j$, the solution of (10) converges to that of (9). Adopting an alternating minimization algorithm for a fixed β can be performed by alternating between two steps, one where we solve for l_j, given values of w_i and vice-versa.

l_j Sub Problem

For a fixed value of w_i, we can minimize l_j by solving the quadratic equation described in (10) as:

$$\min_{l_j} \frac{\mu}{2} \| \hat{k}_j \otimes l_j - b_j \|_2^2 + \frac{\lambda_1 \beta}{2} \sum_{i=1}^{N} \| w_i - \partial_{\phi,i} l_j \|_2^2 + \lambda_2 \sum_{i=1}^{N} \| \partial_{\phi,i} l_j \|_2^2 . \tag{11}$$

We can apply 2D FFTs to find the optimal l_j directly where

$$l_j = F^{-1}\left\{ \frac{\lambda_1 F(\partial_{\phi,i})^* \circ F(w_i) + \frac{\mu}{\beta} F(\hat{k}_j)^* F(b_j)}{(\lambda_1 + \frac{2\lambda_2}{\beta}) F(\partial_{\phi,i})^* \circ F(\partial_{\phi,i}) + \frac{\mu}{\beta} F(\hat{k}_j) \circ F(\hat{k}_j)} \right\} \tag{12}$$

Table 1. Reconstructed numerrable results for NMSE and SSIM

	IDP pattern		USAF test chart	
	NMSE	SSIM	NMSE	SSIM
Richardson-Lucy	0.0217	0.9831	0.0072	0.9828
Wiener filter	0.0344	0.9689	0.0126	0.9645
Shen et al.[8]	0.0210	0.9824	0.0070	0.9819
Tanaka et al.[6]	0.2108	0.8413	0.0311	0.9113
Xiong et al.[7]	0.2451	0.7784	0.0424	0.8691
Levin et al.[5]	0.0147	0.9884	0.0063	0.9855
Krishnan et al.[10]	0.0157	0.9877	0.0066	0.9847
Our proposed	**0.0101**	**0.9916**	**0.0058**	**0.9856**

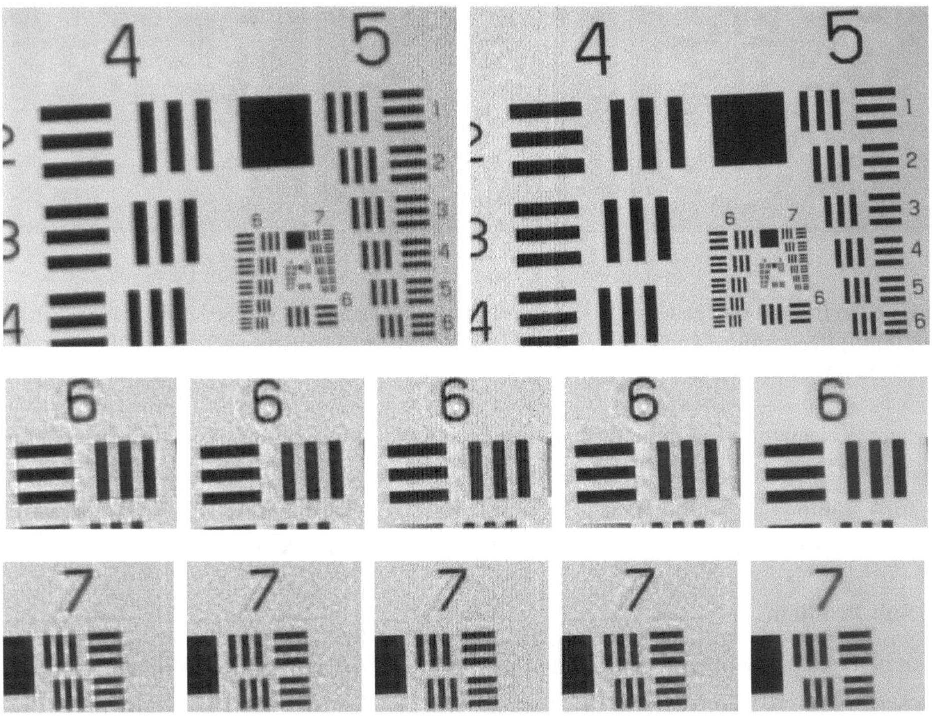

(a) (b) (c) (d) (e)

Fig. 5. Left. Blurred image. ***Right.*** Reconstructed by our proposed (a) Reconstructed by classical RL. (b) Reconstructred by Shen et al.[8] (c) Reconstructed by Levin et al. [5] (d) Reconstructed by Krishnan et al. [10] (e) Reconstructed by our proposed.

where $F(\bullet)$ is the FFT operator, $*$ denotes complex conjugate, $\circ$ denotes component-wise multiplication and the division is also component-wise.

w_i Sub Problem

Table 2. Comparison of the average NMSE of five deblurring algorithms. The best performing algorithm for each image is shown in bold.

NMSE	Lena	Cameraman	Barbara	Baboon	Peppers	Airplane	**Average**
RL	0.0054	0.0209	0.0197	0.0269	0.0049	0.0039	0.01363
[5]	0.0043	0.0176	0.0188	0.0261	0.0035	0.0029	0.01220
[10]	0.0043	0.0170	0.0189	0.0262	0.0034	0.0029	0.01211
[8]	0.0044	0.0168	**0.0186**	0.0256	0.0035	**0.0028**	0.01195
Our proposed	**0.0042**	**0.0160**	**0.0186**	**0.0254**	**0.0033**	**0.0028**	**0.01171**

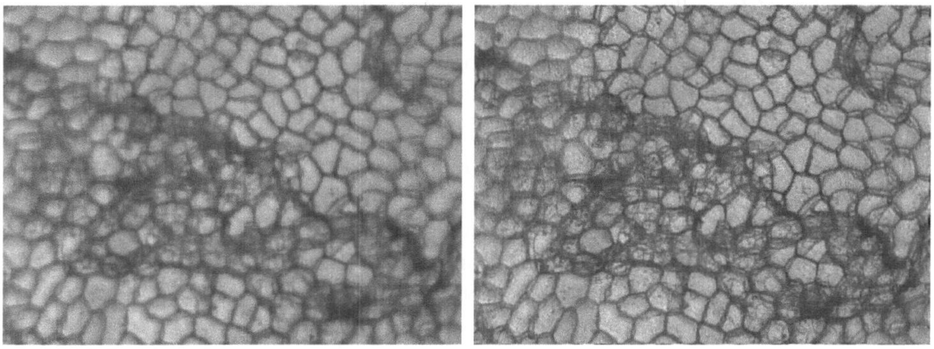

Fig. 6. ***Left***. The coded image. ***Right***. The restored image reconstructed by our proposed algorithm.

Given a fixed l_j, minimizing equation (10) is equivalent to solving the following equation:

$$w_i^\dagger = \min_{w_i} \| w_i \|^\alpha + \frac{\beta}{2}\sum_{i=1}^{N} \| w_i - \partial_{\phi,i} l_j \|_2^2 . \tag{13}$$

According to [10], there are two approaches to find $w_i^\dagger$, LUT and the analytical solution, the detail pseudo algorithm can be found in Table 1. After we have the estimated images l_j, where j = 1, 2, we use the following equation to acquire the final deconvolution image l:

$$\begin{aligned} l_2^* &= l_2 + (b \otimes H_{\text{lowpass}})_{\downarrow 2} \otimes H_{\text{lowpass}} \\ l &= l_1 + (l_2^*)_{\uparrow 2} \otimes H_{\text{lowpass}} \end{aligned} . \tag{14}$$

4 Simulation Results

In this section, we present the experiment results of our proposed system. By using the estimated PSF in Figs. 2(b) and (c), we applied our deconvolution method, along with some classical and state-of-art deblurring algorithms, including the classical Richardson-Lucy, Wiener filter, Shen et al.[8], Tanaka et al. [6], Xiong et al.[7], Levin et al. [5], and Krishnan et al. [10].

In order to test the EDoF system and deblurring ability, we use the IDP pattern and the USAF test chart in our reconstruction scheme. The reconstructed IDP pattern and USAF test chart images recovered by the Richardson-Lucy are shown in Fig. 4(b) and 5(a), and the restored results by Shen et al. [8] are shown in Figs. 4(c) and 5(b). Two state-of-art results are shown, in Figs. 4(d) and 5(c) which are reconstructed by Levin et al. [5] and Krishnan et al. [10] are in Figs. 4(e) and 5(d). Our experiment result for IDP pattern is shown in Fig. 4(f) and the USAF test chart result is shown in Fig. 5(e). The NMSE and SSIM numeral results of the IDP pattern and the USAF test chart can be found in Table 2.

For the classical Richardson-Lucy algorithm, it will cause serious ringing affect which we can find in experiment results figure. Shen et al.[8] use the $l_1 l_2$-norm to the blurred problem which is using the Laplacian model, but it cannot get the sharp image since the Laplacian model is not fit for the image marginal distribution.

Levin et al.[5] and Krishnan et al.[10] both use the hyper-Laplacian distribution model. In fact, the only difference between Levin et al. and Krishnan et al. is the way they solve for the non-convex optimization problem. Levin et al. using the IRLS to solve the non-convex optimization problem which is time-consuming and Krishnan use the fast algorithm to solve the same non-convex problem.

Our proposed method use the hyper-Laplacian distribution model and the l_2-norm prior. By using the L_2-norm, we can significantly suppress the noise but still have the sharp results, and our proposed algorithm is also a time-saving algorithm since we use the LUT and the analytical solution to solve the non-convex optimization problem.

In Table 3, we use the following six images from USC SIPI database as reference images: Lena, Barbara, Cameraman, Baboon, Airplane and Peppers. 2D Gaussian blur with standard deviation set to 3 pixels is applied to the reference images. We can observe that our proposed Hyper-Laplacian L_2 norm priors algorithm outperforms than other compared methods on NMSE. We also show a realistic leaf cell image shooting from the microscopic system in Fig. 6; it is obviously that the blurred coded image became much sharper after our deblurring process.

5 Conclusion

In this paper, we proposed an image reconstruction algorithm for medical microscopic image deblurring and point spread function estimation. We use the Tikhonov regularization to estimate the system PSF. After we acquire the system PSF, we can adopt the system PSF to our proposed deblurring algorithm, a pyramid structure Hyper-Laplacian L_2 norm priors image reconstruction algorithm to reconstruct the blurred medical microscopic images. Since the objective function is no longer a quadratic function or the L_1 norm problem, the previous research methods are no longer applicable. By using the LUT and the analytical solution, the final result can be solved efficiently. On the other hand, we also add the L_2 norm to our objective function, by adding the L_2 norm prior; we can not only get sharp reconstruction images but also have the ability to reduce the noise in smooth region. Experiment results shows that we can restore images with great quality. We successfully establish a deblurring algorithm in the microscopic system and have a state-of-art experiment results for real world shooting images.

Acknowledgement. This work was supported by Industrial Technology Research Institute, Taiwan.

References

1. Asif, M., Malik, A., Choi, T.S.: 3D Shape Recovery from image defocus using wavelet analysis. In: IEEE International Conference on Image Processing, vol. 1, pp. 1025–1028 (2005)
2. Kundur, D., Hatzinakos, D.: Blind Image Deconvolution. IEEE Signal Processing Magazine 13, 43–64 (1996)
3. Joshi, N., Szeliski, R., Kriegman, D.J.: PSF Estimation Using Sharp Edge Prediction. In: IEEE Conference on Computer Vision and Pattern Recognition, pp. 1–8 (2008)
4. Chen, P.C., Liu, C.H., Chang, C.W., Chang, C.C., Angot, L.: Digital decoding design for phase coded imaging system. In: Proc. SPIE, Applications of Digital Image Processing XXXII, vol. 7443, pp. 74431A1–74431A9 (2009)
5. Levin, A., Fergus, R., Durand, F., Freeman, W.T.: Image and Depth from a Conventional Camera with a Coded Aperture. ACM Trans. Graph. 26(3), article number 70 (2007)
6. Tanaka, M., Yoneji, K., Okutomi, M.: High-usability Deblurring Filter. In: IEEE International Symposium on Consumer Electronics, pp. 1–2 (2008)
7. Xiong, R., Ding, W., Ma, S., Gao, W.: A New Image Deblurring Algorithm with Less Ringing Artifacts via Error Variance Estimation and Soft Decision. In: IEEE International Conference on Image Processing, pp. 701–704 (2011)
8. Shen, C.T., Hwang, W.L., Pei, S.C.: Spatially-varying Out-of-focus Image Deblurring with L1-2 Optimization and a Guided Blur Map. In: IEEE International Conference on Acoustics, Speech and Signal Processing, pp. 1069–1072 (2012)
9. Wang, Y., Yang, J., Yin, W., Zhang, Y.: A New Alternating Minimization Algorithm for Total Variation Image Reconstruction. SIAM Journal on Imaging Sciences 1, 248–272 (2008)
10. Krishnan, D., Fergus, R.: Fast Image Deconvolution Using Hyper-Laplacian Priors. In: Advances in Neural Information Processing Systems, vol. 22, pp. 1–9 (2009)
11. Yuan, L., Sun, J., Quan, L., Shum, H.Y.: Progressive Inter-scale and Intra-scale Non-blind Image Deconvolution. ACM Trans. Graph. 27, article number 74 (2008)
12. Ogden, J.M., Adelson, E.H., Bergen, J.R., Burt, P.J.: Pyramid-based Computer Graphics. RCA Engineer 30, 4–15 (1985)

Improved Rotating Kernel Transformation Based Contourlet Domain Image Denoising Framework

Qing Guo[1], Fangmin Dong[1], Shuifa Sun[1,2,*], Xuhong Ren[1], Shiyu Feng[1], and Bruce Zhi Gao[3]

[1] Institute of Intelligent Vision and Image Information, China Three Gorges University, Yichang, 443002, China
[2] Collaborative Innovation Center for Geo-Hazards and Eco-Environment in Three Gorges Area, China Three Gorges University, Yichang, 443002, China
[3] Department of Bioengineering, Clemson University, Clemson, SC, 29634, USA
watersun1977@hotmail.com

Abstract. A contourlet domain image denoising framework based on a novel Improved Rotating Kernel Transformation is proposed, where the difference of subbands in contourlet domain is taken into account. In detail: (1). A novel Improved Rotating Kernel Transformation (IRKT) is proposed to calculate the direction statistic of the image; The validity of the IRKT is verified by the corresponding extracted edge information comparing with the state-of-the-art edge detection algorithm. (2). The direction statistic represents the difference between subbands and is introduced to the threshold function based contourlet domain denoising approaches in the form of weights to get the novel framework. The proposed framework is utilized to improve the contourlet soft-thresholding (CTSoft) and contourlet bivariate-thresholding (CTB) algorithms. The denoising results on the conventional testing images and the Optical Coherence Tomography (OCT) medical images show that the proposed methods improve the existing contourlet based thresholding denoising algorithm, especially for the medical images.

Keywords: Image denoising, Contourlet transform, Direction statistic, Improved Rotating Kernel Transformation.

1 Introduction

The process of image acquisition introduces noise easily, which generates noisy image and have a significant impact on further applications. Lots of useful denoising algorithms have been proposed in past years [1]. In space domain, the denoising methods, such as Lee filter and Rotating Kernel Transformation (RKT) filter, reduce the noise to some extent, but the result is limited [2]. In frequency domain, the denoising algorithm based on wavelet transform has been demonstrated that it can reduce noise

* Corresponding author.

B. Huet et al. (Eds.): PCM 2013, LNCS 8294, pp. 146–157, 2013.

effectively while preserving edges [1]. The literature [3] realizes an efficient retina Optical Coherence Tomography (OCT) image denoising algorithm based on the wavelet transform. In the algorithm, the direction information of the spatial domain is considered by using an artificial setting parameter to control the threshold in vertical subband in wavelet domain. However the algorithm will be invalid while addressing the images of different directions. But its idea is developed further in our work.

Do and Vetterli proposed the contourlet transform to realize the multi-directional and anisotropic representation of image [4]. Since, a series denoising methods based on contourlet transform were proposed. Po and Do [5] proposed the Hidden Markov Tree (HMT) based contourlet denoising method based on the inter-location, inter-scale and inter-direction dependencies of the contourlet coefficients. Guo *.et .al* [6] proposed the bivariate threshold function based contourlet denoising method by taking the inter-direction dependencies into account. Zhou and Shui [7] introduced the local elliptic directional windows to contourlet domain by considering directional information. Nevertheless, this method just modified the shape of window defined by manual operation according to the contourlet transform and ignored the direction information of image. Other more denoising algorithms based on contourlet can be found in [8, 9]. However, the existing contourlet-based threshold denoising algorithms are the extension of the wavelet-based denosing method and their limitations are as follows: 1. in the contourlet transform, the direction information including in space domain is projected into corresponding direction subbands. The noise has no direction and distributes in all subbands uniformly. This will generate the difference between direction subbands. However, the traditional threshold denoising methods don't consider the difference and build the uniform Probability Distribution Function (PDF) in the Bayes framework, which limits the denoising effect. 2. The contourlet transform has no translation invariance, so the denoising result always contains pseudo-Gibbs effect which has a significant impact on the denoising effect.

In order to solve the first problem, a contourlet transform image denoising framework based on IRKT is proposed. Firstly, an Improved Rotating Kernel Transformation (IRKT) technique is proposed to get the direction statistic and edge information of a given image. The IRKT has the capability of noise immunity, which is discussed by performing experiments. Secondly, the direction statistic reflects the difference between subbands in contourlet domain and is introduced into the threshold function in the form of weights to get a satisfactory denoising result.

2 Relationship between the Image Directions and Subbands of Contourlet Domain

This section demonstrates that the direction information of the given image reflects the valid amount of information included in different direction subbands in corresponding contourlet domain. We perform four scales and eight directions contourlet decomposition to four noisy images containing different direction information shown in Fig.1. (a)-(d), and the results of third scale are showed in Fig.1.(e)-(h). According to the theory of contourlet transform, if a image only contains one kind of direction edge, e.g. Fig.1.(a) which only contains horizontal edge, the main edge information will distribute in the corresponding direction subband after contourlet transform, e.g. Fig1.(e). However, the noise has no direction information and distributes in all of the direction subbands. When the number of direction increases, the direction information will be showed in corresponding the sub-bands, e.g. Fig.1(b)-(d) and Fig.1 (f)-(h). The above phenomenon leads to the different coefficient distribution in different direction subband, as it shown in Fig.1.(e)-(h). The direction information of the given image reflects the valid amount of information included in different direction subbands.

In order to utilize this characteristic, in the denoising algorithm a parameter must be introduced to control the threshold for each direction subband. The subband containing less valid edge information should be processed by a larger threshold to get a better denoising result. According to the experiment showed in Fig.1 (a)-(g), the direction information in the spatial domain can be introduced as the parameter. So, the paper tries to find a novel method to count the directions in a given image and introduces the statistic results into the threshold function in different subbands, this will be discussed in section 3 in detail.

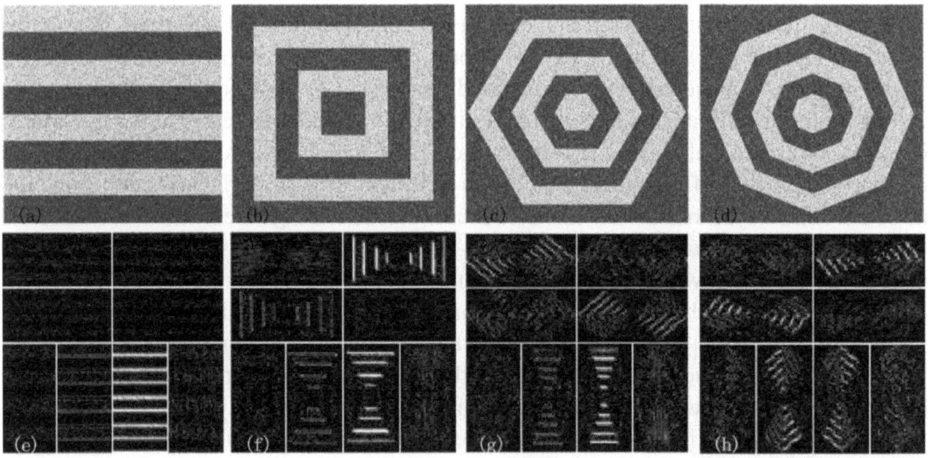

Fig. 1. (a)-(d) are four noisy images containing different number of direction, (e)-(h) are the contourlet decomposition results of (a)-(d)

3 Image Direction Statistic and Weight Model

An Improved Rotating Kernel Transformation (IRKT) is proposed to extract the direction statistic from the given noisy image validly. Various direction kernels are defined to detect the directions of given image. The results of comparing with other conventional edge detection algorithms illustrate that the IRKT can extract the edge information more efficiently, which demonstrate the validity of the direction statistic.

3.1 Improved Rotating Kernel Transformation

Lee and Rhodes proposed the RKT technique which uses several direction kernels to convolve with image and selects the largest convolution of each pixel as the output [10]. Czerwinski *et. al.* use the method to detect the edge of noisy Ultrasound images[11]. Inspired by these works, we propose an Improved Rotating Kernel Transformation (IRKT). The form of kernel is redefined to do better in detecting the edge and counting directions.

Let us define some basic variables. The direction number needing to count is known as *Ndir*; *K* is the size of Kernel, here *K* = (*Ndir* +2)/2. In order to explain the process, we make *Ndir* =8 as the example. The corresponding kernels are shown in Fig.2. I is known as the given image to process. The procedure is as follows:

Step 1: generate the multi direction kernels according to the *Ndir*. Each kernel is defined as k_i (*i*= 1, 2, … , 8) and shown in Fig.2;

Step 2: execute convolutions to I with ki (*i*= 1, 2, … , 8) and get eight matrixes Mi (i = 1,2,…,8) which have the same size with I. So each pixel of I known as I(*x*,*y*) corresponds eight values $M_i(x,y)$ (i =1,2,…,8) in the eight matrices;

$$M_i = k_i * I \qquad i = 1, 2, \ldots Ndir \tag{1}$$

Step 3: calculate the standard variation$\sigma(x,y)$ of the eight values $M_i(x,y)$ (*i* =1,2,…,8). If the$\sigma(x,y)$ is smaller than a threshold T_σ defined in equation (2), the corresponding pixel I(*x*,*y*) will be defined as the pixel in the flat area or noise and not be counted, the corresponding eight values $M_i(x,y)$ will be set to zero. The reason for this step is that the pixels of flat areas or noise have no directions and the standard variation of the convolution results of kernels is small. Actually, the procedure defines the edge very well, as shown in equation (3).

$$T_\sigma = \frac{\max(\sigma_p) + \min(\sigma_p)}{2} \tag{2}$$

$$I_{edge}(x, y) = \begin{cases} 1 & \sigma(x, y) > T_\sigma \\ 0 & \sigma(x, y) \le T_\sigma \end{cases} \tag{3}$$

Step 4: select the largest value *Max*(*x*,*y*) among the eight values $M_i(x,y)$ and count the number of selected of *i* as the frequency of each direction kernel, which generates a 8 dimensional vector known as **Val** and represents the frequency of each direction.

In order to test the validity of this approach, we use the state-of-the-art method, canny and sobel algorithms to detect the edge of four noisy images showed in Fig.1 (a)-(d), and the results are showed in Fig.3.(a)-(l) . The direction statistic results using the proposed method are showed in Fig.3.(m)-(p). The edge detection results indicate that the proposed method can represent edge information more effectively comparing with the canny and sobel algorithm. The statistic results of directions in Fig.3.(m)-(p)

0	1	1	1	1
-1	0	1	1	1
-1	-1	0	1	1
-1	-1	-1	0	1
-1	-1	-1	-1	0

k_1

-1	0	1	1	1
-1	-1	0	1	1
-1	-1	0	1	1
-1	-1	0	1	1
-1	-1	-1	0	1

k_2

-1	-1	0	1	1
-1	-1	0	1	1
-1	-1	0	1	1
-1	-1	0	1	1
-1	-1	0	1	1

k_3

-1	-1	-1	0	1
-1	-1	0	1	1
-1	-1	0	1	1
-1	-1	0	1	1
-1	0	1	1	1

k_4

1	1	1	1	0
1	1	1	0	-1
1	1	0	-1	-1
1	0	-1	-1	-1
0	-1	-1	-1	-1

k_5

1	1	1	1	1
1	1	1	1	0
1	0	0	0	-1
0	-1	-1	-1	-1
-1	-1	-1	-1	-1

k_6

1	1	1	1	1
1	1	1	1	1
0	0	0	0	0
-1	-1	-1	-1	-1
-1	-1	-1	-1	-1

k_7

1	1	1	1	1
0	1	1	1	1
-1	0	0	0	1
-1	-1	-1	-1	0
-1	-1	-1	-1	-1

k_8

Fig. 2. The direction kernels using to count the number of each direction in the image ($NDir = 8$, $K = 5$)

illustrate that the frequency of direction kernels reflect not only the direction information of spatial domain, but also the contourlet decomposition results showed in Fig.1.(e)-(h).

3.2 Direction Weight Model

As the proposed statistic approach can represent the direction information in spatial domain effectively, it is reasonable to use it to build a weight model to represent the valid amount of information in different direction subband. The weight vector is known as $\mathbf{w_v}$ and defined in equation (4).

$$\mathbf{w_v}(1, vi) = \frac{\mathbf{Val}(1, vi)}{\sum_{i}^{Ndir} \mathbf{Val}(1, i)} \qquad vi = 1, \ldots, Ndir \tag{4}$$

The equation (4) illustrates that the larger the frequency of one direction is, the larger the weight of the direction is. As the relationship between subbands and direction information in the contour domain illustrated in section 2, the weight can be explained as: the more the valid amount of information of one direction sub-band contains, the larger the weight of the direction is. So it's reasonable to use the weight vector to control the threshold of direction subbands by considering the different valid information in direction subbands.

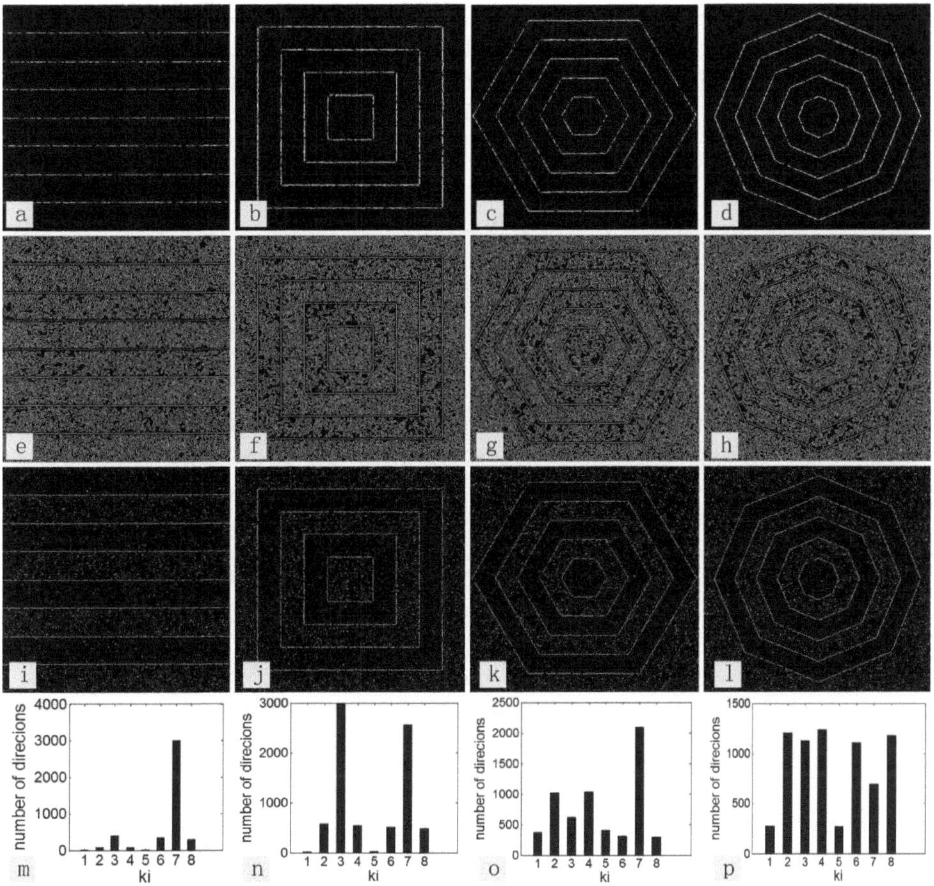

Fig. 3. (a)-(d), (e)-(h), (i)-(l) the edge detection results of the noisy images in Fig.1 by using IRKT, canny and sobel, respectively; (m)-(p) the directions frequency statistic of (a)-(d)

4 Novel Contourlet Transform Denoising Framework Based on IRKT

The basic threshold denoising method for contourlet transform is to deduce a shrinkage function, such as hard threshold function, soft threshold function, bivariate shrinkage function and so on, to handle every coefficient in the transform domain and get the approximation of noiseless coefficients. Then the denoised image can be gotten by contourlet inverse transform. The key of this process is to build an efficient shrinkage function. However, the traditional method only focuses on the characteristic of each direction subband itself, and the difference between subbands is ignored, so the denoising effect is limited.

According to the above introduction, the paper proposes a novel threshold denoising framework for contourlet transform based on IRKT. The process is illustrated as follow and showed in Fig.4 . This framework is used to improve the soft threshold and bivariate threshold denoising method base on contourlet transform. Here, the improved soft and bivariate threshold algorithms based on contourlet transform are labeled as CTSoftDB and CTBDB, respectively.

Step1: perform contourlet transform to the noisy image *I*. *I* can be defined as equation (5).

$$Y=X+N \tag{5}$$

Where Y is noisy image, X is the noiseless pixel, N is the additive noise. The transform result is:

$$y_{i,j} = w_{i,j} + \varepsilon_{i,j} \tag{6}$$

Where y , w and ε is the corresponding noisy coefficient, noiseless coefficient and noise coefficient in contourlet domain, respectively .Here we assume ε is the gaussian noise. j, i represents the decomposition level and the direction subband being discussed, respectively.

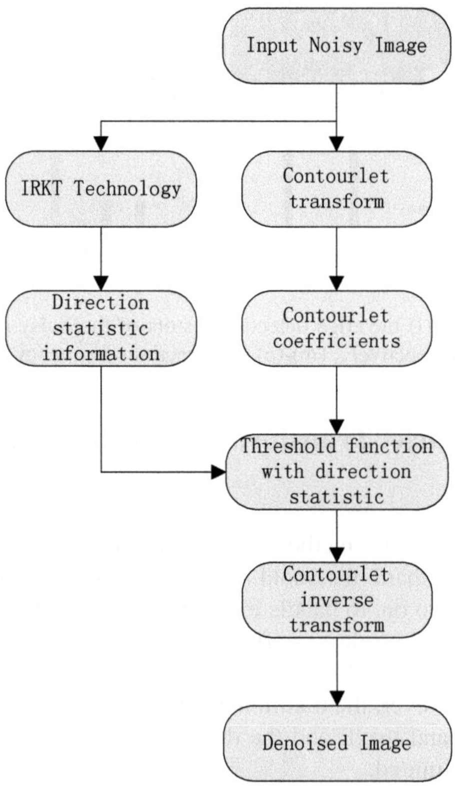

Fig. 4. The proposed threshold denoising framework

Step2: in the decomposition level *j*, the number of direction sub-bands is $Ndir_j$. Use the method proposed in section 3 to get the direction weight $\mathbf{w}_{\mathbf{v,j}}$ of the image in spatial domain with the input $Ndir_j$.

Step3: get the shrinkage function $T(y)$ according to different method mentioned in [4, 6].

Step4: use equation (7) to handle each coefficient in the decomposition level *j* and direction subband *i*, and do the same to other subbands and decomposition levels.

$$\hat{\omega}_{i,j} = \left(\frac{c}{\mathbf{w}_{\mathbf{v,j}}(1,i)+1}\right)\cdot T(y_{i,j}) \tag{7}$$

where $\hat{\omega}_{i,j}$ is the evaluation of the noiseless coefficient, c is a constant to control the extent of the weight.

Step5: perform contourlet inverse transform to the denoised contourlet coefficients and get the denoised image.

5 Experiments

In this section, several conventional testing images and Optical Coherence Tomography(OCT) images are regarded as test images to compare the Wavelet threshold denoising algorithm based on bivariate function (WTB)[12], Contourlet soft threshold denoising algorithm(CTSoft)[4], Contourlet bivariate threshold denoising algorithm (CTB) [6] and two proposed denoising methods (CTSoftDE and CTBDE). The denoising results including metric values illustrate that the proposed denoising framework improves the existing methods. The parameter c in (7) is set to 1.3 for all experiments.

5.1 Metrics

For the standard test images, since the noiseless images are given, the Peak Signal to Noise Ratio (PSNR) and $\beta$are adopted to evaluate the denoising effect and the edge preserving effect [3], respectively. For OCT images, the metrics Signal to Noise Ratio(SNR), Contrast to Noise Ratio (CNR) and Equivalent Number of Look (ENL) are used to evaluate the denoising result[2, 3].

5.2 Experiments and Analysis

Experiment A. conventional testing images denosing experiment

Five algorithms are used to denoise four standard test images showed in Fig.5 (img1-img4). The testing images are added additive gaussian noise with the variance of 0.02 and mean of 0. PSNR and $\beta$are regarded as the metrics to evaluate the denoising results and showed in Table.1. The detected edges are showed in Fig.5.(b).

The denoising results of CTSoftDE and CTBDE are shown in Fig.5.(c) and (d). The experiment indicates that the proposed methods do better in improving the image quality. However the edge information loses to some extent. The reason is that the nature images contain abundant texture information. As a result, the subbands of contourlet domain will also contain amount of texture information, and the uniform lifting threshold will lead to lose some edge information. However, most of the medical images do not have much more detail texture and may mainly contain one direction information. In this case, the proposed framework will do better, as it showed in the B experiment.

Experiment B. OCT Images denoising experiment

In order to test the validity and application of the proposed method, five low SNR OCT images are denoised by using different algorithms. Actually, according to the experiment B, this kind of medical images does not contain the detail texture, so the proposed algorithms will do better. The method proposed in [13] is used to estimate the noise level of the OCT images. The denoising results of five algorithms are shown in Fig.6 and Table.2. The value in the table represents the mean of the selected ROI. The results indicate that the proposed method improves the SNR of the test images, and the CTBDE algorithm based on bivariate threshold function does better than the CTSoftDE for taking the dependency of contourlet coefficients in different scales into account. The CNR has also been improved. The increasing of ENL illustrates that the proposed method smoothes the flat area.

Table 1. PSNR and βvalue of the denoised images in Fig.5

		Noisy	WTB	CTSoft	CTB	CTSoft DE	CTBDE
Img1	PSNR	17.18	26.09	26.45	26.62	26.60	26.65
	β	0.22	0.19	0.29	0.29	0.29	0.28
Img2	PSNR	17.17	26.58	26.50	26.78	26.69	26.89
	β	0.13	0.13	0.17	0. 17	0.16	0.16
Img3	PSNR	17.26	25.68	25.92	26.05	26.07	26.11
	β	0.21	0.17	0.27	0.26	0.26	0.26
Img4	PSNR	17.14	24.86	25.45	25.58	25.56	25.62
	β	0.32	0.18	0.37	0.37	0.38	0.36

Fig. 5. The denoising results of proposed method. Row 1 to 4 is the Img2 to 4, respectively. Column (a) are the noisy images. Column (b) are the dected edges by using the IRKT. Column (c) and (d) are the denoising results of CTSoftDE and CTBDE.

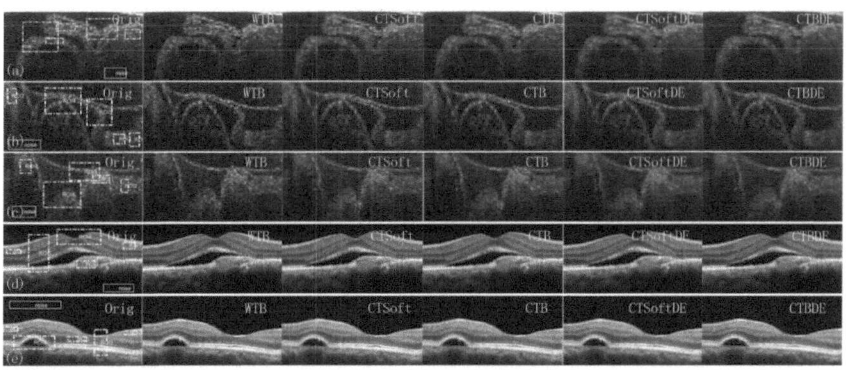

Fig. 6. (a)-(e) are the denoised OCT images based on five algorithms. The first column is the original noisy images, the algorithm used is showed on the top right corner. The images (d) and (c) come from the http://www.optos.com/en/Professionals/Image-library/

Table 2. SNR, CNR and ENL of the denoised images in Fig.6

		Orig	WTB	CTSoft	CTB	CTSoft DE	CTBDE
OCT1	SNR(dB)	17.78	19.72	19.93	20.14	20.85	21.37
	CNR	5.67	6.17	6.21	6.29	6.45	6.61
	ENL	6.47	10.79	14.63	15.74	20.90	25.27
OCT2	SNR(dB)	16.06	18.31	18.46	18.54	19.56	20.18
	CNR	4.42	4.96	5.01	5.08	5.30	5.45
	ENL	7.08	11.889	16.16	17.33	21.10	24.05
OCT3	SNR(dB)	18.97	20.95	21.36	21.69	22.54	23.02
	CNR	6.11	6.50	6.56	6.58	6.74	6.81
	ENL	8. 50	14.93	17.53	19.65	23.98	24.93
OCT4	SNR(dB)	24.08	24.65	24.93	25.06	25.51	25.63
	CNR	8.88	9.13	9.18	9.22	9.33	9.37
	ENL	37.96	56.19	56.31	58.36	61.50	64.71
OCT5	SNR(dB)	21.70	22.13	22.31	22.45	22.75	22.94
	CNR	8.77	9.04	9.09	9.12	9.24	9.33
	ENL	36.95	62.41	64.37	65.97	72.10	75.03

6 Conclusion and Prospect

A novel contourlet transform denoising framework based on IRKT is proposed. The IRKT method is proposed to get the direction statistic representing the difference between subbands and edge information. The direction statistic is added to the existing shrinkage functions. As a result, a novel denoising framework is gotten and used to the soft and bivariate thresholding methods. The experiment results indicate that the proposed methods reduce the noise and improve the SNR efficiently, especially for the medical images, e.g. OCT images. Actually, the paper proposes a novel thresholding denoising framework instead of a novel threshold function to achieve a general improvement of various thresholding denoising methods based on contourlet transform, which may be extent to the denosing methods based on other transformation tools in the future. On the other hand, the edge information exacted in section 3 is not used. The future work will focus on completing the edge exacting process and its use in denoising framework.

Acknowledgements. This project is supported by National Natural Science Foundation of China (61102155, 61272237, 61272236), Outstanding Young and Middle-aged Innovative Research Team Plan of Hubei Province of China (T201002), the Graduate students scientific research innovation fund of the China Three Gorges University(2012CX044), the Graduate students excellent academic dissertation fund of the China Three Gorges University(2013PY038).

References

1. Buades, A., Coll, B., Morel, J.M.: Image Denoising Methods. A New Nonlocal Principle. SIAM Rev. 52(1), 113–147 (2010)
2. Aleksandra, P., Ljubomir, J., Bruno, H., Vladimir, Z., Paul, D., Frans, D., Wilfried, P.: Multiresolution denoising for optical coherence tomography: a review and evaluation. Curr. Med. Imaging Rev. 4(4), 270–284 (2010)
3. Adler, D.C., Ko, T.H., Fujimoto, J.G.: Speckle reduction in optical coherence tomography images by use of a spatially adaptive wavelet filter. Opt. Lett. 29(24), 2878–2880 (2004)
4. Do, M.N., Vetterli, M.: The contourlet transform: a efficient directional multiresolution image representation. IEEE Trans. Image Process. 14(12), 2091–2106 (2005)
5. Po, D.D.Y., Do, M.N.: Directional Multiscale Modeling of Images using the Contourlet Transform. IEEE Trans. Image Process. 6(15), 1610–1620 (2006)
6. Qing, G., Fangmin, D., Shuifa, S., Bangjun, L., Zhi, G.: A Novel Image Denosing Algorithm Based on Contourlet Transform for Optical Coherence Tomography Heart Tube Image. IET Image Process (to be appeared)
7. Zhou, Z.F., Shui, P.L.: Contourlet-based image denoising algorithm using directional windows. Electron. Lett. 43(2), 92–93 (2007)
8. Godzwon, M., Saeed, K.: Biometrics Image Denoising Algorithm Based on Contourlet Transform. In: Bolc, L., Tadeusiewicz, R., Chmielewski, L.J., Wojciechowski, K. (eds.) ICCVG 2012. LNCS, vol. 7594, pp. 735–742. Springer, Heidelberg (2012)
9. Tian, X., Jiao, L., Zhang, X.: Despeckling SAR images based on a new probabilistic model in nonsubsampled contourlet transform domain. Signal Image Video P. (2012), doi:10.1007/s11760-012-0379-0
10. Lee, Y.K., Rhodes, W.T.: Nonlinear image processing by a rotating kernel transformation. Opt. Lett. 15(23), 1383–1385 (1990)
11. Czerwinski, R.N., Jones, D.L., O'Brien, W.D.: Detection of lines and boundaries in speckle images-Application to medical ultrasound. IEEE Med. Imag. 18(2), 126–136 (1999)
12. Sendur, L., Selesnick, I.W.: Bivariate shrinkage functions for wavelet-based denoising exploiting inter-scale dependency. IEEE Trans. Signal Process. 50(11), 2744–2756 (2002)
13. Donoho, D.L., Johnstone, I.M.: Ideal spatial adaptation by wavelet shrinkage. Biometrika 81, 425–455 (1994)

Hand Dorsal Vein Recognition Based on Shape Representation of the Venous Network

Xiangrong Zhu, Di Huang, and Yunhong Wang

Laboratory of Intelligent Recognition and Image Processing
School of Computer Science and Engineering
Beihang University, Beijing, 100191, China
xiangrong.zhu@gmail.com, {dhuang,yhwang}@buaa.edu.cn

Abstract. In recent years, the hand dorsal vein has received increasing attentions in the domain of biometrics. This paper presents a novel approach for hand dorsal vein identification based on shape description of the venous network. In order to generate robust data for shape feature extraction, an effective segmentation method is first proposed which makes use of the enhanced detail of blood vessels. The global and local shape representation are then hierarchically combined according to a coarse-to-fine strategy to encode the distinctiveness. The global one is described as the graph composed by endpoints and crossing points of the skeleton of the vein regions; while the local one is represented by a patch based binary coding scheme. The proposed method was evaluated on the NCUT database that contains 2,040 hand dorsal vein (Near Infrared) images of 102 subjects and achieved a rank-one recognition rate of 98.53%, clearly highlighting its effectiveness.

Keywords: graph matching, vein segmentation, matched filter, binary coding, hand dorsal vein recognition.

1 Introduction

The pattern of hand dorsal vein, as a biometric trait, has received increasing attentions within the research community in recent years due to the property of uniqueness, contactlessness, and robustness. In the literature, many techniques have been investigated, and according to the adopted features, the methods can be mainly classified into two categories, i.e. the shape-based ones and the texture-based ones. The former generally segment the region of interest and the subcutaneous vascular network (represented as its skeleton, Fig. 2(c)) from vein images, and then extract local geometric features for matching, e.g. positions and angles of short straight vectors [2], endpoints and crossing points [3], dominant points [4], vein minutiae, and knuckle shapes [5]. The bottleneck of the shape-based methods lies in that the quality of the generated skeleton image of venous network does not always meet the requirements in the following steps especially when there are some noises in the raw data, and the error will be accumulated and lead to degradation in final performance. While the latter directly extract

B. Huet et al. (Eds.): PCM 2013, LNCS 8294, pp. 158–169, 2013.

features from original hand vein images, leaving out the process of segmentation and thinning. Wang and Li [6] adopted the LBP features to represent local texture variations of hand veins; Tang and Huang [7] explored the SIFT-based feature for the same issue. These techniques achieved comparable results to the shape based ones, but still left space for improvements since the discriminative texture details of hand vein images are not so distinct (as shown in Fig. 1 (a)) that they are commonly not sufficiently exploited.

However, the shape of the venous network is most essential when determining the difference between the hand vein patterns of subjects. Moreover, the shape-based methods tend to accord with the manner that humans distinguish these patterns. In this paper, we present a novel method for dorsal hand vein recognition using shape features, and discuss two critical issues within this domain, i.e. vein region segmentation and shape representation. The contribution of this study is twofold:

- A vein segmentation approach is proposed consisting of image enhancement using the Gaussian matched filter and the extreme points extraction by local curvature analysis.
- A comprehensive and effective shape representation is proposed to distinguish vein patterns of different subjects. It hierarchically combines the global feature described as the graph composed of endpoints and crossing points of vein skeleton as well as the local one embedded in the binary coding feature.

The entire approach is evaluated on the NCUT database, one of the largest hand vein databases publicly available so far known in the literature. The achieved rank-one recognition rate is 98.53%, clearly demonstrating its effectiveness.

The rest of this paper is organized as follows. Section 2 presents the segmentation method, followed by Section 3 describing the shape representation. Section 4 introduces the classification strategy. Experimental results are shown and analyzed in section 5. Finally, Section 6 concludes the paper.

2 Venous Network Segmentation

The primary step of the shape-based methods is the segmentation of the veins from their surrounding corium (as in Fig. 2(b)). Since required by the industrial application, the cost of the imaging devices should be controlled, which brings high noise level and low contrast to the captured images (as shown in Fig. 1 (a)). The back of a hand is a curved surface and the NIR illumination across it is thus not uniform as well [2].

A large number of segmentation techniques have been studied such as local dynamic threshold based [2, 3, 10], directional information of vascular patterns based [11], Mexican hat operator based [5] etc. However, these methods cannot achieve satisfactory results when dealing with poor quality samples. E.g. some thin vessels could be missing, and some dark regions could be considered as vein patterns by mistake [12]. In this paper, we propose a novel segmentation method. Different from the previous ones, it makes use of the properties of blood vessels [13] which can be summarized as:

- Vessels appear darker relative to the background (corium)(Fig. 1 (a)).
- The cross-sectional profile of a vessel has a Gaussian-like shape (Fig. 1 (c)).
- Vessels can be approximately considered as anti-parallel segments and they are locally linear.

Based on the above observations, matched filter is first used to improve the contrast and suppress the noise, then the points near the centerlines are extracted, and after post-processing the final skeleton image is obtained.

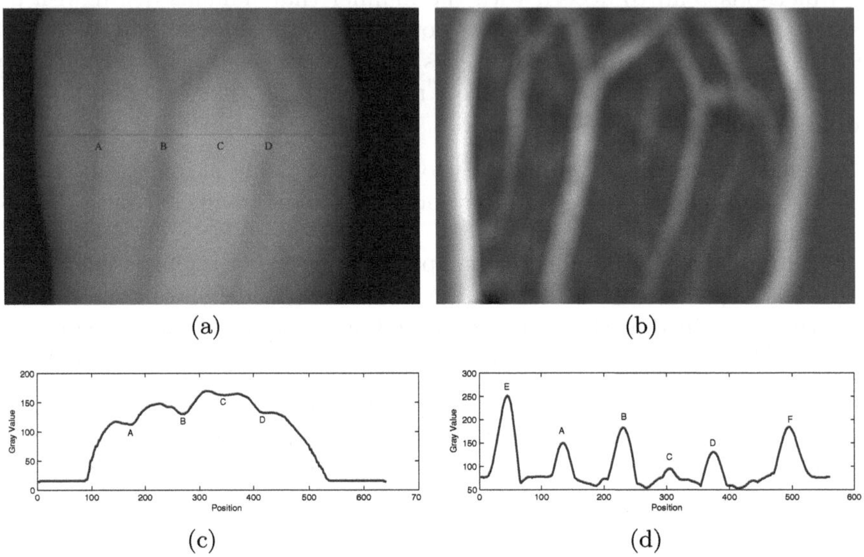

Fig. 1. (a) An one-pixel height line crossing four vessels marked by A, B, C and D. (b) The MFR image. (c) The cross-sectional profile of the line in (a). (d) The response to the Gaussian matched filter of the profile in (c). E and F are the points near the edges of the hand, they have large responses because of the black background but they can be easily removed.

2.1 Matched Filter

Not all the vein areas within the dorsal hand are distinct enough for segmentation due to the low image contrast, and some of them are easily neglected leading to the loss in discriminative power. Therefore, we apply Gaussian matched filter [14] for image enhancement as the first step. It is a convolutional mask designed based on the properties of vessels.

We denote the mask at the orientation θ_i as K_i and N is the neighborhood, which is defined as:

$$N = \{(x, y)||u| \leq 3\sigma, |v| \leq L/2\}, \tag{1}$$

where L is the length of a vessel segment and σ is the width of the Gaussian curve; u and v are given by:

$$[u, v] = [x, y] \begin{bmatrix} \cos\theta & -\sin\theta \\ \sin\theta & \cos\theta \end{bmatrix}. \tag{2}$$

The weight at the position (x, y) is defined as:

$$K_i'(x, y) = -exp\{-u^2/2\sigma^2\}, (x, y) \in N. \tag{3}$$

The convolutional mask K_i is defined as $K_i = K_i' - m_i$ where m_i is the mean value of K_i. Then the response image R_i at the orientation θ_i is obtained by:

$$R_i = f * K, \tag{4}$$

where f is the original hand vein image and $*$ denotes the convolution operation. The final response image R is obtained by selecting the maximal response at all the orientations for each pixel:

$$R(x, y) = max(R_i(x, y)), i = 1, 2 \ldots s, \tag{5}$$

where s is the number of directions.

Because the patterns of blood vessel can 'match' the filter, the corresponding regions thereby give larger responses; while the background and noise cannot 'match' the filter, so they are suppressed. Fig. 1 (d) shows the response values of the profile in (c). Fig. 1 (b) is the image of Matched Filter Response (MFR). As shown in Fig. 1 (b), the contrast between the vessel and the background is obviously enlarged; some tiny vessels even become visible to human eyes.

2.2 Extraction of the Extreme Points

The background of the MFR is not even because of the nonuniform illumination. Besides, the brightness and width of each vessel change greatly. These phenomena make the global and local threshold based approaches not suitable for this case. On the other hand, from Fig. 1 (d), we can see that the points near the centerlines of veins have bigger curvatures although with different gray values and curve widths. The centerlines of the vessels can thus be extracted by localizing the positions with local maximum curvatures [15].

Suppose that z is a point in a profile P and its curvature is denoted as $k(z)$, which can be obtained by (6). If the width of the curve is large, the probability that it's in the vein regions is also high [15]. The score of the point z is then obtained by $s(k) = k(z) \times w$ where w is the width of the curve. The process is conducted at four directions (i.e. 0°, 45°, 90°, and 135°) and the maximum score is selected as the final one.

$$k(z) = \frac{\mathrm{d}^2 P(z)/\mathrm{d}z^2}{\{1 + (\mathrm{d}P(z)/\mathrm{d}z)^2\}^{3/2}} \tag{6}$$

The result of the above operation is shown in Fig. 2 (a). Then, morphological opening and closing operations are utilized to connect broken lines, remove blobs and smooth, Fig. 2 (b) shows the result of this step. The skeleton image (Fig. 2 (c)) is finally achieved through the thinning technique described in [1].

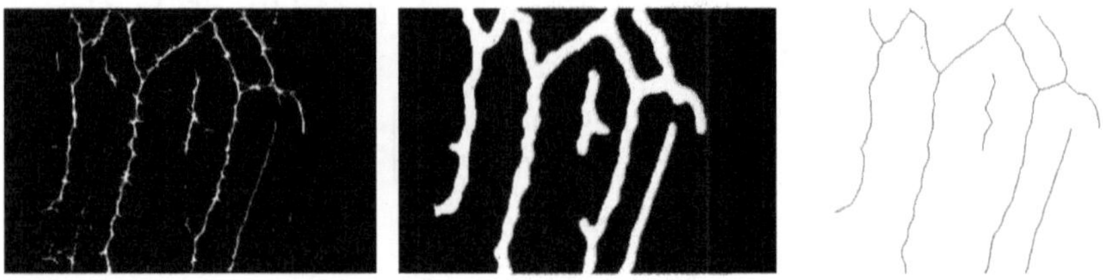

(a) (b) (c)

Fig. 2. The segmentation process. (a) The centerlines of the MFR image. (b) Image after morphological operations. (c) The skeleton image after thinning.

3 Shape Feature Extraction of Hand Vein

To highlight the shape distinctiveness of the hand dorsal vein (expressed by the skeleton image), we propose a comprehensive representation which combines a local shape analysis with a global one. The detail are introduced in the subsequent.

3.1 Binary Coding Based Local Representation

In this part, we propose a new feature, namely Binary Coding (BC), to represent the local shape variations. The proposed BC feature is inspired by Local Binary Patterns (LBP) [16] but more discriminative than LBP when applied to the binary skeleton images.

LBP is a non-parametric kernel which summarizes the local spacial structure of an image. It labels the pixels of a given image by thresholding its neighbors within a circular area and considering the corresponding binary number as a decimal value. LBP is formulated as follows:

$$LBP(x, y) = \sum_{p=0}^{N-1} s(v - v_p) \times 2^p, \tag{7}$$

where v corresponds to the gray value of the center pixel (x, y), v_p to the gray value of the N surrounding neighbors, and function $s(x)$ is defined as:

$$s(x) = \begin{cases} 1 \text{ if } x \geq 0 \\ 0 \text{ if } x < 0 \end{cases}. \tag{8}$$

However, when applied to the skeleton image, LBP tends to make confusion on some patterns commonly appearing. For example, patterns in Fig. 3 (a) and (b) are different but the results achieved by LBP are both 255. To solve this problem, we directly examine each pixel in its neighborhood whether it is a vessel points or not and then label the central pixel as the corresponding decimal number of the binary code. Formally, the decimal number is defined as follows:

$$BC(x, y) = \sum_{p=0}^{N-1} v_p \times 2^p, \quad (9)$$

where the definitions of x,y and v_p are the same with (7). As shown in the right part of Fig. 3, the proposed BC feature can distinguish the patterns (the corresponding BC codes are 182 and 255 respectively) which are confused by LBP.

To make use of spatial information of skeletons, for each skeleton image, we divide it into several rectangle blocks and calculate the histograms of the decimal numbers achieved in the binary coding strategy, and these histograms are orderly concatenated to form the final feature vector. The Chi-square distance is finally used to measure similarities between skeletons.

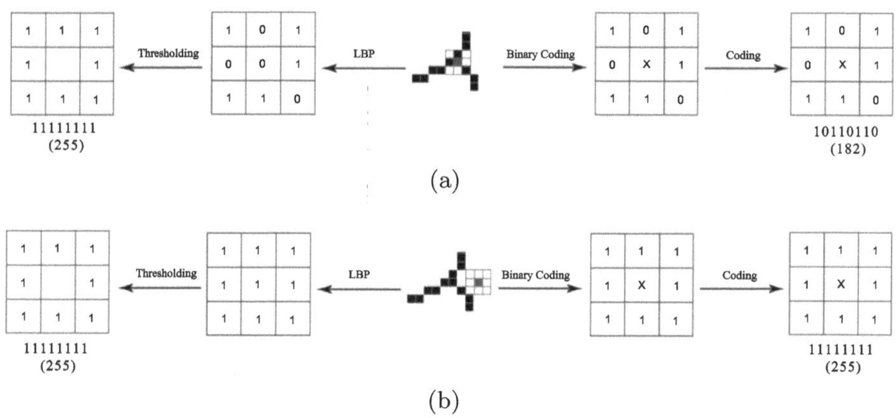

Fig. 3. Comparison between LBP and the proposed BC feature. (a) A pattern from a vein skeleton image and its corresponding results of LBP and binary coding. (b) A different pattern which can be distinguished by the BC feature but confused by LBP.

3.2 Graph Matching Based Global Representation

The BC feature describes local shape variations of vein regions, while in our opinion, it fails to give accurate holistic analysis which is also important in hand dorsal vein based identification. In this subsection, we introduce a graph matching based vein pattern analysis which provides complementary information to the BC based feature.

Intuitively, the relationship of vessel points also contributes to the discrimination; however, the above local statistics of the BC feature cannot take it into account. In order to compensate for such a shortcoming, we propose to encode holistic structures of veins through describing their skeletons as graphs, and graph matching techniques can thus be used to compute their similarities.

Graph matching plays a central role in solving many correspondence problems such as shape matching, character recognition, image analysis (2D and 3D), or action recognition [17, 18]. To the best of our knowledge, graph matching has rarely been investigated in vein recognition. In our previous work [9], we preliminarily proved its effectiveness for such an issue. In order to achieve efficient implementation, we regarded graph matching as a linear assignment problem, and solved it by the Hungarian algorithm which measures the similarity of nodes (defined as edit operation costs) between graphs. However, the amelioration of matching speed comes at the cost of deficiency in performance, because the similarity of edges between graphs is ignored during the process of seeking the optimal node assignment. In this study, we expect to improve the matching accuracy by considering both the relationships of nodes and edges, and thus formulate the pairwise graph matching as the Quadratic Assignment Problem (QAP).

QAP is known to be NP-hard [17] and an exact optimal algorithm can only work for very small graphs. Therefore, the effort in QAP has been made to devise more accurate and faster approximate solvers. To meet these requirements, we employ a recently advanced method, namely Factorized Graph Matching (FGM), showing that affinity matrix can be decomposed as a Kronecker product of smaller ones. On the one hand, it avoids the computation of the cumbersome affinity matrix and hence potentially allows for a more efficient implementation, especially for large graphs. On the other hand, the factorization leads to a better approximation of the graph matching problem that improves state-of-the art approaches [17].

Specifically, a graph $G = (V, E)$ is utilized to describe the structure of the venous network. V is the set of vertices made up by the crossing points and endpoints; E is the set of edges consisting of the vessel lines (as in Fig. 4 (a)). The structure similarity between two skeletons $S1$ and $S2$ can be approximated by the similarity between their corresponding graphs $G1$ and $G2$.

Given a pair of graphs: $G_1 = (V_1, E_1)$ where $|V_1| = n_1$; $|E_1| = m_1$; $G_2 = (V_2, E_2)$ where $|V_2| = n_2$ and $|E_2| = m_2$. To compute the similarity measurement between G_1; G_2, the one-to-one correspondence between their nodes should be obtained in advance. The correspondence matrix is denoted with $X \in \{0,1\}^{n_1 \times n_2}$, where $x_{i_1 i_2} = 1$ if the i_1^{th} node of G_1 corresponds to the i_2^{th} node of G_2. The goal of a graph matching problem is to find the correspondence matrix X that maximize the following function:

$$J(x) = \sum_{i_1 i_2} x_{i_1 i_2} k^p_{i_1 i_2} + \sum_{\substack{i_1 \neq i_2, j_1 \neq j_2 \\ z^1_{i_1 c_1} z^1_{j_1 c_1} = 1 \\ z^2_{i_2 c_2} z^2_{j_2 c_2} = 1}} x_{i_1 i_2} x_{j_1 j_2} k^q_{c_1 c_2}, \tag{10}$$

where K_p and K_q is the similarity matrix of nodes and edges, and Z is the node-edge incidence matrix. $k^p_{i_1 i_2}$ measures the similarity between the node i_1 of G_1 and the node i_2 of G_2, while $k^q_{c_1 c_2}$ measures the similarity between the edge c_1 of G_1 and the edge c_2 of G_2. $Z \in \{0,1\}^{m \times n}$, where $Z_{ic} = Z_{jc} = 1$ if the nodes i and j are connected by the edge c. In our case, $k^p_{i_1 i_2}$ is defined as:

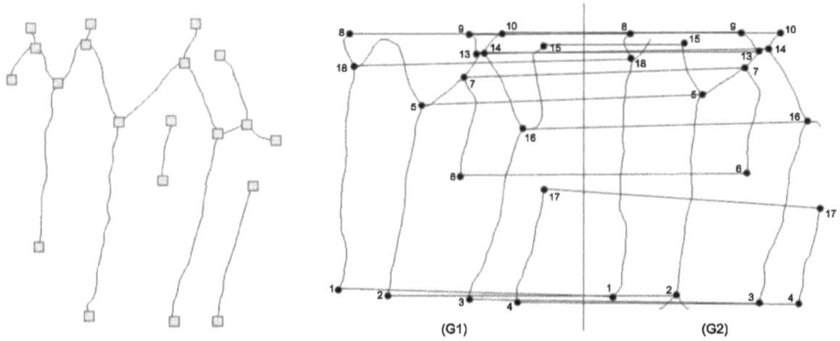

(a) (b)

Fig. 4. (a) Graph based representation of a venous network. (b) Correspondence between two graphs (matched nodes are linked and labeled as the same number).

$$k^p_{i_1 i_2} = exp\{-\sqrt{(x_1 - x_2)^2 + (y_1 - y_2)^2}/k_1\}, \tag{11}$$

where (x, y) is the position of the node in the skeleton image. $k^q_{c_1 c_2}$ is defined as:

$$k^q_{c_1 c_2} = exp\{-\sigma/k_2\} + exp\{-|L_1 - L_2|/k_3\} \tag{12}$$

where σ is the angle between the edge c_1 and c_2, and L_1 and L_2 are the lengths of these edges. In (11) and (12), k_1, k_2, k_3 are calibration factors to balance the data.

To solve the above optimization problem, the nonconvex objective function (10) is relaxed into a convex one $J_{vex}(X)$ and a concave one $J_{cav}(X)$. The details of the two functions can be found in [17]. $J_{cav}(X)$ has a global optimal solution no matter the choice of the initialization and $J_{vex}(X)$ always leads to an integer solution. Inspired by the path following strategy for optimization proposed in [18], the nonconvex optimization problem can be approached by iteratively optimizing a series of the following sub-problems [17]:

$$\begin{aligned} \max_X \quad & J_\alpha(X) = (1-\alpha)J_{cav}(X) + \alpha J_{vex}(X), \\ s.t. \quad & X1_{n_2} \leq 1_{n_1}, X^T 1_{n_1} \leq 1_{n_2}, X \geq 0_{n_1 \times n_2}, \end{aligned} \tag{13}$$

where $\alpha \in [0, 1]$ is a tradeoff between the convex relaxation $J_{vex}(X)$ and the concave one $J_{cav}(X)$. The process starts with $\alpha = 0$ and successively increasing α until 1. In each iteration, the Frank-Wolfe's algorithm [8] is used to optimize $J_\alpha(X)$ and the corresponding maximum points are used as the start searching points in the next iteration. Fig. 4 (b) shows a matching result obtained by the FGM algorithm.

4 Classification

Recall that the binary coding features capture the local distributions of the vessels while the graph based ones describe their holistic structures. The two types of features tend to be complementary to each other in shape representation. Therefore, we combine both of them for a comprehensive analysis. There are several fusion choices, such as at feature level, at matching score level, at decision level, etc. Considering that binary coding runs much (about 50 times) faster than graph matching, to balance the accuracy and the efficiency, we combine them hierarchically at matching score level using a product rule as follows:

$$S = e^{-g/\alpha}c \tag{14}$$

where g is the graph matching based similarity between two venous networks obtained by (10) and c is the Chi-square distance between the two histogram based vectors obtained by binary coding. Specifically, the classification contains the subsequent steps:

1) For a given probe P, list the gallery samples in a descending order according to their Chi-square distances of the BC features.
2) Regard the first N gallery samples as the following candidates, and discard the remaining.
3) Calculate the graph based similarities between the probe P and the N selected gallery candidates.
4) Combine the scores obtained by the BC feature and graph matching according to the product rule in (14).
5) Assign P to the class of its nearest neighbour based on the fused scores.

5 Experimental Results

We evaluate the proposed method on the NCUT dataset. It contains 2,040 images from 102 individuals in which 52 are female and 50 are male. Ten images of the right hand and ten images of the left hand were captured for each individual. To simulate real environment, images were acquired at different times and some of them have shift and rotation [12]. In our experiments, the dataset was divided into a probe set and a gallery set. The gallery consists of the first five images of each hand and the probe uses the others. The left hands and right hands from the same individual are treated as different subjects and there are hence a total of 204 classes.

To demonstrate the effectiveness of our segmentation method, a comparison is made with the Niblack method, which is widely used for hand dorsal vein segmentation [2, 10]. The skeleton images are first generated using the proposed segmentation method and the Niblack method respectively. We then extract the BC features and compute the Chi-square distances between the probe samples and the gallery ones. A probe sample is assigned to the class of its nearest neighbor. To compare the distinctiveness of the BC feature with LBP, the same process is conducted except that the LBP is used for feature extraction instead of BC.

The parameter in the experiments are optimized in the training phase and experimentally set as follows: in matched filter, $L = 9$, $\sigma = 10$ and $s = 12$; in the Niblack method, the only parameter k is 0.01.

The recognition accuracies of different approaches are shown in Table 1. From Table 1, we can see that both LBP and binary coding features perform better on the skeleton images processed by the proposed segmentation method than the ones by the widely used Niblack method [10], indicating that our segmentation is more effective. On the other side, based on both types of segmented data, the binary coding feature outperforms LBP, proving its competency. Additionally, it should be noted that in [10] the Niblack method has proved better than the ones in [2, 3, 11], we thus do not list the comparisons with them due to space limitation.

Table 1. Performance using different methods

Feat./Segm.	This paper	Niblack[10]
Binary Coding	**97.01%**	94.80%
LBP	95.98%	93.82%

As mentioned in section 4, binary coding and graph matching do not match in speed. We combine the similarity measurements of binary coding and graph matching hierarchically. In this experiment, we tune the value of N, i.e. the number of gallery candidates whose Chi-square distances with the probe one in BC feature space are the nearest N ones. Fig. 5 depicts the Cumulative Match Characteristic (CMC) curve, and all the probe samples find their true labels before the rank 50. We thus set N at 50, indicating that no error appears in the first binary coding based matching step. In the following graph matching based step, the number of the gallery samples involved is decreased from 1,020 to 50.

We further compare the rank-one recognition rate of the entire system with the ones of the state of the art. As displayed in Table 2, we can see that our method reaches a recognition rate of 98.53% outperforming all the other methods tested on the NCUT dataset in terms of rank-one recognition rate, clearly highlighting the effectiveness of the entire system.

Table 2. Comparison with the state of the art on NCUT

Methods	Class Num.	Gallery/Probe	Rank-one Recognition Rate
MLBP[8]	15	150	97.30%
LCP[6]	204	1020/1020	90.88%
Nodes Matching[9]	204	1020/1020	96.67%
SIFT[7]	204	1020/1020	98.04%
This paper	204	1020/1020	**98.53%**

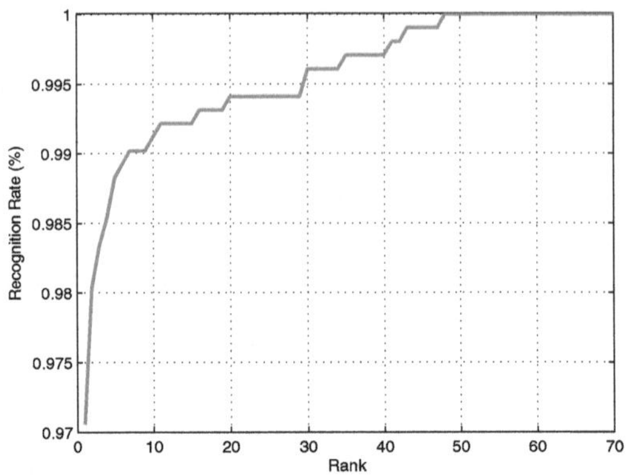

Fig. 5. The CMC curve of the binary coding based matching step

6 Conclusion

This paper proposed a novel approach for hand dorsal vein identification, addressing two key issues within this domain: vein segmentation and shape representation. The segmentation method effectively localized the vein area through operating image enhancement and curvature analysis in the serial way; while the shape representation hierarchically combines the local analysis with a global one. The proposed method was evaluated on the NCUT database and achieved a rank one recognition rate of 98.53%, clearly highlighting its effectiveness. We will investigate the fusion of texture and shape clues for improved performance in further work.

Acknowledgements. This work was supported in part by the National Basic Research Program of China under grant 2010CB327902; the National Natural Science Foundation of China (NSFC) under grant 61202237; and the Fundamental Research Funds for the Central Universities.

References

1. Ding, Y., Zhuang, D., Wang, K.: A Study of Hand Vein Recognition Method. In: IEEE International Conference on Mechatronics and Automation (2005)
2. Cross, J., Smith, C.: Thermographic Imaging of the Subcutane-ous Vascular Network of the Back of the Hand for Biometric Identi-fication. In: Proceedings of International Carnahan Conference on Security Technology (1995)
3. Wang, K., Yuan, Y., Zhang, Z., Zhuang, D.: Hand Vein Recogni-tion based on Multi-supplemental Features of Multi-classifier Fusion Decision. In: Proceedings of IEEE International Conference on Mechatronics and Automation (2006)
4. Lin, C., Fan, K.: Biometric Verification using Thermal Images of Palm-dorsa Vein Patterns. IEEE Trans. on Circuits and Systems for Video Technology 14, 199–213 (2004)

5. Kumar, A., Prathyusha, K.V.: Personal Authentication Using Hand Vein Triangulation and Knuckle Shape. IEEE Trans. on Image Processing 18, 2127–2136 (2009)
6. Wang, Y., Li, K.: Study of Hand-Dorsa Vein Recognition. In: Proceedings of International Conference on Intelligent Computing, pp. 490–498 (2010)
7. Tang, Y., Huang, D., Wang, Y.: Hand-dorsa Vein Recognition based on Multilevel Keypoint Detection and Local Feature Matching. In: Proceedings of International Conference on Pattern Recognition (2012)
8. Frank, M., Wolfe, P.: An Algorithm for Quadratic Programming. Naval Research Logistics Quarterly 3, 95–110 (1956)
9. Zhu, X., Huang, D.: Hand Dorsal Vein Recognition based on Hierarchically Structured Texture and Geometry Features. In: Zheng, W.-S., Sun, Z., Wang, Y., Chen, X., Yuen, P.C., Lai, J. (eds.) CCBR 2012. LNCS, vol. 7701, pp. 157–164. Springer, Heidelberg (2012)
10. Zhao, S., Wang, Y., Wang, Y.: Extracting Hand Vein Patterns from Low-Quality Images: A New Biometric Technique using Low-Cost Devices. In: Proceedings of International Conference on Image and Graphics, pp. 667–671 (2007)
11. Im, S.-K., Choi, H.S., Kim, S.W.: A Direction-Based Vascular Pattern Extraction Algorithm for Hand Vascular Pattern Verification. ETRI Journal 25, 101–108 (2003)
12. Wang, Y., Li, K., Cui, J.: Hand-dorsa Vein Recognition Based on Partition Local Binary Pattern. In: Proceedings of International Conference on Signal Processing, pp. 1671–1674 (2010)
13. Al-Rawi, M., Qutaishat, M., Arrar, M.: An Improved Matched Filter for Blood Vessel Detection of Digital Retinal Images. Computers in Biology and Medicine 37 (2007)
14. Chaudhuri, S., Chatterjee, S., Katz, N., Nelson, M., Goldbaum, M.: Detection of Blood Vessels in Retinal Images using Two-Dimensional Matched Filter. IEEE Trans. on Medical Imaging 8, 263–269 (1989)
15. Miura, N., Nagasaka, A.: Extraction of Finger-Vein Patterns Using Maximum Curvature Points in Image Profiles. IEICE Trans. on Information and Systems E90-D, 1185–1194 (2007)
16. Ojala, T., Pietikainen, M., Maenpaa, T.: Multiresolution Gray-scale and Rotation Invariant Texture Classification with Local Binary Patterns. IEEE Trans. on Pattern Analysis and Machine Intelligence 24, 971–987 (2002)
17. Zhou, F., Torre, F.D.: Factorized Graph Matching. In: Proceedings of IEEE Conference on Computer Vision and Pattern Recognition, pp. 127–134 (2012)
18. Zaslavskiy, M., Bach, F.R., Vert, J.-P.: A Path Following Algorithm for the Graph Matching Problem. IEEE Trans. Pattern Anal. Mach. Intell. 31, 2227–2242 (2009)

A Novel Parallel Video Coding Framework for AVS+ Real Time Encoder

Xiaochen Jiang, Guoping Li, Yake Li, Haiwu Zhao,
Guowei Teng, and Guozhong Wang

Communication and Information Engineering Department,
Shanghai University, Shanghai, 200072, P.R. China
j-xiaochen@hotmail.com

Abstract. Audio Video coding Standard + (AVS+) has made a high progress in coding efficiency, but the algorithm complexity also increased. Parallel coding techniques are efficient solutions to cope with high-complexity coding. A novel parallel video coding framework was proposed to break the traditional video coding back loop and realize efficient parallelism. The framework was applied to AVS+ real time video encoder. Experimental results demonstrate that the proposed framework makes full use of the computing ability of multi-core processor, and improves the code rate greatly, especially in the case of high resolution video.

Keywords: video coding, AVS+, parallel coding, parallel motion estimation.

1 Introduction

AVS+ Promotion Group has formulated the state-of-the-art video coding standard AVS+ that increases four key techniques to further improve the coding efficiency on the basis of AVS [1]. Though AVS+ greatly improves the coding efficiency for the demands of 3D and high definition video coding, computational complexity of the codec has an increment as well. With the development of multi-core hardware, using multi-core processor to parallelly accelerate video coding has become the main trend in standards setting and codec designing. On the condition of fast development of multi-core hardware and parallel processing tools [2], better software and algorithm of parallel coding are needed to realize effective video compress. The research of parallel coding based on H.264 has a focus in recent years [3-5]. For the new generational codec standard High Efficiency Video Coding (HEVC), Joint Collaborative Team on Video Coding (JCT-VC) proposed many valuable proposals for parallel coding, including WPP (Wavefront Parallel Processing) and Tiles [6-7]. These two technologies can only be applied to HEVC standard and simply base on spatial parallelism which has lowered parallelism degree. What's more, the accurate bit rate allocation has been a problem for developers. AVS+ doesn't support of the parallel technologies of WPP and Tiles, only the structure of slices can be used in data parallel computing. Also, the simple use of intra slice parallel coding not only reduces the coding efficiency but brings strip effects [8].

B. Huet et al. (Eds.): PCM 2013, LNCS 8294, pp. 170–179, 2013.

Considering the analysis above, we find that the traditional video coding framework is a main barrier for parallel coding. This paper proposes a novel parallel coding framework to break the parallel limitation of traditional coding framework. The new framework includes three main modules: parallel video preprocessing, parallel motion estimation and parallel coding. The new framework is applied to AVS+ real time encoder and the experimental results demonstrate that this coding framework eliminates the inter-frame dependency and accelerate the encoder efficiently.

2 Traditional Parallel Coding Algorithms

2.1 Intra-frame Parallel Coding

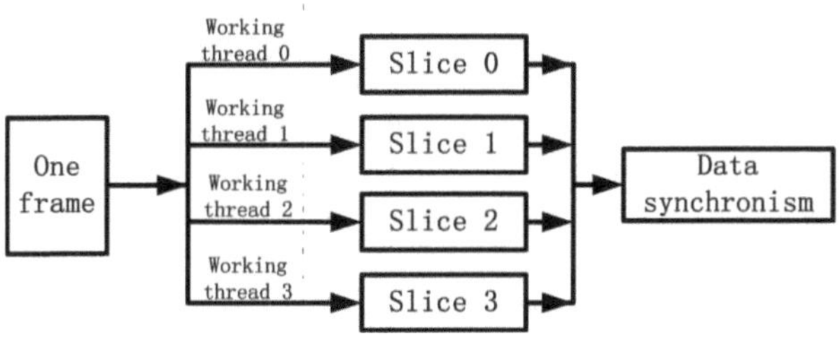

Fig. 1. Slice-level intra-frame parallel coding

Fig.1 shows the fundamental of the intra-frame slice-level parallel coding. One frame is divided into four slices and each slice can be coded within one of the four working threads which work concurrently. There will be a data synchronism after all the threads finish their working task. Though slice-level parallel coding uses the syntactic structural feature to achieve the parallelism and is realized easily, it does not have high coding efficiency and causes strip effects.

2.2 Inter-frame Parallel Coding

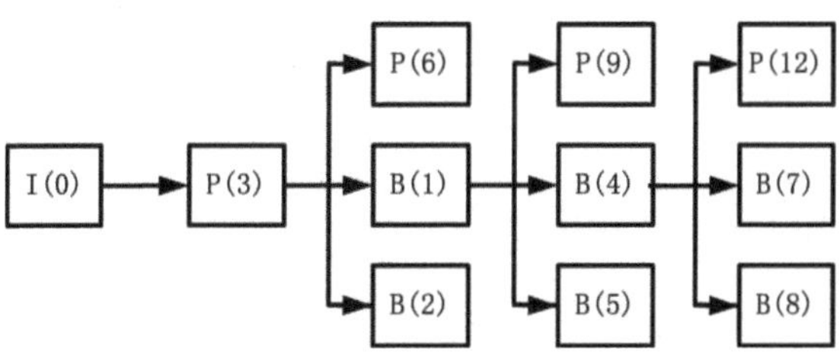

Fig. 2. Inter-frame parallel coding

As shown in Fig. 2, supposing the input sequence type is IPBBP, the first I frame I(0) and P frame P(3) are coded serially. The second P frame-P(6) and the prior two B frames-B(1) and B(2) can be coded concurrently because they don't correlate with each other. For P(6), B(1) and B(2) all refer to I(0) and P(3), three threads are created to process these three frames after I(0) and P(3) are coded parallelly. Then the program returns to the main thread and prepares to code P(9), B(4) and B(5). The main defect of this strategy is that the parallel degree is not high enough for it doesn't eliminate the dependency of inter-frames completely, and may cause system delay.

2.3 Inter-frame Parallel Coding with Reference Dependence

Macroblock's search range is limited because of the continuity of moving object in consecutive frames. The moving object will not beyond a certain range in near reference frames. There is no need for the later frame to wait for the current frame being coded totally. Once a slice of later uncoded frame which refers to the current frame has obtained necessary reference information, the later frame will be coded immediately.

As shown in Fig.3, the input sequence type is IPBBP. Slice-level parallel intra-frame coding is used in intra-frames and the traditional inter-frame coding is employed in inter-frames. Each frame level (a group of frames being coded parallelly in Fig.2) has a time delay, rather than the serial mode of each frame level showed in section 2.2. Fig.3 demonstrates that when the first I (0) is being coded, as long as the slice of P (3) has got the necessary reference information, it can be coded. In a similar way, B (1), B (2) and P (6) as well can start coding, no need to wait for the prior two frames being coded. But this method brings the dependency between reference frames, and reduces the parallel degree.

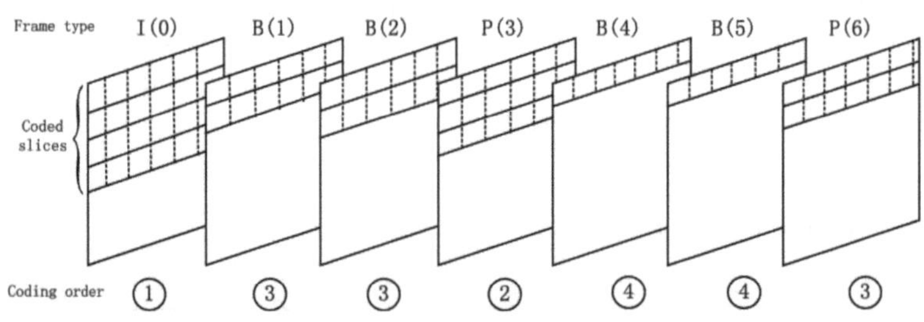

Fig. 3. Inter-frame coding with reference dependence

3 A Novel Parallel Coding Framework

3.1 Limitation of Traditional Video Coding Framework for Parallel Processing

Traditional video coding framework includes two coding paths: forward and reconstructed path. The forward path includes prediction (intra or inter), integer transformation, quantization and entropy coding, and the reconstructed path uses inverse quantization, inverse transformation and motion estimation to reconstruct frames that could be used as the reference frame in prediction. The forward and reconstructed paths constitute the whole coding feedback loop [9].

Owing to this feedback loop, it is difficult to implement the coding parallelism under such structure. A proposed new parallel coding framework breaks this traditional coding framework.

3.2 A New Parallel Coding Framework

As shown in Fig.4, the proposed framework mainly includes three modules: parallel preprocessing module, parallel motion estimation module and parallel coding module. We consider that each module has a time delay of some frames' coding time. In this occasion each module can work concurrently and independently: After the first module has processed the data of M frames, the second module can start to work. After the second module has processed the data of N frames, the third module can start to process.

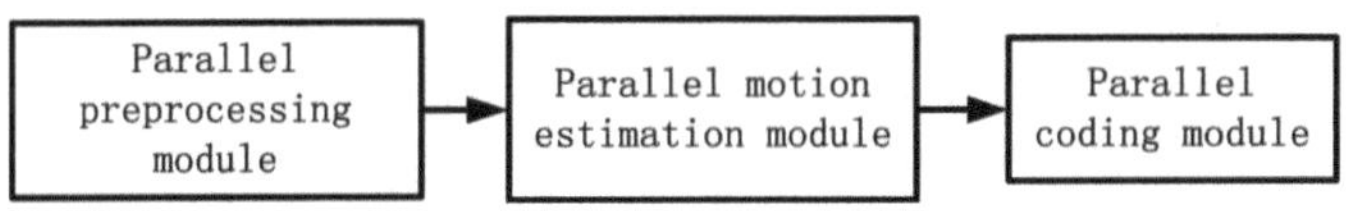

Fig. 4. Proposed parallel coding modules

Fig.5 shows the structure diagram of the entire parallel coding system. The framework breaks the feedback loop and adds the preprocessing module and parallel integer-pixel motion estimation module compared with traditional structure. In parallel coding module, fractional-pixel motion estimation substitutes for the original motion estimation. There are three parts in Fig.5 divided by dotted lines. The three parts are corresponding to the three modules in Fig.4.

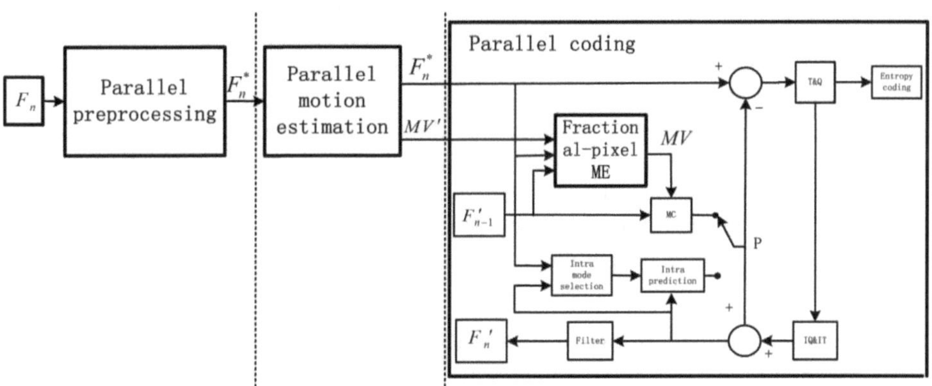

Fig. 5. Structure diagram of the proposed parallel coding framework

(1) Parallel Preprocessing Module

The first part of Fig.5 shows the preprocessing module. Firstly the original video sequence passes through some preprocesses. All preprocesses are conducted based on the original frames spatially and temporally, thus there is no dependency between each other and they can work concurrently. The most important thing in this module is scenes detection and frame types judgment of the sequence by figuring out correlation parameters, which could provide necessary correlation data for the later motion estimation module and parallel coding module.

(2) Parallel Motion Estimation Module

Motion estimation consumes most of time in video coding and integer-pixel motion estimation occupies maximum computational amount and time under the mode of multi references in motion estimation. Moreover, as is mentioned in 3.1, owing to the feedback loop in traditional coding framework, it is hard to make parallel implementation in traditional coding framework. To cope with the problems above, a method that employs the original images as reference images, which separate the most time-consuming integer-pixel motion estimation to conduct the motion estimation parallelly and independently is adopted.

Parallel motion estimation could be conducted under three conditions: (1) There is no dependency between reference frames; (2) Both estimated frame and reference frame could be obtained; (3) The relation between estimated frame and reference frame can be confirmed. The proposed coding framework uses frame types obtained in the preprocessing module and the feature that frames can be buffered by time delay. So it meets all the three conditions listed above and the motion estimation can be performed independently and parallelly in this module. The parallel motion estimation module greatly improves the parallelism of the coding framework and the utilization of multi-core hardware resource.

(3) Parallel Coding Module

Frame type parameters and motion vectors have been obtained through the prior modules, which makes the computational amount in this parallel coding module decrease greatly. We use a method of interlaced coding and each field's process is independent. The inter-frame coding in section 2.3 is adopted. The procedure is showed in the third part of Fig.5. After the necessary selections of macroblock mode, the integer-pixel motion vector MV' is served as the starting point in the fractional-pixel motion estimation. After the fractional-pixel motion estimation we get the final motion vector MV. Next integer transform, quantization and entropy coding are performed to get the AVS stream. This strategy overcomes the disadvantage of method mentioned in section 2.3 and decreases the parallel inter-frame number.

3.3 Parallel Coding Framework for AVS+ Real Time Encoder

AVS+ does not support WPP and Tiles which are parallel techniques in HEVC. Only slice structure coding can be used in parallel coding. AVS+ is mainly applied to 3D, high definition, and interlacing video coding. It can also handle a certain delay time in practical. Considering the features of AVS+ listed above, the new parallel coding framework can be used into AVS+ real time encoder.

Main steps of the parallel real time encoder can be described as follows:

1. Establish L parallel processing threads. Because all preprocessings are implemented using the original frames and the complexity is not high, there is no dependency between preprocessing threads. The thread number commonly sets 2 at most and each thread corresponds to the processing of an original frame.

2. Establish 2×M integer pixel parallel motion estimation threads. After procedure 1, the types of each frame have been confirmed. Parallel integer motion estimation can be conducted according to each original reference frame. AVS+ is mainly applied to interlacing coding. Each frame can be divided into two fields and each of the two fields works as an independent thread when conducting motion estimation.

3. Establish 2×N coding threads. Each thread conducts fractional-pixel motion estimation according to the integer MV from procedure 2. This can efficiently lower the coding complexity. The parallel coding with reference dependency which proposed in 2.3 is adopted in inter-frame coding. Considering AVS+ is mainly applied to interlacing frames coding, each frame can be divided into 2 fields and each field can be treated as a slice coded independently. Thus we not only improve the parallelism, but avoid the loss of coding efficiency and strip effects.

4. Rate control of parallel threads. Because motion vectors and sum of absolute differences (SAD) of coding frames can be obtained after motion estimation in procedure 2, the obtained SADs can be served as the level of complexity of coding frames, which provide basic information for the bit distribution of coding frames.

5. Delay control of the encoder. The time delay of the encoder inferred from above coding procedures is L+M+N frames' coding time.

According to the above procedures and the availability of encoder hardware resources, we should first ascertain time delay of the encoder and the bottleneck of thread management and then distribute L、M、N reasonably.

4 Experimental Results and Analysis

In order to test the performance of the proposed parallel coding structure, a DELL server was used as multi-core test platform. The server has an Intel Xeon CPU with 2×12 cores and each core's basic frequency is 3.06HGz. Six 1920×1080 high definition sequences-parterre, dial, basketball, volleyball, leaf, birdcage-and a spliced sequence by the above 6 sequences are used as test sequences. Coding succession format is IPBB. GOP is 30. Search range is ±32. The testing software is AVS+ high definition real time encoder. The encoder not only sustains the traditional video coding mode (serial coding), but supports the parallel video coding in the framework proposed in this paper. In the experiment, the preprocessing threads parameter L is set 1, parallel motion estimation threads parameter M is set 4, N means parallel inter-frame umber and other parameters' meaning can be seen in Table 1.

Table 1. Descriptions of parameters

Parameter	Description
ME2	UMHexagonS
ME1	Simplified UMHexagonS (remove 5×5 square and big hexogen search)
Fractional-search 1	Sub-pixel MV search 2 times and quarter-pixel MV search 4 times
Fractional-search 2	Sub-pixel MV search 4 times and quarter-pixel MV search 10 times
Fractional-search 3	Sub-pixel MV search 6 times and quarter-pixel MV search 12 times
Parallel inter-frame umber (N)	Number of frames in parallel coding in Fig.3

4.1 CPU Utilization

When computers are doing parallel computation utilizing multi-core processor, CPU utilization is an important feature to reflect the efficiency of algorithm. If the program can make use of all CPU cores to work simultaneously, the speed rate would surely be improved.

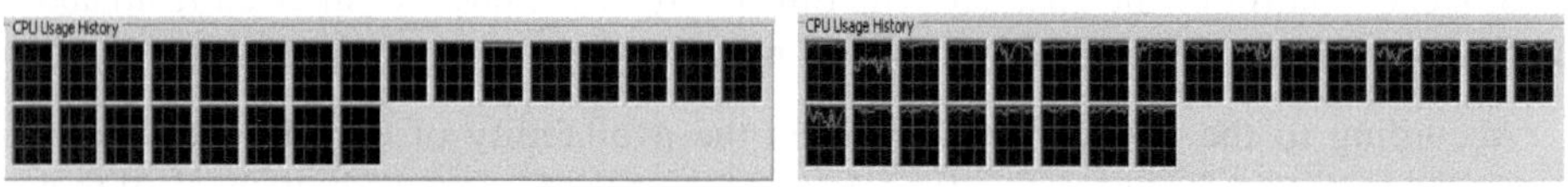

(a) (b)

Fig. 6. Comparison of CPU utilization between serial and parallel coding

Fig.6 shows the comparison of CPU utilization between serial and parallel coding. Figure (a) shows the CPU utilization of serial coding. We can see that only the 11th thread is working, which demonstrates that serial coding cannot make full use of CPU cores. Figure (b) shows the CPU utilization of using the proposed parallel coding algorithm, setting motion estimation parameters ME2 and Sub-search 3, and Parallel inter-frame number 16. The results show that all CPU cores take part in the process and all are of high-efficiently use.

4.2 Speedup Ratio

Speedup ratio is an important index in measuring the performance of the parallel computing. It reflects an increasing degree of the program computing speed from serial process to parallel optimization. In this paper the speedup ratio is defined in formula (1). Coding frame rate can be represented by parameter fps, which also reflects the speed of an encoder.

Table 2. Speedup ratio and fps in different motion estimation

Motion estimation	Fractional-pixel search	Parallel fps	Serial fps	Average speedup ratio
ME1	Fractional-search 1	51.4	4.2	12.24
ME2	Fractional-search 1	43.7	3.7	11.81
ME1	Fractional-search 2	50.4	4.2	12.00
ME2	Fractional-search 2	42.4	3.6	11.78
ME1	Fractional-search 3	50.5	4.2	12.02
ME2	Fractional-search 3	41.3	3.5	11.80

$$speedup = \frac{frame\ rate(fps)\ after\ prallel\ optimization}{frame\ rate\ before\ prallel\ optimization} \tag{1}$$

In formula (1), speedup ratio is presented by speedup, the larger the value, the better the parallelism of the algorithm. Also the coding speed is faster.

Table 2 shows the speedup ratio of testing the 1500 frames of the spliced sequence in ME1, ME2 and Fractional-search 1, 2, 3. It can be seen clearly that encoder employing the proposed parallel framework can get a sufficient speedup ratio that completely meets the practical demands.

4.3 Inter-frame Parallelism

As is mentioned before, Parallel inter-frame number (N) in inter-frame prediction is a key issue to determine coding parallelism. The parameter cannot be too small or too large; otherwise parallelism of the encoder will be negatively affected. What's more, there will be some problems in contributing bit rate. So it is an issue of importance to determine a suitable N in inter-frame prediction.

Table 3. Fps for different index N

N	fps(ME2)	fps(ME1)
4	24.9	24.7
5	29.0	30.4
6	34.0	35.4
8	40.1	43.3
9	40.6	46.1
10	40.8	48.3
11	41.0	49.7
12	41.3	50.5
16	40.9	49.4

Table 3 gives the experimental results for different index N. It is easy to understand that the larger the index N is, the better improvement of the subjective quality of the decoded frames will have. However we should consider that whether the frame rate can improve as well. From Table.3 we can see that when N reaches 8, fps is high enough to meet the practical demands of encoder. Under this circumstances, the delay of encoder is 13 frames' time (L=1, M=4, N=8). Performance of bit-rate control in the encoder will be improved on the condition of low N. It can be seen from Table 3 that fps improves largely as N increases from 4 to 12, but when N arrives 12, fps starts to fluctuate and stops increasing. From the analysis above, we can reduce N if the coding speed is acceptable, which can provide advantages for better realization of rate control.

4.4 Rate-Distortion

After testing 6 high definition sequences, we show rate-distortion curves in different bitrates by serial and parallel coding. As shown in Fig.7, two curves of different color in each figure represent the result of serial and parallel coding. PSNR in different bitrates reflects the quality of decoded frames. The average PSNR loss of the 6 sequences is 0.13 dB and the minimum loss is 0 dB. The largest loss is about 0.2 dB, which can be seen in Fig.7.

We pay more attention on the practical application whose video may be hard to encode, just like volleyball.yuv and basketball.yuv. After testing these two sequences we can see even in the worst condition, there is only a loss about 0.2 dB. Making the original frames as reference frames in motion estimation is feasible and remains little negative impact for encoder. For the quality of decoded frames, using the proposed parallel encoder gets the same satisfactory objective and subjective effects as serial coding does. In addition, the proposed encoder has a remarkable advance in coding rate, which can further display its advantages: good performances and low losses.

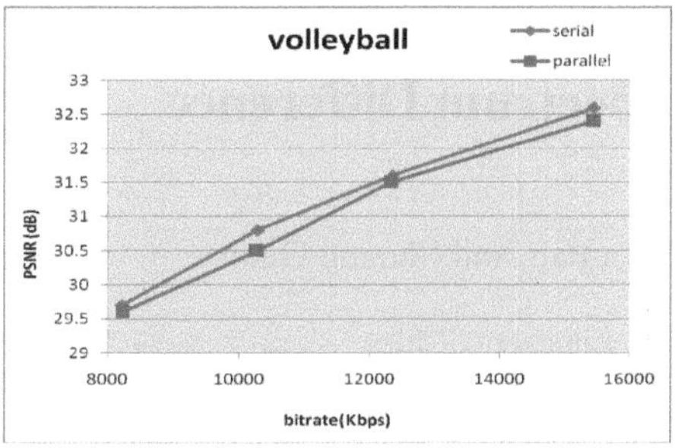

(a)

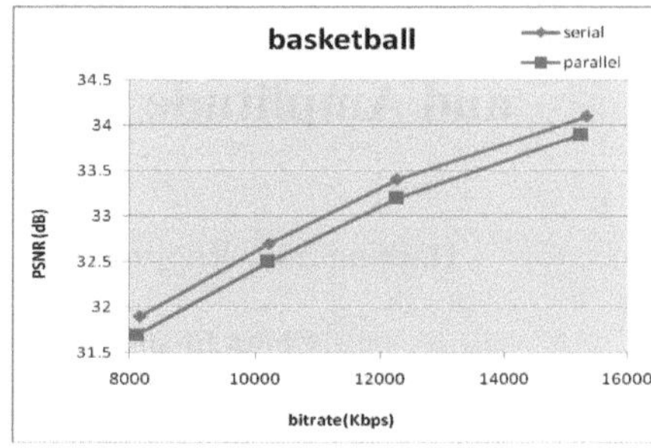

(b)

Fig. 7. The Rate-distortion curves with worst performance in 6 curves

5 Conclusion

We designed a novel parallel coding framework, breaking the traditional video coding structure and making use of the encoder's feature of time delay. Combining the coding framework with AVS+, we proposed a parallel coding framework for AVS+ real time encoder. The experimental results showed that the parallel encoder can make full use of the computational ability of the multi-core processor, and greatly improve the coding speed on the basis of least loss and controllable coding delay (within 0.5 s).

References

1. Zhang, X.L.: The promotion and application of Terrestrial digital TV industrialization. Video Engineering 36(22), 3–4 (2012)
2. Chen, D., Singh, D.: Fractal Video Compression in OpenCL: An evaluation of CPUs, GPUs, and FPGAs AS Acceleration Platforms. In: 18th Asia and South Pacific Design Automation Conference (ASP-DAC), pp. 297–304. IEEE Press, Yokohama (2013)
3. Pang, Y., Zhang, F.Y., Sun, L.F.: Survey of Parallel Acceleration Algorithms of Video Coding on Multi-core Processor. Journal of Frontiers of Computer Science and Technology 3(4), 337–346 (2009)
4. Ge, S., Tian, X., Chen, Y.-K.: Efficient multithreading implementation of H.264 encoder on Intel hyper-threading architectures. In: IEEE Proceedings of the 2003 Joint Conference of the Fourth International Conference, pp. 469–473. IEEE Press (2003)
5. Tian, X., Chen, Y.-K., Girkar, M., et al.: Exploring the use of hyper-threading technology for multimedia applications with Intel OpenMP compiler. In: Proceedings of the Intel Parallel & Distributed Processing Symposium, pp. 36–43. IEEE Press, Nice (2003)
6. JCTVC-G1025: Tiles and Wavefront Parallel Processing. 7th Meeting. Geneva CH, November 21-30 (2011)
7. JCTVC-F274: Wavefront Parallel Processing for HEVC Encoding and Decoding. 6th Meeting. Torino, IT, July 14-22 (2011)
8. Xu, C.M., Li, G.P., Wang, G.Z.: Study and Implementation of AVS Encoding algorithm Based on Slice Parallel. Journal of Image and Graphics 14(6), 1108–1113 (2009)
9. Chiang, C.-K., Lai, S.-H.: Fast H.264 Encoding Based on Statistical Learning. In: Qiu, G., Lam, K.M., Kiya, H., Xue, X.-Y., Kuo, C.-C.J., Lew, M.S. (eds.) PCM 2010, Part II. LNCS, vol. 6298, pp. 179–189. Springer, Heidelberg (2010)

Video Saliency Detection Algorithm Based on Phase and Amplitude Joint Spectrum Difference

Haibing Yin, Jingjing Tan, Chen Pan, and Shuang Guan

China Jiliang University, Hangzhou, China
{yinhb,pc916}@cjlu.edu.cn, yiqi.tj@gmail.com,
guanshuang618@126.com

Abstract. Saliency map can express the position of eye trace, which is a key factor to evaluate the perception quality of video or image. Many saliency detection methods have been proposed based on the human vision system characteristics, but they are time-consuming or with low accuracy. Spectrum residual based algorithm is one typical saliency detection method. It calculates the saliency map of image effectively. The SR method can't detect the saliency in video sequences due to the intrinsic property difference between image and video.

In this paper, we propose a saliency detection method for video. According to the extensive simulation results, we find amplitude difference and phase difference are very important for saliency map detection. On the basis of the spectrum features of the adjacent frames in video, we take spectrum difference between current test frame and its adjacent frame into account to calculate the saliency map. The proposed algorithm using the video sequences with one eye track point. The results indicate that the method proposed in this paper is more effective than others.

Keywords: Spectral, amplitude, phase, saliency map, video.

1 Introduction

With the rapid development of network video application and video monitoring system, high quality videos are demanded increasingly. Research on video quality assessment is very important due to the fact that human eye is the ultimate evaluator of image and video quality. According to the research results, region of interest and eye fixation are very important factors to evaluate the video quality. As a result, saliency map detection is crucial for efficient video quality assessment and perceptual video coding algorithm optimization [1][2].Research on saliency map detection is also an effective method to solve the region of interest coding and transmission in video conference and surveillance. In addition, video saliency research is also related to computer vision, psychology and so on.

Video sequence is nonstationary signal with complex texture or motion edges, especially for the video with rapid scene changes. It was reported in the literatures that the human eyes always focus on a specific area in an image or a video. This is an

B. Huet et al. (Eds.): PCM 2013, LNCS 8294, pp. 180–189, 2013.

intrinsic property of the human visual system. Employing this HVS property and applying it in video quality assessment can greatly assist in perceptual video coding algorithm optimization and video quality improvement.

The research of saliency map started from the last century about the mid 80s. In the literatures, extraction of a significant region is generally concerned by the human visual characteristics, and then can apply it into the visual attention. The region of visual attention can be extracted by analyzing the image information and the feature. Focus of the human eye and the features of image or video are important factors for saliency map detection algorithm modeling. The features such as color, motion, edges are candidate consideration factors for salience map detection. How to create a suitable research model is a significant research in this field.

Currently, the research on the image saliency areas can be grouped into two categories: The one way is based on human visual system characteristics. Another is modeling from a mathematical point of view. Visual attention model was proposed by Koch [3] in 1985, then Itti improved the model (NVT) [4,8,9]. These two articles are both based on the visual physiological and psychological physics experiments, and then fuse the visual features into a saliency model. But the high computational cost is the weakness of these models. Guo, Hou and Li analyzed the amplitude or phase distribution from a mathematical point of view [5-7]. Guo proposed a method named PFT just using the phase information in spectral domain [5]. To some extent, these algorithms reduce the complexity of the former algorithm, and they have better results.

Video has some spatial and temporal correlation; the spatial characteristics of image are obvious. Therefore, these methods of image saliency can't be directly applied in saliency detection for video. Hou proposed an image saliency detection algorithm (SR) [6], which analyzes the difference between the original signal and the smoothed image in the log spectrum, and then transform the difference value into the spatial domain signal. Eventually the algorithm gets the saliency map.

We proposed a video saliency detection algorithm based on spectrum difference (SDASD) between adjacent frames. This paper analyzes the importance of amplitude spectrum and phase spectrum for saliency map detection, and explain the importance spectral characteristic in the adjacent frames. According to these features above, we get the saliency map finally. We test it on a database, named CRCNS [10]. Compared to SR, the result of proposed algorithm increases the detection accuracy of saliency objects especially for the videos with complex image content. The detection accuracy increment is achieved at the cost of slight computation complexity increasement. Nevertheless, the results indicate that the proposed algorithm complexity is moderate and well-suited for real-time implementation.

The following is organized as follows. Spectrum difference of adjacent images is analyzed in section 2. Section 3 describes the proposed algorithm. In section 4, we summarize the experimental results and makes comparison with other algorithms.

2 Analysis of Spatial Spectrum and Spatio-temporal Spectrum

At present, many researchers analyze the saliency region based on the signal spectrum characteristics of the image or the video sequences. Hou proposed a SR algorithm [6], the authors made extensive experiments and found spectrum residuals can express salient areas of the image. As we know, the sum of sine function with different phase values and amplitude values represent image signal. Phase information of the image contains the image texture information while the amplitude information contains chiaroscuro information of image. Phase information is an important feature of the image signal [7]. The frequency of the image is a factor indicating the degree of image grayscale change. Only if the amplitude and phase are all available, it can assess the intensity change of image in the frequency domain. By analyzing the amplitude and phase information of a square wave signal and video sequences with moving objects in the frame, we find that amplitude difference and the phase difference play an important role in detecting salient areas of the video sequence.

2.1 Spectrum Analysis of Square Wave Signal and Image Signal

As analyzed above, the amplitude and phase information of an image or video sequence are important for video saliency detection. Here, we employ one-dimensional square-wave signal and the two-dimensional image signal to analyze the relation among the saliency, the amplitude difference and phase difference. The analysis is achieved by comparison between the original test signal and the reconstructed signal by applying amplitude difference and phase difference to the spectrum in frequency domain.

We take the one-dimensional square wave signal as an example to describe the changes between the original and the reconstructed signal derived from frequency domain processing.

The frequency domain processing may be amplitude spectrum processing, phase spectrum processing, or joint amplitude and phase spectrum processing. Fig.1(a) describes two different reconstructions. Input1 and input2 are the original input signal. We obtain a reconstructed curve in the figure according to the algorithm of SR [6], where the input1 is the original signal. Another reconstructed curve is got by distinct process of amplitude and phase. Amplitude of input1 is invariant. Phase of the reconstruction is the phase difference between input1 and input2. Fig.1(a) indicates that phase difference can express a certain degree of dramatic changes at both signals. Similarly, Fig.1(b) shows that reconstruction by keeping amplitude zero with phase difference of the two-adjacent frame represent salient area are better. Fig.1(c) illustrates that preserving the spectrum difference and phase difference give better representation. Therefore, amplitude difference and phase difference play an important role in the saliency detection.

For the two-dimensional image signal, saliency areas include dramatic changes and the texture area in the image. By analyzing the amplitude and phase information through the reconstructed maps, we can obtain a salience map of a frame in the video.

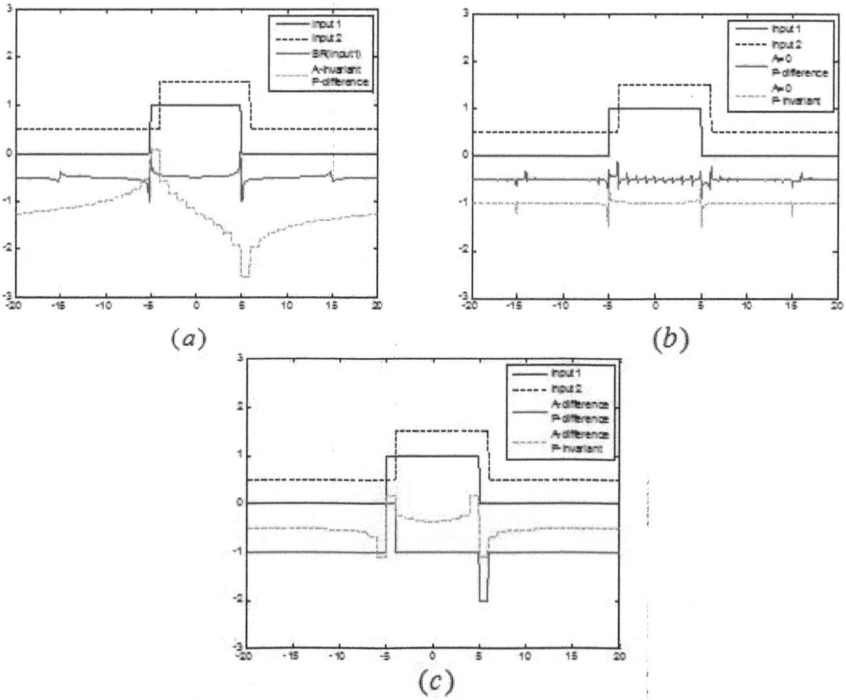

Fig. 1. Reconstructions by changes of two One-dimensional square wave signals

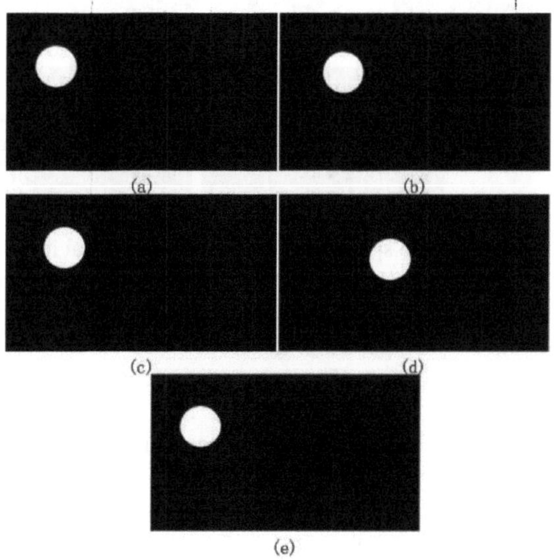

Fig. 2. Two-dimensional original signal (a) test frames (b) (c) (d) (e) may be present in adjacent frames

Fig. 2 is the two-dimensional original signal of the test frames and its adjacent frames. (a) is exactly the same with (e). We test original frame with its adjacent frames by setting up various amplitude and phase. Results are shown in Fig. 3. In the figure, A stands for amplitude and P is phase for short. The observers always pay more attention on the moving object in the adjacent frames. Thus, area of moving objects is of great significance in terms of visual attention and saliency interest. According to the results in Fig.3, we can accurately extract the saliency area through taking the amplitude and phase into consideration. Video Saliency detection Based on Phase and Amplitude Spectrum

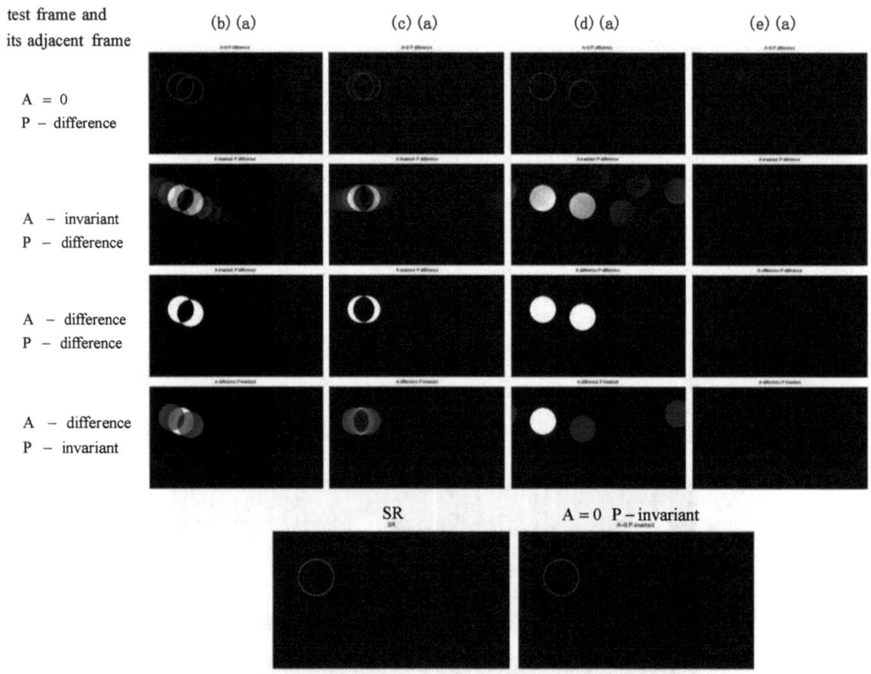

Fig. 3. Changes in the spectrum of the two-dimensional image signal

According to the observation mentioned above, we propose an approach saliency detection algorithm based on spectrum difference (SDASD) which includes phase spectrum and amplitude spectrum. The proposed algorithm can be summarized as follows:

The t-th frame in the test video, denoted as $image_t$, is taken as the current frame for algorithm description. Its previous frame is $image_{t-1}$. The two adjacent frames are as the input image, respectively. After the Fourier transform spectrum of two frames, $\varsigma(u,v)$ and $\eta(u,v)$ were obtained.

$$\varsigma(u,v) = FFT2(image_t(x,y)) \tag{1}$$

$$\eta(u,v) = FFT2(image_{t-1}(x,y)) \tag{2}$$

In equation (1) and (2), $FFT2(\bullet)$ stands for the two-dimensional Fourier transform.

Phase spectrum and amplitude spectrum of the adjacent frames can be calculated as follows:

$$R(u,v) = abs(\varsigma(u,v)) \tag{3}$$

$$R_1(u,v) = abs(\eta(u,v)) \tag{4}$$

$$P(u,v) = phase(\varsigma(u,v)) \tag{5}$$

$$P_1(u,v) = phase(\eta(u,v)) \tag{6}$$

where $abs(\bullet)$ represents a function to obtain the amplitude spectrum, $phase(\bullet)$ is to obtain the phase spectrum. $R(u,v)$ and $R_1(u,v)$ illustrate the amplitude spectrums of the test and previous frames. $P(u,v)$ and $P_1(u,v)$ describe the phase spectrums of the test and previous frames.

Amplitude difference and phase difference are obtained as follows.

$$\alpha(u,v) = R(u,v) - R_1(u,v) \tag{7}$$

$$\beta(u,v) = P(u,v) - P_1(u,v) \tag{8}$$

Then the proposed saliency map is estimated as follows.

$$SM(u,v) = g * \| IFFT2(\exp(\log(\alpha(u,v) + i \cdot \beta(u,v)))) \|^2 \tag{9}$$

Where g is the two-dimensional gaussian filter, the size of filter is set by 11×11. $IFFT2(\bullet)$ means two-dimensional inverse Fourier transform. $SM(u,v)$ is the value of local(u,v).

3 Experimental Results

3.1 Detecting Objects in Saliency Map

Saliency map is an explicit representation of proto-objects. There are many ways to extract the desired objects [6] [8] [9]. In this section, we analyze and extract the target with high saliency weight from the two aspects. First, according to SR method [6], it proposed a simple threshold algorithm to determine the candidate target. Based on the given saliency map $SM(u,v)$, target position image $O(u,v)$ is obtained as follows:

$$O(u,v)=\begin{cases}0 & if\ SM(u,v)>th \\ 1 & otherwise\end{cases} \tag{10}$$

we set $th=E(SM)\times 3$, $E(SM)$ represents the average intensity of saliency map. Because of the existence of false election, the objects image should be subject to further processing. Assuming the position of the candidate image $O(u,v)$ has the largest area O^{max}, the final candidate area $Region_t$ is obtained as follows

$$Region_i=\{(u,v)\,|\,\alpha\cdot O^{\max}<O(u,v)<O^{\max}\} \tag{11}$$

where $i = 1 .. n$, it stands for the number of the candidate regions. α is a threshold to affect the size of region. The bigger the α is, the more precise the selected is. We set α=0.75. We test the objects detection method on videos which are labeled eye-tracking position by CRCNS [10]. When the candidate regions and the eye-tracking position are substantially consistent, we believe that the candidate regions are correct. The number of the candidate area is also an important factor. The video sequence is only one eye tracking, so the fewer the selected area, the more accurate the algorithm.

3.2 Analysis of Result

This database CRCNS, in which eye tracking data were collected from a total of 50 video clips consisting a variety of categories of scenes including indoor scenes, outdoor scenes, television clips, and video games, will be used in analyzing the algorithm. We first adopt some testing sequences listed in the CRCNS. The detailed results of the testing sequences are shown in Figure 4.The follow information is shown: the representative frames of the testing videos (Fig.4(a)), the saliency maps of representative frames (Fig.4(b)), the detected regions of moving objects (Fig.4(c)), the probable interested regions of human vision(Fig.4(d)). From the Figure 4, we can see that the algorithm we proposed is able to detect some regions where interesting objects happen. But the saliency map detecting and objects detecting are not perfect.

As the limitations of the length of an article, we only adopt a video called beverly08.mpg to compare with other algorithms. The sequence may down-sampled to 320 × 240. The video frame rate is 25fps. Figure 5 shows the test results of four methods (NVT, SR, PFT, SDASD) in six frames. SDASD research the saliency maps better, while others will pay some attention to the background. The test results of all frames are shown in Table1.The performance of SDASD is best because it can detect 1.05 objects per frame and is close to the number of eye-tracking position. Objects number of other algorithm is worse than SDASD.

Another important factor to evaluate the algorithm is computation time cost. Table 2 shows the average time cost per frame in the test video. The time cost results of the proposed SDASD algorithm is close to those of the SR and PFT algorithms. It can surely meet real-time requirements.

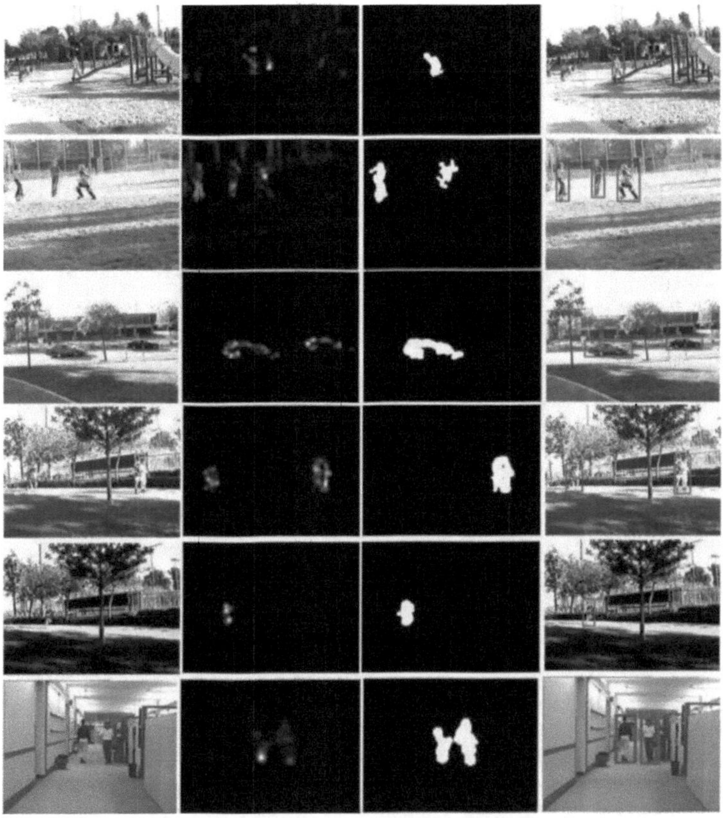

(a) (b) (c) (d)

Fig. 4. Detection results of the testing sequences. (a) shows the representative frames of the testing videos,(b) shows the saliency maps of representative frames, (c) shows the detected regions of moving objects, (d) shows the probable interested regions of human vision.

Table 1. Objects number of five models of the test video

Model	*Selects*	*Number of frames*	*Object number of a frame*
SDASD	127	121	1.05
NVT	183	121	1.51
SR	246	121	2.03
PFT	262	121	2.17

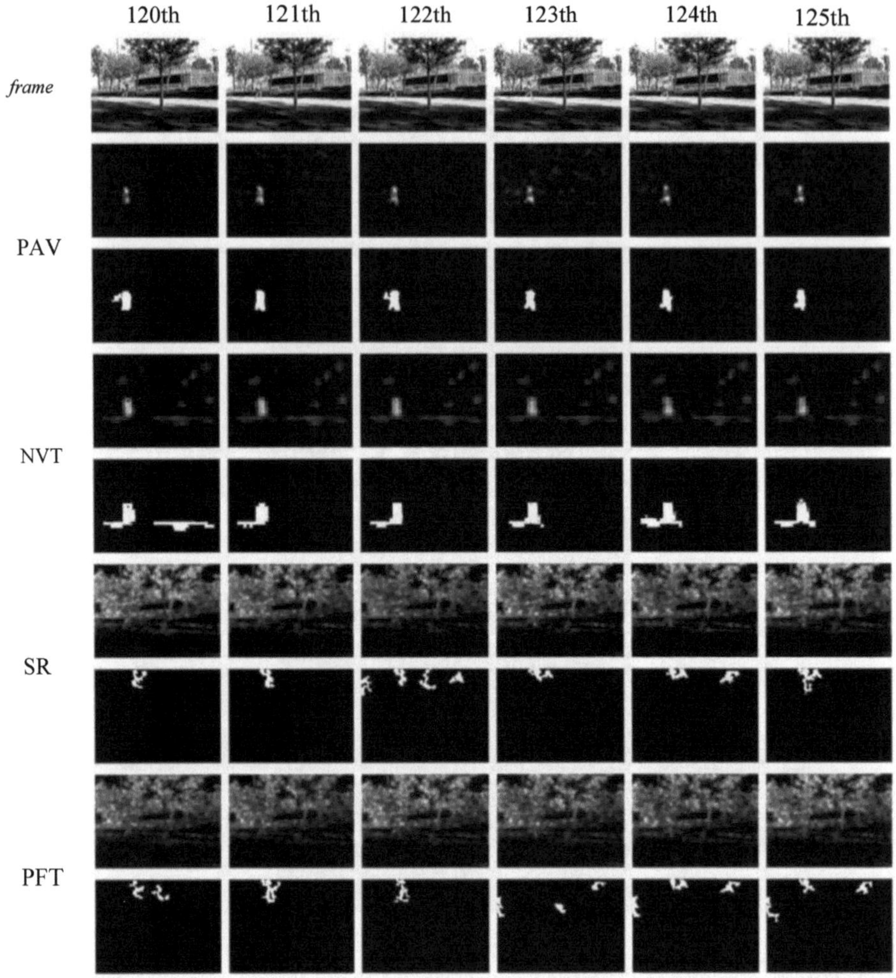

Fig. 5. Saliency map and objects detecting of five models in the test video

Table 2. Average time cost per frame in the test video

Model	*Average time cost(s)*
SDASD	0.064207
NVT	1.769474
SR	0.057834
PFT	0.042874

4 Conclusion

We proposed a method for saliency detecting of video. As we know, SR algorithm shows superior performance in detecting some images [6]. When the test image content is complicated or we detect saliency map of video, SR fails to extract the proto objects. Meanwhile, many effective methods are suffering from huge computation burden. In this paper, we propose a method called SDASD. It can easily detect the objects which are moving and reduce the saliency of background fixed area.

The major contribution of this paper is that we take amplitude and phase difference of the adjacent frames into account, and own low complexity. Compared to the human vision, our method also has some limitations. This algorithm proposed in the paper considers only a phase characteristic and amplitude characteristic of the video frame, and does not take into account the relevant characteristics of the human eye. When the frame and adjacent frame are consistent with the circumstances, the proposed algorithm is not well fixed to the position of eyes. For this, we will further the study to find a solution.

Another potential application of this algorithm is in video quality assessment. Due to simple calculation, it can be good for video quality assessment as a basis for positioning the eye.

Acknowledgments. This work was supported by ZJNSF Y1110114, ZJNSF LY12F01011, NSFC 60802025, and the open project of SKL of Novel Soft Technology, Nanjing University (KFKT2012B09).

References

1. Ninassi, A., Meur, O.L., Callet, P.L., Barba, D.: Does where you gaze on an image affect your perception of quality? Applying visual attention to image quality metric. In: Proc. ICIP, pp. 732–735. IEEE, San Antonio (2007)
2. Lu, Z.K., Lin, W., Yang, X.K., Ong, E., Yao, S.: Modeling visual attention's modulatory aftereffects on visual sensitivity and quality evaluation. IEEE Trans. Image Process. 14(11), 1928–1942 (2005)
3. Koch, C., Ullman, S.: Shifts in selective visual attention: towards the underlying neural circuit. Hum. Neurobiol. 4, 219–235 (1985)
4. Itti, L., Koch, C., Niebur, E., et al.: A Model of Saliency-based Visual Attention for Rapid Scene Analysis. IEEE Transaction on Pattern Analysis and Machine Intelligence 20(11), 1254–1259 (1998)
5. Guo, C., Ma, Q., Zhang, L.: Spatio-temporal Saliency Detection Using Phase Spectrum of Quaternion Fourier Transform. In: Computer Vision and Pattern Recognition, CVPR, pp. 1–8 (2008)
6. Hou, X., Zhang, L.: Saliency Detection: A Spectral Residual Approach. In: Hou, X., Zhang, L. (eds.) Computer Vision and Pattern Recognition, Proc. CVPR, pp. 1–8. IEEE, New York (2007)
7. Li, C., Gao, Y., Lu, K., Qu, Z.: Saliency detection method based on phase spectrum and amplitude spectrum tuning. Journal of Image and Graphics 17(7), 821–827 (2012)
8. Itti, L., Koch, C.: A saliency-based search mechanism for overt and covert shifts of visual attention. Vision Research 40(10-12), 1489–1506 (2000)
9. Itti, L., Koch, C.: Computational Modeling of Visual Attention. Nature Reviews Neuroscience 2(3), 194–203 (2001)
10. Collaborative Research in Computational Neuroscience – Data sharing, `http://crcns.org`

Edge-Respecting Image Smoothing via Extrema Interpolation

Xiaolei Jiang, Hongxun Yao*, and Sicheng Zhao

School of Computer Science and Technology, Harbin Institute of Technology, China
{xljiang,h.yao,zsc}@hit.edu.cn

Abstract. The ability of blurring details while preserving salient edges is valuable for many image processing applications. We propose an extrema interpolation formulation for edge-aware smoothing, in which the extrema of the sought signal are fixed. A set of extrema cannot determine a unique signal, so the final smoothing result is the one that has the least distance with the original signal. Our method can follow significant edges more closely, as can be understood intuitively from the standpoint of monotonic approximation. Experimental results on one- and two-dimensional signals validate the distinctive effects of our method. We also demonstrate the effectiveness of our method for image detail enhancement and image abstraction.

Keywords: Extrema interpolation, edge-preserving smoothing, monotonic regression, quadratic programming.

1 Introduction

The significance of preserving edges while coarsening images has received considerable emphasis in the contexts of base-detail decomposition, image enhancement, tone mapping, and stylization, among many others. Farbman et al. clarify the artifacts resulting from edge blurring and oversharpening for applications in detail enhancement [1]. Other fields, such as image denoising, image restoration, and cartoon-texture decomposition also have some relationship with edge-preserving smoothing, since they all concern not blurring main structures in images. Existing methods can be roughly classified into several classes: global optimization, local methods, and diffusion-based methods.

Global optimization methods find the smoothed image S from the original image I as the solution to the following optimization problem:

$$\min_{S} \left\{ \|S - I\|^2 + \lambda C(S) \right\} , \tag{1}$$

where $\|\cdot\|$ is some kind of norm, and $C(\cdot)$ is a cost. The data (fidelity) term $\|S - I\|^2$ enforces S to be similar in main structures to I, $C(S)$ is a regularization (smoothing) term for penalizing roughness of S, and λ is a smoothing parameter controlling the significance of $C(S)$ relative to $\|S - I\|^2$. Usually, $\|\cdot\|$ is the Euclidean norm, and

* Corresponding author.

B. Huet et al. (Eds.): PCM 2013, LNCS 8294, pp. 190–199, 2013.

$C(S)$ is taken as a functional of gradients of S. Many kinds of $C(S)$ have been studied, including weighted least squares [1], total variation [2], L_0 norm [3], and relative total variation [4].

Local approaches compute each pixel from its neighbors in the input image. In bilateral filtering, every pixel in the filtered result is a weighted average of input pixels, with the weights depending both on spatial distance and difference in intensity [5]. Local histogram filters estimate modes in local image histogram to determine output pixels [6].

Diffusion-based methods employ partial differential equations to describe analogous diffusion or heat transfer processes [7]. Edge preserving is accomplished by setting gradient-dependent conductance.

There are methods that can hardly be seen as belonging to any one of the aforementioned three classes, among them mostly related to our method are [8] and [9]. [8] models detail as oscillations between local maxima and minima, and takes the average of extremal envelopes as the smoothed result. [9] quantifies the smoothness of signals by the notion of locally monotonic signals, which are monotonic within all intervals of a specified length.

In spite of the preceding efforts, edge preservation is still hard to define and quantify. This paper employs monotonic approximation to formalize the intuitive notion of edge preservation, and propose an extrema interpolation formulation for smoothing. By placing constraints on the extrema of resulting images, our method can retain edges to an even higher degree.

2 1D Extrema Interpolation

2.1 Monotonic Regression

Near a significant edge to be preserved, the main feature of the function is monotonically increasing or decreasing. Our key observation is that by enforcing monotonicity, the resulting approximation can follow the main edge fairly well, as shown in Fig. 1 (a). For the original curve I, the smoothed version S is obtained by solving the following monotonic regression problem:

$$\min_S \|S - I\|^2 \quad \text{s.t. } \nabla S_p \geqslant 0, \forall p \ (\text{or } \nabla S_p \leqslant 0, \forall p)\,, \tag{2}$$

where $\|S-I\|^2 = \sum_p (S_p - I_p)^2$ is the square of Euclidean norm, S_p and I_p denote the values taken by S and I at p, and $\nabla S_p = S_p - S_{p-1}$ is the backward difference of S at p. Thus, S is the most similar curve to I, among all monotonic functions. Oscillations violate monotonicity, so the noise-like component of I is suppressed. For the dominant edges, S follows I closely, since monotonic constraint does not penalize large gradients, unless they are opposite in sign. As shown in Fig. 1 (b), monotonic regression can retain significant edges better than other two methods.

For each edge to be preserved, one interval must be specified on which the smoothed curve is constrained to be monotonic. The whole curve, therefore, should be partitioned into segments. This partition induces a cost for measuring smoothness: the number of monotonic segments. We can use this cost as a regularization term and solve (1),

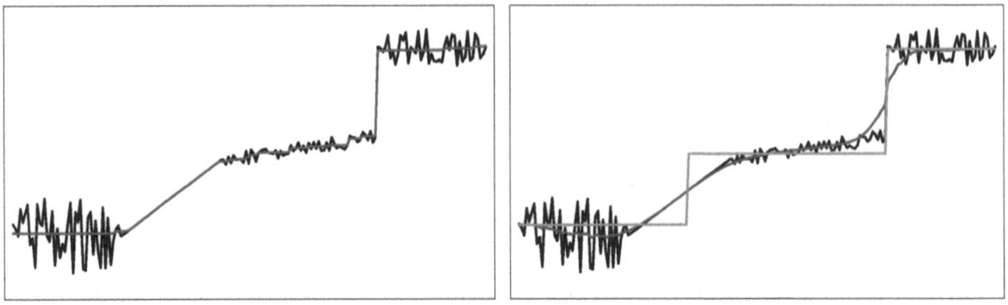

(a) Monotonic regression (b) Other two methods

Fig. 1. Smoothing a signal with details and two significant edges. (a) Result of monotonic regression (red). (b) Results of bilateral filtering [5] (blue) and of L_0 gradient minimization [3] (green).

where $C(S)$ is the smallest number of intervals on each of which S is monotonic. However, this cost is not enough to smooth a curve uniformly, because curves with small costs may have a sharp horn between two monotonic segments. Since first increasing (decreasing) and then decreasing (increasing) means the existence of a local maximum (minimum), $C(S)$ can be replaced by the number of local extrema of S. This leads to an idea of smoothing a signal by controlling the positions and values of extrema. Moreover, conceptually, the perspective of extrema can be generalized more naturally to the 2D case.

2.2 Extrema Interpolation

To penalize sharp corners between monotonic segments, more constraints are needed. One possible way is to prevent the relative magnitudes of extrema from getting too large. This leads us to the following *extrema interpolation* formulation:

For a set of points $\mathbb{E} = \{(p_j, m_j)\}_{j=1}^{J}$, *where* $p_j \in \mathbb{Z}$, $m_j \in \mathbb{R}$, *and* $\{p_j\}_{j=1}^{J}$ *is strictly increasing, find a sequence* S, *such that*
(i) $S_{p_j} = m_j, j = 1, 2, \ldots, J$, *and*
(ii) S *has a local extremum at* p, *if and only if* $p \in \{p_j\}_{j=1}^{J}$.

Let $EI(\mathbb{E})$ denote the set of all interpolating functions satisfying the above two conditions. Intuitively, extrema interpolation means sketching a curve that not only passes through a set of given points but also attains its extrema exclusively at these points. Since extrema provide more information compared with ordinary points, we can roughly sketch a curve similar to the original one from deliberately selected extrema.

For any $S \in EI(\mathbb{E})$, the total number of extrema is controlled by J, the dynamic range of S between p_{j-1} and p_j is limited to lie between m_{j-1} and m_j, and there can be no oscillations between p_{j-1} and p_j, or other extrema will exist. In general, extrema interpolation has infinitely many solutions, so smoothing can be formulated as the following constrained optimization problem:

$$\min_{S \in EI(\mathbb{E})} \|S - I\|^2 . \tag{3}$$

That is, among all interpolating functions, we seek the curve which is the most similar to the original signal. By limiting J and the relative magnitudes of m_j, we can control the smoothness of S, i.e., the smaller we choose J and relative magnitudes of m_j, the smoother will be the resulting interpolation function. To use extrema interpolation for smoothing, we need to specify the positions and amplitudes of all extrema, as discussed in the next subsection.

Efficient numerical solver can be obtained by phrasing the second condition of extrema interpolation in an alternative way:

(ii′) S is monotonic between p_{j-1} and p_j, $j = 2, \ldots, J$.

Since both condition *(i)* and *(ii′)* are linear constraints about S_p, (3) is a quadratic programming problem, which can be solved efficiently. Unfortunately, this is not the case for 2D signals.

Figure 2 shows an example of extrema interpolation for edge-preserving smoothing. Imagine drawing a red curve S which looks most like the original black curve I among all those curves that connect the prescribed blue circles and don't oscillate between any two adjacent blue circles. Differences between adjacent samples shrink to zero after smoothing, if their signs are not consistent with the two enclosing extrema. Samples with amplitudes outside the range of two enclosing extrema have their values clumped in S. However, those samples with both amplitude and difference consistent with two enclosing extrema are basically retained. So smoothing via extrema interpolation can selectively preserve edges and clump humps, under the guidance of extrema. Figure 2 also shows that by changing the positions and values of extrema, a signal can be smoothed to a different degree.

2.3 Determination of Extrema

The ability of extrema interpolation to both preserve edges and suppress oscillations depends on properly prescribed extrema $\mathbb{E}$, which should be compatible with the main structures in the original signal. In addition, mathematically, $\mathbb{E}$ must meet some conditions, e.g. $(m_j - m_{j-1})(m_{j+1} - m_j)$ should not be positive, or $EI(\mathbb{E})$ may be empty. These requirements will be more sophisticated and harder to satisfy for 2D images than for 1D signals. A natural approach is to take $\mathbb{E}$ as the extrema of a smoothed version G, which may not preserve edges well, of the original signal I. Now (3) takes the following form:

$$\min_{S} \|S - I\|^2 \quad \text{s.t.} \begin{cases} S_p = G_p \text{ if } G \text{ has a extremum at } p \\ (\nabla S_p)(\nabla G_p) \geqslant 0 \end{cases}, \tag{4}$$

where the second constraint comes from condition *(ii′)* of extrema interpolation, meaning that S is increasing (decreasing) where G is increasing (decreasing), thus is monotonic between two consecutive extrema of G and attains extrema where G attains extrema. It is understood that $(\nabla S_p)(\nabla G_p) \geqslant 0$ means ∇S_p has the same sign with ∇G_p. If $\nabla G_p = 0$, we let $\nabla S_p = 0$. Obviously, in this case $EI(\mathbb{E})$ is nonempty, since G itself is an interpolating function.

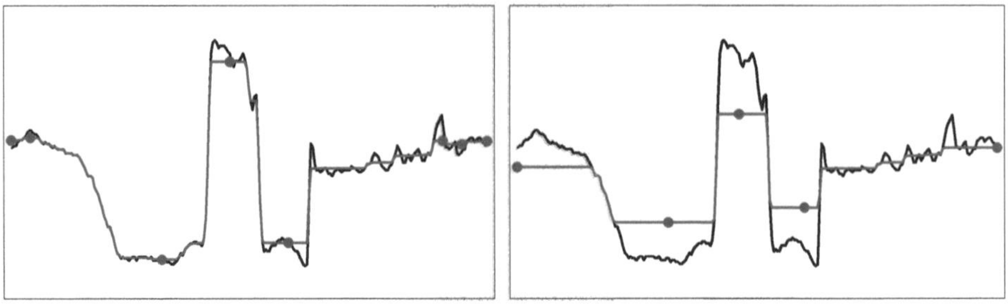

(a) Slightly smoothed (b) Strongly smoothed

Fig. 2. 1D Extrema interpolation. The prescribed extrema are marked by blue circles. These extrema divide the original signal into segments, each of which is smoothed by monotonic regression. Hence the net effects are edge preserving and fluctuation clumping. In (b) the extrema are more restrictive than in (a), thus yielding coarser smoothing result.

Intuitively, we exploit G to extract $\mathbb{E}$, which can be seen as a qualitative trend, and then use extrema interpolation to refine it. In Fig. 3 our method is used to filter a 1D signal I, where G is obtained by WLS [1]. Small fluctuations in I are merged into a longer monotonic interval in G, and then suppressed in S as they are approximated by a monotonic segment. Salient edges in I are smoothed in G, but then their slopes are restored in S, on condition that they are still recognized as having the same monotonicity in G as in I. The dynamic range of restored edges in S is confined by the corresponding two extrema in G. As can be seen in Fig. 3, edges are blurred somewhat by the WLS filter, while our results follow edges more closely.

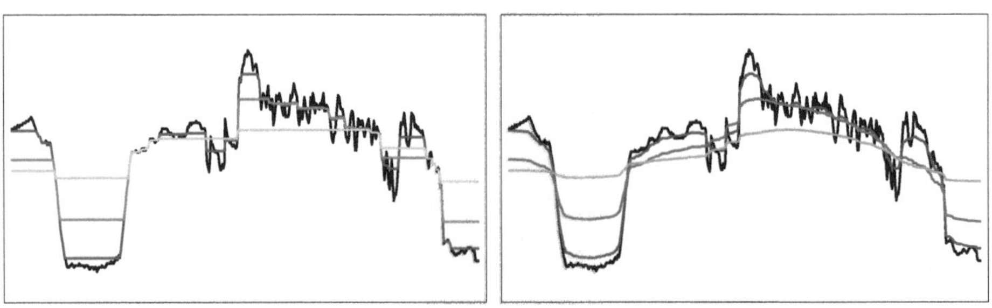

(a) Ours (b) WLS

Fig. 3. Comparison of our method to WLS [1] for a 1D signal. The smoothed versions are shown superimposed on the original signal (using different shades of color: lighter is more strongly smoothed).

2.4 Connection with Total Variation Regularization

Condition (*ii*) in extrema interpolation can be seen as a way to penalize the length of a curve, i.e., to pull a curve taut. Intuitively, the shorter the length, the smaller usually will be the number of extrema. We maintain that by using total variation as a measure of length, condition (*ii'*) can be replaced by

$$\|\nabla S\|_1 = \sum_j |m_j - m_{j-1}| \,, \tag{5}$$

where $\|\nabla S\|_1 = \sum_p |\nabla S_p|$ is the discrete total variation of S. This constraint combined with condition (*i*) implies

$$\sum_{p_{j-1} < i \leqslant p_j} |\nabla S_i| = |S_{p_j} - S_{p_{j-1}}|,\ j = 2, \ldots, J \,, \tag{6}$$

thus condition (*ii'*) holds.

If total variation is employed to obtained a preliminary smoothing result T for extracting extrema $\mathbb{E}$:

$$T = \arg\min_S \left\{ \|S - I\|^2 + \lambda \|\nabla S\|_1 \right\} , \tag{7}$$

then extrema interpolation gives

$$S^T = \arg\min_S \|S - I\|^2 \quad \text{s.t.} \begin{cases} S_p = T_p \text{ if } T \text{ has a extremum at } p \\ \|\nabla S\|_1 = \|\nabla T\|_1 \end{cases} . \tag{8}$$

In this case, extrema interpolation has no effect, i.e., $S^T = T$, otherwise S^T can make the objective function in (7) smaller. Thus total variation smoothing (7) is a special case of extrema interpolation. By the extrema interpolation formulation, we have more flexibility to manipulate extrema, which in turn dominates the resulting smoothing result. For 2D signals, constraining total variation, either isotropic or anisotropic, doesn't have the same kind of influence on extrema as for 1D signals, so total variation regularization loses this extrema interpolation interpretation for images.

3 Smoothing 2D Signals

The extrema interpolation formulation in Sect. 2.2 also applies to images, provided that p and p_j are spatial positions. However, condition (*ii'*) of extrema interpolation doesn't have a corresponding 2D form. In other words, we cannot express the intricate condition (*ii*) in terms of a system of simple inequalities. Fortunately, a substitute can be obtained by following the order relation of neighboring pixels in a preliminarily smoothed version G, that is, extending (4) to 2D by replacing $(\nabla S_p)(\nabla G_p) \geqslant 0$ by $(\partial_x S_p)(\partial_x G_p) \geqslant 0$ and $(\partial_y S_p)(\partial_y G_p) \geqslant 0$, where $\partial_x(\cdot)_p$ and $\partial_y(\cdot)_p$ are differences at p along x and y directions. As in Sect. 2.3, $(\partial_* S_p)(\partial_* G_p) \geqslant 0$ means that $\partial_* S_p$ has the same sign with $\partial_* G_p$ and $\partial_* S_p = 0$ if $\partial_* G_p = 0$.

It should be emphasized that the order relation between any two neighboring pixels cannot be determined from a given set of extrema, so the above formulation is sufficient but not necessary for the nonexistence of extrema besides those of G. This may cause resulting images to have artifacts, such as false ridge-like structures. One natural solution is to add the constraint that magnitudes of differences of images should not become greater after smoothing. Now we are led to:

$$\min_S \|S - I\|^2 \quad \text{s.t.} \begin{cases} S_p = G_p \text{ if } G \text{ has a extremum at } p \\ (\partial_x S_p)(\partial_x G_p) \geqslant 0 \text{ and } |\partial_x S_p| \leqslant |\partial_x I_p| + \varepsilon \\ (\partial_y S_p)(\partial_y G_p) \geqslant 0 \text{ and } |\partial_y S_p| \leqslant |\partial_y I_p| + \varepsilon \end{cases}, \tag{9}$$

where a small positive number ε is introduced to make these requirements not too restrictive. But this presents another problem: the feasible set in (9) may be empty. Although G is a smoothed version of I, the magnitudes of its differences may be larger than that of I somewhere. To guarantee G itself is in the feasible set, the smoothing method for generating G from I must be tailored to this extra constraint on magnitudes of gradients.

We adapt the WLS filtering framework to a constrained optimization formulation:

$$\begin{aligned} &\min_G \sum_p \left\{ (G_p - I_p)^2 + \lambda \left[w_{x,p}(\partial_x G_p)^2 + w_{y,p}(\partial_y G_p)^2 \right] \right\} \\ &\qquad\qquad \text{s.t.} \begin{cases} |\partial_x G_p| \leqslant |\partial_x I_p| + \varepsilon \\ |\partial_y G_p| \leqslant |\partial_y I_p| + \varepsilon \end{cases}, \end{aligned} \tag{10}$$

where $w_{x,p} = (|\partial_x I_p|^\alpha + \varepsilon)^{-1}$, $w_{y,p} = (|\partial_y I_p|^\alpha + \varepsilon)^{-1}$ and λ are from [1]. Here for brevity we use the same ε for both inequality constraints and weights associated with differences. Obviously, the resulting G will satisfy all constraints in (9). Incidentally, constraining magnitudes of gradients, to some extent, may protect edges in G from blurring.

In (9) and (10), the constraints are all linear inequalities and the objective functions are convex quadratic, so they can be solved by standard quadratic programming techniques.

In summary, our method consists of first solving (10) to obtain G, and then solving (9) for the final smoothing result S. Main structures of the original image I are embedded in the extrema of G, which are used in (9) to find S. Whether an edge in I is restored or suppressed in S depends on whether it is consistent with the extrema of G.

4 Experimental Results

All pixel values are normalized to the interval $[0, 1]$. We use the path-following primal-dual interior-point method to solve (9) and (10) [10]. In practice, we find that choosing ε greater than zero can make the final infeasibility in the KKT conditions smaller. We set $\varepsilon = 10^{-4}$ and $\alpha = 2$ by default. To be consistent with (9), 4-connectivity is used when identifying local extrema, that is, a pixel is considered as a local maximum (minimum) if its value is larger (less) than those of the four horizontal and vertical neighbors.

The edge-respecting effect of our method is shown in Fig. 4. As the 1D case, strong edges in I that are retained but somewhat blurred in G are restored in S, and fine details in I that are inconsistent with the positions and amplitudes of extrema in G are subdued in S. As exemplified in Fig. 4, although the constraints in (9) is a restrictive substitute for preserving extrema, our method is effective in preserving main edges. We find that the interior-point method for solving (9) may fail without the constraints in (10) (e.g., when $\lambda = 0.01$ as in Fig. 4 (d)). So the original WLS filter cannot ensure that gradients should not become larger after smoothing. In contrast, our method enforces this requirement explicitly.

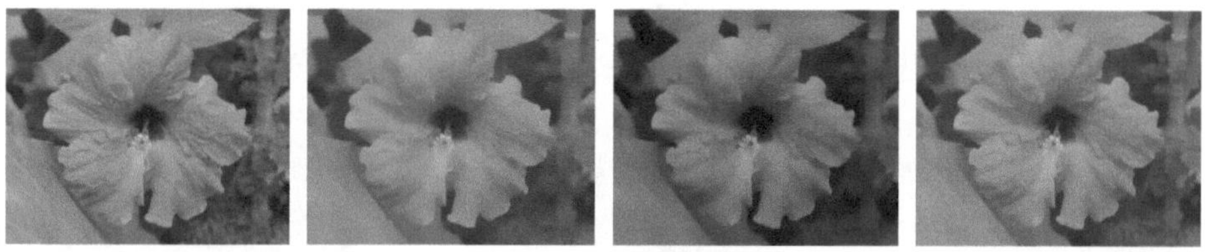

(a) Input I (b) G ($\lambda = 0.01$) (c) (b) with local extrema identified (d) S ($\lambda = 0.01$)

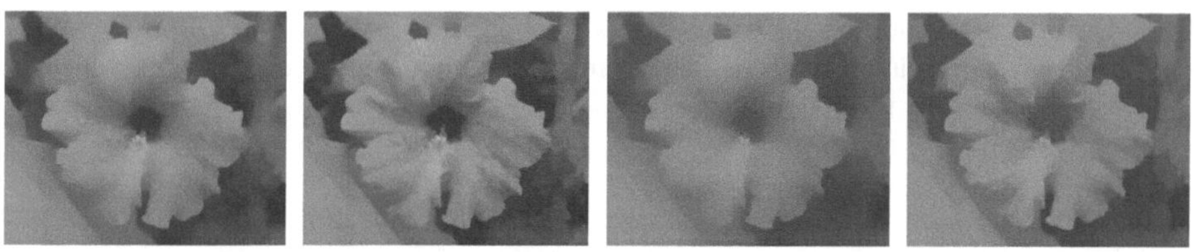

(e) G ($\lambda = 0.05$) (f) S ($\lambda = 0.05$) (g) G ($\lambda = 0.2$) (h) S ($\lambda = 0.2$)

Fig. 4. Image smoothing results. Our method with different λ coarsens an input I to different extents (d)(f)(h). The intermediate results G are also shown (b)(e)(g). All local extrema in (b) are marked in (c), blue for maxima and red for minima. These color markers are much better appreciated when viewed in full size on a computer monitor. By extrema interpolation, the positions and amplitudes of all local extrema in (c) are inherited by (d), this relation also holds for (e)(f) and (g)(h).

Edge-respecting smoothing naturally leads to base-detail decomposition, where the base and detail layers are S and $I - S$, respectively. Detail enhancement can be accomplished by boosting the detail layer. Figure 5 gives enhancement results and comparisons. Our method are performed in the CIELAB lightness channel, leaving the other two color channels unchanged. The WLS filter may cause some artifacts when the boosting multiplier becomes larger, although it performs well in most cases. As can be seen in Fig. 5 (e) and (f), along the silhouettes of distant hills and a sandpit, our results compare favorably with the WLS-based enhancement, yielding less distracting halos and rims.

Fig. 5. Detail exaggeration results (detail layers multiplied by 4, $\lambda = 0.01$). Boosting WLS-based detail layers results in artifacts such as rims along high-contrast edges (b)(e), which are less conspicuous for our method (c)(f).

Fig. 6. Image abstraction results ($\lambda = 0.1$). By our method, details are subdued and main structures are emphasized.

Our method can be used in many other applications that utilize an edge-preserving filter. One application is non-photorealistic stylization, e.g., abstraction. The effect shown in Fig. 6 is produced by superimposing gradient magnitudes on the smoothed image. The three RGB channels are processed independently. Thanks to the nice edge-preserving property of our method, the abstraction results are visually pleasing.

One limitation of our method is about computing time. Most CPU time is for the interior-point method, which requires solving large linear systems of equations at each iteration. For the image shown in Fig. 5, which is of size 533×800, our Matlab implementation takes several minutes for one color channel. To reduce running time, smoothing can be performed in a coarse-to-fine manner in an image pyramid, with results obtained from one resolution as initialization for the next finer resolution.

5 Conclusion

This paper proposes a new method for smoothing images while respecting edges, neither blurring nor sharpening, only with the dynamic range being reduced. Our method is motivated by the observation that monotonic regression can follow significant edges fairly well locally. Extrema interpolation is proposed to employ monotonic regression segment by segment and then glue the resulting locally good segments into a final one. Roughly speaking, the idea is to first identify a set of meaningful local extrema (minima and maxima) of a signal I and then to find the best approximation S of I such that S exclusively has these meaningful extrema.

We apply a constrained version of the WLS filter to obtain a preliminary smoothed image, whose extrema are used by the extrema interpolation formulation to find the final smoothing result. We formulate this problem as a constrained convex quadratic minimization problem, which is therefore solvable with standard tools. For 1D signals, our method has an interesting relation to total variation regularization, which thus obtains an intuitive interpretation for its features. Experiments on detail enhancement show that our method can retain the slope of edges more faithfully, resulting much less objectionable artifacts compared with other methods.

Acknowledgements. This work was supported by the National Natural Science Foundation of China (Grant No. 61071180) and Key Program (Grant No. 61133003).

References

1. Farbman, Z., Fattal, R., Lischinski, D., Szeliski, R.: Edge-preserving decompositions for multi-scale tone and detail manipulation. ACM Transactions on Graphics (TOG) 27(3), 67 (2008)
2. Rudin, L.I., Osher, S., Fatemi, E.: Nonlinear total variation based noise removal algorithms. Physica D: Nonlinear Phenomena 60(1), 259–268 (1992)
3. Xu, L., Lu, C., Xu, Y., Jia, J.: Image smoothing via l 0 gradient minimization. ACM Transactions on Graphics (TOG) 30(6), 174 (2011)
4. Xu, L., Yan, Q., Xia, Y., Jia, J.: Structure extraction from texture via relative total variation. ACM Transactions on Graphics (TOG) 31(6), 139 (2012)
5. Tomasi, C., Manduchi, R.: Bilateral filtering for gray and color images. In: Sixth International Conference on Computer Vision, pp. 839–846. IEEE (1998)
6. Kass, M., Solomon, J.: Smoothed local histogram filters. ACM Transactions on Graphics (TOG) 29(4), 100 (2010)
7. Perona, P., Malik, J.: Scale-space and edge detection using anisotropic diffusion. IEEE Transactions on Pattern Analysis and Machine Intelligence 12(7), 629–639 (1990)
8. Subr, K., Soler, C., Durand, F.: Edge-preserving multiscale image decomposition based on local extrema. ACM Transactions on Graphics (TOG) 28(5), 147 (2009)
9. Acton, S.T.: Locally monotonic diffusion. IEEE Transactions on Signal Processing 48(5), 1379–1389 (2000)
10. Nocedal, J., Wright, S.J.: Numerical optimization. Springer Science+ Business Media (2006)

Image Retargeting by Combining Region Warping and Occlusion

Zhongyan Qiu[1,3], Tongwei Ren[1,2], Yan Liu[4], Jia Bei[1,2], and Muyang Song[2]

[1] State Key Laboratory for Novel Software Technology, Nanjing University, China
[2] Software Institute, Nanjing University, China
[3] School of Software, Nanchang University, China
[4] Department of Computing, The Hong Kong Polytechnic University, China

Abstract. Image retargeting aims to avoid visual distortion while retaining important image content in resizing. However, current image retargeting methods usually fail in difficult situations, such as complex structures and intricate arrangements of objects. In this paper, we propose a novel image retargeting approach, which provides low visual distortion retargeting results by combining region warping and occlusion. We first represent image region with occlusion probability to avoid inaccuracy in image decomposition. Then we combine region warping and occlusion in a unified framework, and transform it to a pixel-level retargeting problem. To verify the performance of our approach, we implement it with two operators, seam carving and pixel warping. Experiments demonstrate the effectiveness of the proposed approach.

Keywords: Image retargeting, visual distortion, region warping, region occlusion.

1 Introduction

The advance of multimedia technology makes it possible to display digital images on mobile devices, such as smart phone and tablet. These devices usually have different screens with arbitrary sizes and aspect ratios. It requires content-aware adaption of image content to satisfy display screens. Such adaptation technology is named as *image retargeting* [1].

While viewing target images, a user prefers to obtain important content for precise understanding and insensible artifacts for good experience [2]. Thus an effective image retargeting method should emphasize important content to provide sufficient information and reduce visual distortion to avoid obvious artifacts [3]. To emphasize important content, current image retargeting methods usually adopt saliency detection technique [4] based on human vision system and/or statistical learning, which can obtain good performance in automatical detection of important image content. Nevertheless, in the aspect of visual distortion reduction, current image retargeting methods still suffer many problems.

B. Huet et al. (Eds.): PCM 2013, LNCS 8294, pp. 200–210, 2013.

For image retargeting attempts to display original image content in a resized screen, visual distortion is inevitably existed in target images, such as changes of region shape or spatial relationships. The essential problem is how to reduce visual distortion and avoid obvious artifacts in target images [5]. In order to reduce obvious visual distortion, current image retargeting methods usually adopt two kinds of strategies, region warping and region occlusion.

Region warping based methods attempt to remove image content from the regions without object or apparent structure. Liu et al. first applied manually selected templates to represent image content, and reduced unimportant regions by fisheye-view warping [6]. Wang et al. represented image content with uniform grids, and warped the low energy regions by optimizing grid vertex positions [7]. Guo et al. utilized content-aware triangle meshes to represent image regions, and resized image by optimizing mesh parametrization [8]. Ren et al. decomposed the image into curve-edge regions and uniformly warped each region with different ratios [3]. Pixel-level image retargeting methods, such as seam carving [1] and pixel warping [9], can be treated as a special kind of region warping based methods. Though these methods do not explicitly represent image content with regions, they focus on removing image content from unimportant regions just as region warping based methods do. Region warping based methods achieve good performance when original image contains inapparent structure and few objects, but suffer serious distortion when obvious structure or many important objects exist.

Region occlusion based methods attempt to change the spatial relationships among regions. For the requirement of determining the overlapping order, these methods usually demand depth information or manual annotated depth order. Dahan et al. enhanced image segmentation by combining depth and color, and occluded the segmented regions by shifting their centers [10]. Mansfiedld et al. decomposed original images into several layers, and re-arranged the layers based on manually specified relative depth [11]. Region occlusion based methods avoid region distortion, but cause serious content loss when too many objects exist and obvious artifacts when the objects have complex structures.

To overcome the above problems, an intuitive solution is using region warping and region occlusion together. Mansfiedld et al. considered combining distortion in scene carving [11]. Lee et al. decomposed stereoscopic images into multiple layers, and warped each layer with its own deformation [13]. Though these methods can improve retargeting performance, they seriously depend on the effect of image decomposition. If image decomposition fails, they may cause serious artifacts. Furthermore, when the objects have complex structures or intricate arrangements, it will be hard to warp or overlap objects. Hence, it cannot obtain good performance by simply using region warping and occlusion together.

In this paper, we propose a novel image retargeting approach, which combines region warping and region occlusion in an unified framework without image decomposition. Fig. 1 illustrates an overview of the proposed approach. We first calculate the occlusion probability of each pixel in the original image to represent

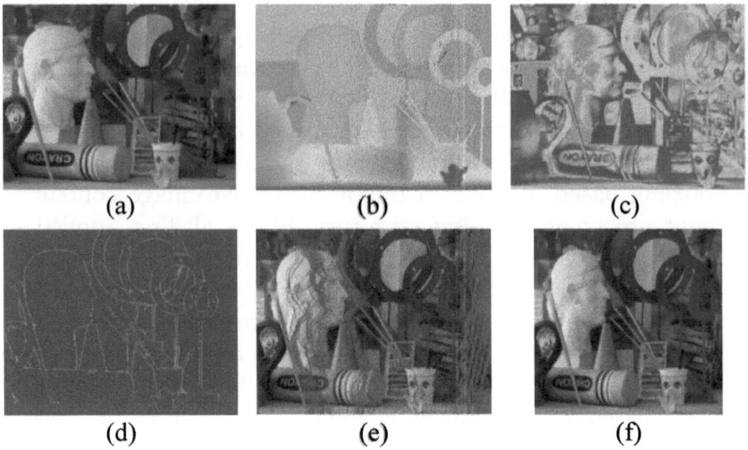

Fig. 1. An overview of the proposed approach. (a) Original image. (b) Depth information, from red to blue denote near to far. (c)-(d) Saliency energy map and occlusion probability map, from blue to red denote zero to one. (e) Retargeting with improved seam carving or other pixel-level retargeting operators. (f) Target image.

image regions implicitly. Then we transform region warping and region occlusion to a unified pixel-level operation, and implement it with pixel-level retargeting operator, such as improved seam carving and pixel warping. Finally, the target image is generated from the result of pixel-level retargeting operation.

2 Image Retargeting by Combining Region Warping and Occlusion

2.1 Implicit Region Representation

Let I denote the original image in size of $m \times n$, p_{ij} denote the pixel with the coordinate (i, j) in I. To carry out region warping and region occlusion on I, we should first represent I as a set of regions $I = \{R_k\}$, where R_k is the kth region in I. For each image region, ignoring the difference in sampling strategy, region warping and region occlusion are only different in the positions of removed pixels. If the removed pixel is on the boundary of the farther image region, its removal will cause region occlusion; otherwise, its removal will cause region warping. Thus, we can represent I as two pixel sets $I = P_W \cup P_O$, where P_W is the set of pixels whose removal will cause region warping and P_O is the set of pixels whose removal cause region occlusion.

Let $\mathbf{\Omega}(i, j)$ denotes whether p_{ij} will cause region occlusion if removed:

$$\mathbf{\Omega}(i,j) = \begin{cases} 0, p_{ij} \in P_W \\ 1, p_{ij} \in P_O \end{cases} . \tag{1}$$

In this way, we represent the image regions in I in an implicit way $I = \{\boldsymbol{\Omega}(i,j)\}$. To calculate $\boldsymbol{\Omega}(i,j)$, I should be segmented into regions. But current automatic image segmentation technique cannot ensure the accuracy of region boundary, especially when the image has complex compositions. To avoid the inaccuracy in image segmentation, we calculate occlusion probability of each pixel, i.e. the probability of its removal may cause region occlusion, instead of $\boldsymbol{\Omega}(i,j)$. The following is the calculation of occlusion probability for horizontal resizing. The occlusion probability for vertical resizing can be calculated similarly.

$$\boldsymbol{\Omega}'(i,j) = \max\{0, \mathbf{D}(i,j) - \mathbf{D}(i,j-1), \mathbf{D}(i,j) - \mathbf{D}(i,j+1)\}, \quad (2)$$

where $\mathbf{D}(i,j)$ is the depth of pixel p_{ij} in the value range [0,1], and larger depth value denotes farther content.

With the representation of occlusion probability, region warping and region occlusion can be unified to pixel-level operation. It will cause region warping when removing the pixels with low occlusion probabilities from the original image, and region occlusion when removing the pixels with high occlusion probabilities.

2.2 Energy Maps Calculation

Energy maps are usually calculated to represent image content characters in retargeting. Considering multiple energy maps provide better performance than single energy map [12], we utilize two kinds of energy maps in our approach, saliency energy map $\mathbf{E}_{sal}$ and distortion energy map $\mathbf{E}_{dis}$. $\mathbf{E}_{sal}$ represents the image content saliency of each pixel, and $\mathbf{E}_{dis}$ represents the brought distortion when removing each pixel.

We use saliency detection algorithm [4] to calculate $\mathbf{E}_{sal}$, in which the pixel with higher saliency value is considered to attract more attention and have higher importance. We calculate $\mathbf{E}_{dis}$ by measuring backward gradient or forward gradient [1] for different retargeting operators as necessary. Equation (3) and (4) show the calculation of backward gradient and forward gradient, respectively. In forward gradient calculation, for the distortion energy of each pixel is influenced by the spatial relationship between the former pixel and current pixel, we use three energy maps, $\mathbf{E}^L_{dis}$, $\mathbf{E}^U_{dis}$ and $\mathbf{E}^R_{dis}$, to represent all three possible cases:

$$\mathbf{E}_{dis}(i,j) = d_{\mathbf{I},\mathbf{D}}(p_{ij}, p_{(i+1)j}) + d_{\mathbf{I},\mathbf{D}}(p_{ij}, p_{i(j+1)}), \quad (3)$$

$$\begin{aligned}
\mathbf{E}^L_{dis}(i,j) &= d_{\mathbf{I},\mathbf{D}}(p_{i(j+1)}, p_{i(j-1)}) + d_{\mathbf{I},\mathbf{D}}(p_{(i-1)j}, p_{i(j-1)}),\\
\mathbf{E}^U_{dis}(i,j) &= d_{\mathbf{I},\mathbf{D}}(p_{i(j+1)}, p_{i(j-1)}),\\
\mathbf{E}^R_{dis}(i,j) &= d_{\mathbf{I},\mathbf{D}}(p_{i(j+1)}, p_{i(j-1)}) + d_{\mathbf{I},\mathbf{D}}(p_{(i-1)j}, p_{i(j+1)}),
\end{aligned} \quad (4)$$

where $\mathbf{E}^L_{dis}$, $\mathbf{E}^U_{dis}$ and $\mathbf{E}^R_{dis}$ are the distortion energy maps to represent the cases that the former pixel is upper-left pixel, upper pixel and upper-right pixel to current pixel, respectively; $d_{\mathbf{I},\mathbf{D}}(,)$ denotes the product of the distances between two pixels in intensity and depth.

When removing a pixel with low occlusion probability, we treat pixel removal as region warping and consider the loss of both saliency energy and distortion energy:

$$e_{ij}^{W} = \alpha \mathbf{E}_{sal}(i,j) + (1-\alpha)\mathbf{E}_{dis}(i,j), \quad (5)$$

where e_{ij}^{W} denotes the energy loss when removing p_{ij} in region warping; α is a weight in range $(0,1)$ to adjust the ratios of saliency energy and distortion energy, which is simply defined as 0.5 in our experiments.

When remove a pixel with high occlusion probability, we treat pixel removal as region occlusion and only consider the loss of saliency energy, for the gradient change on region boundaries does not influence the view of image content:

$$e_{ij}^{O} = \mathbf{E}_{sal}(i,j). \quad (6)$$

We combine the energy losses of region warping and region occlusion by occlusion probability, and calculate the energy loss of removing a pixel as follows:

$$e_{ij} = (1-\boldsymbol{\Omega}'(i,j))e_{ij}^{W} + \boldsymbol{\Omega}'(i,j)e_{ij}^{O}. \quad (7)$$

2.3 Image Retargeting with Energy Optimization

With occlusion probability representation and energy maps, image retargeting by combining region warping and region occlusion is transformed to a pixel-level retargeting problem, which can be implemented with the existing pixel-level retargeting methods. Many pixel-level image retargeting methods have been proposed. To illustrate the generality of our approach, we implement our approach with two popular pixel-level operators, seam carving [1] and pixel warping [9].

Implementation with Seam Carving. Seam carving iteratively removes the seams with the lowest energy till the target image size is satisfied [1]. For retargeting procedure is solved by dynamic programming, we utilize forward gradient in distortion energy map calculation and update energy map after each operation. Based on the energy loss of removing each pixel, we consider vertical seam to reduce image width. Horizontal seam to reduce image height can be handled similarly. A vertical seam $\mathbf{s_v}$ is defined as the eight-connected pixels $\mathbf{s_v} = \{p_{i_j j}\}_{j=1}^{n}$, where $|i_j - i_{j+1}| \le 1$.

We define the energy of seam as the sum of energy of all pixels on it $e_{\mathbf{s_v}} = \sum_{j=1}^{n} e_{i_j j}$, where $p_{i_j j}$ is on $\mathbf{s_v}$, and look for the seam with the lowest energy $\mathbf{s_v}^* = \arg\min_{\mathbf{s_v}} e_{\mathbf{s_v^*}}$. Then, we denote the cumulative minimum energy with $\mathbf{M}$, and look for $\mathbf{s_v^*}$ by dynamic programming:

$$\mathbf{M}(i,j) = \min\left\{\mathbf{M}(i-1,j-1) + e_{ij}^{L}, \mathbf{M}(i-1,j) + e_{ij}^{U}, \mathbf{M}(i-1,j+1) + e_{ij}^{R}\right\}. \quad (8)$$

where e_{ij}^L, e_{ij}^U and e_{ij}^R denote e_{ij} calculated with $\mathbf{E}_{dis}^L(i,j)$, $\mathbf{E}_{dis}^U(i,j)$ and $\mathbf{E}_{dis}^R(i,j)$ in Equation (4), respectively.

Note here, though some previous image retargeting methods have improved seam carving by combining depth, they focused on different aims, such as energy map improvement[16], consistency between binocular images [15] and geometric consistency preservation[14]. Different from them, our approach focuses on implementing the combination of region warping and region occlusion, and reducing visual distortion in retargeting. Mansfiedld et al. also combined object occlusion in seam carving, but their method requires manually specified relative depth map and accurate image decomposition [11].

Implementation with Pixel Warping. Pixel warping reduces the width or height of each pixel to fit target size with the highest retained energy, and generate the target image with pixel fusion [9]. For pixel warping is solved by linear programming, we utilize backward gradient in distortion energy map calculation. We define x_{ij} as the retained width of pixel p_{ij}, and optimized the retained energy:

$$\begin{aligned} &\max \sum_{i=1}^{m}\sum_{j=1}^{n}(e_{ij}x_{ij}),\\ &s.t. \sum_{i=1}^{m} x_{ij} = m', \forall j,\\ &|\sum_{k=1}^{i} x_{kj} - \sum_{k=1}^{i} x_{k(j+1)}| \leq \delta, \forall i \forall j,\\ &0 \leq x_{ij} \leq 1, \forall i \forall j, \end{aligned} \tag{9}$$

where m' is the width of target image and δ is a small positive threshold. In our experiments, δ equals one.

3 Experiments

We verify the proposed approach on Middlebury Stereo Datasets [17], which has 35 subdatasets and each subdataset contains two images with disparity information. We implement our approach with Matlab code, and carry out the experiments on a computer with 2.1 GHz CPU and 2 GB memory. In average, each image requires 2.74 minutes and 14.36 seconds for retargeting with the implementation of seam carving and pixel warping, respectively.

Fig. 2 illustrates an example of generating target images with different resizing orders, horizontal-vertical order and vertical-horizontal order. Experiments show that our approach is robust to resizing order. Furthermore, from the intermediate results, we can find that our approach is suitable for retargeting an image to various aspect ratios.

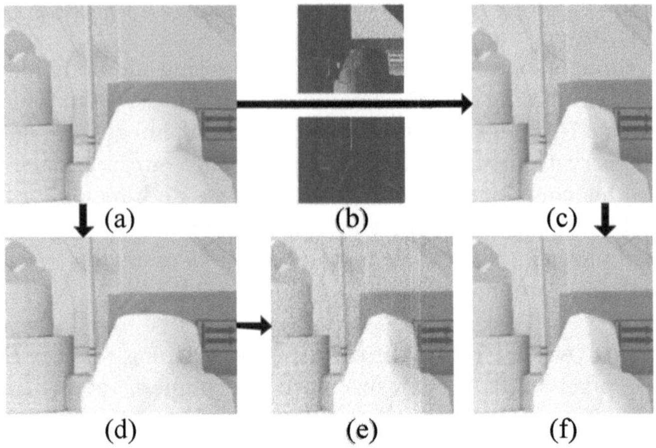

Fig. 2. Results with different resizing orders. (a) Original image. (b) Saliency map (top) and occlusion probability map. (c) and (f) Results of horizontal-vertical resizing. (d) and (e) Results of vertical-horizontal resizing.

To illuminate our performance, we compare our approach with several representative image retargeting methods. We use all 70 images with disparity in Middlebury Stereo Datasets by treating their disparity as depth information, and reduce their width to 1:1 aspect ratio, which is the extreme aspect ratio when the device screens are allowed to be rotated in image viewing. The selected methods for comparison include six representative image retargeting methods based on region warping, region occlusion and their combination, including seam carving [1], curve-edge grid warping [3], scale-and-stretch [7], enhanced segmentation based retargeting [10], scene carving [11], and scene warping [13]. The methods requiring image segmentation all use region growing image segmentation on depth map with manually optimized thresholds. Fig. 3 shows some examples of comparison results. It is found that seam carving, curve-edge grid warping and scale-and-stretch bring in object distortion in target images (Fig. 3(b)-(d)), and enhanced segmentation based retargeting, scene carving and scene warping cause obvious artifacts when image segmentation fails (Fig. 3(e)-(f) and (h)) or subside into cropping (Fig. 3(g)). But our approach can obtain good performance by combining slight region warping and occlusion without image decomposition (Fig. 3(i)-(j)).

To further assess the effectiveness of our approach, we carry out a user study. Thirty-one participants in age of 19 to 58 are invited, including eighteen males and thirteen females with occupation of undergraduate, graduate, worker and officer. During the evaluation, each target image generated by our approach is required to be pairwisely compared with all other corresponding target images by three participants, and the dominant judgment is treated as the final evaluation. If three participants have totally different judgments, we consider our result is similar to the compared target image. Table 1 and 2 show the results of user

(a)

(b)

(c)

(d)

(e)

(f)

(g)

(h)

(i)

(j)

Fig. 3. Examples of comparison results. (a) Original images. (b)-(h) Results of seam carving, scale-and-stretch, curve-edge grid warping, enhanced segmentation based retargeting, scene carving, scene carving without segmentation and scene warping. (i) and (j) Our results implemented by seam carving and pixel warping, respectively.

study, in which “Better”, “Similar” and “Worse” denote the number of original images whose results generated by our approach are better, similar and worse than/to the corresponding methods, respectively. We can find that our method obtains better results than the existing methods on the majority of original images. We can also find the different implementations of our approach are obviously different in each comparison but similar in overall performance. The reason is that the implementation with seam carving can use forward distortion energy but only obtain local optimal result, and the implementation with pixel warping can obtain global optimal result but only use backward distortion energy.

Table 1. User study result of our approach implemented by seam carving

	Better	Similar	Worse
seam carving	36	24	10
curve-edge grid warping	31	24	15
scale-and-stretch	27	25	18
enhanced segmentation	33	26	11
scene carving	26	32	12
scene warping	23	33	14

Table 2. User study result of our approach implemented by pixel warping

	Better	Similar	Worse
seam carving	34	22	14
curve-edge grid warping	29	30	11
scale-and-stretch	32	26	12
enhanced segmentation	40	18	12
scene carving	30	25	15
scene warping	29	32	9

In experiment, we also find some limitations of our approach. For example, it is influenced by the accuracy of depth. When the depth is inaccurate, it cannot avoid visual distortion besides region boundaries (Fig. 4(c)). In addition, our approach is also influenced by the limitation of the utilized pixel-level operation in implementation. For example, if implement our approach with seam carving, it cannot handle edges well in some situations for seam carving has the limitation in edge preservation (Fig. 4(f)).

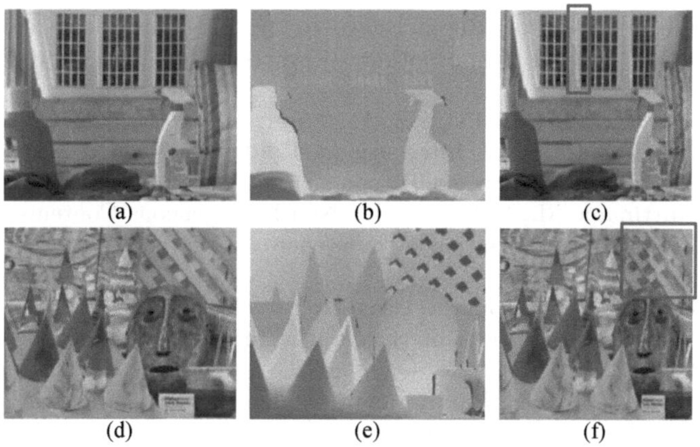

Fig. 4. Examples of bad results. (a) and (d) Original images. (b) and (e) Depth information. (c) and (f) Our results implemented by seam carving.

4 Conclusion

In this paper, we propose an image retargeting approach by combining region warping and region occlusion. Based on the implicit region representation, the proposed approach realize region warping and occlusion with the unique pixel-level operation, and optimize retargeting on multiple energy maps to generate target image. In this way, the proposed approach can effectively avoid artificial and serious overlapping, especially in the situations of complex object structures or intricate object arrangements.

Our future work will focus on improving the proposed approach by handling the inaccuracy of depth and implementing it with other pixel-level operators. We will also consider to extend the approach to video retargeting.

Acknowledgments. The authors want to thank the anonymous reviews for their helpful suggestion. This paper is supported by Natural Science Foundation of China (61202320), Research Project of Excellent State Key Laboratory (61223003), Natural Science Foundation of Jiangsu Province (BK2012304), and National Undergraduate Innovation Program (201210284031).

References

1. Rubinstein, M., Shamir, A., Avidan, S.: Improved Seam Carving for Video Retargeting. In: ACM International Conference on Computer Graphics and Interactive Techniques, New York (2008)
2. Rubinstein, M., Gutierrez, D., Sorkine, O., Shamir, A.: A Comparative Study of Image Retargeting. In: ACM International Conference on Computer Graphics and Interactive Techniques in Asia, Seoul (2010)

3. Ren, T., Liu, Y., Wu, G.: Rapid Image Retargeting based on Curve-edge Grid Representation. In: IEEE International Conference on Image Processing, Hong Kong (2010)
4. Achanta, R., Hemami, S., Estrada, F., Süsstrunk, S.: Frequency-tuned Salient Region Detection. In: IEEE International Conference on Computer Vision and Pattern Recognition, Miami (2009)
5. Wolf, L., Guttmann, M., Cohen-Or, D.: Non-homogeneous Content-driven Video-retargeting. In: IEEE International Conference on Computer Vision, Rio de Janeiro (2007)
6. Liu, F., Gleicher, M.: Automatic Image Retargeting with Fisheye-view Warping. In: ACM Symposium on User Interface Software and Technology, Seattle (2005)
7. Wang, Y.S., Tai, C.L., Sorkine, O., Lee, T.Y.: Optimized Scale-and-stretch for Image Resizing. In: ACM International Conference on Computer Graphics and Interactive Techniques in Asia, Singapore (2008)
8. Guo, Y., Liu, F., Shi, J., Zhou, Z.H., Gleicher, M.: Image Retargeting Using Mesh Parametrization. IEEE Transactions on Multimedia 11(5), 856–867 (2009)
9. Ren, T.: Liu, Y., Wu, G.: Image Retargeting based on Global Energy Optimization. In: IEEE International Conference on Multimedia and Expo, New York (2009)
10. Dahan, M.J., Chen, N., Shamir, A., Cohen-Or, D.: Combining Color and Depth for Enhanced Image Segmentation and Retargeting. The Visual Computer 28(12), 1181–1193 (2011)
11. Mansfield, A., Gehler, P., Van Gool, L., Rother, C.: Scene Carving: Scene Consistent Image Retargeting. In: Daniilidis, K., Maragos, P., Paragios, N. (eds.) ECCV 2010, Part I. LNCS, vol. 6311, pp. 143–156. Springer, Heidelberg (2010)
12. Ren, T., Liu, Y., Wu, G.: Image Retargeting Using Multi-Map Constrained Region Warping. In: ACM International Conference on Multimedia, Beijing (2009)
13. Lee, K.Y., Chung, C.D., Chuang, Y.Y.: Scene Warping: Layer-based Stereoscopic Image Resizing. In: IEEE International Conference on Computer Vision and Pattern Recognition, Providence (2012)
14. Basha, T., Moses, Y., Avidan, S.: Geometrically Consistent Stereo Seam Carving. In: IEEE International Conference on Computer Vision, Barcelona (2011)
15. Utsugi, K., Shibahara, T., Koike, T., Takahashi, K., Naemura, T.: Seam carving for stereo images. In: 3DTV-Conference, Tampere (2010)
16. Ramachandra, V., Zwicker, M., Nguyen, T.Q.: Combined Image Plus Depth Seam Carving for Multiview 3D Images. In: IEEE International Conference on Acoustics, Speech and Signal Processing, Taipei (2009)
17. Middlebury Stereo Datasets, `http://vision.middlebury.edu/stereo/data/`

Automatic 3D Prostate MR Image Segmentation Using Graph Cuts and Level Sets with Shape Prior

Wei Xiong[1], Anthony Lianjie Li[2], Sim Heng Ong[2], and Ying Sun[2]

[1] Institute for Infocomm Research, Singapore
[2] National University of Singapore, Singapore

Abstract. Automatic segmentation for 3D magnetic resonance images of the prostate is a challenging task due to its varying shapes and sizes. Most recent techniques are focused on using variations of the Active Appearance Model (AAM) approach as the main segmentation method. In this paper, an alternative approach using a hybrid of the graph cut technique and the geodesic active contour shape prior level set method is presented. Despite being relatively accurate, level set methods are not commonly used for 3D segmentation purposes because they are computationally expensive. This paper shows that, with 3D graph cut results as initialization for level sets, the processing time for such level set based methods can be substantially reduced while preserving the accuracy of the segmentation.

Keywords: prostate segmentation, level set, graph cut, MRI images.

1 Introduction

Prostate cancer is one of the most lethal cancers amongst males [1]. Magnetic Resonance (MR) imaging is useful in visualizing soft tissue, particularly the apex of the prostate and cancerous tissues [2,3] and has been used widely to treat and diagnose prostate cancer [4]. Determination of prostate volume helps assess the disease stage. Computer aided automatic image segmentation of the images to provide better accuracy and less variability is expected clinically. However, minute differences in grey level intensities between the prostate and surrounding tissues in MR images imposes great challenges.

A possible solution for segmenting 3D prostate MRI is to use a variation of Active Appearance Models (AAM) presented by Cootes et al. [5]. Essentially, these methods place the emphasis on building appearance models that are statistically based on training images, with the prostate subsequently segmented using these models. There are many variants to the AAM method and some of these have even been used for TRUS images [6]. There is also literature to suggest that non-AAM based approach can segment the prostate as well. One such example is Saimee et al.'s [7] approach of incorporating shape prior information with edge based filters to segment the prostate. Dowling et al. [8] has also suggested using a combination of pre-processing methods together with an atlas based method. However, these solutions are inferior to AAM

B. Huet et al. (Eds.): PCM 2013, LNCS 8294, pp. 211–220, 2013.

based methods in terms of accuracy and speed. Most recently, in the MICCAI grand challenge PROMISE 12, which is a 3D prostate MRI segmentation competition, Vincent et al. [9] proposed an AAM based algorithm which topped the competition. As such, AAM based methods could be considered to be the state of the art in the area of prostate image segmentation.

The objective of this paper is to look for a solution that has a faster processing speed with comparable accuracy. Boykov et al.'s [10] efficient N-dimensional graph cuts method is considered for prostate segmentation. However, the outcome of the segmentation is crude and distant from the ground truth since it was treated as a global optimization problem with no shape prior and regional image intensity constraints. We propose that the outcome of the graph cut can be further utilized as an initialization for propagation under the geodesic active contour shape prior level set algorithm proposed by Leventon et al. [11]. In summary, with such modifications and innovations, it is expected that the algorithm proposed in this paper will help to reduce processing time while being comparably accurate with respect to other algorithms that have been proven to be effective. Similar to recent algorithms, this pipeline can also be automated with the necessary pre-processing under a statistical framework.

2 Methodology

An overview of the proposed method is shown in Figure 1 where it will be split into three subsections for further discussion. Subsection 2.1 describes the pre-processing algorithm which produces the relevant models and image maps required by other subsection. Subsection 2.2 describes the graph cut initialization that is necessary to initiate the graph cut method. Finally, subsection 2.3 combines graph cut with shape prior level set method to produce the final segmentation result.

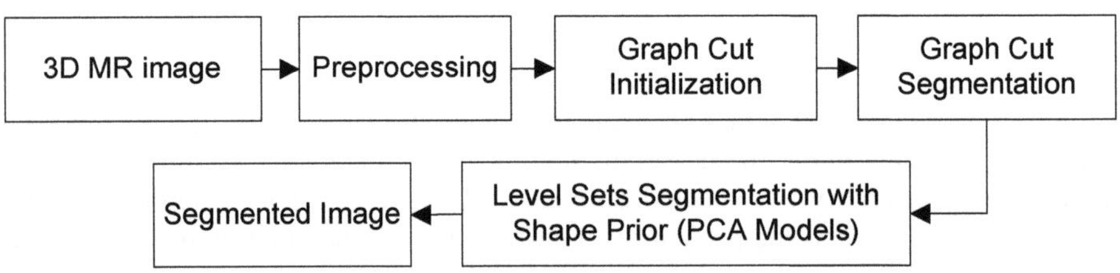

Fig. 1. Overview of processing pipeline

2.1 Pre-processing

The pre-processing algorithm processes a series of binary segmented prostate images that has already been delineated by expert radiologists. This image series is denoted as $I_i^{GT}(x)$ where x refers to the 3D coordinates of the voxel and i refers to the ith set of data and has values in the range $1 \leq i \leq N$. The target image is denoted as $I(x)$.

The MRI data is obtained from different MR imaging scanners with different protocols, the $I_i^{GT}(x)$ will have different spatial resolution, rotation parameters and intensity ranges. The pre-processing algorithm will perform re-sampling to standardize the resolution, spacing and rotation parameters for all $I_i^{GT}(x)$ images with respect to $I(x)$. Subsequently, the images are aligned according to the centroid of the segmented prostate. $\hat{I}_i^{GT}(x)$ will denote the output image series as a result of re-sampling and re-alignment. Principal component analysis (PCA) is then performed on $\hat{I}_i^{GT}(x)$ for the N sets of data. After normalization, the mean image $\mu(x)$ and the eigen-images $\mu_k(x)$ corresponding to the k-th eigenvalue α_k will be produced.

2.2 Graph Cut Initialization and Segmentation

After pre-processing and PCA, further processing will be done to label a suitable number of voxels as foreground and background for the purposes of graph cut algorithm. The foreground refers to the prostate; the background is the rest of the image.

During initialization, voxels which are highly likely to be representative of prostate tissues will be labeled as foreground and voxels which are highly unlikely to be representative of prostate tissues will be labeled as background. The probability of a voxel being representative of prostate tissues is denoted by $P_p(x)$. This probability can be estimated as $\frac{\mu(x)}{max\ (\mu(x))}$ and denoted by $\hat{P}_p(x)$. Here $max\ (\mu(x))$ is the maximum value of all $\mu(x)$ in the image. Note that $\hat{P}_p(x)$ is not correlated with $I(x)$, and $\hat{P}_p(x)$ is not equivalent to $P_p(x)$. Figure 2 shows an example of $\hat{P}_p(x)$ in three views.

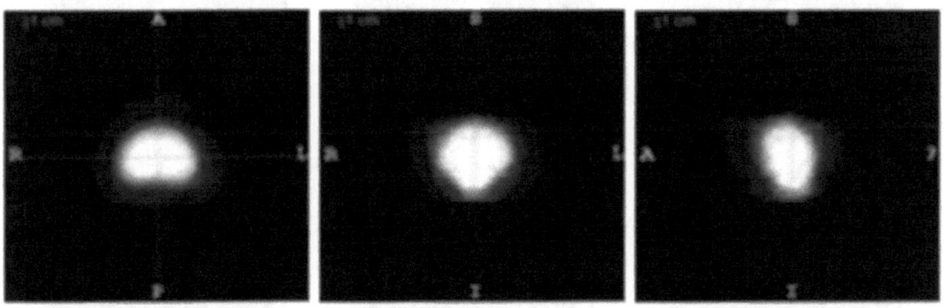

Fig. 2. $\hat{P}_p(x)$: An estimation of the probability of a voxel belonging to true prostate tissues. From left: axial, sagittal and coronal views.

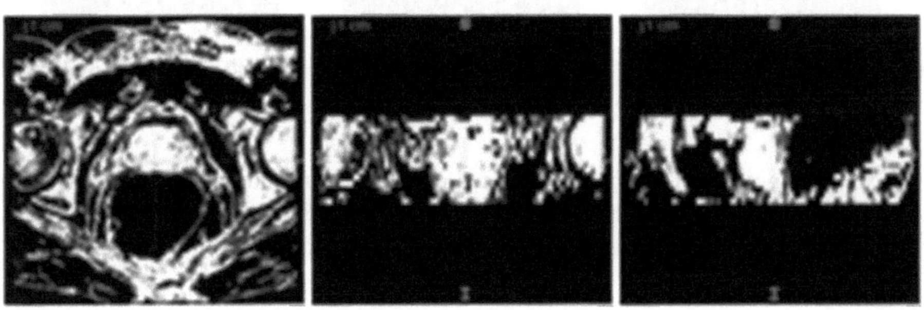

Fig. 3. $P_{kmeans}(x)$: An example of K-means clustering results. From left: axial, sagittal and coronal views.

Next, K-means clustering is performed on $I(x)$ to cluster voxels into K classes. The value of K is set as 3. We have observed that most of the prostate voxels fall within the second label (i.e., prostate) of the classification. An example of $P_{kmeans}(x)$ after K-means clustering result is shown in Figure 3 in different views.

Given that $P_{kmeans}(x)$ denotes the probability of a voxel having an intensity, which is typical of a prostate based on K-means classification, $P_{kmeans}(x)$ can be represented as a binary image map extracted from the output of the K-means classification. Since it is a binary image, $P_{kmeans}(x)$ has either a value of 0 or 1. Subsequently, we compute

$$P_p(x) = \hat{P}_p(x) \times P_{kmeans}(x), \tag{1}$$

which integrates information from $I(x)$ into $\hat{P}_p(x)$. Finally, voxels in $P_p(x)$ with values close to 1 are labeled as the foreground. Similarly, voxels in $P_p(x)$ with probability values close to 0 are labeled as background. Figure 4 illustrates an example of the foreground and background initialization using such a method.

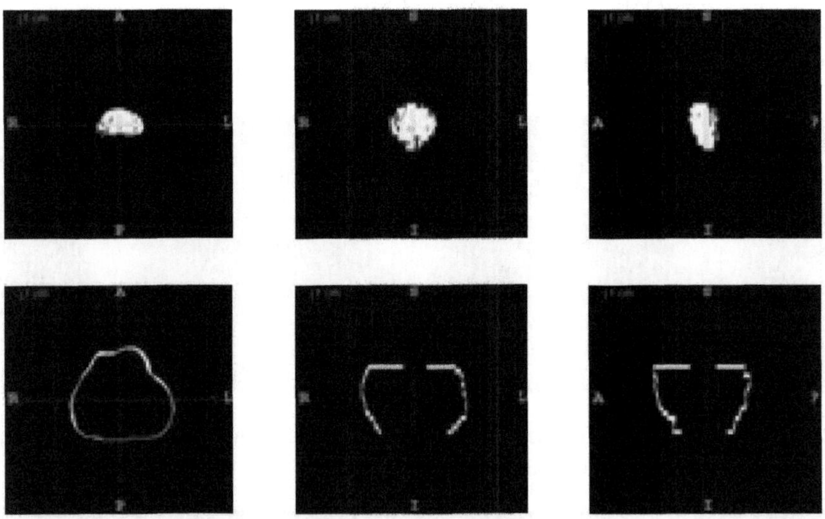

Fig. 4. An example to illustrate foreground (top row) and background (bottom) initializations for graph cuts. From left: axial, sagittal and coronal views.

Next, a 3D graph-cut method [10] based on the foreground and background initializations are performed. As shown in Figure 5, the segmentation is fairly irregular and not smooth, which is typical of graph cut methods applied to MR images.

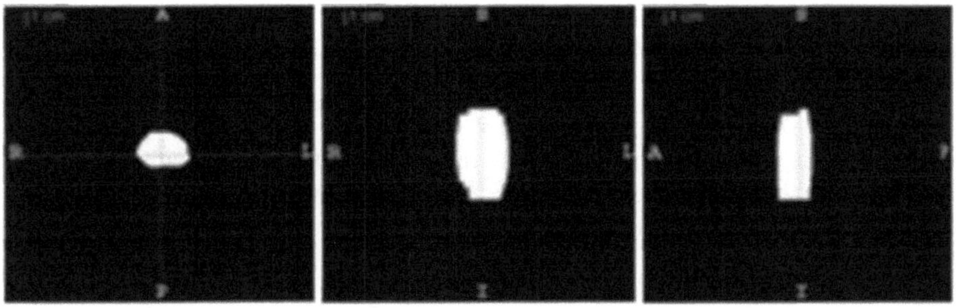

Fig. 5. An example to illustrate graph cut segmentation results based on the foreground and background initialization. From left: axial, sagittal and coronal views.

2.3 Combining Graph Cut and Geodesic Active Contour Shape Prior Level Set Algorithm

Although the segmentation results from graph cut methods are not smooth in the boundaries, they can be used to initialize a level set segmentation procedure. For the sake of completeness, we brief the basic of levels here.

Level sets are typically represented as the signed distance function $\varphi(x,t)$ where t refers to time. An initial level set can be written as $\varphi(x,0)$. The distance of the voxel away from the object boundary determines the absolute value of φ attached to the voxel. The sign of the value will determine whether a voxel is located inside or outside of the object. Values within the object are typically negative. The evolution of $\varphi(x,t)$ involves the use of the following partial differential equation:

$$\frac{d}{dt}\varphi = -\alpha A(x).\nabla\varphi - \beta P(x)|\nabla\varphi| + \gamma Z(x)\kappa|\nabla\varphi|, \tag{2}$$

where $A(x)$, $P(x)$ and $Z(x)$ refer to the advection, propagation and spatial modifier terms respectively, α, β and γ are the weights of the stated terms respectively, and κ is mean curvature. The equations that govern $A(x)$, $P(x)$ and $Z(x)$ are:

$$P(x) = Z(x) = g(x) \tag{3}$$

$$A(x) = -\nabla g(x) \tag{4}$$

This evolution process depends on the gradient image $g(x)$ that can be generated by

$$g(x) = \frac{1}{(1+|(\nabla G)(I(x))|)} \tag{5}$$

where ∇G refers to the Gaussian operator's derivative. $g(x)$ can vary depending on the technique chosen to image $I(x)$. According to (5), $g(x)$ takes on values close to zero in discontinuous regions and values close to 1 in continuous (smooth) regions. The evolution process follows closely the algorithm suggested by Sethian [11]. A modification to the shape guidance term $\varphi(x)$ is required in the event that shape prior information is to be incorporated. As suggested by Leventon et al. [12], the modification will yield the guidance term

$$\xi(\varphi^*(x) - \varphi(x)), \tag{6}$$

where $\varphi^*(x)$ is a function that represents the closest fitting shape within a principal component analysis shape model and ξ represents a variable that determines the weight of the shape constraint. $\varphi^*(x)$ can be obtained by the signed distance

$$\varphi^*(x) = dist(\mu(x) + \textstyle\sum_k \alpha_k \mu_k(x)), \tag{7}$$

where $\mu(x)$ and $\mu_k(x)$ are defined early and α_k refers to the weights of those principal components respectively. The signed distance of the components' sum approximates the value of $\varphi^*(x)$. Constrained by either the maximum number of iterations set by the user or the relative difference of the result between iterations, the final level set function resultant from (2) and (6) is representative of the segmented prostate.

3 Experiment Results and Discussion

The algorithm shown in Figure 1 is implemented in C++ with Insight ToolKit (ITK) [13] libraries. Doria et al.'s [14] 2D graph cut algorithm is extended to 3D for use as part of the algorithm. Algorithms with disabled multi threading capabilities are implemented on a machine that has a clock speed of 2GHz and 1GB RAM with no GPU optimization. No user interaction is required throughout the segmentation. Equation (2)'s parameters k, α, β and γ are fixed constants 1, 1, 20 and 1 respectively.

25 sets of 3D T2 weighted MR images with their respective ground truth are used to build the PCA model described in Section 2.1. Test datasets are extracted from the MICCAI Grand Challenge Promise 2012 competition database [15]. A Leave-One-Out Cross Validation (LOOCV) method is used to validate the performance of the algorithm. Target image $I^{GT}(x)$ is not included in the image series $I_i^{GT}(x)$ used for pre-processing. The segmentation result is then compared with $I^{GT}(x)$ through the calculation of metrics such as the Dice Similarity Coefficient (DSC), Mean Absolute Distance (MAD) and the Hausdorff Distance (HD) [16]. Additionally, the time taken for running the algorithm will be compared with the time taken to run other state of the art algorithms.

The results shown in Table 1 are conclusive that the algorithm proposed within this paper has a substantially shorter segmentation time and similar accuracy when compared to the level set method. On the other hand, it is slower but it produces a much more accurate segmentation compared with the graph cut method. Finally, the algorithm has comparable accuracy when compared with Vincent et al.'s AAM based method. More importantly, it is faster than Vincent et al.'s AAM based method [9] as shown in Table 1 where the algorithm takes a similar amount of time to process despite it being run on a machine with slower processing capabilities. However, it is critical to note that Vincent et al.'s algorithm does not aim for a fast segmentation but rather its focus is on an accurate result.

Table 1. A comparison of algorithm speeds and accuracies

Method	Time taken per 3D image	Clock speed of machine	User in teraction required?	HD (mm)	DSC	MAD (mm)
Graph Cut	1 minute	2.00 GHz	No	2.52 (mean), 7.98 (max)	0.68	3.83
Level Set	30 mi-nutes	2.00 GHz	Yes	1.62 (mean), 7.98 (max)	0.84	1.62
Vincent et al. [11]	8 minutes	2.83 GHz	No	4.17 (95%)	0.89	1.44
Proposed method	8 minutes	2.00 GHz	No	2.33 (mean), 7.75 (max)	0.85	1.66

In Figure 6, an example of the segmented prostate is shown in anterior, posterior and apex views respectively. We also use with pseudo colors to illustrate the Hausdorff Distance values between the segmented prostate and the ground true. Here blue colored areas represent areas of biggest HD; green colored areas have mean HD and red colored areas have zero HD. From left to right: anterior, posterior and apex views of prostate respectively. It is observed that the algorithm has its shortcomings particularly with identifying the apex of the prostate.

To have a better comparison, Figure 7 shows the cross section views (axial and coronal view of the prostate segmentation of three patient data sets (with the first row for the set shown in Figure 6). Here contour in red is the ground truth segmented by the radiologist and the contour in blue is the segmentation result by the proposed algorithm. The axial view of the same patient shows the segmentation is accurate. However the coronal view of the prostate in the top row shows a particularly large deviation at the apex (top part) of the prostate.

Fig. 6. The segmented prostate with pseudo colors to illustrate the Hausdorff Distance values between the segmented prostate and the ground true. Blue colored areas represent areas of biggest HD. Green colored areas have mean HD. Red colored areas have zero HD. From left to right: anterior, posterior and apex views of prostate respectively.

Figures 8 shows segmentation results of two other patients. The top row shows better accuracy results. The bottom row shows deviating substantially from the ground truth. This can happen whenever the intensities and texture of the tissue appears to be relatively similar as can be observed in the images.

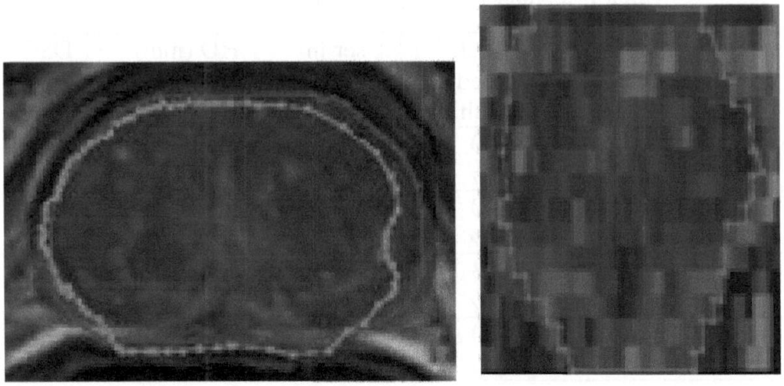

Fig. 7. Prostate segmentation results in axial (left column) and coronal (right column) views. Red contour is the ground truth and the blue is the segmented results.

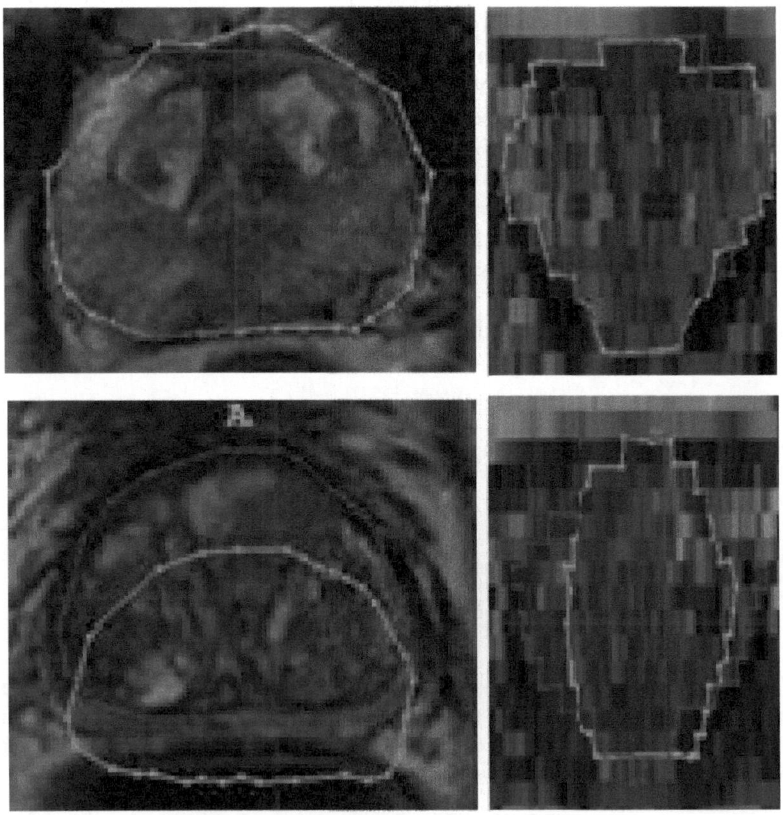

Fig. 8. Two new prostate segmentation results in axial (left column) and coronal (right column) views. Each row is for one patient. Red contour is the ground truth and the blue is the segmented results.

We have tested our algorithm on 25 datasets. 21 of them have shown successful segmentation with reasonable accuracies. The processing time is within 8 minutes per dataset. However, there also four datasets failed to get a reasonable graph cut result. Poor initializations might be the reason. The algorithm took longer than expected to segment the prostate since the initial starting point of level set propagation is heavily dependent on the graph cut getting a reasonable initial estimate.

4 Conclusion

A novel algorithm has been proposed for efficient 3D prostate MR image segmentation by combining 3D graph cuts and 3D geodesic active contour shape prior level set. The 3D graph cuts produced good initializations of the prostates and the level sets refined the prostate segmentation with smoother boundaries. Experimental results have shown that such a combination is viable although certain components of the pipeline require refinement to make the algorithm more robust. More importantly, the algorithm can achieve comparable accuracy with other algorithms. Finally, it is clearly speedier as compared to Vincent et al.'s method. Future work is to integrate the two methods in a single optimization framework and use more data for testing. A recent progress is that MICCAI 2013 has included another prostate segmentation challenges. With more and more research efforts, the challenges in prostate segmentation will be cleared someday.

References

1. Cancer Facts and Figures 2012, American Cancer Society, Atlanta, pp. 19–20 (2012), `http://www.cancer.org`
2. Yan, P., Xu, S., Turkbey, B., Kruecker, J.: Discrete "Deformable Model Guided by Partial Active Shape Model for TRUS Image Segmentation.". IEEE Transactions on Biomedical Engineering 57, 1158–1166 (2010)
3. Villeirs, G.M., Verstraete, K.L., DeNeve, W., et al.: Magnetic Resonance Imaging Anatomy of the Prostate and Per-Prostatic Area: a guide for radiotherapists. Radiotherapy Oncology, 99–106 (2005)
4. Lee, Y.K., Bollet, M., Charles-Edward, C., et al.: Radiotherapy treatment planning of prostate cancer using magnetic resonance imaging alone. Radiotherapy Oncology 66, 203–216 (2003)
5. Cootes, T.F., Petrovi, C.V., Schestowitz, R., Taylor, C.: GroupWise construction of appearance models using piece-wise affine deformations. In: 16th British Machine Vision Conference, vol. 2, pp. 879–888 (2005)
6. Ghose, S., Oliver, A., Martí, R., Lladó, X., Freixenet, J., Mitra, J., Vilanova, J.C., Comet, J., Meriaudeau, F.: Multiple mean models of statistical shape and probability priors for automatic prostate segmentation. In: Madabhushi, A., Dowling, J., Huisman, H., Barratt, D. (eds.) Prostate Cancer Imaging 2011. LNCS, vol. 6963, pp. 35–46. Springer, Heidelberg (2011)

7. Samiee, M., Thomas, G., Fazel-Rezai, R.: Semi-Automatic Prostate Segmentation of MR Images Based on Flow Orientation. In: IEEE International Symposium on Signal Processing and Information Technology, pp. 203–207. IEEE Computer Society Press, USA (2006)
8. Dowling, J., Fripp, J., Chandra, S., Pluim, J.P.W., Lambert, J., Parker, J., Denham, J., Greer, P.B., Salvado, O.: Fast automatic multi-atlas segmentation of the prostate from 3d mr images. In: Madabhushi, A., Dowling, J., Huisman, H., Barratt, D. (eds.) Prostate Cancer Imaging 2011. LNCS, vol. 6963, pp. 10–21. Springer, Heidelberg (2011)
9. Vincent, G., Guillard, G., Bowers, M.: Fully Automated Segmentation of the Prostate using Active Appearance Models. In: MICCAI, United Kingdoms (2012)
10. Boykov, Y.: Graph Cuts and Efficient N-D Image Segmentation. International Journal of Computer Vision (2006)
11. Osher, S., Sethian, J.A.: Fronts Propagating With Curvature-Dependent Speed: Algorithms based on Hamilton-Jacobi Formulation. Journal of Computational Physics 79, 12–49 (1988)
12. Leventon, M., Grimson, W., Faugeras, O.: Statistical shape influence in geodesic active contours. In: Proc. IEEE Conference on Computer Vision and Pattern Recognition (CVPR), vol. 1, pp. 316–323 (2000)
13. Ibanez, L., Schroeder, W., Ng, L., Cates, J.: The ITK Software Guide, 2nd edn. (2012), `http://www.itk.org/ItkSoftwareGuide.pdf`
14. Doria, D., Chen, S.: Interactive Image Graph Cut Segmentation, Release 0.0 (November 2012), `https://github.com/daviddoria/InteractiveImageGraphCutSegmentation`
15. MICCAI 2012 Database, Promise 12 Grand Challenge 2012 (2012), `http://promise12.grand-challenge.org/`
16. Archip, N., Clatz, O., Whalen, S., Kacher, D., Fedorov, A., Chrisochoides, A.N., Jolesz, F., Golby, A., Black, P., Warfield, S.: Non-rigid alignment of preoperative MRI, fMRI, and DT-MRI with intra-operative MRI for enhanced visualization and navigation in image-guided neurosurgery. NeuroImage 35(2), 609–624 (2007)

Improving PMVS Algorithm for 3D Scene Reconstruction from Sparse Stereo Pairs

Bo Li[1], Y.V. Venkatesh[2], Ashraf Kassim[3], and Yijuan Lu[1]

[1] Department of Computer Science, Texas State University, San Marcos, USA 78666
[2] Interactive and Digital Media Institute, NUS, Singapore 119613
[3] Department of Electrical and Computer Engineering, NUS, Singapore 119613
{b_158,lu}@txstate.edu, yv.venkatesh@gmail.com, ashraf@nus.edu.sg

Abstract. 3D scene reconstruction resulting from a limited number of stereo pairs captured by a 3D camera is a nontrivial and challenging task even for current state-of-the-art multi-view stereo (MVS) reconstruction algorithms. It also has many application potentials in related techniques, such as robotics, virtual reality, video games, and 3D animation. In this paper, we analyze the performance of the **PMVS** (Patch-based Multi-View Stereo software) for scene reconstruction from stereo pairs of scenes captured by a simple 3D camera. We demonstrate that when applied to a limited number of stereo pairs, PMVS is inadequate for 3D scene reconstruction and discuss new strategies to overcome these limitations to improve 3D reconstruction. The proposed Canny edge feature-based PMVS algorithm is shown to produce better reconstruction results. We also discuss further enhancements using dense feature matching and disparity map-based stereo reconstruction.

1 Introduction

3D reconstruction of a scene from multiple stereo views is called image-based modeling (**IBM**) technique, which is important for diverse applications ranging from robotics vision, electronic earth maps, and virtual reality to 3D film production, computer games and animation. However, it is known that depth information is lost when 2D images of a 3D world scene are captured by a single camera or multiple cameras. In a single camera image, the depth information is implicit (in the form of shading, shadows, texture and others). In contrast, in two (i.e., binocular) or more (i.e., multi-view) camera images, the same information is explicitly available in what is called *disparity* between pixels in a binocular stereo pair and also in multi-view stereo images, which can be organized in a sequence as stereo pairs. One example demonstrating the disparity information existing in stereos, captured using our 3D camera, is shown in Fig. 1. Therefore, based on the geometry of projection of a 3D scene as a pair of 2D images using a 3D camera, which can be modeled as a pair of pinholes, we can relate the 3D scene point to (i) the disparity between the corresponding points that are projections of the same 3D scene point on the 3D camera images; and (ii) focal lengths (expressed in terms of pixel sizes) of the two lenses and other *intrinsic*

B. Huet et al. (Eds.): PCM 2013, LNCS 8294, pp. 221–232, 2013.

(a) left stereo (b) right stereo (c) overlaid stereo

Fig. 1. An example of disparity information as the foundation of multi-view stereo techniques. Left stereo (a) is coded in red and right stereo (b) is coded into cyan color in (c). 3D depth information can be easily and clearly perceived from the feature stereo pairs by utilizing a cyan-red filter (goggles).

and *extrinsic* parameters of the 3D camera. When multiple 3D cameras are employed (or the same 3D camera is used at multiple locations), this relationship between disparity and 3D scene information can be extended to other binocular pairs. For diagrammatic and theoretical details, please see [6], for instance.

In this paper, our primary goal is reconstruction of a 3D scene from sparsely sampled stereo pairs captured by a single 3D camera in different locations. We present our experimental findings related to the application of the PMVS (Patch-based Multi-View Stereo software) [4] [5] package and its modified version, meant explicitly for *self-calibration* of cameras and scene reconstruction from multiple camera images, to stereo pairs of a laboratory scene having finite depth, as captured by a 3D camera in a limited number of positions. This is an interesting but challenging problem and promising for related applications, i.e., 3D robot vision and real-time video surveillance, which require an efficient and effective 3D scene reconstruction based on only several stereo pairs. Our major findings and contributions are as follows:

- PMVS is inadequate for a complete, accurate and robust 3D scene reconstruction of typical laboratory scenes based on sparse stereo pairs, due to the nature of the extracted Difference-of-Gaussian (DoG) [9] and Harris [7] features used for stereo correspondence.
- For an indirect verification of the relevance of the DoG+Harris features (currently employed in PMVS) to disparity (and hence scene-depth) computation, we have stereo-perceptually analyzed the feature stereo pairs by using the cyan-red filter (goggles) and found that a significant amount of depth perception, and hence of depth information existing in the original stereo pair of images, have been lost.
- In contrast, our experimental results demonstrate that feature stereo pairs obtained from an application of the Canny-operator to the original stereo pairs do retain a significant amount of perceptual depth information of the original stereo pairs. Moreover, these features lead to denser and more accurate feature matches, in contrast with those based on Harris+DoG operator-based features. Better 3D scene reconstruction results also have been obtained and demonstrated on three experimental scenes.

The rest of the paper is organized as follows: Section 2 reviews the related work. An overview of PMVS is presented in Section 3. In Section 4, we present and evaluate the results of PMVS-based scene depth estimation and reconstruction. By way of overcoming the limitations of PMVS, in Section 5 we demonstrate better results arising from a choice of features different from those used in PMVS, that is Canny detector-based features, as well as a proposal of dense matching and disparity-based stereo reconstruction algorithm which is also based on the Canny features for further improvement. We conclude the paper and list the future work in Section 6.

2 Related Work

The literature on 3D scene reconstruction is too vast to be reviewed here satisfactorily. For a taxonomy of algorithms for 3D scene reconstruction from multi-view stereo pairs, please see [12], which builds a benchmark composed of high-quality calibrated multi-view stereo images, ground truth 3D models as well as evaluation methodology. In contrast, without any prior knowledge of the camera or the scene, recent literature deals with the estimation of camera parameters simultaneously with the 3D depth map of the scene for 3D scene reconstruction from one or more image pairs, with or without stereo. This is called the *self-calibration* approach of which bundle adjustment (**BA**) [13] is an example. While this approach has been in use in photogrammetry, its application in computer vision is very recent. It is used to refine scene reconstruction by simultaneously estimating both the 3D structure of the scene and the projection matrices that match its corresponding images in a chosen optimal (i.e., least squares) sense.

Alternative techniques for multi-view 3D scene reconstruction are: (1) the patch-based multi-view (**PMVS**) software [4] [5] which requires camera parameters and outputs a set of oriented points, where both the 3D coordinate and the surface normal are estimated at each oriented point; (2) surface normal estimation and best viewpoint selection as found in [14]. For establishing correspondence among the images, other strategies include photo-consistency [8] as a similarity measure, and correlation-based metrics, which are more robust due to their invariance to global scaling of intensity.

3 Overview of PMVS

PMVS (Patch-based Multi-View Stereo software) has camera parameters for a set of views and possibly view visibility information (about the images that can see common 3D points; used to speed-up computations) as input and reconstruct the 3D structure information based on patch-based multi-view stereopsis. The framework of the PMVS is described in Fig. 2.

PMVS extracts two features using Difference-of-Gaussian (DoG) [9] and Harris [7] operators. The DoG operator performs edge detection by subtracting the results of two Gaussian blurs with different radii. As an approximation of Laplacian of Gaussian, it can filter out high frequency signals such as random noise,

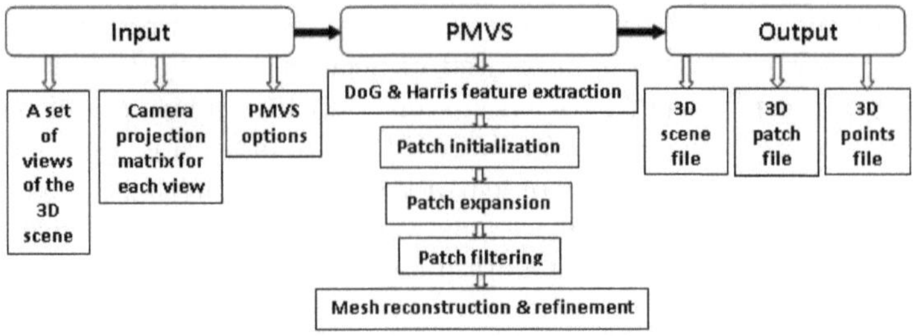

Fig. 2. Frameworks of PMVS

thus making it suitable for processing a highly noisy image. The Harris operator employs local auto-correlation operation to improve the edge consistency by extracting both the edge and corner features of an image. The Harris response is positive in corner regions, negative in edge regions, and small in flat regions. Further, in the PMVS, with image sample points as seeds, epipolar lines are used to decide the corresponding region (within 2x2 pixel) in another image, thereby generating patches (each defined by its center, normal, and visibility) to meet constraints on visibility, and leading to patch-based correspondence between images. Multi-view matching in the PMVS is patch-based and depends on mean photo-consistency of all *visible* pairs. A patch is reconstructed by maximizing the mean value of photo-consistency, and then accepted only if the number of visible images is greater than or equal to three.

PMVS has several input parameters that have different options of choice, such as decomposition level ($level$), cell size ($csize$), threshold value ($threshold$) for photometric consistency measure, window size ($wsize$), and minimum number of images ($minImageNum$) that a 3D point must be visible in. PMVS output comprises the followings: (1) reconstructed 3D scene geometry/model, i.e., 3D coordinates, normal and color data of a set of reconstructed 3D points; (2) patch information (center, normal, and visibility image set); and (3) a 3D point set listing the coordinates and normals of the 3D points, and it is generated as the input of two candidate post-processing algorithms after PMVS: Poisson-based 3D surface reconstruction (PSR) algorithm [5] and Visual Hull (VH) reconstruction [5].

4 PMVS-Based 3D Scene Reconstruction: Evaluations and Limitations

In this section, we target identifying the limitations of PMVS when applied to perform 3D scene reconstruction on a sparse set of stereo pairs captured in a controlled environment.

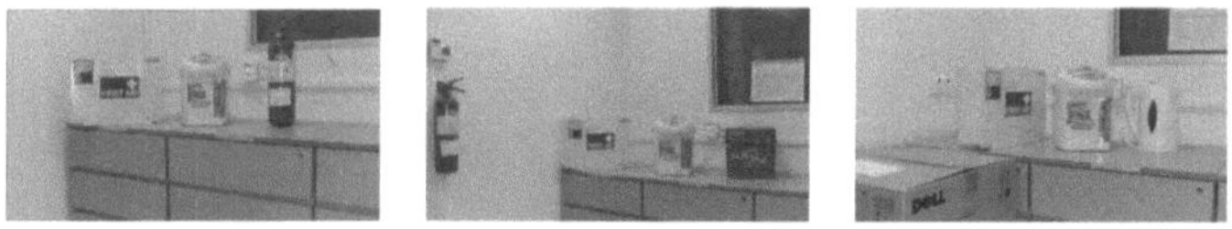

(a) Scene 1 (b) Scene 2 (c) Scene 3

Fig. 3. Sample images of three 3D scenes

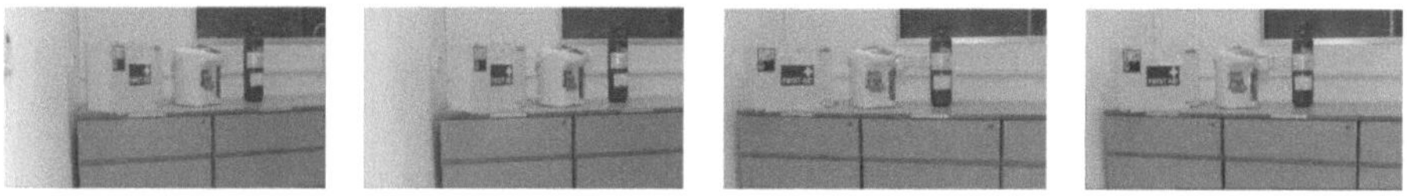

(a) Left stereo (b) Right stereo (c) Left stereo (d) Right stereo

Fig. 4. Sample stereo pairs of Scene 1 from two locations: (a) and (b) are for one pair; (c) and (d) are for the other

4.1 3D Data

We used a 3D camera ("FUJIFILM REAL 3D W1") which has two Fujinon optical zoom lenses (left and right) to capture stereo pairs in a controlled (i.e., measurable) laboratory environment. We captured typical 3D scenes in a laboratory with objects with/without textural information at varying depths in both 2D JPEG image and 3D MPO (Multi Picture Object) file formats. The MPO files are then processed to generate the 3D stereo pairs (left and right). The resulting images obtained for various scenes are shown in Fig. 3. Fig. 4 shows two stereo pairs of Scene 1 from two different locations. Data from Scene 1, 2 and 3 contain 6, 13 and 15 3D stereo pairs sampled roughly in $\frac{1}{4}$ circle, respectively.

4.2 Evaluations

We evaluate PMVS with our captured 3D stereo images with respect to its following two aspects: (1) sparsity degree of 3D stereo pairs; (2) comparison with 2D image-based reconstruction results. Unless stated otherwise, we adopt the PMVS default parameter options in [4] [5], such as *level*, *csize*, *threshold*, *wsize*, and *minImageNum*. We also need to mention that in order to have a comparison based on a common reconstruction process, we only compare the main PMVS reconstruction results till the step of patch filtering (see Fig. 2) and do not consider the final step of mesh reconstruction which still has several options for different types of scenes. We also have found that sometimes the initial scene reconstruction based on the Poisson surface software is unsatisfactory, especially when the reconstructed 3D point set is sparse. It is not easy to tune the parameters as well. It seems to perform well when the surface is a flat plane but not when dealing with curved surfaces.

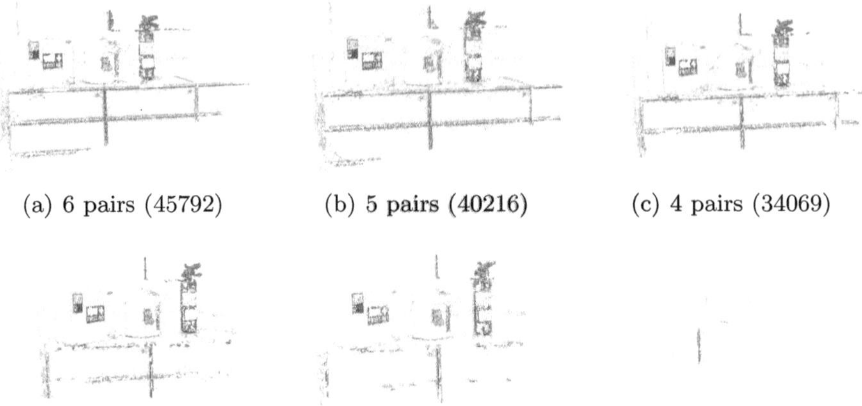

(a) 6 pairs (45792) (b) 5 pairs (40216) (c) 4 pairs (34069)

(d) 3 pairs (23496) (e) 2 pairs (14283) (f) 1 pair (577)

Fig. 5. PMVS reconstruction results comparison on Scene 1 with respect to the sparsity degree of 3D stereo pairs. The digit in a parenthesis is the number of vertices of the 3D reconstruction result. (f) shows the reconstruction result when setting *minImageNum* to be 2 since there is only two images.

Evaluation of PMVS with Respect to the Sparsity Degree of 3D Stereo Pairs. Take Scene 1 as an example, starting from 6 pairs, we sequentially reduce the number of pairs from the end of the view list (Section 4.1) to only 1 pair, that is, testing 6, 5, 4, 3, 2 and 1 pairs. Fig. 5 compares their PMVS reconstruction results. From the above example on Scene 1 as well as on the other two scenes we have tested, we have found that PMVS has apparent limitations in dealing with sparse stereo pairs. For example, it cannot reconstruct a meaningful 3D scene geometry for Scene 1 if only using the first captured pair and for Scene 2, it cannot do that even with first two pairs. At least 4~5 pairs are needed to reconstruct a relatively complete 3D scene for these two cases.

More pairs and denser view sampling will improve the reconstruction quality. However, the computational complexity will also increase rapidly. For example, Table 1 lists the computational time of Scene 1 with respect to different numbers of input pairs using a computer with an AMD Opteron processor 6174, with 2.19 GHz clock and 16.0 GB memory on a Windows 7 64-bit operating system. It is also found that the patch expansion process takes up a large part (about 90%) of the total computational time. We also note that a more efficient algorithm named Clustering Views for Multi-View Stereo (CMVS) [3] based on PMVS and view clustering is available. However, it mainly concentrates on a large number

Table 1. Computational time required for 3D scene reconstruction of Scene 1

Number of stereo pairs	1	2	3	4	5	6
Computational time (sec)	15	155	309	342	350	416

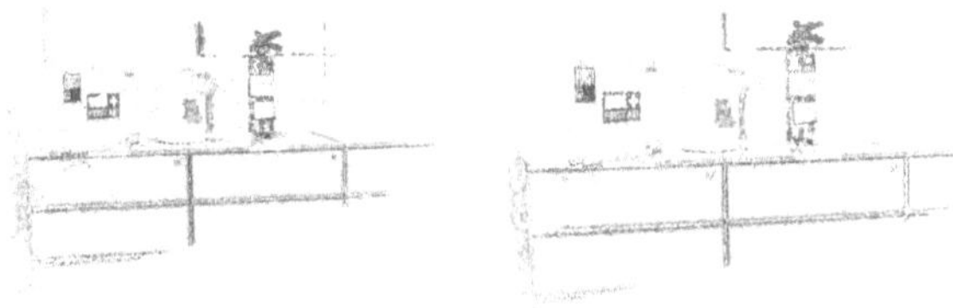

(a) six 3D stereo pairs (b) six 2D images

Fig. 6. Comparison of reconstruction results using of Scene 1

of views and still does not solve the issue of sparse view sampling, which in fact is a general limitation of patch-based reconstruction techniques. Similar situation is with [2] which is also based on PMVS but especially deals with reconstruction of a scene which comprises highly structured objects with little textural information. Therefore, the inadequacy of PMVS to reconstruct a 3D scene using a limited number of 3D camera stereo pairs seems to be *largely* due to the fact that, in the course of patch-based optimization, a patch is accepted only when the number of *visible* views is greater than or equal to three.

Evaluation of PMVS with Respect to the 3D Stereo Pairs versus 2D Images. As mentioned in Section 4.1, at one position of a camera we also obtained one 2D JPEG image. Based on the 2D images of a scene, we perform PMVS-based 3D reconstruction and compare the obtained result with the one based on the 3D pairs. Fig. 6 shows the comparison for Scene 1. It seems that using 2D images directly will obtain at least comparable results as using 3D pairs. However, the number of images employed is reduced by half, thus the reconstruction speed is much faster. We think the main reason for this is because of the small difference in the two baselines of a pair, that is *narrow-baseline*. The 2D correspondence is liable to errors because of the small differences in the coordinates of corresponding feature points in two stereo pair images. In fact, the authors of PMVS [5] also mentioned that one limitation of the algorithm is that the reconstruction is unsatisfactory for narrow-baseline stereo pairs. Therefore, rather than estimating the depth information, we recommend disparity-based reconstruction using a dense stereo matching framework [11].

4.3 Limitations

Though PMVS has demonstrated good performance and ability in their paper [5], it shows several shortcomings when applied to scene reconstruction from a limited number of 3D camera stereo pairs.

(1) PMVS is dependent on the accuracy of camera calibration results. We have found that SIFT feature-based Bundler camera calibration algorithm has a limitation on sparse view sampling and usually the camera estimation results are inaccurate when the input views are sparse or lack texture features. Therefore, a

more robust and accurate camera estimation technique should be considered to provide the important camera parameters input for the PMVS reconstruction.

(2) PMVS employs Harris and DoG operators to extract features which act as seeds for patch definition. However, limited feature representativeness has been found for certain scenes even based on a combinational utilization of these two features. The selection of two Gaussian blur radii for the DoG operator and the response threshold for the Harris operator are subject to the influences of manual settings. A feature extractor (like the Canny edge detection based on automatic thresholding) which avoids manual settings seems to be better suited for implementation in the PMVS package.

(3) PMVS has apparent limitations in dealing with sparse stereo pairs, especially for those captured by a *narrow-baseline* 3D camera. Our finding is that at least four to five pairs are needed to reconstruct a relatively complete 3D scene for these two cases. Compared to 2D images-based reconstruction, 3D stereo pairs-based reconstruction does not achieve an apparently better quality. Both of the above two disadvantages limit the scope of PMVS applications.

5 PMVS Improvement: Canny-Based PMVS

Motivated by the above limitations found during the results analysis of applying PMVS (together with Bundler) to our sparse stereo pairs-based 3D scene reconstruction scenarios, we propose to use Canny features to replace the DoG and Harris features for the feature extraction part of PMVS and have achieved better quality for the reconstruction results for the three sample scenes. Based on this, we further propose a preliminary disparity-based stereo reconstruction algorithm which also utilizes Canny features, and it serves as our next work.

5.1 Disparity Analysis of Different Features

Feature extraction is a very important module of a multi-view stereo reconstruction algorithm. PMVS employs Harris and DoG features. However, it lacks a justification of their feature selection based on a sound analysis of their contributions to the quality of reconstruction results, that is to say, the corresponding relationship between the 2D features selected and the 3D features reconstructed. For example, (1) What 3D features will be highlighted and which 3D features will be missing due to the limitations of the 2D features selected? (2) Whether the 2D features selected are representative enough for the 3D reconstruction? In other words, do the various features contain enough information to facilitate computation of a depth map for the scene and hence the 3D scene reconstruction? (3) Since the 2D feature point sets are extracted and selected as the seeds for the patch definition, then what impact they will have on the accuracy of the patch initialization and expansion till to the final results?

Motivated by the above questions, we compare different candidate features, such as the above mentioned Harris+DoG features, as well as Canny (using automatic thresholding) and SIFT features, in terms of the completeness of the

(a) Harris+DoG features (b) SIFT features (c) Canny features

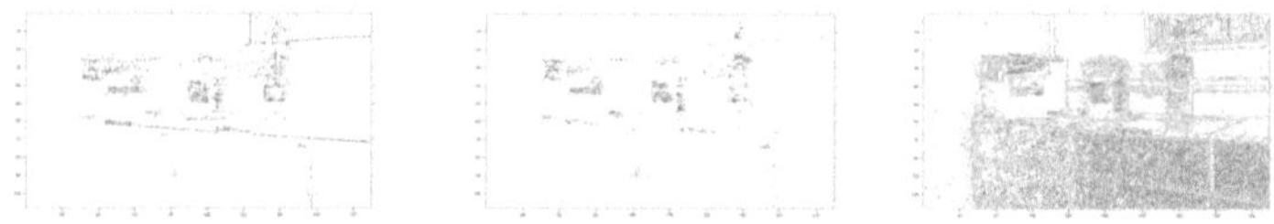

(d) Harris+DoG feature stereo (e) SIFT feature stereo (f) Canny feature stereo

Fig. 7. (a)∼(c): Harris+DoG, SIFT, and Canny features of an example stereo image of Scene 1. (d)∼(f): SIFT-, Harris+DoG- and Canny operator-based overlaid corresponding feature stereo pairs of Scene 1.

disparity information they contain. Please note that *disparity information is the foundation for depth map estimation and hence 3D scene reconstruction.* Our intention to perform the comparison is based on the following assumption: a 2D feature which shows more complete and discernible disparity information for diverse objects in a 3D scene will improve the accuracy and robustness of the 2D feature correspondence, then the 3D triangulation which is directly based on the correspondence results, thus the final reconstructed 3D scene.

Fig. 7 (a)∼(c) compare the three sets of features for the left stereo image of an example stereo pair of Scene 1 and Fig. 7 (d)∼(f) are for a comparison of the overlaid corresponding feature stereo pairs. In fact, we can even view *stereoscopically* the various feature stereo pairs of Fig. 7 (d)∼(f) using the cyan-red filter, and discover that Canny operator-based feature stereo pair contains significantly more depth information of the 3D scene than either the SIFT or Harris+DoG feature stereo pair. More specifically, for the scene under consideration, the number of DoG+Harris feature points in the left image is 1194; and in the right, 1195. For the SIFT operator, the corresponding numbers are 3715 and 3308, respectively; and for the Canny operator, they are 210922 and 194928. Thus, a much more denser feature pair will be obtained if based on Canny operator-based feature extraction. Then, a better disparity map and more accurate reconstruction result can be also expected.

5.2 Canny-Based PMVS

The stereo correspondence algorithm computes disparity at the feature points from which the disparity map can be generated. It has been found that (a) the disparity map computed from matching the Canny operator-based stereo pair is much *denser* than the one obtained from a similar matching operation on either SIFT or DoG+Harris operator-based stereo pair; and (b) the Canny

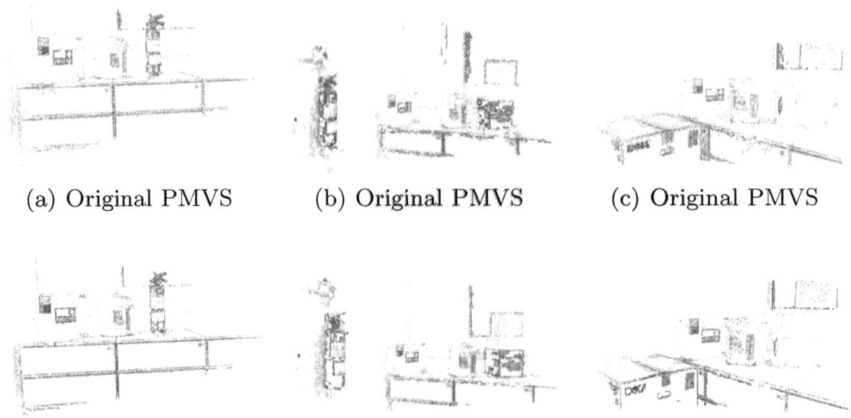

(a) Original PMVS (b) Original PMVS (c) Original PMVS

(d) Canny-based PMVS (e) Canny-based PMVS (f) Canny-based PMVS

Fig. 8. Comparison of reconstruction results, for Scene 1∼3 respectively, using (a) Original PMVS (first row); and (b) Canny features as input to PMVS (second row)

feature-based PMVS algorithm, which we named **Canny-based PMVS**, gives a *denser* depth map and more complete boundaries of objects in the scene, in contrast with the sparse depth maps generated from SIFT- and DoG+Harris operator-based PMVS algorithms.

It should be noted here that, unlike the SIFT or the Harris+DoG operator, the Canny edge detector is based on an automatic threshold selection according to the content of the images, thus it does not involve manual assignment for the parameter selection. We also need to mention that since our Canny edge detection is directly performed on the original images, we need to set the parameter *level* to zero, which means we do not consider the multi-level texture features as in PMVS. However, we will demonstrate that even on a single level for the feature extraction, we can achieve better reconstruction results. Finally, Fig. 8 compares the reconstruction results of the original PMVS and the proposed Canny features-based PMVS for the three scenes.

As can be seen, we have the following findings: (1) Canny-based PMVS algorithm achieves cleaner and more salient edge reconstruction results; (2) More 3D points can be reconstructed if using Canny edge features compared to the SIFT or Harris+DoG features, therefore we have obtained a denser depth map. Please find out the details about the above findings by zooming in on the above figures. For example, stronger and apparent edge information can be found, such as the "***FIRST***" label in Fig. 8 (d)∼(f), the "***DELL***" label and the box object in (f), and the fire extinguisher in (d) and (e). It is likely that a combination of Harris+DOG and Canny features as inputs to PMVS will further improve the results of the Canny-based PMVS algorithm.

Canny Feature and Disparity-Based Stereo Reconstruction Algorithm. As demonstrated above, Canny-based PMVS has achieved better 3D reconstruction quality after utilizing Canny features to replace the Harris+DoG feature. However, as a limitation demonstrated and analyzed before (see Sections 4.2∼4.3), PMVS does not perform comparably well for *narrow-baseline* stereo pairs. Thus, rather than using a patch-based technique PMVS, a dense feature matching and disparity-based technique is more promising to further apparently improve the 3D reconstruction quality and robustness. Motivated by this, we propose an initial Canny feature and disparity-based stereo reconstruction algorithm.

(1) Canny edge feature detection. This is to obtain a set of feature points existing in the left and right images of a pair.

(2) Dense feature matching-based disparity map estimation. We consider all the edge points and utilize a dense feature matching algorithm specially designed for narrow-baseline pair correspondence, such as [10]. Then, we construct a dense disparity map based on the feature correspondence results.

(3) Fundamental matrix computation. Utilizing the matching points and disparity map, we compute the Fundamental matrix **F** based on the RANSAC [1] method (in a loop manner). For implementation, we can refer to the OpenCV function of *cvFindFundamentalMat()*.

(4) Projection matrix computation. We compute the left and right projection matrices **P1** and **P2** from the fundamental matrix **F** [6].

(5) 3D point reconstruction based on triangulation. For this, we can refer to the OpenCV function of *cvTriangulatePoints()*.

6 Conclusions and Future Work

We have first presented results when applying the PMVS package to a limited number of stereo image pairs of typical laboratory scenes captured by a 3D camera. Then, the limitations of PMVS have been identified with respect to the application scenario of sparse stereo pairs-based 3D scene reconstruction. It is demonstrated that the features (Harris+DoG for PMVS, SIFT for Bundler), which are presently extracted as an integral part of the package, lead to not only unsatisfactory camera calibration parameters but also loss of many details in depth map estimates. In contrast, it is shown that Canny operator-based edge features extracted from the stereo pairs retain more depth information (than other features extracted in the PMVS package), and generate denser depth maps, which is important to generate better scene reconstruction. Thus, a Canny feature-based PMVS algorithm has been proposed and better 3D reconstruction results have been achieved on the same example scenes. To further overcome the limitations of the patch-based technique PMVS and SIFT features-based Bundler, we further propose a preliminary disparity map-based stereo reconstruction algorithm based on Canny features and dense feature mapping. Both proposed approaches are promising for related applications which require effective 3D scene reconstruction from a set of sparsely sampled pairs. As the main future work, we plan to further improve the Canny-based PMVS algorithm, as well as its descendent algorithm presented in Section 5.2, for the challenging problem of constrained sparse stereo pairs-based 3D scene reconstruction.

Acknowledgements. This work is partially supported by the Texas State University Research Enhancement Program (REP) and Army Research Office grant W911NF-12-1-0057 to Dr. Yijuan Lu, and the project Grant NRF2007IDM-IDM002-069 on "Life Spaces" from the IDM Project Office, Media Development Authority of Singapore. The authors wish to express grateful thanks to Professor Lawrence Wong (Deputy Director, IDMI) for his keen interest and encouragement.

References

1. Fischler, M.A., Bolles, R.C.: Random sample consensus: a paradigm for model fitting with applications to image analysis and automated cartography. Commun. ACM 24(6), 381–395 (1981)
2. Furukawa, Y., Curless, B., Seitz, S.M., Szeliski, R.: Manhattan-world stereo. In: CVPR, pp. 1422–1429. IEEE (2009)
3. Furukawa, Y., Curless, B., Seitz, S.M., Szeliski, R.: Towards internet-scale multi-view stereo. In: CVPR, pp. 1434–1441. IEEE (2010)
4. Furukawa, Y., Ponce, J.: Accurate camera calibration from multi-view stereo and bundle adjustment. International Journal of Computer Vision 84(3), 257–268 (2009), `http://homes.cs.washington.edu/~furukawa/research/pba/`
5. Furukawa, Y., Ponce, J.: Accurate, dense, and robust multiview stereopsis. IEEE Trans. Pattern Anal. Mach. Intell. 32(8), 1362–1376 (2010)
6. Harltey, A., Zisserman, A.: Multiple view geometry in computer vision, 2nd edn. Cambridge University Press (2006)
7. Harris, C., Stephens, M.: A combined corner and edge detector. In: Proc. of Fourth Alvey Vision Conference, pp. 147–151 (1988)
8. Kutulakos, K.N., Seitz, S.M.: A theory of shape by space carving. International Journal of Computer Vision 38(3), 199–218 (2000)
9. Marr, D., Hildreth, E.: Theory of edge detection 207(1167), 215–217 (1980)
10. Pollefeys, M., Gool, L.J.V.: Visual modelling: from images to images. Journal of Visualization and Computer Animation 13(4), 199–209 (2002)
11. Scharstein, D., Szeliski, R.: A taxonomy and evaluation of dense two-frame stereo correspondence algorithms. International Journal of Computer Vision 47(1-3), 7–42 (2002)
12. Seitz, S.M., Curless, B., Diebel, J., Scharstein, D., Szeliski, R.: A comparison and evaluation of multi-view stereo reconstruction algorithms. In: CVPR (1), pp. 519–528. IEEE Computer Society (2006)
13. Triggs, B., McLauchlan, P.F., Hartley, R.I., Fitzgibbon, A.W.: Bundle adjustment – A modern synthesis. In: Triggs, B., Zisserman, A., Szeliski, R. (eds.) ICCV-WS 1999. LNCS, vol. 1883, pp. 298–372. Springer, Heidelberg (2000)
14. Zabulis, X., Daniilidis, K.: Multi-camera reconstruction based on surface normal estimation and best viewpoint selection. In: 3DPVT, pp. 733–740. IEEE Computer Society (2004)

Fast HEVC Encoding with GPU Assisted Reference Picture Selection

Wei Xiao[1,*], Feng Wu[2], Jizheng Xu[2], and Guangming Shi[1]

[1] School of Electronic Engineering, Xidian University, Xi'an, Shaanxi, China
[2] Microsoft Research Asia, Beijing, China
weixiao@stu.xidian.edu.cn, {fengwu,jzxu}@microsoft.com, gmshi@xidian.edu.cn

Abstract. The emerging video coding standard, HEVC, provides a superior compression efficiency by exploiting flexible motion block partitioning and multiple reference pictures in typical settings. However, to obtain the compression potential of HEVC needs huge amount of computations to get the right motion vectors and associated reference picture indexes for each block, which may take most of the computational resources of encoding. In this paper, we tackle this problem by exploiting the computational capabilities provided by a GPU, to reduce the HEVC encoding time while preserving its high coding efficiency. We analyze the motion estimation module and divide the task into two parts. One favors GPU computing model, which involves lots of calculations without consideration of neighboring states. The other favor CPU computing model to consider the coding results of neighboring block, which can maintain HEVC's high coding efficiency. In such a way of joint CPU-GPU encoding, we show that the encoding time can be significantly reduced. Comparing with the HEVC reference software, experimental results show that our proposed scheme achieves 34.4% encoding time reduction on average while the BD-rate increase is only about 2% for a typical lowdelay setting.

Keywords: High Efficiency Video Coding (HEVC), motion estimation, reference picture selection, Graphic Processing Units (GPU).

1 Introduction

The emerging High Efficiency Video Coding (HEVC) standard[1] defined by the Joint Collaborative Team on Video Coding (JCT-VC) was started in 2010, aiming at an average 50% bitrate decrease in comparison with H.264/AVC (High Profile) when providing the same subjective quality level. In order to reach this goal, various new coding tools have already been aggregated to the current version of the HEVC Test Model (HM) [2], which also increase its computational complexity in comparison with H.264/AVC. Now the standardization has been finalized. On April 13, 2013,

---

* This work was done when W. Xiao was with Microsoft Research Asia as an intern and also supported by the 973 Program of China (No. 2013CB329402), the NSF of China (No. 61227004, 61072104, 61301288, 61003148, 61100155) and the 111 Project (No. B07048).

B. Huet et al. (Eds.): PCM 2013, LNCS 8294, pp. 233–244, 2013.

HEVC was first approved as an ITU-T standard named as H.265/HEVC [3]. The goal of HEVC coding performance has also been achieved, which is verified in [4].

One of the most important part in generations of video standards to improve the coding performance is the more and more sophisticated motion compensation, which can remove the temporal redundancy within video sequences. Compared with H.264/AVC [5], in which only 8 block sizes in 4 types from 16x16 to 4x4 are supported, the emerging standard HEVC supports more flexible motion estimation blocks with 24 different sizes in 7 types from 64x64 to 8x4 and 4x8 though the precision of motion vectors in HEVC is still in quarter pixel level. A longer interpolation filter and elimination of the intermediate rounding error also help to improve the motion compensation precision.

Another important improvement for motion compensation is from the use of more reference pictures. Although multiple reference picture can also be enabled in H.264/AVC, it is not widely used due to its computational burden. With the development of computing technologies and the requirement of high quality video encoding, during the standardization of HEVC, multiple reference pictures has been considered a necessary technique. Thus, even during the call-for-proposal stage of HEVC, multiple reference pictures was defined in the common test condition [6]. Four reference pictures are set for lowdelay configuration for each reference picture list in IPPP or IBBB coding structure and two to four reference pictures for each reference picture list are set in randomaccess configuration for hierarchical B coding structure. Not to mention that with a sophisticated reference picture management, further improvement can be achieved with multiple reference pictures [7].

While the coding performance benefits a lot from the more sophisticated of motion compensation, a heavier load is attached to the complexity in motion estimation to find a more accurate motion vector from a suitable reference picture. It is shown in [8] that motion estimation consumes about 80% encoding time for H.264 encoder. And for the more sophisticated motion estimation in HEVC, the encoding complexity is also one big problem for practical application.

A natural way to accelerate video encoding is through parallel computing. With the development of many multicore parallel computing platforms, e.g. cloud computing, it is promising to find better solutions to speedup encoding. Another cheaper and more convenient local multicore platform is GPU, which has become more and more powerful in computation ability in recent years. For H.264/AVC, many works have proposed different solutions to reduce the computational time of motion estimation (ME) and compensation interpolation. Two examples of H.264/AVC full motion search implementations in GPU are [9] and [10]. There are other works about fast ME algorithm in HEVC. For example, [11] introduces a new fast ME algorithm which outperforms the fast ME algorithm implemented in HEVC reference software HM which brings about 15% overall time reduction.

However, the large time cost spent on motion estimation is not only due to the motion estimation module implementation itself, it is also highly correlated to the times to check this module. For example, if the reference picture number is reduced from 4 to 1, the total time cost spent on motion estimation will be directly reduced to 25% of it was. So a fast reference picture selection scheme is also important in reducing the encoding time or complexity despite what kind of implementation of the sophisticated motion estimation module.

In this paper, we use GPU to do parallel integer motion estimation search to generate best motion vector candidates for the different reference pictures with less information of accurate motion vector prediction and a fast reference picture selection scheme is proposed based on the GPU motion estimation results. With the generated best motion vector candidates from GPU and more accurate motion vector prediction, CPU encoding only needs a cost refinement procedure to finish the final rate-distortion optimization determination.

The rest of the paper is organized as follows. Section 2 presents a brief overview on HEVC motion estimation methods and the selection of best reference picture. The proposed fast reference picture selection scheme based on the GPU parallel motion estimation and CPU cost refinement are described in Section 3. Experimental results in Section 4 give a validation for the proposed scheme. Finally, conclusions are drawn in Section 5.

2 Motion Estimation in the HEVC Test Model

Similar to prior video coding standards, such as H.264/AVC, HEVC still uses the basic hybrid coding architecture while a more generalized and larger quadtree structure based coding block is supported instead of a macroblock in H.264/AVC. Each input video frame is divided into square blocks called Coding Tree Units (CTUs) with the maximum size of 64x64. A CTU can be recursively subdivided into square blocks called coding unit (CU) ranged from 8x8 to 64x64. Each CU is further subdivided into one to four prediction blocks called Prediction Units (PUs) [1]. This flexible and adaptive block partitioning makes HEVC superior to the prior video coding standards.

In HEVC inter prediction, there are 7 types of prediction unit from 2Nx2N to nRx2N, shown in Fig. 1. For CU size 2N larger than 16, all PU types are enabled. As for CU size 8x8, only the first three PU types are allowed. So the PU sizes range from the maximum 64x64 to the minimum 8x4 and 4x8. Motion estimation is the PU level process of finding the best matched block in a search window of the reference picture for each PU in the current picture. The ME algorithm uses Lagrangian cost function (J_{MV}) illustrated in (1),

$$J_{MV} = SA(T)D + \lambda_{MV} \bullet R(MV - MVP) \quad (1)$$

where SA(T)D is the matching distortion function (SAD: Sum of Absolute Difference; SATD: Sum of Absolute Transformed Difference), λ_{MV} is the Lagrangian multiplier, MV is the current motion vector obtained and MVP is the predicted motion vector obtained from the advanced motion vector prediction (AMVP) process[1], one of the most important inter prediction coding tools in HEVC. (MV-MVP) denotes the motion vector difference (MVD) and R(MV-MVP) is the number of bits to represent this MVD. The final motion vector selected for one reference picture is the vector of the block with the minimum cost function.

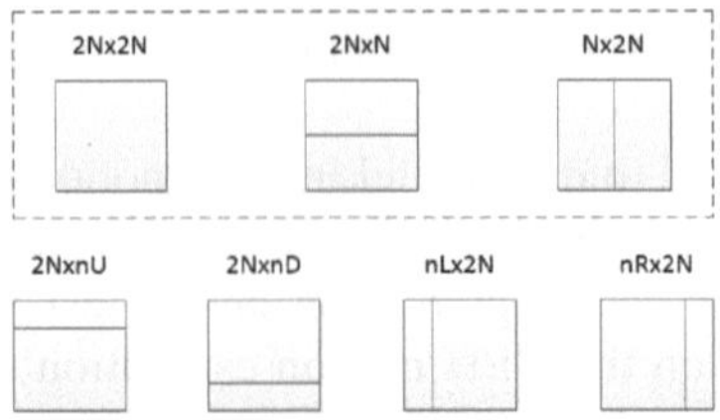

Fig. 1. Inter PU types in HEVC

Same as in H.264/AVC, the precision of motion vectors in HEVC is quarter pixel. Three steps are required to get a motion vector from the course integer pixel search to refined quarter pixel search. In the integer search step, SAD is used as the distortion cost function. The default search window area for the uni-direction inter search is from -64 to +64, which is a 129x129 window with the starting point determined by MVP. In the half pixel and quarter pixel search followed, both 9 refined positions are searched around the best position in the previous step while SATD is the distortion cost function.

When the best MV is found, the best MVP is recalculated accordingly in HEVC to achieve the best rate-distortion performance. Thus the final best MV cost of one PU for one reference picture is obtained. For multiple reference pictures used in HEVC, the encoder will go through the searching areas within all the reference pictures of the current picture and compare the best MV cost of each reference to determine the final best reference picture for current PU.

Two motion estimation integer search algorithms are used in HEVC encoder. One is to search all the blocks in reference picture search window called full search algorithms. The other is the fast ME search algorithm called TZS (Test Zone Search) algorithm, which is originated from H.264 MVC [12]. The fast TZ search is usually used in the common test configuration [6] to avoid suffering from too long encoding time.

3 Proposed Fast Reference Picture Selection Assisted by GPU

3.1 Parallel Motion Estimation Scheme

According to the ME process, the main challenges for an efficient fully parallel ME implementation are the data dependencies in the following:

- Integer search starting position: The selection of a good motion vector search window needs to use the vectors computed in the neighboring blocks as prediction, a fact that introduces a dependency that harms the parallelization.
- Rate distortion optimization (RDO): The MV RD cost computation suffers the same problem of MV prediction. To compute the bit cost of a candidate vector, the final motion vectors of neighboring blocks must be known beforehand.

- Reference picture reconstruction: block motion estimation search process is done in the reference reconstructed pictures, so these pictures should be reconstructed first. It is not a problem in serial processing, but it will restrict the degree of parallel processing.

In our scheme, we don't need GPU to finish a complete motion estimation to get the best reference picture and motion vector. We only target at performing a fast determination of the best reference picture. To remove these dependencies for GPU to achieve the maximum parallel degree, in our proposed scheme, GPU performs a coarse full integer motion estimation for approximation.

For simplicity, we define the motion vector prediction as (0, 0), which means that the motion search starting position is just the collocated position in the reference picture. For RD cost computation, it is not accurate without real MVP information. So we directly use the SAD as the cost to find a best integer motion vector to further accelerate the GPU computation. The last dependency is the reconstruction of reference picture. To make the GPU search independent of the CPU encoding process, we use the original video pictures to approximate the whole motion estimation in GPU, which means the whole GPU motion estimation search process is performed in the original pictures with the same relationship of picture reference. In this way, GPU motion estimation process is totally independent from the CPU encoding in terms of data dependency.

3.2 Motion Estimation Implementations in GPU

NVIDIA Corp. proposes a powerful GPU architecture called Compute Unified Device Architecture (CUDA) [13]. CUDA is basically a Single Instruction Multiple Data (SIMD) computing device. A GPU is built around an array of Streaming Multiprocessors (SMs) consisting of multiple CUDA cores or threads in physical architecture. In the software level, physical CUDA threads can be allocated with scalability to different thread blocks and the threads of a thread block execute the same instruction with multiple data concurrently in one thread block. However, to achieve the best GPU branching performance, we set only one thread block in one SM, which is faithful to its physical architecture.

Suppose the GPU has M SMs with T CUDA threads in one SM. We set it to M thread blocks consisting of T threads for each. An inter picture to be encoded is divided into N CTUs of size 64x64. Then we perform the ME process for M CTUs in M thread blocks in parallel until all the N CTUs are done. For each CTU ME process, we use T threads to do SAD computation for all possible PUs sized from the smallest 8x4 and 4x8 to the largest 64x64, which are listed in Table 1. There are 593 possible PUs (no 4x4) in HEVC in one 64x64 CTU in total. Because the ME process in each search position are independent in the full search, if the thread number T are large enough, we can allocate them to search different positions to further speedup the process. Let t be the thread number to compute the one search position SAD cost for different PUs, then the number of n=T/t positions are searched in parallel.

To make a better use of GPU computing resources to achieve a better performance, some details in implementation must be noticed. First, as shown in Table 1, most of the computation loads are in the first 6 steps, in which the number of PUs to be processed are the usually the multiples of 32. It prefers that the thread number t is also the multiples of 32 to achieve the best GPU performance. Second, all sizes of PU consist of several blocks of 4x4, to avoid multiple integer search and SAD computation, we can only searches SAD for all the 4x4 blocks, and then the SAD costs for larger PUs can be derived from the 4x4 SAD results by combining two smaller PU SADs step by step as described in Table 1. Note that *Ceil(x)* denotes the smallest integer that is greater than or equal to *x*.

The detailed procedures of motion estimation process in GPU are as follows:

1. One CPU thread reads one picture of input video sequence and determine the picture type to be intra picture or inter picture. If the current picture is inter picture, the CPU thread will send the original pixels of the current picture and its relative reference pictures as well as some basic picture parameters to the GPU global memory. The basic picture parameters include CTU number, picture size and reference picture number.
2. For one new inter picture, GPU receives the original current picture data, all reference pictures original data and picture parameters.
3. M thread blocks process M CTUs in parallel. When one thread block finishes its job, it will automatically process the following CTUs in raster scan order.
4. Inside a thread block, T CUDA threads are used to perform motion estimation for one CTU as steps (1) to (3). The GPU will go through all the reference pictures and store the best one set of 593 motion vector candidates for each reference picture of each possible PU in the current CTU to GPU global memory.
5. When one picture finishes the whole motion estimation search for all CTUs in the picture, GPU sends the best motion vectors and their reference picture indexes stored in GPU global memory back to the CPU memory once. In this way, frequent communications between GPU and CPU can be avoided which leads to a negligible communication time.
6. Process for every inter picture of the input video sequence in GPU as the above steps. Note that the CPU memory is limited, we can set a fixed number of data buffer to receive the returned data from GPU. If the buffer is full, GPU will receive a message from the CPU thread and wait for available data buffer.

The algorithm for one CTU ME process in GPU has the following steps:

(1) Each thread loads a part of the CTU (4096/T pixels) in the current picture to shared memory.
(2) Load search window pixels in reference picture from GPU global memory to shared memory and check whether the PU SAD is better than the current best for all possible PUs in the CTU, following the steps (a) to (d).
(a) T threads in one thread block are allocated to check ME cost for T/t different search MV positions in parallel with t threads for each position;
(b) Compute all the 4x4 SADs in one CTU for the position in 256/t loops;

Table 1. Steps to get SADs of PUs in one CTU for one search position

Steps	PU	SAD From	PU Number	Loop Number
Step 1	4x4	GPU SAD search	256	ceil(256/t)
	8x4	4x4 (2)	128	ceil(256/t)
Step 2	4x8	4x4 (2)	128	
	8x8	8x4 (2)	64	
Step 3	16x4	8x4 (2)	32	ceil(128/t)
	4x16	4x8 (2)	32	
Step 4	16x8	8x8 (2)	32	ceil(64/t)
	8x16	8x8 (2)	32	
	16x16	16x8 (2)	16	
Step 5	32x8	16x8 (2)	8	ceil(32/t)
	8x32	8x16 (2)	8	
	16x12	16x16 - 16x4	32	
	12x16	16x16 - 4x16	32	
Step 6	32x16	16x16 (2)	8	ceil(84/t)
	16x32	16x16 (2)	8	
	32x32	16x16 (2)	4	
	32x24	32x32 - 32x8	8	
	24x32	32x32 - 8x32	8	
	64x32	32x32 (2)	2	
Step 7	32x64	32x32 (2)	2	ceil(25/t)
	64x64	32x32 (4)	1	
	64x16	32x16 (2)	2	
	16x64	16x32 (2)	2	
Step 8	64x48	64x64 - 64x16	2	ceil(4/t)
	48x64	64x64 - 16x64	2	

(c) Additional 7 steps to compute the larger PU SADs with smaller PU SAD combinations as in Table 1;

(d) Use all T threads in the thread block to check whether the new derived T/t set of 593 SAD costs are lower than previous best one and update the current best motion vector candidate in the shard memory.

(3) Perform full integer search in the collocated search window in the current reference picture for the current CTU. Get the best motion vector candidate for all 593 PUs in the current reference picture.

Note that when the search range is –R~R, it means that the search window is of (2R+1)x(2R+1). There are (2R+1)x(2R+1) search positions for integer ME. As the maximum PU size is 64x64, the resulting related reference pixel window should be (2R+64)x(2R+64). Limited by the size of the shared memory, it may not be able to load the reference pixel window to the shared memory at once. Thus, multiple times of reference pixel data loading are needed to finish the whole ME search inside the area for one CTU.

3.3 Fast Reference Picture Selection with Motion Vector Cost Refinement in CPU

In GPU motion estimation, the estimated best motion vector selected for each reference picture are returned to CPU. Due to the absence of MVP information in GPU, the cost in GPU is not accurate enough to do reference picture RDO. So the final determination step is operated in CPU. For the CPU side, the whole procedures are in the following steps:

1. Two CPU threads are generated before the whole encoding process, one for the communication with GPU and the other for the normal HEVC encoding.
2. For the first thread, it helps GPU to read original video pictures and determine the picture type according to the encoding configure, the same process as in the normal HEVC encoding thread does. It also sends required data to GPU to do parallel motion estimation as represented in section 3.2 and receives the best motion vector data for every reference picture. As for the encoding thread, it performs the normal encoding process for intra picture, and when it forwards to the inter picture motion estimation module of the first reference picture, it will wait until the best motion vector data of each reference picture have been returned from GPU.
3. When the data are returned for the current ME PU, the encoding thread will do the AMVP process to get accurate MVP information for each reference picture and an integer motion compensation to get the accurate SAD distortion. Combined with the returned best motion vector, the cost as in (1) for each best motion vector candidate can be computed and the motion vector with the minimum cost is selected with its reference picture as the best reference picture for current PU.
4. In the uni-direction ME search, only the best reference picture selected will do the complete motion estimation and motion compensation in CPU to get the best motion vector in quarter pixel accuracy, which saves time in multiple times of complex motion estimation in series for CPU.

4 Experiment Results

The proposed scheme are implemented based on HM 10.0 [2]. Class A (4K, four sequences: Traffic, PeopleOnStreet, Nebuta and SteamLocomotive), Class B (1080p, five sequences: Kimono, ParkScene, Cactus, BasketballDrive and BQTerrace) and Class C (WVGA, four sequences: BasketballDrill, BQMall, PartyScene and RaceHorses) sequences are tested. As the proposed scheme mainly deals with the fast reference picture selection problem in uni-direction motion search. To clearly verify the effectiveness of the scheme, the inter configuration lowdelay P (IPPP coding structure) of main profile are tested under HEVC common test condition [6]. The frame number to be encoded is set to the first one second length of video frames with the frame rate defined in HEVC common test condition. In other words, the test frame number is 24 for 24fps, 32 for 30fps, 48 for 50fps, and 64 for 60fps for lowdelay P configuration for the complete GOP(GOP size = 4). The results are measured in terms of combined BD rate [14], where positive number means bitrate increase (coding performance loss). The experiments are carried on Windows 7 64 bit OS platform with Intel Xeon E5440 @ 2.83 GHz CPU and 8 GB RAM and NVIDIA GeForce GTX 680 GPU.

Note that NVIDIA GeForce GTX 680 has 8 Streaming Multiprocessors (SM) with 192 CUDA cores or threads per Streaming Multiprocessor [15]. And the ME search range in the test is the default range -64~64. So the parameters mapped to Section 3.2

are M=8, T=192 and R=64. Limited by the only 49152 bytes of shared memory per thread block in GTX 680 GPU, the reference pixel window sized of 192x192 are loaded for 8 times with each time a window of 192x80 for the 129x16 search positions. And we set t=64 to use 64 threads to compute one position SAD cost.

Besides the HM10.0 anchor, HM10.0 with only 1 reference (the nearest reference) is also tested as a comparison. The detailed results are shown in Table 2. It can be seen from the table that for the IPPP coding structure, performance loss of only 1 reference is reduced from average 8.1% to only 2.0% with our proposed scheme, which can be compensated quite well for most of the test sequence, especially for larger sizes of video sequences. The RD curves for some sequences are illustrated in Fig.2. From those curves, we can also see that the RD curve of proposed full scheme is quite close to the HM anchor. But there are still some sequences suffer an obvious performance loss, like BQTerrace in class B and PartyScene in class C, especially for the high bitrate when more accurate motion vector plays an important role in performance gain. The two sequences reminds us that the relative coarse integer motion estimation in GPU in our proposed scheme cannot handle all the cases well in reference picture selection.

The encoding time comparison are shown in Table 3 for all the tested sequences. We see that our proposed fast reference picture scheme achieves average 34.4% time reduction than HM anchor, only 3.8% in average less than the HM anchor with only 1 reference, which has huge performance loss. It means that our proposed scheme is able to maintain the high coding performance of HEVC with much less time cost in reference picture selection, which achieves a good balance in encoding time and performance.

Table 2. Coding performance comparison with HEVC Anchor

		HM Anchor_1Ref			Proposed		
Seq Class	Sequences	Y-BD rate	U-BD rate	V-BD rate	Y-BD rate	U-BD rate	V-BD rate
Class A 4K	Traffic	6.7%	7.8%	8.6%	2.4%	1.6%	1.1%
	PeopleOnStreet	2.1%	5.0%	5.0%	1.6%	2.6%	2.2%
	Nebuta	3.6%	16.2%	14.9%	0.7%	1.6%	1.5%
	SteamLocomotive	9.7%	28.8%	27.7%	1.9%	3.2%	2.7%
Class B 1080P	Kimono	0.4%	0.9%	0.8%	0.2%	-0.1%	0.2%
	ParkScene	3.8%	6.2%	7.7%	1.4%	0.8%	0.7%
	Cactus	2.5%	3.4%	3.1%	1.4%	1.2%	0.7%
	BasketballDrive	5.0%	7.5%	6.0%	1.8%	2.4%	2.0%
	BQTerrace	35.3%	57.2%	68.5%	4.4%	2.8%	3.0%
Class C WVGA	BasketballDrill	5.6%	5.9%	5.3%	1.8%	0.4%	1.4%
	BQMall	5.2%	6.0%	5.3%	2.0%	1.7%	1.9%
	PartyScene	19.8%	20.3%	21.8%	5.1%	4.2%	4.6%
	RaceHorses	5.4%	6.4%	6.1%	1.8%	2.2%	1.7%
Average	Class A	5.5%	14.5%	14.0%	1.6%	2.2%	1.9%
	Class B	9.4%	15.0%	17.2%	1.8%	1.4%	1.3%
	Class C	9.0%	9.7%	9.6%	2.7%	2.2%	2.4%
	All	8.1%	13.2%	13.9%	2.0%	1.9%	1.8%

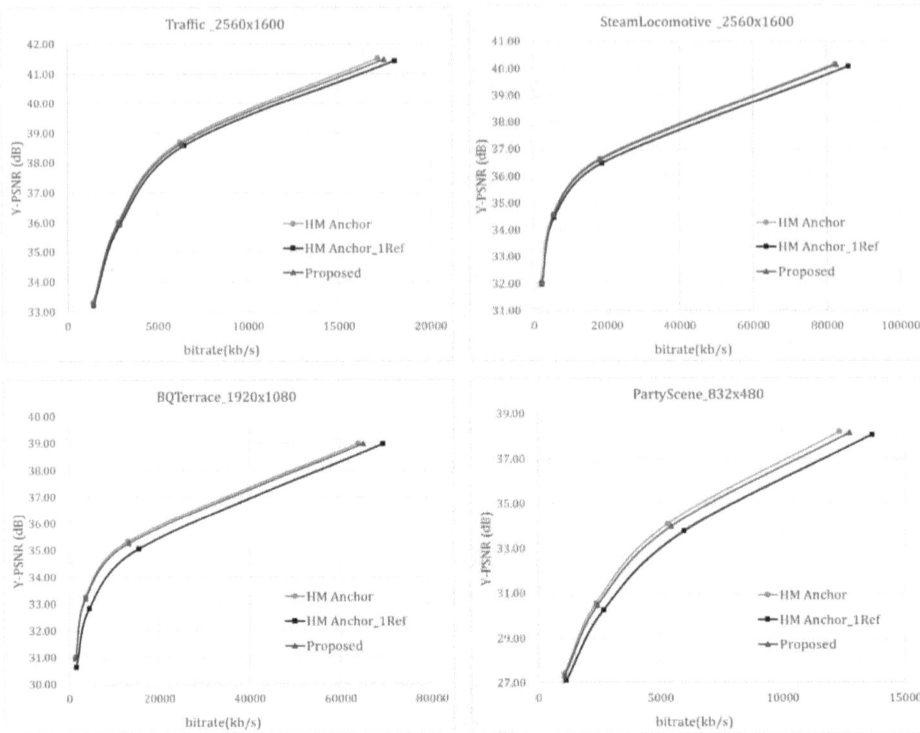

Fig. 2. R-D curves of the lowdelay P coding structure

Table 3. Encoding time reduction comparison with HEVC Anchor

Seq Class	Sequences	HM Anchor_1Ref Encoding Time Reduction(%)	Proposed Encoding time reduction(%)
Class A 4K	Traffic	38.1%	33.8%
	PeopleOnStreet	42.3%	38.4%
	Nebuta	28.9%	25.7%
	SteamLocomotive	37.6%	33.5%
Class B 1080P	Kimono	43.3%	38.5%
	ParkScene	38.0%	32.9%
	Cactus	39.4%	35.3%
	BasketballDrive	45.5%	41.1%
	BQTerrace	32.3%	29.8%
Class C WVGA	BasketballDrill	39.7%	36.2%
	BQMall	38.5%	34.8%
	PartyScene	31.0%	28.8%
	RaceHorses	41.4%	38.0%
Average	Class A	36.8%	32.9%
	Class B	39.7%	35.5%
	Class C	37.7%	34.5%
	All	38.2%	34.4%

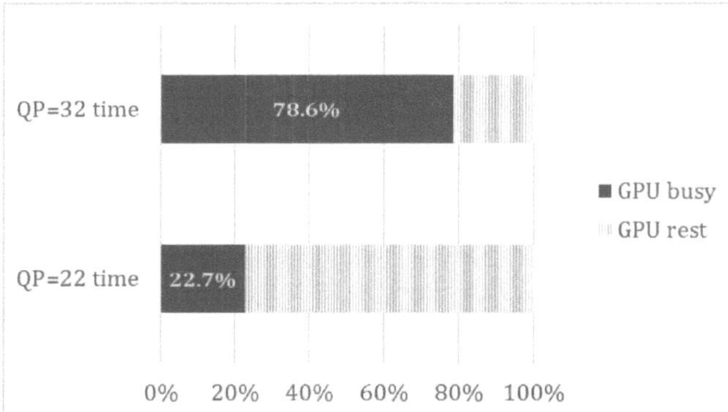

Fig. 3. GPU running time percentage in the whole encoding process of SteamLocomotive at different QPs

To know the GPU performance in our proposed scheme, we also show the running time of GPU percentage compared with the whole encoding time in Fig.3 for different QPs with sequence SteamLocomotive as an example. Note that the running time in GPU excludes the waiting time for available CPU data buffer, it is the running time with effective operations. It shows that GPU running time is shorter than the whole encoding time, especially for smaller QP with higher bitrate. It inspires us that there is a lot of room to improve the proposed scheme if we can make a better use of the remaining GPU computing ability.

5 Conclusions and Future Work

In this paper, we propose a fast reference picture selection scheme for HEVC encoder. The proposed scheme utilizes GPU to fulfill the complex and time consuming motion estimation module in HEVC encoding process and helps CPU to do a fast determination of the best reference picture. The proposed scheme can achieve a huge 34.4% encoding time reduction while the only 2% coding performance loss compared to the HEVC anchor. For those high definition video sequences, whose encoding time are now too long to apply in practice, our scheme may get a better balance in time and performance.

It is also one exploration to one interesting problem, how will the emerging parallel computing platform, e.g. the more and more powerful GPU, help the CPU in one local machine to accelerate the sophisticated HEVC encoding. This is also a very important problem for the practical application of HEVC standard to in our daily life and still worth further investigating. From the experiment results, more gains may be achieved if a more accurate motion estimation search can be implemented in GPU with the rest available GPU computing resources.

References

1. Sullivan, G.J., Ohm, J.-R., Han, W.-J., Wiegand, T.: Overview of the High Efficiency Video Coding (HEVC) standard. IEEE Trans. Circuits Syst. Video Technol. 22(12), 1648–1667 (2012)
2. HM 10.0. HEVC Test Model, https://hevc.hhi.fraunhofer.de/svn/svn-HEVCSoftware/
3. ITU-T and ISO/IEC JTC 1/SC 29/WG 11 (MPEG), H.265: High Efficiency Video Coding (2013), http://www.itu.int/rec/T-REC-H.265-201304-I/
4. Correa, G., Assuncao, P., Agostini, L., da Silva Cruz, L.A.: Performance and computational complexity assessment of High-Efficiency Video Encoders. IEEE Trans. Circuits Syst. Video Technol. 22(12), 1899–1909 (2012)
5. Wiegand, T., Sullivan, G.J., Bjontegaard, G., Luthra, A.: Overview of the H.264/AVC video coding standard. IEEE Trans. Circuits Syst. Video Tech. 13(7), 560–576 (2003)
6. Bossen, F.: Common test conditions and software reference configurations. Document JCTVC-L1100 of JCT-VC, Geneva (2013)
7. Li, H., Li, B., Xu, J.: Rate-Distortion optimized reference picture management for High Efficiency Video Coding. IEEE Trans. Circuits Syst. Video Technol. 22(12), 1844–1857 (2012)
8. Swaroop, K.S., Rao, K.: Performance analysis and comparison of JM 15.1 and Intel IPP H.264 encoder and decoder. In: IEEE Southeastern Symposium on System Theory (SSST) 2010, pp. 371–375 (2010)
9. Kung, M., Au, O.C., Wong, P., Liu, C.H.: Block based parallel motion estimation using programmable graphics hardware. In: International Conference on Audio, Language and Image Processing (ICALIP), pp. 599–603 (2008)
10. Rodríguez-Sánchez, R., Martínez, J.L., Fernández-Escribano, G., Sánchez, J.L., Claver, J.M.: A fast GPU-based motion estimation algorithm for H.264/AVC. In: Schoeffmann, K., Merialdo, B., Hauptmann, A.G., Ngo, C.-W., Andreopoulos, Y., Breiteneder, C. (eds.) MMM 2012. LNCS, vol. 7131, pp. 551–562. Springer, Heidelberg (2012)
11. Purnachand, N., Alves, L.N., Navarro, A.: Fast motion estimation algorithm for HEVC. In: IEEE International Conference on Consumer Electronics - Berlin (ICCE-Berlin), pp. 34–37 (2012)
12. Purnachand, N., Alves, L.N., Navarro, A.: Improvements to TZ search motion estimation algorithm for multiview video coding. In: IEEE International Conference on Systems, Signals and Image Processing (IWSSIP), pp. 388–391 (2012)
13. NVIDIA, CUDA C Programming Guide, PG-02829-001_v5.0. NVIDIA (2012), http://docs.nvidia.com/cuda/pdf/CUDA_C_Programming_Guide.pdf
14. Bjontegaard, G.: Calculation of average PSNR differences between RD-Curves. Document VCEG-M33, Austin (2001)
15. NVIDIA, NVIDIA GeForce GTX 680 Whitepaper, http://international.download.nvidia.com/webassets/en_US/pdf/GeForce-GTX-680-Whitepaper-FINAL.pdf

A Real-Time Mesh Animation Framework Using Kinect

Yirui Wu[1], Tong Lu[1,*], and Jiqiang Song[2]

[1] National Key Lab for Novel Software Technology, Nanjing University, Nanjing, 210046, China
[2] Intel Labs China, 8F, Raycom Infotech Park A, Haidian, Beijing, 100190, China
wuyirui1989@163.com, lutong@nju.edu.cn, jiqiang.song@intel.com

Abstract. This paper proposes a real-time mesh animation framework to animate any kind of virtual characters by using Kinect. A deformation module is first predefined by constructing propagation fields and accordingly the isolines in each field for a character. During animation, motions captured by Kinect are retargeted to movements of users' specified handle constraints, which induce the deformation module to produce natural animation results. Our interactive framework is user-friendly and featured by real-time response for user motions and intuitive manipulations. The experimental animation results and the user study on typical characters demonstrate the effectiveness of our framework.

Keywords: Character animations, Kinect-based motion recognition, Surface-based deformation, Real-time interaction.

1 Introduction

Computer animation is one of the most active areas in computer graphics and multi-media modeling, aiming at producing visually desirable animation results under user specified constraints [1,2]. It has been widely applied in many applications such as art modeling, movie making, game development and virtual reality.

Most existing 3D animation systems [3] employ simplified skeleton representations of characters. The position of a segment on a skeletal model is generally defined by animation variables, named as Avars. By changing values of Avars over time, an animation system performs skeleton-based deformations [4] to compute exact positions and orientations of each mesh frame by frame. However, skeleton-based animation potentially causes many artifacts since the influence of each bone on surface is often defined independently with the local geometry of a mesh. Moreover, rigging numerous controllers on a skeleton model for properly reacting with changed Avars can be quite annoying and inefficient. Instead, mesh-based animation systems [5,6] adopt surface-based deformation algorithms [7,8] to perform the animation due to their capability of preserving geometry details under users' intuitive controls. Unfortunately, most mesh-based animation systems suffer from high computation cost for interactive animation purpose.

* Corresponding author.

B. Huet et al. (Eds.): PCM 2013, LNCS 8294, pp. 245–256, 2013.

This paper proposes a novel mesh-based 3D animation framework, which integrates a Kinect-based motion recognition module and a pre-modeled surface-based 3D deformation module, to perform animation of any virtual character through real-time interactions. The motion recognition module first captures animators' joint motions using Kinect and then transforms joint motions to movements of user-specified handles according to user-customized mapping rules. The deformation module finally produces natural and desirable animation results in real-time by instancing with handle constraints and considering handle movements as inputs.

The main contributions of this paper are:

- The proposed framework only requires a small number of controllers and enables real-time interactivity in most procedures, which results in an efficient and user-friendly animation system, especially for non-professional users.
- According to users' customized mapping rules, any kind of virtual characters can be properly animated to match users' expectation.
- The use of Kinect reduces the system cost, while improving robustness and expansibility. Therefore, the proposed framework can be easily applied to various design scenarios.

The rest of the paper is organized as follows. Section 2 reviews the related work. The proposed animation framework is described in detail in Section 3. Section 4 reports the experiments and discusses the results, and finally Section 5 concludes the paper.

2 Related Work

Animation Methods. There are two approaches in skeleton-based animation for generating Avars, keyframing [11] and motion capture [12]. The former requires animators to set Avars at pre-defined keyframes, and then all intermediate frames are automatically computed by interpolation of keyframes. Unfortunately, complex actions are generally difficult and time-consuming to produce by keyframing methods. The latter makes use of live actions by capturing animator's motions with numerous cameras and sensors, and then motions are applied to animate virtual character directly. Motion capture can produce accurate, fast and smooth movements, which are appropriate for design scenarios that require realistic behaviors and actions. However, such an animation system always requires professional manipulators and expensive hardware.

Former mesh-based animation systems [5,6] perform animation of high-quality surface characters from laser-scanned motions or videos. They generally transform the input motion description into a moving template model, i.e. a mesh or a point cloud. Then the animator specifies a set of markers on the character to map the motion. Finally, their systems automatically generate motion retargeting of characters incorporating with various kinds of surface-based deformation algorithms [7,8]. However, these systems cannot facilitate interactive animation due to their high computation cost and difficult manipulation.

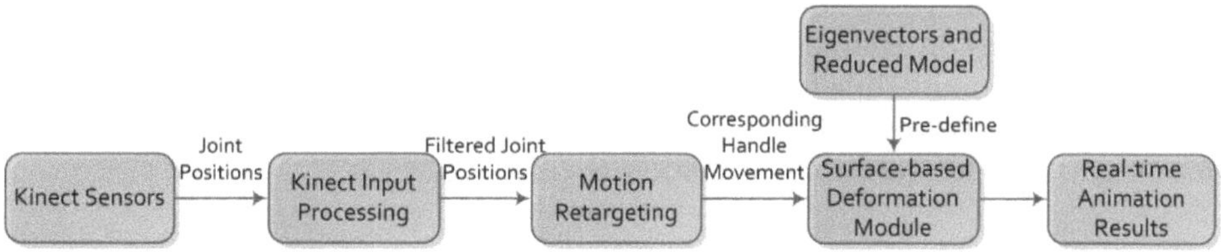

Fig. 1. The run-time work flow of our animation framework

Surface-Based Deformation Methods. In differential mesh editing [9,10], the deformation process is formulated as a global variational minimization problem to preserve local shape characteristics. Laplacian coordinates [13] are often adopted to describe these characteristics and should be properly transformed to match the deformed mesh. Otherwise, distortion like shearing and stretching will occur. After users specify handle constraints, the minimization problem is equivalent to solving the linear system in a least-squares sense:

$$ADG = b(DG), \qquad with$$

$$A = \begin{bmatrix} L \\ \Phi \end{bmatrix} \qquad and \qquad b(DG) = \begin{bmatrix} \sigma(DG) \\ C \end{bmatrix} \tag{1}$$

where L is Laplacian matrix constructed from the original mesh before editing, D is the matrix representing transforming process, G is the transformation set, $\sigma(DG)$ is Laplacian coordinates which are nonlinearly dependent on the deformed vertex positions $X = DG$, and $\Phi G = C$ indicates the constraints of handle positions.

The handle-aware harmonic field method [9] computes transformations through solving a linear system defined by a discrete Laplacian operator with handle positions as its constraints. After sampling the transformations to obtain isoline sets as a reduced model, it updates reduced transformation domain in an iterative Laplacian editing framework [10,14], which achieves fast convergence, low per-iteration cost and stable deformation results. However, it requires a relatively long pre-computing time and high computation cost related to handle numbers if accepted for animation purpose.

3 Our Approach

Fig. 1 shows the run-time work flow of our proposed mesh-based animation framework. Our framework predefines a relatively independent deformation module based on the first few eigenvectors of Laplacian Matrix and the isoline-based reduced model before animation. During animation, we firstly use Microsoft Kinect as the source of Avars and properly filter the captured joint motions to stabilize them. Next, joint motions are transformed to movements of user-specified handles on character according to user-customized mapping rules. Finally, deformation module is instanced by handle constraints and computes exact position of each vertex induced by handles movements. After rendering, we will obtain natural and desirable animation results.

3.1 Predefined Surface-Based Deformation Module

In this subsection, we propose a relatively independent module to pre-model deformation propagation before animation, which consists of two steps, *construct control variables* and *construct interpolation scheme.*

We first compute eigenvectors by eigen-decomposition of Laplacian matrix. Since the first eigenvector is a constant vector, we abandon it and sort the rest eigenvectors in ascending order of the corresponding eigenvalues. In fact, since a Laplacian matrix is mainly dominated by its low-frequency components [17], we can approximately describe a mesh by the first few eigenvectors of Laplacian Matrix. We thus select a fixed set of eigenvectors (vector number $n_e = 8$) as deformation propagation fields $\{e_0, e_1 ... e_7\}$, which describe the global geometric and topological information [15,16]. Then, we adopt an isoline-based reduced model [9] to construct isolines in each field (isoline number $n_{iso} = 10$), which serve as the generalized bones to imitate deformation propagation and will reduce run-time computation complexity. During animation, the transformation assigned to each isoline is regarded as *control variables* of each propagation field.

After constructing control variables, we adopt the same local interpolation scheme as Au. et al [9] to linearly compute transformation of each vertex from control variables in each propagation field. Furthermore, we define a high-level interpolation scheme to integrate transformation of different propagation fields in following three steps. First, we compute a weight factor for each propagation field by projecting one component of the original mesh coordinates to e_i:

$$\xi_{X,i} =< X, e_i >= \sum_{i=1}^{i=m} X \cdot e_i \tag{2}$$

where X represents the x, y, z coordinates of a mesh vertex and $\xi_{X,i}$ denotes the coefficients, which encode how propagation fields can reconstruct the original mesh geometry and more precisely the importance of propagation fields in representing the original shape. Second, we define the interpolation weight for each propagation field as follows:

$$w_i = \sqrt{\xi_{x,i}^2 + \xi_{y,i}^2 + \xi_{z,i}^2} \tag{3}$$

Finally, we normalize the weight set $\{w_i\}$ to make them sum up to 1 for the constancy of volume. The local interpolation scheme and high-level weight set $\{w_i\}$ make up a two-level *interpolation scheme* that provides parameterization for the original mesh space.

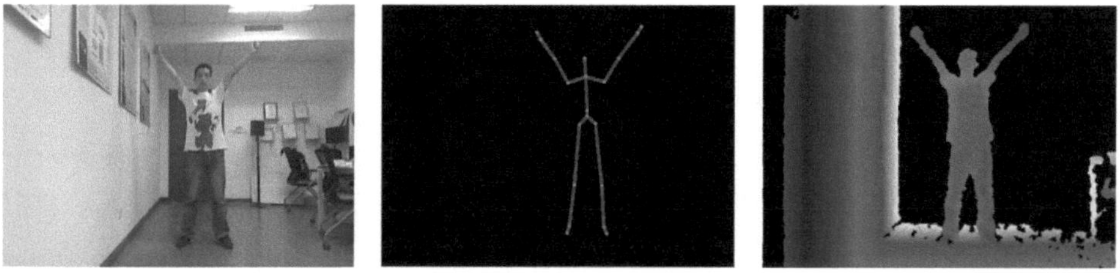

Fig. 2. From left to right: an animator captured by Kinect cameras, the extracted skeleton and the depth image which is transformed to HSL Space for a better view

We can represent the interpolation from the control variable to the vertex as $X = WR$, where R is a column vector constructed from isoline transformation set and W is a matrix constructed from $\{X\}$ and $\{e_i\}$, leading Eq. 1 to be updated with $D = W$ and $G = R$. Since W and L can be pre-computed before animation, we regard it as a relatively independent deformation module to describe the inherent constraints of deformation propagation.

3.2 Kinect Input Processing

Fig. 2 shows examples of a color frame, the extracted skeleton and the depth image captured by Kinect. However, the positions tracked by Kinect may be unstable and accordingly lead to artifacts. To avoid extreme situations and stabilize joint movements, we employ the following filter to process joint positions.

Joint Position Double Exponential Filter: This filter applies Holt-Winters double exponential smoothing [19] to joint positions captured by Kinect with the assumption that there would be a trend in the captured data:

$$\begin{cases} \tilde{p^1} = p^1 & t = 1 \\ b^1 = p^1 - p^0 & t = 1 \\ \tilde{p}^t = \beta p^t + (1-\beta)(p^{t-1} + b^{t-1}) & t > 1 \\ b^t = \gamma(\tilde{p}^t - \tilde{p}^{t-1}) + (1-\gamma)b^{t-1} & t > 1 \end{cases} \tag{4}$$

where p^t represents the original position of one specified joint at time t, b^t represents our estimate of the trend at time t, β is the smoothing factor and γ is the trend smoothing factor. Essentially, this filter adopts the previous historical locations to smooth the current locations. By adopting this filter, our system can remove jitters or noises and predict new joint positions to reduce lags, which leads to stable and desirable animation results.

Furthermore, we transform filtered joint positions to positions of application form based on the proportion between their sizes:

$$\tilde{p}^t = \frac{p^t * w^t}{u} - \frac{r}{2} \tag{5}$$

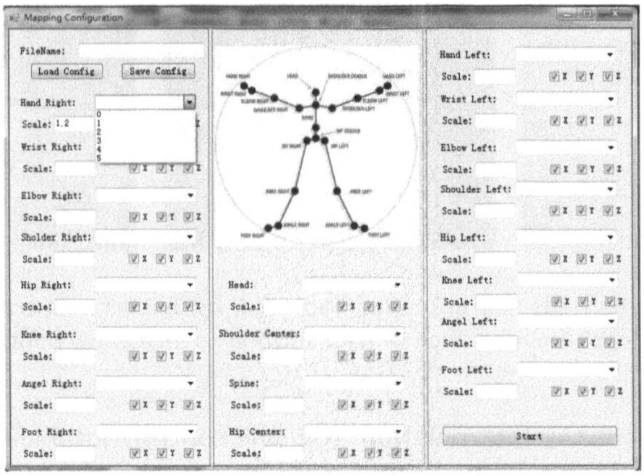

Fig. 3. The configuration panel that allows users to customize the mapping. Joints representing animator's skeleton are demonstrated in the middle to visually assist users.

where p^t represents x or y coordinate of filtered joint positions at time t, w^t and u respectively refer to the width or height of application form and captured frame, and r means the radius of handle point. Note that we accept the mean value of width and height as w^t for z coordinate.

3.3 Motion Retargeting to Handle Movement

In this subsection, we illuminate how to transform the filtered joint positions to the movements of user-specified handles based on user-customized mapping and animation parameters.

Users can intuitively manipulate handles on a mesh for animation. We thus generate a set of handle positions $\{h_0, h_1...h_{m-1}\}$ with handle number m. Considering that there is actually no exact correlations between parts of non-humanoid characters and animators' recognized joints, we allow users to customize the mapping between them in a configuration panel, which is shown in Fig. 3. In particular, the configuration incorporates the possible options related to the handle set $\{h_i\}$ and the joint set $\{j_k\}$. Kinect can recognize up to 20 joints demonstrated in the middle of the panel. Therefore, users' mapping can be formulated as a mapping matrix $M_{m\times 20}$, where the entry from ith row and kth column is set to one only if user defines the mapping between handle h_i and joint j_k, otherwise it's zero.

Furthermore, the configuration panel allows users to customize two parameters to scale joint motion $d_k^t = p_k^t - p_k^{t-1}$, the minus of joint positions between neighbor frames. Specifically, we can discard its x, y or z coordinate when the corresponding checkbox in the panel is unchecked. Therefore, we rewrite the joint movements as

$$d_k^t = f(p_k^t) - f(p_k^{t-1}) \tag{6}$$

where function f denotes the filter discussed in Section 3.2. We further compute the corresponding movement of user-assigned handle in the view of matrix:

$$H^t = M_{m\times 20} S^t_{20\times 3} \tag{7}$$

where S^t is constructed from the joint movements set $\{d^t_k\}$.

3.4 Real-Time Character Animation

Our character animation is based on the deformation module and the handle movements. When handles are specified by users, the deformation module will be instanced as Eq. 1. We solve the corresponding equation in a least square sense:

$$\begin{aligned} \tilde{R} &= (U^T U)^{-1}(W^T L^T \sigma(WR) + W^T \phi^T C) \\ X &= W\tilde{R} \end{aligned} \tag{8}$$

where $U = (W^T L^T \quad W^T \phi^T)^T$. Furthermore, we adopt an iterative Laplacian updating framework [9,10] to achieve the rotation invariant deformation by iteratively and alternatively updating R and X based on the current deformed surface. We also update laplacian coordinates $\sigma(WR^t)$ based on the rotation of triangles crossed by isolines in propagation fields. Therefore, Eq. 8 becomes:

$$\begin{aligned} R^{t+1} &= (U^T U)^{-1}(W^T L^T W_\delta R^t_\delta + W^T \phi^T C) \\ X^{t+1} &= WR^{t+1} \end{aligned} \tag{9}$$

where $\delta = W_\delta R_\delta$. Note that R_δ is a column vector constructed from the normal of the crossed triangles and W_δ is similar to former local interpolation scheme discussed in Section 3.1, i.e., linearly interpolating rotation for rest triangles. It is worth to recall that the constructed matrix W and L can be pre-computed before animation. So does the matrix W_δ. If we store constructed matrices $U^T U$ and $W^T L^T W_\delta$ at the first loading of any new model, and then reuse them for animation, our framework can allow instant animating of any pre-processed mesh.

In animation process, new handle positions are computed by $C^{t+1} = C + H^t$, involving corresponding joint motions to update the instanced deformation module. Since the definition of $\sigma(X)$ is separable in dimension of the vertices, the updating is performed separately for x, y, z coordinates of the deformed vertices. By applying isoline-reduced model, the original system matrix of size $n_{ver} \times n_{ver}$ is reduced to $U^T U$ of size $4n_{iso}n_e \times 4n_{iso}n_e$ after model projection from the original space to the isoline-based reduced space, speeding up the per-iteration updating significantly.

4 Results and Discussion

We animate a virtual character, Armadillo model, to present the effectiveness of our animation framework. Fig. 4 shows both the Kinect captured color frames

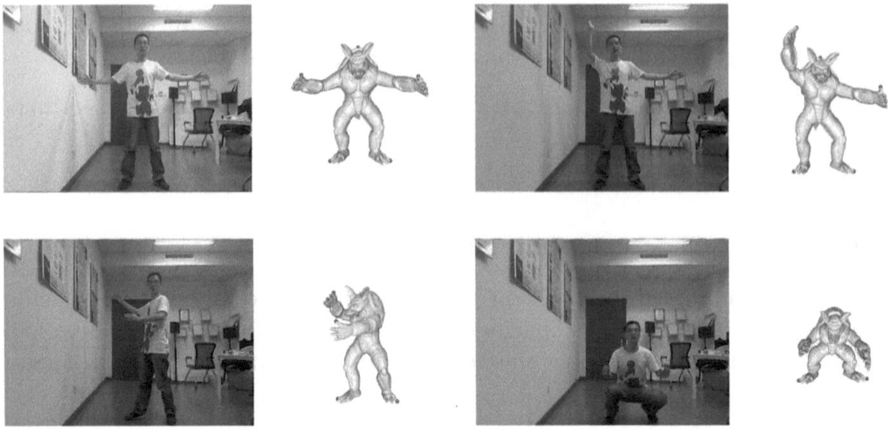

Fig. 4. Animation results of Armadillo model. Left: Kinect captured color frames. Right: Corresponding animation result.

and the corresponding animation frames created by our system. Note that high-frequency details are well preserved by Laplacian coordinates and deformation propagation on the surface is natural due to our well-constructed transformation matrix. In general, our system can capture 30 frames per second which ensures most captured human motions to be reflected in animation.

Table. 1 gives the detailed statistics on character and time performance of our animation system, measured on a 1.7GHz i5 core2 PC with 6GB of RAM. Thanks to the strategy of reusing transformation matrix as discussed in Section 3.4, we can reduce the time of pre-computing deformation module t_1 to a shorter time $\tilde{t_1}$, which is mainly about loading corresponding files and rendering character. We also achieve a rather low computation time $t_{2,3}$ for input processing and retargeting procedure. From Eq. 9, we can see per-iteration updating cost t_4 largely depends on the size of R_δ^t, which is related with the number n_{tri} of the isolines-crossed triangles and is often a relatively small number, resulting in a short enough per-iteration computing time to guarantee real-time response to 30Hz animator's motion input. Moreover, n_{tri} can be a fixed number for one specific model, which leads to a constant per-iteration time in spite of different number of handles and mesh vertices and thus relieves users' concern on causing not real-time response by defining a larger number of handles for animation. Therefore, our animation system achieves nearly instant system computation time $\tilde{t_1}$ and low per-iteration cost $t_{per} = t_{2,3} + t_4$, making the whole system efficient and user-friendly than the existing mesh-based animation systems.

4.1 User Study

We conducted user studies to evaluate the effectiveness of our framework. We observed ten novice animators (male and female, 10-30 years old), equally divided into two groups, Group One is familiar with 3D applications and Group Two

Table 1. Subscripts 1,2,3,4 respectively show time performance of each procedure of our animation system respected to different characters and handle numbers. $\tilde{t_1}(s)$ represents the computing time of the first procedure by reusing transformation matrices, P refers to the proportion of $\tilde{t_1}$ in the original t_1 and t_{per} represents per-iteration computing time

Model	n_{ver}	n_{han}	n_{tri}	$t_1(s)$	$\tilde{t_1}(s)$	$P(\%)$	$t_{2,3}(ms)$	$t_4(ms)$	$t_{per}(ms)$
Dinosaur	14000	6	841	2.917	**0.6132**	**21.02**	0.2353	**9.873**	**10.11**
Dinosaur	14000	10	841	2.915	**0.6411**	**21.99**	0.2185	**9.722**	**9.941**
Armadillo	60000	6	810	15.21	**3.327**	**21.87**	0.2540	**31.91**	**32.16**
Armadillo	60000	10	810	15.22	**3.307**	**21.73**	0.3021	**31.78**	**32.08**
Pegasus	32046	4	837	7.494	**1.704**	**22.74**	0.2785	**18.50**	**18.78**
Dog	18114	8	817	3.611	**0.8725**	**24.16**	0.2203	**11.38**	**11.60**
Human	15154	9	842	3.067	**0.7757**	**25.29**	0.2145	**10.62**	**10.83**

is not, to produce their animations with our system. We began with explaining defining handles and mapping rules with Human character. Next, they were free to play with our system for 3 minutes. After that, they were asked to animate four characters, Armadillo, Dinosaur, Pegasus and Dog. It's worthy noticing that Pegasus is a high-genus character which can not be well animated by the existing skeleton-based systems and the former two characters are humanoid characters which can be animated intuitively. The poses we set for characters are different. Armadillo acts a Kung Fu pose, Dinosaur leans its body, Pegasus lowers down its head, and Dog barks. We show two animation results completed by testers in Fig. 5. Particularly, for the Pegasus character, a user defines its head to correspond to his left hand, its tail to his right hand and the other two handles to two feet. In this way, he can easily achieve the animation in a natural and intuitive manner.

Testers further give a score s ranging from 1 (minimum) to 10 (maximum) to evaluate the accuracy, effectiveness and robustness comparing with the existing keyframing approach, which has been demonstrated to testers with detailed steps by the SketchUp program [20] cooperated with the Ruby plugin [21] and is scored as 5 in all aspects. Table 2 demonstrates the average completion time, total trial times with respect to different groups and testers' subjective evaluation. Although the number of testers involved is relatively small, their objective statistics and subjective impressions are important to understand the current limitation of our system from user's perspective.

The average score of our system 6.475, revealing that the Kinect-based approach do help to improve the productivity during the animation phase from user's perspective. Compared to the long learning time for a special Keyframing program, our system offers a more user-friendly way to perform animation easier and faster with a short completion time t_{avg}. Moreover, it is troublesome for users to control tiny changes of geometric details in Keyframing, while our system provides well-preserved and properly-transformed details automatically. Particularly, our system receives more approval rating, for animating the humanoid characters, Armadillo and Dinosaur, since users find it intuitive to define

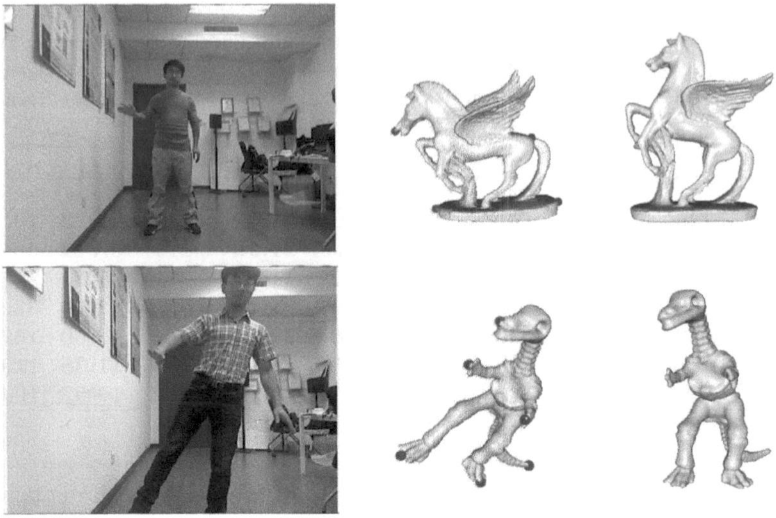

Fig. 5. *The first row* shows animation result of Pegasus character completed by a user from Group One; *The second row* shows result of Dinosaur character animated by a user from Group Two. We show original characters in the right of rows for comparison.

mapping and animation. This is reflected by the much shorter completion times of **1.606** and **1.732** min, and less trial times of **11** and **13** of these two characters. However, nine users try several times (total **35** times) for animating Pegasus, since most of them still apply the rules that map their heads to Pegasus's head, hands and feet to Pegasus's feet, which leads to undesirable animation results. Generally after four times of trials, we will give hints that he/she can control a specific part of a character with any joint and use a relatively small number of handles for animation, since most parts of characters can be animated by our system automatically. Testers can then complete animations following our given hints. Moreover, the tricks they learn from Pegasus can be quickly applied to Dog, which can be proved by a smaller trial times **16** and shorter completion time **2.522** min.

Table 2. Subscripts 1 and 2 show the objective performance in total trial times n and average completion time t (min) of Group One and Group Two. Pre-computing time of deformation module is excluded for fair comparison. We show subjective evaluations for animating compared with the Keyframing [20,21], which is scored as 5 in all aspects.

Model	n_1	n_2	n_{all}	t_1	t_2	t_{avg}	s_1	s_2	s_{avg}
Armadillo	5	6	11	1.476	1.736	1.606	6.5	7.5	7
Dinosaur	5	8	13	1.344	2.120	1.732	6.2	7.2	6.7
Pegasus	15	20	35	6.414	7.376	6.895	5.4	6.4	5.9
Dog	7	9	16	2.394	2.650	2.522	6.0	6.6	6.3

Our system is suitable for inexperienced users, since we find the trial times, completion time of Group One n_1, t_1 and Group Two n_2, t_2 are very close. We contribute this advantage to no further knowledge on 3D model and accordingly users can perform animation in an intuitive manner. For the robustness tested by multiple users, we generally obtain a lower score from Group One, which implies we need further improve our system to make it suitable for experienced users.

4.2 Implementation and Limitation

We use the skeleton and depth API provided by Kinect SDK to capture user motions. We adopt the cotangent weighting scheme to define the discrete Laplace operator, since it gives a better approximation to the normal of triangles. The solver of ARPACK is applied to accomplish the eigen-decomposition efficiently. We solve Eq. 9 by using Cholesky factorization of $U^T U$ for fast computing.

The use of only one Kinect sensor in the current solution could be an obstacle to the precision of our animation system. Particularly, occluded parts or clipped parts of the body can not be tracked accurately, which makes specific motion types hard to catch (e.g. a dancer spins around). Currently, an animator has to follow a few rules to successfully animate such a character.

5 Conclusion

This paper proposes a mesh-based framework for efficient 3D animation. It transforms Kinect captured motions to handle movements, and produces desirable animation results by processing handle movements in a surface-based deformation module. Our framework enables users to customize mapping rules between joints and handles to enable various animation styles. We believe our framework has a large potential for various design scenarios to reduce deployment cost and manipulation complexity, such as family movie-making systems and 3D games. Our further work includes motion recognition by using multiple Kinects.

Acknowledgments. The work described in this paper was supported by the Natural Science Foundation of China under Grant No. 61272218, the 973 Program of China under Grant No. 2010CB327903, and the Program for New Century Excellent Talents under NCET-11-0232.

References

1. Rick, P.: Computer animation - algorithms and techniques, 2nd edn. Morgan Kaufmann, New York (2007)
2. Robert, H., Ankit, G., Brian, C., Maneesh, A.: 3D puppetry: a kinect-based interface for 3D animation. In: UIST, pp. 423–434 (2012)
3. Andrea, S., Fabrizio, L., Gianluca, P., Felipe, D.R.: A kinect-based interface to animate virtual characters. Journal on Multimodal User Interfaces (2012)

4. Xiaohan, S., Kun, Z., Yiying, T., Mathieu, D., Hujun, B., Baining, G.: Mesh puppetry: cascading optimization of mesh deformation with inverse kinematics. ACM Trans. Graph. 26, 81–90 (2007)
5. de Edilson, A., Norimichi, U.: Representing and Manipulating Mesh-Based Character Animations. SIBGRAPI, 198–204 (2012)
6. de Edilson, A., Christian, T., Carsten, S., Hans-Peter, S.: Rapid Animation of Laser-scanned Humans. IEEE Vitual Reality, 223–226 (2007)
7. Mario, B., Olga, S.: On Linear Variational Surface Deformation Methods. IEEE Trans. Vis. Comput. Graph. 14(1), 213–230 (2008)
8. Mario, B., Leif, K.: Multiresolution Surface Representation Based on Displacement Volumes. Comput. Graph. Forum 22(3), 483–492 (2003)
9. Oscar, K.A., Hongbo, F., Chiew-Lan, T., Daniel, C.: Handle-aware isolines for scalable shape editing. ACM Trans. Graph. 26(3), 83 (2007)
10. Oscar, K.A., Chiew-Lan, T., Ligang, L., Hongbo, F.: Dual Laplacian Editing for Meshes. IEEE Trans. Vis. Comput. Graph. 12(3), 386–395 (2006)
11. Burtnyk, N., Wein, M.: Interactive Skeleton Techniques for Enhancing Motion Dynamics in Key Frame Animation. Commun. ACM 19(10), 564–569 (1976)
12. Menache, A.: Understanding motion capture for computer animation and video games. Morgan Kaufmann, New York (2000)
13. Olga, S., Daniel, C., Yaron, L., Marc, A., Christian, R.: Hans-Peter Seidel: Laplacian Surface Editing. In: Symposium on Geometry Processing, pp. 175–184 (2004)
14. Yaron, L., Olga, S., Marc, A., Daniel, C., David, L., Christian, R., Hans-Peter, S.: Laplacian Framework for Interactive Mesh Editing. International Journal of Shape Modeling 11(1), 43–62 (2005)
15. Raif, M.R.: Laplace-Beltrami eigenfunctions for deformation invariant shape representation. In: Symposium on Geometry Processing, pp. 225–233 (2007)
16. Bruno, L.: Laplace-Beltrami Eigenfunctions Towards an Algorithm That "Understands" Geometry. SMI 13 (2006)
17. Zachi, K., Craig, G.: Spectral compression of mesh geometry. SIGGRAPH, 279–286 (2000)
18. Yizhou, Y., Kun, Z., Dong, X., Xiaohan, S., Hujun, B., Baining, G., Heung-Yeung, S.: Mesh editing with poisson-based gradient field manipulation. ACM Trans. Graph. 23(3), 644–651 (2004)
19. Gardner, E.S.: Exponential smoothing: The state of the art. J. Forecast., 4: 1C28 (1985)
20. SketchUp, `http://www.sketchup.com/`
21. Ruby plugin, `http://regularpolygon.org/keyframe-animation/`

Geo-referenced Tourist Attraction Photo Tagging by Mining Community Photo Collections

Xi Liu, Rujie Liu, Qiong Cao, and Fei Li

Fujistu Research and Development Center Co. LTD, Beijing, China
liuxi11@cn.fujitsu.com

Abstract. The advent of photo sharing sites like Flickr has drastically increased the volume of community photo collections on the web. Also the rising popularity of the mobile devices with GPS cameras like iPhone has made most of the photos geo-tagged. These provide new opportunities for automatically tagging the geo-referenced photos such as the tourist attraction photos. In this paper, we propose a framework for automatically tagging geo-referenced tourist attraction photos through mining the community photo collections. The photos collected from social sites are first clustered by fusing several modalities such as GPS and visual features, and then the tags of each cluster are extracted via a simple TF-IDF weighted voting scheme. Finally, for a tourist attraction photo taken with a GPS camera, it is annotated by the tags of the best matched cluster. We download a lot of photos located around some places of interest in Beijing from Flickr and manually construct a geo-referenced photo dataset. Experiment on the dataset shows an overall good performance.

Keywords: Geo-referenced photo, Community photo collections, Tourist attraction, Image clustering, Image tagging, Cluster tagging.

1 Introduction

With the proliferation of digital photo-capture device and the rapid development of Web 2.0 technology, a large number of community contributed photos have been produced and shared on the Web. These photos, especially the user-contributed tourist attraction photos are always associated with GPS information and it is believed to be useful in image tagging by constraining the search space. Besides, the photos, often labeled with keywords or tags by Internet users, provide new opportunities for image tagging since highly labor- and time-consuming manual tagging is no longer required. Thus it is desirable to use community photo collections for image tagging.

However, the GPS information may be inaccurate (e.g. users take photos far away from the sights), and the user-provided tags of the community photo collections are far from perfect due to tag ambiguity(eg. tag "apple" can either refer to a kind of fruit or a computer brand) or the knowledge and terminology limit of users [1, 11] (eg. an image of the Ling Long Pagoda may be tagged

B. Huet et al. (Eds.): PCM 2013, LNCS 8294, pp. 257–268, 2013.

"Pagoda" or "Beijing National Stadium"). Furthermore, it is a nontrivial task to find semantically relevant and visually similar images for query images from the community photo collections. All these issues make it difficult for directly tagging images with the community photo collections.

In this paper, an efficient and effective tagging framework is designed for addressing the above issues. Note that we focus on the tagging of the geo-referenced tourist attraction photos since the GPS coordinates of tourist attraction photos are more easily available and also the GPS coordinates are more meaningful for such tourist attraction photos (e.g. similar flower photos can be taken anywhere, whereas a tourist attraction photo can only be found in the fixed location). As shown in Fig. 1, the framework consists of offline mining process and online tagging process. In the offline process, two major components are designed: (1) Clustering on geo-referenced photos. The meanshift clustering algorithm is first applied to the geo-referenced photos' GPS coordinates to determine the dense photo areas and then the images in each dense area are further clustered by fusing visual features and the link structures. Geographically and visually similar images are grouped together into one cluster which is assumed to represent a specific object. (2) Cluster tagging. For each dense area, a dictionary is constructed by cleaning the tags associated with the images in this area with some word disambiguation rules and the pre-defined visual tag relevance. As to one cluster, it is annotated with the corresponding dictionary and its tags are generated by aggregating the tags of the images in the cluster via a TF-IDF voting scheme. In the online process, given a query image, the dense area where it belongs is first determined and the most similar cluster in the area is then identified. The final tags for the query image are given by the most similar cluster's tags.

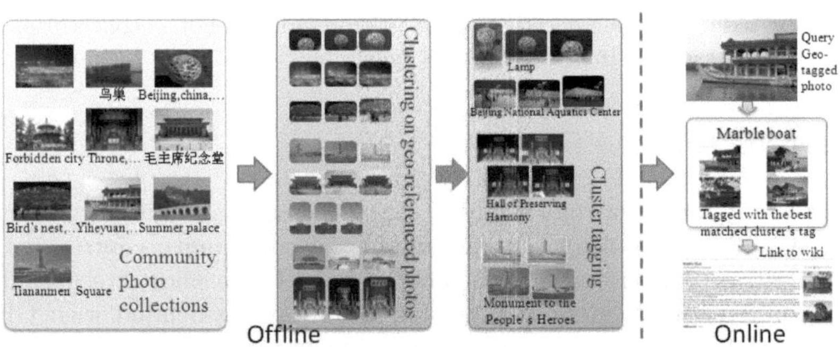

Fig. 1. The overall tagging framework

Several works have used community photo collections to tag images [2, 4–7, 9, 10, 13, 14, 17, 18]. Li et al. [10] propose a search-based tagging approach. For an image, they first find the visually similar images from the user-tagged community photo collections and then learn the tag relevance by accumulating votes from the visually similar neighbors. Quack et al. [13] propose a method for mining objects from the photos. They divide the areas into square tiles and cluster photos in each tile. Each cluster is classified as an object or event and tagged by the most frequent tags. Based on the method introduced in [13], Gammeter et al. [4] first mine the objects and annotate the query image with the most similar object retrieved by use of a scalable object retrieval technique. Jones et al. [7] formulate a two-stage geo-referenced landmark image annotation framework. In the first stage, an input image is matched to a set of community contributed images based on the image's GPS coordinates and a set of SVM classifiers pre-built for the area the input image locates in. In the second stage, the depicted landmark for the input image is inferred from the keyword tags of the matched set.

Compared with the recent works, this paper makes the following contributions. 1) Geographically and visually similar images are clustered together and the cluster's tags are used to annotate the query image. The tag noise and incompletion problems can thus be overcome by using the cluster's tags. Moreover, grouping the similar images can speed up the retrieval process. 2) A novel combination method is presented for fusing two clustering results which are obtained with image visual features and the link structures between images. More consistent clusters are expected by effectively fusing two kinds of clustering results, which provides a good basis for image tagging. 3) We build an online demo for geo-referenced tourist attraction photos tagging over some places of interest in Beijing.

The rest of this paper is organized as follows. In section 2. we briefly introduces the data gathering and image feature extraction. Sections 3 and 4 detail the process of offline community photo collections mining and online automatic geo-referenced photo tagging. The experimental results are reported in Section 5. Lastly, we conclude the paper with future work in Section 6.

2 Data Gathering and Feature Extraction

The community photo collections are collected from Flickr in Beijing city area. The longitude and latitude bounding box of the Beijing city is first manually determined and the geo-tagged photos are then downloaded by querying the Flickr API with the bounding box. About 50,000 photos are obtained, each of which is associated with the metadata, namely the textual descriptions (tags, title, description), user-id, and timestamps.

Three types of visual features are extracted for each image: (1) 225-D block-wise LAB-based color moments extracted over 5*5 grids; (2) 512-D Gist descriptors which is a low-level global image representation capturing the spatial layout; (3) 1000-D bag of visual words based on SIFT descriptor extracted densely on a multi-scale grid.

Let the obtained community photo $P = \{p\}$, where p is one photo denoted by a tuple $(f_{cp}, f_{gp}, f_{sp}, t_p, l_p)$, $\{f_{cp}, f_{gp}, f_{sp}\}$ represents the photo's color feature, gist feature and sift feature respectively, t_p represents the photo's tags, and l_p represents the photo's GPS coordinates.

3 Offline Community Photo Collections Mining

The community photo collections are often quite noisy so that they cannot be directly used for tagging new photos. To solve this problem, we first perform clustering on the geo-referenced community-contributed photos in each dense area and mine the tags for each cluster.

3.1 Clustering on Geo-referenced Photos

For the tourist attractions, many people will take the photos and upload them to the photo sharing sites like Flickr. Therefore, there exist many groups of images that have similar visual content and GPS coordinates in the downloaded community photo collections. This section details the process of finding these groups, which consists of three parts: clustering with GPS coordinates, clustering with visual features and link structures, and fusing clustering results.

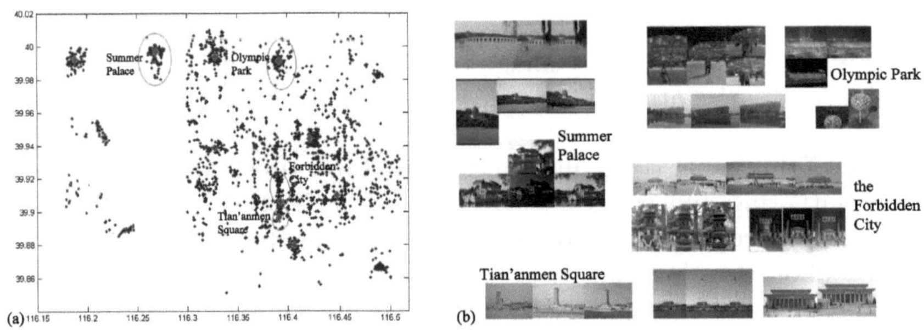

Fig. 2. (a)The GPS distribution and the GPS clustering results. (b) The clusters we get in the four dense areas. The images glued together represent a cluster.

Clustering with GPS Coordinates. The GPS coordinates for each photo are represented by a two dimensional vector of latitude and longitude. Fig. 2(a) shows the distribution of the GPS coordinates of the photos sampled from all the downloaded photos. It can be seen that the geo-referenced photos are not evenly distributed. We argue that the location with high image density is more likely to be a tourist attraction. The meanshift algorithm [3] is employed to cluster the photos based on the GPS coordinates and a flat kernel (Eq. 1) is chosen as follows:

$$g(x) = \begin{cases} 1, if\|x - x_i\|_2 \leq h \\ 0, otherwise \end{cases}, \tag{1}$$

where x is a given data, x_i is x's neighbor. The tolerance h is set as the average distance between each data with its 100 nearest neighbors.

The red diamonds in Fig. 2(a) illustrate the centers of the GPS clusters we find, which signify ten dense areas. By analyzing the locations of these areas, we find that they correspond to the hot tourist attractions such as Tian'anmen

Square, the Forbidden City, Summer Palace, Beijing National Stadium, the Fragrant Hill, which verifies our assumption. In this paper, four dense GPS clusters (marked with red ellipses in Fig. 2(a)) are selected for use. Basically they correspond to the most famous tourist attractions in Beijing, i.e., Beijing Olympic Park, Tian'anmen Square, the Forbidden City, and Summer Palace.

Clustering with Features and Link Structures. Although each dense area corresponds to one tourist attraction, it contains different objects considering that a tourist attraction has many different sights or views. Therefore, the photos in each dense area require to be further clustered based on visual features so that visually similar images are grouped together.

A photo p is represented with three visual features$\{f_{cp}, f_{gp}, f_{sp}\}$ and we can calculate the pairwise distances between photos for each feature. Three pairwise distance matrices D_c, D_g, D_s are created and they are then combined into a final distance matrix D by using the Joint Equal Contribution method proposed in [12].

The hierarchical agglomerative clustering algorithm [15] is applied to the distance matrix D. This clustering approach is chosen since it is rather flexible and very fast once the distance matrix is available. We employ the complete-link criteria for cluster merging, considering that we demand very tight clusters that contain similar views of the same object. Besides, the cut-off distance threshold is set to half of the mean distance.

We also employ a density-based graph clustering (SCAN) algorithm [16] to cluster the images. In contrast to the agglomerative clustering algorithm which directly uses the distance matrix, the graph clustering algorithm uses the link structures between the images. We construct a k-NN graph G based on the distance matrix D and then apply the graph clustering algorithm to G. The k is empirically set to 5.

Fusing Clustering Results. Two clustering results are obtained via the two clustering methods introduced above for each dense area. However, the two clustering results are far from perfect due to several factors such as the difficulty of determining the appropriate cut-off distance threshold or the number of nearest neighbors k.

We consider fusing the two clustering results to obtain more semantically consistent clustering results. Suppose the two clustering results are denoted as C_v and C_l, C_v has n_1 clusters and C_l has n_2 clusters: $C_v = \{c_{v1}, c_{v2}, \ldots, c_{vn_1}\}$, $C_l = \{c_{l1}, c_{l2}, \ldots, c_{ln_2}\}$. The fused clustering result C_f is firstly initialized as C_v. Then we analyze each cluster c_{li} in C_l to decide whether it is directly added to C_f or used to update some clusters in C_f. After analyzing all the clusters in C_l, we get the final fusing results C_f.

The process of analyzing the cluster c_{li} in C_l is as follows: (1) Calculate the overlap between c_{li} with each cluster c_{fj} in C_f. The overlap is calculated by $Overlap(c_{li}, c_{fj}) = \frac{|c_{li} \bigcap c_{fj}|}{min(|c_{li}|, |c_{fj}|)}$. (2) Find the cluster $c_{fj_{max}}$ which has the maximum overlap with c_{li} from C_f. If the maximum overlap is below a pre-specified threshold $Thr_{overlap}$, the cluster c_{li} is directly added to the final cluster results C_f; otherwise, the cluster c_{li} is used to update $c_{fj_{max}}$.

Updating $c_{fj_{max}}$ with c_{li} is illustrated in Alg. 1. The intersection of $c_{fj_{max}}$ and c_{li} forms an initial cluster c_{com}. Each image in $c_{fj_{max}} \bigcup c_{li}$ but not in c_{com} is then added to c_{com} separately to generate a cluster pool. The best cluster in the pool is selected and used to replace c_{com}. The two steps are iteratively performed until no more cluster pool is generated or better cluster cannot be found. To pick out the best cluster, four cluster measures are defined as follows:

For a cluster $c = \{I_1, \ldots, I_{n_c}\}$ with n_c images , suppose there are n images in the dense area where the cluster c locates, D and G are the final distance matrix and kNN graph for the n images respectively. Let $\sigma = mean(D_{ij}), i, j \in \{I_1, \ldots, I_n\}$.

- Global visual coherence $SC_{gv}(c) = mean(exp(-D_{ij}/\sigma)), i, j \in c$.
- Local visual coherence $SC_{lv}(c) = mean(exp(-D_{ij}/\sigma)), i, j \in c, j \in N_{k1}(i)$. $N_{k1}(i)$ is the set of i's k_1 nearest neighbors. k_1 is set to 3 here.
- Global link coherence $SC_{gl}(c) = \sum_{i,j \in c} G_{ij} / \sum_{i \in c \ or \ j \in c} G_{ij}$. G_{ij} is the weight of the edge connecting the images i and j in the k-NN graph G.
- Local link coherence $SC_{ll}(c) = mean(G_{ij}), i, j \in c$.

Algorithm 1. UpdateCluster

Input: cluster c_{li} and the cluster to be updated $c_{fj_{max}}$
Output: the updated cluster $c_{fj_{max}}$

1 Construct a cluster $c_{com} = c_{li} \bigcap c_{fj_{max}}$

2 For each image in c_{li} or $c_{fj_{max}}$ but not in c_{com}, generate a new cluster by adding the image to c_{com}. All the new generated clusters form a cluster pool C_p. If C_p is empty, go to step 5

3 For each cluster c_{pi} in C_p, calculate its score as follows:

$$Score(c_{pi}) = 0.6 * \left[\frac{SC_{gv}(c_{pi}) - SC_{gv}(c_{com})}{SC_{gv}(c_{com})} + \frac{SC_{lv}(c_{pi}) - SC_{lv}(c_{com})}{SC_{lv}(c_{com})} \right] + 0.4 * \left[\frac{SC_{gl}(c_{pi}) - SC_{gl}(c_{com})}{SC_{gl}(c_{com})} + \frac{SC_{ll}(c_{pi}) - SC_{ll}(c_{com})}{SC_{ll}(c_{com})} \right] \quad (2)$$

4 Find the cluster c_{opt} with the largest score from C_p. If the cluster c_{opt}'s score is above Thr_{score}, update c_{com} by replacing it with the cluster c_{opt} and go to step 2; otherwise, go to step 5

5 Replace $c_{fj_{max}}$ with c_{com}

We empirically set the overlap threshold $Thr_{overlap}$ to 0.4 and the score threshold Thr_{score} to 0.05 in our work. Furthermore, to get better clusters we rank the fused clustering results and only retain the top ranked clusters. The ranking method is very simple: four cluster rank lists are created based on the four cluster measures and the final cluster ranking is generated by applying the Borda count rank aggregation algorithm to the four rank lists.

Table 1 gives the precision at top 150 clusters over the four tourist attractions for the two single clustering results with visual features and link structures respectively and the fused clustering results. Each cluster is labeled good or bad by a irrelevant researcher. The precision at top 150 is calculated as #good clusters at top 150/150. From the table, it is easily seen that our fusing method gives the best results. Fig. 2(b) shows several typical clusters we get in the four dense areas and each cluster basically corresponds to an entity of a tourist attraction.

Table 1. Comparison of the precision@150 clusters for different clustering results. Cluster_feat and Cluster_link are the clustering results with visual features and link structures respectively, Cluster_fuse is the fusing results.

	Tian'anmen Square	The Forbidden City	Olympic Park	Summer Palace
Cluster_feat	0.67	0.80	0.91	0.94
Cluster_link	0.64	0.79	0.85	0.89
Cluster_fuse	**0.70**	**0.85**	**0.95**	**0.98**

3.2 Cluster Tagging

In the preceding section, images with similar content and GPS coordinates are grouped into a cluster, which ideally should depict a single entity for a tourist attraction. In this section, the goal is to obtain meaningful tags for each cluster by mining the tags of the images in the cluster.

Considering the tag noises in the community-contributed images, we should perform tag cleaning before performing tag voting to get the cluster tags. We first learn a clean tagging dictionary for each dense area with some word disambiguation rules and the pre-defined visual tag relevance . After that, a cluster is tagged with the dictionary of the dense area that it belongs to by aggregating the tags of the cluster's photos via a TF-IDF voting scheme.

Tag Cleaning. We perform tag cleaning on each dense area. In a dense area, a lot of tags are collected from the photos in this area. These tags are often noisy and need cleaning since they are all given by Internet users.

The tags with low frequency are firstly removed since these tags are always meaningless and helpless for tag voting. Then we consider the following five conditions: 1) abbreviation or acronyms (e.g., Tian'anmen Square vs. T'anmen Sq.); 2) capitalized letters (e.g., grand stage vs. Grand stage); 3) misspelling (e.g., Long Coridor vs. Long Corridor); 4) words with hyphens or dashes (e.g., kun-ming lake vs. kunming lake); 5) plurality (e.g., pagodas, pagoda). Several text processing techniques can be used to analyze these conditions.

Fig. 3. The images for tag "Landscape" and tag "Marble boat". Apparently, "Marble boat" is more visually representative than "Landscape".

Now a list of clean tags is derived. However, not all tags are useful for tag voting and we need to select the tags that are representative of the visual content. As illustrated in Fig. 3, the tag "Marble boat" is apparently more visually representative than the tag "Landscape" and the tag "Marble boat" should be selected. For each tag, we collect all the images with the tag to form a cluster. The cluster's local visual coherence, which is defined in last subsection, is then calculated and used to represent the tag's visual representativeness. The smaller the local visual coherence, the less visual representative the cluster is. Finally, we remove the tags with small visual representativeness and construct a final dictionary, which will be used for tag voting.

Tag Voting. After constructing a dictionary for each dense area, we can tag each cluster in this area with the corresponding dictionary. The tags that are not in the dictionary will be ignored.

For a cluster c, the term frequency for a tag x is given by the number of images tagged with x in the cluster c, i.e. $tf(x,c) = |P_{c,x}|$. The inverse document frequency for a tag x is calculated over all the images from the dense area L where the cluster c locates instead of the cluster. It is calculated as: $idf(x) = |P_L|/|P_{L,x}|$, where $|P_L|$ is the number of the images in L, $|P_{L,x}|$ is the number of the images tagged with the tag x in L. Finally, we multiply the two terms to get the TF-IDF score for each tag, $tfidf(x,c) = tf(x,c) * idf(x)$. The tags with high TF-IDF scores are selected as the cluster c's final tags.

4 Online Automatic Geo-referenced Photo Tagging

Given a geo-referenced photo, the dense area where the photo locates is firstly determined according to the photo's GPS coordinates. Then the most visually similar cluster is found from the clustering results in the dense area. To find the most visually similar cluster for an image, we define an image-to-cluster distance $D_{i-c}(x,c) = \sum_{i=1,\ldots,|c|} D(x,x_i)/|c|$, where x is an image, c is a cluster, x_i is an image in cluster c, $|c|$ is the number of the images in cluster c, $D(x,x_i)$ is the distance between the images x and x_i. The distance $D(x,x_i)$ can be calculated with different visual features. For the sake of tagging speed, we only use SIFT features f_{sp} rather than all the three features to calculate the distance in this work. The smaller the image-to-cluster distance, the more similar the cluster is to the test photo. Finally, the cluster with the smallest image-to-cluster distance is selected and the test photo is tagged with the cluster's tags.

5 Evaluation and Results

Our tagging method is evaluated on the four dense areas mentioned in Section 3.1. Specifically, we first analyze the quality of the cluster tags for each cluster and then evaluate the quality of the recommended tags for each query image. Lastly, an online geo-referenced attraction tourist photo tagging system is presented.

5.1 Evaluation of Cluster Tags

For each dense area, we construct a dictionary with the tag cleaning strategy introduced in section 3.2. Four dictionaries are derived for the four areas. Table 2 gives a brief description of the four dictionaries.

Table 2. Dictionary descriptions for the four dense areas

	Tian'anmen Square	The Forbidden City	Olympic Park	Summer Palace
Rep. Tags	Great Hall of the People, Monument to the People's Heroes, Mausoleum of Mao Zedong	Hall of Supreme Harmony, Hall of Union, Hall of Preserving Harmony	Beijing National Stadium, Beijing National Aquatics Center, Ling Long Pagoda	Kunming Lake, Hall of Dispelling Clouds, Seventeen-Arch Bridge
Dict. Size	66	90	75	163

Table 3. Tagging precision for the clusters in the four dense areas

	Tian'anmen Square	The Forbidden City	Olympic Park	Summer Palace
Tag Prec.	0.80	0.84	0.88	0.85

For each cluster, we use the TF-IDF voting scheme with the corresponding dictionary to obtain the cluster's final tags. We assume that a cluster is correctly tagged if it has at least one relevant tag among the top 3 recommended cluster tags. Table 3 gives the tagging precision for the clusters in the four dense areas. As shown in the table, the precision is satisfying, which provides a good basis for tagging query images. Fig. 4 demonstrates some cluster tags given by our method.

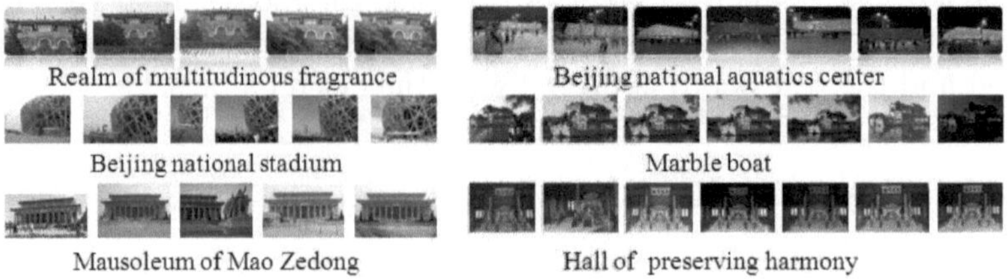

Fig. 4. Some clusters tagged by our method

5.2 Quality of Recommended Tags

To test the quality of the tags recommended by our method for the geo-referenced tourist attraction photos, we manually create the test datasets from the web for the four areas. We query the search engines such as Google with the name of the tourist attraction and select the test images from the search results. We collect 48 photos for Beijing Olympic Park, 71 photos for Tian'an men Square, 44 photos for The Forbidden City and 55 photos for Summer Palace.

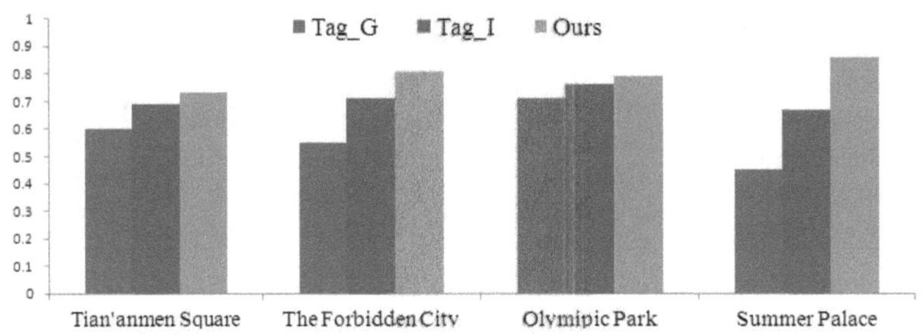

Fig. 5. Comparison of different methods over prec@top1

The recommended tags for test photos are judged by our researchers. A tag is regarded as correct only if it is strongly relevant to the photo. The measure prec@top3, defined as the ratio of the test photos which has at least one correct tag among the top 3 recommended tags, is used for evaluation. Two tagging methods are implemented to make a comparison, one (Tag_G) is tagging by aggregating the tags of the top 30 geographically similar photos, the other (Tag_I) is tagging by aggregating the tags of the top 30 visually similar photos. The results are shown in Fig. 5. Our method performs best for all the tourist attractions. Also note that the performances of Tag_G and Tag_I for Summer Palace are especially low compared with our method. The reason is that there exists large tag noise for images from Summer Palace area and the methods Tag_G, Tag_I are affected a lot by the tag noise while our method can overcome this by using cluster tagging.

5.3 Online Demonstration

The interface of the online system is shown in Fig. 6(a). The user can upload a geo-referenced tourist attraction photo, which is taken when he visits some places of interest. After the photo is uploaded, it will be shown on the corresponding locations of the map according to its GPS coordinates.

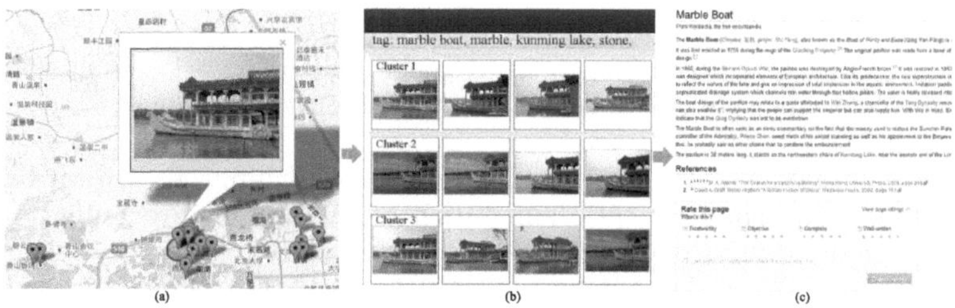

Fig. 6. A demonstration of tagging a geo-referenced photo by our system

Then the system determines the GPS area that the uploaded photo locates in and finds the cluster that is the most visually similar to the photo from the clustering results in this area. Fig. 6(b) shows the retrieved clusters for the uploaded photo. Finally, the uploaded geo-referenced tourist attraction photo is annotated by the tags of the top retrieved cluster and it is tagged “Marble boat”, “Marble” and “kunming lake” in the demo. With these tags, more information can be obtained by using the search engines such as Google or by linking directly to some encyclopedia websites such as Wikepedia. Fig. 6(c) shows the Wikipedia page we get for the uploaded photo.

6 Conclusions

We present a geo-referenced tourist attraction photo tagging system by mining large community photo collections. It first groups the community contributed photos by fusing several modalities and then automatically mines tags for these clusters in an offline process. For a test geo-referenced tourist photo, the system finds the cluster which is visually and geographically most similar to the photo and annotates the test photo with this cluster. We demonstrate our approach on four selected dense areas which correspond to four famous places of interest in Beijing and observe an overall good performance. In the future, we will further make comparisons with other state-of-art methods over our collected datasets and also exploit more tourist attractions.

References

1. Chua, T.S., Tang, J., Hong, R., Li, H., Luo, Z., Zheng, Y.: Nus-wide: a real world web image database from national university of Singapore. In: Proceedings of ACM Conference on Image and Video Retrieval (2009)

2. Fang, Q., Sang, J.T., Xu, C.S., Lu, K.: Paint the City Colorfully: Location Visualization from Multiple Themes. In: International Conference on Multimedia Modeling, pp. 92–105 (2013)
3. Fukunaga, K., Hostetler, L.: The estimation of the gradient of a density function, with applications in pattern recognition. IEEE Transactions on Information Theory 21(1), 32–40 (1975)
4. Gammeter, S., Bossard, L., Quack, T., Gool, L.V.: I know what you did last summer: object-level auto-annotation of holiday snaps. In: International Conference on Computer Vision (2009)
5. Hays, J., Efros, A.: IM2GPS: estimating geographic information from a single image. In: Proceedings of IEEE Conference on Computer Vision and Pattern Recognition (2008)
6. Ivanov, I., Vajda, P., Goldmann, L., Lee, J.S., Ebrahimi, T.: Object-based tag propagation for semi-automatic annotation of images. In: Proceedings of ACM Conference on Multimedia Information Retrieval (2010)
7. Jones, G.J.F., Byrne, D., Hughes, M., O'Connor, N.E., Salway, A.: Automated annotation of landmark images using community contributed datasets and web resources. In: Declerck, T., Granitzer, M., Grzegorzek, M., Romanelli, M., Rüger, S., Sintek, M. (eds.) SAMT 2010. LNCS, vol. 6725, pp. 111–126. Springer, Heidelberg (2011)
8. Joshi, D., Luo, J.: Inferring generic activities and events from image content and bags of geo-tags. In: Proceedings of ACM Conference on Image and Video Retrieval (2008)
9. Kleban, J., Moxley, E., Xu, J.J.: Global annotation on georeferenced photographs. In: Proceedings of ACM Conference on Image and Video Retrieval (2009)
10. Li, X., Snoek, C.G.M., Worring, M.: Learning social tag relevane by neighbor voting. IEEE Transactions on Multimedia 11(7), 1310–1322 (2009)
11. Liu, D., Hua, X., Zhang, H.J.: Image retagging. In: Proceedings of ACM Conference on Multimedia (2010)
12. Makadia, A., Pavlovic, V., Kumar, S.: Baselines for image annotation. International Journal of Computer Vision 90(1), 88–105 (2010)
13. Quack, T., Leibe, B., Gool, L.V.: World-scale mining of objects and events from community photo collections. In: Proceedings of ACM Conference on Image and Video Retrieval (2008)
14. Wang, X.J., Zhang, L., Jing, F., Ma, W.Y.: AnnoSearch: image auto-annotation by search. In: Proceedings of International Conference on Computer Vision and Pattern Recognition (2006)
15. Webb, A.R.: Statistical Pattern Recognition, 2nd edn. John Wiley & Sons (2002)
16. Xu, X., Yuruk, N., Feng, Z., Schweiger, T.A.J.: Scan: a structural clustering algorithm for networks. In: Proceedings of the 13th ACM SIGKDD International Conference on Knowledge Discovery and Data Mining (2007)
17. Yang, Y., Huang, Z., Shen, H.T., Zhou, X.F.: Mining multi-tag association for image tagging. International Journal of World Wide Web 14, 133–156 (2011)
18. Zheng, Y.T., Zha, Z.J., Chua, T.S.: Research and applications on georeferenced multimedia: a survey. Multimedia Tools Appl. 51(1), 77–98 (2011)

Improving Histogram-Based Image Registration in Video Sequences through Warping

Xuesong Le and Ruben Gonzalez

School of Information, Communication, and Technology, Griffith University
Griffith University, Parklands Drive, Southport QLD 4222 Australia

Abstract. This paper presents two computationally efficient dynamic-time warping algorithms for image registration in video sequence through histogram based image segmentation. The key idea is to warp the histogram in an input frame to create an approximation of a reference frame. Any histogram based thresholding method can then be applied to create consistent regions in both the input and the approximated reference frames. Experiments of the proposed algorithm are used to demonstrate that more consistent matches can found after thresholding.

Keywords: histogram-based, dynamic-time warping, consistent regions.

1 Introduction

Compared to numerous feature point based algorithms surveyed in [1-3], the use of image segmentation as a step in image registration for video sequences has been little explored due to the computational requirements and the difficulty in obtaining segmentation consistency. Flusser and Dai [4-5] proposed to extract closed regions by contour tracing detected edges. Goshtasby [6] iteratively split an image into regions recursively until the variation of gray-level histogram in each sub-region is small. These region-based segmentation methods are not widely used in image registration of video sequences due to the computational complexity in part created by the post-processing requirements such as edge linking or boundary closing. In contrast, region-based segmentation methods, such as histogram splitting methods are very efficient because they require only one pass through the pixels in the image to perform segmentation. The key parameter in the histogram thresholding process is the choice of the threshold values.

Histogram threshold selection algorithms can be categorized into two groups: shape-based and clustering–based. The first category of methods achieves thresholding based on the shape properties of the histogram. It comprises three stages: recognizing the modes of the histogram, finding the valleys between the identified modes and finally assigning the valleys as the optimal thresholds. One disadvantage of the mode-seeking method is that it may be difficult to identify significant peaks and valleys in the image. Rosenfeld et al. [7] proposed a valley sharpening technique which restricts the histogram to the pixels with large absolute values of derivatives.

B. Huet et al. (Eds.): PCM 2013, LNCS 8294, pp. 269–280, 2013.

Cheng and Sun [8] proposed a peak finding algorithm to locate the globally significant peaks of the histogram by removing small peaks, peaks which are too close to adjacent peaks, and peaks if valleys between them are not obvious. In the class of clustering methods, gray levels are grouped under various clustering criteria. Otsu [9] selected the global optimal threshold by maximizing the weighted between-class variance or minimizing the weighted within-class variance. Kapur [10] considered a multi-level thresholding scheme where the sum of between-class entropies for two classes is maximized.

Recently, Goncalves [11] proposed an image registration method that uses histogram-based image segmentation but it can only handle translation and rotation because it is not robust to histogram changes and image scaling causes more significant change of the histogram. In this paper, we present a novel method that combines both dynamic time warping (DTW) methods and histogram thresholding techniques to solve this problem. The objective of warping two histograms is to minimize the histogram differences across frames, and thus provide higher correlation between images and creates increased segmentation consistency. In our method, the histogram of the second image is warped first to create the approximation of the histogram in the first image before histogram thresholding. Two efficient DTW algorithms are proposed here to align the histogram sequences. The direct consequence is that each gray level in the second histogram is mapped to its correspondent in the second histogram. Then histogram thresholding techniques in [7-10] can be used to find optimal thresholds for the first histogram. After that, the same thresholds with a lookup table created during the warping can be applied on the second histogram. Finally, regions in both images are found by applying the thresholds. The significance of proposed histogram warping operation is that more consistent regions are found, which reduces the number of outliers in image registration and improves the registration accuracy.

2 Dynamic Time Warping in Histogram Alignment

DTW is a technique that finds the optimal alignment between two time series if one time series is warped non-linearly by stretching or shrinking it along its time axis. This warping operation between two time series can then be used to find corresponding regions between the two time series. In our case, we are dealing with histograms rather than time series and so the DTW problem is expressed as follows: Given two histogram sequences F, and G, both of length 256,

$$F = f_0, f_1, \ldots, f_i, \ldots f_{255} \quad G = g_0, g_1, \ldots, g_j, \ldots g_{255} \tag{1}$$

Construct a warp path W:

$$W = w_0, w_1, \ldots, w_k, \ldots w_n \tag{2}$$

where n is the length of the warp path and the kth element of the warp path is:

$$w_k = (i, j) \tag{3}$$

where i is an index of histogram sequence F, and j is an index of histogram sequence G. The warp path starts at the beginning of each histogram sequence at w_0=(0,0) and finishes at the end of both sequence at w_n=(255,255). An example of warping histogram F to another histogram G is shown in Figure 1. If both of the histograms were identical, the warp path is a straight line joining w_0 ad w_n. There is a constraint on the warp path that forces i and j to be monotonically increasing in the warp path.

$$w_k = (i, j), \quad w_{k+1} = (i', j'), i \le i' \le i+1, j \le j' \le j+1 \tag{4}$$

The minimum distance path, W, between F and G is defined as:

$$Dist(W) = \min(\sum_{k=0}^{k=n} Dist(w_{ki}, w_{kj})) \tag{5}$$

where Dist(w_{ki},w_{kj}) can be any distance metric between two data point indexes (one from F and another from G) in the kth element of the warp path.

2.1 Dynamic Programming Based DTW

A direct solution to find the optimal warp path in equation 5 is through dynamic programming (DP). A 256x256 two dimension cost function matrix D in figure 1, is constructed first where each element, denoted as D(i, j), is defined in equation 6. Each D(i,j) is the sum of the distance d(i,j) found in the current cell and the minimum of the adjacent elements.

$$D(i,j) = \begin{cases} \min[D(i,j-1), D(i-1,j), D(i-1,j-1)] + d(i,j) & if (i>0, \wedge j>0) \\ d(i,j) & else \end{cases} \tag{6}$$

where $d(i,j) = |f(i) - g(j)|$.

After the cost matrix D is filled, backtracking along the minimum cost index pairs $(i, j)_k$ starting from $(255,255)_0$ to $(0,0)_n$ yields the DTW warping path W. A greedy search is performed that evaluates cells to the left, up, and diagonally to the top-left. Figure 1 below shows an example of a cost matrix and minimum-distance warp path traced through it from D(0,0) to D(255,255). If the warp path passes through a cell D(i,j) in the cost matrix, it means that i^{th} point in histogram F is warped to the jth point in another histogram G.

The major disadvantage of the DP-based DTW is that it requires high complexity of $O(N^2)$. A number of complexity reduction methods [12] have been proposed to speed up by a large constant factor, but those methods still runs in $O(N^2)$ time. To avoid the brute-force dynamic programming, a fast approximation of the optimal solution is proposed in section 2.2 which directly aligns two histogram sequences in O(N) time.

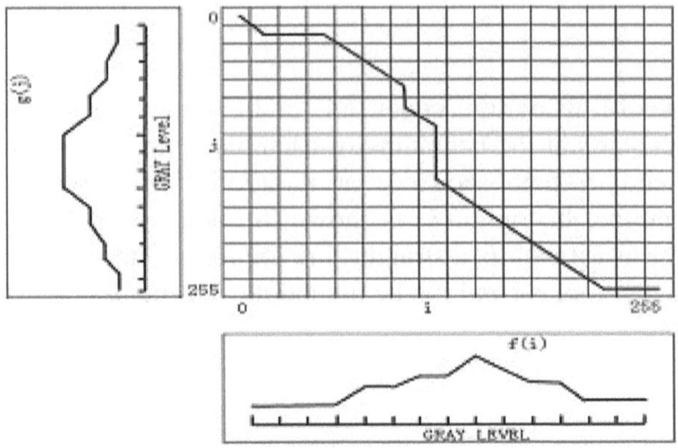

Fig. 1. Illustration of a warping path between histogram F and G

2.2 Integration Based DTW

A faster DTW approach is to use integration. In this approach, the histogram warping problem is described as follows: each gray level i of the second histogram sequence, G, is mapped into a new gray level j in the first histogram sequence, F, such that the quantity:

$$\left|\sum_{k=0}^{j} cF[k] - \sum_{k=0}^{i} cG[k]\right|, \tag{7}$$

$$\sum_{k=0}^{i} cF[k] = \sum_{k=0}^{i-1} cF[k] + f[i] \;, \; \sum_{k=0}^{i} cG[k] = \sum_{k=0}^{i-1} cG[k] + g[i] \;\_ \tag{8}$$

is minimized subject to j.

The mapping table can be constructed by a linear algorithm (O(N)) which does numerical integrations simultaneously on both histogram sequences. For each gray level i in histogram G, we integrate histogram F until a j is reached such that the above condition is satisfied. The integration process exploits the simple recurrence relations as i increases from one to the other. Thus, the overall cost in constructing the mapping table is linear in time.

The C code of warping histogram sequence G with reference to F is essentially a few lines of code:

```
void warp(int cF[], int cG[], int lookupTbl[])
{ int i,j=0;
        for(i=0; i<256; i++)
        {
    while( cF[j]< cG[i] && j<256) j++
    if( j == 0 )  lookupTbl[i] = 0;
    else if( j>255)  lookupTbl[i] = 255;
```

```
        else  //find which one is closer to cG[i]
        lookupTbl[i] = (cF[j] - cG[i]) > (cG[i] - cF[j-1])  ? j-1: j;
            }//End for
}
```

Table 1 lists the average running time on a 600x400 video sequence between DP-warped method and integration-warped method. Experiment runs on a PC with 3.00GHZ Intel Pentium 4 CPU and 1G Memory.

Table 1. Running Time Comparison between Integration-Based DTW and DP-Based DTW

	Integration- based DTW	**DP-based DTW**
comp. time (ms)	0.0287	12.5309

In the above experiment, the DP-based DTW, is 430 times slower than the integration-based method shown in Table 1 since the DP-based DTW runs in $O(N^2)$ time while integration-based method runs in O(N).

3 Proposed Histogram Thresholding Algorithms

After the mapping of gray levels in both histograms is established, a histogram thresholding technique is still required to select optimal levels for the histogram in the first image. We propose three representative thresholding algorithms here and use them during experiments. Two of them are extensions of existing methods.

The problem of multi-level histogram thresholding can be stated as following: It assumes that there are M-1 thresholds, $\{t_1, t_2, \ldots, t_{M-1}\}$, which divide the original histogram into M classes: C_1 for $[0,\ldots, t_1]$, C_2 for $[t_1+1, \ldots, t_2]$, … , C_i for $[t_{i-1}+1, \ldots, t_i]$, …, and C_M for $[t_{M-1}+1, \ldots, 255]$.

3.1 Clustering-Based Thresholding

The proposed clustering algorithm is an extension of OTSU's bi-level thresholding method [9] to a multi-level thresholding method. The cumulative probabilities, ω_k, mean, μ_k, and total variance, σ_k^2, mean for each class C_k is defined as:

$$\omega_k = \sum_{i=C_k} p_i \quad \mu_k = \sum_{i=C_k} ip_i / \omega_k \quad , \quad \sigma_k^2 = \sum_{i=C_k} (i-\mu_k)^2 p_i / \omega_k \tag{9}$$

In [11], the optimal threshold t in any class C_i for $[t_{i-1}+1, \ldots, t_i]$ is the maximum weighted between-class variance, σ_B^2:

$$\sigma_B^2 = \omega_1\omega_2(\mu_1 - \mu_2)^2 \ (\tag{10}$$

ωηερε:

$$\omega_1 = \sum_{j=t_{i-1}}^{t} p_j \ , \quad \omega_2 = \sum_{j=t+1}^{t_i} p_j \ , \quad \mu_1 = \sum_{j=t_{i-1}}^{t} jp_j / \omega_1 \ , \text{ ανδ } \quad \mu_2 = \sum_{j=t+1}^{t_i} jp_j / \omega_2 \ .$$

For this work, we have instead used local optimization to reduce processing time to search for the optimal multi-level thresholds. It is a divisive approach, which starts with a single class and recursively splits the single class into M classes. This algorithm can be described as following:

1. Start with a single class C_0 which has all intensity levels t=0… 255 and add to a list C with count m =0.
2. Compute total variance σ_0^2 in class C_0 with equation 9 and the optimal threshold t with equation 10.
3. Search each candidate class c_i in list C and find the class c_s, which has the largest total variance σ_i^2 and the optimal threshold t_i. Intensity gray level in class c_i ranges from t_s to t_{s+1}, where $0 \le t_1 < ... < t_s < t_{s+1} < ... < L$.
4. Split the class C_s into two new classes C_m, and C_n. Intensity gray level in class c_m ranges from t_s to t_i and ranges from t_i +1 to t_{s+1} in class c_i.
5. Compute total variance σ_m^2, σ_n^2 and the optimal thresholds, t_m, and t_n for two classes.
6. If any new created class c_j has satisfied following criteria $\sigma_j^2 < \sigma_{\min}^2 \vee size(c_j) < \min\_size$, classify this class as a none-divisible class. Otherwise this class is assigned as candidate class.
7. Remove c_s from list C and add two split classes c_m and c_n to list C. Increment m by 1.
8. If the number of classes including all none-divisible classes and candidate classes has reached maximum threshold M or the largest variance is below some threshold, then stop, else go back to step 3.

3.2 Entropy-Based Thresholding

The proposed entropy algorithm is another modification of [13] where the selection of optimal threshold in any class C_i for $[t_{i-1}+1, ..., t_i]$ is defined as the maximum between-class entropy in equation 11:

$$e_B = -\{\sum_{j=t_{i-1}}^{t}((p_j/\omega_1)*\log_2(p_j/\omega_1)) + \sum_{j=t+1}^{t_i}((p_j/\omega_1)*\log_2(p_j/\omega_1))\} \quad (11)$$

where $\omega_1 = \sum_{j=t_{i-1}}^{t} p_j$, $\omega_2 = \sum_{j=t+1}^{t_i} p_j$.

3.3 Peak-Seeking-Based Thresholding

The selection of thresholds in [8] is to seek most significant peaks through several stages of peak detections and is described as the following steps:

1. Find all peaks.
2. Find significant peaks.
3. Remove any small peaks.
4. If two peaks are close, the peak with bigger value is chosen.

5. Remove a peak if the valley between two peaks is not obvious.
6. The minimum values between any adjacent peaks are selected as the thresholds.

After thresholds of the histogram, F, in the first image are selected, every gray level, *i*, is labeled as one of the following:

$$L = \{l_{0,(0,t_0-1}, l_{1,(t_0,t_1-1),\dots}, l_{j,(t_{j-1},t_j-1}, \dots, l_{m,(t_{m1},t_m-1)}\}, \tag{12}$$

$$L_F(i) = l_{j,(t_{j-1},t_j-1}, \quad t_{j-1} \le i \le t_j - 1, \tag{13}$$

where $L_f(i)$ represents the jth label value which gray level, *i*, is assigned to and *i* is within the range $[t_{j-1}, t_j-1]$.

The process of assigning labels to gray levels in the second histogram, G, is a lookup operation, described as follows

$$L_G(j) = L_F(lookupTbl\,[j]), \tag{14}$$

where lookupTbl represents mapping of ith gray level in histogram G to j^{th} gray level in histogram F created in section 2.

Labeling each pixel in the images is another lookup operation which retrieves a label value through its gray level directly. Finally a connected component analysis is performed on both labeled images to extract regions where pixels have been assigned to the same value.

4 Experiment

To test performance of the proposed algorithm in histogram-based segmentation, various video sequences in Figure 2 are used in experiments. Each type has been further tested under varying scaling, rotation, and translation camera transformations.

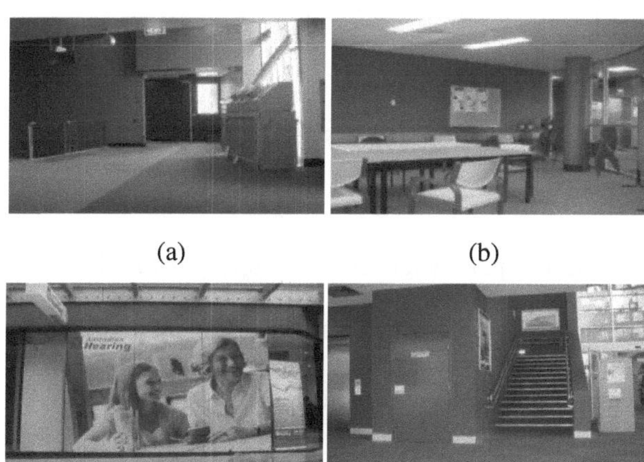

(a) (b) (c) (d)

Fig. 2. (a) Hall (b) Lounge (c) Ladies (d) Stairs

Testing images for rotation tests were selected every 40 frames from the input video sequences. The maximum frame interval range is 200 frames, approximating 10 degree's rotation which is reasonable in video sequences. Images in translation tests were selected every 20 frames among 100 frames as well. The horizontal translation was measured as percentage of amount of translation over the image width. The maximum translation range in the testing sequences was 15%.

4.1 Ground Truth Match Simulation

We used exhaustive RANSAC method [13] to simulate the ground truth matches between region centroids. RANSAC is a non-deterministic algorithm for the estimation of a transformation model from observed data which contains outliers. It is essentially composed of two steps that are repeated iteratively.

Hypothesize: A sample of size m among the N data points is randomly selected. The model parameters are computed from this sample. m is the smallest sufficient cardinality to determine the model parameters.

Test: The hypothesis is verified against the rest of the data by counting the points consistent with the estimated model parameterization.

These two phases are repeated until the iteration finishes.

In video, 2D affine transformation can be modeled as composite matrix operation of scaling, rotation, and translation, which maps a point $p = (x_i, y_i)^T$ to a point $q = (u_i, v_i)^T$ as follows:

$$\begin{bmatrix} u_i \\ v_i \\ 1 \end{bmatrix} = \begin{bmatrix} 1 & 0 & t_x \\ 0 & 1 & t_y \\ 0 & 0 & 1 \end{bmatrix} * \begin{bmatrix} cos\theta & -sin\theta & 0 \\ sin\theta & cos\theta & 0 \\ 0 & 0 & 1 \end{bmatrix} * \begin{bmatrix} s & 0 & 0 \\ 0 & s & 0 \\ 0 & 0 & 1 \end{bmatrix} * \begin{bmatrix} x_i \\ y_i \\ 1 \end{bmatrix}$$
$$= \begin{bmatrix} cos\theta * s & -sin\theta * s & t_x \\ sin\theta * s & cos\theta * s & t_y \\ 0 & 0 & 1 \end{bmatrix} * \begin{bmatrix} x_i \\ y_i \\ 1 \end{bmatrix} \quad (15)$$

The affine matrix $\begin{bmatrix} cos\theta * s & -sin\theta * s & t_x \\ sin\theta * s & cos\theta * s & t_y \\ 0 & 0 & 1 \end{bmatrix}$, in equation 15 can be simplified as $M = \begin{bmatrix} a & -b & c \\ b & a & d \\ 0 & 0 & 1 \end{bmatrix}$ and the parameters in the simplified matrix can be solved by the pseudo-inverse solution. Therefore, the smallest sufficient cardinality to determine the model parameters in RASAC is two in this case.

Given a set of points and its correspondences in the form U = WA.

$$\begin{bmatrix} u_1 \\ \cdots \\ u_n \\ v_1 \\ \cdots \\ v_n \end{bmatrix} = \begin{bmatrix} x_1 & -y_1 & 1 & 0 \\ & \cdots & & \\ x_1 & -y_1 & 1 & 0 \\ y_1 & x_1 & 0 & 1 \\ & \cdots & & \\ y_n & x_n & 0 & 1 \end{bmatrix} * \begin{bmatrix} a \\ b \\ c \\ d \end{bmatrix} \tag{16}$$

The pseudo-inverse solution$A = (W^T W)^{-1} W^T U$ is computed to solve for the four affine coefficients.

U and M are then used for calculating the re-projection error e of the remaining pairs of correspondence. Re-projection error is defined as the Euclidean distance between the transformed point of point p in the previous frame according to the model M and the original correspondence of q in the current frame:

$$e = \left\| \begin{bmatrix} x_i' \\ y_i' \\ 1 \end{bmatrix} - M \begin{bmatrix} x_i \\ y_i \\ 1 \end{bmatrix} \right\| \tag{17}$$

The re-projection error threshold T_{thres} is determined such that among all the initial matches, only those pairs with $e < T_{thres}$ will be regarded as inliers and kept. In our experiment, T_{thres} is set to 2. The process is repeated for different subsets, resulting in different transformations, with the number of inliers recorded for each transformation. The transformation supported by the highest number of inliers among all the iterations is selected to be the winning transformation. In the final step, all inliers from the winning transformation are considered to be the ground truth matches.

4.2 Experiment Results

We extracted the gravity centre points of each image region from two segmented images using both integrated-warped methods, and non-warped methods. In test cases of rotation, and translation, we found that both proposed warped method and non-warped methods were able to provide similar number of matched regions because the change of the histogram is limited. Figure 3 shows number of ground truth matches found in segmented regions using integrated-warped method, and non-warped method for image Hall_r200 and Hall_r400 under 10's degree rotation respectively.

(a) (b) (c)

Fig. 3. (a) segmented regions in Hall_r200 (b) warped segmented regions in Hall_r400 (c) non-warped segmented regions in Hall_r400

Table 2. Matching Results for Scaling

	Ladies			*Lounge*		
	integ	***dp***	***non-warped***	***integ***	***dp***	***non-warped***
scaling ratio 1.3						
Peak-Seeking-Based	*21*	*24*	*20*	*17*	*20*	*10*
Entropy-Based	*28*	*30*	*22*	*18*	*21*	*9*
Clustering-Based	*30*	*31*	*24*	*15*	*23*	*13*
scaling ratio 1.4						
Peak-Seeking-Based	*19*	*23*	*12*	*15*	*20*	*6*
Entropy-Based	*29*	*28*	*15*	*17*	*18*	*10*
Clustering-Based	*27*	*28*	*25*	*17*	*20*	*13*
	Hall			***Stair***		
scaling ratio 1.3						
Peak-Seeking-Based	*21*	*20*	*12*	*12*	*12*	*7*
Entropy-Based	*17*	*16*	*17*	*10*	*13*	*10*
Clustering-Based	*20*	*19*	*16*	*10*	*7*	*10*
scaling ratio 1.4						
Peak-Seeking-Based	*18*	*19*	*9*	*11*	*10*	*4*
Entropy-Based	*17*	*16*	*15*	*10*	*11*	*10*
Clustering-Based	*18*	*20*	*11*	*10*	*7*	*11*

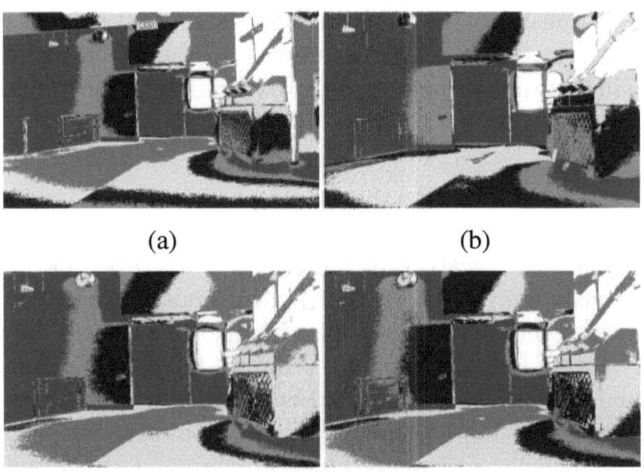

(a) (b) (c) (d)

Fig. 4. (a) segmented regions in Hall_1 (b) non-warped segmened regions in Hall_4 (c) dp-warped segmented regions in Hall_4 (d) integ-warped segmented regions in Hall_4

In test cases of scaling, we selected test images with zooming ratio up to 1.4. Table 2 shows the number of matched regions for Hall sequence under 1.3 and 1.4 scaling. The warping methods performed better than non-warped histogram-based methods 95% of the time. And the warping methods found more number of matches on average 40% higher than those non-warped methods. In these experiments, the DP-based DTW performed only 6% better than the integration-based method as shown in Table 2. Experiments results in Figure 4 show matched regions using DP-warped method, integration warped method, and non-warped method for image Hall_1 and Hall_4 with 1.4 scaling respectively.

5 Conclusion

This paper has presented methods for improving histogram-based segmentation consistency in video sequences for image sequence registration using a DTW approach. The use of DTW is able to improve segmentation consistency by up to 40%. This is necessary for making histogram segmentation based image sequence registration robust against image scaling. While the classical DP-based DTW marginally outperforms the proposed integration-based approach, it is about 430 times slower. Designed for low computation requirements, the proposed technique is effectively able to handle image sequences with significant histogram changes due to either scaling or lighting changes.

References

[1] Brown, L.G.: A survey of image registration techniques. ACM Computing Surveys 24(4), 325–376 (1992)
[2] Zitová, B., Flusser, J.: Image registration methods: a survey. Image Vision Comput. 21(11), 977–1000 (2003)
[3] Maintz, J.B.A., Viergever, M.A.: A survey of medical image registration. Medical Image Analysis 2(1) (1998)
[4] Flusser, J., Suk, T.: A Moment-based Approach to Registration of Images with affine Geometric Distortion. IEEE Trans. on Geoscience and Remote Sensing 32(2) (March 1994)
[5] Dai, X., Khorram, S.: A feature-based image registration algorithm using improved chain-code representation combined with invariant moments. IEEE Trans. Geosci. Remote Sensing 37 (September 1999)
[6] Goshtasby, A., Stockman, G.C., Page, C.V.: A region-based approach to digital image registration with subpixel accuracy. IEEE Trans. Geosci. Remote Sensing 24(3), 390–399 (1986)
[7] Weszka, J.S., Nagel, R.N., Rosenfeld, A.: A threshold selection technique. IEEE Trans. Comput. C-23, 1322–1326 (1974)
[8] Cheng, H.D., Sun, Y.: A hierarchical approach to color image segmentation using homogeneity. IEEE Trans. Image Process. 9(12), 2071–2082 (2000)
[9] Otsu, N.: A threshold selection method from gray-level histograms. IEEE Transactions on Systems Man Cybernet SMC-9, 62–66 (1979)
[10] Kapur, J.N., Sahoo, P.K., Wong, A.K.C.: A New Method for Gray-Level Picture Thresholding Using the Entropy of the Histogram. Graphical Models and Image Processing 29, 273–285 (1985)
[11] Gonçalves, H., Gonçalves, J.A., Corte-Real, L.: HAIRIS: A method for automatic image registration through histogram-based image segmentation. IEEE Trans. Image Process. 20(3), 776–789 (2011)

[12] Chu, S., Keogh, E., Hart, D., Pazzani, M.: Iterative Deepening Dynamic Time Warping for Time Series. In: Proc. of the Second SIAM Intl. Conf. on Data Mining, Arlington, Virginia (2002)
[13] Fischler, M.A., Bolles, R.C.: Random Sample Consensus: A Paradigm for Model Fitting with Applications to Image Analysis and Automated Cartography. Comm. of the ACM 24(6), 381–395 (1981)

Robust Tracking via Hough Forests Based on Weak Classifier Response

Weiyi Chen, Pingyang Dai*, Wei Liu, Cuihua Li, and Yi Xie

Computer Science Department, Xiamen University
Xiamen, 361005, P.R. China
{pydai,chli,csyxie}@xmu.edu.cn,
{weiyichen2013,lwnicky}@gmail.com

Abstract. Discriminative tracking method based on Hough forests has received increasingly attention due to its advantages such as its efficiency and robustness. But Hough forests only use the pixel values in the split test selection, the information of image patches has not been made full use of to discriminate different classes. In this paper, we propose a novel method of Hough forests for visual tracking. This method adopt a new strategy of the split test selection which is based on the weak classifier responses instead of pixel values. Compared with the traditional method of Hough forests, the proposed method can extract more information. Furthermore, this method introduces the LBP feature into Hough forests and then captures both edge-based shape structure and texture information. The proposed method is validated through the experiments on a set of publicly available benchmark dataset. The experiment results have shown that this method outperforms the state-of-the-art algorithms in terms of robustness and effectiveness.

Keywords: visual tracking, Hough forests, multiple features, weak classifier response.

1 Introduction

Visual tracking is a well-known problem in the field of computer vision because of its wide application in surveillance, interactive games, augmented reality and other intelligent vision systems. In unconstrained real world environment, visual tracking is required to deal with practical challenges such as noise, occlusion, varying viewpoints, clutter background and illumination changes. Although many tracking methods have been proposed, it still remains a challenging problem.

Tracking algorithm can be generally categorized as generative or discriminative approaches. Generative method typically learn a model to represent the target object and searching for the region most similar to the target model. The IVT method [1] utilizes an incremental subspace model to adapt the changes of

* Corresponding author.

B. Huet et al. (Eds.): PCM 2013, LNCS 8294, pp. 281–288, 2013.

appearance. This method performs well when target objects encounter illumination changes and pose variations. Adam et al. [2] proposed a fragment-based method to handle occlusion. In this method, the idea of voting has been exploited for tracking where the template of the object is represented by a set of local patches. Each patch votes for the center of the object and is tracked independently. Kwon et al. [3] tracked the target successfully via visual tracking decomposition (VTD) when several types of appearance and motion changes occurred.

Discriminative algorithm formulate the tracking problem as a binary classification problem which aims to separate the target object form the background. The ensemble tracker [4] trained an ensemble of weak classifiers online to distinguish objects from background. Grabner et al. [5] proposed an online boosting method to update discriminative features and a semi-supervised online boosting algorithm was proposed to handle the drift problem [6]. Inspired by the multiple instance learning method which is used to solve the ambiguity problem in face detection, Babenko et al. [7] proposed a tracking method based on the online multiple instance learning (MIL) method. MIL tracker learns a classifier as the appearance model to alleviate the drift problem. The P-N learning algorithm [8] was proposed to exploit the underlying structure of positive and negative samples and learnt an effective classifier for tracking. Recently, Gall et al. propose a Hough forests detector [9]. Hough forests can be regarded as task-adapted codebooks of local appearance and allow fast supervised training and fast matching at test time. Then several tracking methods based on Hough forests have been proposed [10,11].

In this paper, we extend the Hough forests and propose a novel method of Hough forests for visual tracking. Firstly, we propose a new strategy of split test selection which is based on the weak classifier responses instead of pixel values during the training of Hough forests. This new strategy can exploit more information of the target than standard Hough forests. Specially, we integrate the Haar features [12] as weak classifier responses, which are known to be good at capturing geometric structures. Secondly, we introduce the LBP feature [13] into Hough forests combined with color features and orientation features. The integration of multiple features which capturing both edge-based shape structures and texture information makes the tracker more robust and effective.

The rest of paper is organized as follows. The Hough forests integrates LBP feature and the new strategy of split test selection based on weak classifier response are described in Section 2. The details of the proposed method for visual tracking are present in Section 3. Section 4 provides the experimental results. Finally, the conclusion is drawn in Section 5.

2 Hough Forests Based on Weak Classifier Response

In recent years, Hough forests has became a very popular and powerful tool for classification and regression problems by exhibiting many appealing properties like inherent multi-class capability, robustness to label noise and reduced tendencies to overfitting. In this paper, we propose a new strategy of split test selection of Hough forests which based on weak classifier response such haar features. Then we denote our novel Hough forests as Haar Hough forests. The Haar Hough forests are ensembles of Hough decision trees and each tree in the forests is built and tested independently.

2.1 Feature Extraction

As we know, the feature extraction is a prerequisite for the success of a classifier in the training time. It is important to choose appropriate features to train the classifier which is discriminative. In the Hough forests, 16 features channels are extracted including the channels of the Lab color space, the absolute values of the first and second order derivatives , and nine HOG-like channels. Image texture contains information about the spatial arrangement of color or intensities in an image and is very useful for object appearance representation. The LBP feature has been found to be a powerful feature for texture classification. So, we introduce the LBP feature into Hough forests combined with color feature and orientation features. The integration of multiple features which capturing both edge-based shape structures and texture information makes the tracker more robust and effective.

2.2 Building Haar Hough Forests

Hough forests consist of a set of random trees and each tree T is constructed from a set of patches $\{P_i = (I_i, c_i, d_i)\}$, where $I_i = (I_i^1, I_i^2, \cdots, I_i^M)$ is the appearance of the image patch, I_i^j is the j feature channel and M is the number of feature channels. $c_i \in \{0, 1\}$ is the class label of the image patch. A $2D$ offset vector d_i is assigned for each object image patch and d_i is a displacement vector from the centroid of the bounding box to the center of the image patch. The offset vector of a background patch is undefined.

The construction for the tree of Hough forests follows the common random forests framework [14]. Each tree is constructed recursively starting from the root. During construction, each node receives a set of training patches. If the depth of nodes is equal to the maximal one or the number of patches is too small, then the construction node is declared as a leaf. The leaf information (C_L, D_L) is accumulated and stored, where C_L is the proportion of the object patches and $D_L = \{d_i\}$ of the offset vector corresponding to the object patches. Otherwise, a non-leaf node is created and an optimal binary test is chosen from a large pool of randomly tests. The test is defined as:

$$t(I) = \begin{cases} 0 & \text{if } H_1^j < H_2^j + \tau \\ 1 & \text{otherwise} \end{cases} \tag{1}$$

where H_1^j and H_2^j is the Haar feature response in the jth channel, τ is the handicap value, shown as Fig.1. In order to choose the best test in the training time,

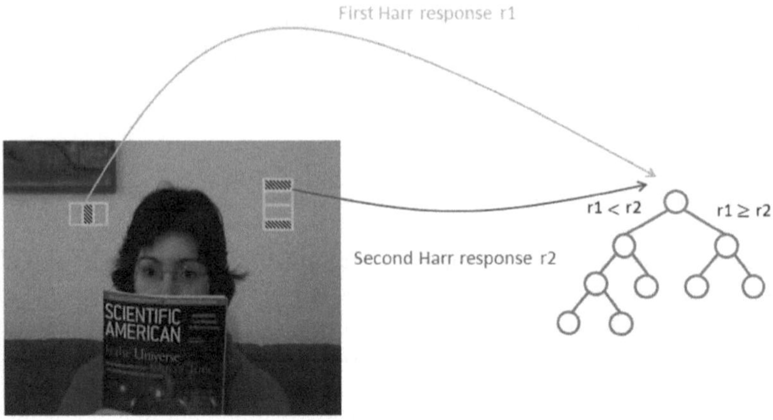

Fig. 1. The split test selection based on Haar feature response

two uncertainties are used. The class-label uncertainties and offset uncertainties are difined as

$$U_c(N) = -|N|(C_N \log(C_N)) + (1 - C_N)log(1 - C_N)) \tag{2}$$

$$U_o(N) = \sum_{d_i \in D_N} (d_i - \bar{d})^2 \tag{3}$$

where C_N is the proportion as defined before, $\bar{d}$ is the mean offset vector over all object patches. Then the randomized decision is choose whether the node should minimize the class-label uncertainty or the offset uncertainty. Finally, we pick the binary test with the minimal sum of the respective uncertainty measures for the two subsets and it split the training set into:

$$\arg\min(U_*(\{p_i|t^k(I_i = 0)\}) + U_*(\{p_i|t^k(I_i = 1)\})) \tag{4}$$

where $*$ indicates the chosen uncertainty measure for the node (U_c or U_o).

3 Visual Tracking with Haar Hough Forests

The Hough forests is a tracking-by-detection and detection-by-tracking method. Given a patch $P(y) = (I(y), c(y), d(y))$ centered at the position y in the image. Here $I(y)$ is the object appearance of the patch, $c(y)$ is the hidden class-label, and $d(y)$ is the hidden offset vector from the center of the object bounding box to y. Furthermore, $E(x)$ denotes the random event corresponding to the existence of the object which is centered at the location x in the image. For a single tree t, the probability estimation can be written as:

$$p(E(x)|I(y);T) = [\frac{1}{|D_L|} \sum_{d \in D_L} \frac{1}{2\pi\sigma^2} \exp(-\frac{\|(y-x)-d\|^2}{2\sigma^2})] \cdot c_L \tag{5}$$

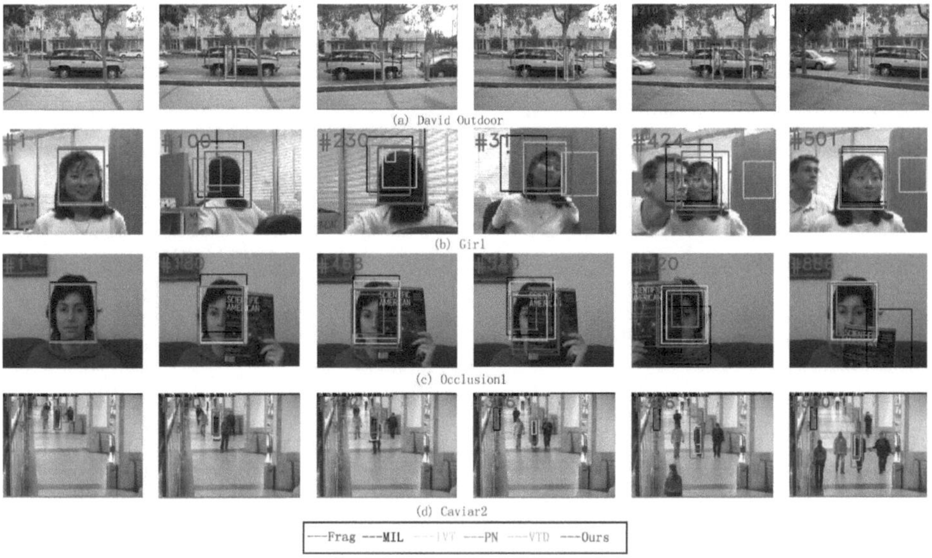

Fig. 2. The tracking result compared with five algorithms

In our Haar Hough forests tracker, the location of the target is given by ground-truth in the initial frame. Then we collect the Hough votes that contributed to the local maximal and the local maximal of the votes is the center of the object.

4 Experiments

We evaluate the performance of the proposed algorithm using several challenging sequences which are available publicly. We compared our tracking algorithm with 5 state-of-the-art methods including incremental visual tracking (IVT) method, fragment-based (Frag) tracking method, visual tracking decomposition (VTD), multiple instance learning (MIL) tracker and P-N learning (PN) tracker.

Our algorithm is implemented in C++. The number of trees used in Hough forests $T = 10$. The depth of each tree $d_T = 15$. The patch size is 16×16 pixels. The search radius $r = 20$. In all of the experiments, the parameters of proposed method are fixed.

4.1 Qualitative Evaluation

The sequence *DavidOutdoor* is very challenging for visual tracking since the tracking object may suffer from occlusion, pose changes and clutter background. Some representative tracking results are shown in Fig.2(a). The indexes of result frames are 1, 60, 140, 183, 210 and 252. When the target object undergoes occlusion and clutter background, other trackers drifts and fails in tracking. Our tracker successfully keeps tracking of the target and performs well throughout the whole sequence.

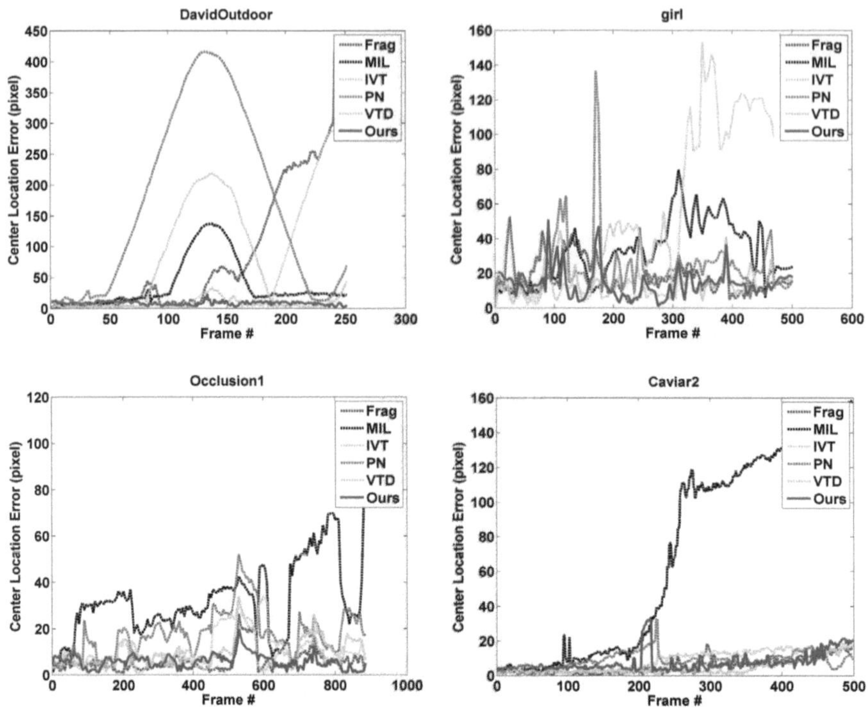

Fig. 3. Quantitative comparison of the trackers in terms of position errors

In the *Girl* sequence, there are out-of-plane rotations and appearance changes. Some representative tracking results are shown in Fig.2(b) with indexes 1, 100, 230, 317, 424 and 501. The IVT method lost the target and other trackers experience drifts. Our method is capable of tracking the target during the entire sequence. Performance on this sequence exemplifiers the robustness of our method to occlusion and out-of-plane rotations.

In the *Occlusion1* sequence, there are frequent heavy occlusions. The tracking results are shown in Fig.2(c). The indexes of representative frames are 1, 180, 463, 540,l 720 and 886. MIL tracker drift away quickly and lost the tracking object at frame 886. Other methods drift in different degrees when the target object is heavily occluded. Our method performs reasonably well in this sequence.

The fourth sequence is *caviar2.* The tracking object is walking person subjected by occlusion by other persons which is similar in color and shape. Fig.2(c) demonstrates some tracking results. The representative frames indexes are 1, 150, 192, 305, 395 and 500. MIL method drift and lost tracking object and other methods drifts respectively. The Frag, VTD and proposed algorithm can track the target in this sequence quite well.

4.2 Quantitative Evaluation

For quantitative evaluation, we use the center location error in pixels as the measurement for our algorithm and all the other tracking methods mentioned above. The results are shown in Fig.3. And Table 1 summarizes the average center location errors in pixels. The quantitative comparison results demonstrate proposed algorithm produce smaller location errors than other trackers in general. This implies that our method can track the target accurately under severe occlusion, pose variations and illumination changes.

Table 1. Average center error(in pixels). The best two are shown in red and blue fonts.

Sequence	Frag	MIL	IVT	PN	VTD	Ours
David Outdoor	90.4	**37.82**	53.12	172.82	62.18	8.54
Girl	**17.94**	31.54	49.12	22.86	21.01	14.06
Occlusion1	5.8	32.39	9.41	17.93	11.19	**5.85**
Caviar2	**5.62**	70.16	8.77	8.67	4.58	7.68
Average	29.94	42.98	30.105	55.57	**24.74**	9.03

5 Conclusion

In this paper, we propose an effective and robust tracking algorithm based on Haar Hough forests. The Haar Hough forests exploits more information of the target by using the strategy of the split test selection which is based on the weak classifier responses instead of pixel values. The integration of multiple features which capturing both edge-based shape structure and texture information also makes the tracker more robust and effective than when only non texture information is used. Experimental results compared with several state-of-the-art algorithms on challenging sequences demonstrate robustness and effectiveness of the proposed algorithm.

Acknowledgements. This research work was supported by the National Natural Science Foundation of China Under Grant No. 61373077, the National Defense Basic Scientific Research Program of China, National Defense Science and Technology Key laboratory Fund, Doctoral Program of Higher Specialized Research Fund (No. 20110121110020), the Natural Science Foundation of Fujian Province of China Under Grant No. 2013J01257 and Fundamental Research Funds for the Central Universities of the People's Republic of China (No. 2010121066), Shenzhen City Special Fund for Strategic Emerging Industries (JCYJ2012).

References

1. Ross, D., Lim, J., Lin, R.-S., Yang, M.-H.: Incremental Learning for Robust Visual Tracking. In: IJCV, pp. 125–141 (2008)
2. Adam, A., Rivlin, E., Shimshoni, I.: Robust Fragments-based Tracking using the Integral Histogram. In: CVPR, pp. 798–805 (2006)
3. Junseok, K., Lee, K.M.: Visual tracking decomposition. In: CVPR, pp. 1269–1276 (2010)
4. Avidan, S.: Ensemble Tracking. PAMI, 261–271 (2007)
5. Grabner, H., Bischof, H.: On-line Boosting and Vision. In: CVPR, pp. 260–267 (2006)
6. Grabner, H., Leistner, C., Bischof, H.: Semi-supervised On-Line Boosting for Robust Tracking. In: Forsyth, D., Torr, P., Zisserman, A. (eds.) ECCV 2008, Part I. LNCS, vol. 5302, pp. 234–247. Springer, Heidelberg (2008)
7. Babenko, B., Yang, M., Belongie, S.: Visual Tracking with Online Multiple Instance Learning. PAMI, 983–990 (2009)
8. Kalal, Z., Mikolajczyk, K., Matas, J.: Tracking-Learning-Detection. PAMI, 1409–1422 (2012)
9. Gall, J., Lempitsky, V.: Class-specific Hough forests for object detection. In: CVPR, pp. 1022–1029 (2009)
10. Gall, J., Yao, A., Razavi, N., Van Gool, L., Lempitsky, V.: Hough Forests for Object Detection, Tracking, and Action Recognition. PAMI, 2188–2202 (2011)
11. Godec, M., Roth, P.M., Bischof, H.: Hough-based tracking of non-rigid objects. In: ICCV, pp. 81–88 (2011)
12. Viola, P., Jones, M.: Rapid object detection using a boosted cascade of simple features. In: CVPR, pp. 511–518 (2001)
13. Ojala, T., Pietikainen, M., Maenpaa, T.: Multiresolution gray-scale and rotation invariant texture classification with local binary patterns. PAMI, 971–987 (2002)
14. Breiman, L.: Random Forests. In: Machine Learning, pp. 5–32 (2001)

Saliency-Based Auditory Detection Method Using Energy Linear Superposition

Yifei Pan[1], Runtian Long[1], Xiaoxue Cheng[1], and Yanxiang Chen[1,2]

[1] School of Computer Science & Information, Hefei University of Technology, Hefei, Anhui 230009, China
[2] Computer Science and Technology Postdoctoral Research Station, Hefei University of Technology, Hefei, Anhui 230009, China
chenyx@hfut.edu.cn

Abstract. In real-world environments, audio signals are usually submerged in background noise originating from various noise sources. In this paper, we present a novel approach to detect salient sound from a mixed sound based on spectrogram saliency map and the energy linear superposition theory. First, an auditory saliency model is used for the spectrogram of the mixed sound to locate the salient part in time dimension. Then we use the energy map of the sound near the salient part to estimate the energy map of background sound in salient part. Finally, we extract salient sound in energy domain based on energy linear superposition theory. Our experiment indicates that the saliency-based auditory attention detection method presented in this paper has a high performance on salient audio signals detection.

Keywords: Auditory attention, Saliency map, Spectrogram, Energy superposition.

1 Introduction

Auditory saliency can help observers sort the information present in a complex auditory scene. From a bottom-up perspective, salient sounds are defined as those sounds that can be determined perceptually what stand out to an observer in a particular scene. Since resources are finite and not all information can be processed equally, we must detect what is most important sound that grabs a listener's attention to give a quick response or decision. This issue is closely related to many areas, including scene analysis.

Numerous research efforts have been undertaken for detection of significant events by means of sounds. Zienowicz [1]applies the spectrographic templates with PCA to detect, classify and model sequences of salient sound events. RuiCai [2]uses HMMs to model the sound effects and a log-likelihood scores based method to locate the highlight sound effects in audio stream. Furthermore, Rapantzikos [3]uses efficient and advanced speech and image processing algorithms to detect key frames that are

B. Huet et al. (Eds.): PCM 2013, LNCS 8294, pp. 289–298, 2013.

acoustically and visually salient for generating video summarization [18]. Zlatintsi [4] approaches the problem of audio summarization by exploring the potential of a modulation model for the detection of perceptually important audio events.

Additionally, in [5] and [6], audio scene analysis is dealt with by classifying the auditory categories, e.g. animals, bells, laughter, mechanical sounds. The results show that the audio levels of the classification have greater impact on the performance. In [7] and [8],the motion-based audio scene analysis is investigated by conducting the perception analysis of audio context.

However, all these methods treat audio signal only from single modality. In fact, spectrogram is the visualization display of audio signal and visual features can be extracted from spectrogram. Inspired by visual saliency map [15] [16] [17], we first locate highlights in the spectrogram and then use energy superposition to process the highlight areas for the audio detection.

The remainder of this paper is organized as follows. In section 2, we briefly review the related work. In section 3, the method for saliency-based auditory detection using energy linear superposition is presented. In section 4, the experiments and results of our method are presented. Conclusions are drawn and future work is suggested in section 5.

2 Related Work

Saliency map is the foundation of the proposed system and the related work is discussed as follows.

The first computational model of bottom-up saliency was proposed by Koch and Ullman in 1985 [9] and further developed by Itti et al. [10]for the visual system. In the auditory modality, only a few such systems have so far been proposed. The first model, by Kayser et al. [11] treats the auditory input spatially by considering the time-frequency representation of an auditory signal ("the auditory image") as the input of the saliency model. The second model, by Kalinli et al.[12] builds on Kayser's model by combining with more features that are extracted from the auditory image, and using a different normalization scheme. Then Duangudom et al. [13] focus on spectro-temporal receptive field (STRF) output computed from the auditory image as the feature base of the system and slightly expands the normalization procedure.

Thereafter, Kaya et al. [14] proposes a novel auditory saliency map that is defined only over time. The proposed method capitalizes on the rich high-dimensional feature space that defines auditory events, where each acoustic dimension is processed across multiple scales. These normalized feature maps are then combined over time into a single temporal saliency map. The peaks of the temporal saliency map indicate the locations of the salient events in the auditory scene.

In the above-mentioned models, the salient points of the audio signal are obtained by using the visual features extracted from the spectrogram. However, to further achieve foreground audio detection based on the salient area of the audio scene, energy linear superposition is implemented in this paper. We distinguish the foreground sound from the mixed sound by using the linear relationship of energy accumulation in the spectrogram. As far as we know, no literature applying energy linear superposition for the saliency-based audio detection has been found.

3 Proposed Method

The proposed method consists of two parts. The first part, we detect and locate the salient area of the audio signal. The second part, we process this salient area to extract target sound base on energy linear superposition. The block diagram of our proposed method is shown in Fig. 1.

For audio signal, short-term energy is represented by intensity in spectrogram, brighter area represent stronger energy. In part one, we detect the salient area in spectrogram by auditory saliency model and locate the time points when salient sound begin and end. The detail of auditory-attention model will be described in Section3.1.

As we know, the energy of a mixed sound is a linear summation of each sound it mixed. In part two, we processed the mixed audio signal in energy domain to extract target sound by subtract energy of background sound from mixed sound. Obviously, the most important in this part is how to estimate energy of background sound when the salient sound occurs. Here, we suppose background sound will not change much in a short time, which is in line with the actual situation. On this premise, we take a period of sound before and after salient part to estimate the energy map of background sound in the salient part. This part is described in Section 3.2.

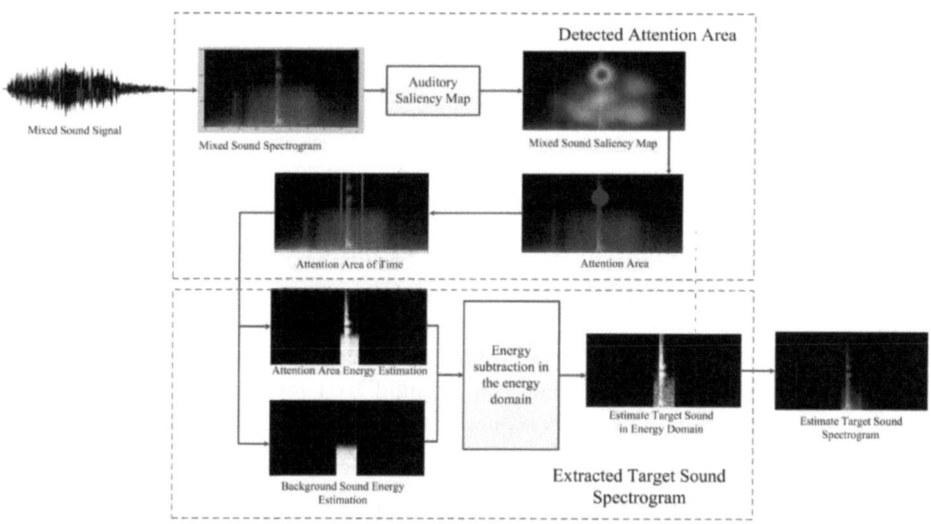

Fig. 1. Salient Audio Event Detection

3.1 Auditory Saliency Model

In spectrogram, the intensity of pixel reflects the energy of signal in corresponding time-point and frequency, the orientation features of local region reflects the characteristic of the corresponding category of sound. We can conclude that the region in the spectrogram where intensity or orientation has significant differences should correspond to the salient part in audio.

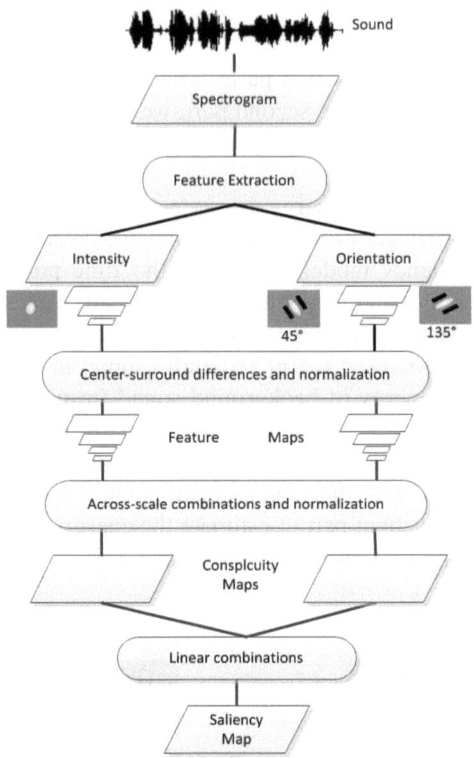

Fig. 2. Auditory Saliency Map

In this paper, our auditory attention model is built up from two angles: intensity and orientation, derived from the Itti model. The diagram auditory attention model is shown in Fig. 2.

Firstly, the intensity feature spectrogram is processed to the Gaussian pyramid structure with eight layers. The Gaussian pyramid $I(\sigma)$ represents intensity pyramid image, where $\sigma \in \{1,..,7,8\}$ is the scale. By the following formula, six strength characteristic figures can be generated.

$$I(c,s) = |I(c) \ominus I(s)| \tag{1}$$

Here, $s = c + k$, $c \in \{2,3,4\}$ and $k \in \{3,4\}$. $I(c)$ and $I(s)$ represent the c and s strength pyramid image respectively, '$\ominus$' is defined as the procedure of generating intensity feature map which is obtained by take the absolute value of corresponding pixel image subtraction $I(s)$ and $I(c)$. In the stage of subtraction, $I(s)$ is enlarged to the scale of $I(c)$. Then the six intensity superposition characteristic diagrams is superimposed and generates the intensity saliency map I^c as follows:

$$I^c = \oplus_{c=2}^{4} \oplus_{s=c+3}^{c=4} N\big(I(c,s)\big) \tag{2}$$

Where, $N(\cdot)$ is a normalization operator, shown in Figure 3. Due to the lack of supervision in t bottom-up model, remarkable error could be made from data. To enhance the overall significance, normalization operator can be the square of each element in the feature map $I(c,s)$.

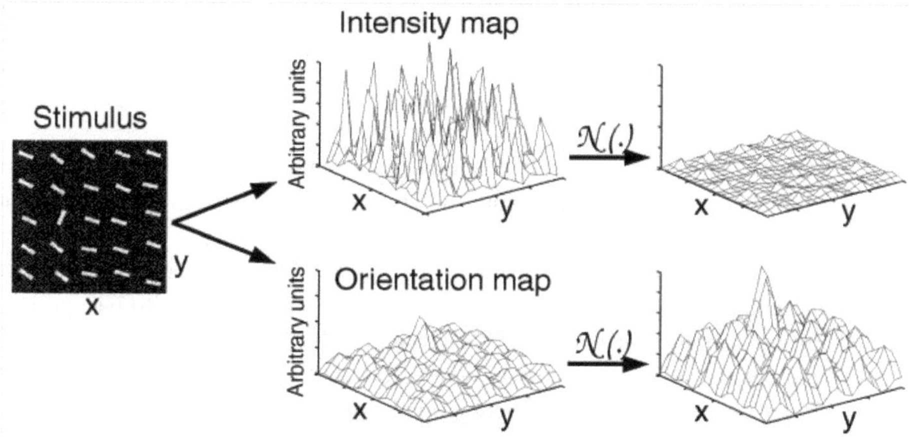

Fig. 3. The normalization operation $N(\cdot)$

Secondly, Gabor pyramid has been used to deal with the direction characteristic spectrogram. Since Gabor kernel has different orientations, four orientations has been used in Itti model. In the auditory model, here we only use $\theta \in \{45°, 135°\}$. For each and every of the orientation from Gabor kernel is used to process the intensity image I and generate a direction Gabor pyramid. The Gabor pyramid image can be represented as $O(\sigma, \theta)$. Similarly, orientation feature maps can be generated by the following formula:

$$O(c,s,\theta) = |O(c,\theta) \ominus O(s,\theta)| \tag{3}$$

Here, $s = c + k, c \in \{2,3,4\}$and $k \in \{3,4\}$.$O(c,\theta)$and $O(s,\theta)$ represents the c and s strength pyramid image from direction θ pyramid respectively. By using the tor'$\ominus$', direction feature map can be generated. We will produce twelve direction feature maps in all. The feature maps are superimposed in all directions, forming direction saliency map, as follows:

$$O^c = \sum_{\theta \in \{45°,135°\}} N\left(\oplus_{c=2}^{4} \oplus_{s=c+3}^{c=4} N\big(O(c,s,\theta)\big)\right) \tag{4}$$

Eventually, we will get three kinds of feature maps based on global and band channel. The final auditory saliency model can be obtained by a superposition process.

3.2 Salient Sound Extraction Method

The saliency map we obtained from section 3.1 pop out the salient part in spectrogram. In this section, we will discuss how to extract spectrogram of salient sound from the salient part.

Spectrogram is an reversible transformation of energy map, so we can obtain spectrogram easily after we get energy map of the signal. For a mixed sound, the energy map of the superimposed area is a linear superposition of these single sounds. With this understanding, we can get the energy map of salient sound as follows:

$$E_{ss} = E_{sp} - E_b \tag{5}$$

Where, E_{ss} is the energy map of salient sound, E_{sp} is the energy map of salient part and E_b is the energy map of background in salient part. The spectrogram of salient sound will be easily achieved after E_{ss} work out.

The only problem left is how to estimate E_b. As the fact what grab our auditory attention in reality is most probably a short and strange sound relative to background sound. Here we suppose the energy of background will not change much during the salient sound occurs and we use energy map before and after salient part to estimate E_b. We extract a period of sound as background sound samples before and after the location of salient part respectively. These two sound samples should have a same length. Evaluate energy maps of these two samples and take average of them, so we get energy map of background sample E_{bs}. We consider the whole energy map of background in salient part is a rank of E_{bs} and the length of the rank is:

$$n = \frac{L_{sp}}{L_{bs}} \tag{6}$$

Where, L_{sp} is length of salient part, L_{bs} is length of background sample.

4 Experiments

To experiment the method, the sound of sea-wave, rain and white noise were chosen as the background and the sound of bird, dog and gun as salient sound, with 10 audio collected for each. L_{sample} is defined as 1/10 of salient part which is pop out from auditory saliency model. An example is shown in Fig. 4.

Since the mixed audio we used is synthesized by software, we can easily compare our result with the standard spectrogram of single sound. We use correlation between our result and standard spectrogram to evaluate the performance of our method. The correlation is computed by the normalized cross-correlation function, defined as follows:

$$\rho = \sum_{n=1}^{N} x(n)y(n) \Big/ \sqrt{\sum_{n=1}^{N} x(n)^2 \sum_{n=1}^{N} y(n)^2} \tag{7}$$

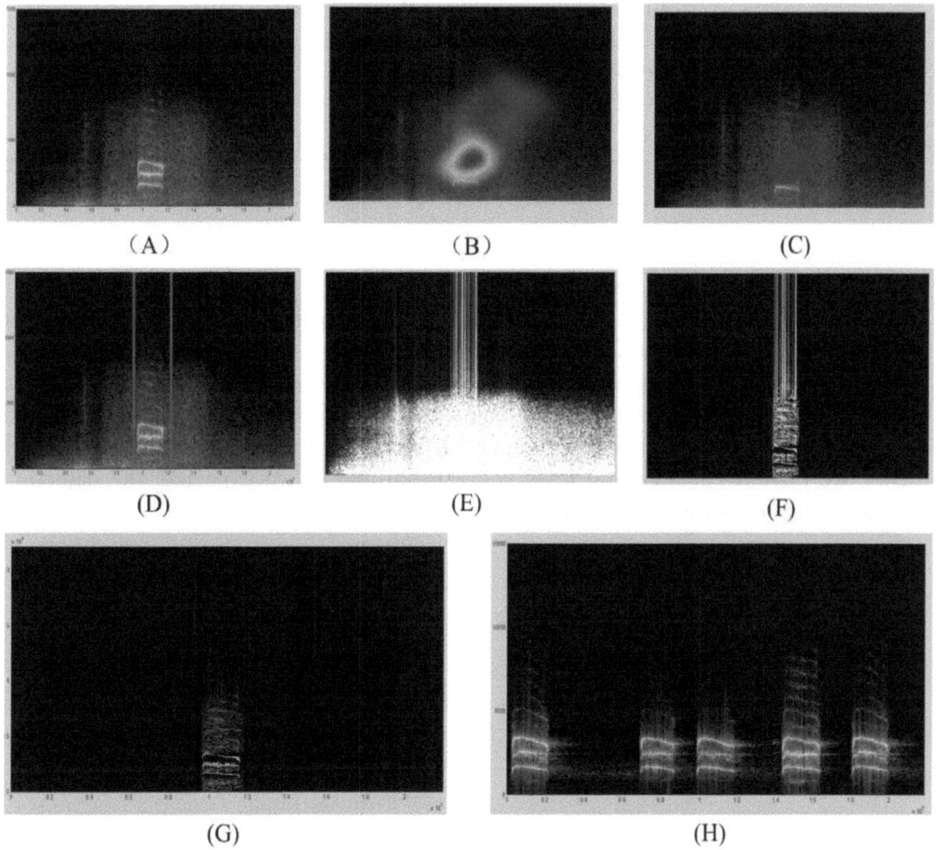

Fig. 4. (A) Spectrogram of mixed audio(sea-wave and bird). (B) Saliency map of mixed audio spectrogram. (C) Locate salient part based on B. (D) Locate salient part in time dimension. (E) Mixed audio in energy domain. (F) Salient sound in energy domain computed by the method we proposed. (G) Salient sound spectrogram we work out. (H) Spectrogram of single bird audio for comparison with G.

Where, x and y are spectrogram we worked out and standard spectrogram respectively. N is the number of pixels in spectrogram. By the Schwarz inequality, ρ belong to interval $[0,1]$. Larger value of ρ indicates higher correlation between x and y, $\rho = 1$ when $x = y$, and $\rho = 0$ when x and y are non-correlation. We assume the detection is successful when $\rho > 0.7$.

As mentioned before, 900 mixed audio were synthesized with the combination of the 30 audio of the background and 30 salient sound. We tested our method on all 900 mixed audio and evaluated by the normalized cross-correlation function. The results was divided into 9 groups, shown in Table 1. The detection rates is the percentage of successful detection in each group. We can see from Table 1 that detection rates of dog and gun beyond 90% for all three conditions, even for the sound of bird, which is changeful and low intensity, our method still performed well.

Table 1. Salient Auditory Detection Rates

	Sea wave	Rain	White noise
Bird	82%	79%	81%
Dog	93%	90%	96%
Gun	96%	92%	97%

It is crucial to note that not like any existing approaches, the performance of our method is not sensitive to noise level. Since what we focus on is irregular part and due to the linear superposition in energy domain, the noise level has tiny influence on either detection or extraction part in our approach. The results of a series repeated experiments with different background level are shown in Fig. 5. In these experiments, we increase background level several-fold and as we expected, the detection rates still satisfactory with the increasing background level.

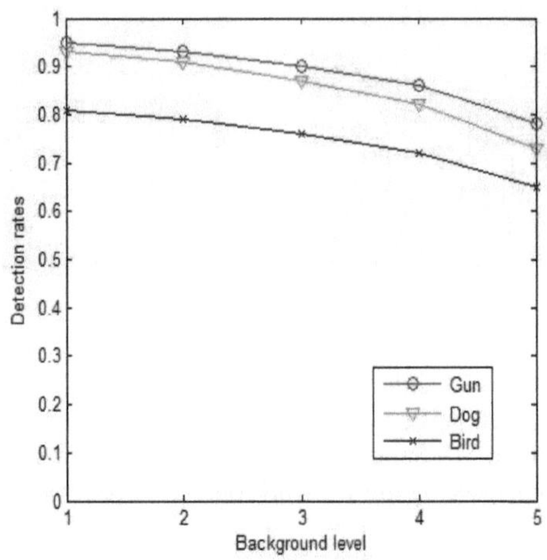

Fig. 5. Average detection rates with increasing background level. Level-1 indicates normal background level.

5 Conclusion and Discussions

In this paper, we present a method to detect auditory attention based on image saliency model and energy linear superposition. The key point of this method is how to locate salient part in spectrogram and estimate E_b accurately. We tested our method on a series of synthetic audio and experimental results proved that the method we proposed has a high performance on detecting salient sound in a smooth and steady background. However, the method proposed here will be high risk for a complex background.

To improve the performance of our method, future work includes taking more audio features into considerations, such as envelope, to assist locating salient part. For extracting salient sound in the audio with complex background, one acceptable method is to use statistical characteristics near salient part to estimate E_b instead of using mean value directly, since background sound should be regular in some way.

Acknowledgments. This work is partially supported by National Nature Science Foundation of China (61105076), China Post-doctoral Science Foundation (2012M511402), Fundamental Research Funds for the Central Universities of China (2012HGCX0001), and MOE project MOE2010-T2-1-087, Singapore.

References

1. Zienowicz, K., Shihab, A.I., Hunter, G.J.: Detecting, classifying and predicting salient events using acoustic signals and Markov models (2008)
2. Cai, R., Lu, L., Zhangand, H.-J., Cai, L.-H.: Highlight sound effects detection in audio stream. In: Proceedings of the 2003 International Conference on Multimedia and Expo, ICME 2003, vol. 3, IEEE (2003)
3. Rapantzikos, K., Evangelopoulos, G., Maragos, P., et al.: An audio-visual saliency model for movie summarization. In: IEEE 9th Workshop on Multimedia Signal Processing, MMSP 2007, pp. 320–323. IEEE (2007)
4. Zlatintsi, A., Maragos, P., Potamianos, A., Evangelopoulos, G.: A saliency-based approach to audio event detection and summarization. In: EUSIPCO 2012, pp. 1294–1298. IEEE (August 2012)
5. Guo, G., Li, S.Z.: Content-Based Audio Classification and Retrieval by Support Vector Machines. IEEE Transactions on Neural Networks 14(1), 209–215 (2003)
6. Liu, L., Zhang, H.J., Jiang, H.: Content Analysis for Audio Classification and Segmentation. IEEE Transactions on Speech and Audio Processing 10(7), 504–516 (2002)
7. Xiong, Z., Radhakrishnan, R., Divakaran, A., et al.: AudioEvents Detection Based Highlights Extraction from Baseball, Golf andSoccer Games in a Unified Framework. In: Proc. of ICASSP, Hong Kong (April 2003)
8. Chu, S., Narayanan, S., Kuo, C.C., et al.: Where am I?Scene Recognition for Mobile Robots using Audio Features. In: Proc. of ICME (July 2006)
9. Koch, C., Ullman, S.: Shifts in selective visual attention: towards the underlying neural circuitry. In: Matters of Intelligence, pp. 115–141. Springer Netherlands (1987)
10. Itti, L., Koch, C., Niebur, E.: A model of saliency-based visualattention for rapid scene analysis. IEEE Transactions on Pattern Analysis and MachineIntelligence 20(11), 1254–1259 (1998)
11. Kayser, C., Petkov, C.I., Lippert, M., Logothetis, N.K.: Mechanisms for allocating auditory attention: An auditory saliency map. Current Biology 15(21), 1943–1947 (2005)
12. Kalinli, O., Narayanan, S.: A saliency-based auditory attention model with applications to unsupervised prominent syllable detection in speech. In: Proc. of Interspeech (August 2007)
13. Duangudom, V., Anderson, D.V.: Using auditory saliency to understand complex auditory scenes. In: EUSIPCO 2007, pp. 1206–1210 (September 2007)
14. Kaya, E.M., Elhilali, M.: A temporal saliency map for modeling auditory attention. In: 46th CISS, pp. 1–6. IEEE (March 2012)

15. Wang, M., Hong, R., Yuan, X.-T., Yan, S., Chua, T.-S.: Movie2Comics: Towards a Lively Video Content Presentation. IEEE Transactions on Multimedia 14(3), 858–870 (2012)
16. Hong, R., Wang, M., Xu, M., Yan, S., Chua, T.-S.: Dynamic Captioning: Video Accessibility Enhancement for Hearing Impairment. In: ACM International Conference on Multimedia, ACM MM (2010)
17. Hong, R., Wang, M., Yuan, X.-T., Xu, M., Jiang, J., Yan, S., Chua, T.-S.: Video Accessibility Enhancement for Hearing Impaired Users. ACM Transactions on Multimedia Computing, Communications, and Applications 7S(1), 24–42 (2011)
18. Wang, M., Hong, R., Li, G., Zha, Z.-J., Yan, S., Chua, T.-S.: Event Driven Web Video Summarization by Tag Localization and Key-Shot Identification. IEEE Transactions on Multimedia 14(4), 975–985 (2012)

Perceptual Based Content Adaptive L_0 Smoothing

Fei Kou[1], Zhengguo Li[2], Changyun Wen[3], Weihai Chen[1], Changchen Zhao[1], and Jianhua Wang[1]

[1] School of Automation Science and Electrical Engineering, Beihang University, Beijing 100191, China
[2] Signal Processing Department, Institute for Infocomm Research, Singapore 639798, Singapore
[3] School of Electrical and Electronic Engineering, Nanyang Technological University, Singapore 138632, Singapore
koufei@yeah.net, ezgli@i2r.a-star.edu.sg, ecywen@ntu.edu.sg, {whchenbuaa,jhwangbuaa}@126.com, zcc8843@gmail.com

Abstract. Edge preserving smoothing is a technique to decompose an image into two layers - one smoothing layer and one detail layer. It is an important image editing tool. The edges are preserved in the smoothing layer and details are decomposed into the detail layer. In this paper, we propose a content adaptive L_0 smoothing method. Unlike common smoothing schemes, we use a perceptual based content adaptive weighted fidelity term. The algorithm gives a larger weight to the region with more information, which is most likely edges, and gives a smaller weight to the region with less information, which is most likely a flat area. So the resulting smoothed image can preserve more edges and smooth the smoothing areas better. Experimental results prove that the proposed method can have better results than existing L_0 smoothing method.

Keywords: image smoothing, SSIM, L_0 sparsity, detail enhancement, perceptual, content adaptive.

1 Introduction

Edge preserving smoothing is an important image editing method, which has wide range of applications in image processing and computer vision. With an edge preserving smoothing scheme, an input image is decomposed into two layers: a base layer and a detail layer. The details are decomposed into the detail layer, while the edges are decomposed into the base layer.

Recently, a lot of edge preserving smoothing schemes has been proposed. The most widely used edge preserving smoothing scheme is bilateral filter [1] involving two filters, one domain filter and one range filter. The range filter can preserve the edges. In [2], Farbman et al. proposed a weighted least squares based edge-preserving decompositions scheme. In [9], He et al. proposed an edge preserving scheme named guided filter that filters the image by considering the content of a guidance image.

B. Huet et al. (Eds.): PCM 2013, LNCS 8294, pp. 299–307, 2013.

Xu et al. proposed an L_0 norm based smoothing scheme [3]. It was shown experimentally in [3] that an L_0 norm based scheme could get better results. Further they developed an improved L_0 norm based scheme which can smooth images with textured surfaces [4]. In [5], Shen et al. proposed an L_0 based smoothing scheme. Different from [3, 4], they used L_1 fidelity instead of L_2 fidelity which was used in [3,4]. In [6], Kou et al. proposed an L_0 smoothing based detail enhancement scheme for fusion of differently exposed images.

Structural similarity (SSIM) is a perceptual based metric for measuring the similarity between two images. Different from the most widely used measuring metric - mean squared error (MSE), SSIM can penalize errors in accordance with their visibility. In [7], Yeo et al. found the relationship between SSIM and MSE and used SSIM to replace MSE in the rate-distortion optimization in video encoding. As a result, they achieved better compression while maintaining the same perceptual quality. Inspired by their scheme, we replace MSE by SSIM in L_0 smoothing. Compared with the original L_0 smoothing optimization problem, our optimization problem gives larger weights to the regions with larger variances, which are most likely edges, while smaller weights to the regions with smaller variances, which are most likely flat areas. The new scheme can preserve more edges and smooth the smoothing areas better in the smoothed image. Experimental results prove that the proposed method can give better results than existing L_0 based smoothing method.

The remainder of this paper is organized as follows. In the next section, we introduce the optimization problem of our L_0 smoothing scheme. In Section 3, we present an approximation solver of the problem in Section 2. In Section 4 our experimental results are illustrated to verify the performance of our proposed schemes, and finally the paper is concluded in Section 5.

2 Using SSIM in L0 Smoothing

SSIM measures the similarity between two images by the following formula [8].

$$SSIM(m,n) = \Big(\frac{2\mu_m\mu_n + c_1}{{\mu_m}^2 + {\mu_n}^2 + c_1}\Big)\Big(\frac{2\sigma_{mn} + c_2}{{\sigma_m}^2 + {\sigma_n}^2 + c_2}\Big) \tag{1}$$

where m and n are two images or two image regions, μ_m and μ_n are the averages of m and n, σ_m^2 and σ_n^2 are the variances of m and n respectively, σ_{mn} is the covariance of m and n, $c_1 = (\kappa_1 L)^2$ and $c_2 = (\kappa_2 L)^2$ are two constants to stabilize the division with weak denominator, $\kappa_1 = 0.01$ and $\kappa_2 = 0.03$, L is the peak value of the two images. For a 8-bit image, L is 256.

In [7], it was concluded that

$$\begin{aligned} dSSIM &= \frac{1}{SSIM} \\ &\approx 1 + \frac{MSE}{2{\sigma_m}^2 + c_2} \end{aligned} \tag{2}$$

where dSSIM is the derivative of SSIM. It is observed from (2) that dSSIM is a weighted MSE, and the weight is $2\sigma_m{}^2 + c_2$, where σ_m is the variances of m.

The original L_0 smoothing optimization problem in [3] is

$$\min_S \left\{ \lambda \cdot C(S) + \sum_p (S_p - I_p)^2 \right\} \tag{3}$$

where λ is a smoothing parameter, which can adjust the weight of the two parts to control the smoothing. Starting from the element at position $(1, 1)$ of a matrix, we can count the elements of the matrix in an order from left to right and then top to bottom. The subscript p denotes the p-th element of the matrix from such counting. S_p is the p-th pixel in image S, I_p is the p-th pixel in image I. $C(S)$ is the L_0 norm of the gradient of S, which is expressed as

$$C(S) = \#\{p|\, |\partial_x S_p| + |\partial_y S_p| \neq 0\} \tag{4}$$

where # stands for the number of p which satisfies $|\partial_x S_p| + |\partial_y S_p| \neq 0$, that is the L_0 norm.

A smaller value of the first term in (3) makes the resulting image S smoother, while a smaller value of the second term makes the image S more similar to the input image I. λ is a smoothing parameter to control the weight of the two terms.

We give different weights to different pixels based on their perceptual importance. Then the optimization problem in (3) is re-written as

$$\min_S \left\{ \lambda \cdot C(S) + \sum_p W_p (S_p - I_p)^2 \right\} \tag{5}$$

where W_p is the weight of pixel p.

$$\begin{aligned} W_p &= \frac{\frac{1}{N} \sum_{p'} \frac{1}{2\sigma^2(p')+c_2}}{\frac{1}{2\sigma^2(p)+c_2}} \\ &= \frac{1}{N} \sum_{p'} \frac{2\sigma^2(p) + c_2}{2\sigma^2(p') + c_2} \end{aligned} \tag{6}$$

where N is the number of pixels in the image. $\sigma^2(p)$ is the variance of the pixels in the neighborhood of the p-th pixel. p' stands for all the p in the image. After numerous experimental test, we use a 9×9 neighborhood in this paper.

3 Solver

Note that there is a discrete counting metric $C(S)$ in (5) and the optimization problem is not convex. So it is very difficult to solve it. As in [3], we change the optimization problem by minimizing the following problem to get an approximation solver

$$\min_{S,h,v}\left\{\lambda \cdot C(h,v) + \sum_p \left\{W_p(S_p - I_p)^2 + \beta\big((\partial_x S_p - h_p)^2 + (\partial_y S_p - v_p)^2\big)\right\}\right\} \tag{7}$$

where $C(h,v) = \#\{p| \, |h_p| + |v_p| \neq 0\}$, β is a parameter controlling the similarity between h, v and $\partial_x S, \partial_y S$. h, v are auxiliary matrices to solve S. h_p and v_p are the p-th element of h and v respectively. When β is large enough, the optimization problem (7) is equivalent to Eq. (5). So we can get the solution of Eq. (5) by solving (7).

Similar to [3], the problem (7) is solved through alternatively minimizing (h, v) and S. In each pass, one set of the variables is fixed with values obtained from the previous iteration. β is set as a small value β_0 at the beginning, and it is multiplied by a constant κ each time. The process ends when β is larger than β_{max}. The details are given as below.

Computing S when h and v are known: The S estimation subproblem corresponds to minimizing

$$\min_{S}\left\{\sum_p \left\{W_p(S_p - I_p)^2 + \beta\big((\partial_x S_p - h_p)^2 + (\partial_y S_p - v_p)^2\big)\right\}\right\} \tag{8}$$

Following similar analysis to [3], the global minimum of (8) is obtained. To accelerate computational speed, we diagonalize the derivative operator after Fast Fourier Transform (FFT) and this gives the solution:

$$S = \mathscr{F}^{-1}\left(\frac{W \cdot \mathscr{F}(I) + \beta(\mathscr{F}(\partial_x)^* \mathscr{F}(h) + \mathscr{F}(\partial_y)^* \mathscr{F}(v))}{W + \beta(\mathscr{F}(\partial_x)^* \mathscr{F}(\partial_x) + \mathscr{F}(\partial_y)^* \mathscr{F}(\partial_y)}\right) \tag{9}$$

where $\mathscr{F}$ is the FFT operator, $\mathscr{F}^{-1}$ is the IFFT operator and * denotes the complex conjugate.

Computing (h, v) when S is known: The objective function for (h, v) is

$$\min_{h,v}\left\{\frac{\lambda}{\beta} \cdot C(h,v) + \sum_p \left\{(\partial_x S_p - h_p)^2 + (\partial_y S_p - v_p)^2\right\}\right\} \tag{10}$$

Similar to [3], the solution is given by

$$(h_p, v_p) = \begin{cases} (0,0) & (\partial_x S_p)^2 + (\partial_y S_p)^2 \leq \dfrac{\lambda}{\beta} \\ (\partial_x S_p, \partial_y S_p) & otherwise \end{cases} \tag{11}$$

By estimating S with equation (9) and h, v with equation (11) alternatively, a smooth image S can be extracted.

4 Experimental Results

Fig. 1. Different selections of λ. Input image courtesy of Li Xu. (a) Input image. (b) Smoothing image obtained by $\lambda = 0.001$ (c) Smoothing image obtained by $\lambda = 0.01$. (d) Smoothing image obtained by $\lambda = 0.05$.

Readers are invited to read the electronic version with full-size figures in order to better appreciate the differences among images. In this section, we first evaluate the choice of λ in (7). As shown in Fig. 1, the smaller λ is, the more similar to the input image the smooth image is, the larger λ is, the less information is decomposed into the smooth layer. This is because if λ is large in (7), the first term will affect more on the result, which results in less none-zero gradients, the result image will be smoother. If λ is small, the second term will effort more on the result and this renders the result image to be more similar to the input image.

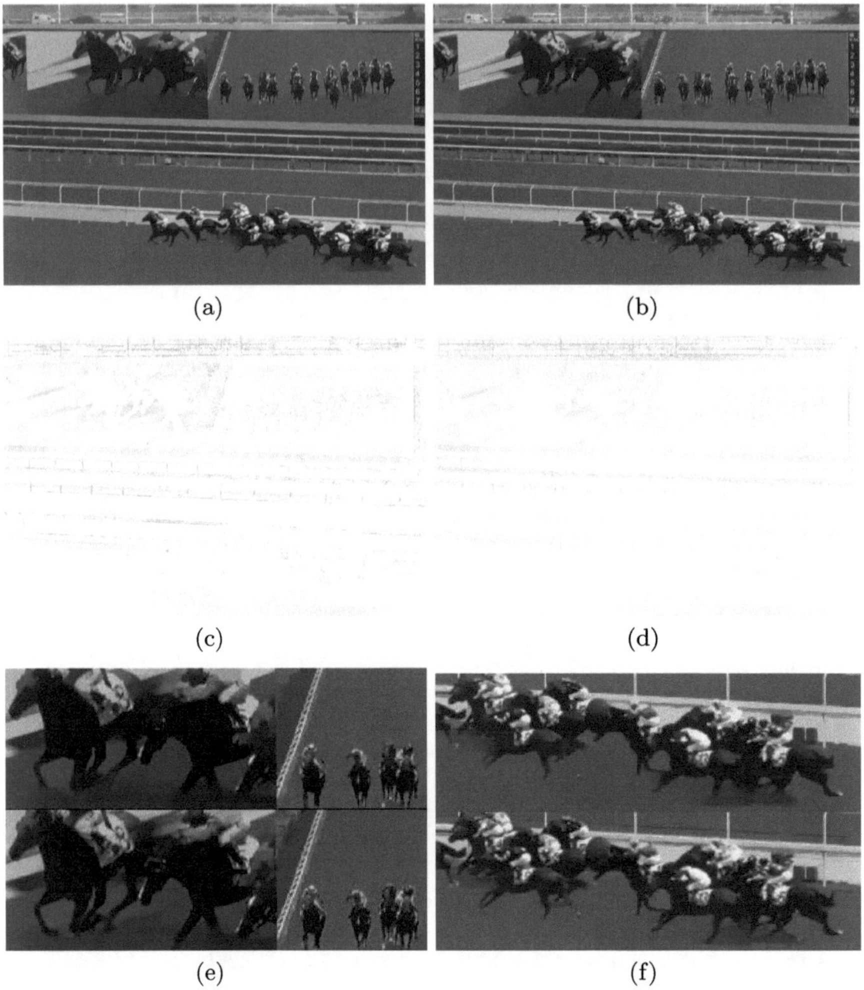

Fig. 2. Comparison of the decomposed results. To better appreciate the difference, the negative images of the detail images are given. Input image courtesy of Li Xu. (a) Smoothing layer obtained by the scheme in [3]. (b) Smoothing layer obtained by the proposed scheme. (c) Detail layer obtained by the scheme in [3]. (d) Detail layer obtained by the proposed scheme. (e) Closeups of (a) and (b). (f) Closeups of (a) and (b).

Then we compare the decomposition result by our scheme with scheme in [3] by choosing the same parameters for both schemes. From Fig. 2, it is seen that with the proposed method, the edges are more clear in the smooth layer than the result in [3] while there are less edges in the detail layer by the proposed method. For example, the horse legs are decomposed into the detail layer by the scheme in [3], while into the smooth layer with the proposed scheme. The horse legs are more clear in the smooth layer based on our scheme. All these can be

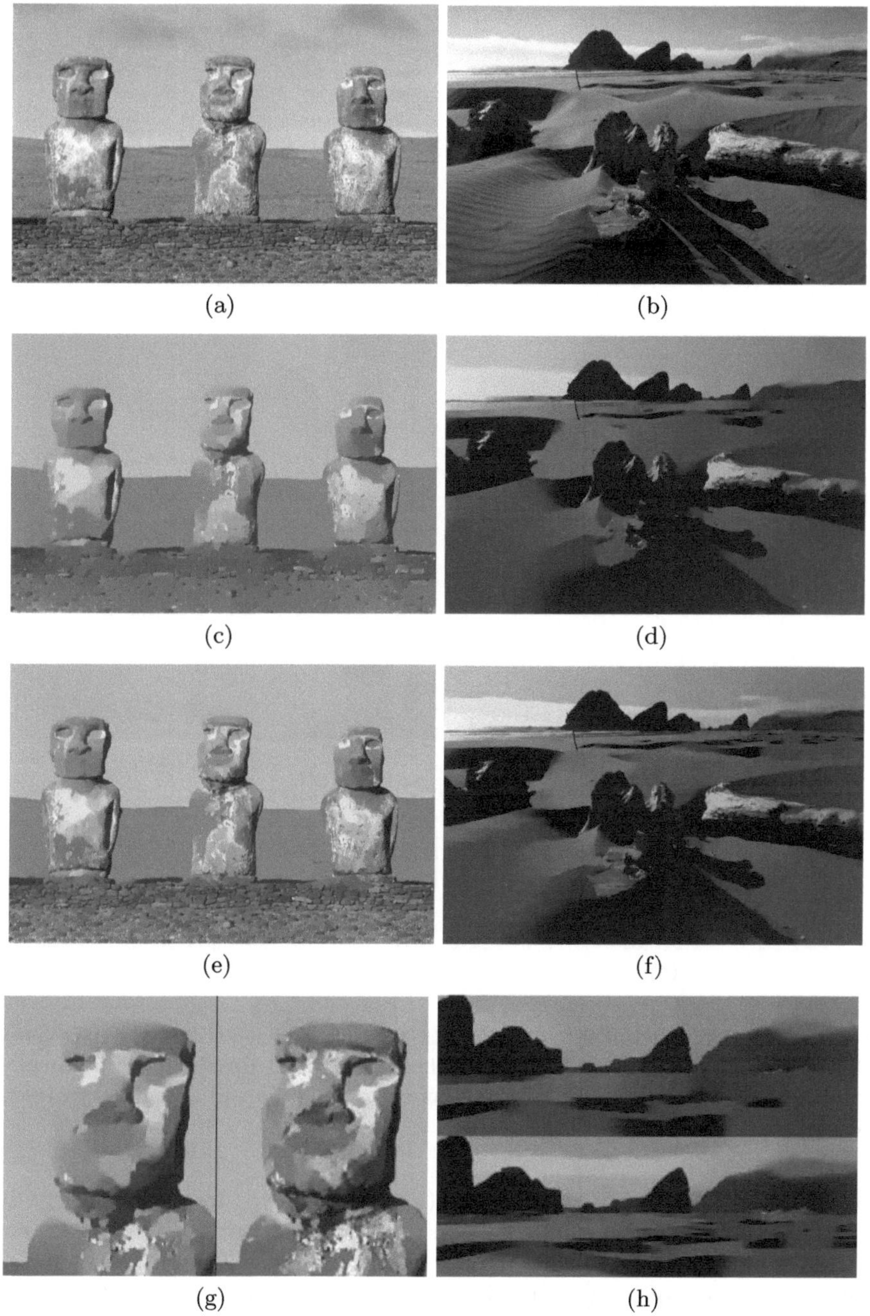

Fig. 3. Comparison of smoothing results. (a) First input image from [2]. (b) Second input image, image courtesy of Norman Koren. (c) Result image obtained by the scheme in [3]. (d) Result image obtained by the scheme in [3]. (e) Result image by the proposed scheme. (f) Result image obtained by the proposed scheme. (g) Closeups of (c) and (e). (h) Closeups of (d) and (f).

Fig. 4. Comparison of detail enhanced results. (a) Result image obtained by the scheme in [3]. (b) Result image obtained by the scheme in [3]. (c) Result image obtained by the proposed scheme. (d) Result image obtained by the proposed scheme. (e) Closeups of (a) and (c). (f) Closeups of (b) and (d).

easily observed in the closeups of Fig. 2(a) and Fig. 2(b) as shown in Fig. 2(e) and Fig. 2(f). From these results, we can conclude that our scheme can preserve edges in the smooth layer better than the existing L_0 based smooth scheme in [3]. Additional two sets of images are also tested and presented in Fig. 3.

Finally, we compare the detail enhance results by our scheme and the scheme in [3]. The detail enhanced images are generated by adding the detail layers to the source images. The original L_0 smoothing scheme in [3] may decompose edges into detail layer. As a result, the result image may be over sharpened around edges. As shown in Fig. 4(a) and 4(c) which are generated by the scheme in [3], several regions are much darker or whiter than those in Fig. 4(b) and 4(d) which are generated by the proposed scheme. A lot of details are missing in Fig. 4(a) and 4(c) in these regions. Based on all the experimental results illustrated, our scheme can give better results.

5 Conclusion

In this paper, a content adaptive L_0 norm based smoothing scheme has been proposed by using a content adaptive weighted fidelity term with larger weights given to the edges and smaller weights to the flat areas. As a result, the resulting image can preserve more edges and smooth the smoothing areas better than existing L_0 norm based smoothing scheme.

Acknowledgment. This work has been supported by National Nature Science Foundation of China under the research project 61075075 and 61175108.

References

1. Tomasi, C., Manduchi, R.: Bilateral filtering for gray and color images. In: Proceedings of the IEEE International Conference on Computer Vision, pp. 839–846 (1998)
2. Farbman, Z., Fattal, R., Lischinshi, D., Szeliski, R.: Edge-preserving decompositions for multi-scale tone and details manipulation. ACM Transactions on Graphics 27(3), 249–256 (2008)
3. Xu, L., Lu, C., Xu, Y., Jia, J.: Image smoothing via L_0 gradient minimization. ACM Transactions on Graphics (SIGGRAPH Asia 2011) 30(6) (2011)
4. Xu, L., Yan, Q., Xia, Y., Jia, J.: Structure Extraction from Texture via Relative Total Variation. ACM Transactions on Graphics (TOG) 31(6), 139 (2012); Proc. ACM SIGGRAPH ASIA 2012 (2012)
5. Shen, C.T., Chang, F.J., Hung, Y.P., Pei, S.C.: Edge-preserving image decomposition using L1 fidelity with L0 gradient. In: SIGGRAPH Asia 2012 Technical Briefs, p. 6 (2012)
6. Kou, F., Li, Z., Wen, C., Chen, W.: L0 Smoothing Based Detail Enhancement for Fusion of Differently Exposed Images. In: in 8th IEEE Conference on Industrial Electronics and Applications (ICIEA 2013), pp. 1398–1403 (2013)
7. Yeo, C., Tan, H.L., Tan, Y.H.: On rate distortion optimization using SSIM. In: IEEE International Conference on Acoustics, Speech and Signal Processing (ICASSP 2012), pp. 833–836 (2012)
8. Wang, Z., Bovik, A.C., Sheikh, H.R., Simoncelli, E.P.: Image quality assessment: From error visibility to structural similarity. IEEE Transactions on Image Processing 13, 600–612 (2004)
9. He, K., Sun, J., Tang, X.: Guided image filtering. In: Daniilidis, K., Maragos, P., Paragios, N. (eds.) ECCV 2010, Part I. LNCS, vol. 6311, pp. 1–14. Springer, Heidelberg (2010)

Fusing Multiple Visual Features for Image Complexity Evaluation

Kunyan Yin, Linbo Wang, and Yanwen Guo*

National Key Lab for Novel Software Technology
Nanjing University, China PR
{yinkunyancs,wanglb.2005}@gmail.com, ywguo@nju.edu.cn

Abstract. In spite of the wide applications in computer vision and cognitive research area, defining an effective complexity measure for color images remains a challenging task. Conventional approaches are generally built upon information theory or a certain visual feature. In this paper, we propose a new method directly exploiting multiple effective visual features including color, clutter and the number of objects to measure the complexity of color images. Furthermore, we present a fuzzy clustering model for combining all the proposed features, which provides specific scores to evaluate image complexity. Experimental results are presented and show good consistency between the proposed objective metric and subjective assessment by human observers.

Keywords: Image complexity, Feature extraction, Gradient orientation, Clustering, Image analysis.

1 Introduction

Evaluating the complexity of an image is very important in the fields of esthetics, visual psychology and cognitive science due to that image complexity reflects a fundamental aspect of human vision system and an effective approach to measure it may be a stepping-stone for better understanding of human recognition process. When it comes to computer science and technologies, the study of image complexity is particularly beneficial to various computer vision applications, including image compression [1], aesthetic evaluation [2], automatic target recognizers [3], watermarking [4], image retargeting [5], image retrieval [6], and photo collage [7]. The psychologists, Oliva et al. [8] launches an investigation on visual complexity and points out that visual complexity originates from color, clutter, symmetry, organization and number of objects and so on. Visual complexity is further closely related to how we perceive and process visual stimuli, which makes it difficult to define image complexity. Early work by Klinger et al. [9] on estimating the image complexity are built upon the content order and disorder in the target image. Salingaros et al. [10] describes the complexity based on the analysis of edge level percentages in the image. Although the order or the

* Corresponding author.

B. Huet et al. (Eds.): PCM 2013, LNCS 8294, pp. 308–317, 2013.

(a) low complexity (b) medium complexity (c) high complexity

Fig. 1. Images of different complexities

number of edge solely plays an important role in image complexity, they both ignore the fact that complexity is dependent on multiple features rather than a single one. Fig. 1 obviously suggests that visual complexity is attributed to multiple features. With the development of information theory, image complexity is directly related to the entropy of image intensity histogram. However, this measure do not take into account the spatial distribution.

In order to solve this problem, an improved method is given in [11], which computes mutual information between local histogram distributions belonging to different regions of the segmented image. In fact, segmentation of nontrivial images itself is one of the most difficult tasks in image processing. In order to provide an objective complexity metric that holds good consistency with the subjective assessment by human perception, we propose a new fuzzy image complexity measure based on visual complexity theory. Our method takes into account multiple fundamental image features, such as color, clutter and the number of objects, leading to visual complexity. We also present a specific model to evaluate the complexity combining these key features.

The rest of this paper is organized as follows. In section 2 , we first introduce three features of image complexity based on color, clutter(gradient orientation) and number of object (edge). Then we present an effective model combining the proposed features using fuzzy clustering theory. In section 3, via extensive experiments, we show that the proposed objective complexity values closely match the subjective ground-truth given by observers, suggesting that our proposed objective metrics are consistent with human perception mechanism. In section 4, we conclude the whole paper and discuss future works..

2 Our Approach

Our basic idea for assessing image complexity is to choose effective features reflecting the complexity of image content and to define a quantitative metric to evaluate the complexity. Following this idea, we propose a solution mainly consisting of two phases. The first phase is to extract effective features based on visual complexity theory, followed by providing a model combining the features and obtaining a specific score to measure image complexity.

2.1 Extraction of Multiple Features

Olives' visual complexity theory [8] suggests that visual complexity originates from the number of objects, clutter, symmetry, organization and variety of colors and so on. It is impractical to extract all features leading to visual complexity. Instead, we resort to exploiting the potential effective features essential for visual complexity. By observation and experimental validation, we select the three key features, color, clutter and the number of objects to evaluate image complexity. In this section, we describe how to extract these features.

Color Complexity. The variety of color is one of the key features which contribute to visual complexity. Fig.1 also shows that vision is sensitive to varied color. To extract this feature, we propose a histogram-based method to define color complexity by color statistics of the input image. Specifically, the complexity of one color channel about an image is defined as,

$$C_{ch} = 1 - \frac{\sqrt{\sum_{i=1}^{n}(f(i) - \frac{1}{n})^2}}{\sigma_{chmax}} \tag{1}$$

where n is the number of bins in histogram, $f(i)$ is the frequency of the i-th bin in histogram, The σ_{chmax} is computed as,

$$\sigma_{chmax} = \sqrt{(1 - \frac{1}{n})^2 + (\frac{1}{n})^2(n-1)} \tag{2}$$

The intuition behind this cue is that the more complex the image looks, the more histogram-equalized the image should be, and hence the less variation the color histogram possesses. σ_{chmax} is an normalization term, which corresponds to the scenario that the image looks extremely simple, consisting of single or slightly varied color. In our application scenario, a color space whose metric has uniform visual perception is desired. We use the HSV space based on opponent-color coordinates, which is more perceptually uniform than several well known color spaces such as RGB, and CIE XYZ. Finally, the color complexity C_{color} is defined as,

$$C_{color} = w_h C_h + w_s C_s + w_v C_v \tag{3}$$

In all experiments in this paper, we use $w_h = w_s = w_v = \frac{1}{3}$, and the number of histograms bins for H, S, V is 32. If the image is gray level, we only statistic the gray value of the input image. Fig.2 shows different complexity values based on HSV and the value of red line is $\frac{1}{n}$ corresponding to the case when the image is painted with a large amount of colors and hence can be regarded as extremely complex. The value of an colorful images histogram bins should be close to the red line.

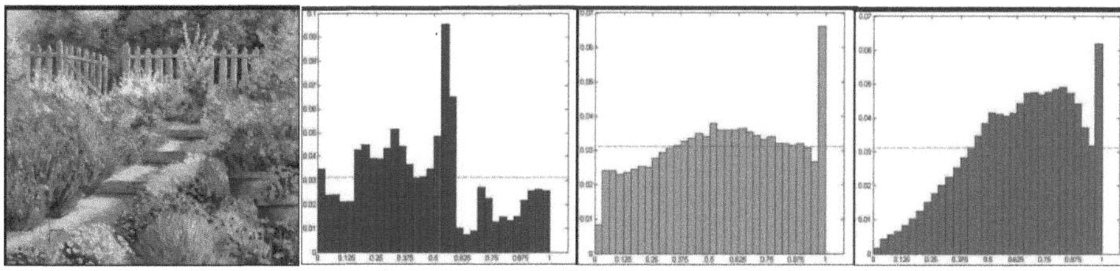

Fig. 2. An example of color complexity evaluation

Gradient Orientation Complexity. As observed from Fig.1, our vision system is also sensitive to the clutter and organization in an image. Intuitively, the perceptual chaos can be caused by the gradient orientation in the sense that the more diverse the orientations of gradient are, the more complex the image looks. Therefore, it should be safe to assume that gradient orientation is another key factor that contributes to image complexity. Following this assumption, we propose the second image complexity feature based on histogram of gradient orientation. More specifically, we measure the complexity by inspecting the variation of the gradient orientation histogram, which is similar to the schema proposed for color complexity. In order to achieve better performance, we investigated various discrete derivative masks for gradient orientation computation, such as various 1-D point derivatives (uncentred [-1,1], centred [-1,0,1]), 3*3 Sobel masks and 2*2 diagonal ones. Among them, we find that the simple 1-D [-1,0,1] mask works best. With the orientation bins evenly spanned over $0^o - 180^o$(unsigned gradient) or $0^o - 360^o$(signed gradient), we define the gradient orientation complexity C_{orient} measure as,

$$C_{orient} = 1 - \frac{\sqrt{\sum_{i=1}^{m}(g(i) - \frac{1}{m})^2}}{\sigma_{gmax}} \tag{4}$$

where m is the number of histogram bins, $g(i)$ is the frequency of the i-th bin in histogram and σ_{gmax} is the max gradient orientation variance which means there is only one gradient orientation in gradient image. σ_{gmax} can be computed as,

$$\sigma_{gmax} = \sqrt{(1 - \frac{1}{m})^2 + (\frac{1}{m})^2(m-1)} \tag{5}$$

The intuitive understanding of the formulation is analogous to the one of color complexity. And as suggested in our experiment, this complexity measure can be very useful in distinguishing the texture and non-texture images. Most texture or ordered images exhibits strong pattern and results in sparse orientation histogram and large covariance, while the non-texture or disordered ones tend to be chaotic and show high complexity. In Fig.3, we illustrate two representative examples and their respective evaluation scores of gradient orientation complexity. As Fig.3 is presented that an image with disordered content are very likely

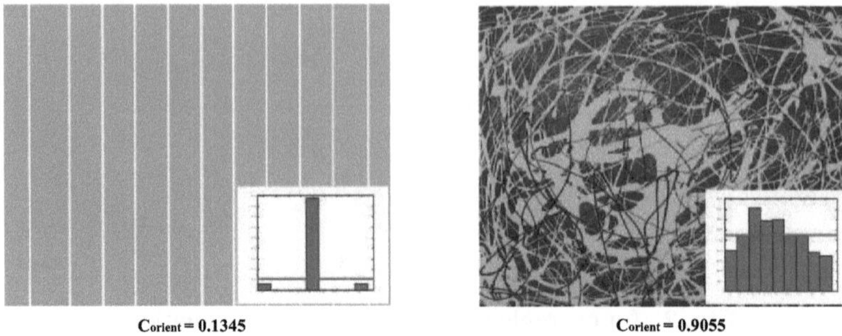

Fig. 3. Comparison results of gradient orientation complexity

to possess gradients with all directions, which further lead to a uniform distribution of orientation, in other words, a flat histogram of orientation with no sharp bins.

Edge Complexity. It is self-evident that the number of objects (the quantity of edge in the image) also has direct influence on visual complexity. The number of strong edges indicate the existence of objects, which leads to our last feature candidate for describing image complexity. Here, we borrowed a fuzzy clustering based approach proposed by Mario et al. [10] to obtain the edge complexity C_{edge}. Specifically speaking, it includes:

1. Compute the gradient map and statics the percentages of different edge levels.
2. Determinate the complexity class(little complex, more or less complex and very complex)using centroids obtained from a fuzzy clustering.
3. obtain a final specific score by a set of interval mapping functions according to three complexity categories above. We regard this score as edge complexity value.

2.2 The Fuzzy Image Complexity Model

Given a test image, each of the complexity measure proposed above can infer a complexity score, which indicates its complexity from different perspective. Therefore, the derived values can be very different from each other and it is more desirable to acquire an overall score by evaluating all the perspectives. To achieve this goal, we provide a comprehensive model to fuse all the complexity measures. More specifically, after computing the color complexity, gradient orientation complexity and edge complexity, each image complexity can be represented by the vector:

$$\mathbf{C} = [C_{color}, C_{orient}, C_{edge}] \tag{6}$$

Using the information of the **C** vectors which are obtained from 900 images, we get three centroids with a fuzzy C-means algorithm. These centroids are

Table 1. Representative vectors of complexity class

Class	Vectors		
	C_{color}	C_{orient}	C_{edge}
V_L	0.2119	0.3721	0.2963
V_M	0.6981	0.6725	0.5297
V_H	0.7826	0.8999	0.8763

the representative vectors for each classes, LC(Low Complexity), MC(Medium Complexity) and HC(High Complexity), represented by V_L, V_M and V_H. The values of these vectors are given in Table 1. Thus, the complexity class $\hat{n}$ of an image can be now determined by the distance between **C** and the centroids,

$$\hat{n} = \underset{n}{argmin} \quad d(\mathbf{C}, \mathbf{V_n}) \tag{7}$$

where,

$$d(\mathbf{C}, \mathbf{V}) = ||\mathbf{C} - \mathbf{V}|| = \sqrt{\sum_{i=1}^{m} (c_i - v_i)^2} \tag{8}$$

With the model, we can only determine the complexity class, but it could not have a comparison with the complexity of images in the same class. We introduce the approach in [10] to acquire a specific score for image complexity. We first map the different complexity classes into different subintervals, as shown in Fig.4. Thereafter, each subinterval is further divided into two segments A and B. A is the region left of the midpoint of the three regions correspond to the subinterval, while region B is the other side. For example, if the distance from **C** to a class centroid is zero, the complexity score equals the midpoint of the subinterval. When it is not zero, the complexity score is acquired by a set of interval mapping functions. The mapping function at subinterval n is defined as,

$$\Psi_n = \begin{cases} M_n & d(\mathbf{C_n}, \mathbf{V_n}) = 0 \\ \max(0, \Psi_n^{'}) \quad AND \quad \min(\Psi_n^{'}, M_n) & |\mathbf{C_n}| < |\mathbf{V_n}| \\ \max(M_n, M_n + \Psi_n^{'}) \quad AND \quad \min(M_n + \Psi_n^{'}, MAX_n) & |\mathbf{C_n}| > |\mathbf{V_n}| \end{cases} \tag{9}$$

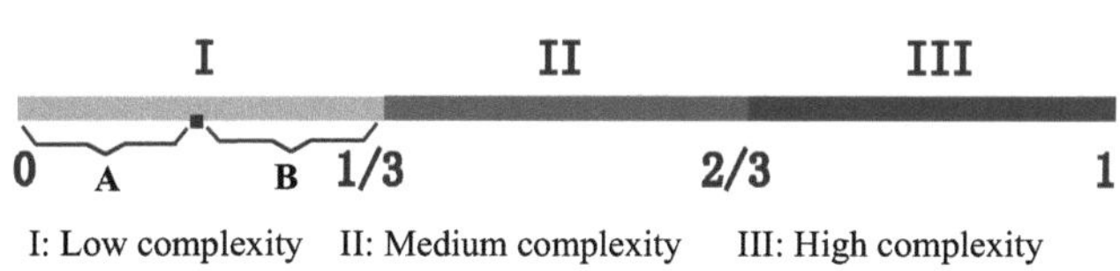

Fig. 4. Complexity region mapping

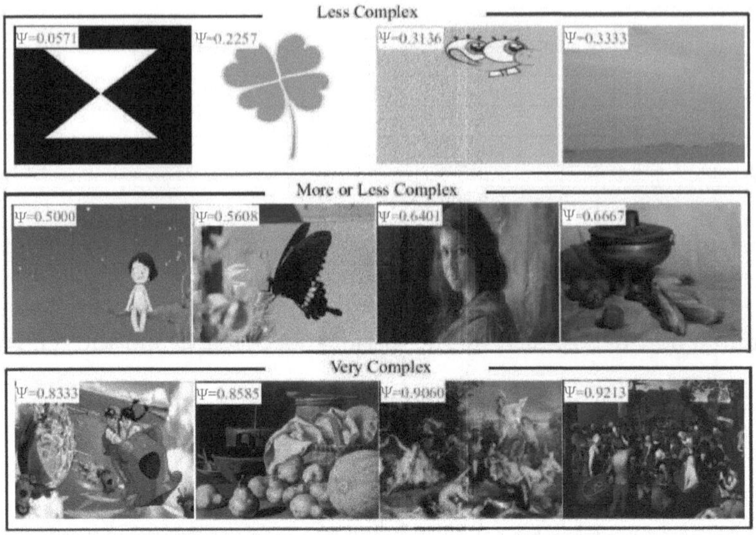

Fig. 5. Complexity scores evaluated for example images in our dataset

For complexity class n, M_n is the value of midpoint in its mapping subinterval, and MAX_n is the max value of its mapping subinterval. $\mathbf{C_n}$ is the complexity vector of an image belonging to complexity class n. $\mathbf{V_n}$ is the centroid of complexity class n. When $|\mathbf{C_n}| < |\mathbf{V_n}|$, the complexity score is on region A, or it is on region B. $\Psi_n^{'}$ can be computed as,

$$\Psi_n^{'} = \frac{1}{6}\frac{d(\mathbf{C_n^{'}}, \mathbf{V_n}) - \mu_n}{\sigma_n} \tag{10}$$

where μ_n is the mean of the distances between the samples' complexity vectors $\mathbf{C_n^{'}}$ and the centroid $\mathbf{V_n}$, and σ_n is the standard deviation.

For an image, we compute its objective complexity score as follows: we first determine its complexity class, then use interval mapping function to obtain a specific score Ψ for the image. For example, if an image belongs to complexity class L, the final complexity result Ψ of the image is Ψ_L.

3 Experiment

In order to validate the effectiveness of the proposed approach, we conduct experiments on a data set with images randomly collected from Internet. We also invite several human testers to score the complexity of the images. By comparing the scores derived with the two methods, we expect to evaluate how consistent with subjective assessment our algorithm is. More specifically, the experiment proceeds as follows:

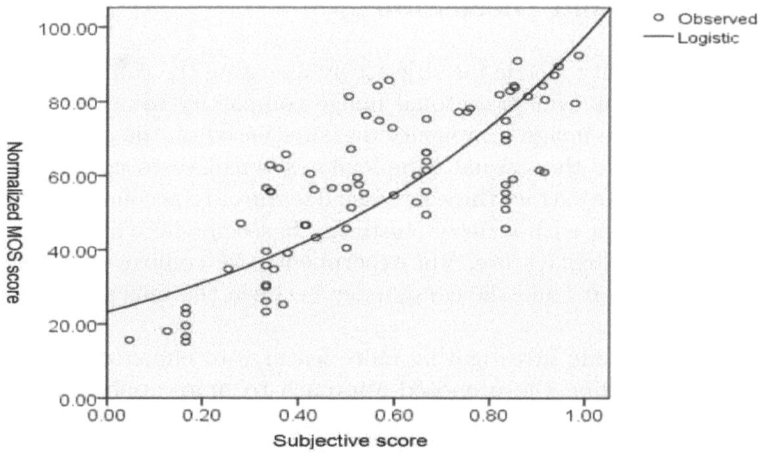

Fig. 6. The fitted curving using logistic function

Participant. 10 university students (5 females and 5 males) with normal visual acuity and normal color vision participated in the study. They were all novel to test. We provided 80 images including 20 natural images, 20 people images, 20 objects images and 20 cartoon images.

Procedure. Participants provided their rating scores according to inter scales of low , medium or high. The rating scores for the three intervals of scale were 1- 5(low), 6-10(medium), 11-15(high). All participants completed 10 trials in the experiment.

Result. The raw scores were firstly normalized by the mean and variance of scores for each participant. Then the score was rescaled to fill the range from 1 to 100. We computed mean opinion scores (MOSs) for each image. Higher MOSs corresponded to higher image complexity.

To measure the performance of image complexity model. the evaluation method proposed by video quality experts group is applied. A nonlinear mapping between the objective/ subjective scores is used with a logistic function:

$$Y = \frac{\beta_1 - \beta_2}{1 + e^{\left(-\frac{X-\beta_3}{|\beta_4|}\right)}} + \beta_2 \tag{11}$$

where the initial estimates of parameters are:

$\beta_1 = max(Y), \beta_2 = min(Y), \beta_3 = mean(X), \beta_4 = 1$

An iterative process is evoked to find optimal β_1, β_2, β_3, β_4 using the SPSS software. The fitting curving is in Fig.6.

Discussion. In this experiment, Spearman rank-order correlation coefficient (r = 0.8698) indicates that there is a high similarity between MOSs and ranking scores of objective image complexity, which means objective computed complexity of image, assessed by the metric proposed in section 2, is consistent with human subjective.

4 Conclusion and Discussion

In this paper, a quality metric for objectively assessing the complexity of image is proposed. Different from traditional image complexity measures, we present a new comprehensive image complexity measure based on the observation from biological perspective that visual complexity is sensitive to color, clutter and amount of objects. We extract three key visual features to account for image complexity and fuse them with a fuzzy clustering based model which also presents a quantitative complexity score. Via experiments, we confirm the effectiveness of the proposed features and the consistency between the final score and human objective assessment.

Future works include investigating more features to enhance the assessment accuracy and extending the proposed approach to more applications, such as image collage, ART and image enhancement.

Acknowledgements. This work was supported in part by the Innovation Foundation for Graduate Students of Jiangsu Province China (CXLX12_0053), the National Science Foundation of China under Grants 61073098, 61021062, 61373059, the National Fundamental Research Program of China (2010CB327903).

References

1. Veisi, H., Jamzad, M.: A complexity-based approach in image compression using neural networks. Int. J. Sign. Process. 5, 82–92 (2009)
2. Guo, Y., Liu, M., Gu, T., Wang, W.: Improving photo composition elegantly: Considering image similarity during composition optimization. In: Computer Graphics Forum., vol. 31, pp. 2193–2202. Wiley Online Library (2012)
3. Peters, R.A., Strickland, R.N.: Image complexity metrics for automatic target recognizers. In: Automatic Target Recognizer System and Technology Conference, pp. 1–17 (1990)
4. Yaghmaee, F., Jamzad, M.: Estimating watermarking capacity in gray scale images based on image complexity. EURASIP Journal on Advances in Signal Processing 2010, 8 (2010)
5. Guo, Y., Liu, F., Shi, J., Zhou, Z.H., Gleicher, M.: Image retargeting using mesh parametrization. IEEE Transactions on Multimedia 11(5), 856–867 (2009)
6. Perkiö, J., Hyvärinen, A.: Modelling image complexity by independent component analysis, with application to content-based image retrieval. In: Alippi, C., Polycarpou, M., Panayiotou, C., Ellinas, G. (eds.) ICANN 2009, Part II. LNCS, vol. 5769, pp. 704–714. Springer, Heidelberg (2009)
7. Yu, Z., Lu, L., Guo, Y., Fan, R., Liu, M., Wang, W.: Content-aware photo collage using circle packing. IEEE Transactions on Visualization and Computer Graphics 99(PrePrints) (2013)
8. Oliva, A., Mack, M.L., Shrestha, M., Peeper, A.: Identifying the perceptual dimensions of visual complexity of scenes. In: Proc. of the 26th Annual Meeting of the Cogn. Sci. Soc., pp. 101–106 (2004)
9. Klinger, A., Salingaros, N.A.: A pattern measure. arXiv preprint arXiv:1108.5508 (2011)

10. Mario, I., Chacon, M., Alma, D., Corral, S.: Image complexity measure: a human criterion free approach. In: Annual Meeting of the North American Fuzzy Information Processing Society, NAFIPS 2005, pp. 241–246. IEEE (2005)
11. Rigau, J., Feixas, M., Sbert, M.: An information-theoretic framework for image complexity. In: Proceedings of the First Eurographics Conference on Computational Aesthetics in Graphics, Visualization and Imaging, Eurographics Association, pp. 177–184 (2005)

An Invisible Gorilla: Is It a Matter of Focus of Attention?

Chen Wang[1], Pablo Cesar[1], and Erik Geelhoed[2]

[1] Centrum voor Wiskunde&Informatica, the Netherlands
{cw,p.s.cesar}@cwi.nl
[2] Falmouth University, UK
erik.geelhoed@falmouth.ac.uk

Abstract. How to evaluate users' attention level in a video task is a challenge. One of the conventional methods is to link users' focus of attention to their performance undertaken in a video. However, this is not always true in a video environment, as users' poor performance may be resulted from some other reasons rather than a lack of focus of attention. In this article, we demonstrated our assumption by using an Electroencephalography (EEG) sensor, which measured the users' attention level in a video task. Our results showed one case that some of the users with a poor performance in the video task had the same level of attention compared to those users with a good performance. In particular, an interference object in the video, aimed for distracting the users' attention, had no impact on some of the users' focus when they were already involved in the video task.

Keywords: EEG Sensors, focus of attention, video training exercises.

1 Introduction

Focus of attention (similar to concentration), in cognitive neuroscience, refers to the individuals' capacity to choose what they pay attention to and what they ignore [10]. It is also known as endogenous attention or executive attention.

Focus of attention is important for learning. The ability to regulate and direct attention makes one being able to guide their attention towards the information-rich areas for learning. Moreover, people believe that the focus of attention has influence on the studying. The higher focus one has during lecture, the higher possibility is to acquire a good score in a test.

Previous studies have proven that multimedia learning resources can be used to enhance users' attention and motivation [12]. Many multimedia training applications are developed to train and improve users' ability of focus. They are considered useful and helpful, especially used as exercises to train users how to concentrate at a certain task. In these training videos, normally there is a task or more than one task included, and users are required to fulfill the task with sufficient attention. Through many training exercises, users may improve their ability of focus which might be benefit for their studies and reduce learning time.

B. Huet et al. (Eds.): PCM 2013, LNCS 8294, pp. 318–326, 2013.

Focus of attention is one of the important factors during studying and it may be intensified through video training exercises. But we cannot simply use the performance results in video training exercises to judge how much attention users have paid. Users may give a poor performance due to a lack of focus or users can not follow the task properly. In the latter situation, the reasons may due to unconsciously users' eyes moved away from the target, but users were still focused at the task they were conducting. Second, some training videos design an interference object in a task. The purpose of using an interference object is to train users how to focus at the task with ignoring the influence of the interference object. Some of the users may alter their visual focus to the interference object when they were involved at the task, while some of the other users may not notice the presence of the interference object as they were concentrated at the task [7]. Therefore, without users' attention feedback, it is difficult to judge whether users have paid different level of focus of attention when they obtained different performance scores in a video task. In particular, there is no proof to conclude that users' attention dropped at the presence of an interference object without users' attention feedback.

In this article, we used an EEG sensor to measure the user s' focus of attention in a training video exercise. With the help of an EEG sensor, we monitored users' attention levels when they were in the task so that we could link the users' attention values to the users' performance results. In addition, we evaluated the effects of an interference object on the users' focus of attention in the training video exercise. Our studies aim to validate the following two hypotheses:

H1: In a video training exercise, users may show the same attention level even though they have different performance scores.

H2: In a video training exercise, users may have the ability to keep attention at the presence of the interference object.

By validating our hypothesis, they may be beneficial for us to learn more knowledge about users' behavior of focus of attention during in a multimedia environment. Our results contribute the further studies on users' focus of attention in terms of multimedia consumption. First, users' attention feedback might be helpful for trainers to objective judge users' performance in a video training exercise. Second, attention feedback may also useful in e-learning environment, e.g., teachers obtain students' feedback from a remote location. Last, it opens a door to the researches on the multimedia design, especially how to design an effective video by detecting users' focus of attention for the purpose of training, teaching and some other applications, for instance advertisement.

This article is structured as follow. The next section describes the related work. Then section 3 describes the experiment design and reports the results. Section 4 discusses the results. And finally, section 5 concludes the article.

2 Related Work

Attention is a complex process. Users' attention level may be affected by the media form, media content and user characteristics [11]. In the previous studies, neuroscientist has found that the users found some tasks difficult done simultaneously if these tasks require the control from the same area of a brain, whereas the users tend to be capable of dealing some of the tasks in the same time if they require the different control parts from a brain. Therefore, it is possible for a user maintains the focus when he or she is conducting multiple tasks.

Recent advancement in the neuroscience technology makes focus of attention measureable. Users' focus of attention is measured by detecting users' EEG electrical activity along the scalp. The EEG sensor is used extensively in neuroscience, cognitive science and cognitive psychology research. The EEG sensor can easily be used without requiring bulky and immobile equipment. Mick et al. [9] studied the accuracy and reliability of the MindSet [2], a type of EEG sensor produced by NeuroSky [8]; Raffaella et al. [3], [4], [5] used this type of (or similar) EEG sensor to detect users' focus of attention in the different applications.

Our research interests were motivated by the availability and convenience of an EEG sensor. An EEG sensor provides the user's attention feedback, which can be used to measure users' attention level in a video consumption. The contributions of this paper are to provide the methodologies to investigate users' behaviour of focus of attention by using an EEG sensor. Educators may use them as a feedback to link users' focus of attention to their performance results in a video training exercise. Teachers may find them beneficial to get an attention feedback from the students sitting in a remote location. Video designers may try to test different design concepts in order to produce the effective videos in the different applications.

3 Experiment Design

3.1 Sensor

The MindWave EEG sensor, a new version of MindSet, processes brainwaves into digital signals to make the measurement available to games and some other applications, for instance attention training. The MindWave uses eSense [6] algorithm to calculate users' engaging attention (similar to concentration), and the algorithm amplifies the raw brainwave signal and removes the ambient noise and muscle movement.

MindWave communicates with the server using a RF dongle. It is attached on the users' forehead, with also reference points located on the ear pad (Fig. 1). The eSense meters (attention) are calculated and being sent to the server. The user' attention value is reported on a relative scale from 1 to 100. A value in the range of 40 to 60 is considered "neutral". A value from 60 to 80 is considered "higher attention", and values from 80 to 100 are considered "highest attention". Similar, on the other end of the scale, a value from 20 to 40 indicates "lower attention", while a value in the range of 1 to 20 indicates "lowest attention"[13].

Fig. 1. MindWave EEG sensor

3.2 Server-Side Component

A laptop is used as the server-side component, and it acts as the monitoring centre. The brain wave sensor communicates with the laptop using a RF dongle. All the data is sampled at 514Hz. The software for monitoring and mixing the data from a brain wave sensor is written in Python and C++ language.

The inputs from a brain wave sensor are being sent to the laptop at the same time. Fig. 2, for example, visualizes the data gathered by the brainwave sensor. In the attention graph, the green line represents the attention value and the red line is the meditation value. These two values reflect the user's internal attention and meditation state.

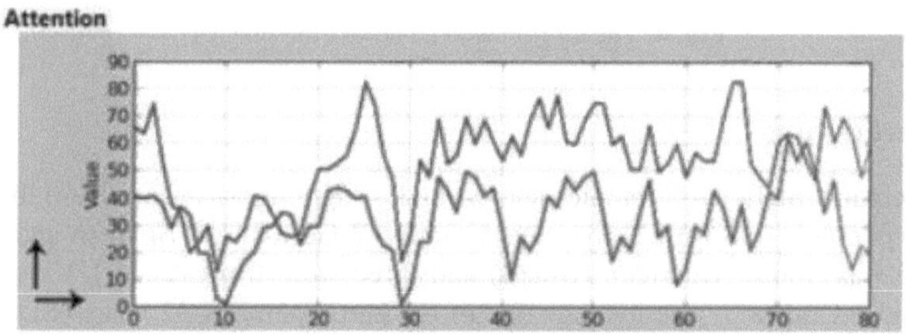

Fig. 2. Real-Time visualization of the attention and meditation

3.3 Video Selection

The video selected in our experiment was produced by the psychologists Daniel Simons and Christopher Chabris [1]. The video showed six people, three in black and three in white, passing a basketball back and forth (Fig. 3). In the meanwhile, a person in a gorilla costume walked across the group of players (Fig. 4). The task in this video was to ask users to pay attention to count passing time of the ball from the team in white. The use of the Gorilla was to test whether users could see the Gorilla or not when they were in the counting task.

Although this video is not special designed for our experiment, it is suitable for our studies. First, it has a counting task, and users may have different counting performance results. Second, users may or may not see the Gorilla when they were busy in the counting task. With the help of an EEG sensor, we can analyse whether different counting scores and the appearance of the Gorilla have effects on the users' attention level.

Fig. 3. The two teams were playing the basketball back and forth

Fig. 4. The Gorilla walked across the players

3.4 Participants

The participants included 8 females and 14 males; the majority were from the Netherlands (10 people), five were European (1 Greek, 2 Germans, 1 Spanish and 1 Bulgarian), one was South American (Brazil) and three were from Asia (1 Vietnamese, 2 Indian, and 1 Chinese). The average age of the participants is 32 years (SD: 6.7) old. All the participants were first time to this video and received a free gift after the experiment.

3.5 Methodology

The EEG sensor data was analyzed using Analysis of Variance (ANOVA). An alpha level of .05 was adopted. For each outcome only F values, p values and Cohen's partial eta squared (η_p^2) are provided.

We performed a two-way between-subjects design in our experiment, including two between-subjects independent variables. The two variables are: counting results and the appearance of the Gorilla. The dependent variable was the focus of attention, which measured by the EEG sensor across all the participants. This design allowed us to analyze three effects on the focus of attention: the main effect of the presence of the Gorilla, the main effect of counting results and the two-way interaction (Gorilla * counting results).

3.6 Procedure

Before watching the video, we helped the users wear the EEG sensor. The users followed the video instruction and fulfilled the counting task in the video. After watching, we took notes about their counting results and whether they saw the Gorilla or not.

3.7 Results

3.7.1 Counting Results Effect

In our experiment there were four different counting results in the answers of the ball passed within the white team: 13 times, 14 times, 15 times and 16 times. The correct answer is 15 times. The users' attention values measured by the EEG sensor at the different counting scores are shown in Table 1:

Table 1. The descriptive statistics on the counting results effect

Score	Mean	Std. Deviation	Number of Participants
13	46.78	8.63	5
14	56.11	12.38	6
15	55.58	12.74	7
16	60.48	10.31	4
Total	54.61	11.59	22

Score: Counting Score, Mean (Std. Deviation): Attention values mean (std. deviation).

The two-way ANOVA analysis showed there was no significant attention difference among the users with the different counting results: $F (3, 14) = 1.18$, $p > .05$, $\eta_p^2 = 0.16$, so that we cannot reject our first hypothesis (H1). This result proved that some users gave a poor performance results (e.g. counting 13 times), who paid the same level of attention as the users counted correctly. The different counting results may result from the users' eyes unconsciously moved to somewhere else in the video, so that they missed counting. Some of the users reported that they were very nervous at the start, so that they missed the one or two counting in the beginning of the task.

3.7.2 Gorilla Presence Effect

Twelve users reported that they saw the Gorilla during counting, while the 10 users did not noticed the appearance of the Gorilla at all. The user' attention values in these two different situations are displayed in the Table 2.

Table 2. The descriptive statistics on the presence of the Gorilla effect

Gorilla	Mean	Std. Deviation	Number of Participants
N	54.13	12.95	10
Y	55.02	10.9	12
Total	54.61	11.59	22

N: the participants not seeing the Gorilla, Y: the participants seeing the Gorilla.

Score: Counting Score, Mean (Std. Deviation): Attention values mean (std. deviation).

The ANOVA analysis results cannot reject the second hypothesis: F (1, 14) = .03, $p > .05$, $\eta_p^2 = .002$, which means that the presence of the Gorilla did not change users' focus of attention when they were busy at the counting task. This again proved our second assumption. Some of the users busying in the counting task were not be interfered by the presence of the Gorilla, or in other words, in this video, the Gorilla designed as an interference object had distracted some of the users' visual attention, but it had the least effect ($\eta_p^2 = .002$) on users' focus of attention.

3.7.3 Two-Way Interaction

There was no interaction found in our analysis results: F (3, 14) = 0.98, p = .43, $\eta_p^2 = 0.16$. It indicates that the simultaneous influence of the two variables (counting results * Gorilla) on the users' focus of attention was not additive. The descriptive statistics on the users' attention value is displayed in Table 3.

Table 3. The descriptive statistics on the interaction

Score	Gorilla	Mean	Std. Deviation	Number of Participants
13	N	48.99	7.48	3
	Y	43.47	12.24	2
	Total	46.78	8.63	5
14	N	58.74	17.66	3
	Y	53.49	7.13	3
	Total	56.11	12.38	6
15	N	49.27	12.24	3
	Y	60.31	12.46	4
	Total	55.58	12.74	7
16	N	70.27		1
	Y	57.21	9.77	3
	Total	60.48	10.31	4
Total	N	54.13	12.95	10
	Y	55.02	10.91	12
	Total	54.61	11.59	22

N: the participants not seeing the Gorilla, Y: the participants seeing the Gorilla.

Score: Counting Score, Mean (Std. Deviation): Attention value mean (std. deviation).

4 Discussion

This paper studied one of the patterns on the user's focus of attention during a training video exercise. An EEG sensor was used to measure the user's attention values when they were at the counting task during the video. The user performance results effect, the Gorilla presence effects and the interaction (counting results * Gorilla) were analyzed on the users' focus of attention.

Our findings showed that focus of attention was not the only one factor which affected users' performance results during a video training exercise. The presence of the Gorilla effect analysis revealed that some of the users' attention levels would not be affected by the presence of the interference object when they were already involved in a task. By employing the EEG sensor and following the methodology mentioned in this article, educators may be more rational assess the users' performance results during a video training exercise. Moreover, in an e-learning environment, although students were in a remote location, their focus of attention may be obtained by using an EEG sensor. Last but not least, based on users' attention feedback, it may be possible to explore the different evaluation methods on the video design, especially a video aimed to improve users' ability of focus.

In this experimental study, it would be interesting to further analyze the attention difference when users obtained the same counting results based on the two different visual attention situations: the users saw the Gorilla and the users did not see the Gorilla. We did not process the analysis due to lack of enough participants in this study, for instance the 4 users saw the Gorilla versus the 3 users did not see the Gorilla when they all counted the 15 times of the ball being passed. Furthermore, it would be more helpful to combine some of the other methods, for instance eye tracking, in our studies to investigate where users looked at during the counting task. In such a way, we may have more arguments to explain why users had different counting results even though they have paid the same level of focus of attention. In the future, we suppose to invite more participants and combine the possible compensative solutions to enrich our studies, in the purpose of exploring users' attention behavior in a video consumption.

5 Conclusion

Based on the results from our user studies, we can conclude that the users' performance results during a video training exercise are not only affected by the users' focus of attention. During the training process, some of the users have the ability to remain their focus at the presence of the interference object, especially when they were already involved in the task.

The results reported in this article are part of a bigger ongoing project, aimed at developing novel video-based communication paradigms. These include remote lecturing and distributed performances. This feedback will be used for the implementation of adaptation mechanisms, which are reactive to the viewers' level of immersion.

Acknowledgements. The research leading to these results has received funding from the European Community's Seventh Framework Program (FP7/2007-2013) under grant agreement no. ICT-2011-287760.

References

1. A selective attention video, http://www.youtube.com/watch?v=vJG698U2Mvo
2. MindSet instruction manual, http://developer.neurosky.com/docs/doku.php?id=mindset_instruction_manual
3. Folgieri, R., Zichella, M.: A BCI-based application in music: Conscious playing of single notes by brainwaves. Comput. Entertain. 10(3), Article 1, 10 pages (2012)
4. Al-Barrak, L., Kanjo, E.: NeuroPlace: making sense of a place. In: Proceedings of the 4th Augmented Human International Conference (AH 2013). ACM, New York (2013)
5. Marchesi, M., Riccò, B.: *BRAVO:* a brain virtual operator for education exploiting brain-computer interfaces. In: CHI 2013 Extended Abstracts on Human Factors in Computing Systems (CHI EA 2013). ACM, New York (2013)
6. eSense, http://company.neurosky.com/files/neurosky_esense_whitepaper.pdf
7. The Invisible Gorilla, http://www.theinvisiblegorilla.com/
8. NeuroSky, http://www.neurosky.com/
9. Grierson, M., Kiefer, C.: Better brain interfacing for the masses: progress in event-related potential detection using commercial brain computer interfaces. In: CHI 2011 Extended Abstracts on Human Factors in Computing Systems (CHI EA 2011). ACM, New York (2011)
10. Attention, http://en.wikipedia.org/wiki/Attention
11. Sheridan, T.B.: Musings on telepresence and virtual presence. Presence: Teleoperations and Virtual Environments 1, 120–125 (1992)
12. Dervan, S., McCosker, C., MacDaniel, B., O'Nuallain, C.: Educational Multimedia Current Developments in Technology-Assisted Education. In: Fourth International Conference on Multimedia and Information and Communication Technologies in Education, Seville, Spain (2006)
13. MindWave User Guide, http://download.neurosky.com/support_page_files/MindWaveMobile/docs/mindwave_mobile_user_guide.pdf

Fine-Grained Image Classification Using Color Exemplar Classifiers

Chunjie Zhang[1], Wei Xiong[1], Jing Liu[2], Yifan Zhang[2], Chao Liang[3], and Qingming Huang[1,4]

[1] School of Computer and Control Engineering, University of Chinese Academy of Sciences, 100049, Beijing, China
[2] National Laboratory of Pattern Recognition, Institute of Automation, Chinese Academy of Sciences, P.O. Box 2728, Beijing, China
[3] National Engineering Research Center for Multimedia Software, Wuhan University, 430072, Wuhan, China
[4] Key Lab of Intell. Info. Process, Institute of Computing Technology, Chinese Academy of Sciences, Beijing, 100190, China
{cjzhang,wxiong,qmhuang}@jdl.ac.cn,
{jliu,yfzhang}@nlpr.ia.ac.cn, cliang@whu.edu.cn

Abstract. The use of local features has demonstrated its effectiveness for many visual applications. However, local features are often extracted with gray images. This ignores the useful information within different color channels which eventually hinders the final performance, especially for fine-grained image classification. Besides, the semantic information of local features is too weak to be applied for high-level visual applications. To cope with these problems, in this paper, we propose a novel fine-grained image classification method by using color exemplar classifiers. For each image, we first decompose it into multiple color channels to take advantage of the color information. For each color channel, we represent each image with a response histogram which is generated by exemplar classifiers. Experiments on several public image datasets demonstrate the effectiveness of the proposed color exemplar classifier based image classification method.

Keywords: Color space, exemplar classifier, fine-grained image classification, structured regularization.

1 Introduction

Recently, fine-grained image classification becomes more and more popular, such as flower classification [1, 2] and bird classification [3]. The goal of fine-grained image classification is to separate images of a basic-level category. Although related with generic image classification, fine-grained classification pays more attention to separate highly similar images which are with subtle differences [4]. The traditional bag-of-visual-words (BoW) model often fails to solve this problem. This is because gray image is used for local feature extraction which ignores the discriminative information within different color spaces. For example, to classify flower images in

B. Huet et al. (Eds.): PCM 2013, LNCS 8294, pp. 327–336, 2013.

Fig. 1. The white daisy can be easily separated with yellow dandelion by color. However, it is relatively more difficult to separate them by using shape features such as SIFT.

Figure 1, the white daisy can be easily separated with yellow dandelion by color. However, it is very difficult to separate them by using local shape features such as SIFT. The same thing also happens when classifying different birds [3].

To combine the color information, researchers have proposed many color based features [5-8]. Weijer *et al.* [5] analyzed the error propagation to the hue transformation and weighted the hue histogram with its saturation. To combine the statistical information of different colors, Mindru *et al.* [6] defined the generalized color moments. Bosch *et al.* [7] used the HSV color model and extracted SIFT features over the three channels which results in a 3×128 dimensional vector. However, this vector has no invariance properties which limit its discriminative power. The OpponentSIFT is generated in a similar way by Sande *et al.* [8] by transforming images into the opponent color space. One problem with these color based descriptors is that they treated the color space jointly without considering the differences within different color spaces. We believe this information should be used to boost the classification performance.

Besides, local features carry little semantic information which is often not discriminative enough for visual classification. To generate more semantically meaningful representation, many works have been done [9-14]. Some researchers [9-12] tried to generate a semantic space with all the training images. Rasiwasia and Vasconcelos [9] proposed a holistic context models for visual recognition. By defining each class as the group of database images labeled with a common semantic label, Carneiro *et al.* [10] proposed a probabilistic formulation for image annotation and retrieval which is conceptually simple and do not require prior semantic segmentation of images. Yao *et al.* [11] proposed a discriminative feature mining method and trained random forest with discriminative decision trees to incorporate the region semantic information and achieved good results. Li *et al.* [12] proposed the Object Bank which represented an image as a response map of generic object detectors. Other researchers [13-15] made use of exemplar image separately. Sparsity constraint [16] is also used by researchers. Malisiewicz *et al.* [13] proposed to ensemble exemplar SVMs for object detection and achieved comparable results with the state-of-the-art methods. Zhang *et al.* [14] used supervised semantic representation for scene classification. Although proven effective, these methods also ignored the color information. If we can combine this semantic based image representation with color information, we will be able to representation images more discriminatively and eventually improve the image classification performance.

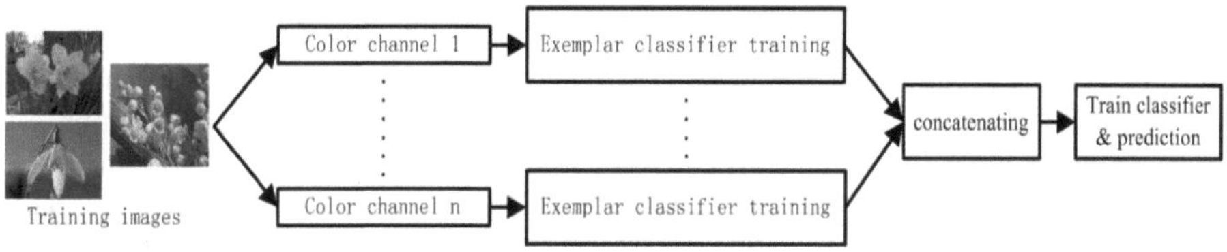

Fig. 2. Flowchart of the proposed method

In this paper, we propose a novel fine-grained image classification method using the color exemplar classifiers. For each channel, we first decompose each image into different color channels (*e.g.* RGB, HSV). Then we extract SIFT local features and use the sparse coding along with max pooling technique to represent images of this channel. Exemplar SVM classifiers are trained and we use the outputs of each exemplar classifier as the final image representation for this channel. Each image is represented by concatenating the final image representation of all color channels. Experimental results on several public datasets demonstrate the effectiveness of the proposed method. Figure 2 gives the flowchart of the proposed method.

The rest of this paper is organized as follows. Section 2 gives the details of the proposed color exemplar classifier based fine-grained image classification method. We give the experimental results and analysis in Section 3. Finally, we conclude in Section 4.

2 Color Exemplar Classifier Based Fine-Grained Image Classification

In this section, we give the details of the proposed color exemplar classifier based fine-grained image classification algorithm. We first decompose images into multiple color channels. For each channel, exemplar classifiers are trained and we use the output of these exemplar classifiers for image representation. We then concatenate the image representation for each channel as the final image representation. Finally, we learn one-vs-all SVM classifiers to predict the categories of images.

2.1 Color Channel Decomposition and Local Feature Extraction

The color information plays an important role for visual application. The traditional BoW model often extracts local features on gray images without considering the color information. Besides, the state-of-the-art color descriptors [5, 8] jointly consider the color spaces without making good use of the information within each color channel. Figure 3 shows an example of this problem. Different color channels review different aspects of the images and should be treated separately.

To make full use of the color information as well as take advantage of different color channels, we propose to first decompose images into different color channels, such as RGB, HSV. For each channel, we extracted SIFT features and use these features to represent images. This is different from traditional color descriptors [5, 8] which concatenate the local features of all the color channels to form a long descriptor. We treat each channel separately to preserve the discriminative information of each color channel. Besides, to represent images for each channel, we use the popular sparse coding with max pooling strategy [17] as it has been proven very effective. In this way, we are able to combine the color information for better image representation which will eventually help to improve the final image classification performance.

Fig. 3. The decomposition of one image into different color channels (RGB, HSV). It is best viewed in color.

2.2 Image Classification with Color Exemplar Classifier

After representing images for each channel, we are able to take advantage of this color representation. However, the local feature carries little semantic information. To get a more semantic image representation, we use the exemplar classifier based method. For each color channel, we train exemplar classifier for each image and use the output of these exemplar classifiers as the new image representation. We choose linear SVM as the exemplar classifier. This is because it is very effective and computational efficient.

Formally, let $p_{i,j}^k \in \mathbb{R}^{M\times 1}, i=1,2,...,N, j=1,2,..., M, k=1,2,...,K$ be the predicted values of the i-th image with the j-th class of the k-th color channel, where N is the number of training images and M is the number of training image categories, K is the number of color channels. For color channel k, the exemplar classifier based image representation is $h_i^k = [h_{i,1}^k;...;h_{i,j}^k;...;h_{i,M}^k]$. To combine the image representation of multiple color channels, we concatenate the h_i^k for each channel k, where $k=1,2,...,K$, as $h_i = [h_i^1, h_i^2, ..., h_i^K]$ for the i-th image.

To predict the categories of images, we use the one-vs-all strategy and train linear SVM classifiers. Let $y_i \in Y = \{1,2,...,M\}$ be the corresponding labels of image i, we try to learn M linear functions such that:

Fig. 4. Example images of the Oxford Flower 17 dataset

$$y = \max_{c \in Y} w_c^T h \tag{1}$$

This is achieved by solving the following optimization problem as:

$$\min_{w_c} \|w_c\|^2 + \lambda \sum_{i=1}^{N} \ell(w_c, y_i, h_i) \tag{2}$$

Where $\ell(\bullet)$ is the loss function. We choose to use the quadratic hinge loss as the loss function which has the form of:

$$\ell(w_c, y_i, h_i) = [\max(0, w_c^T h_i y_i - 1)]^2 \tag{3}$$

We adopt the LBFGS algorithm [17] to solve the optimization problem (2). After learning the classifier parameters, we are able to predict the image categories with Eq. (1).

3 Experiments

We conduct experiments on the Oxford Flower datasets [1, 2] and the Scene-15 dataset [18] to evaluate the performance of the proposed color exemplar classifier for fine-grained image classification method. We densely extract local features of 16×16 pixels with an overlap of 6 pixels. The RGB, HSV, C-invariant, and opponent color spaces are used, as [8] did. To combine the spatial information, the spatial pyramid matching (SPM) approach [18] is adopted with $2^l \times 2^l, l = 0,1,2$. The codebook size is set to 1,024 for each channel. Sparse coding with max pooling [17] is used to extract the BoW representation of images for each channel. The one-versus-all rule is used for multi-class classification. We use the average of per-class classification rates as the quantitative performance measurement method. This process is repeated for several times to get reliable results.

Table 1. Performance comparison on the Oxford Flower 17 dataset

Methods	Performance
Nilsback and Zisserman [1]	71.76 ± 1.76
Varma and Ray [19]	82.55 ± 0.34
LP-B [20]	85.40 ± 2.40
KMTJSRC-CG [21]	88.90 ± 2.30
Ours	**91.53 ± 1.39**

3.1 Oxford Flower 17 Dataset

The first flower dataset we consider is a smaller one with 17 (*buttercup, colts' foot, daffodil, daisy, dandelion, fritillary, iris, pansy, sunflower, windflower, snowdrop, lily valley, bluebell, crocus, tigerlily, tulip* and *cowslip*) different flower categories [1]. This dataset has 1,360 images with 80 images for each class. For fair comparison, we follow the same experimental setup as [1] did and use the same training/validation/test (40/20/20) split. Figure 4 shows some example images of the Oxford Flower 17 dataset.

We compare the proposed color exemplar classifier based image classification method with [1, 19-21]. Nilsback and Zisserman [1] tried to learn the vocabularies for color, shape and texture features respectively. Varma and Ray [19] chose the most discriminative features by optimization. A boosting procedure is used by Gehler and Nowozin [20] to combine multiple types of features while Yuan and Yan [21] combined multiple types of features using a joint sparse representation approach.

Table 1 gives the performance comparison results. We can see from Table 1 that the proposed color exemplar classifier based image classification method outperforms other methods. This demonstrates the effectiveness of the proposed method. By combining color information per color channel, we are able to use the color information more efficiently. The effectiveness of considering each color channels separately can be seen from Table 1. The proposed method outperforms KMTJSRC-CG [21] by about 2.7 percent while [21] treated different color channels jointly. Besides, the use of exemplar classifier based semantic representation helps to represent image with more semantically meaningful histogram which helps to alleviate the semantic gap.

3.2 Oxford Flower 102 Dataset

The Oxford Flower 102 dataset is an extension of the Oxford Flower 17 dataset, hence is more difficult to classify than the Flower 17 dataset. This dataset has 8,189 images of 102 classes. Each class has 40-250 images. For each class, we use 10 images for training, 10 images for validation and the rest for testing, as [2] did for fair comparison. Figure 5 shows some example images of the Oxford Flower 102 dataset.

Fig. 5. Example images of the Oxford Flower 102 dataset

Table 2. Performance comparison on the Oxford Flower 102 dataset

Methods	Performance
Nilsback and Zisserman [1]	72.8
KMTJSRC-CG [21]	74.1
Ours	**76.3**

We compare the proposed color exemplar classifier based image classification method with [1, 21]. Table 2 gives the quantitative comparison results. We can have similar conclusions as on the Oxford Flower 17 dataset. The proposed color exemplar classifier based method outperforms the baseline methods [1, 21]. This again shows the effectiveness of the proposed method. As the Flower 102 dataset has more flower classes and large inter-class variation, a well chosen image representation is vital for the final image classification. This problem can be solved by using the proposed color exemplar classifier based representation hence helps to improve the classification performance.

3.3 Scene-15 Dataset

The last dataset we consider is the Scene-15 dataset [18]. This dataset consists of 15 classes of images (*bedroom, suburb, industrial, kitchen, livingroom, coast, forest, highway, insidecity, mountain, opencountry, street, tallbuilding, office* and *store*). Figure 6 shows some example images of this dataset. Each class has different sizes ranging from 200 to 400 images with an average of 300×250 pixel size. For fair comparison, we follow the same experimental procedure as [18] and randomly choose 100 images per class for classifier training and use the rest of images for performance evaluation.

We give the performance comparison of the proposed method with [17, 18, 22, 23, 24] in Table 3. We can see from Table 3 that the proposed method outperforms the baseline methods. Compared with exemplar based method [24], the use of color information can further improve the semantic representativeness of the exemplar based methods. These results demonstrate the proposed method's effectiveness.

Fig. 6. Example images of the Scene 15 dataset

Table 3. Performance comparison on the Scene 15 dataset

Methods	Performance
SPM [18]	81.40 ± 0.50
SPC [23]	81.14 ± 0.46
ScSPM [17]	80.28 ± 0.93
KC [22]	76.67 ± 0.39
WSR-EC [24]	81.54 ± 0.59
Ours	**83.75 ± 0.52**

4 Conclusion

This paper proposes a novel fine-grained image classification method using color exemplar classifiers. To combine the color information, we decompose each image into different color channels. We also train exemplar SVM classifiers for each channel and use the output of exemplar classifiers as the image representation for this channel to take advantage of the effectiveness of semantic based image representation. The final image representation is obtained by concatenating the representation of different channels. Experimental results on the Oxford Flower 17 and 102 datasets and the Scene-15 dataset demonstrate the effectiveness of the proposed method.

In our future work, we will consider how to encode the local features with smooth constraints [25, 26].

Acknowledgement. This work is supported in part by National Basic Research Program of China (973 Program): 2012CB316400; National Natural Science Foundation of China: 61303154, 61025011, 61272329, 61202325; the President Fund of UCAS; the Open Project Program of the National Laboratory of Pattern Recognition (NLPR); China Postdoctoral Science Foundation: 2012M520434, 2013T60156, 2013M530350. We thank Weigang Zhang and Qi Tian for their helpful suggestions.

References

1. Nilsback, M., Zisserman, A.: A visual vocabulary for flower classification. In: Proc. CVPR, pp. 1447–1454 (2006)
2. Nilsback, M., Zisserman, A.: Automated flower classification over a large number of classes. In: Proc. CVPR, pp. 722–729 (2008)
3. Yao, B., Bradski, G., Fei-Fei, L.: A codebook-free and annotation-free approach for fine-grained image categorization. In: Proc. CVPR, pp. 3466–3473 (2012)
4. Rosch, E., Mervis, C., Gray, W., Johnson, D., BoyesBraem, P.: Basic objects in natural categories. Cognitive Sci. 8(3), 382–439 (1976)
5. van de Weijer, J., Gevers, T., Bagdanov, A.: Boosting color saliency in image feature detection. IEEE Transactions on Pattern Analysis and Machine Intelligence 28(1), 150–156 (2006)
6. Mindru, F., Tuytelaars, T., Gool, L., Moons, T.: Moment invariants for recognition under changing viewpoint and illumination. Computer Vision and Image Understanding 94(1-3), 3–27 (2004)
7. Bosch, A., Zisserman, A., Muoz, X.: Scene classification using a hybrid generative/discriminative approach. IEEE Transactions on Pattern Analysis and Machine Intelligence 30(4), 712–727 (2008)
8. Sande, K., Gevers, T., Snoek, C.: Evaluating color descriptors for object and scene recognition. IEEE Transactions on Pattern Analysis and Machine Intelligence 32(9), 1582–1596 (2010)
9. Rasiwasia, N., Vasconcelos, N.: Holistic context models for visual recognition. IEEE Transactions on Pattern Analysis and Machine Intelligence 34(5), 902–917 (2012)
10. Carneiro, G., Chan, A., Morena, P., Vasconcelos, N.: Supervised learning of semantic classes for image annotation and retrieval. IEEE Transactions on Pattern Analysis and Machine Intelligence 29(3) (2007)
11. Yao, B., Khosla, A., Fei-Fei, L.: Combining randomization and discrimination for fine-grained image categorization. In: Proc. CVPR, pp. 1577–1584 (2011)
12. Li, L., Su, H., Xing, E., Fei-Fei, L.: Object bank: A high-level image representation for scene classification & semantic feature sparsification. In: Proc. ECCV (2010)
13. Malisiewicz, T., Gupta, A., Efros, A.: Ensemble of exemplar-SVMs for object detection and beyond. In: Proc. ICCV (2011)
14. Zhang, C., Liu, J., Liang, C., Tang, J., Lu, H.: Beyond local image features: Scene classification using supervised semantic representation. In: Proc. ICIP (2012)
15. Jiang, W., Chang, S.-F., Loui, A.: Context-based concept fusion with boosted conditional random fields. In: Proc. ICASSP (2007)
16. Yang, Y., Huang, Z., Yang, Y., Liu, J., Shen, H., Luo, J.: Local image tagging via graph regularized joint group sparsity. Pattern Recognition 46(5), 1358–1368 (2013)
17. Yang, J., Yu, K., Gong, Y., Huang, T.: Linear spatial pyramid matching using sparse coding for image classification. In: Proc. CVPR (2009)
18. Lazebnik, S., Schmid, C., Ponce, J.: Beyond bags of features: Spatial pyramid matching for recognizing natural scene categories. In: Proc. CVPR (2006)
19. Varma, M., Ray, D.: Learning the discriminative power invariance trade-off. In: Proc. ICCV (2007)
20. Gehler, P., Nowozin, S.: On feature combination for multiclass object classification. In: Proc. ICCV (2009)
21. Yuan, X., Yan, S.: Visual classification with multi-task joint sparse representation. In: Proc. CVPR (2010)

22. Gemert, J., Veenman, C., Smeulders, A., Geusebroek, J.: Visual word ambiguity. IEEE Trans. Pattern Anal Machine Intell. 32(7), 1271–1283 (2010)
23. Zhang, C., Wang, S., Huang, Q., Liu, J., Liang, C., Tian, Q.: Image classification using spatial pyramid robust sparse coding. Pattern Recognition Letters 34(9), 1046–1052 (2013)
24. Zhang, C., Liu, J., Tian, Q., Liang, C., Huang, Q.: Beyond visual features: A weak semantic image representation using exemplar classifier for classification. Neurocomputing (2012), doi:10.1016/j.neucom.2012.07.056
25. Gao, S., Tsang, I., Chia, L.: Lappacian sparse coding, hypergraph laplacian sparse coding, and applications 35(1), 92–104 (2013)
26. Zhang, C., Wang, S., Huang, Q., Liang, C., Liu, J., Tian, Q.: Laplacian affine sparse coding with tilt and orientation consistency for image classification. Journal of Visual Communication and Image Representation 24(7), 786–793 (2013)

Smart Thumbnail: Automatic Image Cropping by Mining Canonical Query Objects

Yang Yang[1], Linjun Yang[2], and Gangshan Wu[1]

[1] State Key Laboratory for Novel Software Technology, Nanjing, 210046, P.R. China
[2] Microsoft Corporation, One Microsoft Way, Redmond, WA, 98052, USA
charlie.yang.nju@gmail.com, linjuny@microsoft.com, gswu@nju.edu.cn

Abstract. In this paper, we present a query-dependent thumbnailing approach for web image search. Motivated by the fact that uniform down-sampling cannot emphasize query objects while saliency-based methods may present incorrect foreground objects, we propose to employ common object discovery (COD) algorithms to mine the underlying canonical query objects from the result image collection and adopt the detected object regions of interest (ROIs) as a guide for image cropping. To make the employed COD approach more adaptive to our scenario, we enhance it by introducing text-based search rankings. We then decide for each image whether it should be cropped and determine the final cropping boundary by expanding the detected bounding box, so that the produced thumbnails are of proper appearances. The experimental results demonstrate that our method can outperform down-sampling and saliency-based methods on both object localization accuracy and general thumbnail quality.

1 Introduction

Web image search, which serves as an important business for most search engines, enables the users to access relevant images using key words, and thumbnailing techniques aim to adapt large quantity of result images into a limited space such as a web page. In this paper, we focus on a special category of queries, by which the users attempt to search for objects such as "BMW Q5", "Pearl Tower" and "Avocado". For these queries, only some parts of the image are relevant to the query. Therefore, they have specific preference for thumbnails that the relevant parts should be emphasized in the thumbnail while the other parts should be either abandoned or suppressed.

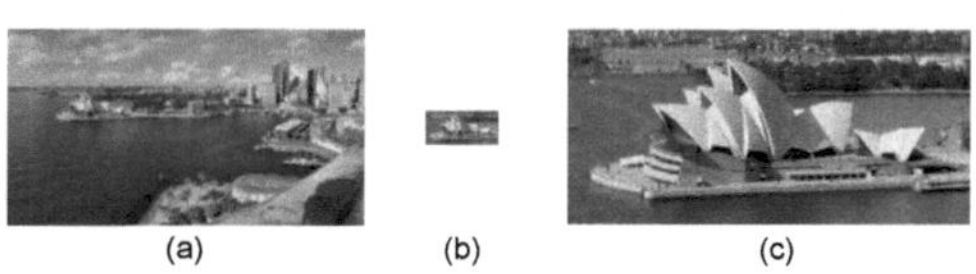

Fig. 1. An example of "Sydney opera". (a) thumbnail by down-sampling, (b) region of the opera, (c) thumbnail by cropping.

B. Huet et al. (Eds.): PCM 2013, LNCS 8294, pp. 337–349, 2013.

Uniformly down-sampling the image to thumbnail size has low time consumption but may probably give up object details. For example, Fig. 1.(a) shows a thumbnail of Sydney opera produced by down-sampling. After resizing the object to thumbmail size, the building of Sydney opera only occupies a very small region and can hardly be recognized by the users (see Fig. 1.(b)). Recently, some saliency-based methods [1, 2] propose to find the foreground object of each image and crop the image based on the detected bounding box. However, saliency-based cropping may probably produce incorrect thumbnails. First, saliency fails to indicate the correct object region when the background content is cluttered. We can see from Fig. 2.(a) the detected bounding box using saliency also includes the trees of Champs Elysee. Second, saliency-based methods are query-independent, thus may produce irrelevant thumbnail when the image multiple foreground objects. For example, Fig. 3.(a) shows an image of multiple sport brand logos where the "Adidas" logo is the most salient. Since this image is also relevant to the queries such as "Nike" and "Reebok", saliency-based algorithms still presents the "Adidas" logo as an irrelevant thumbnail.

We argue that the problem is caused by the fact that existing cropping methods only consider the appearance of each single image and do not figure out what the query object is. For object queries, we believe that the query's meaning is reflected by a few object categories underlying the result images, called "Canonical Query Objects". In this paper, the canonical query objects are mined using common object discovery (COD) with the assumption that the relevant object regions are visually consistent. We employ a COD algorithm based on Link-Analysis (LA) due to its time efficiency, which meets the requirements of web image search. Because the algorithm is not specifically designed for result image collection, we modify it by introducing text-based search ranking as a priori so that the algorithm can discover query-relevant object categories from noisy image search results. Then, for each image, we decide whether it needs to be cropped according to image relevance and object appearance. Finally, we determine the cropping boundary for each image by expanding the detected bounding box so as to include some background context that better fits the user's expectation. In convenience, the term "canonical query object" is called "query object" in the following descriptions for short. The proposed approach is regarded as query-dependent because it produces different thumbnails for one image when it is retrieved by different queries. The main contributions of this paper are summarized as follows:

- We propose to discover the underlying canonical query objects as a guide of result image cropping.
- We enhance the common object discovery algorithm in [3] to improve the relevance and completeness of the detected objects.
- We propose to crop the image based on an expanded object bounding box for a proper appearance of thumbnail.

The proposed method is tested on a dataset comprising 50 web queries and 5000 images, where an evaluation of object localization is applied as well as an

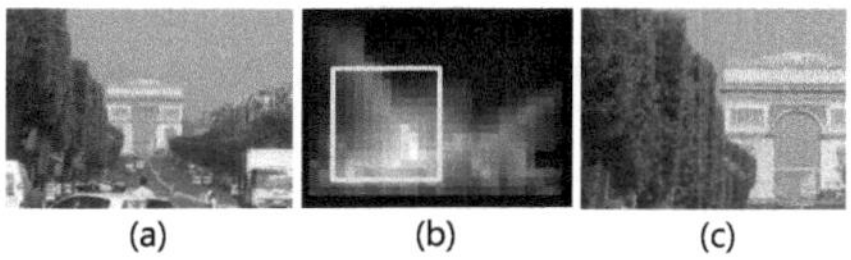

Fig. 2. An example of "Champs Elysee": (a) original image, (b) saliency map with localized foreground object, (c) produced thumbnail

Fig. 3. An example of multiple logos": (a) original image, (b) produced thumbnail in the result list, (c) saliency map with localized object

user study. The experimental results suggest that our method can outperform other baseline methods on the quality of thumbnails.

The rest of this paper is organized as follows: After reviewing the related approaches in Sect. 2, we describe the details of the proposed method in Sect. 3. In Sect. 4, we present two experiments settings and analyze the experimental results. Finally, we conclude this work in Sect. 5 with some remarks on the future work.

2 Related Work

The studied approaches on thumbnails can be generally classified into two categories: resizing-based approaches and cropping-based approaches. Other than uniform down-sampling, advanced resizing-based approaches [4, 5, 6, 7, 8] apply different sampling rates on different parts of the image according to their visual importance such as saliency, which is also known as "image re-targeting". For example, [4] removes eight-connected pixels with the lowest energy in each iteration while in [5] the authors decompose the image into curve-edge grids and optimize their shapes using max-flow-min-cost model. Image re-targeting approaches do not explicitly drop any image content but non-uniform pixel sampling may probably destroy the image structure.

Cropping-based approaches only keep the foreground object in the thumbnail, while the other parts of the image are abandoned. For example, [1] expands the foreground bounding-box by applying a greedy algorithm on the saliency map. Instead of focusing on saliency, [9] proposed an attention driven foreground detection method in localizing foreground objects. Recently, some advanced salient object detection methods [10, 11, 12, 2, 13] succeed to localize foreground objects accurately, and we believe these approaches can also be adopted in producing thumbnails. In [12], the author generates saliency maps by analyzing the frequency domain while [11] integrates multiple psychological evidence in saliency calculation, such as color, contrast, visual frequency, layout and human faces. In [2], the saliency map is calculated by comparing the content inside and outside the hypothesis bounding box, meanwhile in [13], the author generates saliency map using background priori on boundary and connectivity, and calculates saliency using geodesic distance. As mentioned on Sect. 1, saliency-based methods only consider the object appearance in a single image, leading to incorrect thumbnails.

Unsupervised object detection approaches are also known as "Common Object Discovery". According to different forms of output, they can be generally divided into two categories. The approaches in the first category provide pixel level segmentation of objects, called "co-segmentation". For example, in [14] the author simultaneously segments the foreground objects from multiple images using an expansion function analogue to heat spreading while [15] categorizes the pixels from different images through a discriminative clustering framework. The second category represents an image as a bag of ROIs and discovers the objects by finding common ROIs in different bags. For example, in [16] a conditional random field (CRF) is built for all the candidate ROIs and the common object discovery problem is transformed as finding an optimal configuration in the CRF. Guided by salient object detection result, [17] trains the object detection model through bottom-up multi-class learning and discovers object categories using discriminative Expectation-Maximization (EM) framework. In this paper, we adopt an efficient COD approach [3], where the author alternately detects the foreground ROI from each image and builds foreground model by finding representative ROIs using link analysis method. Since this method is not specifically designed for result image collection, we modify it to fit our application.

3 Approach

The proposed thumbnailing approach consists of the following three steps. We first apply the modified common object discovery algorithm to find the canonical query objects from the result image collection. Then, we decide for each image whether it should be cropped. At last, we expand the detected foreground bounding box to include some parts of background.

3.1 Canonical Query Object Discovery

At the very beginning, 100 ROIs with the highest saliency in each image are detected by the salient object detection method proposed in [2]. Each image is thus represented as a bag of ROIs. Since some images contain only foreground, the image itself is also added into the bag as an ROI. The employed COD algorithm is of an alternative optimization framework. In each iteration, the algorithm first finds a set of diverse and representative ROI exemplars. Then, the foreground ROI of each image is refined based on these selected exemplars. The process of exemplar discovery and foreground refinement are alternatively executed until convergence.

ROI Exemplar Discovery. We denote the foreground ROI of the i_{th} image selected in the last step as $r_i^{(t-1)}$, and denote $\boldsymbol{R}^{(t-1)} = \{r_i^{(t-1)}\}$, where t denotes current number of iterations. The initial ROI set $\boldsymbol{R}^0$ is selected as the ROIs presenting the entire image. Our goal of this step is to find N most representative ROIs called "exemplars". Other than simple object detection, the application of thumbnailing requires the exemplars to be as relevant to the query as possible so as to avoid the ambiguity of multiple foregrounds mentioned in Sect. 1.

With $\boldsymbol{R}^{(t-1)}$ and a similarity measure s, we apply Affinity-Propagation (AP) [18] to vote for qualified exemplars. Initially, all $r \in \boldsymbol{R}^{(t-1)}$ are considered as potential exemplars and represented as nodes in the network, and the voting process is done by recursively passing messages between all node pairs. There are two kinds of messages being transmitted over the network. 1) "Responsibility". Let $p(r_i, r_j)$ denote the "responsibility" transmitted from r_i to r_j. $p(r_i, r_j)$ reflects how well-suited for r_j to represent r_i as its exemplar, considering other potential exemplars for r_i. 2) "Availability". Denoting $a(r_i, r_j)$ as the "availability" transmitted from r_j to r_i, it indicates how appropriate for r_i to choose r_j as its exemplar, taking into account other ROIs that may choose r_j as their exemplar. $p(r_i, r_j)$ and $a(r_i, r_j)$ are recursively updated by follows:

$$\begin{aligned} p(r_i, r_j) &\leftarrow s(r_i, r_j) - \max_{j' != j}\{s(r_i, r_{j'}) + a(r_i, r_{j'})\}, \\ a(r_i, r_j) &\leftarrow \min\{0, p(r_i, r_j)\} + \sum_{i' \notin \{i,j\}} \max\{0, p(r_{i'}, r_j)\}. \end{aligned} \tag{1}$$

The "self-availability" is updated as follows:

$$a(r_j, r_j) = \sum_{j' \notin j} \max\{0, p(r_j, r_{j'})\}. \tag{2}$$

In order to improve the relevance of the exemplars, we assume that ROIs from top images are more likely to be exemplars than those from the bottom. In Affinity Propagation, the above assumption can be read as follows: for a pair of ROIs, r_i and r_j, r_j should have higher "Responsibility" to be the exemplar of r_i if r_j's parent image has higher ranking than r_i's. Here, we achieve this assumption by modifying the similarity matrix s. Originally, matrix s is calculated by comparing visual feature vectors of image pairs, such as color histogram and histogram of oriented gradient (HOG) [19]. Therefore, s is a symmetric matrix, $s(r_i, r_j) = s(r_j, r_i)$. To implement the ranking preference, we embed a non-symmetric term s_r to matrix s, according to image rankings. The enhanced similarity matrix s' is defined as follows:

$$\begin{aligned} s'(r_i, r_j) &= s(r_i, r_j) + \eta s_r(r_i, r_j), \\ s_r(r_i, r_j) &= \log(rpos(r_i) + 1) - \log(rpos(r_j) + 1), \end{aligned} \tag{3}$$

where $rpos(r)$ denotes the ranking position of the images where r is from. If r_j got a higher ranking than r_i, thus $s'(r_i, r_j) > s(r_i, r_j)$ and r_j obtains higher "responsibility" to serve as the exemplar of r_i than using the original measure of s.

The above messages keep passing over the network until convergence. Finally the exemplar collection E^t is selected by solving

$$\arg \max_{\substack{R' \subset R^{(t-1)} \\ \|R'\| = N}} \sum_{r \in R'} (p(r, r) + a(r, r)). \tag{4}$$

Foreground ROI Refinement. The goal of this step is to chose an ROI from each image bag, which is regarded as its foreground in the condition of the current

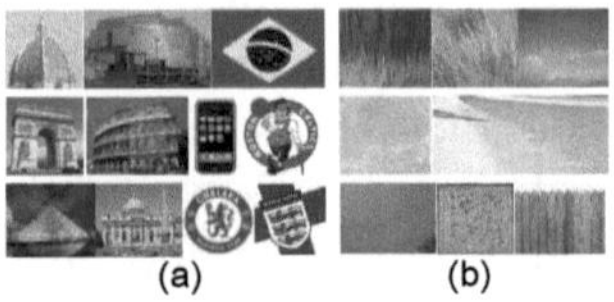

Fig. 4. Two cases where images shall not be cropped: (a) the foreground object occupies most of the image area, (b) there's no recognizable foreground object

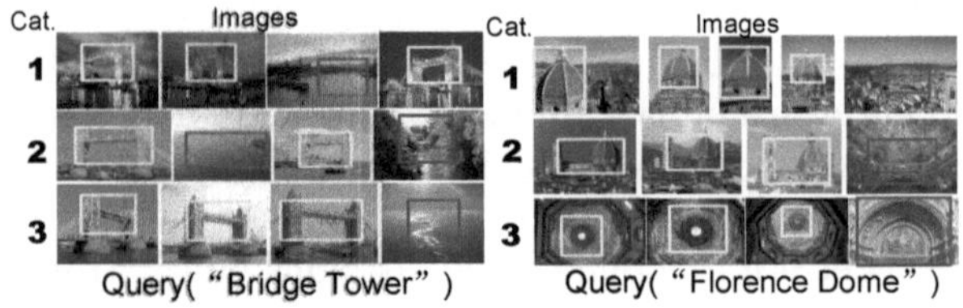

Fig. 5. The object categories from the query "Tower Bridge" and "Florence Dome" discovered by the proposed COD approach, where the correctly localized foreground ROIs are marked yellow. We can see that the localized ROIs of "foreground-less" images are playing as outliers in each category.

exemplars. The chosen foreground ROIs from all the images are then adopted as the input of exemplar selection process of the next iteration. The result of the final iteration is consequently taken as the query-relevant object for each image. In this paper, we follow the ROI selection step described in [3], where for image i, the author constructs an augmented bipartite graph W_i^* between $\boldsymbol{G}^i$ and $\boldsymbol{E}^t$. The matrix form of W_i^* is expressed as follows:

$$W_i^* = \begin{bmatrix} \alpha W_i & (1-\alpha)W_i' \\ W_i' & 0 \end{bmatrix}, \tag{5}$$

where W_i is the self-similarity graph constructed by pairwise comparison on $\boldsymbol{G}^i$, and W_i' is a bipartite graph constructed between $\boldsymbol{G}^i$ and $\boldsymbol{E}^t$. After that, we apply PageRank [20] algorithm on the constructed bipartite graph and obtain a score vector $\boldsymbol{p}$, then the foreground ROI for current iteration is selected by $\arg\max_{g_j \in \boldsymbol{G}^i} \boldsymbol{p}(j)$.

3.2 Cropping Image Selection

In our experiment, we observe that cropping is not necessary for two kinds of images. Motivated by this, we make a decision for each image whether it should be cropped after the query-relevant foreground object is localized. Figure 4.(a) shows some examples of the first kind, whose foreground objects occupy most of the image area and appear as high-quality thumbnails without any cropping. In this paper, we ignore those images in the cropping step whose foreground objects occupy more than $\frac{3}{4}$ of the image size.

The second kind of images are of an opposite situation, in which there's no recognizable foreground object. As shown in Fig 4.(b), these images only present simple colors or textures such as "blue sky" or "green grass". Although the employed COD approach assigns each image with a foreground ROI no matter there is a foreground object or not, we can observe from Fig. 5 that the so-called foreground ROIs of the "foreground-less" are usually playing as outliers within the object collection. To identify these outliers, we first cluster all foreground

ROIs by assigning each ROI to its nearest exemplar and then eliminate the ROIs with low visual densities. The evaluation of visual density is done by applying the PageRank algorithm within each category, where we calculate the self-similarity matrix by pairwise comparing the belong ROIs. Assume we obtain a vector $\boldsymbol{p}^c$ of PageRank scores from categroy c, the outliers of c are selected as $\{i|\boldsymbol{p}^c(i) < \delta \max_j\{\boldsymbol{p}^c(j)\}\}$. In our experiment, δ is set to 0.5.

3.3 Bounding Box Expansion

The second row of Fig. 6 shows some localized objects from the query "Triumphal Arch" and "Big Ben", from which we observe that most of the objects are tightly bounded by the detected bounding boxes. The thumbnails produced by cropping the ROIs straightly are illustrated in the third line of Fig. 6, and obviously, no user could accept such kind of thumbnails because the image content is incomplete without any background context. Besides, the thumbnail looks more attractive when the object is set off by some background. Motivated by this, we first expand the detected bounding box according to a conducted user study and then refine the expanded bounding box if it exceeds the image boundary or it is of an unacceptable aspect ratio.

Study of Expansion Distances. To investigate how much background should be included from the users' perspectives, 500 images of various topics are crawled from the Bing[21] search engine. We invite 5 people including 4 males and 1 female in the study, and each participant is asked to manually crop the images into thumbnails which he thinks are the best. For each image, we calculate the distances between the detected object bounding box and the manually labeled thumbnail boundary, and the average expansion distances of all samples on the four directions are shown in Fig 7. To make it fair for bounding boxes of different sizes, the horizontal and vertical expansion distances are respectively normalized by the widths and heights of the foreground ROIs. According to Fig. 7, expanding the detected bounding boxes by 41% horizontally and 36% vertically is the best expansion strategy from the users' perspectives. The last line of Fig. 6 shows the thumbnails based on the expanded bounding box. We can see that the query objects are of better appearances with the presence of proper background.

Bounding Box Refinement. The expansion distances from the above study are just a rough guide for image cropping. In the following cases, the bounding box should be refined after the expansion so as to produce the correct thumbnail. First, the expanded bounding box sometimes exceeds the image boundary as demonstrated in Fig. 8. Since the query object is away from the image center after cropping, the thumbnail may have a weird appearance. To deal with this problem, we suggest to uniformly scale-down the expanded bounding box until all edges are inside the image boundary, such that the expansion distances on the four directions remain consistent. Second, some of the object ROIs are of very high/low aspect ratios, such as the "Big Ben" towers shown in Fig. 6. In such circumstances, the expanded bounding box would follow the shape of the tower and produce a narrow thumbnail if the above expansion distances are adopted. To control the non-reasonable expansion, we bring in a threshold σ for the thumbnail aspect ratios, and truncate the expanded bounding box if the aspect ratio is higher than σ or lower than $\frac{1}{\sigma}$. In this paper, σ is empirically set to $\frac{5}{3}$.

4 Experiment

In this section, we first demonstrate an experiment on object localization to test whether our approach can present the correct objects. Then, we present an user study to show the effectiveness of our approach on the quality of produced thumbnails.

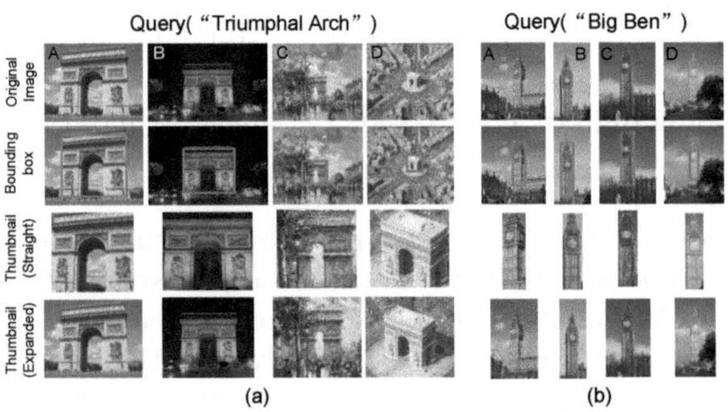

Fig. 6. Two examples of the query "Triumphal Arch" and "Big Ben". Lines from top to bottom: the original images, the detected object bounding boxes, the thumbnails by straight cropping and the thumbnails based on the expanded bounding box.

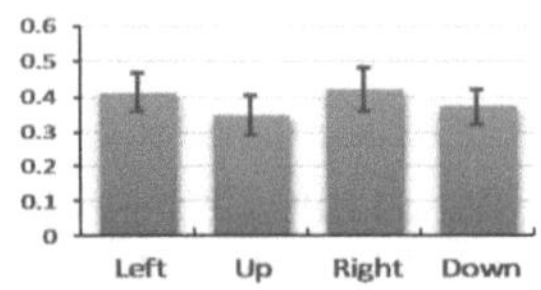

Fig. 7. The average expansion distances on the 4 directions. The corresponding variances are marked red.

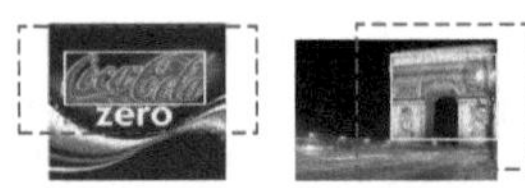

Fig. 8. Two examples when the expanded bounding box exceeds the image boundary

4.1 Dataset and Settings

The proposed approach is tested on 50 queries selected from the "Web Queries" dataset[1] [22], comprising object queries such as landmarks, vehicles, logos and productions. Since "Web Queries" dataset only provides result images of very low resolutions, we re-issue these queries to the Bing search engine [21] and download the top 200 result images. For all the approaches evaluated in this paper, including the proposed one, the visual similarity between the ROIs are calculated using a visual feature called "Pyramid Histogram Of visual Words (PHOW)" [23], which is a histogram of densely sampled visual words. The total dimensionality of this feature is 16000.

[1] The dataset is available at
`http://lear.inrialpes.fr/~krapac/webqueries/webqueries.html`

4.2 Object Localization Evaluation

Evaluation Steps. The motivation of this experiment is to test whether the proposed approach can present the correct objects in the produced thumbnails. Before the evaluation, each image is manually labeled with an object bounding box as the ground-truth of object location. For each image i, the detected bounding box is denoted as D_p^i while D_t^i denotes the ground-truth. The localization accuracy is calculated based on the overlapping area of the two bounding boxes, expressed as follows:

$$Accuracy = \frac{1}{|\mathcal{C}|} \sum_{i \in \mathcal{B}} \frac{Area(D_p^i \cap D_t^i)}{Area(D_p^i \cup D_t^i)}, \tag{6}$$

where $\mathcal{C}$ stands for the result image collection.

The proposed is compared with 3 baseline methods:

- ***FREQ***[12] and ***SAL_COMP***[2]. These two baselines are typical methods in salient object detection, in which *SAL_COMP* serves as the state-of-the-art. They are regarded as the representation of query-independent methods.
- ***LA***[3]. This is the original version of the employed COD approach. It is adopted to show the effectiveness of the modification we made in this paper.

Towards the query dependency problem illustrated in Fig 3, we pick up a subset of 120 images from the dataset with multiple distinct foreground objects. For these images, only the locations of the query-relevant objects are labeled as ground-truth.

Result Analysis. The localization accuracies of the 3 evaluated methods are shown in Table 1. We can observe from Table 1 that all COD-based methods achieve higher localization accuracy than saliency based methods. Especially for those images with multiple objects, the improvement is more obvious. For example, the basic *LA* method outperforms *SAL_COMP* by 6.53% while the proposed approach outperforms it by 19.19%. It suggests that the problems mentioned in Sect. 1 can be better handled by mining the canonical query objects. We can also observe that the proposed methods outperforms *SAL_COMP* by 8.64%, implying that the foreground objects can be more accurately localized with the help of text-based rankings.

Table 1. The localization accuracies of the proposed method and some baseline methods

Accuracy	*FREQ*[12]	*SAL_COMP*[2]	*LA*[3]	*Proposed*
All	41.11%	46.43%	52.96%	60.60%
Multi-Foregrounds	37.96%	39.93%	45.42%	59.12%

4.3 User Study

Evaluation Steps. In order to evaluate the proposed approach using quantified image quality, we invite 9 participants to take part in our user study, including 6 males and 3 females with various professional backgrounds. Due to the workload limit of the participants, the participants are only asked to rate the top 50 images in the search result list. The thumbnails produced by 4 different approaches are compared, including 1) uniform down-sampling (*DW_SAM*), 2) *SAL_COMP*[2], 3) *LA* and 4) the proposed approach. To make it close to a real search engine, all thumbnails are scaled to a height of 120 pixels. In the user study, each image is labeled into 3 quality levels: **Good, Fair, Bad**, according to the users' own perspectives. After above process, the ratings from the users on each level are accumulated and the proportion of each quality level by the different approaches are shown in Fig. 9.

Result Analysis. From Fig. 9, we can see that the proposed approach produces 38.84% more good thumbnails than down-sampling and 11.79% than *SAL_COMP*. It indicates that mining the canonical objects is generally effective in producing high-quality thumbnails. However, our approach produces 2.2% more bad thumbnails than down-sampling. These low-quality thumbnails should be regarded as the risk of adopting automatic cropping techniques. Once the foreground object of an image is incorrectly detected, the thumbnail probably appears to be content-less. On the other hand, down-sampling trends to produce thumbnails with small foreground objects thus obtains large quantity of "Fair" ratings. We are still glad to see that our approach produces 3.82% less low-quality thumbnails than saliency-based cropping because of the raise on localization accuracy.

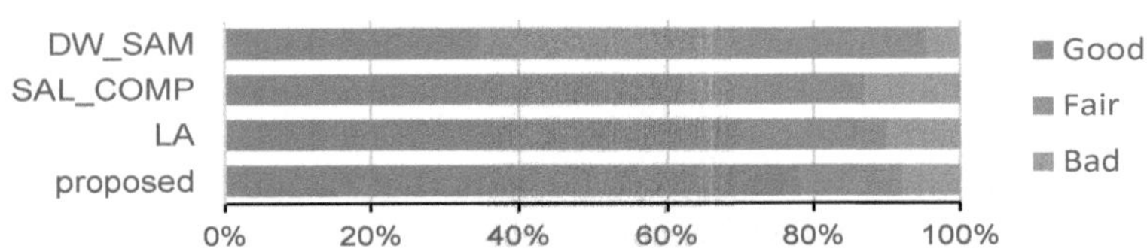

Fig. 9. The comparison of the five evaluated approaches on user ratings: 1) uniform down-sampling, 2) *SAL_COMP*, 3) *RNK_PREF*, 4) *SZ_PREF*, 5) proposed method

Fig. 10 shows a dozen of thumbnail examples produced by down-sampling, *SAL_COMP* and the proposed method. We can observe that *SAL_COMP* detects the incorrect foreground ROIs in many images, while the proposed method finds the right ones. For example, in image **A** of "Pizza Tower" and image **D** of "Arsenal Logo", *SAL_COMP* presents incorrect foreground objects while in image **A** of "Forbidden City" and image **C** of "Hollywood Sign", it presents some

[2] To make it fair for *SAL_COMP*, the detected bounding boxes are also expanded by our expansion method.

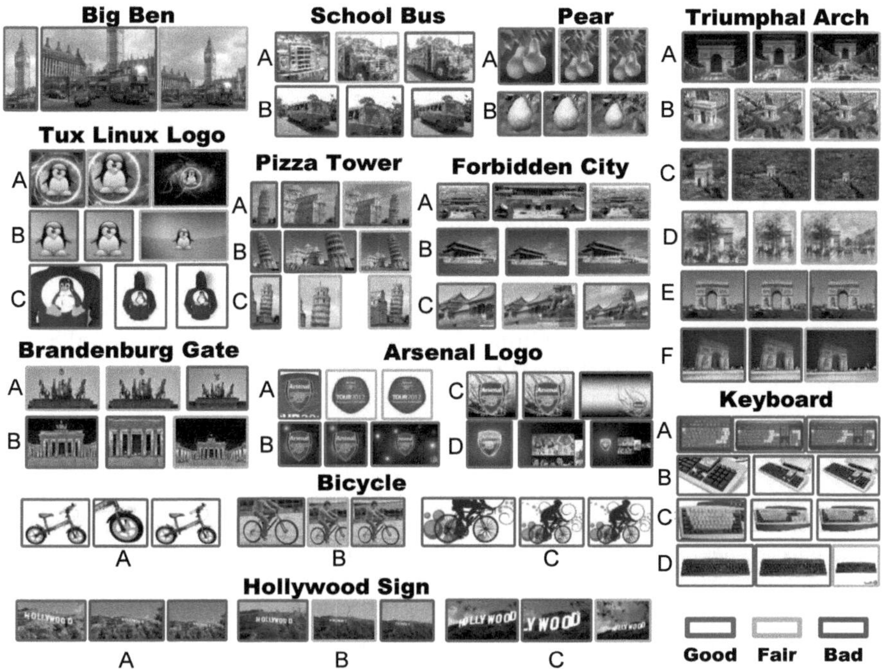

Fig. 10. Comparison of thumbnails produced by the three approaches. Each group from left to right: 1) proposed method, 2) *SAL_COMP*, 3) uniform down-sampling.

incomplete object parts. Our approach performs the worst of all in the query "Keyboard" By exploring the result image collection, we find that none of the keyboards are of a consistent gesture, causing the COD approach fail to discovery correct canonical query objects. For image **B** and **C** in the query "Bicycle", all participants give "bad" ratings to our approach, because it only discovers the region of bikes and ignores the people riding them. Although the cyclists on the bikes are not relevant to the query, all users insist them to be kept in the thumbnails. This problems is rare but typical, we decide to investigate it in our following research.

5 Conclusion and Future Work

This paper presented a novel thumbnailing approach for web image search. Motivated by the fact that saliency-based cropping can not always present correct objects in thumbnails, we suggested to mine the canonical query objects via common object discovery and use them as a guide for image cropping. We then brought in two measurements to test whether cropping is necessary for each image and proposed to obtain the cropping boundary by expanding the detected bounding box so as to include some background. The evaluation on object localization suggested that the proposed method is more likely to detect correct objects than the baseline methods while the user study reported that our approach can provide the higher thumbnail quality. Our future work will concentrate on following two directions: First, the paper only presents a simple strategy for bounding box expansion. In the future, we are going to consider some other information such as image layout and object composition. Second, we need to further improve the efficiency so that it can be integrated into an existing search engine as an online component.

Acknowledgement. This work is supported by the NSFC of China (No. 61021062), the 863 Program of China (No. 2011AA01A202) and National Special Fund (No 2011ZX05035-004-004HZ).

References

[1] Suh, B., Ling, H., Bederson, B., Jacobs, D.: Automatic thumbnail cropping and its effectiveness. In: ACM Symposium on User Interface Software and Technology, pp. 95–104. ACM (2003)
[2] Feng, J., Wei, Y., Tao, L., Zhang, C., Sun, J.: Salient object detection by composition. In: ICCV, pp. 1028–1035. IEEE (2011)
[3] Kim, G., Torralba, A.: Unsupervised detection of regions of interest using iterative link analysis. In: NIPS (2009)
[4] Avidan, S., Shamir, A.: Seam carving for content-aware image resizing. In: TOG, vol. 26, p. 10. ACM (2007)
[5] Ren, T., Liu, Y., Wu, G.: Rapid image retargeting based on curve-edge grid representation. In: ICIP, pp. 869–872 (2010)
[6] Ren, T., Liu, Y., Wu, G.: Image retargeting based on global energy optimization. In: ICME, pp. 406–409. IEEE (2009)
[7] Ren, T., Liu, Y., Wu, G.: Image retargeting using multi-map constrained region warping. In: ACM Multimedia, pp. 853–856. ACM (2009)
[8] Liu, F., Gleicher, M.: Automatic image retargeting with fisheye-view warping. In: Proceedings of the 18th Annual ACM Symposium on User Interface Software and Technology, pp. 153–162. ACM (2005)
[9] Amrutha, I., Shylaja, S., Natarajan, S., Murthy, K.: A smart automatic thumbnail cropping based on attention driven regions of interest extraction. In: ICIS, pp. 957–962. ACM (2009)
[10] Alexe, B., Deselaers, T., Ferrari, V.: What is an object? In: CVPR, pp. 73–80. IEEE (2010)
[11] Goferman, S., Zelnik-Manor, L., Tal, A.: Context-aware saliency detection. PAMI 34(10), 1915–1926 (2012)
[12] Achanta, R., Hemami, S., Estrada, F., Susstrunk, S.: Frequency-tuned salient region detection. In: CVPR, pp. 1597–1604. IEEE (2009)
[13] Wei, Y., Wen, F., Zhu, W., Sun, J.: Geodesic saliency using background priors. In: Fitzgibbon, A., Lazebnik, S., Perona, P., Sato, Y., Schmid, C. (eds.) ECCV 2012, Part III. LNCS, vol. 7574, pp. 29–42. Springer, Heidelberg (2012)
[14] Kim, G., Xing, E., Fei-Fei, L., Kanade, T.: Distributed cosegmentation via submodular optimization on anisotropic diffusion. In: ICCV, pp. 169–176 (2011)

[15] Joulin, A., Bach, F., Ponce, J.: Discriminative clustering for image co-segmentation. In: CVPR, pp. 1943–1950. IEEE (2010)
[16] Deselaers, T., Alexe, B., Ferrari, V.: Localizing objects while learning their appearance. In: Daniilidis, K., Maragos, P., Paragios, N. (eds.) ECCV 2010, Part IV. LNCS, vol. 6314, pp. 452–466. Springer, Heidelberg (2010)
[17] Zhu, J.Y., Wu, J., Wei, Y., Chang, E., Tu, Z.: Unsupervised object class discovery via saliency-guided multiple class learning. In: CVPR, pp. 3218–3225. IEEE (2012)
[18] Frey, B., Dueck, D.: Clustering by passing messages between data points. Science 315(5814), 972–976 (2007)
[19] Dalal, N., Triggs, B.: Histograms of oriented gradients for human detection. In: CVPR, vol. 1, pp. 886–893. IEEE (2005)
[20] Brin, S., Page, L.: The anatomy of a large-scale hypertextual web search engine. Computer networks and ISDN systems 30(1-7), 107–117 (1998)
[21] Bing, `http://www.bing.com`
[22] Krapac, J., Allan, M., Verbeek, J., Juried, F.: Improving web image search results using query-relative classifiers. In: CVPR, pp. 1094–1101 (2010)
[23] Bosch, A., Zisserman, A., Muoz, X.: Image classification using random forests and ferns. In: CVPR, pp. 1–8 (2007)

An Efficient Zigzag Scanning and Entropy Coding Architecture Design*

Tongbing Cui, Chuang Zhu, Yangang Cai, Meng Li,
Huizhu Jia, Don Xie, and Wen Gao

National Engineering Laboratory for Video Technology, Peking University Beijing, China
{tbcui,czhu,ygcai,limeng,hzjia,xdxie,wgao}@jdl.ac.cn

Abstract. Rate distortion optimization (RDO) technique is the best known mode decision method in recent video coding standard, such as H.264 and AVS. However, the unbearable computational burden limits its application. According to the proposed block-level pipeline architecture of RDO-based MD, we find that zigzag scanning and entropy coding are the bottlenecks. In our paper, we firstly analyze the time consumption of the bottlenecks, and then we propose our efficient zigzag scanning and entropy coding architecture. Finally, our enhanced architecture is implemented in AVS encoder. The experimental results show that 20% throughput can be increased compared with the 4-way parallel scanning and entropy coding. With the proposed architecture, the real time RDO-based MD processing of 1080P@30fps can be supported. And our design is realized in high-level Verilog/VHDL hardware description language and implemented under SMIC 0.18μm CMOS technology with 50K logic gates and 6 KB SRAMs at 237MHZ operation frequency.

Keywords: VLSI architecture, RDO, zigzag scan, entropy, AVS.

1 Introduction

The RDO-based MD is adopted in many video coding standards, such as H.261, MPEG-1,2,4, H.264/AVC and AVS, etc. However, dramatic data processing throughput and computation complexity in genuine RDO-based mode decision is a big challenge [1].

To solve the problem of the unbearable computational burden in RDO-based MD, some algorithms, like [2], try to do a rough estimate on distortion (D) and rate (R) instead of using the real ones. The others, like [3], cut down the number of the candidate modes to alleviate the computation burden. The work [4-5], which belong to the second category of the algorithms above, adopted highly efficient pipeline structure to increase the throughput of RDO-based MD, as shown in Fig.1. To get the rate

* This work is partially supported by grants from the Chinese National Natural Science Foundation under contract No.61171139 and No. 61035001, and National High Technology Research and Development Program of China (863 Program) under contract No.2012AA011703.

B. Huet et al. (Eds.): PCM 2013, LNCS 8294, pp. 350–358, 2013.

distortion cost (RDcost) of every candidate mode, we need to perform discrete cosine transform (DCT), quantization (Q), inverse quantization (IQ), inverse discrete cosine transform (IDCT), reconstruction (REC) functions, zigzag scanning (ZIGZAG), and entropy coding (ENTROPY) for every block, in series. Among the processing units above, ZIGZAG and ENTROPY are always the bottlenecks.

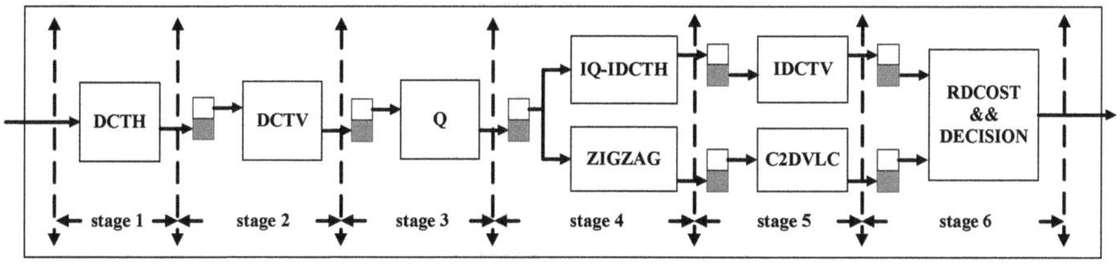

Fig. 1. Block-level pipeline structure

In [6], the 64 coefficients of one 8x8 block are segmented into eight groups along the scanning trace. The scanning process of the proposed parallel scan strategy is divided into 2 pipeline stages, and the total scanning cycles are different with different number of nonzero coefficients. Work [7] integrated the scanning unit into the entropy coding unit, the total scanning time consumption is equal to the total number of coefficients. But with the increase of coefficients, the total processing time of zigzag scanning and entropy coding will be very long. So, how to reduce the processing time when the coefficients are many will be a key issue.

In this paper, we propose an efficient zigzag scanning and entropy coding architecture, which is suitable for the genuine RDO-based MD. And our architecture is verified through AVS high definition video encoder. The rest of this paper is organized as follows. In section 2, we analyze the time consumption of ZIGZAG and ENTROPY in detail. Then the universal parallel ZIGZAG and ENTROPY architecture will be given in section 3. And we will realize our proposed architecture in AVS encoder in section 4. At last, the experiment result and conclusion of our paper will be presented.

2 Time Consumption Analysis

2.1 Time Consumption Analysis of ZIGZAG

For an M×M block in the traditional serial scan strategy, we need one cycle to judge whether each coefficient is zero or nonzero, thus M^2 cycles are totally required for an M×M block. To reduce the time consumption of ZIGZAG, some parallel scan algorithms [5][6] have been proposed. Generally, these algorithms used multiple scan engines to accelerate the scanning process. But how many engines should be used to get the minimum time consumption will be a crucial problem.

To scan the coefficients in parallel manner, we firstly divided M^2 coefficients into several groups. In each group, we use one scan engine to detect the (run, level) pairs in reverse zigzag order. After the detection, the (run, level) pairs are generated from each group. We know that the run of (run, level) pair is the number of zeros preceding each non-zero coefficient and the continuous zeros may be interrupted at the end of every group, so the first and last (run, level) pair of each group may be wrong because of the interruption, as the red points shown in Fig.2. So, we need to fix those wrong (run, level) pairs of each group. It is clear to see that the more groups we divide, the more (run, level) pairs are to be fixed.

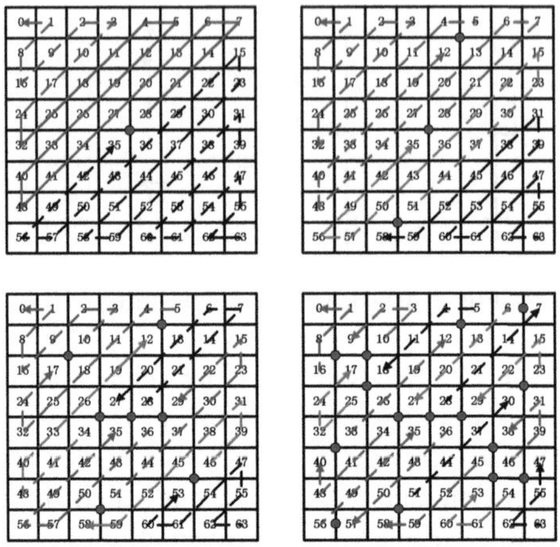

Fig. 2. Scanning diagrams of different segmentation

Assuming the number of groups is N and the block size M×M equals 8×8. Fig.2 shows different scanning diagrams when N equals 2, 4, 8, and 16. The number of the red points represents the number of (run, level) pairs need to be fixed. For N = 2, 4, 8, and 16, the corresponding number of (run, level) pairs need to fixed are 1, 3, 7 and 15.

Let T_{ZIGZAG} the total time consumption of ZIGZAG, $T_{DETECTION}$ the detection time of each group, T_{FIX} the total time of fixing all the possible wrong (run, level) pairs. Then, we can get function (1).

$$T_{ZIGZAG} = T_{DETECTION} + T_{FIX}. \quad (1)$$

In Function (1), $T_{DETECTION}$ is directly proportional to the total number of coefficients in each group. In fact, we need one cycle to detect every coefficient, and the number of groups is N, so, $T_{DETECTION}$ can be described as function (2).

$$T_{DETECTION} = \mathrm{ceil}(M^2 / N). \quad (2)$$

Here, ceil(M^2 / N) returns the smallest integer greater than M^2 / N. In addition, T_{FIX} is proportional to the number of (run, level) pairs which need to be fixed. It is clear that the number of (run, level) pairs need to be fixed is N - 1. Due to the parallel detection, the run of (run, level) pair in the interface between every 2 groups may be interrupted. And, we can not fix all the (run, level) pairs at the same time but fix them one by one in the zigzag order. For every (run, level) pair, we need α cycles to fix it. Then, we can get the following function (3).

$$T_{FIX} = \alpha \times (N - 1). \quad (3)$$

So, function (1) can be re-written as function (4),

$$T_{ZIGZAG} = T_{DETECTION} + T_{FIX}$$
$$= \text{ceil}(M^2 / N) + \alpha \times (N - 1). \quad (4)$$

2.2 Time Consumption Analysis of Entropy

In order to generate the total bits of one block, we need to look up the codenums of all the (run, level) pairs according to the appropriate variable length coding tables (such as C2DVLC in AVS, CAVLC in H.264) and then generate bits through Exp-Golomb coding. We use pipeline structure to generate the total bits of one coding unit. Therefore, for ENTROPY, the total time consumption is composed of three parts. The first part, T_{SETUP} is the pipeline setup time. The second part, T_{PRO} is the processing time of the (run, level) pairs. Consider the worst case, $T_{PRO} = \text{ceil}(M^2 / N)$. The last part of the time consumption, T_{BIT} comes from the total bits accumulating, and the time consumption will increase with the increase of N. It is because that the number of addend increases with the increase of N (for every group, there is one number of bits, called an addend).

So, we can get the total time consumption of ENTROPY, which is described as function (5).

$$T_{ENTROPY} = T_{SETUP} + \text{ceil}(M^2 / N) + T_{BIT}. \quad (5)$$

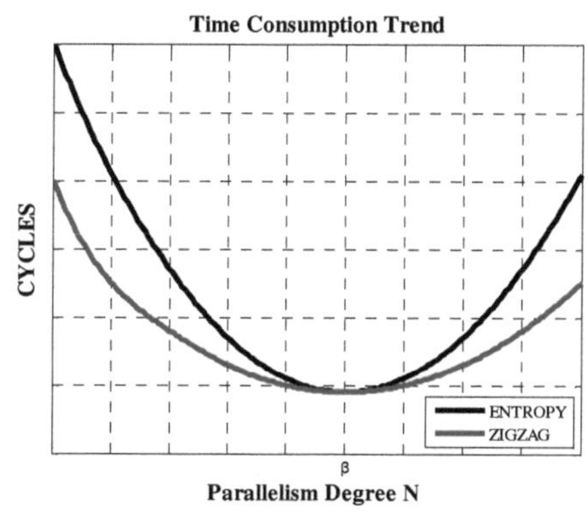

Fig. 3. Time consumption trend of ENTROPY and ZIGZAG

3 The Proposed Universal Architecture

Based on the analysis above, the method of solving the bottleneck of ZIGZAG and ENTROPY is to find the best solution of function(6). In other words, we should find the best degree of parallelism in ZIGZAG and ENTROPY.

$$T_{BEST} = \min\{\max\{ T_{ZIGZAG}, T_{ENTROPY} \}\}$$
$$= \min\{\max\{ \mathrm{ceil}(M^2 / N) + \alpha \times (N - 1), T_{SETUP} + \mathrm{ceil}(M^2 / N) + T_{BIT} \}\}. \quad (6)$$

Assuming the best degree of parallelism is β. Fig.3 shows the time consumption trend of ZIGZAG and ENTROPY. For the ZIGZAG, when N is less than β, due to $T_{DETECTION}$ is the major part of T_{ZIGZAG}, total scan time dropped significantly with the growth of N. But with the increase of N, T_{FIX} gradually becomes the major part of T_{ZIGZAG}, when N is more than β, the scan time begin to increase. It is similar for the ENTROPY, when N is less than β, T_{PRO} is the major part of $T_{ENTROPY}$, and T_{BIT} will be the major part after N is more than β.

We can get the best degree of parallelism β by optimizing (6). So the enhanced parallel ZIGZAG and ENTROPY architecture is proposed, as shown in Fig.4. β detection engines detect the coefficients belonging to the different groups of one block in parallel. After detection, β groups of (run, level) pairs generated by the β detection engines are stored in the (run, level) buffer array. Because the (run, level) pairs from the β detection engines may be wrong, they will be fixed by the fix engine. To fix them, the fix engine examines the boundary (run, level) pairs between two groups. If the level of the last (run, level) pair (in reverse zigzag order) in previous group equals 0, then we modify the run of the first (run, level) pair in current group to run_previous + run_current.

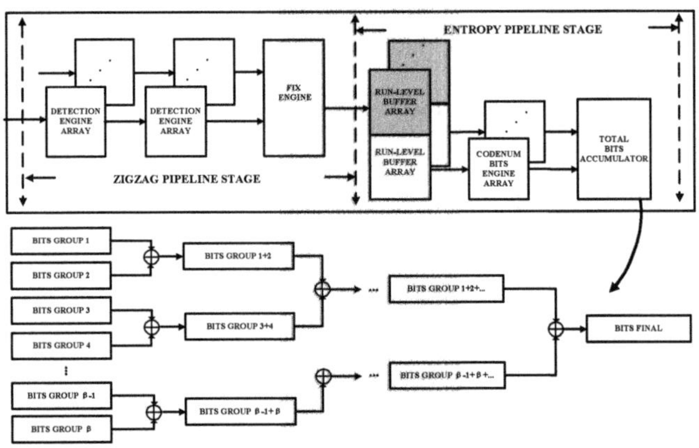

Fig. 4. Parallel ZIGZAG and ENTROPY coding architecture

While at the same time, the (run, level) pairs, which have been fixed in the previous block-level pipeline period, residing in the other buffer array are transferred to codenum engine array of VLC to generate the codenums of the (run, level) pairs. After that, the codenums generated are used to produce bits of every group. At last, all the bits from different groups are added together to form the final bits (R) for an M×M block.

4 Implementation

4.1 MD Pipeline

Our MD module is integrated to an AVS encoder system, as shown in Fig.5, in which MB_CTRL is a central control unit to synchronize all the processing modules in concert and to configure all the MB-level pipelining modules. We adopted a 5-stage MB-level pipelining structure, and they are: fetch unit (FETCH), integer motion estimation (IME), fractional motion estimation (FME), MD, bit stream generating unit (BG) and de-blocking unit (DBK). Besides, we used Multi-stage Motion Vector Prediction (MVP) Schedule Strategy for the AVS HD Encoder, to solve the data dependency problem [8].

The MD module, which belongs to the fourth stage of the MB-level pipelining, gets the original and predicted pixels from FME and IP, and then chooses the best mode based on RDcost. After that, MD transmits the codenums to BG, which are the the values of encoded coefficients generated by C2DVLC, and reconstructed pixels of the best mode to DBK, which filters the reconstructed pixels to filter out the block effect, respectively.

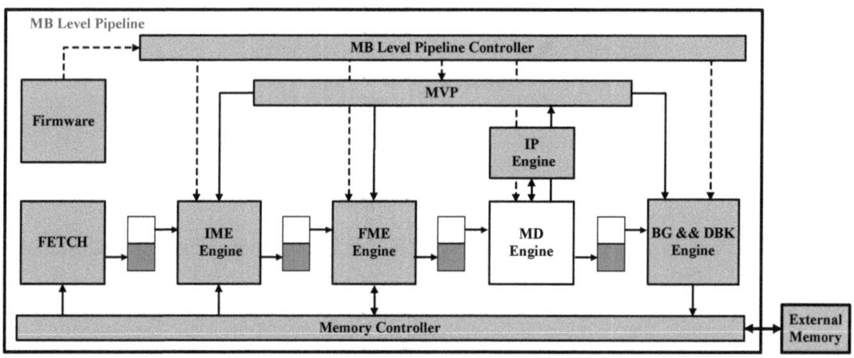

Fig. 5. MB-level pipeline structure

Just like the work [5], we selected block-level 6-stage pipeline to get the RDcost for every mode. As Fig.1 shows, the first stage is DCTH and the second stage is DCTV. The third stage is Q and we adopted a 8-way quantization. After quantization, the pipeline is divided into 2 branches. IQ-IDCTH and ZIGZAG both belong to the fourth stage, while IDCTV and C2DVLC are at the same stage, the fifth stage. At the last stage, the 2 branches of the pipeline again merge into one part, and the RDcost can be then calculated.

4.2 Architecture Realization

In section 3, we have given the universal method to solve the bottleneck of ZIGZAG and ENTROPY. Now, we need to determine related parameters for the special realization in AVS.

For T_{ZIGZAG}, the size of base block is 8 × 8, so M = 8, and we need 2 cycles to fix every (run, level) pair (one cycle to read out the (run, level) pair from buffer, the other to fix it), so $\alpha = 2$. According to Function (4), we can get function (7).

$$T_{ZIGZAG} = \text{ceil}(64 / N) + 2 \times (N - 1). \tag{7}$$

So, we can get the time consumption of ZIGZAG in different degree of parallelism, as Fig.6 shows.

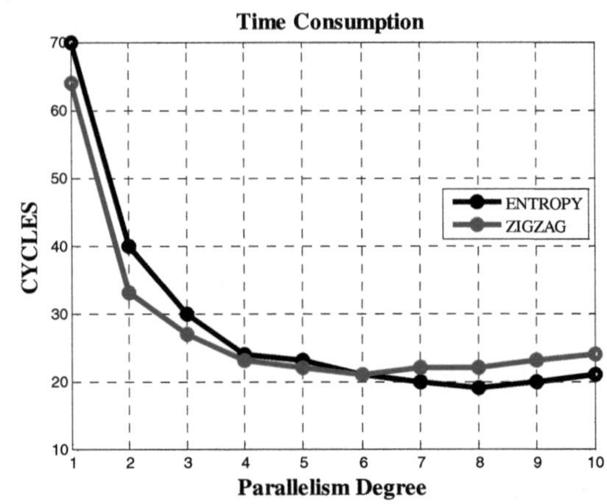

Fig. 6. Time consumption of C2DVLC and ZIGZAG in different degree of parallelism

From Fig.6, we can see that when N equals 5 or 6, the time consumption is the least.

For $T_{ENTROPY}$, we need 6 cycles to setup the pipeline in our implementation, so $T_{SETUP} = 6$. And Table 1 shows the T_{BIT} with different N.

Table 1. Time consumption T_{BIT} with different N

N	1	2	3	4	5	6	7	8	9
T_{BIT}	1	2	3	3	4	4	4	4	5

According to Table 1 and Function (5), we can get function (8).

$$T_{ENTROPY} = 6 + \text{ceil}(M^2 / N) + T_{BIT} \tag{8}$$

So, we can get C2DVLC part in Fig.6, which depicts the time consumption of ENTROPY coding.

From the figure above, it is obvious that the time consumption of ENTROPY coding declines with the increase of N. When N equals 7, the time consumption of ENTROPY will be 20 cycles, which is less than the time consumption, when N is 6. But if we choose N = 7, ZIGZAG will be the bottleneck of the MD pipeline, and the number of cycles is 22. However, we should make good balance of ZIGZAG and ENTROPY. Through careful observation of Fig.6, when N equals 6, the bottleneck of MD pipeline will be the minimum, and the corresponding processing time is 21 cycles. So we adopted 6-way ZIGZAG and ENTROPY coding architecture to get the best performance.

5 Experimental Results

We implemented our proposed 6-way ZIGZAG scanning and ENTROPY coding design based on SMIC 0.18-μm CMOS technology, the on-chip memory is 6K bits and the gate count is 50K. And the working frequency is 237MHZ, with which the system can support real time 1080p@30fps.

Table 2. Time consumption of every processing unit

Processing Unit	Time Consumption(cycles)
DCT-H	18
DCT-V	18
Q(8-way)	20
IQ && IDCTH	21
ZIGZAG(4-way)	22
IDCTV	18
C2DVLC(4-way)	26(MAX)
RDCOST && DECISION	17

Compared with the previous work as shown in Table 2, which adopted 4-way ZIGZAG and ENTROPY coding. We reduced the bottleneck of MD pipeline from 26 cycles to 21 cycles. If we use RDO-based MD, for I frame, there are 28 block-level tasks entering into the MD pipeline (In AVS-P2, for every intra-luma block, there are 5 different modes; and for every intra-chroma block, 4 different modes in total). While for P frame, {P16x16, P16x8, P8x16, P8x8, Intra8x8} will enter into the MD pipeline. And for B frame, {Bdirect, B16x16, B16x8, B8x16, B8x8, Intra8x8}, 6 MB-level modes will enter into the pipeline. Then, we can get the number of cycles saved in I, P and B frame respectively, shown in Table 3.

Table 3. Cycles saved in proposed architecture

PICTURE-TYPE	Saved cycles for one MB(cycles)
I	28×(26-21)=140
P	30×(26-21)=150
B	36×(26-21)=180

6 Conclusion

To solve the bottleneck of RDO-based MD pipeline, we analyzed the time consumption of ZIGZAG and ENTROPY coding in detail respectively. Based on the analysis, we proposed an enhanced parallel ZIGZAG and ENTROPY coding architecture, which can be used to genuine RDO-based MD. The implementation result shows that the proposed architecture can support real time 1080p@30fps encoding when integrated to the encoder system. Compared with the previous work, we reduce the bottleneck of MD pipeline from 26 cycles to 21 cycles, and the number of cycles saved in one MB of I, P and B frame are 140, 150 and 180 respectively when using RDO-based MD.

References

1. Zhang, T., Li, S., Tian, G., Ikenaga, T., Goto, S.: High throughput VLSI architecture of a fast mode decision algorithm for H.264/AVC intra prediction. In: International Conference on Communications, Circuits and Systems, ICCCAS (May 2008)
2. Wang, Q., Zhao, D., Gao, W., Ma, S.: Low complexity RDO mode decision based on a fast coding-bits estimation model for H.264/AVC. In: IEEE International Symposium on Circuits and Systems, ISCAS 2005 (May 2005)
3. Pan, F., Lin, X., Susanto, R., Lim, K.P., Li, Z.G., Feng, G.N., Wu, D.J., Wu, S.: Fast mode decision algorithm for intraprediction in H.264/AVC video coding. IEEE Trans. Circuits Syst. Video Technol. 15(7), 813–822 (2005)
4. Yin, H., Wang, X., Zhu, X., Qi, H.: Hardware Friendly Mode Decision Algorithm for High Definition AVS Video Encoder. In: 2nd International Congress on Image and Signal Processing, CISP 2009 (October 2009)
5. Wang, X., Zhu, C., Yin, H., Gao, W., Xie, X., Jia, H.: Fast Mode Decision Based on RDO for AVS High Definition Video Encoder. In: Qiu, G., Lam, K.M., Kiya, H., Xue, X.-Y., Kuo, C.-C.J., Lew, M.S. (eds.) PCM 2010, Part II. LNCS, vol. 6298, pp. 62–72. Springer, Heidelberg (2010)
6. An, D., Tong, X., Zhu, B., He, Y.: A novel fast DCT coefficient scan architecture. In: Picture Coding Symposium, pp. 1–4 (2009)
7. Huang, Y.-W., Hsieh, B.-Y., Chen, T.-C., Chen, L.-G.: Hardware Architecture Design for H.264/AVC Intra Frame Coder. In: Proceedings of ISCAS 2004, May 23-26, vol. 2, pp. II-269–II-272 (2004)
8. Yang, W., Yin, H., Gao, W., Qi, H., Xie, X.: Multi-stage motion vector prediction schedule strategy for AVS HD encoder. In: 2010 Digest of Technical Papers International Conference on Consumer Electronics, ICCE (January 2010)

COGE: A Novel Binary Feature Descriptor Exploring Anisotropy and Non-uniformity

Zhendong Mao[1], Yongdong Zhang[1], and Qi Tian[2]

[1] Institute of Computing Technology, Chinese Academy of Sciences, Beijing, China
[2] University of Texas at San Antonio, United States
{maozhendong,zhyd}@ict.ac.cn, qitian@cs.utsa.edu

Abstract. Matching keypoints across images is the base of numerous Computer Vision applications, which is often done with local feature descriptors. Hand-crafted descriptors such as SIFT and SURF are still established leaders in the field since they are discriminative as well as robust.

In this paper, we introduce a novel COGE descriptor, a simple yet effective method for keypoint description. By exploiting the anisotropy and the non-uniformity of the underlying gradient distributions, the proposed COGE is highly discriminative and robust. In addition, COGE contains only 480/240/120 bits and can be matched by using Hamming distance, making it ideal for mobile applications. To evaluate the performance of COGE, a comprehensive comparison against SIFT, SURF, ORB and BRISK is performed on three benchmark datasets: the dataset of Mikolajczyk, the INRIA Holidays and the UKbench. Experimental results show that our proposed COGE descriptor significantly outperforms existing schemes.

1 Introduction

The deployment of cameras on mobile phone and the broadband wireless network enable a new class of applications such as mobile image retrieval. Many such applications are constrained by the memory capacity of their platforms and the speed of wireless link, which demand higher requirements for local descriptors: smaller to store, faster to transmit and easier to compare.

Since the publication of Lowe's article in 1999 [11], the SIFT descriptor has proven remarkably successful in a number of applications using local features. Inspired by SIFT, Bay *et al.* proposed the SURF descriptor [2]. SURF descriptor exhibits comparable results with respect to robustness and distinctiveness, but at the same time is less expensive to compute. However, the size of SIFT/SURF descriptors from an image are typically larger than the size of the JPEG-compressed image itself. Therefore, SIFT and SURF impose large burdens in feature storage, feature transmission, and feature matching, making them unsuitable for mobile applications.

To best address the above issues, various binary descriptors have been recently proposed in literature, including BRIEF [3], ORB [15], BRISK[10], and FREAK [1]. These schemes encode the intensity comparison results of sample points, making descriptors faster to compute and more compact. Another advantage of binary descriptors is that

B. Huet et al. (Eds.): PCM 2013, LNCS 8294, pp. 359–371, 2013.

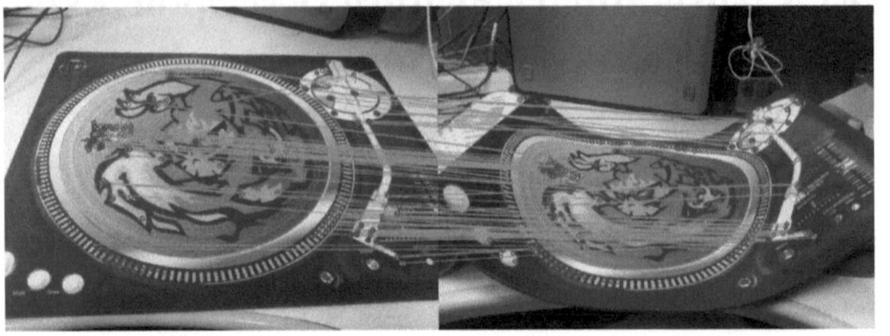

Fig. 1. Typical matching result using COGE on real-world images with viewpoint change and surface deformation. Red circles indicate keypoints; Green lines show the first 100 matches considering the nearest neighbor matching strategy. We can observe that false positives are very few in number.

the Hamming distance, which is very fast to compute, can be used for feature matching. However, despite of the clear advantage in storage, transmission and matching, the binary approaches suffer the loss in terms of discriminative power and robustness. On one hand, the intensities of sample points are not very stable. On the other hand, the intensity comparison results of sample points provide limited information.

In this paper, we propose a novel Combinatorial Comparison of Gradient Element based descriptor (COGE). The COGE descriptor consists of two parts. The first part exploits the anisotropy of the underlying gradient distributions, while the second part exploits the non-uniformity of the underlying gradient distributions. By using simple binary tests between gradient statistics in a carefully selected pattern, the proposed COGE descriptor is very compact while remaining robust and discriminative.

To validate COGE, experiments are performed on three benchmark datasets of small, medium and large size, respectively. We test the properties of COGE relative to SIFT, SURF, ORB and BRISK, for raw matching ability and performance in image retrieval. Experimental results reveal that 480-bit length COGE and 240-bit length COGE significantly outperform the state-of-the-art methods. Surprisingly, even 120-bit length COGE achieves comparable performance to SIFT on all three datasets. Regards the computational complexity, COGE is more expensive than SURF, ORB and BRISK, but comparable to SIFT.

2 Relative Work

Local feature is the base of vast applications including image retrieval, object recognition and stereo matching [6] [7] [18]. A comprehensive review of local features is referred to [17], and [12]. The state-of-the-art local features usually consist of two parts: feature detector and feature descriptor. In this section, we mainly focus on descriptors.

2.1 Statistics of Gradient Based Descriptors

A number of descriptors are based on the statistics of gradient within a keypoint patch, aiming at obtaining highly discriminative power. A prime example is SIFT descriptor, which is invariant to image scaling and rotation, and partially invariant to viewpoint changes. SIFT is applied in many fields and a number of its variants have been developed. Other examples are DAISY descriptor by Tola *et al* [16], gradient location and orientation histogram (GLOH) by Mikolajczyk and Schmid [12]. Typically, the statistics of gradient based descriptors are discriminative and robust, but not compact, making them unsuitable for real-time applications. Several schemes have been proposed to reduce the length of descriptors. They can be categorized into schemes based on dimensionality reduction (PCA-SIFT) [9], encoding (CHOG)[4], and vector quantization [8]. However, despite of the clear advantage in storage, the performances of compressed descriptors are typically worse than SIFT.

2.2 Binary Descriptors

Obtaining highly compact descriptors is important in many applications, especially for applications on handheld devices. Therefore, a number of binary visual descriptors have been recently proposed to meet the need of memory efficiency and fast computing. Calonder *et al.* in [3] propose to compute a binary descriptor directly on image patches. The binary descriptor, named BRIEF, is obtained by comparing the intensity of pre-selected pixel pairs. Although BRIEF descriptor is simple and efficient to compute, it is very sensitive to in-plane rotation. In view of this, Rublee *et al.* in [15] propose the Oriented Fast and Rotated BRIEF (ORB) descriptor, which computes the orientation of each keypoint . Compared to BRIEF, ORB is much robust to image rotation. While BRIEF and ORB simply use random pairs to generate descriptor, a method called BRISK [10] makes a judicious selection of pairs to improve discriminative power. It uses a circular pattern where points are equally spaced on concentric circles. Alahi *et al.* [1] propose FREAK to further improve BRISK. They believe that redundancy brings more discriminative power. The sampling pattern of FREAK is nearly the same as BRISK but having higher density of points near the center. Heinly *et al.* evaluate the performance of the different binary descriptors in [5].

3 COGE Descriptor

Our approach is inspired by earlier works [3], [15], [10], [1] that showed simple brightness comparison of several pairs of sampling points can be good enough to describe and match image patches. The brightness of a sampling point represents the intensity statistics of a cell centered at it, since Gaussian smoothing is pre-processed. The brightness comparison of two sampling points can be regarded as the intensity statistics comparison of two cells, namely inter-cell intensity comparison.

The comparison responses based idea has been demonstrated to be very effective, however in this paper we employ it in a more qualitative manner. The improvement lies on two folds. First, with the aim of achieving more discriminative power, we compute not only inter-cell but also intra-cell comparisons. The inter-cell comparison exploits the non-uniformity of the low-level feature (intensity, color, gradients *etc.*) distributions, while the intra-cell comparison exploits the anisotropy of the low-level feature distributions. All above binary features focus on non-uniformity information, yet the discriminative power of anisotropy information is not fully explored. Second, we use quantized statistics of gradient locations and orientations instead of intensity. One reason is that intensity is not suited for the intra-cell comparison. They are very close in the same cell and hence the comparison results are too sensitive to be robust. The other reason is various schemes have demonstrated that the quantized statistics of gradient locations and orientations are highly discriminative, and very robust to image transformations. We show that our intra- and inter- cell comparisons based descriptor significantly outperforms the state-of-the-arts descriptors in Section 4.

3.1 Preliminary Procedures

We first apply an Gaussian smoothing to the keypoint patch to reduce the noise sensitivity. For a detected keypoint at scale σ, the keypoint patch is a $55\sigma \times 55\sigma$ window and the smoothing kernel is 2σ. These parameters are chosen from our experiments and are different from previous binary features since intra-cell comparisons are involved. The big patch size ensures that each cell is informative and the small smoothing parameter will not accidentally distort the information content of intra-cell comparisons.

Next, the orientation of the patch is computed and the patch is then rotated to this orientation. Following ORB, the orientation is determined by the moments of the patch:

$$\theta = \arctan\left(\sum_{(x,y)} y \cdot I(x,y) / \sum_{(x,y)} x \cdot I(x,y)\right) \tag{1}$$

where $I(x,y)$ is the intensity of the pixel (x,y).

Then, we compute the quantized statistics of gradient locations and orientations. This step is similar to SIFT, each patch is divided into a square 4×4 grid with 16 cells and 8 orientation components of gradients are extracted separately for each cell (see Figure 2). We number cells from left to right and from top to bottom, and then the j-th orientation component of i-th cell can be formulated as:

$$\delta(i,j) = \sum_{(x,y)\in C_i,(d_x,d_y)\in \Omega_j} \sqrt{d_x^2 + d_y^2} \tag{2}$$

where C_i is the i-th cell ($i = 1...16$) and $\Omega_j = \{(d_x,d_y) | \frac{\pi(j-1)}{4} \leq \arctan(\frac{d_y}{d_x}) < \frac{\pi j}{4}, j = 1...8\}$.

3.2 Building the Descriptor

We create a bit vector out of the orientation components comparison responses. The comparison is a binary test:

$$T_k = T(i_1,j_1,i_2,j_2) = \begin{cases} 1 \; if \; \delta(i_1,j_1) > \delta(i_2,j_2) \\ 0 \qquad otherwise \end{cases} \tag{3}$$

The descriptor is defined as a vector of n test responses:

$$f_n = \sum_{1\leq k<n} 2^{k-1} T_k \tag{4}$$

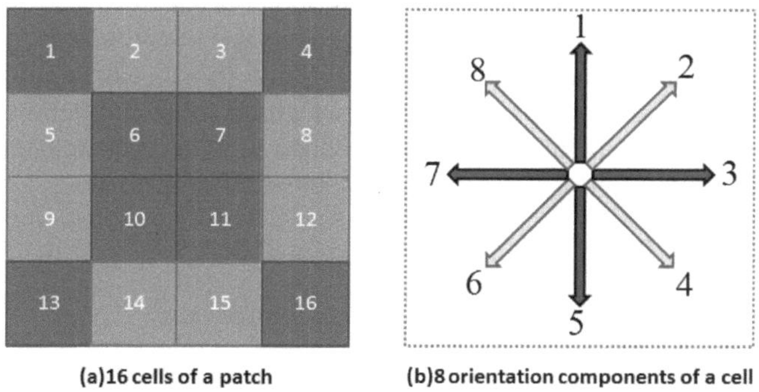

Fig. 2. Illustration of the gradient statistics of a keypoint patch (similar to SIFT). (a) shows the cells, which are divided into red and green groups. (b) shows the gradient components, which are divided into blue and yellow groups.

According to section 3.1, i_1, i_2 range from 1 to 16 and j_1, j_2 range from 1 to 8, resulting in $16 \times 16 \times 8 \times 8 = 16384$ possible tests. However, not all of the possible tests are valuable. Firstly, some tests such as $T(i, j, i, j)$ are useless. Secondly, some tests have sensitive results. Lastly, some tests are significantly correlated. A real challenge is to efficiently select the tests T_k, resulting compact, discriminative and robust descriptor. We classify the tests into intra-cell comparisons and inter-cell comparisons, and select 256 and 224 tests from the former and the latter, respectively. The first part and the second part of our descriptor are formed by the responses of the selected intra-cell comparisons and the selected inter-cell comparisons, respectively. As a result, it contains 480 bits ($n = 480$). To validate our selection, various selection strategy are compared in section 4.4.

Intra-cell Comparisons Selection. Intra-cell comparisons are comparisons between two different orientation components of the same cell. It exploits the anisotropy of the underlying gradient distributions. Given a cell, there are eight orientation components and each component can be compared with the others (See Figure 2 (b)). For one cell, we have $\binom{8}{2} = 28$ intra-cell comparisons.

A selection strategy is desired for choosing intra-cell comparisons. Take component 1 for instance, it can be compared with components from 2 to 8 (See Figure 2 (b)). We seek to select a good subset of components set $\{2, 3, 4, 5, 6, 7, 8\}$ for comparing with component 1. All possible subsets are tested and the results show that the subset $\{2, 4, 6, 8\}$ performs the best in matching precision. This result is in accordance with the continuity of gradient distributions. Components 2 and 8 are the neighboring components of component 1, which means the comparison responses of component 1 and these two components are sensitive, and hence discriminative. The results also show that the subset $\{2, 8\}$ performs better than other 2-elements subsets, which demonstrates the above analysis. It is worth noting that the comparison responses of two neighboring components are not only discriminative, but also robust. The reason is that the selected angle ($\pi/4$) between two neighboring components results the right trade-off between discriminative power and robustness. An angle interval less than $\pi/4$ will make the comparison response of two neighboring components too sensitive to be robust.

As a result, we classify the components of a cell into {north, east, south, west} group and {northeast, southeast, southwest, northwest} group (blue and yellow components in Figure 2 (b)). Each element of a group is compared with each element of the other group. The selection of intra-cell comparisons can be formalized as:

$$\{T_k | T_k = T(i, j_1, i, j_2), i = 1...16, j_1 = 1, 3, 5, 7, j_2 = 2, 4, 6, 8\} \quad (5)$$

Therefore, we select $16 \times 4 \times 4 = 256$ intra-cell comparisons.

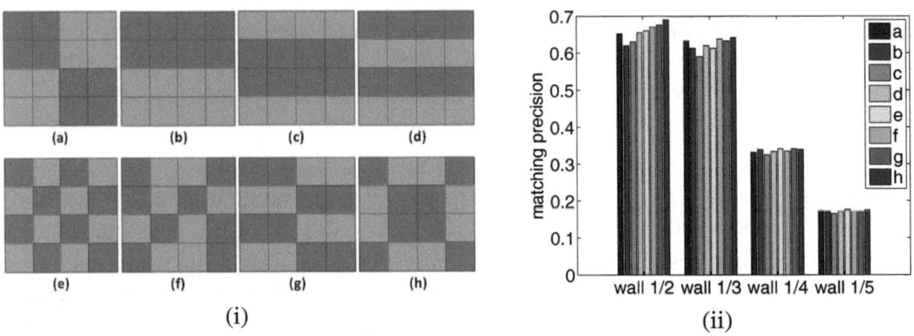

Fig. 3. (i) shows eight investigated grouping geometries for choosing inter-cell comparisons and (ii) shows their matching performances. In each grouping geometry, cells are divided into red and green groups. The selected inter-cell comparisons are comparisons between two cells of the same group. We see that grouping geometry (h) leads to best performance.

Inter-cell Comparisons Selection. Inter-cell comparisons are comparisons between the corresponding orientation components of two different cells. It exploits the non-uniformity of the underlying gradient distributions. Generating a compact, discriminative and robust bit vector leaves many options for selecting cell pairs. We evenly divide the 16 cells into 2 groups G_1 and G_2, and the selected cell pairs are:

$$\begin{aligned} &\{(C_i, C_j) | C_i \in G_1, C_j \in G_1, C_i \neq C_j\} \\ \cup &\{(C_i, C_j) | C_i \in G_2, C_j \in G_2, C_i \neq C_j\} \end{aligned} \quad (6)$$

Note that comparisons are performed within groups but not between groups, since the former generates $2 \times \binom{8}{2} = 56$ cell pairs while the latter generates $8 \times 8 = 64$ cell pairs.

We experiment with a number of grouping geometries and eight of them are illustrated in Figure 3i. The matching precision of these grouping geometries are shown in Figure 3ii. As it can be seen, grouping geometry (h) leads to best performance, and hence identified. Although the grouping geometry is heuristic, it seems to match our understanding of the model of the gradient distribution. The comparison responses of short-distance cell pairs are sensitive and thus discriminative, while the comparison responses of long-distance cell pairs are stable and thus robust. Grouping geometry (h) is a good combinations of cell pairs of two types, which balance the invariance and the discriminative power. We further observe that comparing the corresponding 1,3,5,7 or 2,4,6,8 components of two cells is equivalent to comparing all components of them. Therefore, the cell pairs of one group compare their 1,3,5,7 components while the cell pairs of the other group compare their 2,4,5,6 components. To summarize, the selection of inter-cell comparisons can be formalized as:

$$\begin{aligned} &\{T_k | T_k = T(i_1, j, i_2, j), i_1, i_2 \in G_1, i_1 \neq i_2, j = 1, 3, 5, 7\} \\ \cup &\{T_k | T_k = T(i_1, j, i_2, j), i_1, i_2 \in G_2, i_1 \neq i_2, j = 2, 4, 6, 8\} \end{aligned} \tag{7}$$

where $G_1 = \{1, 4, 6, 7, 10, 11, 13, 16\}$ and $G_2 = \{2, 3, 5, 8, 9, 12, 14, 15\}$ (G_1 and G_2 are the red and the green cells in Figure 3i (h), respectively). Therefore, we select $2 \times \binom{8}{2} \times 4 = 224$ inter-cell comparisons.

Dimensionality Reduction. According to section 3.2.1 and 3.2.2, $480(256 + 224)$ comparisons are selected and hence a 480-bit descriptor is introduced. However, in many applications such as mobile image retrieval, more compact descriptor is desired. Here we aim to seek some variants of the proposed 480 dimensional descriptor, with greatly reduced dimensionality and slightly decreased performance.

One possible strategy is to use PCA or some other dimensionality-reduction method, but the new descriptor would be less effcient to compute. For simplicity, we search for a fixed subset of uncorrelated dimensions rather than find a mapping that uses all of the dimensions. The steps of our dimensionality reduction algorithm are summarized:

Step 1: Set up a training set of 5 million keypoints, which are extracted from the UK-bench dataset.

Step 2: Build a 480 dimension vector for each keypoints by using the selected comparisons. Note that the elements of these vectors are floating values denoted by $\delta(i_1, j_1) - \delta(i_2, j_2)$, not binary values denoted by $T(i_1, j_1, i_2, j_2)$.

Step 3: For each dimension, we compute the absolute correlation between it and the others.

Step 4: Top m most uncorrelated dimensions are selected.

In this paper we consider $m = 120, 240$, and 480, and will show in the results section that these yield good compromises between storage efficiency and matching performance. In the remainder of the paper, we will refer to COGE descriptors as COGE-m.

4 Experimental Results and Discussion

In this section, we present extensive experimental results obtained on three benchmark datasets of small, median and large size: the dataset of Mikolajczyk *et al.* [12], the INRIA Holidays dataset [8] and the University of Kentucky Benchmark (UKbench) [13]. We compare COGE descriptors with SIFT, SURF, ORB and BRISK descriptors (BRIEF and ORB are similar, we only select ORB for comparison since it performs better than BRIEF in our experiment. Likewise we select BRISK but not FREAK). We use multi-scale FAST detector[14] for COGE (same as ORB), and use their own detectors for the other descriptors. The thresholds of different detectors are set to proper values to output an approximately equal number of keypoints per image. Except COGE, all detectors and descriptors are use the OpenCV2.4.3 implementations with default parameters.

4.1 Evaluation on Mikolajczyk Dataset

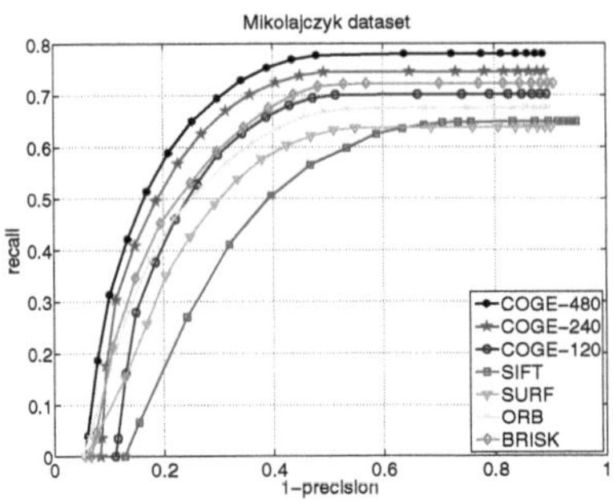

Fig. 4. Recall vs 1-precision curves of different descriptors on Mikolajczyk dataset

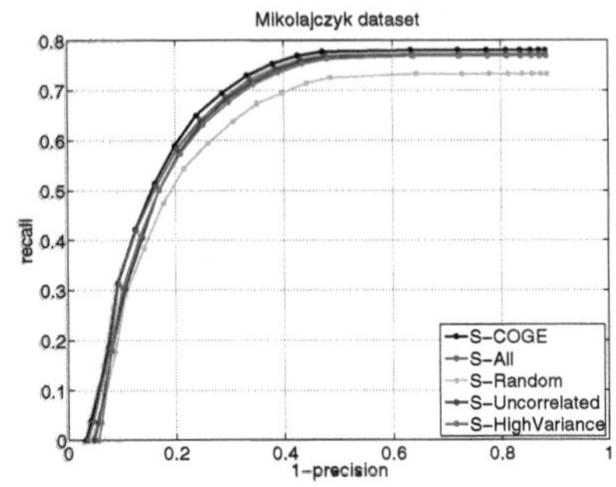

Fig. 5. Recall vs 1-precision curves of five selection strategies on Mikolajczyk dataset

This image dataset consists of eight categories and each category contains a sequence of six images exhibiting an increasing amount of transformation such as viewpoint change, light change, blur, rotation, zoom + rotation, and JPEG compression. The ground truth homographies are provided for every image transformation with respect to the first image of every sequence. Following [12], the raw matching ability of descriptors is evaluated by means of precision-recall curves, which are shown in Figure 4. The matching strategy is nearest neighbor matching. The curves clearly show the robustness and the discriminative power of descriptors. Note that these curves are the average results of all categories of this dataset. It is clear that COGE-480 and COGE-240 consistently outperform the others, yielding higher precision and better recall rates. The performance of COGE-m drops as m decreases. COGE-480 shows best performance, indicating robust and distinctive information is most contained in the selected intra-cell and inter-cell comparisons. COGE-240 yields good compromises between storage and matching performance. COGE-120 drops a bit but still compares well to other descriptors, i.e. lower than BRISK, but better than most of the others. The daisy sampling pattern with various smoothing kernels makes BRISK very robust to image blur transformations. We notice that 25% of the test images exhibit blur transformation(2 categories of the image dataset), resulting in the high score of BRISK. In addition, the performance of ORB and BRISK are better than SIFT and SURF. A possible reason is that the 0-1 quantization of intensity difference makes ORB and BRISK more robust, while the discriminative power of SIFT and SURF is not fully explored due to the small size of this dataset.

4.2 Evaluation on Holidays Dataset

The Holidays dataset contains 1491 holiday images, which are divided into 500 groups. The first image of each group is the query image and the correct retrieval results are the other images of the group. The evaluation criteria on this data set are different from previous dataset since the ground truth homographies are not available. Descriptors are evaluated by image retrieval precision:

Image Retrieval Precision: Two descriptors a and b are matched if a is the nearest neighbor to b. Each descriptor from a query image A is compared with each descriptor from the rest images to find its match. It is a brute-force search. The image B is the retrieval result if B has most matches of A's descriptors. The image retrieval precision is: correct retrieved results / retrieved results (*i.e.*, 500).

Image retrieval precision depicts how descriptors perform in a strict image retrieval application, but the ratio of descriptors contributing to the correct retrieval results is still unknown. Typically, a query image has several correct retrieval results exhibiting different level of transformations on this dataset. More robust approaches find more correct matches on relevant images with large transformations, but the retrieval result may remain the same as the less robust approaches. To further investigate the performance of different approaches, we present feature voting precision:

Feature Voting Precision: Given two descriptors a from a query image A and b from one of the rest images B, if b is the nearest neighbor to a and B is one of the correct retrieval results of A, a is considered as a correct vote. The feature voting precision is: correct votes / descriptors of all query images.

We extract 1 million keypoints from Holidays dataset to extensively evaluate the matching and the retrieval performance. Figures 6i presents the quantitative results. As illustrated in Figure 6i, COGE-480 and COGE-240 obtain superior results in both two criteria. The image retrieval precision and the feature voting precision of COGE-480, as compared with SIFT, increase 10.6% and 126%, respectively. The increases are even more remarkable compared with SURF. COGE-120 performs comparablely with SIFT and SURF in image retrieval precision but outperforms them in feature voting precision, which indicates more COGE-120 descriptors of query images find their matches in correct images. Note that the image retrieval precision is not exactly in the same order of the feature voting precision. As could be expected, the feature voting precision increases as robustness increases for all the descriptors but the image retrieval precision may remain unchanged. Figure 6i clearly shows that the feature voting precision of all COGE descriptors significantly outperforms the others, indicating COGE is more robust. The ranking of the approaches is different from previous dataset. We can observe that SIFT and SURF obtain better results than ORB and BRISK on this dataset. This result reveals that the ranking of ORB and BRISK is affected by the size of the dataset, indicating that they are more robust than SIFT and SURF, but less discriminative. In addition, the feature matching precision of ORB and BRISK is even higher than SIFT, which also demonstrate that they are more robust.

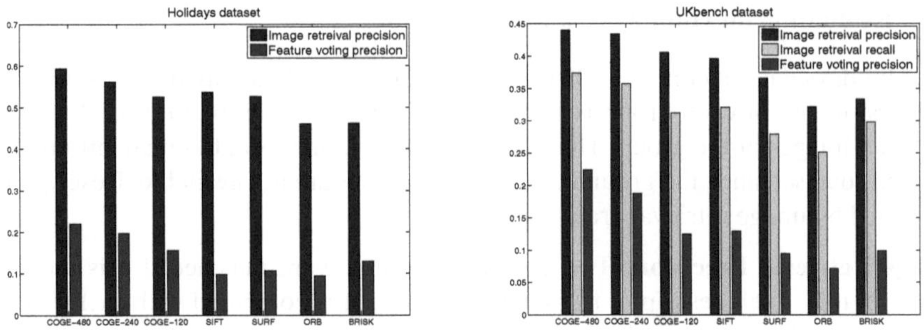

(i) (ii)

Fig. 6. Image retrieval precision and feature voting precision of different methods on Holidays dataset are shown in (i). Image retrieval precision, image retrieval recall and feature voting precision of different methods on UKbench dataset are shown in (ii).

4.3 Evaluation on UKbench Dataset

The UKbench dataset is a collection of 10200 object images, which are evenly divided into 2550 groups. The first image of each group is selected as the query image and we compute image retrieval precision and feature voting precision. Note that these two criteria are using the same definitions of section 4.2. Since all query images have 3 correct retrieval results, image retrieval recall is also provided to further evaluate descriptors.

Image Retrieval Recall: number of relevant images in top 3 retrieved results of all query images / number of all relevant images (*i.e.*, 3×2550).

We extract 5 million keypoints from UKbench dataset to further evaluate the discriminative power of descriptors. The results are shown in Figure 6ii. As it can be seen, COGE-480 and COGE-240 give the highest scores. Compared to SIFT, the image retrieval precision and recall of COGE-480 are increased 11.1% and 16.4% respectively, and that of COGE-240 are increased 9.6% and 11.2%. The feature voting precision is exactly in the same order of the image retrieval recall, which justifies the introduced feature voting precision. COGE-120 performs slightly worse than SIFT but better than the others. We observe that the differences of image retrieval precision between COGE descriptors and the other descriptors increase as the size of dataset(e.g. from 10.6% to 11.1%), indicating COGE is more discriminative. We also notice that ORB and BRSIK perform worse than SIFT and SURF in this large database, which further demonstrates that they are less discriminative.

4.4 Selection Strategy Validation

In this section, we validate our selection strategy of intra-cell and inter-cell comparisons. According to section 3.2, intra-cell comparisons are comparisons between two different orientation components of the same cell and inter-cell comparisons are comparisons between the corresponding orientation components of two different cells.

In this section, various selection strategies of comparisons are applied to generate different descriptors and the performance of them are compared on Mikolajczyk dataset. We consider the following selection strategies:

COGE Strategy (S-COGE): Select 256 intra-cell and 224 inter-cell comparisions according to section 3.2.1 and section 3.2.2.

Select All Strategy (S-All): All intra-cell comparisons ($\binom{8}{2} \times 16 = 448$), and all inter-cell comparisons ($\binom{16}{2} \times 8 = 960$). S-All generates 1408-bit descriptors.

Random Strategy (S-Random): Select 256 intra-cell and 224 inter-cell comparisons randomly.

Uncorrelated Strategy (S-Uncorrelated): Select 256 intra-cell and 224 inter-cell comparisons that are most uncorrelated by using the algorithm of section 3.2.3.

High Variance Strategy (S-HighVarience): Sort intra-cell and inter-cell comparisons in order from highest variance to lowest variance, pick the top 256 intra-cell comparisons and the top 224 inter-cell comparisons.

The results of the five selection strategies are presented in Figure 5. We can observe that COGE strategy (S-COGE) yields best result, validating the proposed selection strategy of section 3.2.1 and 3.2.3. Select all strategy (S-All) ranks second, since it preserves most information. The reason for S-COGE outperforms S-All is S-COGE removes sensitive comparisons and redundancies. Uncorrelated strategy (S-Uncorrelated) performs better than High variance strategy (S-HighVarience) and random strategy (S-Random), which demonstrates our dimensionality reduction algorithm.

4.5 Computational Cost

Table 1. Description time per point (in milliseconds)

COGE-480	SIFT	SURF	ORB	BRISK
0.863	0.812	0.144	0.065	0.029

In this section we perform a timing comparison of descriptors. Our COGE code is implemented in C++ based on OpenCV data structures and other descriptors are using OpenCV source code. All descriptors are running on an Intel quad-core of 2.83 GHZ using a single core. 5000 keypoints are extracted from the first image of Holidays dataset and the average description times of one keypoint are present in Table 1. The description time of COGE-240 and COGE-120 are almost the same as COGE-480, and thus not shown. We see that the description time of COGE is computationally more expensive than SURF, ORB and BRISK, but comparable to SIFT. Nevertheless, an advantage of COGE, as well as other binary descriptors, is that the Hamming distance can replace the usual Euclidean distance. Therefore, COGE can speed-up the computation of feature matching which is the most time consuming step in large-scale image retrieval.

5 Conclusion

In this paper, we present a novel description method named COGE, which exploits the anisotropy and the non-uniformity of the underlying gradient distributions. The proposed COGE exhibits better results than state-of-the-art methods with respect to distinctiveness and robustness, but at the same time contains only 480/240/120 bits and can be fast matched by Hamming distance. We perform a comprehensive performance comparison with SIFT, SURF, ORB and BRISK on three benchmark datasets of different sizes. Even though our descriptor are more expensive to compute than SURF, ORB and BRISK , but comparable to SIFT, our results reveal a significant improvement in performance.

Acknowledgments. This work is supported by National Natural Science Foundation of China (61273247, 61271428,61173054), National Key Technology Research and Development Program of China (2012BAH39B02). This work was supported in part to Dr. Qi Tian by ARO grant W911NF-12-1-0057, NSF IIS 1052851, Faculty Research Awards by Google, FXPAL, and NEC Laboratories of America, and 2012 UTSA START-R Research Award respectively. This work was supported in part by NSFC 61128007.

References

1. Alahi, A., Ortiz, R., et al.: Freak: Fast retina keypoint. In: CVPR (2012)
2. Bay, H., Tuytelaars, T., Van Gool, L.: Surf: Speeded up robust features. In: Leonardis, A., Bischof, H., Pinz, A. (eds.) ECCV 2006, Part I. LNCS, vol. 3951, pp. 404–417. Springer, Heidelberg (2006)
3. Calonder, M., Lepetit, V., Strecha, C., Fua, P.: BRIEF: Binary robust independent elementary features. In: Daniilidis, K., Maragos, P., Paragios, N. (eds.) ECCV 2010, Part IV. LNCS, vol. 6314, pp. 778–792. Springer, Heidelberg (2010)
4. Chandrasekhar, V., Takacs, G.: et al. Compressed Histogram of Gradients: A Low-Bitrate Descriptor. IJCV (2012)
5. Heinly, J., Dunn, E., Frahm, J.-M.: Comparative Evaluation of Binary Features. In: Fitzgibbon, A., Lazebnik, S., Perona, P., Sato, Y., Schmid, C. (eds.) ECCV 2012, Part II. LNCS, vol. 7573, pp. 759–773. Springer, Heidelberg (2012)
6. Hui Tang, J., Cheng Yan, S., Chang Hong, R., Qi, G.-J., Chua, T.-S.: Inferring Semantic Concepts from Community-Contributed Images and Noisy Tags. ACM Multimedia (2009)
7. Hui Tang, J., Jie Li, H., Qi, G.-J., Chua, T.-S.: Image Annotation by Graph-based Inference with Integrated Multiple/Single Instance Representations. In: TMM (2010)
8. Jegou, H., Douze, M., Schmid, C.: Hamming Embedding and Weak Geometric Consistency for Large Scale Image Search. In: Forsyth, D., Torr, P., Zisserman, A. (eds.) ECCV 2008, Part I. LNCS, vol. 5302, pp. 304–317. Springer, Heidelberg (2008)
9. Ke, Y., Sukthankar, R.: PCA-SIFT: A More Distinctive Representation for Local Image Descriptors. In: CVPR (2004)
10. Leutenegger, S., et al.: Brisk: Binary robust invariant scalable keypoints. In: ICCV (2011)
11. Lowe, D.: Distinctive image features from scale-invariant keypoints. In: IJCV (2004)
12. Mikolajczyk, K., Schmid, C.: A performance evaluation of local descriptors. PAMI (2005)
13. Nistr, D., Stewnius, H.: Scalable Recognition with a Vocabulary Tree. In: CVPR (2006)

14. Rosten, E., Drummond, T.W.: Machine learning for high-speed corner detection. In: Leonardis, A., Bischof, H., Pinz, A. (eds.) ECCV 2006, Part I. LNCS, vol. 3951, pp. 430–443. Springer, Heidelberg (2006)
15. Rublee, E., Rabaud, V., et al.: Orb: an efficient alternative to sift or surf. In: ICCV (2011)
16. Tola, E., Lepetit, V., Fua, P.: Daisy: An efficient dense descriptor applied to wide-baseline stereo. PAMI (2010)
17. Tuytelaars, T., Mikolajczyk, K.: Local invariant feature detectors: a survey. Foundations and Trends® in Computer Graphics and Vision 3(3), 177–280 (2008)
18. Wang, M., Hua, X.-S., Hong, R.-C., Tang, J.-H., Qi, G.-J., Song, Y.: Unified Video Annotation Via Multi-Graph Learning. TCSVT (2009)

A Complementary Aggregation Approach for Local Stereo Matching Using Color and Correlation Cues

Ran Ju, Yang Yang, Xiangyang Xu, Chunrong Xia, and Gangshan Wu

State Key Laboratory for Novel Software Technology, Nanjing University, China
juran@smail.nju.edu.com, {charlie.yang.nju,chunrong.xia}@gmail.com,
xiangyang_xu@smail.nju.edu.cn, gswu@nju.edu.com

Abstract. Existed local stereo methods usually choose the support region for aggregation using correlation or color information independently. The correlation cue works well with high textures but has a poor performance near depth discontinuities while the color cue plays the complementary role. In this paper we first propose a new soft segmentation approach for correlation-based aggregation. Then we make a combination of the two cues and adopt the advantages of them to overcome the limitation of each other. Our approach performs a two stage aggregation based on correlation and color respectively. Each stage is operated by a bilateral filter on the cost volume. The combination is simple and effective, which enables our approach to achieve a better performance in both highly textured areas and depth discontinuities than existed methods. The experimental results conform to our expectation and do make improvements to state-of-the-art methods.

Keywords: stereo matching, cost aggregation, adaptive weight, bilateral filter, cost volume.

1 Introduction

Stereo matching tries to find corresponding pixels in two or more images to reconstruct the 3D structure of a scene. The result is usually given in the form of disparity map, where the intensity of each pixel indicates the horizontal displacement to its corresponding pixel in the other view. The process of searching corresponding pixels may be confused by noise, poor texture etc. To overcome such difficulties there are generally two categories of solutions [1]: global and local methods. Global methods [2,3,4] use explicit smoothness assumptions to get global optimal solutions, which can make very accurate results but have a very high computational cost. On the contrary, local methods only rely on local information and thus work much faster than global methods. However, the accuracy and robustness can't fulfill the requirement of real applications and still need to be improved.

Generally, local methods need to take an aggregation, which sums or averages the matching cost over a local support region such that the ambiguity can be

B. Huet et al. (Eds.): PCM 2013, LNCS 8294, pp. 372–383, 2013.

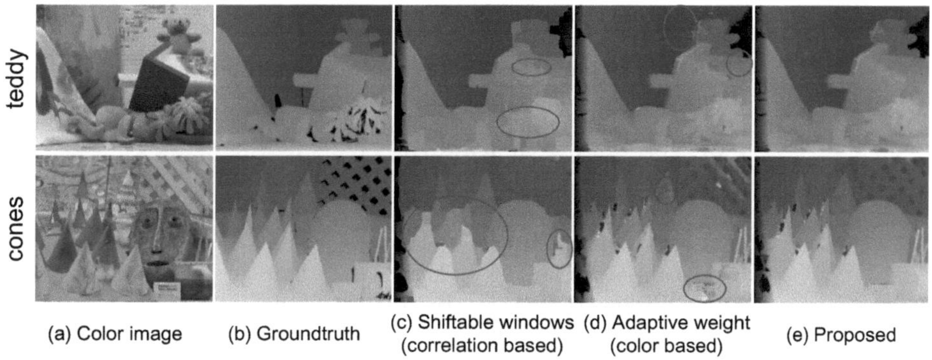

Fig. 1. Correlation-based and color-based methods. (a) (b) The two images with groundtruth disparity maps are from the Middlebury stereo benchmark [5]. (c) (d) The disadvantages in dealing with depth borders and high textures of correlation and color based methods respectively are emphasized using red ellipses. (e) Our approach makes a combination of the two approaches and thus achieves a satisfactory balance between smoothness in highly textured areas and edge-preserving property in depth discontinuities.

reduced effectively. However, how to choose the size and shape of the support region is a challenging problem and may influence the matching results directly. Correlation-based methods used to be the most successful solution. They search for the surrounding area with highest correlation on all disparity hypothesizes as the support region [1,6,7]. The basic prior is, a local set of pixels with similar disparities get the highest correlation value (or lowest matching cost) given the correct disparity hypothesis. These methods work well in highly textured areas but has a poor performance near low textured depth borders. Besides, existed correlation-based methods are constrained by the shape of rectangle windows, which may lead to an edge-fattening problem. Recently, color-based methods have made great advances in local stereo matching and thus represent state-of-the-art. They [8,9,10,11] choose the support region based on color similarity and space proximity. The basic prior is, pixels with similar colors tend to have similar disparities. These methods are good at handling depth discontinuities but easily confused in highly textured areas for insufficient aggregation support. The disadvantages of the two kinds of methods can be seen in Fig. 1. Obviously, correlation and color based methods play complementary roles and thus can be combined to cover the shortage of each other.

Our contribution lies in two aspects. First, we propose a new correlation-based aggregation method using soft segmentation to overcome the edge-fattening problem. Second, we make a combination of correlation and color cues by a two stage aggregation. The proposed approach is evaluated using the Middlebury dataset [5], computer-generated stereo images and photos taken in real life. The results show that our approach works well in both highly textured areas and depth discontinuities. The rest of this paper is organized as follows: First, we give a review of related works in Sect. 2. Then in Sect. 3 we describe the proposed

Table 1. Typical local stereo methods

Algorithm	Basis	Segmentation
Shiftable window [1]	Correlation	Hard
Multiple windows [6]	Correlation	Hard
Variable window [7]	Correlation	Hard
Adaptive weight [8]	Color	Soft
Segment support [9]	Color	Soft
Geodesic support [10]	Color	Soft
Cost-volume filter [11]	Color	Soft
Proposed	Both	Soft

approach. After that, we present our experiment and analysis in Sect. 4. At last, we make a conclusion and give some remarks on the future work in Sect. 5.

2 Related Works

Table 1 lists some typical local methods and their basic attributes. Adaptive window methods like [1,6,7] choose the sub window with best matching cost or some best surrounding windows around the central pixel as support region. However, besides the inherent disadvantage in low textured depth borders, these methods are constrained by the shape of rectangle windows, which is not suitable for irregular borders. The color-based methods use the local color distribution to guide aggregation. Typically, the adaptive weight method [8] employs the color similarity and geometric proximity to measure the contribution of each pixel while [9] and [10] take connectivity into consideration. To compute the aggregation cost more efficiently, some latest methods use edge-preserving filters on the cost volume. For example, Yang et al. [12] use a joint bilateral filter and Rhemann et al. [11] use the guided image filter instead. These methods work very well near depth discontinuities. However, they are easily confused by high textures for insufficient support. Generally speaking, color-based methods are unsuitable for the case that color distribution is inconsistent with disparity distribution. An extreme example is the random-dot stereogram. In this case color-based methods will get very poor results because color cue fails to work.

In the last column of Table 1, we list the segmentation type of the mentioned methods. Hard segmentation indicates the pixels are simply accepted or rejected for aggregation, while to soft segmentation the aggregation support of a pixel can be partially accepted. For two reasons soft segmentation is more suitable than hard segmentation. First, in depth borders the pixels may be mixed by foreground and background. Second, soft segmentation works in a probabilistic way which is more robust than hard constraints. The performances of the listed methods also show that soft segmentation is preferable and thus taken by our approach.

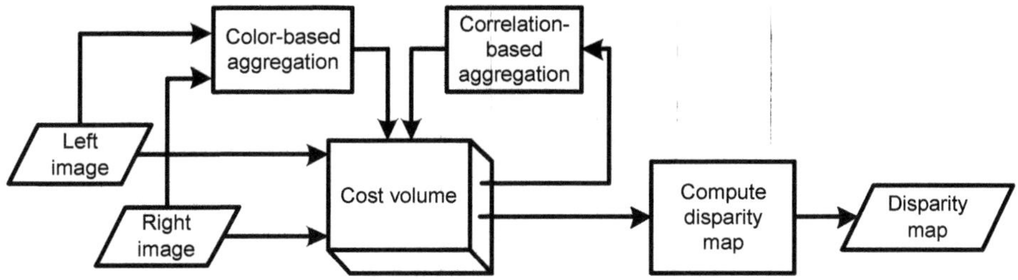

Fig. 2. Overview of the proposed approach

3 Approach

We show an overview of our approach in Fig. 2. First we use the two color images (also called the left and right image) to compute the cost volume. Then we utilize an independent bilateral filter slice by slice on the cost volume as a correlation-based aggregation. Next we use a joint bilateral filter to perform a color-based aggregation. At last we generate the disparity map by trivially choosing for each pixel the disparity with minimum aggregated matching cost.

3.1 Construction of the Cost Volume

We first construct the cost volume by computing the matching cost of each pixel at all disparity hypothesizes. There have been a lot of studies on stereo matching cost [13,14] and Hirschmuller et al. made an excellent survey [15]. However up to now there still doesn't exist a perfect matching cost function. For example, NCC (normalized cross correlation) is insensitive to illumination changes but time-consuming, while AD (absolute difference) is simple and fast but easy to confused by illumination changes. In this paper, as we focus mainly on aggregation approaches, we choose the absolute difference of color intensities as our matching cost function for simplicity. Note for the other methods we also use AD to compute matching cost to give a fair comparison on aggregation. Suppose p is a pixel in the left image. Given a disparity hypothesis d, the corresponding pixel in the right image is $p - d$. The matching cost of pixel (p, d) is:

$$C(p,d) = \frac{1}{3} \sum_{c \in \{R,G,B\}} |I_l^c(p) - I_r^c(p-d)| \tag{1}$$

where $I_l^c(p)$ indicates the intensity of pixel p in the left image, and c is the color channel. As d varies from 0 to a maximum value d_{MAX}, we get a 3 dimensional cost volume, the intensity of each voxel is assigned with $C(p, d)$. In Fig. 3 we show a few slices of the cost volume for the image called "Teddy" (as shown in the first row of Fig. 1). The sequence number of each slice indicates the disparity value.

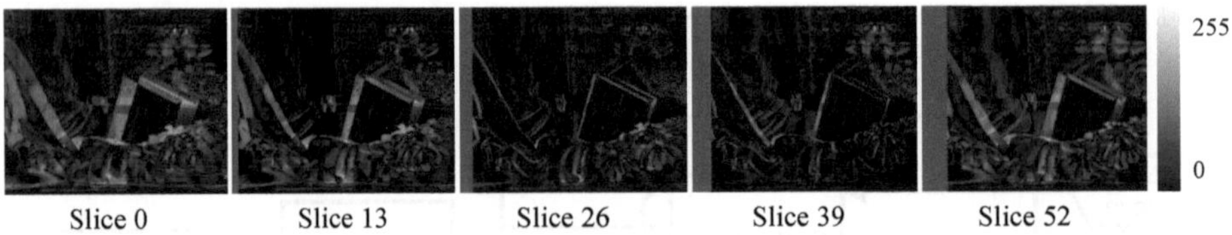

Fig. 3. A few slices of the cost volume for Teddy

3.2 Correlation-Based Aggregation

To improve the performance of classical adaptive window methods [1,6,7] near depth discontinuities, we use soft segmentation in aggregation. Generally, in the cost volume the pixels with lower matching costs are more likely to be correctly matched and thus should be assigned with higher weights for aggregation. Meanwhile, the pixels with higher matching costs should also be highlighted for rejection. The bilateral filter [16] achieved this goal very well. First, it adaptively aggregates the matching costs according to the cost distribution. Second, it performs a soft segmentation, which is more robust than hard segmentation and also handles depth discontinuities finely. Furthermore, the bilateral filter can also eliminate noises and strengthen edges. Thus we utilize a bilateral filter on the cost volume to break through the limitation of window shape constraints. The matching cost of each pixel (p, d) in the cost volume is updated as:

$$C_A(p,d) = \frac{\sum_{q \in N(p)} w_o(p,q) C(q,d)}{\sum_{q \in N(p)} w_o(p,q)} \tag{2}$$

where $N(p)$ indicates the neighboring pixels of p, usually a square window centered at p. $C(q, d)$ is the pixel-wise matching cost. $w_o(p, q)$ is the support weight of pixel q for p, which can be expressed as:

$$w_o(p,q) = \exp\left(-\left(\frac{|C(q,d) - C(p,d)|}{\gamma_o} + \frac{\|p-q\|_2}{\eta_o}\right)\right) \tag{3}$$

where $|C(q, d) - C(p, d)|$ indicates the intensity difference and $\|p - q\|_2$ is the Euclidean distance between the two pixels. γ_o and η_o are two parameters can be adjusted by users to control the power of color and space influence respectively.

To show the effect of the proposed correlation-based aggregation, we compute the intermediate disparity results by choosing the disparity with minimum aggregated matching cost for each pixel. We compare the results with the shiftable windows method [1] and adaptive support weight method [8], which represents the correlation-based and color-based method respectively. As shown in Fig. 4, adaptive window method generates smoother disparity maps but has serious edge-fattening problem. On the contrary, adaptive weight method has very accurate results near depth discontinuities but gets a poor performance in highly textured areas. The proposed approach shows more balanced results in both the two areas. However, there still exists edge-fattening problem due to the inherent disadvantage of correlation cue. Thus, we follow with a color-based aggregation.

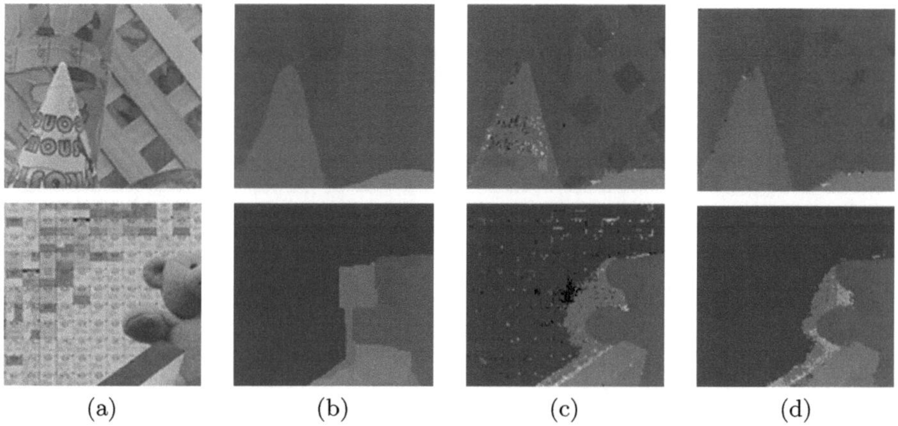

Fig. 4. The close-up disparity results. (a) Left color image. (b) Shiftable windows [1]. (c) Adaptive weight [8]. (d) Proposed method.

3.3 Color-Based Aggregation

According to the adaptive support weight method [8], the aggregated matching cost is computed as:

$$C'_A(p,d) = \frac{\sum_{q\in N(p),q_d\in N(p_d)} w_c(p,q)w_c(p_d,q_d)C_A(q,d)}{\sum_{q\in N(p),q_d\in N(p_d)} w_c(p,q)w_c(p_d,q_d)} \tag{4}$$

where p and q are the pixels in the left image and p_d and q_d are the corresponding pixels in the right image. $w_c(p,q)$ is the support weight of pixel q for p. It is expressed as:

$$w_c(p,q) = \exp\left(-\left(\frac{\Delta_{pq}}{\gamma_c} + \frac{\|p-q\|_2}{\eta_c}\right)\right) \tag{5}$$

Δ_{pq} is the Euclidean distance between pixel p and q in the RGB color space. $\|p-q\|_2$ is the Euclidean distance between the two pixels.

We give a comparison of the two kinds of aggregation in Fig. 5. It is obviously that the correlation-based aggregation has more sufficient supports in highly textured areas than the color-based aggregation. Meanwhile near depth discontinuities the color-based aggregation is more accurate. As we make a combination, the goal of dealing with more complicated scenes is achieved by the complementary approach.

3.4 Disparity Computation and Time Complexity

We use the WTA (winner-take-all) strategy to compute final disparity map, where the disparity for each pixel is computed as:

$$d(p) = \arg\min_{0\le d\le d_{MAX}} C'_A(p,d) \tag{6}$$

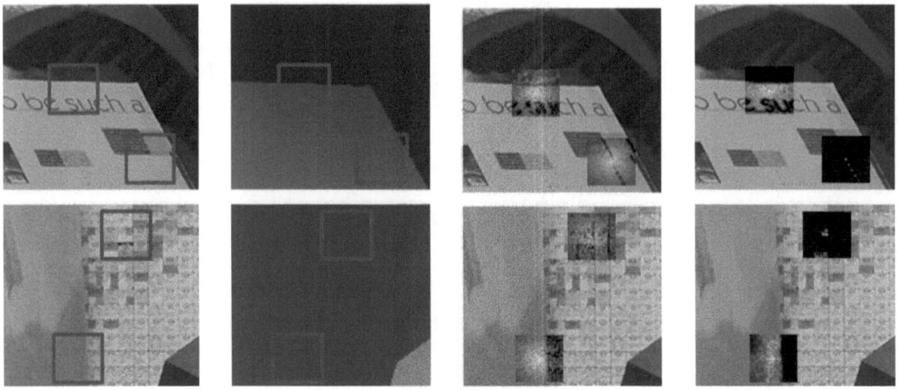

Fig. 5. Local supports of correlation-based and color-based methods. (a) Color image. (b) Truth depth map. (c) Local supports of correlation-based aggregation. (d) Local supports of color-based aggregation. The red windows in the first and second column indicate example aggregation windows. The gray windows in the third and fourth column indicate local support weights corresponding to the red windows. The pixels appear brighter indicate higher support weights and vice versa.

where d_{MAX} is maximum disparity hypothesis. It can be seen that the time complexity for each pixel is $O(d_{MAX}W)$, where W is the size of the local support window. The complexity is the same with [8] and the approach can be easily accelerated by parallel processing.

4 Experiments

4.1 Datasets and Parameter Settings

We test our algorithm using the Middlebury dataset [5], computer-generated stereo images and real world stereo photos. Four images used as benchmark in the Middlebury website are shown in the first row of Fig. 6. The benchmark uses percent of bad pixels as evaluation standard and we also take this way. To show the improvement of our approach we compare with state-of-the-art methods. The parameters in our algorithm are set as $\{\gamma_o, \eta_o, \gamma_c, \eta_c\} = \{10, 24, 15, 50\}$. The window size for correlation-based aggregation and color-based aggregation are 13×13 and 35×35 respectively. The parameters of the competitors are set according to their papers.

4.2 Experimental Results and Analysis

We choose four state-of-the-art methods for comparison, i.e. "AdaptWeight" [8], "SegmentSupport" [9],"GeoSup" [10] and "CostFilter" [11]. In their papers, the results have preformed different matching cost computation and post-processings.

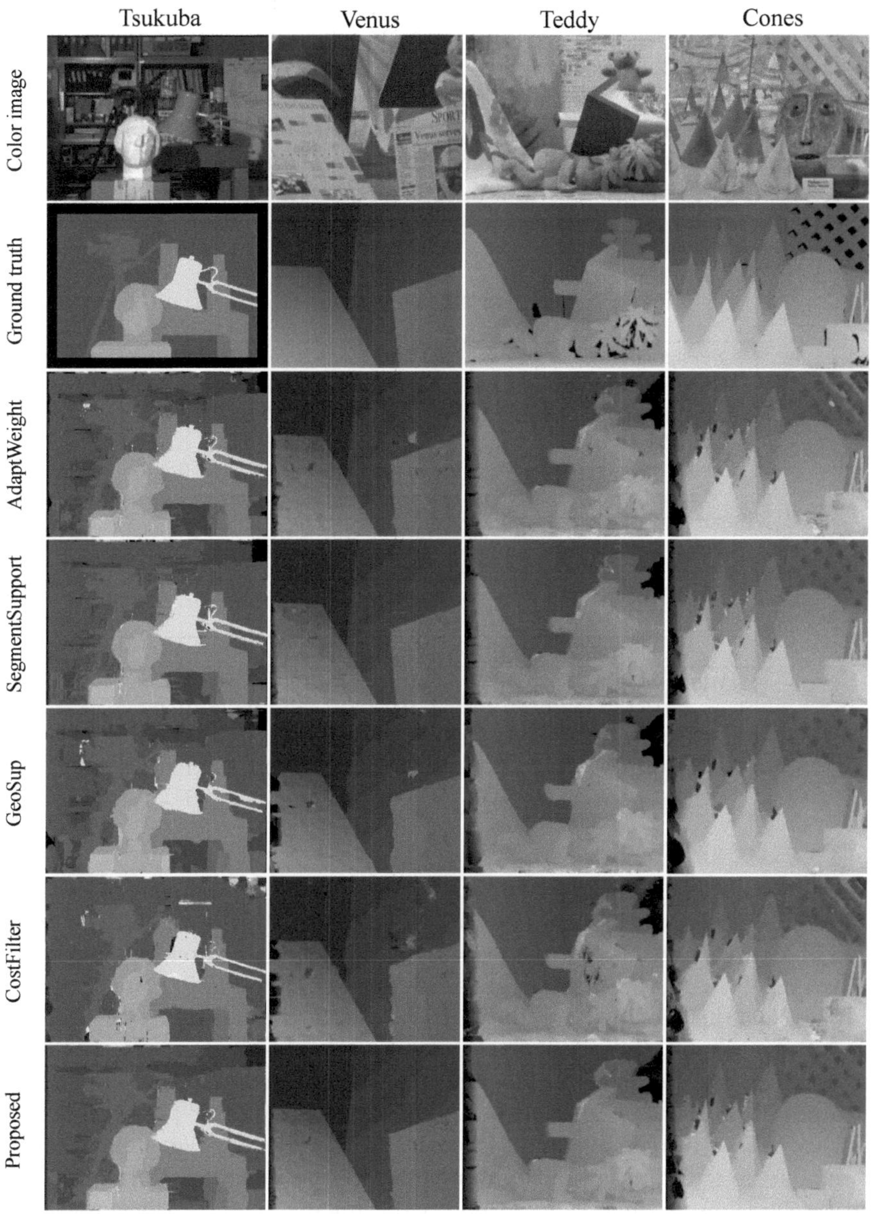

Fig. 6. Results comparison on the Middlebury dataset

To give a fair comparison, we use TAD (truncated absolute difference) as matching cost function and compute the results without any post-processing for all methods. The results on the Middlebury dataset are shown in Fig. 6. It can be seen that for the images with larger area of high textures like Venus and Cones our algorithm has an obvious improvement compared to the other competitors. AdaptWeight and CostFilter generate false matches in highly textured areas for insufficient aggregation support. GeoSup has edge-fattening problems, especially for the image named "Venus". SegmentSupport also has an excellent performance in highly textured areas because it aggregates more support by clustering small color pieces. However, it is still strongly dependent on color distribution while our algorithm uses correlation cue, which is more insensitive to color distribution and thus performs more robust as will be shown in the following.

Table 2. Bad pixels in non-occluded regions

Algorithm	Tsukuba	Venus	Teddy	Cones	Average
AdaptWeight [8]	2.82	2.76	12.1	9.66	6.84
SegmentSupport [9]	2.05	1.47	10.8	5.08	4.85
GeoSup [10]	3.16	2.74	11.6	5.11	5.65
CostFilter [11]	3.23	4.53	13.7	15.3	9.19
Proposed	2.01	1.25	11.1	4.93	4.82

Table 3. Bad pixels in discontinuous regions

Algorithm	Tsukuba	Venus	Teddy	Cones	Average
AdaptWeight [8]	7.38	10.4	21.4	15.9	13.77
SegmentSupport [9]	7.14	10.5	21.7	12.5	12.96
GeoSup [10]	10.1	17.3	22.9	12.3	15.65
CostFilter [11]	9.16	20.8	24.1	22.1	19.04
Proposed	7.07	5.86	21.2	11.4	11.38

We also use the evaluation standard by calculating the percent of bad pixels as the Middlebury benchmark does. In Table 2 and Table 3 we give the percent of bad pixels in non-occluded regions and near depth discontinuities respectively. The bad pixels in non-occluded regions give an overall evaluation on the disparity results except the occluded area because we didn't perform any post-processing to handle occlusion. The bad pixels in discontinuous regions give a targeted evaluation on the performance of handling depth discontinuities. As shown in the tables, in both the two areas our algorithm has a better performance than the other competitors in average. However, the Middlebury benchmark contains four images is still limited and may not reflect the overall performance of the local methods. And thus we give more experiments to show the power of our approach.

We give a few more examples in Fig. 7. As shown in the first row, we first use computer-generated random-dot stereo images to test the performance of handling the case that color distribution is independent of disparity distribution. The competitors all generate poor results because they all depend on the color prior, that is, pixels with similar colors tend to have similar disparities. Although the prior works well in most cases, it is not a "one size fits all" standard.

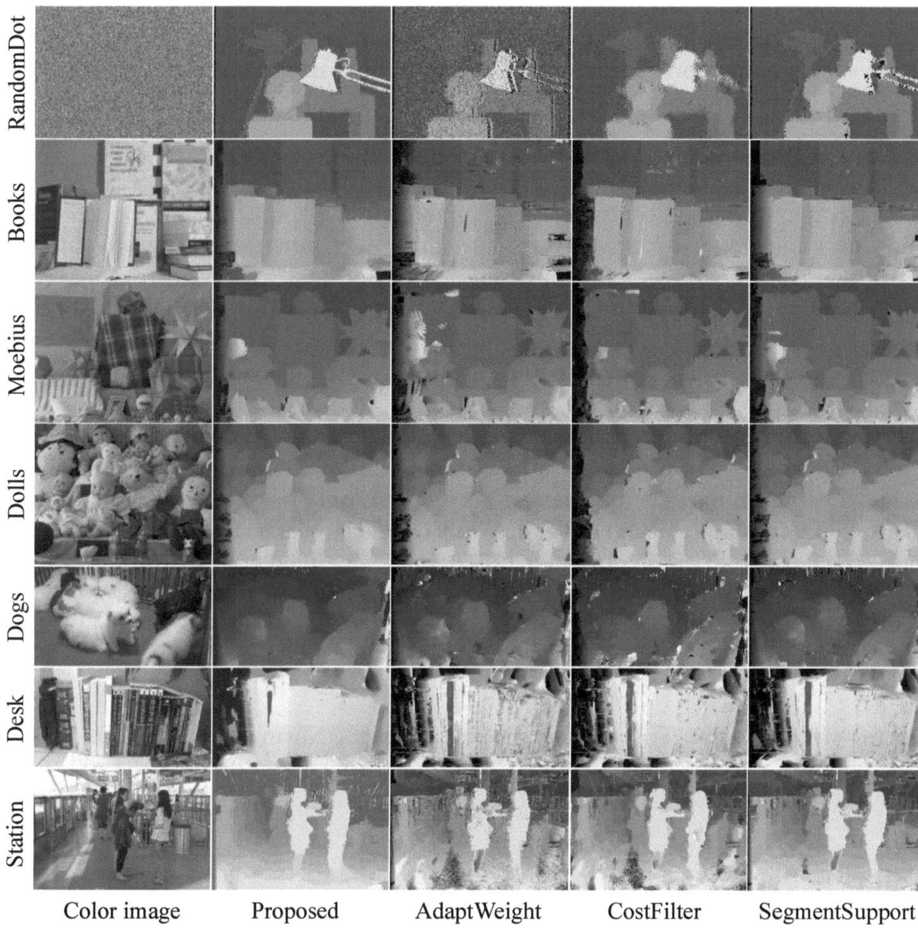

Fig. 7. More results tested on random-dot images (1st row), the Middlebury dataset (2nd to 4th row) and photos taken by a Fujifilm 3D camera (5th to 7th row). Color images (1st column), disparity maps generated by our approach (2nd column), AdaptWeight (3rd column), CostFilter (4th column) and SegmentSupport (5th column).

This can also be seen in the other results in Fig. 7. More or less, an image contains the parts do not meet the color prior. In comparison, our approach still works well even for random-dot stereo images because we combined the correlation cue for aggregation. The other images are from the Middlebury dataset (from the 2nd row to the 4th row) and taken by a FujiFilm W3 stereo camera (from the 5th row to the 7th row). It can be seen that our algorithm performs more robust than the other algorithms, especially for the real world stereo photos. This can be explained that the Middlebury images are taken under ideal illumination and of high quality, also the scenes are artificially designed.

In comparison, the real world photos contain more noise and the power of color prior may be reduced for natural scenes. However, our approach made a combination of color and correlation cues and thus works more robust.

5 Conclusion and Future Work

In this paper we have proposed a new aggregation approach using both correlation and color cues for local stereo matching. By integrating the advantages of the two cues, the proposed approach can easily deal with high textures and depth borders. The experimental results indicate our approach gives a better solution for local stereo matching no matter the color distribution of the image is consistent or inconsistent with the disparity distribution and performs more robust than state-of-the-art.

In the future, we plan to make efforts in the following two directions. First, we will incorporate our approach with more robust matching cost measurement [15] to deal with illumination changes and non-ideal Lambertian reflectance. Second, we are going to supply a high level description of the disparity map, e.g. we may apply the surface fitting technique [17] to get a few geometric primitives of the disparity map. We believe the disparity maps generated by our approach can be very useful for further applications like reconstruction, object segmentation, instance recognition and scene analysis etc.

Acknowledgments. This work is supported by the Natural Science Foundation of China (No. 61021062), the 863 Program of China (No. 2011AA01A202) and National Special Fund (No 2011ZX05035-004-004HZ).

References

1. Scharstein, D., Szeliski, R.: A taxonomy and evaluation of dense two-frame stereo correspondence algorithms. International Journal of Computer Vision 47(1), 7–42 (2002)
2. Klaus, A., Sormann, M., Karner, K.: Segment-based stereo matching using belief propagation and a self-adapting dissimilarity measure. In: 18th International Conference on Pattern Recognition, ICPR 2006, vol. 3, pp. 15–18. IEEE (2006)
3. Yang, Q., Wang, L., Yang, R., Stewénius, H., Nistér, D.: Stereo matching with color-weighted correlation, hierarchical belief propagation, and occlusion handling. IEEE Transactions on Pattern Analysis and Machine Intelligence 31(3), 492–504 (2009)
4. Wang, Z.F., Zheng, Z.G.: A region based stereo matching algorithm using cooperative optimization. In: IEEE Conference on Computer Vision and Pattern Recognition, CVPR 2008, pp. 1–8. IEEE (2008)
5. Scharstein, D., Szeliski, R.: Middlebury stereo vision page, `http://vision.middlebury.edu/stereo`
6. Hirschmüller, H., Innocent, P., Garibaldi, J.: Real-time correlation-based stereo vision with reduced border errors. International Journal of Computer Vision 47(1), 229–246 (2002)

7. Veksler, O.: Fast variable window for stereo correspondence using integral images. In: Proceedings of the 2003 IEEE Computer Society Conference on Computer Vision and Pattern Recognition, vol. 1, p–556. IEEE (2003)
8. Yoon, K., Kweon, I.: Adaptive support-weight approach for correspondence search. IEEE Transactions on Pattern Analysis and Machine Intelligence 28(4), 650–656 (2006)
9. Tombari, F., Mattoccia, S., Di Stefano, L.: Segmentation-based adaptive support for accurate stereo correspondence. In: Mery, D., Rueda, L. (eds.) PSIVT 2007. LNCS, vol. 4872, pp. 427–438. Springer, Heidelberg (2007)
10. Hosni, A., Bleyer, M., Gelautz, M., Rhemann, C.: Local stereo matching using geodesic support weights. In: 2009 16th IEEE International Conference on Image Processing (ICIP), pp. 2093–2096. IEEE (2009)
11. Rhemann, C., Hosni, A., Bleyer, M., Rother, C., Gelautz, M.: Fast cost-volume filtering for visual correspondence and beyond. In: 2011 IEEE Conference on Computer Vision and Pattern Recognition (CVPR), pp. 3017–3024. IEEE (2011)
12. Yang, Q., Yang, R., Davis, J., Nistér, D.: Spatial-depth super resolution for range images. In: IEEE Conference on Computer Vision and Pattern Recognition, CVPR 2007, pp. 1–8. IEEE (2007)
13. Birchfield, S., Tomasi, C.: A pixel dissimilarity measure that is insensitive to image sampling. IEEE Transactions on Pattern Analysis and Machine Intelligence 20(4), 401–406 (1998)
14. Heo, Y.S., Lee, K.M., Lee, S.U.: Robust stereo matching using adaptive normalized cross-correlation. IEEE Transactions on Pattern Analysis and Machine Intelligence 33(4), 807–822 (2011)
15. Hirschmuller, H., Scharstein, D.: Evaluation of stereo matching costs on images with radiometric differences. IEEE Transactions on Pattern Analysis and Machine Intelligence 31(9), 1582–1599 (2009)
16. Tomasi, C., Manduchi, R.: Bilateral filtering for gray and color images. In: Sixth International Conference on Computer Vision, pp. 839–846. IEEE (1998)
17. Cohen-Steiner, D., Alliez, P., Desbrun, M.: Variational shape approximation. ACM Transactions on Graphics (TOG) 23, 905–914 (2004)

GhostMesh: Cloud-Based Interactive Mesh Editing

Zhong Zhou*, Fei Dou, Yi Li, and Lin Zhang

State Key Laboratory of Virtual Reality Technology and Systems
Beihang University, Beijing, China
{zz,zhanglin}@vrlab.buaa.edu.cn

Abstract. Interactive deformation of complex meshes is a challenging task in modeling and animation, due to high computation and storage requirements. The emerging cloud computing technology, as well as cloud storage, provides a possible consumer-level solution so that interactive mesh editing could work in light-weighted clients with public cloud services. In this paper we present a system for efficient interactive editing of complex models. Building over a client-server-based distributed architecture, our system transfers computationally heavy tasks (e.g., computing of blending weights, construction of handles) and storage from the client to a cloud computing environment, and allows the user to edit a complex mesh that stored in the cloud server by interactively modifying a simplified mesh locally. When the user specifies coarse deformations on the client, user manipulations are sent to the cloud server, and the original complex mesh is deformed accordingly in the background. Experiments conducted on various complex models show that our system achieves very efficient performance, and the most important, remains fully compatible with the classical WYSIWYG deformations.

Keywords: interactive graphics, shape deformation, 3D modeling.

1 Introduction

The modeling and animation of complex geometric models has become a hot topic today. Although much progresses have been made over the years on 3D modeling systems, effective interactive deformation of complex objects is still a challenging task, due to limitations of computation capability and storage resources.

In previous research, a fair amount of techniques have been developed for the purpose of overcoming these challenges. Among them, space deformation and multi-resolution shape deformation are most representative. To reduce the computational cost of shape deformation, space deformation methods deform a low polygon-count polyhedron in which a shape is embedded, instead of explicitly deforming the shape itself. The main advantages of space deformation techniques lie in the simplicity and speed. Multi-resolution shape deformation is

* Corresponding author.

B. Huet et al. (Eds.): PCM 2013, LNCS 8294, pp. 384–395, 2013.

another preferable choice for interactive mesh editing. Different from space deformation methods that introduce an additional ambient space, multi-resolution techniques, such as subdivision surfaces, organize the objective meshes at different levels of detail and perform hierarchical editing to achieve efficiency. However, both space deformation methods and multi-resolution deformation techniques still strongly depend on the local computing power, meaning that deformations of large complex objects would be still not available on a device with low computing power or storage.

Cloud computing has been a hot topic of computing paradigm in the last years. It transfers heavy computing and storage from the client to a cloud computing environment. Enterprises currently employ cloud services to improve the scalability of their services and to deal with bursts in resource demands. In this work, we are interested in an interactive mesh editing system based on cloud services. In particular, we aim at enabling the user to edit large complex models with the help of cloud server even using a terminal with low computing power and storage. Furthermore, we desire our system to provide interactive manipulation for users.

Our solution is a client-server-based system that transfers heavy computing and storage from the client to the cloud server. Specifically, a coarse deformation is specified on the client side by interactively editing a simplified model instead of the original complex model, while computationally heavy tasks, such as the computing of blending weights for handles and the deformation of the original complex model, are performed on the cloud server. We propose efficient strategies to classify and stream blending weights and user manipulations. As a result, the system allows the user to deform a large complex mesh only by editing a small simplified mesh interactively in local.

2 Related Work

In the past decades, a fair amount of research on surface deformation has been developed. Variational methods [1,2] require post time optimization, which are still too slow to deform high-resolution objects at high framerate. Variational harmonic maps [3] restrict the degrees of freedom to harmonic deformations of a specified cage although they are faster. Other methods using a weighted blend of handle transformations such as least squares [4], dual quaternions [5], and linear blend skinning (LBS) [6] which is most frequently used for real-time character deformation, are fast at pose time. Bounded biharmonic weights [7] is one of the techniques using LBS for efficient pose-time computation with great deformation result. Each vertex is transformed according to the affine transformations of the handles, which are linearly averaged with different weights.

Multi-resolution shape deformation is a very effective way to perform interactive mesh editing. B-spline is presented as a pioneering work on hierarchical editing by Forsey and Bartels [8], in which the user should decide where to add detail and to manipulate the corresponding controls. Wavelets enable multi-resolution editing on objects that can be represented by the hierarchical basis functions [9]. Subdivision surfaces [10] is a preferable choice for interactive mesh editing, due to the possibility to edit mesh at varying levels of detail. For purpose of higher efficiency, a fair amount of research on simplification techniques has long been developed. Progressive mesh, first proposed by Hoppe [11], is a multi-resolution technique using edge collapse, which simplifies models by iteratively contracting edges with precise results and high performance. Among numerous edge collapse algorithms, quadric error metric (QEM) [12], which estimates the error introduced by a pair collapse as the distance from a vertex to a quadratic surface, is the one with high performance and usually the most precise results.

There have been numerous shape editing systems for designing of 3D models. iWIRES [13] and LineFFD [14] are analyze-and-edit approaches. They detect important features of a model to extract a descriptive set of lines and then users can utilize the feature lines to deform the model. Systems like Teddy [15] and FiberMesh [16] for quickly and easily designing freeform surfaces with a collection of 3D curves provide convenience for users. A user draws several 2D freeform strokes interactively on the screen and the system automatically convert them to 3D curves and constructs plausible 3D polygonal surfaces afterwards. When using Wires [17] or SilSketch [18], the user applies the deformation by giving nothing but the 'over-sketch' of feature lines to the system. The lines have to be drawn manually and the system will deform the shape according to them.

3 User Interface

The user interface of GhostMesh consists of a button panel and a rendering window (Fig. 1). The rendering window is embedded with controls for navigation and the capability of drawing viewport-aligned strokes. By clicking option buttons on the button panel and holding some meta keys, the user can drag the mesh along the horizontal and the vertical axies, rotate it and scale the current projection by clicking (left or right) and dragging the mouse.

To edit a large mesh, the system first streams a simplified version of the original large mesh from the cloud server to the client. The user determines an appropriate view, places control handles on the surface. Then the system binds the mesh to these handles. The user interactively manipulates these handles and the system deforms the mesh accordingly.

The system has three main tools for the mesh editing: handle tool, ROI tool, and deformation tool. The user switches between these tools via menu selection or a keyboard shortcut.

3.1 Handle Tool and ROI Tool

The handle tool lets the user construct a number of handles on the mesh. We aim to provide users a flexible and interactive way to manipulate large 3D mesh on the client side. In our system, three common handle types are supported: point, skeleton and control curve. A control curve is described as a set of points.

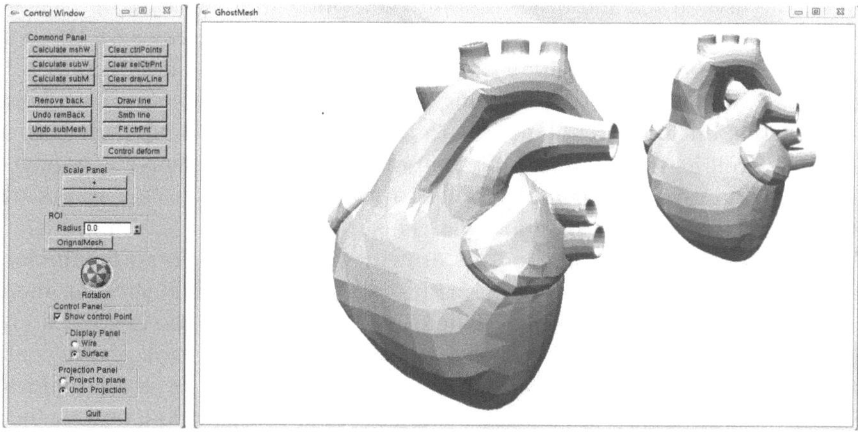

Fig. 1. User interface of our system

Fig. 2. Our mesh editing tool in action. GhostMesh supports points, skeletons, and control curves arranged in an arbitrary configuration. The user can specify translations, rotations and scales at handles (middle left). A sequence of point handles can be selected automatically by drawing a stroke on the surface (middle right), which would be deformed by drawing another control curve (rightmost).

To construct handles, the user clicks or drags on the simplified surface to specify a number of points. Fig. 2 shows illustration of our handle tools.

The ROI tool lets the user specify a certain part of mesh to conduct deformations on. After constructing handles, an optional ROI around specified handles can be computed by diffusion on the surface, and the ROI can be interactively adjusted to fit the target deformation.

3.2 Deformation Tool

The deformation tool lets the user deform the mesh by manipulating handles. One of our user interface for mesh deformation is a usual direct manipulation method: the user grabs and drags a point (i.e., point handle, point of a skeleton and point on a curve), specifies translations, rotations and scales on the point, and the mesh deforms smoothly within the specified ROI. Generally, interactive manipulations of the handles (dragging a point or drawing a control stroke) are usually carried out on a 2D plane due to the limitation of 2D input devices and displays, as well as the 2D depth perception. Therefore, the current implementation always moves the points parallel to the screen. If the user wants more control, new handles must be added on the surface.

Another of our deformation tools is drawing a control stroke to manipulate a curve handle. Using this tool, the user draws a stroke which suggests a modification of the curve, and the shape is deformed according to the deformed curve handle. It is worth noting that because of the sketching strokes can contain noise, our system allows the user to smooth a control stroke or the deformed curve handle directly by brushing on the target curve with the mouse, which is intuitive and convenient.

By manipulating handles with our deformation tools, the simplified mesh is deformed in real-time, showing the user a coarse deformation. At the same time, the system streams the manipulation to the server side, and the original large mesh is deformed accordingly in the background.

4 Algorithm

To implement our interactive mesh editing and the described interfaces, we propose a cloud service which consists of four main parts: handle optimization, weight computing, data transmission, and deformation computation.

Before describing our algorithms, we introduce notations. Let $M = (V, E, F)$ be a given triangular mesh. V denotes the set of vertices, E denotes the set of edges and F denotes the set of faces. Let $\mathbf{e}_{ij}$ be the edge linking two distinct vertices $\mathbf{v}_i$ and $\mathbf{v}_j$. Let H_j $(j = 1, ..., m)$ be the control handles.

4.1 Handle Selection and Optimization

In current implementation we choose some of the mesh vertices as handle members. To select handle members, the user clicks or drags on the simplified mesh M_S to specify a number of points.

It is obvious that handles constructed on the original large mesh are more close to the user intention (see Fig. 3). Therefore, when the user picks 3D points on the simplified mesh M_S, the system streams these points to the cloud server, constructing handles on the complex surface.

When click or drag on M_S, 3D intersection points p_i are selected on faces f_{p_i} of the simplified mesh (3D mouse picking). In current implementation, any two of points p_i are guaranteed not in the same face of the simplified mesh M_S, avoiding redundancy. A path line L formed by a discrete set of points $\{p_i\}$ is constructed (we take the single point selected by clicking as a line with size 1). L is usually not a subset of the vertices of mesh M_S and the original large mesh

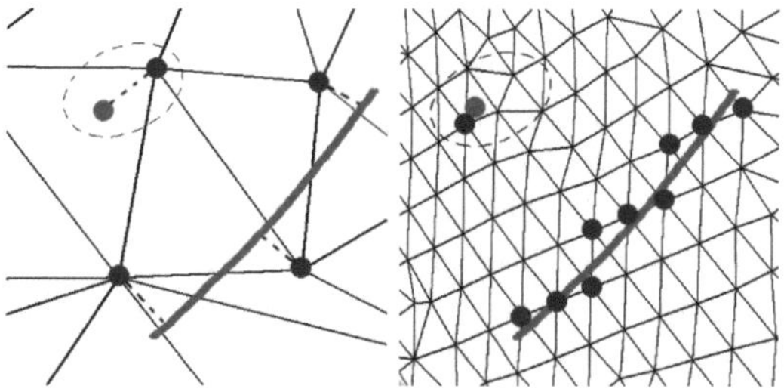

Fig. 3. Comparison between constructing handles on simplified (left) and original (right) surface. In both left and right figures, red shapes are user interactions, and blue points are the possible result handles.

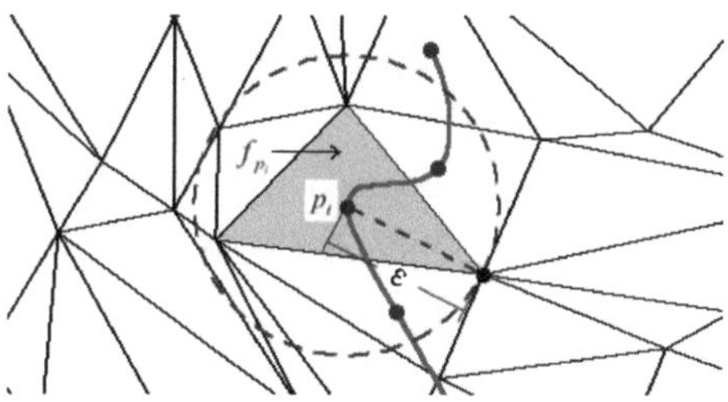

Fig. 4. Initial parameters of collecting p_i'

M. We define a line $L' = \{p_i', p_i' \in V\}$ as minimizers of a shape-aware functional subject to:

$$\underset{p_i', i=1,\ldots,|L|}{\arg\min} \sum_{i=1}^{|L|} \sum_{p_j \in \mathcal{N}_i^+} \|p_i' - p_j\|_2 \tag{1}$$

with $|L|$ the size of set $L = \{p_i\}$, and $\mathcal{N}_i^+$ the set of p_i and its geodesic neighbors in L. In practice we compute line L' by collecting p_i' through a flood filling process performed on M, starting at the first vertex of face f_{p_i} and restricted to the euclidean distance ε from it to p_i (Fig. 4).

When the user draws a stroke on the surface, the sampling density of handle members depends on that of the 3D mouse picking during dragging, and can be controlled according to the practical requirement. A skeleton point is constructed by moving a 3D intersection point inside into the volume.

4.2 Weight Computing

Real-time performance of weight computing for blending is critical for user experience as well. The less the user has to wait, the better the user experience will be. In our system, to shorten time spent waiting for binding, weights are computed by the powerful cloud server. In current implementation, we use bounded biharmonic weights [7], which produces smooth and intuitive deformations, for real-time blending-based deformations. Since handles are selected on the original mesh M, handle members are usually not a subset of the vertices of the simplified mesh M_S. As discussed in [7], this is not consistent with the requirement of computation of bounded biharmonic weights. A direct solution is recomputing a new collapse list for PM that keeps all p'_i during edge collapse, meaning more computation and re-streaming of a new simplified mesh. Therefore, in current implementation, we directly discretize face f_{p_i} by adding edges between each vertex of f_{p_i} and the handle member p'_i.

As will be discussed further below, to make our system more efficient, some of the weights are transmitted over unreliable channels. Then missing weight of vertex v_i is estimated by interpolation as the formula below:

$$w_j(i) = \sum_{k \in \mathcal{N}(i)} \mu_{ik} \cdot w_j(k), \tag{2}$$

where w_j is the weight function associated with handle H_j, the function $\mathcal{N}(i)$ is the 1-ring neighbors of vertex v_i, and $\sum \mu_{ik} = 1$. In current implementation, the choice of neighbor weights

$$\mu_{ik} = \begin{cases} \frac{1}{|\mathcal{N}(i)|-1} \cdot \left(1 - \frac{\|v_i - v_k\|_2}{\sum_{l \in \mathcal{N}(i)} \|v_i - v_l\|_2}\right) \\ 0 & \text{if } w_j(k) \text{ is missing} \end{cases} \tag{3}$$

defines the nature of w_j.

4.3 Data Transmission

Since manipulations are carried out on the simplified mesh on the client, all of them should be sent to the cloud server for the desired corresponding deformation on the original mesh. In order to facilitate data transmission and maintenance, we divide user manipulations into two types: intermediate and key manipulation. There are mainly two cases producing intermediate manipulations. The first case is the user chooses a subset of handles and specifies a same kind of manipulation (i.e., translations, rotations or scales) on them *multiple times* to get a desire deformation. The second case is the system records continuous changes of handles caused by just one user manipulation, such as the trajectory of moving a point by dragging. The relative key manipulation is the final result of a series of intermediate manipulations. A single manipulation is also considered a key manipulation. In fact, the same result of shape editing can be obtained by only computing these key manipulations in order, and the intermediate manipulations are only details of user editing process. Fig. 5 shows an illustration of the user's manipulations.

In our system, data transmission between the cloud server and the client includes: progressive meshes, creation of handles, ROI, blending weights, manipulations of handles and other auxiliary information such as weight computing

n new handle   t translation   r rotation   s scale
key manipulation   intermediate manipulation

Handle	User Interaction													
	1	2	3	4	5	6	7	8	9	10	11	12	13	...
H_1	n	r	r					t	t	t	s	s		...
H_2	n			t	t	t	t				s	s		...
⋮	... ...													
H_m	n			t	t	t	t				s	s		...
H_{m+1}													n	...

Fig. 5. Illustration of different types of manipulations. A key manipulation is a single manipulation or the final result of a series of intermediate manipulations of a subset of handles.

progress. In order to make our system more efficient, we send these data differently on hybrid transmissions to take the advantages of both TCP and UDP. The progressive meshes, creation of handles, ROI parameters, greater blending weights, and key manipulations of handles are encoded and transmitted over reliable channels, while the smaller blending weights and intermediate manipulations are delivered over unreliable channels.

We use QEM [12] for streaming progressive meshes, a simplified mesh is firstly streamed from the cloud server to the client. The user specifies handle selecting and ROI parameters on this simplified mesh, and the system sends them to the cloud server. Then the server constructs handles, computes blending weights accordingly, and returns weights to the client.

In a typical mesh editing system, each handle has the maximum effect on its immediate region and its influence disappears in distant parts of the object, in other words, weights decay quickly. According to this phenomenon we send weights greater than an user-defined threshold δ over reliable channels, otherwise over unreliable channels (Fig. 6). There is no need to request retransmission of lost weight data. In practice we interpolate the missing weights with formula 2 through a simple dynamic process starting at the vertex with the most neighbors that have weight.

During editing, the client encodes the user manipulation, and sends them to the server. The encoded manipulation data includes the following information:

- The manipulation ID (numbered in ascending order);
- The manipulation type (i.e., intermediate or key);
- IDs of manipulated handles;
- The result parameters of each handle.

Since the key and intermediate manipulations are transmitted via reliable and unreliable channels respectively, two sorted lists are maintained on the server, and they can be easily retrieved and traversed as one according to the manipulation IDs. The cloud server processes only the key manipulations in order, or

(a) plot of weights (b) plot of weight groups

Fig. 6. 3D plot of weights of a point handle on the tip of the Knife model and their groups. Note that weights are smooth and local (a). Those smaller weights (in yellow) are transmitted over unreliable channels, and the others over reliable channels. Parameter value $\delta = 0.2$.

the intermediate manipulations as well, to computes the desired deformations of the complex original mesh.

5 Evaluation

In this section, we introduce the experiments results to evaluate GhostMesh. We implement the real-time 3D interactions and the proposed data transmission algorithm on a Windows XP PC with Intel Core2 Xeon 3.07GHz CPU and 4GB DDR3 RAM, and a Core2 Duo Laptop with a 2.5GHz CPU and 4GB RAM. Clients are tested on the Laptop. Laptop client communicates with the PC server through Wi-Fi.

Table 1. Statistics for the various examples in the paper. The binding time is measured in seconds. LOD is the percentage of visible vertices after mesh simplification.

Model	# of vertices	# of faces	Binding time per handle in different LODs			
			25%	50%	75%	100%
Heart	7349	14676	0.296	0.755	1.403	2.241
Sword	18001	35966	0.785	3.978	6.143	9.277
Dinosaur	56194	112384	3.446	11.167	19.681	30.193
Neptune	99996	200000	9.453	23.738	45.664	69.157

We list in Table 1 the binding time measurements of our unoptimized code. The computational overhead of blending weights can be reduced significantly by using a simplified mesh. The data size of weights is determined by the number of vertices and handles. Take the simplified *Neptune* model with 50K vertices and 100K triangles (LOD=50%) as an example, the data size of weights is about 0.19MB per handle without compression, and it takes an average of 0.561s to send that of 25 handles from the server to the client in our experiment.

Fig. 7 gives a comparison of deformation results between the one using the original weights and the other one using the interpolated weights. As is shown, only small distortions can be seen at a loss ratio even as high as 50% (review

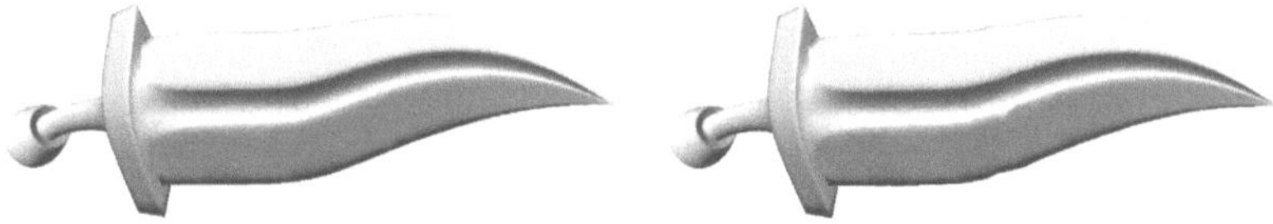

Fig. 7. Comparison between deformation results of using the original weights (left) and the interpolated weights (right). For parameters value of $\delta = 0.2$ and the loss ratio is 50%.

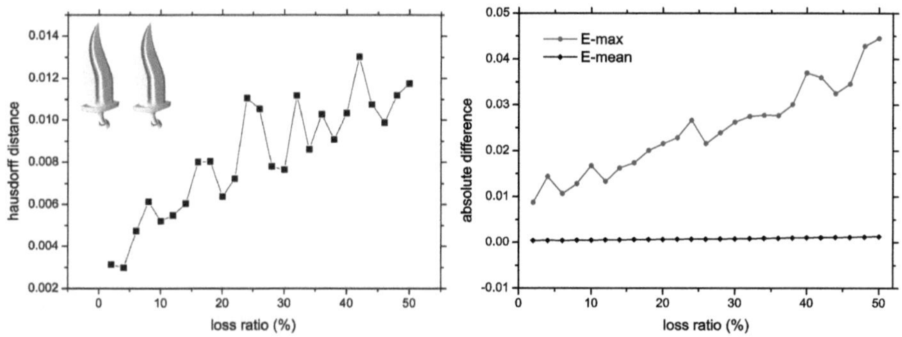

Fig. 8. The geometric distortion of the deformed Sword model and weight interpolation error with different data loss ratios. The geometric distortion is measured in Hausdorff Distance (left). E-max and E-mean are respectively the max and mean absolute difference between our interpolation and the original bounded biharmonic weights (right).

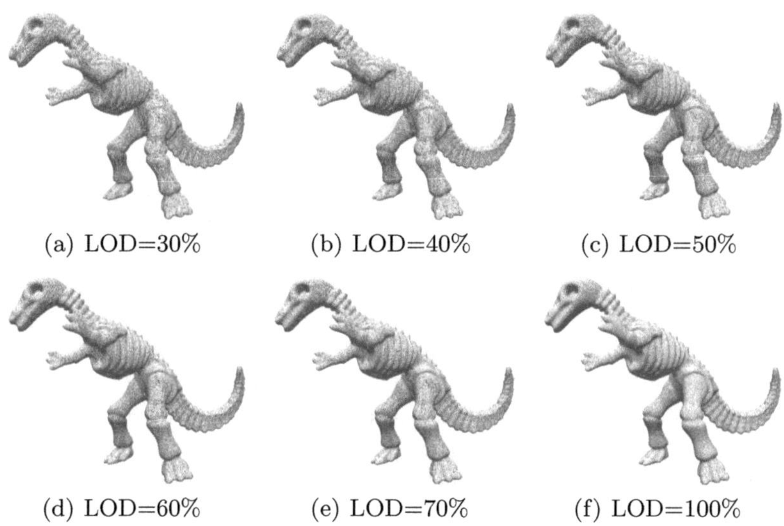

(a) LOD=30% (b) LOD=40% (c) LOD=50%

(d) LOD=60% (e) LOD=70% (f) LOD=100%

Fig. 9. Editing results of simplified mesh with different LOD (a-e), and the corresponding deformation result computed by the cloud server

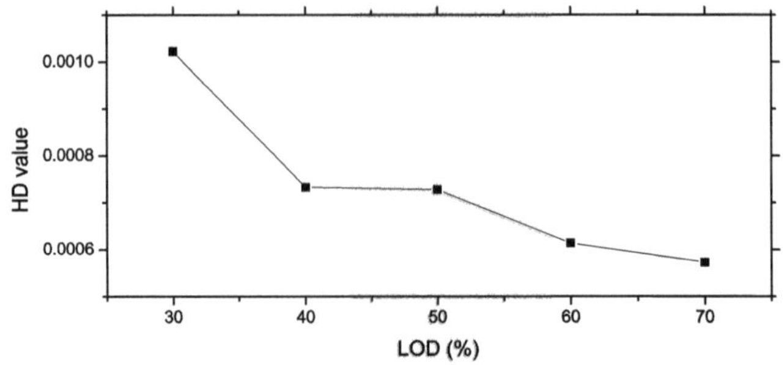

Fig. 10. The geometric difference between deformations of simplified and the original meshes with different LOD, measured in Hausdorff Distance

Fig. 6(a) for the plot of weights). Fig. 8 shows our analysis results on the weight interpolation and the geometric distortion according to different loss ratios. As expected, errors and distortions are very limited, even though with the increasing of data loss, the error and distortion raises as well.

Fig. 9 shows the editing results of simplified mesh with different LOD (a-e), and the corresponding deformation result computed by the cloud server. As we can see, deformation results are very similar. Furthermore, we show our analysis results on the geometric difference between the client (Fig. 9(a-e)) and the server (Fig. 9(f)) according to different LOD in Fig. 10. Consistent with Fig. 9, the geometric differences are very small, meaning the system deforms the original complex mesh in a way very close to user's desire.

6 Conclusion and Future Work

We have presented a system for cloud-based interactive mesh editing, namely GhostMesh. The system allows the user to edit a complex mesh that stored in the cloud server by interactively modifying a simplified mesh locally. We proposed efficient strategies to classify and stream blending weights and user manipulations. As a result, the system can transfer heavy computing and storage from the client to a cloud computing environment, allowing the user to edit a large complex model on the remote cloud server even using a lightweight terminal with low computing power and small storage space. Mobile-based augmented reality is an emerging technology that provides immersive experiences over wireless networks. In future work, we wish to extend our system to include multiple mobile users in a collaborative virtual environments. We also plan to compare with other applications and perform user studies in the future.

Acknowledgments. This work is supported by the National 863 Program of China under Grant No.2012AA011801, the Natural Science Foundation of China under Grant No.61170188, and the Specialized Research Fund for the Doctoral Program of Higher Education of China under Grant No.20121102130004.

References

1. Igarashi, T., Moscovich, T., Hughes, J.F.: As-rigid-as-possible shap manipulation. ACM Trans. Graph. 24(3), 1134–1141 (2005)
2. Botsch, M., Sorkine, O.: On linear variational surface deformation methods. IEEE TVCG 14(1), 213–230 (2008)
3. Ben-Chen, M., Weber, O., Gotsman, C.: Variational harmonic maps for space deformation. ACM Trans. Graph. 28(3) (2009)
4. Schaefer, S., Mcphail, T., Warren, J.: Image deformation using moving least squares. ACM Trans. Graph. 25(3), 533–540 (2006)
5. Kavan, L., Collins, S., Zara, J., OSullivan, C.: Geometric skinning with approximate dual quaternion blending. ACM Trans. Graph. 27(4) (2008)
6. Magnenat, N., Laperriére, R., Thalmann, D.: Joint-dependent local deformations for hand animation and object grasping. In: Graphics Interface, pp. 26–33 (1988)
7. Jacobson, A., Baran, I., Popović, J., Sorkine, O.: Bounded biharmonic weights for real-time deformation. ACM Trans. Graph., 78:1–78:8 (2011)
8. Forsey, D.R., Bartels, R.H.: Hierarchical B-spline refinement. ACM SIGGRAPH Computer Graphics 22(4), C205–C212 (1988)
9. Gortler, S.J., Cohen, M.F.: Hierarchical and variational geometric modeling with wavelets. In: Proceedings of ACM Symposium on Interactive 3D Graphics, pp. 43–54 (1995)
10. Kobbelt, L.P., Bareuther, T., Seidel, H.P.: Multiresolution shapedeformations for meshes with dynamic vertex connectivity. Computer Graphics Forum 19(3), 249–260 (2000)
11. Hoppe, H.: Progressive meshes. In: Proc. of SIGGRAPH 1996, pp. 99–108 (1996)
12. Garland, M., Heckbert, P.: Surface Simplification Using Quadric Error Metrics. In: Proceeding of SIGGRAPH 1997, pp. 209–216 (1997)
13. Gal, R., Sorkine, O., Mitra, N.J., Cohen-Or, D.: iWIRES: an analyze-and-edit approach to shape manipulation. ACM Trans. Graph. 28(3) (2009)
14. Schemali, L., Thiery, J.M., Boubekeur, T.: Automatic Line Handles for Freeform Deformation. Eurographics (Short), 81–84 (2012)
15. Igarashi, T., Matsuoka, S., Tanaka, H.: Teddy: a sketching interface for 3D freeform design. In: Proceedings of SIGGRAPH 1999 (1999)
16. Nealen, A., Igarashi, T., Sorkine, O., Alexa, M.: FiberMesh: designing freeform surfaces with 3D curves. ACM Trans. Graph. 26(3) (2007)
17. Singh, K., Fiume, E.: Wires: a geometric deformation technique. In: Proceedings of SIGGRAPH 1998, pp. 405–414 (1998)
18. Zimmermann, J., Nealen, A., Alexa, M.: SilSketch: automated sketch-based editing of surface meshes. In: Proceedings of ACM Eurographics Workshop on Sketch-Based Interfaces and Modeling, pp. 23–30 (2007)

Efficient Sketch-Based 3D Shape Retrieval via View Selection

Dawei Lu, Huadong Ma, and Huiyuan Fu

Beijing Key Lab of Intelligent Telecomm. Software and Multimedia,
Beijing University of Posts and Telecomm., Beijing 100876, China
ludawei19@gmail.com, mhd@bupt.edu.cn,
fu_hui_yuan@yahoo.com.cn

Abstract. Since the available 3D models are becoming more and more popular, there is an increasing need for effective methods of retrieving 3D models. For this purpose, we propose a sketch-based 3D model retrieval approach using a hand-drawn sketch as a query input. In our approach, we use a combined line rendering technique to represent project views of each 3D model from several predefined viewpoints. Then we compute a descriptor based on the orientation of the feature lines for both the project views and sketches. Our method consists of two processing stages: offline and online. In the offline stage, we use a sketch dataset that corresponding to the query 3D model dataset for each class to measure similarity between the sketches and the view images. With the measurement the preference viewpoints of the 3D model in each class can be selected. In the online stage, we match the query sketch to the selected view images and retrieve the query results. Evaluative experiments demonstrate that our approach is robust against variations of shape, pose and rotation. We compare our method with an approach (DTF-SC) by the first-tier precision, and the results show higher precision for most classes of 3D models.

Keywords: Sketch-based 3D shape retrieval, combined line rendering, preference viewpoint.

1 Introduction

Recently, with the rapid increase in the number of 3D geometric models, how to accurately and effectively search for 3D model has been an important issue for many applications such as industrial design, engineering, and entertainment area.

The most direct way for retrieving 3D models is using textual annotations or keywords to describe the 3D objects. However, for many 3D models, there might be some different labels attached to the model and many models do not carry any labels. This fact leads to the development of the research on content-based 3D model retrieval in recent years [1], [20], [21]. Content-based shape retrieval takes the shape itself as a query, thus user requires a 3D model as an example query object. However, the application of these approaches is also difficult, since a good example is not always at hand. This fact clearly limits the 3D model retrieval development and application.

B. Huet et al. (Eds.): PCM 2013, LNCS 8294, pp. 396–407, 2013.

Instead of above approaches, sketch-based retrieval has been proposed, where users draw 2D strokes and lines to depict the important features of the model as seen from one or more viewpoints. Although the sketch does not include color and texture, it could be a clear and accurate representation of the projection of the model. However, there are also some problems in this approach. One important issue is how to represent the two-dimensional projections of the 3D model and match them against the query sketch lines. Next, the preference viewpoint of the sketch is unknown, so handling all possible view directions may be an inevitable choice. Finally, sketches are almost always not realistic depictions of the original shapes, and this could leads to incorrectly recognition by the system and unexpected retrieval results.

In order to handle above issues, several approaches have been proposed. For example, Funkhouser et al. [2] and Chen et al. [3] both develop a content-based retrieval system that also supports sketch as input, their approaches only consider the contours of sketches and the project views. Yoon et al. [4] propose an approach using the HOG descriptors in diffusion tensor fields for effective retrieval. Recently, a novel sketch-based retrieval approach has been described by Eitz et al. [5] using Bag–of-features method. It considers the predicted best user sketch views and good local features for better results, but the choice of the project views is also not sufficient for effective retrieval.

In this paper we propose an approach for sketch-based 3D shape retrieval that takes into account user viewpoint preference. We project the 2D view images of each 3D model from some predefined viewpoints and use combined lines to render these images. In this way the rendering images resemble the user hand-drawn sketches very closely. For the rendering images and the sketches we compute a feature vector based on the orientation of the feature lines. Furthermore, we divide our approach into two stages: offline and online. In the offline stage, we use a sketch dataset that corresponding to the query 3D model dataset for each class to measure similarity between the sketches and the view images. By this way we could select the preference viewpoints of the 3D model in each class. In the online stage, we match the query sketch to the selected view images and get the retrieval results. The effectiveness and robustness of our approach are demonstrated by comparative experiments.

The rest of this paper is organized as follows. In section 2 we review the related work in the area of sketch-based 3D model retrieval. Details in our approach are presented in section 3. Section 4 describes the experiments and the analysis of the results. Finally, section 5 concludes the paper.

2 Related Work

Since there exists a lot of work on content-based 3D model retrieval [15], [16], [17], [18], [19], sketch-based retrieval is only an additional approach in the example-based retrieval system. One of these kind approaches is a web-based 3D search engine that supports 3D sketches, 2D sketches, 3D model examples, and keywords as input(s) described by Funkhouser et al. [2]. In the part of sketch queries, they extract the boundary contours of each 3D model from 13 orthographic view directions and represent each view and sketch by a rotation invariant boundary descriptors that modified by the Fourier Descriptors. Then they index these descriptors using the nearest neighbor method for the developed matching algorithm to compare the views to the input sketch. Another example is a system developed by Chen et al. [3]. They use a Lightfield descriptor to represent sampling views of each 3D model. For each view and the sketch, closed contours are extracted by a hybrid of Zernike moments and Fourier descriptors.

The above two approaches only use sketch-based retrieval as an additional way so that they do not consider much difficulty in this area. After these researches, there are several approaches proposed to handle the important issues for sketch-based shape retrieval. Since we use 2D project views from 3D models to compare to the query sketches, an appropriate representation for each view is needed. Yoon et al. [4] describe an approach to efficiently solve this problem. They use suggestive contours [6] to depict the 2D view from the 3D model, which render the occluding contours and the important areas of the shape naturally. Then the project views of the shape could be represented as sparse lines similar with the input sketches so as to make discrimination between different views. With the above representations they use the HOG [7] in the diffusion tensor field as the descriptor for the 2D views and the sketches and match them to achieve the retrieval procedure.

Besides the feature measure and matching ways from the content-based retrieval, the approaches in image retrieval area can also be used for the sketch-based 3D model retrieval. Recently Eitz et al. [5] develop a bag-of-features (BOG) approach for sketch-based 3D shape retrieval, which are well known in the image retrieval community [8], [11]. They also use the line rendering techniques to represent the project views of the shape, and use Gabor filters to extract features. With the BOG approach, they extract the local features as local descriptors and quantize them for matching phase based on histograms.

In summary, previous sketch-based 3D model retrieval approaches measure the similarity between a query sketch and a 3D model by directly computing the distances between the sketch and several project views of the 3D model and then choose the most similar one. However, these approaches have a shortcoming of the cost for matching all viewpoints, which motivates us to develop a sketch-based retrieval method by selecting views of 3D models that are similar with the viewpoints of user sketch preference.

3 Our Approach

In this section we describe our proposed sketch-based 3D shape retrieval approach. Since the user input is a 2D sketch image, we need to get 2D representation from 3D models using several line rendering techniques. For image features extraction and matching we choose a descriptor that could catch the spare lines information effectively, the details are presented in section 3.2. In our approach, the main property is that we use a sketch dataset to match the 3D models and select the project images of user preference viewpoints in the preprocessing phase. In this way, we could make the 3D shape retrieval more accurately and effectively.

Our proposal is composed of two stages: offline stage and online stage. In the offline stage, 3D models are represented as 2D lines rendering images, and then match the sketches of training sketch dataset by extracted features. Finally the shape view images of user preference viewpoint are selected for the user search. In the online stage, similarity measurement is implemented between input sketches and viewpoint preference images, and then the query results are obtained. These two stages are shown in Figure 1.

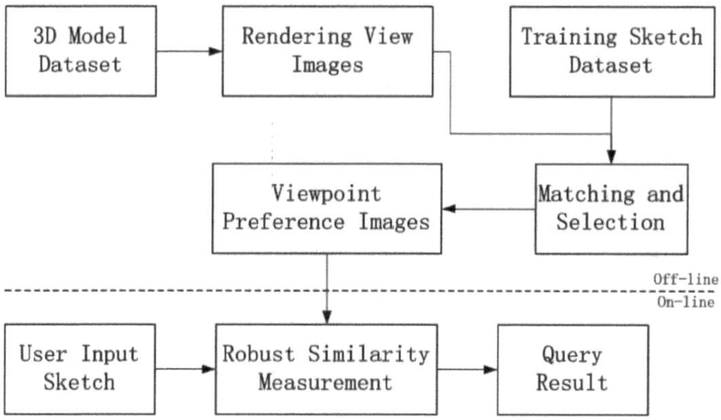

Fig. 1. Overview of our 3D shape retrieval approach

3.1 3D Model Representation in 2D Views

Since the sketch as a query input is a rough description of a project view of the 3D model, for sketch-based 3D model retrieval we could represent the 3D model as 2D projections from different viewpoints. In order to improve the visual similarity between the user sketches and the project views that could make the matching phase more effectively, good choices for appropriate projection and rendering techniques are needed.

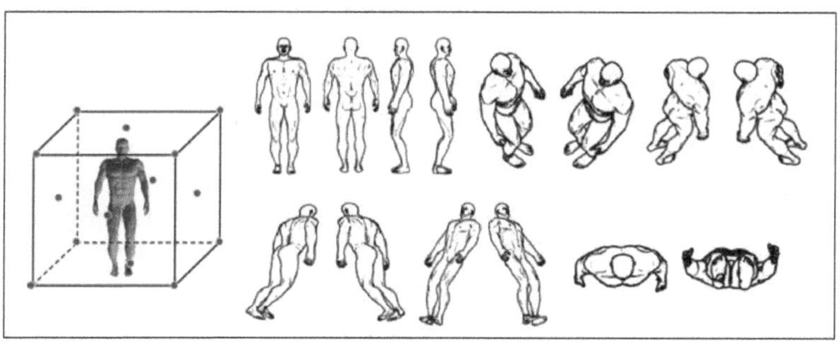

Fig. 2. A 3D model and its rendering lines from 14 different view directions

A simplest way for depicting the 2D views of a 3D model is the occluding contours. They could catch the view-dependent information of the object with sparse outlines that resemble rough user sketches. However, most natural 3D models contain bumps, dips, and undulations of varying geometry, the features could not be depicted by these lines easily.

To handle this issue, we extract both suggestive contours [6] and apparent ridges [9] to render the project views of the 3D model. Suggestive contours are extensions of occluding contours. They are locations which are almost contours and correspond to true contours in nearby viewpoints. Apparent ridges are extensions of the valleys and ridges, but they are view-dependent with a more sophisticated curvature metric. These two lines both have drawbacks and neither of them is better in all conditions, so we use a hybrid of them. Besides these lines we use the boundary lines as additional step for the occlusion of the shape. In this way we can get a minimalistic representation of the project views of the 3D models.

For better matches between the user sketches and the 3D models, we render the combined lines of each model from 14 different view directions. As shown in Figure 2, we take viewpoints at the center of the six faces and the eight corners of a bounding cube. For each category of 3D models datasets that we have used in the experiments, we align them with consistent orientation for the work on estimation of the viewpoint preference that users sketch the model more likely. Details of the work are presented in the section 3.3.

3.2 Feature Extraction and Matching

In a sketch or a line image, the key information is the direction of the stroke relative to its position, and this information relates best to the direction of gradients. So we analyze the features in the diffusion tensor fields [10] and extract the SHOG features [11] as the descriptors for the sketches and the line drawing views.

First, we need preprocessing of the combined line drawings from a 3D model and the input sketches to extract the features in the diffusion tensor fields. Then we organize the features into a histogram as a feature vector according to their orientation and their magnitude. The feature vector could be used as a HOG descriptor for the sketches and the line drawing images, but in our approach we use the SHOG descriptor to represent them. SHOG descriptor improve the histogram of oriented gradient descriptor by enhance only the sketch feature lines in the histogram gradients, this could help to make descriptors extracted from user sketches and the view images more similar. The descriptor is also invariant to geometric and photometric transformations of the features, so it is very robust in the 3D model retrieval procedure.

So given an input sketch I_s and a line drawing view image I_v, we could normalize them to a fixed size and compute feature vectors H_s and H_v by the above approaches. Finally we normalize the two feature vectors of histogram orientation into unit vectors, and then define the similarity between them as

$$S(I_v, I_s) = \frac{H_v \cdot H_s}{\|H_v\| \|H_s\|} \tag{1}$$

Then the similarity between a query sketch and a 3D model can be defined by the maximum of the comparing results between the sketch and each line drawing view of the 3D model.

3.3 Handling Viewpoint Preference

In our approach we represent a 3D model as 2D project views from 14 different viewpoints, and these views could contain almost all possible viewpoints. However, it is clear that users do not draw the shapes with equal probability. For example, if people draw a car they usually choose a side as the viewpoint rather than the top or bottom. So we do not need to compares all the view images with the query sketches because of the user viewpoint preference.

Unlike some computation model for viewpoint estimation [5], [22], [23], we predict the good viewpoints by comparing user sketches in a collected dataset with all rendering views of the 3D models. In this way we use the sketch dataset from [12] and the 3D model dataset from [13]. The 2D sketch dataset contains 20,000 sketches, categorized into 250 classes. It is collected by crowd–sourcing from non-expert participants and it covers most object categories in everyday life. Furthermore it has equal sketches in each class, so we could use this dataset corresponding to the 3D model dataset in our approach for good realization. The 3D model dataset contains a train dataset and a test dataset, which each of them has 907 models with 97 and 90 classes, respectively.

Here we select sketches from the sketch dataset and models from the 3D model dataset to form two target datasets as in [14], and then we have 7200 sketches and 1258 relevant models in 90 classes. So given the two datasets and classes correspondence, we propose an approach to select the images of viewpoint preference.

Fig. 3. Viewpoint selection of 3D models in each class

For each class we compare m sketches with all the 14 views of n models, and the similarity score S of an image pair could be computed by Eq.1. So for one viewpoint, the mean similarity score V of all matches can be given by

$$V = \sum_{i}^{k} S_i \Big/ k, k = m \times n \tag{2}$$

Then for the mean similarity scores of all 14 viewpoints, we array them in descending order, and the first one corresponds to the viewpoint of user selection most likely. By this order we could select several rendering images of viewpoint preference per class. The procedure of view selection is shown in Figure 3. In this way the matches between input sketches and views of 3D models can be reduced, thus the shape query procedure could be more effectively and the results can be more accurately. Now we summarize the preprocessing algorithm in the block below.

Algorithm 1. Preprocessing algorithm for sketch-based 3D shape retrieval.

Input: Training sketch dataset Q, 3D model dataset P and C classes
Output: The top n viewpoints for each class.

Initialization: Divide Q and P into datasets in each class: $Q^c = \{q_i\}_{i=1}^{M}$, a sketch dataset in class c; $P^c = \{p_j\}_{j=1}^{N}$, a 3D model dataset in class c; $K = 14$, all viewpoints of a 3D model.

repeat
  for $k = 1$ to K **do**
    for $i = 1$ to M **do**
      Compute feature vector H_{q_i} for q_i.
      for $j = 1$ to N **do**
        Render combined lines for p_j.
        Compute feature vector H_{p_j} for p_j.
        Compute the similarity S_j^i between H_{q_i} and H_{p_j}.
      end for
    end for
    Computer the average similarity V_k for a viewpoint.
  end for
  Sort the average similarities for all viewpoints.
  Select the top n viewpoints of user preference.
until All the C classes have been handled.

4 Experiments

In this section we perform comparative and evaluative experiments of our sketch-based 3D shape retrieval approach. First of all, the datasets used in two stages are presented, and then we analyze the retrieval results and the comparison of retrieval performance.

4.1 Dataset Description

For the two stages of our approach, we use different sketch datasets in our experiments. In the offline stage, the training sketch dataset and 3D model dataset we use has been presented in section 3.3. Figure 4 shows examples of the 3D model dataset and the sketch dataset for the offline stage.

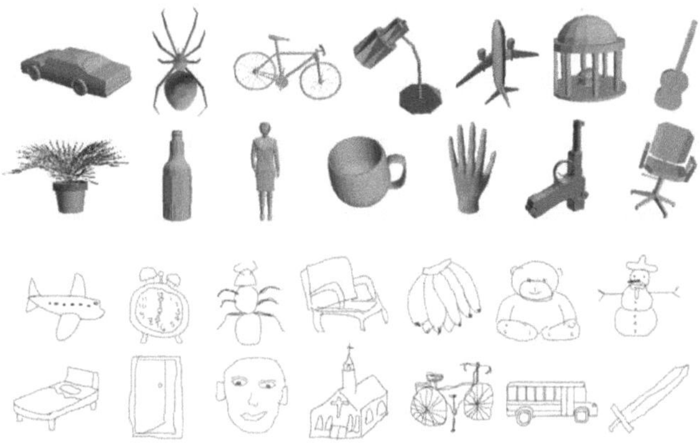

Fig. 4. Examples of the 3D models and the sketches

In the online stage we use the same 3D model dataset as our query shape dataset, but we use our hand-drawn sketches as the query input for the search procedure.

To measure the retrieval accuracy, we use the performance metric of First Tier (FT) that could define the percentage of a class has been retrieval among the first M retrieved models. For our case, M = 20. So we need to make the dataset fit our evaluation approach.

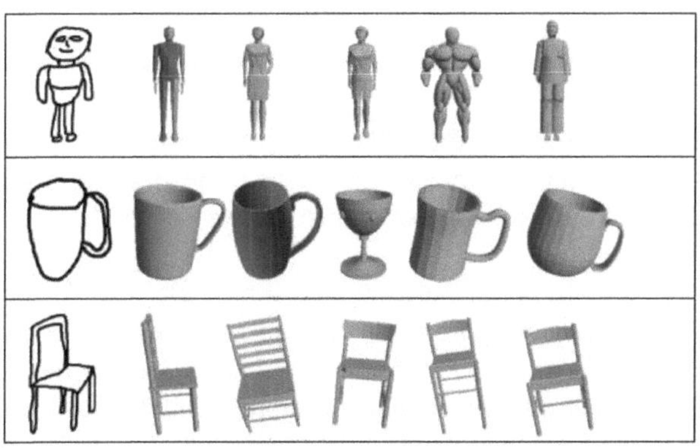

Fig. 5. Examples of the retrieval approach with the first five results

Here we use the test dataset including 907 models of 90 classes and choose 300 models (15 classes, 20 models each) as target test 3D model dataset. We collect a sketch dataset corresponding to the categories of test 3D model dataset from non-expert, untrained users in the experiments. We ask the users to draw one sketch for a class name with only two requirements: (1) the sketch should be as simple as possible; (2) other humans can recognize the sketch as belonging to a class easily. After removing the hand-drawn sketches that do not follow our requirements, the final test sketch dataset contains 106 sketches and each class has 5 sketches at least.

In our experiments we use the same two test datasets for all approaches and conditions, and we will give the performance analysis in the next section.

4.2 Result Analysis and Evaluation

In this section, we show the results of the query of 3D models from our test sketch dataset and the comparison to another approach.

Since the 3D model dataset has different numbers of models for different classes, there is a class bias for retrieval performance. In other words, for some classes of only several models the retrieval results may not be very good. Figure 5 shows the examples of the retrieval results, but here we only depict the first 5 retrieved models.

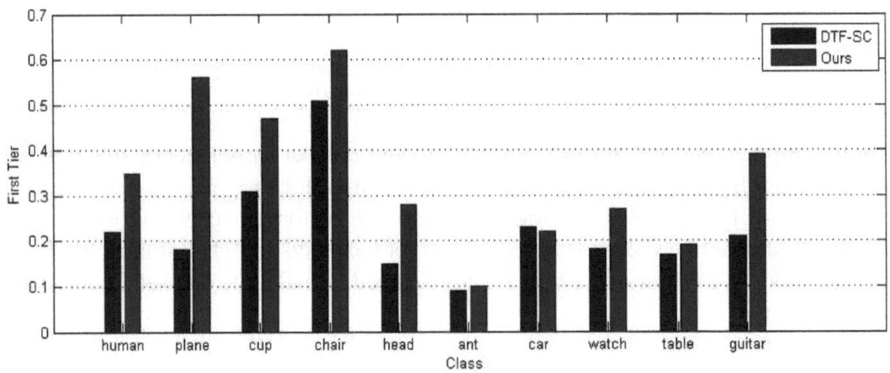

Fig. 6. First Tier Performance comparison between our method and DTF-SC [4]

In our experiments we compare the First Tier performances of our approach with the method DTF-SC [4] on each class. We retrieve the first 20 models for one input sketch by the two methods, and in our approach we select the first 7 views by the algorithm in section 3.3.

Figure 6 shows the comparing performance for 10 classes. We could see that our approach increases the retrieval precision for most classes, as the average First Tier performances of the classes with DTF-SC is 22.4% and our approach is 35.2%. So for respective classes and overall performance our approach both has the better results.

To have a comprehensive evaluation of our approach, we further provide the comparing performance for the different numbers of selected views in our method. Here we select the first 3 views, the first 5 views and the first 7 views respectively, and the performance is shown in Figure 7.

Similarly, we compute the average First Tier performances of selection views. The result in 3 views is 32.2%, in 5 views is 33.4% and 7 views is 35.2%. We can see that there is just a little increasing in precision from 3 views to 7views. The results demonstrate that our retrieval method is robust with different numbers of selected views. However, for some classes the First Tier with fewer views may be higher than more views. The reason is that more views means more matches between the input sketch and view images, and then some irrelevant models may be more similar to the sketch by matching algorithm. Thus the wrong results are retrieved more likely.

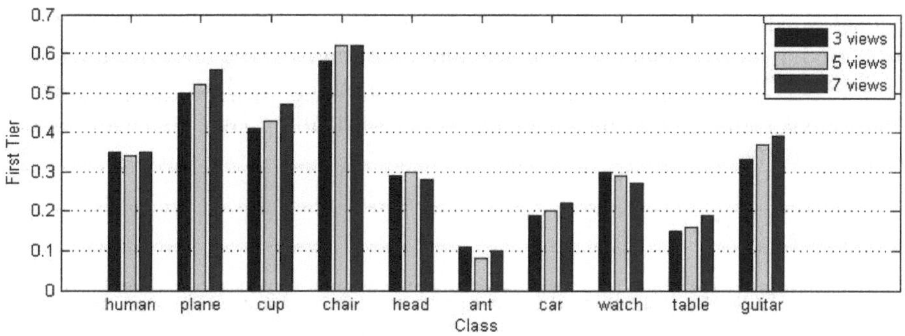

Fig. 7. First Tier Performance comparison using different selections of views

4.3 Limitations

Our approach has a good accuracy in terms of sketched-based 3D model retrieval that depends on the viewpoint preference. However, our view selection strategy can not cover all the categories in our daily life, and sometimes users do not draw the sketches from the usual viewpoints. So the results are not better than other regular approaches, sometimes may be worse, in these conditions.

Moreover, the matching algorithm in our experiments is based on geometric similarity, which is efficiency but simple for the retrieval procedure. Thus without semantic behaviors our approach may get some results from the wrong class among all the retrieve models. The tradeoff of the matching algorithm between the accuracy and efficiency is an important issue that does not be well handled in our approach.

5 Conclusions

In this paper, we have presented a novel approach for sketch-based 3D shape retrieval by view selecting. Considering the user viewpoint preference, we use a sketch dataset that corresponding to the query 3D model dataset for each class to measure similarity between the sketches and the project view images of 3D models, then we select the preference viewpoints of the 3D model in each class in the preprocessing phase. In the query phase, we match the query sketch to the selected view images by a feature vector and return to the user the retrieved results. Experiments based on hand-drawn sketches and the project view images of line rendering demonstrate our approach is robust against variations of shape, pose and rotation. Our method achieves superior performance in precision increasing as well as retrieval time reducing. For the future work, the sketch recognition will be considered to make the 3D model retrieval more accurately and effectively and we will extend our approach for sketch-based 3D modeling as well as sketch-based 3D scene reconstruction.

Acknowledgment. The research reported in this paper is supported by The National Key Technology R&D Program of China under Grant No.2012BAH39F07; the China National Funds for Distinguished Young Scientists under Grant No. 60925010; the Specialized Research Fund for the Doctoral Program of Higher Education under Grant No.20120005130002; China Education and Research Grid – ChinaGrid; The Cosponsored Project of Beijing Committee of Education; The Funds for Creative Research Groups of China under Grant No.61121001.

References

1. Tangelder, J.W.H., Veltkamp, R.C.: A survey of content based 3D shape retrieval methods. Multimedia Tools and Applications 39(3), 441–471 (2008)
2. Funkhouser, T., Min, P., Kazhdan, M., Chen, J., Halderman, A., Dobkin, D., Jacobs, D.: A search engine for 3D models. ACM Transactions on Graphics 22(1), 83–105 (2003)
3. Chen, D.-Y., Tian, X.-P., Shen, Y.-T., Ouhyoung, M.: On visual similarity based 3D model retrieval. Computer Graphics Forum 22(3), 223–232 (2003)
4. Yoon, S.M., Scherer, M., Schreck, T., Kuijper, A.: Sketch-based 3D model retrieval using diffusion tensor fields of suggestive contours. ACM Multimedia, 193–200 (2010)
5. Eitz, M., Richter, R., Boubeker, T., Hildebrand, K., Alexa, M.: Sketch-based shape retrieval. ACM Transactions on Graphics 31(4), 31:1–31:10 (2012)
6. DeCarlo, D., Finkelstein, A., Rusinkiewicz, S., Santella, A.: Suggestive contours for conveying shape. ACM Transactions on Graphics 22(3), 848–855 (2003)
7. Dalal, N., Triggs, B.: Histograms of oriented gradients for human detection. In: IEEE International Conference on Computer Vision and Pattern Recognition, pp. 886–893 (2005)
8. Sivic, J., Zisserman, A.: Video Google: a text retrieval approach to object matching in videos. In: IEEE International Conference on Computer Vision, pp. 1470–1477 (2003)
9. Judd, T., Durand, F., Adelson, E.H.: Apparent ridges for line drawing. ACM Transactions on Graphics 26(3), 19:1–19:8 (2007)
10. Yoon, S.M., Kuijper, A.: View-based 3D model retrieval using compressive sensing based classification. In: International Symposium on Image and Signal Processing and Analysis, pp. 437–442 (2011)
11. Eitz, M., Boubeker, T., Hildebrand, K., Alexa, M.: Sketch-based image retrieval: benchmark and bag-of-features descriptors. IEEE Transactions Visualization and Computer Graphics 17(11), 1624–1636 (2011)
12. Eitz, M., Hays, J., Alexa, M.: How do humans sketch objects? ACM Transactions on Graphics 31(4), 44:1–44:10 (2012)
13. Shilane, P., Min, P., Kazhdan, M., Funkhouser, T.: The Princeton shape benchmark. Shape Modeling International, pp. 167–178 (2004)
14. Li, B., Lu, Y.J., Godil, A., Schreck, T., Aono, M., Johan, H., Saavedra, J.M., Tashiro, S.: Eurographics Workshop on 3D Object Retrieval. In: SHREC 2013 Track: Large Scale Sketch-Based 3D Shape Retrieval, pp. 1–9 (2013)

15. Li, P.J., Ma, H.D., Ming, A.L.: Combining topological and view-based features for 3D model retrieval. Multimedia Tools and Applications 65(3), 335–361 (2012)
16. Li, P.J., Ma, H.D., Ming, A.L.: Non-rigid 3D model retrieval using multi-scale local features. ACM Multimedia, 1425–1428 (2011)
17. Li, P., Ma, H., Ming, A.: 3D model retrieval using 2D view and transform-based features. In: Qiu, G., Lam, K.M., Kiya, H., Xue, X.-Y., Kuo, C.-C.J., Lew, M.S. (eds.) PCM 2010, Part I. LNCS, vol. 6297, pp. 449–460. Springer, Heidelberg (2010)
18. Li, P.J., Ma, H.D., Ming, A.L.: View-based 3D model retrieval with topological structure. In: IEEE International Conference on Multimedia and Expo, pp. 1–6 (2011)
19. Li, P.J., Ma, H.D., Ming, A.L.: View-based 3D model retrieval using two-level spatial structure. In: IEEE International Conference on Image Processing, pp. 3657–3660 (2011)
20. Bustos, B., Keim, D., Saupe, D., Schreck, T.: Content-based 3D object retrieval. IEEE Computer Graphics and Applications 27(4), 22–27 (2007)
21. Bimbo, A.D., Pala, P.: Content-based retrieval of 3D models. ACM Transactions on Multimedia Computing, Communications, and Applications 2(1), 20–43 (2006)
22. Gao, Y., Wang, M., Zha, Z.J., Tian, Q., Dai, Q.H., Zhang, N.Y.: Less is More: efficient 3D object retrieval with query view selection. IEEE Transactions on Multimedia 13(5), 1007–1018 (2011)
23. Secord, A., Lu, J.W., Finkelstein, A., Singh, M., Nealen, A.: Perceptual models of viewpoint preference. ACM Transactions on Graphics 30(5), 109:1–109:12 (2011)

Guided Depth Enhancement via Anisotropic Diffusion*

Junyi Liu and Xiaojin Gong**

Dept. of Information Science and Electronic Engineering
Zhejiang University, Hangzhou, 310027, P.R. China
{shshzaa,gongxj}@zju.edu.cn

Abstract. In this paper, we propose to conduct inpainting and upsampling for defective depth maps when aligned color images are given. These tasks are referred to as the guided depth enhancement problem. We formulate the problem based on the heat diffusion framework. The pixels with known depth values are treated as the heat sources and the depth enhancement is performed via diffusing the depth from these sources to unknown regions. The diffusion conductivity is designed in terms of the guidance color image so that a linear anisotropic diffusion problem is formed. We further cast the steady state problem of this diffusion into the famous random walk model, by which the enhancement is achieved efficiently by solving a sparse linear system. The proposed algorithm is quantitatively evaluated on the Middlebury stereo dataset and is applied to inpaint Kinect data and upsample Lidar's range data. Comparisons to the commonly used bilateral filter and Markov Random Field based methods are also presented, showing that our algorithm is competent.

Keywords: Depth Inpainting, Depth Upsampling, Anisotropic Diffusion.

1 Introduction

Although recent revolutionary developments in the range sensing technologies have enabled rapid progress in many computer vision applications, the data captured by state-of-the-art range sensors are still inherently defective. For instance, Microsoft's Kinect [16], a type of indoor ranging sensors, produces depth maps containing numerous invalid regions due to occlusions and other factors. Velodyne HDL-64E Lidars [25], which are applicable to both indoor and outdoor environments, provide range information at a much lower resolution than images. These defects nowadays lead to a variety of research on depth inpainting and upsampling.

This paper aims to address both inpainting and upsampling problems for depth maps when aligned color images are given. Such tasks are referred to as guided depth enhancement in this work in that the color images are taken as guidance [9]. Previous works on guided depth enhancement mainly employ filtering techniques [7] [26] [12] [10] or rely on Markov Random Fields (MRFs) [18] [28] [31] [6].

In this paper, we propose to conduct the guided depth enhancement based on the anisotropic diffusion framework. The pixels with known depth values are treated as

* This research work was supported in parts by the National Natural Science Foundation of China via grants 61001171, 61071219, 90820306, and Chinese Universities Scientific Fund.

** Corresponding author.

B. Huet et al. (Eds.): PCM 2013, LNCS 8294, pp. 408–417, 2013.

heat sources. The depth diffuses from the sources to unknown regions, guided by the color information provided by the reference image. Although PDE-based anisotropic diffusions are extensively used in image smoothing and inpainting [20] [4] [2] [3], to the best of our knowledge, none of them integrates extra images for guidance.

Moreover, instead of conducting the depth diffusion iteratively, we cast the steady state problem of the guided anisotropic diffusion into the well-known random walk model [8] [30] [11]. This model is prevalent in image segmentation [8] [30] [11], but it is seldom used in image enhancement applications. With this model, the guided depth enhancement problem is able to be addressed by solving a sparse linear system.

2 Related Work

2.1 Guided Depth Upsampling

It aims to generate high-resolution range maps by fusing low-resolution range data with high-resolution visual images. To this end, Markov Random Fields (MRFs) are firstly employed in the work [6]. It infers depth by optimizing an energy function that consists of a term evaluating depth consistency of known measurements and a term regularizing depth smoothness. A couple of MRF based methods are proposed later on, which extend the pioneering work [6] by further taking into account new regularizations, for instance, nonlocal smoothness constraints [28] [18] and temporal coherence [31]. Another group of methods for this problem are based on filtering techniques. They mostly use joint bilateral filters or their variations [1] [29] [7] to integrate color information for depth upsampling.

2.2 Guided Depth Inpainting

This problem is mainly intrigued by the development of Kinect-like sensors. In the last two years, various methods have been proposed to fill-in invalid regions on depth maps with reference to aligned visual images. Some researches extend traditional inpainting algorithms [2] [5] [23] to the guided case. Particularly, new inpainting models [14] [21] and propagation orders [14] [15] are devised in order to incorporate guidance information. These methods are limited to the scenarios where invalid regions are surrounded by valid areas, so that they are not applicable for guided upsampling. In contrast to them, another group of researches adopt filters [19] [10], which work for upsampling as well, for guided depth inpainting.

2.3 Anisotropic Diffusion

Anisotropic diffusion here mainly refers to the heat diffusion framework that is described by partial differential equations (PDEs). The PDE-based anisotropic diffusion is extensively used in image or surface smoothing [20] [4], inpainting [2] [3], segmentation [24] [11], and other image processing applications [27]. The most related works are the smoothing and inpainting techniques. To the best of our knowledge, none of these techniques integrates high-resolution reference images to guide the diffusion process.

In this work, we formulate guided depth enhancement, including both inpainting and upsampling, as a linear anisotropic diffusion problem. Instead of solving the problem iteratively as traditional techniques do, our diffusion is conducted by solving a large sparse linear system that is derived from the steady state.

3 Methodology

3.1 Problem Formulation

Suppose we are given an aligned depth-color image pair. The depth map either contains invalid regions or is of lower resolution than the paired color image. The color image is assumed to be indefective. We denote the depth and the color of a pixel $\mathbf{p} = (x, y)$, respectively, by $D(\mathbf{p})$ and $\mathbf{I}(\mathbf{p})$. $\mathbf{p}$ is in an image domain Ω. The pixels with known depth values are referred to as seeded pixels, which form a set S. If we treat the seeded pixels as the heat sources, then depth enhancement can be conducted via diffusing the heat, i.e. depth, on the image starting from these sources. The diffusion procedure is represented by the following partial differential equation:

$$\begin{aligned} &\frac{\partial D(\mathbf{p};t)}{\partial t} = \mathrm{div}\left(\mathbf{W}(\mathbf{p};t)\nabla D(\mathbf{p};t)\right) \\ &s.t.\ \ D(\mathbf{p};0) = D_0(\mathbf{p}), \end{aligned} \tag{1}$$

where $\mathbf{W}(\mathbf{p};t)$ is the diffusion conductance at location $\mathbf{p}$ at time t and $D_0(\mathbf{p})$ is the depth at the initial state. The conductance is a spatial varying term that depends on the given color guidance image, leading Eq.(1) to be an anisotropic diffusion.

The above PDE is an initial value problem [13]. In the diffusion system represented by this PDE, we are interested in the steady state in which the depth goes stable [8] [30] [11]. Hence, without ambiguity, we drop t in our notation and obtain the following equation:

$$\begin{aligned} &\mathrm{div}\left(\mathbf{W}(\mathbf{p})\nabla D(\mathbf{p})\right) = 0 \\ &s.t.\ \ D(\mathbf{p}) = D_0(\mathbf{p}) \quad \text{for} \quad \mathbf{p} \in S. \end{aligned} \tag{2}$$

The solution of Eq.(2) corresponds to the enhanced depth map that we expect to achieve. When the initial depth map is given, the enhancement quality highly depends on the choice of $\mathbf{W}(\mathbf{p})$. Therefore, we explore the design of the diffusion conductance and present the way of solving the steady state problem in the followings.

3.2 Numerical Scheme

We have formulated the depth enhancement problem as an anisotropic diffusion process. The diffusion conductance is a term depending on the color guidance image, but independent of the depth values. Therefore, our guided depth enhancement is a linear anisotropic diffusion problem [11]. The linearity means that $\mathbf{W}$ is not a function of D or ∇D.

Now, let us consider the steady state of the linear anisotropic diffusion as described in Eq(2). We first construct the discrete depth map as a graph $\mathcal{G} = (\mathcal{V}, \mathcal{E})$, in which $\mathcal{V}$ is a set of vertices, i.e. pixels, and $\mathcal{E}$ contains the edges representing the 4-connected neighborhood. The diffusion conductance between two vertices is devised by Gaussian similarity in color. That is

$$w(\mathbf{p}, \mathbf{q}) = \begin{cases} \exp\left(-\frac{||\mathbf{I}(\mathbf{p})-\mathbf{I}(\mathbf{q})||^2}{2\sigma^2}\right) & \text{if}(\mathbf{p}, \mathbf{q}) \in \mathcal{E} \\ 0 & \text{otherwise} \end{cases} \tag{3}$$

This conductivity implies that the depth diffuses fast when two vertices are similar in color.

Once we have built up the graph, the discrete formulation of the diffusion process therefore reduces to the well-known random walk model [8] [30] [11]

$$D(\mathbf{p}) = \frac{1}{a_{\mathbf{p}}} \sum_{(\mathbf{p},\mathbf{q})\in\mathcal{E}} w(\mathbf{p}, \mathbf{q}) D(\mathbf{q}) \tag{4}$$

$$s.t. \;\; D(\mathbf{p}) = D_0(\mathbf{p}) \quad \text{for} \quad \mathbf{p} \in S,$$

where

$$a_{\mathbf{p}} = \sum_{(\mathbf{p},\mathbf{q})\in\mathcal{E}} w(\mathbf{p}, \mathbf{q}). \tag{5}$$

If we treat the depth map as an one column vector $\mathbf{D}$, Eq(4) turns to be the following equation:

$$(\mathbf{L} - \mathbf{w})\mathbf{D} = 0, \tag{6}$$

where $\mathbf{L}$ is an identity matrix and $\mathbf{w}$ is an $N \times N$ matrix whose entries are the pairwise affinities. N is the total number of the pixels of the entire image. Given the initial depth map D_0, the nontrivial solution of $\mathbf{D}$ can be computed by solving a linear system $\mathbf{AD} = \mathbf{b}$, with

$$\mathbf{A}(p, q) = \begin{cases} \mathbf{L}(p, q) & \text{if } \mathbf{p} \in S \\ (\mathbf{L} - \mathbf{w})(p, q) & \text{otherwise} \end{cases}, \tag{7}$$

and

$$\mathbf{b}(p) = \begin{cases} D_0(\mathbf{p}) & \text{if } \mathbf{p} \in S \\ 0 & \text{otherwise} \end{cases}. \tag{8}$$

Here, p and q correspond to the indices of the pixels $\mathbf{p}$ and $\mathbf{q}$ in the column vector $\mathbf{D}$. Note that $\mathbf{A}$ is a sparse and positive-definite matrix. Thus, the guided depth enhancement is turned to be a sparse linear system which can be solved efficiently.

4 Experiments

In order to validate the proposed method, we have conducted a series of experiments. Quantitative evaluations and comparisons to the bilateral filter and the MRF approach are first made on the Middlebury stereo dataset [22]. Then, we apply the algorithms to an inpainting application for Kinect sensors, as well as an upsampling application for a Velodyne 64E Lidar. Details of the experiments are presented below.

4.1 Experiments on Middlebury Dataset

This experiment aims to quantitatively evaluate the performance of our algorithm. A set of depth-image pairs is taken from the Middlebury stereo dataset. The depth maps in the dataset are served as the ground truth. The maps are downsampled by different factors, $5\times, 10\times, 15\times, 20\times$, respectively, to create the low-resolution depth maps. The color images are taken as the guidance for depth upsampling.

Throughout all the experiments, we set the σ in the diffusion conductivity to be 10. We implement the naive bilateral filter by ourselves which do not sacrifice performance for speed. In the bilateral filter, the neighborhood size is 41×41, $\sigma_d = 2$, and $\sigma_s = 10$. The MRF method in [6] is implemented for comparison as well, in which the weighting coefficient is 1, regularizing the weight between the data and smoothness terms.

The results obtained by each algorithm under each upsampling factor are evaluated in terms of the root mean square error (RMSE) against the ground truth. Table 1 summarizes the results, from which we get several observations. When the upsampling factor is small, the proposed method and the bilateral filter are comparable. When the factor goes up, our method performs best. And in all cases, MRF has the largest errors due to its over-smoothing effect. Fig. 1 also presents the upsampling results under $10\times$ upsampling factor. From the images we can see that the bilateral filter gets sharp boundaries. Both the proposed method and the MRF approach get blurred boundaries. MRF blurs most. Moreover, both the proposed method and the MRF approach perform better on thin-structured regions than the bilateral filter does.

Table 1. Quantitative evaluation results on the Middlebury stereo dataset [22]. The error is measured in RMSE.

Method \ Scenario	BF				MRF				AD			
	5×	10×	15×	20×	5×	10×	15×	20×	5×	10×	15×	20×
Art	**5.501**	**9.007**	11.716	13.917	6.636	10.013	12.037	13.746	5.786	9.152	**11.148**	**12.598**
Books	2.794	4.783	5.857	6.374	2.811	4.377	5.498	6.251	**2.459**	**3.829**	**4.835**	**5.204**
Laundry	**3.167**	5.372	**6.873**	8.967	3.875	6.007	8.244	9.898	3.432	**5.346**	7.338	**8.865**
Moebius	2.427	**3.638**	4.805	6.057	2.912	4.555	5.726	6.621	**2.358**	3.681	**4.756**	**5.486**
Reindeer	**3.667**	**5.077**	**6.330**	8.081	4.437	6.260	7.882	9.371	3.723	5.425	6.732	**7.963**

4.2 Experiments on Kinect Dataset

In this experiment, we apply the algorithms to inpaint depth maps captured by Kinect sensors. A Kinect sensor produces aligned depth and image pairs at a high frame rate.

(a) Color image (b) Ground truth (c) BF (d) MRF (e) AD

Fig. 1. The results of the guided depth upsampling conducted on the Middlebury dataset. (a) are the high-resolution color images. (b) are the ground truth depth maps. (c), (d), and (e) are, respectively, the results obtained by the bilateral filter, the MRF, and the proposed method under 10× upsampling factor.

However, the produced depth maps usually contain numerous invalid regions where range measurements are missing, due to occlusions, reflective or dark surfaces, or the limit of depth of field. Meanwhile, the alignment of the depth map to the displaced color image leads to large invalid regions along the map's boundary.

We present some inpainting results in Fig. 2. The data are either taken by ourselves or chosen from the published Kinect dataset [10]. From the results, we see that the bilateral filter is not able to fill-in all unknown regions due to the fixed window size we set. It means that an adaptive scheme should be designed in order to use it efficiently while applicable to all scenarios. Another observation is quite the same as we see from the previous experiments, the bilateral filter gets sharp edges and the other two get blurred boundaries. The results obtained by the MRF and the proposed method visually are almost the same.

(a) Color image (b) Depth (c) BF (d) MRF (e) AD

Fig. 2. The results of the guided depth inpainting conducted on the Kinect dataset. (a) are color images. (b) are depth maps which contain numerous unknown regions. (c), (d), and (e) are the inpainting results obtained by the bilateral filter, the MRF and the proposed method respectively.

4.3 Experiments on Lidar Data

In the last experiment, we apply the algorithms to upsample range data obtained by a Velodyne 64E Lidar, which is used in conjunction with a camera. The images captured by the camera are taken as the guidance. Different from the upsampling experiments on the Middlebury dataset, the range data we obtain here are of non-uniform resolutions. Fig. 3 presents several examples. From them, we can also observe that the proposed method preserve thin-structures such as the fence best.

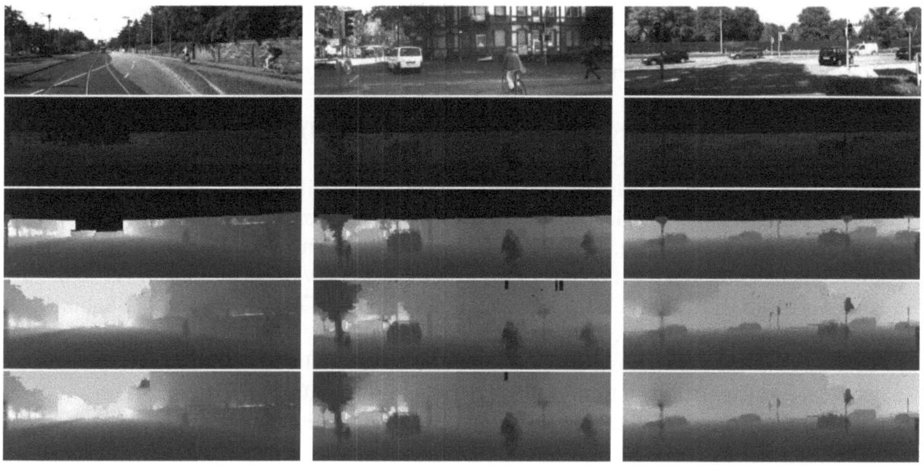

Fig. 3. The results of the guided depth upsampling conducted on the Velodyne 64E Lidar dataset. The first row presents color images. The second row contains the range data registered on the images. The bottom three rows are the upsampling results obtained by the bilateral filter, the MRF and the proposed method respectively.

5 Conclusions

In this paper, we have presented a novel method for guided depth enhancement. The method is based on the heat diffusion framework and addressed efficiently by using the random walk model. Thus, it can be applied for inpainting degraded depth maps produced by Kinect-like sensors, but also for upsampling for range data obtained by Lidars. Both these two applications are prevalent nowadays.

Meanwhile, in order to investigate the essential relations from our algorithm to the bilateral filter and the MRF based methods, we have not considered complicated diffusion conductivities and the variants of the BF and MRFs in this work. A better design of the conductivity will improve the diffusion around boundaries. It will be our future work.

References

1. Barash, D.: Bilateral Filtering and Anisotropic Diffusion: Towards a Unified Viewpoint, HP Laboratories Israel, HPL-2000-18(R.1) (2000)
2. Bertalmio, M., Sapiro, G.: Image Inpainting. SIGGRAPH (2000)
3. Chan, T., Shen, J.: Mathematical Models for Local Non-texture Inpaintings. UCLA CAM TR 00-11 (2001)
4. Clarenz, U., Diewald, U., Rumpf, M.: Anisotropic Geometric Diffusion in Surface Processing. In: IEEE Conference on Visualization (2000)
5. Criminisi, A., Perez, P., Toyama, K.: Region filling and object removal by exemplar-based image inpainting. IEEE Transactions on Image Processing 13(9), 1–13 (2004)
6. Diebel, J., Thrun, S.: An Application of Markov Random Fields to Range Sensing. In: NIPS (2005)
7. Dolson, J., Baek, J., Plagemann, C., Thrun, S.: Upsampling Range Data in Dynamic Environments. In: CVPR (2010)
8. Grady, L.: Random Walks for Image Segmentation. IEEE Transactions on Pattern Analysis and Machine Intelligence 28(11), 1–17 (2006)
9. He, K., Sun, J., Tang, X.: Guided Image Filtering. In: Daniilidis, K., Maragos, P., Paragios, N. (eds.) ECCV 2010, Part I. LNCS, vol. 6311, pp. 1–14. Springer, Heidelberg (2010)
10. Janoch, A., Karayev, S., Jia, Y., Barron, J.T., Fritz, M., Saenko, K., Darrell, T.: A category-level 3-d object dataset: Putting the kinect to work. In: ICCV Workshops (2011)
11. Kim, G., Xing, E.P., Fei-Fei, L., Kanade, T.: Distributed Cosegmentation via Submodular Optimization on Anisotropic Diffusion. In: ICCV (2011)
12. Kim, S.-Y., Cho, J.-H., Koschan, A., Abidi, M.A.: Spatial and temporal enhancement of depth images captured by a time-of-flight depth sensor. In: ICPR (2010)
13. Larsson, S., Thomee, V.: Partial Differential Equations with Numerical Methods. Springer, Heidelberg (2003)
14. Liu, J., Gong, X., Liu, J.: Guided Inpainting and Filtering for Kinect Depth Maps. In: ICPR (2012)
15. Gong, X., Liu, J., Zhou, W., Liu, J.: Guided Depth Enhancement via A Fast Marching Method. Image and Vision Computing 31, 695–703 (2013)
16. Microsoft (2010), `http://www.xbox.com/en-US/kinect/`
17. Paris, S., Durand, F.: A fast approximation of the bilateral filter using a signal processing approach. International Journal of Computer Vision 99(3), 24–52 (2009)
18. Park, J., Kim, H., Tai, Y.-W., Brown, M.S., Kweon, I.: High Quality Depth Map Upsampling for 3DTOF Cameras. In: ICCV (2011)
19. Parker, R.: Kinect depth inpainting and filtering (2011), `http://www.radfordparker.com/papers`
20. Perona, P., Malik, J.: Scale-Space and Edge Detection Using Anisotropic Diffusion. IEEE Transactions on Pattern Analysis and Machine Intelligence 12(7), 629–639 (1990)
21. Qi, F., Han, J., Wang, P., Shi, G., Li, F.: Structure Guided Fusion for Depth Map Inpainting. In: Pattern Recognition Letters (2012) (in press)
22. Scharstein, D., Pal, C.: On Learning conditional random fields for stereo. International Journal of Computer Vision 99(3), 319–337 (2007)
23. Telea, A.: An image inpainting technique based on the fast marching method. Journal of Graphics Tools 9(1), 25–36 (2004)
24. Tschumperle, D., Deriche, R.: Anisotropic Diffusion PDE's for Multi-Channel Image Regularization: Framework and Applications. In: Advances in Imaging and Electron Physics, pp. 1–47 (2007)
25. Velodyne. Velodyne HDL-64E (2012), `http://velodynelidar.com/lidar/`

26. Wasza, J., Bauer, S., Hornegger, J.: Real-time Preprocessing for Dense 3D Range Imaging on the GPU: Defect Interpolation, Bilateral Temporal Averaging and Guided Filtering. In: Workshop in ICCV (2011)
27. Weickert, J.: Anisotropic Diffusion in Image Processing. ECMI Series. Teubner-Verlag (1998)
28. Wong, U., Garney, B., Whittaker, W., Whittaker, R.: Camera and lidar fusion for mapping of actively illuminated subterranean voids. In: Howard, A., Iagnemma, K., Kelly, A. (eds.) Field and Service Robotics. STAR, vol. 62, pp. 421–430. Springer, Heidelberg (2010)
29. Yang, Q., Yang, R., Davis, J., Nister, D.: Spatial-Depth Super Resolution for Range Images. In: CVPR (2007)
30. Zhang, J., Zheng, J., Cai, J.: A Diffusion Approach to Seeded Image Segmentation. In: CVPR (2010)
31. Zhu, J., Wang, L., Gao, J., Yang, R.: Spatial-Temporal Fusion for High Accuracy Depth Maps Using Dynamic MRFs. IEEE Transactions on Pattern Analysis and Machine Intelligence 32(5), 899–909 (2010)

Stereo GrabCut: Interactive and Consistent Object Extraction for Stereo Images

Ran Ju, Xiangyang Xu, Yang Yang, and Gangshan Wu

State Key Laboratory for Novel Software Technology, Nanjing University, China
{juran,xiangyang_xu}@smail.nju.edu.cn, gswu@nju.edu.cn,
charlie.yang.nju@gmail.com

Abstract. This paper presents an interactive object extraction approach for stereo images. The extraction task on stereo images has two significant differences compared to that on monoscopic images. First, the segmentation for both images should be consistent. Second, stereo images have implicit depth information, which supplies an important cue for object extraction. In this paper, we generate consistent segmentation by putting the correspondence relationship in a graph cut framework. Besides, we leverage depth information, which is obtained by stereo matching, to give a pre-estimation of foreground and background. The pre-estimation is then used to generate accurate color models to perform a graph cut based segmentation. To simplify the user interaction, we supply an interface similar to GrabCut, which only needs the user to drag a compact rectangle in most cases. The experiments show our approach works fast and produces more satisfactory results than state-of-the-art.

Keywords: object extraction, graph cut, stereo matching, saliency.

1 Introduction

Object extraction (segmentation) is a classical and widely studied problem which aims at finding an optimal binary segmentation for pixels in an image. The two categories of pixels are labeled as "Foreground (Object)" and "Background" respectively. The technique has a large number of applications in object tracking, instance recognition, image editing and so on. A lot of excellent methods have been proposed to solve the problem. And today as the world is generating more and more stereo images by mobile phones, stereo cameras and so on, it is urgently required to develop object extraction methods for stereo images. However, the traditional methods designed for monoscopic images can not be directly applied to stereo images for two reasons.

First, a stereo image is made up of two slightly different images and the segmentation should be consistent for both images. However, if we use traditional methods for each image separately, the consistency can't be guaranteed due to the difference between the two images. Also the workload will be doubled. We show an example in Fig. 1 (a).

B. Huet et al. (Eds.): PCM 2013, LNCS 8294, pp. 418–429, 2013.

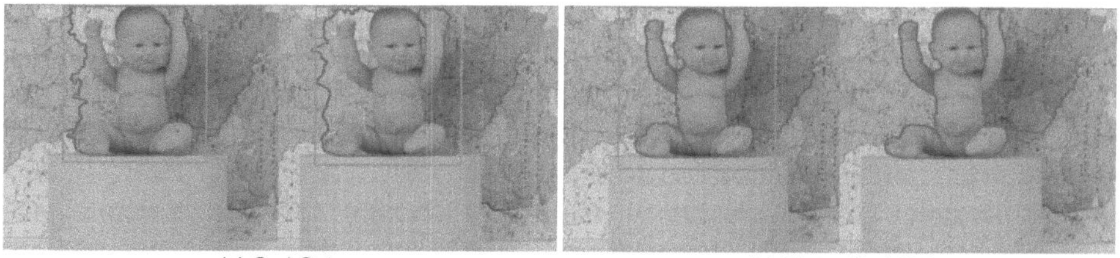

Fig. 1. Differences between object extraction on monoscopic images and stereo images. (a) GrabCut [1] is applied to the two images separately. The consistency can't be guaranteed due to the differences between the images and the user's inputs. (b) Stereo GrabCut is only applied to one image and produces consistent results. The segmentation results are improved owing to the depth guidance.

Second, stereo images have implicit depth information, which can be used as an important cue for segmentation. However, how to utilize the depth information effectively has not yet been well studied before. Additionally, the disparity map may be inaccurate and contain noise, which makes it difficult to analyze.

Our approach—Stereo GrabCut, takes the two points into consideration. For the first one, we perform stereo matching to establish a correspondence relationship between the two images to guarantee consistency. For the second point, we use depth saliency analysis to get a pre-estimation of foreground and background to improve the segmentation results. We show an example in Fig. 1.

To make the interaction easier for users, we design a user interface similar to GrabCut [1]. First, the user drags a compact rectangle around an object then we automatically generate an initial extraction result. Then, if the result needs to be improved, the user can scribble with a foreground and background brush for further editing. Generally our approach generates satisfactory results in the first step. Besides, the operating time is acceptable for interactions. Typically, computing the segmentation for a $320 \times 240 \times 2$ stereo image takes 0.9 second.

2 Previous Work

For monoscopic images there have been a lot of successful interactive methods [2,3,4,1,5]. The graph cut based methods [4,1,5] model the segmentation task as a energy minimization problem. These methods make a best balance between data observation and smoothness. However, as we illustrated before, these methods can't be directly applied to stereo images.

Recently a few works have focused on consistent segmentation for stereo images [6,7,8]. Price et al. [6] proposed a consistent interactive object selection framework based on graph cut [4]. The method enforces consistency using dense stereo correspondence probability distributions. In [7] Tasli et al. used sparse corresponding to enforce consistency. These methods depend on the selection of initial seeds and thus are hard to master for amateur users. Besides,the depth information has not been used.

Leveraging depth information for object extraction has been introduced to some special purpose applications [9,10,11]. These methods focus more on automatic segmentation in particular domains and produce segmentation results for only one view. In contrast, our approach focus more on general purpose segmentation and preserving consistency.

The methods based on saliency analysis [12,13] for object extraction have attracted a large attention because the results are more natural and perceptual acceptable. While previous methods focus on color images, Niu et al. [13] proposed two methods to use stereo information for saliency analysis. One is based on global contrast [12] and the other is based on stereoscopic photography knowledge. However the priors are not usually satisfied which makes it limited in general purpose applications. To overcome the limitations, we utilize the users' inputs to guide depth saliency analysis and use the results to generate color models for graph cuts, which is similar to [14].

3 Approach

We give an overview of our approach in Fig. 2. We assume the input stereo images are rectified so that the corresponding pixels in both images are constrained in the same horizontal line. In step 1, the user drags a rectangle around the object. Then we perform stereo matching to get the disparity map. We compute the saliency map in the user selected area based on the disparity information and then give a pre-estimation of foreground and background. The pre-estimation is used to create the color models. Next we construct the graph model using the stereo image, the disparity map and the color models. At last we optimize the model by a max-flow/min-cut algorithm [15] to get initial segmentation results. If it is needed, in step 2, the user improves the details using a foreground and background brush. The color models, graph model and optimal flow are re-computed. Step 2 can be repeated until the user gets the satisfactory results.

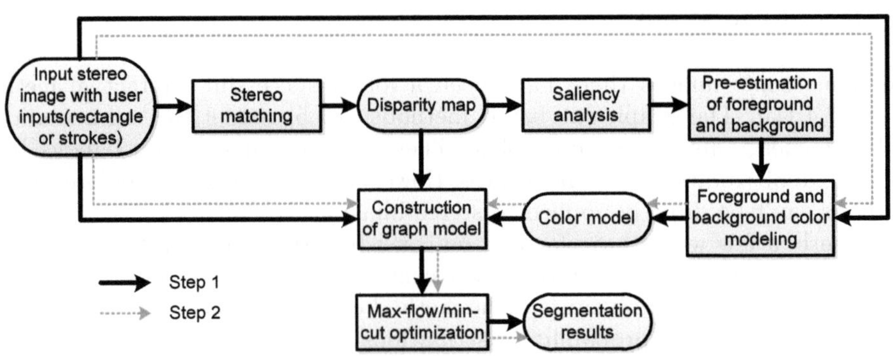

Fig. 2. Overview

3.1 Stereo Matching and Consistent Segmentation

The disparity map plays a key role in consistent segmentation and pre-estimation of foreground and background. However, today there still doesn't exist a perfect solution for stereo matching [16]. To overcome the problem, first we choose a robust and fast stereo matching algorithm called ELAS [17] to get the disparity map. The algorithm has been tested on the KITTI vision benchmark [18] which contains lots of challenging stereo pairs for autonomous driving task and performs excellent. Second, we don't require very accurate disparity maps for the following processes, which will be showed in the following.

Next we add the correspondence constraint to the global energy function:

$$\begin{aligned} E(A) = & \sum_{p \in P_l \cup P_r} R_p(A_p) + \lambda_B \sum_{\{p,q\} \in N_B} B_{\{p,q\}} \mid A_p - A_q \mid \\ & + \lambda_C \sum_{\{p_l,q_r\} \in N_C} C_{\{p_l,q_r\}} \mid A_{p_l} - A_{q_r} \mid \end{aligned} \tag{1}$$

where $E(A)$ is the global energy.

In the right of the formulation, the first two terms R and B are almost the same with [4]. The first term (region term) reflects the penalty for assigning "Object" or "background" to pixel p. This can be interpreted as how the intensity of a pixel p fits into a color model:

$$R_p(A_p) = -\log P(A_p|c_p) \tag{2}$$

where $P(A_p|c_p)$ is the probability of p assigned with A_p given its color c_p. In our approach, we use the Gaussian mixture model (GMM) for color modeling. $p \in P_l \cup P_r$ indicates all pixels in the left and right image. The second term measures the boundary properties between two neighboring pixels p and q. $\{p,q\} \in N_B$ indicates all neighboring pixels in the left and right image. The term indicates when p and q have similar colors they should be assigned with the same label and vice versa. The coefficient $B_{\{p,q\}}$ is defined as $\exp(-\beta \| c_p - c_q \|^2)$ in [4], where β is a constant controls the smoothing strength. The coefficient λ_B stands for a relative importance of boundary term to the other two terms.

The last term in (1) reflects the correspondence property between the two images of a stereo pair. Obviously, corresponding pixels should be assigned with the same label. And thus the correspondence term can be interpreted as a penalty for the mismatch between a pixel p_l in the left image and p_r in the right. In [6] $\{p_l, p_r\} \in N_C$ indicates all possible corresponding pixels. This adds to the graph model $N \cdot D$ links where N is the number of all pixels and D is the maximum disparity. In our approach we reduce the links by only building correspondence edges for consistently matched pixels satisfying $p_l - d_l(p_l) = q_r$ and $q_r + d_r(q_r) = p_l$, where $d_*(p)$ indicates the disparity of p. This improvement reduces the link number to less than N. However, the consistency will not be sacrificed for two reasons. First, the consistently matched pixels plays a role of "seeds" to propagate the segment label from one image to the other with high confidence. Second, small errors in correspondence due to the inaccuracy of the disparity map will be

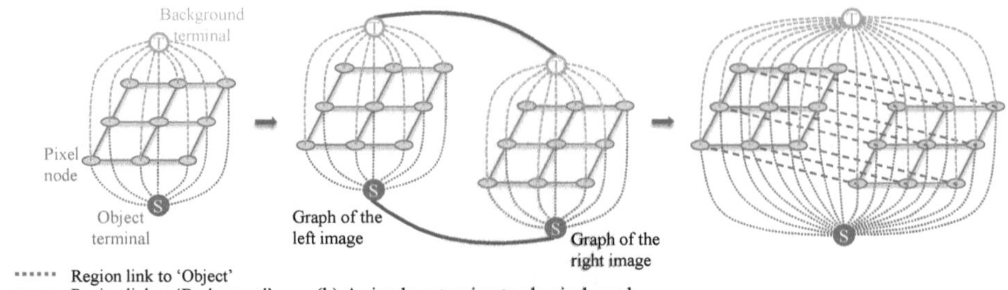

(a) Classical graph cut model

(b) A simple extension to classical graph cut model. The graph is constructed by simply linking the graphs of the left image and the right image at the terminal nodes.

(c) Consistent graph cut model. The graph extends (b) by adding correspondence edges.

Fig. 3. Graph model comparison

recovered by neighboring pixels. The coefficient $C_{\{p_l,q_r\}}$ is defined as the support of p_l corresponding to q_r:

$$C_{\{p_l,q_r\}} = -\log E(p_l, q_r) \tag{3}$$

where $E(p_l, q_r)$ is the matching energy between p_l and q_r generated by ELAS. A lower energy indicates a higher confidence of correspondence. λ_C is a weight factor to balance the relative importance of correspondence term.

Now we construct the graph model of the energy function. The process may be interpreted as a transformation from the classical graph cut model [4] to our consistent graph cut model as shown in Fig. 3. First we construct a classical graph cut model for each image (see Fig. 3 (a)). The vertices are all pixels plus two terminal nodes called "Object terminal" and "Background terminal". Edges between pixel nodes are assigned weights with $B_{p,q}$. Edges linking pixel nodes with object or background terminal nodes are assigned weights with R_p("*Object*") or R_p("*Background*"). Then we link the two graphs of the left and right image by connecting the terminal nodes (see Fig. 3 (b)). At last, we link the corresponding pixels between the two images and assign the linking edges with weights C_{p_l,q_r} (see Fig. 3 (c)). The graph model is optimized by a max-flow algorithm [15], which indicates a minimization to Equation (1). In step 2, the graph model is updated only for the region term and the user marked seed pixels are set as hard constraints [4], e.g. for pixels marked as "foreground", the weights of the edges linking to background terminal are set as 0 and those linking to object terminal are assigned with K ($K \gg B_{\{p,q\}}$).

3.2 Pre-estimation of Foreground and Background

The user input rectangle provides a rough segmentation of foreground and background named S_O and S_B. Next we use saliency analysis to give a finer segmentation because salient regions are more likely to attract the user's interests

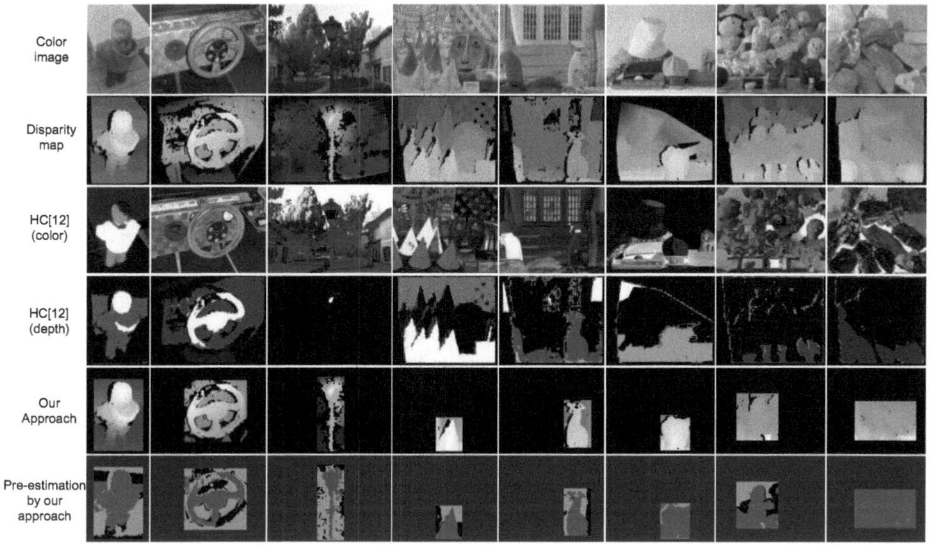

Fig. 4. Depth saliency and pre-estimation of foreground and background. The HC method [12] is applied to color and depth image separately (the 3rd and 4th row). The last row shows the pre-estimation of foreground and background. Red indicates probable "Object". Green indicates probable "Background". Black indicates "Unsure" and won't contribute to color modeling. Blue indicates stable "Background".

perceptually. We only use depth saliency analysis because color is easily confused by textures and thus unsuitable for general purpose segmentation.

We define the saliency value of a pixel as:

$$S(p_i) = \sum_{p_k \in S_B} |d(p_i) - d(p_k)|, \quad p_i \in S_O \tag{4}$$

where $d(p_i)$ indicates the disparity at p_i. The basic assumption is obvious: salient regions are more likely to differ from background. The assumption works well in most cases although it is not always true. Given a counter example, in the case the object is similar to background in depth, the saliency of the object is low everywhere. This degenerates to that using user selected rectangle as object segmentation without saliency analysis, which comes to no harm. Another counter case is that the interesting object is occluded by another object. However, in the following color modeling process the GMM can tolerate some noise so if the occluded region is small we can still produce acceptable results.

Our saliency analysis approach is similar to [12]. The difference is that the saliency value of a pixel is computed as the contrast over all image in their work, but in our approach the contrast is computed over the background, which is more targeted. The computation can be speeded up by a disparity histogram:

$$S_D(d(p_i)) = \sum_{0 < d_k \leq D_{max}} f_{d_k} |d_k - d(p_i)|, \quad p_i \in S_O \tag{5}$$

where f_{d_k} is the number of pixels with disparity d_k. The time complexity can be reduced to $O(N_O)$ where N_O is the number of pixels inside the object rectangle. Note unstable pixels (zero disparity) are not computed. We show a few examples of the saliency map in Fig. 4. It can be seen that depth saliency maps are more robust than color saliency maps. Also our approach is more suitable for depth images owing to the utilization of users' guidance.

Next we use the saliency map to make a pre-estimation of foreground and background. We define the top 50% salient pixels as probable "Object" and bottom 20% as probable "Background". The rest are labeled as "Unsure" and not used in color modeling. The pixels outside the object rectangle are stable "Background". We show a few examples of pre-estimation in the last row of Fig. 4. The estimation is rough due to the inaccuracy of the depth map. Also a false example of our depth saliency assumption is shown in the last column. However, the pre-estimation gives a better training set for color modeling than the original rectangle segmentation and thus leads to a considerable improvement for the segmentation results.

We use GMM (Gaussian mixture model) for color modeling, which is the same as [1]. The parameters are initialized using K-means. Then we use the color model, the original color images and the disparity map to create a consistent graph model and perform segmentation as shown in Fig. 2 and Fig. 3.

4 Experiments and Analysis

Our experiment is made up of three parts. First we focus on consistency and accuracy. Next we demonstrate some user interaction examples. At last we show the running time. We compare our approach with two methods: GrabCut [1] and StereoCut [6], which represents the state-of-the-art method for monoscopic and stereo images respectively.

4.1 Dataset, Input Type and Parameter Settings

We use the open dataset introduced in [6] for evaluation. The dataset consists of 31 stereo images with groundtruth segmentations, 14 of them are introduced in [16] and have groundtruth disparity maps. We extend the dataset to 100 images covering more natural and daily life scenes. The images are downloaded from `http://www.flickr.com` and taken by a Fuji W3 stereo camera.

We define three kinds of input, rectangle, stroke, and rectangle plus stroke. The rectangle and rectangle plus stroke is used for GrabCut and our approach. Stroke is used for StereoCut. Each kind of input is performed both by machine and by user. In machine type, the input rectangle is a minimum window covering the object generated from groundtruth. The stroke is generated by skeletonization of the object and background. The two types are shown in Fig. 5. In user type, the users freely use the tools to finish the segmentation. Throughout our experiment the parameters are set as: $\{\lambda_B, \lambda_C, \beta, K\} = \{39, 50, 0.001, 1000\}$. Our approach is implemented in C++ and tested on a 2.4GHz Intel T8300 CPU with 2GB RAM.

(a) Stereo image (b) Segmentation (c) Rectangle (d) Stroke

Fig. 5. Input types. For both rectangle and stroke input, the white pixels indicate "Object" and black ones indicate "Background".

Table 1. Comparison of average consistency (in percentage)

Approach	Input by machine			Input by user		
	Rect	Stroke	Rect+Stroke	Rect	Stroke	Rect+Stroke
GrabCut	95.93	—	98.33	85.72	—	97.20
StereoCut	—	99.13	—	—	99.12	—
Stereo GrabCut	99.44	—	99.38	99.25	—	99.27

Table 2. Comparison of average accuracy (in percentage)

Approach	Input by machine			Input by user		
	Rect	Stroke	Rect+Stroke	Rect	Stroke	Rect+Stroke
GrabCut	81.52	—	94.85	74.01	—	95.02
StereoCut	—	91.49	—	—	94.44	—
Stereo GrabCut	87.36	—	96.62	85.89	—	98.16

4.2 Consistency and Accuracy Evaluation

Consistency is evaluated as $\frac{|C_l|+|C_r|}{|N_l|+|N_r|}$, where N_* is the set of pixels labeled as "Object" in the left or right image, and C_* indicates the set of pixels consistently labeled in N_*. $| * |$ indicates the number of pixels in the set. The pixels without corresponding pixels are not counted to both C_* and N_*.

We show the consistency evaluation in Table 1. The value is averaged over all images and all users. It can be seen that no matter what input type is, our approach and StereoCut performs much better than GrabCut. Especially in the user input type, as users' input rectangles usually differ between the two images of a pair, the consistency value of GrabCut drops down to only 85.72%. StereoCut can also produce highly consistent results but is much more time-consuming, which will be shown in Sect. 4.4. To demonstrate the consistency more clearer, we give a few examples of segmentation results in Fig. 6. It is very interesting that even small inconsistency in statistics appears very sharp in visual perception e.g. "Bowling" and "Dolls" using GrabCut with rectangle inputs (in the 5th row of Fig. 6). They get consistency value of 91.33% and 93.94% respectively but appear to be obviously inconsistent.

Next we evaluate accuracy using the intersection over union metric [19], which is defined as the ratio of intersection to union between the results and ground

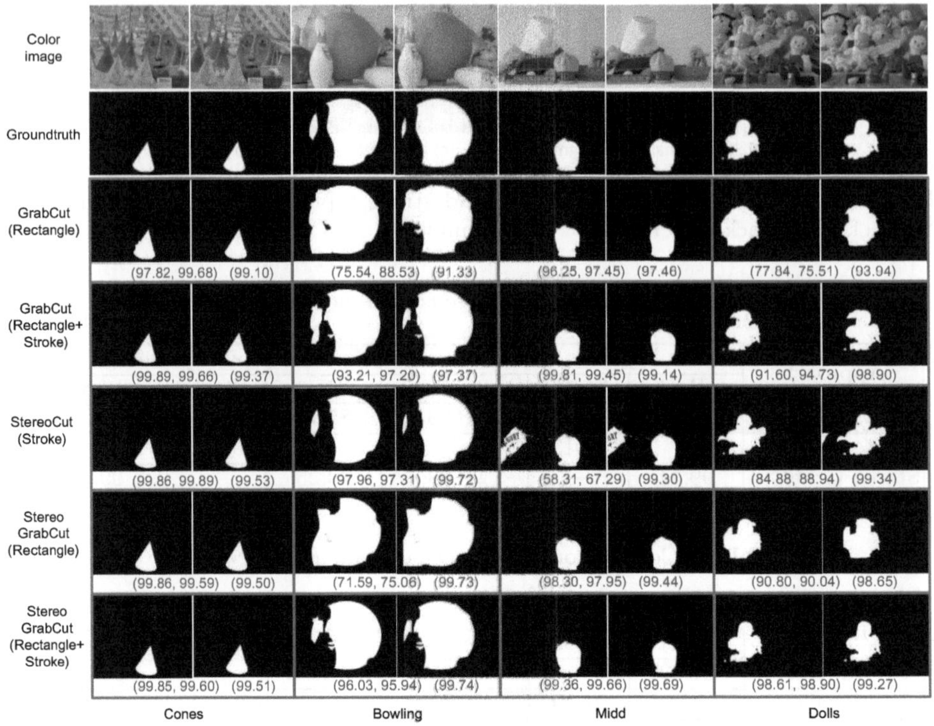

Fig. 6. Examples of GrabCut, StereoCut and our approach. The results are all generated in machine input type. The accuracy and consistency values are also shown below each image. Red numbers within parentheses indicate accuracy values for the left and right segmentation results. Blue numbers indicate consistency values.

truth. As shown in Table 2, in machine input type, when using stroke input all the three methods get high accuracy values. However, GrabCut and our approach both have initial segmentations and thus perform better than StereoCut. In user input type, our approach has made significant improvements to GrabCut. Especially in the first segmentation, our approach makes an improvement of 11.88% to GrabCut because users generally can't draw a minimum rectangle around the object. This is very meaningful for simplifying user interactions because our approach obtains good results at the first segmentation and thus users can focus more on improving small details, which makes our approach perform better than the other two (as shown in the last two columns of Table 2). A few examples are shown in Fig. 6. The stereo images named "Bowling" and "Dolls" are two failure examples of our depth saliency assumption. For "Bowling" the object is occluded by another object and for "Dolls" the object is mingled in background. However for "Dolls" our approach still produces suboptimal results because the GMM can tolerate some noise. For "Bowling" with very little interactions the error can be removed.

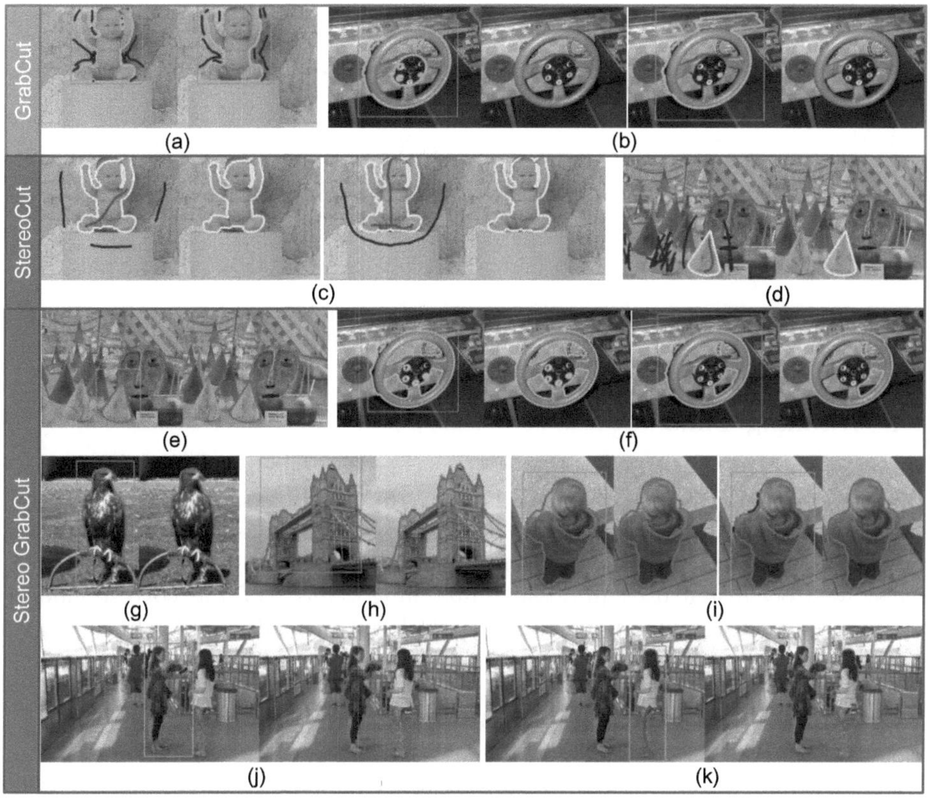

Fig. 7. User interaction examples

4.3 User Interaction

We asked 10 amateur users to participate in our experiments. We taught them how to use the rectangle and stroke tools, and then show the groundtruth object of each image. Then each user finishes the tasks including 100 stereo images freely. Stereo GrabCut gets the best results with minimum time, which takes about 85 minutes totally. The total time increased to 120 minutes for StereoCut because it is hard to master and responses slowly. For GrabCut it increased to 190 minutes because the workload is doubled. We found the users tend to lose patience when repeating the jobs, especially in GrabCut (Fig. 7 (a)). Besides, users like to scribble a large area when the results are unsatisfactory (Fig. 7 (d)). In comparison, our approach doesn't require the users to draw professional seed labels as StereoCut (Fig. 7 (c)). Also our results do not have high dependence on the input rectangles as GrabCut (Fig. 7 (b) (f)). For most images our approach generates satisfactory results in the first step (see Fig. 7 (e) (g) (h)), which makes the users focus more on refining details (Fig. 7 (i)). Our approach can also handle multi-object extraction (Fig. 7 (j) (k)).

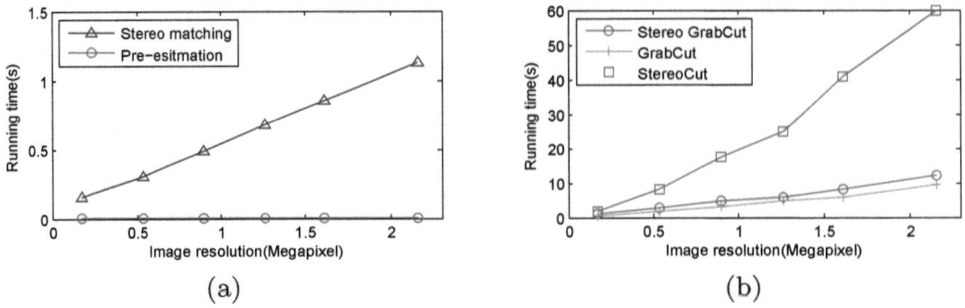

Fig. 8. Running time. (a) Running time of stereo matching and pre-estimation. (b) Running time comparison of GrabCut [1], StereoCut [6] and our approach.

4.4 Running Time

We show the running time in Fig. 8. The time complexity of stereo matching and pre-estimation is almost linear. For a 2.16 Megapixels stereo image, stereo matching takes 1.13s and pre-estimation takes only 0.008s. The total running time of GrabCut, StereCut and our approach is shown in Fig. 8 (b). StereoCut is the slowest because it uses all possible matching pixels (ND) as correspondence terms. The running time of our approach is close to GrabCut but a little slower. This is mainly due to the expense of consistency forcing.

5 Conclusion and Future Work

We presented an interactive and consistent object extraction approach for stereo images based on graph cut, robust stereo matching and depth saliency analysis. Experiments show that our approach makes promising results in acceptable interaction time and outperforms the other two state-of-the-art methods.

In the future we plan to make efforts in the following directions. First, we will investigate more effective object estimation algorithms using depth information. We have shown the power of leveraging depth for object extraction by saliency analysis. We believe progress in depth analysis can effectively improve the object extraction results. Second, we will apply Stereo GrabCut to further applications such as object-based image editing etc. Also, we will extend our stereo image dataset and use Stereo GrabCut to generate groundtruth, which will be surely valuable for object extraction and recognition researches.

Acknowledgments. This work is supported by the Natural Science Foundation of China (No. 61021062), the 863 Program of China (No. 2011AA01A202) and National Special Fund (No 2011ZX05035-004-004HZ).

References

1. Rother, C., Kolmogorov, V., Blake, A.: Grabcut: Interactive foreground extraction using iterated graph cuts. ACM Transactions on Graphics (TOG) 23, 309–314 (2004)

2. Kass, M., Witkin, A., Terzopoulos, D.: Snakes: Active contour models. International Journal of Computer Vision 1(4), 321–331 (1988)
3. Mortensen, E.N., Barrett, W.A.: Intelligent scissors for image composition. In: Proceedings of the 22nd Annual Conference on Computer Graphics and Interactive Techniques, pp. 191–198. ACM (1995)
4. Boykov, Y.Y., Jolly, M.P.: Interactive graph cuts for optimal boundary & region segmentation of objects in nd images. In: Proceedings of the Eighth IEEE International Conference on Computer Vision, ICCV 2001, vol. 1, pp. 105–112. IEEE (2001)
5. Li, Y., Sun, J., Tang, C.K., Shum, H.Y.: Lazy snapping. ACM Transactions on Graphics (ToG) 23(3), 303–308 (2004)
6. Price, B.L., Cohen, S.: Stereocut: Consistent interactive object selection in stereo image pairs. In: 2011 IEEE International Conference on Computer Vision, ICCV, pp. 1148–1155. IEEE (2011)
7. Tasli, H.E., Alatan, A.A.: User assisted stereo image segmentation. In: 3DTV-Conference: The True Vision-Capture, Transmission and Display of 3D Video, 3DTV-CON, pp. 1–4. IEEE (2012)
8. Lo, W.Y., van Baar, J., Knaus, C., Zwicker, M., Gross, M.: Stereoscopic 3d copy & paste. ACM Transactions on Graphics (TOG) 29, 147 (2010)
9. Kolmogorov, V., Criminisi, A., Blake, A., Cross, G., Rother, C.: Bi-layer segmentation of binocular stereo video. In: IEEE Computer Society Conference on Computer Vision and Pattern Recognition, vol. 2, pp. 407–414. IEEE (2005)
10. Harville, M., Gordon, G., Woodfill, J.: Foreground segmentation using adaptive mixture models in color and depth. In: Proceedings of the IEEE Workshop on Detection and Recognition of Events in Video, pp. 3–11. IEEE (2001)
11. Ahn, J.H., Kim, K., Byun, H.: Robust object segmentation using graph cut with object and background seed estimation. In: 18th International Conference on Pattern Recognition, ICPR 2006, vol. 2, pp. 361–364. IEEE (2006)
12. Cheng, M.M., Zhang, G.X., Mitra, N.J., Huang, X., Hu, S.M.: Global contrast based salient region detection. In: 2011 IEEE Conference on Computer Vision and Pattern Recognition, CVPR, pp. 409–416. IEEE (2011)
13. Niu, Y., Geng, Y., Li, X., Liu, F.: Leveraging stereopsis for saliency analysis. In: 2012 IEEE Conference on Computer Vision and Pattern Recognition (CVPR), pp. 454–461. IEEE (2012)
14. Fu, Y., Cheng, J., Li, Z., Lu, H.: Saliency cuts: An automatic approach to object segmentation. In: 19th International Conference on Pattern Recognition, ICPR 2008, pp. 1–4. IEEE (2008)
15. Boykov, Y., Veksler, O., Zabih, R.: Fast approximate energy minimization via graph cuts. IEEE Transactions on Pattern Analysis and Machine Intelligence 23(11), 1222–1239 (2001)
16. Scharstein, D., Szeliski, R.: A taxonomy and evaluation of dense two-frame stereo correspondence algorithms. International Journal of Computer Vision 47(1-3), 7–42 (2002)
17. Geiger, A., Roser, M., Urtasun, R.: Efficient large-scale stereo matching. In: Kimmel, R., Klette, R., Sugimoto, A. (eds.) ACCV 2010, Part I. LNCS, vol. 6492, pp. 25–38. Springer, Heidelberg (2011)
18. Geiger, A., Lenz, P., Urtasun, R.: Are we ready for autonomous driving? the kitti vision benchmark suite. In: Conference on Computer Vision and Pattern Recognition, CVPR (2012)
19. Lee, W., Woo, W., Boyer, E.: Silhouette segmentation in multiple views. IEEE Transactions on Pattern Analysis and Machine Intelligence 33(7), 1429–1441 (2011)

A Novel Hardware-Based UHD Video Up-Scaler Based on Local Structure Estimation

Hang Sun[1], Shengfu Dong[1,2,*], Xiaodong Xie[1], Meng Li[1], Xiaofeng Huang[1], and Wen Gao[1]

[1] National Engineering Laboratory for Video Technology, Peking University, Beijing, China
[2] Peking University Shenzhen Graduate School, Shenzhen, China
sunhang618@126.com, dongsf@pkusz.edu.cn, {xdxie,limeng,xfhuang,wgao}@jdl.ac.cn

Abstract. A novel HW-based video up-scaling method proposed in this paper performs the video image up-scaling. Based on the geometry similarity between the original image and the target image, the proposed algorithm produces new pixels by exploiting the directional weights from the errors estimated by a preset interpolation model in the local area, and using the weights to combine two pixel values predicted by the same model. The algorithm can well preserve the local structure, the edge sharpness in particular, and effectively reduce artifacts such as jaggies and ringings. Based on this method, an efficient HW architecture is developed which is capable of processing at real-time up-scaling to HD or UHD@30fps. Experimental results demonstrate that the proposed method even outperforms some methods using more complex algorithms in both objective and subjective image qualities. In addition, the low complexity of the algorithm simplifies the hardware implementation, which makes real-time enlargement of video image practical.

Keywords: local structure based image interpolation, video scaler, lowpass filter, VLSI design.

1 Introduction

Image up-scaling technique aims to reconstruct high resolution (HR) images from the original low resolution (LR) ones. This is a very prevalent technique in digital image applications, ranging from consumer electronics to professional display devices. Video scaler is a crucial component of these applications for resolution format conversion, for example, images enlarging from QCIF to CIF, or from standard definition (SD) video to high definition (HD) video or more recently to ultra high definition (UHD) video. The more advanced method in preserving the image quality usually increases the computational burden at the same time. But it is agreed in studies that for real applications, the criteria should be put not only on the perceptual quality of images, but also on the computational/implementation cost [1, 2].

* Corresponding author.

B. Huet et al. (Eds.): PCM 2013, LNCS 8294, pp. 430–441, 2013.

Classical linear interpolation methods, such as bilinear [3] or bicubic [4] interpolation, have their benefits on low computational cost and are hence widely used as scaling methods in early years. But these methods face difficulties in preserving the fast-changing geometry structures, such as sharp edges and fine textures. Unfortunately, this problem could not be solved by linear interpolation methods because of the unsuitability of their inherent space invariant model for a variety of natural geometry structures.

Non-linear interpolation method is the primary domain of research in recent years. The most outstanding advantage of non-linear interpolation is its possibility to preserve the sharpness of edges and textures, which is highly sensitive to human visual system. So a wide variety of methods [5-11] have been devoted toward this topic. The methods in [5, 6, 7] quantize the edge orientation into finite models and a VLSI architecture has been implemented in [7]. But the process of identifying edge orientations is quite complicated and may result in image distortion if the identification for orientation goes wrong. In order to adjust to an arbitrarily oriented edge, the methods in [8, 9, 10] apply auto-regression model to the estimation of the local interpolation coefficients. NEDI [8] proposed a method to estimate the covariance matrix of the high resolution image based on the corresponding low resolution one, and NLEDI [9] extends the choice of estimating window to larger areas. However, auto-regression model always introduces huge computation time by solving a set of linear equations: it is cited in [8] itself that the computation of each pixel will require about 1300 multiplications when the local window size is 8, which is obviously unacceptable for hardware design. In [11], two pre-estimated samples are combined by using the statistics obtained by linear minimum mean square-error estimation (LMMSE). However, the estimation is only taken in the pixel interpolation itself that ignores its connection with its surrounding pixels, which is not enough to get a satisfactory result.

In this paper, we propose a new image up-scaling method that exploits the local geometry structure by applying a preset interpolation model to calculate orientation estimation errors of four local neighboring pixels from LR image and interprets the structure into directional weights for HR image pixel interpolation. Due to the simplicity of this method, we propose a possible hardware architecture that accomplishes real-time HD to UHD image enlargement with limited resource consumption.

The rest of the paper is organized as follows. In Section 2, the proposed method is introduced. Section 3 gives the experimental results in objective and subjective quality. Section 4 proposes an architecture for implementation of our algorithm and shows the main resource consumption. The conclusion is provided in Section 5.

2 The Proposed Image Interpolation Method

Without losing generality, we first consider the case of scaling factor equal to 2, i.e. the problem of the LR image of size H W scaling up to the HR image of size 2H 2W. After this, we then consider the more general case where the scaling factor is an arbitrary number.

Let I_h be the HR image that is intended to be interpolated by the LR image, I_l. As in Fig.1, the black dots represent the I_l pixels and the gray and white pixels represent the missing ones in I_h that need to be interpolated. In our method, white pixels are able to be calculated only after all of the gray ones are obtained, hence the whole process is divided into two stages: first, gray pixels are interpolated by black pixels. Second, white pixels are interpolated by black and gray pixels together.

At the first stage, let's focus on P(2x+1, 2y+1) interpolation process as an example.

According to the assumption that HR image and its associated LR image have geometric similarity in the local area, we consider that the HR area centered at P(2x+1, 2y+1) and its four nearest surrounding LR areas centered at P(2x, 2y), P(2x+2, 2y), P(2x, 2y+2), P(2x+2, 2y+2) have the same geometric characteristics. In this paper, we mark the areas with two mutually orthogonal lines illustrated in Fig.1. The solid lines cover an HR local area, and the dotted lines cover one of the surrounding LR areas.

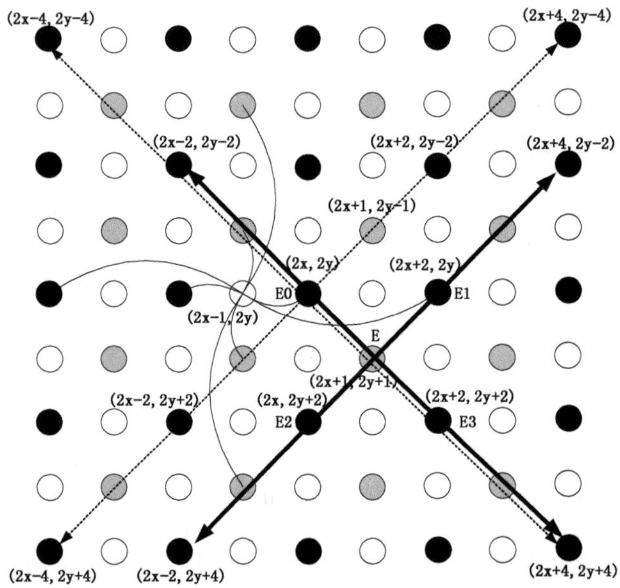

Fig. 1. HR image filled with LR image pixels and the missing pixels. The black dots represent the LR image pixels and the gray and white dots represent the missing HR pixels.

We find that P(2x+1, 2y+1), the center pixel of the HR local area, can be interpolated using the information extracted from the already existed LR pixels - P(2x, 2y), P(2x+2, 2y), P(2x, 2y+2) and P(2x+2, 2y+2). The question emerges naturally now: since the local areas share the same structure, is it possible to exploit the structure characteristics by the existing LR pixels, and to use this local structure to interpolate the missing ones? NEDI [8] proposed a possible method: it gets interpolation coefficients according to a principle of minimum mean square error in the local area, and then uses these coefficients to interpolate the missing samples. The disadvantage of this method is the high complexity of solving a set of linear equations for generating interpolation coefficients. Meanwhile, inspiration for an alternative method comes up in the opposite way: to save the computation time, we can prepare a preset interpolation model (two mutually orthogonal lines in Fig.1) and use the same model to interpolate P(2x+1, 2y+1) and its surrounding pixels- P(2x, 2y), P(2x+2, 2y), P(2x, 2y+2), P(2x+2, 2y+2). Since the surrounding pixels are already existing, prediction errors can be easily calculated between the true pixel values and their predicted ones, then the errors are used for generating a set of weighting factors to derive the more accurate P(2x+1, 2y+1). At the meantime, we introduce an edge Sensitive Factor (SF) to better represent the sharpness of edges and textures, which will be discussed in detail later. The procedure of our algorithm is described as follows:

At first, for missing pixel P(2x+1, 2y+1), we interpolate its two predicted values E_{45} and E_{135} along 45° and 135° respectively by using the eight nearest pixels on the two diagonal lines (4 for each). Then we fuse the two predicted values into the true value of P(2x+1, 2y+1) by a set of weighting factors W_{45} and W_{135}, which remain to be determined.

$$E_{45} = \mathrm{C0} \times \mathrm{P}(2\mathrm{x}-2{,}2\mathrm{y}+4) + \mathrm{C1} \times \mathrm{P}(2\mathrm{x}, 2\mathrm{y}+2) + \mathrm{C2} \times \mathrm{P}(2\mathrm{x}+2{,}2\mathrm{y}) + \mathrm{C3} \times \mathrm{P}(2\mathrm{x}+4{,}2\mathrm{y}-2) \tag{1}$$

$$E_{135} = \mathrm{C0} \times \mathrm{P}(2\mathrm{x}-2{,}2\mathrm{y}-2) + \mathrm{C1} \times \mathrm{P}(2\mathrm{x}, 2\mathrm{y}) + \mathrm{C2} \times \mathrm{P}(2\mathrm{x}+2{,}2\mathrm{y}+2) + \mathrm{C3} \times \mathrm{P}(2\mathrm{x}+4{,}2\mathrm{y}+4) \tag{2}$$

$$\mathrm{P}(2\mathrm{x}+1{,}2\mathrm{y}+1) = W_{45} \times E_{45} + W_{135} \times E_{135} \tag{3}$$

The values C0, C1, C2 and C3 (1, 2) are coefficients derived using a typical low-pass filter model ($-\frac{1}{8}, \frac{5}{8}, \frac{5}{8}, -\frac{1}{8}$ for example). And the second step is to use the same model to interpolate predicted values for surrounding pixels, which are $E0_{45}$, $E0_{135}$ for P(2x, 2y), $E1_{45}$, $E1_{135}$ for P(2x+2, 2y), $E2_{45}$, $E2_{135}$ for P(2x, 2y+2) and $E3_{45}$, $E3_{135}$ for P(2x+2, 2y+2). Now taking the predicted values $E0_{45}$ and $E0_{135}$ as an example:

$$E0_{45} = \mathrm{C0} \times \mathrm{P}(2\mathrm{x}-4{,}2\mathrm{y}+4) + \mathrm{C1} \times \mathrm{P}(2\mathrm{x}-2{,}2\mathrm{y}+2) + \mathrm{C2} \times \mathrm{P}(2\mathrm{x}+2{,}2\mathrm{y}-2) + \mathrm{C3} \times \mathrm{P}(2\mathrm{x}+4{,}2\mathrm{y}-4) \tag{4}$$

$$E0_{135} = \mathrm{C0} \times \mathrm{P}(2\mathrm{x}-4{,}2\mathrm{y}-4) + \mathrm{C1} \times \mathrm{P}(2\mathrm{x}-2{,}2\mathrm{y}-2) + \mathrm{C2} \times \mathrm{P}(2\mathrm{x}+2{,}2\mathrm{y}+2) + \mathrm{C3} \times \mathrm{P}(2\mathrm{x}+4{,}2\mathrm{y}+4) \tag{5}$$

Then, we regard the sum of four prediction errors along a particular direction as the local structure estimation error along this direction:

$$Err_{45} = |\mathrm{P}(2\mathrm{x}, 2\mathrm{y}) - E0_{45}| + |\mathrm{P}(2\mathrm{x}+2{,}2\mathrm{y}) - E1_{45}| + |\mathrm{P}(2\mathrm{x}, 2\mathrm{y}+2) - E2_{45}| + |\mathrm{P}(2\mathrm{x}+2{,}2\mathrm{y}+2) - E3_{45}| \tag{6}$$

$$Err_{135} = |\mathrm{P}(2\mathrm{x}, 2\mathrm{y}) - E0_{135}| + |\mathrm{P}(2\mathrm{x}+2{,}2\mathrm{y}) - E1_{135}| + |\mathrm{P}(2\mathrm{x}, 2\mathrm{y}+2) - E2_{135}| + |\mathrm{P}(2\mathrm{x}+2{,}2\mathrm{y}+2) - E3_{135}| \tag{7}$$

Err_{45} implies the possibility that we get the true value of P(2x, 2y) if we only make interpolation along 45° direction. The greater Err_{45} is, the smaller this possibility will be, and vice-versa.

However, the experimental results (Fig.4 (b)) show that these prediction errors are not sufficient to reflect the sharpness of local structure. So before generating directional weighting factors, we introduce an edge Sensitive Factor (SF) to revise two prediction errors, which will make our algorithm more sensitive to edge structure:

$$ESF_{45} = (Err_{45})^{SF} \tag{8}$$

$$ESF_{135} = (Err_{135})^{SF} \tag{9}$$

Then, the weighting factors for both directions can be calculated as follows:

$$W_{45} = \frac{ESF_{135}}{ESF_{45}+ESF_{135}} \tag{10}$$

$$W_{135} = 1 - W_{45} \tag{11}$$

Finally, we can obtain P(2x+1, 2y+1) by substituting W_{45} and W_{135} to (3).

Once all the gray pixels in the entire image are interpolated in the first stage, they will be treated as available pixels in the second stage. Then the white pixels (for example, P(2x-1, 2y) covered by curved lines in Fig.1) can be obtained in the same fashion after rotating the image plane by π/4 clockwise.

3 Experimental Result

We firstly down-sample HR images to LR images by eliminating even position pixels both in horizontal and vertical directions, then conduct different methods to reconstruct the LR images to their original size, and compare them with their original HR images in term of objective and subjective qualities. Extensive experiments are

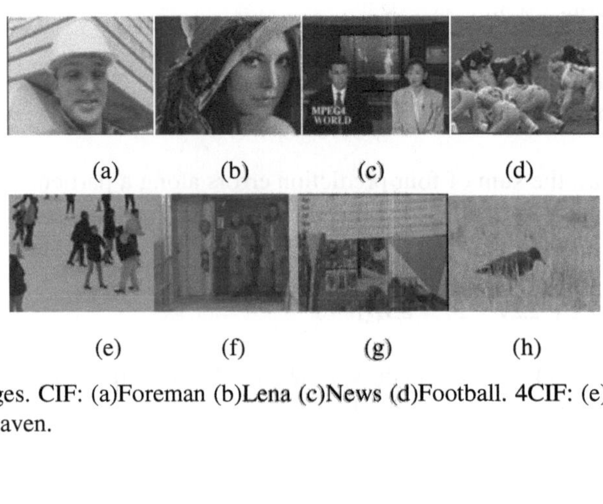

Fig. 2. Test images. CIF: (a)Foreman (b)Lena (c)News (d)Football. 4CIF: (e)Ice (f)Crew. HD: (g)Calendar (h)Raven.

conducted to evaluate our interpolation technique. Without the loss of generality, we choose test images with a variety of structures including parallel lines in Foreman and Crew, large curved edges in Lena and Football. High contrast contents including texts, individuals and big areas of still background are also presented in News and Ice. Calendar and Raven contain fine textures with plenty of details. Moreover, several universal resolutions are taken into consideration, including CIF (352×288), 4CIF (704×576) and HD (1280×720). Fig.2 depicts all these images.

To verify the vital role of Sensitive Factor introduced in Section 2, we choose integers 1~8 as the values of SF and substitute them into (8) and (9) respectively, then we compare the reconstructed images with their original ones and plot PSNR values in Fig.3. The PSNR results prove that the image quality is changing with the value of Sensitive Factor. To be more specific, in the interval between SF=1 and 8, an almost linear increase is detected for SF=1~3 and a decline trend is found for higher SF values from 4 to 8. The changing trend implies that the PSNR curves always have a peak point appearing around values of 3 and 4.

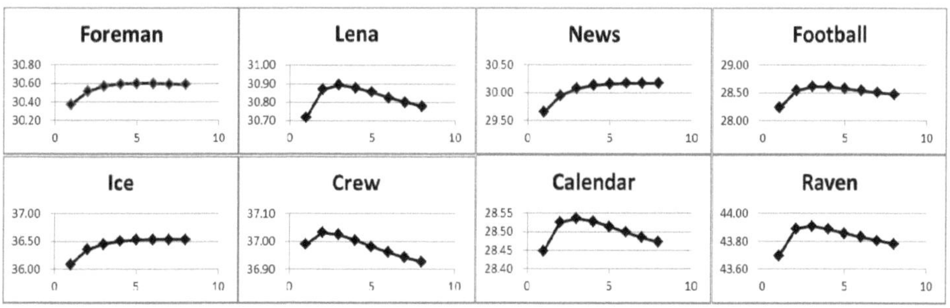

Fig. 3. The PSNR change trend from SF=1 to 8 for test images

The subjective quality measurements also verified this interesting phenomenon, and an example for enlarged Forman is illustrated in Fig.4. Not surprisingly, bicubic method introduces apparent jaggies along the sharp lines (Fig.4 (a)), whereas these artifacts can be amended by our method when SF=1 (b), but not satisfactory enough. Huge quality improvement is observed in Fig.4 (c) for SF=3, because the increase of Sensitive Factor gives more priority to the interpolation process that matches local fine structures. However, with the increase of SF value, the differences between Fig.4 (c) and (d) become almost invisible. In fact, the phenomena mentioned above are also observed in other test images.

We compared our method with other four methods including two classical methods of bilinear and bicubic, and two competitive methods in [8] and [11]. These different methods are applied to the eight test images of Fig.2 and PSNR results are tabulated in Table 1. Although a variety of the structures and resolutions are involved, the proposed method consistently ranks the first in terms of PSNR performance.

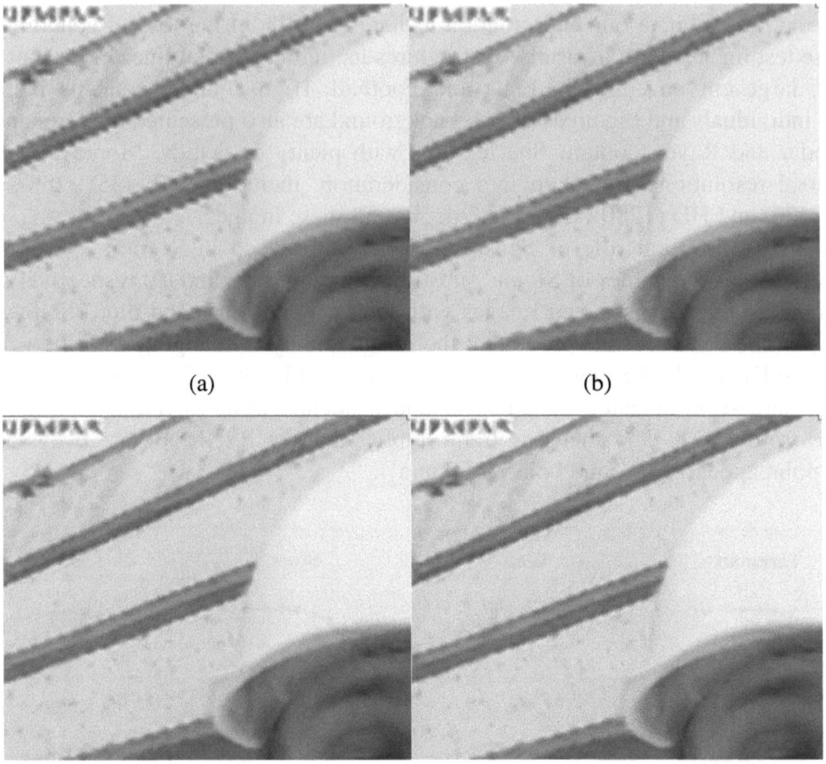

(a) (b) (c) (d)

Fig. 4. Reconstructed images enlarged to compare different SF values. (a)bicubic (b)SF=1 (c)SF=3 (d)SF=8.

Table 1. PSNR results of the reconstructed HR image by different methods

Images	bilinear	bicubic	Method in[8]	Method in [11]	Proposed
Foreman 352x288	29.82	30.01	30.49	30.46	**30.60**
Lena352x288	30.13	30.42	30.53	30.27	**30.89**
News 352x288	28.84	29.38	28.73	29.16	**30.17**
Football 352x288	27.42	28.04	27.38	27.38	**28.61**
Ice 704x576	35.36	35.74	35.92	36.42	**36.54**
Crew704x576	36.31	36.98	35.81	36.56	**37.03**
Calendar 1280x720	28.10	28.25	27.64	28.32	**28.54**
Raven 1280x720	42.52	43.68	42.63	43.03	**43.91**
Average	32.31	32.81	32.39	32.70	**33.29**

The subjective qualities of test images for these methods are demonstrated by News and Football in Fig.5 and Fig.6. The performance of using bilinear interpolation is similar to that of using bicubic interpolation as shown in (b) and (d). Both of the methods tend to blur the details and generate jaggies throughout the whole image, so the visual qualities are inferior to others.

(a) (b)

(c) (d)

(e) (f)

Fig. 5. Comparison of methods on News. (a)original image (b)bilinear (c)method in [8] (d)bicubic (e)method in [11] (f)proposed.

In Fig.5, differences among (c), (e) and (f) are easily discovered by evaluating the text quality of "MPEG4 WORLD" in the bottom left of the images. The text deformation in (c) is probably caused by the incorrect estimation of covariance [9, 10], which may generate speckle artifacts around high contrast curvatures. That is the reason that the PSNR measure for method [8] is the second lowest on average, next

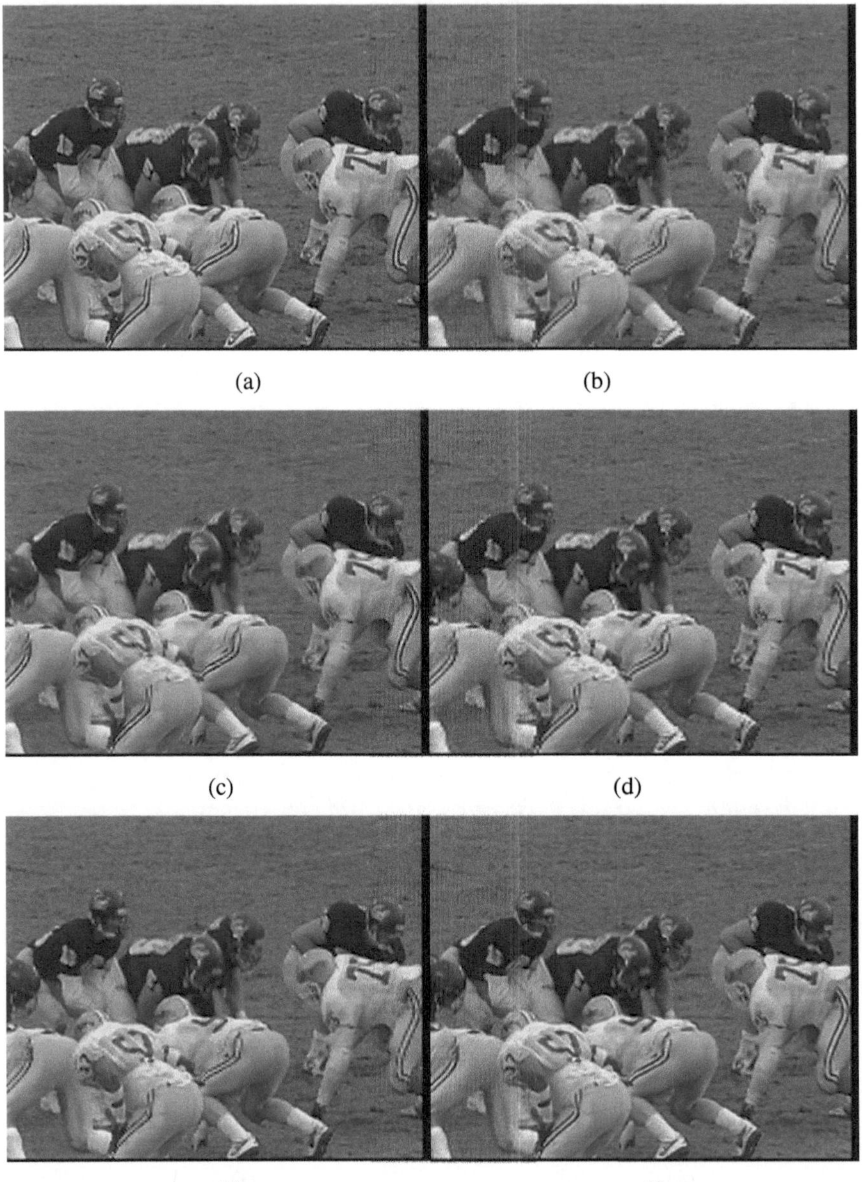

Fig. 6. Comparison of different methods on Football. (a) original image (b) bilinear (c) method in [8] (d) bicubic (e) method in [11] (f) proposed.

only to the bilinear interpolation. Fig.5 (e) presents "ringing" around the texts and the contents are discontinuous. A possible explanation for the undesired artifacts is that the method of [11] fails to catch the changing trend of nearby area structure as a whole since its LMMSE estimation is only effective in limited area. As a result, its PSNR ranks in between bicubic and the method of [8], while the same conclusion is also drawn in [10]. However, our proposed method produces favorable image (f) with less artifacts, the most noticeable places are in texts and female's collar.

The same situation occurs in Fig.6 as well. The reconstruction for parallel stripes on the sport pants is a good case to evaluate different interpolation methods, because the parallel stripes consist of multiple characteristics such as thin lines, step edges and non-directional curves, any of which is a serious challenge for image interpolation. The proposed method succeeds in reconstructing a fine image (f) with continuous lines and smooth edges rather than other images (a~e) consisting of jaggies, ringings and distortion artifacts.

4 Hardware Architecture

Except for the high reconstruction quality achieved by our method, low complexity is also an important factor. Based on our method, a novel hardware architecture is proposed in Fig.7. As mentioned in Section 2, the whole process is divided into two stages: calculating all the gray pixels (Fig.1) in the first stage and completing the others in the second stage. However, unlike the stage separating method in our algorithm, extra memory for frame buffering between the two stages is not necessarily required. The first pipeline stage processes pixels one by one and directly stores them in Parallel Memory Unit in Second-stage; the Second-stage starts its process while all the required pixels arrive.

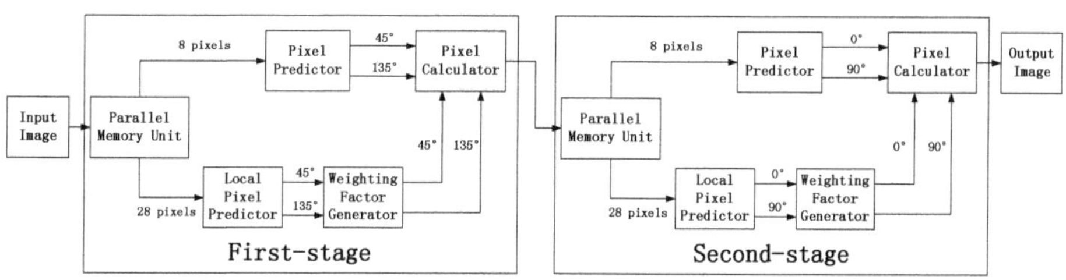

Fig. 7. Block diagram of proposed hardware architecture

The First-stage is composed of five blocks: Parallel Memory Unit, Missing Pixel Predictor, Local Pixel Predictor, Weighting Factor Generator, and Pixel Calculator. According to the data dependency deduced from Fig.1, Parallel Memory Unit composes of a six-line-buffer array for buffering input image pixels line by line. And the pixels are pre-fetched one at a time from each buffer by the Missing Pixel Predictor to warrant no conflict in memory accessing, then the predicted values for both directions are generated and stored in registers temporarily. At the same time,

Local Pixel Predictor generates two groups of predicted values and sends them to the next unit for weighting factor generation. The Pixel Calculator produces final pixel value by the time that both the weights and the predictions for current missing pixel are available. For each new pixel processing, twenty eight original pixels are required in total except the overlapped ones. The Pixel Calculator requires two multipliers for weighted summation operation (3). Weighting Factor Generator (Fig.8) needs four multipliers for SF enhancement (8, 9) when SF=3 (recommended value referring to PSNR change trend) and one divider for the following weight assignment (10). In addition, the consumption of multipliers and divider can be totally avoided by configuring a lookup table directly indexed by prediction errors. As an optimized result, main resources consumed in the First-stage are only two multipliers in Pixel Calculator since the rest modules are only composed of shifters and adders. It is estimated that the proposed method consumes about 1% of hardware resources that would otherwise required in method [8].

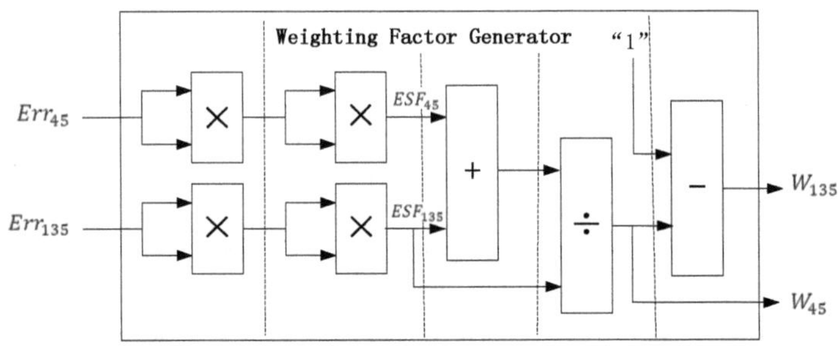

Fig. 8. Block diagram of Weighting Factor Generator

The architecture and operation process of Second-stage are quite similar to that of First-stage, which are only distinguished by the memory accessing pattern if we rotate the interpolation model by $\pi/4$ clockwise (curved lines in fig.1).

Because of the intrinsic pipeline nature of the architecture, it is found that our proposed design can easily support the up-scaling to the UHD@30fps resolution, which would require the operating clock frequency of 297MHz.

5 Conclusion

A HW efficient image up-scaling method is introduced to solve the contradiction between algorithm complexity and visual quality. Specifically, we exploit local structure by a preset interpolation model to eliminate the computation brought by either edge identification or auto-regression. The edge Sensitive Factor performs well in edge regions while suppressing the artifacts and preserving the sharpness. The experimental results have proven an overall superiority through objective and subjective measurements. Finally, the advantage in much reduced HW resource and efficient architecture makes the proposed method appropriate for real-time imaging applications, even for UHD applications.

Acknowledgements. This work is partially supported by grants from the Chinese National Natural Science Foundation under contract No.61171139 and No. 61035001, and National High Technology Research and Development Program of China (863 Program) under contract No.2012AA011703.

References

1. Wünschmann, J., Zanker, S., Günter, C., Rothermel, A.: Reduction of Computational Cost for High Quality Video Scaling. IEEE Trans. on Consumer Electronics 56(4), 2584–2591 (2010)
2. Wang, Z., Zhai, J., Zhou, M.: A Fast Autoregression Based Image Interpolation Method. In: IEEE International Conference on Networking, Sensing and Control, pp. 1400–1404 (2008)
3. Gonzalez, R.C., Woods, R.E.: Digital Image Processing. Publishing House of Electronics Industry, Reading (2010)
4. Hou, H.S., Andrews, H.C.: Cubic splines for image interpolation and digital filtering. IEEE Trans. on Signal Processing 26, 508–517 (1978)
5. Jensen, K., Anastassiou, D.: Subpixel edge localization and the interpolation of still images. IEEE Trans. on Image Processing 4(3), 285–295 (1995)
6. Chen, M., Huang, C., Lee, W.: A fast edge-oriented algorithm for image interpolation. Image and Vision Computing 23, 791–798 (2005)
7. Raghupathy, A., Chandrachoodan, N.: Algorithm and VLSI Architecture for High Performance Adaptive Video Scaling. IEEE Trans. on Multimedia 5(4), 489–502 (2003)
8. Li, X., Orchard, M.T.: New edge-directed interpolation. IEEE Trans. on Image Processing 10(10), 1521–1527 (2001)
9. Zhang, X., Ma, S., Zhang, Y., Zhang, L., Gao, W.: Nonlocal Edge-directed Interpolation. In: Muneesawang, P., Wu, F., Kumazawa, I., Roeksabutr, A., Liao, M., Tang, X. (eds.) PCM 2009. LNCS, vol. 5879, pp. 1197–1207. Springer, Heidelberg (2009)
10. Zhang, X., Wu, X.: Image Interpolation by Adaptive 2-D Autoregressive Modeling and Soft-Decision Estimation. IEEE Trans. on Image Processing 17(6), 887–896 (2008)
11. Zhang, L., Wu, X.: Image interpolation via directional filtering and data fusion. IEEE Trans. on Image Processing 15(8), 2226–2238 (2006)

Binary Adaptive Luminance Mapping for Motion Estimation

Xianghu Ji, Jie Liu, Chuang Zhu, Huizhu Jia, Xiaodong Xie, and Wen Gao

National Engineering Laboratory for Video Technology, Peking University
Beijing 100871, P.R. China
{xhji,liuzimin,czhu,hzjia,xdxie,wgao}@jdl.ac.cn

Abstract. Integer Motion Estimation (IME) for block-based video coding introduces significant challenges in power consumption and silicon area usage with the adoption of more complex coding tools and higher resolution. To conquer these problems, this paper proposes an Binary Adaptive Luminance Mapping (BALM) algorithm by exploiting the local correlation in image and give a Very-large-scale Integration (VLSI) architecture for its implementation. We test the algorithm performance with different Number of Truncated Bits (NTB). And, the experimental results show that our proposed BALM achieves higher rate-distortion (RD) performance compared with previous proposed bit reduction approach and PSNR degradation is relatively small when NTB≤5 using our scheme. And, the NTB4 BALM can achieve 37.3% silicon area saving and power consumption reduction with just PSNR loss of 0.1 dB in our proposed IME architecture.

Keywords: integer motion estimation, binary adaptive luminance mapping, VLSI, video coding.

1 Introduction

The H.264/AVC video coding standard [1][2] can save 25%-45% and 50%-70% of bit rate while maintaining the same video quality compared with MPEG-4 advanced simple profile and MPEG-2, respectively [3][4]. The high coding efficiency comes from the new features of the standard such as variable block sizes (VBS) and multiple reference frames (MRF) [2] in Integer Motion Estimation (IME). However, these new techniques also increase the computation complexity and power consumption dramatically in a video encoder.

Among all the signal processing techniques required for video compression, IME is the most computational intensive and hence consumes the largest amount of hardware resource and power. Therefore, low complexity motion estimation that reduces the computation complexity and power consumption with reasonable rate-distortion (RD) performance are required, especially for Very-large-scale Integration (VLSI)-based implementations.

There are many techniques to reduce the complexity and the power consumption of the IME. Bit reduction is frequently used optimization technique. Bit Truncation (BT)

B. Huet et al. (Eds.): PCM 2013, LNCS 8294, pp. 442–452, 2013.

[5][6][7] that drops some least significant bits of 8-bit pixels in Sum-of-Absolute-Difference (SAD) calculation (used in block matching) is one type of bit reduction method. BT includes both the fixed bit truncation [5] and the adaptive bit truncation [6][7], but in all these methods, the least significant bits are truncated directly for each pixel. The number of bits that can be truncated is very limited and it usually causes higher performance degradation with the increase of Number-of-Truncated-Bits (NTB). Recently some novel bit reduction methods [8]-[11] are proposed for ME hardware implementation. In [8], current and reference frames are filtered with a multi band-pass filter. Then the filtered frame is compared with original frame to get the one bit-depth frame. Finally, low cost one bit matching criterion using Exclusive-OR (EX-OR) operations is adopted in the ME searching. In [9], the multi band-pass filter adopted in [8] is modified to avoid multiplication operations. In [10], a two bit-depth frame representation method is presented to improve the accuracy of motion estimation. However, excess computation is brought by low bit-depth generation method. A two bits ME using constrained one-bit transform is proposed in [11]. It employs simple operations to improve performance of one bit ME. And, it can generally provides better motion estimation performance compared to two bit-depth based ME. In this paper, we will introduce a novel bit reduction method by using locality property of an image which can be easily applied to VLSI-based ME architecture for silicon area reduction and power saving with less performance deterioration.

The remainder of this paper is organized as follows. Section 2 gives an introduction to the proposed Binary Adaptive Luminance Mapping algorithm. In Section 3, Binary Adaptive Luminance Mapping architecture and its application on a typical motion estimation are presented. Performance analysis as well as the comparisons with other works are discussed in Section 4. Finally, some concluding remarks are drawn in Section 5.

2 Adaptive Binary Luminance Mapping Algorithm

In this section, a novel adaptive binary luminance mapping algorithm is presented. By the nature of a typical video sequence, the spatially adjacent luminance data are highly correlated [5]. Generally, luminance pixels in a video are represented in 8~12 bits. The most significant bits are typically important in describing the skeleton of the image while the least significant bits play roles in describing the details. Intuitively, dropping some least significant bits of 8-bit luminance pixels during block matching using SAD calculation may not affect the ME performance. Reducing the bits of operands in SAD calculation can decrease the width of data path logic of a ME search engine which translates into less gate count, smaller routing area and power consumption. Less gate count and smaller routing area can greatly reduce challenges in the chip back-end design.

Bit Truncation (BT) [5][6][7] in which the least significant bits are truncated directly for each luminance pixel in an image is the common adopted algorithm for ME bit reduction. The N-bit BT means luminance dynamic range is reduced to 2^{8-N} levels in which the N least significant bits of pixels are dropped. By using BT, full luminance dynamic range (typically from 0 to 255) is directly mapped to the low dynamic range from 0 to $2^{(8-N)}-1$. The problem of this method is that the number of bits that can be reduced is very limited and it usually causes higher performance degradation when large NTB is used.

Towards conquering the problem, our proposed new algorithm is based on the observation that the luminance values in a certain area (a 8x8 block, for example) have a low relative dynamic range according to the local luminance correlation of an image. It is also noted that the block-based motion estimation uses a block as a basic unit for matching. By taking advantage of these observations, a more precise algorithm can be proposed by using adaptive luminance mapping for each block, as illustrated in Fig.1. This implies that different block may have different dynamic range centered at a different median value and hence the pixels values in a search window (reference picture) have to be adaptively mapped to this range before matching is performed.

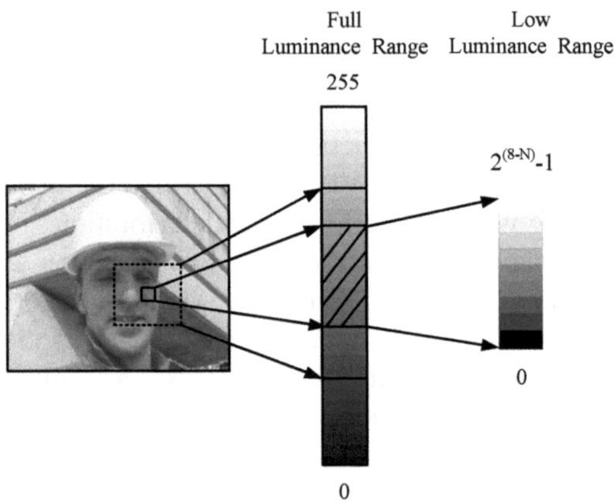

Fig. 1. Motivation of Adaptive Luminance Mapping Methods

The proposed Adaptive Luminance Mapping (ALM) is also illustrated in Fig.2.

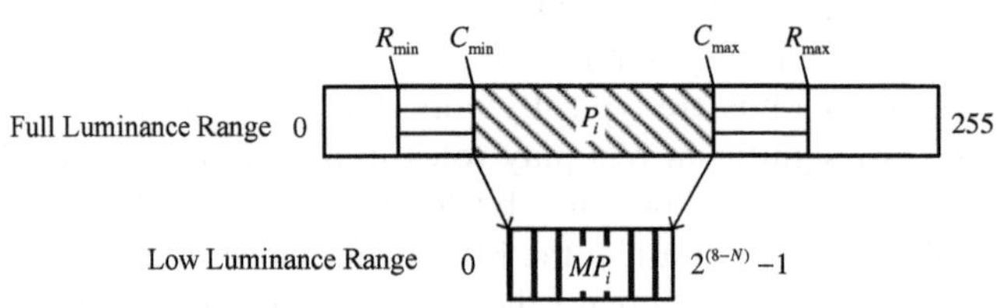

Fig. 2. Adaptive Luminance Mapping

Here C_{max} and C_{min} are respectively the minimum and maximum luminance values in the current block. And R_{max} and R_{min} are the maximum and minimum luminance values in the reference window. P_i is an arbitrary pixel in the current block (or in the reference window). MP_i denotes the luminance value of P_i after applying the proposed mapping algorithm. N is the number of reduced bits. The ALM can be defined by:

$$MP_i = \begin{cases} 0 & P_i < C_{min} \\ \dfrac{P_i - C_{min}}{C_{max} - C_{min} + 1} \times 2^{(8-N)} & C_{min} \le P_i \le C_{max} \\ 2^{(8-N)} - 1 & P_i > C_{max} \end{cases} \quad (1)$$

To simplify the mapping for hardware implementation, some adjustments are employed for the mapping algorithm, as is shown in Fig.3.

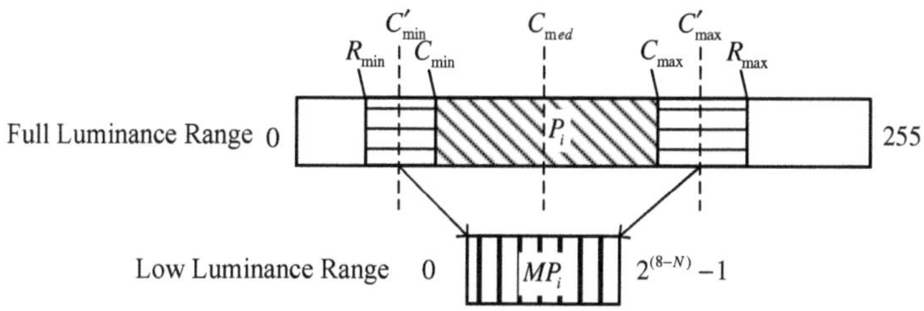

Fig. 3. Binary Adaptive Luminance Mapping

Here, $C_{med} = \frac{C_{min}+C_{max}}{2}$. C_{min} and C_{max} are respectively extended to C'_{min} and C'_{max} ($C'_{min} \le C_{min}$, $C'_{max} \ge C_{max}$) which is also centered at C_{med} to satisfy the following constraint (rounding to the nearest integer bit).

$$\begin{cases} 2^{(M-1)} \le C_{max} - C_{min} + 1 < 2^M \\ C'_{max} - C'_{min} + 1 = 2^M \end{cases} \quad (2)$$

M is the number of bits used to represent the difference between C'_{min} and C'_{max}. We can get two kinds of ALM as shown in Fig.4 (a) and Fig.4 (b) based on the value of M.

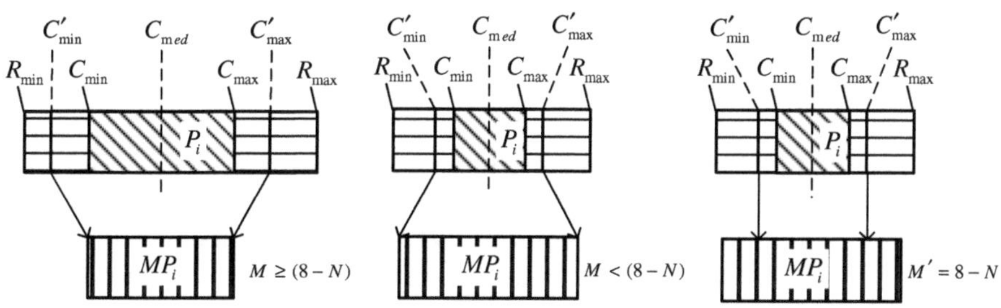

(a) ALM for $M \ge (8-N)$ (b) ALM for $M < (8-N)$ (c) Revised ALM for $M < (8-N)$

Fig. 4. Two kinds of ALM

As P_i is expressed in integer values, it is unnecessary to perform the mapping which will also deteriorate the distortion for pixels in reference window as shown in Fig.4 (b). To minimize the distortion for reference pixels, M is revised to M' by Eq.(3).

$$M' = \text{Max}(M, 8 - N) \tag{3}$$

Fig.4 (c) shows the revised ALM using M' when M is less than $8 - N$.

And, C'_{max} and C'_{min} are updated using the revised M'. Hence, the ALM can be described as follows:

$$MP_i = \begin{cases} 0 & P_i < C'_{min} \\ \left\lfloor \dfrac{P_i - C'_{min}}{C'_{max} - C'_{min} + 1} \times 2^{(8-N)} \right\rfloor & C'_{min} \le P_i \le C'_{max} \\ 2^{(8-N)} - 1 & P_i > C'_{max} \end{cases} \tag{4}$$

So we can get our scheme, Binary Adaptive Luminance Mapping (BALM) using only shift and subtract operation:

$$BMP_i = \begin{cases} 0 & P_i < C'_{min} \\ (P_i - C'_{min}) >> K & C'_{min} \le P_i \le C'_{max} \\ 2^{(8-N)} - 1 & P_i > C'_{max} \end{cases} \tag{5}$$

Here, $K = M' - (8 - N)$ and it is a constant value for each current block. BMP_i is the luminance pixel value after the BALM. The sum of absolute differences (SAD) for the $N \times N$ blocks can be defined by:

$$SAD = \sum_{i=0}^{N} \sum_{j=0}^{N} |CMP_{ij} - RMP_{ij}| \tag{6}$$

Here, CMP_{ij} and RMP_{ij} are BALM pixels in the current and reference blocks respectively. Finally, we can get the Rate-Distortion cost function for the reduced-bit motion estimation using BALM:

$$COST = SAD \ll K + \lambda \times MVD\_COST \tag{7}$$

3 The VLSI Architecture

The proposed BALM VLSI structure includes two types of mapping unit: binary adaptive luminance mapping unit (MU) and simplified binary adaptive luminance mapping unit (SMU). Fig.5 (a) presents the VLSI implementation of the BALM (Eq.(5)). Here, $MP_{max} = 2^{(8-N)} - 1$. Reference pixels in the search window can be mapped by MU. When BALM is performed on the original pixels, the original el P_i always satisfy the constraint of $C'_{min} \le P_i \le C'_{max}$. Hence, the compare operation in Eq.(5) can be simplified for the mapping of original pixels. And, a simplified VLSI architecture (SMU) including only one subtracter and one shift as shown in Fig.5 (b) is proposed for the original pixels.

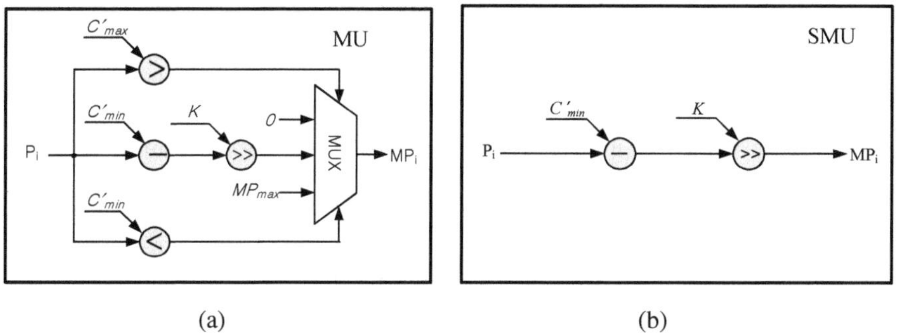

Fig. 5. (a) binary adaptive luminance mapping unit (b) Simplified binary adaptive luminance mapping unit

Fig.6 shows the architecture diagram of a typical ME engine adopting the proposed BALM for silicon area and power reduction. The binary mapping parameters generation module (BALM Gen) generates the C'_{min}, C'_{max}, K and MP_{max} after the fetch operation of current micro-block. Pixels in the current micro-block are mapped by SMU Array and sent to the current buffer at the beginning of search. Reference pixels read from on-chip SRAMs are mapped by MU Array before shifted into Reference Buffer. And, the Rate-Distortion cost is calculated in the Processing Element Array (PEA) for each search.

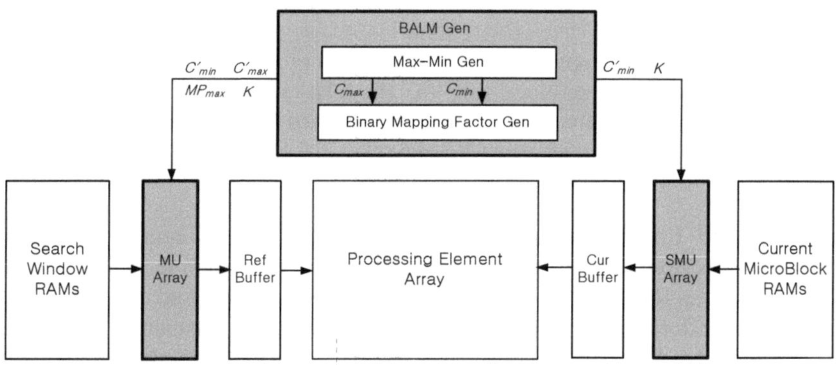

Fig. 6. Typical ME Engine Adopting the Proposed BALM

4 Experimental Results

In this section, we will give out the experimental results of the proposed BLAM and the comparisons with other works. Fig.7 shows the comparison between the ME reference window for BT [5][6][7] and the proposed BALM algorithms. The best matching positions for the current block in the reference window are marked as red block. The BT and BALM error maps contribute to the luminance distortion of the reference window for BT and BALM algorithms, respectively. It can be seen that luminance distortion for the BT is quite irregular and the best matching position suffers from large luminance distortion which will greatly deteriorate the ME searching performance. The BALM behaves a more 'smart' distortion pattern. Reference region in which pixel is similar to the luminance value in the current block will get small distortion. And, large distortion in the BALM usually occurs at the most unlikely matching position.

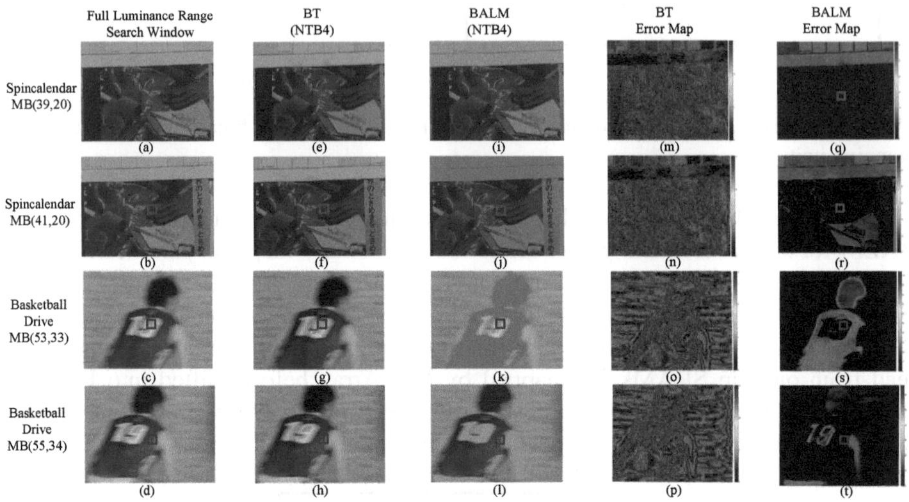

Fig. 7. Reference Window Samples of BT and BALM. (a) - (d) The full luminance range reference window for different micro-block. (e) – (h) The NTB4 reference window using BT algorithm. (i) – (l) The NTB4 reference window using proposed BALM algorithm. (m) – (p) The NTB4 reference window error map using BT algorithm. (q) – (t) The NTB4 reference window error map using BALM algorithm.

Four 720P and 1080P video sequences *City, Cyclists, Kimono and Spincalendar* are used for the RD performance evaluation. Fig.8 shows the simulation results of BT and our proposed BALM on Rate-Distortion for the four video sequences. And different NTBs are chosen for the performance evaluation. And the performance comparison between previous proposed bit reduction MF1BT[9] and C1BT[11] approaches is shown in Fig.9.

Contrary to the BT approach, our proposed BALM achieves higher RD performance. The PSNR degradation is relatively small when NTB ≤ 5 using our scheme. For the 1 bit-depth and 2 bit-depth ($NTB = 7$ and $NTB = 6$) ME, the proposed BALM performs better than the previous MF1BT and C1BT approach. And, The VLSI implementation of BALM is as easy as the implementation of BT approach as result of simple operation. Finally, both BT and BALM are implemented in our proposed IME architecture [12]. We implemented different bit truncation schemes (for comparison purpose) on the proposed IME architecture. The silicon areas of different bit truncation schemes are measured by gate counts. The results are given in TABLE I. The NTB4 BALM can achieve 37.3% silicon area saving which can translate into proportionate reduction in power consumption of IME with just PSNR loss of 0.1 dB. And, it is finally adopted in our proposed IME architecture.

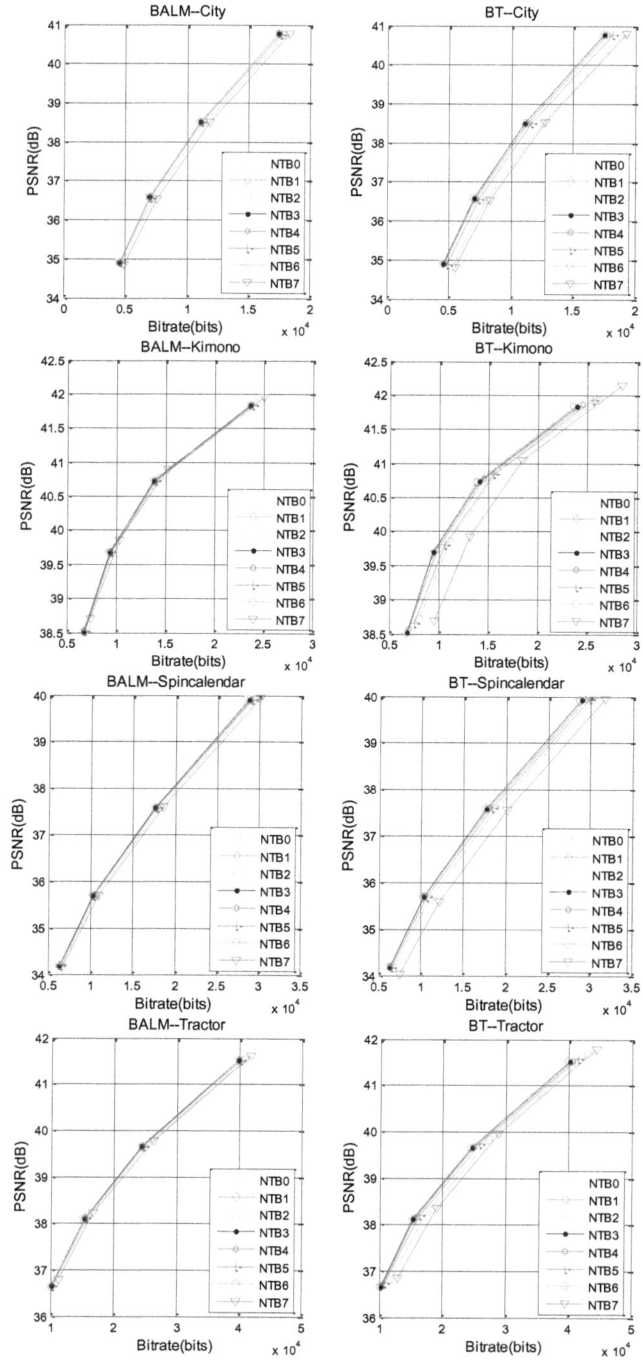

Fig. 8. Simulation Results for BALM and BT, With Hierarchical Search (FME ON) AND Search Range of 128×64 Pixels

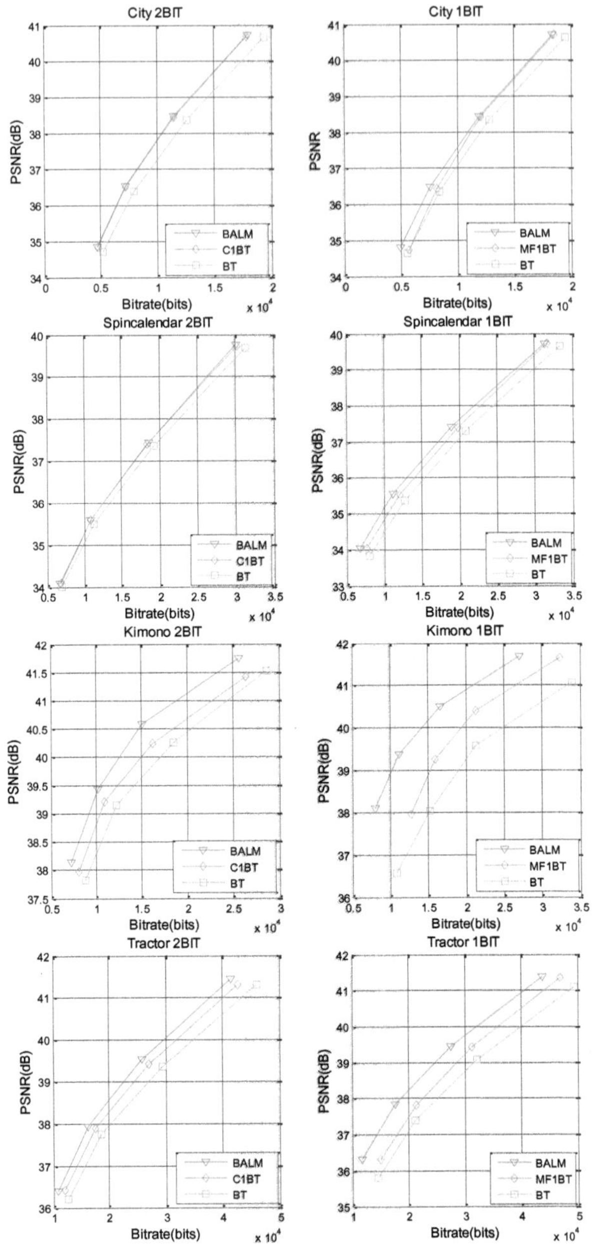

Fig. 9. Simulation Results for BALM, MF1BT and C1BT, With Hierarchical Search (FME OFF) AND Search Range of 128×64 Pixels

Table 1. Hardware ImplementationResults Of Different Bit Reduction Schemes Results

NTB	BALM		BT	
	Gate Count (K)	Gate Count Saving	Gate Count (K)	Gate Count Saving
0	260	0.0%	260	0.0%
1	233	10.2%	227	12.5%
2	216	16.9%	209	19.5%
3	190	26.9%	178	31.7%
4	163	37.3%	141	45.7%
5	139	46.6%	113	56.5%
6	119	54.1%	97	62.6%

5 Conclusion

In this paper, we proposed a Binary Adaptive Luminance Mapping algorithm by using a dynamic mapping for each block which is based on the local pixel correlation of an image and give an architecture for its VLSI implementation. The proposed BALM VLSI architecture can be implemented using only small amount of integer arithmetic subtracters and shifts for silicon area and power reduction. Experimental results show that our proposed BALM achieves higher RD performance compared with previous algorithms and PSNR degradation is relatively small when NTB ≤ 5 using our scheme. And, the NTB4 BALM can achieve 37.3% silicon area saving and reduction in power consumption with just PSNR loss of 0.1 dB in our proposed IME architecture.

References

1. Draft ITU-T Rec. Final Draft Int. Standard of Joint Video Specification, ITU-T Rec. H.264 and ISO/IEC 14496-10 AVC (May 2003)
2. Wiegand, T., Sullivan, G.J., Bjøntegaard, G., Luthra, A.: Overview of the H.264/AVC video coding standard. IEEE Trans. Circuits Syst. Video Technol. 13(7), 560–576 (2003)
3. Iain, E.G., Richardson, H.: 264 and MPEG-4 Video Compression: Video Coding for Next-Generation Multimedia. ch. 7, pp. 225–267. Wiley, New York (2003)
4. Wiegand, T., Schwarz, F.K.H., Joch, A., Sullivan, G.: Rate constrained coder control and comparison of video coding standards. IEEE Trans. Circuits Syst. Video Technol. 13(7), 688–703 (2003)
5. Moshnyaga, V.G.: An MSB truncation scheme for low-power video processors. In: Proc. IEEE Int. Symp. Circuits Syst., vol. 4, pp. 291–294 (1999)
6. He, Z.-L., Tsui, C.-Y., Chan, K.-K., Liou, M.L.: Low-Power VLSI Design for Motion EstimationUsing Adaptive Pixel Truncation. IEEE Trans. Circuits Syst. Video Technol. 10(5), 669–678 (2000)
7. Bahari, A., Arslan, T., Erdogan, A.T.: Low power variable block size motion estimation using pixel truncation. In: Proc. IEEE Int. Symp. Circuits Syst., pp. 3663–3666 (2007)

8. Natarajan, B., Bhaskaran, V., Konstantinides, K.: Low-complexity block-based motion estimation via one-bit transforms. IEEE Trans. Circuits Syst. Video Technol. 7(4), 702–706 (1997)
9. Ertürk, S.: Multiplication-free one-bit transform for low-complexity block-based motion estimation. IEEE Signal Process. Lett. 14(2), 109–112 (2007)
10. Kim, N.J., Ertürk, S., Lee, H.Y.: Two-Bit Transform Based Block Motion Estimation using Second Devivaties. IEEE Trans. Consumer Electron 55(2), 902–910 (2009)
11. Urhan, O., Ertürk, S.: Constrained one-bit transform for low complexity block motion estimation. IEEE Trans. Circuits Syst. Video Technol. 17(4), 478–482 (2007)
12. Yin, H.B., Jia, H.Z., Qi, H.G., Ji, X.H., Xie, X.D., Gao, W.: A Hardware-Efficient Multi-Resolution Block Matching Algorithm and Its VLSI Architecture for High Definition MPEG-Like Video Encoders. IEEE Trans. Circuits Syst. Video Technol. 20(9), 1242–1254 (2010)

Adaptive Video Super-Resolution Based on Superpixel-Guided Auto-Regressive Model

Yanming Zhu[1], Kun Li[1,⋆], Jianmin Jiang[1], and Jingyu Yang[2]

[1] Tianjin Key Laboratory of Cognitive Computing and Application,
School of Computer Science and Technology,
Tianjin University, Tianjin 300072, China
[2] School of Electronic Information Engineering,
Tianjin University, Tianjin 300072, China
lik@tju.edu.cn

Abstract. This paper proposes an adaptive video super-resolution (SR) method based on superpixel-guided auto-regressive (AR) model. The key-frames are automatically selected and super-resolved by a sparse regression method. The non-key-frames are super-resolved by simultaneously exploiting the spatiotemporal correlations: the temporal correlation is exploited by a GPU-based optical flow method while the spatial correlation is modelled by a superpixel-guided AR model. Experimental results show that the proposed method outperforms the existing benchmark in terms of both subjective visual quality and objective peak-to-peak signal-to-noise ratio (PSNR). At the same time, the running time of the proposed method is the shortest in comparison with the state-of-the-art methods, which makes the proposed method suitable for practical applications.

1 Introduction

Video super-resolution (SR) is a classic and hot issue in computer vision and video processing in the last few decades [6, 13, 14, 21], which can be classified into three categories: 1) using the single image SR approaches to super-resolve each video frame independently; 2) using motion compensation techniques and information extracted from multiple views of the same object along frames to construct the high-resolution (HR) frames (usually known as reconstruction-based method) [3,19,22,23]; and 3) using codebooks derived from key-frames in mixed-resolution-video to generate the HR frames (usually known as learning-based method or example-based method) [5,11,14,21].

The methods of super-resolving each frame independently can use the newest and the most effective single image SR approaches to get high quality results. But these methods do not take advantage of the great similarity and interrelationships between adjacent frames. In addition, they usually need to take a very long time when the video contains a large number of frames. Therefore, the reconstruction-based video SR methods are proposed to overcome

⋆ Corresponding author.

B. Huet et al. (Eds.): PCM 2013, LNCS 8294, pp. 453–464, 2013.

this problem in the last few decades. Among the reconstruction-based methods, most are formulating the process of frame information with a generic smoothness prior, and finding the best solution by using maximum a posteriori [2,3,20], projections onto convex sets [24], iterative back-projection [15,25], or a combination of these methods [22]. In these methods, registration is a very important step to the success of HR frames reconstruction. As general registration methods are using motion models to represent multiple object motion, accurate motion estimation between adjacent frames should be performed for successful registration. However, this requires considerable computational complexity. Moreover, existing registration methods become invalid when the motion between two frames are complex. Therefore, learning-based or example-based methods are more widely and successfully applied to video SR in recent years [6,7,16,18]. Latest works [14,21] are two masterpieces in this area. In [21], Song et al. propose a video SR algorithm using bi-directional overlapped block motion compensation and on-the-fly dictionary training, which achieves good results. However, the essence of this method is block matching, which may incur blocking artifacts and information missing. In [14], Huang et al. propose a method using codebooks derived from key-frames, and get better results. It trains codebooks which contain large number of HR and LR patch pairs, and recovers the HR non-key-frame by retrieving the best matching HR patch corresponding to the LR patch. However, this method is computational complex because of the huge dictionary training and the best match retrieving. Furthermore, the results of learning-based or example-based method largely depend on the similarity between the input frame and the samples in the dictionary, new noises may be introduced from the training set.

From the above analysis, the performance of video SR can be improved: 1) by fully exploiting the temporal correlation between frames and 2) by accurately modeling the spatial correlation within video frames. In this paper, we propose a new video SR method addressing the two aspects for better SR performance. In the proposed method, key-frames are upsampled by an image SR method based sparse regression. Then, the resolution of non-key-frames are enhanced by simultaneously exploiting the spatiotemporal correlations: the temporal correlation is exploited by a GPU-based optical flow method while the spatial correlation is modelled by a superpixel-guided AR model. Experimental results prove that our method obtains both better visual quality and higher PSNR than the existing methods. At the same time, the running time of our method is the shortest among the state-of-the-art methods.

2 The Proposed Method

In this section, we present the details of our method as depicted in Fig. 1. To avoid error propagation, the video SR process is refreshed at some time instants (the corresponding frames are named as key-frames). For an input LR video, the key-frames (frames with the red border in the figure) are first automatically selected and super-resolved with SRNIP (Section 2.1). Then, the

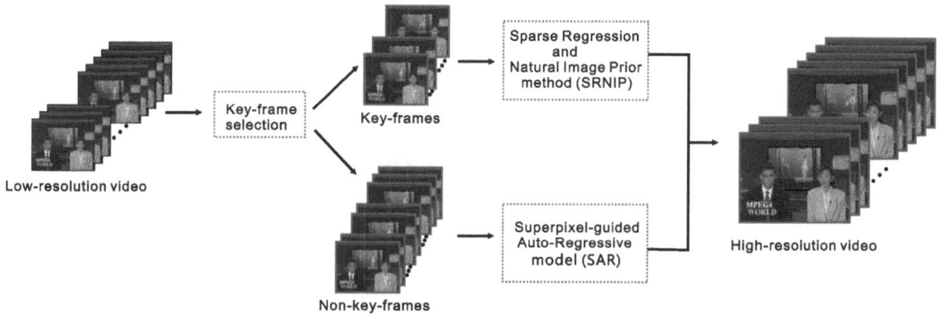

Fig. 1. The framework of the proposed method

non-key-frames (frames with the blue border in the figure) are super-resolved based on superpixel-guided AR model (SAR) with respect to the previous nearest key-frame (Section 2.2).

2.1 The Key-Frame Selection and Super-Resolution

The key-frames are automatically selected according to the motion error with respect to the previous nearest key-frame. For a LR video, we set the first frame as the initial key-frame, and super-resolve it using a method based on sparse regression and natural image prior (SRNIP) [17]. Then, the motion error is calculated for each LR frame. If the motion error is larger than a threshold T, this frame is selected as a new key-frame and super-resolved by the SPNIP method [17]. Otherwise, it will be treated as a non-key frame and super-resolved by the SAR method (Section 2.2).

To calculate the motion error, we first up-scale the current frame by bicubic method [12] and subtract this frame from the previous nearest key-frame. Then we get a difference matrix which represents the difference between the two frames. The average of the non-zero entries in the difference matrix is calculated as the motion error, which is formulated as:

$$e = \frac{f_{NZ}\left(P(y) - P(y_{pre})\right)}{k}, \tag{1}$$

where y and y_{pre} are the current frame and the previous nearest key-frame, respectively. $P(\cdot)$ is an operator that converts the RGB image into gray image. f_{NZ} is an function which selects non-zero entries from matrix, and k is the number of the non-zero entries in the difference matrix.

We use an iterative method to select the threshold T automatically. First, we define two constants c_1 and c_2, which represent the lower and upper bounds of the threshold T, respectively. The proportion of non-key-frames in the video sequence is represented by constant P_0. The mean value $\bar{e}$ of motion errors between each adjacent frame can be calculated by

$$\bar{e} = (\sum_{i=1}^{N-1} e_{i,i+1})/(N-1), \tag{2}$$

where $e_{i,i+1}$ is the motion error between frame i and frame $i+1$, and N is the number of frames in the video. The iterative process is run in the following manner. If the mean error $\bar{e} < c_1$, we set $T = \bar{e}$ as an initial value of the threshold T and execute the key-frames selection process based on this threshold. Then, we examine the number n of key-frames and calculate the proportion p of the non-key-frames by $p = (N-n)/N$. If p is smaller (or larger) than the initial setting value P_0, we set the threshold $T = T + (or-)0.1$, and repeat the above process until $p \geq (or \leq) P_0$. Finally, we get the ultimate threshold T subject to $p \approx P_0$ by this iterative process. Otherwise, if $c_1 \leq \bar{e} \leq c_2$, we set the ultimate threshold $T = \bar{e}$. The initial definition of the parameter c_1 uses a statistical approach. We set c_1 to be 3.0 based on large amounts of experimental data analysis and statistics. Parameter c_2 is decided by the maximum value of all the motion errors between adjacent frames. The definition of parameter P_0 is dependent on the request of SR quality and time cost. Obviously, when P_0 becomes smaller, the video SR quality becomes higher and the running time becomes longer.

The SR of the selected key-frame in essence is a single image SR. Many approaches have been developed to generate high quality scaled images in recent years. The image SR that is applied to the video SR, need have two main advantages: 1) low computation complexity and short running time; 2) easy to accomplish and can generate high quality result. Therefore, in this paper, we adopt the method of Kim et al. [17] to super-resolve the key-frames. The reasons are that: 1) it uses sparse regression to reduce the computation complexity which satisfies the time requirement of the video SR; 2) it uses a natural image prior model to deal with the discontinuity of images which satisfies the quality requirement of video SR.

2.2 The Non-Key-Frame Super-Resolution

The framework of the non-key-frame SR is described in Fig. 2. For the input LR non-key-frame, we first initialize it into the desired scale using bi-cubic method [12]. Then, superpixel and geometric flow information is calculated to generate the AR predictor. At the same time, the previous nearest HR key-frame is warped to the current non-key-frame, and corrected based on the upscaled non-key-frame to construct the data term. Finally, the SAR model is used to generate the HR non-key-frame.

Key-Frame Warping. As the displacement of the adjacent frames is not large, we use a GPU-based optical flow method [8] to warp the previous nearest HR key-frame to the current non-key-frame. Denote $O_{i,j}$ is the optical flow at point

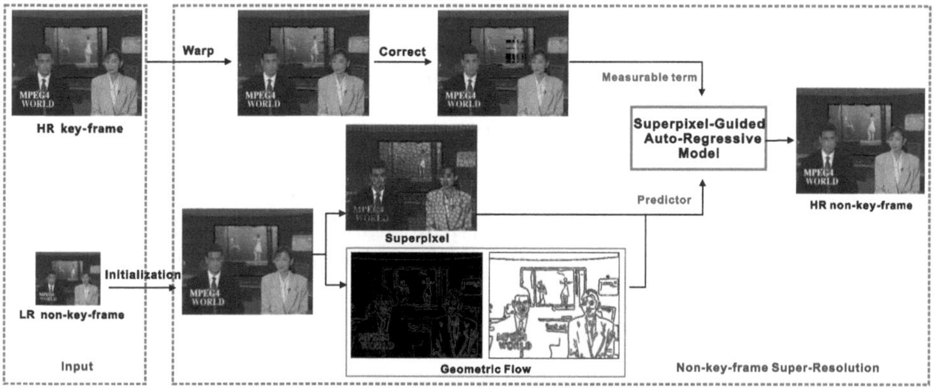

Fig. 2. The framework of the non-key-frame SR

(i, j). Warping the previous nearest HR key-frame with the optical flow O to current non-key-frame is formulated as:

$$F^{\dagger}{}_{k,l} = K_{i,j}, \quad (i \in (1, m), j \in (1, n)), \tag{3}$$

with

$$\begin{aligned} k &= i + O^1_{i,j} \\ l &= j + O^2_{i,j}, \end{aligned} \tag{4}$$

where K is the previous nearest HR key-frame, m and n are the height and width of K, $F^{\dagger}$ is the warped version of the current non-key-frame, and O^1, O^2 are the first and second dimension of the optical flow O, respectively. We ignore the pixel values of $F^{\dagger}$, when its subscripts k and l are beyond the image boundary.

Although the adopted GPU-based optical flow can warp the key-frame to non-key-frame accurately, possible deviation may exist in the warped non-key-frame compared to the original non-key-frame I. Therefore, we correct the warped non-key-frame by:

$$F^0_{(p,q)} = \begin{cases} 0 & R(p,q) > t \\ F^{\dagger}_{(p,q)} & R(p,q) \leq t \end{cases}, \tag{5}$$

with

$$R = \sum_{u \neq v} |(F^{\dagger} - I)^u - (F^{\dagger} - I)^v|, \tag{6}$$

where F^0 is the corrected version of the current non-key-frame, p and q are the height and width of frame $F^{\dagger}$, u and v are the indexs of color channel $\mathcal{C}$, and t is a threshold used to prevent large deviation.

Non-Key-Frame Super-Resolution Based on AR Model. Denote by Ω the set of pixels with corrected non-key-frame values. Given the corrected non-key-frame F^0, we proposed the following non-key-frame SR model based on AR:

$$\min_F E_{data}(F, F^0) + \lambda E_{AR}(F), \tag{7}$$

where $E_{data}(F, F^0)$ is the data term to make the super-resolved non-key-frame F consistent with the corrected non-key-frame, and $E_{AR}(F)$ is the AR term to impose AR model on the super-resolved non-key-frame. The data term and the AR term are weighted by λ.

The data term is expressed as

$$E_{data}(F, F^0) \triangleq \sum\nolimits_{x\in\Omega} \left(F_x - F_x^0\right)^2, \tag{8}$$

and an AR model is incorporated into the non-key-frame SR as the AR term

$$E_{AR}(F) \triangleq \sum\nolimits_{x} \left(F_x - \sum\nolimits_{y\in\mathcal{N}_x} c_{x,y} F_y\right)^2, \tag{9}$$

where the AR coefficient $c_{x,y}$ is defined according to the upscaled non-key-frame.

The AR model is powerful in describing non-key-frame only when the AR coefficients are properly designed. However, an accurate AR model is difficult to infer from only the corrected non-key-frame F^0. Noting that the original LR non-key-frame has strong structural information, the information loss in the corrected frame can be complemented by the accompanied structural information of original non-key-frame. To achieve high quality super-resolved non-key-frame, we design pixel-wise adaptive AR predictors in the following section using both corrected non-key-frame and the original non-key-frame.

Superpixel-Guided AR Model. An AR predictor $c_{x,y}, y \in \mathcal{N}_x$ is constructed for each pixel x by considering both superpixel and geometric flow information of the original non-key-frame. Represent the non-key-frame with $I = \{I^i, i \in \mathcal{S}\}$, where I^i is the intensity of the color channel with index i and $\mathcal{S}$ is a certain color space. In our method, we use the *YCbCr* and *RGB* color space, i.e., $\mathcal{S} = \{Y, Cb, Cr\}$ and $\mathcal{S}' = \{R, G, B\}$. The AR coefficient $c_{x,y}$ consists of two terms:

$$c_{x,y} = \frac{1}{N_x} c_{x,y}^B c_{x,y}^S, \tag{10}$$

where N_x is the normalization factor, $c_{x,y}^B$ and $c_{x,y}^S$ are the bilateral term and superpixel term, respectively.

The bilateral term $c_{x,y}^B$ is defined on the upscaled non-key-frame $\hat{I}$ by a range filter:

$$c_{x,y}^B = \exp\left(-\frac{\|x-y\|_2^2}{2\sigma_1^2}\right) \exp\left(-\frac{\sum_{i\in\mathcal{S}'} \left(\hat{I}_x^i - \hat{I}_y^i\right)^2}{2\times 3\times \sigma_2^2}\right) \times C^Y, \tag{11}$$

where σ_1 and σ_2 are parameters of the bilateral filter to adjust the importance of the spatial distance and intensity. As the human eye is more sensitive to the luminance domain, in our method, we add filter C^Y on the luminance Y in the color spaces $\{Y, Cb, Cr\}$ defined as:

$$C_{x,y}^{Y} = \exp\left(-\frac{(\hat{I}_x^Y - \hat{I}_y^Y)^2}{2\sigma_3^2}\right), \tag{12}$$

where σ_3 is parameter to adjust the importance of the luminance.

The superpixel term $c_{x,y}^S$ is designed to take benefit of the prevention for the AR model from predicting across non-key-frame discontinuities. Based on this purpose, we propose the following super-pixel term:

$$c_{x,y}^{S} = \begin{cases} 1 & SP_x = SP_y \\ 0.3 & SP_x \neq SP_y \end{cases}, \tag{13}$$

where SP_x and SP_y are the superpixel value at point x and y ,respectively. In our method, the superpixel is calculated by SLIC method [1].

The proposed SR model is equivalent to solve a linear system, say $Cf = f_0$, where f is the vector form of the non-key-frame to be super-resolved, f_0 is the vector form of the corrected non-key-frame, and the AR predictor for each pixel is associated with each row of C.

3 Experimental Results

In this section, we evaluate the performance of the proposed video SR method with four publicly available YUV video sequences: *Container*, *News*, *Mobcal* and *Shields*. In order to compare with other methods, the same as method [21], we also use the decompressed HR key-frames to super-resolve the LR non-key-frames. In the following experiments, we first down-sample the original HR non-key-frame by the factor of 1/2 and then super-resolve the LR version using the proposed method. The parameters are set as: $\lambda = 0.02$, $t = 80$, $\sigma_1 = 4$, $\sigma_2 = 14$, $\sigma_3 = 4$. We investigate the influence of the parameters on the SR performance, and find that the performance is stable when the parameters are within certain ranges: $\lambda \in [0.01, 0.20]$, $\sigma_1 \in [3.0, 7.0]$, $\sigma_2 \in [3.8, 19.0]$, $\sigma_3 \in [3.0, 10.0]$. The neighborhood $\mathcal{N}_x$ is chosen as a 11×11 patch with x as its center. All the experiments are run on a laptop with Intel Core i3 CPU at 2.5 GHz and 2.0 GB RAM using MATLAB(R2011a).

We compare the performance of the proposed method with six existing methods: the bi-cubic method [12], and the methods in Ref. [4,9,10,14,21]. For the method in Ref. [21], we compare with its LSR version, MSR version and Hybrid SR version, respectively. Table 1 shows the PSNR improvement of the 15th frame from the bi-cubic interpolation, where the values in red are the highest of all methods for each sequence and the values in blue indicate the second highest. As shown in Table 1, our method has obvious better performance than all the

Table 1. Quantitative evaluation of different algorithms: improvement of PSNR values from the bi-cubic interpolation

	[10]	[9]	[4]	MSR [21]	LSR [21]	Hybrid SR [21]	[14]	Our Method
Container	1.4	-0.1	0.9	4.0	2.7	5.3	8.1	11.2
News	2.1	1.0	1.0	2.5	4.5	6.7	9.4	10.7
Mobcal	1.6	-0.8	0.1	3.2	1.1	3.3	7.3	10.5
Shields	0.2	-1.1	-0.2	0.3	1.2	1.6	4.9	2.5

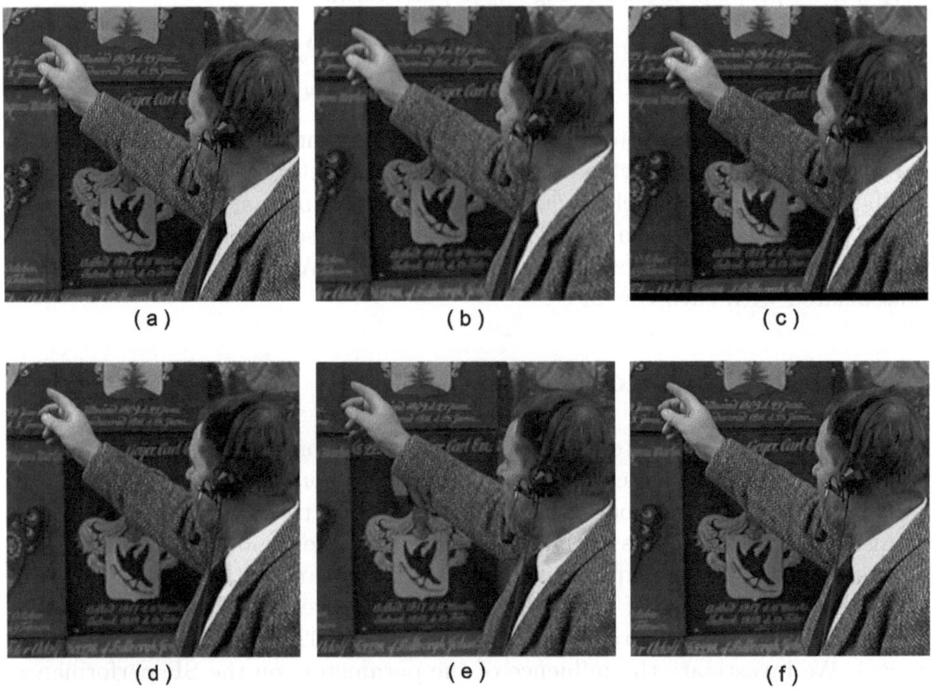

Fig. 3. Super-resolution results for a center region of *Shields*. (a) Original. (b) Bi-cubic. (c) MSR in Ref. [21]. (d) Hybrid SR in Ref. [21]. (e) Method in Ref. [14]. (f) Our Method.

methods for the sequences *Container*, *News* and *Mobcal*, and the PSNR value is even 11.2 dB higher than bi-cubic method. For the sequence *Shields*, the performance of our method is better than other methods except Hung's method [14]. Although the PSNR value is lower than Hung's result, the visual quality of our result is significantly better than it. Fig. 3 gives the qualitative evaluation of the 15th frame for the *Shields* sequence. We compare the result of our method with the results of the bi-cubic method [12], the MSR version and Hybird SR version

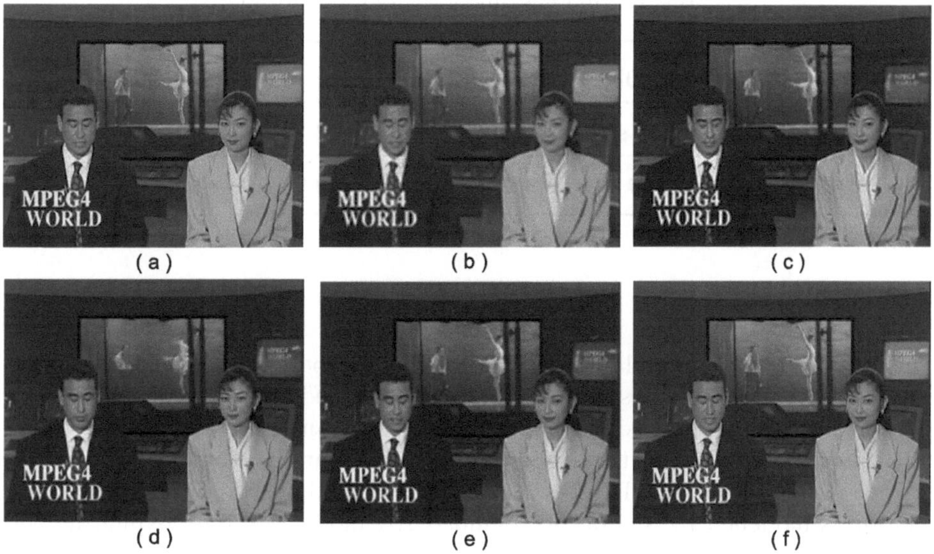

Fig. 4. Super-resolution results for *News*. (a) Original. (b) Bi-cubic. (c) Brandi's in Ref. [4] (d) MSR in Ref. [21]. (e) Hybrid SR in Ref. [21]. (f) Our Method.

in method [21], and the method in Ref. [14]. As shown in Fig. 3, the result of bi-cubic method is very blurred in the whole image, even though its PSNR value is not low. Obvious errors such as the logo in the middle of the image and information missing such as the bottom of the image, can be seen in the result of MSR method in Ref. [21]. Hybrid SR method in Ref. [21] performes slightly better than MSR method, but the blocking artifacts are still apparent, and the edges and cloth textures are also blurred. The result of Hung's method [14] also has error regions, such as, the logo in the middle of the image. On the contrary, our method provides the best visual quality with sharper edges and richer textures.

Fig. 4 gives the SR results of the 15th frame for the *News* sequence. We compare the result of our method with the results of the bi-cubic method [12], the method in Ref. [4], the MSR version and Hybrid SR version in method [21]. As shown in Fig. 4, the results of bi-cubic and the method in Ref. [4] are very blurred. In the result of the MSR method in Ref. [21], the information of dancers in the background is wrong and missing, and the hair details of the female anchor's forehead is also missing. The result of the Hybrid SR method in Ref. [21] has the same problem in the hair area, and the words on the background LED screen are blurred. Our method overcomes all the above mentioned problems and gets the significantly better visual quality than all the other methods.

Table 2 gives the running time for sequences with various resolutions. The running time is measured by seconds. As shown in the table, although the running time will increase with the improvement of the video resolution, the time of our method is shortest in all the methods.

Table 2. Running time for various algorithms

	Bi-cubic	[10]	[9]	[4]	[21]	Accelerated [21]	Our Method
1280×720	3.6 s	84 s	724 s	27.9 s	33.5 s	12.7 s	9.1 s
1920×1080	9.5 s	156 s	998 s	56.3 s	65.2 s	23.2 s	14.1 s

4 Conclusion

This paper proposes an adaptive video SR method based on superpixel-guided AR model. The key-frames are automatically selected and super-resolved by a method based on sparse regression. The non-key-frames are super-resolved by simultaneously exploiting the spatiotemporal correlations: the temporal correlation is exploited by a GPU-based optical flow method while the spatial correlation is modelled by a superpixel-guided AR model. By fully exploiting the temporal correlation between frames and accurately modeling the spatial correlation within video frames, the proposed method achieves high performance. As shown in the above experiments, the proposed method outperforms the existing benchmark in terms of both subjective visual quality and objective PSNR. At the same time, the running time of our method is the shortest among the state-of-the-art methods. The proposed method is applicable to various video sequences, which is a significant progress in video SR.

Acknowledgment. This work is supported by the NSF of China (61072062), Ph.D. Programs Foundation of Ministry of Education of China (20120032120040), the Tianjin Research Program of Application Foundation and Advanced Technology (12JCYBJC10300, 13JCQNJC03900), and the Research Academy of China Space Technology.

References

1. Achanta, R., Shaji, A., Smith, K., Lucchi, A., Fua, P., Süsstrunk, S.: SLIC superpixels. École Polytechnique Fédéral de Lausssanne (EPFL), Tech. Rep. 149300 (2010)
2. Altunbasak, Y., Patti, A.J., Mersereau, R.M.: Super-resolution still and video reconstruction from MPEG-coded video. IEEE Trans. Circuits and Systems for Video Technology 12(4), 217–226 (2002)
3. Belekos, S., Galatsanos, N.P., Katsaggelos, A.K.: Maximum a posteriori video super-resolution using a new multichannel image prior. IEEE Trans. Image Processing 19(6), 1451–1464 (2010)
4. Brandi, F., Queiroz, R., Mukherjee, D.: Super-resolution of video using key frames and motion estimation. In: Proc. IEEE International Conference on Image Processing, pp. 321–324 (October 2008)

5. Brandi, F., de Queiroz, R., Mukherjee, D.: Super resolution of video using key frames. In: Proc. IEEE International Symposium on Circuits and Systems, pp. 1608–1611. IEEE (2008)
6. Brandi, F., de Queiroz, R., Mukherjee, D.: Super-resolution of video using key frames and motion estimation. In: Proc. IEEE International Conference on Image Processing, pp. 321–324 (October 2008)
7. Choi, B.D., Han, J.W., Kim, C.S., Ko, S.J.: Motion-compensated frame interpolation using bilateral motion estimation and adaptive overlapped block motion compensation. IEEE Trans. Circuits and Systerms for Video Technology 17(4), 407–416 (2007)
8. Eisemann, M., De Decker, B., Magnor, M., Bekaert, P., De Aguiar, E., Ahmed, N., Theobalt, C., Sellent, A.: Floating textures. In: Proc. Computer Graphics Forum, vol. 27, pp. 409–418. Wiley Online Library (2008)
9. Fan, W., Yeung, D.Y.: Image hallucination using neighbor embedding over visual primitive manifolds. In: Proc. IEEE Conf. Computer Vision and Pattern Recognition, pp. 1–7 (2007)
10. Farsiu, S., Robinson, M.D., Elad, M., Milanfar, P.: Fast and robust multiframe super resolution. IEEE Trans. Image Processing 13(10), 1327–1344 (2004)
11. Freeman, W.T., Jones, T.R., Pasztor, E.C.: Example-based super-resolution. IEEE Computer Graphics and Applications 22(2), 56–65 (2002)
12. How, H.S., Andrews, H.C.: Cubic spline for image interpolation and digital filtering. IEEE Trans. Acoustics, Speech and Signal Proc. 26(5), 508–517 (1978)
13. Hung, E.M., de Queiroz, R.L.: Blocking-effect reduction in a reversed-complexity video codec based on a mixed-quality framework. In: Proc. IEEE Int. Telecommun. Symp., pp. 1–6 (September 2010)
14. Hung, E.M., de Queiroz, R.L., Brandi, F.: Video super-resolution using codebooks derived from key-frames. IEEE Trans. Circuits and Systerms for Video Technology 22(9), 1321–1331 (2012)
15. Jiang, Z., Wong, T.T., Bao, H.: Practical super-resolution from dynamic video sequences. In: Proc. IEEE Conference on Computer Vision and Pattern Recognition, vol. 2, pp. 549–554. IEEE (2003)
16. Kawano, H., Suetake, N., Cha, B., Aso, T.: Super-resolution via matching from self-decomposed codebook with local distance measure incorporating pixel correlation. In: Proc. Biennial Conference of the Australian Pattern Recognition Society on Digital Image Computing Techniques and Applications, pp. 317–323 (2007)
17. Kim, K.I., Kwon, Y.: Single-image super-resolution using sparse regression and natural image prior. IEEE Trans. Pattern Analysis and Machine Intelligence 32(6), 1127–1133 (2010)
18. de Oliveira, K.F., Brandi, F., Hung, E.M., de Queiroz, R.L., Mukherjee, D.: Bipredictive video super-resolution using key-frames. In: Proc. SPIE Symposium on Electronic Imaging, Visual Information Processing and Communication (2010)
19. Protter, M., Elad, M., Takeda, H., Milanfar, P.: Generalizing the nonlocal-means to super-resolution reconstruction. IEEE Trans. Image Processing 18(1), 36–51 (2009)
20. Segall, C.A., Katsaggelos, A.K., Molina, R., Mateos, J.: Super-resolution from compressed video. Kluwer International Series in Engineering and Computer Science, pp. 211–242 (2001)
21. Song, B.C., Jeong, S.C., Choi, Y.: Video super-resolution algorithm using bi-directional overlapped block motion compensation and on-the-fly dictionary training. IEEE Trans. Circuits and Systerms for Video Technology 21(3), 274–285 (2011)

22. Su, H., Tang, L., Tretter, D., Zhou, J.: A practical and adaptive framework for super-resolution. In: Proc. IEEE International Conference on Image Processing, pp. 1236–1239. IEEE (2008)
23. Takeda, H., Milanfar, P., Protter, M., Elad, M.: Super-resolution without explicit subpixel motion estimation. IEEE Trans. Image Processing 18(9), 1958–1975 (2009)
24. Tappen, M.F., Russell, B.C., Freeman, W.T.: Exploiting the sparse derivative prior for super-resolution and image demosaicing. In: Proc. IEEE Workshop on Statistical and Computational Theories of Vision, Citeseer (2003)
25. Tom, B.C., Katsaggelos, A.K.: Resolution enhancement of monochrome and color video using motion compensation. IEEE Trans. Image Processing 10(2), 278–287 (2001)

Motion Estimation for Video Coding Based on Subspace Pursuit

Yanfei Shen[1,2], Jintao Li[1], Zhenmin Zhu[1], and Yongdong Zhang[1]

[1] Institute of Computing Technology, Chinese Academy of Sciences, Beijing, P.R. China
[2] University of Chinese Academy of Sciences, Beijing, P.R. China
{syf,jtli,zmzhu,zhyd}@ict.ac.cn

Abstract. In this paper, a motion estimation algorithm is proposed based on subspace pursuit, which can be used in video coding to reduce the temporal redundancy. The main idea of our proposed algorithm is that the process of motion estimation will be imagined as sparse representation, that is, the predicted block is grouped as an observation vector and the corresponding reference blocks in previous reconstructed frame are used to construct a sparse dictionary, then the sparse coefficients are calculated by subspace pursuit algorithm. In order to reduce the transmission of sparse coefficients, the idea of template matching is adopted. Moreover, our proposed method can be combined with traditional motion estimation algorithm to further enhance the inter prediction accuracy. Simulation results show that our proposed method can outperform the decoder motion vector derivation method in term of the peak signal to noise ratio (PSNR).

Keywords: motion estimation, sparse representation, subspace pursuit.

1 Introduction

In video coding techniques, including the current state-of-the-art H.264/AVC [1] and high efficiency video coding (HEVC) [2], motion estimation and compensation are key components for reducing temporal redundancy between target frame and reference frame. There are two groups of motion estimation algorithms used in inter frame prediction. The first group is forward motion estimation in which motion vectors are needed to be transmitted in bit stream. Such as in H.264/AVC and HEVC, a frame is divided into non-overlapping rectangular blocks with different sizes and every block is assigned with one or more motion vectors which represent the displacement in the reference frame relative to the location of the block in the current frame. The block size and motion vectors are selected based on the well-known Lagrangian rate-distortion optimization theory [3] [4]. The prediction operation then simply consists of copying the reference block to the block in the current frame. The second group is backward motion estimation in which motion vectors is recursively estimated based on the reconstructed pixels adjacent to the current block as shown in the literature [5-10]. Therefore, in the second group, no motion vector needs to be transmitted to video decoder. Our proposed method belongs to the second group.

B. Huet et al. (Eds.): PCM 2013, LNCS 8294, pp. 465–474, 2013.

In the forward motion estimation, an analysis of the bit rate consumption for different types of coded syntax elements reveals that a significant portion of the rate is spent for motion vector information [5], especially in some low bit rate video coding cases, the encoded motion vectors consume up to 40% of the total frame bit rate. In the past decade, several methods have been proposed for reducing the amount of bits necessary to represent motion vectors in the bit stream, such as advanced motion vector prediction in HEVC [11], it improves the motion vector (MV) prediction by applying an MV prediction competition, where multiple MVs of spatial and temporal neighbor blocks are used as candidates, and the difference between the best predictor and the actual motion vector and the index of the candidate are transmitted to the decoder. Moreover, motion vectors with fractional pixel units are also supported to improve the efficiency of motion estimation and motion compensation, where interpolation filters are used to generate the prediction pixels in non-integer positions, such as HEVC uses an eight-tap filter for the half-sample positions and a seven-tap filter for the quarter-sample positions. However, it is impossible for such simple and fixed interpolation filter to adapt all the complex texture and motion in video sequences.

In addition, the motion estimation algorithms always assume that the motion of objects in video sequences is rigid and generally pure translation motion model is used, but in fact the objects in the real world undergo more complicated motions, including zooming, rotating and shearing etc, it is difficult for a simple translation model to adapt the complex motions in natural video content.

In fact, the process of motion estimation can be imagined as a solution for sparse representation. In particular, it is equal to seek sparse representation for the current block in the space which is constructed by the same shaped reference blocks in reference frame. It assumes that the current block is generated by some affine transforms of reference blocks, so there is no restriction of translation motion model and no explicit interpolation filter for calculating the pixel value in fractional pixel position. Because generally the current block is only relevant to several reference blocks in the reference frame, the representation is sparse and can be calculated by some greedy methods or convex optimization methods [12]. In this paper, the subspace pursuit algorithm (SP) [13] will be used to calculate the sparse representation coefficients for the current block, because the SP algorithm has two important characteristics: low computation complexity, comparable to that of orthogonal matching pursuit techniques when applied to sparse signals reconstruction, and representation accuracy of the same order as that of linear programming optimization methods.

Based on the above observation, in this paper we will propose a motion estimation algorithm based on subspace pursuit to improve the inter frame prediction efficiency. Similar to decoder-side motion vector derivation (DMVD) algorithm [5] and template matching prediction algorithm (TMP) [8] which aim at the reduction of motion parameter bit rate, the main idea is to firstly calculate the sparse coefficients by the reconstructed pixels adjacent to the current block in the same shaped reference block space, then the current block is predicted by the same sparse coefficients in its corresponding space. The major difference of our proposed algorithm with DMVD or TMP is that our proposed algorithm needs not to interpolate the sub-pixel value in the process of motion estimation, so it can better adapt to the texture characteristic than the fixed interpolation filters as used in traditional video coding techniques. Furthermore, it is very easy to extend our proposed algorithm to multihypothesis inter prediction by adding the reference blocks to the representation space or sparse dictionary in motion estimation [15-16].

The rest of this paper is organized as follows. In section 2, we will introduce the concept of subspace pursuit, and its relationships with traditional ME algorithm. Our proposed ME algorithm based on SP (SPME) will be detailed in section 3. Section 4 gives the experimental results in terms of prediction quality and results of comparison with DMVD and TMP. Finally, we conclude this paper in section 5.

2 Subspace Pursuit and Motion Estimation

Subspace pursuit algorithm is early proposed to reconstruct sparse signal in compressive sensing domain, that is, if a large but sparse signal is encoded by a relatively small number of linear measurement, and then the original signal can be recovered from the encoded one. Let $x \in \mathbb{R}^N$ be an arbitrary signal vector which sparse degree is K, and let $y \in \mathbb{R}^m$ be an observation of x via m linear measurement as

$$y = \Phi x \tag{1}$$

where $\Phi \in \mathbb{R}^{m \times N}$ is referred to as the measurement matrix. The compressive sensing theory guarantees that the original signal x can be recovered from the measurement y. A natural formational of the recovery problem is within a ℓ_1 norm minimization framework, which seeks a solution to the problem as

$$\min \|x\|_1 \quad \text{s.t.} \quad y = \Phi x \tag{2}$$

where

$$\|x\|_1 = \sum_{i=1}^{N} |x_i| \tag{3}$$

Recently, a family of iterative greedy algorithm has received significant attention to solve the above problem due to their low complexity and simple geometric interpretation, especially for SP algorithm, because it has provable reconstruction capability comparable to that of LP methods, and low reconstruction complexity of matching pursuit techniques for very sparse signals. The main steps of the SP algorithm are listed in Table 1.

In the viewpoint of sparse representation, SP algorithm can also be used to solve the following sparse approximation problem.

$$\alpha = \min \|x - D\alpha\|_2^2 \quad \text{s.t.} \quad \|\alpha\|_0 \leq K \tag{4}$$

where x is a signal vector similar to observation value, D is a overcomplete dictionary corresponding to measurement matrix, α is sparse coefficient vector, K is the sparse degree of α, which is the number of nonzero elements in vector α.

Motion estimation algorithm in video coding can be imagined as a special case of sparse representation, where the predicted current block is the signal vector x, all reference blocks in the search window of previous reconstructed reference frame can be viewed as atoms in overcomplete dictionary D. The process of motion estimation is to seek K atoms which are the most relevant to the current predicted block, and the indices of K atoms are equivalent to the motion vectors. For example, in the case of integer pixel accuracy ME, the sparse degree K is equal to one, and in the case of sub-pixel accuracy ME, the sparse degree K is equal to the tap of interpolation filters. Let bilinear interpolation filter $f = \{0.5,0.5\}$ be used in half pixel ME, the sparse coefficient α will be equal to $\{0.5,0.5\}$ and sparse degree K is 2, that is, the most two relevant atoms will be selected. Of course, the motion estimation algorithm based on sparse representation can simulate more complex models, such as polynomial affine motion model. Another advantage of this algorithm is that no explicit interpolation operation for sub-pixel is needed and it can adapt to various texture context in video sequences.

Table 1. SP Algorithm

Input: K, Φ, y
Initialization:
1) $\theta = \Phi^* y$, Φ^* is the transposition of Φ
2) $\Gamma^0 = \{K^*$ indices corresponding to the largest magnitude entries in $\theta\}$
3) $y_p = \Phi_\Gamma \Phi_\Gamma^\dagger y \qquad \Phi_\Gamma^\dagger = (\Phi_\Gamma^* \Phi_\Gamma)^{-1} \Phi_\Gamma^*$
4) $y_r^0 = y - y_p$
Iteration: at the ℓth iteration
1) $\theta = \Phi^* y_r^{\ell-1}$
2) $\Gamma^\ell = \{K^*$ indices corresponding to the largest magnitude entries in $\theta\}$
3) $\Gamma = \Gamma^\ell \cup \Gamma^{\ell-1}$
4) $x_p = \Phi_\Gamma \Phi_\Gamma^\dagger y \qquad \Phi_\Gamma^\dagger = (\Phi_\Gamma^* \Phi_\Gamma)^{-1} \Phi_\Gamma^*$
5) $\Gamma^\ell = \{K^*$ indices corresponding to the largest magnitude entries in $x_p\}$
6) $y_p = \Phi_{\Gamma^\ell} \Phi_{\Gamma^\ell}^\dagger y \qquad \Phi_{\Gamma^\ell}^\dagger = (\Phi_{\Gamma^\ell}^* \Phi_{\Gamma^\ell})^{-1} \Phi_{\Gamma^\ell}^*$
7) $y_r^\ell = y - y_p$
8) If $\|y_r^\ell\|_2 > \|y_r^{\ell-1}\|_2$, quit the iteration
Output: the estimated signal x

3 Motion Estimation Algorithm Based on SP

The process of motion estimation is to seek sparse representation for the current block in the dictionary constructed by the reference blocks from the previous reconstructed frame. In order to guarantee the same prediction between the video encoder and decoder, the sparse coefficients should be transmitted, which is similar to the motion vector in traditional motion estimation. However, on the one hand, the sparse coefficients are higher precision, and it is usually represented in float point format, so it will consume more bits compared with traditional motion vector if it is explicitly transmitted; on the other hand, the positions of sparse coefficients are also needed to encoded because they are variable block by block. As a result, although the motion estimation prediction accuracy is improved, the side information with sparse representation will consume more bits and reduce the video coding efficiency.

To tradeoff the higher prediction accuracy and the more side information, the idea from template matching prediction algorithm will be used, in particular, the reconstructed pixels adjacent to the current block are grouped as support region and used to firstly calculated the spare coefficients, then the current block is predicted by the sparse coefficients based on the same dictionary. As shown in the Fig.1.

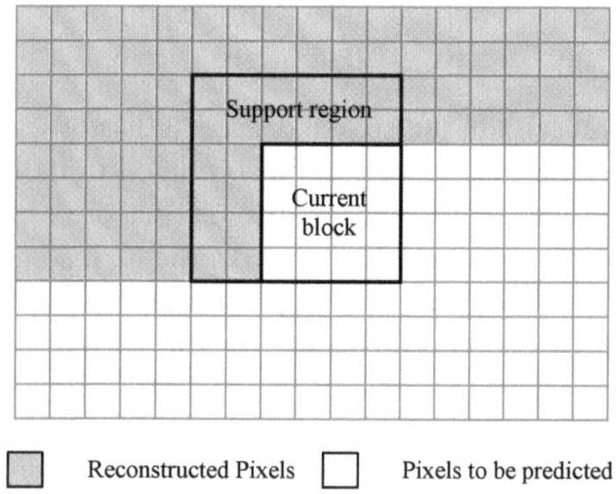

Fig. 1. Current block and its support region

Let the pixel value in the support region and in the current block be arrayed as column vector y_s and y_c, the vector y comprises of y_s and y_c as follows. It notes that y_s is the reconstructed pixels and y_c is the predicted pixels which are not known now.

$$y = \begin{bmatrix} y_s \\ y_c \end{bmatrix} \tag{5}$$

Let D denote the sparse dictionary which is a matrix of size $N \times M$, N is equal to the number of elements in vector y and M is decided by the size of search windows in reference frame. The columns in dictionary D are constructed by the same method as vector y. In general, the dictionary D can also be divided two sub-matrix D_s and D_c as follows.

$$D = \begin{bmatrix} D_s \\ D_c \end{bmatrix} \tag{6}$$

The problem of motion estimation can be formulated as the following sparse representation problem.

$$\alpha = \min \|y - D\alpha\|_2^2 \quad \text{s.t.} \ \|\alpha\|_0 \le K \tag{7}$$

That is

$$\alpha = \min \left\| \begin{bmatrix} y_s \\ y_c \end{bmatrix} - \begin{bmatrix} D_s \\ D_c \end{bmatrix} \alpha \right\|_2^2 \quad \text{s.t.} \ \|\alpha\|_0 \le K \tag{8}$$

If the pixels in the support region and in the current block share the same motion, the sparse coefficient α calculated by the support region pixels can be used to predict the pixels y_c in the current block. Therefore, the sparse coefficient α is firstly calculated by SP algorithm as follows.

$$\alpha = \min\|y_s - D_s\alpha\|_2^2 \quad s.t. \ \|\alpha\|_0 \leq K \tag{9}$$

Once the sparse coefficient α has been calculated, the pixels y_c in the current block can be predicted as

$$y_c = D_c\alpha \tag{10}$$

In our proposed ME algorithm, the sparse degree K is not only related to the true motion of the objects in video sequences, but also related to the texture characteristic. In addition, it is possible that there is inconsistency between the support region and the current block. To select the optimal sparse degree K, the sparse coefficient α_k is calculated for different sparse degree K in support region, and then the final optimal K is selected based the prediction performance for the current block by sparse coefficient α_k as follows

$$K = \min_k\left\|y_{org} - D_c\alpha_k\right\|_2^2 \tag{11}$$

where y_{org} is the original pixel value in the current block. However, because the pixel value y_{org} is not available in video decoder, the sparse degree K should be transmitted to the video decoder. Our proposed algorithm is detailed in Table 2.

Table 2. Motion estimation algorithm based on SP

Input: y_s, y_{org}, D_s, D_c, K_{max}
Step 1 Calculate the sparse coefficient α_k by SP in support region for k = 1 to K_{max} $\alpha_k = \min\|y_s - D_s\alpha_k\|_2^2 \quad s.t. \ \|\alpha\|_0 \leq k$ **Step 2** Select the optimal sparse degree K_{opt} $K_{opt} = \min\left\|y_{org} - D_c\alpha_k\right\|_2^2$ **Step 3** Predict the current block by the sparse coefficient α_{opt} corresponding the sparse degree K_{opt} $y_c = D_c\alpha_{opt}$
Output: the predicted current block y_c and sparse degree K_{opt}

4 Experimental Results

In this section, in order to evaluate the performance of our proposed method, the proposed method was tested by different standard video sequences, including foreman, coastguard, Waterfall etc with CIF resolution. The traditional translational motion estimation (TME) with quarter-pixel motion vector accuracy is used and compared. The block size is set 8x8 pixels and the search range of motion estimation is ± 16 pixels. The fractional sample interpolation for luma samples uses separable application of an eight-tap filter for the half-sample positions and a seven-tap filter for the quarter sample positions [17] as shown in Table 3.

Table 3. Filter coefficients for fractional sample interpolation

tap	-3	-2	-1	0	1	2	3	4
half	-1	4	-11	40	40	-11	4	1
quarter	-1	4	-10	58	17	-5	1	

In order to perform comparison, another decoder-side motion vector derivation algorithm (DMVD) for hybrid video coding is also implemented [5], which is based on template matching and aims to reduce motion parameter bit-rate by re-estimating the motion parameter at the decoder side. The motion estimation search method is full search algorithm.

The reference frame used in our experiment is the original frame and the extrapolation pixel beyond the image edges is repeated by the nearest pixel. The size of the current block and support region is shown in Fig.2. DMVD and our proposed algorithm use the same block structure, that is, the current block is 8×8 pixels and the adjacent template or support region is 4 pixels width from the current block.

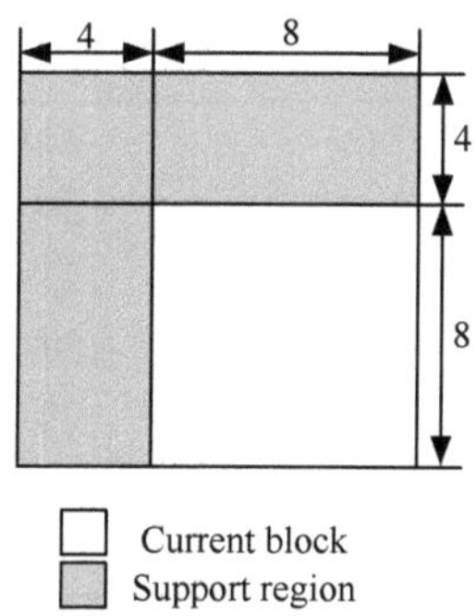

Fig. 2. The size of current block sand support region

The performance of motion estimation algorithm is measured by PSNR between the current original frame and the reconstructed frame, as shown in Table 4. It is demonstrated that our proposed motion estimation algorithm can outperform DMVD method in all video sequences, which is one of the most competitive proposals subjected to HEVC video coding standard [5]. The temporal redundancy can be efficiently reduced by our proposed SPME algorithm in video coding. Although the prediction efficiency of our proposed algorithm is lower than the TME method, no more side information except sparse degree needs not to be transmitted. Moreover, it is very easy for our proposed method to extent multihypothesis inter prediction. For video sequences with strong edge, such as bus and tempete, because the texture property of the support region and current block may be inconsistent, SPME algorithm is less efficient to use sparse coefficients trained from support region to predict the current block than for other video sequences, this is also the main drawback of DMVD method. But the performance of our proposed motion estimation can be enhanced by using mixed ME model, for example, if the disagreement between the current block and the support region is detected, our proposed algorithm can be switched to TME algorithm, it is similar to the switching from inter prediction to intra prediction when the inter prediction is less efficient than intra prediction.

To guarantee the same prediction for the current block between video encoder and decoder, the sparse degree K for every block is needed to be transmitted to video decoder as side information and it can be very efficiently coded by arithmetic coder or variable length coder, as can be seen from the distribution of sparse degree K in Fig.3. If sparse degree is one, our proposed motion estimation algorithm corresponds to traditional translational motion model with full pixel accuracy, otherwise, it corresponds to fractional pixel motion estimation algorithm, but the difference between them is that the coefficients of interpolation filters used to generate fractional pixel value in our proposed method are calculated in real time and can be adapt to real motion model hidden in video sequence. In addition, the bit overhead of transmitting this side information is lower than motion vectors. Fig.3 shows that the major of sparse degree is less than eight, that is, the current block is predicted by linear combination of less eight reference block, this is coincide with the length of interpolation filter defined in video coding standard HEVC.

Table 4. PSNR performance of DMVD, TME and SPME (dB)

sequence	TME	DMVD	SPME
Forman	37.69	33.00	36.84
Flower	33.41	30.77	32.97
Bus	29.30	23.81	26.66
Container	42.97	34.79	41.20
Coastguard	33.86	29.48	33.51
hall	38.84	33.49	37.28
Waterfall	40.36	39.07	39.81

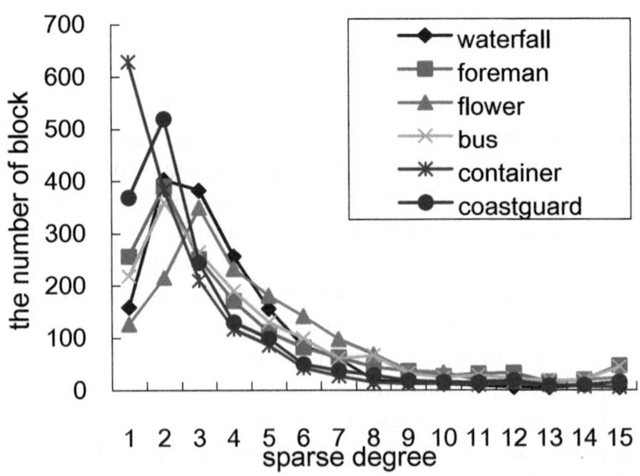

Fig. 3. The distribution of sparse degree K

In traditional translational motion model, the coefficients of interpolation filter are fixed and there are many fast motion estimation algorithms, so its computational burden is lower. The coefficients of sparse representation are calculated in real time for our proposed motion estimation algorithm, so it has more computation complexity.

5 Conclusion

In this paper, a motion estimation algorithm is proposed based on subspace pursuit, which can be used in video coding to reduce the temporal redundancy. The process of motion estimation is imagined as sparse representation, that is, the reference blocks in previous reconstructed frame are constructed as sparse dictionary, by which the current block is sparse represented. In order to reduce the transmission of sparse coefficients side information, the idea of template matching prediction is used. The sparse coefficients of the support region are firstly calculated by SP algorithm, and then the current block is predicted by the same sparse coefficients. The key benefit of our proposed algorithm is that it can model all kinds of motion, such as rotate motion and shear motion etc. and no motion vector side information is transmitted and no explicit fractional pixel interpolation is performed. Simulation results show that our proposed method can outperform the decoder motion vector derivation method in term of the peak signal to noise ratio (PSNR).

Acknowledgement. This work described in this paper was supported by the Nation Natural Science Foundation of China (Grant No.61001123, 61379084, 61272323) and National Key Technology Research and Development Program of China (2012BAH06B01).

References

[1] Wiegand, T., Sullivan, G.J., Bjøntegaard, G., Luthra, A.: Overview of the H.264/AVC video coding standard. IEEE Trans. Circuits Syst. Video Technol. 13(7), 560–576 (2003)
[2] Bross, B., Han, W.-J., Ohm, J.-R., Sullivan, G., Wiegand, T.: High Efficiency Video Coding (HEVC) Text Specification Draft 6, document JCTVC-H1003. In: Joint Collaborative Team on Video Coding (JCT-VC), San Jose, CA (February 2012)
[3] Sullivan, G.J., Wiegand, T.: Rate-Distortion Optimization for Video Compression. IEEE Signal Processing Magazine, 74–90 (November 1998)
[4] Wiegand, T., et al.: Rate-Constrained Coder Control and Comparison of Video Coding Standards. IEEE Trans. Circuits and Systems for Video Technology 13(7), 688–703 (2003)
[5] Kamp, S., Wien, M.: Decoder-Side Motion Vector Derivation for Block-Based Video Coding. IEEE Trans. Circuits and Systems for Video Technology 22(12), 1732–1746 (2012)
[6] Klomp, S., Munderloh, M., Vatis, Y., Ostermann, J.: Decoder-side block motion estimation for H.264/MPEG-4 AVC based video coding. In: Proc. IEEE Int. Symp. Circuits Syst., pp. 1641–1644 (May 2009)

[7] Sugimoto, K., Kobayashi, M., Suzuki, Y., Kato, S., Boon, C.S.: Inter frame coding with template matching spatio-temporal prediction. In: Proc. IEEE Int. Conf. Image Process., vol. 1, pp. 465–468 (October 2004)
[8] Suzuki, Y., Boon, C.S., Kato, S.: Block-based reduced resolution inter frame coding with template matching prediction. In: Proc. IEEE Int. Conf. Image Process., pp. 1701–1704 (October 2006)
[9] Suzuki, Y., Boon, C.S., Tan, T.K.: Inter frame coding with template matching averaging. In: Proc. IEEE Int. Conf. Image Process., pp. III-409–III-412 (2007)
[10] Shen, Y., Li, J., Zhu, Z.: Motion Estimation for Video Coding Based on Sparse Representation. In: Proc. IEEE Int. Conf. Acoust. Speech Signal Process. (to appear, 2013)
[11] Helle, P., Oudin, S., et al.: Block Merging for Quadtree-Based Partitioning in HEVC. IEEE Trans. Circuits and Systems for Video Technology 22(12), 1720–1731 (2012)
[12] Boyd, S., Vandenberghe, L.: Convex Optimization. Cambridge University Press (March 2004)
[13] Dai, W., Milenkovic, O., et al.: Subspace Pursuit for Compressive Sensing Signal Reconstruction. IEEE Trans. Information Theory 55(5), 2230–2249 (2009)
[14] Flierl, M., Wiegand, T., Girod, B.: Rate-constrained multihypothesis prediction for motion-compensated video compression. IEEE Trans. Circuits Syst. Video Technol. 12(11), 957–969 (2002)
[15] Girod, B.: Efficiency analysis of multihypothesis motion-compensated prediction for video coding. IEEE Trans. Image Process. 9(2), 173–183 (2000)
[16] Karczewicz, M., Nieweglowski, J., Haavisto, P.: Video coding using motion compensation with polynomial motion vector fields. Signal Process.: Image Commun. 10, 63–91 (1997)
[17] Sullivan, G.J., et al.: Overview of the high efficiency video coding (HEVC) standard. IEEE Trans. Circuits and Systems for Video Technology 22(12), 1649–1668 (2012)

Graph-Guided Fusion Penalty Based Sparse Coding for Image Classification

Xiaoshan Yang[1,2], Tianzhu Zhang[1,2], Changsheng Xu[1,2], and Min Xu[3]

[1] Institute of Automation, Chinese Academy of Science, Beijing, China
[2] China-Singapore Institute of Digital Media, Singapore
[3] iNEXT, Faculty of Engineering and IT, University of Technology, Sydney, Australia
{xiaoshan.yang,tzzhang,csxu}@nlpr.ia.ac.cn, Min.Xu@uts.edu.au

Abstract. In image classification, conventional sparse coding only encodes local features independently. As a result, the similar local features may be encoded into code vectors with large discrepancy. This sensitiveness has became the bottleneck of the traditional sparse coding based image classification methods. In this paper, we propose a novel graph-guided fusion penalty based sparse coding method. To alleviate the sensitiveness of the traditional sparse coding, our approach constrains that the similar local features are encoded into similar code vectors. To achieve this goal, we add the popular graph-guided fusion penalty term into the traditional $l1$-regularized sparse coding formulation. Finally, we adopt the multi-task form of the smoothing proximal gradient method to solve our optimization problem efficiently. Experimental results on 3 benchmark datasets demonstrate the effectiveness of our improved sparse coding method in image classification.

Keywords: image classification, sparse coding, smoothing proximal gradient.

1 Introduction

Image classification is one of the most challenging research tasks in computer vision [13,9]. The improvement on image classification can also benefit other useful applications, such as image search and retrieval [14]. A lot of research efforts devoted to image classification have led to significant progress, and the accuracies on several extremely difficult data sets have been increasing year by year. However, compared with the human recognition abilities, there is still a long way to go for a practical image classification algorithm.

Most of the currently used image classification methods are based on Bag-of-Words [19,18,13,9]. The idea behind these methods is borrowed from text retrieval. In text retrieval, words which represent the minimum semantic units are extracted firstly. Then a quantized code vector which is computed as the frequency of the extracted semantic words is used to represent the text semantic information of a document. In computer vision, the semantic words are replaced by visual words which are comprised of clustered centers in local feature space [21,22]. To be simply, there are five main steps for the conventional Bag-of-Words based image classification methods. **(1) Feature extraction.**

B. Huet et al. (Eds.): PCM 2013, LNCS 8294, pp. 475–484, 2013.

Local features, such as SIFT [15] and SURF [3], are densely extracted for a given image. Actually, each local feature is a statistic histogram of gradients in a local image patch. **(2) Codebook design.** k-means clustering is the most widely used method for constructing codebook. Local features extracted from training images are sampled and clustered into thousands of visual words. Many improved methods have been proposed recently. For example, on-line clustering and mean-shift are combined to create much more effective codebook in [10]. **(3) Feature coding.** Different kinds of coding methods were proposed in the past several years. Yang *et al.* [19] proposed a method called ScSPM which introduces sparse regularization to the soft-assignment coding method. This results in competitive image classification accuracies by only using linear SVM [6]. Wang *et al.* [18] further improved ScSPM with a locality constraint, which leads to an analytical solution to the coding problem and a fast approximated solving method. Liu *et al.* [13] proposed a localized soft-assignment coding(LSC) method which adds a locality constraint to the distance function in traditional soft-assignment coding method, and this idea is similar to the salient coding proposed by Huang *et al.* [9]. **(4) Pooling and concatenating.** This is another key step for extracting high level sematic information from local features. There are three mostly used pooling methods, sum-pooling, average-pooling, and max-pooling. The max-pooling method was proved to be much more effective for sparse feature coding [19,18]. Liu *et al.* [13] proposed a new "mix-order" max-pooling method with max-pooling as its special case. Lazebnik *et al.* [11] proposed a concatenating method called SPM(Spatial Pyramid Matching) which takes advantages of global geometric correspondence among code vectors in a given image. **(5) Classifier construction.** This is a canonical problem in machine learning. Most of the sate-of-the-art image classification methods use SVM classifier due to its efficiency and simplicity. There are also some other methods for classification. Yao *et al.* [20] proposed a combined method of random forest and SVM where a linear SVM is used to replace the ordinary weak classifier at each node of random trees.

In the remainder of this paper, we mainly focus on feature coding in image classification. As the most popularly used feature coding method, sparse coding has improved the image classification accuracies tremendously in the past few years. However, there are still many problems in conventional sparse coding [2]. As we know, the bottleneck of the conventional sparse coding is that a small change of the local feature will leads to really large variation of the code vector. A principal reason is that the traditional sparse coding only minimize the reconstruction error of each local feature with a sparsity regularization item independently. In other words, conventional sparse coding can not guarantee that the similar local features are encoded into similar code vectors.

In this paper, in order to break through the bottleneck of the conventional sparse coding mentioned in above paragraph, we propose a graph-guided fusion penalty based sparse coding (GFP-SC) algorithm. We introduce the graph-guided fusion penalty into the traditional sparse coding formulation. Thus the code vectors encoded from similar local features are constrained to be similar.

Then we adopt the smoothing proximal gradient method to solve our optimization problem efficiently. In our implementation, we construct a graph with 4-neighborhood structure to describe the contextual structure information among local features in a given local region. Finally, the linear SVM [6] is applied to perform classification. Extensive experimental results on several popular benchmarks show that our GFP-SC outperforms several recently published methods.

This paper is organized as follows. In Section 2, we summarize the works most related to ours. Our improved sparse coding approach is presented in Sections 3. In Section 4, we report and analyze extensive experimental results. At last, a conclusion to our method is given in Section 5.

2 Related Work

In this section, we review 6 mostly used feature coding methods for image classification. Let $\mathbf{b}_i$ ($\mathbf{b}_i \in \Re^d$) denote a visual word or a basis vector, where d is the same as the dimensionality of the local feature. Matrix $\mathbf{B} = [\mathbf{b}_1, \mathbf{b}_2, \cdots, \mathbf{b}_m]$ denotes the visual codebook. The total number of visual words contained in $\mathbf{B}$ is m. Generally, the codebook is trained in advance by clustering. There are also methods which optimize the codebook in coding process [18,8]. Let $\mathbf{x}_i$ ($\mathbf{x}_i \in \Re^d$) be the ith local feature for an image $\mathbf{X} = [\mathbf{x}_1, \mathbf{x}_2, \cdots, \mathbf{x}_n]$. Let $\mathbf{v}_i$ ($\mathbf{v}_i \in \Re^m$) be the encoded code vector of $\mathbf{x}_i$, each element $\mathbf{v}_{ij}$ is the coefficient with respect to one visual word $\mathbf{b}_j$ contained in codebook $\mathbf{B}$. Define $\mathbf{V} = [\mathbf{v}_1, \mathbf{v}_2, ..., \mathbf{v}_n]$. With these denotations, 6 related coding methods can be uniformly written as follows.

Hard-Assignment Coding: In original Bag-of-Words method [5], the simplest hard-assignment or vector quantization was used to construct a bag of keypoints(original name of visual words). The main idea is to count the number of local features assigned to each clustered keypoint. Given a query local feature $\mathbf{x}_i$, the formally expression can be written as

$$\mathbf{v}_{ij} = \begin{cases} 1 & if \quad j = \arg\min_{j=1,\ldots,m} \|\mathbf{x}_i - \mathbf{b}_j\|_2^2, \\ 0 & otherwise. \end{cases} \tag{1}$$

Soft-Assignment Coding: A local feature $\mathbf{x}_i$ is assigned to just a single visual word by hard-assignment coding while soft-assignment coding assigns it to all visual words in the codebook $\mathbf{B}$ [17] with different weights,

$$\mathbf{v}_{ij} = \frac{\exp(-\alpha \|\mathbf{x}_i - \mathbf{b}_j\|_2^2)}{\sum_{k=1}^{m} \exp(-\alpha \|\mathbf{x}_i - \mathbf{b}_k\|_2^2)}. \tag{2}$$

Sparse Coding: This method [19] is more complicated than Soft-assignment. An optimization problem with sparse regularization constraint needs to be solved for computing the correspondent weight to each visual word in codebook $\mathbf{B}$. If the codebook $\mathbf{B}$ is trained in advance, the optimization problem can be simplified as

$$\mathbf{v}_i = \arg\min_{\mathbf{v}_i \in \Re^m} \|\mathbf{x}_i - \mathbf{B}\mathbf{v}_i\|_2^2 + \lambda \|\mathbf{v}_i\|_1. \tag{3}$$

Locality-Constrained Linear Coding: In the method [18], a locality constraint is introduced to the sparse coding method [19] by multiplying a Euclidean distance weight vector to $\mathbf{v}_i$ in the optimization equation,

$$\begin{aligned} \mathbf{v}_i = \arg\min_{\mathbf{v}_i \in \Re^m} \|\mathbf{x}_i - \mathbf{B}\mathbf{v}_i\|_2^2 + \lambda \|\mathbf{d}_i \odot \mathbf{v}_i\|_2^2 \\ s.t. \quad \mathbf{1}^T \mathbf{v}_i = 1. \end{aligned} \tag{4}$$

where $\mathbf{d}_i = exp\left(dist\left(\mathbf{x}_i, \mathbf{B}\right)/\delta\right)$, and $\odot$ represents element-wise multiplication.

Localized Soft-Assignment Coding: As proposed by Liu *et al.*[13], localized soft-assignment coding method mainly focuses on the k nearest visual words for each local feature $\mathbf{x}_i$. Here $\mathbf{N}_k\left(\mathbf{x}_i\right)$ is the k-nearest neighbor of $\mathbf{x}_i$.

$$\mathbf{v}_{ij} = \frac{\exp(-\alpha \|\mathbf{x}_i - \mathbf{b}_j\|_2^2)}{\sum_{\mathbf{b}_m \in \mathbf{N}_k(\mathbf{x}_i), \mathbf{b}_m \neq \mathbf{b}_j} \exp(-\alpha \|\mathbf{x}_i - \mathbf{b}_m\|_2^2)}. \tag{5}$$

Laplacian Sparse Coding: Obviously, all the previously mentioned 5 feature coding methods carry out the coding process independently for each local feature. In contrast, as proposed by Gao *et al.*[8], structure information among the local features is combined in the Laplacian sparse coding method. The structure information is used to construct a Laplacian matrix which is added to the optimization problem of the conventional sparse coding method.

$$\begin{aligned} &\arg\min_{\mathbf{B},\mathbf{V}} \|\mathbf{X} - \mathbf{B}\mathbf{V}\|_F^2 + \lambda \|\mathbf{V}\|_1 + \frac{\beta}{2}\sum_{i,j} \|\mathbf{v}_i - \mathbf{v}_j\|^2 w_{ij} \\ &= \arg\min_{\mathbf{B},\mathbf{V}} \|\mathbf{X} - \mathbf{B}\mathbf{V}\|_F^2 + \lambda \|\mathbf{V}\|_1 + \beta\, Tr(\mathbf{V}\mathbf{L}\mathbf{V}^T) \\ &subject\ to: \|\mathbf{b}_m\|^2 \leq 1, \end{aligned} \tag{6}$$

where $\mathbf{L} = \mathbf{D} - \mathbf{W}$ is the Laplacian matrix, $\mathbf{W}$ is the similarity matrix comprised of w_{ij}, w_{ij} denotes the relationship among local features $\mathbf{x}_i$ and $\mathbf{x}_j$. $\mathbf{D}_{ii} = \sum_j w_{ij}$, $w_{ii} = 0 (i = 1, 2, ..., n)$, $\|.\|_1$ is the matrix entry-wise $l1$-norm. An approximate method was adopted to solve $\mathbf{V}$ and $\mathbf{B}$.

3 Graph-Guided Fusion Penalty Based Sparse Coding

In this section, we first give the details of our graph-guided fusion penalty based sparse coding (GFP-SC) for image classification. Then we make a discussion about some advantages of our GFP-SC method [8]. Here, we still use the uniform notation in Section 2.

3.1 Our Formulation

The sparse coding method shown in Equation (3) can be equivalently rewritten as

$$\arg\min_{\mathbf{V} \in \Re^{m \times n}} \|\mathbf{X} - \mathbf{B}\mathbf{V}\|_F^2 + \lambda \|\mathbf{V}\|_1, \tag{7}$$

where $\|.\|_F$ is the matrix Frobenius norm(entry-wise $l2$-norm), and $\|.\|_1$ is the matrix entry-wise $l1$-norm. As discussed in Section 2, the contextual structure information among neighborhood local features is not considered in traditional sparse coding. In order to make the full use of the contextual structure information among local features, we add a graph-guided fusion penalty to Equation (7). Then, we get Equation (8)

$$\arg\min_{\mathbf{V}\in\Re^{m\times n}} \|\mathbf{X}-\mathbf{B}\mathbf{V}\|_F^2 + \gamma\Omega_G(\mathbf{V}) + \lambda\|\mathbf{V}\|_1, \tag{8}$$

where γ is the regularization parameter for structured sparsity, the graph penalty is defined as

$$\Omega_G(\mathbf{V}) = \sum_{e=(i,j)\in E} w_{ij}\|\mathbf{v}_i - \mathbf{v}_j\|_1, \tag{9}$$

here, the weight w_{ij} is computed according to the visual similarity among two query local features

$$w_{ij} = \exp\left(-\frac{\|\mathbf{x}_i - \mathbf{x}_j\|_2^2}{2\delta^2}\right), \tag{10}$$

e is one edge which connected node i and j in the graph, and the edges set of the graph is defined as

$$E = \Big\{(i,j) \mid (i<j) \wedge \big(\mathbf{p}_i \in N_k(\mathbf{p}_j) \vee \mathbf{p}_j \in N_k(\mathbf{p}_i)\big), i \text{ or } j \in \{1,...,n\}\Big\}, \tag{11}$$

where $\mathbf{p}_i$ and $\mathbf{p}_j$ denote the spatial coordinates of local features $\mathbf{x}_i$ and $\mathbf{x}_j$ in an image respectively, $N_k(\mathbf{p})$ denotes the spatial k nearest neighbors of the $\mathbf{p}$. This graph-guided fusion penalty gives a structure regularization using contextual information among the local features.

3.2 Solutions

Equation (9) can be rewritten as the matrix form

$$\Omega_G(\mathbf{V}) = \|\mathbf{C}\mathbf{V}^T\|_1, \tag{12}$$

where $\mathbf{C} \in \Re^{|E|\times n}$ is the incident matrix defined as

$$\mathbf{C}_{(i,j),k} = \begin{cases} w_{ij}, & if\ \ k=i \\ -w_{ij}, & if\ \ k=j \\ 0 & otherwise. \end{cases} \tag{13}$$

As mentioned in [4], the Equation (12) can be further rewritten as

$$\Omega_G(\mathbf{V}) = \max_{\mathbf{A}\in\mathcal{Q}} \langle \mathbf{C}\mathbf{V}^T, \mathbf{A}\rangle = \max_{\mathbf{A}\in\mathcal{Q}} Tr(\mathbf{V}\mathbf{C}^T\mathbf{A}), \tag{14}$$

where $\mathbf{A} \in \mathcal{Q} = \{\mathbf{A} \mid \|\mathbf{A}\|_\infty \leq 1, \mathbf{A} \in \Re^{|E|\times m}\}$ is a matrix of auxiliary variables. The difficulty to solve the optimization problem in Equation (8) is the non-smoothness of the graph penalty $\Omega_G(\mathbf{V})$. Here we introduce smooth approximation to $\Omega_G(\mathbf{V})$ as follows

$$f_u(\mathbf{V}) = \max_{\mathbf{A}\in\mathcal{Q}} \langle \mathbf{C}\mathbf{V}^T, \mathbf{A}\rangle - ud(\mathbf{A}), \tag{15}$$

where u is the smoothness parameter and $d(\mathbf{A})$ is defined as $\frac{1}{2}\|\mathbf{A}\|_F^2$. It is easy to prove that $f_u(\mathbf{V})$ is convex and smooth. Replace $\Omega_G(\mathbf{V})$ in Equation (8) with $f_u(\mathbf{V})$, we get the final optimization equation

$$\arg\min_{\mathbf{V}\in\Re^{m\times n}} \|\mathbf{X} - \mathbf{B}\mathbf{V}\|_F^2 + \gamma f_u(\mathbf{V}) + \lambda\|\mathbf{V}\|_1. \tag{16}$$

Then, above optimization Equation (16) can be solved by smoothing proximal gradient method.

3.3 Discussion

In this subsection, we give a detailed comparison between our coding method and the traditional sparse coding. Besides, we give the difference between our coding method and the Laplacian sparse coding which is, to our best knowledge, the most similar published method to ours.

Our GFP-SC vs. Traditional Sparse Coding: The basic idea of our method is shown in (a) of Figure 1. Naturally, the densely sampled SIFT features in an image are connected with each other by their spatial and visual information. Traditional sparse coding discards the contextual structure information while our coding method effectively retains all this kind of information. Besides, (b) of Figure 1 shows a toy example about the detailed difference between traditional sparse coding and our coding method. What is noteworthy is that local features x, y and z are three adjacent local features sampled in (a), but z is much more similar with y than x in visual space. In traditional sparse coding, coding processes for x, y and z are carried out independently. Without loss of generality, we assume that visual words $\{b_1, b_2, b_3\}$ are assigned to x, $\{b_1, b_2, b_4\}$ are assigned to y and $\{b_4, b_5, b_6\}$ are assigned to z. In this case, we can see that the visual words $\{b_4, b_5, b_6\}$ assigned to z are much more different from $\{b_1, b_2, b_4\}$ assigned to y than $\{b_1, b_2, b_3\}$ assigned to x, though z is much more similar with y than x in visual space. However, if we consider the constraint of graph-guided fusion penalty, the visual words $\{b_4, b_5, b_6\}$ assigned to local feature y are the same as visual words of z. Thus, our GFP-SC method will definitely alleviate the sensitiveness of the traditional sparse coding.

Our GFP-SC vs. Laplacian Sparse Coding: From Equation (8) and Equation (6), it is easy to find that both Laplacian sparse coding and our coding methods introduce new regularization items to take the contextual structure information into account. Here, we show the differences between our graph-guided fusion penalty based sparse coding method and the Laplacian sparse coding method. **(1)** From equations (6) and (9), we can see that the $l1$ norm is used in our method to construct the graph-guided fusion penalty while $l2$ norm is

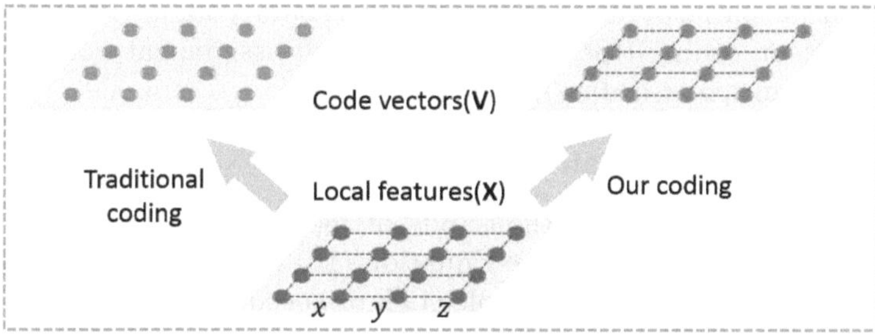

(a)

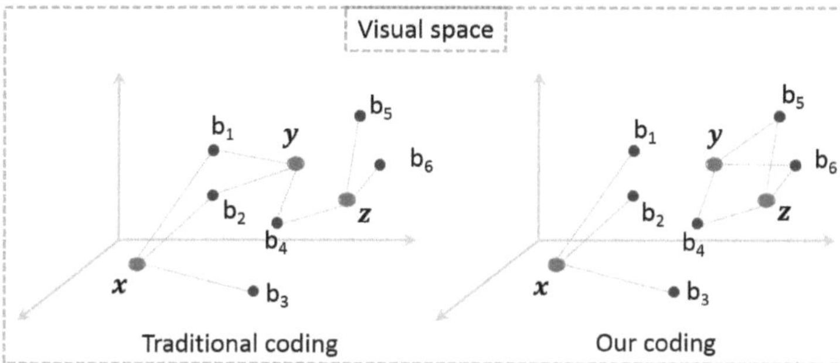

(b)

Fig. 1. Illustration of the difference between traditional sparse coding and our improved sparse coding method (GFP-SC). **(a)** Traditional sparse coding discards the contextual structure information while our coding method retains all this kind of information. **(b)** Note that x, y, z are adjacent local features sampled in (a), but z is much more similar with y than x in visual space. Our method is more likely to assign similar visual words (purple points) to similar local features in visual space.

used in the Laplacian sparse coding. In this case, code vectors computed from similar local features by our method are almost identical except several elements where related visual words are different. **(2)** We construct a sparse matrix $\mathbf{C}$ to regularize the code vectors other than the Laplacian matrix $\mathbf{L}$ in the Laplacian sparse coding method. **(3)** Only the structure information in local feature space is considered in Laplacian sparse coding method while our method combines the structure information in both feature space and spatial space as shown in Equations (9), (10) and (11) respectively. **(4)** Due to the time complexity, only an approximate solver was given in [8]. The contextual structure information is introduced from a subset of sampled local features. In contrast, our formulation can be solved using the smoothing proximal gradient method efficiently, where all the local features are encoded simultaneously. Thus the more complete contextual structure information can be retained to the code vectors after coding.

4 Experiments

To validate the effectiveness of our GFP-SC method for image classification, we compare it with 4 state-of-the-art local feature coding methods which are denoted as: ScSPM(linear spatial pyramid matching using sparse coding)[19], LLC(locality-constrained linear coding)[18], LSC(localized soft-assignment coding)[13] and LScSPM(laplacian sparse coding)[8].

4.1 Implementation Details

Since there are always tens of thousands of local features contained in each image, it is time consuming to solve Equation (16) for all SIFT features in each image. In this paper, we adopt a simple trick to speed up our structured coding algorithm. We use SLIC (Simple Linear Iterative Clustering) algorithm [1] to segment each image into smaller regions. Then, our graph-guided fusion penalty based sparse coding (GFP-SC) algorithm is applied in each image region. We choose the parameter of the initial image region size for the SLIC method to be 60. Bigger initial region size for the SLIC method leads to higher accuracy but also more computation time [1].

To fairly compare with ScSPM, LLC and LSC, we use the publicly available source codes provided by the authors. For these three feature coding methods and our GFP-SC, we use the same experimental setup in all 5 steps (mentioned in Section 1) of the image classification except the feature coding step. We use same image size and same training/testing splits on each dataset. In feature extraction, single scale SIFT features (16×16) with step size 4 are extracted. We also use the same codebook trained by k-means clustering in advance. Max-pooling is used for pooling and a 3 levels SPM with 1×1, 2×2, 4×4 cells is used for concatenating. For the linear SVM used in ScSPM, LLC, LSC and our GFP-SC methods, we adopt the same code from Liblinear [6] instead of the different implementations provided by authors of the three baseline methods.

To compare with LScSPM, we borrow the results reported in [8], because the authors do not share their codes for comparison.

4.2 Results and Analysis

The experiments are conducted on three popularly used datasets in the literatures: Caltech101 [7], UIUC 8-Sports [12], 10 class Corel dataset [16]. On each dataset, all the results shown in Table 1 are calculated by averaging the accuracies of 10 random training/testing splits.

Following are details of the three datasets. **Caltech101** [7] contains 9144 images of objects with 101 classes. Each image is resized to be no bigger than 300 in height and width. We use 30 images for training and remaining images for testing in each split. The accuracies are shown in the first column of Table 1. **UIUC 8-Sports** dataset contains 1579 images of 8 category sport events [12]. Each image is resized to be no bigger than 400 in height and width due to the high resolution of images in this dataset. We use 70 images for training and 60 images for testing in each split. The accuracy results are shown in the second column of Table 1. **10-Corel** dataset contains 1000 images with 10 categories [16]. Each image is resized to be no bigger than 300 in height and width. We use 50 images for training and 50 images for testing in each split. The accuracy results are shown in the third column of Table 1.

Table 1. Accuracies on 3 data sets

Method	Caltech101	Sports8	Corel10
ScSPM [19]	73.0±1.1	86.0±0.8	87.8±1.0
LLC [18]	72.3±0.8	86.4±1.6	88.4±1.3
LSC [13]	73.2±1.2	86.2±1.5	88.3±0.8
GFP-SC	**74.1±1.0**	**87.9±1.4**	**89.1±1.4**
LScSPM [8]	–	85.3±0.5	88.4±0.8

From Table 1, we can see that the classification accuracies for the first three baseline coding methods are almost the same. Compared with these three baseline methods, our coding method makes 1% increases of the accuracies on all three datasets. Due to the difficulty of these datasets, a slightly even smaller than 1% improvement is acceptable as reported by recent top-level literatures in computer vision. Thus we believe our graph-guided fusion penalty based sparse coding method is effective by incorporating the structured contextual information ignored in other coding methods. Authors of [8] only give an approximate algorithm where the codebook $\mathbf{B}$ and $\mathbf{V}$ are optimized iteratively. Nevertheless, our coding method still shows competitive results on the last two datasets by only using a fixed codebook $\mathbf{B}$.

5 Conclusions

In this paper, we propose a novel graph-guided fusion penalty based sparse coding for image classification. Combined with the contextual structure information among local features, our algorithm tends to encode the similar local features into similar code vectors. Our method is implemented based on the the smoothing proximal gradient method. Experimental results on three popularly used datasets demonstrate the effectiveness of our coding algorithm compared with four state-of-the-art feature coding methods.

Acknowledgments. This work is supported by National Natural Science Foundation of China (61225009, 61003161), and Beijing Natural Science Foundation (4131004), also by the Singapore National Research Foundation under International Research Centre@Singapore Funding Initiative and administered by the IDM Programme Office.

References

1. Achanta, R., Shaji, A., Smith, K., Lucchi, A., Fua, P., Süsstrunk, S.: Slic superpixels compared to state-of-the-art superpixel methods. IEEE Trans. Pattern Anal. Mach. Intell. 34(11), 2274–2282 (2012)

2. Bao, B.-K., Zhu, G., Shen, J., Yan, S.: Robust image analysis with sparse representation on quantized visual features. IEEE Transactions on Image Processing 22(3), 860–871 (2013)
3. Bay, H., Tuytelaars, T., Van Gool, L.: Surf: Speeded up robust features. In: Leonardis, A., Bischof, H., Pinz, A. (eds.) ECCV 2006, Part I. LNCS, vol. 3951, pp. 404–417. Springer, Heidelberg (2006)
4. Chen, X., Lin, Q., Kim, S., Carbonell, J.G., Xing, E.P.: Smoothing proximal gradient method for general structured sparse learning. In: UAI, pp. 105–114 (2011)
5. Csurka, G., Dance, C.R., Fan, L., Willamowski, J., Bray, C.: Visual categorization with bags of keypoints. In: Workshop on Statistical Learning in Computer Vision, ECCV, pp. 1–22 (2004)
6. Fan, R.-E., Chang, K.-W., Hsieh, C.-J., Wang, X.-R., Lin, C.-J.: Liblinear: A library for large linear classification. Journal of Machine Learning Research 9, 1871–1874 (2008)
7. Fei-Fei, L., Fergus, R., Perona, P.: Learning generative visual models from few training examples: an incremental Bayesian approach tested on 101 object categories. In: Workshop on Generative-Model Based Vision, CVPR (2004)
8. Gao, S., Tsang, I.W.-H., Chia, L.-T., Zhao, P.: Local features are not lonely - laplacian sparse coding for image classification. In: CVPR, pp. 3555–3561 (2010)
9. Huang, Y., Huang, K., Yu, Y., Tan, T.: Salient coding for image classification. In: CVPR, pp. 1753–1760 (2011)
10. Jurie, F., Triggs, B.: Creating efficient codebooks for visual recognition. In: ICCV, pp. 604–610 (2005)
11. Lazebnik, S., Schmid, C., Ponce, J.: Beyond bags of features: Spatial pyramid matching for recognizing natural scene categories. In: CVPR (2006)
12. Li, L.-J., Li, F.-F.: What, where and who? classifying events by scene and object recognition. In: ICCV, pp. 1–8 (2007)
13. Liu, L., Wang, L., Liu, X.: In defense of soft-assignment coding. In: ICCV, pp. 2486–2493 (2011)
14. Liu, S., Feng, J., Song, Z., Zhang, T., Lu, H., Xu, C., Yan, S.: Hi, magic closet, tell me what to wear? In: ACM Multimedia (2012)
15. Lowe, D.G.: Distinctive image features from scale-invariant keypoints. International Journal of Computer Vision 60(2), 91–110 (2004)
16. Lu, Z., Ip, H.H.-S.: Image categorization with spatial mismatch kernels. In: CVPR, pp. 397–404 (2009)
17. van Gemert, J.C., Geusebroek, J.-M., Veenman, C.J., Smeulders, A.W.M.: Kernel codebooks for scene categorization. In: Forsyth, D., Torr, P., Zisserman, A. (eds.) ECCV 2008, Part III. LNCS, vol. 5304, pp. 696–709. Springer, Heidelberg (2008)
18. Wang, J., Yang, J., Yu, K., Lv, F., Huang, T.S., Gong, Y.: Locality-constrained linear coding for image classification. In: CVPR, pp. 3360–3367 (2010)
19. Yang, J., Yu, K., Gong, Y., Huang, T.S.: Linear spatial pyramid matching using sparse coding for image classification. In: CVPR, pp. 1794–1801 (2009)
20. Yao, B., Khosla, A., Li, F.-F.: Combining randomization and discrimination for fine-grained image categorization. In: CVPR, pp. 1577–1584 (2011)
21. Zhang, T., Liu, J., Liu, S., Ouyang, Y., Lu, H.: Boosted exemplar learning for human action recognition. In: ICCV Workshop on Video-oriented Object and Event Classification (2009)
22. Zhang, T., Liu, J., Liu, S., Xu, C., Lu, H.: Boosted exemplar learning for action recognition and annotation. IEEE Transactions on CSVT 21(7), 853–866 (2011)

Image Classification by Iterative Semi-Supervised Sparse Coding

Haixia Zheng and Horace H.S. Ip

Centre for Innovative Applications of Internet and Multimedia Technologies,
Department of Computer Science, City University of Hong Kong, Hong Kong
Haixia.Zheng@my.cityu.edu.hk, cship@cityu.edu.hk

Abstract. In this paper, we propose a novel algorithm, *Iterative Semi-supervised Sparse Coding(ISSC)*, which combines sparse coding and graph-based semi-supervised learning into a unified framework, in order to learn the discriminative sparse codes as well as an effective classification function. The ISSC algorithm fully exploits initial labels and the subsequently predicted labels for sparse codes learning. At the same time, during the graph-based semi-supervised learning stage, the similarity matrix is firstly adjusted through the latest learned sparse codes, and then is utilized to obtain a better classification function. In particular, by solving quadratic optimization, the ISSC approach can give rise to closed-form solution for learned sparse codes. We have extensively evaluated the proposed ISSC method over the widely used datasets for image classification task. The experimental results in terms of classification accuracy demonstrate that the proposed ISSC method is robust and can also achieve significant performance improvements with respect to the state-of-the-arts.

Keywords: Sparse coding, graph-based semi-supervised learning, image classification.

1 Introduction

Image classification assigns an unknown image to a category of interest based on its visual and/or semantic content. It can be applied in image retrieval (searching through an image dataset to retrieve those images with particular visual content), image annotation (attaching textual meta-information or semantic linkage to images), and so on. Although image classification experienced fast development in the past decade, it is still one of the key challenges in computer vision research community. There are still two central issues in image classification. One is how to make the best use of the limited supervisory information, since only a few data are labeled and large amounts of data remain unlabeled. The other issue is the very high dimensionality in multimedia data.

Semi-supervised learning [1] has been applied to cope with the very situations of limited labeled data and abundant unlabeled data. It is able to incorporate knowledge about unlabeled data into the training process in order to learn a better classification function. The typical semi-supervised learning methods include

B. Huet et al. (Eds.): PCM 2013, LNCS 8294, pp. 485–496, 2013.

co-training [2] and graph-based semi-supervised learning. The graph-based semi-supervised learning has attracted increasing attentions recently, because it can generate elegant mathematical formulation and is easy to be implemented [3]. The graph-based semi-supervised learning approaches treat the whole dataset as a graph where the vertices represent the data and each edge reflects the similarity between two respective data samples.

Moreover, due to the high dimensionality of multimedia data, efficiency concerns naturally lead to sparsely decomposing data, rather than directly utilizing the original high-dimensional data. Sparse coding approaches can provide a sparse representation for original high-dimensional data. Furthermore, different from the general decomposition methods, sparse coding does not require the basis vectors to be orthogonal, therefore, provides a higher flexibility for the data representation. Sparse coding approaches have been shown very useful in many applications, such as color image restoration [4], image denoising and demosaicking tasks [5], image classification [6] [7] etc.

However, graph-based semi-supervised learning approaches do not utilize the advantage of sparse coding; the sparse coding methods neither exploit the prior information in the labeled data, nor consider the manifold structure presented in the unlabeled data. Therefore, the motivation of our approach is to jointly explore the merits of both graph-based semi-supervised learning and sparse coding methods.

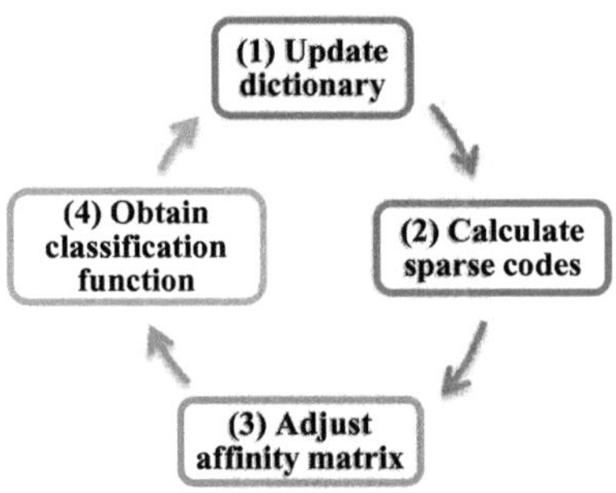

(a) Iterative semi-supervised sparse coding (ISSC)

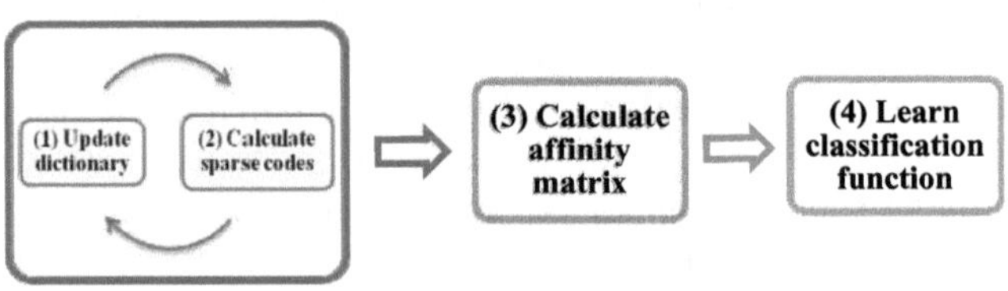

(b) The naive algorithm of first sparse coding and then graph-based semi-supervised learning without iterative step

Fig. 1. The framework comparison between two algorithms

In this paper, we presented a novel learning algorithm called Iterative Semi-supervised Sparse Coding (ISSC), which couples sparse coding and graph-based semi-supervised learning into a unified iterative framework, as shown in Fig. 1(a). A joint cost function is optimized to learn the smooth and discriminative sparse codes as well as a better classification function in a semi-supervised manner. Thus, the local neighborhoods relationships of both labeled and unlabeled data can be maintained while the discriminative property of labeled data is exploited.

Different from the naive algorithm of first sparse coding and then graph-based semi-supervised learning without iterative step [8] in Fig. 1(b), the proposed ISSC approach iterates between the two stages: sparse coding stage and graph-based semi-supervised learning stage. In the sparse coding stage we make good use of the initial labels and the subsequently predicted labels to learn sparse codes. During the graph-based semi-supervised learning stage, firstly the similarities between the data are adjusted based on their latest learned sparse codes, and then the updated similarity matrix is used to obtain a better classification function. In particular, through solving the quadratic optimization, the proposed ISSC approach can give rise to closed-form solution for the learned sparse codes.

In order to evaluate our proposed algorithm, extensive experiments are carried out over the well-known UIUC-Sport [9] and LabelMe [10] datasets for image classification task. We also compare our algorithm with several related supervised and semi-supervised learning approaches. The experimental results illustrate that our proposed learning framework outperforms other state-of-the-art image classification methods.

The remainder of this paper is organized as follows. Section 2 gives a brief review of some previous works that are closely related to this paper. In section 3, we present the iterative semi-supervised sparse coding algorithm. Experimental results are shown in section 4. Section 5 draws the conclusion on this paper.

2 Related Work

In this section, we will briefly introduce the work that is related to the approach proposed in this paper.

2.1 Sparse Coding

Sparse coding represents the data as a linear combination of a few basis vectors in the dictionary. Given only unlabeled input data, sparse coding algorithm tries to learn a suitable subspace to capture the underlying data manifold over which distinct class become more separable. One of the most important properties of sparse coding is that it usually produces a sparse representation for the original data. Such a sparse representation encodes the data using few basis vectors, which makes the encoding easy to interpret and at the same time saves the storage and computation time.

Give a dataset $X = \{x_1, \cdots, x_n\} \subset R^m$, the optimization formula to learn sparse codes is:

$$\min_{A,B} \sum_{i=1}^{n} \left(\|x_i - B\alpha_i\|^2 + \lambda\varphi(\alpha_i) \right) \quad \text{s.t. } \|b_j\|^2 \le 1, \forall j = 1, \cdots, p \tag{1}$$

where, $B = [b_1, \cdots, b_p] \in R^{m\times p}$ is the basis set, also called dictionary; $A = [\alpha_1, \cdots, \alpha_n] \in R^{p\times n}$ is the corresponding sparse codes set for X, and most entries in A are zeros; λ is regularization parameter; $\varphi(\alpha_i)$ is regularization function.

Usually, $\varphi(\alpha_i)$ is set to be L_1 penalty function $\|\alpha_i\|_1$, because L_1 penalty yields a sparse solution for α_i and can be robust to irrelevant features [11]. But it does not consider the inherent domain knowledge embedded in the data itself.

A more efficient regularization function $\varphi(\alpha_i)$ is using the locality-constrained linear coding [7]:

$$\varphi(\alpha_i) = \alpha_i^T E_i \alpha_i \tag{2}$$

where, E_i is a $p \times p$ diagonal matrix with its (j, j)-element equal to $\exp(\|x_i - b_j\|_2 / \delta)$ and δ is used to adjust the weight decay speed for the locality constraint.

The optimization problem in Eq.(1) is not jointly convex for basis set B and sparse codes A simultaneously. But it is convex for B when A is fixed, and is also convex for A when B is fixed. Therefore, the problem in Eq.(1) is usually solved by minimizing the objective function with respect to A and B alternatively.

2.2 Graph-Based Semi-Supervised Learning

Semi-supervised learning is designed to fully utilize the large amount of unlabeled data together with the limited labeled data. The key to semi-supervised learning is based on the prior assumption of consistency, also called cluster assumption, which states that: (1) nearby points are likely to have the same label; and (2) points on the same structure (typically referred as a cluster or a manifold) are likely to have the same label [3]. The basic idea behind the graph-based semi-supervised learning is label propagation on graph with the cluster consistency, which means two data points on the same structure (cluster or manifold) are likely to have the same class label.

Graph has proved to be successful in characterizing pairwise data relationship and manifold exploration. For a dataset $X = \{x_1, \cdots, x_n\} \subset R^m$, a weighted undirected graph G=(V, E, W) can be constructed to characterize the pairwise similarities among data points, where V is the vertices set and each node v_i corresponds to a data point x_i; E is the set of edges; W is the set of weights measuring the strength of the pairwise similarity.

Given a dataset $X = \{x_1, \cdots, x_\ell, x_{\ell+1}, \cdots, x_n\} \subset R^m$ and a label set $\mathcal{L} = \{1, \cdots, c\}$, the first ℓ data $x_i (i \le \ell)$ are labeled as $y_i \in \mathcal{L}$ and the remaining data $x_i (\ell + 1 \le i \le n)$ are unlabeled. To predict the label of the unlabeled data through graph-based semi-supervised learning, the cost function associated with classification function F can be defined as:

$$\min_{F} \sum_{i,j=1}^{n} W_{ij} \|F_i - F_j\|^2 + \gamma \sum_{i=1}^{n} \|F_i - Y_i\|^2 \tag{3}$$

where, $\gamma > 0$ is the regularization parameter; Let $\mathcal{F}$ denote the set of $n \times c$ matrices with nonnegative entries. Matrix $F = [F_1^T, \cdots, F_n^T]^T \in \mathcal{F}$ corresponds to a classification on the dataset X by labeling each point x_i as the label $y_i = \arg\max_{j \le c} F_{ij}$. The $n \times c$ matrix $Y \in \mathcal{F}$ is defined with $Y_{ij} = 1$ if x_i is labeled as $y_i = j$ and $Y_{ij} = 0$ otherwise. Clearly, Y is consistent with the initial labels according to decision rule [3].

3 Iterative Semi-Supervised Sparse Coding

The proposed Iterative Semi-supervise Sparse Coding (ISSC) couples sparse coding and graph-based semi-supervised learning into a unified framework in order to learn discriminative sparse codes as well as an effective classification function. It makes the best use of the merits of both graph-based semi-supervised learning and sparse coding.

For the dataset $X = \{x_1, \cdots, x_\ell, x_{\ell+1}, \cdots, x_n\} \subset R^m$, our goal is to learn the sparse codes and predict the label of the unlabeled data. Therefore, the joint cost function is defined by synergistically combining locality constraint sparse coding in Eq. (1) & Eq.(2) and discriminative cost function in Eq.(3) as follows:

$$\begin{aligned} \min_{A,B,F} & \sum_{i=1}^{n} (\|x_i - B\alpha_i\|^2 + \lambda \alpha_i^T E_i \alpha_i) + \\ & \sum_{i,j=1}^{n} W_{ij} \|F_i - F_j\|^2 + \gamma \sum_{i=1}^{n} \|F_i - Y_i\|^2 \\ \text{s.t. } & \|b_j\|^2 \le 1, \forall j = 1, \cdots, p \end{aligned} \tag{4}$$

As introduced in above Section 2, in Eq.(4), $B = [b_1, \cdots, b_p] \in R^{m \times p}$ is the dictionary; $A = [\alpha_1, \cdots, \alpha_n] \in R^{p \times n}$ is the sparse codes set for X; $\lambda > 0, \gamma > 0$ are the regularization parameters; E_i is a $p \times p$ diagonal matrix with its (j, j)-element equal to $\exp(\|x_i - b_j\|_2 / \delta)$ and δ is used to adjust the weight decay speed for the locality constraint; F is a classification function; Y is the initial label matrix with $Y_{ij} = 1$ if x_i is labeled as $y_i = j$ and $Y_{ij} = 0$ otherwise; The similarity matrix W is constructed using the well-known Gaussian kernel, which is usually defined by Eq.(5):

$$k_{ij} = \begin{cases} \exp(-\|x_i - x_j\|^2 / 2\sigma^2), & i \ne j \\ 0, & i = j \end{cases} \tag{5}$$

In Eq.(4), the first term denotes the reconstruction error; the second term is the locality constraint regularization; the third term represents the smoothness constraint, which means that a good classification function should not vary too much between nearby points. The fourth term stands for the fitting constraint, which means a good classification function should not vary too much from the initial label assignment [3]. Note that the fitting constraint term contains labeled and unlabeled data.

The Eq. (4) is not jointly convex for sparse codes A, dictionary B, and classification function F simultaneously. But it is convex for one of parameters when the other two are fixed. Similar to [11], we can minimize the above objective with respect to A, B and F alternatively when holding the other two fixed.

The following describes the four basic steps in the iterative learning process:

(I) When A and F are fixed, the joint cost optimization problem with respect to dictionary B is transformed into Eq.(6). It can be solved using incremental codebook optimization method [7]:

$$\begin{aligned} &\min_B \sum\nolimits_{i=1}^{n} (\|x_i - B\alpha_i\|^2 + \lambda \alpha_i^T E_i \alpha_i) \\ &\text{s.t. } \|b_j\|^2 \leq 1, \forall j = 1, \cdots, p \end{aligned} \tag{6}$$

(II) When B and F are fixed, Eq.(4) can be reduced to:

$$\min_A \sum\nolimits_{i=1}^{n} (\|x_i - B\alpha_i\|^2 + \lambda \alpha_i^T E_i \alpha_i) + \sum\nolimits_{i,j=1}^{n} W_{ij} \|F_i - F_j\|^2 \tag{7}$$

We can optimize the Eq.(7) with respect to each α_i alternatively and do not consider all the sparse codes $A = [\alpha_1, \cdots, \alpha_n]$ simultaneously. That is, when we focus on α_i, the other sparse codes $\alpha_j (j \neq i)$ are fixed. Then we can get the analytical closed-form solution of Eq. (7) with respect to α_i as follows:

$$\begin{aligned} \alpha_i^* =& \arg\min_{\alpha_i} \|x_i - B\alpha_i\|^2 + \lambda \alpha_i^T E_i \alpha_i + \sum\nolimits_{j=1}^{n} W_{ij} \|F_i - F_j\|^2 \\ =& (2B^T B + 2\lambda E_i - \frac{1}{\sigma^2} \sum\nolimits_{j=1}^{n} (W_{ij} \|F_i - F_j\|^2) I)^{-1} \cdot \\ & (2B^T x_i - \frac{1}{\sigma^2} \sum\nolimits_{j=1}^{n} (W_{ij} \|F_i - F_j\|^2 \alpha_j)) \end{aligned} \tag{8}$$

where, I is $p \times p$ identity matrix.

(III) During the semi-supervised learning stage of the iterative framework, we will first adjust the affinity matrix $W = [W_{ij}]_{n \times n}$ based on the latest sparse codes $A = [\alpha_1, \cdots, \alpha_n]$ by the following Eq. (9). Then the updated affinity matrix W is used to learn classification function F.

$$W_{ij} = \begin{cases} \exp(-\|\alpha_i - \alpha_j\|^2 / 2\sigma^2), & i \neq j \\ 0, & i = j \end{cases} \tag{9}$$

(IV) When A and B are fixed, the problem with respect to F is:

$$\min_F \sum\nolimits_{i,j=1}^{n} W_{ij} \|F_i - F_j\|^2 + \gamma \sum\nolimits_{i=1}^{n} \|F_i - Y_i\|^2 \tag{10}$$

The optimal classification function F in Eq. (10) can be obtained through the following Algorithm 1.

Algorithm 1. Graph-based Label Propagation

Input: Initial label matrix Y; Affinity matrix W.
Output: The optimal classification function F.

(i) Construct the propagation matrix $P = W$.
(ii) Iterate $F_{t+1} = \beta P F_t + (1 - \beta)Y$ until convergence.

return the optimal classification function $F = \lim_{t\to\infty} F_t$.

The above four steps will be repeated until convergence. Based on the previous analysis, the Iterative Semi-supervised Sparse Coding (ISSC) algorithm is summarized in Algorithm 2.

Algorithm 2. Iterative Semi-supervised Sparse Coding

Input: Dataset $X = \{x_1, \cdots, x_\ell, x_{\ell+1}, \cdots, x_n\} \subset R^m$. Label set $\mathcal{L} = \{1, \cdots, c\}$. The first ℓ data $x_i (i \leq \ell)$ are labeled as $y_i \in \mathcal{L}$ and the remaining data $x_i(\ell + 1 \leq i \leq n)$ are unlabeled.
Output: Labels for the unlabeled data in the dataset X.
Repeat
(i) Compute dictionary $B = [b_1, \cdots, b_p] \in R^{m\times p}$ in Eq.(6) using incremental codebook optimization method [7].
(ii) Learn each sparse codes α_i through Eq.(8).
(iii) Adjust affinity matrix $W = [W_{ij}]_{n\times n}$ for semi-supervised learning using latest sparse codes by Eq.(9).
(iv) Calculate the classification function F using Algorithm 1.
Until convergence
return the label of each unlabeled data $x_i(\ell + 1 \leq i \leq n) : y_i = \arg\max_{j\leq c} F^*_{ij}$.

In the ISSC algorithm, the analytical closed-form solution for sparse codes α_i in Eq. (8) includes the terms related to classification function F, which provide the supervisory information about the initial labels and the subsequently predicted labels. In Eq. (9), the affinity matrix $W = [W_{ij}]_{n\times n}$ is first updated using the latest learned sparse codes α_i and then used in graph-based semi-supervised learning. Therefore, in the jointly learning framework, sparse coding and graph-based semi-supervised learning are effectively integrated. In particular, the analytical closed-form solution for sparse codes α_i can be generated through solving the quadratic optimization.

4 Experimental Results

In this section, the proposed ISSC method will be evaluated on the widely used datasets: UIUC-Sport [9] and LabelMe [10] datasets. We first describe the experimental setup, including information about the two image datasets and implementation details. Furthermore, our method is compared with other related supervised and semi-supervised learning methods on the two datasets.

4.1 Experimental Setup

UIUC-Sport dataset contains totally 1792 images falling into 8 different sports: badminton, bocce, croquet, polo, rock climbing, rowing, sailing and snow boarding. The number of images ranges from 137 to 250 per category. In the LabelMe dataset, to be consistent with the work of Wang et al. [12], we used the images from the following 8 classes: highway, inside city, tall building, street, forest, coast, mountain, and open country. We also only utilized images with 256×256 pixels. The total image number is 1600.

Our implementations utilized a single descriptor type, the popular SIFT descriptor, as in [6][13]. The SIFT descriptors extracted from 16×16 pixel patches were densely sampled from each image in the whole dataset on a grid with step size of 8 pixels. In order to train the codebooks, we employed the standard k-means algorithm to quantize the features. The spatial pyramid representation [13] was utilized to preserve the spatial information. The images were all preprocessed into gray scale, and were resized to be no larger than 300×300 pixels with preserved aspect ratio.

In the experiments, we set $\beta = 0.01$, locality constraint parameter $\lambda = 0.15$. The other parameters are set to be their respective optimal values according to the 3-fold cross validation. We employ classification accuracy to evaluate the performance of these approaches. Following the common benchmarking procedures, the final classification accuracy is obtained by averaging the results over 10 different trials.

4.2 Performance Evaluation

UIUC-Sport Dataset. Following the standard evaluation procedure [14], we randomly select 70 images from each category as labeled data, and randomly select another 60 images from each category as unlabeled data. The average classification accuracies computed on the unlabeled images over UIUC-Sport dataset are shown in Table 1.

LabelMe Dataset. Similar to [15], we randomly select 100 images per category as labeled data and randomly select another 100 images per category as unlabeled data. The average object recognition accuracies computed on the unlabeled images over LabelMe dataset are reported in Table 2.

Moreover, we also list the confusion matrices of the UIUC-Sport and LabelMe datasets in Fig. 2 and Fig. 3, respectively. In each confusion matrix, the entry located in i-th row, j-th column stands for the percentage of class i being misclassified to class j. From Table 1 and Table 2, it can be observed that the proposed ISSC approach outperforms other existing state-of-the-art approaches. The reason is that, by simultaneously discovering sparse data manifold and balancing

Table 1. Performance comparison on UIUC-Sport dataset (%)

Method	Classification Accuracy
Li [9]	73.40
MlGSC [15]	76.63
AC [16]	81.53
ScSPM [6]	82.74
HIK+ocSVM [17]	83.54
KSRSPM [14]	84.92
LScSPM [18]	85.31
LLC [7]	85.34
ISSC	**86.50**

Table 2. Performance comparison on LabelMe dataset (%)

Method	Classification Accuracy
SGLIBA	75.13
ScIBA	75.24
McsLDAA	76.00
MlGSC	76.24
KmlGSC	82.94
ISSC	**85.75**

class discrimination, the local neighborhoods relationships of both labeled and unlabeled data can be better maintained in the proposed ISSC approach.

To further evaluate the effectiveness of the proposed ISSC, we also compare ISSC with the naive approach of first sparse coding and then graph-based semi-supervised learning without iterative step [8]. The average classification results are demonstrated in Fig. 4. Compared with the naive approach of first sparse coding and then graph-based semi-supervised learning without iterative step, the ISSC approach can achieve up to 4% (on UIUC-Sport) and 6% (on LabelMe) of performance improvement, respectively. That is, owing to fully exploiting the prior information of labeled and unlabeled data in an iterative learning framework, ISSC can perform better in discriminative manifold and classification function learning than the naive approach of first sparse coding and then graph-based semi-supervised learning without iterative step.

4.3 Convergence

We also evaluate the convergence of the ISSC algorithm on UIUC-Sport dataset. According to the experimental results, the ISSC algorithm converges rapidly in less than six iterations on average. One of the results during the convergence test of the ISSC algorithm is shown in Fig. 5.

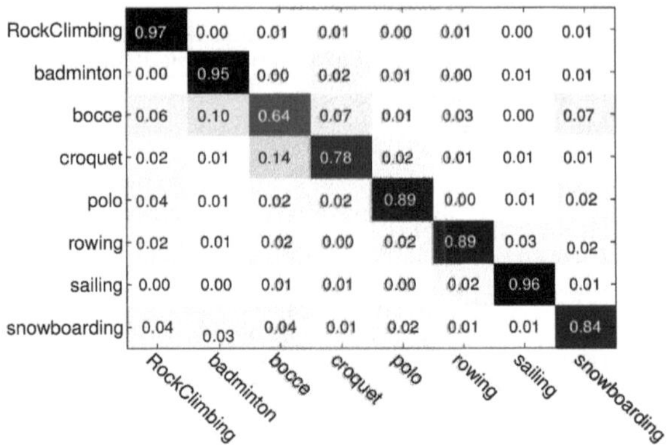

Fig. 2. Confusion matrix on UIUC-Sport dataset

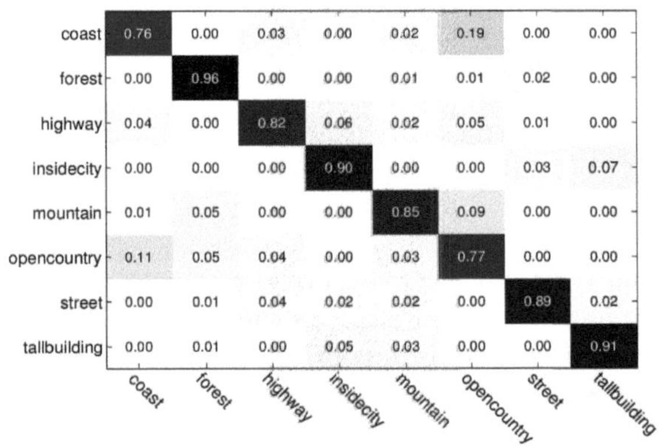

Fig. 3. Confusion matrix on LabelMe dataset

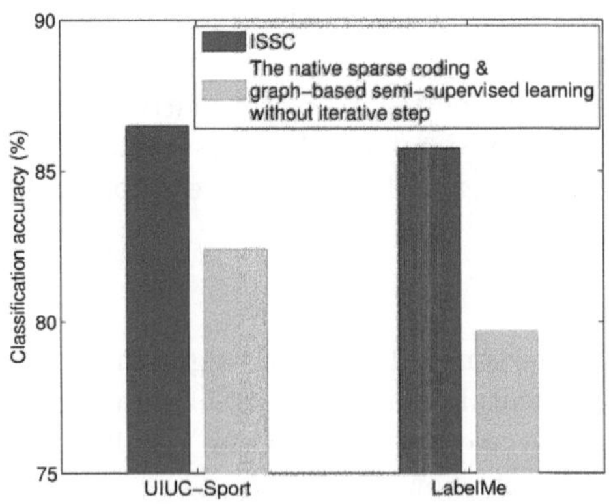

Fig. 4. Performance comparison between the proposed ISSC and the native sparse coding and graph-based semi-supervised learning without iterative step

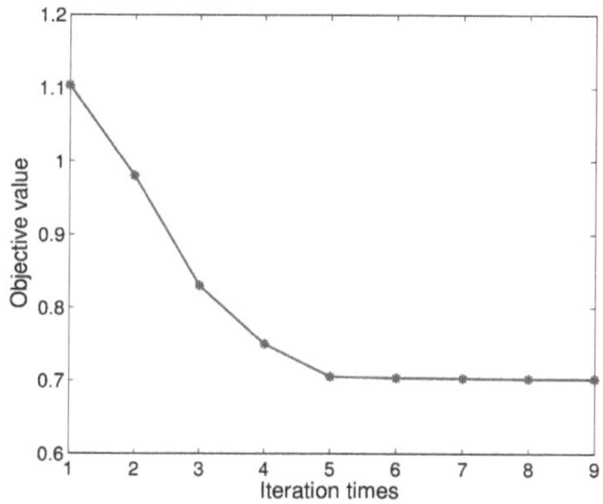

Fig. 5. Convergence test of the proposed ISSC algorithm on UIUC-Sport dataset

5 Conclusions

The scarcity of labeled data and the high-dimensionality of features are the major obstacles for image classification. Semi-supervised learning leverages the large amount of unlabeled data in developing effective classification function. Sparse coding provides an optimal sparse representation for data to help classification. This paper proposed a novel algorithm, Iterative Semi-supervised Sparse Coding (ISSC), which utilizes the advantages of both graph-based semi-supervised learning and sparse coding in order to learn effective sparse codes as well as a discriminative classification function in a joint learning framework. The promising experimental results evaluated in the well-known datasets clearly demonstrate the effectiveness of the proposed ISSC method. Furthermore, although the proposed ISSC algorithm is evaluated on image classification task in this paper, we can also extend it to other image and video tasks such as annotation and retrieval.

References

1. Olivier, C., Bernhard, S., Alexander, Z.: Semi-Supervised Learning. MIT Press (2006)
2. Blum, A., Mitchell, T.: Combining labeled and unlabeled data with co-training. In: Proceedings of the Eleventh Annual Conference on Computational Learning Theory, pp. 92–100. ACM, New York (1998)
3. Dengyong, Z., Olivier, B., Thomas, N.L., Jason, W., Bernhard, S.: Learning with Local and Global Consistency. In: Advances in Neural Information Processing Systems, vol. 16. MIT Press (2003)
4. Mairal, J., Elad, M., Sapiro, G.: Sparse Representation for Color Image Restoration. Trans. Img. Proc. 17, 53–69 (2008)
5. Shi, Y., Qi, N., Yin, B., Ding, W.: Two dimensional K-SVD for the analysis sparse dictionary. In: Lin, W., Xu, D., Ho, A., Wu, J., He, Y., Cai, J., Kankanhalli, M., Sun, M.-T. (eds.) PCM 2012. LNCS, vol. 7674, pp. 861–871. Springer, Heidelberg (2012)

6. Jianchao, Y., Kai, Y., Yihong, G., Thomas, S.H.: Linear spatial pyramid matching using sparse coding for image classification. In: Computer Vision and Pattern Recognition, pp. 1794–1801 (2009)
7. Jinjun, W., Jianchao, Y., Kai, Y., Fengjun, L., Thomas, S.H., Yihong, G.: Locality-constrained Linear Coding for image classification. In: The Twenty-Third IEEE Conference on Computer Vision and Pattern Recognition, pp. 3360–3367. IEEE Press (2010)
8. Zheng, H., Ip, H.H.S., Tao, L.: Adjacency Matrix Construction Using Sparse Coding for Label Propagation. In: Fusiello, A., Murino, V., Cucchiara, R. (eds.) ECCV 2012 Ws/Demos, Part III. LNCS, vol. 7585, pp. 315–323. Springer, Heidelberg (2012)
9. Li-Jia, L., Fei-Fei, L.: What, where and who? Classifying events by scene and object recognition. In: ICCV, pp. 1–8 (2007)
10. Aude, O., Antonio, B.T.: Modeling the Shape of the Scene: A Holistic Representation of the Spatial Envelope. International Journal of Computer Vision 42, 145–175 (2001)
11. Honglak, L., Alexis, B., Rajat, R., Andrew, Y.N.: Efficient sparse coding algorithms. In: Proceedings of the Twentieth Annual Conference on Neural Information Processing Systems, pp. 801–808. MIT Press (2006)
12. Chong, W., David, M.B., Fei-Fei, L.: Simultaneous image classification and annotation. In: IEEE Computer Society Conference on Computer Vision and Pattern Recognition, pp. 1903–1910 (2009)
13. Lazebnik, S., Schmid, C., Ponce, J.: Beyond Bags of Features: Spatial Pyramid Matching for Recognizing Natural Scene Categories. In: Computer Vision and Pattern Recognition, pp. 2169–2178. IEEE Computer Society, Washington (2006)
14. Shenghua, G., Tsang, I.W., Liang-Tien, C.: Sparse Representation With Kernels. IEEE Transactions on Image Processing 22, 423–434 (2013)
15. Shenghua, G., Liang-Tien, C., Tsang, I.W.H.: Multi-layer group sparse coding-For concurrent image classification and annotation. In: CVPR, pp. 2809–2816 (2011)
16. Subhransu, M., Alexander, C.B.: Max-margin additive classifiers for detection. In: IEEE 12th International Conference on Computer Vision, pp. 40–47 (2009)
17. Jianxin, W., Rehg, J.M.: Beyond the Euclidean distance: Creating effective visual codebooks using the Histogram Intersection Kernel. In: IEEE 12th International Conference on Computer Vision, pp. 630–637 (2009)
18. Shenghua, G., Wai-Hung, I.T., Liang-Tien, C.: Laplacian Sparse Coding, Hypergraph Laplacian Sparse Coding, and Applications. IEEE Trans. Pattern Anal. Mach. Intell. 35, 92–104 (2013)
19. Jerome, F., Trevor, H., Robert, T.: A note on the group lasso and a sparse group lasso. arXiv:1001.0736v1 (2010)
20. John, W., Allen, Y.Y., Arvind, G., Shankar, S.S., Yi, M.: Robust Face Recognition via Sparse Representation. IEEE Trans. Pattern Anal. Mach. Intell. 31, 210–227 (2009)

Cross-Codebook Image Classification

Shuai Liao[1,2], Xirong Li[1,2], and Xiaoyong Du[1,2,3]

[1] Key Lab of Data Engineering and Knowledge Engineering, MOE, China
[2] School of Information, Renmin University of China, Beijing 100872, China
[3] State Key Lab of Software Development Environment, Beijing 100191, China
{leoshine,xirong,duyong}@ruc.edu.cn

Abstract. Representing images by bag of visual codes (BoVC) features has been the cornerstone of state-of-the-art image classification system. Since the BoVC features depend on a precomputed codebook in use, when the codebook applied to test images differs from the codebook of an existing image classification system, the system becomes inapplicable. To resolve the codebook incompatibility problem, we propose in this paper *cross-codebook* image classification. This is achieved by transforming BoVC features derived from one codebook to make them compatible with another codebook. Two BoVC transform methods, i.e., code-reassignment and least squares, are studied. Experiments on a popular image classification benchmark set show that both methods are better than random guess when crossing the codebooks. In particular, when the BoVC features are transformed from a higher dimension to a relatively small dimension, cross-codebook image classification has a similar performance compared to within-codebook image classification, with a relative performance loss of 1.3% only. The results justify the feasibility of the proposed cross-codebook image classification.

Keywords: Image classification, bag of visual codes, cross-codebook.

1 Introduction

Automatically classifying images into a set of predefined visual categories such as 'beach', 'building', and 'food' is crucial for organizing and retrieving the ever-growing amounts of images on personal devices and the Internet. Image classification is thus a focal topic in multimedia computing and applications.

The state-of-the-art in image classification is to learn Support Vector Machines classifiers from manually labeled examples represented by bag of visual codes (BoVC) features [1–3]. Since the seminal work by Csurka *et al.* [4] in 2004, BoVC features have been the *de facto* choice for image classification.

To extract a BoVC feature vector for a given image, a number of local descriptors, e.g., SIFT [5], are first extracted from image patches. These descriptors are then quantized by a precomputed *codebook*. The codebook is constructed by conducting k-means clustering on many local descriptors, with each cluster center corresponding to a specific visual code. Consequently, the given image is represented by a histogram with its dimensionality equal to the size of the codebook.

B. Huet et al. (Eds.): PCM 2013, LNCS 8294, pp. 497–504, 2013.

Each bin of the histogram corresponds to a visual code, and its value is the accumulated frequency of that code. Therefore, the extraction of BoVC features depend on the codebook in use.

When the codebook applied to extract BoVC from a test image set is different from the codebook with which a classification system is built, the existing system is inapplicable to classify the test images. We term this problem as *codebook incompatibility*, and conceptually illustrate it in Fig. 1.

One might argue to avoid the codebook incompatibility problem by sticking to one codebook. For instance, use the codebook of the existing classification system to re-extract BoVC features of the test data. However, this solution is unfeasible when the original test images are inaccessible because of privacy or copyright concerns, or they are simply no longer available due to varied reasons including the limit in storage. Consider for instance the NUS-WIDE [6], a popular dataset for consumer photo classification. Due to the copyright concern, the dataset releases BoVC features, without providing the original images. As a consequence, one cannot directly use this dataset to test the effectiveness of an image classification system at hand. In addition, as re-extracting BoVC features for large-scale datasets is cumbersome [3], a lightweight solution to codebook incompatibility is desirable.

In order to make BoVC features derived from one codebook compatible with another codebook, we propose in this paper cross-codebook image classification, as shown in Fig. 1. In particular, we study BoVC transform methods, given that the local descriptors are of the same type. Cross-codebook image classification is, to the best of our knowledge, non-existing in the literature.

The rest of the paper is organized as follows. We elaborate in Section 2 the proposed BoVC transform methods, followed by cross-codebook image classification experiments in Section 3. Conclusions are given in Section 4.

2 BoVC Transform for Cross-Codebook Classification

Given an image classification system based on one codebook C_s and a set of test images with their BoVC features w.r.t another codebook C_t, we cannot directly employ the system to classify the test data due to the incompatibility between C_s and C_t. For varied reasons as discussed in Section 1, re-extracting BoVC features of the test data using C_s or re-training the system with C_t might be infeasible. In that regard, we study how to transform the BoVC features of the test data into new features that are compatible with C_s, and thus can be accepted by the existing classification system.

2.1 Problem Formalization

To make our discussion more formal, we first introduce a few notations. Let $C_t = \{c_{t1}, c_{t2}, \ldots, c_{tk_t}\}$ be the codebook with which BoVC features of the test data are extracted, where c_{ti} denotes the vector of the i-th code, and k_t is the size of the codebook. In a similar fashion, we denote the codebook for the existing image classification system as $C_s = \{c_{s1}, c_{s2}, \ldots, c_{sk_s}\}$.

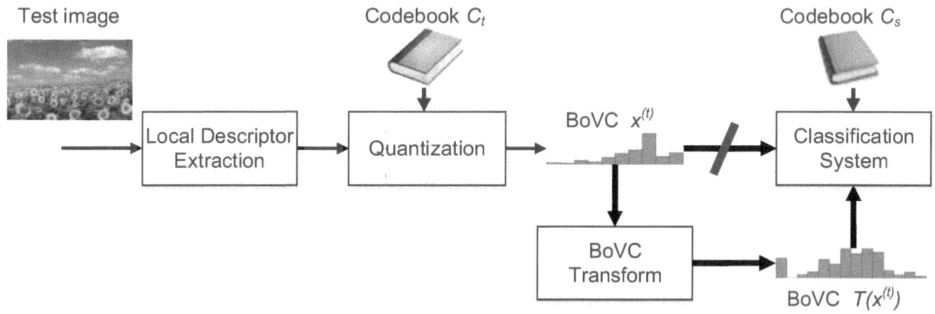

Fig. 1. Illustrating the codebook incompatibility problem in bag-of-visual-codes (BoVC) based image classification. When the codebook (C_t) applied to a test image differs from the codebook (C_s) upon which a classification system is built, the existing system is inapplicable to classify the test image. In order to make BoVC features w.r.t C_t compatible with C_s, we propose in this paper cross-codebook image classification.

Let x be a specific image. Its BoVC feature quantized by C_t is denoted by $x^{(t)} = (x_1^{(t)}, \ldots, x_{k_t}^{(t)})$, and $x^{(s)} = (x_1^{(s)}, \ldots, x_{k_s}^{(s)})$ indicates the BoVC feature quantized by C_s. For a given category ω, let $H(x^{(s)}, \omega)$ be its classifier. As aforementioned, this classifier is built on the base of the codebook C_s, so it cannot handle $x^{(t)}$.

To resolve the codebook incompatibility problem, we look for a *transform* function $T : \mathcal{R}^{k_t} \rightarrow \mathcal{R}^{k_s}$ such that classification with the new feature $T(x^{(t)})$ would have similar performance compared to classification with $x^{(s)}$, i.e.,

$$\arg_T sign(H(T(x^{(t)}), \omega)) = sign(H(x^{(s)}, \omega)), \tag{1}$$

where $sign : \mathcal{R} \rightarrow \{1, -1\}$ is a decision function.

Notice that Eq. (1) states an ideal case of BoVC transform. In this paper we instantiate the transform function T as a linear mapping for its simplicity and efficiency. In particular, we define

$$T(x^{(t)}) := x^{(t)} A, \tag{2}$$

where A is a $k_t \times k_s$ matrix. By doing so, cross-codebook image classification boils down to finding the matrix A.

Next, we give two BoVC transform methods, which compute A by visual code re-assignment and by least squares, respectively.

2.2 BoVC Transform Methods

As we have discussed in Section 1, if the codebook C_t is simply a re-order of the codebook C_s, codebook incompatibility is resolved by re-assigning the i-th bin of $x^{(t)}$ to the bin corresponding to c_{sj}, the code in C_s that is the closest to c_{ti}.

In this case, A in Equation (2) is a binary matrix. In practice, however, such a hard re-assignment method would lose much information. To that end, we first introduce a *soft re-assignment* method which generates the new feature $T(x^{(t)})$ by allocating each dimension of $x^{(t)}$ in terms of visual code similarities.

Method I: Code Re-assignment. Given two visual codes c_{ti} and c_{sj}, let $d(c_{ti}, c_{sj})$ be the Euclidean distance. To convert the distance into a similarity score, we use a Gaussian function, and consequently the similarity function between the two codes is defined as $\mathcal{K}(c_{ti}, c_{sj}) = \exp(-\frac{d(c_{ti}, c_{sj})}{\rho})$. The bandwidth parameter ρ is empirically set to be the mean of the distances of all code pairs.

In the code re-assignment method, all dimensions of $x^{(t)}$ can contribute to each dimension of $T(x^{(t)})$, with more importance given to closer codes. In particular, we compute the j-th dimension of the transformed feature as

$$T_j(x^{(t)}) = \sum_{i=1}^{k_t} x_i^{(t)} A_{ij}, \tag{3}$$

where the matrix element A_{ij}, computed as

$$\frac{\mathcal{K}(c_{ti}, c_{sj})}{\sum_{l=1}^{k_s} \mathcal{K}(c_{ti}, c_{sl})}, \tag{4}$$

determines the amount of c_{ti} that is assigned to the bin corresponding to c_{sj}.

As shown in Eq. (4), the code re-assignment method only utilizes the two codebooks to compute A, without explicitly considering the connection between $T(x^{(t)})$ and $x^{(s)}$. Given that extracting $x^{(s)}$ from the test data is unfeasible, as an alternative, we leverage an auxiliary image set for which BoVC features quantized by distinct codebooks can be extracted. In that regard, we propose the second *least square* method as follows.

Method II: Least Squares. In order to distinguish the auxiliary set from the test set in consideration, we use $\mathbf{z}$ to denote an auxiliary image, and let $\{\mathbf{z}_1, \ldots, \mathbf{z}_m\}$ be the set of m auxiliary images. Assuming that the BoVC feature $\mathbf{z}^{(s)}$ can be fully reconstructed from $\mathbf{z}^{(t)}$ by A, we have a system of m linear equations, i.e.,

$$\mathbf{z}_i^{(s)} = A \cdot \mathbf{z}_i^{(t)}, \qquad i = 1, \ldots, m. \tag{5}$$

When $m > k_t$, Eq. (5) is mostly an overdetermined system, meaning none exist of an exact solution. Therefore, we resort to least squares, a classical approach to approximated solutions of overdetermined systems [7]. Accordingly, we define the objective function as

$$\underset{A}{\operatorname{argmin}} \sum_{i=1}^{m} ||A \cdot \mathbf{z}_i^{(t)} - \mathbf{z}_i^{(s)})||^2. \tag{6}$$

Notice that the least squares method requires no labeled examples, so the auxiliary set can be obtained, e.g., from the Internet, with ease.

3 Empirical Study

3.1 Experimental Setup

Data sets. We use the ImageCLEF 2010 development set [8], a leading benchmark for image classification. The set consists of two prespecified subsets, one subset with 5,000 images for training, and the other subset with 3,000 images for testing. All the images were collected from Flickr, with ground truth available for 93 visual concepts. To resolve the optimization problem as defined in Equation (6), we downloaded a number of Flickr images as the source of *unlabeled* auxiliary data, without using its annotations.

BoVC feature Extraction. To extract the BoVC feature for each image, we use the color descriptor software [3] to extract the SIFT descriptor in a dense manner. In order to simulate a cross-codebook scenario, the descriptors of a training image are quantized by the codebook C_s, while the descriptors of a test image are quantized by the codebook C_t different from C_s.

Image Classification. We adopt the fast intersection kernel SVMs (FikSVMs) for its effectiveness and efficiency [9]. The key technique for FikSVMs comes from the observation that the decision function of intersection kernel SVMs can be expressed as the sum of decision functions with respect to individual feature dimensions. For each dimension, its decision function can be efficiently computed by linear interpolation on a limited set of precomputed points. As a consequence, classification boils down to a few table-lookup operations, and thus becomes very fast. In this work, we use the implementation of FikSVM by Li *et al.* [10].

Notice that for most of the 93 concepts, their positive training examples are in minority. Hence, to address class imbalance for training SVMs, the positive class and the negative class are assigned with different cost parameters in light of the reciprocal of their distribution in the training data.

Evaluation Criteria. For each concept, we report the widely used Average Precision. To measure the overall performance, we use mean Average Precision (mAP).

3.2 Experiments

To verify the effectiveness of the two BoVC transform methods, the codebook C_t and the codebook C_s were independently generated by performing k-means clustering on random subsets of SIFT descriptors extracted from the training set. For comparison, we build a within-codebook image classification system using C_s only. We also build a random guess run, which ranks the test set at random. We take the mean score of multiple random guess runs. While random guess with an mAP of 0.129 is a performance lower bound, the within-codebook system establishes a performance upper bound on cross-codebook image classification.

We construct a number of distinct codebooks, by executing k-means clustering multiple times on different sets of local descriptors. Depending on whether the size of C_t is equal to the size of C_s, we divide our experiments into two parts, that is, BoVC transform with $k_t = k_s$ and BoVC transform with $k_t \neq k_s$.

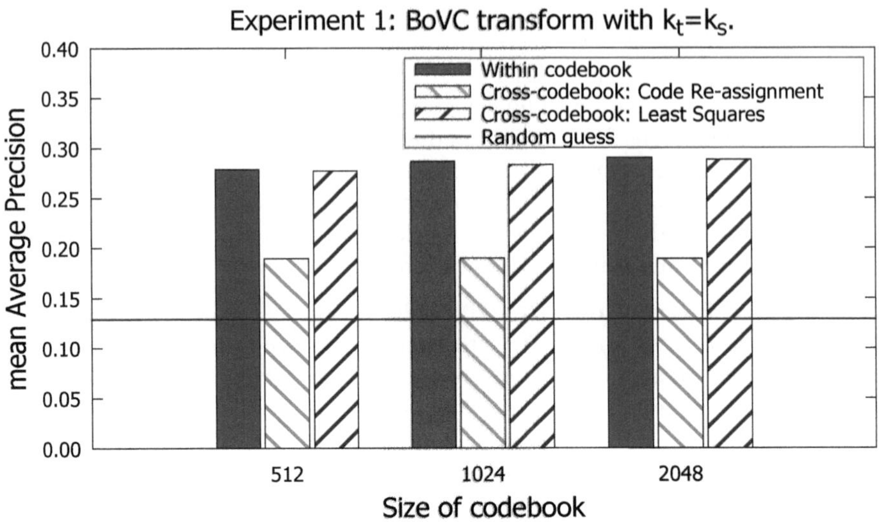

Fig. 2. Experiment 1: BoVC transform with $k_t = k_s$. With the least squares BoVC transform method, the cross-codebook image classification system is comparable to the within-codebook system.

Experiment 1: BoVC Transform with $k_t = k_s$. In this experiment, BoVC features derived from the two distinct codebooks are of the same dimensionality. We try codebooks of varied sizes by choosing k_t and k_s from {512; 1,204; 2,048}.

As shown in Fig. 2, both the code re-assignment method and the least squares method outperform the random guess run. Comparing the two methods, least squares is clearly better than code re-assignment. We attribute this to the fact that by exploiting extra data, the least squares method finds a better mapping between BoVC features derived from C_s and C_t. Moreover, using the mapping found by solving least squares, the cross-codebook classification system is comparable to the within-codebook system, with a relative performance loss of 1.3% at most. From the above results we conclude that when the two codebooks are of equal size, the codebook incompatibility problem can be well resolved by the least squares method, and consequently cross-codebook image classification is doable.

Experiment 2: BoVC Transform with $k_t \neq k_s$. To study BoVC transform between codebooks of different sizes, we fix k_s to be 512, and choose k_t from {64; 256; 1,024; 2,048; 4,096}.

As shown in Fig. 3, the least squares method is again better than the code re-assignment method. When $k_t < k_s$, the performance of both methods improves as k_t increases, but is lower than the within-codebook run, which has an mAP of 0.289. This is because we are mapping BoVC features of the test data from

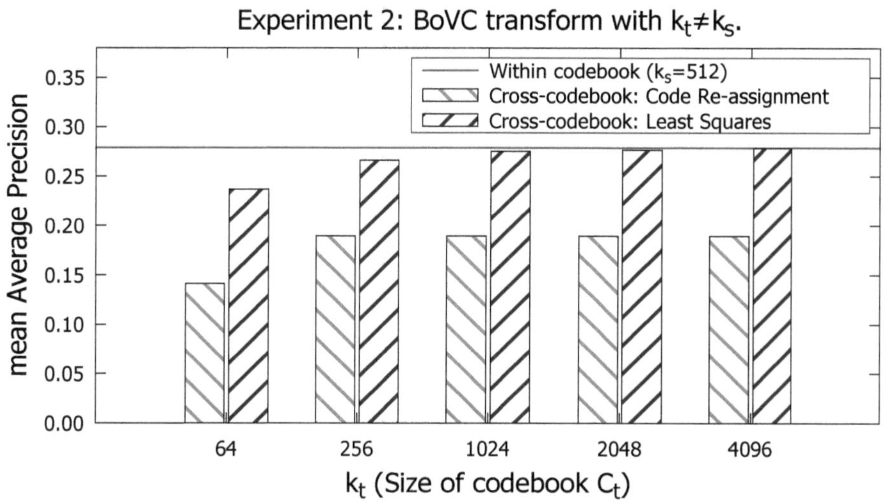

Fig. 3. Experiment 2: BoVC transform with $k_t \neq k_s$. We fix the size of C_s to be 512, and vary the size of C_t. When the size of C_t is larger than the size of C_s, the codebook incompatibility problem can also be well addressed by least squares BoVC transform.

a lower dimensions, e.g., 64 or 256, to the higher dimension, i.e., 512, but the transform methods do not bring in novel information about the original images.

When $k_t > k_s$, while the performance of the code re-assignment method remains lower than the performance of the within-codebook run, the cross-codebook runs with the least squares method are close to the within-codebook run. Our explanation is that while mapping BoVC features of the test data from a higher dimension, e.g., 2,048 or 1,024 to the lower dimension, i.e, 512, also causes information loss, the transformed BoVC features seem adequate for the existing image classification system. Therefore, when the size of C_t is larger than the size of C_s, the codebook incompatibility problem can also be well addressed by least squares BoVC transform.

4 Conclusions

In this paper we extend the research on bag of visual codes (BoVC) based image classification in a new dimension, namely cross-codebook image classification. When the codebook applied to test images differs from the codebook of an existing image classification system, the system is inapplicable to classify the test images. To tackle the codebook incompatibility problem, we study two BoVC transform methods, i.e., code re-assignment and least squares.

From experiments on the ImageCLEF 2010 development set we draw the following conclusions. While both methods are better than random guess, the least squares method is superior to its code re-assignment counterpart. When the size of the codebook of the test data is no smaller than the size of the codebook of the existing classification system, cross-codebook classification with BoVC transformed by the least squares method is found to be comparable to within-codebook classification, with a relative performance loss of 1.3% at most. Given local descriptors of the same type, cross-codebook image classification is feasible, and the least squares method is our recommendation.

For future work, it will be interesting to evaluate the effectiveness of the proposed BoVC transformation methods for other image analysis tasks such as image retrieval and copy detection. Another challenging extension is to consider cross-codebook image classification for local descriptors of different types.

Acknowledgments. This research was supported by the Basic Research funds in Renmin University of China from the central government (13XNLF05), State Key Laboratory of Software Development Environment Open Fund (SKLSDE-2012KF-09), and National Natural Science Foundation of China (No. 61303184).

References

[1] Perronnin, F., Akata, Z., Harchaoui, Z., Schmid, C.: Towards good practice in large-scale learning for image classification. In: CVPR, pp. 3482–3489 (2012)

[2] Jiang, Y.G., Yang, J., Ngo, C.W., Hauptmann, A.: Representations of keypoint-based semantic concept detection: A comprehensive study. IEEE Transactions on Multimedia 12(1), 42–53 (2010)

[3] van de Sande, K., Gevers, T., Snoek, C.: Empowering visual categorization with the gpu. IEEE Transactions on Multimedia 13(1), 60–70 (2011)

[4] Csurka, G., Dance, C., Fan, L., Willamowski, J., Bray, C.: Visual categorization with bags of keypoints. In: ECCV Workshop on Statistical Learning in Computer Vision, vol. 1, p. 22 (2004)

[5] Lowe, D.: Distinctive image features from scale-invariant keypoints. International Journal of Computer Vision 60(2), 91–110 (2004)

[6] Chua, T.S., Tang, J., Hong, R., Li, H., Luo, Z., Zheng, Y., Zheng, Y.: Nus-wide: a real-world web image database from national university of singapore. In: CIVR, p. 48 (2009)

[7] Hastie, T., Tibshirani, R., Friedman, J.: The Elements of Statistical Learning: Data Mining, Inference, and Prediction. Springer (2001)

[8] Nowak, S., Huiskes, M.: New strategies for image annotation: Overview of the photo annotation task at Image CLEF 2010. In: CLEF (2010)

[9] Maji, S., Berg, A.C., Malik, J.: Classification using intersection kernel support vector machines is efficient. In: CVPR, pp. 1–8. IEEE (2008)

[10] Li, X., Snoek, C., Worring, M., Koelma, D., Smeulders, A.: Bootstrapping visual categorization with relevant negatives. IEEE Transactions on Multimedia 15(4), 933–945 (2013)

Spatio-temporal SURF for Human Action Recognition

Sameh Megrhi, Wided Souidène, and Azeddine Beghdadi

L2TI, Institut Galilée, Université Paris 13, Sorbonne Paris Cité
{sameh.megrhi,wided.mseddi,azeddine.beghdadi}@univ-paris13.fr

Abstract. In this paper, we propose a new spatio-temporal descriptor called ST-SURF. The latter is based on a novel combination between the speed up robust feature (SURF) and the optical flow. The Hessian detector is employed to find all interest points. To reduce the computation time, we propose a new methodology for video segmentation into Frame Packets (FPs), based on the interest points trajectory tracking. We consider only moving interest points descriptors to generate robust and powerful discriminative codebook based on K-mean clustering. We use a standard bag-of-visual-words Support Vector Machine (SVM) approach for action recognition. For the purpose of evaluation, the experimentations are carried out on KTH and UCF sports Datasets. It is demonstrated that the designed ST-SURF shows promising results. In fact, on KTH Dataset, the proposed method achieves an accuracy of 88.2% which is equivalent to the state-of-the-art. On the more realistic UCF sports Dataset, our method surpasses the performance of the best results of space-time descriptors/Hessian detector with 80.7%.

Keywords: Action recognition, SURF, optical flow, spatio-tempral feature, group of interest points, frames packets.

1 Introduction

In user-generated video footage, the quantity of video data containing human actions and scenes is growing exponentially (about 48 hours of video uploaded per minute on YouTubeTM) [1]. With this growth the demand for action and scene recognition or content-based video data retrieval is, certainly, colossal. Usually, considerable events are characterized by actions, for example, boxing, kissing, or some stealthy actions or behavior in a surveillance video, examples of which are shown in Fig. 1.

Recognizing human actions from videos is receiving increasing attention due to its wide range of applications such as video indexing and retrieval [2], human-computer interaction, digital entertainment, surveillance videos [3] etc. However, action recognition is usually confronted to many issues, including the necessity of handling considerable occlusions, scale changes, illumination, and the existence of background clutter, as well as viewpoint changes. In the context of action recognition in video, the representation of video objects as a bag of visual words through a histogram has become a very active research field [4].

B. Huet et al. (Eds.): PCM 2013, LNCS 8294, pp. 505–516, 2013.

Fig. 1. Samples representing three classes of human actions: boxing, kissing and running

This histogram can be used in classifier framework to make the difference between object's classes. However, the main weakness of a given bag of visual words is that, not all words will be informative, accurate and objective in terms of describing actions. Consequently, the selection of the most informative words is required. The most used methods to select visual words are using Machine learning techniques Boosting [5] or adaptation process such as Multiple Instance Learning (MIL) [6] or many other State-of-the-art algorithms [7], [8]. As long as these approaches proved significant results for action recognition, they need to be adapted to be applied into the temporal domain, for action recognition or data retrieval. Recent studies in both the spatial [9] and temporal [10] domains explore the descriptive and discriminative performances of these features. In particular, spatio-temporal local features have been widely studied as visual features to detect human actions, objects and events in videos. Although, video analysis with spatio-temporal features is not new, but has not been much explored yet [11], [12]. To extract spatio-temporal features, one of the most used methods is local cuboid. Dollar et al. [13] and Laptev et al. [14] extract Histogram of Gradient (HoG) and Histogram of Flow (HoF) respectively from a cuboid. In this context, we propose in this paper a novel spatio-temporal feature based on the SURF [15] local descriptor. The proposed method is based on detecting spatio-temporal interest points. We extract the descriptor by extending the original SURF descriptor to a 3D spatio-temporal space. These descriptors are then quantized by K-means clustering and each Video clip is represented as a histogram with K bins. SVM is then used for classification. We propose a novel codebook based on spatio-temporal descriptors called Bag of Spatio-temporal Visual Words BoSTVW. We prove experimentally that this contribution outperforms other state-of the-art approaches on the increasingly complex and popular KTH Dataset [16] and UCF sports Dataset [17].

2 Related Works

Inspired from the text retrieval community, the 'bag of words' BoW, has recently become popular for image [18] and video analysis [19]. In action recognition's state-of-the-art, the BoW models were widely used since they have shown the effectiveness of local appearance based descriptors [20]. However, in comparison with other approaches, Bag of visual word selection is still in its infancy. To extract video descriptors, many researchers have been investigating in tracking major parts of human bodies then extracting features from these regions [21].

However, they need to setup many hypothesis. These considerations and hypothesis are often difficult. So that, methods based on spatio-temporal features are promising for action recognition. Some of them are based on the extraction of low-level optical flows from cuboids [22] this method gives good results in terms of feature selection and a good classiffications accuracy [22]. But they presents limits concerning the long computational time they require. Dollar et al. detect local cuboids to apply 1-D Gabor filters in the temporal direction and 2-D Gaussian kernels in the spatial space [13], and they produce video visual words based on vector-quantizing in the same way as bag-of-visual-words for object recognition [18]. In the same direction, Laptev et al proposed STIP (Spatio-Time Interest Points) to detect cuboids [14]. This method is considered as an extension of Harris detector. Nevertheless, the limits of the aforementioned methods not only concerns the hardness of finding the best cuboid size, but also the high computational requirements. To overcome these problems, we propose to detect interest points using SURF/Hessian [15]. Then we segment the videos into Groups of interest points (GIP) and Frame Packets (FPs) to reduce the computation time. We use Sun, D at al. [23] optical flow detection methods which allows to extract spatio-temporal SURF by tracking interest points instead of cuboids.

Fig. 2. Example of SURFs found using Hessian detectors on diffrent frames from UCF sports dataset

3 The Proposed Approach for Action Recognition

The proposed method aims at detecting human actions, to reach this goal, first video sequences are segmented in Frame Packets (FPs) and Group of Interest Points (GIP). Second, based on a novel combination between optical flow computed by [23] and **S**patio-**T**emporal SURF (Speeded-Up Robust Feature) [15], the interest points **ST-SURF** are localized and extracted, from all training video FPs. Then, the extracted ST-SURFs are clustered using K-means clustering algorithm. The video clips are represented as a K-bins histogram of the quantized descriptors 'bag of spatiotemporal visual words' BoSTVW. Finally, an SVM classfier is trained using these histograms *(One vs all)*.

3.1 Frame Packets (FPs) and Group of Interest Points (GIP) Segmentation

To be able to achieve accurate and fast computation, our algorithm does not use all the frames available in a video in order to extract its descriptors.

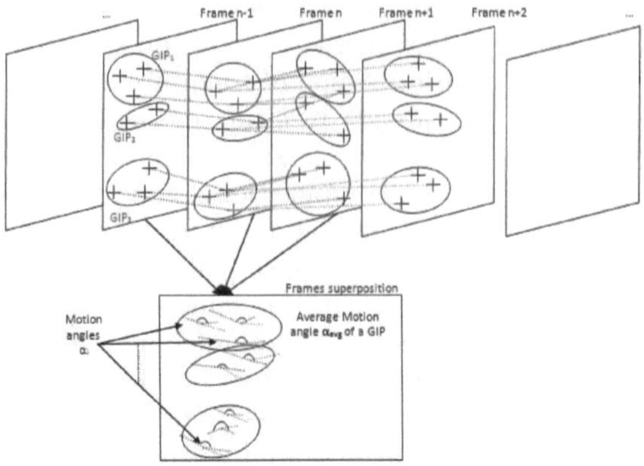

Fig. 3. IPs trajectory tracking for FPs segmentation

Instead, we have created and used the concepts of Frame Packets (FPs) and Group of Interest Points (GIP). We assume that, between three successive frames (n-1, n and n+1), an interest point (from one picture to another) can have three possible states: still, moving and disappear. The first and last states are not treated because ,in the first case, no motion is detected. In the last case the interest point IP disappeares and cannot be tracked anymore. The second state is the one that concerns us the most, since there is a displacement and we can track the displacement angle. From now on, we assume that α is the angle between the lines segments supporting the motion of an IP from the couple of frames (n-1, n) and (n, n+1), see Figure 3 for details. By comparing α to α_{max} (a parameter fixed experimentally at the beginning of the processing) we are able to segment a succession of frames, that we call here Frame Packets (FPs), in which each IP has an α lower than α_{max}. By calibrating this angle of tolerance, we are able to certify that, within this FP, all IPs movements are within this tolerance parameter. This means that we cannot miss any significant movement likely to influence the remaining computing. In order to be able to have more control over the size (in number of frames) of the FP, we introduce, the concept of Group of Interest Points (GIP). In fact, a NGIP is a parameter defining the number of IP that must be grouped together (this constant has to be experimentally finetuned to find the most suitable N). This grouping is performed over successive IP in a frame. By defining this number NGIP we can compute an average angle (α_{avg}) for a certain GIP and compare it to the α_{max}. The higher NGIP is the less the α_{avg} will be sensitive to motion and the more the FP will contain frames. Here are the steps of our segmentation algorithm. Let us suppose that we are beginning the computation of a new FP:

- We extract the IP of the frames one and two.
- We define the GIPs based on the NGIP parameter fixed at the beginning of the algorithm.

- We compute the line supporting the motion for each corresponding IP within these two frames.
- We apply the above three steps to the frames two and three.
- We compute the angle between each motion line and we extract the average angle for each GIP.
- We compare each average angle to the α_{max} (fixed at the beginning of the algorithm).
- We continue performing the above six steps over the next frames (taking, always, the first motion direction as reference to all remaining comparisons) until finding an average angle of a GIP higher than the maximum angle. In this case we can define the FP and assume, with confidence, that the first and last frames of this FP can fully describe the motion within.

3.2 Interest Points Extraction

In the following, we present the used interest point detector followed by a description of the feature we extract. For interest points detection we choose the Hessian detector [24]. It searches for image locations that exhibit strong derivatives in two orthogonal directions. It is based on the matrix of second derivatives, the so-called Hessian [24]. In 2004, Lowe [15], presented SIFT for extracting invariant features from images that can be robust against image scale and rotation. Then it was widely used in image, recognition and retrieval etc. However, extracting robust features approaches are very slow. Bay et al. speeded up robust features by using integral images for image convolutions and Fast-Hessian detector [15]. Their experiments turned out that SURF was faster and it works well. We use the extraction solution given by [15] to extract interest point feature. This choice is motivated by the robustness, the smaller size of this feature and their excellent performances attested in various datasets for action recognition [25]. The SURF feature is a 64-D vectors that describes spatial patterns around detected points. Refer to [15]for the detail.

3.3 Surf Tracking into 3D Feature Space

Features' tracking is performed by estimating optical flow. To increase optical flow estimation accuracy, many researches are inspired from the Horn and Schunck (HS) Optical flow formulation [23], [26]. In fact, they focuses on optimizing an objective function which combines the image's properties and its spatial motion prediction. Sun end al. proposed a new algorithm to approximate an optimized computationally tractable objective function, based on the original HS formulation. They first, use median filtering to denoise the flow, Exploiting connections between median filtering and L1-based denoising. They proved that algorithms relying on a median filtering step are approximately optimizing a different objective that regularizes the flow over a large spatial neighbourhood [23]. The resulting algorithm ranks 1st in both angular and end-point errors in the Middlebury evaluation in March 2010 [23]. In our work, we considered every Frame Packet as a volume of frames in the 3D space called FP Volume (FPV), this cubic volume is characterized by its frames' number(FN) going from 1 to t_{max}, its frames' surfaces dimensions (FS) and its center ($FPVc$). A given interest point $IP = (x, y, t)$ is defined by its position (x, y) and its frame t. In frame $(t+n)$, the IP moves by a displacements u in the x direction, and v in the y direction. IP becomes, $IP(t+n) = (x+u, y+v, t+n)$. In all our experiments, unless mentioned otherwise, we assume that due to the video segmentation into FPs, the motion vectors trajectory remain stable. For stagnant interest points $u = v = 0$. Thus, in the FPV, the 3D direction (u, v, n) represent the direction of the IP motion. The motion vector is calculated by the Sun et al. [37] optical flow approach. Our contribution consists on the use of motion orientation and position to characterize the motions, instead of using the direction vector (u, v, n) generated from optical flow computation. We suppose that the motion vector in the 3D space can be defined as the intersection of two planes perpendicular respectively to the plane (t, x) and the plane (t, y). This parameterization is one among several possible representations of 3D lines [27]. To extract IP orientation, we project its motion vectors onto the planes (t, x) and (t, y) of the FPV to define an angle for each projection, the first angle α_x between optical flow and the plane (t, x), the angle α_y between the plane (t, y) and the motion vector.

$$\propto_x = 90 - \frac{180}{\Pi} arctan(u), \propto_y = 90 - \frac{180}{\Pi} arctan(v). \quad (1)$$

For each IP, we project its motion vector onto the planes (t, x) and (t, y) and obtain two lines L_x and L_y. The orthogonal projection of $FPVc_x$ and $FPVc_y$ onto the lines L_x and L_y allows the computing of both distances D_x and D_y between the cube center and the lines supporting the motion vectors (L_x and L_y).

For an IP located at (x, y, t):

$$D_x = D_{xu} - D_{tv}, D_y = D_{yv} - D_{tu} \quad (2)$$

where

$$D_{xu} = (x - x_{max}/2)cos(180/\Pi arctan(u)) \quad (3)$$

$$D_{tv} = (t - t_{max}/2)sin(180/\Pi arctan(v)) \quad (4)$$

$$D_{yv} = (y - y_{max}/2)cos(180/\Pi arctan(v)) \quad (5)$$

$$D_{tu} = (t - t_{max}/2)sin(180/\Pi arctan(u)) \quad (6)$$

where t_{max}, x_{max} and y_{max} are the dimensions of the Frame Packet volume with t_{max} depend on the number of the frames into a segmented (FPV). In the following, D_x and D_y describe the motion distances of a given interest point. Fig. 4, is a graphical illustration of the cube center and its projection into the planes (t, x) and (t, y).

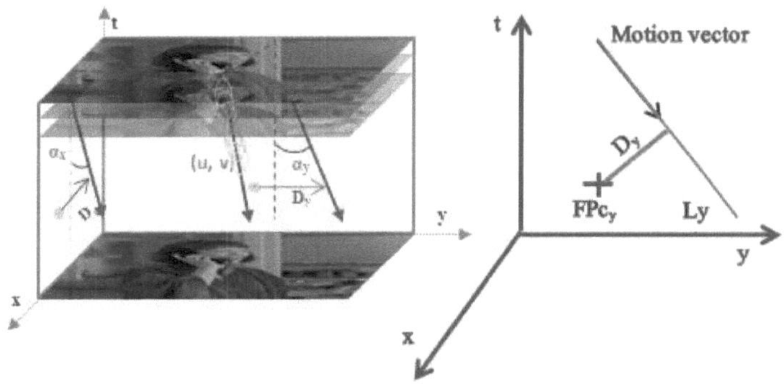

Fig. 4. The projection of a motion vector in the adjacents planes

ST-SURF Extraction: This step consists in the generation of the novel ST-SURF that we designed. This descriptor is represented by spatial feature 64-D vector, and temporal 4-D feature, we concatenate both vectors into one 68-D spatiotemporal descriptor vector, and thus we extend the image SURF descriptor [15] to videos. These features will tracks the interest point through time in each FP we defined. The size of a FP depends on its average frames number. In our work we consider only moving interest points (where $\propto_x \neq 0$ and $\propto_y \neq 0$).

ST-SURF Training Pipeline: After the extraction of ST-SURF descriptors, we define a spatio-temporal words dictionary. The basic idea is to assign a set of objects into groups so that the objects of similar type will be in one cluster, in order to construct a visual codebook, which can be used to represent an action, a scene or en object. Recently, the K-means algorithm has been widely used to construct the visual codebook because of its high performances and simplicity. Fig. 5, illustrate the training steps of UFC sports dataset's videos.

Evaluation Pipeline: After the extraction step, the generated ST-SURFs are quantized into visual words using k-means clustering. Each video sequence can then be represented as the frequency histogram over the visual words. Generally, using a large-sized codebook allows to obtain high recognition accuracy, yet an oversized codebook leads to high quantization errors. The resulting histograms of visual word occurrences are used as classification inputs. We use a non linear support vector machine to classify human actions.

4 Experiments

In the following we describe the datasets used for the evaluation of the proposed work. We evaluate the ST-SURF in a bag-of-features based action classification task and compare our approach to the state-of-the-art employ.

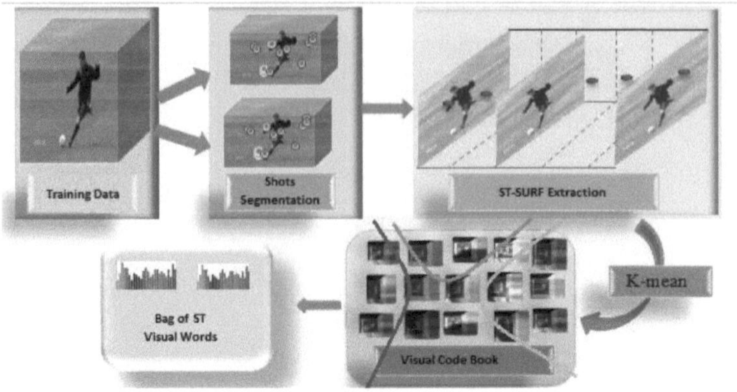

F

Fig. 5. Training pipeline

4.1 Experimental Setups and Data

Dataset: The proposed framework is tested on the KTH dataset [16] and UCF sports Dataset [17]. The KTH dataset is commonly used as a public benchmark test of spatio-temporal features [25]. This dataset contains six kinds of actions such as walking, running, jogging, boxing, hand waving and hand clapping. We consider 6 action classes by 25 persons in 4 different scenarios with a total of 2391 video samples. The average length of videos in the KTH dataset is about 20 second long. The second one is the UCF sports dataset, more realistic and challenging data obtained from broadcast sport videos by Ahmed et al. [17]. The collection represents a natural pool of actions featured in a wide range of scenes and viewpoints. The publicly available part of this dataset contains nine actions namely diving, golf, swinging, kicking, lifting, horseback riding, running, skating, swinging and walking. This dataset contains close to 200 video sequences at a resolution of 720x480 [17].

Parameter Settings: In all our experiments, we explored optimal parameter settings. We evaluate the classification rates of both KTH and UCF datasets while changing the codebook and the FPs sizes. The results shows that the empirically optimal size book is k = 4000 with $\alpha_{max} = 42\circ$ and NGIP = 38. These settings gave us empirically satisfactory results.

4.2 Experimental Results

KTH Dataset: From the recently reported results of the state-of-the-art, we can clearly conclude that using Hessian detector, Laptev and al. [10] obtained 88.7% using a combination of HOG (histograms of gradient orientations) and HOF (histograms of optical flow) descriptors, 88.6% using HOF, and 77.7% with HOG . We note that Kläser and al. achieved an accuracy of 84.6% using HOG3D descriptor, which is a comparable results with HOG/HOF [10]. The combination SURF/Hessian detector gives 84.6% for williems and 86% for Noguchi [25]. In the table. 1, the first row compares the **b**est **a**verage **a**ccuracy (**BAA**) for the differents detector/descriptor combinations reported by other researchers, on the KTH dataset. The **a**verage **a**ccuracy for **H**essian (**HAA**) detector/descriptor combinations on the KTH dataset are drawn in the second row.

UCF Sports Dataset: We note that Kläser and al. achieved an accuracy of 85% using HOG3D/Gabor descriptor, Laptev and al. [10] obtain 81.6%using HOG/HOF, 82.6% using HOF, and 77.4% with HOG. The first row, in Table. 2, compares the **b**est **a**verage **a**ccuracy (**BAA**) for the differents detector/descriptor combinations reported by other researchers, on the UCF sports dataset. The **a**verage **a**ccuracy for **H**essian (**HAA**) detector/descriptor combinations on the UCF sports dataset are drawn in the second row.

Table 1. Average accuracy for various detector/descriptor combinations on the KTH dataset

.	HOG3D	HOG/HOF	HOG	HOF	E-SURF	t-SURF	ST-SURF
BAA	90%	91.8%	82.3%	92.1%	81.4%	86%	**88.2%**
HAA	84.6%	88.7%	77.7%	88.6 %	81.4%	86%	**88.2%**

Table 2. Average accuracy for various detector/descriptor combinations on the UCF sports dataset

.	HOG3D	HOG/HOF	HOG	HOF	E-SURF	t-SURF	ST-SURF
BAA	85%	81.6%	77.4%	82.6%	77.3%	-	**80.7%**
HAA	78.9%	79.3%	66.0%	75.3%	77.3%	-	**80.7%**

Table 3. SURF confusion matrix action recognition on the KTH dataset

KTH	Boxing	clapping	Waving	Walking	Jogging	Running
Boxing	0.6	0.01	0.03	0.13	0.1	0.1
clapping	0.06	0.58	0.25	0.04	0.02	0.05
Waving	0.05	0.08	0.74	0.03	0.01	0.09
Walking	0.01	0	0	0.7	0.16	0.13
Jogging	0	0.01	0	0.12	0.58	0.29
Running	0	0	0	0.1	0.21	0.59

Tables 3-4 are the confusion matrices of the actions classification results based on two type of features. the first matrix describle the classification result for the visual SURF feature reported in [25]. The result among a single visual feature give bad classification results for all the actions. Based on Table. 4, confusion matrix show that the original combination, that we proposed, of both visual and motion features (ST-SURF) boosted significantly the classification accuarcy. Regarding the average result over each of the six actions' KTH dataset, ST-SURF

Table 4. ST-SURF confusion matrix action recognition on the KTH dataset

KTH	Boxing	clapping	Waving	Walking	Jogging	Running
Boxing	0.9	0.07	0	0.03	0	0
clapping	0. 07	0.9	0.03	0	0	0
Waving	0.01	0.06	0.95	0	0	0
Walking	0	0	0	0.91	0.06	0.3
Jogging	0	0	0	0.04	0.85	0.1
Running	0	0	0	0.06	0.14	0.8

Table 5. ST-SURF confusion matrix action recognition on the UCF sports dataset

UCF	Dive	Golf	Kick	Lift	Ride	Run	Skate	Swing	Walk
Dive	0.8	0.17	0	0	0	0	0	0.03	0
Golf	0	0.78	0.2	0	0	0	0	0	0.02
Kick	0	0	0.9	0	0	0.07	0	0	0.03
Lift	0	0	0	0.92	0	0	0	0.08	0
Ride	0	0	0.2	0	0.62	0.18	0	0	0
Run	0	0	0.02	0	0	0.88	0	0	0 .1
Skate	0	0	0.08	0	0	0	0.6	0	0.32
Swing	0	0	0	0	0	0	0.21	0.79	0
Walk	0	0	0	0	0	0.04	0	0	0.96

produced good result, however, less accuracy is observed in the jogging and running actions because thes actions are almost similar. lastly but not least, comparing with results driven by the best result of the state-of-the-art, our method achieve 88.2% better than the 86% reported by Noguchi and al. [25] using Spatio-temporal SURF. Outperforming the results of the Cuboids/HOG combinaison obtained by [14] 82.3% and the 81.4% reported by Willem and al. [28]. Based on the confusion matrix of UCF sports Dataset given in Table. 2, the ST-SURF outperform the best result driven by the state-of-the-art using Hessian detector and achieves 80.7% of accuracy. We note that ST-SURf/Hessian gave better results in realistic videos. We are still below the results driven by Laptev and al. with 85% using the HOG3D/Gabor combination, and their 91.8% reached using Harris3D/(HOG/HOF), this can be due to different codebook generation and the use of different interest points detectors. This motivates further investigations of different interest points detectors and realistic vidoe settings. Regarding all these results our method is equivalent to the state of the art, and shows significantly better performance, outperforming many results driven in the same setup.

5 Conclusion

In this paper, We have investigated a novel scheme to efficiently segment video sequences into a new concept we called Fram Packets. Then we proposed a novel spatio-temporal descriptor based spatio-temporal interest points. The designed descriptor is an extension of the SURF to the temporal domain. The proposed feature extraction consists on detecting of the Surf points and mapping them into a 3D feature space based on an original exploitation of the optical flow orientation and position. Only the moving SURF are then selected. The extracted features are embedded into a bag of visual word pipeline, to finally classify six actions from KTH Dataset and then nine actions from the UCF sport dataset. Furthermore, the proposed framework demonstrate promising recognition performance on tow standard benchmarks with the accuracy about 88.2% in KTH and 80.7% in UCF sports. We consider many perspectives for the future, the most important includes applying the same work to tackl more complex and realistic actions. We also plan to improve our ST-SURF and consider combining it with variety of low level features and use these data to video retrieval. We also look to investigate diffeterent interest points detectors and larger datasets in bag-of-visual-words based representations. Finally, the results we obtained demonstrate the viability of our approach and proove that we already are equivalent to the state-of-the-art performances.

References

1. Youtube: Statistiques@ONLINE (June 2009)
2. Jiang, Y.G., Ngo, C.W., Yang, J.: Towards optimal bag-of-features for object categorization and semantic video retrieval. In: Proceedings of the 6th ACM International Conference on Image and Video Retrieval, CIVR, pp. 494–501. ACM, New York (2007)
3. Hu, W., Tan, T., Wang, L., Maybank, S.: A survey on visual surveillance of object motion and behaviors. IEEE Transactions on Systems, Man, and Cybernetics, Part C: Applications and Reviews, 334–352
4. Willamowski, J., Arregui, D., Csurka, G., Dance, C.R., Fan, L.: Categorizing nine visual classes using local appearance descriptors. Illumination 21 (2004)
5. Viola, P., Jones, M.: Rapid object detection using a boosted cascade of simple features. In: Proceedings of the IEEE Computer Society Conference on Computer Vision and Pattern Recognition,CVPR, p–511. IEEE (2001)
6. Maron, O., Lozano-Pérez, T.: A framework for multiple-instance learning. In: Advances in Neural Information Processing Systems, 570–576 (1998)
7. Kim, T.K., Wong, S.F., Cipolla, R.: Tensor canonical correlation analysis for action classification. In: IEEE Conference on Computer Vision and Pattern Recognition, CVPR, pp. 1–8. IEEE (2007)
8. Lin, Z., Jiang, Z., Davis, L.S.: Recognizing actions by shape-motion prototype trees. In: IEEE 12th International Conference on Computer Vision, pp. 444–451. IEEE (2009)
9. Mikolajczyk, K., Tuytelaars, T., Schmid, C., Zisserman, A., Matas, J., Schaffalitzky, F., Kadir, T., Van Gool, L.: A comparison of affine region detectors. International Journal of Computer Vision 65(1-2), 43–72 (2005)
10. Wang, H., Ullah, M.M., Klaser, A., Laptev, I., Schmid, C.: et al.: Evaluation of local spatio-temporal features for action recognition. In: BMVC British Machine Vision Conference (2009)
11. Sameh, M., Wided, S., Beghdadi, A., Amar, C.B.: Video indexing using salient region based spatio-temporal segmentation approach. In: International Conference on Multimedia Computing and Systems, pp. 170–173. IEEE (2012)
12. Megrhi, S., Souidene, W., Beghdadi, A.: Spatio-temporal salient feature extraction for perceptual content based video retrieval. In: The Colour and Visual Computing Symposium 2013. IEEE (2012)

13. Dollár, P., Rabaud, V., Cottrell, G., Belongie, S.: Behavior recognition via sparse spatio-temporal features. In: IEEE International Workshop on Visual Surveillance and Performance Evaluation of Tracking and Surveillance, pp. 65–72. IEEE (2005)
14. Laptev, I., Lindeberg, T.: Local descriptors for spatio-temporal recognition. In: MacLean, W.J. (ed.) SCVMA 2004. LNCS, vol. 3667, pp. 91–103. Springer, Heidelberg (2006)
15. Bay, H., Tuytelaars, T., Van Gool, L.: SURF: Speeded up robust features. In: Leonardis, A., Bischof, H., Pinz, A. (eds.) ECCV 2006, Part I. LNCS, vol. 3951, pp. 404–417. Springer, Heidelberg (2006)
16. Schuldt, C., Laptev, I., Caputo, B.: Recognizing human actions: a local svm approach. In: Proceedings of the 17th International Conference on Pattern Recognition, ICPR, pp. 32–36. IEEE (2004)
17. Rodriguez, M.D., Ahmed, J., Shah, M.: Action mach a spatio-temporal maximum average correlation height filter for action recognition. In: IEEE Conference on Computer Vision and Pattern Recognition, CVPR 2008, pp. 1–8. IEEE (2008)
18. Csurka, G., Dance, C., Fan, L., Willamowski, J., Bray, C.: Visual categorization with bags of keypoints. In: Workshop on Statistical Learning in Computer Vision, ECCV, p. 22 (2004)
19. Brandão Lopes, A.P., Alves do Valle Jr., E., Marques de Almeida, J., Albuquerque de Araújo, A.: Action recognition in videos: from motion capture labs to the web. arXiv preprint arXiv:1006.3506 (2010)
20. Niebles, J.C., Wang, H., Fei-Fei, L.: Unsupervised learning of human action categories using spatial-temporal words. International Journal of Computer Vision (3), 299–318 (2008)
21. Mojarrad, M., Dezfouli, M.A., Rahmani, A.M.: Feature extraction of human body composition in images by segmentation method. World Academy of Science, Engineering and Technology (2008)
22. Fathi, A., Mori, G.: Action recognition by learning mid-level motion features. In: IEEE Conference on Computer Vision and Pattern Recognition, CVPR, pp. 1–8. IEEE (2008)
23. Sun, D., Roth, S., Black, M.J.: Secrets of optical flow estimation and their principles. In: IEEE Conference onComputer Vision and Pattern Recognition, CVPR, pp. 2432–2439. IEEE (2010)
24. Beaudet, P.R.: Rotationally invariant image operators. In: Proceedings of the International Joint Conference on Pattern Recognition, pp. 579–583 (1978)
25. Noguchi, A., Yanai, K.: A surf-based spatio-temporal feature for feature-fusion-based action recognition. In: Kutulakos, K.N. (ed.) ECCV 2010 Workshops, Part I. LNCS, vol. 6553, pp. 153–167. Springer, Heidelberg (2012)
26. Beghdadi, A., Mesbah, M., Monteil, J.: A fast incremental approach for accurate measurement of the displacement field. Elsevier Image and Vision Computing 21, 383–399 (2003)
27. Dementhon, D., Doermann, D.: Video retrieval of near-duplicates using κ-nearest neighbor retrieval of spatio-temporal descriptors. Multimedia Tools and Applications (3), 229–253 (2006)
28. Willems, G., Tuytelaars, T., Van Gool, L.: An efficient dense and scale-invariant spatio-temporal interest point detector. In: Forsyth, D., Torr, P., Zisserman, A. (eds.) ECCV 2008, Part II. LNCS, vol. 5303, pp. 650–663. Springer, Heidelberg (2008)

Human Action Recognition by Mining Discriminative Segment with Novel Skeleton Joint Feature

Weijia Zou[1], Baoyuan Wang[2], and Rui Zhang[1]

[1] Institute of Image Communication and Network Engineering, Shanghai Jiao Tong University, Shanghai, P.R. China
zwj.sjtu@gmail.com, zhang_rui@sjtu.edu.cn,
[2] Microsoft Research Asia, Beijing, P.R. China
baoyuanw@microsoft.com

Abstract. In this paper, we present a "key segment" mining approach for human action recognition. Our model is able to locate discriminative segments for action samples via multiple instance learning. Moreover, we propose a dynamic pooling approach to automatically find the optimal length of segment for each action sample. In addition, an effective feature is proposed for action recognition with 3D skeleton joints. It can effectively capture informative motion and shape cues of skeletons, and leads to a compact and discriminative representation. The experimental results validate the effectiveness of the proposed human action recognition method on two benchmark datasets (i.e., MSR Action3D and UCF-Kinect). Moreover, our method demonstrates superior accuracy than previous methods of using only skeleton data on MSR Action3D, and achieves the state-of-the-art performance on UCF-Kinect.

Keywords: action recognition, multiple instance learning, skeleton feature, dynamic pooling.

1 Introduction

Human action recognition plays an important role in many computer vision and multimedia applications such as video surveillance, human computer interfaces, sports video analysis and video retrieval. In this paper, we consider the problem of recognizing human actions from pre-segmented video sequences. Previous methods of addressing this problem can be divided into two major categories: One is to treat the sequence as a whole unit and use off-the-shelf classifiers to do classification after gaining a holistic representation of it [1,2]. The other one is to extract key pose frames by manual selection or automatic learning [3,4]. Both of these methods have some limitations. Considering the fact that many action classes share similar motions during some phases such as the onset phase, treating the sequence as a whole unit is not necessarily the optimal choice for gaining discriminative representation. On the other hand, it is usually difficult to distinguish some similar actions via only one or several key frames.

B. Huet et al. (Eds.): PCM 2013, LNCS 8294, pp. 517–527, 2013.

In this paper, in order to obtain a discriminative representation for action recognition, we regard the video clip as "bag of segments" instead of "bag of frames" through looking backward for a number of previous frames for each frame. We typically cast this problem as a multiple instance learning problem. Specifically, it's up to our learner to choose the most discriminative segment within the positive bag rather than manually annotated. Moreover, we propose a dynamic pooling approach to automatically determine the best segment length for each action sample rather than using fixed length.

Another contribution of our paper lies in a novel feature extracted from skeleton data. As the release of depth cameras such as Microsoft Kinect sensor, it becomes possible to use 3D depth data for action recognition. Based on the estimated 3D joint positions, we propose a novel skeleton joint feature which can effectively capture informative shape and motion cues of skeletons in video segments and leads to a compact and discriminative representation for human action recognition.

The rest of the paper is organized as follows. Section 2 presents the related work. Section 3 elaborates our method of mining discriminative segment, including the scheme on extracting feature representation of video frames from skeleton data and dynamic pooling approach. Experimental results and discussions are given in Section 4. Finally, Section 5 concludes this paper.

2 Related Work

Human action recognition has been extensively studied in the past decades and the recent survey of [5] provides a detailed overview of current approaches to our discussion. Many prior work [3, 4] have been dedicated to extract key pose frames in video sequences for action recognition. Our work can be viewed as an extension of key pose frames methods for a more discriminative representation. Hoai's work [6] is similar with our research on mining discriminative segments. But they focused on mining non-contiguous discriminative segments for weakly labeled time series, and the performance on human action recognition has not been evaluated. Schindler and Van Gool [7] investigate how many frames are required to make sense for action recognition and find that very short snippets (e.g., 1-7 frames) are sufficient for basic action recognition.

With the release of the Microsoft Kinect sensor and other depth cameras, many efforts [1, 8, 9] have been made to utilize the depth information and skeleton data for action recognition. Li *et al.* [8] characterize postures through sampling the bag of 3D points from the depth maps. Wang *et al.* [1] utilize fourier transform to extract feature insensitive to temporal misalignment and propose a local occupancy patters feature to capture the human-object interactions.

The systems presented in [2, 4, 10] also attempt to find discriminative features based on skeleton data for action recognition. [2] and [4] aim to capture the information of offset, posture, and motion. However, [2] uses all pairwise joint differences which result in redundant information and extremely high-dimensional feature vectors. [4] utilizes distances of all joint pairs between current frame x_t

and frame x_{t-10} to capture motion information which may be more suitable for key pose representation, but not the segment-based representation used in this paper, as shown in our experiments.

3 Mining Discriminative Action Segment

Similar to the approach for finding key pose frames in [4], we find the key segments based on a model obtained by multiple instance learning [11]. For each frame of action samples, we get a segment of length l through looking backward for previous $l-1$ frames. In the case that the number of pervious frames is less than $l-1$, we add proper amount of the action sample's first frame to the beginning of the segment to ensure its length. Then we regard each action sample as a bag of segments, and the amount of segments is equal to the length of action sample if we fix the segment length l. Here we start with fixed segment length, and discuss the case of dynamic length in section 3.2.

For solving this multiple instance learning problem, we introduce a latent variable h which indicates the end point's position of segments. Then an action sample x can be scored as follow:

$$f_\omega(x) = \max_h \ \omega^\top \Phi(x,h) \tag{1}$$

where ω is a vector of model parameters and $\Phi(x,h)$ is the segment representation given the latent variable h.

Let $D = (< x_1, y_1 >, ..., < x_n, y_n >)$ be a set of labeled training examples, where $y_i \in \{-1, 1\}$, we train ω by solving the following optimization problem:

$$\min_\omega \frac{1}{2} \parallel \omega \parallel^2 + C \sum_{i=1}^{n} \max(0, 1 - y_i f_\omega(x_i)) \tag{2}$$

where C is a predefined regularization parameter. We solve the problem of Equ. 2 via a "coordinate descent" approach [12] that alternates between inferring the latent variable h and optimizing the weight vector ω. We use the Liblinear toolbox [13] to solve the quadratic program in each iteration. For multi-class recognition, we train a model for each action class according to "one-versus-all" manner. For testing an action sample, we compute its score of learned model according to Equ. 1 for each candidate class, and then classify it into the one with the highest score.

For each instance of a bag (i.e. a segment of frames), we utilize "*max pooling*" operation, which has shown its significance for robust representation in image classification [14–16], to get its representation from the features of frames. Suppose an action sample x_i with m frames $\{x_i^1, ..., x_i^m\}$ has a feature vector $\phi(x_i^j)$ for each frame x_i^j $(j = 1, 2, ..., m)$. Given the segment length l and the end point of the segment h, we can obtain the segment-level feature representation as:

$$\Phi(x_i, h) = \max\{\phi(x_i^{h-l+1}), ..., \phi(x_i^h)\}. \tag{3}$$

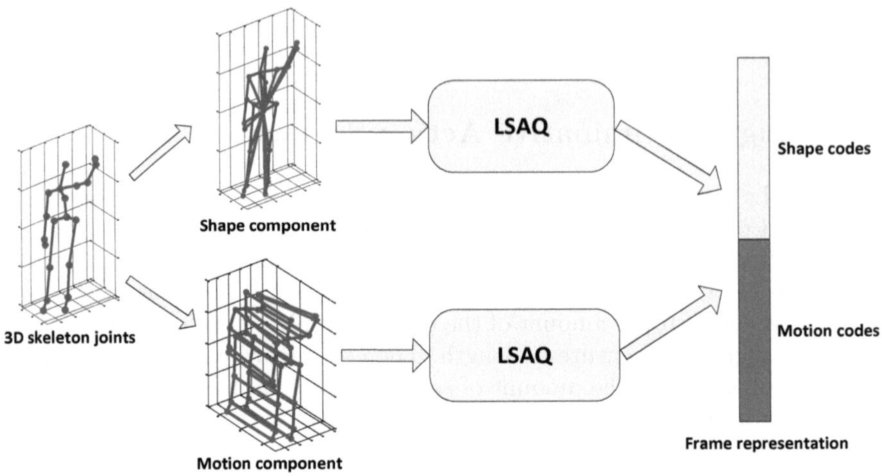

Fig. 1. Illustration on the scheme of computing frame representation based on the proposed skeleton joint feature (Best viewed in color)

3.1 Frame Representation

Given the 3D coordinates of joints, we propose a novel skeleton joint feature by jointly considering the motion and shape cues, which can produce a discriminative representation of video frame for human action recognition. Fig. 1 illustrates the scheme of computing frame representation based on the proposed skeleton-based feature.

Motion Component. Suppose each frame of an action sample has n joints, we can get the motion feature of current frame by subtracting the coordinates of the corresponding joints in previous frame from the coordinates of joints in current frame. As shown in Fig. 1, the resultant $3 * n$ dimensional vector, which is the concatenation of n 3-D displacement vectors, can capture the relative motion between current frame and previous one.

Shape Component. To capture the shape information, we first compute the center point's position through averaging the coordinates of n joints. Then, by subtracting the coordinates of the center point from n joints, we can get a $3 * n$ dimensional vector as the shape feature of current frame. It roughly describes the shape of skeleton by considering the relative positions of all joints w.r.t. object center.

After extracting motion and shape components of skeleton joint features, we build a codebook for each component respectively. Then we adopt the localized soft assignment coding (LSAQ) [15] to perform feature coding for motion and shape components respectively. Finally, the feature representation of each frame can be obtained by concatenating the code vectors of both components.

3.2 Dynamic Pooling

In this subsection, we extend our framework from the fixed length pooling to a dynamic length pooling. As shown in Fig. 2, the same motion procedure in different action samples may have distinct lengths due to people's different speeds when performing the action. Therefore, the fixed length of segment may not be capable of locating optimal temporal extension of object for action recognition. On the basis of above considerations, we propose a dynamic-length pooling method for mining more discriminative segment, which is adaptive to different action samples in practice.

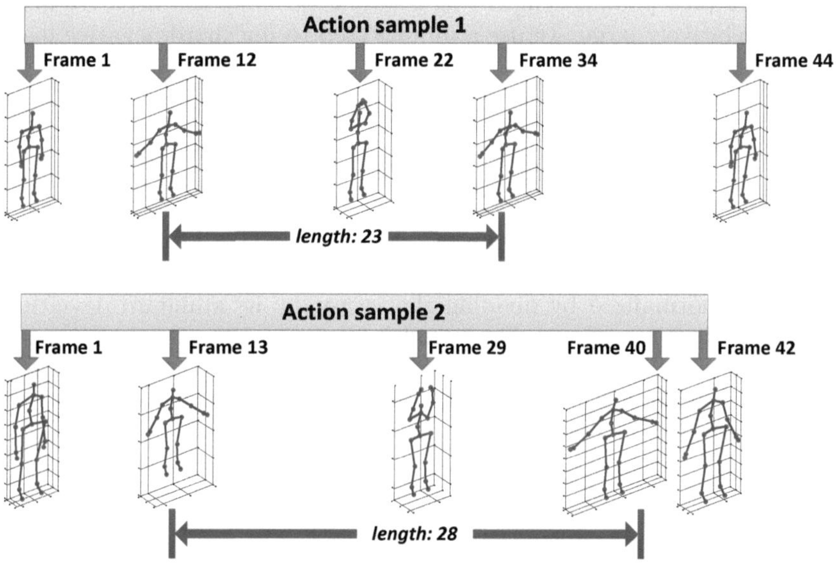

Fig. 2. This figure shows two action samples of "Two hand wave" performed by different subjects. The discriminative motion procedures of rising and falling of hands in these two samples have different lengths. (Best viewed in color)

Given an action sample of length n, for fixed length segment, we only need to choose h from n candidate segments:

$$h^* = \underset{h \in \{h_1, \dots h_n\}}{\arg\max} \ \omega^\top \Phi(x, h) \tag{4}$$

As we allow the dynamic length of the segment, it needs to select a best segment length l. To automatically determine the best pooling length l, we regard pooling length l as a latent variable. Then Equ. 1 changes to the form:

$$f_\omega(x) = \max_{h,l} \ \omega^\top \Phi(x, h, l) \tag{5}$$

where $l \in [l_{min}, l_{max}]$ and the form of $\Phi(x, h, l)$ is the same with formula (3).

To solve this problem, we first select the best pooling length l_i for each h_i:

$$l_i = \underset{l \in [l_{min}, l_{max}]}{\arg\max} \quad \omega^\top \Phi(x, h_i, l), \quad \forall i = 1, ...n \tag{6}$$

then we choose the optimal h_i from $\{h_1, ...h_n\}$ with corresponding l_i, i.e.

$$i = \underset{i \in \{1,...n\}}{\arg\max} \ \omega^\top \Phi(x, h_i, l_i). \tag{7}$$

Finally, we obtain the optimal segment position h_i and pooling length l_i for action sample x.

As we introduce pooling length l as a latent variable, the negative instance set become relatively large, we use a mining step to find hard negative instance as [12] in training stage.

4 Experiments

In this section, we demonstrate the superiority of our proposed method through a series of experiments on two datasets—MSR Action3D dataset [8] and UCF-Kinect dataset [4]. In our experiments, both the motion component and the shape one are normalized by dividing the vector by its standard deviation. In the feature coding step, we build a codebook with 500 visual words via K-means for each component respectively. The β in LSAQ is fixed to 0.1 for the motion component and 1 for the shape component, while the number of nearest neighbors is set as 5 for both of them.

4.1 MSR Action3D Dataset

The MSR Action3D is a public dataset recorded by a depth sensor similar to the Kinect device. It contains 20 action classes performed by 10 subjects, and a total of 557 depth map sequences. Each frame in the dataset has 20 skeleton joints. This dataset contains both skeleton and depth information, but we only use skeleton data to extract our feature as discussed in section 3.1. In order to provide a fair comparison, we follow [8] to divide the 20 actions into three subsets (see table 1) and adopt cross-subject test setting where a half of the subjects are used for training and the rest ones are used for testing. The C value of SVM is fixed to 10 for all the experiments on this dataset. For the fixed pooling method, the pooling length is set by 30, while the possible pooling length range is [20, 40] for the dynamic pooling method.

At first, we evaluate the performance of our feature through comparison with the feature used in [4]. Follow the settings in [4], we use current frame x_t, the frame captured 10 frames previously, x_{t-10}, and the frame captured 30 frames previously, x_{t-30} to get the feature of current frame x_t. Then we build a codebook with 2000 visual words and set the nearest neighbor number in LSAQ coding as 20. As shown in Fig. 3, in which "no pooling(a)" and "fixed pooling(a)" are the results of feature used in [4], our feature (which corresponds to "no pooling(b)"

Table 1. The three subsets of actions used for the MSR Action3D dataset

ActionSet 1	ActionSet 2	ActionSet 3
Horizontal arm wave(How)	High arm wave(HiW)	High throw(HT)
Hammer(H)	Hand catch(HC)	Forward kick(FK)
Forward punch(FP)	Draw x(DX)	Side kick(SK)
High throw(HT)	Draw tick(DT)	Jogging(J)
Hand clap(HC)	Draw circle(DC)	Tennis swing(TSw)
Bend(B)	Two hand wave(HW)	Tennis serve(TSr)
Tennis serve(TSr)	Forward kick(FK)	Golf swing(GS)
Pickup & Throw(PT)	Side boxing(SB)	Pickup & throw(PT)

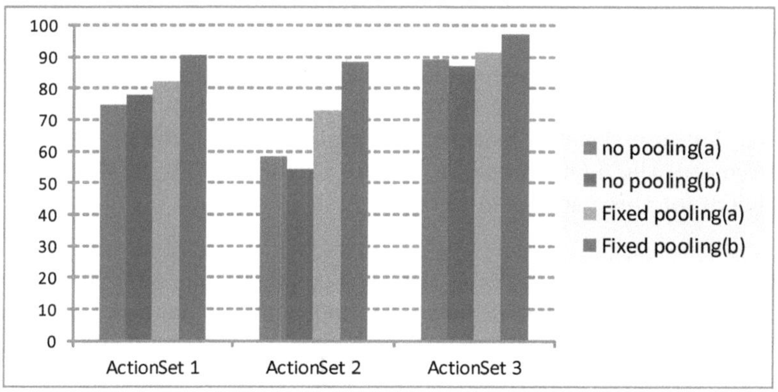

Fig. 3. Performance comparison on different features ((a) indicates feature of [4], (b) indicates our feature. Best viewed in color)

and "fixed pooling(b)" in Fig. 3) gains a slightly worse performance than the feature of [4] in the case of "no pooling" method (i.e., segment length $l = 1$). However, when we adopt the fixed pooling method, our feature outperforms [4] by an obvious margin. It implies that our feature is more suitable for the action segment representation. On the other side, both the feature of [4] and ours gain significant performance improvement via pooling, which demonstrates segment representation is more discriminative than a single frame[1].

Moreover, we demonstrate the superiority of dynamic pooling relative to fixed length pooling. As shown in Fig. 4, dynamic pooling outperforms fixed pooling on action set 1 and action set 2, and gains a similar performance on action set 3. Given the fact that the actions in action set 1 and action set 2 are with similar motions while action set 3 groups complex but pretty distinct actions, it implies dynamic pooling can better deal with similar actions.

[1] Although the feature in [4] is extracted from three frames, we can regard it as a representation for a single frame.

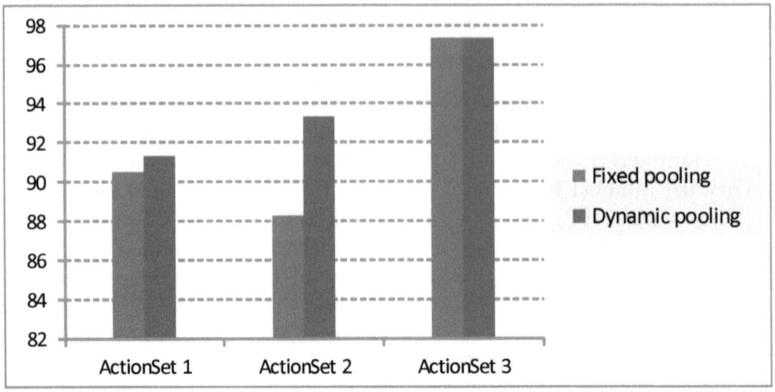

Fig. 4. Performance comparison on different pooling methods (Best viewed in color)

Finally, we compare the recognition accuracy of our method with previous results in literature. As shown in table 2, our method outperforms all methods which only use skeleton feature. The method in [9] which outperforms our method requires the depth maps for feature extraction. Fig. 5 shows the confusion matrices for three action subsets.

Table 2. Experimental results from the MSR-Action3D dataset ("S" indicates the methods which only use skeleton data, "D" indicates the method which only uses depth data, "S+D" indicates that it uses both skeleton data and depth data)

Method	Accuracy
(S) Ellis *et al.* [4]	65.7%
(D) Action Graph on Bag of 3D Points [8]	74.7%
(S) Actionlet Ensemble [1]	88.2%
(S) Our method	**94.0%**
(S+D) Joint Angles Similiarities [9]	94.8%

4.2 UCF-Kinect Dataset

The UCF-Kinect dataset [4] was gathered from 16 subjects using a Kinect sensor. It contains 16 actions and a total of 1280 action samples. The 3-dimensional coordinates of 15 joints are available for each frame. Following the experimental setting used in [4], all of our experiments on this dataset are implemented using 4-fold cross-validation. The C value of SVM is fixed to 1 for all the experiments on this dataset.

To evaluate the performance w.r.t different length of observations, we create new datasets by varying a parameter termed *maxFrames* as [4]. New dataset is created by selecting the first *maxFrames* frames from the original sequence.

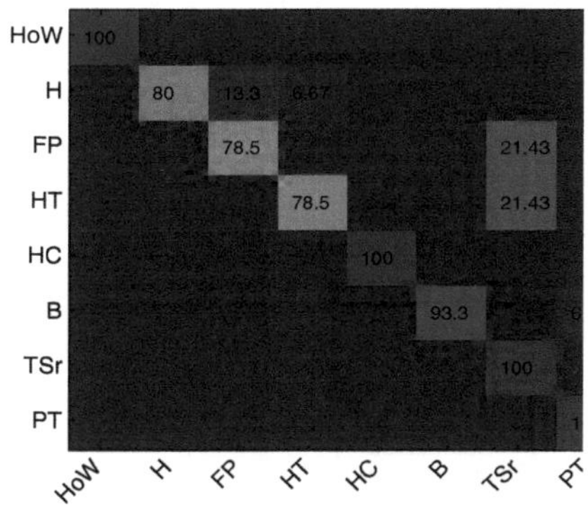

ActionSet 1

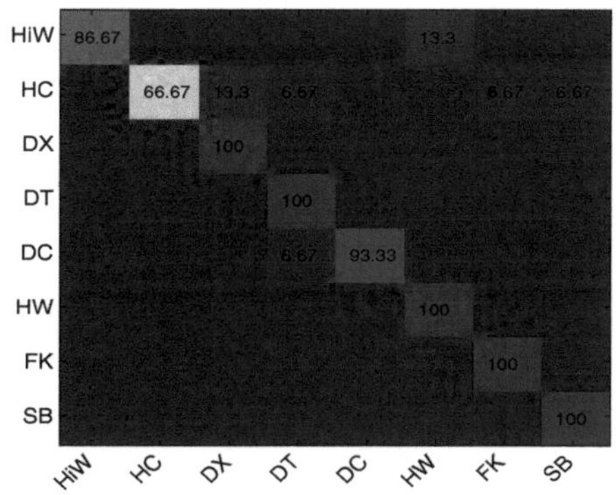

ActionSet 2

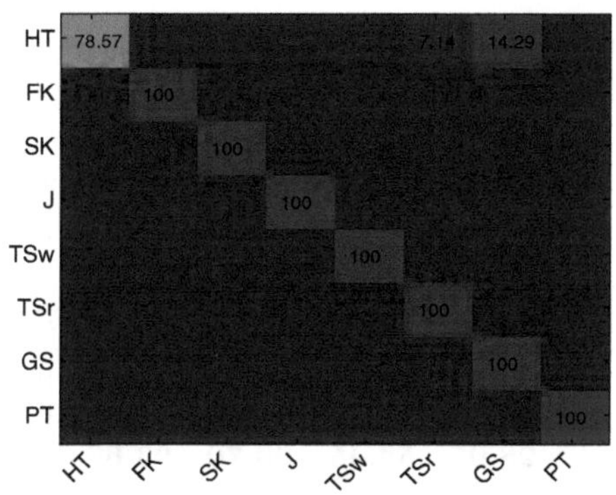

ActionSet 3

Fig. 5. Confusion matrices of MSR Action3D's three action subsets (Best viewed in color)

The entire video is used when the original sequence is shorter than *maxFrames*. We compare the performance of our method with existed methods [4] and [9]. For our method, we set the pooling length range to $[maxFrames - 10, maxFrames]$ when *maxFrames* is less then 40, otherwise the range is set to [20, 40]. As shown in Fig. 6, our method gets a much higher accuracy than [4] and [9], especially when *maxFrames* is small. Table 3 shows numerical results at several points along the curves in Fig. 6. The accuracy of our method outperforms other two methods by at least 10% when *maxFrames* equals to 10, 15 and 20, which demonstrates the superiority of our method for mining discriminative information.

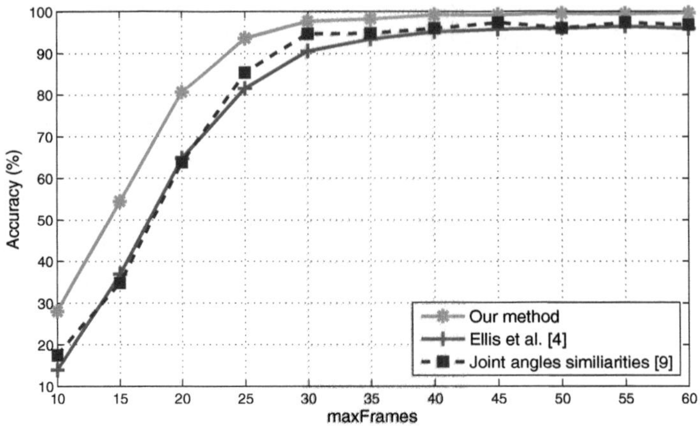

Fig. 6. Accuracy curves w.r.t. different *maxFrames*

Table 3. Accuracy (%) of UCF-Kinect dataset with different *maxFrames*

Method \ Frames	10	15	20	25	30	40	60
Ellis *et al.* [4]	13.91	36.95	64.77	81.56	90.55	95.16	95.94
Joint Angles Similiarities [9]	17.50	34.80	63.85	85.45	94.72	95.80	96.83
Our method	**28.05**	**54.45**	**80.70**	**93.67**	**97.73**	**99.22**	**99.69**

5 Conclusions

In this paper, we propose a discriminative segment mining approach for human action recognition, and the length of key segment for each action sample can be automatically determined via multiple instance learning and dynamic pooling. In addition, we present a novel skeleton joint feature extracted from 3D joints which is capable of capturing both motion and shape cues. To the best of our knowledge, our method achieves the state-of-the-art performance on UCF-Kinect dataset, and shows superior accuracy than previous methods of using only skeleton data on MSR Action3D dataset.

Acknowledgments. This work is funded by the Project NSFC 61071155. We would like to thank Jun Zhu for valuable comments on this paper.

References

1. Wang, J., Liu, Z., Wu, Y., Yuan, J.: Mining actionlet ensemble for action recognition with depth cameras. In: 2012 IEEE Conference on Computer Vision and Pattern Recognition (CVPR), pp. 1290–1297. IEEE (2012)
2. Yang, X., Tian, Y.: Effective 3d action recognition using eigenjoints. Journal of Visual Communication and Image Representation (2013)
3. Shao, L., Ji, L.: Motion histogram analysis based key frame extraction for human action/activity representation. In: Canadian Conference on Computer and Robot Vision, CRV 2009, pp. 88–92. IEEE (2009)
4. Ellis, C., Masood, S.Z., Tappen, M.F., LaViola Jr., J.J., Sukthankar, R.: Exploring the trade-off between accuracy and observational latency in action recognition. International Journal of Computer Vision 101(3), 420–436 (2013)
5. Poppe, R.: A survey on vision-based human action recognition. Image and Vision Computing 28(6), 976–990 (2010)
6. Nguyen, M.H., Torresani, L., De la Torre, F., Rother, C.: Weakly supervised discriminative localization and classification: a joint learning process. In: 2009 IEEE 12th International Conference on Computer Vision, pp. 1925–1932. IEEE (2009)
7. Schindler, K., Van Gool, L.: Action snippets: How many frames does human action recognition require? In: IEEE Conference on Computer Vision and Pattern Recognition, CVPR 2008, pp. 1–8. IEEE (2008)
8. Li, W., Zhang, Z., Liu, Z.: Action recognition based on a bag of 3d points. In: 2010 IEEE Computer Society Conference on Computer Vision and Pattern Recognition Workshops, CVPRW, pp. 9–14. IEEE (2010)
9. Ohn-Bar, E., Trivedi, M.M.: Joint angles similiarities and hog2 for action recognition (2013)
10. Xia, L., Chen, C.C., Aggarwal, J.: View invariant human action recognition using histograms of 3d joints. In: 2012 IEEE Computer Society Conference on Computer Vision and Pattern Recognition Workshops (CVPRW), pp. 20–27. IEEE (2012)
11. Andrews, S., Tsochantaridis, I., Hofmann, T.: Support vector machines for multiple-instance learning. Advances in Neural Information Processing Systems 15, 561–568 (2002)
12. Felzenszwalb, P.F., Girshick, R.B., McAllester, D., Ramanan, D.: Object detection with discriminatively trained part-based models. IEEE Transactions on Pattern Analysis and Machine Intelligence 32(9), 1627–1645 (2010)
13. Fan, R.E., Chang, K.W., Hsieh, C.J., Wang, X.R., Lin, C.J.: LIBLINEAR: A library for large linear classification. Journal of Machine Learning Research 9, 1871–1874 (2008)
14. Yang, J., Yu, K., Gong, Y., Huang, T.: Linear spatial pyramid matching using sparse coding for image classification. In: IEEE Conference on Computer Vision and Pattern Recognition, CVPR 2009, pp. 1794–1801. IEEE (2009)
15. Liu, L., Wang, L., Liu, X.: In defense of soft-assignment coding. In: 2011 IEEE International Conference on Computer Vision (ICCV), pp. 2486–2493. IEEE (2011)
16. Zhu, J., Zou, W., Yang, X., Zhang, R., Zhou, Q., Zhang, W.: Image classification by hierarchical spatial pooling with partial least squares analysis. In: Proceedings of the British Machine Vision Conference, pp. 102.1–102.11. BMVA Press (2012)

Multi-cue Based Multi-target Tracking with Boosted MHT

Long Ying[1,2], Tianzhu Zhang[1,2], Shengsheng Qian[1,2], and Changsheng Xu[1,2]

[1] Institute of Automation, Chinese Academy of Science, Beijing, China
[2] China-Singapore Institute of Digital Media, Singapore

Abstract. Tracking multiple objects is critical to automatic video content analysis and virtual reality. The major problem is how to solve data association problem when ambiguous observations are caused by objects in close proximity or occlusion. To tackle this problem, we propose a boosted multiple hypotheses tracking (BMHT) algorithm for multi-object tracking. Here, on-line boosting learning is adopted to enhance the discriminative property and enlarge search space of the generative tracker MHT. To make the tracker be more reliable, a multi-cue integration strategy is adopted to consider different kinds of features under the on-line boosting framework. In this paper, we integrate both appearance and motion pattern information. For simplicity, Haar-like features and optical flow are adopted. We test our BMHT tracker on several challenging video sequences that involve heavy occlusion and pose variations. Experimental results show that the proposed BMHT achieves good performance.

Keywords: Multi-object Tracking, MHT, Association, Online Boosting.

1 Introduction

Object tracking is an important subject in computer vision with a wide range of applications [24,23] such as automatic surveillance, video content analysis, virtual reality, etc. For a visual tracking algorithm to be useful in real-world scenarios, it should be designed to handle and overcome cases where the target's appearance changes over time. Significant and rapid appearance variations due to noise, occlusion, varying viewpoints, scale changes, background clutter and illumination pose major challenges to any tracker. Over the years, numerous tracking algorithms have been proposed to overcome these challenges. For a survey of many of these algorithms, we refer the reader to [17].

Generally speaking, object tracking methods can be categorized as either generative or discriminative. In generative tracking methods, a generative appearance model (possible with updating strategy) is used to represent target observations explicitly. Here, the tracker searches for a potential target location that has the most similar appearance to the generative model. Popular generative trackers include eigentracker [6], mean shift tracker [9], sparse trackers [21,20,22], and incremental tracker [15]. For multi-object tracking, the major

B. Huet et al. (Eds.): PCM 2013, LNCS 8294, pp. 528–537, 2013.

problem is how to do data association between the known possible objects and the observations [18]. To deal with this problem, many data association methods have been proposed in the literature, such as Joint Probabilistic Data Association Filter (JPDA) [3] and methods base on particle filter [12]. However, considering information of future feedback and past simultaneously is usually more effective to overcome the ambiguities in tracking process, which is the main conception of current methods, such as multiple hypotheses tracking (MHT) [10,7], MCMC data association [18], network flow graph model [19], and hierarchical association [8]. Most of the existing methods are based on the similar formulation and conception with MHT. The efficient MHT is the most classical approach for data association in tracking by detection framework, which is widely and successfully used for point tracking in radar and sonar signal processing formerly. It is practical in real-time tracking with moderate computational complexity, and performs robustly for application. However, for all of the existing generative methods, they rely on the knowledge of the foreground only. Consequently, as the appearance of the object changes over time, a generative tracker is prone to generate inaccurate association in many scenarios.

Instead of building a model to describe the appearance of an object, discriminative trackers formulate the tracking problem as a binary classification problem. In this case, the tracker finds the target location that best separates the target from the background. Popular discriminative methods include online boosting [11], ensemble tracking [1], and online multiple instance learning tracking [2]. In the OLB tracker [11], online boosting is adopted to select useful features for object tracking. Its performance is affected by background clutter, and the tracker can easily drift. In addition, the online boosting tracker uses limited features, and is difficult for robust object tracking. The ensemble tracker [1] formulates the tracking task as a pixel based binary classification problem. Although this method is able to differentiate between target and background, the pixel-based representation is rather limited and thereby constrains its ability to handle heavy occlusion and clutter. In the MIL tracker [2], the multiple instance learning method is extended to an online setting for object tracking. While it is capable of reducing tracker drift, this method is unable to handle large nonrigid shape deformation. The discriminative trackers [11,1,2] have achieved a great success in visual tracking by considering the discriminative information between foreground and background. However, most of the discriminative algorithms treat tracking as the binary classification problem, making themselves unsuitable for tracking multiple objects.

Inspired by the previous work, in this paper, we propose a boosted MHT for multi-object tracking to address the above issues. The main contributions of this work are the following two points. **(1)** To enhance the discriminative property of MHT for multi-object tracking, we adopt the on-line boosting learning framework to improve the likelihood calculation within MHT, which can make the generative tracker MHT be discriminative to distinguish the foreground and background. Moreover, multiple measurements are generated for each object to augment the search space of MHT, conducting more robust data association.

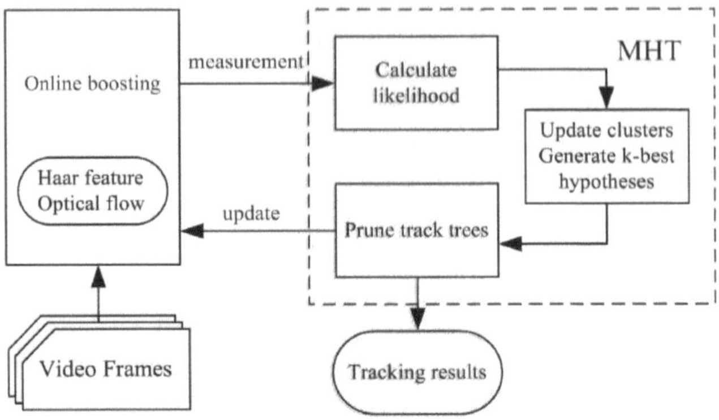

Fig. 1. The framework of our proposed boosted MHT multi-object tracking method

(2) To make our tracker be more reliable, we adopt a multi-cue integration strategy to consider different kinds of features for the on-line boosting classifier. Here, we integrate both appearance and motion pattern information. Specially, Haar-like features which are well known to be good at capturing geometric structures are adopted to describe the appearance, and optical flow feature which is calculated between two frames is used to model the object motion pattern. The proposed tracker is evaluated using several challenging video sequences and excellent performances are observed both qualitatively and quantitatively.

2 Overview

The proposed object tracking framework is shown in Figure 1, which includes two different stages: **(1)** online boosting for measurement generation and similarity calculation; **(2)** MHT tracking framework for multi-object association. In the first stage, online boosting classifiers based on Haar-like features and optical flow feature are used to incorporate discriminative information to generate multiple measurements (detection results). Here, for each object, a specific online boosting classifier is built and updated when the states of multiple best assignment results obtained by MHT at previous frame are similar enough. In addition, we also consider the occlusion between objects. When the distance between two objects are close enough, patches around them are extracted as negative samples to update corresponding classifiers respectively. After online boosting learning, multiple measurements are generated for each object by clustering in corresponding search window. The second stage is the MHT framework for multiple objects association. Given the measurements for each object, the corresponding classifier confidence is incorporated with dynamic similarity calculated by Kalman filter to enhance the discriminative power of likelihood. Then, track tree clusters in previous hypotheses are updated and k-best hypotheses are generated. After track trees pruning, the multi-object tracking results can be obtained.

3 The MHT Algorithm

Our work is based on the MHT algorithm, which is efficiently constructed and implemented by Cox [10]. To make this paper self-contained, we review MHT algorithm in this section. For details, please refer to the reference [10]. In tracking by detection framework, new measurements can be received at every instance. Each measurement may either **(1)** belong to a previously known object, **(2)** be the start of a new object, **(3)** be a false positive. For objects that are not assigned measurements, there is the possibility of **(4)** termination, or **(5)** continuation of an object.

Let Z^k be the set of all measurements until time k. A single object trajectory hypothesis τ_j is defined as a list of measurements at different time. A specific partition of the measurement set, named association hypothesis Θ_l^k, where l is index, is defined as a set of hypothesis trajectories, in which an additional trajectory τ_0 contains all false positives. When measurements $Z(k)$ at time k are received, a particular global hypothesis Θ_l^k can be derived from certain hypothesis Θ_m^{k-1} at time $k-1$ by the specific set of assignments of the origins of all measurements received at time k with all objects assumed by the parent hypothesis Θ_m^{k-1}. Let $\theta_l(k)$ denotes this set of assignments which is defined to consist of γ measurements from known objects, υ measurements from new objects, ϕ spurious measurements, and χ terminated objects. A constraint that one observation originates from at most one object and one object has at most one associated observation is imposed to reduce the size of the search space.

The probability of an association hypothesis Θ_l^k can be calculated according to parent hypothesis Θ_m^{k-1} using Bayes' rule:

$$\begin{aligned} P\{\Theta_l^k|Z^k\} = \tfrac{1}{c}P\{Z(k)|\theta_l(k), \Theta_m^{k-1}, Z^{k-1}\} \\ P\{\theta_l(k)|\Theta_m^{k-1}, Z^{k-1}\}P\{\Theta_m^{k-1}|Z^{k-1}\} \end{aligned} \tag{1}$$

where c is a normalization constant. Multi-object tracking aims to find out the MAP hypothesis. It is assumed that a measurement with index i obeys uniform distribution if it is considered as a false positive, and obeys Gaussian distribution if it is associated with trajectory object τ_{j_i} whose index is j_i:

$$P\left\{z_i(k)|\tau_{j_i}(k-1)\right\} = N_{j_i}\left[z_i(k)\right] \triangleq N[z_i(k); \hat{z}_i(k|k-1), S^{j_i}(k)] \tag{2}$$

where $\hat{z}_i(k|k-1)$ denotes the predicted measurement for object τ_{j_i}, and $S^{j_i}(k)$ is the innovation covariance [14].

If the probabilities of detection and termination of object τ_{j_i} are P_D^j and P_χ^j respectively, and the numbers of false positives and new objects are assumed to obey Poisson distributed with densities λ_F and λ_N respectively, taking all these settings and Eq.(2) into Eq.(1), the posterior probability of an association hypothesis can be expressed as:

$$\begin{aligned} P\{\Theta_l^k|Z^k\} = \tfrac{1}{c}\prod_{i=1}^{m_k}\left[N_{j_i}[z_i(k)]\right]^{\gamma_i} \\ \left\{\lambda_F^{\phi}\lambda_N^{\upsilon}\prod_j \left(P_D^j\right)^{\delta_j}\left(1-P_D^j\right)^{1-\delta_j}\left(P_\chi^j\right)^{\chi_j}\left(1-P_\chi^j\right)^{1-\chi_j}\right\}P\{\Theta_m^{k-1}|Z^{k-1}\} \end{aligned} \tag{3}$$

where γ_i, δ_j and χ_j are respectively indicator variable of measurement originating from known object, detected object or terminated object. ϕ and υ are the total number of false positives and new objects respectively.

In practical application, tracks that do not compete for common measurements can be partitioned into separate clusters. The k-best hypotheses are generated directly to achieve more efficient performance, which is implemented by optimizing Murty's algorithm [13] in $O(kN^3)$. Additionally, pruning stage is executed to reduce complexity.

4 Our Boosted MHT Algorithm

The MHT algorithm, which provides a rigid probabilistic formulation of multi-object tracking problem in tracking by detection framework, is successfully used in radar and sonar signal processing. When this kind of methods are applied in image patches tracking such as pedestrian, the similarity metric and the strategy of measurement generation should be adapted to the task.

4.1 Our Formulation

The assignment likelihood calculation in classical generative methods such as MHT depends on the knowledge of the target appearance. As the appearance of the object changes over time, it is easy to incorrectly associate with measurements of other objects or background clutter in many scenarios. The discriminative tracker, online boosting tracker, is realized by selecting useful features to distinguish the foreground and background for object tracking, and it can help to deal with the problem of MHT.

In online boosting framework, a set of weak classifiers are initialized with randomly selected features subset. When a new sample arrives, all weak classifiers are trained with respect to the importance weight of the sample, and the weak classifier with the smallest error is selected. Then the corresponding voting weight is calculated and the importance weight is updated. This process is repeat n times to form strong classifier. In each iteration, selected weak classifiers are excluded. Finally, The worst weak classifier is replaced with a new randomly initialized one. The strong classifier can give a label confidence score for a candidate patch.

Due to considering the foreground and background distribution, the label confidence output by online boosting classifier is discriminative to the patches around specific object, which is suitable similarity metric for data association. In our method, specific online boosting classifier is built for each object. Classifier updating is conducted when the states of pruned multiple best assignment results obtained by MHT at previous frame are similar enough. If the distance between two objects is smaller than a threshold, they are treated as negative sample for each other, enhancing discrimination between different objects which may overlap. Multiple best measurements are generated for each object by k-means clustering in corresponding search window. The false positive density which is related with these generated measurements and used in posterior probability calculation in MHT, is estimated using kernel clutter intensity estimation method proposed in [16].

The value of label confidence ranges from −1 to 1. To incorporate it under the probability framework of MHT, we scale this value to the range [0, 1] for a particular object (namely category confidence) by confidence transformation. Thus, likelihood of the assignment of an measurement in Eq.(2) can be improved as follows:

$$P\left\{z_i(k)|\tau_{j_i}(k-1)\right\} = (1-\alpha)N_{j_i}\left[z_i(k)\right] + \alpha P_b\left\{z_i(k)|\tau_{j_i}(k-1)\right\} \quad (4)$$

where $N_{j_i}[z_i(k)]$ is the dynamic confidence of the assignment of measurement $z_i(k)$ to object trajectory τ_{j_i}; $P_b\{\bullet\}$ is the on-line boosting category confidence to the assignment. The parameter α is tradeoff factor, which can be dynamically set depending on tracking situations, including isolation, inter-occlusion and clutter. Note that the adopted boosting confidence probability consider the discriminative property of the measurement. Therefore, it is robust to calculate the likelihood probability.

4.2 Multiple Cues Fusion

The traditional on-line boosting tracker is adopted to select useful features for object tracking. But only one kind of feature can not represent the objects very well, it is difficult for robust multi-object tracking. Therefore, a natural way is to use multiple cues, making object tracking more robust and accurate. In this work, two kinds of features, Haar-like features and optical flow, are employed in our tracker. The reasons why we select these two kinds of features lie in two folds. First, Haar-like features are the region-like features, which can describe the appearance effectively. Second, the optical flow can represent the motion pattern of object. Therefore, they are complementary to a certain extent.

For Haar-like features, we implement as similar as in on-line boosting tracker [11]. Optical flow is the pattern of apparent motion of objects in a visual scene caused by the relative motion between an observer and the scene [5], which is discriminative when objects move in different direction and have obviously different velocity. Optical flow vectors of pixels in image regions of measurements are calculated by Lucas-Kanade algorithm, which are then split into horizontal and vertical channels and the effect of noise is reduced. The image regions of measurements are divided into 18 pie slices. The values of each direction channel are integrated over the domain of every slice, which are concatenated into a feature vector with 36 dimensional.

To integrate multiple features, we construct two on-line boosting classifiers using Haar-like features and optical flow feature separately. Each classifier is first used as an individual discriminative tracker, then we integrate the two trackers into the final one. For simplicity, the final classifier confidence is calculated based on their linear combination.

4.3 Discussion

Based on our formulation as in Eq.(4), with different α and different measurements selection strategy, our tracker has different property. When $\alpha = 0$ and only global detected measurements are considered, our algorithm reduces to the MHT [10]. And it is on-line boosting tracker [11] if $\alpha = 1$ and only the best measurements generated by each on-line classifier are taken. However, our proposed tracker should be much better than both of them. The reasons are the following:

(1) Due to embedding the discriminative property for the appearance and motion pattern modeling via on-line boosting tracker, our proposed boosted MHT should be much better than its old version in [10]. Moreover, multiple detection candidates are generated by k-means clustering for each object, enlarging the search space of MHT for potential optimal association.

(2) The performance of on-line boosting tracker is affected by background clutter, and it can easily drift. That's because the patch with the highest confidence of classifier is not always the target, which leads to drift. However, in our implementation, multiple detection candidates are obtained, and input to the MHT framework to calculate the best one using multi-cue in time window. As a result, it can help the on-line boosting tracker solve the drift problem. Thus, our proposed boosted MHT is much better than on-line boosting tracker [11].

5 Experimental Results

In this section, we present experimental results that validate the effectiveness of our boosted MHT method. We also conduct a thorough comparison between our boosted MHT and state-of-the-art tracking methods.

5.1 Dataset and Baseline methods

Dataset: In out experiments, 5 surveillance video sequences from TRECVID 2011, which are taken in an airport hall, with resolution 720×576 and 1000 frames respectively. Each sequence contains about 20 objects (pedestrians) on average, containing frequently occlusion of moving or static travelers with luggage are used. The ground truths of these videos are manually labeled.

Baseline Methods: We compare our boosted MHT (BMHT) with two recent and commonly used trackers. The effect of online boosting algorithm [11] is illustrated, which is denoted as OLB in the following paragraphs. MHT algorithm [10] with appearance model which is commonly incorporated in many multi-object tracking algorithms is tested in the experiments and denoted as MHT. We implemented these trackers using publicly available source codes or binaries provided by the authors.

5.2 Results and Analysis

In Fig. 2 and Fig. 3, we show qualitative tracking results of these three methods respectively on video1 and video3 in the form of sampling frames. To obtain

Fig. 2. The qualitative tracking results on video1 for three methods: OLB (first row), MHT (second row) and BMHT (third row)

Fig. 3. The qualitative tracking results on video3 for three methods: OLB (first row), MHT (second row) and BMHT (third row)

quantitative evaluation of the overall performances of the 3 trackers on each video sequences, we adopt the most commonly used CLEAR metrics for evaluation [4]. CLEAR metrics consist of average rate of object misses, rate of false positives, rate of identity switches, and MOTA which evaluates the overall situation of object identities by considering misses, false positives and identity switches together. The quantitative results are shown in Fig. 4.

From the results, we can see that BMHT consistently outperform the other trackers in all sequences. As shown in Fig. 2, when OLB or MHT suffer from drift or identity switch in frame 215, 323 and 485, our BMHT can successfully maintain the trajectories, such as object 1, 4, 7, 8 and 9 in the third row. The same situation also occurs in Fig. 3 in frame 113, 512, 600 and 915, such as object 1, 9, 10, 14 and 15. The quantitative evaluation in Fig. 4 shows that BMHT performs much better than OLB and MHT overall. The proposed method has higher MOTA, lower miss rate and almost the same false and ID switch rate compared with the other two methods. This is because OLB is effected by background clutter and easily drifts from the target. For OLB tracker, it selects the best one as the tracking result and may be trapped in local minima.

Videos	Methods	MOTA	Miss	False	Switch
Video1	OLB	0.2926	0.5955	0.1108	0.0010
	MHT	0.4903	0.4591	0.0475	0.0031
	BMHT	**0.5684**	0.3657	0.0638	0.0020
Video2	OLB	0.1447	0.6011	0.2516	0.0027
	MHT	0.2841	0.5414	0.1710	0.0035
	BMHT	**0.4492**	0.4319	0.1171	0.0018
Video3	OLB	0.3260	0.3853	0.2883	0.0005
	MHT	0.3279	0.5685	0.1017	0.0019
	BMHT	**0.4431**	0.4219	0.1335	0.0010
Video4	OLB	0.3283	0.4884	0.1826	0.0007
	MHT	0.3005	0.5862	0.1113	0.0020
	BMHT	**0.4566**	0.4357	0.1057	0.0020
Video5	OLB	0.3472	0.4193	0.2326	0.0009
	MHT	0.3501	0.4987	0.1483	0.0029
	BMHT	**0.4715**	0.3852	0.1412	0.0021

Fig. 4. The quantitative comparison for baseline methods and the proposed method

MHT is affected by the fluctuant appearance and motion pattern of objects and may loss discriminative capability in some situations. The search space of MHT is also restrict by inaccurate and insufficient measurements.

6 Conclusions

We propose a boosted MHT method for multi-object tracking, which has the advantages of both online boosting and MHT. Two kinds of features, Haar-like features and optical flow, are adopted to improve the robustness of the online boosting method for measurements detection. Then, multiple best measurements are input to the probabilistic framework of MHT to decide the best solution in time window. The category confidence of online boosting classifier on each measurement is integrated to likelihood calculation for more discrimination. Extensive experimental results on 5 challenging video sequences with two related baseline methods demonstrate the effectiveness of the proposed method.

Acknowledgments. This work is supported by National Natural Science Foundation of China (61225009), National Basic Research Program of China (973 Program) under Grant 2012CB316304, and Beijing Natural Science Foundation (4131004), also by the Singapore National Research Foundation under International Research Centre@Singapore Funding Initiative and administered by the IDM Programme Office.

References

1. Avidan, S.: Ensemble tracking. In: CVPR, pp. 494–501 (2005)
2. Babenko, B., Yang, M.-H., Belongie, S.: Visual tracking with online multiple instance learning. In: CVPR (2009)
3. Bar-Shalom, Y., Daum, F., Huang, J.: The probabilistic data association filter. IEEE Control Systems Magazine 29(6), 82–100 (2009)

4. Bernardin, K., Stiefelhagen, R.: Evaluating multiple object tracking performance: the clear mot metrics. Journal on Image and Video Processing, 1–8 (2008)
5. Black, M., Anandan, P.: The robust estimation of multiple motions: Parametric and piecewise smooth flow fields. Computer Vision and Image Understanding 63(1), 75–104 (1996)
6. Black, M.J., Jepson, A.D.: Eigentracking: Robust matching and tracking of articulated objects using a view-based representation. In: IJCV, pp. 63–84 (1998)
7. Breitenstein, M.D., Reichlin, F., Leibe, B., Koller-Meier, E., Gool, L.V.: Robust tracking-by-detection using a detector confidence particle filter. In: Proceedings of ICCV, pp. 1515–1522 (2009)
8. Brendel, W., Amer, M., Todorovic, S.: Multiobject tracking as maximum weight independent set. In: Proceedings of CVPR, pp. 1–8 (2011)
9. Comaniciu, D., Ramesh, V., Meer, P.: Kernel-Based Object Tracking. TPAMI 25(5), 564–575 (2003)
10. Cox, I., Hingorani, S.: An efficient implementation of reid's multiple hypothesis tracking algorithm and its evaluation for the purpose of visual tracking. IEEE Trans. on PAMI 24(6), 843–854 (1979)
11. Grabner, H., Grabner, M., Bischof, H.: Real-Time Tracking via On-line Boosting. In: BMVC (2006)
12. Knan, Z., Balch, T., Dellaert, F.: Mcmc-based particle filtering for tracking a variable number of interacting targets. IEEE Transcactions on PAMI 27(11), 82–100 (2005)
13. Miller, M., Stone, H., Cox, I.: Optimizing murty's ranked assignment method. IEEE Transactions on Aerospace and Electronic Systems 33(3), 851–862 (1997)
14. Reid, D.: An algorithm for tracking multiple targets. IEEE Trans. on Automatic Control 24(6), 843–854 (1979)
15. Ross, D., Lim, J., Lin, R.-S., Yang, M.-H.: Incremental Learning for Robust Visual Tracking. IJCV 77(1), 125–141 (2008)
16. Song, T.L., Musicki, D.: Adaptive clutter measurement density estimation for improved target tracking. IEEE Transactions on Aerospace and Electronic Systems 47(2), 1457–1466 (2011)
17. Yilmaz, A., Javed, O., Shah, M.: Object tracking: A survey. ACM Comput. Surv. 38(4), 13+ (2006)
18. Yu, G.M.: Multiple-target tracking by spatio-temporal monte carlo markov chain data association. IEEE Trans. on PAMI 31(12), 91–110 (2009)
19. Zhang, L., Li, Y., Nevatia, R.: Global data association for multi-object tracking using network flows. In: Proceedings of CVPR, pp. 1–8 (2008)
20. Zhang, T., Ghanem, B., Liu, S., Ahuja, N.: Low-rank sparse learning for robust visual tracking. In: Fitzgibbon, A., Lazebnik, S., Perona, P., Sato, Y., Schmid, C. (eds.) ECCV 2012, Part VI. LNCS, vol. 7577, pp. 470–484. Springer, Heidelberg (2012)
21. Zhang, T., Ghanem, B., Liu, S., Ahuja, N.: Robust visual tracking via multi-task sparse learning. In: IEEE Conference on Computer Vision and Pattern Recognition (2012)
22. Zhang, T., Ghanem, B., Liu, S., Ahuja, N.: Robust visual tracking via structured multi-task sparse learning. International Journal of Computer Vision 101(2), 367–383 (2013)
23. Zhang, T., Liu, S., Xu, C., Lu, H.: M^4l: Maximum margin multi-instance multi-cluster learning for scene modeling. Pattern Recognition 46(10), 2711–2723 (2013)
24. Zhang, T., Liu, S., Xu, C., Lu, H.: Mining semantic context information for intelligent video surveillance of traffic scenes. IEEE Transactions on Industrial Informatics 9(1), 149–160 (2013)

Haar-Like Feature Based On-Line Boosting Tracking Algorithm with Directional Texture Entropy

Song Hu[1], Shui-fa Sun[1,2,*], Bang-jun Lei[1,2], and Zhi-ping Dan[1,2]

[1] Institute of Intelligent Vision and Image Information, China Three Gorges University, HuBei, YiChang, 443002

[2] Collaborative Innovation Center for Geo-Hazards and Eco-Environment in Three Gorges Area, Hubei province, YiChang, 443002

watersun1977@hotmail.com

Abstract. Large numbers of features are needed for tracking in the conventional Haar-like feature based on-line boosting methods (HBT). The vast amount guarantees the existence of useful features. But it leads to expensive computing and memory requirements. On the other hand, the coexisted useless features may bias the tracking towards random influences. To address those two problems, in this paper, we propose a new method named HBTDTE (Haar-like feature based on-line Boosting Tracking with Directional Texture Entropy). In this novel algorithm, the texture direction and richness information are extracted by the entropy of gray level co-occurrence matrix (GLCM). Then Haar-like features are only calculated along directions with richest texture. And the employed feature number is also governed by the texture richness. In this way, not only the number of useless features is largely reduced, but also the total size of the feature pool. Experiments reveal the higher robustness and performance of this new HBTDTE method. We show the superiority of our method over several current state-of-the-art tracking methods in several experiments on publicly available data.

Keywords: object tracking, on-line boosting, Haar-like feature, GLCM.

1 Introduction

Video object tracking has been extensively applied to intelligent traffic control, advanced medical treatment and national defense etc. Feature-based tracking methods need to extract object features and follow these features through every frame. It is a process of feature matching and therefore the selection of correct features is very important. For the tracking purpose, there exist many kinds of features, such as the color histogram that is frequently employed in mean-shift or particle filter based tracking [1][2][13]. Color histogram is robust against occlusion, non-rigid motion and the change of viewpoint. However, when an object has similar appearance to the background, color histogram will not be able to distinguish it. Haar-like features

* Corresponding author.

B. Huet et al. (Eds.): PCM 2013, LNCS 8294, pp. 538–549, 2013.

have been widely applied in the field of pattern recognition and surveillance systems for its strong classification capability and the fast calculation [3][4]. Whereas, Haar-like features are lack of texture information compared with other features, such as the histogram of oriented gradient (HOG) [5]. Most recently, a lot of other features, such as SIFT [6], LBP [7] and texture feature from gray level co-occurrence matrix (GLCM) [8], have been proposed. Among them, one of the most stable features is SIFT, which is invariant to scale, translation, rotation and even affine transformation. SIFT has been widely used in image matching. But its computation is high and it is a sparse representation. LBP [9] can be applied to every pixel. But it is sensitive to noise. The gray level co-occurrence matrix (GLCM) model was first proposed by R.Haralick in early 1970s [10]. An object's texture information can be obtained and various feature vectors are formed through the statistics of GLCM.

The quality of object features is critical to the tracking in feature based object tracking. Only if the selected features are effective enough can a high quality tracking be expected. Avidan [11] adopted off-line learning of SVM to track an object by classify it out of the surrounding background. But its main drawback is the sensitiveness to the possible posture change of the object and the background. Therefore, on-line learning to reflect the object current state is needed. For this purpose, Grabner et al. designed an on-line boosting classifier which segments an object out of the background through adaptive feature selection [12].

Haar-like features are based on the gray intensity and insufficient in describing an object's texture. On the other hand, GLCM can describe an object's texture effectively. Combining GLCM and Haar-like principle will improve the descriptive capability. The study of existing Haar-like feature templates reveals that the templates are directional and templates' directions are associated with object's texture directions. In this paper, a novel feature generation method is introduced. It improves the conventional Haar-like based on-line boosting tracking algorithm (HBT). Because the HBT generates a large number of Haar-like features randomly that consist of also many redundant features and have low computational efficiency, we propose an approach that builds the Haar-like feature pool by combining the HBT with the directional texture entropy (HBTDTE). In HBTDTE, the size of the feature pool is also made adaptive to an object's texture richness.

2 Tracking Framework and Feature Selection

2.1 On-line Boosting Tracking Algorithm [12]

On-line boosting algorithm is made up of weak classifier, selector and strong classifier. Each weak classifier consists of one Haar-like feature based basic classifier whose recognition rate is higher than 50%. Each selector consists of one weak classifier and all selectors share one feature pool, namely Haar-like feature pool. In the process of update, every selector will be updated in order. The weak classifier that has the lowest classification error rate and has not been used by previous selectors is selected by the current selector as its basic classifier. Finally, the algorithm can obtain a strong classifier by weighted summation on all selectors. The weight of each selector is determined by the classification error rate of the corresponding weak classifier. The higher the error rate is, the lower the weight of the selector.

The number of the weak classifiers is a key factor to the algorithm efficiency. A large number of redundant features will be generated for the HBT algorithm. During update, every selector will be updated and every feature in the feature pool will be sequentially considered. Therefore the more selectors the algorithm has, the less efficient it will be. If a compact feature pool can be built in the initialization, the efficiency and accuracy of the tracking will be greatly improved.

2.2 Entropy from GLCM

Gray level co-occurrence matrix (GLCM) is defined on top of joint probability density of two pixels. Suppose (x1, y1) and (x2, y2) are two pixels in the image, in order to determine their positional relationship, two parameters are calculated including the distance d between them and the angle θ between the x axis and the line connecting them. Thus GLCM can be described as P(i, j, d, θ) (i and j are the gray values of those two pixels respectively).

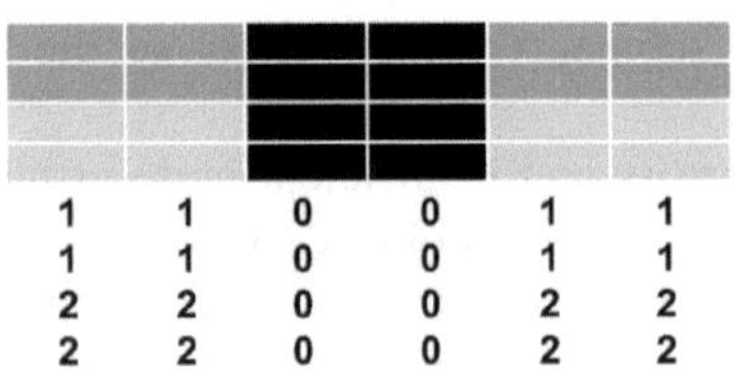

Fig. 1. Texture in vertical direction

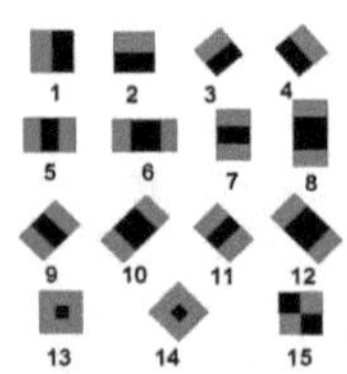

Fig. 2. Haar-like feature template

An integer specifies the number of gray-levels to use when scaling the grayscale values in image. For example, if this integer is 8, GLCM scales the values in image so they are integers between 1 and 8. The number of gray-levels determines the size of the GLCM. By scanning through the object area, GLCMs of horizontal, vertical, 45° and 135° direction can be calculated respectively. In practice, symmetrization and normalization are performed subsequently on the obtained GLCM. The frequency G(i, j) of grayscale pair (i, j) is gotten. Then the entropies at the four directions are calculated independently. The entropy at each direction reflects the texture complexity in that direction. If a GLCM has a uniform distribution, the entropy will have a large value. The formula for calculating entropy is as follows:

$$ENT = -\Sigma_{i=1}^{k}\Sigma_{j=1}^{k} G(i,j)\log G(i,j) \quad . \tag{1}$$

For an area with regular texture, the entropy value is small in the texture direction and the perpendicular direction. And it is relatively larger in the direction of 45° and 135° to the texture (see Fig. 1). To further clarify this, we learn about entropy change first. Entropy change is frequently mentioned in chemistry and it describes a process of chemical reaction from ordered to unordered. After a chemical reaction, the entropy value of this chemical system increased. Analogously, the texture direction and its perpendicular direction is ordered in general, but the other directions, like the direction of 45° and 135° to the texture, is unordered. So the entropy value is larger in an unordered direction, like the direction which has a uniform distribution.

Then four directions' GLCM of data in Fig. 1 are presented in Table 1(d=1, four different directions have been calculated respectively and the gray-level is 3):

Table 1. GLCM and its symmetrization and normalization

Angle	GLCM	Symmetrization	Normalization
	4 2 2	8 4 4	1/5 1/10 1/10
0°	2 4 0	4 8 0	1/10 1/5 0
	2 0 4	4 0 8	1/10 0 1/5
	3 2 1	6 3 3	1/5 1/10 1/10
45°	1 2 0	3 4 2	1/10 2/15 1/15
	2 2 2	3 2 4	1/10 1/15 2/15
	6 0 0	12 0 0	1/3 0 0
90°	0 4 4	0 8 4	0 2/9 1/9
	0 0 4	0 4 8	0 1/9 2/9
	3 1 2	6 3 3	1/5 1/10 1/10
135°	2 2 2	3 4 2	1/10 2/15 1/15
	1 0 2	3 2 4	1/10 1/15 2/15

The entropy value of each direction is as follows: S0=1.8867, S45=2.1413, S90=1.5230, S135=2.1413. This simple experiment proved the above statement.

2.3 The Entropy Direction of Haar-Like

In order to generate Haar-like features guided by the entropy information, a new concept named Entropy Direction of Haar (EDH) is proposed. For Haar-like features, EDH is defined as the two directions which have the higher entropy values than those of the other two directions. As clarified in Section 2.2, we can view the orientation of rectangles in Haar-like feature as its texture direction. Fifteen kinds of Haar-like feature templates that used in the paper are presented in Fig. 2 and EDHs of them are listed in Table 2.

Table 2. Entropy Direction of Haar-like

EDH	TypeNumber
0° 90°	3 4 9 10 11 12 14
45° 135°	1 2 5 6 7 8 13 15

Note that, in Table 2, Type 13, Type 14, Type 15 are all centrosymmetric. They generate similar entropy values in all four directions. Type 13 and Type 15 are classified as 45° and 135°, and Type 14 is classified as 0° and 90°. Classification in this way does not affect the descriptive capability of the features.

3 Feature Pool Construction and Update Strategy

3.1 Feature Pool Initialization

To get rid of the deficiency in the traditional algorithm, our method extracts entropies at the four directions from GLCM. The larger the entropy value at each direction is, the richer the detailed texture information is and the object needs more features to be described. The number of features in each direction is influenced by the value of entropy in that direction. If the object is simple and has a low entropy value in every direction, the algorithm will generate a relatively small number of features. Otherwise, it will generate a large number of features. Here the exponential distribution function is used to determine the feature number in each direction. The density function of exponential distribution is formed as formula (2).

$$p(x)=\begin{cases}\lambda e^{-\lambda x} & x>0\\ 0 & x\le 0\end{cases}\ . \tag{2}$$

Let λ=1 and replace x with Ei*ER. Ei represents the entropy value of direction i and i∈{0, 45, 90, 135}. ER represents entropy ratio that is used to adjust the value of entropy. Usually, Ei*ER is greater than zero and formula (2) can be simplified into the form of formula (3).

$$p(Ei) = \mathrm{EXP}(-Ei*ER)\ . \tag{3}$$

The smaller p(Ei) is, the larger the number of features generated. Through our method, every type of features can be found probably in the initial feature pool. So the flexibility of the boosting framework which put emphasis on the fusion of multiple weak classifiers is not degraded. But the total number of features is decreased and the distribution of features is corresponding with the texture information of object.

3.2 Self-adaptive Update of Feature Pool

In order to get a better tracking when the object state is changing during tracking, the feature pool must be updated partially to address this issue by removing the worst features and adding features with better classification ability to the pool. Specifically, the proposed HBTDTE algorithm introduces an update strategy based on the object's texture richness: the new method selects feature type with the biggest value of the entropy. The update of feature pool during tracking is depicted in Fig. 3.

From Fig. 3 we can see that the new generated feature is corresponding with the current texture direction of the object when applying our update strategy. The generated feature is added in the reserve features, so the feature pool can always be updated properly with the current state of the object when it is necessary. Fig. 4(a) and (b) show the features generated during training.

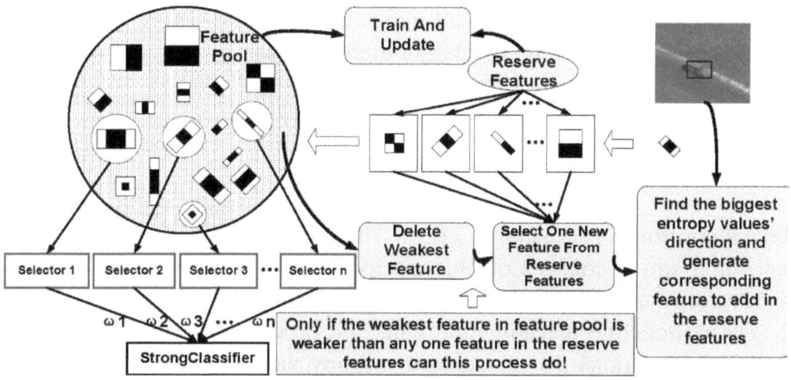

Fig. 3. Update of feature pool during tracking

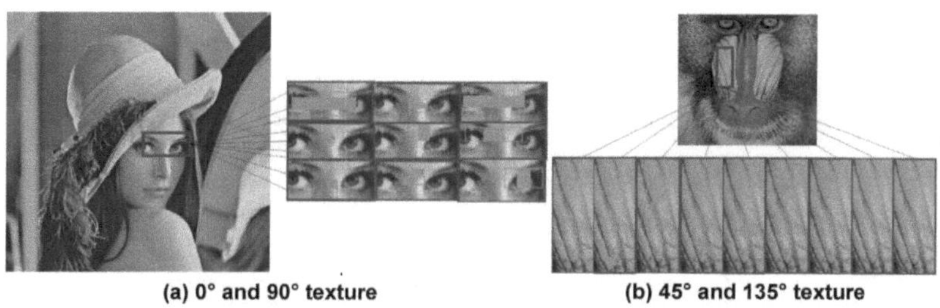

Fig. 4. Features generated during training

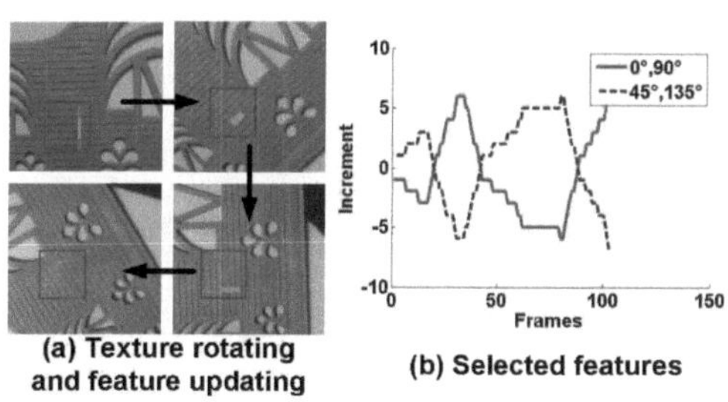

Fig. 5. Relationship between the texture direction and the feature updating procedure

The following experiment with rotating the texture illustrates the relationship between the texture richness and the feature update. In Fig. 5(a), the object in experiment video was rotated slowly and the texture's direction was therefore changing. As a result, the feature type during updating also changed slowly (see Fig. 5(b)). The features of EDH 0° and 90°(red solid) increases while the features of EDH 45° and 135°(blue dot) decreases and vice versa. The vertical coordinate indicates the increment of specified type of features relative to feature pool. The horizontal ordinate shows the frame number.

The results of the above experiment indicate that HBTDTE can generate new features corresponding to the object's texture direction and therefore guarantee an effective tracking.

4 Algorithm

The whole algorithm is presented in Fig. 6. The solid line at left represents the process of initialization which consists of object selection and feature number determination in each direction based on the texture direction information from the entropy of the GLCM. The dotted line at right represents the process of tracking, and extracting GLCM in four directions and calculating entropy values to find the maximum. The feature pool is then updated based on the entropy information.

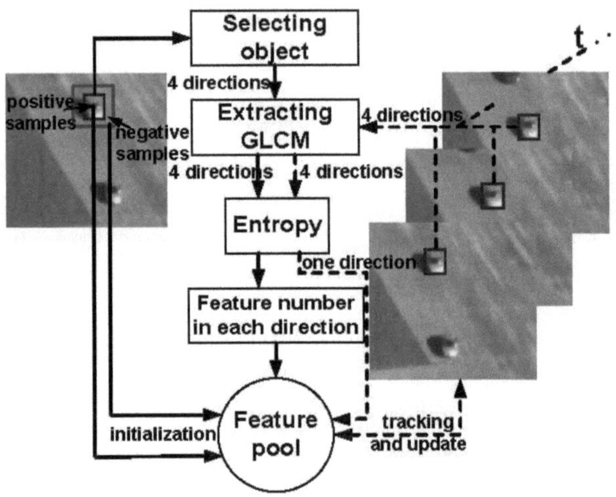

Fig. 6. Whole algorithm process

5 Experiments

5.1 Test of Basic Performance: Tracking Accuracy

The experimental videos come from VIVID data set[1]. Experimental results show that tracking effect of HBT and HBTDTE on EgTest01 and RedTeam have no obvious difference. But in 310th frame of EgTest03, tracking lost occurred in HBT. The lost happened when the vehicle made a turn. At that time, vehicle's texture does not stay parallel with the road edge any more. The angle between the vehicle and the road edge was about 45°. Texture direction changed from 135° to 90°. Tracking results of

[1] VIVID dataset web site: `http://vision.cse.psu.edu/data/vividEval/datasets/datasets.html`

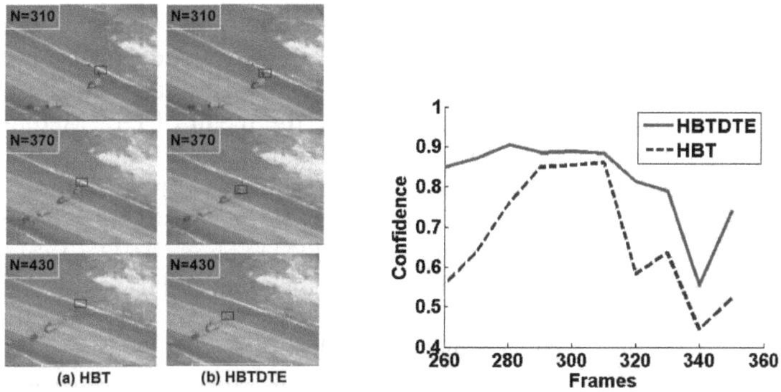

Fig. 7. Comparison of the tracking effect between HBT and HBTDTE (*left*) and distribution of the tracking confidence (*right*)

three frames (310th, 370th and 430th) on EgTest03 by HBT and HBTDTE respectively are shown in the left of Fig. 7(a) (b). We can see that the object was lost by HBT, but was continuously tracked by HBTDTE.

The tracking results from 260th to 350th frames are collected for close investigation. Right of Fig. 7 describes distribution of tracking confidence. Tracking confidence is a metric which is obtained through calculating weighted average of all selectors' weight and their results of selecting(-1 or 1). Table 3 lists the update times during tracking and the feature type distribution of selected features during update. As shown in the right of Fig. 7, tracking confidence in HBT is less than that in HBTDTE as a whole. This indicates that the feature pool of HBT had weaker classification ability than that of HBTDTE. Table 3 also indicates that there exist a lot of features with weaker classification ability in the feature pool of HBT. So update times of HBT is higher than those of HBTDTE. Most features generated during update have the EDH of 45° and 135° because when the vehicle made a turn, the texture changed to be at vertical direction. Therefore the best features to describe object were also changed. This shows that there is a strong relationship between the direction of the object texture and the EDH of Haar. As there were a small number of Haar-like features with the EDH of 45° and 135° in the feature pool of HBT, so more features of that type should be added through update from the reserved features. When the texture direction changes, HBTDTE can adapt to the change and generate proper features. But HBT generates features randomly and when these features cannot describe an object's texture effectively, HBT will fail. See Fig. 7, after the 310th frame, the tracking box stopped at the direction of original road edge and lost the object.

Table 3. Features update state

Me-thod	FeatureUpdate-Times	Type 0_90	Type 45_135
HBT	58	14	44
HBT DTE	17	7	10

In order to further analyze the two tracking algorithms, we propose to use the evaluation metric DE (Displacement Error) [13]. DE is the Euclidean distance between the object center and the GT (Ground Truth). Three metrics are used to evaluate the tracking performance statistically. Dmn stands for the mean of DE. Median value of DE is Dmd and Dmm=0.5Dmn +0.5Dmd. The smaller these three metrics are, the better the performance of the object tracking algorithm is. The statistical results of EgTest01, RedTeam and the continuous tracking part of EgTest03 are shown in Table 4, and PI is the percentage improvement of HBTDTE above HBT. Table 4 clearly indicates that tracking effect of HBTDTE is better than that of HBT.

Table 4. Statistical results of DE

Video	Method	Dmn	PI	Dmd	PI	Dmm	PI
EgTest01	HBT	11.16		10.84		11.00	
	HBTDTE	8.96	19.71%	9.43	13.01%	9.19	16.45%
RedTeam	HBT	4.34		4.12		4.23	
	HBTDTE	3.39	21.89%	3.00	27.18%	3.19	24.59%
EgTest03	HBT	2.96		2.62		2.79	
	HBTDTE	2.22	25.00%	2.24	14.50%	2.23	20.07%

5.2 Tracking Speed and Effectiveness

We further analyze the tracking speed and effectiveness. Table 5 lists the feature pool size (FPSize), mean of tracking confidence (MeanConf) and mean of FPS (Frame per Second) of above videos through the two algorithms.

Table 5. Analysis of tracking speed and effectiveness

Video	Method	FPSize	MeanConf	PI	MeanFPS	PI
EgTest01	HBT	350	0.922		18.78	
	HBTDTE	222	0.951	3.21%	29.90	59.21%
RedTeam	HBT	350	0.747		23.82	
	HBTDTE	225	0.766	2.46%	25.97	9.03%
EgTest03	HBT	350	0.850		28.80	
	HBTDTE	150	0.903	6.22%	37.08	28.75%

From Table 5 we can see that the feature pool size generated by HBTDTE was obviously smaller than HBT. But the tracking performance was not affected. HBTDTE not only improves the tracking speed, but also the tracking confidence.

Table 6. Comparison of mean confidence with the same feature pool size

Video	Method	MeanConf	PI
EgTest01	HBT	0.921538	
	HBTDTE	0.959205	4.09%
RedTeam	HBT	0.747342	
	HBTDTE	0.829618	11.01%
EgTest03	HBT	0.849696	
	HBTDTE	0.934743	10.01%

We let two algorithms' feature pool size be the same through adjusting ER. Their mean confidences are listed in Table 6.

As shown in Table 6, the mean confidence of HBTDTE is obviously higher than that of HBT and higher than that of HBTDTE in Table 5. The statistical data indicates that the classification capability of weak classifiers with the feature pool generated by HBTDTE is better than that of HBT when the feature pool size is the same.

5.3 Comparison to Other Tracing Algorithms

Previously, we have compared our new method with the original method HBT in both accuracy and efficiency. Now we further compared it with other state-of-the-art approaches. HBTDTE was compared with results reported in [14] which reports on performance of 4 trackers (On-line Random Forests (ORF) [15], Fragments-based Tracking (FT) [16], Multiple Instance Learning (MIL) [17], and Parallel Robust On-line Simple Tracking (PROST) [14]) on 4 sequences. It was also compared with results reported in [18] which reports on performance of 3 trackers(Color Histograms (CH) [19], Multi-Component-Descriptor (MCD) [20], and Haar-like Center-surround Feature based Classifiers (HCFC) [18]) on 9 sequences. The first four videos come from web site[2] and the last nine videos come from web site[3].

The performance was reported using Recall - number of true positives divided by the length of the sequence (true positive is considered if the overlap with ground truth is > 50% (the first four videos) or > 33.33% (the last nine videos)). Table 7 shows the performance measured by Recall.

Table 7. Percentage of frames tracked correctly

Sequence	Frames	ORF	FT	MIL	PROST	CH	MCD	HCFC	HBTDTE
Board	698	10.0	67.9	67.9	**75.0**	-	-	-	71.06
Box	1161	28.3	61.4	24.5	**91.4**	-	-	-	78.12
Lemming	1336	17.2	54.9	83.6	70.5	-	-	-	**87.23**
Liquor	1741	53.6	79.9	20.6	**83.7**	-	-	-	80.13
BoBoT-A	601	-	-	-	-	70.73	63.24	59.35	**96.01**
BoBoT-B	628	-	-	-	-	67.02	50.73	77.38	**89.83**
BoBoT-C	403	-	-	-	-	47.58	63.71	91.33	**100.00**
BoBoT-D	946	-	-	-	-	63.35	76.39	75.21	**80.48**
BoBoT-E	304	-	-	-	-	78.21	77.42	86.32	**98.36**
BoBoT-F	452	-	-	-	-	44.43	40.02	**68.32**	37.37
BoBoT-G	715	-	-	-	-	46.27	49.62	71.16	**74.16**
BoBoT-H	411	-	-	-	-	62.19	86.50	94.47	**100.00**
BoBoT-I	1016	-	-	-	-	**68.94**	47.63	56.33	50.93
Mean	-	27.3	66.0	49.2	80.2	60.97	61.70	75.54	**80.28**

[2] PROST dataset web site: `http://www.gpu4vision.org`.
[3] BoBoT dataset web site: `http://www.iai.uni-bonn.de/~kleind/tracking/index.htm`

Fig. 8. Two frames of each sequence during tracking

The best result is printed in bold faced letter; the second best result is underlined. HBTDTE scored best in 8/13 outperforming the other algorithms. Fig. 8 shows two frames of each sequence during tracking.

6 Conclusion and Future Work

In this paper, a novel object tracking algorithm that combines the texture information with the Haar-like feature is proposed on top of the conventional HBT algorithm. Our algorithm absorbs the texture description characteristic of the entropy extracted from GLCM into the Haar-like feature. But the robustness of the occlusion handling problem has yet to be improved. If the background is treated as the positive sample during training when there is occlusion, the error is accumulated that leads to a failed tracking. Semi-supervised learning can effectively solve this problem to some extent. In addition, we can consider the texture information of background and object together to guide the construction and update of the feature pool.

Acknowledgments. This project is supported by National Natural Science Foundation of China (61102155, 61272237, and 61272236), the Young and Middle-aged Science Funding of Hubei Provincial Department of Education (Q20111205).

References

1. Nummiaro, K., Koller-Meier, E., Van Gool, L.: An adaptive color-based particle filter. Image and Vision Computing 21(1), 99–110 (2003)
2. Comaniciu, D., Ramesh, V., Meer, P.: Real-time tracking of non-rigid objects using mean-shift. In: Proc. CVPR, vol. 2, pp. 142–149 (2000)

3. Lienhart, R., Maydt, J.: An extended set of Haar-like features for rapid object detection. In: Proc. ICIP, vol. 1, pp. 900–903 (2002)
4. Viola, P., Jones, M.: Robust real-time object detection. International Journal of Computer Vision 57(2), 137–154 (2004)
5. Dalal, N., Triggs, B.: Histograms of oriented gradients for human detection. In: Proc. CVPR, vol. 1, pp. 886–893 (2005)
6. Lowe, D.: Distinctive image feature from scale-invariant keypoints. International Journal of Computer Vision 60(2), 91–110 (2004)
7. Ojala, T., Pietikainen, M., Maenpaa, T.: Multiresolution gray scale and rotation invariant texture classification with local binary patterns. IEEE Transactions on Pattern Analysis and Machine Intelligence 24(7), 971–987 (2002)
8. Wu, G., Tang, Z.-M., et al.: Object Tracking Method Based on Gray Level Co-occurrence Matrix Texture Characteristic. Journal of Nanjing University of Science and Technology (Natural Science) 34(4), 459–463 (2010)
9. Heikkila, M., Pietikainen, M.: A texture based method for modeling the background and detecting moving objects. IEEE Transactions on Pattern Analysis and Machine Intelligence 28(4), 657–662 (2006)
10. Jafari-Khouzani, K., Soltanian-Zadeth, H.: Radon transform orientation estimation for rotation invariant texture analysis. IEEE Transactions on Pattern Analysis and Machine Intelligence 27(6), 1004–1008 (2005)
11. Avidan, S.: Support vector tracking. IEEE Transactions on Pattern Analysis and Machine Intelligence 26(8), 1064–1072 (2004)
12. Grabner, H., Bischof, H.: On-line boosting and vision. In: Proc. CVPR, pp. 260–267 (2006)
13. Shen, H., Sun, S.: Comparative study of color feature for particle filter based object tracking. In: Proc. ICMLC, vol. 3, pp. 1104–1110 (2012)
14. Santner, J., Leistner, C., Saffari, A., et al.: PROST:Parallel Robust Online Simple Tracking. In: Proc. CVPR, pp. 723–730 (2010)
15. Saffari, A., Leistner, C., Santner, J., Godec, M., Bischof, H.: On-line random forests. In: 3rd IEEE ICCV Workshop on On-line Comput. Vision (2009)
16. Adam, A., Rivlin, E., Shimshoni, I.: Robust fragments-based tracking using the integral histogram. In: Proc. CVPR (2006)
17. Babenko, B., Ming-Hsuan, Y., Belongie, S.: Robust Object Tracking with Online Multiple Instance Learning. IEEE Transactions on Pattern Analysis and Machine Intelligence 33(8), 1619–1632 (2011)
18. Klein, D.A., Schulz, D., Frintrop, S., Cremers, A.B.: Adaptive Real-Time Video-Tracking for Arbitrary Objects. In: Proc. IROS, pp. 772–777 (2010)
19. Pérez, P., Hue, C., Vermaak, J., Gangnet, M.: Color-Based Probabilistic Tracking. In: Heyden, A., Sparr, G., Nielsen, M., Johansen, P. (eds.) ECCV 2002, Part I. LNCS, vol. 2350, pp. 661–675. Springer, Heidelberg (2002)
20. Frintrop, S.: General object tracking with a component-based target descriptor. In: Proc. ICRA, pp. 4531–4536 (2010)

Robust Object Tracking Using Motion Context in Crowded Scenes

FeiMing Xu, Tong Lu*, and Yirui Wu

National Key Lab for Novel Software Technology
Dept. of Computer Science and Technology, Nanjing University, 210046, China
{xfm18901105,wuyirui1989}@163.com, lutong@nju.edu.cn

Abstract. Tracking objects in a crowded scene with occlusions has been a challenge in computer vision and multimedia in the past years. This paper presents a novel framework to track any arbitrary object through modeling its coupled motion context. For a scene which is densely packed, an individual movement is restricted into a specific pattern to make it regular to be detected. Moreover, members in a crowded scene are clustered into groups and thereby modeled as crowded scene contexts of a tracking object. Accordingly, we present a novel framework to track motions for any object in a crowded scene even with occlusions by using the modeled contexts. Experiments on a number of real-life surveillance videos illustrate the effectiveness and robustness of our method especially in handling occlusions in crowded scenes.

Keywords: object tracking, crowded scenes, occlusion, scene context.

1 Introduction

Object tracking in crowded scenes plays a crucial role for a wide range of computer vision and multimedia applications such as daily surveillance, behavior modeling and abnormal detection. Although many object tracking algorithms have been investigated and a significant progress has been made in the past few years, reliable object tracking in a crowded scene still remains a challenge due to the complexity of noises, varying viewpoints, clutter backgrounds and even illumination changes. Most seriously, a tracking object is always easily to lose when occlusions exist, which are very common and sometimes unavoidable in real-life crowded scenes.

In this paper, we propose a novel algorithm to track any arbitrary object through modeling its coupled motion context in a crowded scene. Essentially, the aim of the proposed method is to continually track an object even when it becomes unseen, which is in general brought by occlusions in a crowded scene. To achieve this, we hypothesize that when a scene is densely packed, individual movements will be restricted into a coupled pattern and thereby makes themselves relatively regular to be described. Moreover, members in a crowded scene

* Corresponding author.

B. Huet et al. (Eds.): PCM 2013, LNCS 8294, pp. 550–560, 2013.

are clustered into groups, which essentially constituting scene contexts for a tracking object. Accordingly, we present a novel object tracking framework to predict motions for continually tracking in a crowded scene with occlusions.

The main contributions of this paper are as follows:

1. We exploit a hybrid representation by simultaneously modeling the motions of an expected object and its coupled environmental contexts. Comparing with the existing object-centric representations, our method is more efficient to characterize a crowded scene, and
2. A novel object tracking algorithm is accordingly presented based on the representation, where specific occlusions can be detected and further well handled for a continuous and relatively long-time tracking. In this way, objects can be robustly tracked especially in crowded scenes.

The rest of the paper is organized as follows. Section 2 introduces the related works. Section 3 discusses the details of our object tracking framework by modeling motion context in crowded scenes. Experimental results and discussions are illustrated in Section 4. Finally, Section 5 concludes the paper and gives our future work.

2 Related Work

Object tracking has always been one of the interests in computer vision and multimedia in the past decade [4][10]. Many efforts have been paid to solve the challenges and the existing methods can be roughly classified into three categories: appearance modeling, motion modeling, and scene context modeling.

The appearance modeling approach avoids the problems of tracking drifts and occlusions by improving appearance templates. For instance, adaptive appearance modeling techniques track objects by indirectly employing statical analysis [7][9][17]. However, their appearance models are susceptible to be contaminated by long-time occlusions due to their blind update strategies. In [12], a tracking object is divided into regular grid cells and occlusion states are accordingly determined for every cell using a classifier. However, the classifiers have to be trained manually.

Considering the complexity of a crowded scene, the motion modeling approach analyzes spatial or temporal motion patterns to assist object tracking. Ali *et al.* [1] track objects in extremely crowded scenes and propose floor fields to predict target motions. [16] models various motion modalities at different locations inside a scene by employing the Correlated Topic Model [13]. Similarly, [11] captures the spatial and temporal variations in crowded scenes by training a hidden Markov model to model motion pattern. Unfortunately, occlusions remain a difficulty during tracking objects.

The context modeling approach makes use of additional scene information to aid object tracking. [6] employs networking targets in the same scene to estimate the location of an tracking object especially when it is unseen. But the relationship between the target and its supporters still faces difficulties in handling complex situations. [18] mines auxiliary objects from scene background and collaboratively uses them to track objects as a strong verification of expected locations. Unfortunately, it is uncertain to verify the relationship between an auxiliary object and the tracking object in the method.

3 The Proposed Tracking Framework

We propose a novel Bayesian tracking framework to realize the mentioned steps inspired by [8], where the tracking problem is formulated by maximizing the posterior distribution of state x_t at time t with the following available measurement:

$$p(x_t|z_{1:t}) \propto p(z_t|x_t) \int p(x_t|x_{t-1})p(x_{t-1}|z_{1:t-1})dx_{t-1} \tag{1}$$

z_t is the frame at time t and $p(z_t|x_t)$ is its likelihood. The state x_t for frame z_t is modeled as a vector describing the location, height and width of a tracking object.

Essentially, our framework consists of the following three modules.

3.1 Modeling Motion Transitions

We represent the motion transition from frame $t-1$ to frame t by $f(x_t|x_{t-1})$, which can be modeled by a Gaussian distribution of $N(\mu, \sigma)$, where μ is the mean of optical flows to describe object displacements and σ is its covariance matrix. Based on the observation that pedestrians in a crowded scene tend to move in a coupled manner, we hypothesize that an individual pedestrian can not only be tracked according to his/her own motion characterizations, but also the accompanied motion constraints from his/her neighbors to improve the robustness. It can be described as follows

$$p(M|z_t) = w_{indi} \cdot p(M_{indi}|z_t) + w_{group} \cdot p(M_{group}|z_t) \tag{2}$$

We model the latter item as an *object group* to characterize the coupled motions between a specific tracking object and its neighbors, facilitating predicting the trajectory when occlusions are detected. The neighbors here are simultaneously referred as *supporters* for the tracking object.

3.2 Appearance and Motion Updating during Tracking

To well measure the likelihood distribution $p(z_t|x_t)$, we associate an appearance template together with motion characteristics in our method, which are helpful to distinguish a specific object from its visually similar backgrounds or motion similar couplers.

Specifically, we first filter the extracted optical flows which are too large or too small and hash them into the polar coordinates to obtain a group of motion histograms. Since our observation likelihood is essentially decided by an appearance term and a motion term, it can be computed as follows:

$$p(z_t|x_t) = \alpha_{app,t} Simi(x_t, x_{obs}) + \alpha_{mo,t} Simi(x_t, x_{obs}) \tag{3}$$

where $Simi(x_t, x_{obs})$ is a Bhattachayya distance measure to evaluate the similarity of two vectors. The coefficients of the appearance term and the motion term of $Simi(x_t, x_{obs})$ are decided by their abilities to distinguish an object from its surroundings, respectively. We use the score of the particle template calculated in a previous frame to iteratively update the appearance and motion coefficients of $\alpha_{app,t}$ and $\alpha_{mo,t}$ by

$$\alpha_{app,t} = \alpha_{app,t-1} \cdot \frac{var_{app}^2}{var_{app}^2 + var_{mo}^2} \tag{4}$$

$$\alpha_{mo,t} = \alpha_{mo,t-1} \cdot \frac{var_{mo}^2}{var_{app}^2 + var_{mo}^2} \tag{5}$$

The detection of an occlusion or a drift during object tracking is as follows. We store the motion and appearance templates for a tracking object in the beginning k frames of $T = T_1, .., T_k$ as an initialization for tracking. Then for a new frame $t(t > k)$, we compare the tracking result with the stored templates in T as follows: if the matching score is greater than an empirical threshold, we update the stored templates with the latest tracking result from frame t, else an occlusion or a drift is accordingly detected and we thereby stop updating the appearance and motion templates to avoid unexpected results.

3.3 Tracking Objects through a Supporter Model

After modeling motion transitions and giving motion updating strategies, a specific object in a crowded scene with occlusions can be continually tracked as follows. We first track an object using the particle filter method [14] until there is an occlusion or a drift is detected. Then we use the supporters of the tracking object, which are essentially the motion contexts inside the same crowded scene, to assist predicting the trajectory and further verifying whether the object can be tracked continually. These two stages are alternately switched during the whole tracking process in a dynamic way.

Obviously, it can be seen that the key to track an object through its contexts is to detect and model the relationship between a tracking object and its supporters. To search for proper supporters as scene contexts for a specific tracking object, we consider the following two properties are important:

1. A supporter should move regularly to make itself consistent in predicting another tracking object.
2. A supporter should be easy to match, thereby an object together with its supporters will be tracked in a coupled and robust way by propagating the motion details from the supporters to the tracking object as a prediction basis.

A tracking object together with its supporters forms the mentioned object group, which is generally characterized by similar motion patterns and composed of spatially close scene objects. We denote such a group by a state vector which describes its motion and spatial properties as (L, G, M, t), where L is the 2-D location of the group, G denotes the spatial range of the group which is represented as a variance Σ of the distance from its boundary to the center, m denotes its motion characterization, l is the supporter number, t is its existing time. Then the details to search for supporters of a tracking object are shown by the following steps.

Segment a Crowded Scene. We first segment a crowded scene based on the motions characterized by optical flows. Due to the fact that the extracted feature points from a scene image may be too sparse, we employ the Large Displacement Optical Flow (LDOF) [3] to obtain dense optical flows. We segment every crowded scene frame according to the extracted optical flow vectors. Specifically, instead of clustering the vectors using K-means or kNN algorithms which do not consider spatial constraints, we employ the graph-based method [5] to obtain scene segmentation results (see the example results in Fig. 1). We use $G = (V, E)$, which is an undirected graph, to describe the segmentation results. V is the set of segmented regions where every vertex $v \epsilon V$ denotes a set of optical flow vectors, and $(v_i, v_j) \epsilon E$ corresponds to a pair of neighboring vertices. Note that an edge $(v_i, v_j) \epsilon E$ has a weight of $w(i, j)$, which is a non-negative measure of the dissimilarity between the two neighbors of v_i and v_j. The weight function is measured by $w(i, j) = |Op(p_i) - Op(p_j)|$, where $Op(p_i)$ is the optical flow of pixel p_i.

Fig. 1. Example results of graph-based segmentation for relatively crowded scenes

Search for Supporters from the Segmented Regions. After segmenting a crowded scene into several regions, we search for the supporters for a tracking object from the segmented regions which have regular motion descriptions. We employ the method in [14][15] to extract the attention regions (ARs) from the segmentation results, which we consider can be a good choice as motion supporters since the defined ARs are relatively easy to match and less error-prone. Moreover, we can narrow the searching space for more efficient computing since the potential max range of possible ARs in a new frame will be accordingly predicated.

Specifically, we consider a detected AR as a supporter instead of directly using a segmented scene region. Suppose the motion of a detected AR can be predicted by optical flows as a Gaussian distribution (μ, σ), we set the searching space for a new frame as $(-3\sigma, 3\sigma)$ and the local discrimination score is accordingly defined as $\rho_L(x) = min(D(x), D(y))$. Then in a new frame, we use the gradient decent method to search for the most discriminative regions. An expected AR supporter is defined as a state vector of $Sm_i = \{T(i), S(i)\}$, where $T(i)$ denotes the templates describing the color histogram and the motion pattern of an AR, while $S(i)$ is its state vector characterizing the width, height, 2-D position and its duration time. Moreover, we categorize every supporter into two types according to whether it together with the tracking object exist in the same group, due to the fact that the objects inside the same group tend to have relatively closer motion correlations.

Track an Object Using Supporters in a Crowded Scene. In a crowded scene, we assume a tracking object and its detected supporters have regular relative movements, and therefore motions of a pedestrian are generally easy to be interfered by their surroundings. Specifically, we describe the relationship between a tracking object and its detected supporters by $y = f(x) + \epsilon$, where $f(x)$ is the mapping from the supporters to the object, ϵ denotes Gaussian noise with zero mean and variance σ^n, y and x are the states of the tracking object and its supporters. We model the states of x by the width, height, 2-D location of a supporter within frame t and its duration time t_s, respectively. Accordingly, the mapping from the supporters to the tracking object can be defined as $f(x) = w^T \cdot x$.

Then we need learn the function $f(x)$ which transforms the states from the supporters to the tracking object as $y = f(x) + \epsilon$. This is essentially a Gaussian linear regression problem. Assume for frame t, we obtain a matched supporter of x^* and calculate the function f^* by

$$p(f^*|x^*, X, y) = \int f(x^*|w)p(w|X, y)dw = N(\frac{1}{\sigma^2}(x^*)^\top A^{-1}Xy, (x^*)^\top A^{-1}x^*) \tag{6}$$

where $A = \Sigma_p^{-1} + \frac{1}{\sigma^2}XX^\top$, and w is the Gaussian prior over the parameter $w \sim (0, \Sigma_p)$. The X and y are the matching results of the supporters and the tracking object in $(t - t_s, ..., t - 1)$. Next, after predicting f through Gaussian linear regression, we combine the two types of supporters S to reduce the uncertainly of the position for the tracking object as follows:

$$P(y|S) = \lambda \sum_{S_i \in G_{indi}^t} w_i p(f_i^*) + \lambda_{context} \sum_{S_j \in G_{context}^t} w_j p(f_j^*) \tag{7}$$

Accordingly, the prediction result during object tracking is predicted by the coupled motions of the tracking object when occlusions or drifts are detected.

4 Experiments and Discussion

To demonstrate both the effectiveness and the robustness of our method, we test our algorithm on several challenging two real-world crowded scenes of a campus street and campus gate. The pedestrains within the scene walking together and

Fig. 2. We use two real-world crowded scenes containing a large amount of individuals to evaluate our method. The Campus Street(left) contains pedestrains and bicycles resulting in severe occlusions. The Campus Gate scene(right) has a simpler background and more regular motion, but it's easy to drift since the individuals walking closely.

Fig. 3. Comparing dense optical flows (left) with sparse optical flows (right) to characterize motions. We note that the detected feature points of sparse optical flows are sometimes even unable to see. Obviously, dense optical flows help analyze motion characteristics inside a crowded scene.

cover each other that result in severe occlusions(see Fig. 2). For every scene image, we zoom it to calculate LDOF features, which are found better than the sparse optical flows as a motion descriptor (see Fig. 3).

We first test our algorithm in the Campus Street scene, where there are many vehicles and pedestrians and thereby is challenging since a tracking object is easy to be partially or even completely occluded by other pedestrians and vehicles. We set the cuboid size of a supporter as 20×20, the initial coefficient of a motion template as 0.3 and the coefficient of appearance template as 0.7. Fig. 4 shows some detected supporters which help predict the estimated position of a tracking object (see the green bounding box in Fig. 4). A samller green bounding box in Fig. 4 indicates a single detected supporter, and the opposite end of the line connecting with the supporter is the calculated prediction indicating the possible position of the tracking object. Once the tracking object is occluded ,its position still can be predicted by its supporters(the red point means the supporter's prediction of the target's position when the occlusion or drift occurs). Accordingly, after the object appears again, we can keep tracking it again(see Fig. 4).

Fig. 5 shows more object tracking examples in a crowded scene. Every object is successfully tracked before occlusion occurs. Once an occlusion occurs, the tracking result will deviate from the ground truth. As illustrated, the OF

Fig. 4. Object tracking examples. The largest bounding box indicate the tracking target and the small green bounding box indicates a detected supporter, while the end of a connecting green line indicates the prediction of the tracking object from a supporter, and our algorithm can track a specific object even when it temporarily disappears.

algorithm [2] successfully handle weak occlusions (see Frame 2 in the first row of Fig. 4), but fail to deal with serious occlusions (see Frame 4 in the first row). The tracking object thereby shifts to the vehicle which is passing by the tracking object. Comparatively, our method successfully tracks the object (see the second row of Fig. 5 as a comparison). More object tracking examples can refer Fig. 5.

To quantitatively compare our algorithm with the existing object tracking methods of PF [15] and OF [2], we measure the results against the ground truth as follows. We first manually label 20 objects in the Campus Street scene and 18 objects in the Campus Gate scene, respectively. Then we define the error measure as the average distance $\|y_t - x_t\|$, where y_t is the ground truth and x_t is a tracked result. Fig. 6 shows the distribution of the errors, which are the means and the standard errors computed by trajectories analysis. The finial mean error of our method is 5.3403 in the Campus Street scene, which is lower than that of PF 8.0440 and OF 6.2129. In the Campus Gate, the mean error of our method is 13.244, which is lower than PF 16.2854 and OF 15.6778. We find the mean error is near to the quarter of object width, indicating that the tracking error is smaller than the size of the tracking object. It can be found that motion supporters help object tracking even under specific heavy occlusions and drifts. Error occurs when the pedestrains unexpectedly changed the motion and our method could not capture the multiple motion since the occluding individual changing the motion in different direction.

Fig. 5. Example object tracking results when occlusions occur. The green bounding box is the ground truth, the purple bounding boxes are the detected results of the OF method, while the red bounding boxes are the tracking results of our method. The red point indicates the centroid of the tracking object when occlusions or drifts are detected. Row 1 and Row 2, Row 3 and Row 4, Row 5 and Row 6 illustrate object tracking results using the OF method and our method for comparison.

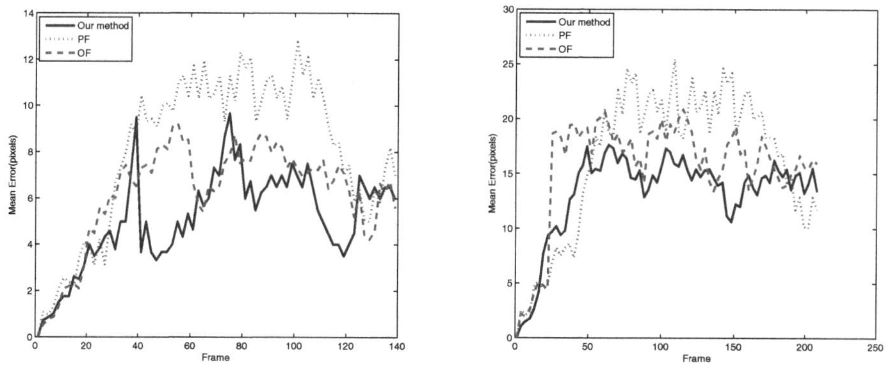

(a) (b)

Fig. 6. The mean distance between the manually and automatically tracked trajectories of Street scene (a) and Gate scene (b)

5 Conclusion

This paper proposes a novel algorithm to track individual objects in a crowded scene with occlusions. We restrict an individual movement into a specific pattern to make it regular to be detected. Moreover, members in a crowded scene are clustered into groups and thereby modeled as crowded scene contexts of a tracking object. Accordingly, we present a novel framework to track motions for any object in a crowded scene even with occlusions by using the modeled contexts. The experimental results show that our method can guarantee a relatively long-term tracking even with specific heavy occlusions. We believe crowded motions can be further leveraged to help track irregular objects in our future work.

Acknowledgments. The work described in this paper was supported by the Natural Science Foundation of China under Grant No. 61272218 and 61021062, the 973 Program of China under Grant No. 2010CB327903, and the Program for New Century Excellent Talents under NCET-11-0232.

References

1. Ali, S., Shah, M.: Floor fields for tracking in high density crowd scenes. In: Forsyth, D., Torr, P., Zisserman, A. (eds.) ECCV 2008, Part II. LNCS, vol. 5303, pp. 1–14. Springer, Heidelberg (2008)
2. Baker, S., Matthews, I.: Lucas-kanade 20 years on: A unifying framework. International Journal of Computer Vision 56(3), 221–255 (2004)
3. Brox, T., Bregler, C., Malik, J.: Large displacement optical flow. In: IEEE Conference on Computer Vision and Pattern Recognition, CVPR 2009, pp. 41–48. IEEE (2009)

4. Comaniciu, D., Meer, P.: Mean shift: A robust approach toward feature space analysis. IEEE Transactions on Pattern Analysis and Machine Intelligence 24(5), 603–619 (2002)
5. Felzenszwalb, P.F., Huttenlocher, D.P.: Efficient graph-based image segmentation. International Journal of Computer Vision 59(2), 167–181 (2004)
6. Grabner, H., Matas, J., Gool, L.V., Cattin, P.: Tracking the invisible: Learning where the object might be. In: 2010 IEEE Conference on Computer Vision and Pattern Recognition (CVPR), pp. 1285–1292. IEEE (2010)
7. Han, B., Davis, L.: On-line density-based appearance modeling for object tracking. In: Tenth IEEE International Conference on Computer Vision, ICCV 2005, vol. 2, pp. 1492–1499. IEEE (2005)
8. Isard, M., Blake, A.: Condensation conditional density propagation for visual tracking. International Journal of Computer Vision 29(1), 5–28 (1998)
9. Jepson, A.D., Fleet, D.J., El-Maraghi, T.R.: Robust online appearance models for visual tracking. In: Proceedings of the 2001 IEEE Computer Society Conference on Computer Vision and Pattern Recognition, CVPR 2001, vol. 1, pp. I–415. IEEE (2001)
10. Khan, Z., Balch, T., Dellaert, F.: An mcmc-based particle filter for tracking multiple interacting targets. In: Pajdla, T., Matas, J(G.) (eds.) ECCV 2004. LNCS, vol. 3024, pp. 279–290. Springer, Heidelberg (2004)
11. Kratz, L., Nishino, K.: Tracking with local spatio-temporal motion patterns in extremely crowded scenes. In: 2010 IEEE Conference on Computer Vision and Pattern Recognition (CVPR), pp. 693–700. IEEE (2010)
12. Kwak, S., Nam, W., Han, B., Han, J.H.: Learning occlusion with likelihoods for visual tracking. In: 2011 IEEE International Conference on Computer Vision (ICCV), pp. 1551–1558. IEEE (2011)
13. Lafferty, J.D., Blei, M.D.: Correlated topic models. In: Proceedings of the 2005 Conference Advances in Neural Information Processing Systems, pp. 147–155. Citeseer (2006)
14. Nummiaro, K., Koller-Meier, E., Van Gool, L.: An adaptive color-based particle filter. Image and Vision Computing 21(1), 99–110 (2003)
15. Pérez, P., Hue, C., Vermaak, J., Gangnet, M.: Color-based probabilistic tracking. In: Heyden, A., Sparr, G., Nielsen, M., Johansen, P. (eds.) ECCV 2002, Part I. LNCS, vol. 2350, pp. 661–675. Springer, Heidelberg (2002)
16. Rodriguez, M., Ali, S., Kanade, T.: Tracking in unstructured crowded scenes. In: 2009 IEEE 12th International Conference on Computer Vision, pp. 1389–1396. IEEE (2009)
17. Ross, D., Lim, J., Yang, M.-H.: Adaptive probabilistic visual tracking with incremental subspace update. In: Pajdla, T., Matas, J(G.) (eds.) ECCV 2004. LNCS, vol. 3022, pp. 470–482. Springer, Heidelberg (2004)
18. Yang, M., Wu, Y., Hua, G.: Context-aware visual tracking. IEEE Transactions on Pattern Analysis and Machine Intelligence 31(7), 1195–1209 (2009)

Person Re-identification by Smooth Metric Learning

Yimin Wang[1], Ruimin Hu[1], Chao Liang[1], Chunjie Zhang[2], and Qingming Leng[1]

[1] National Engineering Research Center for Multimedia Software, School of Computer, Wuhan University, Wuhan, 430072, China
[2] School of Computer and Control Engineering, University of Chinese Academy of Sciences, Beijing, 100190, China

Abstract. Person re-identification, which is to match people across areas covered by multiple non-overlapping surveillance cameras, has drawn great research interests in the video surveillance domain. Previous research mainly focus on feature extraction and distance measure, where the former aims to find robust feature representation while the latter seeks to learn an optimal metric space. Because of the introduction of supervised information, metric learning methods can usually achieve better performance over feature based methods. As one of the most representative metric learning method, the Large Margin Nearest Neighbor (LMNN) algorithm was recently applied in person re-identification task and achieved satisfactory results [1]. However, LMNN uses a standard hinge loss function, which is neither differentiable everywhere nor time-efficiency due to the usage of all training samples. In this paper, we propose to replace hinge loss function with the logistic loss function, which transforms LMNN to a smooth unconstrained convex optimization problem easily solved with gradient descent algorithm. Thereafter, we further design a stochastic sampling scheme to accelerate the optimization process of the above problem with randomly selected training samples. Extensive comparative experiments conducted on two standard datasets have shown the effectiveness and efficiency of the proposed algorithm over a series of standard baseline methods.

Keywords: Person re-identification, Surveillance video, Multiple cameras, Metric learning.

1 Introduction

Recently, increasing number of surveillance camera networks have been set up for monitoring pedestrian activities over a large public areas, such as airports, metro stations and parking lots. In these scenarios, matching person across non-overlapping cameras in a surveillance camera network, a.k.a, person re-identification, is becoming a hot research spot in the computer vision community. During the past five years, a lot of research effort has been devoted to this field [2–4, 1, 5]. However, due to significant changes of view angles, scales

B. Huet et al. (Eds.): PCM 2013, LNCS 8294, pp. 561–573, 2013.

Fig. 1. Some typical samples from VIPeR dataset [6] where appearance changes are mainly caused by variant views and lighting conditions. Each column shows two images of the same person captured by two different cameras.

and illumination conditions (see Fig. 1), different persons captured by the same camera usually appear more alike than the same person across different cameras, making person re-identification remain a unsolved and challenging problem.

Generally speaking, person re-identification can be considered as a visual retrieval problem [5], i.e. given a query person image taken in one camera, the algorithm is expected to search images of the same person captured by other cameras. Typically, it consists of two stages: feature extraction and distance measure. The former aims to seek a discriminative and robust feature representation which can easily separate different persons in various cameras. Wang et al. [7] used a co-occurrence matrix to capture the spatial distribution of the appearance relative to each of object parts to modeling the appearance of people. Farenzena et al. [2] divided the image of person into 5 regions by exploiting symmetry and asymmetry perceptual principles, and then combine multiple color and texture features to represent the appearance of people. However, in the condition of significant intra-object appearance variation cased by severe viewing changes, computing a set of features that are both distinctive and stable is extremely difficult [4].

In contrast, distance learning based methods focus on the second stage, distance measure, and aim to seek a suitable distance metric function through supervised learning. Gray and Tao [8] transformed the matching problem into a classification problem, in which they assigned a pair of images belonging to the same individual with positive sample or negative sample otherwise and learn the classifier using AdaBoost. The essence of the method is learning a weight vector assigning higher weight to the more discriminative feature. Recently, metric learning methods targeting for a matrix rather than a vector are exploited for person re-identification problem and are reported better performance. Hizer et al. [1] learned an optimal distance metric through LMNN algorithm, which was designed for k Nearest Neighbor classification [9]. The core idea of LMNN algorithm is that the feature distance of the same individual is small while that of different persons is large under the learned metric. However, standard LMNN algorithm uses a hinge loss function that is not differentiable everywhere resulting in difficulty of solution. Instead, Zheng et al. [4] used a logistic loss function and got a smooth objective function to learn the metric, in which a probabilistic relative distance comparison (PRDC) constrain was exploited to learn a low rank projection matrix. Recently, Kostinger et al. [5] developed a simpler algorithm to achieve an analytical form of the learned metric with the assumption that all samples coming from the Gaussian distribution.

In this paper, we focus on the distance learning based methods. Our idea is motivated from LMNN algorithm, which is known as the-state-of-the-art metric learning algorithm. However, according to the above discussion, LMNN is difficult to solve due to the usage of a hinge loss function, which is not differentiable everywhere. Instead, we use a logistic loss function, a soft approximation to original hinge loss function, to improve the standard LMNN algorithm, which transforms the original problem to a smooth unconstrained convex optimization problem. In addition, the new model is more flexible and achieves better performance with seeking a suitable approximation parameter. Besides, the LMNN algorithm is time consuming, even though an active set strategy was exploited in [9] to improve the efficiency. Motivated by Stochastic Gradient Descent (SGD) algorithm [10], we further propose a stochastic sampling based gradient descent algorithm. In particular, for each positive examples, we randomly selected some negative samples to consist real triple training samples. Extensive experiment results show that the proposed method is superior than the standard LMNN method in both accuracy and efficiency.

2 The Approach

This section presents our approach. We begin with terms and notations used in metric learning based person re-identification. Then, technical details of improved LMNN using logistic loss function is discussed and a stochastic sampling strategy is given to accelerate the algorithm implementation.

2.1 Person Re-identification Based on Metric Learning

For the convenience of presentation, we consider a pair of cameras C_a and C_b covering different area without overlapping, and $O_a = \{o_a^1, o_a^2, \cdots, o_a^m\}$ and $O_b = \{o_b^1, o_b^2, \cdots, o_b^n\}$ are images of persons captured by C_a and C_b, respectively. The task of person re-identification is that finding images of the same person from O_b for each instance o_a in O_a. Usually, the algorithm executes as follows: First, a d-dimensional feature vector is extracted to represent each instance. Second, for each pair images, a specialized distance function, such as Euclidean distance, is defined to computing the distance between the feature vectors. Finally, a ranking list is obtained with top positions corresponding to more alike the query person images.

Specially, let $\mathbf{x} = (x_1, \cdots, x_d)$ represent the d-dimensional feature vector of of a person o, then the Euclidean distance between o_a^i and o_b^j can be formulated as

$$D(o_a^i, o_b^j) = D(\mathbf{x}_a^i, \mathbf{x}_b^j) = \left\|\mathbf{x}_a^i - \mathbf{x}_b^j\right\|_2^2 = (\mathbf{x}_a^i - \mathbf{x}_b^j)^\top (\mathbf{x}_a^i - \mathbf{x}_b^j) \tag{1}$$

where $(\cdot)^\top$ represents the transpose of a vector or matrix. With the above notation, the formulation of metric learning can be denoted as:

$$D_{\mathbf{M}}(\mathbf{x}_a^i, \mathbf{x}_b^j) = (\mathbf{x}_a^i - \mathbf{x}_b^j)^\top \mathbf{M} (\mathbf{x}_a^i - \mathbf{x}_b^j) \tag{2}$$

where $\mathbf{M}$ is a positive semi-definite matrix for the validity of metric. In this case, $\mathbf{M}$ can be factored into real-valued matrices as $\mathbf{M} = \mathbf{L}^\top \mathbf{L}$ [11], and the Eq. (2) can be rewritten as

$$D_{\mathbf{L}}(\mathbf{x}_a^i, \mathbf{x}_b^j) = \left\|\mathbf{L}(\mathbf{x}_a^i - \mathbf{x}_b^j)\right\|_2^2 \tag{3}$$

The purpose of metric learning is to learn the positive semi-define matrix $\mathbf{M}$ or the real-valued matrix $\mathbf{L}$.

2.2 The Smooth LMNN Method

Before going further, we first briefly review the classic LMNN metric learning algorithm [9]. LMNN learns a real-valued matrix $\mathbf{L}$ under two constrain terms. One pulls the same labeled examples closer together, the other pushes examples with different labels further apart.

The *pull* term penalizes large distances between each instance and its target match, the formulation of which is given by:

$$\varepsilon_{pull}(\mathbf{L}) = \sum_{i,j \rightsquigarrow i} D_{\mathbf{L}}(\mathbf{x}_a^i, \mathbf{x}_b^j) \tag{4}$$

where $j \rightsquigarrow i$ means that $\mathbf{x}_b^j$ has the same label as $\mathbf{x}_a^i$.

The *push* term penalizes small distances between differently labeled examples. In particular, the term penalizes the differently labeled examples that invalidate the inequality in Eq. (5), named as *imposter*:

$$D_{\mathbf{L}}(\mathbf{x}_a^i, \mathbf{x}_b^l) \geq D_{\mathbf{L}}(\mathbf{x}_a^i, \mathbf{x}_b^j) + 1 \tag{5}$$

where $\mathbf{x}_a^i$ and $\mathbf{x}_b^j$ have the same label (belong to the same person), while $\mathbf{x}_b^l$ has another label. The formulation of *push* term is given by:

$$\varepsilon_{push}(\mathbf{L}) = \sum_{i,j \rightsquigarrow i}^{m} \sum_{l} (1 - y_{il})[1 + D_{\mathbf{L}}(\mathbf{x}_a^i, \mathbf{x}_b^j) - D_{\mathbf{L}}(\mathbf{x}_a^i, \mathbf{x}_b^l)]_+ \tag{6}$$

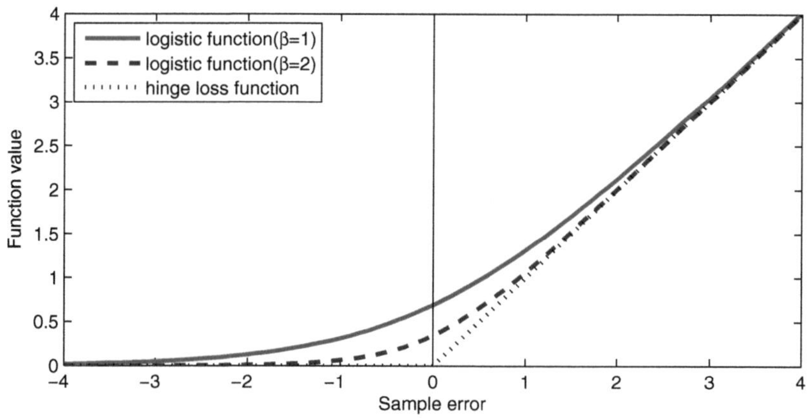

Fig. 2. Explanation for that the logistic loss gives a soft approximation to hinge loss, where the β is larger, the logistic loss is more near to hinge loss

where y_{il} is an indicator variable that equals to 1 if and only if $\mathbf{x}_a^i$ and $\mathbf{x}_b^l$ having the same label, and 0 otherwise; $[z]_+ = max(z, 0)$ denotes the standard hinge loss function.

Finally, the two terms $\varepsilon_{pull}(\mathbf{L})$ and $\varepsilon_{push}(\mathbf{L})$ are combined into an integrated loss function for distance metric learning:

$$\varepsilon(\mathbf{L}) = (1-\mu)\varepsilon_{pull}(\mathbf{L}) + \mu\varepsilon_{push}(\mathbf{L}) \tag{7}$$

where $\mu \in [0, 1]$ is the balance parameter.

Note that, the hinge loss $[z]_+$ is not differentiable at $z = 0$, which can be replaced with a smooth logistic loss function as shown in Fig. 2. The new loss function can be rewritten as

$$\varepsilon_{\text{LMNN-S}}(\mathbf{L}) = (1-\mu)\varepsilon_{pull}(\mathbf{L}) + \mu\varepsilon^*_{push}(\mathbf{L}) \tag{8}$$

$$\varepsilon^*_{push}(\mathbf{L}) = \sum_{i,j\rightsquigarrow i}^{m} \sum_{l} (1-y_{il})\ell_\beta(1 + D_{\mathbf{L}}(\mathbf{x}_a^i, \mathbf{x}_b^j) - D_{\mathbf{L}}(\mathbf{x}_a^i, \mathbf{x}_b^l)) \tag{9}$$

where $\ell_\beta(x) = \frac{1}{\beta}\log(1 + e^{\beta x})$ is the generalized logistic loss function. It is easy to see from the Fig. 2 that the logistic loss gives a soft approximation to hinge loss. The parameter of logistic loss β is the approximation parameter. As larger as the β is, the logistic is more near to hinge loss. The LMNN-S improves the performance with seeking a more suitable β (see the experiments results in Sec. 3.4).

Since the logistic loss function is convex and differentiable everywhere, Eq. (9) is a smooth convex optimization problem with respect to $\mathbf{L}$ and hence Eq. (9) is smooth convex optimization problem as quadratic term is smooth convex.

Algorithm 1. Learning the matrix $\mathbf{L}$

Input: The training set data: Positive samples with pair form $S_p = \{(\mathbf{x}_p^i, \mathbf{x}_p^j)_k\}$, Negative Samples with triple form $S_n = \{(\mathbf{x}_n^i, \mathbf{x}_n^j, \mathbf{x}_n^l)'_k\}$

1: Initialize $\mathbf{L}_0$ as identical matrix;
2: **for** $i = 1$ to $MaxIter$ **do**
3: Compute $\nabla\varepsilon(\mathbf{L}) = \nabla\varepsilon_{pull}(\mathbf{L}) + \nabla\varepsilon_{push}(\mathbf{L})$ as Eq.(10-12)
4: Choose a proper step λ
5: Compute $\mathbf{L}_{i+1} = \mathbf{L}_i - \lambda\nabla\varepsilon(\mathbf{L})$ as Eq.(13)
6: **if** converge **then**
7: break;
8: **end if**
9: **end for**

Output: The optimal matrix $\mathbf{L}^*$

2.3 Stochastic Sampling Based Optimization Algorithm

As the above discussion, we transform the problem to a smooth convex optimization problem, which can be solved with a simple gradient descent method. However, it is time consuming with massive training samples. In particular, assume there are N different persons, the size of positive samples is $O(N)$. In addition, for each positive samples, there are $O(N-1)$ negative samples and hence the size of total training samples is $O(N^2)$. In [9], an active set strategy was exploited to improve the efficiency. The active set consists of the *imposter* samples which is relatively little. Even so, it is time consuming for using the k closet within-class samples [12].

SGD algorithm was designed to solve large scale linear prediction [10] where the gradient was compute on randomly selected samples. Motivated by SGD algorithm, we propose a stochastic sampling based gradient descent algorithm. In particular, for each positive examples, we randomly selected $k \ll N$ negative samples, with which the size of training set reduce to $O(kN)$ from $O(N^2)$.

With stochastic sampling strategy, a simple gradient-descent method is exploited to learn the projection matrix $\mathbf{L}$. The detail is as follows: The gradient of the loss function $\varepsilon(\mathbf{L})$ is given as

$$\frac{\partial\varepsilon_{\text{LMNN-S}}(\mathbf{L})}{\partial\mathbf{L}} = \frac{\partial\varepsilon_{pull}(\mathbf{L})}{\partial\mathbf{L}} + \frac{\partial\varepsilon_{push}(\mathbf{L})}{\partial\mathbf{L}} \tag{10}$$

$$\frac{\partial\varepsilon_{pull}(\mathbf{L})}{\partial\mathbf{L}} = 2L\sum_{i,j\rightsquigarrow i}^{m}(\mathbf{x}_a^i - \mathbf{x}_b^j)(\mathbf{x}_a^i - \mathbf{x}_b^j)^\top \tag{11}$$

$$\frac{\partial\varepsilon_{pull}(\mathbf{L})}{\partial\mathbf{L}} = 2L\sum_{i,j\rightsquigarrow i}^{m}\sum_{l} g(\varepsilon_{pull}(\mathbf{L}))((\mathbf{x}_b^j)(\mathbf{x}_b^j)^\top - (\mathbf{x}_b^l)(\mathbf{x}_b^l)^\top - 2(\mathbf{x}_b^j - \mathbf{x}_b^l)(\mathbf{x}_a^i)^\top) \tag{12}$$

where $g(x) = (1 + e^{-\beta x})^{-1}$ is the derivative of logistic loss function $\ell_\beta(x)$.

With the gradient, an iterative optimization algorithm is used. Starting from an initial identical matrix, which means no projection, $\mathbf{L}$ is optimized iteratively as follows

$$\mathbf{L}_{i+1} = \mathbf{L}_i - \lambda \cdot \frac{\partial \varepsilon(\mathbf{L}_i)}{\partial \mathbf{L}_i} \tag{13}$$

where $\lambda > 0$ is a step length automatically determined at each updating step using a similar strategy in [4]. The iteration is terminated when the updating times is greater than the maximum iterative times (i.e. 1000 in this work) or the following criterion is met.

$$| \varepsilon_{i+1} - \varepsilon_i | < \epsilon \tag{14}$$

where ϵ is a small positive value and fixed to 10^{-9} in this paper. The pseudo-code of algorithm is given in Algorithm 1.

3 Experimental Results

In this section, we evaluated the proposed approach for person re-identification on two widely used datasets, the VIPeR dataset [6] and the i-LIDS Multiple-Camera Tracking Scenario (MCTS) dataset [3]. First, we compared our method with several state-of-the-art person re-identification methods on both datasets, under the similar experiment settings. Moreover, we validate the influence of performance with different algorithm parameters on VIPeR dataset.

3.1 Datasets

The VIPeR dataset is collected by Gray et al. [8] and contains 632 persons with two images for each person taken from two non-overlapping outdoor cameras. All images of individuals are normalized to a size of 128×48 pixels. Example images are shown in Fig. 1, in which each column contains two images of the same person under different cameras. As shown, view changes are the most significant cause of appearance change, and most of the examples contain a viewpoint of 90 degrees. Besides, other variations are also considered, such as illumination conditions.

The i-LIDS MCTS dataset [3] is taken at a busy airport hall with several disjoint indoor cameras. There are 119 individuals with 476 images in all. As VIPeR dataset, all images are normalized to 128×48 pixels. The differences are that the light variances are more and some examples subject to large occlusion.

3.2 Image Representation

A combination feature descriptor consisting of color and texture features is used to represent images of individuals. Specifically, for each image, the RGB and HSV color histograms and LBP descriptor are extracted from overlapping blocks of size 16×16 and stride of 8×8, i.e. 50% overlap in both directions. RGB and HSV

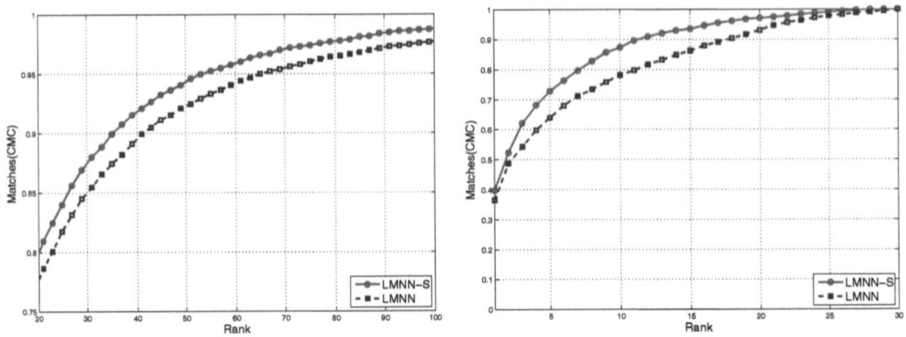

(a) On VIPeR dataset (b) On i-LIDS MCTS dataset

Fig. 3. Comparative results with LMNN algorithm on VIPeR and i-LIDS MCTS datasets

histograms encode the different color distribution information in the RGB and HSV color space, respectively. LBP descriptors, encoding the texture feature, are extracted in grayscale images. All of the features are then put together to concatenated to a vector. To accelerate the learning process and reduce noise, we conducted principle component analysis (PCA) to obtain a low-dimension representation.

3.3 Baselines and Settings

To evaluate the effectiveness of the proposed LMNN-S, we compare with several representative person re-identification methods, including the Adaboost based feature selection method (Adaboost) [8], Rank support vector machine (RandSVM) [3], symmetry-driven accumulation of local feature (SDALF) [2] and probabilistic relative distance comparison (PRDC) [4], and metric learning methods, containing large margin nearest neighbor (LMNN) [9], information theoretical metric learning (ITML) [13] and KISSME [5]. As LMNN and ITML are not development for person re-identification, we use codes of these methods provided by their authors and report their results under the optimal parameter configurations.

The experiments are conducted as follows: for each dataset, we randomly selected p people as training set and the rest as testing set. For training set, positive samples consist a pair of images from the same person, which used to pull the the same labeled persons together closer, meanwhile, k images with different label are randomly selected for each positive samples to form a triple of images making up negative samples, which used to push the different labeled people further apart. Then, each testing set is divided into a probe and gallery set. The gallery set consists of one image for each person in testing set, and the remaining images are set as the probe set. Finally, each image in the probe set is matched with all images of the gallery set. The procedure was repeated 20

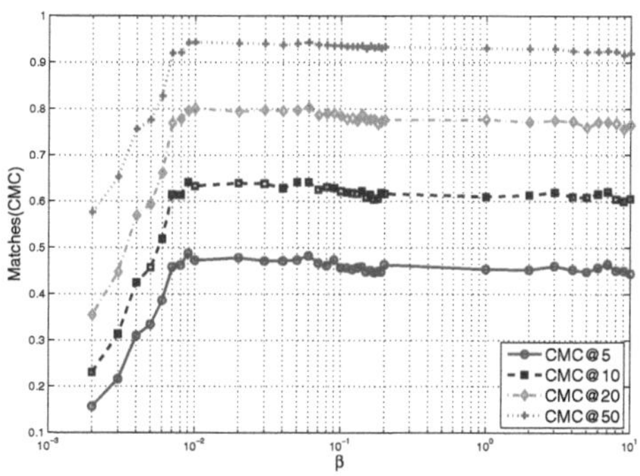

Fig. 4. Comparative results of different β

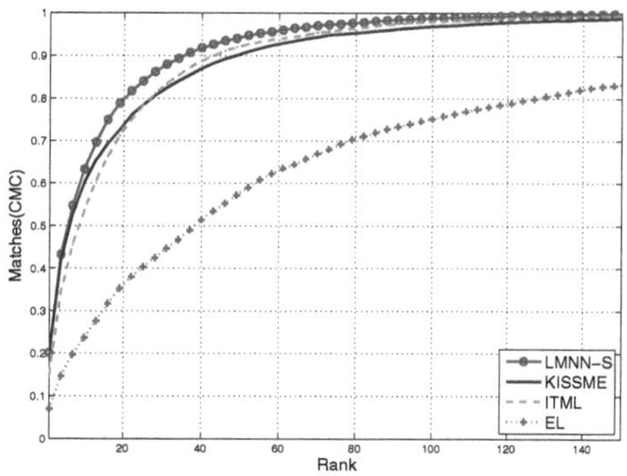

Fig. 5. Comparative results of metric learning

times, and the average of the Cumulative Matching Characteristic (CMC) curve, which is suggested in [7], was reported.

The CMC curve is exploited by most papers on the person re-identification problem [2, 3, 8, 5, 4] as well as. The value of CMC@k indicates the percentage of the real match ranked in the top k. In particular, let $P = \{p_1, \cdots, p_{|P|}\}$ be a probe set, where $|P|$ is the size of P, and $G = \{g_1, \cdots, g_n\}$ be a gallery set. For each probe image $p_i \in P$, all gallery images $g_j \in G$ are ranked based a defined distance function. The correct match is denoted as g_{p_i}, and the rank index of which is denoted as $r(g_{p_i})$. The CMC@k is defined as

$$CMC_k = \frac{\sum_{i=1}^{|P|} \mathbf{1}(r(g_{p_i}) \leq k)}{|P|} \tag{15}$$

where $\mathbf{1}(\cdot)$ is the indicator function.

3.4 Results and Analysis

Comparing to LMNN. We first compare LMNN-S with the LMNN method used in [11] on both VIPeR and i-LIDS MCTS datasets with settings in above subsection. As can be seen from Fig. 3, obviously, the LMNN-S is outperform LMNN on both datasets. The maximum improvement achieves 10% at CMC@5 on i-LIDS MCTS dataset. These results empirically validate the previous discussion in Sec. 1.

In addition, we evaluate the influence of the parameter of logistic loss β. We change the β from 0.001 to 10 and the results are shown in Fig. 4. The performance is going straight up when β change from 0.001 to 0.01, then reduce slightly and nearly stable when $\beta > 1$. It is consistent with the discussion in Sec. 1.

Comparing to the State-of-the-art Person Re-identification Methods. The comparison results with the state-of-the-art methods for person

Table 1. Comparative results with state-of-the-art person re-identification methods on top ranked matching rate(%)

Methods	VIPeR Dataset					iLIDS MCTS			
	r=1	5	10	20	50	1	5	10	20
SDALF [2]	**20**	40	50	65	85	28.5	48.0	57.5	68.0
Adaboost [8]	8.2	24.2	36.6	52.1	90.1	35.6	66.4	79.9	93.2
RankSVM [3]	16.3	38.2	53.7	69.9	85	43.0	71.3	85.2	97.0
PRDC [4]	15.7	38.4	53.9	70.1	87	**44.1**	72.7	84.7	96.3
KISSME [5]	**20.0**	46.0	60.1	74.2	90.5	-	-	-	-
ITML [13]	15.0	38.5	53.7	72.8	92.3	36.4	68.0	83.1	95.55
LMNN [9]	18.7	47.4	62.6	77.8	92.3	36.5	64.0	78.2	93.1
Ours	19.9	**49.3**	**64.3**	**79.8**	**94.1**	39.8	**72.9**	**87.3**	**97.1**

Table 2. Comparative results of computing time with varying numbers of negative samples on VIPeR dataset

Methods	1	2	3	4	5	6	7	8	9	10	LMNN
Time(s)	1.36	2.89	4.11	5.6	6.72	7.94	9.51	11.66	13.12	14.56	20.1
(STD)	(0.53)	(1.53)	(0.91)	(1.19)	(1.55)	(0.84)	(1.15)	(1.50)	(1.80)	(2.24)	(0.61)

re-identification are shown in Table 1. For a fair comparison, the results of Adaboost, RankSVM, SDALF, PRDC and KISSME are directly taken from their original papers. As shown in Table 1, our approach obtains competitive results across all ranks. Besides, the metric learning based methods, such as PRDC and KISSME, outperforms the non-metric learning methods, such as SDALF. As the discussion in Sec. 1, the metric learning seeks a optimal distance function with the supervise of the training samples. The learned distance function is more suitable for the problem than standard distance function, such as Euclidean distance. Moreover, we conducted several experiments with several popular metric learning methods with the same feature representation. The Fig. 5 gives a summery of comparison results. EL indicates the Euclidean distance without learning. Obviously, the metric learning significantly improve the performance and our approach outperform others metric learning methods.

Influence of the Number of Negative Samples. As discussion in Sec. 2.3, the number of negative samples is much larger than positive samples, and we use a stochastic sampling based algorithm. Therefore, we conducted experiments where the number of negative samples is set from one to twenty for each positive sample. As shown in Fig. 6, the effectiveness is about the same when the number

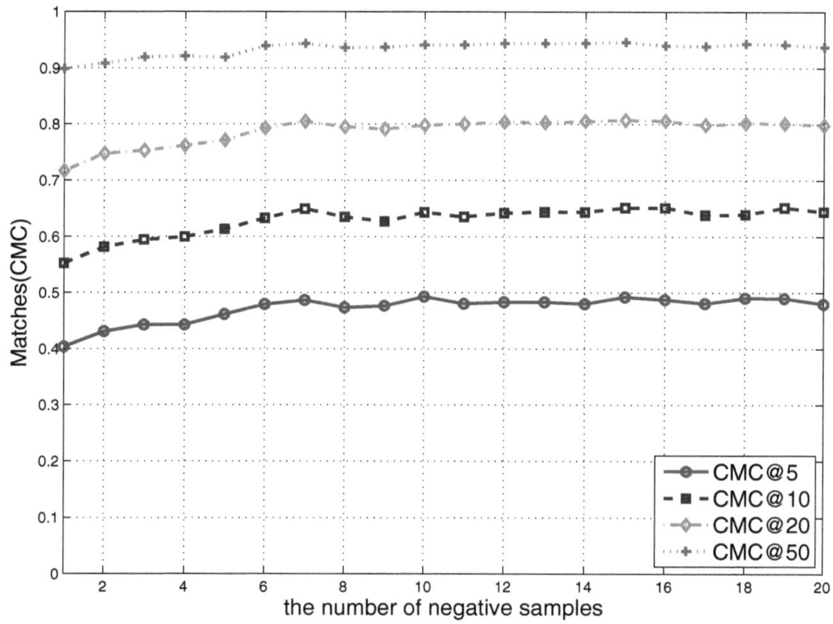

Fig. 6. Comparative results with different size of negative samples on VIPeR dataset

is larger than eight. It is very important for the real system, as the total size of training set can be reduced to $O(n)$ rather than $O(n^2)$.

Computational Cost. Finally, we conducted our experiments on Intel dual-core 2.59GHz CPU PC to evaluate the efficiency of the proposed method, LMNN-S, and compare it with LMNN as well as. The VIPeR dataset was used to evaluate, with which the training size is 316 and dimension of feature vector is 50. Then, we count the computation time of the provided algorithm with varying numbers of negative samples from 1 to 10. For each setting, we repeated 20 times, and the average time was reported in Table 2. As shown in Table 2, we observed that LMNN-S is more efficient than LMNN even though we did not employ the active set method which was designed to improve efficiency of LMNN [9]. Besides, as increasing negative samples, the computational cost descend significant. It is a good selection that improve our implementation with the active set technology at processing large number of training samples. Fortunately, it achieves an enough good effectiveness selecting relatively little negative samples.

4 Conclusions and Future Work

In this paper, we exploit logistic loss function to improve a popular metric learning method, LMNN, and hence transform the problem to a smooth convex optimization problem. Then, we provide a stochastic sampling based optimization algorithm to solve the problem motivated by SGD algorithm. Extensive comparative experimental results reported in Sec. 3 show that the proposed approach superior than the standard LMNN method in both accuracy and efficiency and outperforms the state-of-the-art person re-identification methods and several popular metric learning methods on two challenging public datasets, VIPeR and i-LIDS MCTS. In the future, we plan to further investigate the metric learning methods with the environment in which the instance of each person consist of a sequence of images and metric learning methods can not be directly used.

Acknowledgement. The research was supported by the major national science and technology special projects(2010ZX03004-003-03, 2010ZX03004-001-03), the National Basic Research Program of China (973 Program) (2009CB320906), the National Natural Science Foundation of China (60970160, 61070080, 61003184, 61172173, 61170023, 61231015), the China Postdoctoral Science Foundation funded project (2013M530350), the President Fund of UCAS, the Open Project Program of the National Laboratory of Pattern Recognition (NLPR).

References

1. Hirzer, M., Beleznai, C., Kstinger, M., Roth, P.M., Bischof, H.: Dense appearance modeling and efficient learning of camera transitions for person re-identification. In: ICIP (2012)
2. Farenzena, M., Bazzani, L., Perina, A., Murino, V., Cristani, M.: Person re-identification by symmetry-driven accumulation of local features. In: Computer Vision and Pattern Recognition (CVPR) (2010)
3. Prosser, B., Zheng, W.S., Gong, S., Xiang, T.: Person re-identification by support vector ranking. In: British Machine Vision Conference (BMVC) (2010)
4. Zheng, W.S., Gong, S., Xiang, T.: Person re-identification by probabilistic relative distance comparison. In: Computer Vision and Pattern Recognition (CVPR) (2011)
5. Kostinger, M., Hirzer, M., Wohlhart, P., Roth, P., Bischof, H.: Large scale metric learning from equivalence constraints. In: Computer Vision and Pattern Recognition (CVPR) (2012)
6. Gray, D., Brennan, S., Tao, H.: Evaluating appearance models for recognition, reacquisition, and tracking. In: IEEE International Workshop on Performance Evaluation of Tracking and Surveillance (PETS) (2007)
7. Wang, X., Doretto, G., Sebastian, T., Rittscher, J., Tu, P.: Shape and appearance context modeling. In: International Conference on Computer Vision (ICCV) (2007)
8. Gray, D., Tao, H.: Viewpoint invariant pedestrian recognition with an ensemble of localized features. In: Forsyth, D., Torr, P., Zisserman, A. (eds.) ECCV 2008, Part I. LNCS, vol. 5302, pp. 262–275. Springer, Heidelberg (2008)
9. Weinberger, K.Q., Blitzer, J., Saul, L.K.: Distance metric learning for large margin nearest neighbor classification. Journal of Machine Learning Research 10, 207–244 (2009)
10. Zhang, T.: Solving large scale linear prediction problems using stochastic gradient descent algorithms. In: Proceedings of the Twenty-First International Conference on Machine Learning (ICML) (2004)

11. Dikmen, M., Akbas, E., Huang, T.S., Ahuja, N.: Pedestrian recognition with a learned metric. In: Kimmel, R., Klette, R., Sugimoto, A. (eds.) ACCV 2010, Part IV. LNCS, vol. 6495, pp. 501–512. Springer, Heidelberg (2011)
12. Tao, D., Jin, L., Wang, Y., Yuan, Y., Li, X.: Person re-identification by regularized smoothing kiss metric learning. IEEE Trans. on Circuits and Systems for Video Technology (2013)
13. Davis, J.V., Kulis, B., Jain, P., Sra, S., Dhillon, I.S.: Information-theoretic metric learning. In: International Conference on Machine learning (ICML) (2007)

Venue Semantics: Multimedia Topic Modeling of Social Media Contents

Weizhi Nie[1], Xiangyu Wang[2], Yi-Liang Zhao[2], Yue Gao[2], Yuting Su[1], and Tat-Seng Chua[2,*]

[1] Tianjin University of China
[2] National University of Singapore

Abstract. With the rapid development of location-based social networks (LBSNs), multimedia topic modeling on location-related user generated contents (UGCs) for venues is strongly desired. However, most of the previous topic modeling approaches only work on single modality data, or correlated multimodal data. The intrinsic property of UGCs in LBSNs that the heterogeneous UGCs are generally independent makes these approaches unsuitable for multimedia venue topic modeling. In this paper, we propose a novel multimedia topic modeling approach for extracting venue semantics from heterogeneous location-related UGCs. The approach relates multimedia UGCs by leveraging on multiple data sources. Furthermore, a graph clustering method is proposed to detect the topics which are considered as the dense subgraphs. Based on the multimedia venue topic modeling, we further propose the semantic based venue summarization, which verifies the effectiveness of the proposed framework. The integration of these heterogeneous UGCs into semantic topics provides users with an easier way to understand the venues, and therefore enriches the user experience. Extensive experiments have been conducted on a cross-platform dataset and promising results have been obtained.

Keywords: Location, Multimedia Topic Modeling, Social Media.

1 Introduction

In the era of Web 2.0, the rapid development of location-based social networks (LBSNs) such as Foursquare, Jiepang, etc., has been generating huge amount of location-oriented user generated contents (UGCs). When users are physically visiting some venues, they can check in at these places to let the world know their current presences, or share location related multimedia contents. The fast-growing location-oriented UGCs have given rise of many opportunities and challenges [8,9].

There are plenty of UGCs at different venues in LBSNs. In LBSNs, a venue is different from a location. A location is a geographical region represented by latitude and

* This research is supported by the Singapore National Research Foundation under its International Research Centre @ Singapore Funding Initiative and administered by the IDM Programme Office and A*STAR Project Geographical Information Retrieval via Spatial Annotation of Web Media.

B. Huet et al. (Eds.): PCM 2013, LNCS 8294, pp. 574–585, 2013.

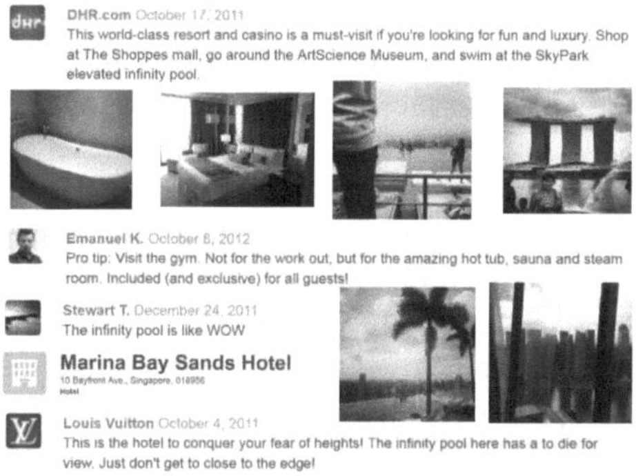

Fig. 1. Examples of UGCs at Marina Bay Sands Hotel in Singapore

longitude, while a venue is a specific place (*e.g.*, Starbucks) within a location. A location can have thousands of venues. These UGCs describe different aspects of venues, and semantics about the venues can be extracted from these rich sources. Venue semantics is the venue content that is identified to understand the meaning of a venue, such as the expression of venue functions and activities. Venue semantics characterize the venues and can be beneficial to various location-based services such as venue summarization and recommendation. For example, given the venue semantics, we can measure the semantic similarity between different venues. If a user likes a venue, the venues with similar semantics can be recommended to him/her.

However, the venue UGCs are unstructured, and usually include different modalities and various topics. Figure 1 presents a sampled subset of UGCs at Marina Bay Sands Hotel from Foursquare. As can be seen, the UGCs contain both texts and images and cover various topics. Among the textual descriptions, people not only talked about the facilities such as the gyms and pools, but also talked about the surrounding attractions. Among the photos, people not only took photos for the bed rooms and various views, but also uploaded their portraits for moments of pleasure. Moreover, different from Flickr, Foursquare does not provide annotation facilities, which makes contents for the venues hard to browse and represent due to the sheer volume of user contributed photos.

An intuitive solution to modeling multimedia topics from the Foursquare UGCs is to directly fuse them together using approaches proposed in [4,5]. However, this simple approach faces the following three challenges. First, the UGCs from Foursquare consist of short texts and images. They are of heterogeneous modalities, which lead to non-uniformity of data. Second, the textual contents and visual contents have little correspondence. The textual descriptions may not come with photos, and vice versa. Even if the images have surrounding text, it is quite common in social networks that the text does not describe the image content. It makes the links among UGCs difficult to construct. Thus, the previous multimodal topic models, which assume the correspondence between texts and images, are also not suitable for location-related UGCs. Third, there exists much noise in the UGCs, which is a common problem in social media. For example, the texts in Foursquare are usually very short and some may be of low relevance to the venue. Due to these intrinsic challenges, the traditional textual and multimodal topic modeling methods are not suitable for this task.

To solve these issues, we propose a novel *multimedia venue topic modeling* approach to simultaneously integrate the independent multimodal UGCs. The goal of multimedia venue topic modeling is to discover the various visual and textual topics from heterogeneous UGCs from LBSNs. In this work, a multimedia dataset from Foursquare is collected. External text information from Wikitravel [1] and TripAdvisor [2] about the venues is collected and used to supplement the short texts in Foursquare. Moreover, we combine cross-platform resources from other social networks such as Flickr to bridge the gap between independent multimodal UGCs. The Flickr data usually has photos and associated texts (tags), and thus can be used to link the texts and images from Foursquare. Then, a multi-modality graph is constructed and the topics in the UGCs are detected as the dense subgraphs by the proposed graph clustering method. Figure 2 illustrates the system flow. It is worth noting that our approach can be applied to other LBSNs though Foursquare is chosen as the experimental dataset in this work.

As a result, venue semantics can be represented as multimedia topics. We can then summarize the venue based on their semantic topics. Effective venue topic modeling provides a more comprehensive understanding of venues. Given a venue, we are able to both know the textual topics and have a visual representative view through multimedia venue topic modeling. Different from the traditional text topic models, such as Latent Dirichlet Allocation (LDA) [14], multimedia venue topic modeling deals not only with texts but also visual information. Compared to the traditional visual based venue summarization [6], the integration of these heterogeneous UGCs into topics provides users with an easier way to understand the venues semantics, and therefore can enrich the user experience.

The main contributions of our work are summarized as follows.

- To the best of our knowledge, this is the first work that targets venue semantics based on multimedia venue topic modeling from UGCs.
- We propose a unified framework to solve the novel application of multimedia venue topic modeling by leveraging on multiple social network resources, which greatly enhances various location-based services.
- We propose the semantic based venue summarization application, which demonstrates the effectiveness of the proposed method in real applications.

2 Related Work

One relevant research is data integration in database, but it focuses on merging data from heterogeneous sources into a common schema, which generally do not deal with topics or multimodal data. The most related research direction is topic modeling, which has been studied for decades and many effective methods have been proposed [2,14,17,10]. Anaya-Sánchez *et al.* [2] introduced a clustering algorithm to discover the topics comprised in text collections, which relies on the term pairs. Blei [14] introduced LDA, which is a statistical graphical model of document collections. Later, Pan *et al.* [13]

[1] `http://wikitravel.org`

[2] `http://www.tripadvisor.com`

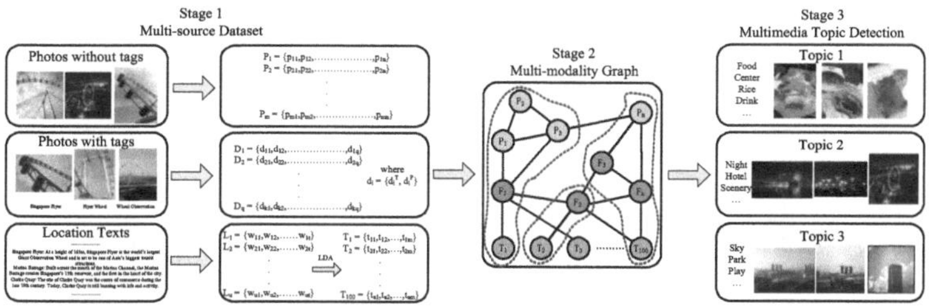

Fig. 2. The framework of multimedia venue topic modeling. It consists of three stages. First, we crawl data from multiple sources and generate heterogeneous data representations. Second, we construct the multi-modality graph. Third, we apply graph clustering approach to detect the dense subgraphs and generate multimedia topics.

proposed an improved LDA model with temporal and spatial clustering for topic detection. Basically, these models are designed for text collections. More recently, Wang *et al.* [18] proposed a Spatial LDA topic model, and applied it to visual application. Although the methods work well, they generally have been proposed for single modality, and are not suitable for venue topic modeling because the UGCs are usually multimodal.

In recent years, multimodal topic modeling methods are proposed for multimodal data, such as [3,16]. These models consider data with multiple types where the instance of one type (such as a caption) serves as a description of the other type (such as an image). However, the heterogeneous UGCs in LBSNs are usually independent. Recently, an effective multi-clue fusion approach for web video topic detection [5] is proposed by utilizing the video and its related tag information. Zhang *et al.* [20] proposed a cross-media topic detection framework that requires some documents involving multiple modalities. They are also not suitable for venue topic modeling because the tags for photos in LBSNs are usually missing. To the best of our knowledge, there are no multimodal topic mining methods that can be directly applied in our scenario. The most recent multimodal topic mining work is [15] which only utilizes text information.

3 Multimedia Venue Topic Modeling

As mentioned in the introduction, heterogeneous UGCs in LBSNs are not directly correlated due to the lack of photo annotations. Thus, we propose a novel multimedia venue topic modeling framework to construct the correlations among different types of UGCs by leveraging on cross-platform data sources. The framework is shown in Figure 2. The scheme consists of three main stages. The first stage (Data Enrichment) crawls data from multiple sources and generates heterogeneous data representations. The second stage (Multi-modality Graph Construction) constructs a multi-modality graph by measuring the similarities among different multimedia documents. The third stage (Topic Modeling in Multi-Modality Graph) detects the multimedia topic by finding the dense subgraphs. Each stage is described in details in the following subsections.

Fig. 3. A motivational example of introducing Flickr data. The photo from Flickr in the middle connects the photo and text from Foursquare.

3.1 Data Enrichment

In this paper, the main data source related to venues is Foursquare. However, the textual and visual information are usually contributed separately and have no clear correlations. In order to link the textual and visual information, we leverage on the images from Flickr, since the Flickr images usually contain correlated visual information and textual tag information. For each venue in Foursquare, we retrieve the relevant images from Flickr by using the venue name and the corresponding geographical bounding box. We only keep photos with at least one tag after stop word filtering. To this end, we are able to indirectly relate the textual documents and photos in Foursquare with the help of Flickr images. On one hand, the relations between the Flickr images and Foursquare texts can be built via the similarities of text. On the other hand, the Foursquare images and Flickr images can be related via visual similarities. Figure 3 gives a motivational example. Here, the photo on the left is from Foursquare and has no textual information. The photo in the middle is from Flickr with the tags (Hotel and Bathroom). These two photos are visually similar. In addition, the tags are correlated with the textual descriptions of the venue on the right. In this way, the photo from Flickr is able to indirectly relate the texts and images in Foursquare.

Moreover, the textual descriptions of venues in Foursquare may be too short and incomplete to describe all aspects of the venues. In order to handle the problem of possible incomplete coverage of venue aspects, we select descriptions for each venue from Wikitravel and TripAdvisor as supplements.

3.2 Textual Information Processing

The textual information for each venue includes venue description, a set of tips and descriptions from Wikitravel and TripAdvisor. While descriptions are generally relevant to venues, tips are of varying qualities. Some users may write information unrelated to venues in tips, or some suspicious users may leave spams. Figure 4 shows examples of tips with high/low relevance to the venue: Singapore Flyer. These non-relevant tips degrade the quality of venue semantics.

In order to filter out the non-relevant tips, we remove all the stop words and links such as urls and email addresses. After that, the textual information, including venue descriptions, tips from Foursquare, descriptions from Wikitravel and TripAdvisor, and

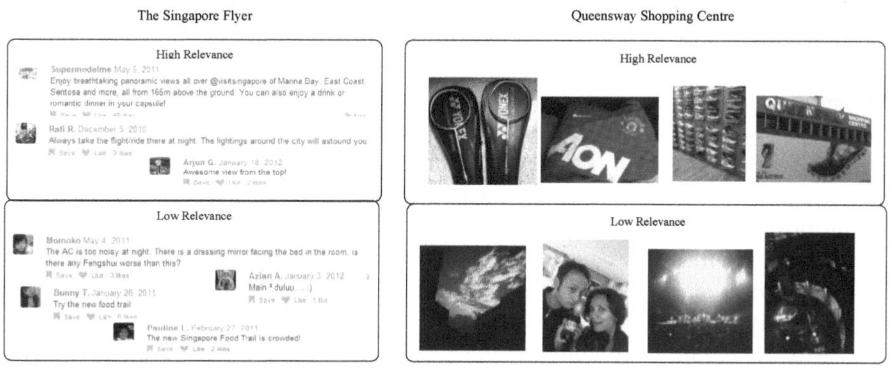

Fig. 4. Example tips and photos with high and low relevance. The left is the example tips with high/low relevance of the Singapore Flyer. The right is the example photos with high/low relevance of Queens Way Shopping Centre.

tags from Flickr, is treated as a document for each venue. A set of latent textual topics from the venue documents is then generated by LDA, which could further remove noisy information from the venue documents as is reported in previous studies [14,13].

3.3 Visual Information Processing

Similar to the textual information, the photos uploaded by users may also have varying relevance. Figure 4 gives examples of photos with high/low relevance to the venue: Queesway Shoppting Centre. We can see that the photos of high relevance are thematically related to the venue, while those of low relevance do not help to reveal the themes of the venue.

From careful observations, we find that the photos including people usually have low relevance to the venue. This is particularly so for portrait-type photos where the size of people/faces is large. In order to remove these low relevance photos, we apply the Part-based detection model [7] to detect faces and people. Part-based detection model is reported to have achieved good results for different objects and has been utilized in many research areas [7]. In our work, we apply the person detection model trained using PASCAL dataset for people detection. Those photos in which the area of people is larger than 45% of the photo size are removed. To remove the photos with faces, the Harr face detection model is used. The filter threshold is also empirically set to be 45%. The body and face detection model is quite mature to achieve good results. Though the performance is not perfect, high accuracy is not critical for noise removal purpose. After these steps, most of the noisy photos can be removed.

3.4 Multi-modality Graph Construction

In this section, we describe the construction of the multi-modality graph. Considering a geographical region, let $\mathbb{T} = \{t_1, \ldots, t_m\}$ denote the Foursquare textual topic set, and $\mathbb{P} = \{p_1, \ldots, p_n\}$ denote the Foursquare photo set, respectively for venue L. Let

$\mathbb{D} = \{d_1, \ldots, d_q\}$ denote the Flickr photo set, where $d_i = (d_i^T, d_i^P)$ includes text information d_i^T and the visual information d_i^P.

Then, we build a graph to represent the relations between the multimedia documents. Each mutlimodal document, including textual topic and images, is considered as a node in the graph. Thus, there are three types of vertices in the heterogeneous graph $G = (\mathbb{V}, \mathbb{E}, w)$, where $\mathbb{V} = \{v_1, \ldots, v_n\}$ is the vertex set, $\mathbb{E} \subset \mathbb{V} \times \mathbb{V}$ is the edge set, and w is the weighting function. A vertex v may be t_i, p_j or d_u. Edges represent neighborhood relationships, and edge-weight reflects similarity between a pair of linked vertices. It is worth mentioning that our focus here is multimedia topic modeling for venues and the proposed multi-modality pairwise graph is actually the most natural modeling in this case, though hypergraph model may be used.

For textual similarity, we compute the similarities between the latent topic t and a tagged photo d as follows:

$$S_{text}(t, d) = \max\left(h(u_i^d, u_j^t) \cdot p(u_j^t|t)\right), \tag{1}$$

where $u_i^d \in \mathbb{W}^d$, $u_j^t \in \mathbb{Z}^t$. $\mathbb{W}^d$ is the set of words from the tags of photo d, and $\mathbb{Z}^t$ is the set of words from the topic t. u_i^d is the i-th word in $\mathbb{W}^d$, u_j^t is the j-th word in $\mathbb{Z}^t$, $h(u_i^d, u_j^t)$ is the similarity between different words computed according to the approach proposed by [19], and $p(u_j^t|t)$ is the posterior probability of word u_j^t in topic t obtained from LDA.

For visual similarity, we use LBP [1] as the visual feature. The input images are first resized into 752×576 and segmented into a set of 64×64 cells. The LBP feature is extracted for each cell, and a 5,012-dimension feature vector is obtained for each image. The similarities between different feature vectors are calculated using cosine similarity as:

$$S_{visual}(v_i, v_j) = \frac{v_i \cdot v_j}{||v_i|| \cdot ||v_j||}, \tag{2}$$

Then, we represent the multi-modality graph G with the corresponding weighted adjacency matrix $A \in \mathbb{R}^{n \times n}$, where $a_{ij} = w(v_i, v_j)$ is computed as follows:

$$w(v_i, v_j) = \begin{cases} S_{visual}(v_i, v_j) & \text{if } v_i, v_j \in \mathbb{D} \cup \mathbb{P} \\ S_{text}(v_i, v_j) & \text{if } v_i, v_j \in \mathbb{D} \cup \mathbb{T} \end{cases} \tag{3}$$

As long as v_i, v_j both contain visual information, the adjacency is based on the visual similarity. Otherwise, it is calculated using textual similarity. Figure 5 gives an illustration of multi-modality graph construction.

3.5 Topic Modeling in Multi-Modality Graph

Given the constructed multi-modality graph, the dense subgraphs are considered as the multimedia topics. To detect the dense subgraphs, we extend the pair-wise clustering method of graph shift (GS) [12]. The graph shift method is an effective method in detecting dense subgraph. However, it assumes that the nodes are of the same modality, thus does not allow for subgraphs that contain different types of nodes. We extend GS to deal with multi-modality data and ensure that each subgraph contains a sufficient

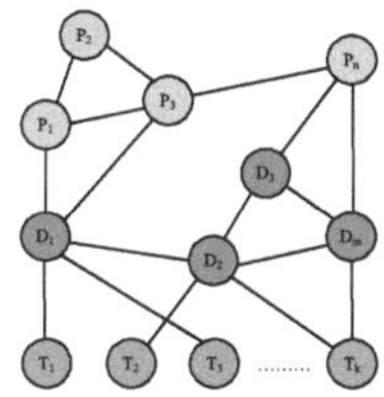

Fig. 5. Illustration of multi-modality graph construction. The yellow nodes denote the photos from $\mathbb{P}$, the green nodes denote the external data from $\mathbb{D}$, the blue nodes denote the topic data from $\mathbb{T}$.

number of multimedia nodes by intuitively set a constraint $x_i \geq c_i$. Here, c_i is the lower bound of the existence probability of vertex i within the final subgroup. The input of the method is the adjacency matrix $\mathbf{A}$ of the multi-modality graph. A subgraph of the multimedia graph is represented by a probabilistic cluster $\mathbf{x} \in \Delta^n$, where $\Delta^n = \{\mathbf{x}|\mathbf{x} \in \mathbf{R}^n, \mathbf{x} \geq 0, |\mathbf{x}|_1 = 1\}$. In fact, $\mathbf{x}$ is a unit mapping vector, which is the probability that the subgraph contains each vertex. Particularly, $x_i = 0$ means that the i-th vertex is not included in the dense subgraph. The GS method measures the average connection strength of subgraph $\mathbf{x}^*$ by $g(\mathbf{x}) = \mathbf{x}^T\mathbf{A}\mathbf{x}$ and efficiently finds all local maximums $\mathbf{x}^*$ of $g(\mathbf{x})$. Each local maximum indicates a dense subgraph of the graph, which is defined as a topic in our method.

Finding the dense subgraph is equivalent to maximizing $g(\mathbf{x})$ $s.t. \sum x_i = 1$, $x_i \geq c_i$. Since our aim is to find the dense subgraph that contains one textual topic and its corresponding visual information, hence each time we only consider one topic node and all the image nodes. Without loss of generality, we denote the topic node as x_1. Thus, c_1 is set to be a positive number so that the dense subgraph always includes the topic node, and $c_i = 0$, for $i = 2, ..., n$.

Since the problem is a constrained optimization problem, by adding Lagrangian multipliers λ, α_i for all $i = 1, \dots, n$, we obtain its Lagrangian function:

$$L(\mathbf{x}) = g(\mathbf{x}) - \lambda(\sum_{i=1}^{n} x_i - 1) + \sum_{i=1}^{n} \alpha_i(x_i - c_i), \tag{4}$$

Any local maximizer $\mathbf{x}^*$ must satisfy the Karush-Kuhn-Tucker (KKT) condition. That is:

$$\begin{aligned} f_i(\mathbf{x}^*) - \lambda + \alpha_i &= 0, \\ \sum_{i=1}^{n} \alpha_i(x_i^* - c_i) &= 0, \end{aligned} \tag{5}$$

where $f_i(\mathbf{x}^*) = \frac{\partial g(\mathbf{x}^*)}{\partial x_i}$. As pointed out by [11], we can optimize the problem in the pairwise manner described by Algorithm 1.

Input: Weighted adjacency array A and an initialization $x(0)$;
Set $\mathbf{x} = x(0)$;
Repeat:
    Update the partial derivative $g(\mathbf{x})$ with respect to each variable x_i;
    Find a pair (v_i, v_j) and compute the best step size to maximize the increase in the objective function;
Until: $\mathbf{x}$ is a local maximizer;
Output: A KKT point $\mathbf{x}^*$

Algorithm 1. Dense subgraph detection

3.6 Semantic Based Venue Summarization

Based on the proposed approach, the venue semantics can be extracted as multimedia topics, and a semantic based venue summarization can be achieved by selecting the top relevant topics. Compared to the traditional visual based location summarization [6,21], the integration of these heterogeneous UGCs into topics provides users with an easier way to understand the venues semantics. The previous works generate distinct visual themes, but they are visual only and did not consider venue semantics. For example, the landmarks such as Marina Barrage in Singapore may have different appearance from different angle of views, and the traditional methods will generate different themes for these views. However, by considering the venue semantics, we may merge them into one semantic scenery topic, and detect some other topics, such as picnic.

4 Experiments

In this section, we first collect a cross-platform dataset for the multimedia venue topic modeling. These platforms include Foursquare, Flickr, Wikitravel, and TripAdvisor. Then, we conduct several experiments to evaluate the performance of our multimedia topic modeling.

4.1 Dataset

We evaluate our method on a Foursquare dataset enriched with multiple social network sources. We crawled the venue homepages of Singapore venues from Foursquare. We started with several venues as seeds, and expanded to other venues through the first order links in the homepage. Each venue homepage contains all the user generated photos, venue tips and descriptions. Moreover, some related textual venue descriptions from other media platforms such as Wikitravel and TripAdvisor were also collected. Here, we manually collected some venue descriptions from Wikitravel and the top 100 tips for each venue in TripAdvisor website by querying with the venue names. As a result, the data for each venue includes both enriched textual and visual information. Eventually, we collected 81,165 venues in Singapore. A subset was then selected based on the number of check-in, photos and tips. The selection criteria are as follows: (1) the

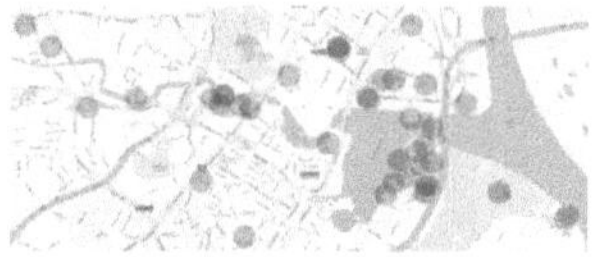

Category	Number	Example
Hotel	6	Gardens By The Bay
Park	3	Sands SkyPark
Food	17	Trasapura Masters Food Court
Shopping	8	Marina Square
Landmark	7	The Singapore Flyer

Fig. 6. The left is the venue geographical distribution, and the right is the venue category distribution

number of check-in is more than 1,000; (2) the number of photos is more than 1,000; and (3) the number of tips is more than 100 as at the crawling time. Finally, we selected 41 popular venues in Singapore that are close to Merlion Park. Figure 6 shows the geographical distribution and the category distribution of the 41 venues.

As mentioned in previous session, all the tips and description for one venue are treated as a single text document. From these 41 venues, we obtained 41 documents and 10,631 photos. The visual and textual information from Foursquare is intrinsically independent. We collected multimedia data from Flickr. In the collection procedure, we used the theme of venue as queries to search in Flickr and crawled the top 60 images for each searching results. In total, we collected 2,080 photos from Flickr.

4.2 Semantic Based Multimedia Venue Summary

With the cross-platform dataset, we first apply LDA on the textual venue documents to generate $N = 100$ latent topics, *i.e.*, the set $\mathbb{T}$ in Section 3.4. After the multimedia topic modeling, we obtain the venue semantics, represented by a set of latent multimedia topics. Each topic includes a textual topic and some corresponding photos. Since the textual topic is generated by LDA, we can also obtain the correlation between the textual topics and venues. This correlation helps to get the topic rank list for each venue. We select the top 3 relevant topics for each venue and use them to illustrate the venue semantics.

The results are illustrated in Figure 7. As can be seen, for each venue, the venue semantics are summarized into different topics and relevant words and photos in each topic are shown (words in the same topic may vary among venues). For example, in Marina Barrage, people not only enjoy the beautiful views, but also enjoy the delicious food and relax by flying the kites. The Esplanade not only contains the landmark topics, but also other topics, such as books. Based on the venue semantics, a more diversified and complete venue summarization is achieved.

To evaluate the performance, we have conducted a user study, where a group of users are invited to judge the multimedia topics for each venue. The judgements are based on three criteria: (1) the level of consistency of the multimedia topics for each venue; (2) the level of completeness of the multimedia topics for each venue; and (3) how satisfactory are the multimedia topics. 20 people from different countries comprising 11 male and 9 female were invited to evaluate the venue summary performance from both textual and visual aspects. The results are shown in Figure 8(a). We can see that the average scores are generally over 7, which means that the summarization is satisfactory in different aspects. We further compared the multimedia topic summarization with text

Fig. 7. Examples of semantic based venue summarization. The topics for each venue are ranked and the top 3 topics are selected. These examples show that the proposed multimedia venue topic modeling helps to extract venue semantics and generate better summarization.

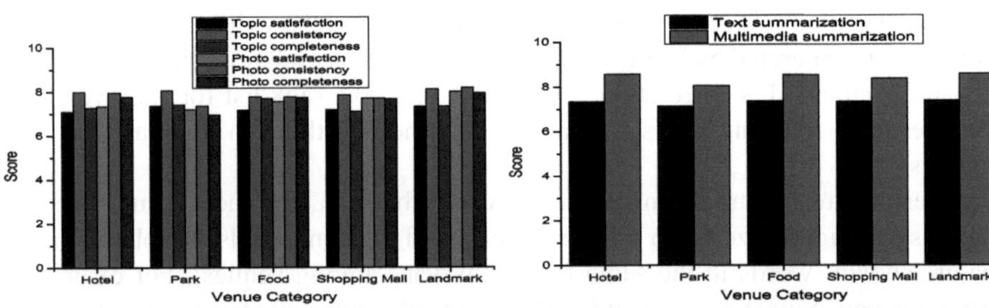

(a) Results of summarization evaluation (b) Results of summarization comparison

Fig. 8. Results of user study for semantic based venue summarization

only summarization as shown in Figure 8(b). The average score of multimedia summarization (8.5) is significantly better than that of text summarization (7.4). In general, our method can effectively summarize the multimedia venue topics.

5 Conclusions

In this paper, we identified a novel problem of multimedia venue topic modeling for venue semantics and proposed a novel framework, which is able to relate heterogeneous multimedia documents leveraging on multiple data sources. We extended the graph shift algorithm by adding constraints to ensure the availability of multimedia descriptions of each venue topic. Semantic based multimedia venue summary have been proposed to further verify the effectiveness of the proposed framework. Experimental results showed that our proposed approach was able to generate satisfactory multimedia semantic summarizations. In the future, we will look into refined multimedia venue topic modeling by introducing user influence.

References

1. Ahonen, T., Hadid, A., Pietikainen, M.: Face description with local binary patterns: Application to face recognition. IEEE TPAMI 28(12), 2037–2041 (2006)
2. Anaya-Sánchez, H., Porrata, A.P., Llavori, R.B.: A document clustering algorithm for discovering and describing topics. Pattern Recognition Letters (2010)
3. Blei, D.M., Jordan, M.I.: Modeling annotated data. In: SIGIR (2003)
4. Chen, K., Luesukprasert, L., Chou, S.T.: Hot topic extraction based on timeline analysis and multidimensional sentence modeling. In: IEEE TKDE (2007)
5. Chen, T., Liu, C., Huang, Q.: An effective multi-clue fusion approach for web video topic detection. In: ACM MM (2012)
6. Fang, Q., Sang, J., Xu, C., Lu, K.: Paint the city colorfully: Location visualization from multiple themes. In: Li, S., El Saddik, A., Wang, M., Mei, T., Sebe, N., Yan, S., Hong, R., Gurrin, C. (eds.) MMM 2013, Part I. LNCS, vol. 7732, pp. 92–105. Springer, Heidelberg (2013)
7. Felzenszwalb, P.F., Girshick, R.B., McAllester, D.A., Ramanan, D.: Object detection with discriminatively trained part-based models. IEEE TPAMI 32(9) (2010)
8. Gao, Y., Tang, J., Hong, R., Dai, Q., Chua, T.-S., Jain, R.: W2go: A travel guidance system by automatic landmark ranking. In: MM, pp. 123–132 (2010)
9. Ji, R., Gao, Y., Zhong, B., Yao, H., Tian, Q.: Mining flickr landmarks by modeling reconstruction sparsity. TOMCCAP 7S(1), 31:1–31:22 (2011)
10. Kurashima, T., Iwata, T., Hoshide, T., Takaya, N., Fujimura, K.: Geo topic model: joint modeling of user's activity area and interests for location recommendation. In: WSDM (2013)
11. Liu, H., Yan, S.: Robust clustering as ensembles of affinity relations. In: NIPS (2010)
12. Liu, H., Yan, S.: Robust graph mode seeking by graph shift. In: ICML (2010)
13. Pan, C., Mitra, P.: Event detection with spatial latent dirichlet allocation. In: JCDL (2011)
14. Putthividhya, D., Attias, H.T., Nagarajan, S.S.: Topic regression multi-modal latent dirichlet allocation for image annotation. In: CVPR (2010)
15. Sang, J., Xu, C.: Browse by chunks: Topic mining and organizing on web-scale social media. TOMCCAP 7S(1), 30:1–30:18 (2011)
16. Virtanen, S., Jia, Y., Klami, A., Darrell, T.: Factorized multi-modal topic model. CoRR, abs/1210.4920 (2012)
17. Wang, Q., Xu, J., Li, H., Craswell, N.: Regularized latent semantic indexing: A new approach to large-scale topic modeling. ACM TOIS (2013)
18. Wang, X., Grimson, E.: Spatial latent dirichlet allocation. In: NIPS (2007)
19. Wu, Z., Palmer, M.S.: Verb semantics and lexical selection. In: ACL (1994)
20. Zhang, Y., Li, G., Chu, L., Wang, S., Zhang, W., Huang, Q.: Cross-media topic detection: A multi-modality fusion framework. In: ICME (2013)
21. Zhao, Y.-L., Zheng, Y.-T., Zhou, X., Chua, T.-S.: Generating representative views of landmarks via scenic theme detection. In: Lee, K.-T., Tsai, W.-H., Liao, H.-Y.M., Chen, T., Hsieh, J.-W., Tseng, C.-C. (eds.) MMM 2011 Part I. LNCS, vol. 6523, pp. 392–402. Springer, Heidelberg (2011)

Conditional Random Fields for Image Region Labeling with Global Observation

Zhe Lin[1], Wen Chan[1], Kai He[1], Xiangdong Zhou[1], and Mei Wang[2]

School of Computer Science, Fudan University, Shanghai, China
{11210240019,11110240007,xdzhou}@fudan.edu.cn
School of Computer Science and Technology, Donghua University, Shanghai, China
wangmei@dhu.edu.cn

Abstract. Region (pixel) labeling has attracted increasing attentions from both research and industry communities. In this paper, we present a new approach based on Conditional Random Fields (CRF) to assign the semantic labels to the corresponding regions of images. Different from previous work, our model incorporates the global observation into the region labeling framework with the harness of spatial context modeling of CRF model. The experimental results with two commonly used datasets demonstrate that our method achieves significant improvement on region labeling tasks compared with the strong baselines.

Keywords: Image Region Labeling, Global Observation, Conditional Random Fields, Image Annotation.

1 Introduction

Labeling local region (pixel) of images is very important in many computer vision applications, such as robot vision, scene understanding, and semantic based image retrieval. A straightforward manner is to regard it as the multi-class classification problem, that is with the region (pixel) input, the classifier picks a keyword (class label) from a preset vocabulary and assigns it to the region (pixel). The stat-of-the-art computer vision and machine learning techniques have been exploited to address the problem, for instance, bi-layer sparse coding formulation is used to transfer the labels to the image region by reconstructing them using segmented image patches[1]. A nonparametric scene parsing system is proposed to transfer the labels of exiting examples to test images [10], just named a few.

An interesting idea is to explore the spatial contextual information for region or pixel labeling. It is believed that neighbor pixels (regions) in an image often have the same labels, so called "label smoothing" which is illustrated in Figure 1 (b). The most successful approach for spatial context modeling is the Conditional Random Fields (CRF) [2]. Kumar and Hebert [12] present an enhancement CRF model for detecting man-made structures in natural images. They employ discriminative models to capture the individual pixel observation and the interactions between neighbor pixels. However, the pixel discriminative

B. Huet et al. (Eds.): PCM 2013, LNCS 8294, pp. 586–597, 2013.

model is still difficult to predict the semantic label by only viewing a pixel (local region) of the object or the whole semantic region. Some researchers attempt to integrate the local image feature and the global image feature to deal with the region labeling problem[17,18]. He et al. [17] propose a method based on Constraint Boltzmann Machine (CBM) for region labeling with the observations of multi-scale image features. However, the philosophy of "label smoothing" is not apparently adopted in their model.

In this paper, we propose a new method based on Conditional Random Fields (CRF) to solve the region labeling problem by utilizing visual feature of both the local region and the whole image. Our motivation has two aspects: the whole image recognition is incorporated into the site potential with the prior probability of the semantic label occurring in the image; the "label smoothing" of the neighbor pixels is also adopted in our model. We conduct experiments on the MSRC[21] and LabelMe Outdoors[10] datasets to evaluate the performance our proposed model. In the experiment, we partition the image into identical grids. We train CRFs for 18 categories on the MSRC dataset and 8 categories on the LMO dataset. The experimental results shows that our model achieves the region labeling accuracy about 0.62 on both datasets. It demonstrates that the combination of whole image observation and "label smoothing" improves the performance of the CRF compared with other strong baselines and state-of-the-art methods. The rest of the paper is organized as follows: Section 2 introduces some related works. Section 3 presents the model of Conditional Random Fields with global observation. Section 4 describes details of parameters estimation of the proposed algorithm. Section 5 presents the model inference and label assignment. Section 6 presents the experimental setup, and section 7 provides the experimental results and discussion on it. Section 8 presents the conclusion on this paper.

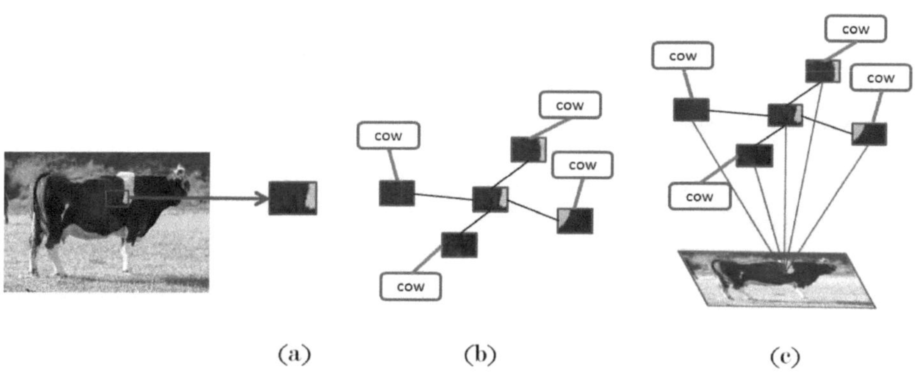

Fig. 1. Left: an image from the MSRC dataset. (a) Observation of a singular grid. (b) Observation of the grid and its neighborhood grids. (c) Observation of the local area and the global image. The global observation of the whole image makes it clear that the grid is part of a cow in the image.

2 Related Work

Region labeling can be regarded as one sub topic of image annotation[7][8] [9][20]. It is an area of research that has something in common with image segmentation and object recognition[3,4]. A context-based region labeling method has been proposed for simultaneous image segmentation and region labeling[5]. Alexander Vezhnevets et al. [6] propose a multi-image model to predict the labels of the pixels of the image by segmenting the image into semantic regions. Some researchers also consider the region labeling to be a problem coupled segmentation and classification[13]. They propose an approach based on energy minimization algorithm to obtain the labels of image regions. Label transferring methods are also proposed, Liu et al. propagate the labels at the image level to the corresponding regions by reconstruction coefficients[1]. The method solve the region labeling problem by exploiting the consistency of the regions with the identical labels. Rubinstein et al. [11] propagate the label map of the training images to the test images via dense correspondence.

It is beneficial to exploit the information of local area when labeling the regions. Sanjiv Kumar and Martial Hebert [12] apply the CRF into classifying image regions by considering the site visual content and its interaction with neighbor sites. However, it could be inaccurate when classifying the regions just by employing the observation of the local area. Some researchers propose the algorithms based on CRF for employing the observation from different scales[17,18]. The incorporation of observation from larger scale could reduce the misclassification of visual similar regions with different labels.

In this paper, we incorporate both the local and global observation into the region labeling framework: the whole image recognition is incorporated into the site potential with the prior probability of the semantic label occurred in the image; The "label smoothing" is exploited in our model, which is a power ability the CRF model provided.

3 Multiple Conditional Random Fields Based Image Region Labeling

In this section, we give a brief introduction of CRF theory, and then explain the details of construction of CRFs for region labeling.

3.1 Conditional Random Fields

Conditional Random Fields (CRF) uses discriminative potentials for the singular sites and the interactions between sites. Let G=(S,E) be a graph with site set $S = \{1, 2, ..., m\}$ and edge set E. Let $y = \{y_1, y_2, ..., y_m\}$ be a set of variables indexed by S, and $X \in \chi$be the feature vector of observed data. The pair (y, X) is said to be conditional random field if, when conditioned on X, the random variable y_i obey the Markov property with respect to the graph: $P(y_i|X, y_{s-\{i\}}) = P(y_i|X, y_{N_i})$ where $s - \{i\}$ is the set of all nodes in G except node i, N_i is the set of neighbors of node i in G. The conditional distribution over the labels y given X is defined as

$$P(Y^k|X) = \frac{1}{Z}exp(\sum_{i\in S} A(y_i^k, X) + \sum_{i\in S}\sum_{j\in N_i} I(y_i^k, y_j^k, X)) \quad (1)$$

where Z is the normalizing constant called partition function, $-A$ and $-I$ are the site potential and edge potential respectively. For the image region labeling application, the sites of the random field are grids of the images, and the neighborhoods of the grid are the adjacent 4-nearest grids in space. As the label of a grid could be one of the keyword set $W = \{w_1, w_2, ..., w_K\}$, we construct K different CRFs for labeling the image regions. Each CRF is concerned with one keyword by binary classification. Y^k indicates that the CRF is constructed for the keyword w_k.

3.2 Site Potential with Local and Global Observation

In our CRF framework, site potential function $A(y_i^k, X)$ is used to model the association between the singular site and the keyword w_k given the observed features of the local region and the whole image. In particular, $X_I \in X$ is the feature vector of the images. We define the site potential as:

$$A(y_i^k, X) = y_i^k(\lambda^k + \alpha^k P(y_i^k = 1|x_i) + u\delta^k P(y_I^k = 1|X_I)\phi(k)) \quad (2)$$

where λ^k , α^k and δ^k are the parameters on the site potential and u is the smoothing parameter. $\phi(k)$ is the occurring prior probability related to the keyword w_k.

Local Observation. For the region labeling task, we use the random variable $y_i{}^k$ which takes value from $\{-1, +1\}$ to indicate the absence or presence of keyword w_k for grid i. The probability that the keyword presents in the local site i is defined as

$$P(y_i^k = 1|x_i) = \frac{1}{1 + exp(\beta_{k0} + \beta_k^T x_i)} \quad (3)$$

where $x_i \in X$ is the observed region features of grid i. We employ the logistic regression to generate $P(y_i^k = 1|x_i)$, which is the prior probability of that the local region is labeled with the keyword w_k. We adopt the LIBLINEAR toolbox [19] for training the logistic regression classifiers.

Global Observation. Compared with the image region, observation of the whole image is the global description relevant to the local area. Incorporated with the image observation, the site potential could be boosted by the knowledge from the larger scale and higher layer. It's the effective complement to the limit and deficiency of the local observation. The random variable $y_I^k \in \{-1, +1\}$

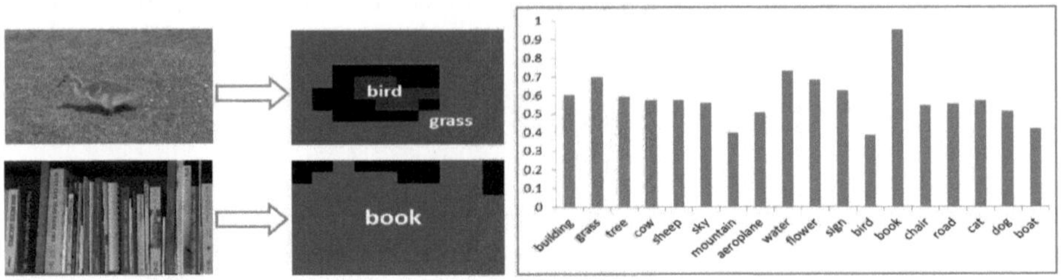

(a) (b)

Fig. 2. The illustration for probability estimation of label occurring. (a) The left figure shows that the regions of different categories have different sizes in the image. (b) The occurring prior probability of 18 categories on MSRC dataset

is the label of the whole image related to the keyword w_k. Similar with local observation, we also employ the logistic regression to generate the probability $P(y_I^k = 1|X_I)$ as defined in Equ. 3, which denotes the probability of the label occurs in the whole image. We consider the site data to be the observation from different scales. If we are faced with some part of an object, the overview of the whole object could be the valuable information for recognizing it. Different from the multi-scale based method[17], we just employ the information related to the whole image, avoiding the reduplicate observation and increasing computational complexity.

In addition to taking the probability of the label occurring in the whole image into account, we also exploit the prior probability of the label occurring. Figure 2 (a) gives some hints of the prior probability. Obviously, the regions labeled with " bird" and regions labeled with "book" have different prior occurring probability. The heuristic is that in an image, different semantic label covers different amount of image regions (pixels), and more pixels a keyword covers, more opportunity it occurs in the image. Therefore, with a training image database, we can estimate the prior probability of label occurring based on counting their corresponding pixels (regions) in an image database. Specifically, for each image in the training set, we segment the image into many grids and the each grid is annotated with one keyword. We can obtain the grid counting distribution related to the keyword by calculating the corresponding number of the grids. Suppose there are m_k images is labeled with the keyword w_k, where image i has N_i grids and n_i^k grids are related to w_k. The label occurring prior probability estimation is defined as

$$\phi(k) = \frac{1}{m_k} \sum_{i=1}^{m_k} \frac{n_i^k}{N_i} + l \tag{4}$$

where l is the default label occurring prior probability. As each image has certain probability to be relevant with the keyword w_k, we set default prior probability $l = 0.3$. Figure 2 (b) shows the estimation of label occurring prior distribution in MSRC dataset. The label occurring probability of the whole image is vulnerable to the complex contents, thus we employ the smoothing parameter u for adjusting the influence of the global observation.

3.3 Edge Potential

The edge potential in CRF is based on the labels of current site and the neighborhood sites. Since the regions extracted from the images are just part of the whole object, it should be favored that the adjacent regions have the identical labels. The edge potential is defined as:

$$I(y_i^k, y_j^k, X) = \beta_{ij}^k y_i^k y_j^k P(y_j^k = 1|x_j) sim(x_i, x_j) \tag{5}$$

where

$$sim(x_i, x_j) = exp(-dist(x_i, x_j)) \tag{6}$$

where β_{ij}^k is the parameter on the edge potential, $x_j \in X$ is the observed data and $y_j^k \in \{-1, +1\}$ is the label of neighborhood grid j. $P(y_j^k = 1|x_j)$ is the conditional probability of label y_j^k for keyword w_k given observation x_j. $sim(x_i, x_j)$ is used to measure the similarity of the regions given their observation data. We employ chi-square distance which is represented as the function $dist(x_i, x_j)$ to calculate the distance of the two feature vectors.

4 Parameters Estimation

4.1 Pseudo-likelihood

Maximum likelihood method is widely used to estimate the parameters of CRF. While the computation of the partition function is a generally NP-hard problem. We adopt the approximation scheme called pseudo-likelihood to avoid computing the partition function. The pseudo-likelihood is defined as

$$PL(Y^k) = \prod_{i \in S} P(y_i^k | y_{N_i}^k, X) = \prod_{i \in S} \frac{e^{-U_i^k(y_i^k, y_{n_i}^k)}}{\sum_{y_i^k} e^{-U_i^k(y_i^k, y_{n_i}^k)}} \tag{7}$$

where

$$U_i^k(y_i^k, y_{n_i}^k) = A(y_i^k, X) + \sum_{j \in N_i} I(y_i^k, y_j^k, X) \tag{8}$$

is the potential of the site i. Let

$$\theta_k = (\lambda^k, \alpha^k, \delta^k, \beta_{ij \forall j \in N_i}^k)^T \tag{9}$$

$$v_i^k = (1, P(y_i^k = 1|x_i), uP(y_I^k = 1|X_I)\phi(k), y_j^k P(y_j^k = 1|x_j) sim(x_i, x_j)_{\forall j \in N_i})^T \tag{10}$$

then we can rewrite Equ. 8 to

$$U_i^k(y_i^k, y_{n_i}^k) = y_i^k \theta_k^T v_i^k \tag{11}$$

where θ_k is the parameters to be estimated of the CRF for keyword w_k, and v_i is the training data of the site i. Substituting Equ. 11 into Equ. 7, we can rewrite the pseudo-likelihood to

$$PL(Y^k) = \prod_{i \in S} \frac{e^{-y_i^k \theta_i^T v_i^k}}{e^{\theta_i^T v_i^k} + e^{-\theta_i^T v_i^k}} \tag{12}$$

The parameters $\theta = (\theta_1^T, \theta_2^T, ..., \theta_K^T)^T$ are related to different CRFs for different keywords. We estimate the parameters by maximizing the pseudo-likelihood with regularization.

4.2 Maximum Pseudo-likelihood with Regularization

We have constructed the training set $T_k = \{(v_i^k, y_i^k)\}_{i=1}^{M_k}$ for the CRF related to keyword w_k. We penalize the parameters of CRF with a default l_2 regularization in order to avoid over-fitting. Then the log pseudo-likelihood with regularization on the training set T_k is:

$$L_k(\theta_k) = \sum_{i=1}^{M_k} \{(1 - y_i^k)\theta_k^T v_i^k - log(1 + e^{2\theta_k^T v_i^k})\} - \frac{||\theta_k||^2}{2\sigma^2} \tag{13}$$

To maximize the above equation, we adopt the Newton-Raphson method to solve optimization problem. The Newton-Raphson algorithm will converge as the penalized log pseudo-likelihood is convex with respect to the parameters.

5 Model Inference and Label Assignment

5.1 Model Inference

The inference problem in CRF is to find the most probable configuration of the sites:

$$Y^{k^*} \leftarrow \underset{Y^k}{argmax}\, P(Y^k|X) \tag{14}$$

where $P(Y^k|X)$ is defined in Equ. 1. The iterative conditional modes (ICM) algorithm is employed for inference which maximizes the local conditional probabilities sequentially. In the $(t+1)^{th}$ iteration step, given the observation X and labels on neighboring sites ${y_{N_i}^k}^{(t)}$ obtained in the last iteration, the algorithm sequentially updates each $y_i^{k(t)}$ to $y_i^{k(t+1)}$ that yields the maximal conditional probability $P({y_i}^k|{y_{N_i}^k}^{(t)}, X)$ defined in Equ. 7. The update rule can be written as follows

$$y_i^{k(t+1)} = \begin{cases} +1, & \text{if } P(y_i^k = 1|{y_{N_i}^k}^{(t)}, X) > P(y_i^k = -1|{y_{N_i}^k}^{(t)}, X) \\ -1, & \text{otherwise} \end{cases} \tag{15}$$

The ICM algorithm starts with an initial configuration and iteratively obtains the labels of the sites in the CRF. The algorithm will stop if the outputs of two consecutive iterations are identical. Otherwise, it will stop after a certain number iterations. We run the ICM algorithm on all the CRFs constructed for different keywords.

5.2 Label Assignment

As the regions are partitioned from the images, each image region is related to only one keyword. So the final label assignment of the image regions is a multi-class classification according to the posterior probability generated from the CRF. We compute the posterior probability $P(y_i^k|y_{N_i}^k, X)$ defined in Equ. 7 given the observed data X and the final labels of the neighboring sites for keyword w_k ,which is obtained from the inference of the CRF. The rule of the label assignment is as follows

$$k \leftarrow \text{argmax}_k P_{k,k=1,\dots,K} \quad (16)$$

where $P_k = P(y_i^k|y_{N_i}^k, X)$ is the posterior probability.

6 Experimental Setup

In this section, we give the details of the dataset used in our experiments and describe the features that we extract from the image regions and images.

6.1 Data Sets

MSRC. We evaluate the performance of the model on the MSRC [21] dataset where each image is densely labeled. The MSRC dataset has 591 images and 23 keywords. We get 491 images from the dataset for training the model and the left 100 images for testing. We resize all the images into the resolution of 320*240 and segment one image into 100 grids with the same size. Since some grids may have pixels with different labels, we assign the label of maximun pixels to the grid. We remove the set of keywords which have less than 10 positive sample images for training the related CRF. Finally, we construct the CRFs of 18 categories: 'building', 'grass', 'tree', 'cow', 'sheep', 'sky', 'mountain', 'aeroplane', 'water', 'flower', 'sign', 'bird', 'book', 'chair', 'road', 'cat', 'dog', 'boat'.

LabelMe Outdoors. LabelMe Outdoors (LMO)[10] is a subset of the dataset SUN[15]. The LMO database consists of 2,688 outdoor images and 33 keywords which have been randomly split into 2,488 training and 200 test images. All of the images have been densely labeled with the 33 categories. We obtain the label of the image regions with the same method on MSRC dataset. We randomly select 100 training images and 50 test images from the original training and test sets. We finally construct the CRFs for 8 categories: 'sky', 'tree', 'building', 'road', 'car', 'sea', 'sign', 'grass'.

6.2 Feature Extraction

In order to train the model, we extract the features for both image regions and entire images. We extract HSV and SIFT descriptors from the whole images for training the logistic regression classifiers for different categories. It's difficult to capture the details of the characteristics of the small image regions. In addition, the visual content of an object or semantic region could be quite different from separate part. The diversity of the categories could be enhanced at the region level. In summary, we use the following features to capture the characteristics of the regions.

SIFT. Scale-Invariant-Feature-Transform (SIFT) [16] is a strong feature descriptor which is widely applied for object recognition and image classification. We adopt the Bag-of-Words method for dealing with SIFT descriptors. We extract the SIFT feature points from the images and construct a visual dictionary by k-means clustering algorithm with k=500. Then we quantify the feature descriptors into the uniform vector with the same length 500.

Region Moments. Region Moments is a class of appearance descriptors based on image moments which is firstly used in vehicle detection. The special design of the moments and the image features makes the descriptors scale and rotation invariant. It is especially effective to capture the appearance information of small strucures.

Color Descriptors and Coordinates. We also extract HSV color descriptors and location information of the image regions. As the grids are segmented equably, each grid has a corresponding location information in the image. We employ the location information of grids to distinguish the visual similar region which usually appear at different part of the image such as regions of 'sky' and 'water', 'grass' and 'tree'.

7 Experimental Results and Discussions

We implement our method on MATLAB and C++ platform and run the algorithm on the MSRC and LMO dataset. In order to get the image regions, we partiton the images into 100 grids on MSRC dataset and 64 grids on LMO dataset. All the grids have the identical width and height. The global observation of the whole image is the prior probability generated by logistic regression. In the training process, we just incorporate the prior probability in case that the image is labeled with the keyword. While in the test process, we set a threshhold to incorporate the prior probability if it is larger than the threshhold 0.5.

We implement three algorithms as benchmark baselines and the proposed algorithm for image region labeling task. (1) Binary logistic regression (BLR), which converts the m-class classification into m binary classification problems.

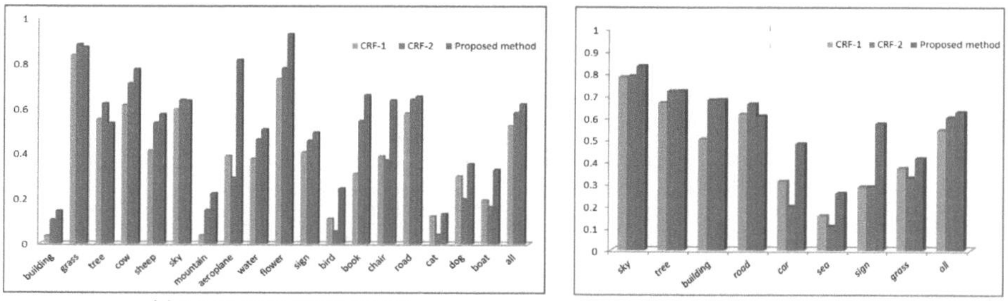

(a) Accuracies on MSRC dataset

(b) Accuracies on LMO dataset

Fig. 3. The region labeling accuracies for (a) MSRC dataset and (b) LabelMe Outdoors (LMO) dataset. The horizontal axis shows the name of each class and the vertical axis demonstrates the region labeling accuracy. The CRF-1 method is the CRF with site potential defined in Equ. 2 only. The CRF-2 method is the CRF without global observation.

Table 1. Region Labeling accuracy comparison on MSRC and LMO datasets. Numbers are given in percentages.

Dataset	BLR	CRF-1	CRF-2	Ours
MSRC	46.14	52.47	58.35	62.14
LMO	45.36	54.05	59.72	62.36

(2) Conditional Random Fields with site potential only (CRF-1), defined in Eque. 2. (3) Conditional Random Fields without global observation (CRF-2), which just employs the site and edge potentials. Table 1 shows the accuracy comparison between the baseline algorithms and our proposed algorithm on the MSRC and LMO datasets. From the results, we have the following observations: (1) The proposed algorithm achieves accuracies of 62.14% and 62.36% on the MSRC and LMO datasets repectively. It demonstratee the effectiveness of our algorithm for region labeling. (2) The proposed algorithm outperforms the baseline algorithms on both datasets. CRF-1 incorporates the global information of the whole image , while CRF-2 adopts the information of adjacent neighborhoods. It demonstrates that the integration of both local and global observation effectively improves the performance of image region labeling task. (3) The experimental results also demonstrate the CRF-2 algorithm outperforms than CRF-1, it is noted that the multi-scale model is not effective due to the lose of "label smoothing". (4) Figure 3 shows the detailed accuracies of different categories on the MSRC and LMO dataset. We notice that the proposed algorithm obtains lower accuracy on labeling some objects such as 'building','mountain' and 'cat'. As the over-partitioned based approach could damage the original whole appearance of objects, it is difficult for labeling the regions which has blurry and limited visual information. Some results of the region labeling are displayed in Figure 4 and Figure 5.

Fig. 4. Examples of region labeling results on MSRC dataset

Fig. 5. Examples of region labeling results on LMO dataset

8 Conclusion

We have proposed a model based on the Conditional Random Fields for assigning the labels to the corresponding image regions. It takes consideration with both local and global observation of images as well as harnesses the benefit of spatial context modeling of CRF model. Experimental results on the MSRC and LMO datasets demonstrate the significant improvement and effective performance of the proposed algorithm.

Acknowledgments. This work was partially supported by Natural Science Foundation of China under grant 61073002 and Natural Science Foundation of China under grant 61103046.

References

1. Liu, X., Cheng, B., Yan, S.: Label to region by bi-layer sparsity priors. In: MM (2009)
2. Lafferty, J.D., McCallum, A., Pereira, F.C.N.: Conditional Random Fields: Probabilisitc Models for Segmenting and Labeling Sequence Data. In: ICML (2001)
3. Huang, Q., Han, M., Wu, B., Ioffe, S.: A hierarchical Conditional Random Field model for labeling and segmenting images of Street Scenes. In: CVPR (2011)
4. Kohli, P., Ladicky, L., Torr, P.H.S.: Robust higher order potentials for enforcing label consistency. In: CVPR (2008)

5. Athanasiadis, T., Mylonas, P., Avrithis, Y.: A context-based region labeling approach for semantic image segmentation. In: Avrithis, Y., Kompatsiaris, Y., Staab, S., O'Connor, N.E. (eds.) SAMT 2006. LNCS, vol. 4306, pp. 212–225. Springer, Heidelberg (2006)
6. Vezhnevets, A., Ferrari, V., Buhmann, J.M.: Weakly supervised semantic Segmentation with a multi-image model. In: ICCV (2011)
7. Tang, J., Zha, Z.-J., Tao, D., Chua, T.-S.: Semantic-Gap Oriented Active Learning for Multi-Label Image Annotation. IEEE Tran. on Image Processing 21(4) (2012)
8. Tang, J., Hong, R., Yan, S., Chua, T.-S., Qi, G.-J., Jain, R.: Image Annotation by kNN-Sparse Graph-based Label Propagation over Noisily-Tagged Web Images. ACM Tran. on Intelligent Systems and Technology 2(2) (2011)
9. Liu, J., Xu, C., Lu, H.: Cross-media retrieval: state-of-the-art and open issues. International Journal of Multimedia Intelligence and Security 1(1) (2010)
10. Liu, C., Yuen, J., Torralba, A.: Nonparametric Scene Parsing via Label Transfer. IEEE Tranactions on Pattern Analysis and Machine Intelligence 33(12) (2011)
11. Rubinstein, M., Liu, C., Freeman, W.T.: Annotation Propagation in Large Image Databased via Dense Image Correspondence. In: Fitzgibbon, A., Lazebnik, S., Perona, P., Sato, Y., Schmid, C. (eds.) ECCV 2012, Part III. LNCS, vol. 7574, pp. 85–99. Springer, Heidelberg (2012)
12. Kumar, S., Hebert, M.: Discriminative Fields for Modeling Spatial Dependencies in Natural Images. In: NIPS (2003)
13. Corso, J.J., Yuille, A., Tu, Z.: Graph-shifts: Natural Image Labeling by Dynamic Hierarchical Computing. In: CVPR (2008)
14. Galleguillos, G., Rabinovich, A., Belogie, S.: Objet categorization using co-occurrence, location and appearance. In: CVPR (2008)
15. Russel, B.C., Torralba, A., Murphy, K.P., Freenam, W.T.: Labelme: a database and web-based tool for image annotation. International Journal of Computer Vision 77(1-3), 157–173 (2008)
16. Lowe, D.: Distinctive image features from scale-invariant keypoints. International Journal of Computer Vision 60(2), 91–110 (2004)
17. He, X., Zemel, R.S., Carreira-Perpinan, M.A.: Multiscale Conditional Random Fields for Image Labeling. In: CVPR (2004)
18. Toyoda, T., Hasegawa, O.: Random Field Model for Integration of Local Information and Global Information. IEEE Transactions on Pattern Analysis and Machine Intelligence 30(8), 1483–1489 (2008)
19. Fan, R., Chang, K., Hsieh, C., Wang, X., Lin, C.: LIBLINEAR: A library for large linear classification. Journal of Machine Learning Research 9, 1871–1874 (2008)
20. Xiang, Y., Zhou, X., Chua, T.-S., Ngo, C.-W.: A revisit of generative model for automatic image annotation using markov random fields. In: CVPR 2009 (2009)
21. `http://research.microsoft.com/en-us/projects/objectclassrecognition`

Multi-feature Subspace Learning via Sparse Correlation Fusion and Embedding

Hong Zhang[1,2,3] and Yanpeng Zhang[1]

[1] College of Computer Science & Technology, Wuhan University of Science & Technology, China
[2] Hubei Province Key Laboratory of Intelligent Information Processing and Real-time Industrial System, China
[3] State Key Laboratory of Software Engineering, Wuhan University, 430072, China
zhanghong_wust@163.com

Abstract. Subspace learning is most traditional and important in multimedia analysis. Numerous researches have focused on how to introduce machine learning and statistical methods to multimedia subspace learning for semantic understanding and denoising, and have gained remarkable achievement in different multimedia applications, such as content-based retrieval, data clustering, face recognition, etc. However, most of these researches are based on multimedia data of single modality. Nowadays, with the rapid development of multimedia and information technology, multimedia data of different modalities often coexist, and the presence of one has a complementary effect on the other to some extent. Because different multimedia data are usually represented with heterogeneous low-level features and there exists the well-known semantic gap, it is interesting and challenging to learn multimedia semantics by multi-feature subspace learning of different modalities. In this paper, we analyze sparse canonical correlation between feature matrices of different multimedia data, construct an isomorphic sparse multi-feature subspace; moreover, we propose subspace optimization strategy with correlation fusion, which explores both geometrical-based content correlation and graph-based semantic correlation. Our algorithm has been applied to content-based multimodal retrieval and data classification. Comprehensive experiments have demonstrated the superiority of our method over several existing algorithms.

Keywords: multi-feature subspace, multimedia analysis, correlation fusion.

1 Introduction

Subspace learning plays an important role in many machine learning tasks and applications, such as CBMR (Content-based Multimedia Retrieval) [1][2][3][7][8], data clustering [4][5], face recognition [14][15] and cartoon generation [6][13]. Most of these researches focus on how to analyze low-level multimedia features and find a low-dimensional semantic subspace with minimum noise and underlying semantic correlation discovered. So far considerable subspace learning methods are based on

B. Huet et al. (Eds.): PCM 2013, LNCS 8294, pp. 598–607, 2013.

multimedia data of single modality, such as image subspace learning for retrieval and recognition [19][20][21][22][29], audio subspace learning for semantic understanding [23], etc.

Recently, with the rapid development of multimedia and information technology, multimedia data of different modalities usually coexist and the presence of one has a complementary effect on the other to some extent [16]. Some researches have shown that underlying correlations among different modalities help improve the efficiency of multimedia analysis [1][9][10][17][18][24]. It is important and interesting to utilize multi-feature correlation mining of different modalities in the process of subspace learning. However, most of traditional subspace learning methods, which are based on multimedia data of single modality, are hardly usable for different modalities.

In this paper we propose a novel multi-feature subspace learning method via sparse correlation detection and fusion for multimedia analysis. Our method is formulated based on two typical modalities, i.e., image and audio. First, we analyze visual-auditory multi-feature correlation and build the Sparse Multi-feature Subspace (SMFS) based on sparse canonical correlation analysis. Secondly, we further explore underlying content and semantic correlation in the SMFS: the content correlation is explored with geometrical motivated local linear regression model, and semantic correlation is analyzed with graph-based nonlinear learning to bridge the semantic gap. Thirdly, both of above content and semantic learning results are fused into an objective function to calculate a global optimized solution. The efficiency and superiority of our approach is tested and demonstrated with several multimedia applications: content-based multimodal retrieval and data classification.

The rest of this paper is organized as follows. Section 2 describes sparse multi-feature subspace construction for image and audio data. Section 3 gives details of the subspace optimization with correlation fusion. Section 4 presents experimental results and comparisons. Concluding remarks are in section 5.

2 Sparse Multi-feature Subspace

Since multimedia data of different modalities are initially represented with heterogeneous low-level features, in this section we construct a Sparse Multi-feature Subspace (SMFS) where different multimedia data all reside and canonical correlation among original features are furthest preserved.

In our previous work, Canonical Correlation Analysis (CCA) was used to find a map to the isomorphic subspace [10]. CCA is a classical method to explore statistical correlation between two sets of variables. The underlying ideas of CCA are as follows: it looks for two basis vectors for two sets of variables such that the correlation between the projections onto the basis vectors is mutually maximized. However, when the dimension of low-level features is very high it is important to preserve meaningful correlations instead of all of them. Therefore, in this section, we focus on how to find sparse mapping vectors during feature correlation analysis. In the following description our method is formulated based on two typical modalities: image and audio.

Formally, let $X \in R^{n \times p_1}$ denote image feature matrix, and $Y \in R^{n \times p_2}$ denote audio feature matrix. We assume that the columns of X and Y have been standardized to have mean zero and standard deviation one. CCA calculates linear combinations of the variables in X and Y that are maximally correlated with each other.

Let $u \in R^{p_1}$ and $v \in R^{p_2}$ denote canonical vectors which maximize the correlation between Xu and Yv, CCA is to solve the following extremum problem:

$$\begin{aligned} &\max\ u^T X^T Y v \\ &s.t.\ u^T X^T X u = 1,\ v^T Y^T Y v = 1 \end{aligned} \tag{1}$$

The main challenging issue of CCA is that it is not appropriate when $p_1, p_2 \approx n$ or $p_1, p_2 \gg n$. Then Sparse Canonical Correlation Analysis (SCCA) was proposed to address the problem [12], and its objective function takes the following form:

$$\begin{aligned} &\max\ u^T X^T Y v \\ &s.t.\ \|u\|_2 \le 1,\ \|v\|_2 \le 1,\ P_1(u) \le c_1,\ P_2(v) \le c_2 \end{aligned} \tag{2}$$

where P_1, P_2 are convex and non-smooth sparsity-inducing penalty functions that yield sparse u, v. And the constraints $\|u\|_2 \le 1,\ \|v\|_2 \le 1$ are convex relaxations of the quality constraints. Paper [12] studied two specific forms of the penalty P_1, P_2 with structure of L1-norm penalty and the chain-structured fused lasso penalty, which resulted in unique canonical vectors, even when $p_1, p_2 \gg n$. With u fixed, the criterion is convex in v, and with v fixed, it is convex in u.

To solve image canonical vector u and audio canonical vector v, the objective function is used the same as function (2). Here canonical vectors u and v define a linear combination of visual features in X that is correlated with a linear combination of audio features in Y. Elements of u and that equal zero indicate features in X and Y that are not involved in the linear combinations. In this paper we impose sparsity on v. For the ease of illustration, we assume:

$$P_1(u) = \|u\|_1 \tag{3}$$

$$P_2(v) = \sum_j |v_j| + \sum_j |v_j - v_{j-1}| \tag{4}$$

where $P_1(u)$ is an L_1 penalty and $P_2(v)$ is a fused lasso penalty. To substitute Eq.(3) and Eq.(4) for $P_1(u)$ and $P_2(v)$ in the objective function (2), the optimized sparse canonical vectors u and v can be calculated with the algorithm proposed in paper [12]. Accordingly, image feature matrix and audio feature matrix can be mapped to the Sparse Multi-feature Subspace (SMFS).

3 Subspace Optimization by Correlation Fusion

Although image and audio samples are represented with isomorphic dimensionality in the SMFS where sparse canonical correlation is preserved, the SMFS isn't well consistent with high-level semantics because of the well-known semantic gap. Thus, in this section, we propose subspace optimization strategy by correlation fusion, which explores content correlation with local linear regression and utilizes semantic correlation based on a weighted k-nearest neighbor graph. Moreover, both content and semantic correlation we learned are fused into an overall objective function; the optimum solution of this function is therefore the Optimized Sparse Multi-feature Subspace (OSMFS)

In the following descriptions, $z_i, (i \in [1, 2n])$ denote a sample (which could be image or audio) in the SMFS, and $m_i, (i \in [1, 2n])$ denote the corresponding coordinate vector in the OSMFS after optimization, $Z = \{z_1, z_2, ..., z_{2n}\}$ denote all of the samples in the SMFS.

3.1 Content Correlation Analysis with Local Linear Regression

To predict the value of m_i, we use a local linear regression model of $\psi_i^T z_i + \xi_i$ where ψ_i is a matrix and ξ_i is a vector of bias term. The definition of local means k-nearest neighbors of z_i (including z_i itself), denoted as $\mathbb{N}^k(z_i)$. The regression parameters ψ_i and ξ_i are common for all the samples in $\mathbb{N}^k(z_i)$. Based on the local linear regression model, we define a prediction error for each z_i as:

$$\sigma(z_i) = \left\|\psi_i^T z_i + \xi_i - m_i\right\|_F^2 \tag{5}$$

Then, by summing up all prediction errors for samples in $\mathbb{N}^k(z_i)$ we obtain the local prediction error as below:

$$\sigma_L = \sum_{z_j \in \mathbb{N}^k(z_i)} (\left\|\psi_i^T z_j + \xi_i - m_j\right\|_F^2 + \alpha\left\|\psi_i\right\|_F^2) \tag{6}$$

where the second term is added as a regularizer to avoid overfitting. We minimize the local error and get the objective function

$$\arg\min_{m_i, \psi_i, \xi_i} (\left\|\psi_i^T Z_i + \xi_i \xi_i^T - M_i\right\|_F^2 + \alpha\left\|\psi_i\right\|_F^2) \tag{7}$$

where $Z_i = [z_i, z_i^1, z_i^2, ..., z_i^k]$ is a feature matrix of samples in $\mathbb{N}^k(z_i)$, and $M_i = [m_i, m_i^1, m_i^2, ..., m_i^k]$ is a matrix of the coordinate vectors for the samples in $\mathbb{N}^k(z_i)$. It is easy to find that the global projection error is to sum (7) on all the samples in training set. Therefore the objective function can be rewritten as:

$$\arg\min \sum_{i=1}^{n} (\left\|\psi_i^T Z_i + \xi_i \xi_i^T - M_i\right\|_F^2 + \alpha\left\|\psi_i\right\|_F^2) \tag{8}$$

By setting the derivatives to be zero with respect to ψ_i and ξ_i, we have:

$$\begin{cases} \psi_i = (Z_i H Z_i^T + \alpha I)^{-1} Z_i H M_i \\ \xi_i = (k+1)^{-1}(M_i \theta_{k+1} - Z_i^T \psi_i \theta_{K+1}) \end{cases}, H = I - (k+1)^{-1}\theta_{k+1}\theta_{k+1}^T \in \mathbb{R}^{(k+1)\times(k+1)} \tag{9}$$

where $\theta_{k+1} \in \mathbb{R}^{k+1}$ is a column vector with all ones, and H is the centering matrix. With (8)(9), the objective function becomes:

$$\arg\min_{M_i} \sum_{i=1}^{n} M_i L_i M_i^T \tag{10}$$

$$L_i = H - HZ_i^T (Z_i H Z_i^T + \alpha I)^{-1} Z_i H \quad , \quad L_i \in \mathbb{R}^{n\times(k+1)} \tag{11}$$

We define $M_i = MS_i$ where matrix M consists of all coordinate vectors in the OSMFS and $S_i \in \mathbb{R}^{n\times(k+1)}$ is a selecting matrix made up of 1 and 0. To be specific, the cell value of $S_i(r,t)=1$ when both z_r and z_t are in the k-nearest neighbor set of $\mathbb{N}^k(z_i)$, and otherwise $S_i(r,t)=0$. Therefore, the objective function can finally be written as:

$$\arg\min_{M} tr(MLM^T) \tag{12}$$

where $L=[S_1,S_2,...,S_n]\begin{bmatrix} L_1 & & \\ & ... & \\ & & L_n \end{bmatrix}[S_1,S_2,...,S_n]^T$ is name as content correlation Laplacian matrix.

3.2 Semantic Correlation Analysis Based on Graph Model

When the query example r is inside training set, the system finds its k-nearest neighbors in the MSS, ranks them by distance in ascending order, and returns to the user as query results. On the other hand, when the query example r is out of training set it needs to be mapped into the MS, then the retrieval process is the same as before. Moreover, we use weighted k-nearest neighbor graph to explore underlying semantic correlation. Since it is difficult to find large amount of labeled image and audio samples for supervised analysis, we utilize both labeled and unlabeled datasets for semantic exploration. Formally, let $G_s(V,E)$ denote a weighted k-nearest neighbor graph with its vertex set V being the set of Z, and the corresponding semantic weight matrix $A=[a_{ij}]$ is defined by

$$a_{ij} = \begin{cases} 1, & if\ z_i \in \mathbb{N}^k(z_j)\ or\ z_j \in \mathbb{N}^k(z_i) \\ 0, & otherwise \end{cases} \tag{13}$$

where $\mathbb{N}^k(z_j)$ denotes the set of k-nearest neighbors of z_j. Note that: if z_i and z_j are both labeled and belong to the same class then definitely there is

$z_i \in \mathbb{N}^k(z_j)$, $z_j \in \mathbb{N}^k(z_i)$; if one of them is unlabeled or both of them are unlabeled we use the Euclidean distance in the SMFS to calculate whether they are k-nearest neighbors to each other. The normalized semantic correlation Laplacian matrix is then defined as

$$L_s = I - B^{-1/2} A B^{1/2} \tag{14}$$

where I is a $n \times n$ identity matrix and B is an $n \times n$ diagonal matrix with its i-th diagonal element being the sum of the i-th row of A . Then, similar to above subsection, we have the following objective function according to graph-based semantic learning result of L_s :

$$\arg \min_{M^T M = I} tr(M L_s M^T) \tag{15}$$

So far, we obtain two objective functions (12) and (15). Therefore, the overall objective function to calculate a good map to the OSMFS is defined as:

$$\arg \min_{M^T M = I} tr(M (L + L_s) M^T) \tag{16}$$

Therefore, the optimum solution of (16) can be obtained by eigen-decomposition of $(L + L_s)$. In summary, the construction of SMFS and its optimization via content and semantic correlation fusion are stated below:

Step 1. Extract low-level features of images and audio clips, and formulate visual feature matrix $X \in R^{n \times p_1}$ and auditory feature matrix $Y \in R^{n \times p_2}$;

Step 2. Analyze sparse canonical correlation between matrices X and Y , compute SMFS where the correlation learned is furthest preserved, and map all samples into the SMFS;

Step 3. Compute k-nearest neighbors $\mathbb{N}^k(z_i)$ for each sample z_i in the SMFS, calculate sub-matrices S_i, L_i $(i \in [1, n])$ in (11), then get content correlation Laplacian matrix L in (12);

Step 4. Construct the k-nearest neighbor graph $G_s(V, E)$ with labeled and unlabeled data, calculate corresponding semantic weight matrix $A = [a_{ij}]$ in (13), then get the normalized semantic correlation Laplacian matrix L_s in (14);

Step 5. Compute $V = [v^1, v^2, ... v^c]$ in which $v^1, v^2, ... v^c$ are eigenvectors obtained from c minimum non-zero eigenvalues of (16), then for sample z_i its coordinate vector in the OSMFS is calculated by $m_i = (v_i^1, v_i^2, ..., v_i^c)$ where v_i^j is the ith entry of eigenvector v^j .

Fig. 1. The construction and optimization of SMFS

4 Experiments

4.1 Dataset

To evaluate the effectiveness of the proposed approach, we experimented with an image-audio dataset consisting of 12 semantic categories, including dog, car, bird, explosion, tiger, train, dolphin, drum, piano, plane, zither, lightning. There are 2000 images and 1200 audio clips in total, which are divided equally into the 20 categories. 85% of the image-audio dataset are used as labeled data and the rest 15% are used as unlabeled. We test our proposed algorithm with content-based multimodal retrieval and data classification.

The extracted visual features include Color Histogram, CCV, and Tamura Texture. Auditory features are made up of Centroid, Rolloff, Spectral Flux and Root Mean Square. Since audio is a kind of time series data, the dimensionalities of combined auditory feature vectors are inconsistent. We require collected audio clips not exceed 7 seconds and employ Fuzzy Clustering [10] on auditory features in preprocessing to get index vectors.

4.2 Performance Comparison

Content-based multimodal retrieval is performed based on the Euclidean distance in the subspace. In our experiments, if a returned result and the query example are in the same semantic category, it is regarded as a correct result. Precision is defined as the percentage of correctly retrieved samples in the top-k-returned results.

To evaluate the efficiency of our method, k-nearest neighbors, which could be image and audio, are calculated for each query example according to the Euclidean distance in the SMFS and in the OSMFS respectively. The SMFS is obtained with SCCA which explores original multi-feature canonical correlation, while the OSMFS is obtained with further optimization of the SMFS. Therefore, our experiments compare the retrieval performance before and after subspace optimization, together with the correlation ranking algorithm in [11]. In all the following figures, the precision is the averaged precision values from all the query samples.

Figure 2 shows the comparison results. From Figure 2 we have the following observations: after subspace optimization the retrieval performance is improved, and the SMFS slightly outperform the correlation ranking method in [11]. The phenomenon is attributed to the following reasons:

(1) Sparse correlation is explored from a global perspective and preserved in the SMFS, while in the OSMFS the knowledge learned from both content correlation analysis and semantic correlation analysis further helps bridge the semantics gap.

(2) Paper [11] estimated multimodal similarity based on correlation ranking in the text-image-audio graph, in which text data worked as a bridge to propagate ranking scores among image and audio; when text data is not included the correlation ranking method in [11] doesn't work well, while our multi-feature subspace learning method are based on intrinsic sparse canonical correlation learned from low-level visual and auditory features.

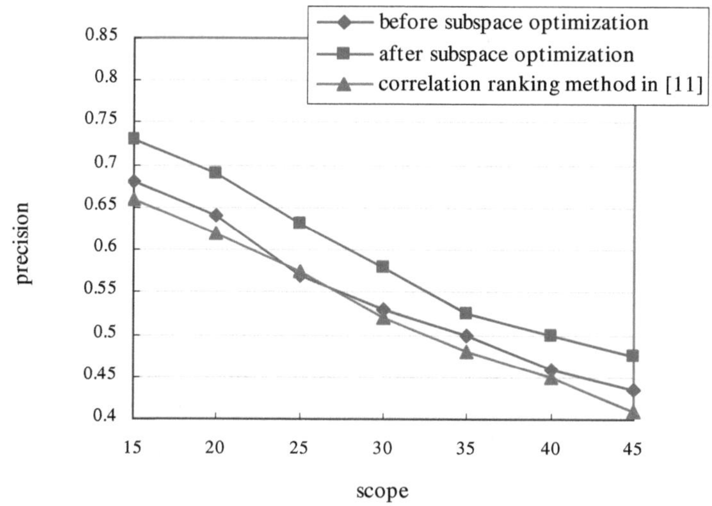

Fig. 2. Comparison of before and after subspace optimization and the correlation ranking algorithm in [11] for multimodal retrieval

To further evaluate the efficiency of our algorithm, data classification is performed in the OSMFS. We use images and audio clips from all 12 categories for performance evaluation respectively. Table1 shows ACC (Accuracy) and AUC(Area Under Curve) results for image classification and audio classification.

Table 1. ACC & AUC performance for image clustering and audio clustering

	Image clustering		Audio clustering	
	ACC	AUC	ACC	AUC
Dog	0.412	0.512	0.397	0.503
Car	0.426	0.533	0.422	0.522
Bird	0.453	0.549	0.389	0.512
Explosion	0.445	0.538	0.431	0.534
Tiger	0.493	0.563	0. 418	0. 518
Train	0.408	0.522	0. 392	0. 508
Dolphin	0.447	0.557	0. 442	0. 537
Drum	0.398	0.515	0. 429	0. 521
Piano	0.456	0.573	0. 425	0. 528
Plane	0.461	0.562	0. 413	0. 514
Zither	0.458	0.519	0. 399	0. 509
Lightning	0.418	0.487	0.408	0.503

In our experiments KNN (k-nearest neighbor) method is used to execute classification process on image and audio data located in the OSMFS space. We observe that our algorithm works well on the whole. This demonstrates that the OSMFS space learns underlying correlation among image and audio data, and is basically in accordance with high-level semantics.

5 Conclusions

Different from most existing subspace learning methods which focus on single modality feature analysis and denoising, this paper proposes a novel multi-feature subspace learning method, which not only maps multimedia data of different modalities into the sparse multi-feature subspace where original canonical correlation is furthest preserved, but also optimizes the subspace with correlation fusion which explores content correlation with local linear regression and utilizes semantic correlation based on a weighted k-nearest neighbor graph. The experimental results are promising and advantageous; also show that our approach is effective on several multimedia applications. To the best of our knowledge, it is hard to find a public multimodal database as a benchmark. The collected image-audio database is relatively small. Therefore, future work includes further study on large-scale multimodal data analysis and semantic understanding.

Acknowledgment. This work is supported by National Natural Science Foundation of China (No.61003127), State Key Laboratory of Software Engineering (SKLSE2012-09-31).

References

[1] Yang, Y., Nie, F., Xu, D., Luo, J., et al.: A Multimedia Retrieval Framework based on Semi-Supervised Ranking and Relevance Feedback. IEEE Transactions on Pattern Analysis and Machine Intelligence (TPAMI) 34(4), 723–742 (2012)
[2] Datta, R., Joshi, D., Li, J., Wang, J.Z.: Image retrieval: Ideas, influences and trends of the new age. ACM Computing Surveys 40(2) (2008)
[3] Zhang, R., Zhang, Z.: Effective Image Retrieval based on Hidden Concept Discovery in Image Database. IEEE Transactions on Image Processing 16(2), 562–572 (2007)
[4] Nie, F., Xu, D., Tsang, I., Zhang, C.: Spectral Embedded Clustering. In: International Joint Conference on Artificial Intelligence (IJCAI), California, pp. 1181–1186 (2009)
[5] Ye, J., Zhao, Z., Wu, M.: Discriminative k-means for clustering. Advances in Neural Information Processing Systems 20, 1649–1656 (2008)
[6] Liang, D.W., Liu, Y., Huang, Q.M., et al.: Video2Cartoon: Generating 3D Cartoon from Broad-cast Soccer Video. In: Proceedings of ACM Multimedia (2005)
[7] Typke, R., Wiering, F., Veltkamp, R.: A survey of music information retrieval systems. In: Proceedings of ISMIR, pp. 153–160 (2005)
[8] Lew, M., Sebe, N., Djeraba, C., Jain, R.: Content-based multimedia information retrieval: state-of-the-art and challenges. ACM Transactions on Multimedia Computing, Communication, and Applications 2(1), 1–19 (2006)

[9] Yang, Y., Xu, D., Nie, F., Luo, J.: Ranking with local regression and global alignment for cross-media retrieval. ACM Multimedia (2009)
[10] Zhang, H., Zhuang, Y., Wu, F.: Cross-modal correlation learning for clustering on image-audio dataset. ACM Multimedia (2007)
[11] Zhang, H., Meng, F.: Multi-modal Correlation Modeling and Ranking for Retrieval. In: IEEE Pacific-Rim Conference on Multimedia, pp. 637–646 (2009)
[12] Witten, D.M., Tibshirani, R.: Extensions of sparse canonical correlation analysis, with applications to genomic data. Statistical Applications in Genetics and Molecular Biology 8(1) (2009)
[13] Yang, Y., Zhuang, Y., Xu, D., Pan, Y., Tao, D., Maybank, S.: Retrieval Based Interactive Cartoon Synthesis via Unsupervised Bi-Distance Metric Learning. ACM Multimedia, 311–320 (2009)
[14] Turk, M.A., Pentland, A.P.: Face Recognition using Eigenface. In: Computer Vision and Pattern Recognition, pp. 586–591 (1991)
[15] Guo, G., Li, S.Z., Chan, K.: Face Recognition by Support Vector Machines. In: IEEE Intl. Conf. on Auto. Face and Gesture Recognition, pp. 196–201 (2000)
[16] McGurk, H., MacDonald, J.: Hearing Lips and Seeing Voices. Nature 264, 746–748 (1976)
[17] Zhang, H., Liu, Y., Ma, Z.: Fusing inherent and external knowledge with nonlinear learning for cross-media retrieval. Neurocomputing (2013), doi:10.1016/j.neucom.2012.03.033
[18] Ma, Q., Akiyo, N., Katsumi, T.: Complementary Information Retrieval for Cross-media News Content. In: Proceedings of Information Systems, vol. 31(7), pp. 659–678 (2006)
[19] Joliffe: Principal component analysis. Springer, New York (1986)
[20] He, X.F., Yan, S.C., Hu, Y.X., et al.: Face recognition using laplacianfaces. IEEE Transactions on Pattern Analysis and Machine Intelligence 27(3), 328–340 (2005)
[21] Tenenbaum, J.B., Silva, V.D., Langford, J.C.: A Global Geometric Framework for Nonlinear Dimensionality Reduction. Science 290, 2319–2323 (2000)
[22] Hansen, L., Larsen, J., Kolenda, T.: On Independent Component Analysis for Multimedia Signals. In: Multimedia Image and Video Processing, pp. 175–200. CRC Press (2000)
[23] Guo, G., Li, S.Z.: Content-based Audio Classification and Retrieval by Support Vector Machines. IEEE Transactions on Neural Networks 14(1), 209–215 (2003)
[24] Slaney, M., Covell, M.: FaceSync: A linear operator for measuring synchronization of video facial images and audio tracks. In: NIPS, pp. 814–820 (2000)
[25] Belhumeur, P., Hespanha, J., Kriegman, D.: Eigenfaces vs. Fisherfaces: recognition using class specific linear projection. IEEE Transactions on Pattern Analysis and Machine Intelligence 19(7), 711–720 (1997)
[26] Zhang, H., Yu, J., Wang, M., Liu, Y.: Semi-supervised Distance Metric Learning based on Local Linear Regression for Data Clustering. Neurocomputing 93, 100–105 (2012)
[27] Lovasz, L., Plummer, M.: Matching Theory, pp. 307–349. Akadémiai Kiadó, North Holland (1986)
[28] Cai, D., He, X., Han, J.: Semi-supervised Discriminant Analysis. In: IEEE 11th International Conference on Computer Vision, pp. 1–7 (2007)
[29] Ma, Z., Yang, Y., Nie, F., Uijlings, J., Sebe, N.: Exploiting the entire feature space with sparsity for automatic image annotation. In: Proceedings of the 19th ACM International Conference on Multimedia, pp. 283–292

Auditory Context Recognition Combining Discriminative and Generative Models

Feng Su* and Li Yang

State Key Laboratory for Novel Software Technology,
Nanjing University, Nanjing 210023, China
suf@nju.edu.cn

Abstract. The paper considers the task of recognizing the category of a context surrounding an audio sensor. Due to the unstructured and diverse nature of the auditory context and constituent environmental sounds, which differs from the usual structured audio data like speech or music, the recognition of auditory context faces many difficulties and relatively fewer researchs have addressed it. In this paper, we propose an ensemble recognition scheme based on the Hough forest framework for unstructured auditory contexts, which combines the discriminative and generative modeling of the context. We learn the effective audio feature representation for environmental sounds in the context with the LDB algorithm, and recognize the context using the Hough forest based ensemble classifier, which aggregates both the segmental and the contextual probabilistic votes on the context category by the segments of the auditory context. The experimental results demonstrate the effectiveness of the proposed approach for auditory context recognition.

Keywords: auditory context, environmental sound, hough forest, local discriminant bases, HMM.

1 Introduction

Nowadays, more and more smart mobile or embedded devices are equipped with acoustic sensors and powerful CPUs with rapidly increasing speed and capacity. This allows the inclusion of sound as a common type of data in many exciting applications such as content-based indexing and retrieval of multimedia databases, robotics systems and various context-aware applications. Since compared with vision-based approaches, audio signals are robust in situations like visual obstruction or weak lighting and relatively inexpensive in storing and computing, audio data are widely exploited either independently or as an important component in multi-modal processing.

For context aware processing specifically, it's very useful for machines to automatically recognize the category of the surrounding auditory context where the auditory sensor is currently located and, furthermore, to understand the high-level semantics of the context. Such information, for example, may be used to

* Corresponding author.

B. Huet et al. (Eds.): PCM 2013, LNCS 8294, pp. 608–619, 2013.

automatically switch the operational mode of one mobile phone or provide the user with customized information as the circumstance changes.

The term auditory context here refers to the acoustic modeling of a specific location or site such as restaurant, bus station or conference hall, which contains and is usually characterized by various auditory events occurring in it, such as laughing, whistling and gunfire, as illustrated in Fig. 1. Such auditory events are short audio segments corresponding to specific object or event in the context and with distinct acoustic patterns.

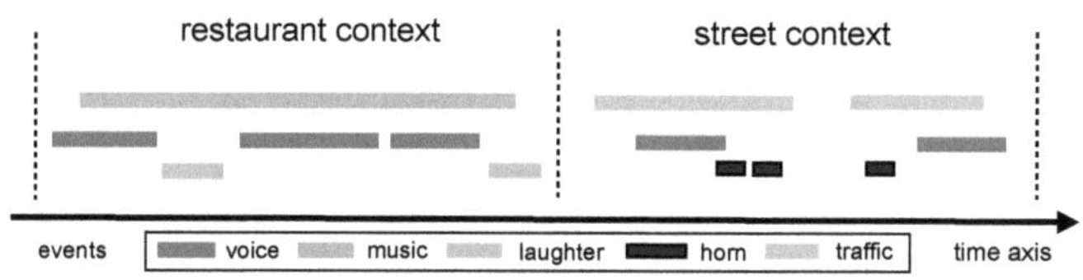

Fig. 1. Illustration of auditory contexts

The difficulty for auditory context recognition lies in two aspects. First, most auditory events can be categorized to the so-called *environmental sounds*. Different from speech or music, which can be categorized to structured sounds due to their formantic or harmonic structure characteristic, environmental sounds are typically unstructured, which have a broad noise-like flat spectrum and inherently diverse variety of signal composition [1] and are difficult to build models.

Research in general audio signal analysis and recognition has received long time interests in the past years [2,3,4,5] and a variety of audio features and processing models have been proposed in temporal or spectral domain such as the popular Mel-frequency cepstral coefficients (MFCCs). However, most of previous efforts deal with structured sounds like speech or music, relatively fewer research address the unstructured environmental sounds. In [1], the author proposed to use the matching pursuit (MP) algorithm to select effective time-frequency features as the supplementation to MFCC for audio environment characterization. In [6], the performances of three new types of transforms, Chirplet, curvelet, and Hilbert transforms, are investigated for environmental sound classification. In [7], a hybrid SVM/KNN classifier is used for environmental sound classification based on MPEG-7 audio low-level descriptors.

As the second aspect of auditory context recognition, the loose and possibly ambiguous connection between audio signal components (events) and the context brings considerably difficulties to the modeling of auditory context. Similar audio patterns may appear in different auditory contexts, and the samples of one context category may exhibit variant composition of characteristic events, or even lack of such distinct event. For these reasons, sophisticated context models on the basis of audio event or characteristic audio fragments are usually needed to capture the correlation between the audio components and the context. For example, In [8], a set of key audio effects are modeled with HMMs and a Bayesian network-based approach is proposed to discover the high-level semantics of an

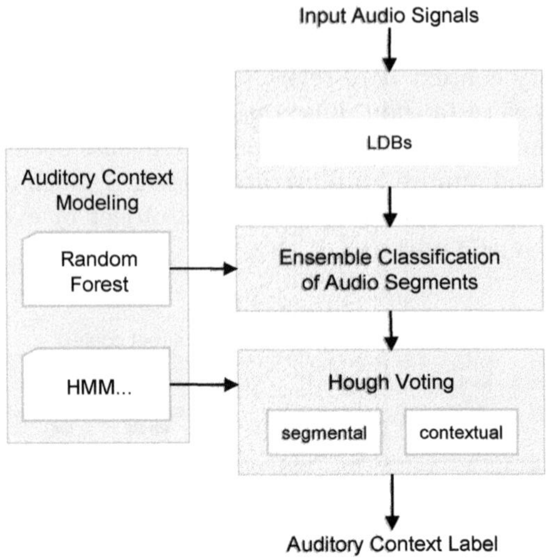

Fig. 2. Block diagram of the proposed auditory context recognition algorithm

auditory context embedded in key effect sequences. In [9], the ergodic hidden Markov model and support vector machine (SVM) approaches are investigated to fuse the characteristics and correlations among various audio events at the semantic context level. In [10], Eronen et al. developed a system to evaluate the recognition accuracy of various audio features (ZCR, MFCC, Band-energy, Spectral flux, etc.), feature transforms (PCA, ICA, LDA), classifiers (k-NN and HMM) on 24 audio contexts and compared with human listeners. In [1], on the other hand, the author considers the sound events as properties of each environment, hence characterize the acoustic environment of one context of specific category as a whole with Gaussian Mixture Model (GMM), rather than a collection of discrete sound events pre-extracted.

In this paper, on the basis of our previous work [11] that describes the environmental sounds with local discriminant bases (LDB), we propose an ensemble recognition scheme based on the Hough forest framework for unstructured auditory contexts, which combines the discriminative and generative modeling of the context. We learn the effective audio feature representation for environmental sounds in the context with the LDB algorithm, and recognize the context using the Hough forest based ensemble classifier, which aggregates both the segmental and the contextual probabilistic votes on the context category by the segments of the auditory context. The block diagram of the proposed algorithm is outlined in Fig. 2.

The remainder of the paper is organized as follows. Section 2 describes the audio feature extraction. Section 3 describes the Hough forest based recognition model for auditory contexts. In Section 4, we present the experimental results of the proposed method and some discussions.

2 Audio Feature Extraction

Due to the unstructured and diverse nature of environmental sounds, we exploit the LDB algorithm proposed by Saito and Coifman [12] to learn the effective feature representation for auditory contexts from training samples. LDB is basically one kind of "best-basis" method to select an orthonormal optimum basis or subspace from a large collection of bases, which minimizes entropy or maximizes a certain discriminant measure among classes. Once such a basis is selected, a small number of most significant coordinates (features) can be used to enhance the performance of the classifier without losing important information of the problem.

To exploit the LDB algorithm to identify the feature components that exhibit high discrimination between different auditory contexts, like in [5], we decompose an audio signal into a binary wavelet packets tree and use LDB as a pruning algorithm to obtain a set of tree nodes (subspaces) that provides maximum dissimilarity information between different signal classes. The block diagram of the LDB algorithm used in our work is depicted in Fig. 3.

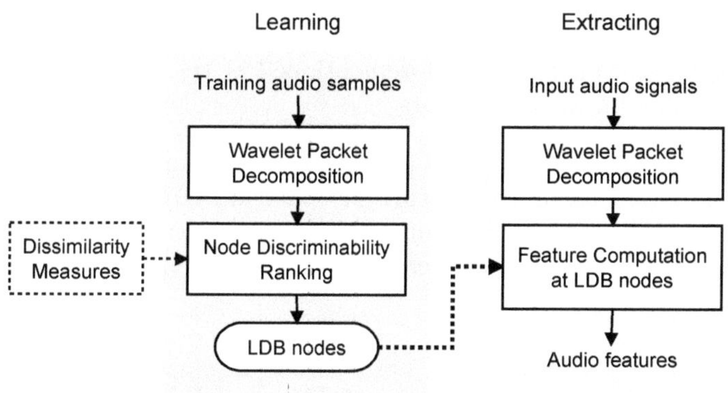

Fig. 3. Block diagram of the LDB-based audio feature extraction

Specifically, an audio signal x_i is decomposed into a binary wavelet packets tree $\{\Omega_{j,k}\}$, where j indicates the level of the tree, k represents the node index in level j, $\Omega_{j,k}$ denotes the subspace spanned by a set of wavelet packet basis vectors $\{w_{j,k,l}\}_{l=0}^{l=2^{u-j}-1}$ and 2^u corresponds to the length of the signal x_i. The signal x_i can then be expressed as

$$x_i = \sum_{j,k,l} [a_{j,k,l}]_i \cdot w_{j,k,l} \tag{1}$$

where, $a_{j,k,l}$ are the basis vector coefficients at node (j, k).

Next, from all 2^J mutually orthogonal subspaces (J is the number of decomposition levels), we adopt the following steps to prune the wavelet packet tree based on the discriminative capability of a wavelet packet node (subspace) (j,k) on P signal classes:

1. Compute the dissimilarity values $M_\beta^{(j,k)}$ of node (j,k) on the $\beta^{th} \in \binom{P}{2}$ combination of two signal classes, which is formulated as the mean of the dissimilarity values $\{D^{(j,k)}\}_{t\in[1..T]}$ between a set of T sample signal pairs extracted randomly from the training sets of two classes. As in [5], $D^{(j,k)}$ is computed as the difference of the normalized energy, $E_1^{(j,k)} - E_2^{(j,k)}$, or the difference of temporal variance of local piecewise variances of wavelet packet basis coefficients, $\frac{var[v^1_{(j,k)}]}{var[v^2_{(j,k)}]}$, on node (j,k) of two classes' samples being compared, which reveal the variance of energy concentration locations on the TF plane or the nonstationarity for these two audio classes.
2. Choose Q LDB nodes with consistently high values of $M_\beta^{(j,k)}$ over the random trials of training samples.
3. From the pool of $Q \times \binom{P}{2}$ LDB nodes collected from all class combinations, choose the H most frequently occurring LDB nodes to constitute the final feature representation of audio samples, where H is the required dimension of LDB features.
4. For an input audio segment, we extract the wavelet packet basis coefficients or the energy/variance values aforementioned at the H learned LDB nodes to form the H-dim feature vectors.

The intrinsic multi-scale property of the wavelet packet decomposition of signals and the adaptive learning of feature subspace in LDB are particularly useful in identifying and describing discriminative characteristics between signal components of the auditory context, which vary significantly in duration and spectral composition.

3 Ensemble Auditory Context Recognition

Given the sequence of short-time audio features extracted from one audio stream of certain context, in this section, we propose a Hough forest-based ensemble recognition model to classify it to one of the predefined auditory context categories. The main differences of the proposed method from other existing ones including our previous work [11] are in two aspects: 1) We omit the modeling of specific categories of audio events, so that employ a one-layer processing model that directly operates on the auditory context samples. 2) Different from the HMM-based modeling of auditory contexts adopted by most works, we propose a Hough forest-based model for auditory contexts, which combines the discriminative classification of individual audio features and the HMM-based generative modeling of the correlations between various audio features.

The reasons for 1) is mainly due to the complexity and variety of the compositions of auditory contexts, in which a predefined set of audio event categories usually cannot cover all discriminative components of one context. On the other side, to automatically discover representative patterns from contexts will increase the adaptibility of the algorithm and allow it to be expanded incrementally.

For 2), considering the unstructured characteristics of environmental sound components and their unstructed occurrences in complex auditory contexts, we employ an ensemble voting scheme, instead of explicit temporal modeling at the context level, for auditory context recognition, in which the votes for the potential context category casted by individual segments of the input audio stream and their contextual combinations are aggregated by a generalized Hough transform.

3.1 Hough Forest Classification Model

Hough forest [13] is basically a random forest [14], which in turn is a collection of binary decision tree predictors, each classifying the input feature vector and outputting its own vote on the possible class label, and the final output of the whole forest is then the integration of the individual votes. The binary test at each non-leaf node of decision trees performs thresholding on one feature channel to determine the branch the input feature vector goes down, where the choice for feature channel and thresholds are learned in the training stage. In Hough forest, a generalized Hough transform is exploited to aggregate votes on parameters of interest (e.g. the symbol category in our case) casted by individual samples of the object. Compared to a single decision tree or other simple classifier, Hough forest assembles together several trees trained in a randomized way and usually achieves superior generalization and stability.

In our work, a Hough forest model is used to learn the mapping between features of individual audio segments and their probabilistic votes on potential categories of the auditory context. The training process, given samples of different auditory contexts, is as follows:

1. Initialize a set of L decision trees $\{T_1, T_2, \ldots, T_L\}$ with required parameters (maximum depth, minimum sample count per node, etc.).
2. Divide each training audio sample into sequence of segments with sufficient overlapping, extract the aforementioned feature vector $\{\boldsymbol{v}_i\}$ for each segment, and attach the auditory context label r_i ($r_i \in [1..K]$, K is the number of context categories) of the training sample to each $\boldsymbol{v}_i$.
3. Train the random forest $\{T_1, T_2, \ldots, T_L\}$ with all $\{< \boldsymbol{v}_i, r_i >\}$. Once trained, the leaf nodes of forest trees encode the characteristic features of audio segment for specific context classes.

Given an unknown testing audio stream s, we send its feature vectors corresponding to all its segments $\{s_j\}$ down the forest and collect the sets of training samples R_j contained in the predicted leaf nodes of the forest for all s_j. The probabilistic votes on context category casted by all segments $\{s_j\}$ are then aggregated with the generalized Hough transform. We consider two forms of votes, the segment-specific vote $H_s(k)$ and the contextual (cross-segment) vote $H_c(k)$, and both $H_s(k)$ and $H_c(k)$ are 1-D Hough matrice ($k = 1..K$, K is the number of context categories) with each element containing the accumulated votes for the corresponding context category.

3.2 Segmental Hough Aggregation

Considering each sample $< \boldsymbol{v}_i, r_i >$ in the training sample set R_j of segment s_j makes a contribution $u(j)$ to the hypothesis that it belongs to an auditory context of category r_i, thus the corresponding bin of the segment-specific Hough matrix $H_s(r_i)$ is incremented as:

$$H_s(r_i) = H_s(r_i) + u(j), \quad u(j) = \frac{w(j)}{|R_j|} \tag{2}$$

where, $w(j)$ is certain weight associated with the segment s_j, denoting the importance or the confidence of its votes. In this work, we fix $w(j)$ to 1.0 for simplicity.

3.3 Contextual Hough Aggregation

The aforementioned segment-specific vote $H_s(k)$ relies solely on the characteristic of one segment of the input audio stream, however, similar acoustic patterns may appear in several context categories, while the combination of multiple segment, a.k.a. the context, will usually reduce such ambiguity, especially for auditory context with stronger correlations (or less unstructed in certain senses) between its components.

To measure the contextual contribution of all the segments $\{s_j\}$ of the input audio stream to the potential context category, we define the contextual Hough voting $H_c(k)$ as follows:

1. For each auditory context category $k \in [1, K]$ (K is the number of context categories), learn one HMM model HMM_k on the corresponding training samples using the Baum-Welch algorithm to capture the probabilistic correlations between them. In this work, we employ HMMs with discrete states and continuous observations modeled by mixtures of Gaussians. The number of states in the HMM varis with different context categories and is determined experimentally.
2. Group all the training samples in $\{R_j\}$ of an input audio stream according to the category label $r_i \in [1, K]$ into K sets $\{\bar{R}_k\}_{k=1..K}$, and sort the samples in each $\bar{R}_k$ according to their temporal order in the original training audio streams.
3. Send the samples in each $\bar{R}_k$ into the corresponding HMM_k, and use the (normalized) output likelihood l_k by the Forward algorithm as the vote for the context category k. The corresponding bin of the contextual Hough matrix $H_c(k)$ is then incremented as:

$$H_c(k) = H_c(k) + l_k \tag{3}$$

The contextual correlation between audio segments can also be described using the latent Dirichlet allocation (LDA) model [15], in which the topic can be mapped to the auditory event (or some other latent parameter of the feature set), the words can be mapped to the discrete cluster labels of audio features obtained by applying certain clustering algorithm like the k-means or spectral clustering on all training audio features, and the document can therefore be mapped to one audio stream. Specifically, we train one LDA model LDA_k for each auditory context category k with its number of topics (i.e. constituent event categories) be presumed large enough. Then, for each set $\bar{R}_k$ of training samples triggered by the input audio stream, we use the normalized probability of the document $\bar{R}_k$ on the corresponding LDA_k as the vote l_k for the context category k, and then proceed as in the HMM scheme. Because this LDA-based generative contextual voting scheme performs not as good as the aforementioned HMM-based model in experiments on our current feature representation and model configuration, however, in this paper we still focus on the HMM model, letting the former explored in more depth in succeeding work.

3.4 Combination of Segmental and Contextual Votes

Similar to the potential function of the Markov Random Field model that integrates the first-order unary cliques and higher order cliques involving multiple sites, we combine Hough votes from the discriminative random forest classifier and the generative HMM model by the weighted sum of both:

$$H(k) = \beta H_s(k) + (1.0 - \beta)H_c(k) \tag{4}$$

By adjusting the value of the weight β, we can balance the contribution of the segmental and the contextual Hough votes to the final recognition result.

After all the segments $\{s_j\}$ of the input audio stream are processed, the maxima $\arg\max_{k=1...K} H(k)$ indicates the MAP estimate for the potential context category.

4 Experimental Results

To evaluate the proposed auditory context recognition method, we collect test data for 10 auditory contexts from Internet and some movie/TV clips, including 6 outdoor and 4 indoor contexts: inside vehicle, beach, train station, street, restaurant, auditorium, forest, raining, playground and war field. The training and testing set for each context category contain around 100 and 150 mono channel audio samples, respectively, with typical sample length of 15s-2min and 44.1kHz sampling rate.

In experiment, we use a 1s analysis window with 50% overlapping on the audio stream for generating signal segments and extracting features. We learn a 40-dim LDB from the 6-level wavelet packets tree (totally 127 nodes). The Hough forest model for context classification has 30 decision trees with max depth of 15. The number of states of the contextual HMM is learned by the clustering algorithm on training samples and set to 4, 4, 6, 5, 4, 5, 3, 3, 5, 4 for the 10 context categories respectively, while the number of Gaussian components for the observation is experimentally set to 4.

Fig. 4 shows the confusion matrix of the proposed Hough forest based method on a typical trial of the test set with β set to 0.2. The average accuracy on 10 context categories is 80.2%. For comparison, Fig. 5 shows the confusion matrix of the method that employ the single HMM context model to capture the temporal correlations on audio features of individual segments at the context level, though some auditory contexts are typically unstructured. The average accuracy on 10 context categories is 75.1%. The result shows that the proposed scheme precedes overall the one-layer HMM-based context model, while the two models exhibit variant discriminability on different specific context categories.

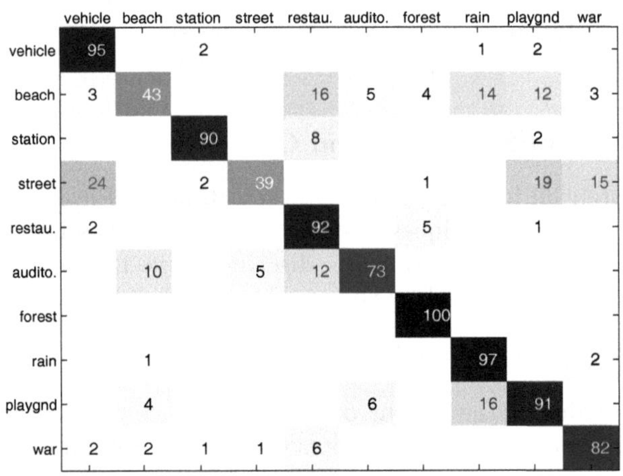

Fig. 4. Confusion matrix of the proposed Hough forest-based method (with $\beta = 0.2$) for 10 auditory contexts

In Fig. 6, we compare the overall recognition accuracy for the 10 considered context categories using the proposed method, either with or without ($\beta = 1$) the contextual Hough voting scheme, and our previous method with explicit HMM-based modeling of 21 audio events in the context (engine, horn, gun-shot, explosion, cutlery, music, thundering, talking, applause, laughter, cheer, etc.), as well as the GMM-based context models like in [1] but with LDB audio features.

The experiment results show that the proposed contextual Hough voting helps to improve the accuracy of the Hough forest recognition framework on the auditory contexts with relatively strong correlations between its components, such as inside-vehicle and restaurant, while for contexts with loose but distinctive components, like beach and street, the segment-specific voting performs better, showing the complementary aspects of both schemes for the final recognition rate. On the other hand, both schemes perform significantly better than

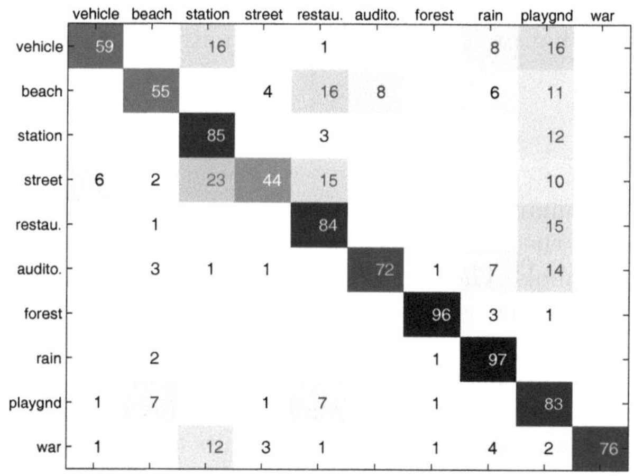

Fig. 5. Confusion matrix of the HMM-based context model of audio segments for 10 auditory contexts

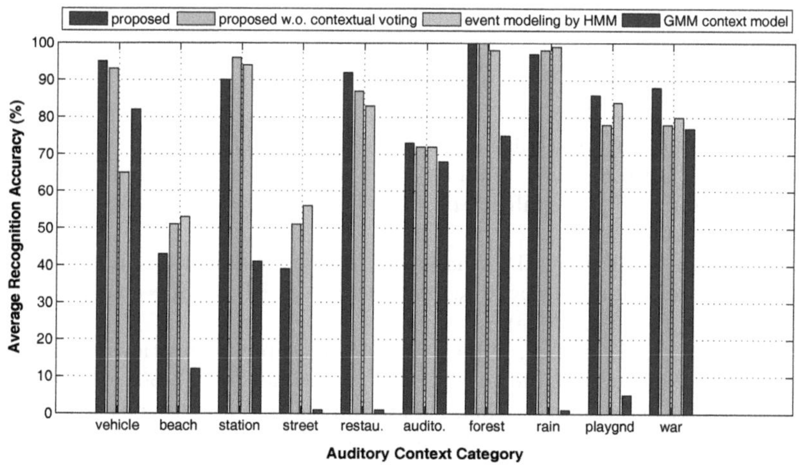

Fig. 6. Overall recognition accuracy by the proposed method with / without contextual Hough voting, the explicit modeling of audio events by HMM, and the GMM-based context model

the one-layer GMM-based context model, which employs a simpler GMM-based description of the context composition and similarly without supervised event modeling.

Moreover, the results also show that, in spite of the absence of explicit modeling of various audio events in the context, the proposed method achieves the comparable recognition accuracies to the method with HMM-based event models, while the proposed method can be more easily adapted to new context categories and those that are difficult to define limited prominent components.

5 Conclusion

We propose a discriminant subspace learning and part weighting scheme that consistently enhances the performance of the parts-based ensemble recognition model for graphic symbols. The main contribution of the paper is twofold: First, we propose to combine the generative HMM-based contextual Hough voting with the Hough forest-based recognition framework, which helps to resolve the segment-specific ambiguity of similar audio components in different auditory contexts. Second, we evaluate the one-layer auditory context model without explicit audio event modeling and obtain comparable results to our previous work, while the former is more flexible and extensible in handling various context categories.

Acknowledgments. Research supported by the National Science Foundation of China under Grant Nos. 61003113, 61272218 and 61021062.

References

1. Chu, S., Narayanan, S., Kuo, C.C.J.: Environmental sound recognition with time-frequency audio features. IEEE TASLP 17(6), 1142–1158 (2009)
2. Kiranyaz, S., Qureshi, A.F., Gabbouj, M.: A generic audio classification and segmentation approach for multimedia indexing and retrieval. IEEE TASLP 14(3), 1062–1081 (2006)
3. Lin, C., Chen, S., Truong, T., Chang, Y.: Audio classification and categorization based on wavelets and support vector machine. IEEE T. Speech and Audio Processing 13(5), 644–651 (2005)
4. Umapathy, K., Krishnan, S., Jimaa, S.: Multigroup classification of audio signals using time-frequency parameters. IEEE T. Multimedia 7(2), 308–315 (2005)
5. Umapathy, K., Krishnan, S., Rao, R.K.: Audio signal feature extraction and classification using local discriminant bases. IEEE TASLP 15(4), 1236–1246 (2007)
6. Han, B., Hwang, E.: Environmental sound classification based on feature collaboration. In: ICME 2009, pp. 542–545 (2009)
7. Wang, J., Wang, J., He, K., Hsu, C.: Environmental sound classification using hybrid SVM/KNN classifier and mpeg-7 audio low-level descriptor. In: IJCNN 2006, pp. 1731–1735 (2006)
8. Cai, R., Lu, L., Hanjalic, A., Zhang, H., Cai, L.: A flexible framework for key audio effects detection and auditory context inference. IEEE TASLP 14(3), 1026–1039 (2006)
9. Chu, W., Cheng, W., Wu, J.: Generative and discriminative modeling toward semantic context detection in audio tracks. In: MMM 2005, pp. 38–45 (2005)
10. Eronen, A.J., Peltonen, V.T., Tuomi, J.T., Klapuri, A.P., Fagerlund, S., Sorsa, T., Lorho, G., Huopaniemi, J.: Audio-based context recognition. IEEE TASLP 14(1), 321–329 (2006)

11. Su, F., Yang, L., Lu, T., Wang, G.: Environmental sound classification for scene recognition using local discriminant bases and HMM. In: ACM Multimedia 2011, pp. 1389–1392 (2011)
12. Saito, N., Coifman, R.R.: Local discriminant bases and their applications. J. of Mathematical Imaging and Vision 5(4), 337–358 (1995)
13. Gall, J., Lempitsky, V.: Class-specific hough forests for object detection. In: CVPR 2009, pp. 1022–1029 (2009)
14. Breiman, L.: Random forests. Machine Learning 45(1), 5–32 (2001)
15. Blei, D.M., Ng, A.Y., Jordan, M.I.: Latent dirichlet allocation. Journal of Machine Learning Research 3(4-5), 993–1022 (2003)

An Integrated Model for Bayesian Learning of Sparse Representation and Classifier Training

Jin Li[1], Gang Hua[2], Xuguang Lan[1], and Nanning Zheng[1]

[1] Institute of Artificial Intelligence and Robotics, Xi'an Jiaotong University
[2] Department of Computer Science, Stevens Institute of Technology
{xglan,nnzheng}@mail.xjtu.edu.cn
ghua@stevens.edu
lij.09054137@stu.xjtu.edu.cn

Abstract. We present an integrated model for Bayesian learning of sparse representation and classifier training, and apply it for the task of visual recognition. Most previous work learns the sparse representation and trains the classifier on top of it in two separate steps. We cast these two into a unified probabilistic model. This way, the supervised labels can effectively affect the learning of the sparse representation. In the training phase, the inference of the joint expectation for dictionary, code, classifier and other variables under the observation of descriptors and labels is carried out by Gibbs Sampling. In the testing phase, based on the learned parameters, the sparse code and the class label of the image are obtained by Bayesian inference. The proposed model is evaluated on Caltech 101 dataset and its efficacy is demonstrated by a careful analysis of the experimental results.

1 Introduction

The sparse coding algorithms attempt to automatically find a small number of representative patterns when given a large set of input patterns. This technique has been successfully applied to many problems in computer vision such as denoising, restoration and compressive sensing [1-4]. For object classification, many works [5, 6, 7] have shown that sparse representation can be more discriminative in differentiating different visual categories. Specially, Yang et al. [5] introduced the sparse code to reconstruct features before spatial pyramid pooling and Wang [6] added a local constraint on the sparse code to improve the performance further. Jia et al. [12], Feng et al. [16] and Zhu et al. [17] focused on ameliorating the pooling step, especially, Geometric ℓ_p-norm feature pooling method proposed by Feng et al. [16] holds a leading performance for image classification in many datasets. However, sparse representation is separated from the classifier training in the above-mentioned methods, so the dictionary will not benefit from the label information in supervised recognition task. To this end, discriminative SVD method [14] is proposed as an available solution, in which the dictionary and classifier are calculated together in one equation. In addition, the label information is associated with each dictionary item to

B. Huet et al. (Eds.): PCM 2013, LNCS 8294, pp. 620–628, 2013.

enforce discriminability in sparse codes during the dictionary learning process in the LC-KSVD [15]. Both of them show better performances when the sparse representation and classification are pulled together.

In this paper, an integrated model is proposed based on beta process factor analysis (BP-FA), which combines the dictionary learning and classifier training in a Bayesian framework. Compared with the optimization model such as K-SVD and LC-KSVD, a non-parametric Bayesian model has several advantages for estimation of parameters. For example, the parameters can be computed directly as the joint expectation. In addition, the non-parametric technique may also ascertain the proper size of parameters.

The main contribution of this paper is a unified Bayesian model which jointly learns the sparse representation and trains the classifier, where all the parameters can be conveniently estimated by Bayesian inference.

The remainder of this article is arranged as follows. Section 2 describes the beta process factor analysis for sparse representation; Section 3 introduces our non-parametric model and then the inference algorithms are derived; Section 4 describes the settings of the experiments along with the experimental results and discussions; Section 5 concludes the model and discusses future work.

2 BP-FA for Sparse Representation

The beta process factor analysis [8] is a new method to sparsely represent input signals, which is different from the traditional methods that always focus on minimizing the reconstruction error. In BP-FA, the dictionary and sparse vectors are random variables and the calculation is inferring the expectation of variables.

Original beta process is defined in measurable space [9]. Nevertheless Paisley and Carin give a finite approximation form and furthermore develop a BP-FA model to reconstruct the input signals.

Based on reference [8], the finite approximation form is written as

$$\begin{aligned} H(\omega) &= \sum_{k=1}^{\infty} \pi_k \delta_{\omega_k}(\omega) \\ \pi_k &\sim Beta\left(a/K, b(K-1)/K\right) \\ \omega_k &\sim H_0 \end{aligned} \tag{1}$$

Where $\omega \in \Omega$ is a random variable in a high-dimensional space, H_0 is a base distribution.

π_k is the weight and $\delta_{\omega_k}(\omega)$ is a basis vector, accordingly, $H(\omega)$ stands for a linear combination of items which draw from space Ω. When beta process is used for sparse representation, the input signals can be regarded as drawn from a beta process, and the item $\delta_{\omega_k}(\omega)$ of the dictionary is also considered to be drawn from a base distribution. However, π is a weight vector rather than the sparse code in expression (1). In order to let the coefficient vector be sparse, the equation can be modified as

$$H(\omega) = \sum_{k=1}^{\infty} z_k \delta_{\omega_k}(\omega), \tag{2}$$

where z_k is the weight drawn from a Bernoulli distribution with parameter π_k, thus the sparsity of z depends on the value of π_k. To avoid the restrict of the weight, a weight vector is further introduced, and the reconstruction step can be represented as

$$H(\omega) = \sum_{k=1}^{\infty} s_k z_k \delta_{\omega_k}(\omega), \tag{3}$$

where the value of s_k is between 0 and 1.

BP-FA is proposed to reconstruct the input signal, its algebra model is

$$X = D(W \circ Z) + E, \tag{4}$$

where $X \in \mathbb{R}^{P \times N}$ is a set of P-dimensional N input signal, D is the dictionary with P rows and K columns. W is the weight matrix and Z is a sparse binary one, they both have K rows and N column. E is an error matrix, and the symbol $\circ$ represents the Hadamard multiplication of two matrices. The details of BP-FA can be found in reference [8].

3 The Non-parametric Bayesian Model for Classification

In this section, a mathematic model is modified based on BP-FA, in which the variables for label and classifier are added. The label vector describes the class of image for supervised learning, and the parameters of the classifier form the variable that connects the label and sparse code.

3.1 The Non-parametric Model

In order to depict the relationship among sparse code, classifier and label, a linear expression is introduced

$$C = L(W \circ Z) + E_c, \tag{5}$$

where $C = [c_1, c_2, \ldots, c_N] \in \mathbb{R}^{M \times N}$ is the set of labels of the N input feature x_i extracted from the original image; $L \in \mathbb{R}^{M \times K}$ is a multi-classifier matrix where the jth row represents a binary classifier for class j, $j = 1, \ldots, M$; E_c is an error matrix as E in Equation (4).

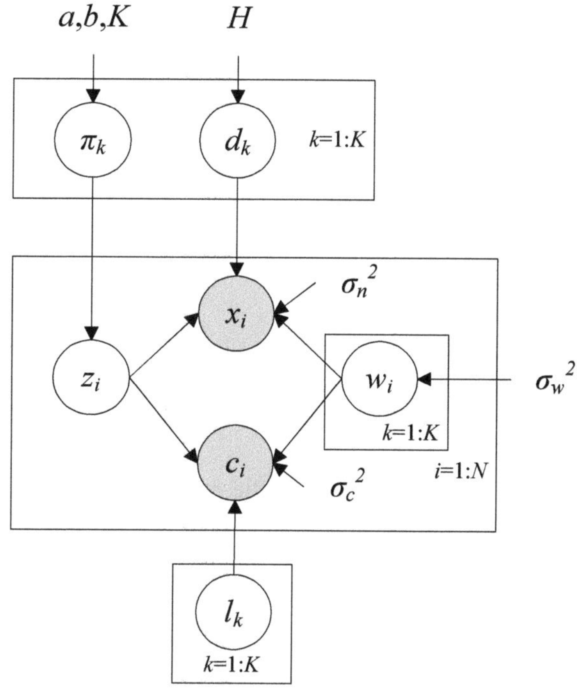

Fig. 1. The graphics of classification model

Figure 1 illustrates the graphical model of the proposed non-parametric Bayesian model, in which σ_c^2 is the variance of E_c that also follows the normal distribution. If feature x_i belongs to class j, the jth element of vector c_i is one and others equal to zero. When the classifier is added to graphical model, the dictionary and classifier are trained together.

The joint probability density of all nodes is

$$\begin{aligned} P(X,C,\Psi) = &\prod_{i=1}^{N} \mathrm{N}(x_i; D(w_i \circ z_i), \sigma_n{}^2\mathrm{I})\, \mathrm{N}(c_i; L(w_i \circ z_i), \sigma_c{}^2\mathrm{I})\, \mathrm{N}(w_i; 0, \sigma_w{}^2\mathrm{I}) \\ &\prod_{k=1}^{K} \mathrm{N}(d_k; 0, P^{-1}\mathrm{I}) Beta(\pi_k; a, b) \prod_{i=1}^{N}\prod_{k=1}^{K} Bernoulli(z_{ik}; \pi_k) \\ &\Gamma(\sigma_w{}^{-1/2}; c, d)\Gamma(\sigma_n{}^{-1/2}; e, f)\Gamma(\sigma_c{}^{-1/2}; e, f) \end{aligned} \tag{6}$$

where $\Psi = \{D, L, S, Z, \pi, \sigma_n^2, \sigma_w^2, \sigma_c^2\}$ is the set of parameters that need to be inferred, and a, b, c, d, e, f are the hierarchy parameters. When calculating the expectation in Equation (6), the label can affect the sampling of dictionary indirectly via sparse code.

In addition, the non-parametric method is also used for inference to find a proper size of the dictionary [10]. When sampling the matrix Z, the kth item in dictionary and classifier is removed if $z_{ik} = 0$ $(i = 1, \dots, N)$, thereby a proper size of dictionary can be inferred in the training phase. In testing step, the new features are sparse represented with the discriminative dictionary D, and the product of the sparse vectors and the linear classifier is used to predict the label of each feature.

We design a contrast experiment in Section 4 to evaluate the improvement of classification accuracy when the dictionary is influenced by classifier and label.

3.2 Gibbs Sampling Inference

The Gibbs sampling is carried out to infer the expectation of the joint probability density in Equation (6).

Sample for d_k

According to Equation (6), the posterior of d_k is easily established as

$$P(d_k \mid -) \propto \prod_{i=1}^{N} \mathrm{N}\left(x_i; D(w_i \circ z_i), \sigma_n^2 \mathrm{I}\right) \mathrm{N}\left(d_k; 0, P^{-1}\mathrm{I}\right). \tag{7}$$

It follows a normal distribution because the prior and likelihood of d_k are conjugate. The mean μ_{d_k} and covariance Σ_{d_k} can be expressed respectively

$$\Sigma_{d_k} = (P\mathrm{I} + \sigma_n^{-2} \sum_{i=1}^{N} w_{ik}^{\ 2} z_{ik}^{\ 2}) \tag{8}$$

$$\mu_{d_k} = \sigma_n^{-2} \Sigma_{d_k} \sum_{i=1}^{N} w_{ik} z_{ik} \left(x_i - B\right), \tag{9}$$

where $B = D(w_i \circ z_i) - d_k w_{ik} z_{ik}$ and does not contain the item d_k. The posterior of w_{ik} is similar to d_k since its prior and likelihood are also normal distributions.

Sample for z_{ik}

$$P(z_{ik} \mid -) \propto \mathrm{N}\left(x_i; D(w_i \circ z_i), \sigma_n^{-2}\mathrm{I}\right) \mathrm{N}\left(c_i; L(w_i \circ z_i), \sigma_c^{-2}\mathrm{I}\right) \\ Bernoulli\left(z_{ik}; \pi_k\right) \tag{10}$$

When z_{ik} =1 or z_{ik} =0, the probability yields

$$P_1 = P(z_{ik} = 1) \propto \pi_k \exp\{-\frac{1}{2}[w_{ik}^{\ 2}(\sigma_n^{-2} d_k^{\ T} d_k + \sigma_c^{-2} l_k^{\ T} l_k) - \\ 2 w_{ik}(d_k + l_k)^T (x_i - B)]\} \tag{11}$$

and

$$P_0 = P(z_{ik} = 0) = 1 - \pi_k. \tag{12}$$

For satisfying the definition of a distribution, z_{ik} may be drawn from the Bernoulli distribution as follows

$$z_{ik} \sim Bernoulli(\frac{P_1}{P_1 + P_0}). \tag{13}$$

Sample for π_k

$$P(\pi_k \mid -) \propto Beta(\pi_k; \frac{a}{K}, \frac{b(K-1)}{K}) \prod_{i=1}^{N} Bernoulli(z_{ik}; \pi_k). \tag{14}$$

The posterior is also a beta distribution

$$\pi_k \sim Beta(\frac{a}{K} + \sum_{i=1}^{N} z_{ik}, \frac{b(K-1)}{K} + N - \sum_{i=1}^{N} z_{ik}). \tag{15}$$

Sample for σ_w^2

Let $\gamma_w = \sigma_w^2$, which is from gamma distribution $\Gamma(\gamma_w; c, d)$. The posterior can be written as

$$P(\gamma_w \mid -) \propto \Gamma(\gamma_w; c, d) \prod_{i=1}^{N} N(w_i; 0, \gamma_w^{-1} \mathrm{I}). \tag{16}$$

And a gamma distribution is deduced as

$$\gamma_w \sim \Gamma(c + \frac{NK}{2}, d + \frac{1}{2} \sum_{i=1}^{N} w_i^T w_i). \tag{17}$$

In a similar way, the posterior of σ_n^2 and σ_c^2 can be derived.

4 Experiments

The above model is evaluated on Caltech 101 [11] and the results are compared with some well-known method. And like LC-KSVD [15], the spatial pyramid feature x_i of each image is input as observation, and the details are shown as follows. 1) the SIFT descriptors are extracted from dense 16×16 patches; 2) a codebook is trained with k-means clustering for spatial pyramid; 3) the histograms for descriptors are calculated and then the descriptors are pooled by spatial pyramid with 1×1, 2×2 and 4×4 grid; 4) PCA is used to reduce the dimension of features.

In the experiment, for 9144 features from 102 categories, we train on 5, 15, 20 and 30 random samples per class. The initial size of dictionary is set to 1024, which can be reduced to an appropriate value in the training phase.

The experimental results show that the classification accuracy of 9 classes achieves 100% in our evaluation with 30 training images for each category, and seven out of them contain at least 10 testing images, which are partly shown in Figure 2. And when training sample is 5, 15, 20 and 30, the size of dictionary is reduced to K= 923, 1011, 1000 and 1018.

Then we design a comparative experiment to discuss if the dictionary is more discriminative when it is influenced by label and classifier. In this experiment, the settings are the same as the ones above, but the dictionary learning is separate from classifier training like the most previous methods did. In other words, learning the dictionary did not leverage any information from the label.

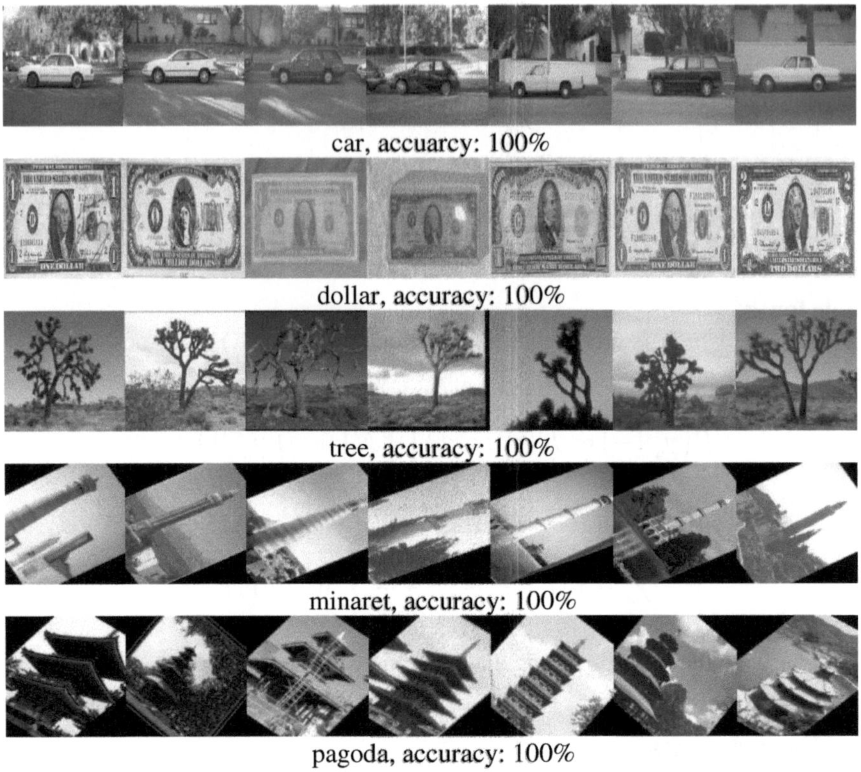

Fig. 2. Examples with high accuracy in Caltech 101

Table 1 displays the improvement when the dictionary is combined with classifier. Despite the accuracy of our incorporated model is not higher than many existing classification method, the comparison shows large potential when the sparse code is influenced by class index and classifier.

Table 1. Comparison between combined and separate dictionary learning

Number of train. sample	5	15	20	30
Combined classification model	49.1%	66.4%	68.7%	73.2%
Baseline	28.6%	41.9%	48.5%	57.2%

Table 2 gives accuracies of classification calculated by several well-known approches. The results indicate that our model is compared to the state-of-the-art.

Table 2. Performance in other method (30 training sample)

method	accuracy	method	accuracy
ScSPM[5]	73.2%	LC-KSVD[15]	73.6%
LLC[6]	73.4%	GLP[16]	82.6%
Jia et al.[12]	75.3%	Zhu et al.[17]	77.7%
SRC[13]	70.7%	Boureau et al.[18]	77.3%
D-KSVD[14]	73.0%	**Ours**	**73.2%**

5 Conclusions

A new non-parametric classification model is established for image object recognition, which presents several contributions: first, the beta process factor analysis is applied to train a discriminative dictionary for classification, and the non-parametric inference is used to reduce the size of dictionary. In addition the label is added to affect dictionary learning while the sparse reconstruction and classifier training steps are combined together, therefore the code is more discriminative. Our experiment shows that the accuracy is significantly improved when dictionary learning is combined with classifier training. The proposed model is also comparable to the state-of-the-art image recognition algorithms. Notice that we just use a linear classifier with a simple probability model, therefore one of the future work is to explore to extend it with non-parametric kernel machines such as Gaussian processes classifiers.

Acknowledgements. This work was supported in part by the Program 973 No. 2012CB316400, No. 2010CB327902, NSFC No.61175010, No.61228303 and International Cooperation Research funding

References

1. Elad, M., Aharon, M.: Image denosing via sparse and redundant representations over learned dictionaries. IEEE Trans. Img. Proc. 54(12), 3736–3745 (2006)
2. Marial, J., Elad, M., Sapiro, G.: Sparse representation for color image restoration. IEEE Trans. Img. Proc. (2008)
3. Ranzato, M., Poultney, C., Chopra, S., LeCun, Y.: Efficient learning of sparse representations with an energy-based model. In: NIPS (2006)
4. Duarte-Carvajalino, J.M., Sapiro, G.: Learning to sense sparse signals: Simultaneous sensing matrix and sparsifying dictionary optimization. IMA Preprint Series 2211 (2008)
5. Yang, J., Yu, K., Gong, Y., Huang, T.: Linear spatial pyramid matching using sparse coding for image classification. In: CVPR (2009)
6. Wang, J., Yang, J., Yu, K., Lv, F., Huang, T., Gong, Y.: Locality-constrained linear coding for image classification. In: CVPR (2010)
7. Bradley, D., Bagnell, J.: Differential sparse coding. In: NIPS (2008)

8. Paisley, J., Carin, L.: Nonparametric factor analysis with beta process priors. In: Proceedings of the 26th Annual International Conference on Machine Learning, pp. 777–784. ACM (2009)
9. Hjort, N.: Nonparametric Bayes estimators based on beta processes in models for life history data. The Annals of Statistics, 1259–1294 (1990)
10. Zhou, M., Chen, H., Paisley, J., et al.: Non-parametric Bayesian dictionary learning for sparse image representations (2009)
11. Li, F., Fergus, R., Perona, P.: Learning generative visual models from few training examples: an incremental bayesian approach tested on 101 object categories. In: CVPR Workshop on Generative-Model Based Vision (2004)
12. Jia, Y., Huang, C., Darrell, T.: Beyond spatial pyramids: Receptive field learning for pooled image features. In: Proc. CVPR (2012)
13. Wright, J., Yang, M., Ganesh, A., Sastry, S., Ma, Y.: Robust face recognition via sparse representation. TPAMI 31(2), 210–227 (2009)
14. Zhang, Q., Li, B.: Discriminative k-svd for dictionary learning in face recognition. In: Proc. CVPR (2010)
15. Jiang, Z., Lin, Z., Davis, L.: Learning a discriminative dictionary for sparse coding via label consistent K-SVD. In: 2011 IEEE Conference on Proc. CVPR, pp. 1697–1704. IEEE (2011)
16. Feng, J., Ni, B., Tian, Q., Yan, S.: Geometric ‘p-norm feature pooling for image classification. In: Proc. CVPR (2011)
17. Zhu, J., Zou, W., Yang, X., et al.: Image Classification by Hierarchical Spatial Pooling with Partial Least Squares Analysis. In: Proc. BMVC (2012)
18. Boureau, Y.L., Roux, N.L., Bach, F., Ponce, J., LeCun, Y.: Ask the locals: multi-way local pooling for image recognition. In: Proc. ICCV (2011)

Generalized N-Dimensional PCA for Compact Representation of Medical Volumes and Effective Mode Selection by Adaboost

Junping Deng, Rui Xu, Xianhua Han, and Yen-Wei Chen

College of Information Science and Engineering, Ritsumeikan University
gr0119xv@ed.ritsumei.ac.jp

Abstract. The paper addresses the issues of compact representation of medical volumes and effective mode selection. Medical volumes are unlike traditional two dimensional images in pattern recognition which in spatial domain has three dimensions. Meanwhile it often meets small sample problem. Thus common compact representation methods, such as PCA, do not fit medical volumes. Because the number of eigenvector is so little that it will lose much useful information. In previous work, we proposed a Generalized N-dimensional principal component analysis (GND-PCA) for reconstruction of medical volumes with only few samples. The core tensor of GND-PCA can keep most useful information. However, making diagnosis using the core tensor is difficult due to most modes very general for all samples. It will affect finally diagnosis. To resolve this problem, Adaboost is used as classifier in diagnosis because it can choose distinctive mode according to its definition. The proposed method was evaluated using a medical volume database. In our experiment, we compare Adaboost with SVM and KNN. The classification accuracy of Adaboost is slightly better than that of SVM and KNN,meanwhile time consuming of classification of Adaboost is greatly less than those of other two classification methods.

Keywords: GND-PCA, Adaboost, distinctive mode, classification accuracy.

1 Introduction

In recently years, the research of medical volume processing is a hot topic. One of the most important work is the representation of medical volume which is a key part of computer-assisted diagnosis (CAD). While due to inter and intra personal variability of anatomic structures, it makes medical volume representation to be a difficult task.

The representation of medical volume can be mainly categorized as shape based methods and texture based methods. Statistical shape model (SSM) can construct the generic structure (mean structure) and deformation for a shape ensemble[1]. It is widely utilized in medical image processing, such as medical image registration and segmentation[2][3], because for most diseases the organ shape has the deformation. 3D active shape model which inspired by the work of active shape model was proposed for construction of 3D statistical models for segmentation of the left ventricle of the heart [4].SSM also show its good performance for distinguish the abnormal liver from the normal one in reference[7]. But many diseases has not only shape variations, but also texture variations. So the active

B. Huet et al. (Eds.): PCM 2013, LNCS 8294, pp. 629–636, 2013.

appearance model (AAM) is proposed which can represent both shape and texture information. In reference[5], 3D active appearance model is used for segmentation of cardiac MR and ultrasound images. For example, Mitchell et al.[6] used a combination of ASM and AAM to segment cardiac images. At each iteration, the two models ran independently to compute new estimates of the pose and shape parameters. For diagnosis assistance, making an accuracy diagnosis decision of liver is important for patient. Radiologists are mainly depending on the intensity variations (texture information) in livers on medical images to identify nodules or tumors and make a diagnostic decision. However, there has been little research on applications of digital atlas to CAD.

Compared to statistical shape modeling, statistical texture modeling usually faces overfitting problems, and the statistical texture modeling for medical volumes is a challenging task because medical volumes are unlike traditional two dimensional images in pattern recognition which in spatial domain has three dimensions. Meanwhile it often meets small sample problem. Thus common compact representation methods, such as PCA, do not fit medical volumes. Because the number of eigenvector is so little that it will lose much useful information. In previous work, we proposed a Generalized N-dimensional principal component analysis (GND-PCA)[8] for reconstruction of medical volumes with only few samples. The core tensor of GND-PCA can keep most useful information. However, making diagnosis using the core tensor is difficult due to most modes very general for all samples. It will affect finally diagnosis. To resolve this problem, Adaboost is used as classifier in diagnosis because it can choose distinctive mode according to its definition. The proposed method was evaluated using a medical volume database.In our experiment, we compare Adaboost with SVM and KNN. The classification accuracy of Adaboost is slgihtly better than that of SVM and KNN, meanwhile time consuming of classification of Adaboost is greatly less than those of other two classification methods.

This paper is organized as follows: In section 2, a brief review GND-PCA is made. Adaboost is introduced in section 3. Section 4 is the experimental part. Section 5 summarizes the key pints of this paper.

2 GND-PCA

Because medical volume is treated as tensor in this paper. A simple explanation of tensor will be given firstly. A tensor is a multidimensional array. More formally, an Nth-order tensor is an element of the tensor product of N vector spaces, each of which has its own coordinate system.The order of a tensor in the number of dimensions, as known as ways or modes. An N-th order tensor $\mathcal{A}$ is defined as a multi-array with N indices, where $\mathcal{A} \in \mathbb{R}^{I_1 \times I_2 \times \ldots \times I_N}$ and $\mathbb{R}$ is the real manifold. Elements of the tensor $\mathcal{A}$ are denoted as $a_{i_1 \ldots i_n \ldots i_N}$, where $1 \leqslant i_n \leqslant I_n$.From the perspective of $\mathcal{A}$, scalars, vectors and matrices can be seen as zeroth-order, first order and second order tensors, respectively. Unfolding of tensor is the process of reordering the elements of an Nth-order array into a matrix. The mode-n matricization of a tensor is denoted by and arranges the mode-n fibers to be the columns of the resulting matrix. Fig. 1 illustrates a example of matricization for a 3rd-order tensor.More details of tensor please see reference [11]..

And the notations in this paper are given below. Scalers are denoted by italic-shape letters, i.e. $(a, b, \ldots)$ or $(A, B, \ldots)$. Bold lower case letters, i.e. $(\mathbf{a}, \mathbf{b}, \ldots)$, are used to

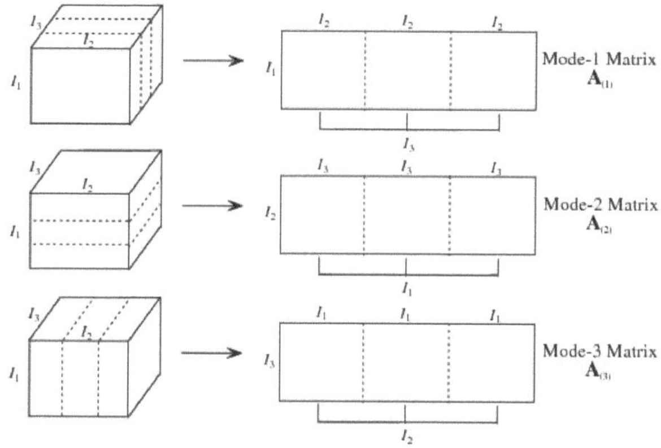

Fig. 1. Matricization of a 3rd-order tensor

represent vectors. Matrices are denoted by bold upper case letters, i.e. $(\mathbf{A}, \mathbf{B}, ...)$; and higher-order tensors (more than third order tensor) are denoted by calligraphic upper case letters, i.e. $(\mathcal{A}, \mathcal{B}, ...)$.

GND-PCA was proposed by Xu and Chen et al. for statistical appearance modeling of medical volumes with few samples[8]. The medical volume is treated as a 3rd-order tensor, and the optimal subspace on each mode is calculated simultaneously by minimizing of the square error between the original volumes and reconstructed ones. In the following part of this section, we will briefly review the algorithm of GND-PCA.

Given a series of the N-order tensors with zero-means $\mathcal{A}_i \in \mathbb{R}^{I_1 \times I_2 \times ... \times I_N}$, $i = 1, 2, \ldots, M$, M is the number of samples. We aim to get another series of low rank $\{J_1, J_2 \cdots J_N\}$ tensors $\hat{\mathcal{A}}_i$ which accurately approximate the original tensors, where $J_n \leq I_n$. The new series is decomposed by the matrices $\mathbf{U}^{(n)} \in \mathbb{R}^{I_n \times J_n}$ with orthogonal columns according to the Turker Model[10] which is shown by

$$\hat{\mathcal{A}}_i = \mathcal{B}_i \times_1 \mathbf{U}^{(1)} \times_2 \mathbf{U}^{(2)} \times \cdots \times_n \mathbf{U}^{(n)} \times \cdots \times_N \mathbf{U}^{(N)} \tag{1}$$

where $\mathcal{B}_i \in \mathbb{R}^{J_1 \times J_2 \times \cdots \times J_N}$ are the core tensors. The product operator in Eq.4 is matrix product, tensor $\mathcal{B}_i$ will unfold to matrix in each mode and then product with orthogonal matrix in each mode. The orthogonal matrices $\mathbf{U}^{(\mathrm{n})}$ can be determined by minimizing the cost function as:

$$S = \sum_{i=1}^{M} \left\| \mathcal{A}_i - \mathcal{B}_i \times_1 \mathbf{U}^{(1)} \times_2 \mathbf{U}^{(2)} \times \cdots \times_N \mathbf{U}^{(N)} \right\|^2 \tag{2}$$

If the tensor $\mathcal{B}_i$ is chosen as:

$$\mathcal{B}_i = \mathcal{A}_i \times_1 \mathbf{U}^{(1)^T} \times_2 \mathbf{U}^{(2)^T} \times ... \times_N \mathbf{U}^{(N)^T} \tag{3}$$

Minimization of Eq.2 is equal to the maximization of the following equation:

$$S' = \sum_{i=1}^{M} \left\| \mathcal{A}_i \times_1 \mathbf{U}^{(1)^T} \times_2 \mathbf{U}^{(2)^T} \times ... \times_N \mathbf{U}^{(N)^T} \right\|^2 \tag{4}$$

Algorithm 1. GND-PCA

IN: a series of N-th order tensors, $\mathcal{A}_i \in \mathbb{R}^{I_1\times I_2\times\ldots\times I_N}, i = 1, 2, ..., M$.

OUT: N Matrices $\mathbf{U}^{(n)}_{opt} \in \mathbb{R}^{I_n\times J_n}$ $(J_n = 1, n = 1, 2, ..., N)$ with orthogonal column vectors.

1. Initial values: $k = 0$ and $\mathbf{U}^{(n)}_0$ whose columns are determined as the first J_n leading eigenvectors of the matrices $\sum_{i=1}^{M} (\mathbf{A}_{i(n)} \cdot \mathbf{A}^T_{i(n)})$.
2. Iterate for k until convergence

   - Maximize $S' = \sum_{i=1}^{M} \| C_i \times_1 \mathbf{U}^{(1)^T} \|^2, C_i = \mathcal{A}_i \times_2 \mathbf{U}^{(2)^T}_k \times ... \times_N \mathbf{U}^{(N)^T}_k$
     Solution: $\mathbf{U}^{(1)}$ whose columns are determined as the first J_1 leading eigenvectors of $\sum_{i=1}^{M} (\mathbf{C}_{i(1)} \cdot \mathbf{C}^T_{i(1)})$
     Set $\mathbf{U}^{(1)}_{k+1} = \mathbf{U}^{(1)}$.
   - Maximize $S' = \sum_{i=1}^{M} \| C_i \times_2 \mathbf{U}^{(2)^T} \|^2, C_i = \mathcal{A}_i \times_1 \mathbf{U}^{(1)^T}_{k+1} \times_3 \mathbf{U}^{(3)^T}_k \times ... \times_N \mathbf{U}^{(N)^T}_k$
     Solution: $\mathbf{U}^{(2)}$ whose columns are determined as the first J_2 leading eigenvectors of $\sum_{i=1}^{M} (\mathbf{C}_{i(2)} \cdot \mathbf{C}^T_{i(2)})$
     Set $\mathbf{U}^{(2)}_{k+1} = \mathbf{U}^{(2)}$.

     ...
   - Maximize $S' = \sum_{i=1}^{M} \| C_i \times_n \mathbf{U}^{(n)^T} \|^2, C_i = \mathcal{A}_i \times_1 \mathbf{U}^{(1)^T}_{k+1} \times ... \times_{n-1} \mathbf{U}^{(n-1)^T}_{k+1} \times_{n+1} \mathbf{U}^{(n+1)^T}_k \times ... \times_N \mathbf{U}^{(N)^T}_k$
     Solution: $\mathbf{U}^{(n)}$ whose columns are determined as the first J_n leading eigenvectors of $\sum_{i=1}^{M} (\mathbf{C}_{i(n)} \cdot \mathbf{C}^T_{i(n)})$
     Set $\mathbf{U}^{(n)}_{k+1} = \mathbf{U}^{(n)}$.

     ...
   - Maximize $S' = \sum_{i=1}^{M} \| C_i \times_N \mathbf{U}^{(N)^T} \|^2, C_i = \mathcal{A}_i \times_1 \mathbf{U}^{(1)^T}_{k+1} \times ... \times_{N-1} \mathbf{U}^{(N-1)^T}_{k+1}$
     Solution: $\mathbf{U}^{(N)}$ whose columns are determined as the first J_N leading eigenvectors of $\sum_{i=1}^{M} (\mathbf{C}_{i(N)} \cdot \mathbf{C}^T_{i(N)})$
     Set $\mathbf{U}^{(N)}_{k+1} = \mathbf{U}^{(N)}$.

   $k = k + 1$
3. Set $\mathbf{U}^{(1)}_{opt} = \mathbf{U}^{(1)}_k, \mathbf{U}^{(2)}_{opt} = \mathbf{U}^{(2)}_k, \ldots, \mathbf{U}^{(N)}_{opt} = \mathbf{U}^{(N)}_k$.

There is no close-form solution to simultaneously resolve the matrices for the Eq.4; however the explicit solution for one matrix can be obtained if the other matrices are fixed. This is expressed by Lemma 1, and is explained below.

Lamma 1. Given the fixed matrices, $\mathbf{U^{(1)}}, \mathbf{U^{(2)}}, \ldots, \mathbf{U^{(n-1)}}, \mathbf{U^{(n+1)}}, \ldots, \mathbf{U^{(N)}}$, if the columns of the matrix $\mathbf{U^{(n)}}$ are selected as the first J_N leading eigenvectors of matrix $\sum_{i=1}^{M} (C_{i(n)} \cdot C_{i(n)}{}^T)$, $C_{i(n)}$ is the mode-n matrix of the tensor $C_i = \mathcal{A}_i \times_1 \mathbf{U}^{(1)} \times_2 \mathbf{U}^{(2)} \times \cdots \times_{n-1} \mathbf{U}^{(n-1)} \times_{n+1} \mathbf{U}^{(n+1)} \times \cdots \times_N \mathbf{U}^{(N)}$, the cost function S' can be maximized.

The proof of Lemma 1 is given in reference[8], here it will not be given again. According to Lemma 1 we can use an iteration algorithm to get the N optimal matrices, $\mathbf{U}^{(1)}_{opt}, \mathbf{U}^{(2)}_{opt}, \ldots, \mathbf{U}^{(N)}_{opt}$, which are able to maximize the cost function S'. This algorithm is summarized by Algorithm 1.

Using the calculated matrices $\mathbf{U}^{(n)}_{opt}, n = 1, 2, ..., N$, each of the volume $\mathcal{A}_i$ with least errors, where $\hat{\mathcal{A}}_i = \mathcal{B}_i \times_1 \mathbf{U}^{(1)} \times_2 \mathbf{U}^{(2)} \times \cdots \times_n \mathbf{U}^{(n)} \times \cdots \times_N \mathbf{U}^{(N)}$. The approximation can be

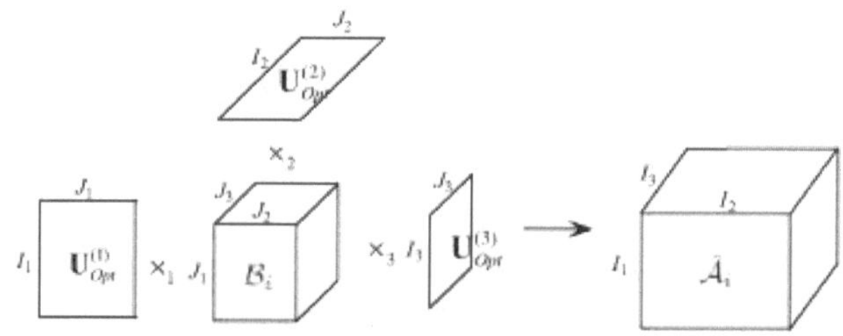

Fig. 2. Illustration of reconstructing a third-order tensor by the principal component $\mathcal{B}_i$ and the three orthogonal bases of mode subspaces $\mathbf{U}_{opt}^{(1)}, \mathbf{U}_{opt}^{(2)}, \mathbf{U}_{opt}^{(3)}$

illustrated by Fig.2 for the 3D case. The core tensor $\mathcal{B}_i$ are the principle components. The core tensor can keep most useful information and are useful for reconstruction. However, making diagnosis using the core tensor is difficult due to most modes very general for all samples. It will affect finally diagnosis. We need to find a method to select more distinctive modes. Thus Adaboost which will be introduced in next section is adapted as classifer.

3 Adaboost

The AdaBoost algorithm was introduced in 1995 by Freund and Schapire [12]. The algorithm takes as input a training set $(x_1; y_1), \cdots, (x_m; y_m)$where each x_i belongs to some domain or instance space X, and each label y_i is in label set $Y = \{-1, 1\}$. AdaBoost calls a given weak learning algorithm repeatedly in a series of rounds $t = 1, \cdots, T$. One of the main ideas of the algorithm is to maintain a distribution or set of weights over the training set. The weight of this distribution on training example i on round t is denoted $D_t(i)$. Initially, all weights are set equally, but on each round, the weights of incorrectly classified examples are increased so that the weak learner is forced to focus on the hard examples in the training set. The distribution D_t is next updated using the rule shown in the Fig.3. The effect of this rule to increase the weight of examples misclassified by h(t), and to decrease the weight of correctly classified examples. Thus, the weight tends to concentrate on hard examples. The final hypothesis H is a weighted majority vote of the T weak hypotheses where t is the weight assigned to h(t). Pseudocode for AdaBoost is given in Fig. 3.

As the definition of Adaboost, it can find more distinctive weak learners among all learners. The effective elements in core tensor can be selected by founding more distinctive weak learners among all learners. This advantage of Adaboost just fits our requirement of diagnosis. In the next section, we will show Adaboost is better than other classifiers in the diagnosis.

Input: Given labeled examples $\mathbf{x}$;

Set the initial w_1 to the uniform distribution;

For $t = 1, \ldots, T$:

1. Train a weak classifier $h_j : x \rightarrow \{-1, +1\}$;
2. Calculate w_t-weighted error
$$e_j = P[h_j(x_i) \neq y_i \mid w_t];$$
3. Choose $h_k(x)$, such that $e_k < e_j, \forall j \neq k$;
4. Let $e_t = e_k$.
5. Choose $\alpha_t = \frac{1}{2}\log\left(\frac{1-e_t}{e_t}\right)$;
6. Update $w_{t+1,i} \leftarrow w_{t,i}\exp(-\alpha_t y_i h_i(x_i))$;
7. Normalize w_{t+1} to $\sum_i w_{t+1,i} = 1$;

Output $H(x)$ as
$$H(x) = \text{sign}\left(\sum_{t=1}^{T} \alpha_t h_t(x)\right)$$

Fig. 3. The algorithm of AdaBoost

4 Experimental Results

The proposed method is evaluated by using a liver database[9]. In this database, there are 10 normal healthy ones and 10 abnormal ones. The size of each sample is $256 \times 256 \times 79$. In the experiment, the size of core tensor is $20 \times 20 \times 3$. So the dimension of the samples is 1200. In this work, we mainly considered the texture information.Thus before extracting the core tensor, non-rigid registration method was adapted to remove the shape information.Fig.4 shows some original data and morphed data respectively. The first column is original data and the second column is morphed data. The first row is one samples of the normal ones, the second row is one sample of the abnormal ones. This illustrates that all the samples have familiar shapes, so the shape information does not effect experimental results.

Table 1 is the classification accuracy of original data and compressed data using different classifiers by leave-one-out method. Compared with these two results, we can find that the result of Adaboost is better than that of SVM and KNN, also the result of compressed data is better than that of original data.Meanwhile, Adaboost only used a few basis for classification, thus the time consuming of classificaiton is less than other two classification methods. Table 2 is the time consuming using different classification methods at the same condition. It shows that at the same condition, the speed of Adaboost is faster than other two methods.

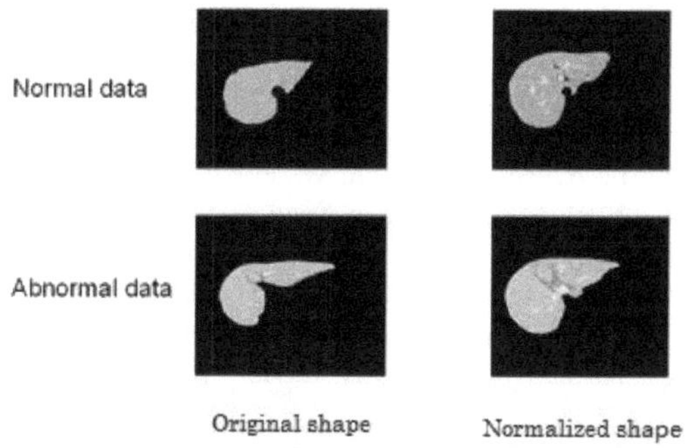

Fig. 4. Original shape (left) and normalized shape (right)

Table 1. The classification accuracy of original data and compressed data using different classifiers

	SVM	KNN	Adaboost
Orignal	10/20	10/20	11/20
GND-PCA	11/20	11/20	13/20

Table 2. The time consuming using different classification methods at the same condition

	SVM	KNN	Adaboost
Time Consuming(s)	1.5	0.025	0.0007

5 Conclusion

In this paper, we addresses the issues of compact representation of medical volumes and effective mode selection. GND-PCA is used for compact representation. The core tensor of GND-PCA can keep most useful information. Due to most modes very general for all samples, it will affect finally diagnosis. To resolve this problem, Adaboost is used as classifier in diagnosis because it can choose distinctive mode. The proposed method was evaluated using a medical volume database. In our experiment, we compare Adaboost with SVM and KNN. The classification accuracy of Adaboost is better than that of SVM and KNN.

References

1. Cootes, T.F., Taylor, C.J., Cooper, D.H., Graham, J.: Active shape models-their training and application. Computer Vision and Image Understanding 61(1), 38–59 (1995)
2. Huang, H., Makedon, F., McColl, R.: High dimensional statistical shape model for medical image analysis. In: IEEE International Symposium on Biomedical Imaging: From Nano to Macro (ISBI 2008), pp. 1541–1544 (May 2008)

3. van Assen, H.C., Danilouchkine, M.G., Behloul, F., Lamb, H.J., van der Geest, R.J., Reiber, J.H.C., Lelieveldt, B.: Cardiac LV segmentation using a 3D active shape model driven by fuzzy inference. In: Ellis, R.E., Peters, T.M. (eds.) MICCAI 2003. LNCS, vol. 2878, pp. 533–540. Springer, Heidelberg (2003)
4. Kaus, M.R., von Berg, J., Weese, J., Niessen, W., Pekar, V.: Automated segmentation of the left ventricle in cardiac MRI. Medical Image Analysis 8(3), 245–254 (2004)
5. Cootes, T.F., Edwards, G.J., Taylor, C.J.: Active appearance models. IEEE Transactions on Pattern Analysis and Machine Intelligence 23(3), 681–685 (2001)
6. Mitchell, S., Lelieveldt, B., van der Geest, R., Schaap, J., Reiber, J., Sonka, M.: Segmentation of cardiac mr images: An active appearance model approach. In: SPIE Medical Imaging (February 2000)
7. Chen, Y.-W., Uetani, M., Kohara, S., et al.: Application of statistical shape model of the liver in classification of cirrhosis. International Journal of Digital Content Technology and its Applications (in press, 2013)
8. Xu, R., Chen, Y.W.: Generalized N-dimensional principal component analysis (GND-PCA) and its application on construction of statistical appearance models for medical volumes with fewer samples. Neurocomputing 72, 2276–2287 (2009)
9. Qiao, X., Chen, Y.W.: A Statistical Texture Model of the Liver Based on Generalized N-Dimensional Principal Component Analysis (GND-PCA) and 3D Shape Normalization. International Journal of Biomedical Imaging (2011)
10. Lieven, D.L., Bart, D.M., Joseph, V.: A multilinear singular value decomposition. SIAM Journal of Matrix Analysis and Application 4(21), 1253–1278 (2001)
11. Kolda, T.G.: Orthogonal Tensor Decompositions. SIAM Journal on Matrix Analysis and Applications 3(1), 243–255 (2001)
12. Freund, Y., Schapire, R.E.: A decision-theoretic generalization of on-line learning and an application to boosting. Journal of Computer and System Sciences 55(1), 119–139 (1997)

Constructing Hierarchical Visual Tree for Discriminative Image Representation and Classification

Hao Lei[1], Ning Zhou[3], Kuizhi Mei[1], Peixiang Dong[1], and Jianping Fan[2]

[1] Institute of Artificial Intelligence and Robotics, Xi'an Jiaotong University, Xi'an 710049, PR China
[2] School of Information Science and Technology, Northwest University, Xi'an 710069, PR China
[3] Department of Computer Science, University of North Carolina, Charlotte, NC 28223 USA

Abstract. To support large-scale visual recognition, it is critical to train a large number of classifiers with high discrimination power. To achieve this task, in this paper a hierarchical visual tree is constructed for organizing a large number of object classes and image concepts according to their inter-concept visual correlations. Based on the hierarchical visual tree, a novel approach is proposed for learning multi-scale group-based dictionary to support discriminative bag-of-visual-words (BoW-based) image representation. In addition, a structural learning approach is developed to enable large-scale classifier training over such hierarchical visual tree. We have also compared the performance of our hierarchical visual tree with traditional label tree over large-scale image collections.

Keywords: Hierarchical Visual Tree, Dictionary Learning, Discriminative Image Representation, Structural Learning, Image Classification.

1 Introduction

Image classification becomes increasingly important and necessary to support automatic image annotation, so that large-scale image retrieval can be made more intuitively by using the adequate keywords [23]. However, the performance of image classification largely depends on two inter-related issues: (a) discriminative representation for visual content of images, and (b) effective algorithm for multi-class classifier training.

For the first issue, bag-of-visual-words (BoW) has become one of popular methods for visual content representation of images due to its effectiveness and flexibility. BoW-based approach represents an image as a histogram based on the frequencies of a set of "visual words", which is known as visual dictionary [4]. Learning effective visual dictionary is a crucial issue for supporting discriminative BoW-based image content representation. For the second issue, when large-scale image categories come into view, large amount of classifiers should be effectively learnt for bridging the semantic gap successfully by mapping the

B. Huet et al. (Eds.): PCM 2013, LNCS 8294, pp. 637–648, 2013.

low-level visual features onto high-level image concept (human interpretation of image semantics).

In addition, when a large number of object classes (i.e., image semantics are interpreted by the visual content of object regions) and image concepts (i.e., image semantics are interpreted by the visual content of entire images) come into view, some of them are strongly inter-related (visually similar) because their relevant images may share some similar or even common visual properties (i.e., strong inter-concept visual correlations) [8,9]. In view of huge inter-concept visual correlations, to address the above two issues, robust techniques should leverage the inter-concept visual correlations for discriminative dictionary learning and effective large amount of classifiers training.

In this paper, a hierarchical visual tree has been constructed to organize a large number of object classes and image concepts according to their inter-concept visual correlations. We construct the hierarchical visual tree directly in the visual feature space rather than in the label space, because image representation and classifier training indeed happen in the visual feature space. On the one hand, we use the hierarchical visual tree to determine the groups of visual similar object classes and image concepts at different levels, and then jointly learn a discriminative dictionary for each group at different levels. On the other hand, we take advantage of the hierarchical visual tree to discover the inter-related classifiers for different object classes and image concepts, and develop a novel structural learning algorithm for effectively training the inter-related classifiers.

2 Related Work

In this section, we briefly review some related work on dictionary learning for BoW-based image representation and multi-class classifier training for image classification.

Most of prevailing algorithms have learnt a universal visual dictionary [17,21,27] for BoW-based image representation, where the same bases (the same set of visual words in the universal dictionary) are used to obtain the BoW histograms for all the images in the database. It is worth noting that the images from different object classes and image concepts may have diverse visual properties, thus such universal dictionary may not be optimum for all the object classes and image concepts. To improve the performance, Nistér et al. [18] developed a vocabulary tree to allow a larger and more discriminatory vocabulary, [14,28,30] learned the universal dictionary combined with the classifier training. However, these existing algorithms only considered intra-class (category) or intra-image visual correlations, while completely ignored inter-concept (category) visual correlations. Pioneering work proposed by Zhou et al. [31] have leveraged the inter-concept (category) visual correlations to learn multiple inter-related dictionaries jointly for the visually similar object classes and image concepts, where a common dictionary (which is shared among multiple visually similar object classes and image concepts) and multiple individual dictionaries (which are class-specific for each object class or image concept) are learnt jointly.

For multi-class classifier training, rather than providing an exhaustive and comprehensive review of all related works on image classification, we focus on giving a brief overview of the work on hierarchical image classification. In order to support hierarchical image classification, Barnard et al. [1], Vasconcelos et al. [24], and Fan et al. [8,9] have incorporated hierarchical mixture models and concept ontology to leverage the hierarchical inter-concept semantic similarity contexts for training multiple inter-related classifiers jointly. Fei-Fei et al. [12] have also incorporated prior knowledge of object parts and their locations to improve hierarchical image classification. The major problem with the hierarchical approach is that the classification errors may be propagated among the classifiers for the relevant image concepts (i.e., inter-concept error propagation) [10].

Some previous work have recently been proposed to characterize the inter-concept correlations for image classifier training. Marszalek et al. [15] and Fei-Fei et al. [5] have used WordNet [16] to find the semantic relationships between the labels and combined discriminative classifiers through the semantic hierarchies. Wang et al. [26] have utilized the normalized Google distance (NGD) [3] as the inter-word contextual potential for multi-label image annotation. However, these previous work measure the inter-concept correlations by using semantic information in the label space, it is not quite reasonable for determining the inter-related classifiers. Because classifier training and image classification are performed in the feature space rather than in the label space. Pioneering work proposed by Dong et al. [7] and Fan et al. [11] have utilized the inter-concept visual network for determining the inter-related classifiers and training them jointly.

3 Automatic Construction of Hierarchical Visual Tree

The images from the inter-related object classes and image concepts may share some similar or even common visual properties (e.g., visual features), i.e., these inter-related object classes and image concepts may have huge inter-concept visual correlations. In this paper, a hierarchical visual tree is constructed to organize the object classes and image concepts from multi-levels and determine the inter-related (visually similar) object classes and image concepts directly in the visual feature space.

To construct the hierarchical visual tree, we first propose a new method to characterize the inter-concept (category) visual correlations explicitly. We extract dense SIFT features at multi-scales for each image instance in the database, and then use spatial pyramid histograms [13] to represent each image instance. Let H_x and H_y respectively denote the spatial histogram descriptors of a pair of image instances x and y. A kernel function $k(x, y)$ is then defined to characterize their visual similarity relationship as follows

$$k(x, y) = \sum_{l=0}^{L} \alpha_l I_l(H_x^l, H_y^l) \tag{1}$$

Table 1. Inter-concept visual correlations

Concept pair	φ	Concept pair	φ	Concept pair	φ
crab & box turtle	0.96	monitor & microwave	0.62	canoe & shoji	0.38
deerhound & lynx	0.82	sky & cloud	0.62	bookcase & dune	0.37
go-kart & horse cart	0.78	clog & accordion	0.57	flat bench & sky	0.36
bookcase & cabinet	0.78	tiger cat & wave	0.51	microwave & crab	0.35
limousine & cab	0.74	cab & oilskin	0.44	lynx & website	0.33

where $I_l(H_x^l, H_y^l)$ is the histogram intersection of H_x and H_y at pyramid level l, $l = 1, ..., L$ is the pyramid level, and α_l is the weight at level l, which is 2^{-L} for $l = 0$ and 2^{l-L-1} for others.

According to the image visual similarity relationship, for two given conceptual image sets C_i and C_j, we characterize their inter-concept visual correlation $\varphi(C_i, C_j)$ as follows:

$$\varphi(C_i, C_j) = \frac{1}{|C_i||C_j|} \sum_{x \in C_i} \sum_{y \in C_j} k(x, y) \tag{2}$$

where $|C_i|$ and $|C_j|$ are the total numbers of image instances in the conceptual image sets C_i and C_j respectively, $k(x, y)$ is the visual similarity between two image instances x and y from conceptual image sets C_i and C_j. If relevant image instances from C_i and C_j share more similar visual properties (i.e., they are visually similar), their inter-concept visual correlation $\varphi(.,.)$ should have larger value. Table 1 gives some experimental results for the inter-concept visual correlations $\varphi(.,.)$ in our test dataset. It can be seen that our proposed method can well characterize the inter-concept visual correlations, e.g., relevant images from the image classes "bookcase" and "cabinet" look visually similar, and thus their inter-concept visual correlation has lager value in Table 1.

In the following, we use the top-down method to construct the hierarchical visual tree according to the inter-concept visual correlations. At the first level, we construct a visual concept network for all the object classes and image concepts in the database, where each object class or image concept is linked with multiple visually similar (inter-related) object classes and image concepts with larger values of the inter-concept visual correlations $\varphi(.,.)$. To determine the groups of visually similar object classes and image concepts at the first level, the visual concept network is treated as a weighted undirected graph, where each object class or image concept is treated as a node of the graph and the value of the inter-concept visual correlation $\varphi(.,.)$ is taken as the weight of the corresponding edge. And then, we partition this graph (visual concept network) into a set of disjoint subgraphs (visual concept sub-networks), and the Normalized cut (Ncut) [22] algorithm is employed to address this partition problem. These sub-networks constitute the second level of the hierarchical visual tree. Subsequently, we recursively do the above steps until reaching the maximum level L. Consequently, the hierarchical visual tree with L levels is constructed, which determines the multi-scale sub-networks of inter-related object classes and image

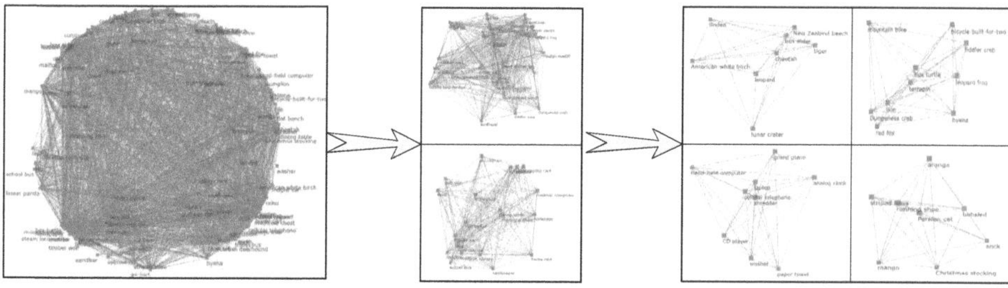

(a) Network at the first level (b) Sub-networks at the second level (c) Sub-networks at the third level

Fig. 1. Some visual concept networks at different levels of the hierarchical visual tree for our test dataset

concepts at different levels. Fig. 1 demonstrates some visual concept networks at different levels of the hierarchical visual tree for our test dataset (100 most popular object classes and image concepts from ImageNet [6] image set).

4 Multi-scale Group-Based Dictionary Learning

In Section 3, we have developed a new method for constructing the hierarchical visual tree directly in the visual feature space. The hierarchical visual tree is composed of a set of multi-scale visual concept sub-networks at the different levels, and each sub-network consists of a set of visually similar object classes and image concepts. We refer to such sub-network as a visually similar group. The feature vectors for such visually similar object classes and image concepts may overlap in the visual feature space, thus it is harder to distinguish them effectively. In this paper, we develop a group-based approach for leveraging the inter-concept visual correlations to jointly learn discriminative dictionary for all the object classes and image concepts in the same group. Our motivation is to learn a discriminative group-based dictionary for each group at different levels in order to enhance its discrimination power on distinguishing the visually similar object classes and image concepts in the same group, while different groups with weak inter-concept visual correlations will have different group-based dictionaries. In addition, because of different diversities of visual properties for different-level sub-networks, we learn different-scale dictionaries (i.e., multi-scale dictionaries) for different sub-networks (groups).

In the following, we discuss our algorithm for learning multi-scale group-based dictionary in details. We use the bottom-up approach to learn the dictionary for the visually similar groups at every level over the hierarchical visual tree. Because the inter-group visual correlations are weaker than the inner-group visual correlations at each level, so we can use the hypothesis that different groups at the same level are independent (i.e., ignoring weak inter-group correlations) in our algorithm. For legible expression, we first define some notations. Denote $l = 1, ..., L$ is the level of the hierarchical visual tree (first level is the top level and L-th level is the bottom level). Let G_i^l is the i-th ($i = 1, ..., M^l$)group at the l-th level, D_i^l is the dictionary corresponding to the group G_i^l, and the size of D_i^l is all K^l. First, we learn the dictionary for the groups at the L-th level (bottom level). Let X_i is the set of training samples from the object classes and image concepts in the group G_i^L, x_j, $j = 1, ..., N_i$ is a training sample from the set X_i, and w_k, $k = 1, ..., K^L$ be a visual word in the group dictionary D_i^L. We learn the $\{D_i^L\}$ by solving the following optimization problem:

$$\arg\min_{\{D_i^L\}} \sum_{i=1}^{M^L} \sum_{\substack{j=1 \\ x_j \in X_i}}^{N_i} ||x_j - \mathcal{V}(x_j, D_i^L)||^2 \tag{3}$$

where $\mathcal{V}(x_j, D_i^L)$ is the representation of the training sample x_j by using the group dictionary D_i^L. Here we utilize the vector quantization to accomplish the representation. Then $\mathcal{V}(x_j, D_i^L)$ is formulated as

$$\mathcal{V}(x_j, D_i^L) = \arg_{w_k} \min_{\substack{w_k \in D_i^L \\ k=1,\dots,K^L}} ||x_j - w_k||^2 \tag{4}$$

We apply the iterative scheme to optimize the formula (3). At the beginning, we randomly select K^L training samples from each X_i to initialize the corresponding group dictionary D_i^L, and then we separately optimize each D_i^L iteratively. After we have learned the dictionary $\{D_i^L\}$ for the L-th level (bottom level), we then treat $\{D_i^L\}$ as the training samples and recursively learn the dictionary for other higher levels. It is worth noting that the parent groups from the high level only use the training samples from their children groups of the adjacent lower level. Because the more visual diversities for the group at higher level (such group consists of large amount of object classes and image concepts), we learn larger-scale dictionary for the group at higher level to enhance the discrimination power, while we choose the same scale for the the groups at the same level. As a result, the size of dictionary for the group at the l-th level is set to $2^{(L-l)} \times K^L$. Once multi-scale group-based dictionaries have been learned, we can use the BoW-based method to represent the images. It is worth noting that the children groups at low level are represented only by using the dictionary belong to their parent group at adjacent higher level, and images at different levels have different-scale representations.

5 Structural Classifier Training for Image Classification

In this section, we proposed a novel structural learning approach for effectively training multi-class classifiers over the hierarchical visual tree. We use support vector machine (SVM) with χ^2-based kernel [29] as the basic classifier to design our structural classifiers. Our structural classifiers consist of two components: intra-group classifiers for distinguishing the visually similar object classes and image concepts in the same group, and inter-group classifiers for discriminating those from different groups. For the intra-group, we only consider the groups at L-th level (bottom level), because the object classes and image concepts in such groups are most visually similar. We use the pairwise method to train the intra-group classifiers for these visually similar object classes and image concepts. For the inter-group, we treat the visually similar object classes and image concepts in the same group as a whole group concept at different levels. We learn inter-related classifiers for group concepts (image concepts for L-th level) in the same parent group at different levels, and only consider the adjacent high level groups from the same ancestry. Subsequently, we combine the intra-group and inter-group classifiers according to the hierarchical visual tree.

For a given object class or image concept C_j, X denotes the input test sample. We use $f(C_+, C_-, X)$ to denote the basic pairwise discriminant function, where C_+ denote the positive training concept(s) and C_- denote the negative training concept(s). The value of $f(C_+, C_-, X)$ is designed to 1 for the positive samples and 0 for the negative samples. Then our structural classifier $F(C_j, X)$ respect to $C_j (C_j \in G_i^L)$ can be defined as follows:

$$F(C_j, X) = \beta_0 \overbrace{\sum_{C_k \in G_i^L, k \neq j} f(C_j, C_k, X)}^{intra-group} + \beta_1 \overbrace{\sum_{G_k^L \bowtie G_i^L, k \neq i} f(C_j, G_k^L, X)}^{bottom\ level\ inter-group} + \overbrace{\sum_{l=2}^{L} \beta_l \sum_{G_k^l \not\subseteq G_n^{l-1}} f(G_k^l, G_n^{l-1}, X)}^{high\ level\ inter-group} \tag{5}$$

where $G_k^L \bowtie G_i^L$ denote the group G_k^L is inter-related with the group G_i^L; G_k^l and G_n^{l-1} respectively denote the group at l and $l-1$ level both including object class or image concept C_j. And β are the weighted coefficients, which determine the contribution ratio of the intra-group and inter-group classifiers to the final decision, and they are subjected to $\sum_{l=0}^{L} \beta_l = 1$. Once the structural classifiers have been trained, the automatic image classification is achieved by computing

$$C_{(X)} = arg \max_{C_j} F(C_j, X) \tag{6}$$

where $C_{(X)}$ is the predicted label for X. Fig. 2 demonstrates a simple example with 3 levels for our structural learning approach, where the intra-group, bottom level inter-group and high level inter-group are indicated expressly.

In the following, we discuss the computational complexity of our structural learning approach. Let N is the number of total object classes and image concepts. Without loss of generality, we suppose each group has M ($M \ll N$) children groups (M_i^L is the number of object classes and image concepts in the i-th group at level L). Therefore, the total number of binary classifiers for our structural approach is determined as follows

$$\sum_{i=1}^{M^L} \frac{M_i^L(M_i^L - 1)}{2} + N(M-1) + \sum_{l=2}^{L-1} M^l(M-1) \tag{7}$$

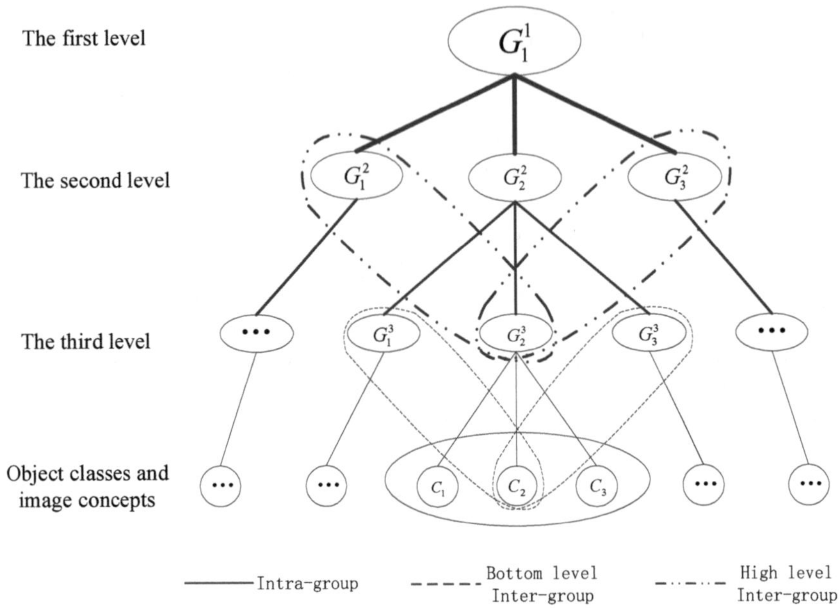

Fig. 2. A simple example for our structural learning approach

where the first item is the total number of intra-group classifiers, the second item is the total number of inter-group classifiers for the bottom level, and the third item is the total number of high level inter-group classifiers. Typically, $M^l \ll N$, $M \ll N$ and $M_i^L \ll N$, therefore our structural classifier training algorithm can achieve sub-quadratic complexity with the number of object classes and image concepts N, much less than the traditional pairwise method with the quadratic complexity of $O(N^2)$. Especially for large scale classification task (N is large), our structural learning algorithm is more efficient.

6 Algorithm Evaluation

6.1 Experimental Setup

Test Dataset: We chose the ImageNet [6] image set as our test dataset, because ImageNet is widely used for large scale image classification task and the images in ImageNet are more complex and have more inter-concept visual correlations. In our experiments, we selected 100 image concepts (most popular real-world object classes and scene categories) from different semantic levels and randomly selected 200 images for each image concept, which were used as the test bench for evaluating our algorithms. The selected subset from ImageNet image set is referred to as ImageNet100 in the following experiments. For ImageNet100 dataset, we randomly selected 100 images from each image concept as training data and the rest images as testing data.

Features Extraction: The images used in the evaluations are all preprocessed into gray scale, and we resize all the training and testing images to a max side length of 300 pixels (width or height) without changing their aspect ratios. SIFT descriptors are then computed at the points on a regular grid with spacing of 3 pixels for each image. In addition, in order to allow scale variation between images, multiple descriptors over supporting regions with different sizes (4, 6, 8, and 10 pixels) are computed at each grid point. We use the publicly available VLFeat toolbox (version 0.9.14) [25] to compute these SIFT descriptors.

Hierarchical Visual Tree: In the following experiments, we constructed the hierarchical visual tree with 3 levels according to our proposed inter-concept visual correlations for ImageNet100 dataset, and each parent group was partitioned into 4 children groups. It is worth noting that some groups at high level maybe stop growing in advance and become a leaf node of the visual tree.

Binary SVM Classifier: The SVM with χ^2-based kernel is used as the basic classifier, and the LIBSVM software [2] is employed to train the basic SVMs. For each basic SVM classifier, we use the same penalty parameter determined by searching over 7 values $(1, 2^2, ..., 2^{12})$ and the same kernel parameter determined by searching over 7 values $(2^{-9}, 2^{-7}, ..., 2^3)$ with five-fold cross validation on the training data.

6.2 Image Classification Evaluation

Performance evaluation for our algorithms and comparison with other algorithms: According to the algorithm proposed in Section 4, we have learned the multi-scale group-based dictionaries, and used the dictionaries to represent the images in ImageNet100 dataset. In the experiment, we evaluated three different sizes for K^L: 256, 512 and 1024. And then we trained the structural classifiers according to the hierarchical visual tree for multi-class image classification. We compared our algorithm with the traditional pairwise method and the hierarchical classification method over traditional label tree, the dictionary size used in these algorithms is set to 1024. We also compared our algorithm with some other inter-related classifier training algorithms. Fig. 3 demonstrates the classification accuracy of our algorithm and the hierarchical classification method [15] over all the object classes and image concepts on ImageNet100. The average classification accuracy for ImageNet100 dataset is shown in Table 2. From the experimental results, it can be seen that our algorithm outperforms all other compared methods. The reasons for this outcome are: (a) our algorithm can determine the inter-related object classes and image concepts in the visual feature space and effectively train the inter-related classifier for them; and (b) our structural learning approach can restraint the error propagation through the hierarchical tree.

Performance Comparison under various Measurements for Inter-concept Correlation Characterization: We have evaluated our correlation characterization approach compared with two popular measurement

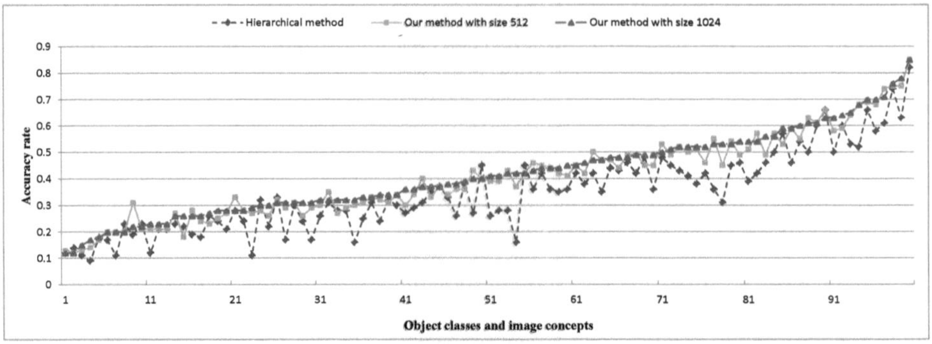

Fig. 3. Classification accuracy for our algorithm and the hierarchical classification method on ImageNet100 dataset

Table 2. Average classification accuracy comparison on ImageNet100 dataset

Algorithms	Average accuracy
Pairwise	0.4037
Hierarchical method [15]	0.3489
Inter-related method [7]	0.3696
Our method ($K^L = 256$)	0.3866
Our method ($K^L = 512$)	0.4031
Our method ($K^L = 1024$)	**0.4159**

Table 3. Inter-concept correlation measurement methods comparison on ImageNet100 dataset

Correlation Measurement	Dictionary Size (K^L)		
	256	512	1024
WordNet similarity [19]	0.3720	0.3833	0.3863
NGD [3]	0.3693	0.3836	0.3906
Our method	**0.3866**	**0.4031**	**0.4159**

methods which used WordNet distance and Google distance respectively. In our experiment, we used the method in [19] to compute WordNet distance between two object classes or image concepts, and the software tool provided by [20] (based on WordNet 3.0) was employed. Normalized Google distance (NGD) [3] is computed by incorporating the results of Google search engine to measure the inter-concept similarity between two object classes or image concepts. For given two image concept x and y, NGD for x and y is defined as follows

$$NGD(x,y) = \frac{\max\{\log f(x), \log f(y)\} - \log f(x,y)}{\log N - \min\{\log f(x), \log f(y)\}} \tag{8}$$

where $NGD(x, y)$ represents the normalized Google distance between the image concepts x and y. $f(x)$, $f(y)$ and $f(x, y)$ denotes the number of web pages containing x, y, both x and y, separately reported by Google. N is the total number of web pages indexed by Google. The experimental results of the average classification accuracy for different inter-concept correlation measurement approaches are shown in Table 3. One can observe that our inter-concept correlation measurement approach performs better than others.

7 Conclusions and Future Work

In this paper, a new method is developed for constructing the hierarchical visual tree automatically, where the inter-concept visual correlations are precisely characterized directly in the visual feature space. Based on the hierarchical visual tree, we propose a group-based approach for leveraging the inter-concept visual correlations to learn multi-scale dictionaries for discriminative image representation. In addition, a novel structural learning approach is also developed to effectively train inter-related classifiers for image classification. Our experiments over a large number of object classes and image concepts have approved the effectiveness of our algorithm for image classification task.

In the future, we will focus on how to determine the proper number of levels and groups at each level (i.e., the structure of the hierarchical visual tree). We believe that the more proper structure can provide better performance.

Acknowledgments. This work is partially supported by National Science Foundation of China under Grant No. 61272285, Ministry of education program for New Century Excellent Talents under NCET-11-0427, and Doctoral Program of Higher Education of China under Grant No. 20126101110022.

References

1. Barnard, K., Forsyth, D.: Learning the semantics of words and pictures. In: ICCV, vol. 2, pp. 408–415 (2001)
2. Chang, C.C., Lin, C.J.: LIBSVM: A library for support vector machines. ACM Transactions on Intelligent Systems and Technology 2, 27:1–27:27 (2011)
3. Cilibrasi, R., Vitanyi, P.: The google similarity distance. IEEE Trans. on Knowledge and Data Engineering 19(3), 370–383 (2007)
4. Csurka, G., Dance, C., Fan, L., Willamowski, J., Bray, C.: Visual categorization with bags of keypoints. In: ECCV, vol. 1, p. 22 (2004)
5. Deng, J., Berg, A.C., Li, K., Fei-Fei, L.: What does classifying more than 10,000 image categories tell us? In: Daniilidis, K., Maragos, P., Paragios, N. (eds.) ECCV 2010, Part V. LNCS, vol. 6315, pp. 71–84. Springer, Heidelberg (2010)
6. Deng, J., Dong, W., Socher, R., Li, L.J., Li, K., Fei-Fei, L.: Imagenet: A large-scale hierarchical image database. In: CVPR, pp. 248–255 (June 2009)
7. Dong, P., Mei, K., Zheng, N., Lei, H., Fan, J.: Training inter-related classifiers for automatic image classification and annotation. Pattern Recognition 46(5), 1382–1395 (2013)
8. Fan, J., Gao, Y., Luo, H.: Integrating concept ontology and multitask learning to achieve more effective classifier training for multilevel image annotation. IEEE Trans. on Image Processing 17(3), 407–426 (2008)
9. Fan, J., Gao, Y., Luo, H., Jain, R.: Mining multilevel image semantics via hierarchical classification. IEEE Trans. on Multimedia 10(2), 167–187 (2008)

10. Fan, J., Luo, H., Hacid, M.S.: Mining images on semantics via statistical learning. In: ACM SIGKDD, pp. 22–31 (2005)
11. Fan, J., Shen, Y., Yang, C., Zhou, N.: Structured max-margin learning for inter-related classifier training and multilabel image annotation. IEEE Trans. on Image Processing 20(3), 837–854 (2011)
12. Fei-Fei, L., Perona, P.: A bayesian hierarchical model for learning natural scene categories. In: CVPR, vol. 2, pp. 524–531 (2005)
13. Lazebnik, S., Schmid, C., Ponce, J.: Beyond bags of features: Spatial pyramid matching for recognizing natural scene categories. In: CVPR, vol. 2, pp. 2169–2178 (2006)
14. Mairal, J., Ponce, J., Sapiro, G., Zisserman, A., Bach, F.R.: Supervised dictionary learning. In: NIPS, pp. 1033–1040 (2008)
15. Marszalek, M., Schmid, C.: Semantic hierarchies for visual object recognition. In: CVPR, pp. 1–7 (2007)
16. Miller, G.A.: Wordnet: a lexical database for english. Commun. ACM 38(11), 39–41 (1995)
17. Moosmann, F., Triggs, B., Jurie, F.: Fast Discriminative Visual Codebooks using Randomized Clustering Forests. In: NIPS, pp. 985–992 (2007)
18. Nistér, D., Stewénius, H.: Scalable recognition with a vocabulary tree. In: CVPR, vol. 2, pp. 2161–2168 (2006)
19. Patwardhan, S.: Incorporating dictionary and corpus information into a context vector measure of semantic relatedness. Master thesis, University of Minnesota, Duluth (2003)
20. Pedersen, T., Patwardhan, S., Michelizzi, J.: Wordnet: similarity - measuring the relatedness of concepts. In: AAAI, pp. 1024–1025 (2004)
21. Ries, C., Romberg, S., Lienhart, R.: Towards universal visual vocabularies. In: ICME, pp. 1067–1072 (July 2010)
22. Shi, J., Malik, J.: Normalized cuts and image segmentation. IEEE Trans. on PAMI 22(8), 888–905 (2000)
23. Smeulders, A., Worring, M., Santini, S., Gupta, A., Jain, R.: Content-based image retrieval at the end of the early years. IEEE Trans. on PAMI 22(12), 1349–1380 (2000)
24. Vasconcelos, N.: Image indexing with mixture hierarchies. In: CVPR, vol. 1, p. I (2001)
25. Vedaldi, A., Fulkerson, B.: VLFeat: An open and portable library of computer vision algorithms (2008), `http://www.vlfeat.org`
26. Wang, Y., Gong, S.: Refining image annotation using contextual relations between words. In: CIVR, pp. 425–432 (2007)
27. Winn, J., Criminisi, A., Minka, T.: Object categorization by learned universal visual dictionary. In: ICCV, vol. 2, pp. 1800–1807 (October 2005)
28. Yang, L., Jin, R., Sukthankar, R., Jurie, F.: Unifying discriminative visual codebook generation with classifier training for object category recognition. In: CVPR, pp. 1–8 (2008)
29. Zhang, J., MarszaLek, M., Lazebnik, S., Schmid, C.: Local features and kernels for classification of texture and object categories: A comprehensive study. International Journal of Computer Vision 73, 213–238 (2007)
30. Zhang, Q., Li, B.: Discriminative k-svd for dictionary learning in face recognition. In: CVPR, pp. 2691–2698 (June 2010)
31. Zhou, N., Shen, Y., Peng, J., Fan, J.: Learning inter-related visual dictionary for object recognition. In: CVPR, pp. 3490–3497 (2012)

Group Sparse Ensemble Learning for Visual Concept Detection

Yongqing Sun, Kyoko Sudo, and Yukinobu Taniguchi

NTT Media Intelligence Laboratories
1-1 Hikarinooka Yokosuka-Shi Kanagawa 239-0847 Japan
yongqing.sun@lab.ntt.co.jp

Abstract. To exploit the hidden group structures of data and thus detect concepts in videos, this paper proposes a novel group sparse ensemble learning approach based on Automatic Group Sparse Coding (AutoGSC). We first adopt AutoGSC to learn both a common dictionary over different data groups and an individual group-specific dictionary for each data group which can help us to capture the discrimination information contained in different data groups. Next, we represent each data instance by using a sparse linear combination of both dictionaries. Finally, we propose an algorithm to use the reconstruction errors of data instances to calculate the ensemble gating function for ensemble construction and fusion. Experiments on the TRECVid 2008 benchmark show that the ensemble learning proposal achieves promising results and outperforms existing approaches.

Keywords: Ensemble learning, Visual concept detection, Semantic indexing, Sparse representation.

1 Introduction

With the advent of the big data era, the explosive growth of visual contents on the Internet presents a challenge in how to manage the ever-growing size of the multimedia collections, particularly in how to extract sufficiently accurate semantic metadata (concepts) to make them searchable [5,7,17]. Visual concept detection is essentially a classification task in which classifiers are learned with various features extracted from training samples to predict the presence of a certain concept in a video shot or keyframe (image) [17,19]. Ranging from objects such as *"boat"* and *"car"* to scenes such as *"sky"* and *"sea"*, semantic concepts can serve as good intermediate semantic metadata for video content indexing and understanding [17]. Establishing a large set of robust concept detectors will yield significant improvements in many challenging applications, such as image/video search and summarization [19]. Due to the existence of the well-known semantic gap [12] between the low level visual features and the users's semantic interpretation of diversified visual data, concept detection is a challenging yet essential task that has attracted the attention of many researchers.

Studies on pedestrian classification [4,8] and visual concept detection [5] indicate that the data matters most; this was highlighted recently by machine

B. Huet et al. (Eds.): PCM 2013, LNCS 8294, pp. 649–657, 2013.

learning researchers who stated that "the quickest path to success is often to just get more data" [3]. Due to the low efficiency and unscalability of the classical methods based on global classification such as Support Vector Machine (SVM) [2], Gaussian Mixture Model [1], Hidden Markov Model [10], etc., various ensemble learning methods such as LDA-SVM [14–16], multi-kernel ensemble learning [13] and sparse ensemble learning [17], etc., were developed for visual concept detection; they exploit the "divide and conquer" strategy to train large amounts of samples both effectively and efficiently.

In particular, [17] proposed an efficient sparse ensemble learning method by exploiting a sparse non-negative matrix factorization process for ensemble construction and fusion. It was shown to achieve state-of-art performance on the TRECVid 2008 dataset. However, this approach adopts traditional sparse coding and so treats each data instance as an individual and no data group information is considered. It considers each visual feature such as Bag of Visual Word (BoVW) of an image as a separate coding problem and does not take into account the fact that the sparse coding of each feature does not guarantee the sparse coding of all images in the dataset.

Each dataset usually consists of many categories, and is assured of having hidden group structures. Once a dictionary atom has been selected to represent an image, it may as well as be used to represent other images of a given category without much additional regularization cost [11]. Therefore, it makes more sense to learn a group level sparse representation [11]. To exploit the group structures hidden in the data set, [11] proposed Group Sparse Coding (GSC), which learns a sparse representation on the group level as well as a shared dictionary. However, GSC assumes that the data group identities are pre-given, even though they are often hidden in many real world applications, and it can only learn a common dictionary [18]. To discover the hidden data groups, [18] proposed Automatic Group Sparse Coding (AutoGSC) by learning both a common dictionary over different data groups and an individual group-specific dictionary for each data group which can help us to capture the discrimination information contained in different data groups.

Inspired by the sparse ensemble learning work [17] and the advantages of AutoGSC [18] in discovering hidden structures of data, in this paper, we propose a novel group sparse ensemble learning approach based on automatic group sparse coding to exploit the hidden group structures of data.

2 Proposed Approach

2.1 Problem Formulation

Ensemble learning refers to the process of combining multiple classifiers to provide a single and unified classification decision [17]. Recent research has demonstrated that a good ensemble of localized classifiers can outperform a single (best) classifier learned over the entire dataset [3, 17]. Additionally, learning a set of "smaller" localized classifiers is usually more efficient in terms of algorithmic complexity than a global classifier, which has motivated researchers to adopt the ensemble learning approach for concept detection [17]. [3] advocates to "learn many models not just one".

In visual concept detection, an image or a keyframe of a video shot is processed to detect the presence of a set of predefined concepts. Without loss of generality, we assume that the data instances are represented as vectors, such as the visual feature vectors of keyframes. Mathematically, we denote the observed data matrix as $\mathbf{X} = [x_1, x_2, \ldots, x_n] \in \Re^{d \times n}$, where $x_i \in \Re^d$ represents the i-th data instance vector with dimensionality d. For each concept, we have the label $Y = \{y_i \in \pm 1, i = 1, 2, \ldots, n\}$ for the training set matrix $\mathbf{X}$. Consequently, with the binary classification in the framework of SVM, the ensemble discriminant function $F(x_t)$ for a given test sample x_t is [17]:

$$F(x_t) = \sum_{c=1}^{k} \Psi_c(x_t) \cdot \left(\sum_{i \in \pi_c} \beta_i y_i \langle \Phi(x_t), \Phi(x_i) \rangle + b_c \right) \tag{1}$$

where k localized classifiers are built on instance localities π_c, and $\Psi_c(x_t)$ are the gating functions that govern how localized classifiers are coordinated for the final classification of test sample x_t. Learning the ensemble discriminant function $F(x_t)$, i.e., ensemble construction, can be decomposed into two steps [17]: (1) learning the instance localities π_c and gating function $\Psi_c(x_t)$, and (2) training the individual classifiers to estimate the kernel classifier parameters such as the optimal classification hyperplane parameters β_i, b_c.

2.2 Framework

Here, we propose to construct the ensemble through AutoGSC [18] to take advantage of the hidden group structures of data. The overall framework is illustrated in Fig.1.

Specifically, as shown in the figure, after feature extraction, we use AutoGSC to learn both a common shared dictionary $\mathbf{D}^S$ over different data groups and the k individual group-specific dictionaries $\{\mathbf{D}_c^I\}_{c=1}^k$ which can help us to capture the discrimination information contained in the different data groups. We then represent each data instance x_i by using a sparse linear combination of both dictionaries, i.e., get the shared sparse code matrices $\{\mathbf{G}_c^S\}_{c=1}^k$ from the common shared dictionary $\mathbf{D}^S$ and the individual sparse code matrices $\{\mathbf{G}_c^I\}_{c=1}^k$ from the individual group-specific dictionaries $\{\mathbf{D}_c^I\}_{c=1}^k$ for data matrix $\mathbf{X}$. Finally, we compute the reconstruction errors of the group sparse coding for each data instance, and use them to calculate the gating functions $\Psi_c(x_t)$ which are used for instance grouping during ensemble construction and individual classification result fusion during testing.

The following subsections will detail the gating function calculation with AutoGSC, ensemble construction and fusion, and complexity analysis.

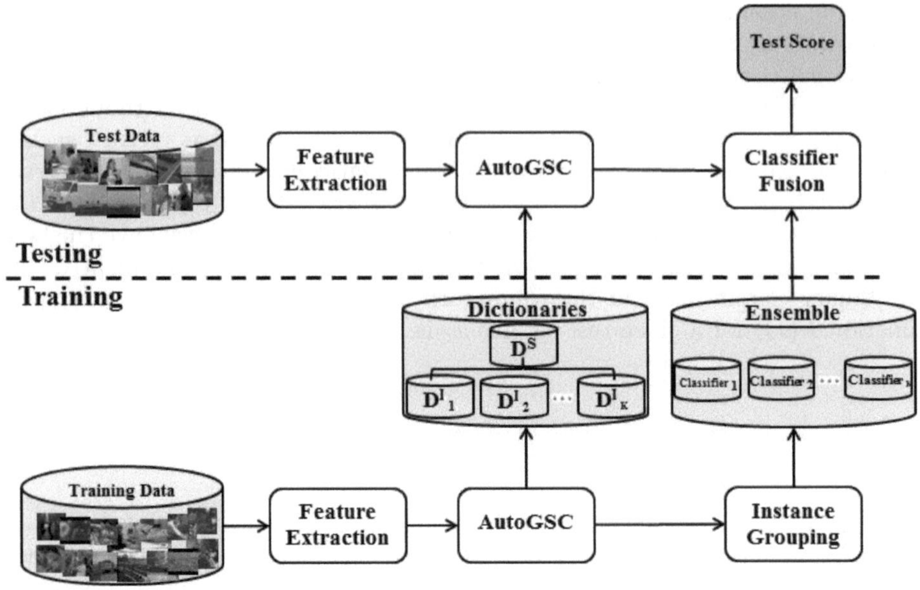

Fig. 1. Proposed Framework

2.3 Gating Function Calculation with AutoGSC

AutoGSC tries to discover the hidden group structures of data by solving the optimization problem under the following non-negativity constraints [18]:

$$\begin{aligned} &\min\sum_{c=1}^{k}\left\|\mathbf{X}_c-\mathbf{D}^S\mathbf{G}_c^S-\mathbf{D}_c^I\mathbf{G}_c^I\right\|_F^2+\sum_{c=1}^{k}\left[\gamma_I\phi\left(G_c^I\right)+\gamma_S\phi\left(G_c^S\right)\right] \\ &s.t.\mathbf{D}^S\geq 0,\forall c=1,2,\ldots,k,\mathbf{D}_c^I\geq 0,\mathbf{G}_c^I\geq 0 \end{aligned} \tag{2}$$

where the normalized matrices $\mathbf{D}^S$ and $\{\mathbf{D}_c^I\}_{c=1}^k$ to be solved are the common shared dictionary and k individual group-specific dictionaries on each group locality π_c respectively. The matrices $\{\mathbf{G}_c^S\}_{c=1}^k$ and $\{\mathbf{G}_c^I\}_{c=1}^k$ to be solved are the sparse code (i.e., reconstruction coefficient) matrices decoded by the two dictionaries correspondingly. In order to achieve group sparsity, the second term is introduced to impose some regularization on the sparse code where function ϕ is used to compute the ℓ_1-norms of the row vectors of the input matrix.

After merging the two kinds of dictionaries and corresponding sparse codes by:

$$\begin{aligned} \mathbf{D}_c &= \left[\mathbf{D}^S,\mathbf{D}_c^I\right] \\ \mathbf{G}_c &= \left[\mathbf{G}_c^S,\mathbf{G}_c^I\right] \end{aligned} \tag{3}$$

the objective function $\hat{\mathcal{J}}_0$ of the optimization problem (2) can be rewritten as:

$$
\begin{aligned}
\hat{\mathcal{J}}_0 &= \sum_{c=1}^{k} \left[\left\| \mathbf{X}_c - \mathbf{D}_c \mathbf{G}_c^\top \right\|_F^2 + \gamma \phi(\mathbf{G}_c) \right] \\
&= \sum_{c=1}^{k} \sum_{x_i \in \pi_c} \left[\left\| x_i - \sum_j G_{cij} \mathbf{D}_{c\cdot j} \right\|_F^2 + \gamma \sum_j |G_{cij}| \right]
\end{aligned} \tag{4}
$$

where $\mathbf{D}_{c\cdot j}$ is the j-th column of $\mathbf{D}_c$, and G_{cij} is the (i, j)-th entry of $\mathbf{G}_c$.

AutoGSC uses a Lloyd style algorithm [18] to solve to the problem by alternating between the dictionary and sparse code.

Since AutoGSC searches each locality π_c to obtain the group identity of a data instance x_i with the minimum reconstruction error, we propose to use the reconstruction error to calculate the gating function vector $\Psi(x_i) = \{\Psi_c(x_i)\}_{c=1}^{k}$ in the four steps shown in Fig.2. This makes our proposed algorithm very different from the gating function calculation method proposed in [17], which uses the sparse code to obtain the gating function directly.

To achieve high efficiency, especially testing efficiency, we adopt the Instance-Locality Assignment Algorithm proposed in [17] in our proposed algorithm, and restrict the maximum number of groups to which data instance x_i can be assigned with the input replication parameter l, i.e., an instance can only be allocated to at most l group localities. After calling the algorithm, the returned gating function $\Psi_c(x_i)$ has only r ($r \leq l \ll k$) non-zero elements, which means x_i can be allocated into only r groups at the same time.

2.4 Ensemble Construction and Fusion

For ensemble construction, we use $\Psi_c(x_i)$ to allocate each training data instance x_i into multiple r group localities π_c among all k hidden groups, and use SVM to train the individual classifiers for estimating the kernel classifier parameters such as the optimal classification hyperplane parameters β_i, b_c.

Whereas for ensemble fusion, with Equation (1), we use $\Psi_c(x_t)$ to coordinate the classification results of the related r group localities to obtain the ensemble discriminant function $F(x_t)$ for a given test sample x_t.

2.5 Complexity Analysis

Similar to the computational complexity analysis in [17], since the theoretical computational complexity of SVM training is between $O(n^2)$ and $O(n^3)$ (n is the number of training samples) depending on the value of the hyper-parameter C [17], we can greatly improve the training efficiency after we use AutoGSC by partitioning a given training sample into at most l hidden groups. Specifically, the number of training samples in a given group locality π_c is nl/k on average, and we need to train k individual classifier in the ensemble. Therefore, the complexity is greatly reduced to only l^2/k to l^3/k^2 ($l \ll k$) times that of global classification. While for testing, since we only invoke at most l individual classifiers in the

Algorithm: Gating Function Calculation

Input: Instance x_i, Dictionaries $\{\mathbf{D}_c\}_{c=1}^k$, and replication parameter l.

Output: Gating function vector $\Psi(x_i)$.

1: Measure the reconstruction error vector $\{\varepsilon_{ic}\}_{c=1}^k$ for quantizing x_i with dictionary $\mathbf{D}_c$ by:

$$\varepsilon_{ic} = \|x_i - \mathbf{D}_c \cdot g_{ci}\|^2 + \gamma \left|g_{ci}\right|_1 \qquad (5)$$

where g_{ci} is the corresponding sparse code of x_i with dictionary $\mathbf{D}_c$, i.e., i-th column of $\mathbf{G}_c$.

2: Compute the grouping weight w_{ic} of assigning x_i to the group locality π_c as the reciprocal of the error, i.e., $w_{ic} = 1/\varepsilon_{ic}$. Thus, we obtain the weight vector $w_i = \{w_{ic}\}_{c=1}^k$ over all groups.

3: Call the Adaptive Instance-Locality Assignment Algorithm proposed in [17] with the input parameter l to get the membership vector $\alpha_i = \{\alpha_{ic}\}_{c=1}^k$ from w_i.

4: ℓ_1-normalize membership vector to return the gating function vector $\Psi(x_i) = \{\alpha_{ic}/\left|\alpha_i\right|_1\}_{c=1}^k$.

Fig. 2. Gating Function Calculation Algorithm

ensemble for a given test sample, and the classifiers are very compact with many fewer support vectors than those used by a global classifier, the testing efficiency can be greatly improved compared with global classification. Thus we conclude that our proposed group sparse ensemble learning method is very efficient in both training and testing.

3 Experiments and Results

To evaluate the performance of our proposed group sparse ensemble learning method, we selected the same TRECVid 2008 video benchmark collection [9] as [17] to conduct comparison experiments with the sparse ensemble learning (SEL) method proposed by [17]. TRECVid is now widely regarded as the actual standard for evaluation the performance of concept based video retrieval systems [17]. Refer to [9, 17] for more details about the dataset. We also used the same VIREO-374 BoVW features released by [6] as [17] to train and test our system. Additionally, we used the same parameters such as number of locality $k = 800$, replication parameter $l = 20$, RBF kernel of SVM, and same evaluation criteria InfAP as [17] for direct comparison.

First, to verify the effectiveness of our ensemble fusion method based on the proposed gating function calculation algorithm, we compared it with average fusion. InfAP of each concept and MAP yielded by the two fusion schemes are

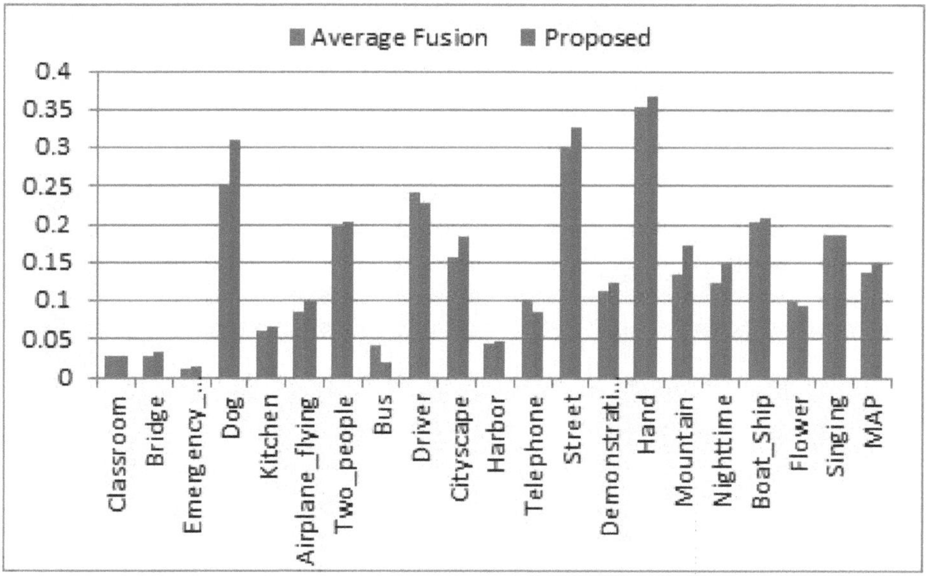

Fig. 3. Fusion comparison with average fusion

shown in Fig.3. It shows that our fusion method, based on the proposed gating function calculation algorithm, yields MAP of 0.147, 6.5% higher than average fusion (MAP=0.138). We can also see that the proposed method clearly outperforms average fusion on most of the concepts, such as "Dog", "Airplane-flying", "Cityscape", "Street", "Hand", "Mountain", "Nighttime". This superiority does not extend to rare concepts with too few positive training samples such as "Bus" and "Emergency-Vehicle"; their detection rates are low and unstable. Thus we can conclude that our fusion method is effective.

Next, we compared our method with the SEL method [17], global SVM classification (Single-SVM). Ours yielded a MAP of 0.147, an improvement of 4.3% and 12.2% over SEL (MAP=0.141) and Single-SVM (MAP=0.131), respectively. Fig.4 compares InfAP for each concept. From the figure, we can see that the proposed method clearly outperforms SEL and single SVM on 11 out of 20 concepts, including both scene concepts like "Cityscape", "Harbor", "Street", "Mountain", "Nighttime" and object concepts like "Dog", "Airplane-flying", "Two-people", "Telephone", "Hand", "Flower". For the remaining concepts such as "driver" and "Demonstration-Or-Protest", we can found that the instances of these concepts in which our proposed method performs best are not too diversified, and often have good tendency or consistence on being grouped. This reflects the advantages of group sparse coding in discovering the hidden group structures.

The above results show that the ensemble learning proposal has achieved promising results and can outperform existing approaches.

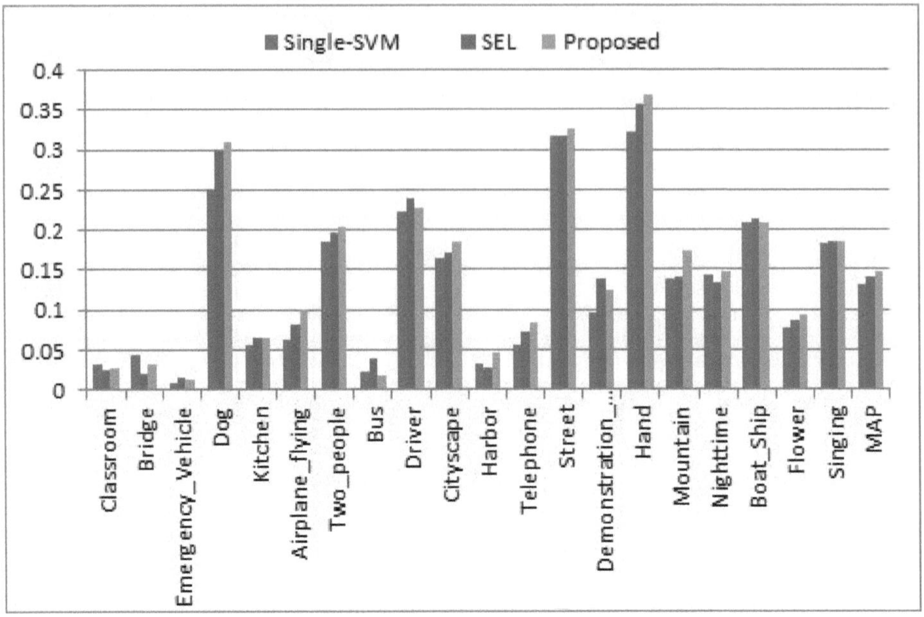

Fig. 4. Comparison with Single-SVM and SEL

4 Conclusion

In this paper, we propose a novel group sparse ensemble learning approach for detecting concepts in videos. To exploit the hidden group structures of data, we first adopt Automatic Group Sparse Coding (AutoGSC) to learn both a common dictionary over different data groups and an individual group-specific dictionary for each data group. Next, we represent each data instance by using a sparse linear combination of both dictionaries, and use the reconstruction errors of data instances for ensemble construction and fusion.

Experiments on the TRECVid 2008 benchmark verified the effectiveness of the proposed ensemble learning approach.

References

1. Amir, A., Berg, M., Chang, S.-F., et al.: Ibm research trecvid-2003 video retrieval system. In: NIST TRECVID Workshop (November 2003)
2. Cao, J., Lan, Y., Li, J., et al.: Intelligent multimedia group of Tsinghua University at trecvid 2006. In: NIST TRECVID Workshop (November 2006)
3. Domingos, P.: A few useful things to know about machine learning. Commun. ACM 55(10), 78–87 (2012)
4. Enzweiler, M., Gavrila, D.M.: Monocular pedestrian detection: Survey and experiments. IEEE Trans. Pattern Anal. Mach. Intell. 31, 2179–2195 (2009)

5. Huiskes, M.J., Thomee, B., Lew, M.S.: New trends and ideas in visual concept detection: the mir flickr retrieval evaluation initiative. In: MIR 2010, New York, NY, USA, pp. 527–536 (2010)
6. Jiang, Y.-G., Yang, J., Ngo, C.-W., Hauptmann, A.G.: Representations of keypoint-based semantic concept detection: A comprehensive study. IEEE Transactions on Multimedia 12, 42–53 (2010)
7. Li, H., Wang, X., Tang, J., Zhao, C.: Combining global and local matching of multiple features for precise retrieval of item images. ACM/Springer Multimedia System Journal 19(1), 37–49 (2013)
8. Munder, S., Gavrila, D.: An experimental study on pedestrian classification. IEEE Transactions on Pattern Analysis and Machine Intelligence 28, 1863–1868 (2006)
9. Over, P., Awad, G., Rose, R.T., Fiscus, J.G., Kraaij, W., Smeaton, A.F.: Trecvid 2008 - goals, tasks, data, evaluation mechanisms and metrics. In: NIST TRECVID Workshop (2008)
10. Pytlik, B., Ghoshal, A., Karakos, D., Khudanpur, S.: TRECVID 2005 Experiment at Johns Hopkins University: Using Hidden Markov Models for Video Retrieval. In: NIST TRECVID Workshop (November 2005)
11. Samy Bengio, D.S., Pereira, F., Singer, Y.: Group Sparse Coding. In: Neural Information Processing Systems - NIPS (2009)
12. Smeulders, A.W.M., Worring, M., Santini, S., Gupta, A., Jain, R.: Content-based image retrieval at the end of the early years. IEEE Transactions on Pattern Analysis and Machine Intelligence 22(12), 1349–1380 (2000)
13. Song, Y., Zheng, Y.-T., Tang, S., et al.: Localized multiple kernel learning for realistic human action recognition in videos. IEEE Transactions on Circuits and Systems for Video Technology 21(9) (2011)
14. Tang, S., Li, J.-T., Li, M., Xie, C., Liu, Y., Tao, K., Xu, S.-X.: Trecvid 2008 high-level feature extraction by MCG-ICT-CAS. In: NIST TRECVID Workshop (November 2008)
15. Tang, S., Li, J.-T., Yong-Dong Zhang, E.: Pornprobe: an lda-svm based pornography detection system. In: ACM Multimedia 2009 (October 2009)
16. Tang, S., Zheng, Y.-T., Cao, G., Zhang, Y.-D., Li, J.-T.: Ensemble learning with lda topic models for visual concept detection. In: Multimedia - A Multidisciplinary Approach to Complex Issues, pp. 175–200 (2012)
17. Tang, S., Zheng, Y.-T., Wang, Y., Chua, T.-S.: Sparse ensemble learning for concept detection. IEEE Transactions on Multimedia 14(1), 43–54 (2012)
18. Wang, F., Lee, N., Sun, J., Hu, J.: Automatic Group Sparse Coding. In: Twenty-Fifth AAAI Conference on Artificial Intelligence (August 2011)
19. Zhu, S., Wang, G., Ngo, C.-W., Jiang, Y.-G.: On the sampling of web images for learning visual concept classifiers. In: CIVR 2010, New York, NY, USA, pp. 50–57 (2010)

High-Level Video Semantic Concept Detection Based on Multi-level Feature Representations

Lijuan Liu[1], Haojie Li[1,*], Fuming Sun[2], Yaomin Yin[3], and Chenxin Liu[1]

[1] Dalian University of Technology
[2] Liaoning University of Technology
[3] Agricultural Bank of China
hjli@dlut.edu.cn, sunfm@mail.ustc.edu.cn,
yinyaomin@abchina.com

Abstract. Semantic concept detection is a fundamental problem with many practical applications such as concept-based video retrieval. The major challenge of concept detection lies in the existence of the well-known semantic gap between the low-level visual features and the user's semantic interpretation of visual data. To bridge the semantic gap, we propose to promote the low-level visual feature to the middle-level representation, expecting that the underlying semantic aspects of image data can be discovered, and such latent aspects can better model the semantic of images. Specifically, latent Dirichlet allocation (LDA) is adopted to cluster the image data into semantic topics and the distributions on such topics are used as the middle-level feature vectors of the image. Meanwhile, a recently developed more efficient probabilistic representation of low-level features, i.e., Fisher Vector (FV) is used to complement the LDA representation for video concept detection. The experimental results on the TRECVID 2013 Semantic Indexing dataset have demonstrated the effectiveness of the proposed approach.

Keywords: Concept Detection, Fisher Vector, LDA.

1 Introduction

With the amount of huge multimedia collections increasing at a tremendous speed, there is an urgent need to develop automatic video retrieval systems. Traditional content-based retrieval methods cannot satisfy people and have shown their limitations [1]. In recent years semantic concept-based video retrieval has been proposed and attracted attentions of many researchers [2-5]. Here, semantic concept detection, which aims to detecting the presence or absence of high-level concepts such as car, mountain, sky in video shots, is the key task. However, though lots of research efforts have been dedicated to the concept detection, it is still a challenging problem within the multimedia communities [6].

Essentially, we can regard concept detection as a classification task, in which a binary classifier is usually learned to predict the presence of a certain concept in a video

* Corresponding author.

B. Huet et al. (Eds.): PCM 2013, LNCS 8294, pp. 658–668, 2013.

shot or keyframe based on the input feature descriptors. Bag-of-visual-words (BOV) [7-9], which transforms local image descriptors into image representations, is the widely used approach for concept detection. A baseline BOV approach employs k-means clustering of local descriptors from a training dataset and assigning each descriptor to the nearest cluster. There are also many variation methods of BOV such as the soft-weighting assigning scheme [13] or modeling the vocabulary using GMM [11], [12], et al. However, BOV only considers the occurring number of local descriptors assigned to each visual word. Due to the high degree of the diversity in creator, content, style and production qualities about the video data, we need to model the original video frames naturally. Therefore, we generate the low-level features as dense-SFIT [14] instead of SIFT [15], and in order to avoid the increasing of the dimension of the BOV representation, we propose to use Fisher Vector [16], [17], which is measured by Fisher kernel, to represent the low-level features. As showed in [18], [19], as a generative probabilistic model, Fisher vector can describe the video frames naturally. Recently, by assuming that the samples follow a parametric generative model estimated on a training set, Fisher vector is represented as the gradient of the sample's likelihood with respect to the parameters of distribution, has achieved state-of-the-art performance in many tasks [20], [21]. In our work, we use Fisher vectors to represent keyframes, expecting that these vectors can also gain effective information for concept detection.

One more important problem with the traditional methods is that they just used the low-level features extracted from the frames, and then described them with researcher's purpose, which often leads to the so-called semantic gap and it has been the bottleneck in this research [6]. Various methods have been proposed to narrow the semantic gap. Early work on this problem focused concept-specific handcrafted decision rules which mapped restricted sets of low-level visual features to a single high-level concept [22]. [23] proposed a Bayesian probabilistic framework for mapping multimedia representation to high-level semantics, and recently G. E. Hinton et al [24] proposed to use convolutional neural network to learn the features of images automatically based on GPU. In this work, we propose to introduce a middle-level representation, specifically, Latent Dirichlet Allocation (LDA) [25], to encode low-level image features to bridge the semantic gap in concept detection. In recent research, as a topic model inspired from text mining, LDA has been widely used in vision processing in various ways [10], [26], [28]. LDA is a probabilistic generative model derived from low-level local features, such as SIFT; however, it could describes the co-occurrence between low-level features which indicates some higher-level implicit semantic over the underlying image data. Just like the specificity of an object might rely on the geometrical configuration of a limited number of visual patterns, in our approach we expect that LDA can capture the specificity of a particular concept greatly through modeling the particular co-occurrences of a large number of visual components. We regard the co-occurrences of visual components under a concept as latent semantic contextual information. It is noted that in our work, to reduce the computational cost in low-level feature processing, we use only one kind of image feature, namely, we extract the dense SIFT of video keyframes and then generate two kinds of representations, Fisher vector and LDA.The contributions of our work are two-fold. First, we propose to use milddle-level representation to encode low-level features to narrow the semantic gap for concept detection, and the experimental results on TRECVID 2013 Semantic Indexing dataset demonstrate its practicality and effectiveness. Second, by fusing the results of Fisher vector and LDA approaches, we show that the descriptors from different probabilistic models could capture the different aspects of images, thus improve the final detection performance.

The rest of the paper is organized as follows. We first introduce the overall framework of our approach in Section 2 and then detail the approach in Section 3. The experiments are conducted in Section 4 and we conclude in Section 5.

2 The Proposed Framework

In this section, we present the proposed framework of our work. As described in Fig. 1, we propose an approach to using a middle-level representation to encode low-level image features. Here we propose to use LDA to indicate some higher-level implicit semantic information over the underlying image data. As a complement, we also use a recently developed more efficient probabilistic representation of low-level features referred as Fisher vector to quantize the extracted image features. We fuse the two level features generated from different probabilistic models, and expect that these vectors can describe the original frame effectively.

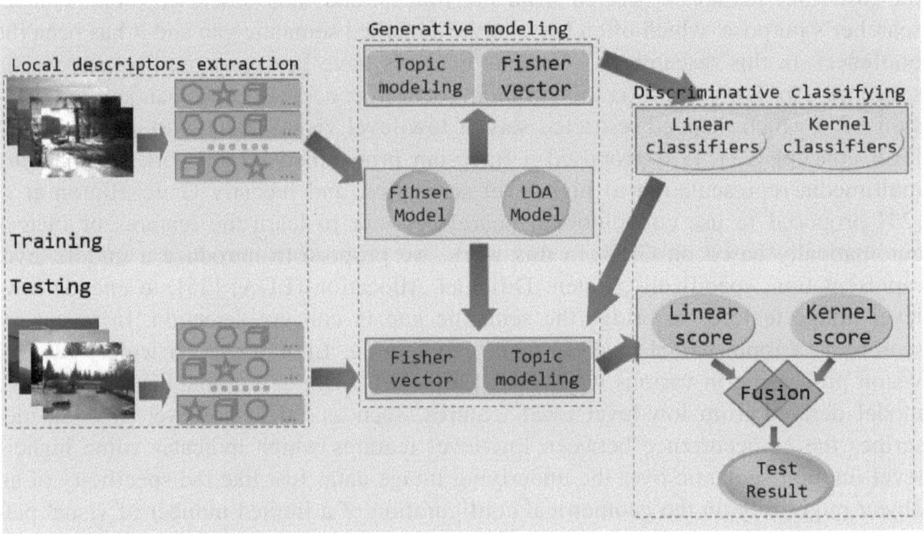

Fig. 1. The proposed Framework of our work

As the dataset comes from the web video with huge variations, we propose to use the densely sampled SIFT descriptors in low-level feature extraction. While in quantization phrase, we model the original frame with generative probabilistic methods, and we expect to generate the representation of frame naturally and effectively. In such phase, we quantize the descriptors in different views in which the low-level quantization using Fisher vector [16] to estimate the parametric generative model on the training set and characterize the distribution of the vectors with respect to the model, and the intermediate vector which is generated over the low-level descriptors using LDA [25] to model the association between varying visual patterns under one semantic concept. In the last phrase, we propose to train different classifiers for each representation, and then fuse the predict results in an unsupervised way [27]. In our experiment, we choose support vector machine to train models, and because of the high dimension of the Fisher vector, we prefer the linear SVM rather than kernel SVM.

3 The Approach

In this section, we briefly review two popular approaches that quantize local descriptors into a vector representation of a frame. Then we propose our method to combine these two representations.

3.1 Low-level Feature Representation: Fisher Vector

The Fisher Vector is an extension of BoV representation. Instead of characterizing each frame by the number of occurrences of each visual word, it is characterized by a gradient vector derived from a generative probabilistic model. In our work, we use Fisher Kernel based on Gaussian Mixture Model to gain frame representation which is referred as Fisher Vector.

Fisher kernels [16], [17] have been introduced to combine the benefits of generative and discriminative approaches. For two data items, we use Fisher kernel to calculate and compare the gradient of the log-likelihood of the data item with respect to the model with a given set of parameters. Let U_x be a probability model, the Fisher score U_x is defined as the gradient of the log-likelihood of x with respect to the model with a given set of parameters λ:

$$U_x = \nabla_\lambda \log P(x \mid \lambda) \tag{1}$$

We then use it as frame representation, as it gains enough information with lower computational complexity. The Fisher score gives us an embedding into the feature space. The Fisher kernel refers to the inner product in this space, and is defined as

$$K(x_i, x_j) = U_{x_i}^T F_\lambda^{-1} U_{x_j}^T \tag{2}$$

Where F_λ is the Fisher information matrix:

$$F_\lambda = E_{x \sim p(x|\lambda)}[(\nabla_\lambda \log p(x \mid \lambda))(\nabla_\lambda \log p(x \mid \lambda))^T] \tag{3}$$

As F_λ is symmetric and positive define, F_λ^{-1} can be decomposed as

$$F_\lambda^{-1} = L_\lambda^T L_\lambda \tag{4}$$

Then we can rewrite Fisher kernel as

$$K(x_i, x_j) = (L_\lambda U_{x_i})^T (L_\lambda U_{x_j}) \tag{5}$$

$\Gamma_\lambda = L_\lambda U_x$ is referred as Fisher Vector. Because of the high cost of computation of inverse, in [21] F_λ is regarded as an identity matrix and use the Fisher score as an approximation. In fact, if Fisher scores are similar in the sense that they would require similar adaptations to the parameters.

In the case of generative approach, the probability density function (pdf) – denoted by p - can be used to model the emission of the low-level descriptors in the frame. Assuming that we generate the local descriptors $\{x_t, x_t \in R^D, t = 1 \dots T\}$ of frame I, we model the visual vocabulary with a Gaussian Mixture Model (GMM) [11][12] where $\lambda = \{w_i, \mu_i, \Sigma_i, i = 1 \dots N\}$ is the input parameter in which w_i, μ_i and Σ_i denote respectively the weight, mean vector and covariance matrix, and each Gaussian corresponds to a visual word. Let p_i be the distribution of Gaussian i and we get:

$$p(x \mid \lambda) = \sum_{i=1}^{N} w_i p_i(x \mid \lambda) = \sum_{i=1}^{N} w_i N(x \mid \mu_i, \Sigma_i) \tag{6}$$

We denote by $\gamma_i(x_t)$ the probability for x_t to have been assigned to the $i-th$ Gaussian. By Bayes formula, we get:

$$\gamma_i(x_t) = \frac{w_i p_i(x_t \mid \lambda)}{\sum_{j=1}^{N} w_j p_j(x_t \mid \lambda)} \tag{7}$$

In the following, we can use the gradient of the log-likelihood to describe the contribution of the parameters to the process. As it was shown in [19] that the gradient with respect to the weights w_i generally bring little additional information, we only consider the gradients with respect to the mean and standard deviation parameters. By assuming the covariance matrices are diagonal, and while given the sample $X = \{x_t, x_t \in R^D, t = 1 \dots T\}$, we denate the $d-th$ element of u_i as u_i^d, and the $d-th$ element of Σ_i as $(\sigma_i^d)^2$, hence, we have the following formulas for $\Gamma_{u_i^d}(X)$ and $\Gamma_{\sigma_i^d}(X)$:

$$\Gamma_{u_i^d}(X) = \sum_{t=1}^{T} \gamma_t(i) \left[\frac{x_t^d - u_i^d}{\left(\sigma_i^d\right)^2} \right] \tag{8}$$

$$\Gamma_{\sigma_i^d}(X) = \sum_{t=1}^{T} \gamma_t(i) \left[\frac{\left(x_t^d - u_i^d\right)^3}{\left(\sigma_i^d\right)^3} - \frac{1}{\sigma_i^d} \right] \tag{9}$$

Hence, we get the Fisher Vector representation with the dimension of $2KD$, in which, K represents the number of the Gaussian in GMM, and D represents the dimension of each lower feature. Actually, it characterizes the distribution of the vector with respect to the center.

3.2 Middle-level Feature Representation: LDA

As shown in [3], traditional methods which use low-level features only could not gain better result, we propose to use topic model to capture co-occurrence information between elements. In our work, we propose Latent Dirichlet Allocation (LDA) [25] model for that purpose. LDA, proposed by Blei et al., is a probabilistic generative model that can be used to estimate the properties of multinomial observations by unsupervised learning. LDA is originally proposed to model text problem and in our work we use it to address the problem of latent semantic modeling in video concept detection. In our approach we represent each frame with a topic representation assignment, which can deal with the polysemy and synonymy issue [9], which suffered in BOV, very well.

We perform this model based on frame features, and the following frame-text analogy is thus considered for the proposed processing:

- Each low-level feature $(w_i)_{i=1...N}$ of the frame corresponds to a word, as the basic unit of discrete data, where N is the number of the low-level features for one frame;
- Each visterm $(v_j)_{j=1...V}$ of the vocabulary corresponds to a term under the topic, where V is the size of the vocabulary;
- Each frame $(W_m)_{m=1...M}$ in the dataset is regarded as a document in the corpus, where M is the number of frames in the dataset;
- We regard the frame dataset as the corpus.

Under the assumption that the occurrence of a visterm v_j is independent of the frame W_m it belongs to, given an aspect z_i, the generative process within LDA will follow the next steps to construct corpus:

1. For each topic $k \in [1, K]$, we sample a mixture component $\vec{\varphi}_k \sim Dir(\vec{\beta})$, here each $\vec{\varphi}$ is a v-dimensional vector which is a distribution of terms under the topic;
2. For each frame $m \in [1, M]$, we sample a mixture proportion $\vec{\theta}_k \sim Dir(\vec{\alpha})$ firstly, here each $\vec{\theta}$ is a k-dimensional vector which is a distribution of the topics for the current frame;
   - For all low-level features $n \in [1, N]$ in current frame, we first sample a topic index $z_{m,n} \sim Mult(\vec{\theta}_m)$, in which $z_{m,n}$ is topic of the current feature; then sample a visterm for the feature $w_{m,n} \sim Mult(\vec{\varphi}_{z_{m,n}})$;
   - Repeat the previous step until we get all features for the current frame;
3. Repeat the previous step until we get all frames.

Here, $\vec{\alpha}$ and $\vec{\beta}$ are the priori parameters for each Dirichlet distribution. From the topology of the Bayesian network, we can get the joint distribution of all known and hidden variables given the hyperparameters for one frame:

$$p(\vec{w}_m, \vec{z}_m, \vec{\theta}_m, \phi \mid \vec{\alpha}, \vec{\beta}) = \prod_{n=1}^{N} p(w_{m,n} \mid \vec{\varphi}_{z_{m,n}}) p(z_{m,n} \mid \vec{\theta}_m) \cdot p(\vec{\theta}_m \mid \vec{\alpha}) \cdot p(\phi \mid \vec{\beta}) \tag{10}$$

According to the model, the probability that a feature $w_{m,n}$ instantiates a particular visterm v_j given the LDA parameters is:

$$p(w_{m,n} = v_j \mid \vec{\theta}_m, \phi) = \sum_{k=1}^{K} p(w_{m,n} = v_j \mid \vec{\varphi}_k) p(z_{m,n} = k \mid \vec{\theta}_m) \tag{11}$$

Here $w_{m,n} = v_j$ means that low-feature $w_{m,n}$ is assigned to the visterm v_j through the similarity measurement. As showed in Eq. (11), we can generate a feature from the topic distribution based on the vocabulary, and generate the frame dataset we expected on a proper topic distribution. Inversely, we can infer a probabilistic distribution which is referred as a topic structure over the vocabulary if we have a steady dataset. For each frame we can have a topic assignment from its BOV representation over the topic structure. This is used as the frame representation to describe the associations between latent topics under one semantic concept in our work.

4 Experiments

In this section, we first test the approaches using Fisher vector and LDA separately to evaluate their validation for concept detection. Then we evaluate the proposed combined approach using a late-fusion fashion for Fisher vector and LDA.

4.1 Datasets and Experimental Setups

Datasets: To evaluate the performance of our approach, we use videos selected from the entire Trecvid 2013 Semantic Indexing development dataset [29], which includes approximately 3200 internet archive videos (50 GB) with creative commons licenses in MPEG-4/H.264 format with durations between 3.6 and 4.1 min. The concepts can be basically categorized as follows.

- **Object**, e.g. Airplane, Boat_Ship, Bus, Computers;
- **Scene**, e.g. Hills, Oceans, Forest, Nighttime;
- **Site**, e.g. Bridges, Highway, Office, Studio_with_Achorperson;
- **Activity**, e.g. Bicycling, Meeting, Running, Skating;
- **Event**, e.g. Demonstration_Or_Protest, Explosion_Fire, Press_Conference, Perople_Marching.

In our experiments, we divide the keyframe set of each concept into two parts, one is the training set and the other is the testing set. The evaluation criteria used here is the inferred mean average precision (mAP) [29] on the top 2000 returned detection results.

Experimental Setups: Our experiments can be divided into three steps briefly: 1) local feature extraction; 2) feature representation under two probabilistic models; and 3) training classifiers and fusing the prediction results.

In first step, we use densely sampled SIFT implementation, with the parameters setting as step = 6 pixels, scale = 1.2 and radius = 10 pixels, and at last we gain a 128-dimension descriptor by default.

In the image representation phase, we generate image descriptors both with Fisher vector and latent semantic aspects. As the dimension of Fisher Vector is linear to the dimension of local features, we use PCA to reduce the SIFT features into a lower dimension. Hence we get a 64-dimension vector for each sampled location.

We randomly select about 1M features from the training set and use them to train the PCA projection matrix and then generate the codebooks with GMMs. We use 256 Gaussians in our experiment and only the mean vector denoted by $\Gamma_{u_i^d}(X)$ according to Eq. (8) are kept as the Fisher vector. While in topic modeling, we first use the randomly selected sample features to build a vocabulary of 300 visual words and then map each feature with the soft-weighting scheme [13]. We learn 150 topics using LDA approach and describe each image with the distribuions on the learned topics.

Finally, classifiers are trained using SVM for each representation of each concept. As mentioned above, we train linear classifiers for Fisher vector and RBF classifiers for LDA model. We use 5-fold cross validation for parameter tuning. To aggregate the scores of Fisher vector and LDA results, we use the weighted fusion method.

4.2 Results and Discussions

Figure 2 shows the results for methods using *Fisher vector only*, *LDA representation only* and our proposed approach respectively.

Overall, the method of *LDA representation only* even significantly outperforms over *Fisher vector only* with about 10%. This observation suggests that as a probabilistic model under Bayesian framework, LDA topical model can capture more semantics than Fisher vector by effectively modeling the co-occurrence between low-level features, validating the practicality of LDA for concept detection. On the other hand, the inferior performance of Fisher vector representation maybe caused by the overfitting in training classifiers due to the high dimension of Fisher vectors.

From Fig. 2, we can also observe that the proposed combined approach achieves about 5% improvement over the *LDA representation only* approach. This suggests that even only one kind of low-level features are extracted, when encoded with different representations, we can gain more information. This maybe explained from the multi-view perspective, that is, the different representations reflect the different or even complementary aspects of the data, thus improve the final detection performance.

We also conduct experiments to analyze the influences of the different weights on the performance of late fusion. As Table 1 shows, our proposed approach can achieve a relatively steady and satisfying performance when the weight lies in a wide range (i.e., from 0.2~0.6). We can also notice that by incorporating the results of LDA, the final fusion results can always outperform the Fisher vector approach.

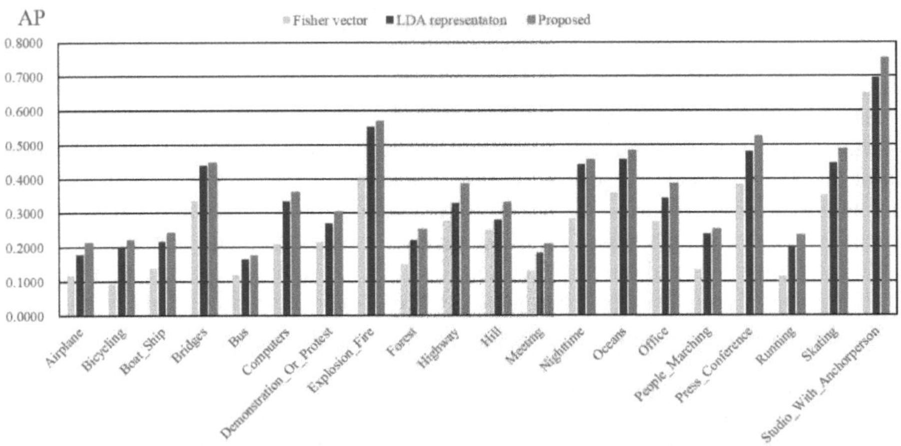

Fig. 2. Comparison of AP for each concept by different methods

Table 1. mAP for *Fisher vector only*, *LDA representation only* and different weights in the late fusion of both. (0.1..0.9 indcates the occupy of Fihser vector in late fusion).

FV	LDA	0.2	0.3	0.4	0.5	0.6	0.7	0.8
0.249	0.333	0.365	0.364	0.367	0.363	0.351	0.334	0.308

5 Conclusions

In this paper we proposed a novel high-level video concept detection method based on multi-level representation of image features, aiming at bridging the semantic gap between low-level festures and high-level semantic concept. Fisher vector and LDA are adopted as the low-level and middle-level representations respectively. Our experiments show that LDA, which can discover immediate semantics from image data, is practical for video concept detection and satisfying performance has been achieved. Moreover, by fusing Fisher vector and LDA, we have obtained superior performance than using either single representation, revealing that the low-level and middle-level representations derived from same low-level feature can gain more and complementary information. In the future, we will investigate the influence of different number of semantic topics of LDA on the detection performance, and will also study more fusion mechanisms under this promising framework.

Acknowledgements. This work was supported by National Natural Science Funds of China (61173104, 61272214) and the Open Projects Program of National Laboratory of Pattern Recognition.

References

1. Smeulders, A.W.M., Worring, M., Santini, S., Gupta, A., Jain, R.: Content Based Image Retrieval at The End of The Early Years. J. IEEE Trans. PAMI 22, 1349–1380 (2000)
2. Tang, J., Yan, S., Hong, R., Qi, G.J., Chua, T.S.: Inferring Semantic Concepts from Community-Contributed Images and Noisy Tags. In: Proc. ACM Conf. Multimedia, pp. 223–232 (2009)
3. Snoek, C.G.M., Worring, M., Gemert, J.C., Geusebroek, J.M., Smeulders, A.W.: The Challenge Problem for Automated Detection of 101 Semantic Concepts in Multimedia. In: Proc. ACM Conf. Multimedia, pp. 421–430 (2006)
4. Wang, M., Hong, R., Li, G., Zha, Z.J., Yan, S., Chua, T.S.: Event Driven Web Video Summarization by Tag Localization and Key-Shot Identification. J. IEEE Trans. on Multimedia 14, 975–985 (2012)
5. Tang, S., Zheng, Y.T., Wang, Y., Chua, T.S.: Sparse Ensemble Learning for Concept Detection. J. IEEE Trans. on Multimedia 14, 43–54 (2012)
6. Lew, M.S., Sebe, N., Dheraba, C.: Content-Based Multimedia Information Retrieval: State of the Art and Challenges. J. TOMCCAP 2, 1–19 (2006)
7. Sivic, J., Zisserman, A.: Video Google: A Text Retrieval Approach to Object Matching in Videos. In: ICCV, vol. 2, pp. 1470–1477 (2003)
8. Csurka, G., Dance, C.R., Fan, L., Willamowski, J., Bray, C.: Visual Categorization with Bags of Keypoints. In: ECCV Workshop, pp. 1–22 (2004)
9. Quelhas, P., Monay, F., Odobez, J.M.: Modeling Scenes with Local Descriptors and Latent Aspects. In: ICCV, vol. 1, pp. 883–890 (2005)
10. Fei-Fei, L., Perona, P.: A Bayesian Hierarchical Model for Learning Natural Scene Categories. In: Proc. CVPR, vol. 2, pp. 524–531 (2005)
11. Farquhar, J., Szedmak, S., Meng, H., Shawe-Taylor, J.: Improving "Bag-of-Keypoints" Image Categorisation. J. Technical report, University of Southampton (2005)

12. Perronnin, F., Dance, C.R., Csurka, G., Bressan, M.: Adapted vocabularies for generic visual categorization. In: Leonardis, A., Bischof, H., Pinz, A. (eds.) ECCV 2006. LNCS, vol. 3954, pp. 464–475. Springer, Heidelberg (2006)
13. Jiang, Y.G., Ngo, C.W., Yang, J.: Towards Optimal Bag-of-Features for Object Categorization and Semantic Video Retrieval. In: CIVR 2007, pp. 494–501 (2007)
14. Bosch, A., Zisserman, A., Munoz, X.: Image Classifcation Using Random Forests and Ferns. In: Proc. ICCV, pp. 1–8 (2007)
15. Li, H., Wang, X., Tang, J., Zhao, C.: Combining Global and Local Matching of Multiple Features for Precise Retrieval of Item Images. J. ACM/Springer Multimedia System Journal 19, 37–49 (2013)
16. Jaakkola, T.S., Haussler, D.: Exploiting Generative Models in Discriminative Classifiers. Advances in Neural Information Processing Systems 11, 487–493 (1999)
17. Jaakkola, T.S., Haussler, D.: Probabilistic kernel regression models. In: Proceedings of the 1999 Conference on AI and Statistics (1999)
18. Jegou, H., Douze, M., Schmid, C., Perez, P.: Aggregating local descriptors into a compact image representation. In: CVPR, pp. 3304–3311 (2010)
19. Perronnin, F., Dance, C.: Fisher Kernels on Visual Vocabularies for Image Categorization. In: CVPR 2007, pp. 1–8 (2007)
20. Sun, C., Nevatia, R.: Large-scale Web Video Event Classification by use of Fisher Vectors. In: WACV, pp. 15–22 (2013)
21. Csurka, G., Perronnin, F.: Fisher Vectors: Beyond Bag-of-Visual-Words Image Representations. In: Richard, P., Braz, J. (eds.) VISIGRAPP 2010. CCIS, vol. 229, pp. 28–42. Springer, Heidelberg (2011)
22. Vailaya, A., Figueiredo, M.A.T., Jain, A.K., Zhang, H.J.: Image Classification for Content-based Indexing. J. IEEE Transactions on Image Processing 10, 117–130 (2001)
23. Naphade, M.R.: A Probabilistic Framework for Mapping Audio-visual Features to High-Level Semantics in Terms of Concepts and Context. Dissertation of the University of Illinois at Urbana-Champaign (2001)
24. Hinton, G.E., Krizhevsky, A., Sutskever, L.: ImageNet Classification with Deep Convolutional Neural Networks. J. NIPS, 1106–1114 (2012)
25. Blei, D.M., Ng, A.Y., Jordan, M.I.: Latent Dirichlet Allocation. J. JMLR, 993–1022 (2003)
26. Lienou, M., Maitre, H., Datcu, M.: Semantic Annotation of Satellite Images Using Latent Dirichlet Allocation. J. IEEE 7, 28–32 (2010)
27. Snoek, C., Worring, M., Smeulders, A.: Early Versus Late Fusion in Semantic Video Analysis. In: Proc. ACM Int. Conf. Multimedia, Singapore, pp. 399–402 (2005)
28. Tang, S., Li, J.T., Li, M., Xie, C., Liu, Y.Z.: TRECVID 2008 High-Level Feature Extraction By MCG-ICT-CAS. In: Proc. TRECVID 2008 Workshop, Gaithesburg, USA (2008)
29. TREC Video Retrieval Evaluation, `http://trecvid.nist.gov/`

The Impact of Global and Local Features on Multiple Sequence Alignment Clustering-Based Near-Duplicate Video Retrieval

Yandan Wang[1], Guojun Lu[2], Mohammed Belkhatir[3], and Christopher H. Messom[1]

[1] Faculty of Information Technology, Monash University, Sunway Campus, Malaysia
{yandan.wang,c.h.messom}@monash.edu
[2] Faculty of Information Technology, Monash University, Gippsland Campus, Australia
guojun.lu@monash.edu
[3] Faculty of Computer Science, University of Lyon, Campus dela Doua, France
mohammed.belkhatir@univ-lyon1.fr

Abstract. Traditionally, the performance of Near-Duplicate Video Retrieval (NDVR) is enhanced through different video features, matching scheme and indexing methods. The video features have been intensively investigated and it has been shown that local features outperform global features in terms of accuracy. However, local features have the expensive computational problem. Therefore, indexing structure is introduced to assist in scaling up, whilst the accuracy will drop slightly or dramatically in most time by using indexing approaches. Recent progress shows that NDVR based on clustering could reduce searching space while maintains equivalent retrieval accuracy compared to that of non-clustering based. In this paper, we will continue to evaluate clustering based NDVR, but using popular global and local features. Before conducting NDVR, dataset will be pre-processed offline into groups by using clustering algorithm that near-duplicate videos (NDVs) are assembled in the same cluster. Each cluster will be represented by member video or the centroid. The query video will then be compared to the representative videos instead of all videos in database (non-clustering based). Our experiment shows that clustering-based NDVR using global and local features outperforms than that of non-clustering based in terms of both retrieval accuracy and speed.

Keywords: Clustering, Near-duplicate video retrieval, Visual words.

1 Introduction

With the advancement of the internet technology, web videos are uploaded all over the world with exponential growth. Meanwhile, Near-duplicate videos are collected and rising rapidly due to the flooding video edit softwares. Authors in [3] reported that based on a sample of 24 popular queries from YouTube, Google Video and Yahoo! Video, on average there are 27% redundant videos that are duplicates or near duplicates to the most popular version of a video in the search results. Based on this observation, it is important to investigate approaches to manage video collections, and take the advantage of video management to improve the performance of NDVR. Authors in [5] developed a

B. Huet et al. (Eds.): PCM 2013, LNCS 8294, pp. 669–677, 2013.

clustering framework that is based on Multiple Sequence Alignment (MSA) to group NDVs in the same cluster and the centroid is computed to represent each cluster. Then a query video is compared to all centroids instead of all videos in the collection. Videos are pre-clustered offline as prior dataset process step. In such a process, the retrieval speed is obviously enhanced while maintaining the retrieval accuracy compared to that of non-clustering based NDVR. Compared to the traditional approach that improves NDVR performance through video feature representations [2] [7], matching schemes [2] [3], and indexing algorithms [2] [3] [7], pre-process videos offline is simpler and less complicated. The advantage of clustering-based NDVR is that the pre-clustering is processed offline and the complexity of clustering algorithm will not affect the retrieval speed.

In [5], the authors show that clustering-based NDVR outperforms that of non-clustering based. However, the evaluation result sounds only by using the ordinal features, which divides each keyframe into 4 blocks and forms video histogram by considering the blocks ranking permutations as histogram bins. Thus, we have no knowledge whether clustering-based NDVR outperforms that of non-clustering based sounds for some popular features. Therefore, in this paper, we will continue evaluate clustering-based NDVR compared to that of non-clustering based by using most popular visual global and local features. We will show the impact of most popular global and local features on clustering-based NDVR compared to that of non-clustering based. The comparison is under the same condition for both clustering and non-clustering based, for example using the same threshold, same matching method etc. The results show us that clustering-based NDVR outperforms that of non-clustering based by using popular global and local features. Thus, it is promising and feasible to apply clustering algorithm as pre-process step and database management tool for improving the performance of NDVR.

The rest of the paper is organized as following. Section 2 presents the clustering-based NDVR framework. The global and local features are introduced in Section 3. Section 4 describes how to represent the cluster. Distance/similarity measurement is detailed in Section 5, and Section 6 presents the experimental results. Section 7 concludes the paper.

2 Clustering-Based NDVR Evaluation Framework

Fig.1 shows the clustering-based NDVR framework. The difference between clustering-based and non-clustering based is that clustering-based NDVR will pre-process videos offline and group NDVs in the same cluster by using a clustering algorithm, which we use MSA clustering method in this paper. Instead of comparing the query video to all videos in database in non-clustering based framework, we only have to compare query video to the cluster representations. The number of comparisons is reduced from the number of videos in database to the number of clusters. Therefore, searching space is obviously reduced. However, it is important for us to know whether it could maintain or enhance retrieval accuracy while retrieval speed is significant faster than that of non-clustering based. Our focus of this paper is to evaluate the impact of features on clustering-based NDVR compared to that of non-clustering based, rather than how to do the clustering. We will evaluate directly on the clusters processed by the MSA clustering method introduced in [5].

In next section, we will describe global and local features that we are going to evaluate for clustering-based NDVR.

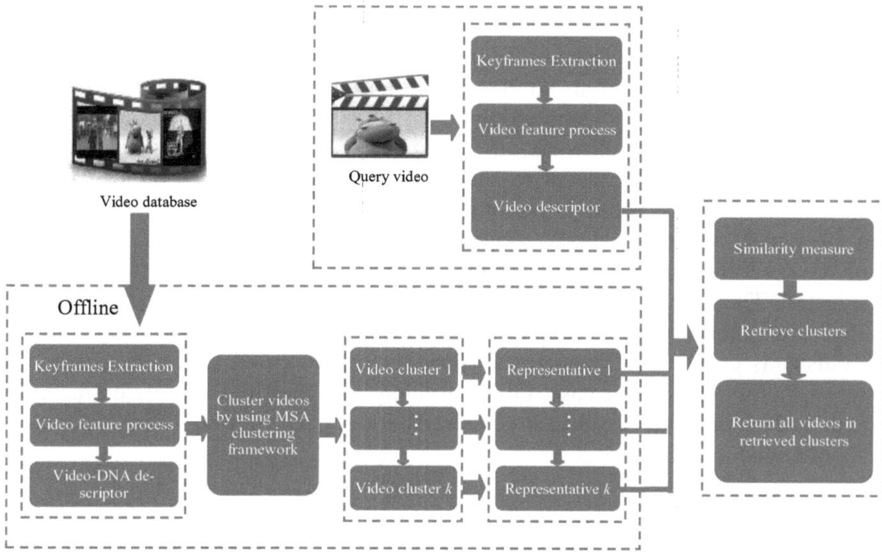

Fig. 1. MSA Clustering-based NDVR framework

3 Video Features Representation

Color and texture are the most common global features, and Scale-Invariant Feature Transform (SIFT) is the most popular local feature. In this paper, we will evaluate scalable color, edge histogram and SIFT for clustering-based NDVR.

3.1 Global Features

Color and texture are the most common global visual features. There are various color and texture types [4]. In this paper, we will evaluate scalable color and edge histogram. Thus each keyframe could be represented by either a scalable color descriptor or an edge histogram. Therefore, a video is described as a sequence of vector descriptors.

Scalable color is a color histogram that is quantized in the HSV color space and then encoded by Haar transform. City-block distance also known as Manhattan distance is the best metric to measure the distance between two color histograms [4].

Edge histogram describes the spatial distribution of five types of directional and non-directional edges. Edge histogram outperforms color descriptors. Metric such as Euclidean distance and cosine similarity could be used to measure the distance/similarity between two edge histograms.

3.2 Local Features

Local scale-invariant features (SIFT) [1] was proposed in 1999 by Lowe, D.G. It is still the most popular local feature. Local features outperform global features in terms of accuracy. However, it suffers from expensive computational complexity problem. Accordingly, Sivic and Zisserman [2] proposed text retrieval method by using SIFT, which using k-means to group similar keypoints in the same cluster and mapping keypoints in image/video to closest cluster to form the bag of words (BoW) to describe the image/video. Following this paper, there are many other papers [3] proposed to apply the BoW concept to solve the real-time NDVR in large-scale video collections problem and achieves rather good performance.

In this paper, we employ BoW concept to process the SIFT feature to video histogram. We randomly select one representation video for each video cluster. 708 videos are selected. Then we randomly select 40000 keypoints descriptor from the 708 videos. K-means then applied to cluster 40000 keypoints with k=200. The centroid of each cluster is computed. The size of video histogram equals the size of keypoints clusters which is 200 in our case. Given a video with the SIFT feature extracted, we map each SIFT point to the closest cluster. The number of points fall in each cluster bin is computed.

4 Cluster Representations

The quality of cluster representation is connected to the NDVR performance in terms of accuracy. Two ways are always used to represent the cluster: 1) Select one or more than one videos to represent the cluster. 2) Compute centroid of the cluster to represent the cluster.

4.1 Representative Video Selection

In our experiment, for global features (color and textures), we select one single video to represent the cluster. For each video in the cluster, we compute the number of nearest videos that meet the required distance/similarity threshold. The video with the greatest number of nearest videos will be selected as the representative video of the cluster.

4.2 Cluster Centroid Representation

If the video is represented in the form of single vector (BoW of SIFT feature in our experiment), the cluster centroid is computed to represent the cluster.

5 Video Similarity Measurement

In this section, we will describe the similarity measure between two videos. Jaccard Similarity is employed to measure videos described by global features and cosine similarity is applied to score the similarity between two videos represented by local features (video histrogram of BoW SIFT).

5.1 Global Features Similarity Measure

We use Jaccard similarity (Equation 1) to measure similarity between videos represented by color and texture features. Each keyframe is represented by a color/texture vector, thus video is described by a sequence of color/texture vectors. Absolute city-block distance is used to score the distance between two color vectors (scalable color) and cosine similarity is applied to score the similarity between two texture vectors (edge histogram). We detail the measurement in algorithm 1.

$$Sim(V_1, V_2) = \frac{|V_1 \cap V_2|}{|V_1 \cup V_2|} \tag{1}$$

Algorithm 1. Jaccard similarity measure between videos

INPUT: two sets of video color/texture vectors descriptors V1 and V2 with length L1 and L2 respectively.

Let variable *X* be the number of keyframes intersection in two videos, *T* be the threshold of city-block distance or cosine similarity.

```
LOOP all of keyframes in V1
                // score = 100 (if distance metric applied);
    Var score=-1(similarity metric applied)
    LOOP rest of keyframes in V2
                // (Var dis=city-block(f1, f2) if scalable color)
        Var sim=cosineSimilarity(f1, f2)
                // (score>dis && dis<=T) when color distance applied
        IF (score<sim && sim>=T)
                score=sim/dis and record the current keyframe index in V2
        ENDIF
    ENDLOOP
                // score1!=100 (if distance)
    IF (score!=-1) {
            X++; remove the keyframe descriptor at the preserved index
    ENDIF
ENDLOOP
get intersection=X; and union= L1 + L2 –intersection
apply the jaccard equation (1) and get the similarity
```

OUTPUT: intersection/union, which is similarity between V1 and V2

5.2 Local Features Similarity Measure

We employ SIFT for our local feature evaluation and process SIFT to video histogram by using BoW. Thus distance/similarity metric such as Euclidean distance, cosine similarity could be simply applied to measure the distance/similarity between two video histograms. In our case, we employ cosine similarity metric. Given two video histograms V^1 and V^2, the cosine similarity is measured by the equation (2), where k is the length of the video vector.

$$Similarity(V^1, V^2) = \frac{\sum_{i=0}^{k-1} V_i^1 \times V_i^2}{\sqrt{\sum_{i=0}^{k-1} \left(V_i^1\right)^2} \times \sqrt{\sum_{i=0}^{k-1} \left(V_i^2\right)^2}} \quad (2)$$

6 Experiment

We select 5000 videos and 120 query videos from CC_WEB_VIDEO [6] for our experiment. The 5000 experimental dataset consists of 3789 near duplicate videos and 1211 noise videos [5]. The example videos could be found in [6]. We use the MSA-based clustering technique to group videos into 807 clusters by using video-DNA feature [5]. Then we represent clusters by using global and local features as described in Sections 4.1 and 4.2 respectively. The global and local features are also used to evaluate the NDVR based on clustering compared to that of non-clustering based. Given a query video, instead of comparing to all 5000 videos (non-clustering based NDVR), we only have to compare to 807 cluster representations (clustering based NDVR). We can clearly see that the clustering-based physically reduces the size of dataset offline and slim down the searching space. We measure the similarity between videos depend on the usage of global and local features as described in Sections 5.1 and 5.2 respectively. Three experiments are conducted, which compares the performance of clustering based NDVR to that of non-clustering based by using color, texture, and SIFT respectively. Each comparison is fulfilled fairly by following conditions: i) same video feature processing (color, texture and SIFT); ii) same similarity measure method with same threshold; iii) 120 video queries are processed sequentially in the same manner; iv) the experiment is conducted on a standard PC with an Intel(R) Core(TM) 2.67GHz processor and 2 GB of RAM.

In Section 6.1, we discuss the evaluation results of NDVR based on clustering by using global features edge histogram and scalable color. In Section 6.2, we present the evaluation result of NDVR based on clustering by using local features SIFT compared to that of non-clustering based.

6.1 MSA Clustering-Based NDVR with Global Features

In this section, we show the NDVR performance based on clustering compared to that of non-clustering based by using texture and color respectively. 120 query videos are submitted and the average precision-recall retrieval curves are presented.

Fig.2 shows the performance by using texture (edge histogram) in terms of retrieval accuracy and speed. The left figure compares the average precision-recall curve of NDVR based on clustering (with clusters) to that of non-clustering based (no clusters). The figure clearly shows that the clustering based performs much better than non-clustering based before the recall 0.9. The curve is superimposed after the recall 0.9. Overall, clustering-based NDVR is in much better retrieval accuracy than that of non-clustering based. Meanwhile, the figure on the right shows time consumed corresponding to the number of query copies processed. The clustering based (blue curve) NDVR cost only little time. The red curve of non-clustering NDVR consume more time. With the number of query videos are processed, the time consumed by non-clustering based NDVR climbs much faster than that of clustering based. When 120 videos are process, clustering based NDVR performs more than 6 times faster than that of non-clustering based. In summary, by using texture feature, clustering based NDVR performs much better than that of non-clustering based in terms both of retrieval accuracy and speed.

Fig.3 shows NDVR performance by using color (scalable color) in terms of retrieval accuracy and speed. Same as Fig.2, the left figure shows the average NDVR precision-recall curve based on clustering and non-clustering. Clearly, the blue curve of clustering-based NDVR significantly outperforms the red curve of non-clustering based NDVR before the recall 0.8. However, after the recall 0.8, the two curves found no significant difference and almost superimposed. Overall, clustering based stands significant better retrieval accuracy. Meanwhile, the figure in the right presents the retrieval speed, which shows the time consumed corresponding to the number of query videos processed. We can see that NDVR based on clustering (blue curve) consumes little time only when all 120 query videos are processed and the curve is fairly flat, which explains with the number of query videos increased, the consumed time climbs much slower than that of non-clustering based. By observing the slope of two curves, we can see that the non-clustering based NDVR, the incensement of time consumed of processing every single query video is more than 2 times than that of clustering based NDVR. When all 120 query videos are processed, the clustering based NDVR performs around 6 times faster than that of non-clustering based. In summary, by using the color feature, NDVR based on clustering also significantly outperforms non-clustering based in terms of retrieval accuracy and speed.

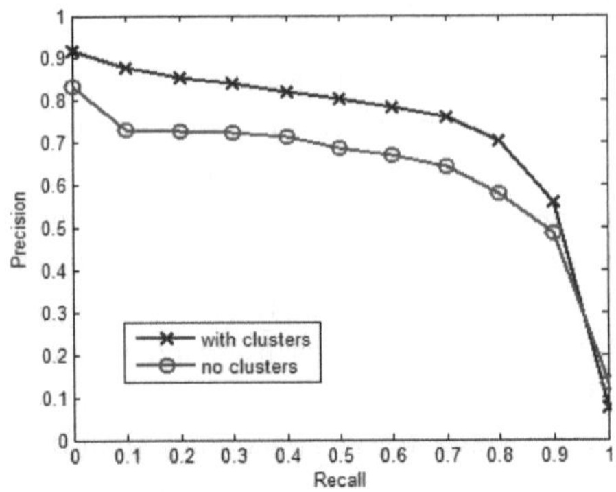

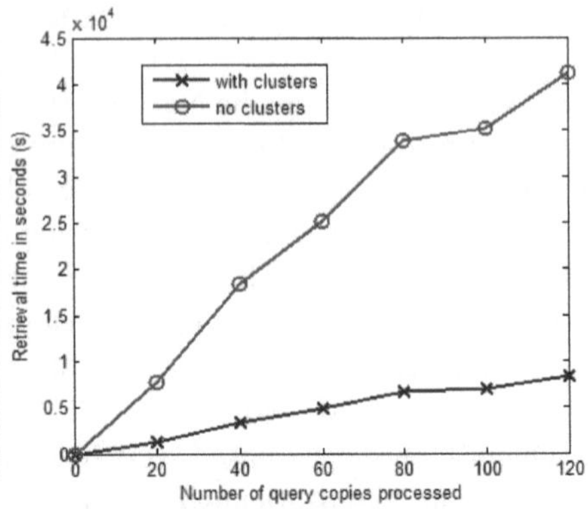

Fig. 2. Precision-recall of NDVR by using edge histogram and time consumed corresponding to the number of query copies processed

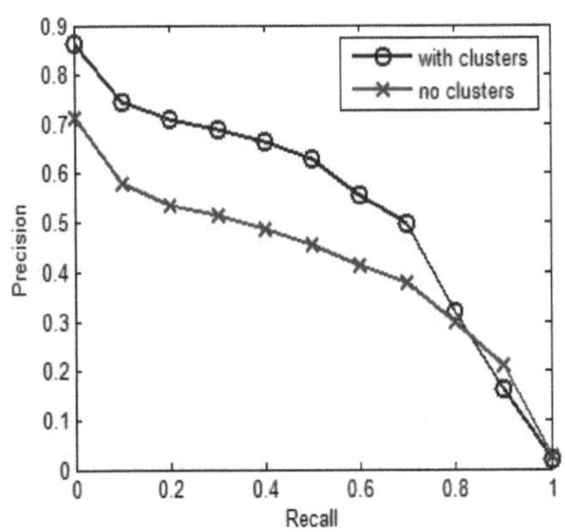

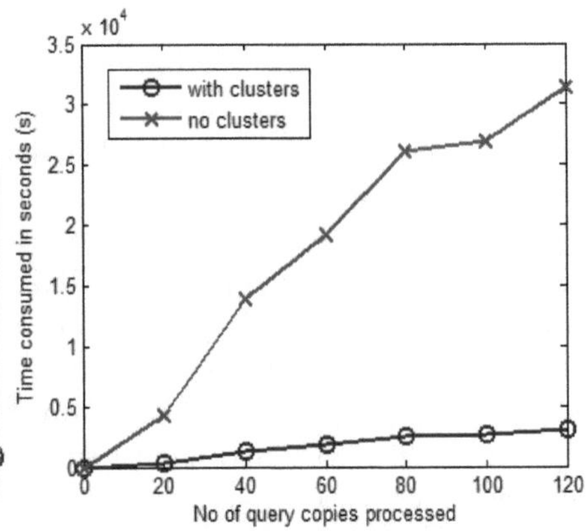

Fig. 3. Precision-recall of NDVR by using scalable color and time consumed corresponding to the number of query copies processed

6.2 MSA Clustering-Based NDVR with Local Features

In this section, we will present the NDVR performance in terms of accuracy and speed by using sift feature. Same as last section, 120 query videos are submitted for evaluation. The cosine similarity is employed to measure the similarity between videos (Section 5.2).

Fig.4 shows the NDVR performance in terms of precision-recall accuracy and speed in left and right figures respectively in the same showing manner in Section 6.1. As we can see in the right figure, the time consumed by non-clustering based (red) is still greater than that of clustering based (blue), while the accuracy of two precision-recall curves remains at least equivalent. Overall, two precision-recall curves are found no significant difference. However, the retrieval speed is dramatically enhanced by using clustering. When 120 query videos are processed, clustering based NDVR is around 4 times faster than that of non-clustering based.

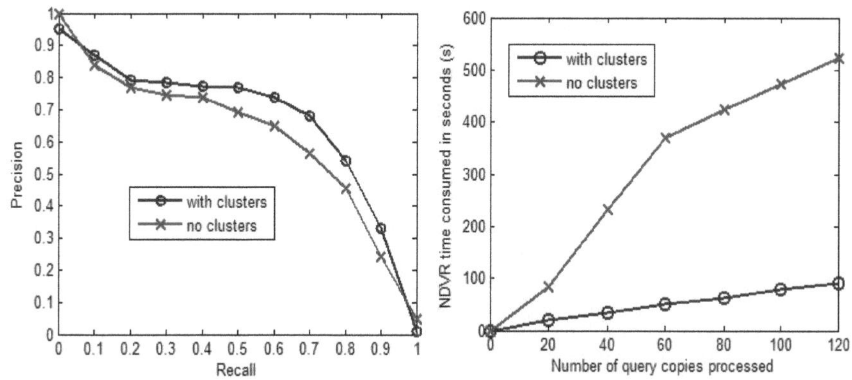

Fig. 4. Precision-recall of NDVR by using sift and time consumed corresponding to the number of query copies processed

7 Conclusion

In this paper, we showed that NDVR based on clustering outperforms that of non-clustering based in terms of retrieval accuracy and speed by using popular global and local features. By showing this, we clearly see that enhancing NDVR through clustering by using popular global and local features is feasible. By using clustering to pre-process video dataset offline, we can reduce the size of database offline, which greatly enhance NDVR performance. Accordingly, researchers could consider pre-clustering video collections offline first before the common NDVR process (matching and indexing process), since pre-process the dataset could greatly enhance NDVR performance in terms of both accuracy and speed. In future works, we will continue evaluating the impact of different clustering algorithms on clustering-based NDVR.

References

1. Lowe, D.G.: Object recognition from local scale-invariant features. In: The Proceedings of the Seventh IEEE International Conference on Computer Vision, pp. 1150–1157 (1999)
2. Sivic, J., Zisserman, A.: Video Google: A text retrieval approach to object matching in videos (2003)
3. Wu, X., Hauptmann, A.G., Ngo, C.W.: Practical elimination of near-duplicates from web video search. In: Proceedings of the 15th International Conference on Multimedia, pp. 218–227 (2007)
4. `http://mpeg.chiariglione.org/standards/mpeg-7/visual`
5. Wang, Y., Belkhatir, M., Tahayna, B.: Near-duplicate video retrieval based on clustering by multiple sequence alignment. In: Proceedings of the 20th ACM International Conference on Multimedia, pp. 941–944 (2012)
6. `http://vireo.cs.cityu.edu.hk/webvideo/`
7. Cai, Y., Yang, L., Ping, W., Wang, F., Mei, T., Hua, X.S., Li, S.: Million-scale near-duplicate video retrieval system. In: Proceedings of the 19th ACM International Conference on Multimedia, pp. 837–838 (2011)

Saliency Detection-Based Mixture of Reality and Non-Photorealistic Rendering Effects for Artistic Visualization

Zhipeng Wu and Kiyoharu Aizawa

Dept. of Information and Communication Eng., The University of Tokyo, Tokyo, Japan
{zhipengwu,aizawa}@hal.t.u-tokyo.ac.jp

Abstract. In this paper, we bring the novel idea to automatically combine Non-Photorealistic Rendering (NPR) effects with real-world images based on saliency detection. Noticing that the key idea of NPR is to focus on enabling a wide variety of expressive styles for digital visual art (e.g. painting, sketch, and cartoon), such mixture of Reality and NRP effects always provides an extremely intriguing sense of art beyond the original content. Technically, given an input image, we devote a fast approach to convert it into manga or pencil sketch on-the-fly. Moreover, guided by a hierarchical saliency detection strategy, the mixture of NPR effects and Reality can be finished in the most effective way. On the other hand, the proposed 'RealMeetsArt' system also provides the function to let user manually select interested foreground regions. User can easily select a fine-grained area with only several stroke drawings.

Keywords: Non-Photorealistic Rendering, Saliency Detection, RealMeetsArt.

1 Introduction

Image is the core of multimedia. As old saying says, "a picture is worth a thousand words", comparing with plain words, a piece of image always encodes more information and becomes better acceptable to users. Meanwhile, a lot of research endeavors focused on enhancing the expressive power of image, adding a sense of artistic feelings, and even to make the image "speak". Recent research progress in Non-Photorealistic Rendering (NPR) amazingly enables us to automatically convert an image into various kinds of art forms [1], [2].

While standing on the bridge of NPR, our left side is the real-life image, and right side is the artistic painting. According to recent news from The Daily Mail, rather than completely adopting real images or art paintings, the so-called "half-real image composition", which is the combination of real and artificial effects, sometimes yields a surprisingly intriguing feeling [3] (Fig. 1).

Inspired by this idea, we proposed the 'RealMeetsArt' system to generate artistic images as well as create 'half-real compositions'. Our contribution is three-fold:

1. We propose fast algorithms to convert input image into high-quality sketch or manga. Our approach outperforms previous works by its higher conversion quality and real-time processing capability, which provides better user experience.

B. Huet et al. (Eds.): PCM 2013, LNCS 8294, pp. 678–689, 2013.

2. Motivated by the hierarchical saliency detection strategy [4], we propose an enhanced multi-layer saliency model with special focus on high-level attention priors. The introduction of face and skin prior makes our model particularly capable when coping with images contain human.
3. We provide a user-interactive mode to help the user manually label foreground regions. With the help of super-pixel segmentation, the user can easily select a fine-grained area with only several stroke drawings.

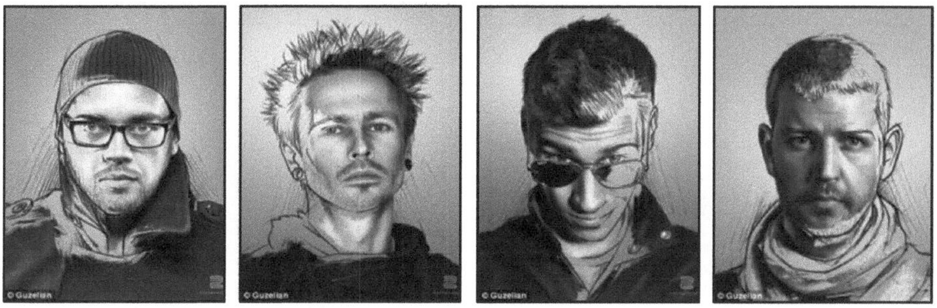

Fig. 1. Artists took self-portraits and sketched themselves into the pictures. The mixture of art and reality creates an extremely intriguing composite effect [3].

Fig. 2 shows a snapshot of the proposed approach. In this case, the image is automatically converted into manga and further mixed with the foreground pixels returned by the saliency detector. Please refer to Section 3 for more details.

Our paper is organized as the follows. In Section 2, we briefly review the Non-Photorealistic Rendering (NPR) module which functions to add a visually amiable artistic effect (pencil sketch or manga) for the original image. Section 3 and Section 4 further describe the saliency guided half-real image composition and its alternative: user-interactive image composition. The details about the hierarchical saliency model and high-level attention priors are discussed as well. In Section 5, the results with respect to processing time and visualized examples are shown to the reader.

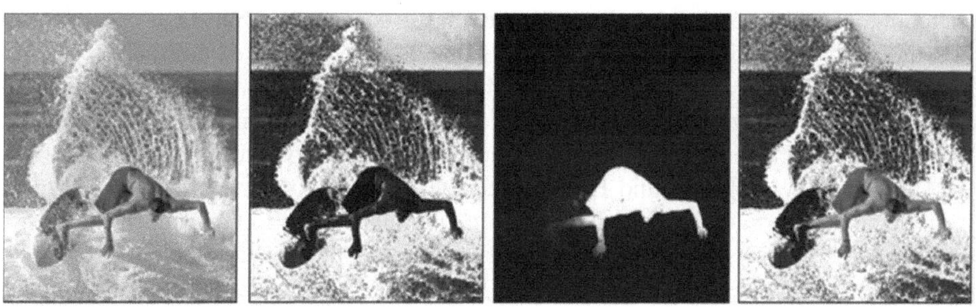

(a) (b) (c) (d)

Fig. 2. Automatically combine real images and non-photorealistic rendering effects. (*a*): Input image. (*b*): Manga rendering result. (*c*) Saliency map. (*d*) Half-real image composition.

(a) (b)

(c) (d)

Fig. 3. Non-Photorealistic Rendering results. (*a*): Input image. (*b*): Manga rendering result. (*c*) Pencil sketch rendering result. (*d*) Color pencil sketch rendering result.

2 Non-Photorealistic Rendering

In visual art, pencil sketch and manga play important roles to abstract human perception of natural scenes (Fig. 3). Previous works for NPR-based sketch synthesis generally fall into two categories: model-based and image-based sketching. In model-based sketching, researchers simulated pencil drawing by studying the physical properties of 3-D model. They draw outlines on every visible edge [5], detected contours from a 3-D surface, perturbed them and generated textures with directional strokes to express shading [6]. For image-based sketching, Mao et al. detected local structure orientation from the input image and incorporated Linear Integral Convolution (LIC) to produce the shading effect [7]. Li and Huang used the feature geometric attributes obtained from local-region moments and textures to simulate pencil drawing [8]. As we know, sketch drawing is an inherently difficult problem because not only structural information should be selectively preserved but as well the appearance should approximate that of real pencil drawings. In this paper, we follow Lu's recent work [1], which proposed a two-stage system combining both line and tonal drawing.

2.1 Sketch Drawing

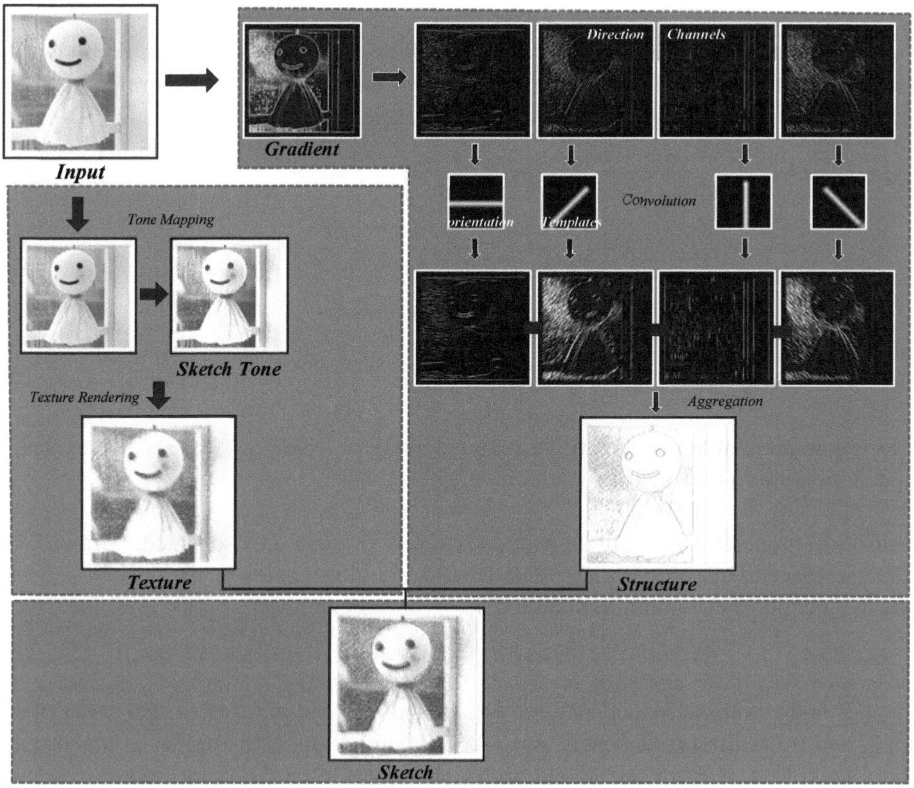

Fig. 4. Sketch drawing

Structure. The extraction of image structure is done in an efficient way. We first compute the gray-scale of the input image, yielding magnitude:

$$G = ((\partial_x I)^2 + (\partial_y I)^2)^{\frac{1}{2}} \tag{1}$$

Then the gradient map is divided into different channels $\{C_1, C_2, \ldots, C_8\}$ according to the local direction. Finally, we get the image structure $\boldsymbol{S}$ by aggregating the convolution between direction channels and their corresponding orientation templates $\{T_1, T_2, \ldots, T_8\}$ (Fig. 4, blue region):

$$S = I - \sum_{i=1}^{8} (T_i \otimes C_i) \tag{2}$$

Tone. The tone distribution of a gray-scale photo generally differs significantly from that of the pencil sketch. By learning the tonal characteristics from real pencil sketches, [1] proposed a parametric model P to map real-world photos to sketch-tonal images.

$$P(v) = \frac{1}{Z}\sum_{i=1}^{3} \omega_i p_i(v) \tag{3}$$

where v is the tone value and Z is the normalization factor. The three components $p_i(v)$ account for the three tonal channels (bright, middle, dark) and ω_i refers to their weightings. For [0, 255] gray-scale image, we define them as:

Bright Layer

$$p_b = \frac{1}{z_b} exp(\frac{i-255}{9}) \tag{4}$$

Middle Layer

$$p_m = \begin{cases} \frac{1}{z_m}(u_a - u_b) & \text{if } u_a \le i \le u_b \\ 0 & \text{otherwise} \end{cases} \tag{5}$$

Dark Layer

$$p_d = \frac{1}{z_d} exp(\frac{-i}{9}) \tag{6}$$

In our implementation, weighting parameters $W = \{\omega_b{:}52, \omega_m{:}37, \omega_d{:}11\}$ gives a good simulation to real-world sketch drawings.

Texture. In human drawing, tonal pencil texture is generated by repeatedly drawing at the same place. This process is simulated using multiplication of strokes, which results in an exponential combination model:

$$H(x)^{B(x)} \approx J(x) \tag{7}$$

where H is an example pencil pattern, $B(x)$ stands for the number of repeats at position x, and J is the result pencil tone. To efficiently apply this model to our framework, we collect H from real-world art works, resize it to the input image size, and approximate matrix B by:

$$B(x) = \frac{In(I(x))}{In(H(x))} \tag{8}$$

where I is the result image after tone mapping the original photo. The combination of structure $\boldsymbol{S}$ and texture $\boldsymbol{T}$ finally yields out the result for sketch drawing $\boldsymbol{R}$:

$$\boldsymbol{R} = \boldsymbol{S} \cdot \boldsymbol{T} \tag{9}$$

2.2 Manga Rendition

Originated in the late 19th century, as a unique form of art, Manga has won a worldwide popularity. Recently, with the rapid progress of science and technology, computational manga quickly won significant reputation. Qu et al. proposed a novel approach for generating manga-style screen patterns with input color images [9]. Their goal is to preserve the visual richness in the original photograph by utilizing not only screen density, but also the variety of screen patterns. The combination of tone similarity, texture similarity, and chromaticity distinguishability ensures the high quality of image to manga conversion, especially the rendition of manga screening. However, restricted to several time-consuming procedures (segmentation, texture matching, color-to-pattern matching, tone matching, and minimal user control), this work is not feasible for real-time applications. Given an 800×1000 input image, it takes around four minutes on a P4 3.2GHz CPU [9].

While taking both of the efficiency and effectiveness into consideration, we propose a lightweight manga rendition framework which provides high-quality rendering results. As shown in Figure 5, the core Image2Manga module is a combination of structure and texture processing. The structure extraction step aims to provide a visual abstraction of the image content (flow-based DoG) as well as filter out noise and trivial details (vectorization). In order to reproduce the bitonal screen pattern, we propose a multi-layer tone mapping and halftoning strategy. The manga texture is further obtained by merging multi-layer results with respect to manga images' characteristics.

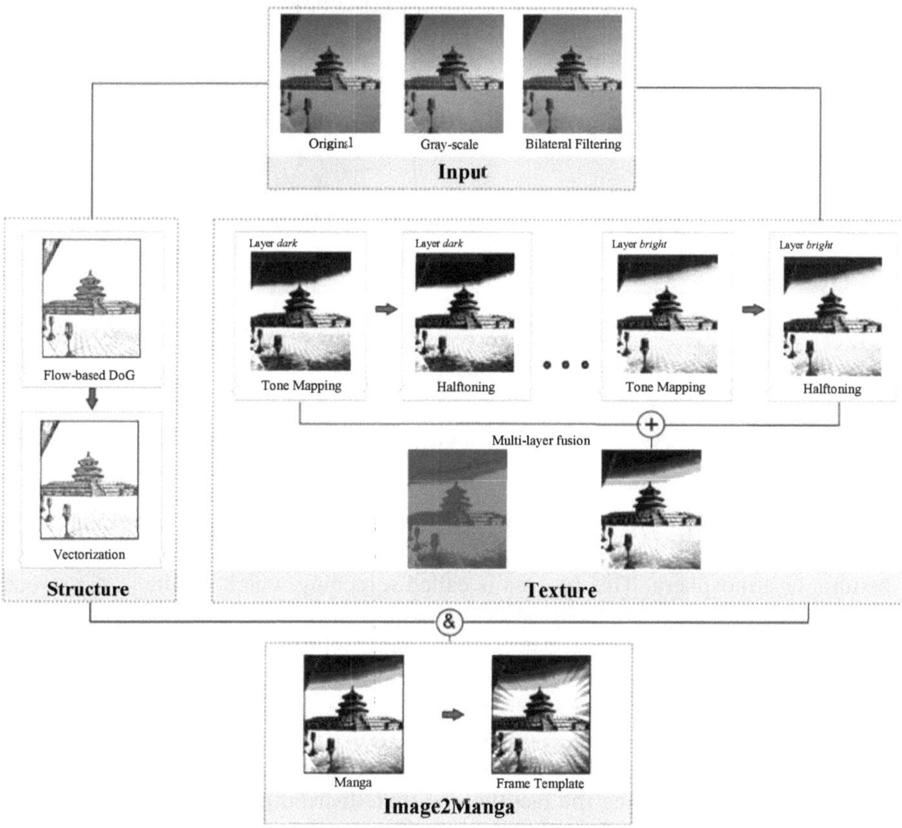

Fig. 5. Manga Rendition

Structure. In general, manga production starts with drafting outlines, and further deepens the reader-impression by summarization, generalization, and exaggeration of daily scenes. Likewise, our structure extraction scheme should not only keep the major components of the original image but also eliminate the noises and trivial details. Comparing with the classical edge detection problem, our task requires two factors: (1) Filtering non-obvious lines to simplify the visual cues. (2) Enhancing major contours to convey scene aspects. Line drawing has drawn a lot of attention in recent NRP research. As proposed in [10], the main idea of line drawing was to take into account the 'flow' of edges. Rather than conducting classical 2-D DoG operations to extract edge responses, it was shown that we could receive the biggest contrast while applying 1-D DoG operator along the perpendicular direction to local 'edge flow'. Then, after flow-based DoG filtering, the edge line drawing is further vectorized to remove noise and trivial details.

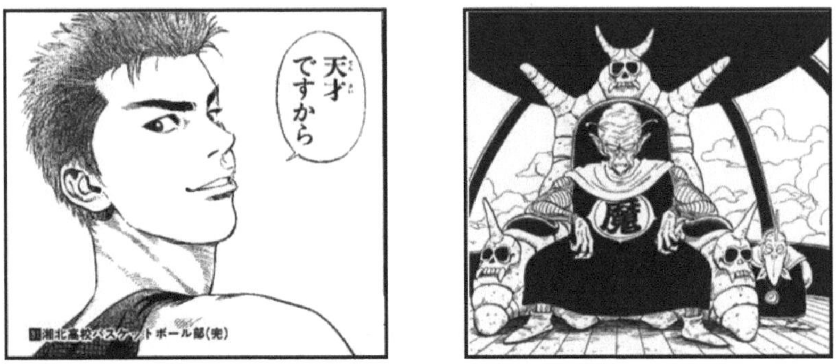

(a) Manga Image from *SLAMDUNK* (b) Manga Image from *DRAGON BALL*

Fig. 6. Genuine manga images

Texture. Comparing with other non-photorealistic art forms, manga has two common characteristics:

- *Screentone Pattern*: After drawing structure lines, manga artists usually select appropriate pre-print screen sheets to fill regions in order to express shading, tone, texture, or atmosphere. This process is called screening which results in the screentone pattern in manga productions. In digital processing, the conversion from grayscale image into binary image, namely halftoning, reproduces the screentone texture pattern. We employ Bayer templates [11] as a fast halftoning implementation.
- *White/Black Regions*: Another key observation of manga images is the prevalent existence of solid white/black regions (Figure 6). Since manga is a highly abstract art form, these 'Flat Regions' are used to express not only full white/black but also bright/dark areas. Noticing the fact that the tone distribution of manga images generally differ significantly from daily photos, we employ a multi-layer tone mapping and halftoning method which efficiently yields stable manga-like texture patterns.

The final manga $\boldsymbol{M}$ is simply the pixel-wise '&' combination of structure $\boldsymbol{S}$ and Texture $\boldsymbol{T}$ (Fig. 5). We can add some pre-defined frame templates as well.

3 Saliency Guided Image Composition

Striking the balance between NPR artistic effects and real-world image content has never been a trivial problem. Inspired by Fig. 1, we tend to keep the foreground, or salient regions unchanged while convert the background areas into visual art (sketch, manga). Therefore, the problem is degenerated into a classical topic in multimedia and computer vision – saliency detection [12, 22].

Saliency detection, which is closely related to selective processing in human visual system [13], aims to locate important regions or objects in images. Early works of saliency analysis focused on local information [14]. Researchers directly computed center-surrounding color difference in a fixed neighborhood for each pixel [15]. On the other hand, global methods mostly consider color statistics [16]. Recently, hierarchical saliency model was proved to be extremely effective when dealing with objects with complex structures [4]. Their basic idea is to analyze saliency cues from multiple levels of structure, and then integrate them to infer the final saliency map. Such a hierarchy model finds foundation from studies in psychology, which believes that the human attention system operates from more than one layer and has complex interactions between layers. Based on this novel saliency hierarchy, we propose robust saliency detection with special focus on high-level priors.

3.1 High-level Priors

Previous works prove the feasibility for using high-level priors in saliency detection [4, 17, 18]. For instance, image center is more likely to contain salient objects than border regions. And such prior can be effectively modeled, e.g., using radial basis function. Return back to our problem, when we are trying to mix visual artistic effects with genuine photo pixels, we believe the following high-level priors are indispensible.

(a) Input Image (b) Face Prior [18] (c) Our Face Prior (d) Our Skin Prior

Fig. 7. High-level priors

1) ***Face Prior.*** As shown in Fig. 1, facial area always drives people's attention because of the critical information it contains (e.g. 'Who is in the photo?'). Previous works such as [18] performed face detection on the images, and assigned semantic priors using $p_f(x) = \exp(-d(x, f_c)/\sigma^2)$, where f_c is the center of face (Fig. 7 (b)).

A more intuitive estimation of face priors is proposed in this paper. Our basic assumption is based on the co-occurrence of face and human body. We hold that not only the facial areas are salient regions, the human bodies, especially the upper bodies are with high attention as well. As shown in Fig. 7 (c), the body area is model by drawing a rectangle under the detected face box. Its width is set to 3 times the face width and the height is set to 9 times the face height. The values inside the body box are measured by: $P_{face}(x) = e^{(-L_2(x', f_c'))}$, where x' and f_c' are normalized between 0 and 1.

2) Skin Prior. Comparing with face and apparel, naked skin areas are not detect as salient regions because they do not contain complex patterns and high local contrasts. However, when we are trying to create a half-real image composition, it is expected that the foreground should be compact and connected, elaborately located at the center or any side of the image. In order to effectively connect the salient regions, we estimate skin prior based on the skin map obtained by [19] (Fig. 7 (d)).

3) Location Prior. Psychophysical study shows that human attention favors central regions. We simply use the same location prior estimation as [4] and [18].

3.2 Saliency Hierarchy

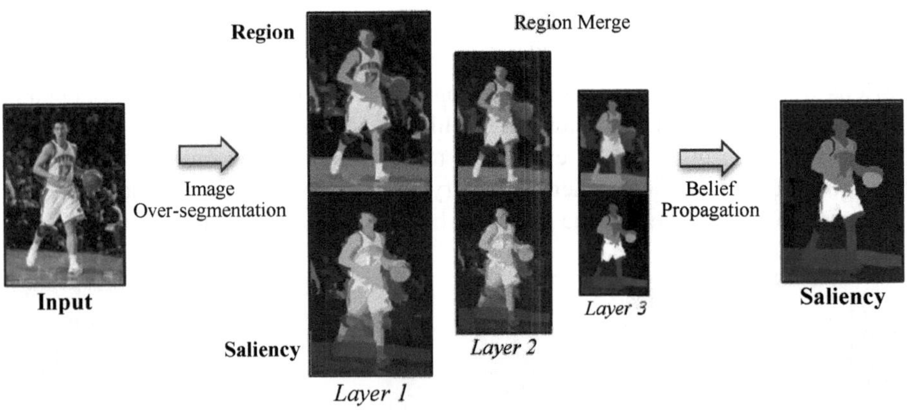

Fig. 8. Saliency hierarchy

The saliency hierarchy is built based on image over-segmentation. We use a 3-layer hierarchy, where each of the upper layer l_k is generated by merging neighbor regions in previous layer l_{k-1}. The saliency value S_i^l for region i in layer l_k is the combination of local contrast C_i and high-level priors P_i:

$$S_i^l = C_i^l \times P_i^l \tag{10}$$

$$C_i = \sum_{j=1}^{n} w(R_j)\phi(i,j)\left\|c_i - c_j\right\|_2 \tag{11}$$

$$P_i = \frac{1}{w(R_i)} \sum_{x \in R_i} Max(P_f(x), P_s(x), P_l(x)) \quad (12)$$

where $w(R_j)$ is the number of pixels in region R_j. $\phi(i, j)$ is the Gaussian weighting which controls the spatial distance influence between region i and j. c_i and c_j are average colors in CIELUV color space. P_f, P_s, P_l stand for face prior, skin prior, and location prior, respectively. The final saliency map is built by an efficient hierarchical inference algorithm based on belief propagation. Noticing that the modeled saliency hierarchy is actually a loop-free Bayes network, the iterations can be finished in one simple bottom-up and top-down message propagation.

After saliency detection, an adaptive saliency threshold t is adopted for foreground/background segmentation [20]. As we mentioned above, the background pixels are further converted into a sketch or manga. To let the user easily control the composition effect, a drag bar which controls the threshold t is also added to the system's user interface.

4 User-Interactive Image Composition

An alternative way for half-real image composition is by user-interaction. To be more specific, we let the user to select the foreground/background regions. With the help of superpixel segmentation [21], user only needs to draw several strokes on the selected areas.

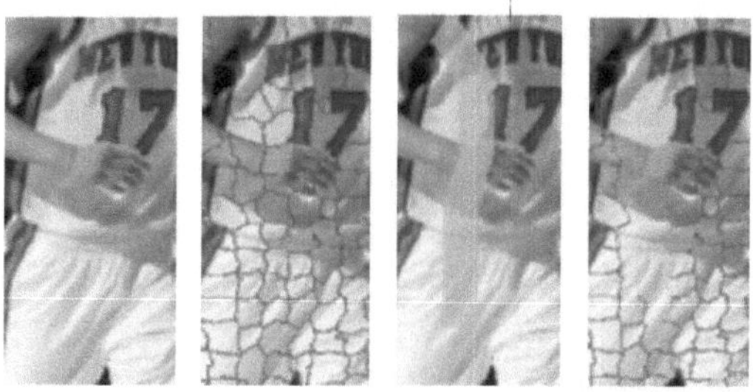

(a) Input image.

(b) Super-pixel segmentation.

(c) User stroke drawing

(d) Selected foreground regions.

(a) (b) (c) (d)

Fig. 9. User-interactive image composition

5 Results

Table 1. Average processing time for image to sketch/manga conversion

Resolution	VGA(640×480)	WVGA(800×480)	XGA(1024×768)	HD(1280×720)
Avg. Manga	0.184 s	0.234 s	0.487 s	0.577 s
Avg. Sketch	0.272 s	0.359 s	0.772 s	0.907 s

5.1 Processing Time

We first evaluate the processing time for image to sketch/manga conversion. Table 1 shows the average result of 100 runs by single thread on a Laptop equipped with Intel Core i7 CPU. For hierarchical saliency detection and super-pixel segmentation, the input image is resized to 300-pixel according to its short edge. According to our observation, hierarchical saliency detection with face and skin priors takes 0.4 to 0.6 seconds. And super-pixel segmentation usually costs 0.2 seconds.

5.2 Visualized Examples

Fig. 10 illustrates the results for our half-real image composition (row 1: manga-based composition, row 2: sketch-based composition). For more examples, please view out video demo (http://youtu.be/m4KtY8JKjf4) or visit the online image gallery (http://www.flickr.com//photos/95064380@N07/sets/72157633275818101/show/).

Input Image **Automatic Composition**

Fig. 10. Visualized examples

6 Conclusion

Inspired by the intriguing feeling when we mix real images with artistic effects (sketch, manga), in this paper, we propose an automatic way to generate the so-called 'half-real image composition'. Guided by an efficient hierarchical saliency detection strategy, the detected background pixels are filtered into sketch/manga while the foreground pixels are kept unchanged. Experimental results show that the proposed 'RealMeetsArt' system can generated high-quality image compositions on-the-fly.

References

1. Lu, C., Xu, L., Jia, J.: Combining sketch and tone for pencil drawing production. In: NPAR, pp. 65–73 (2012)
2. Winnemöller, H., Olsen, S.C., Gooch, B.: Real-time video abstraction. ACM TOG 25(3), 1221–1226 (2006)
3. That's one way to draw attention to yourself: Artist who takes self-portraits and sketches himself into the pictures. The Daily Mail (March 28, 2013)
4. Yan, Q., Xu, L., Shi, J., Jia, J.: Hierarchical Saliency Detection. In: CVPR (2013)
5. Sousa, M.C., Buchanan, J.W.: Computer generated graphite pencil rendering of 3d polygonal models. Comput. Graph. Forum 18(3), 195–208 (1999)
6. Lee, H., Kwon, S., Lee, S.: Real-time pencil rendering. In: NAPR, pp. 37–45 (2006)
7. Mao, X., Nagasaka, Y., Imamiya, A.: Automatic generation of pencil drawing from 2d images using line integral convolution. In: CAD/Graphics, pp. 240–248 (2001)
8. Li, N., Huang, Z.: A feature-based pencil drawing method. In: GRAPHITE, pp. 135–140 (2003)
9. Qu, Y., Pang, W.-M., Wong, T.T., Heng, P.-A.: Richness-preserving manga screening. ACM ToG 27(5), 155 (2008)
10. Kang, H., Lee, S., Chui, C.K.: Flow-based image abstraction. IEEE Transactions on Visualization and Computer Graphics 15(1), 62–76 (2009)
11. Bayer, B.: An optimum method for two-level rendition of continuous-tone pictures. In: IEEE International Conference on Communications, pp. 11–15 (1973)
12. Cerf, M., Harel, J., Einhauser, W., Koch, C.: Predicting human gaze using low-level saliency combined with face detection. Advances in Neural Information Processing Systems, 241–248 (2008)
13. Palmer, S.: Vision science: Photons to phenomenology (1999)
14. Itti, L., Koch, C., Niebur, E.: A model of saliency-based visual attention for rapid scene analysis. IEEE TPAMI 20(11), 1254–1259 (1998)
15. Ma, Y.-F., Zhang, H.: Contrast-based image attention analysis by using fuzzy growing. In: ACM Multimedia, 374–381 (2003)
16. Cheng, M.-M., Zhang, G.-X., Mitra, N.J., Huang, X., Hu, S.-M.: Global contrast based salient region detection. In: CVPR, pp. 409–416 (2011)
17. Goferman, S., Zelnik-Manor, L., Tal, A.: Context-aware saliency detection. In: CVPR, pp. 2376–2383 (2010)
18. Shen, X., Wu, Y.: A unified approach to salient object detection via low rank matrix recovery. In: CVPR (2012)
19. Jones, M.J., Rehg, J.M.: Statistical color models with application to skin detection. IJCV 46(1), 81–96 (2002)
20. Achanta, R., Hemami, S., Estrada, F.J., Susstrunk, S.: Frequency-tuned salient region detection. In: CVPR, pp. 1597–1604 (2009)
21. Achanta, R., Shaji, A., Smith, K., Lucchi, A., Fua, P., Süsstrunk, S.: SLIC Superpixels Compared to State-of-the-art Superpixel Methods. IEEE TPAMI, 2274–2282 (2012)
22. Lang, C., Liu, G., Yu, J., Yan, S.: Saliency detection by multi-task sparsity pursuit. IEEE TIP 21(3), 1327–1338 (2012)

Hand Tracking with Mid-Level Visual Cues and Ensemble Learning

Yikai Fang and Baocai Yin*

Beijing Key Laboratory of Multimedia and Intelligent Software Technology
College of Computer Science and Technology
Beijing University of Technology, 100124, Beijing, China
{fyk,ybc}@bjut.edu.cn

Abstract. Hand tracking is a crucial step in vision based gesture interaction. Due to illumination and hand shape variation, tracking of such object is a difficult task. In this paper, an ensemble tracking framework integrated with superpixels for hand tracking is proposed. With treating tracking as a binary classification problem, ensemble tracking utilizes weak classifiers to get strong classifier and get robustness to illumination and appearance change. However, outliers in confidence map caused by irregular object shape bring ill effects to tracking performance. As hand area has similar color and texture, superpixel method can be used to cluster the pixels in hand area and provide solution to exclude outliers. Experiments show that with superpixe integrated, the proposed method get more accurate tracking result.

Keywords: hand tracking, ensemble learning.

1 Introduction

There are more and more emerging interaction applications using hand gestures. Among them vision based gesture interaction is the most favorable way as it does not need extra hardware and is convenient in practice. In visual gesture interaction, hand tracking is an important step. Although visual tracking has been an important topic in computer vision and is intensively investigated in academic and industrial research, the tracking of hand-like objects with irregular shape or shape variations are still regarded as very difficult tasks.

Unlike mostly rigid objects that can be tracked with very limited shape modeling efforts, hand is articulated object and most conventional tracking methods fails to deal with it. With rigid hand models, condensation may work well for a few selected hand configurations and relatively static lighting conditions[1].

Many approaches utilize shape-free color information or background differences[3][4]. These methods are vulnerable to unimodal failure modes, caused by camera movement or skin-like objects in the background. Other multi-cue

* Supported by Beijing Postdoctoral Research Foundation, NSFC(No.61133003, 61370119) and BJNSF(No.4132013, KZ201310005006).

B. Huet et al. (Eds.): PCM 2013, LNCS 8294, pp. 690–697, 2013.

methods integrate texture and color information and can then recognize and track a small number of fixed shapes despite arbitrary backgrounds [5] also integrate color information with motion data. In their work, a particle filtering method is optimized for speed with deterministic mean shift, and dynamic weights determine the blend of color with motion data. Depth information combined with color[2] also yields a robust hand tracker, yet stereo cameras are more expensive and cumbersome. KLT features are also used to track hand[6][7]but it is easy to drift to other skin-like areas such as arms.

In recent years, there are many attempts to use on-line learning in visual tracking and ensemble tracking[8] is an efficient representation. With given video sequence, it firstly trains a set of weak classifiers. A strong classifier is then computed with Adaboost, which classifies the pixels in the next frame and generate a confidence map for further analysis by the Mean Shift algorithm[9] and finally determine the new position of target. The strong classifier is trained with a set of weak classifiers handling a single feature vector. One of the the drawbacks of ensemble tracking is that if the object to be tracked is not a pure rectangle then the bounding box that we use for tracking will include some pixels that are labeled as positive, while in fact they should be labeled negative. And it was shown that AdaBoost is sensitive to outliers[10]. Ensemble tracking uses a simple approach to treat too difficult examples as outliers and change their label. In the proposed method, a more effective way with segmentation is used for that. The rest of this paper is organized as follows. section 2 describes related work on superpixels. The proposed method is depicted in section 3 and experiments are shown in section 4. The last section gives conclusion and direction of future work.

2 Related Work

Superpixel is a kind of macro pixel. It provides a convenient primitive for computation of local image features. With capture of redundancy in the image, it greatly reduce the complexity of subsequent image processing tasks. They have proved increasingly useful for applications such as depth estimation, image segmentation, skeletonization, body model estimation and object localization. The advantage of superpixels is obvious- well adherence to image boundaries, which is an important mid-level visual cue and can be used to simplify some vision task.

According to EPFL's work on superpixels[15], algorithms for generating superpixels can be broadly categorized as either graph-based or gradient ascent methods. Graph-based approaches to superpixel generation treat each pixel as a node in a graph. Edge weights between two nodes are proportional to the similarity between neighboring pixels. Superpixels are created by minimizing a cost function defined over the graph. Such approaches includes Normalized Cuts[11], Graph Segmentation[12] and Superpixel lattices[13].

Starting from a rough initial clustering of pixels, gradient ascent methods iteratively refine the clusters until some convergence criterion is met to form superpixels.Turbopixel, Quick shift and SLIC are in this category.

Here in this paper, we use the implementation of SLIC for outlier removal. As there is no obvious edge and rich texture inside hand area, it does not matter that SLIC is not the best one in boundary adherence. Besides it performs better in computational efficiency.

3 The Proposed Method

The proposed framework mainly consists of three parts, initialization with detection, tracking and hand area refinement with superpixels. An initialization or given position of target is indispensable for tracking of any objects. From practical perspective, hand tracking in this paper is triggered by a detection of specified hand posture. Empirically an "Open" posture with five fingers extended is selected as trigger gesture. Then the bounding box of hand and its surrounding area are extracted as positive and negative area respectively for initial training in the stage of ensemble tracking. To remove outliers in positive examples, superpixels of image patch inside bounding box area are generated. Different superpixels are clustered into two categories of hand and background.

3.1 Specified Posture Detection

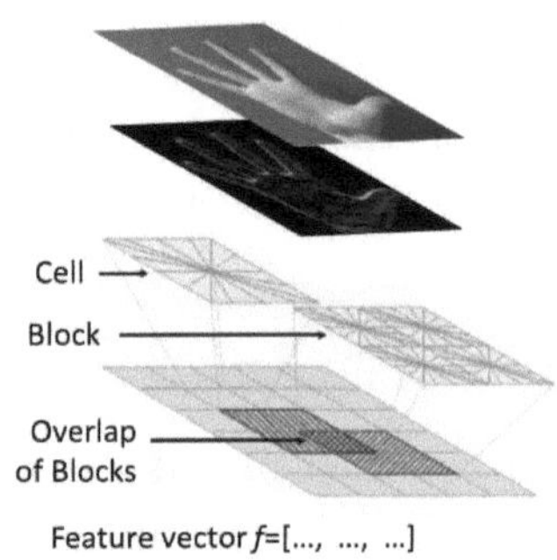

Fig. 1. HOG hand posture detection

The histogram of oriented gradients (HOG)[14] 1descriptor operates on the entire hand image window It. Similarly to the SIFT feature, the image window It is partitioned into cells, and for each cell, a histogram of gradient orientations is computed. The feature is the concatenation of the gradient orientation histograms from all cells. The size of the cells and the granularity of the histograms affect the generalization capabilities of the feature. A more detailed discussion on how different parameters of the HOG affect human detection can be found in [14]. Due to its spatially coarser binning, the HOG descriptor is expected to be less discriminative than shape context. However, the robustness of the feature will probably lead to a very smooth mapping with good generativity C the feature value can be predicted in an accurate and deterministic manner from a hand pose.

3.2 Outliers Removal with Superpixel

It is obvious that hand has an irregular shape. For such object, the bounding box contains not only positive examples in ensemble tracking, but also negative examples. So it is necessary to refine positive area.

Fig. 2. Superpixels analysis in hand area

With hand posture detection result, the bounding box (green rectangle) contains positive example and an extension to bouding box (blue rectangle) contains negative examples. As the detection bouding box cannot be an exact boundary of positive and negative area. Superpixes in blue rectangle are generated and different superpixels in hand area are connected as they share similar color and texture. Thus the outliers in positive area are removed as shown in Fig.2.

3.3 Ensemble Tracking

In tracking applications, the target and its surrounding background can be viewed as two different classes. Thus tracking can be formulated as a binary classification problem and be handled with learning-based approaches. The ensemble tracking framework, introduced by Avidan [8], works by constantly updating a collection of weak classifiers to separate the foreground object from the background. The AdaBoost algorithm, widely applied in machine learning tasks, is also used in the ensemble tracking framework. A set of weak classifiers are trained on increasingly more difficult examples and combined to produce a strong classifier.

The object being tracked is not represented explicitly, instead, an ensemble of weak classifiers are used to determine if a pixel belongs to the object or not. The classifiers can be added or removed during tracking to reflect changes in object appearance or incorporate new information about the background.

As in [8], the example extracted from a pixel that we use in AdaBoost is an 11D feature vector that consists of an 8- bin local histogram of oriented gradients as well as R, G, and B values of the pixel. The 8-bin local histogram of oriented gradients is calculated on a 5x5 window centered at the pixel. The histogram of oriented gradients (HOG), first used for detection purposes, conveys rich information and is easy to compute. In addition, the computation of HOG can be further accelerated by using integral histograms as efficient data structures. Other features, such as local binary patterns can be used as well.

In the initialization stage, a rectangle is given by detection in the first video frame showing where the object is. Examples coming from the rectangle of object are used as positive examples, while examples coming from the background as negative examples, and then T weak classifiers are trained using AdaBoost and added to the ensemble. When the ensemble of weak classifiers are ready, the following tracking process is repeated for every new incoming frame (see [8] for more details):

1) Test all pixels of interest in the new frame using the current strong classifier and create a confidence map of those pixels.

2) Run mean shift on the confidence map and report the new object rectangle.

3) Label pixels inside the new rectangle as object and all those outside it as background.

4) Keep K best weak classifiers and remove those that perform badly.

5) Train $T - K$ new weak classifiers on the current frame and add them to the ensemble.

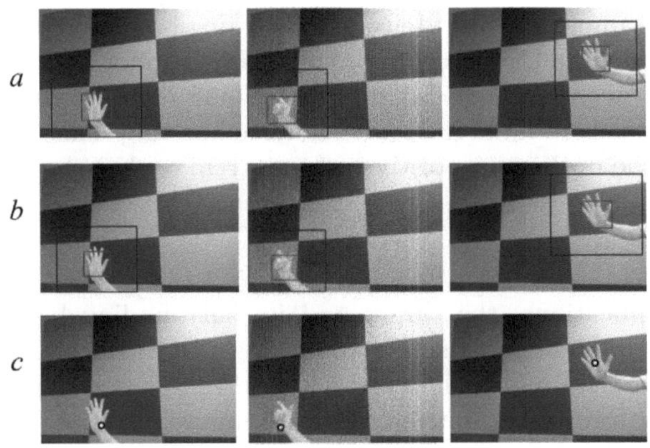

Fig. 3. Similar results. (simple background).

4 Experiments

To validate the framework proposed in this paper, We implement the proposed method as well as the original ensemble tracking method in [8].We also tried the the KLT tracking with flock of features described in [6] and replace the detection with mentioned HoG detection as the initialization of tracking step. The original ensemble tracker was equipped with proposed superpixel method to form an integrated tracking system and demonstrate it on different video sequences.

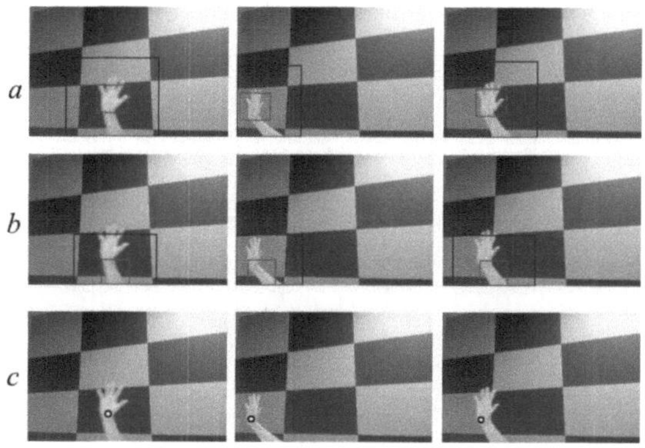

Fig. 4. Different results. (simple background).

Fig. 5. Different results. (cluttered background).

In our experiments, the initial object rectangle is obtained with above detector automatically. Video sequences including different backgrounds are tested with KLT tracking[6], original ensemble tracking and the proposed method. Part of a

b

Fig. 6. Similar results. (cluttered background).

results on 2 video sequences are in this section. Fig. 3 and Fig. 4 show tracking results under simple background. Row a, b and c contain tracking position of proposed method, ensemble tracking and KLT respectively. The results in row a and b are marked with red rectangle while in row c the results are represented with white dots which are the mean position of all flock features. In some cases, different methods give similar results (Fig.3). But it is obvious that the proposed method achieves better accuracy in general.

Fig. 5 and Fig. 6 shows results with natural and cluttered background. As KLT often missed object and cannot give comparable results, it was excluded from our comparison. Fig.6 lists similar results obtained by ensemble tracking(row b) and our method(row a). It seems that fist and blurred object caused by motion are easier to track. But with different hand shapes, the proposed method get more accurate tracking results under such circumstance. It is not easy to be distracted by artifacts as shown by row a and c (like red themos and the wall clock) in Fig.5. With comparison, it is obvious that original tracking method (in row b and d) misses part of the object.

5 Conclusion

In this paper, an ensemble tracking framework integrated with superpixels for hand tracking is proposed. The outliers in confidence map caused by irregular object shape bring ill effects to tracking performance. As hand area has similar color and texture, superpixel method can be used to cluster the pixels in hand area and provide solution to exclude outliers. Experiments show that with such mid-level visual cues integrated, our method obtained more accurate tracking result. The main problem in this method lies in the dependency on target. Hand is special object with uniform color and superpixel based method can deal with it easily. For objects with complicated color and texture, it is necessary to expand such method.

References

1. Isard, M., Blake, A.: A mixed-state CONDENSATION tracker with automatic model-switching. In: ICCV 1998, pp. 107–112 (1998)
2. Segen, J., Kumar, S.: GestureVR: Vision-Based 3D Hand Interface for Spatial Interaction. In: The Sixth ACM Intl. Multimedia Conference (September 1998)
3. Kurata, T., Okuma, T., Kourogi, M., Sakaue, K.: The Hand Mouse: GMM Hand-color Classification and Mean Shift Tracking. In: Second Intl. Workshop on Recognition, Analysis and Tracking of Faces and Gestures in Real-time Systems (July 2001)
4. Grange, S., Casanova, E., Fong, T., Baur, C.: Vision-based Sensor Fusion for Human-Computer Interaction. In: Intl. Conference on Intelligent Robots and Systems (October 2002)
5. Bretzner, L., Laptev, I., Lindeberg, T.: Hand Gesture Recognition using Multi-Scale Colour Features, Hierarchical Models and Particle Filtering. In: Proc. IEEE Intl. Conference on Automatic Face and Gesture Recognition, Washington D.C., pp. 423–428 (2002)
6. Kolsch, M., Turk, M.: Fast 2D Hand Tracking with Flocks of Features and Multi-Cue Integration. In: Proc. IEEE Workshop on Real-Time Vision for Human-Computer Interaction (July 2004)
7. Liu, H., Cui, W., Ding, R.: Robust Hand Tracking with Hough Forest and Multi-cue Flocks of Features. In: Bebis, G., Boyle, R., Parvin, B., Koracin, D., Fowlkes, C., Wang, S., Choi, M.-H., Mantler, S., Schulze, J., Acevedo, D., Mueller, K., Papka, M. (eds.) ISVC 2012, Part II. LNCS, vol. 7432, pp. 458–467. Springer, Heidelberg (2012)
8. Avidan, S.: Ensemble Tracking. IEEE Trans. Pattern Anal. Mach. 29(2), C261–C271 (2007)
9. Comaniciu, D., Meer, P.: Mean shift: a robust approach toward feature space analysis. IEEE Trans. Pattern Anal. Mach. Intell. 24(5), C603–C619 (2002)
10. Freund, Y.: An adaptive version of the boost by majority algorithm. Machine Learning 43(3), 293–318 (2001)
11. Shi, J., Malik, J.: Normalized cuts and image segmentation. IEEE Transactions on Pattern Analysis and Machine Intelligence (PAMI) 22(8), C888–C905 (2000)
12. Felzenszwalb, P., Huttenlocher, D.: Efficient graph-based image segmentation. International Journal of Computer Vision (IJCV) 59(2), C167–C181 (2004)
13. Moore, A., Prince, S., Warrell, J., Mohammed, U., Jones, G.: Superpixel Lattices. In: IEEE Computer Vision and Pattern Recognition (CVPR) (2008)
14. Dalal, N., Triggs, B.: Histograms of oriented gradients for human detection. In: CVPR (2005)
15. Achanta, R., Shaji, A., Smith, K., Lucchi, A., Fua, P., Ssstrunk, S.: SLIC Superpixels Compared to State-of-the-art Superpixel Methods. IEEE Transactions on Pattern Analysis and Machine Intelligence 34(11), 2274–2282 (2012)

Real-Time Head Pose Estimation by RGB-D Camera

Fan Liu[1], Jinhui Tang[1], Yan Song[1], Xinguang Xiang[1], Ting Rui[2], and Zhenmin Tang[1]

[1] Nanjing University of Science and Technology, Nanjing, China
[2] PLA University of Science and Technology, Nanjing, China
fanliu.njust@gmail.com

Abstract. Conventional RGB image-based head pose estimation methods encounter many difficulties due to pose variation and illumination. In this paper, we present a real-time 3D head motion estimation method using both RGB image data and depth data. Head is detected from depth data in a simple way. Based on a rigid-body motion model, we derive the linear depth and optical flow constraint equations respectively. These constraints are combined into a single linear system, from which head motion vector is recovered by minimizing a least-squares. Experimental results have shown that the use of both depth data and RGB data in our method overcomes the shortcomings of single depth or RGB data. In addition, it's still robust when there is only one type of data reliable.

Keywords: head detection, depth rate constraint, optical flow constraint, rigid head motion, head pose estimation, least-squares.

1 Introduction

Head pose estimation has become one of the most active research areas in computer vision, due to the large number of potential applications. For example, head pose is critical in driving applications, allowing the driver to receive cues based on his/her awareness and visual attention. In addition, it's also beneficial to human-computer interaction.

Researches on head pose estimation can be conveniently categorized by the type of input data they rely on, i.e., RGB image data or depth data. Similar to the categorization in the survey paper [1], the algorithms based on RGB image data can be further divided into appearance-based methods, which analyze the whole face region, and feature-based methods, which rely on the localization of specific facial features. A common appearance-based approach is to discretize the head poses and learn a separate detector for each pose [2], [3]. Feature-based methods need either the same facial features to be visible in all poses [4, 5, 6], or use pose-dependent features. Generally, appearance-based technique performs better, since feature-based one highly depends on the accuracy of feature extraction. However, all RGB image-based methods suffer from several problems, in particular changes in illumination, lack of features and partial occlusion. Moreover, the annotation of head poses from RGB images is an error-prone task itself.

B. Huet et al. (Eds.): PCM 2013, LNCS 8294, pp. 698–707, 2013.

With the development and availability of depth-sensing technologies, researchers have focused on using depth information to solve the problem of head pose estimation, either as unique cue [7, 8, 9], or in combination with RGB image data [10, 11]. Although a few notable works [7, 8, 9] have shown the usefulness of depth information, they require either a training phase [7, 8], or parallel computation power of GPU [9]. In addition, both [7] and [9] rely on the high quality depth data. Therefore, Kondori et al. [14] proposed to employ a simple technique, the least-squares, for head pose estimation. It recovers head motion from a sequence of low quality range images taken by the Miscrosoft Kinect. A similar approach has been presented with a time-of-flight sensor using the constancy of 2D motion fields and depth from optical flow [15]. However, it only uses depth information in optical flow constraint. Any noise from depth or RGB image data will degrade its performance.

In this paper, we propose a simple, yet effective head pose estimation method to overcome the aforementioned limits. It not only takes full use of both RGB image data and depth data from the Kinect, but also guarantees the real-time performance by using the least-squares technique. The depth information enables us to easily segment and locate head from the scene. According to a rigid-body motion model, we derive the linear depth and optical constraint equations based on depth and RGB data respectively. The two constraints are integrated jointly over the head region, yielding a single linear system. The head motion vector is eventually recovered from the linear system by minimizing the error in least-squares. Experimental results show that the use of the combined RGB image data and depth data outperforms the use of either independently. In addition, the weights of RGB image data and depth data in the system can be adjusted according to their quality in real conditions. Therefore, even if there is only depth or RGB image data available, the system is still robust.

The remainder of this paper is organized as follows: Section 2 gives a description of the head localization algorithm. Section 3 describes our 3D head pose estimation method. Experimental results are included in Section 4 and finally we present our conclusion in Section 5.

2 Head Detection

Head detection is a key component of head pose estimation. Although many sophisticated head detection methods are available, they usually use intensity images as the input to the detection system [16] [17] and are easily affected by illumination changes. In this paper we perform a simple head detection method by using depth data, which is robust against illumination variation. First, we extract the background depth data, which can be regarded as an initialization step. Then a difference matrix is computed by subtracting background from a new depth array. If the difference is below a threshold, the pixel is set zero, otherwise it will be retained, resulting in a matrix containing the depth information of the foreground. According to the prior knowledge about human body, a pixel and its neighbors are considered to be on the same object if the depth difference between them is less than a threshold (0.1-0.2m). Any segmented object that contains fewer pixels than a particular number is considered as non-human and discarded.

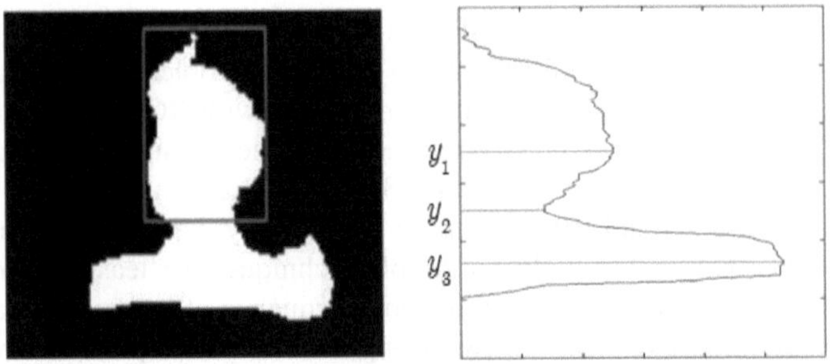

Fig. 1. Head detection: left is the binary image; right is the horizontal projection histogram

To get more correct head detection result, we perform a simple head/shoulder separation based on vertical projection signature operation by efficient use of depth image. A similar method has been used in [15]. As shown in the Fig 1, the binary image is projected to the vertical direction. The bottom boundary of the head area is segmented by finding the minimal value between the two peaks of projection histogram. Then the head area is eventually located by finding the topmost, leftmost and rightmost pixel in the area above bottom boundary. The four pixels constitute the boundaries of the rectangle containing the head.

For head tracking, we use the same way in [14] due to its simplicity. The mean x, y and depth values for head area in one frame are sorted and compared with those in the next frames. If a similar segment is found between frames, they are regarded as the same segment.

3 Head Pose Estimation

In order to recover the head pose, we could assume that the sensor is rigid and the global motion of the head is a kind of rigid motion. Then the motion of the head relative to the sensor can be parameterized by a rigid motion vector μ, which includes the 3D rotation angles $\left(W_X, W_Y, W_Z\right)$ and the 3D translation vector $\left(T_X, T_Y, T_Z\right)$. As shown in Figure 2, when the 3D coordinate of a head point is $A=\left(X_A, Y_A, Z_A\right)^T$ in the camera-centered 3D coordinate system, then the new location of the point A transformed by the rigid motion vector μ is $B=\left(X_B, Y_B, Z_B\right)^T$.The relationship of A and B can be described by

$$B = RA + T \tag{1}$$

Where the 3D translation vector $T \in \mathbb{R}^{3\times1} = \left(T_X, T_Y, T_Z\right)^T$ and the 3D rotation matrix $R \in \mathbb{R}^{3\times3}$ corresponds to $\left(W_X, W_Y, W_Z\right)$. When the rotation is small, the rotation matrix can be approximated as

$$R = \begin{bmatrix} 1 & -W_Z & W_Y \\ W_Z & 1 & -W_X \\ -W_Y & W_X & 1 \end{bmatrix} \tag{2}$$

using the exponential map [12].

Then the 3D rigid motion between point A and B can be described by

$$\begin{bmatrix} \Delta X_{AB} \\ \Delta Y_{AB} \\ \Delta Z_{AB} \end{bmatrix} = \begin{bmatrix} T_X \\ T_Y \\ T_Z \end{bmatrix} + \begin{bmatrix} 0 & -W_Z & W_Y \\ W_Z & 0 & -W_X \\ -W_Y & W_X & 0 \end{bmatrix} \begin{bmatrix} X_A \\ Y_A \\ Z_A \end{bmatrix} \tag{3}$$

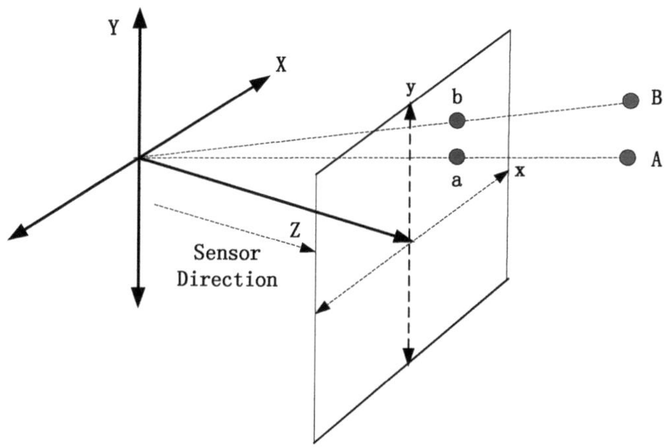

Fig. 2. Relationship between 3D motion from A to B and 2D motion from a to b

3.1 Depth Constraint

The time-varying depth map from the Kinect can be viewed as a function of the form $Z(X,Y,t)$. Taking a full time derivative of Z via the chain rule, the following equation is obtained:

$$\frac{dZ}{dt} = \frac{\partial Z}{\partial X}\frac{dX}{dt} + \frac{\partial Z}{\partial Y}\frac{dY}{dt} + \frac{\partial Z}{\partial t} \tag{4}$$

The above equation will be called the *depth rate constraint equation* [14], where the three partial derivatives of Z are denoted by

$$p = \frac{\partial Z}{\partial X}, \quad q = \frac{\partial Z}{\partial Y}, \quad Z_t = \frac{\partial Z}{\partial t} \tag{5}$$

Where the $\left(\frac{dX}{dt},\frac{dY}{dt},\frac{dZ}{dt}\right)$ denote the differentiation with respect to time, which can be computed by

$$\begin{aligned}\frac{dX}{dt}&=\frac{\Delta X_{AB}}{\Delta t}\\ \frac{dY}{dt}&=\frac{\Delta Y_{AB}}{\Delta t}\\ \frac{dZ}{dt}&=\frac{\Delta Z_{AB}}{\Delta t}\end{aligned} \tag{6}$$

Then (4) can be written in the form:

$$\Delta Z_{AB}=p\Delta X_{AB}+q\Delta Y_{AB}+Z_t\Delta t \tag{7}$$

By substituting $\Delta X_{AB},\Delta Y_{AB},\Delta Z_{AB}$ of (3) into (7), we obtain the following equation:

$$-pT_X-qT_Y+T_Z+rW_X+sW_Y+uW_Z=Z_t\Delta t \tag{8}$$

Where $r=Y_A+qZ_A$, $s=-X_A-pZ_A$, $u=pY_A-qX_A$. If there are n pixels in the head area, the resulting n equations can be written in a matrix form as:

$$\underbrace{\begin{pmatrix}-p_1 & -q_1 & 1 & r_1 & s_1 & u_1\\ -p_2 & -q_2 & 1 & r_2 & s_2 & u_2\\ \vdots & \vdots & \vdots & \vdots & \vdots & \vdots\\ \vdots & \vdots & \vdots & \vdots & \vdots & \vdots\\ \vdots & \vdots & \vdots & \vdots & \vdots & \vdots\\ -p_n & -q_n & 1 & r_n & s_n & u_n\end{pmatrix}}_{\mathbf{G}}\underbrace{\begin{pmatrix}T_X\\ T_Y\\ T_Z\\ W_X\\ W_Y\\ W_Z\end{pmatrix}}_{\mathbf{w}}=\underbrace{\begin{pmatrix}(Z_t\Delta t)_1\\ (Z_t\Delta t)_2\\ \vdots\\ \vdots\\ \vdots\\ (Z_t\Delta t)_n\end{pmatrix}}_{\mathbf{c}} \tag{9}$$

3.2 Optical Flow Constraint

The 3D head rigid motion also forms motion field in the head region of the RGB image plane from Kinect. Optical flow, based on constant image brightness constraint over time, is used to estimate the motion filed between successive frames. If the time-varying RGB image from Kinect is viewed as a function of the form $I(x,y,t)$, the well-known Optical Flow Constraint Equation [13] can be described as:

$$\frac{\partial I}{\partial x}\frac{dx}{dt}+\frac{\partial I}{\partial y}\frac{dy}{dt}+\frac{\partial I}{\partial t}=0 \tag{10}$$

Where the three partial derivatives of I can be denoted by

$$I_x=\frac{\partial I}{\partial x},\quad I_y=\frac{\partial I}{\partial y},\quad I_t=\frac{\partial I}{\partial t} \tag{11}$$

Based on the pinehole camera model, there is a following equation:

$$x=\frac{F}{Z}X,\quad y=\frac{F}{Z}Y \tag{12}$$

Where F is the camera focus length. By taking the derivative with respect to time, we have:

$$\begin{cases}\dfrac{dx}{dt}=\dfrac{F}{Z}\dfrac{dX}{dt}-\dfrac{FX}{Z^2}\dfrac{dZ}{dt}\\ \dfrac{dy}{dt}=\dfrac{F}{Z}\dfrac{dY}{dt}-\dfrac{FY}{Z^2}\dfrac{dZ}{dt}\end{cases} \tag{13}$$

By substituting (3), (6) and (12) into (13), we obtain the relationship between instantaneous 3D motion vector and 2D motion vector:

$$\begin{cases}\Delta x=\dfrac{F}{Z}T_X-\dfrac{x_A}{Z}T_Z-\dfrac{x_Ay_A}{F}W_X+\left(F+\dfrac{x_A^2}{F}\right)W_Y-y_AW_Z\\ \Delta y=\dfrac{F}{Z}T_Y-\dfrac{y_A}{Z}T_Z-\left(F+\dfrac{y_A^2}{F}\right)W_X+\dfrac{x_Ay_A}{F}W_Y+x_AW_Z\end{cases} \tag{14}$$

By substituting (12) and (14) into (10), we get the equation (15):

$$pT_X+qT_Y+kT_Z+rW_X+sW_Y+uW_Z=I_t\Delta t \tag{15}$$

Where

$$p=-\frac{I_xF}{Z}\quad q=-\frac{I_yF}{Z}\quad k=\frac{I_xx_A-I_yy_A}{Z}$$
$$r=\frac{x_Ay_A}{F}+\left(F+\frac{y_A^2}{F}\right)\quad s=-\frac{x_Ay_A}{F}-\left(F+\frac{x_A^2}{F}\right)\quad u=y_A-x_A$$

As there are n pixels in head area, the resulting n equations could be written in a matrix form as:

$$\underbrace{\begin{pmatrix} p_1 & q_1 & k_1 & r_1 & s_1 & u_1 \\ p_2 & q_2 & k_2 & r_2 & s_2 & u_2 \\ \vdots & \vdots & \vdots & \vdots & \vdots & \vdots \\ \vdots & \vdots & \vdots & \vdots & \vdots & \vdots \\ \vdots & \vdots & \vdots & \vdots & \vdots & \vdots \\ p_n & q_n & k_n & r_n & s_n & u_n \end{pmatrix}}_{\mathbf{H}} \underbrace{\begin{pmatrix} T_X \\ T_Y \\ T_Z \\ W_X \\ W_Y \\ W_Z \end{pmatrix}}_{\mathbf{w}} = \underbrace{\begin{pmatrix} (I_t \Delta t)_1 \\ (I_t \Delta t)_2 \\ \vdots \\ \vdots \\ \vdots \\ (I_t \Delta t)_n \end{pmatrix}}_{\mathbf{d}} \tag{15}$$

3.3 Head Pose Estimation

The two linear systems according to depth and optical flow constraints can be combined into a single linear system:

$$\mathbf{y} = \mathbf{X}\mathbf{w},\ where\ \mathbf{X} = \begin{bmatrix} \lambda\mathbf{G} \\ \mathbf{H} \end{bmatrix}, \mathbf{y} = \begin{bmatrix} \lambda\mathbf{c} \\ \mathbf{d} \end{bmatrix} \tag{17}$$

The scaling factor, λ, controls the weighting of the depth constraints relative to the optical flow constraints. When depth data is more reliable than RGB image data, λ is set to a value higher than 1. When depth data is much noisier than RGB image data, λ is set to a small value. In addition, the depth and optical flow constraints can also be used independently for pose estimation. It usually happens when one kind of data, depth or RGB image data, is not valid any more. The parameter $\mathbf{w}$ can be determined by solving the following optimization problem:

$$\min_{\mathbf{w}} \left\| \mathbf{y} - \mathbf{X}\mathbf{w} \right\|^2 \tag{18}$$

The least squares solution to the above equation is given by

$$\hat{\mathbf{w}} = (\mathbf{X}^T\mathbf{X})^{-1}\mathbf{X}^T\mathbf{y} \tag{19}$$

Consequently, the six DOF (degree of freedom) of the head motion will be recovered. After 3D head motion between two successive image frames is solved by (19), cumulative 3D head motion over time from the starting status can be calculated by summing up all the rotation angles between every two successive image frames.

4 Experimental Results

An experiment is designed to evaluate the system performance. In this experiment, the subject rotates his head as smooth as possible. The method is tested on a sequence of RGB and depth data with a core Intel T9300 at 2.0 GHz. As both RGB and depth data are reliable, we set λ to 1 which means that they have the same importance for pose estimation. The total processing speed is about 20 frames per second.

Consequently our system is fast enough operate in real time applications. The user must maintain a frontal pose for initialization before the system has begun. The six DOF of the head motion are recovered and used to manipulate a 3D head model. As shown in Fig 3, the position and orientation of the head model is updated whenever the user moves his head.

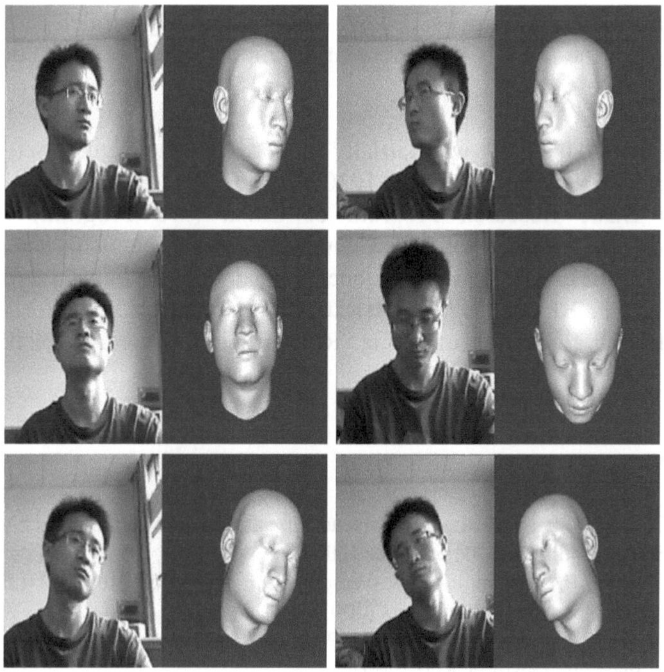

Fig. 3. The three rows show the rotation in yaw, pitch and roll respectively

We also conduct experiments in which head pose estimation just relies on the only one of depth and optical flow constraints. The experimental results show that depth constraint is more robust than optical flow constraint as depth data is less sensitive to illumination variation and shading effects. But the valid sensing distance of depth sensor of Kinect is only from 0.6 up to 6 meters. The RGB image data doesn't have this limitation. In addition, head pose estimation based on depth constraint is also influenced by self-occlusion (e.g. the neck by the chin) as it creates large depth gradients which do not remain consistent with head pose estimation during motion. Therefore, the use of combined RGB image data and depth data is necessary and performs better.

5 Conclusions

In this paper, we propose a real time 3D head pose estimation method that uses both the RGB data and depth data obtained from the Kinect. The linear depth constraint and optical flow constraint are incorporated into a single linear system for pose estimation. The two constraints correspond to depth data and RGB data respectively and their weight in the system is adjustable. The experimental results reveal that the use of combined RGB image data and depth data is more reliable and robust than the use of single one. This occurs not only because the two kinds of data provide more data for estimation, but also because each type of data helps offset the shortcomings of the other. Although the efficiency and robustness of the system have been demonstrated, our method has high dependency on high accurate head detection. It must be reinitialized when the track is lost. We plan to improve the head detection algorithm in the future work.

Acknowledgment. This work was partially supported by the Natural Science Foundation of China (NSFC) under Grant 61103059, 61202133, 61301106 and 61305134, the National Basic Research Program of China (973 Program) under Grant 2012CB316304, the Program for New Century Excellent Talents in University under grant NCET-12-0632, The Natural Science Foundation of Jiangsu Province under Grant BK2012033, BK2012397 and BK2011700, Research Fund for the Doctoral Program of Higher Education of China (RFDP) under Grant 20113219120022 and 20123219120024.

References

1. Murphy-Chutorian, E., Trivedi, M.M.: Head Pose Estimation in Computer Vision: A Survey. IEEE Trans. on PAMI 31, 607–626 (2009)
2. Jones, M., Viola, P.: Fast multi-view face detection. Technical Report TR2003-096, Mitsubishi Electric Research Laboratories (2003)
3. Morency, L.-P., Sundberg, P., Darrell, T.: Pose estimation using 3d view-based eigenspaces. In: Aut. Face and Gestures Rec. (2003)
4. Vatahska, T., Bennewitz, M., Behnke, S.: Feature-based head pose estimation from still images. In: Workshop on Subspace Methods (2009)
5. Yang, R., Zhang, Z.: Model-based head pose tracking with stereovision. In: Aut. Face and Gestures Rec. (2002)
6. Matsumoto, Y., Zelinsky, A.: An algorithm for real-time stereo vision implementation of head pose and gaze direction measurement. In: Aut. Face and Gestures Rec. (2000)
7. Fanelli, G., Gall, J., Van Gool, L.: Real time head pose estimation with random regression forests. In: Proc. of IEEE CVPR (2011)
8. Fanelli, G., Weise, T., Gall, J., Van Gool, L.: Real time head pose estimation from consumer depth cameras. In: DAGM (2011)
9. Breitenstein, M.D., Kuettel, D., Weise, T., Van Gool, L.: Real-time face pose estimation from single range images. In: Proc. of IEEE CVPR (2008)
10. Cai, Q., Gallup, D., Zhang, C., Zhang, Z.: 3D deformable face tracking with a commodity depth camera. In: Daniilidis, K., Maragos, P., Paragios, N. (eds.) ECCV 2010, Part III. LNCS, vol. 6313, pp. 229–242. Springer, Heidelberg (2010)
11. Seemann, E., Nickel, K., Stiefelhagen, R.: Head pose estimation using stereo vision for human-robot interaction. In: Aut. Face and Gesture Rec. (2004)
12. Bregler, C., Vetter, T.: Tracking people with twist and exponential maps. In: Proc. of IEEE CVPR (1998)

13. Horn, B.K.P., Schunck, B.G.: Determining Optical Flow. Artificial Intelligence 17, 185–203 (1981)
14. Kondori, F.A., Yousefi, S., Li, H., Sonning, S.: 3D head pose estimation using the Kinect. In: Wireless Communication and Signal Processing (WCSP) 2011 International Conference on Digital Object Identifier (2011)
15. Zhu, Y., Fujimura, K.: 3D Head Pose Estimation with Optical Flow and Depth Constraints. In: Fourth International Conference on 3-D Digital Imaging and Modeling (2003)
16. Nguyen, H., Smeulders, A.: Robust Tracking Using Foreground-Background Texture Discrimination. International Journal of Computer Vision 69, 277–293 (2006)
17. Zhao, T., Nevatia, R.: Tracking Multiple Humans in Complex Situations. IEEE Trans. on PAMI 26, 1208–1221 (2004)

Real-Time Multi-pedestrian Tracking Based on Vision and Depth Information Fusion

Shan Gao, Zhenjun Han, Ce Li, and Jianbin Jiao*

University of Chinese Academy of Sciences, Beijing, China
{gaoshan10,lice09}@mails.ucas.ac.cn,
{hanzhj,jiaojb}@ucas.ac.cn

Abstract. Visual object tracking plays an essential role in vision based applications. Most of the previous Multi-pedestrian Tracking has limitations due to considering each pedestrian with the same motion and appearance model in a uniform observation space, leading to tracking failures in complex occlusions. To address this problem without losing real-time performance, we propose a graph based approach for multi-pedestrian tracking using fused vision and depth data in this paper, where one main contribution is devoted in terms of the consideration of pedestrians with different priori probability in distinguishing observation space divided based on vision and depth information. Then we formulate the tracking model using an Improved Bipartite Graph (IBG), which is then optimized with a heuristic algorithm. Experiments on three datasets of fused vision and depth data demonstrate robust tracking results of the proposed approach.

Keywords: Multi-pedestrian Tracking, Bipartite Graph, Data Fusion.

1 Introduction

Multi-pedestrian tracking can provide fundamental information to lots of applications including video content analysis and driving assistant systems, which makes it one of the most active topics in the areas of computer vision and image processing in recent years [1, 2].

Given a video sequence, the task of multi-pedestrian tracking is to locate all pedestrians in each frame and correctly associate pedestrians over frames. However, the task is still an open problem for the following challenges. First, the outputs from a pedestrian detector are not always reliable, especially when vision data is captured with a moving camera [10, 13, 14]. Second, various occlusions among pedestrians are common [3, 6, 16].

From the perspective of image based tracking methods [11, 17, 18], people tend to find proper appearance models that can distinguish one object from others or the background [4, 5]. Some association based methods focus on tracking multiple pedestrians of a pre-known class simultaneously [6, 7]. They typically associate

* Corresponding author.

B. Huet et al. (Eds.): PCM 2013, LNCS 8294, pp. 708–719, 2013.

detection responses produced by a pre-trained detector into long tracks, and then pursuit a global optimal solution for all targets. Appearance models are often pre-defined [8] or online learned to distinguish multiple targets [9]. In addition, motion models, such as linear models on tracklet pairs [10, 16], are also adopted to constrain smoothness of motion.

It is noticed that most previous methods assume that all the track associations are independent. However, in a context of serious occlusion, the observations for pedestrians are not precise and trajectories can be mixed with each other. This make the independent assumption does not hold in a tracking process of serious occlusion. So in this paper, we divide the trajectories into distinguishing observation space, and give them different priori probability.

In recent years, benefiting from the reliable depth information of a laser rangefinder, tracking of pedestrians with occlusions achieves more promising results. However, there are also limitations when using depth information from rangefinder. In common, a laser rangefinder cannot discriminate different pedestrians of same depth. Furthermore, when there is inter-pedestrian occlusion, the mixture of depth information makes it difficult to discriminate pedestrians in the depth domain, and consequently tracking failure happens.

In this paper, considering the mutual complementary of vision and depth data, fusion of different sensing devices helps to overcome the limitation of each sensing technology, leading to enhance the performance of multi-pedestrian tracking. We formulate the data association progress from the perspective of Maximum A Posteriori (MAP) by proposing an IBG matching approach. We also propose an occlusion model cooperated with the depth information from laser rangefinder in consequent frames, which dynamically outputs the discriminated weight cost in IBG. The proposed approach can process independent observations and trajectories in a long-term tracking with complicated occlusion problem. Consequently, the robustness and accuracy of multi-pedestrian tracking using vision and depth can be improved.

The remainder of this paper is organized as follows: the proposed IBG matching approach is presented in section 2. Experiments are presented in section 3. We conclude the paper in section 4.

2 IBG Matching

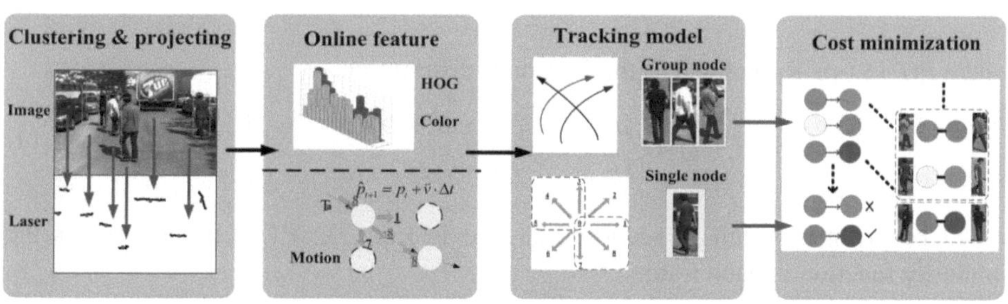

Fig. 1. Framework of our method

Framework: Given an input video sequence with the pedestrians' initialization after laser pattern classifier, clustering and projecting job, we can obtain the candidate regions both laser and image planes. And then the features are extracted to describe the pedestrians in terms of their appearances and motion characteristics, which include their position, velocity and moving orientation information. An online adaptive feature pool, named adaptive HOGC [19], is used to represent pedestrians' temporal association, considering the temporal consistence property of tracking. After that we reformulate the MAP estimation problem as a new tracking model with fused online updating information, and divide the observation space into two parts: group node and single node. Finally, we minimize the cost function using a heuristic searching algorithm to near optimality enlightened by IBG matching, in which the cost edges can be dynamically updated by the spatial and temporal relation from group and single nodes, and then we track each pedestrian and keep its identity.

2.1 Cost Function

Like some previous MAP work formulation [20, 21], observations of the i-th pedestrian at frame *t* are formulated as $T_{i_t} = \{pos_i, ori_i, vel_i, f_i\}$, indicating its position, orientation, velocity and appearance feature. Therefore the whole trajectory hypothesis of the i-th pedestrian can be written as $s_i = (T_{i_1}, T_{i_2}, \cdots, T_{i_n})$. Association hypothesis is a set $S = \{s_i\}$ composed by all the single trajectory hypotheses. The objective of data association is to maximize the posteriori probability of *S*, given the observation set $T = \{T_{i_t}\}$:

$$S^* = \arg\max_S P(S \mid T) \propto \arg\max_S P(T \mid S)P(S). \tag{1}$$

Assuming that motion of each pedestrian is independent, we can decompose Eq. (1) as

$$S^* \propto \arg\max_S \prod_i P(T_i \mid S) \prod_{s_k \in S} P(s_k). \tag{2}$$

Supposing that $S = -\log S^*$, the Eq. (2) can be equivalent to rewrite as

$$S = \arg\min_S \sum_{s_k \in S} -\log P(s_k) + \sum_i -\log P(T_i \mid S), \tag{3}$$

where $P(T_i \mid S)$ is formulated as the Bernoulli distribution.

$$P(T_i \mid S) = \begin{cases} p_i & if\ s_k \in S, T_i \in s_k \\ 1 - p_i & otherwise \end{cases}, \tag{4}$$

p_i will be given in detail in section 2.2. And in Eq. (3), $P(s_k)$ is modeled as the similarity function in each frame:

$$P_k(T_{i_{k-1}} \mid T_{i_k}) = P_{app}(T_{i_{k-1}} \mid T_{i_k}) P_{motion}(T_{i_{k-1}} \mid T_{i_k}) P_{ori}(T_{i_{k-1}} \mid T_{i_k}). \tag{5}$$

where P_{app}, P_{motion} and P_{ori} represent the similarity of the appearance, motion model and moving orientation respectively. And the similarity function is modeled as a Gaussian function.

The observation space S can be divided into two parts $S = \{S^{\alpha} + S^{\beta}\}$ by the affinity relation of trajectory hypothesis among pedestrians, and then a new priori probability $P(S)$ can be calculated in a new observation space, which is used to measure the trajectory relation among all the pedestrians. S^{α}, called group nodes, denotes the trajectories of pedestrians with the priori probability $P(S^{\alpha})$, who walk close each other (shown in the red rectangle in Fig.2a), and S^{β} denotes the trajectories of isolate pedestrians with the priori probability $P(S^{\beta})$ (shown in the blue rectangle in Fig.2a), known as single nodes. Based on this reasonable division, we rewrite Eq. (1) as:

$$S^* \approx \arg\max_S \prod P(S^{\alpha}) \prod P(T \mid S^{\alpha}) + \prod P(S^{\beta}) \prod P(T \mid S^{\beta}). \tag{6}$$

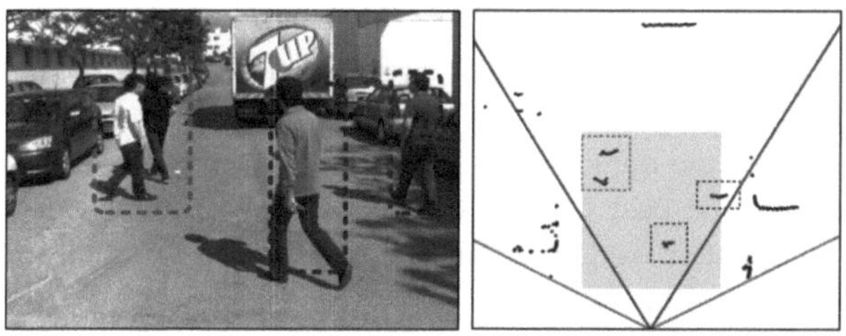

(a) (b)

Fig. 2. (a) Pedestrians in the image domain; and (b) pedestrians in depth domain

In Eq. (6), there are four terms in total. Since the values of $P(S^{\alpha})P(T \mid S^{\beta})$ and $P(S^{\beta})P(T \mid S^{\alpha})$ are small, these two terms can be omitted in our MAP formulation. To couple the non-overlap constraints with the objective function (7), then we define two 0-1 indicator variables $t_{j,i}$ and t_i as

$$t_{j,i} = \begin{cases} 1 & \text{if } T_j \text{ is right after } T_i \\ 0 & \text{otherwise} \end{cases}, \tag{7}$$

$$t_i = \begin{cases} 1 & \text{if } T_i \in S_k \\ 0 & \text{otherwise} \end{cases}. \tag{8}$$

It's easy to see that these variables are determined for a given association hypothesis S, and vice versa. S is non-overlap if and only if

$$t_i = \sum_j t_{j,i} \leq 1. \tag{9}$$

Based on two 0-1 variables, a new objective cost function is obtained when substituting Eq. (3) into Eq. (6), as follows:

$$S = \arg\min_{S^{\alpha}} \sum_{i,j} C_{j,i} t_{j,i} + \sum_{i} C_i t_i + \arg\min_{S^{\beta}} \sum_{i,j} C_{j,i} t_{j,i}. \tag{10}$$

Subject to:

$$C_{j,i} = -\log P_k \left(T_{j_{k-1}} \mid T_{i_k} \right), \tag{11}$$

$$C_i = -\log P(T_i \mid S) = -\log \frac{p_i}{1-p_i} = \log \frac{1-p_i}{p_i}. \tag{12}$$

To solve the objective function Eq.(10), subjecting to the constrains Eq.(7-9), is equivalent to find a min-cost path in a fully connected graph model: The object candidate is represented by the node of the graph in successive frames, C_i and $C_{j,i}$ donate the cost of edges in graph, and the 0-1 indicator t_i and $t_{j,i}$ donate the connection relation of the nodes. In this paper, our goal is to find the min-cost path combination among the different weighed edges of graph model.

2.2 Occlusion Model

The cost of edge C_i and $C_{j,i}$ in the graph play a key role in graph model, which leads the discriminating capacity among objects candidates (nodes). In our method, an occlusion model along the depth dimension is defined to discriminate pedestrians by their depth ordering. The cost C_i can be dynamically updated, which are defined in Eq. (12). Then an updated weight factor as a cost C_i can be obtained in each frame. We definite the factor in Eq. (12) as

$$p_i = \left[1 + \exp(\bar{Z} - Z_i) \right]^{-1}. \tag{13}$$

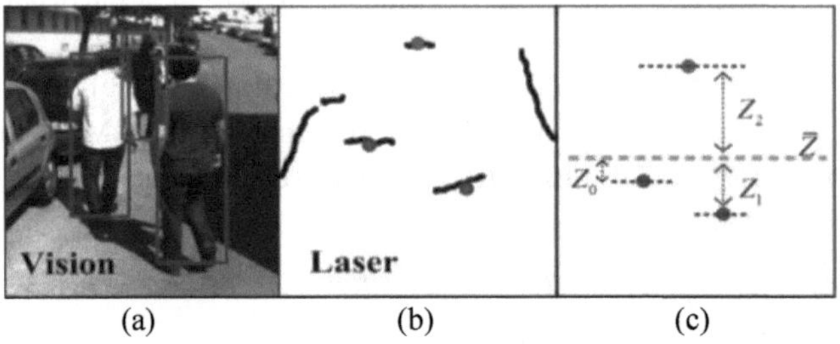

Fig. 3. (a) Pedestrians in image domain; (b) pedestrians in depth domain; (c) occlusion measure with depth information

Eq. (13) can be solved by depth value from the laser rangefinder. As shown in Fig.3a, three pedestrians can be divided into S^{α}, also known as the group nodes. Firstly, we calculate the average depth $\bar{Z}$ (the green line in Fig.3c) of the objects (red and blue points in Fig.3c). Then we can obtain the relative depth Z_i between each object and $\bar{Z}$. Finally, we can get the probability p_i, which can be invoked by Eq. (12) and then we obtain C_i. Based on the definition, the pedestrians with bigger C_i is more likely to be occluded, such as the red point with positive C_i is apt to be occluded by the blue points with negative C_i. Certainly, a single pedestrian in S^β always holds the zero relative depth and then has a zero C_i.

2.3 Optimization of IBG

The formulation in Eq. (10) can be mapped into an improved bipartite graph G, as shown in Fig.4. Each space-time location j, or equivalently pedestrian observation at time t-1, corresponds to a pair of nodes $(n_{t-1,j}, n'_{t-1,j})$ connected by an edge of cost C_i. Each putative pedestrian observation at time t, corresponds to a node $n_{t,j}$ .The transition from $n'_{t-1,j}$ to $n_{t,j}$ in two frames is an edge $(n'_{t-1,j}, n_{t,j})$ with cost $C_{j,i}$.

To optimize the IBG, we adopt an iterative strategy. It is known that a good initial solution can yield convergence to a better optimum. Therefore, we use a greedy algorithm to get initial solution. The greedy algorithm will find out the shortest path on the IBG and remove the selected cost edge from the graph until there is no C_i edge in the residue graph. Given an initial solution, the cost minimization algorithm is described in Algorithm 1.

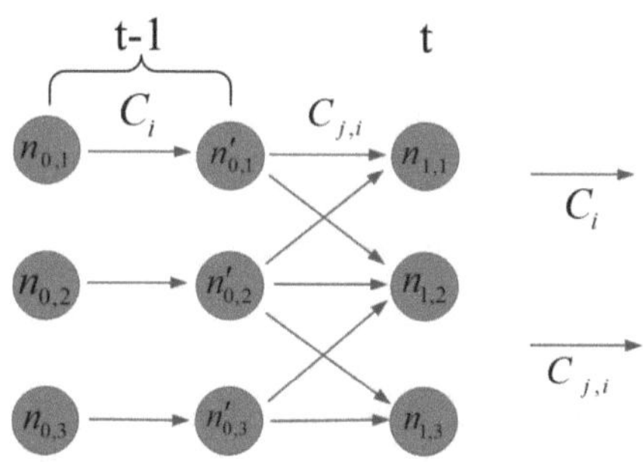

Fig. 4. Improved bipartite graph (IBG)

Algorithm 1. Switching labels with low cost

Input: IBG G; cost$\{C_i\}$, $\{C_{j,i}\}$; observation S;
Initial: Finding the label set L' of lowest cost by a greedy algorithm and evaluate its overall cost S^* by (10);

For $i<n$ **do**
  Set maximum cost $S' = +\infty$;
  For $j=i,\cdots,n$ that $C_{i,j} \in G$ **do**
    -switch label of $n'_{t-1,j}$ and $N_{t,j}$ under constrains in Eq.10 and evaluate new cost S'';
    -if $S'' < S'$, $S' = S''$;
  end
  If $S' < S^*$, $S^* = S'$, update L with this switch;
end
Output: Label set L of G.

3 Experiments

To demonstrate the effectiveness of the proposed approach, we performed comprehensive experiments on three public datasets. One is ISR-UC-imglidar-sync dataset [11]. The other two are SDL-1 and SDL-2 datasets [12], which are collected on a robot platform with a camera and a laser rangefinder. The platform moves in a campus, captures two video sequences of pedestrians occluded with each other. To evaluate the tracking performance, we adopt evaluation metrics defined in Table 1.

Table 1. Evaluation metrics

Items	Definition
GTP	Number of positions in ground truth
LF (↓)	Label failure, different with the ground truth
PF (↓)	Position failure, different with the ground truth
GTI	Number of ID in ground truth
MT(↑)	Mostly tracked trajectories, the number of trajectories that are tracked for more than 80%
ML(↓)	Mostly lost trajectories, the number of trajectories that are tracked for less than 20%
PL (↓)	Partially lost trajectories, the number of trajectories that are tracked in 20-80%
Sp (↑)	Speed, frame per second (fps)

For items with ↑, higher scores indicate better results, for those with ↓, lower scores indicate better results. The first three items (**GTP**, **LF**, **PF**) measure the accuracy between tracking result and ground truth in each frame. The following four items (**GTI**, **MT**, **ML**, **PL**) reflect the overall discrimination capacity in the occurrence period of targets. **Sp** shows the calculating speed of different algorithms, which includes observation detecting time in laser coordinates.

Table 2. Comparisons of four methods on three datasets

Dataset	Method	GTP	PF	LF	GTI	MT	ML	PL	Sp(fps)
SYNC	Dep (NN)	2360	612	566	10	4	2	4	**60-80**
	Dep+Fea	2360	229	312	10	7	1	2	25-30
	Dep+Mot	2360	268	347	10	6	2	2	25-35
	Dep+IBG	2360	**0**	**0**	10	**10**	**0**	**0**	20-25
SDL-1	Dep (NN)	384	189	189	4	1	1	2	**60-80**
	Dep+Fea	384	68	68	4	2	1	1	25-30
	Dep+Mot	384	72	72	4	2	1	1	25-35
	Dep+IBG	384	**6**	**6**	4	**4**	**0**	**0**	20-25
SDL-2	Dep (NN)	1954	512	495	31	13	8	10	**55-70**
	Dep+Fea	1954	274	294	31	18	6	7	25-30
	Dep+Mot	1954	240	286	31	17	7	7	25-30
	Dep+IBG	1954	**43**	**46**	31	**25**	**3**	**3**	18-25

Four relevant methods are evaluated on these datasets for comparison and results are listed in Table 2. "Dep(NN)" denotes the "nearest neighbor" method only applying depth information from laser rangefinder. "Dep+Fea" means the combination of the first method and the feature matching. "Dep+Mot" is the combination of the first method and motion estimate and "Dep+IBG" is the ultimate results of our method. It can be seen in Table 2 that our method has the best performance in all of the three datasets.

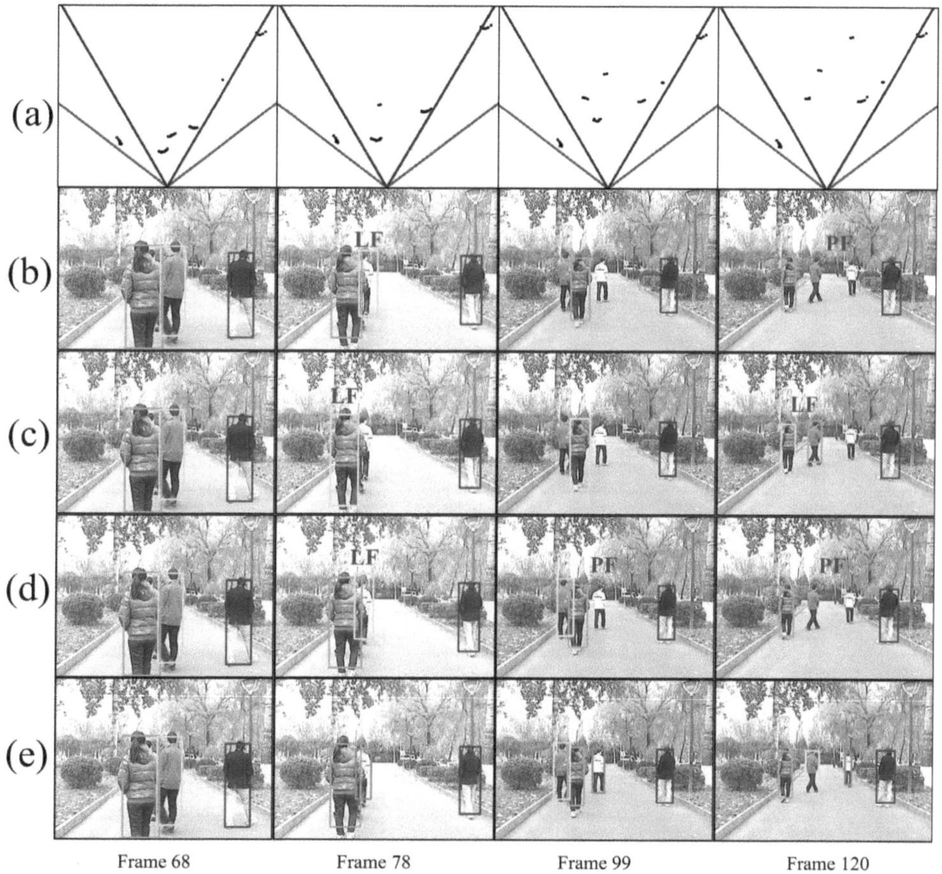

Fig. 5. Tracking examples on the dataset SDL-1. The row (a) denotes "depth picture" obtained by laser rangefinder; the row (b)-(e) show the results of "Dep (NN)", "Dep+Fea", "Dep+Mot" and "Dep+IBG" based tracking methods respectively. The **LF** and **PF** mean "label failure" and "position failure".

SDL Datasets: Fig.5 and Fig.6 show the tracking results by our online tracking model approach. The first rows (a) in both figures are "depth picture" obtained by our laser rangefinder. Traditional methods with motion and appearance model or just the simple combination of these two models can only get the cost edge of "single-node" in the improved bipartite graph. Regardless of the affinity relation among the group pedestrians, the other three methods ("Dep (NN)" in row (b), "Dep+Fea" in row (c), "Dep+Mot" in row (d)) result in mislabeling and losing pedestrians (**LF**, **PF**) during long-term and complex occlusion. However, by considering the group-node in tracking model, the pedestrians can get greatly different cost edges in our objective function, which can explain the conspicuous improvements on **MT**. Particularly, when the number of pedestrian increasing, our tracking model can keep great accuracy and robustness in the last row (e)of Fig.5 and Fig.6.

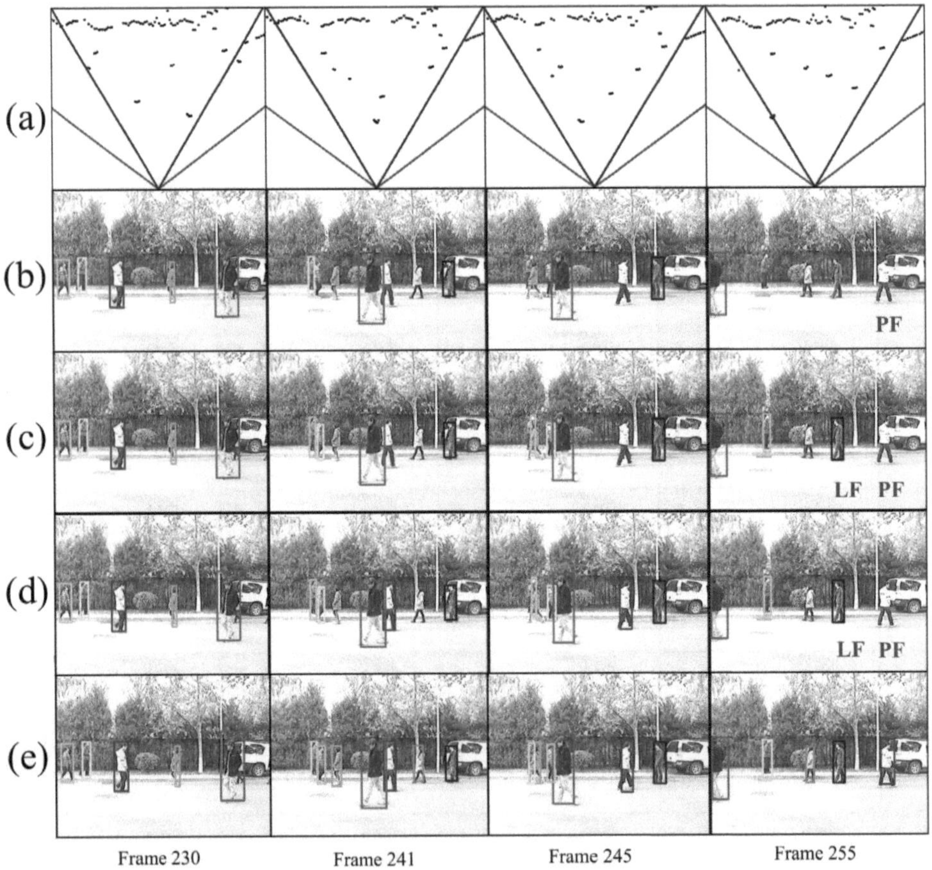

Fig. 6. Tracking examples on the dataset SDL-2. The row (a) denotes "depth picture" obtained by laser rangefinder; the row (b)-(e) show the results of "Dep (NN)", "Dep+Fea", "Dep+Mot" and "Dep+IBG" based tracking methods respectively. The **LF** and **PF** mean "label failure" and "position failure".

SYNC Dataset: as shown in Fig.7, only using the position information of laser rangefinder can not discriminate the pedestrians, especially when the merge and split happened in pedestrians' moving process, as shown in row (b), frame 351, Fig.7.

In the laser level, pedestrians and cars are treated as the same cluster when the pedestrian is so close to the car. As the result, the algorithm of "Dep (NN)" makes the **PF** and **LF** errors increase at the same time. As the appearance model added, we can finish the online feature extracting work using the "Dep+Fea" method in row (c) of Fig.7. Mostly like the "Dep+Fea" method, the motion model "Dep+Mot" can predict the motion state when pedestrians get out of occlusion. So the **PF** and **LF** errors in these two methods given in row (c) and (d) decrease greatly compared with the method of "Dep (NN)". Then we further apply tracking model "Dep+IBG" to associate the candidates in observations, the errors have dropped greatly compared with the first three methods. Because our tracking model makes full use of the fused depth and vision information among tracked pedestrians as discriminated costs in objective function, which make it easier to find the min-cost path in IBG.

(a) (b) (c) (d) (e)

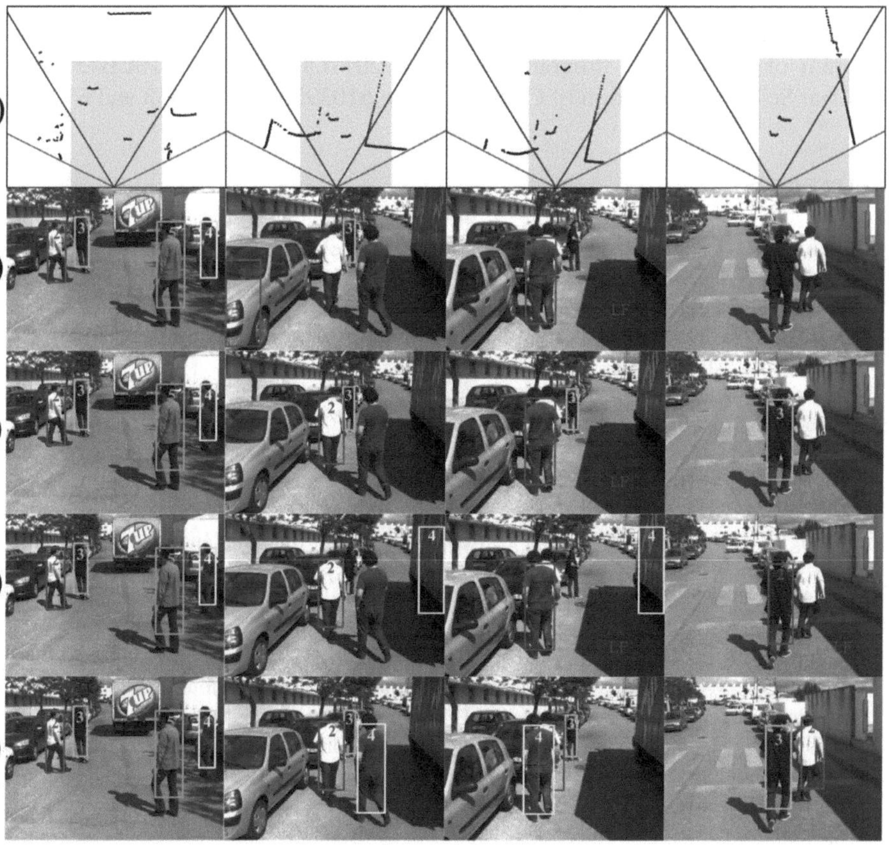

Frame 159 Frame 334 Frame 351 Frame 582

Fig. 7. Tracking examples on dataset SYNC. The row (a) denotes "depth picture" obtained by laser rangefinder; the row (b)-(e) show the results of "Dep (NN)", "Dep+Fea", "Dep+Mot" and "Dep+IBG" based tracking methods respectively. The **LF** and **PF** mean "label failure" and "position failure".

4 Conclusions

Data association is a primary issue of pedestrians tracking. In this paper, we implemented a real-time pedestrians tracking approach by integrating the depth and vision information into an IBG, which was formulated as the association cost function, and then solved by a heuristic algorithm. The extensive evaluation based on the public data sets demonstrates that our approach is powerful for complex tracking problems compared with the existing work. However, there is a limitation in our work. We use three kinds of patterns in the laser coordinates to classify the objects. Even though our data association is accurate and efficient, the unstable detection responses probably bring in more false detection observation. We will cope with this limitation in the future work.

Acknowledgments. This work is supported in part by National Basic Research Program of China (973 Program) with No. 2011CB706900, 2010CB731800, and National Science Foundation of China with No. 61039003, 61271433 and 61202323.

References

1. Okuma, K., Taleghani, A., de Freitas, N., Little, J.J., Lowe, D.G.: A boosted particle filter: Multitarget detection and tracking. In: Pajdla, T., Matas, J(G.) (eds.) ECCV 2004. LNCS, vol. 3021, pp. 28–39. Springer, Heidelberg (2004)
2. Zhao, T., Nevatia, R.: Tracking multiple humans in crowded environment. In: Proc. IEEE Int. Conf. CVPR (2004)
3. Andriyenko, A., Roth, S., et al.: An Analytical Formulation of Global Occlusion Reasoning for Multi-Target Tracking. In: Proc. IEEE Int. Conf. ICCV (2011)
4. Li, Y., Ai, H., Yamashita, T., et al.: Tracking in low frame rate video: A cascade particle filter with discriminative observers of different lifespan. In: Proc. IEEE Int. Conf. CVPR (2007)
5. Wang, S., Lu, H., Yang, F., Yang, M.-H.: Superpixel tracking. In: Proc. IEEE Int. Conf. ICCV (2011)
6. Song, B., Jeng, T.-Y., Staudt, E., Roy-Chowdhury, A.K.: A stochastic graph evolution framework for robust multi-target tracking. In: Daniilidis, K., Maragos, P., Paragios, N. (eds.) ECCV 2010, Part I. LNCS, vol. 6311, pp. 605–619. Springer, Heidelberg (2010)
7. Andriyenko, A., Schindler, K.: Multi-target tracking by continuous energy minimization. In: Proc. IEEE Int. Conf. CVPR (2011)
8. Pirsiavash, H., Ramanan, D., et al.: Globally-optimal greedy algorithms for tracking a variable number of objects. In: Proc. IEEE Int. Conf. CVPR (2011)
9. Shitrit, H.B., Berclaz, J., Fleuret, F., Fua, P.: Tracking multiple people under global appearance constraints. In: Proc. IEEE Int. Conf. ICCV (2011)
10. Breitenstein, M.D., Reichlin, F., Leibe, B., et al.: Online multi-person tracking-by-detection from a single, uncalibrated camera. IEEE Transactions on PAMI 33(9), 1820–1833 (2011)
11. Oliveira, L., Nunes, U.: Semantic fusion of laser and vision in pedestrian detection. In: Pattern Recognition, pp. 3648–3659 (2010)
12. SDL Dataset: `http://www.ucassdl.cn/resource.asp`

13. Ess, A., Leibe, B., Schindler, K., Van Gool, L.: A mobile Vision System for Robust Multi-Person Tracking. In: Proc. IEEE Int. Conf. CVPR (2008)
14. Wojec, C., Walk, S., Roth, S., et al.: Monocular Visual Scene Understanding: Understanding Multi-Object Traffic Scenes. IEEE Transactions on PAMI 35(4), 882–896 (2013)
15. Andriyenko, A., Schindler, K., Roth, S.: Discrete-Continuous Optimization for Multi-Target Tracking. In: Proc. IEEE Int. Conf. CVPR (2012)
16. Yang, B., Nevatia, R.: An online learned CRF model for multi-target tracking. Proc. IEEE Int. Conf. CVPR (2012)
17. Ess, A., Leibe, B., Van Gool, L.: Depth and Appearance for Mobile Scene Analysis. In: Proc. IEEE Int. Conf. CVPR (2008)
18. Bajracharya, M., Moghaddam, B., et al.: A Fast Stereo-based System for Detecting and Tracking Pedestrians from a Moving Vehicle. Journal of Robotics Research 2009 (2009)
19. Han, Z.J., Jiao, J.B., Zhang, B.C., Ye, Q.X., Liu, J.Z.: Visual Object Tracking via Sample-Based Adaptive Sparse Representation (AdaSR). Pattern Recognition (44), 2170–2183 (2011)
20. Zhang, L., Li, Y., Nevatia, R.: Global data association for multi-object tracking using network flows. In: Proc. IEEE Int. Conf. CVPR (2008)
21. Berclaz, J., Fleuret, F., Turetken, E., Fua, P.: Multiple Object Tracking Using K-Shortest Paths Optimization. IEEE Transactions on PAMI 33(9), 1806–1819 (2011)

Traffic Sign Detection Based on Camera Imaging Apriority

Weiwei Wu, Yanyun Qu*, Xiaoqing Yang, and Li Lin

Computer Science Department, Xiamen University, 361005, P.R. China
{wuww0528,quyanyun}@gmail.com

Abstract. Traffic sign detection is very important to the vehicle intelligent auxiliary driving system and the driverless system. However, traffic sign detection is still a challenging problem, and there is not a satisfactory solution until now. In this paper, we aim at improving the speed and accuracy of traffic sign detection. In order to improve the detection speed, we use the image-forming principle to select the scale of sliding windows instead of the standard sliding window scheme. This operation will reduce the computational complexity from $O(N^4)$ to $O(N^2)$. In order to improve the detection accuracy, we adopt the hierarchical detection scheme. In the first stage, we use the cascade GentleAdaBoost classifier combined with the Haar-like features; in the second stage, we use the GentleAdaboost classifier combined with the multiple features fusing the color cues. The hierarchical detection scheme greatly reduces the false positive rate. We implement our approach on the Swedish Traffic Signs Dataset, the experimental results demonstrate our approach is effective and our approach could greatly reduce the false positive rate while keeping the detection rate.

Keywords: traffic sign detection, camera imaging apriority, multiple features, cascade classifier, hierarchical detection, sliding window scheme.

1 Introduction

With the quick increase of the amount of vehicles, the study of intelligent auxiliary driving system and driverless vehicles attracts more and more attentions. As we know, traffic signs give important information of road condition. In this paper, we aim at improving the speed and accuracy of traffic sign detection. The driver could be reminded to pay attention to the road condition by the traffic-sign-detection system and avoid the dangers.

However, traffic sign detection is still a challenging problem. There are two difficulties: 1) the searching scheme which can achieve real-time detection; 2) the design of the detector, which can accurately detect traffic signs. The sliding window scheme is widely used in object detection. However, it is time-consuming, because it needs to search the object at all locations and scales. Its computational complexity is $O(N^4)$, where N is the width of an image. In the aspect

* Yanyun Qu is the corresponding author.

B. Huet et al. (Eds.): PCM 2013, LNCS 8294, pp. 720–729, 2013.

of the detector design, it is very difficult to design a perfect detector, because the appearances of traffic signs vary with the changes of illumination, weather condition, and viewpoint and so on.

In order to overcome the two difficulties, many researchers discussed the problem of traffic sign detection. Karla [1] gave an overview of the methods of traffic sign detection, and he divided these methods into three classes: the color-based methods, the shape-based methods and the machine learning based methods. In this paper, we are concerned about machine learning based methods. As we mentioned above, two critical factors greatly influence the performance of traffic sign detection: the detector design and the searching scheme. Xie [2]used HOG [3] to represent traffic signs, and design a SVM classifier to distinguish traffic signs from non-traffic-sign object. Karla [1] used Haar-like features [4] to represent traffic signs and designed an Adaboost classifier to determine the traffic signs. In the aspect of the searching scheme, the methods can be divided into two classes: the sliding window methods and the saliency methods. As we know, the sliding window strategy is time consuming. The saliency methods solve the problem by mimicking visual attention. Thresholding based methods [5,6,7] can be regarded as belonging to saliency methods. Kamada [5] proposed using thresholding method in RGB space, and Ohta [6] improved this method. Gavrila [7] used shape information to detect traffic signs. However, saliency methods are not reliable.

In this paper, motivated by Hoiem [8], who proposed to put objects into perspective and model the relationship between the scale and location variances in an image, we use the camera imaging principle to determine the scale of the sliding windows at a given position so as to reduce the computational cost of the sliding window strategy. We derive the interplay among the traffic signs in the real world, the camera and the projection of the traffic signs in an image. Furthermore, we design a hierarchical detection scheme to filter the non-traffic-sign regions. In the first stage, we use the cascade GentleAdaBoost classifier combined with the Haar-like features to find the candidate traffic-sign regions; in the second stage, we use the GentleAdaboost classifier combined with the multiple features fusing color cues to filter the false positive regions. The hierarchical detection scheme could greatly reduce the false positive rate. Therefore, the contributions of our work are as follows: 1) using camera imaging apriority to select the scale of sliding windows; 2) designing a hierarchical detection strategy to distinguish traffic-sign regions from non-trafffic-sign regions.

The rest of this paper is organized as follows: Section 2 briefly introduces camera image apriority. Section 3 introduces the framework of our method, and Section 4 details the implementation of our approach. The experimental results are shown in Sections 5. In section 6 we give the conclusions.

2 Camera Imaging Apriority of Traffic Signs

Hoiem [8] proposed putting local objects in the context of the overall $3D$ scene. He modeled the interdependence of objects, surface orientations and camera

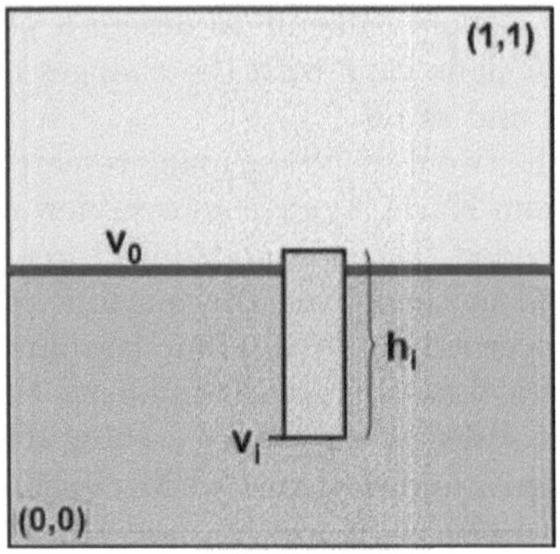

Fig. 1. The illustration of the parameters used in camera imaging principle

viewpoints. In this paper, we would use his method to derive the project information of traffic signs in an image in order to determine its scale in an image.

Now we give a brief introduction of the camera imaging principle. We assume that objects stand upright on the ground plane. In our approach, we regard a traffic sign and its stick as a whole. If we know the horizon line and the camera height, we will derive the relationship between the object's world height y and the object image height using the following notation: world coordinates (x, y, z) with y being height and z being depth which is the distance between the object and the camera; camera tilt θ_x and focal length f; camera height y_c; pixel coordinate (u, v) in an image ranging from $(0, 0)$ at the bottom-left and $(1, 1)$ at the top-right; camera optical center (u_c, v_c); the horizon position v_0 as the vanishing line of the ground plane in image coordinate and we assume that the ground plane is at $y = 0$ and its location in an image is v_0. According to the camera imaging principle, the camera tilt angle can be approximately computed as $\theta_x \approx 2arctan\frac{v_c - v_0}{2f}$. Furthermore, the relationship between the pixel coordinates (u, v) and the object's world coordinates (x, y, z) is given

$$\begin{bmatrix} u \\ v \\ 1 \end{bmatrix} = \frac{1}{z} \begin{bmatrix} f & 0 & u_c \\ 0 & f & v_c \\ 0 & 0 & 1 \end{bmatrix} \begin{bmatrix} 1 & 0 & 0 & 0 \\ 0 & cos\theta_x & -sin\theta_x & y_c \\ 0 & sin\theta_x & cos\theta_x & 0 \end{bmatrix} \begin{bmatrix} x \\ y \\ x \\ 1 \end{bmatrix} \tag{1}$$

Inferring Equ. (1), we can obtain the object's height y:

$$y = \frac{z(fsin\theta_x - (v_c - v)cos\theta_x) - fy_c}{(v_c - v)sin\theta_x + fsin\theta_x} \tag{2}$$

We let v_t and v_b denote the heights of the top and bottom of the object in an image. When the height of the object is zero, that is, $y = 0$, from Equ. (2), we can solve the object depth z as

$$z = \frac{fy_c}{fsin\theta_x - (v_c - v_b cos\theta_x)}.$$

We assume that the camera tilt is small, that is,

$$cos\theta_x = 1, sin\theta_x \approx \theta_x, \theta_x \approx \frac{v_c - v_c}{f};$$

the object's world height y can be inferred as

$$y = \frac{y_c \frac{v_t - v_b}{v_0 - v_b}}{1 + \frac{(v_c - v_0)(v_c - v_i)}{f^2}}$$

As we assumed before, the horizon position is near the camera optical center, that is

$$\frac{(v_c - v_0)(v_c - v_i)}{f^2} \approx 0,$$

so we can simplify Equ. (2) as follows,

$$y = y_c \frac{v_t - v_b}{v_0 - v_b}. \tag{3}$$

In this paper, we modify Equ. (3) as follows,

$$y_i = k y_c \frac{h_i}{v_0 - v_b} \tag{4}$$

where the h_i, y_i, v_i are the object's image height, world height, and bottom position respectively of object i. Thus, we can solve for object image height h_i

$$h_i = \frac{(v_0 - v_b) y_i}{k y_c} \tag{5}$$

where the threshold $k = 1.5$. The explanation of the parameters used in this section is shown in Figure 1. As mentioned before, we regard a traffic sign and its stick as a whole, so h_i represents the height of the traffic-sign object. In the real world, the size of traffic signs is about $0.5m \times 0.5m$ and the possible height of the whole is $1.0m, 2.5m, 4.0m$. So from Equ. (5), we can limit the size of the sliding windows as follows,

$$\alpha \frac{(v_0 - v_b) y_i}{k y_c} - \varepsilon < h_i < \alpha \frac{(v_0 - v_b) y_i}{k y_c} + \varepsilon \tag{6}$$

3 Framework

Figure 2 shows the framework of our approach. For a query image, we firstly sample the patches by the sliding window scheme which selects the scale of sliding windows based on the camera imaging apriority. After that, we determine if a sampled patch belongs to the traffic-sign class by a cascade binary classifier in which each classifier is an Adaboost classifier with Haar-like features. Furthermore, we filter the candidate regions by the second-level Adaboost classifier with multiple features fusing the color cues. The second level operation can effectively reduce the false positive rate.

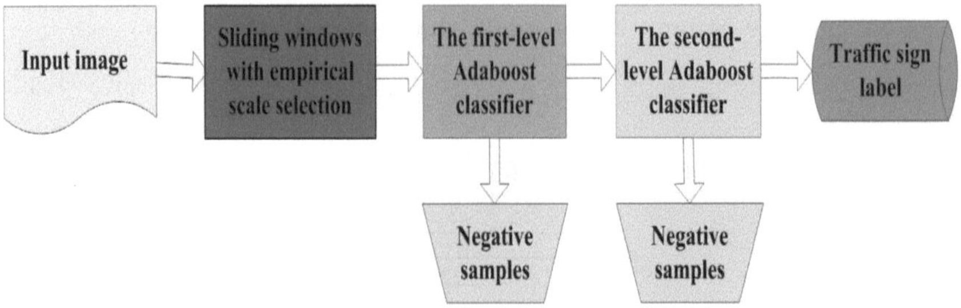

Fig. 2. The framework of our approach

3.1 Implementation Details

In our approach, we first select the scale of the sliding windows via camera imaging apriority. As shown in Figure 3, we show an example how the sliding windows vary with the location. The camera imaging apriori knowledge implies that the size of traffic-sign objects is among a range. In our experiment, the width of sliding windows is selected between 20 pixels and 300 pixels.

In the first stage, we use the cascade GentleAdaBoost classifier [9] combined with the Haar-like features [4]. Figure 4 shows the flowchart of the cascade classifier. This classifier has a high detection rate and the lower false detection rate. Consider that the results obtained from the first level detection contain many false positive patches, we implement the second level classifier to refine the traffic sign regions. In the second stage, we use the GentleAdaboost classifier combined with the multiple features fusing the color cues and it was shown effective in traffic sign detection and recognition in our previous work [10].

In this section, we design the color fused feature which is concatenated by HOG, LBP and Hue histogram. The HOG (Histogram of Oriented Gradients) was proposed by Dalal [3] for pedestrian detection. In this paper, we normalize

Fig. 3. The scales of sliding windows based on camera imaging apriority

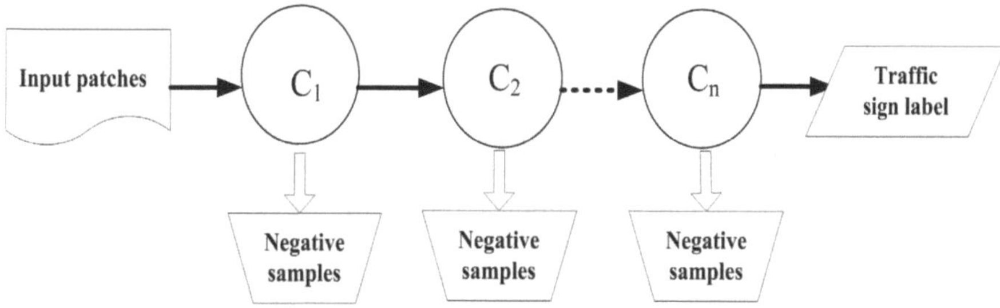

Fig. 4. The cascade classifier

a traffic-sign window to 40 × 40. Then, it is divided into some block and each block is divided into some cells. In our experiment, we use 49 blocks, each block includes 2 × 2 cells, the size of a cell is 5 × 5 , in which a gradient histogram of 9 bins is obtained. Thus,we can obtain a HOG feature vector with 1764 dimensions.

LBP feature is proposed by Ojala [11], and it is a popular texture descriptor. The basic operator assigns a label to every pixel of an image by thresholding its 3 × 3-neighborhood with the center pixel value and considering the result as a binary number. Then the histogram of the labels can be used as a descriptor. In our approach, we normalized the labeled patch to 40 × 40. The normalized patch is tiled by a 4 × 4 grid. The descriptors are then concatenated to form a global description of the labeled region.

Both of HOG features and LBP features neglect the color cues, but the color cues are important to traffic sign detection. So we use the HOG and LBP features fusing the color cues. Our previous work has shown that the concatenation of HOG features, LBP features and Hue histogram can achieve better detection performance. Figure 5 shows the multiple features which fuse the hue histogram. In our approach, we convert images from RGB to HSV, and the Hue histogram is obtained by dividing Hue to 256 bins. Thus, the color descriptor is a vector of 256 dimensions.

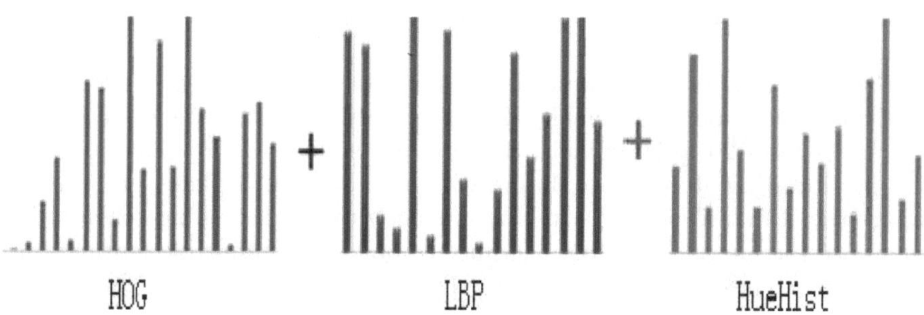

Fig. 5. The color cue fused multiple features

4 Experimental Results

In this section, we implement our approach on the STSD dataset (Swedish Traffic Signs Dataset). The STSD contains more than 20000 pictures and the size of each picture is 1986×960. In our experiments, we use 1970 labeled pictures from the STSD as our testing set. From the remaining pictures, we construct the Swedish30 set as our training set which contains 3192 positive samples belonging to 30 classes of traffic signs. Figure 6 a) shows a typical labeled picture in the STSD, and Figure 6 b) shows the 30 classes of traffic signs in Swedish30.

a) b)

Fig. 6. An example and the traffic signs of STSD dataset. a) An example of STSD dataset, b) 30 classes of traffic signs in STSD dataset.

In the first experiment, we evaluate how the camera imaging apriority is reasonable to traffic sign detection. We select 100 pictures from the testing set randomly. The 100 pictures contain 159 traffic signs. And we use the camera imaging apriority to sample. In the samples, there are 156 traffic signs. The result demonstrates that the camera imaging apriori information could keep about 98% of traffic signs detected. Figure 7 shows some regions generated by camera imaging apriority.

In the second experiment, we evaluate how the camera imaging apriority reduce the computational complexity. In our experiments, the size of sliding windows ranges from 20 to 300 pixels, the size of a sliding window increases by 10 pixels per iterating step and a sliding window moves at 5 pixels per time. We compare the standard sliding window scheme with our searching scheme. Both of them use the GentleAdaboost classifier combined the multiple features fused with the color cues to filter the sampled patches. Table 1 shows the comparison results in a query image. It demonstrates that our approach could reduce the computational cost theoretically from $O(N^4)$ to $O(N^2)$. The real consuming time could be reduced from 40000 seconds to 50 seconds in a query image.

In the last experiment, we evaluate the performance of our hierarchical detection method. We implement our detection approach in the testing set and

a) b)

Fig. 7. The reasonableness of scale selection via the camera imaging apriority. a) A query image; b) The sampled regions based on camera imaging apriority.

Table 1. The comparison of two searching schemes

	Standard sliding window scheme	Our approach
Computation complexity	$O(N^4)$	$O(N^2)$
Time(s)	40000	50

Table 2. The comparison in terms of the false detection rate (%)between the first level and second level detections

Scales of windows	10×10	20×20	30×30	40×40	50×50	70×70	100×100
The first level	99.23	97.73	92.79	89.46	87.47	89.36	81.03
The second level	88.18	73.34	53.31	43.39	36.94	37.42	26.66

Table 3. The comparison in terms of the detection accuracy (%)between the first level and second level detections

Scales of windows	10×10	20×20	30×30	40×40	50×50	70×70	100×100
The first level	89.4	90.69	90.47	88.66	86.96	81.52	69.23
The second level	89.3	90.66	90.46	88.62	86.95	81.52	69.23

we use the detection accuracy and the false detection rate as our criterion. The detection accuracy is the ratio of the number of correctly detected regions and the number of ground truth regions. A region is called to be correctly detected if the ratio of the intersection area and the union area between the detected region and the ground truth region is greater than 50%. The false detection rate is the ratio of the number of false positive samples and the total number of detected regions. Considering that the pictures in STSD are complex and the scales of traffic signs change greatly, we count our results according to scales of traffic signs in the testing set. As shown in Tables 2 and 3, the results in the column of 10×10 (the false positive rate or the detection accuracy) are obtained on all the samples whose size is greater than 10×10 and the results in other columns are explained in the same way. Table 2 shows the false detection rates. We can see that the second level detection can effectively reduce the false positive rate. In the aspect of the detected samples, the false detection rate is only 26.6%, which reduces about 55% compared with the first level detection. Moreover, Table 3 shows that the detection accuracy after the second level detection is almost unchangeable compared with the first level detection. Figure 8 shows the visual results of the first level and second level detections. We can see that most of the obtained ROI from the first level detection can be filtered in the second level detection. Thus, the proposed hierarchical detection method could effectively reduce the false detection rate while keeping the detection rate.

Fig. 8. The comparison of the first level detection and the second level detection. a) The result obtained by the first level detection, b) The result obtained by the second level detection.

5 Conclusions

In this paper, we design a prototype system of traffic sign detection. The experimental results have demonstrated that the camera imaging apriority is effective for reducing the computational complexity of the sliding window searching, and the hierarchical detection scheme has a good performance in traffic sign detection, which greatly reduces the false positive rate. In the future, we would optimize the algorithm in order to implement it on real videos in real time.

Acknowledgments. This research work was supported by the National Natural Science Foundation of China Under Grant No. 61373077, the Natural Science Foundation of Fujian Province of China Under Grant No. 2013J01257, Research Funds for the Central Universities Under Grant No.2010121067 and the 2013 national college students' innovative and entrepreneurial training project.

References

1. Brkic, K.: An overview of traffic sign detection methods (2010)
2. Xie, Y., Liu, L., Li, C., Qu, Y.: Unifying Visual Saliency with HOG Feature Learning for Traffic Sign Detection. In: IVS 2009, pp. 24–29 (2009)
3. Dalal, N., Triggs, B.: Histograms of oriented gradients for human detection. In: IEEE Computer Society Conference on Computer Vision and Pattern Recognition, vol. 1, pp. 886–893 (2005)
4. Viola, P., Jones, M.: Rapid object detection using a boosted cascade of simple features. In: Proceedings of the 2001 IEEE Computer Society Conference on CVPR 2001, vol. 1, pp. I-511–I-518 (2001)
5. Kamada, H., Naoi, S., Gotoh, T.: A compact navigation system using image processing and fuzzy control. In: Proceedings of the Southeastcon 1990, pp. 337–342. IEEE (1990)
6. Ohta, Y.I., Kanade, T., Sakai, T.: Color information for region segmentation. Computer Graphics and Image Processing 13, 222–241 (1980)
7. Gavrila, D.M., Philomin, V.: Real-time object detection for "smart" vehicles. In: The Proceedings of the Seventh IEEE International Conference on Computer Vision, pp. 87–93 (1999)
8. Hoiem, D., Efros, A.A., Hebert, M.: Putting objects in perspective. In: IEEE Computer Society Conference on Computer Vision and Pattern Recognition, vol. 2, pp. 2137–2144 (2006)
9. Friedman, J., Hastie, T., Tibshirani, R.: Additive logistic regression: a statistical view of boosting (With discussion and a rejoinder by the authors). The Annals of Statistics 28, 337–407 (2000)
10. Yang, X., Qu, Y., Fang, S.: Color fused multiple features for traffic sign recognition. In: Proceedings of the 4th International Conference on Internet Multimedia Computing and Service, pp. 84–87 (2012)
11. Ojala, T., Pietikäinen, M., Harwood, D.: A comparative study of texture measures with classification based on featured distributions. Pattern Recognition 29, 51–59 (1996)

Recommend Social Network Users Favorite Brands

He Feng and Xueming Qian

SMILES LAB, Xi'an Jiaotong University, China
fenghe7658@stu.xjtu.edu.cn, qianxm@mail.xjtu.edu.cn

Abstract. With the development of social network and image sharing websites, users are willing to upload their favorite photos on the websites and assign them some texts to describe the image content. Thus we can capture their interest by these photos and corresponding texts, and recommend relevant brands based on user's interest. This paper proposes a novel brands recommendation approach for social network users based on their browsing images and labeled texts. Firstly, we enrich the uploaded image's texts by image annotation approach. Secondly, we build brand tree from the collected datasets. And then, we recommend brands by scalable brand mining based on tree structure. Finally, we conduct a series of experiments on real Flickr users. The experiment results show the effectiveness of our approach.

Keywords: User's Interest, Brands Recommendation, Brand Tree, Scalable, Social Network.

1 Introduction

Nowadays, we are moving forward into a new era with the rapid development of network. It almost becomes an essential part of daily life, and an important way to obtain and disseminate information. Especially, the birth of web2.0 improves user's initiative in Internet greatly, and then brings volume of social network such as Flickr, Facebook, Twitter etc. The booming of information and user's invaluable resource on the Internet can bring huge economic benefits. Advertising is one of the important ways. It is not only a way of businessmen to release goods information but also an approach of users to get services in network economy.

If ads can meet people's need, then they become useful information to them and can facilitate their life. For example, for the social user shared photos, Qian et al., proposed to model users' tagging behavior and then recommend user preferred vocabularies for user by fusing the photo taken time, visual information, and photo taken locations [8]. How to make ads meet user's need? We think that it is necessary to advertise from user's interest. It is the first step to attract user's attention. Furthermore, some users have their favorite brands, which are the more accurate interests. If we can mine user's favorite brands information from their browsed images and labeled texts, then we will recommend more user-targeted ads to get the favor of users. So, to make ads meet user's need and clicked, we should solve brand recommendation problem firstly.

B. Huet et al. (Eds.): PCM 2013, LNCS 8294, pp. 730–739, 2013.

Previous works [5],[6],[9] paid more attention on produces advertising based on user's interest. They learnt and matched the topics of user's photos and products based on ODP tree, which did well in semantic gap and vocabulary impedance problem of the vocabulary of users' photos and products, but little attention on brands recommendation, because both products' description and ODP's webs content have less relevance to brand. We think user's favorite brand is more accurate user interest. So we build brand tree including more brand information offline to handle semantic gap and vocabulary impedance problem like ODP tree of Argo [6]. However, Argo mapped user photos tags (user labeled and annotated) to all leaf nodes of ODP tree to get user photos' topic distribution to represent user's interest. It did solve textual ambiguities and semantic mismatch problem, but make user's interest less prominent and susceptible to noise cause of the large number of leaf nodes. Some of the leaf nodes' topic has little relevance to user photos' topic and little effect on textual ambiguities and semantic mismatch problem. And different users correspond to different leaf nodes. So, to make user's interest prominent and robust, we need get candidate leaf nodes relevant to user photos' topic, which will also shorten match time.

Motivated by the necessity of user favorite brand mining and the potential improvement of Argo, we propose a novel approach scalable brands recommendation to match user interested brand products from one picture and corresponding texts of user browsed. The contribution of this paper is twofold: Firstly, we could get user's more accurate interest by brand mining. Secondly, our scalable approach recommends brands layer by layer, which firstly detect user interested categories to get candidate brands (leaf nodes) relevant to user photos' topic and then rank them. We believe that user's interest is relevant to several categories usually. For example, user's interested in sports, and he may be interested in clothes of Adidas.

2 The Approach

With the rage of the social network, users' browsed photos and corresponding texts reflect their interest and even brand preference.

To recommend more user-targeted ads, we need to mine user's favorite brand from their browsed photos and corresponding texts. So we should get keywords to represent user's interest and then match brands effectively. The system overview of scalable brand recommendation for social network users is shown as Figure 1. Firstly, one of user's browsing photo and corresponding texts are system input. Secondly, we enrich picture's texts with image annotation approach in [7], [10]. Here we denote user's label texts and generated annotations as user tags (UT). Thirdly, we use scalable brand mining to get brand distribution to represent user's brand preference based on brand tree, which can be seen as the constraint condition to brand recommendation online. To express our scalable approach better, we detail the brand tree in next part firstly, and then scalable brand mining. Fourth, we rank brands based on brand mining. At last, the targeted brands are the real results from our approach.

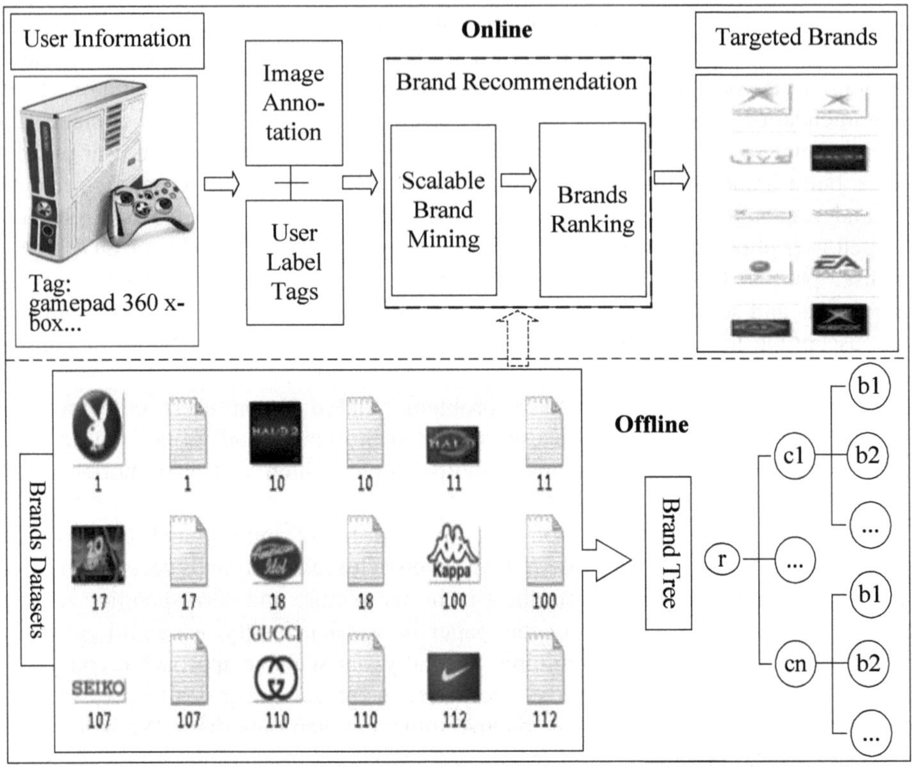

Fig. 1. System overview of scalable brands recommendation for social network users

2.1 Brand Tree

Argo's topic distribution model solved vocabulary impedance problem well, but little attention on brands mining, because both products' description and ODP's webs content have less relevance to brand. So, we build brand tree (Figure 2) from brands datasets to get brand distribution like ODP tree of Argo. However, there are two main difference: 1) brand tree is build based on volumes of brands' description [4] including brand category, brand name, products' description of the brand, and some keywords of the products' description as shown in Table 1. This information can be mined from Internet; and 2) Argo recommends products based on ODP tree without consideration of brand, while ours recommends user interested brands from brand tree. There are two layers of brand tree structure. The first layer is products category layer, and the second is brand layer. We put the same brand category of brands datasets into one big category, which build the first layer of brand tree. In our test brand datasets, we get 20 big categories which almost cover brand category of all products. They are Arts & Entertainment, Automobile, Clothes & Fashion & Beauty product, Community & Government, Computers & Electronics, Education, Food &

Dining, Health & Medicine, Home & Family, Hotel, Industry & Energy, Legal & Financial, Media & Communications, Miscellaneous, Office Products, Other, Pets, Sports, Tools & Hardware, Travel & Transportation.

Different categories of the first layer have different brand nodes of the second layer. For example, the Sports category has brand nodes of ADIDAS, NIKE, and YONEX etc.

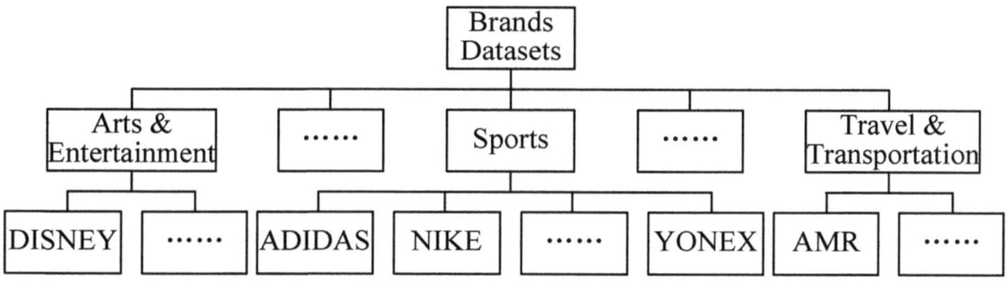

Fig. 2. Brand tree

Table 1. An example of brand description

Nike	
brand category	Sports
brand name	Nike
products' description	Nike is a major American supplier of athletic shoes, apparel and sports equipment.
products' keywords	athletic shoes, sport

2.2 Scalable Brand Mining

To learn efficient user interest, we remove the high-noise term of UT. And then, we detect user interested categories (UIC) to filter irrelevant brand nodes. Finally, we score candidate brands' similarity to user's interest to get brand distribution.

Noise Tag Filter

Usually, the labeled texts for user shared photo and even the annotations are not all relevant to photo topic. Some of texts are noise which will give negative results for detecting user interested category.

Many previous works used TF-IDF to evaluate term's importance of a document in set of files. It can be also used to reduce noise terms of a document out set of files. We reduce noise of UT by IDF as Eq. (1). Here UT is the document out set of files including 20 categories' description of first layer in brand tree. Because noisy terms have global distribution (the term appears in most categories of first layer) while theme-clear terms have localized distribution (the term appears in a few categories of first layer) [6], we can discriminate them by IDF. Let $T=\{t_1,t_2,...,t_M\}$ denotes UT, and $\Gamma=\{\tau_1,\tau_2,...,\tau_m\}$ denotes the filtered texts, where M is the total number of terms in UT, m is the number of lower-noise terms in UT, and $m \le M$, $\tau_i \in T$.

$$IDF_i = \sum_{j=1}^{n} D(d_j, t_i), i \in \{1, \cdots, M\} \tag{1}$$

where $D(x, y)$ denotes whether term y appears in document x or not. We have $D(x, y) = 1$ if y appears in x, and $D(x, y) = 0$ otherwise. Here $y = t_i$ is the i-th term of T, $x = d_j$ is the document of all brands product description of the j-th category of first layer. And n=20 is the number of categories in first layer.

Thus IDF_i is the times that t_i appears in all categories of first layer. The higher the IDF_i, the more likely the term t_i is a noise term. If IDF_i is large enough then we consider the term t_i has less contribution for detecting user interested category. So, it is settled as user interest irrelevant term.

Detect UIC

We use α_j to denote the possibility of the j-th category to be user's interest as Eq. (2).

$$\alpha_j = \sum_{i=1}^{m} F(d_j, \tau_i), j \in \{1, ..., n\} \tag{2}$$

where $F(x, y)$ is frequencies that term y appears in document x. Here $y = \tau_i$ is the i-th term of Γ, $x = d_j$ is the document of all brands product description of the j-th category of the first layer.

Thus the higher the α_j, the more likely the category is user's interest, and the more likely the brands belonging to the category are user interested brands. Then we rank these categories by α_j to get top k user's interest relevant categories to filter irrelevant brand nodes. And we compare the performance with different k in Section 3.1.

We weight top k categories in the following two ways: 1) put $c_j = 1, j \in \{1, \ldots, k\}$. We call it the unweighted scale method; and 2) normalize α_j of the top k categories as Eq. (3):

$$c_j = \alpha_j / \sum_{j=1}^{k} \alpha_j, j \in \{1, ..., k\} \tag{3}$$

And we compare the performance of these two weighted ways in Section 3.2.

Brand Distribution

After detecting UIC, we consider the k categories' brands as candidate. We use β_j^l to denote the possibility of the l-th brand of j-th category to be user interested brand as Eq. (4).

$$\beta_j^l = \sum_{i=1}^{m} F(b_j^l, \tau_i), j \in \{1, ..., k\}, l \in \{1, ..., N\} \tag{4}$$

where τ_i is the i-th term of Γ, b_j^l is the document of the l-th brand node description of j-th category. And we will normalize β_j^l. It's worth noting that different category has different brand node number N of brand tree.

We use s_j^l to denote similarity between the l-th brand of j-th category to user's interest as Eq. (5), and then we get brand distribution to represent user's brand preference.

$$s_j^l = c_j \cdot \beta_j^l, j \in \{1, ..., k\}, l \in \{1, ..., N\} \tag{5}$$

We can see that not all of brand nodes are taken into consideration and we add a new factor (category weight) to make suggested brands more prominent.

2.3 Brands Ranking

We rank brands by the value of s_j^l. Then, We have two ranking results (unweighted scale and scalable approach) corresponding to two category weight ways. From Eq. (5), we can see that Argo is a special case of our approach when k equals n (the number of categories of first layer) and categories are unweighted.

3 Experiments

Experiments are conducted to evaluate the performance of our proposed scalable brands recommendation based on tree structure approach.

We use 125 users' pictures and corresponding labeled text from Flickr [1] to test our scalable approach. And the test brand datasets contains 7,284 brands logo and corresponding brands description [3][4].

We invite three volunteers to evaluate top 10 targeted brands for each of the Flickr user as irrelevant, relevant or perfect like Argo [6], by browsing the shared photos and the textual descriptions. "irrelevant" means the recommended brand is a false-alarm. "relevant" means the recommended brand is somewhat relevant, and "perfect" means strong relevance. And then, AP (average precision) and WAP (weighted average precision) [6] defined as Eq. (6) and Eq. (7) were used for the evaluation.

$$\mathrm{AP} = (p+r)/(p+r+i) \tag{6}$$

$$\mathrm{WAP} = (p+0.5*r)/(p+r+i) \tag{7}$$

where p, r, i denote the number of "perfect", "relevant", and "irrelevant" brands respectively.

3.1 Performance Comparison with Different *k* Categories

Here we conducted experiment to compare our approach scalable brands recommendation based on tree structure with different k categories, and the categories are unweighted.

Figure 3 illustrates the AP and the WAP of top 10 recommended brands to the 125 users of three different k values. We can see that the performance of k=2 has the highest AP and WAP at almost all top 10 results, because user's interest is usually relevant to a few categories.

3.2 Performance Comparison between Scalable Brand Ranking and Argo

Here we conducted experiment to compare our scalable approach with non-scale approach Argo [6] based on brand tree. And we make k=3 in detect UIC part.

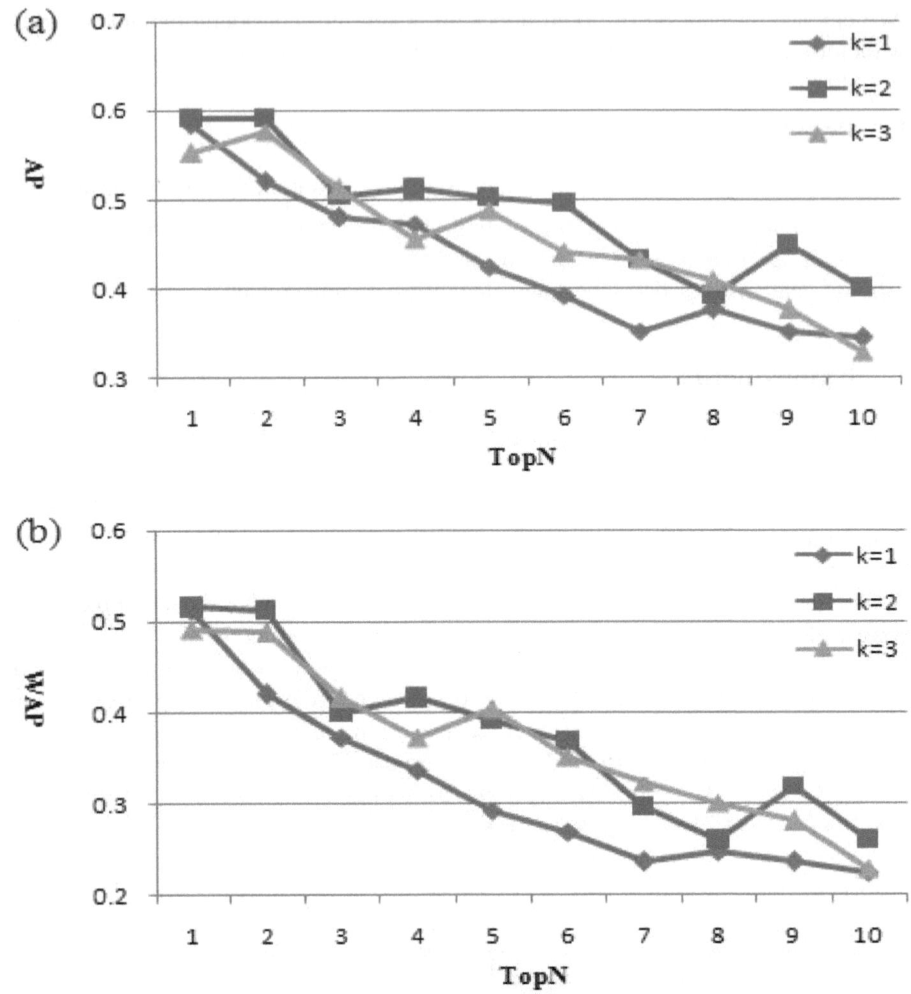

Fig. 3. Performance of scalable brands recommendation with different *k* categories

Figure 4 illustrates the AP and the WAP of top 10 recommended brands to the 125 users of three approaches (Scalable brand ranking, unweighted scalable approach and Argo). From Figure 4, we can see that the AP and WAP value are declining overall from the top1 to top10 result. The weighted scalable approach has the highest AP and WAP values. The AP value ranged from 0.648 to 0.376, the WAP value ranged from 0.576 to 0.3 for the top ranked 10 products. And all of the WAP value is under the AP value from the top1 to top10 result. Because not all of the recommended brands are perfect result, some of them are only relevant.

Table 2 shows the execution time and average AP of top 5 results of Argo, unweighted scale and scalable approach. We can see that execution time of scalable approach is lower than Argo, and average performance is higher than Argo.

Table 2. Execution time and average performance of Argo and scalable approach

Comparison	Time(ms)	AP
Argo	225.128	0.402
Unweighted Scale	17.683	0.517
Scalable	17.128	0.587

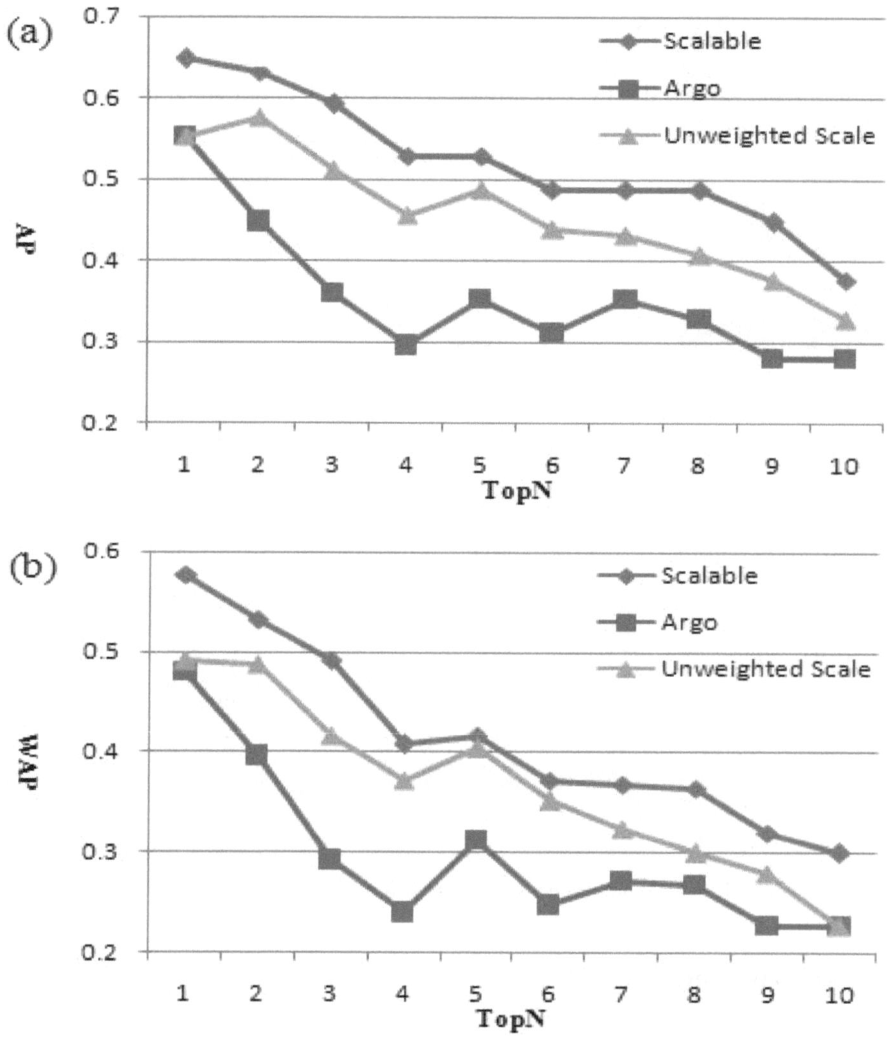

Fig. 4. Performance of scalable brand ranking with weighted and unweighted categories and Argo

3.3 Example Cases

Figure 5 illustrates a few examples of the system output. Each row corresponds to a Flickr user and recommended brands. In the first row, top 6 results are high relevant to Sony products. In the second and third row, only the top 1 or 2 results are relevant to user interest because it's the only relevant brands in brand datasets. In the fourth and fifth row, although users didn't label their favorite brand, our approach still recommends high relevant brands to their interest.

We consider that some Flickr users' photos are mainly products, and it is unreasonable to suggest relevant products to them. Thus, we assured that the 125 tested users' photos could reflect their interest rather than commercial intentions.

Fig. 5. Examples of flickr users and recommended brands

4 Conclusions

We expand brand recommendation for social network users by this work. Firstly, we built the two layer brand tree from brands datasets to match relevant brand efficiently. Further, we recommend brands layer by layer to filter irrelevant brand nodes. And, we noticed the importance of the connection between the two layers by scalable brand mining. At last, real Flickr users' experiments proved the efficiency of our approach.

At present, our approach only recommends the relevant brands from user's submission but not takes the advertiser into consideration. Our approach can be expanded in following aspects: 1) Competitive bid from advertiser when we recommend brands. 2) Recommend products based on user's favorite brand.

Acknowledgments. This work is supported partly by NSFC No. 60903121, No.61173109, Microsoft Research Asia.

References

[1] Flickr, `http://www.flickr.com/`
[2] ODP, `http://www.dmoz.org/`
[3] Mei, T., Hua, X.S., Li, S.: Contextual in-image advertising. In: ACM Multimedia, pp. 439–448 (2008)
[4] Mei, T., Hua, X.-S.: Contextual Internet Multimedia Advertising. In: Proceedings of the IEEE, pp. 1416–1433 (2010)
[5] Wang, X.-J., Yu, M., Zhang, L., Ma, W.-Y.: Advertising based on users' photos. In: ICME 2009, pp. 1640–1643 (2009)
[6] Wang, X.-J., Yu, M., Zhang, L., Cai, R., Ma, W.-Y.: Argo: Intelligent Advertising by Mining a User's Interest from His Photo Collections. In: ACM Data Mining and Audience Intelligence for Advertising, pp. 18–26 (2009)
[7] Wang, X.-J., Zhang, L., Li, X.-R., Ma, W.-Y.: Annotating Images by Mining Image Search Results. IEEE Transactions on Pattern Analysis and Machine Intelligence, 1919–1932 (2008)
[8] Qian, X., Liu, X., Zheng, C., Du, Y., Hou, X.: Tagging photos using users' vocabularies. Neurocomputing 111, 144–153 (2013)
[9] Feng, H., Qian, X.: Recommendation via user's personality and social contextual. In: ACM CIKM 2013 (2013)
[10] Li, Q., Gu, Y., Qian, X.: LCMKL: Latent-community and multi-kernel learning based image annotation. In: ACM CIKM 2013 (2013)

Event Representation and Visualization from Social Media

Xueliang Liu[1,*] and Benoit Huet[2]

[1] Hefei University of Technology, Hefei, China
[2] EURECOM, Sophia Antipolis, France
liuxueliang@hfut.edu.cn, huet@eurecom.fr

Abstract. The user generated content, available in massive amounts in social media, is receiving increased attention due to its many potential applications. One of such applications is the representation of events with multimedia data. This paper addresses the problem of retrieving and summarizing events on a given topic, and propose a novel and original framework for leveraging social media data to extract and illustrate social events automatically on any given query. The problem is tackled in three steps. First, the input query is parsed semantically to identify the topic, location, and time information related to the event of interest. Then, we use the parsed information to mine the latest and hottest related events from social news web services. In the end, for each event, we retrieve both relevant tweets on Twitter and compelling images from Google image search. The resulting documents are shown within a vivid interface featuring both event description, tag cloud and photo collage.

1 Introduction

The amount of social media data posted daily by online users has grown rapidly in recent years and continues to do so. How to leverage the explosion of this vast amount of data to benefit web users at large is, however, still an open and challenging problem with numerous potential application.

An event is one of the most important cues for people to recall past moments of one's life. The reminder value of an event makes it extremely helpful in organizing data. Thus, events-based media analysis has recently drawn much attention within the multimedia research community.

In this paper, we address the problem of retrieving and summarizing events from social media data, and propose a novel and original framework to extract and illustrate social events automatically on a given query. To solve the problem, we use natural language processing algorithm to parse the input query semantically and extract the most popular events from social news web service. Then, the obtained events are depicted through a multi-modal faceted representation composed of a textual description, a tag cloud and a photo collage, providing

* The work was performed while the first author was working towards his PhD at EURECOM.

B. Huet et al. (Eds.): PCM 2013, LNCS 8294, pp. 740–749, 2013.

Fig. 1. The snapshot of a query example results

the viewer with a rich informative panel about events. The novelty of our proposed framework is twofold. Firstly, as opposed with other methods [15] that mine events from social media data directly, we extract events from social news website Digg.com. This has considerable advantages: no additional time cost, computation and storage are needed and the results can be achieved online. Furthermore, the built-in collective intelligence of social news web services can be used to rank the hottest events. Secondly, we collect relevant media documents from different media source and use tag cloud and photo collage to provide vivid interface illustrating as visually as possible the events. This enables content from different view to be shown at the same time. The result is generated automatically on any given query. A subset of the events for an example query can be found in Figure 1.

This paper is organized as follows: we review the related work in Section 2, and describe our proposed method in Section 3. Experimental results are presented in Section 4. Finally, we summarize the paper and propose future work in Section 5.

2 Related Work

Nowadays, Twitter has become one of the most important tools for people to share their interest, their personal issues, their views and their experiences as well as comment on other's. Some research has been done to find events directly from Twitter post [15,14]. In [15], the authors studied how to employ a wavelet-based techniques to detect events from Twitter stream. A similar method can be found in [4] to detect events from Flickr time series data. In [14], the authors investigate how to filter the tweets to detect seismic activity as it happens. In [2], a system is proposed to detect emerging topics from social streams and illustrate the topics with the corresponding information in multiple modalities. Quack *et al.* [13] presented methods to mine events and object from community photo collections by clustering approaches. In [4], a wavelet based approach is proposed to detect events from social media data. Pan and Mitra [12] developed a system to combine the popular LDA model with temporal segmentation and spatial clustering for automatically identifying events from a large document collection. In [7], the authors focused on building a Naive Bayes event models which classify photos as either relevant or irrelevant to given events.

Illustrating events with media addresses the problem of how to leverage vivid multi-modal content to share experience. In [10], the authors studied users' uploading behaviors on Flickr and matched concert events with photos based on different modalities; such as text/tags, time, and geo-location. It results in an enriched photo set which better illustrates events. A similar framework involving more modalities is proposed in [3] for enriching event descriptions. In [6], to improve the users' attention when reading news articles, a system was proposed to help people reading news by illustrating the news story. In [8], an unsupervised approach was presented to describe stories with automatically collected pictures. In this framework, semantic keywords are extracted from the story, and used to search an annotated image database. Then a novel image ranking scheme automatically choose the most important images. In [17], a Text-to-Picture system was developed that synthesizes a picture from natural language text without limitation. The system firstly identified "picturable" text units by natural language processing, then searched for the most likely image parts conditioned on the text, and finally optimized picture layout conditioned on both the text and image parts.

A web service with similar illustration functionalities can be found in EventBurn[1]. It creates a summary of a given hot event from popular services like Twitter, Facebook, and Flickr, but fails to extract events automatically from social media streams.

3 Our Proposal

Our framework, shown in Figure 2, extracts and illustrates public events by leveraging the social media data directly. Since events can be defined as something happening at a given location and time, we start by parsing the query to identify its topic, location and time information with natural language processing algorithm. Rather than detecting events from Twitter data directly, events can be obtained by crawling and scraping social news web service. This saves time, computation and storage compared to alternative detection processes. To provide a vivid illustration for each event, we retrieve the relevant tweets from Twitter, and show the collected data with tag cloud. In addition, we also retrieve

[1] `http://www.eventburn.com/`

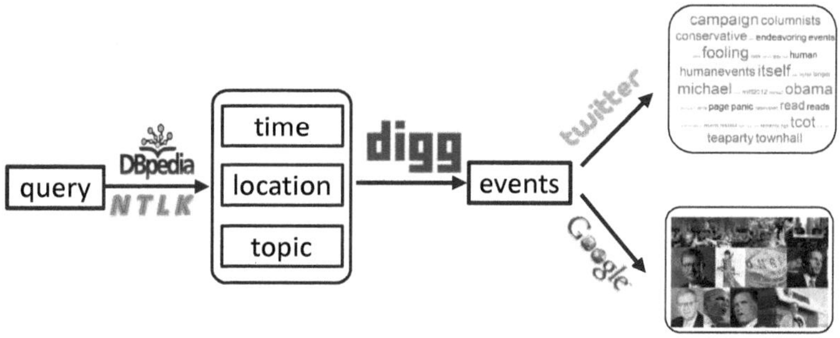

Fig. 2. Overview of the proposed framework

photos with Google image search engine, and summarize the results with photo collage/montage.

3.1 Semantic Query Parsing

The three basic properties of events are location, time, and topic, as stated in [10]. To identify the meaning of a given query input, we would like to extract the information in the three dimensions. Here, we assume that the query input is a noun phrase headed or tailed with complements, such as "New York in the past three days". We extract the structured data from this noun phrase, where there is a predictable organization of entities and relationships. The question of extracting structured data from text has been well studied in Natural Language Processing (*NLP*). This process, composed of 3 steps is performed using *NLTK*[2], a well-known *NLP* package. First, the input text is segmented into words using a tokenizer. Then, each word is tagged with part-of-speech tag (*POS*), which provides the lexical categories for words. With the *POS* tag, we use the *Regexp-Parser chunker* in *NLTK* to create the chunker tree and identify each sub noun phrases in the input string. The process is depicted in Figure 3

Then, to determine the semantic meaning of each noun phrase, different techniques are employed to extract the location, time, and topic information from the parsed noun phrases. The location is obtained through a query of the DBpedia[3] database, which provides structured information extracted from Wikipedia. If geographical metadata, such as "geo:point", "geo:area", are found for a noun phrase, the corresponding location is kept as reference for the events to collect. For determining the temporal information, we develop a script which parses and converts the human readable string, such as "tomorrow","last week", "Monday" to a time structure. We use the DBpedia API and our script to parse the sub noun phrases in order to obtain the location and time information addressed by the query. Since it is hard to model the topic in sophisticated way, we assign the

[2] http://nltk.org/
[3] http://dbpedia.org

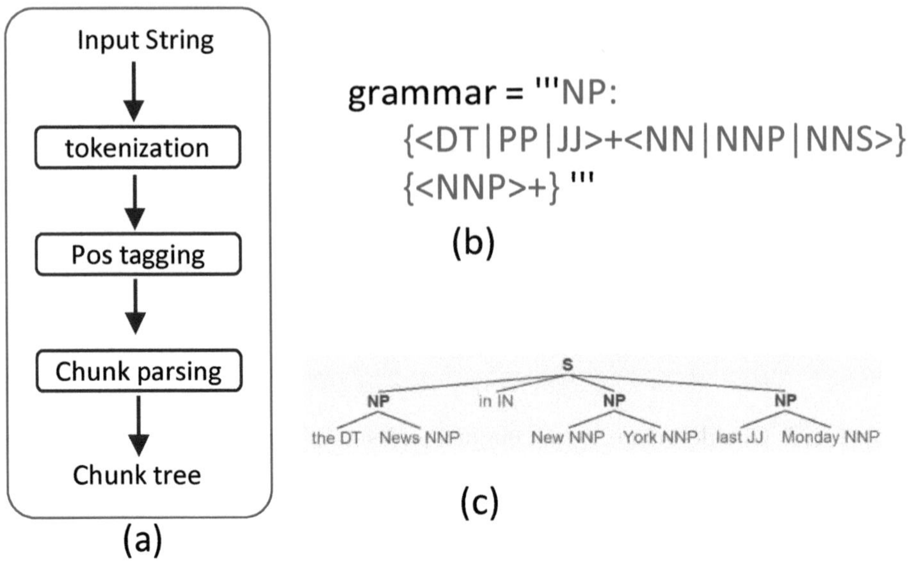

Fig. 3. Semantic parsing using NLP. (a) the flowchart; (b) the grammar used for chunker parsing; (c) an example for input query "the News in New York last Monday"

nouns as the topic keywords of the event to search for, if neither time or location can be determined from it.

3.2 Events Extraction

Recently, there has been some research focusing on detecting events from social media data, such as detecting events from Twitter stream [15], or Flickr [4,9]. Indeed, useful information can be mined from community-contributed data. However, in these methods, huge amount of data have to be downloaded from the web service. Therefore, it takes lots of resources to store the data and lot of time to process it.

Currently, the world's happenings are collected as News by communication services and broadcasted publicly to the mass audience through various channels. Some web services, such as Google News[4] or Digg[5] have been developed to organize the news data in a structured data. In this paper, rather than detecting events from social streams, we aim at querying the events from the social news web service Digg. Digg features user-posted news that are ranked based on the user popularity and comments. With the ranking, popular news can easily be identifyed, leveraging from users' collective intelligence. We take the time,

[4] `http://news.google.com`
[5] `http://digg.com/`

location and topic keywords as the query parameters to retrieve popular events, ranked according to their popularity. In the returned results, many useful metadata can be found for each event, such as "title", "submit_date" to describe the property of events. We also extract the "link" metadata that refers the original content of the news, so that the original content could be retrieved and be used in the text description in the final result.

Although the events is extracted from Digg in this paper, the whole framework is flexible and can be extended to handle more events directories, such as Google News or any such public API easily.

3.3 Media Illustration

Years of multimedia research have shown that it is easier and more accurate for the computer to identify specific pattern compared with abstract concept. To find media illustrating events, a query is specifically tailored for each event. To illustrate events using text and images, we retrieve the original news content, related tweets from Twitter and images from Google image search, and show them in the proper format.

The event is fully described in the original news web page. For each event, we extract the "link" metadata in Digg and crawl the original web page. To extract the the main body of the web page, the method proposed in [5] is employed. It recognizes the main content based on character number under the assumption that characters accumulate more in the main content than in other part of web page.

For each event, we would like to collect other textual description and comments from different users. The textual data is retrieved from Twitter, the famous online microblog service that enables the users to send instance posts known as "tweets". As there are hundreds of thousands of tweets posted every minute in a city, tweets are also studied as the data source to mine events/trends [15]. In this paper, we retrieve related tweets to illustrate events. The event title is often the most useful information for describing/retrieving an event. We perform full text search queries on Twitter based on the event title extracted from Digg. The tweets retrieved are also filtered using a time interval from the time of posting the event in Digg to the end of querying period.

To provide a nice and meaningful visualization, we use a tag cloud to organize the textual data. Tag cloud is a form of histogram which can represent the frequency of over a hundred items. In tag cloud, the importance of each tag is shown with different format. This format is useful for quickly perceiving the most prominent terms. For each event, we segment each tweets into tags, count the frequency of each tag and generate the tag cloud with tags in different font size and color.

Visual summerization is fundamental tool to provide vivid description on image/video collections [16]. Besides tag cloud, we also employ such techniques to illustrate event. Photo collage [11] is a visual clustering technique that can depict the event in different points of view. To generate the photo collage, we retrieve photos from the Google image search engine with the event title as parameter and filter out those photos for which the cosine distance between its textual metadata and the event title is below a given threshold. We decided, after experimentation, not use Flickr as photo source because most of photos are taken during tours/vacations and the dominant concepts are scene, landmark, and building. With the selected photos, the methods proposed in [11] is used to create the photo collage for each events.

Last but not least, to assist the users quickly grasp preponderant information, we would like to show the events according their importance. In details, we measure the importance of each event by the entropy of its tag cloud [1]. After ranking, the events with more information will be shown first and events with less information shown later.

Table 1. Events found for query "New York in the last 3 days", performed on June 2rd, 2012

Date	Event Title
01/06/2012	Companies fire back at proposed NYC big soda ban
01/06/2012	Motorcyclist clocked at 193 mph? in the rain
01/06/2012	New York Bill Proposes Mandatory Wearable ID Tags @ skewnews.com
31/05/2012	Carmelo Anthony Becomes VP Of PowerCoco Energy Drinks
01/06/2012	Cheap Dresses Online Australia - On Sale From $10
02/06/2012	A Powerful Interview with Former Guant?namo Prisoner Lakhdar Boumediene
01/06/2012	NYPD vs. CPD: How Police Deal With Press and Protesters - New York - Slideshows
31/05/2012	Obama Campaign May Be Fooling Itself
03/06/2012	6 New York cops claim they were demoted for being Republicans
02/06/2012	Mets Pitch First No-Hitter in 50-Year History
31/05/2012	2013 Mercedes-Benz GL63 AMG Makes Debut
01/06/2012	Shadows Meet The Clouds, Gray On Gray, Like Dusty Charcoal On An Ashen Brow
31/05/2012	Scientific Proof That Men Have the Dirtiest Desks
03/06/2012	Space shuttle hardware is on the move in Houston and NYC
31/05/2012	Obama Presides Over Secret 'Kill List'
01/06/2012	Facebook Forced To Let You Vote On Privacy Changes
02/06/2012	Paul Krugman Pompously Insults Ron Paul And His Supporters Mediaite
02/06/2012	NBA Arrested For Marijuana Possession
31/05/2012	Bar Refaeli gets close to Olympic star

4 Experimental Results

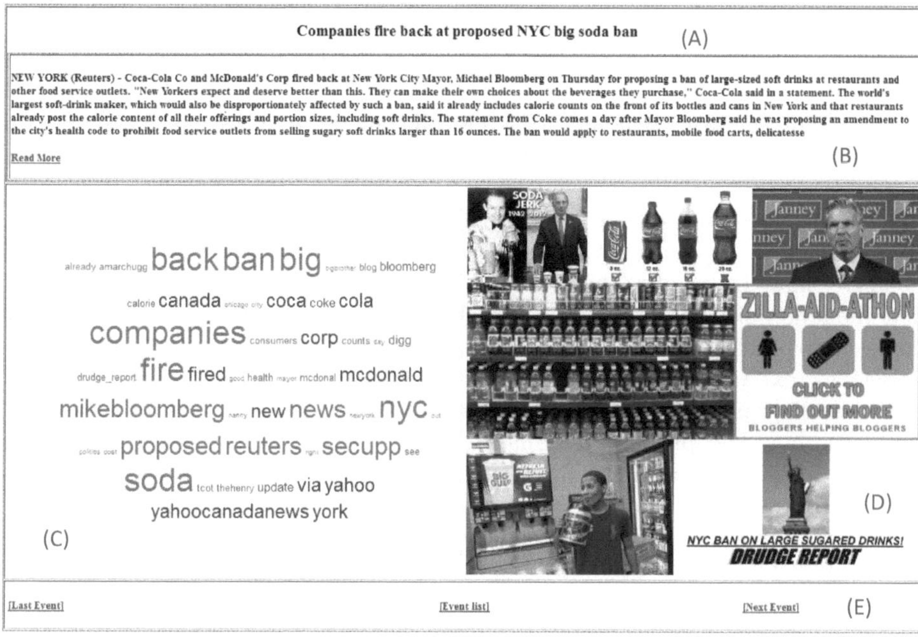

Fig. 4. An example of event visualization, to illustrate the event "Companies fire back at proposed NYC big soda ban", (A) Event title; (B) Event abstract, a link to the original news; (C) Tag cloud; (D) Photo collage; (E) Navigation

Our system can be used to extract and illustrate events from a query input without limitation. For a given query, we first parse it to find out the topics, location, and time information. It should be noticed that for some queries, one or more semantic dimension is missing, and a default parameter will be set. For example, in query " the Olympics in London", no time information is provided. The default value are "in the last 7 days" for time, to ensure the timeliness of events, and "worldwide" for location, and "" for topic, to reduce the limitation of event querying. The motivation for using "in the last 7 days" as default value for time, is that recent news are more likely to be the topic of interest when not specified. We then use the parsed topic, time, and location to query events from Digg. The top 20 events are kept as hot candidates according their ranking promoted by the users.

For example, from the query "New York in the last 3 days", the following events in Table 1 are found.

For all of the queried events, we provide an indexing interface as shown in Figure 1, that provides the thumbnails for all of the events with the combination of photo and event title. All of the events are ranked by their importance as described in Section 3.3. When navigating to a specific event, the event illustration as shown in Figure 4 is provided, giving the user information related on that event solely. Although currently limited to a simple layout, which can easily be improved by a graphic designer, it is composed of five individual parts to help people understand the event well.

The event title is shown in part A, which is the highest level description for the events.

To provide the users with detailed information about the event, we also parse the original news web page, and mine the main textual content part, as shown in part B.

We parse the title and time metadata from the obtained events, use the parameters to query from Twitter the comments from different users, then use the tag clouds to show the results. The tag cloud is presented in part C. We can clearly see that the larger size content such as "back", "ban", "companies", "fire", matches the event topic "companies fire back at the big soda ban" very well.

Besides text visualization, we also retrieve photos with Google image search, and filter the ones that can not be matched based on their textual metadata. All of the matched photos are collaged in the same layout. The photo collage is shown in part D. From the figure, it could be found that most of photos are relevant to the event; The ban of large-sized soft drinks at food service outlets.

To assist users navigating between events, the link to the previous/next event and the index page is located at the bottom of the webpage, as shown in part E.

To conclude, while tag cloud and photo collage in part C and D provide attractive interface with abstract content, the textual content in part B gives concrete description that would assist the users to understand the event well.

5 Conclusion

We proposed an original framework leveraging on social media data (News, Media Sharing Platform and Microblog) to extract and to depict public events. The process is done without any human assistance on a given textual query. At first, natural language processing is employed to parse and make sense of the input query. Then, a social news web service is employed to extract relevant events (with respect to the query) and the metadata from selected events is used to query textual and visual content from different online sources (Twitter and Google). We present the results with an attractive visual format combining tag clouds and photo collage. We demonstrated how our novel approach addresses the event retrieval and illustrating problem and showed its effectiveness on a real world example.

In future work, we will study how to model the relation between multi-modal content by exploiting multiple features, such as text, tag, visual content to help remove noisy data and therefore improve the quality of the presentation of events.

Acknowledgments. This work was supported by the European Commission under contracts FP7-287911 LinkedTV and FP7-318101 MediaMixer.

References

1. Aouiche, K., Lemire, D., Godin, R.: Web 2.0 OLAP: From data cubes to tag clouds. In: Cordeiro, J., Hammoudi, S., Filipe, J. (eds.) WEBIST 2008. LNBIP, vol. 18, pp. 51–64. Springer, Heidelberg (2009)
2. Bao, B.-K., Min, W., Sang, J., Xu, C.: Multimedia news digger on emerging topics from social streams. In: Proceedings of the 20th ACM International Conference on Multimedia, MM 2012, pp. 1357–1358 (2012)
3. Becker, H., Iter, D., Naaman, M., Gravano, L.: Identifying content for planned events across social media sites. In: ACM Conference on WSDM (2012)
4. Chen, L., Roy, A.: Event detection from flickr data through wavelet-based spatial analysis. In: ACM Conference on CIKM (2009)
5. Chengjie, S., Yi, G.: A statistical approach for content extraction from web page. Journal of Chinese Information Processing 18(5), 17–22 (2004)
6. Delgado, D., Magalhães, J., Correia, N.: Assisted News Reading with Automated Illustrations. In: ACM Conference on Multimedia, pp. 1647–1650 (2010)
7. Firan, C.S., Georgescu, M., Nejdl, W., Paiu, R.: Bringing order to your photos: Event-Driven Classication of Flickr Images Based on Social Knowledge. In: Proceedings of the 19th ACM International Conference on Information and Knowledge Management, New York, USA, p. 189 (October 2010)
8. Joshi, D., Wang, J.Z., Li, J.: The Story Picturing Engine—a system for automatic text illustration. ACM Transactions on Multimedia Computing Communications and Applications 2(1), 68–89 (2006)
9. Liu, X., Huet, B., Troncy, R.: Eurecom@ MediaEval 2011 social event detection task. In: Proceedings of the MediaEval 2011 Workshop (2011)
10. Liu, X., Troncy, R., Huet, B.: Finding Media Illustrating Events. In: ACM International Conference on ICMR, Trento, Italy (2011)
11. Mei, T., Yang, B., Yang, S.-Q., Hua, X.-S.: Video Collage: Presenting a Video Sequence Using a Single Image. The Visual Computer, 39–51 (2009)
12. Pan, C.-C., Mitra, P.: Event detection with spatial latent Dirichlet allocation. In: Proceeding of the 11th Annual International ACM/IEEE Joint Conference on Digital Libraries, New York, USA, p. 349 (June 2011)
13. Quack, T., Leibe, B., Van Gool, L.: World-scale mining of objects and events from community photo collections. In: Proceedings of the 2008 International Conference on Content-Based Image and Video Retrieval, New York, USA, p. 47 (July 2008)
14. Sakaki, T., Okazaki, M., Matsuo, Y.: Earthquake shakes twitter users: real-time event detection by social sensors. In: International Conference on World Wide Web, Raleigh, North Carolina, USA (2010)
15. Weng, J., Lee, F.: Event detection in twitter. In: AAAI Conference on Weblogs and Social Media, Barcelona, Spain (2011)
16. Yahiaoui, I., Merialdo, B., Huet, B.: Comparison of multiepisode video summarization algorithms. EURASIP J. Appl. Signal Process. 2003, 48–55 (2003)
17. Zhu, X., Goldberg, A.B., Eldawy, M., Dyer, C.R., Strock, B.: A Text-to-Picture Synthesis System for Augmenting Communication. In: Proceedings of the 22nd National Conference on Artificial Intelligence, vol. (2), pp. 1590–1595 (2007)

Online Web-Video Topic Detection and Tracking with Semi-supervised Learning

Guorong Li[1], Weigang Zhang[3], Junbiao Pang[4], Qingming Huang[1,2], and Shuqiang Jiang[2]

[1] Graduate University, Chinese Academy of Sciences (CAS), Beijing, China
[2] Key Lab of Intell. Info. Process., Inst. of Comput. Tech., CAS, Beijing, China
[3] School of Computer Science and Technology, Harbin Inst. of Tech., China
[4] Beijing Municipal Key Lab. of Multimedia and Intelligent Software Tec., Beijing Univ. of Tech, China
{grli,wgzhang,jbpang,qmhuang,sqjiang}@jdl.ac.cn

Abstract. With the rapid growth of web data, a large amount of web videos are available online. However, how to organize them for facilitating users' experience and government supervision remains a problem yet to be seriously investigated. Topic detection and tracking, which has been a hot research topic for decades, could cluster web videos into different topics according to their semantic content. However, how to online discover topic and track them from web videos and images has not been fully discussed. In this paper, we formulate topic detection and tracking as an online tracking, detection and learning problem. First, by learning from historical data including labeled data and plenty of unlabeled data using semi-supervised multi-class multi-feature method, we obtain a topic tracker which could also discover novel topics from the new stream data. Second, when new data arrives, an online updating method is developed to make topic tracker adapt to the evolution of the stream data. We conduct experiments on public dataset to evaluate the performance of the proposed method and the results demonstrate its effectiveness for topic detection and tracking.

Keywords: Topic Detection and Tracking, web video, Multi-feature fusion, semi-supervised learning.

1 Introduction

With the development of network technology, many forms of social media such as video, twitter, blogs spring up. The openness, immediacy and reality of the Internet information made it an important indication of the trend of the society. More and more people use them to share their experiences or to find out what was happening outside and the web data is increasing at a unprecedented rate. According to the report, only in 2011 we had 1.8 trillion GB data and the data volume will grow with the speed of 50%. It is estimated that in 2020 global data will reach 35.2ZB, 90 percent of which are images or videos.

B. Huet et al. (Eds.): PCM 2013, LNCS 8294, pp. 750–759, 2013.

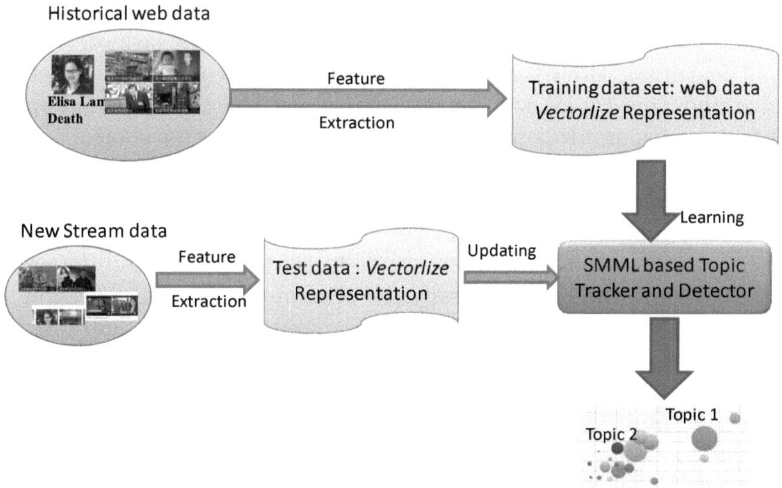

Fig. 1. The framework for topic tracking and detection from web data

Facing so large volume video or image data, it is inconvenient for users to get the main contents and find out what they are interested with. Therefore how to organize those video and image data is an important but very difficult problem. Topic detection and tracking [1](TDT) aims to discover topical related data and thus could classify web data by its semantic content. This brings at least three advantages:

(1) For users, it is easy to quickly understand what happens from the Internet and find out the materials related to the topic they are concerned with.

(2) For administrators, it is able to organize different modal data according to their contents. Furthermore, they can recommend topics which are related with that users are taking part in.

(3) For governments, they can discover hot or sensitive topics such as "watch" quickly and understand the peoples' necessities. Moreover, for some topics which are rumors but becoming more and more attractive, such as "the salt is polluted because after the earthquake shook Japan on March 11, 2011", it is possible for governments to make timely response and prevent its further spread.

During the past years, as text is the main form of the media data in the past, researchers focus on TDT from a collection of documents and have achieved great success [2] [3–5]. However, directly apply those methods to other types of media data (e.g. video, microblog) often generates poor results because of their different characteristics. For example web video contains short text tags and rich visual information. Intuitively, different modal features could complement with each other, but how to fuse them is still an open problem. In [6], a new Co-clustering method is developed to detect emerging topics from social streams and could leverage three sources of data including news article, micro-blog and picture. Recently, some works such as [7–10] are developed to mine topics from web videos through fusing multi-modal features, but most of them are based on clustering method and perform topic detection in an off-line fashion (mainly rely on textual features). However, the number of topics could not be determined in advance and new topics will be revealed along the time, so perhaps online learning feature fusion for topic detection would be a better choice.

As web videos or images usually arrives in streams over time, their content contains strong correlations. Historical data is very useful to better understand and predict the underlying topics in the future data. Therefore, instead of analyzing them in an off-line manner, we propose to learn topic detector from historical data and process and classify the stream of video data as soon as they are available. In this way, it is possible to track the topics of interest, predict their evolutions and discover the emerging trends in time. First, provided historical data including labeled and unlabeled samples, we learn a classifier using semi-supervised multi-feature multi-class learning method[11]. This method can fuse different modality features without destroying the local structure of individual feature. Second, when new data arrives, identify whether they are new topics or which topic they are belong to. Then updating classifier to adapt to the evolution of the streaming data. This approach has the added advantage that it could build models for the interested topics and track them.

2 The Proposed Algorithm

In this section, we will describe our algorithm for topic detection and tracking in detail. Here we need to introduce some notations.

Table 1. Notations of variables

X	the labeled sample set at time t
Y	the corresponding labels of X
x_i	the ith sample
x_i^g	the gth feature representation for sample x_i
X^g	the gth feature representation for labeled sample set
N_i^g	the k nearest neighbors of sample x_i according to the gth feature
U	unlabeled sample set at time
h_i^g	the local classifier learned from N_i^g
b^g	the global classifier learned from X^g

2.1 Semi-supervised Multi-feature Multi-class Learning Problem

Provided the historical data including labeled samples $\mathcal{X}$ and unlabeled samples $\mathcal{U}$, we want to learn a multi-topic classifier which could be used to track known topics and detect new topics. To exploit the structure information implied in the training data, for each sample x_i we assume there exists local classifiers h_i^g, which can classify N_i^g into c classes. Let f_i^g denote the expected predicting result of x_i using according to the gth feature. Then the loss function of local classifiers derived from g^{th} feature is defined as following:

$$L_h^g = \sum_{i=1}^{n} \sum_{x_j^g \in N_i^g} (loss(h_i^g(x_j^g), f_j^g)) + \mu_1 \parallel h_i^g \parallel \tag{1}$$

where $\parallel h_i^g \parallel$ denotes the model complexity of h_i^g and λ is a parameter. Then through minimizing the loss function Equ. 2, the v features are combined together.

$$L_h = \sum_{g=1}^{v} L_h^g \tag{2}$$

To classify samples outside the training set, we assume there exist global classifies b_g whose expected predicted results are denoted by F. Similarly, the loss function of the global classifier is defined as:

$$L_b = \sum_{g=1}^{v} \sum_{i=1}^{n} loss(b_g(x_i), F_i) + \mu_1 \parallel b_g \parallel \tag{3}$$

Finally the consistency loss function among f, F and Y is simply formulated by Equ. (4).

$$L_c = \sum_{g=1}^{v} \parallel F - f_g \parallel_2^2 + \mu_3 \sum_{i=1}^{n_L} \parallel F_i - y_i \parallel_2^2 \tag{4}$$

where μ_3 is a parameter. Then by summing over the above three loss functions, we obtain the final objective function:

$$L(F, f_g, h_g^i, b_g) = L_h + \alpha L_b + \beta L_c \tag{5}$$

where α and β are parameters. There are many selections for the form of local and global classifiers, such as multiclass boosting[12], SVM. We select the most common construction: linear classifiers for its generalization[13]. Let $h_i^g(x_i^g) = (w_i^g)^T x_i^g + c_i^g$, $b_g(x_i^g) = (W^g)^T x_i^g + C^g$ and the complexity of the linear classifier is defined as the norm of its coefficient. The objective loss function can be rewritten as:

$$\begin{aligned} L(w_i^g, c_i^g, W^g, C^g) = &\sum_{g=1}^{v} \sum_{i=1}^{n} \sum_{x_j^g \in N_i^g} \parallel (w_i^g)^T x_j^g + c_i^g - f_j^g \parallel^2 \\ &+ \alpha \sum_{g=1}^{v} \sum_{i=1}^{n} \parallel (W^g)^T x_i^g + C^g - F_i \parallel^2 + \gamma_1 \sum_{g=1}^{v} \parallel F - f^g \parallel^2 \\ &+ \beta \sum_{i=1}^{n_L} \parallel F_i - Y_i \parallel^2 + \gamma_2 \sum_{i=1}^{n} \sum_{g=1}^{v} \parallel w_i^g \parallel^2 + \gamma_3 \sum_{g=1}^{v} \parallel W^g \parallel^2 \end{aligned} \tag{6}$$

where the γ_1, γ_2 and γ_3 are parameters. To obtain the minimum value of L, similar to [11], we set the derivative L with respect to variables c_i^g, w_i^g, C^g, W^g to be zero, we have

$$\begin{cases} c_i^g = \frac{\sum_{x_j^g \in N_i^g} f_i^g - (w_i^g)^T x_j^g}{k+1} \\ w_i^g = \left(\sum_{x_j^g \in N_i^g} x_j^g (x_j^g - \bar{N_i^g})^T + \gamma_2 I\right)^{-1} \sum_{x_j^g \in N_i^g} x_j^g (f_j^g - \bar{f_i^g})^T \\ C^g = \frac{\sum_{i=1}^n (F_i - (W^g)^T x_i^g)}{n} \\ W^g = \left[\sum_{i=1}^n x_i^g (x_i^g - \bar{X^g})^T + \gamma_3 I\right]^{-1} \sum_{i=1}^n x_i^g (F_i - \bar{F})^T \end{cases} \tag{7}$$

where $\bar{N_i^g} = \frac{\sum_{x_j^g \in N_i^g} x_j^g}{k+1}$, $\bar{f_i^g} = \frac{\sum_{x_j^g \in N_i^g} f_j^g}{k+1}$, $\bar{F} = \frac{\sum_{i=1}^n F_i}{n}$ and $\bar{X^g} = \frac{\sum_{i=1}^n x_i^g}{n}$. To simplify the equation, we introduce some variables $D_i^g = \left[\sum_{x_j^g \in N_i^g} x_j^g (x_j^g - \bar{N_i^g})^T + \gamma_1 I\right]^{-1}$, $B^g = \left[\sum_{i=1}^n x_i^g (x_i^g - \bar{X}_g)^T + \gamma_3 I\right]^{-1}$, $M_i^g(j) = \begin{cases} 1, x_j^g \in N_i^g \\ 0, otherwise \end{cases}$, $S_i^g(j,j) = \begin{cases} 1, x_j^g \in N_i^g \\ 0, otherwise \end{cases}$ then w_i^g and W^g become

$$\begin{cases} c_i^g = \frac{(f^g - (w_i^g)^T X^g) M_i^g}{k+1} \\ w_i^g = D_i^g X^g S_i^g \left[I - 1_n \frac{(M_i^g)^T}{k+1}\right] (f^g)^T \\ C^g = \frac{(F - (W^g)^T X^g) 1_n}{n} \\ W^g = B^g X^g (I - \frac{1_n (1_n)^T}{n}) F^T \end{cases} \tag{8}$$

where $f^g = [f_1^g, f_2^g, \cdots, f_n^g]$, $X^g = [x_1^g, x_2^g, \cdots, x_n^g]$. Similarly, through setting the derivative L with respect to variables f_i^g, F_i to be zero and substituting c_i^g, w_i^g into the derivation, we have

$$f^g \left\{ \sum_{x_i^g \in N_j^g} \left[(\Omega_j^g)^T (X_i^g - \frac{X^g M_i^g}{k+1}) + \frac{M_i^g}{k+1} - P_i \right] + \gamma_1 P_i \right\} = -\gamma_1 F_i \tag{9}$$

where $\Omega_j^g = D_j^g X^g S_j^g (I - \frac{M_j^g (1_n)^T}{k+1})$. Let $T_i^g = \sum_{x_i^g \in N_j^g} ((\Omega_j^g)^T (X_i^g - X^g M_i^g) + M_i^g - P_i)$, $T^g = \{T_1^g, T_2^g, \cdots, T_n^g\}$, easily, we derive

$$f^g = -\gamma_1 F (T^g + \gamma_2 I)^{-1}. \tag{10}$$

Algorithm 1. Semi-supervised Multi-Feature Multi-class learning

Input: training examples: X, Y and stream web data $x_t, x_{t+1}, \cdots$, parameters α, γ_1, γ_2, γ_3

Output: paratmers of the learned global and local classifiers:
$\{< W^1, C^1 >, < W^2, C^2 >, \cdots, < W^v, C^v >\}$
$\{< w_1^1, c_1^1 >, < w_2^1, c_2^1 >, \cdots, < w_n^1, c_n^1 >, < w_1^2, c_1^2 >, \cdots, < w_n^2, c_2^n >, \cdots, < w_1^v, c_1^v >, \cdots, < w_n^v, c_n^v >\}$

for $g = 1; g \leq v; g++$ **do**
  for $i = 1; i \leq n; i++$ **do**
    Compute M_i^g, S_i^g, N_i^g;
    Compute inverse of the D_i^g;
    Calculate Ω_i^g;
  end
  Calculate B^g;
  for $i = 1; i \leq n; i++$ **do**
    Calculate T_i^g;
  end
end
Compute F;
for $g = 1; g \leq v; g++$ **do**
  Compute f^g, W^g and C^g;
  for $i = 1; i \leq n; i++$ **do**
    Calculate w_i^g, c_i^g;
  end
end

Next substituting C^g, W^g and f^g into $\frac{\partial L}{\partial F_i}$, we obtain the optimal solution for F is:

$$\begin{aligned} F = {} & diag(I(1 \leq n_L), \cdots, I(n \leq n_L))Y \\ & \left\{ \left[\alpha \sum_{g=1}^{v} (B^g X^g (I - \frac{1_n(1_n)^T}{n}))^T (X^g - \frac{X^g 1_n (1_n)^T}{n}) \right] \right. \\ & + \frac{1_n(1_n)^T - I}{n} + \left[\gamma_1 \sum_{g=1}^{v} I - \gamma_1 (T^g + \gamma_3 I)^{-1} \right] \\ & \left. + diag(I(1 \leq n_L), \cdots, I(n \leq n_L)) \right\}^{-1} \end{aligned} \tag{11}$$

After obtaining the solution, we could compute f^g, W^g, w_i^g, C^g and c_i^g sequtially according to Equ. (10),Equ. (8). Alg. 1 provides the details of semi-supervised multi-feature multi-class learning procedure.

2.2 Online Topic Tracking and Detection

Provided a new data $x_t = \{x_t^1, x_t^2, \cdots, x_t^v\}$, we first use the learned global classifier to predict its label

Algorithm 2. Online updating method for Topic Tracker

Input: test examples: $x_{n+1}, \cdots, x_{n+p}$, $\{< M_i^g, D_i^g, \omega_i^g, T_i^g >\}_{i=1,g=1}^{i=n,g=v}, < iB^g >$, F

Output: parameters of the learned global and local classifiers:
$\{< W^1, C^1 >, < W^2, C^2 >, \cdots, < W^v, C^v >\}$
$\{< w_1^1, c_1^1 >, < w_2^1, c_2^1 >, \cdots, < w_n^1, c_n^1 >, < w_1^2, c_1^2 >, \cdots, < w_n^2, c_2^n >$
$, \cdots, < w_1^v, c_1^v >, \cdots, < w_n^v, c_n^v >\}$

for $g = 1; g \leq v; g++$ **do**
    Select k neighbors from $\tilde{N}_{n+1}^g$ to compose N_{n+1}^g; Calculate $N_{n+1}^{\bar{g}}$ and set M_{n+1}^g;
    $D_{n+1}^g = \left[N_{n+1}^g * (N_{n+1}^g)^T - (k+1) N_{n+1}^{\bar{g}} (N_{n+1}^{\bar{g}})^T + \gamma_1 I\right]^{-1}$;
    $\Omega_{n+1}^g = D_{n+1}^g X^g S_{n+1}^g (I - \frac{M_{n+1}^g (1_{n+1})^T}{k+1})$;
    for $i = 1; i \leq n; i++$ **do**
        if $M_{n+1}^g(i) == 1$ **then**
            $T_i^g += (\Omega_{n+1}^g)^T (x_i^g - \bar{N}_i^g) + M_i^g - P_i$;
        end
    end
    $T_{n+1}^g = 0$;
    $\bar{X}_g = \frac{n * \bar{X}_g + x_n^g}{n+1}$;
    $B_g = \left[X_g * (X_g)^T + x_n^g * (x_n^g)^T - (n+1) \bar{X}_g * (\bar{X}_g)^T + \gamma_3 I\right]^{-1}$;
end
Compute F;
for $g = 1; g \leq v; g++$ **do**
    Compute f_g, W^g and C^g;
    for $i = 1; i \leq n; i++$ **do**
        Calculate w_i^g, c_i^g;
    end
end

$$\tilde{F}_t = \frac{1}{v} \sum_{g=1}^{v} (W^g)^T x_t^g + C^g \tag{12}$$

$$c_t = max_c \{F_t^c\}. \tag{13}$$

Then we want to verify whether x_t belong to the cth topic. Generally, if x_t belong to the c_t^{th} topic, its feature vector should be similar to those training samples coming from c_t^{th} topic. Therefore, the local classifiers $h_c(x_t) = \{h_i^g : y_i(c_t) = 1\}$ are likely to classify x_t correctly. We define credible local classifiers set of x_t $CLC(x_t) = \{h_i^g : y_i(c_t) = 1\}$. If the ratio of the credible local classifiers which classifies x_t into the cth topic is larger than the threshold $thresh$, $y(x_t) = e_{c^{th}}$ (e_c denote unit vector where the cth element is one); otherwise, x_t is assumed to come from a novel topic and $y(x_t) = e_{M+1}$ and the set of known topic classes is updated accordingly, $Y = \begin{pmatrix} Y \\ 0 \end{pmatrix}, y(x_t) \Big)$.

2.3 Online Updating

To be adaptive to the evolution of the web data, when new data x_{n+1} arrives the topic tracker and detector should be updated. However, it is time consuming if we re-minimize the loss function considering into the new sample. Therefore, we select an approximate method to online update the variables of the local classifier and global classifier. The k-nearest neighbors of $x_i, i = 1, 2, \cdots, n$ are considered to remain unchanged, while that of x_{n+1} are selected from historical data. Then M^g_{n+1}, N^g_{n+1} and $N^{\bar{g}}_{n+1}$ could be easily obtained. The detail procedures for online updating are shown in Algorithm. 2.

3 Experiments

Table 2. F-measure of the best 10 topics which are tracked successfully by SMMLTDT/MMLTDT

best topic	1	2	3	4	5	6	7	8	9	10	average
SMMLTDT	0.9637	0.9616	0.9549	0.9509	0.9235	0.9117	0.8833	0.8718	0.8458	0.8276	0.9095
MMLTDT	0.8758	0.8754	0.8709	0.8678	0.8500	0.8353	0.8287	0.8197	0.7845	0.7822	0.8390

The experiments are mainly performed on public dataset MCG-WEBV[14], which includes 248,887 videos downloaded from YouTub. The CoreData are 14,473 videos, which are the "Most Viewed" videos of "This Month" from Dec. 2008 to Nov. 2009. Different types of visual feature such as Color Histogram, Edge Histogram, Texture Concurrence, and textual features including title, tag, description etc. are provided. There are 73 topics annotated as ground truth. Each key frame of a video is considered as an example. The label of video is obtained by voting of the images belong to it. Moreover, to verify whether the unlabeled data is useful, we test the proposed method without using unlabeled data (MMLTDT). $k = 20$, $\alpha = \beta = \gamma_1 = \gamma_2 = \gamma_3 = 1.0$, $thresh = 0.5$ in the following experiment.

3.1 Online Topic Tracking

We select 10 hottest topics from the annotated 73 topics, the web data from the other topics are considered as from 11th class. The data are divided into two parts according to their upload time: training data and testing data. Test data are used as unlabeled samples. Moreover, to test whether unlabeled data could be useful, we set $\mathcal{U} = NULL$ and SMMLTDD degrades into supervised multi-class multi-feature learning (referred as MMLTDT). Precision, Recall and F-measure defined in [9] are used for quantitative evaluations. Table. 2 shows the top 10 F-measure of the topics that are successfully tracked by MMLTDT and SMMLTDT respectively. Figure. 2 provides comparisons of SMMLTDT, MMLTDT and [9] in terms of F-measure. We can see that MMLTDT could integrate multi-modal features to build effective classifier for target topics and mining information implied in the unlabeled data SMMLTDT could further improve the performance.

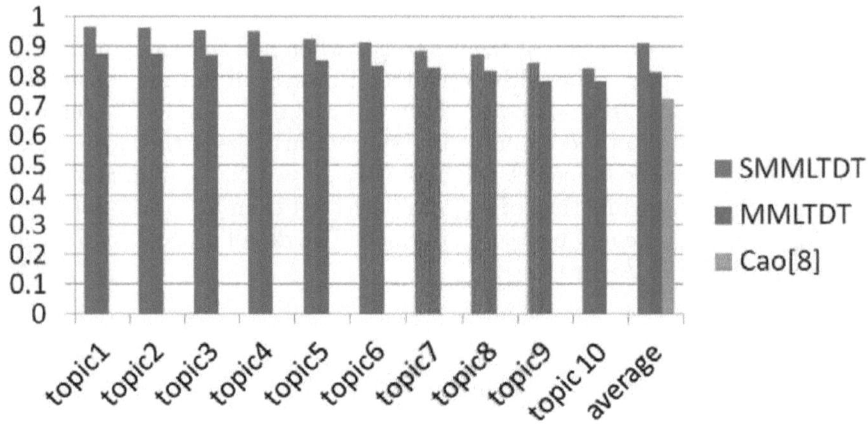

Fig. 2. The F-measure of 10 topics respectively and average F-measure of MMLTDT, SMMLTDT and Cao[9]

3.2 Online Topic Discovery

The data of the earliest one month is used as historical data. For each topic, we randomly choose part of samples belong to it as unlabeled samples. If the F-measure of a discovered topic is larger than 0.5, we considered this topic as successfully detected and tracked (referred as Topic Successfully Detected and Tracked TSDT). The number of TSDT is 30.

4 Conclusions

To analyze the massive web video streaming data, we propose an online topic tracking and detection method, which fuses different types of features through multi-class multi-feature learning algorithm. To be adaptive to the dynamic of streaming data, we develop an online updating method to update our topic tracker and detector. New topics could be incremental revealed as learning and updating go on. Experiments on public dataset show that the proposed method can not only track the interested topics but also can discover new topics and build efficient classifiers for them.

Acknowledgments. This work was supported by China Postdoctoral Science Foundation: 2012M520436, in part by National Basic Research Program of China (973 Program): 2012CB316400, National Natural Science Foundation of China: 61025011 and 61202322.

References

1. Allan, J., Carbonell, J., Doddington, G., Yamron, J., Yang, Y.: Topic detection and tracking pilot study final report (1998)
2. Chen, K., Luesukprasert, L., Chou, S.: Hot topic extraction based on timeline analysis and multi-dimensional sentence modeling. IEEE Transactions on Knowledge Data Engeering 19(8), 1016–1025 (2007)
3. Sun, A.X., Hu, M.: Query-guided event detection from news and blog streams. IEEE Transactions on Systems, Man and Cybernetics 41(5), 834–839 (2011)
4. Zhai, Y., Shah, M.: Tracking news stories across different sources. In: Proceedings of the 20th ACM International Conference on Multimedia, MM 2005, pp. 2–10. ACM (2005)
5. Kasiviswanathan, S.P., Melville, P., Banerjee, A., Sindhwani, V.: Emerging topic detection using dictionary learning. In: Proceedings of the 20th ACM International Conference on Information and Knowledge Management, CIKM 2011, pp. 745–754. ACM (2011)
6. Bao, B.K., Min, W., Sang, J., Xu, C.: Multimedia news digger on emerging topics from social streams. In: Proceedings of the 20th ACM International Conference on Multimedia, MM 2012, pp. 1357–1358. ACM, New York (2012)
7. Shao, J., Ma, S., Lu, W., Zhuang, Y.: A unified framework for web video topic discovery and visualization. Pattern Recognition Letters 33(4), 410–419 (2012)
8. Hong, R., Tang, J., Tan, H., Ngo, C., Yan, S., Chua, T.: Beyond search: event driven summarization for web videos. ACM Transactions on Multimedia Computing, Communications and Applications 33(4), 410–419 (2011)
9. Cao, J., Ngo, C.W., Zhang, Y.D., Li, J.T.: Tracking web video topics: Discovery, visualization, and monitoring. IEEE Transactions on Circuits and Systems for Video Technology 21(12), 1835–1846 (2011)
10. Chen, T., Liu, C., Huang, Q.: An effective multi-clue fusion approach for web video topic detection. In: Proceedings of the 20th ACM International Conference on Multimedia, MM 2012, pp. 781–784. ACM, New York (2012)
11. Yang, Y., Song, J., Huang, Z., Ma, Z., Sebe, N., Hauptmann, A.: Multi-feature fusion via hierarchical regression for multimedia analysis. IEEE Transactions on Multimedia 15(3), 572–581 (2013)
12. Freund, Y., Schapire, R.: A decision theoretic generalization of on-line learning and an application to boosting. Journal of Computer and System Sciences 55(1), 119–139 (1997)
13. Hastie, T., Simard, P.: Models and metrics for handwritten character recognition. Statistical Science 13(1), 54–65 (1998)
14. Cao, J., Zhang, Y., Song, Y., Chen, Z., Zhang, X., Li, J.: Mcg-webv: A benchmark dataset for web video analysis (2009)

Temporal Context Analysis for Action Recognition in Multi-agent Scenarios

Yifan Zhang[1], Chunjie Zhang[2], Zhiqiang Tang[1], and Hanqing Lu[1]

[1] National Laboratory of Pattern Recognition,
Institute of Automation, Chinese Academy of Sciences, Beijing 100190, China
{yfzhang,hqlu}@nlpr.ia.ac.cn, zqtang2013@gmail.com
[2] School of Computer and Control Engineering,
University of Chinese Academy of Sciences, Beijing 100049, China
cjzhang@jdl.ac.cn

Abstract. In multi-agent scenarios such as sports videos, multiple actions are played by different players. Such actions do not necessary appear strictly sequentially but can happen in parallel. Approaches which only consider a single stream of actions are not competent to handle such scenarios. The temporal and causal relationships between the action streams such as "concurrence", "mutually exclusion" and "triggering" need to be captured so as to correctly recognize the actions. In this paper, a novel method is presented for action recognition in multi-agent scenarios leveraged by analyzing the relationships among the temporal contextual actions. The multi-streams of actions are modeled by a Dynamic Baysian Network (DBN) containing several temporal processes corresponding to each type of action. Comparing to the Coupled Hidden Markov Model (CHMM), only the necessary interlinks between the temporal processes are built by a structure learning algorithm to capture the salient relationships. Empirical results on real-world video data demonstrate the effectiveness of our proposed method.

Keywords: Multi-agent action recognition, graphical model, structure learning.

1 Introduction

Recognizing human actions from videos is a task of obvious scientific and practical importance. In this paper we consider the problem of recognizing actions in multi-agent scenarios, which is challenging due to that the interactions between actions would lead to large state spaces and complicate the already uncertain low-level visual processing. In multi-agent scenarios, group or interactive actions from multiple players may occur sequentially or in parallel. It is not uncommon to have primitive actions with parallel streams. One action does not need to be completed before continuing on to the later one. For example, in the case of basketball, "the offensive player is *shooting* while the defender is *blocking* him" (shown in Fig. 1), which includes parallel actions. However, most of the existed

B. Huet et al. (Eds.): PCM 2013, LNCS 8294, pp. 760–768, 2013.

Fig. 1. Interactive actions in multi-agent scenario. The action *shooting* and *blocking* are highly likely to co-occur together. The frame is captured from the OSUPEL basketball dataset [2].

approaches such as Finite State Machines (FSMs) [5] or Hidden Markov Models (HMMs) [8] consider the scenario as a temporally ordered single stream of actions.

In multi-agent scenarios such as sports videos, one action could maintain certain relationships with the contextual ones which are governed by the domain knowledge and certain rules of thumb. These relationships may impose a temporal structure on constituent primitives, which can be leveraged to correctly recognize the actions. Besides the "concurrence" relationship between the parallel actions described above, there are two other important relationships: "mutually exclusion" and "triggering". If two actions are mutually exclusive, it means that their concurrence is inhibited. For example, "if one player is *dribbling*, he cannot be *holding* the ball at the same time". The "triggering" relationship indicates that one action is caused by another, such as "one player *passing* the ball leads to the other one *catching* the ball".

To capture these kinds of relationships, we propose to use a Dynamic Baysian Network (DBN) to represent the temporal structure on top of the actions. A DBN is a directed acyclic graphical model, which models the temporal evolution of a set of random variables X over time. In the DBN, each variable corresponds to one type of action, and its evolution forms a temporal process. To model the interactions between the actions, instead of being fully connected between the temporal processes, the interlinks are discovered by a structure learning algorithm so as to capture the salient temporal relationships while still controlling the complexity of the network.

Our main contribution is providing a system that efficiently and robustly recognizes actions in multi-agent scenarios by using (1) temporal context analysis for capturing meaningful temporal and causal relationships to disambiguate amongst noisy visual observations and (2) structure learning for discovering salient interlinks in the DBN model instead of fully connected and make it computationally tractable.

2 Related Work

There has been considerable research exploring how to represent and model multiple action interactions. The typical representations are Finite State Machines (FSMs) [5] or Hidden Markov Models (HMMs) [8], in which actions cause state transitions in a strictly sequential order and a successful transition through the stream implies the recognition of the action. Unfortunately, these type of approaches may not handle the scenarios with multiple streams of actions. A main difficulty with these approaches are that the system can only be in one state at a time, they cannot well represent parallel actions. Coupled Hidden Markov Model (CHMM) [1,7] was presented by Brand et al., which factorize the actions into two parallel transition processes to deal with the complexity in highly coupled T'ai Chi hand movements. However, the transition model structure of the CHMM is fully connected. Direct extensions of this model to multi-agent scenarios which contains multiple transition precesses may encounter computationally tractability issues.

Expert domain knowledge can be leveraged to create models for multi-agent actions. Ryoo and Aggarwal [9] model two person interactions by a context-free grammar (CFG), where high-level interactions are defined hierarchically using logical spatial and temporal predicates on sub-actions. The CFG parsing is not probabilistic and can be sensitive to low-level failures. Brendel et al. proposed Probabilistic event logic (PEL) [2] to capture temporal relationship between actions based on confidence weighted formulas. However, the temporal relationships must be known in advance in order to manually encode them into the event logic formulas.

Based on the limitations of the approaches mentioned above, we need to find a method which can automatically discover interactive relationships so as to model multiple streams of actions. In addition, it should be a probabilistic model which can handle the uncertainty brought by the low-level visual processing.

3 Relationships Modeling Using a DBN

We propose to use a DBN to model and learn the relationships among the actions. Let X^t represent a set of random variables at a discrete time slice t. A DBN is defined as $B = (G, \Theta)$, where G is the model structure, and Θ represents the model parameters, i.e., the conditional probabilistic distributions for all nodes. A DBN is a time-sliced model, as shown in Fig. 2(a), where each time slice is used to represent the snapshot of certain evolving temporal processes at a time instant. The neighboring time slices are interconnected by the links joining certain temporal nodes from two consecutive slices. Hence, to specify a DBN, we need to define the intra-slice topology (within a slice), the inter-slice topology (between two slices), as well as the parameters for the first two slices. Such a two-slice temporal Bayesian network is often called a 2TBN, as shown in Fig. 2(b). A 2TBN is used as a template which can be unrolled to the whole DBN.

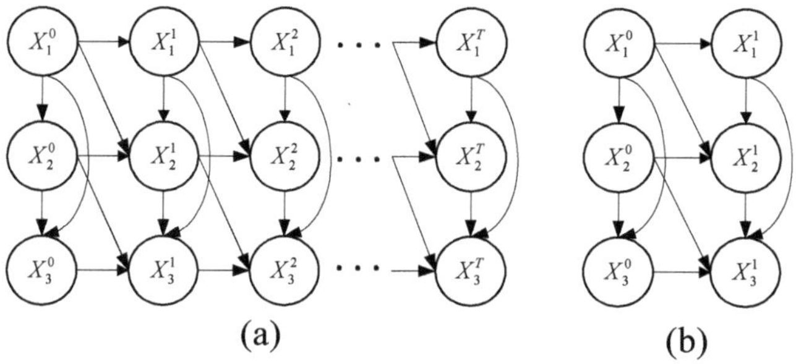

Fig. 2. (a) The DBN for $T + 1$ time slice, and (b) the corresponding 2TBN representation

Given a DBN model, the joint probability over all variables $X^0, ..., X^T$ is computed as follows:

$$P(x^0, ..., x^T) = P(x^0) \prod_{t=0}^{T-1} P(x^{t+1}|x^t), \tag{1}$$

where x^t represents the sets of values taken by the random variables X at time t, $P(x^0)$ captures the joint probability of all variables in the first slice, and $P(x^{t+1}|x^t)$ represents the transition probability.

3.1 Structure Initialization

The DBN model learning is started by structure initialization. An initial 2TBN structure is derived by analyzing the relationships among temporal contextual actions in the training data. Based our definition, there are mainly three types of relationships for actions at two adjacent time slices: concurrence, mutually exclusion and triggering. Concurrence and mutually exclusion can be captured in the intra-slice topology, while triggering can be captured in the inter-slice topology.

In the training data, the "concurrence" relationship is discovered by the statistic of the concurrences between each pair of actions. The pairwise concurrence dependency between two actions is computed as follows:

$$P(A_i^t = 1 | A_j^t = 1) = \frac{N_{A_i^t + A_j^t}}{N_{A_j^t}}, \tag{2}$$

where $N_{A_i + A_j}$ is the total number of concurrences of action A_i and A_j in the same time slice, and N_{A_j} is the total number of occurrences of A_j in the whole database.

The "mutually exclusion" relationship is captured by the statistic of the absence of one action given the presence of the other one. The pairwise negative dependency between two actions is computed as follows:

$$P(A_i^t = 0 | A_j^t = 1) = \frac{N_{\neg A_i^t + A_j^t}}{N_{A_j^t}}, \tag{3}$$

where $N_{\neg A_i^t + A_j^t}$ is the total number of absences of A_i^t given the presence of A_j^t, and $N_{A_j^t}$ is the total number of occurrences of A_j^t in the whole database.

The "triggering" relationship indicates that one action's presence at the current time slice is caused by another one's presence at the previous time slice. We obtain the statistic of such situation in the training data. The pairwise causal dependency between two actions is computed as follows:

$$P(A_i^t = 1 | A_i^{t-1} = 0, A_j^{t-1} = 1) = \frac{N_{A_i^t + \neg A_i^{t-1} + A_j^{t-1}}}{N_{\neg A_i^{t-1} + A_j^{t-1}}}, \tag{4}$$

where $N_{A_i^t + \neg A_i^{t-1} + A_j^{t-1}}$ is the total number of the presences of A_i at the current time slice given its absence and the presence of A_j at the previous time slice, and $N_{\neg A_i^{t-1} + A_j^{t-1}}$ is the total number of the absence of A_i and the presence of A_j at the same time slice in the database.

We have obtained the three types of pairwise dependency for each pair of actions by temporal contextual analysis. The first two types of dependency are represented within a time slice. If they are higher than predefined thresholds, we assume that the two actions have strong dependency and build a link between the two nodes in the intra-slice topology. The last type of dependency is represented between two adjacent time slices. Similarly, if it is higher than a predefined threshold, we build a link between the two nodes in the inter-slice topology. This way, an initial 2TBN structure has been constructed.

3.2 Model Learning

After analyzing the contextual relationships, we obtain an initial DBN structure. Although it is our best guess based on the contextual analysis, it is necessary to use training data to refine it with a structure learning algorithm. The structure learning algorithm first defines a score that describes the fitness of each possible structure G to the observed data, and then, the best fitted network structure is identified with the highest score. The fitness score is defined as:

$$Score(G) = logP(D, G) = logP(G) + logP(D|G), \tag{5}$$

where $logP(G)$ is the log prior probability of the DBN structure. We do not give an equal prior to all possible structures. Instead, we assign a higher probability to the prior structure we initialized. $logP(D|G)$ is the log likelihood of the training data which can be approximated by the Bayesian information criterion (BIC) as follows:

$$logP(D|G) \approx logP(D|G, \hat{\Theta}) - \frac{d}{2}log(K), \tag{6}$$

where $\hat{\Theta}$ is the set of parameters of G which maximizes the likelihood of the training data D, d is the number of free parameters in G, and K is the number of training data D. The first term is used to measure how well the model fits the data, and the second term is a penalty term to punish the structure complexity. To obtain the model parameters $\hat{\Theta}$, we maximize the posterior distribution $p(\Theta|D, G)$ (MAP), given the training data D and the current structure G:

$$p(\Theta|D, G) = \prod_{i=1}^{n}\prod_{j=1}^{m_i} p(\theta_{ij}|D, G), \tag{7}$$

where n is the number of variables in G, and m_i is the number of all the parent instantiations for variable X_i. Since the training data set is complete, each parameter θ_{ij} can be calculated by a counting process. After we define the fitness score of the model, we can use an iterative way to learn the structure and parameters of the DBN. We firstly start with the initial DBN structure G_0, learn the parameters based on G_0 and compute the fitness score. Secondly, we generate the nearest neighbors of G_0 by adding, deleting, or reversing a single link, subjecting to the acyclicity constraint. Then we update G_0 with the structure which has the maximum score among the neighbors. The iteration process will be terminated until the score converges or the maximum iteration time is reached.

3.3 DBN Inference

Once the DBN model has been learned, each node is attached with an observation node so as to form a two-layer model. The top layer encodes the actions and their temporal and causal relationships. The bottom layer comprises a set of observation nodes that ingest the preliminary detection from low-level features. During action recognition, the node in the top layer are hidden and must be inferred from the observations in the bottom layer. The inference is conducted by finding the most probable explanation (MPE) of the evidence.

Let $A^t_{1:n}$ represents all the nodes for actions at time t, where n is the number of action nodes. Given the available evidence until time t: $O^{1:t}_{A_{1:n}}$, the action nodes are inferred over time by maximizing the probability $p(A^t_{1:n}|O^{1:t}_{A_{1:n}})$. The probability can be factorized by performing the DBN updating process as described in [6], and the inference can be solved by the widely used junction tree inference algorithm.

By DBN inference, we can employ the contextual information and mutual dependencies in a holistic way to refine the isolated preliminary detection results, and thus correctly recognize the actions in the multiple action streams.

4 Experimental Results

To evaluate the performance of our action recognition methods in multi-agent scenario, we choose the OSUPEL basketball dataset [2]. The OSUPEL basketball dataset is publicly available and it consists of multiple players playing against each other in a real basketball court. There are six primitive actions in the dataset: Pass, Catch, Hold Ball, Shoot, Jump and Dribble. This dataset is suitable for evaluating localization and recognition of multiple primitive actions characterized by rich spatiotemporal constraints.

Before discussing the recognition results in the multiple action streams, we first briefly describe the method on how to get the visual evidence of the actions from low-level features. The computed tracks of the players in the videos have been already provided in the dataset. We extract features from the bounding box of the computed tracks and use an HMM to detect each action separately. The features are derived from the histogram of oriented gradients (HOG)[3] and histogram of oriented optical flow (HOF) [4]. We train HMMs for each action class and use them to detect actions in the video separately. Please note that the preliminary detection results are not satisfying. The performance of the noisy detectors on the OSUPEL basketball data are summarized in Table 1. They can only be considered as noisy evidences to infer the true occurrence of each action.

Table 1. Preliminary action detection performance

	Dribble	Jump	Shoot	Pass	Catch	Hold
Recall	0.52	0.33	0.20	0.25	0.24	0.49
Precision	0.86	0.63	0.43	0.74	0.67	0.64

We feed the preliminary detection results as the observation values into the bottom layer of the DBN model so as to infer the state of the hidden nodes in the top layer. Fig. 3 shows the inference results on an example sequence with 10000 frames in the the OSUPEL basketball dataset. It contains 6 plots corresponding to 6 action classes. In each plot, the ground truth of the action occurrences is shown in the first row; the inference results are shown in the second row; the observations from the low-level detectors are shown in the third row. It is clear that using the holistic inference based on the contextual information, our model can correct the missing errors and false alarms in the preliminary detection results, and thus improve the action recognition accuracy.

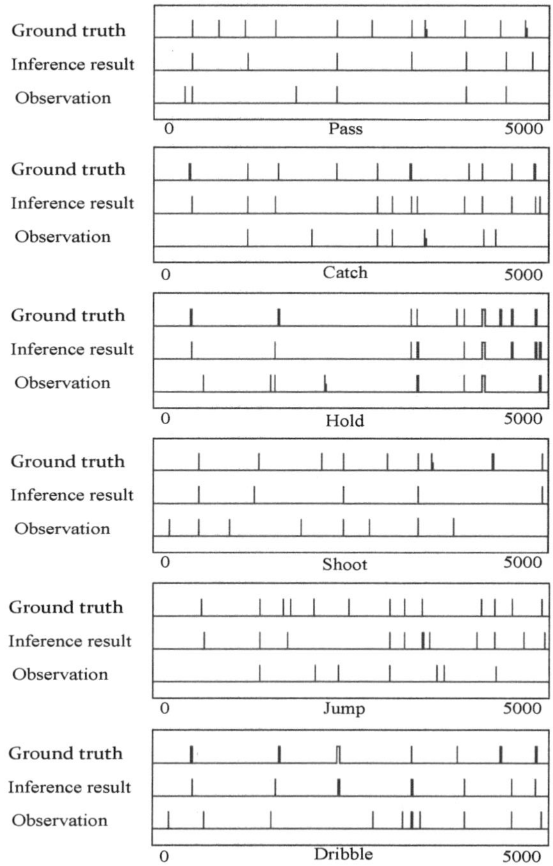

Fig. 3. Comparison of inference results with ground truth and observations from low-level detector on an example sequence in the OSUPEL basketball dataset

The performance of our model on the whole dataset is demonstrated in Table 2. Comparing to Table 1, we can find that both the precision and recall have been improved. To compare with our method, we also implement a CHMM [1,7] model, the structure of which is fully connected between all the temporal processes. The performance of CHMM model on action recognition is also shown in Table 2. Since the CHMM model cannot learn an adequate transition model, it performs worse than our model as expected. This demonstrates that the structure learning reduces the number of unnecessary parameters and caters for better network structure discovery.

5 Conclusions

We have proposed a DBN model for action recognition in multi-agent scenarios. Three kinds of mutual relationships between the interactive actions, concurrence,

Table 2. Inference performance using CHMM and our method

		Dribble	Jump	Shoot	Pass	Catch	Hold
CHMM	Recall	0.53	0.32	0.23	0.29	0.36	0.53
	Precision	0.83	0.61	0.44	0.72	0.69	0.66
Our method	Recall	0.54	0.36	0.27	0.32	0.34	0.56
	Precision	0.86	0.60	0.44	0.73	0.70	0.67

mutual exclusion and triggering, can be successfully captured in our model, so as to compensate the limitation of the low-level visual detectors. An advanced structure learning algorithm has been presented to discover meaningful and salient dependencies in order to construct a computationally tractable network. Currently, our model is time slice based. In the following work, we intend to extent our model to temporal interval based to make it more expressive and be able to capture more complex relationships.

Acknowledgement. This work was supported in part by the National Natural Science Foundation of China under Grant No. 61202325, the President Fund of UCAS, the Open Project Program of the National Laboratory of Pattern Recognition, China Postdoctoral Science Foundation:2012M520434, 2013T60156.

References

1. Brand, M., Oliver, N., Pentland, A.: Coupled hidden markov models for complex action recognition. In: CVPR Conference Proceedings (1997)
2. Brendel, W., Fern, A., Todorovic, S.: Probabilistic event logic for interval-based event recognition. In: CVPR Conference Proceedings (2011)
3. Dalal, N., Triggs, B.: Histograms of oriented gradients for human detection. In: CVPR Conference Proceedings, pp. 886–893 (2005)
4. Dalal, N., Triggs, B., Schmid, C.: Human detection using oriented histograms of flow and appearance. In: Leonardis, A., Bischof, H., Pinz, A. (eds.) ECCV 2006. LNCS, vol. 3952, pp. 428–441. Springer, Heidelberg (2006)
5. Hongeng, S., Nevatia, R.: Multi-agent event recognition. In: ICCV Conference Proceedings, pp. 84–91 (2001)
6. Korb, K., Nicholson, A.: Bayesian Artificial Intelligence. Chapman and Hall/CRC (2004)
7. Kratz, L., Nishino, K.: Anomaly detection in extremely crowded scenes using spatio-temporal motion pattern models. In: CVPR Conference Proceedings (2009)
8. Lv, F., Nevatia, R.: Recognition and segmentation of 3-D human action using HMM and multi-class adaBoost. In: Leonardis, A., Bischof, H., Pinz, A. (eds.) ECCV 2006. LNCS, vol. 3954, pp. 359–372. Springer, Heidelberg (2006)
9. Ryoo, M.S., Aggarwal, J.K.: Recognition of composite human activities through context-free grammar based representation. In: CVPR Conference Proceedings (2006)

Target Detection Based on Kernel Density Estimation Combined with Correlation Coefficient

Ting Rui[1], Zijian Yang[1], You Zhou[2], Husheng Fang[1], and Jingwei Zhu[1]

[1] PLA University of Science and Technology, Nanjing, China
[2] Jiangsu Institute of Economic & Trade Technology, Nanjing, China
rtinguu@sohu.com

Abstract. Kernel density estimation algorithm is an effective method to segment fore/background, but the computation of kernel density estimation is complexity. The conditions of dynamic scene of backgrounds and light mutational change make the robustness of the method is not good enough. Correlation coefficient is an effective method to describe the similarity of images, and the method is not sensitive when the image changes and has small differences in the intensity. A hierarchical block detection mechanism is proposed. First of all, the disturbance of the dynamic background scene and the light mutation are filtered by the correlation coefficient. And then, the block with fore-ground is segmented by kernel density estimation. Experiments confirmed that the proposed method is effective under the dynamic background disturbance.

Keywords: intelligent video, background modeling, kernel density estimation, correlation coefficient.

1 Introduction

Intelligent Video Surveillance (IVS) is a research, which originated from Computer Vision (CV) and Artificial Intelligence (AI). It aims at building a mapping relationship between images and incidents by the aid of massive computer process ability to screening the interference of useless information. With the help of AI, key objectives can be identified and automatic analyze from numerous and complicated video images.

The typical intelligent vision analysis process includes: Environment Modeling, Motion Segmentation, Object Tracking, Object Classification, Behavior understanding and Description. Environment Modeling, as the first step for most visional analysis tasks, which plays an important role of foundation to Motion Segmentation, Object Tracking, Object Classification, Behavior understanding.

The difficulty of Environment Modeling lies in the background disturbance. Generally, background may produce the following typical disturbance [1] [2]: change of scene light, which includes gradual change and sudden change; change of state, which including objective into and out of background. The characteristics of background changes put forward very restricted requirements to the dynamic update of background model. Traditional methods include Kalman Filtering and the Single Gaussian

B. Huet et al. (Eds.): PCM 2013, LNCS 8294, pp. 769–778, 2013.

Model and Gaussian Mixture Model. Background estimation model, which proposed by Koller, is based on the theory of Kalman Filter[3]. This method can track the changes of scene light, but not the changes of background objectives in time, therefore, it can lead to mistakenly identified goals and targets. Stauffer and Grimson proposed a Gaussian mixture model for background extraction [4][5], through fitting of distribution to single pixel value changes, background changes can be tracking timely. But the amount of calculation of algorithm is massive, real-time monitoring and control system relies on expensive special hardware.

In recent years, many new methods of background extraction have been developed, Elgammal and Duraiswami presents a nonparametric Kernel density estimation method [6], which need no presupposes the distribution of pixel values, but can directly estimate the probability density function from the continuous variation of the pixels value.

At present, there are many researches to the method of target detection which bases on background modeling [7-10]. Both Gaussian mixture model and Kernel density estimation model takes computational cost, "real-time" cannot be guaranteed in high resolution intelligent video analysis system. At the same time, the two models are based on the current frame pixel which belongs to the probability of foreground/background, so that to realize target detection as well as the model update, so it possesses a gradient background with good adaptability, a model collapse may be caused when there is a camera shake or sudden light change. Tracking method to correlation detection is based on the images similarity measure, which tracking the most close to the template image area in the current frame, the correlation coefficient is a kind of effective method to describe the correlation[11], it is not sensitive to the change of light gray which caused by small scale camera shake, translation, or sudden change of light , but the way of point by point scanning also takes computational complexity, real-time detection cannot be guaranteed.

This paper proposes a method which combines Kernel density estimation and correlation coefficient, for fulfilling the purpose of background modeling and moving objective detection. This modeling method is a kind of multilevel detection system, firstly, partitioning the current frame and background. Secondly, executing the instruction of correlation detection in the corresponding area. Thirdly, starting the Kernal density estimation method for the foreground/background segmentation if the correlation coefficient is less than the threshold value, otherwise, target object does not exist and update the model directly. Such comprehensive utilization embodies the adaptability of mutation to the environment of coefficient of correlation detection method and the precision of Kernel density estimation to target detection. Besides, this method greatly reduces the computational complexity, and ensures the real-time performance of the moving target detection, the validity of the method has been proved by experimental test.

2 The Background Modeling of Kernel Density Estimation

Kernel Density Estimation (KDE), is an important tool of mathematical statistics and related applications. It uses analytical method to solve the distribution density function, which is random variable for a distribution of the sample of unknown given data, so that to construct the data samples in a specific space distribution model.

In a complex environment, the background are not statics, but the density distribution of background is usually unpredictable,therefore, nonparametric multimodal background model is very suitable for dealing with this kind of density distribution of the unknown phenomenon. Kernel density estimation method does need to assume the background characteristics of probability distribution form, also do not need to set parameters for the model and the parameter optimization, but the probability density estimation function, which can obtained directly from the sample. As long as there is sufficient sample, Kernel density estimation can gradually converge to any a probability density function.

Though the weighted average of the center which located in the sampling value of the local function, Kernel density estimation can estimate the unknown density distribution. A pixel characteristic space of a given sample set $S=\{x_1, x_2, ..., x_N\}$, observed value x_t density distribution $p(x_t)$ is used to estimate

$$p(x_t) = \sum_{i=1}^{N} \alpha_i K_h(x_t - x_i) \tag{1}$$

the weights α_i=1/N; $K_h = (1/h)K(t/h)$ are the Kernel function. h is the width of the Kernel function. If a pixel consider m features, choose Gaussian function K_h as Kernel function, σ_m represents the feature of Kernel bandwidth, then type (1) can be stated as:

$$p(x_t) = \alpha_i \sum_{i=1}^{N} \prod_{m=1}^{d} \frac{1}{\sqrt{2\pi\sigma_m^2}} \exp(-\frac{1}{2}\left[\frac{(x_{tm} - x_{im})^2}{\sigma_m^2}\right]) \tag{2}$$

In Kernel density estimation, the bandwidth selection of σ is much more important than Kernel function itself. Bandwidth selection will bring about probability density estimation to owe a smooth phenomenon, the opposite will lead to smooth, both of which will adversely affect the estimate result. Assuming x_i is normal distribution $N(\mu,\sigma^2)$, then $(x_i - x_{i+1})$ is normal distribution $N(0,2\sigma^2)$, m is median of absolute value of eigenvalue difference of two adjacent pixels, namely $m = median(|x_i - x_{i+1}|)$, By the normal distribution and median symmetry properties get:

$$P[N(0,2\sigma^2 > m] = 0.25 \tag{3}$$

Known by the normal distribution:

$$\sigma = \frac{m}{0.68\sqrt{2}} \tag{4}$$

It is proved that the calculation of pixel values in the sample frame sequence will take a lot of time, when a longer sample frame sequence estimated used by Kernel density estimation. In order to improve the real-time performance, we can use average values in the sample sequence approximation, this way does not impact on the estimated result significantly.

Background modeling Kernel density estimation is essentially an extension of Gaussian mixture background modeling method, compared with the Gaussian mixture model in only 3-5 Gaussian background model method, it can describe the sample density distribution more accurately, thus has stronger robustness. Due to the advancement of Kernel density estimation method in theory and good effect in the practical use, but it also features in two deficiencies: 1.the complicated calculation, since the beginning of the algorithm is proposed, Kernel density estimation calculation complexity problems has caused widespread concern of the researchers, especially in video analysis, and such other kind type of application which requires high standard for real-time, the improvement of calculating efficiency has became important problem to the application for Kernel density estimation method. 2.the adaptability to dynamic background, in order to improve the real-time performance of the algorithm, when we carries on the Kernel density estimation, the sample frame sequence cannot be too long, usually in 30-50 frames, the segmentation error may appear foreground/background in great disturbance if a smaller sample frame sequence itself is very smooth. In order to reduce the complexity of Kernel density estimation method and improve the disturbance of adaptability to the background, this article put forward that we can use the advantage of correlation coefficient test method is not sensitive to dynamic background disturbance and the relatively small calculating amount of blocked area detection, carries on the correlation detection of the current frame and background model before using Kernel density estimation fore/background segmentation to the current frame.

3 Foreground Area Detection Based on Correlation Coefficient

Correlation coefficient is a good criterion for identifying the similarity of image, which has a good adaptability to the disturbance such as noise, gray circumstances. But in traditional target tracking method which based on correlation algorithm, it is rarely used in the correlation coefficient as a test criterion, this mainly because of the correlation coefficient of the computational cost is larger, point by point by point calculation is needed in target tracking method for the current frame and template of the correlation coefficient, which cannot guarantee real-time tracking.

This paper uses a new thought, which concerns the current frame and background of block, using the corresponding between correlation detection of current frame and background model. If the correlation coefficient is larger than the set threshold, suggests that this block does not have foreground objective, then updating Kernel density estimate the background model directly, if the correlation coefficient is less than the set threshold, and suggests the corresponding block in foreground objective, starting the Kernel density estimation method of the fore/background segmentation, positioning the foreground objective accurately.

3.1 Background Model Corresponding to the Background

In the Kernel density estimation method, the background model is consisted of updated dynamic background of sequence and Kernel density estimation function. In this article, background is defined as a sequence corresponds to the background of actual background. It can be simply obtained by average background sequences, when the dynamic update can be obtained by type (5) recursively

$$\mu_{t+1} = (1-\alpha)\mu_t + \alpha X_t \tag{5}$$

μ is the background gray value, X is the division of the current frame as the background pixel gray value, α is the update, constant for less than 1 If the background sequence elements on the background of the contribution of the same, then $\alpha = 1/n$ (n is the sequence length), taking into account the dynamic background segmentation, background pixels in the current frame with the most "fresh" information on the background should a greater contribution to so take $\alpha = (2\sim3)1/n$.

To the prospected pixels which have been determined, keep the same original background picture. In case of the original background objects movement become into foreground object, or the original foreground object keeps stationary for too long term, and converted into background objects, we can set up a counter, such as the background is the current frame pixel, $counter = 0$, such as the foreground object is the current frame pixel, $counter = counter + 1$, when $counter$ is greater than the counter threshold $counterthresh$, the pixels of the state did not make change for a long time, the pixel as the background, which should be regarded as a part of the of Kernel density model updating, at the same time, the background should be updated too.

3.2 The Detection to Foreground of Correlation Coefficient

The correlation coefficient is defined as two pairs of images

$$Corr(f,g) = \frac{\sum_{i=1}^{M}\sum_{i=j}^{N}(f_{i,j}-\bar{f})(g_{i,j}-\bar{g})}{\sqrt{\sum_{i=1}^{M}\sum_{j=1}^{N}(f_{i,j}-\bar{f})^2}\sqrt{\sum_{i=1}^{M}\sum_{j=1}^{N}(g_{i,j}-\bar{g})^2}} \tag{6}$$

$$\overline{f}=\frac{1}{MN}\sum_{i=1}^{M}\sum_{j=1}^{N}f_{i,j}\text{ , }\overline{g}=\frac{1}{MN}\sum_{i=1}^{M}\sum_{j=1}^{N}g_{i,j}$$

f is a block in current frame, g is a block of background frame, $\overline{f}$ is the average in current frame, $\overline{g}$ is the average of the block which in the background frame, M,N are the dimension of the block.

Correlation coefficient meet $|Corr|\le 1$, to describe similarity of two images within the scope of[-1,1], the $Corr$ more closer to 1,the two images has more similarity. As a result, in the correlation detection to the corresponding block of the current frame and background, if the correlation coefficient of the corresponding block is close to 1, the more similarity does exists between the current frame and background in the area of the block, the foreground object does not show up. If the correlation coefficient of the corresponding block is small, which indicates in the area of the block, a big difference does exists between the current frame and background, and this difference is caused by the existence of foreground object, by this way , we can fulfill the purpose of foreground object detection.

Such a mechanism is adopted to accomplish the multilevel detection of background model, the first level is coarse detection of block, which is accomplished by correlation coefficient analysis. This method has a good adaptability to dynamic background and the gray change. The second level is taking precise segmentation to the block which has foreground object within by the way of Kernel density estimation method, because the target only possible appears in a limited number of blocks, so the computational complexity of Kernel density estimation could be greatly reduced, the requirement of real-time can be fulfilled at the same time.

4 Analysis of Algorithm and Experiment Result

4.1 Algorithm Steps

In this paper, the process of algorithm is shown in figure 1.Firstly, set up Kernel density estimation background model by the number of initial frame with Gaussian Kernel function, and obtains background by background frames. Then the current frame and background figure divides the block and the corresponding block do coefficient of correlation detection, if the correlation coefficient is greater than the threshold, the corresponding block indicates the current frame. The block area with no vision, can update background model directly, if the corresponding block of the correlation coefficient is less than the threshold, indicates the current frame appear prospects that are within the area of targets, the use of Kernel density estimation method of the area fore/background segmentation, part prospects ground model update will not be involved.

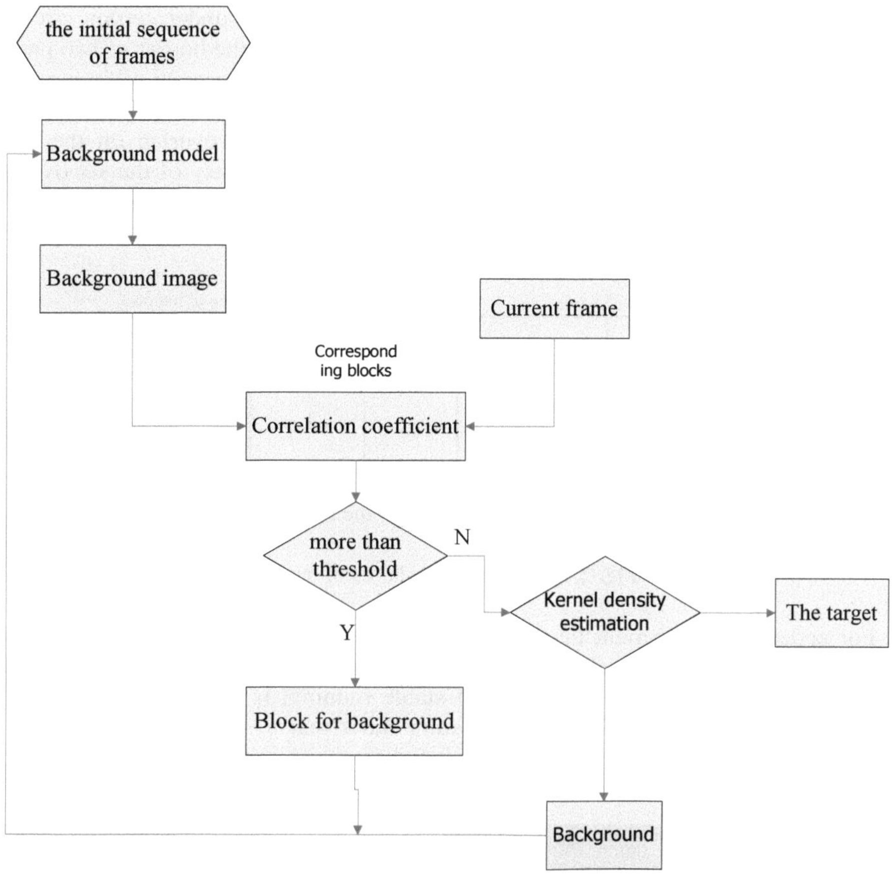

Fig. 1. This algorithm processes

4.2 Experiment Results

To verify the algorithm adaptability to disturbance in natural conditions, the test select outdoor natural environment, the target monitoring scene camera around 50m, a proportion of the target dimension in the frame is small, mainly to test the effectiveness of the algorithm of target detection and adaptive disturbance to the environment. Test Equipment used for the 2.4G frequency CPU notebook, Algorithm based on the VC + +6.0 achieve OpenCV1.0,Video resolution 640 × 480 pixels.

Kernel density estimation model made up of 30 frames of images sequences and Gaussian Kernel function, the image block, according to 35x35pixels experiment found there is no vision of cases, the current frame and background corresponds to each piece of the correlation coefficient of stability of keep at around 0.95, in the event of a future goal, including target block in the current frame and background of the correlation coefficient is less than 0.8.

To the blocks does not overlap, residual error may occur under certain circumstances, when the scale of smaller object happen to appear on the border of two pieces, for each block, only a few pixels in the current frame and background differences, the correlation coefficient is greater than 0.95 at this time, will cause the miss the target.

For solving this error, when the block divided with the overlap on the block boundary, as shown in figure2, make each piece of the boundary of the six overlapping areas of a pixels, so as to ensure that when the target is in two piece of border, can at least make one piece of correlation coefficient is less than the threshold value detection.

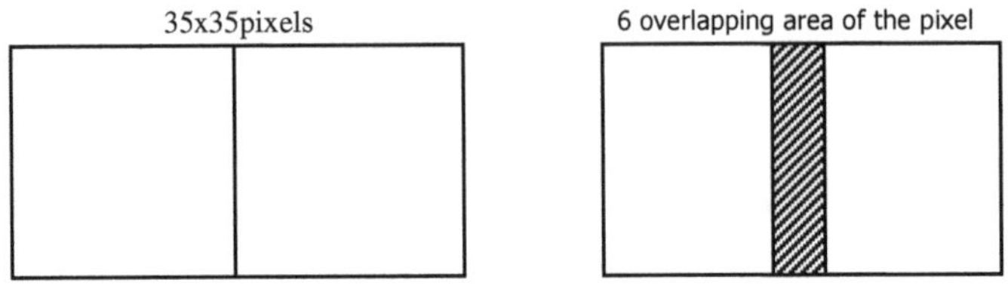

(a) the blocks does not overlap (b) the block divided with the overlap

Fig. 2. The division methords of module

For pedestrian movement target test, test result as shown in figure3 ,we can see complete algorithm to extract the small target, does not appear the mistake area, at the same time the test measured in hours of stable running, is not affected by ambient light and other disturbances, the test results show that system uses the method is effective.

Fig. 3. The results of Pedestrian motion target test

5 Conclusion

As a kind of effective nonparametric estimation method, Kernel density estimation dose not need to assure the density distribution forms, it can estimated the unknown density function from the data directly, which avoids the complex and difficult problems such as model form assumptions and parameters optimization of, it is very suitable for the background modeling of intelligent video analysis.. But in practice, this method has two drawbacks, one is Kernel density estimation method relies on the data sequence of huge computation, it cannot meet the requirement of real-time video analysis. The other is its poor adaptability to disturbance of environment mutations, when light change occurred, the model is easy to collapse.

This paper presents a joint block multilevel object detection estimation of correlation coefficient, which is only in view of the corresponding block, it greatly reduced the computational complexity, at the same time, provided by the coefficient of correlation detection to the environment disturbance adaptive mutation, only exists in the correlation coefficient test indicate that some piece of foreground object, only by the Kernel density estimation for these blocks fore/background segmentation. This, comprehensive utilize the respective advantages of the Kernel density estimation and correlation coefficient test, while reducing the work of complexity computation significantly, the robustness is improved obviously through a long term of experiments which proved the effectiveness of the method.

Acknowledgment. This work is supported by the National Natural Science foundation under Grants 61070108 and 61070104. It is also funded by 2011BAK07B00.

References

1. Mittaland, A., Huttenlocher, D.: Scene modeling for wide area surveillance and image synthesis. In: IEEE Conf. Computer Vision and Pattern Recognition, vol. 2, pp. 160–167 (2000)
2. Huang, S.S., Fu, L.C., Hsiao, P.Y.: Region-level motion-based background modeling and subtraction using MRFs. IEEE Transactions on Image Processing 16(5), 1446–1456 (2007)
3. Koller, D., Weber, J., Huang, T., et al.: Towards robust automatic traffic scene analysis in real-time. In: Proceedings of International Conference on Pattern Recognition, Jerusalem, Palestine, pp. 1262–1311 (1994)
4. Stauffer, C., Grimson, W.E.L.: Adaptive background mixture models for real-time tracking. In: Proceedings of IEEE Computer Society Conference on Computer Vision and Pattern Recognition, Fort Collins, Colorado, USA, pp. 2462–2521 (1999)
5. Stauffer, C., Grimson, W.E.L.: Learning patterns of activity using real-time tracking. IEEE Trans. PAMI 22(8), 474–757 (2000)
6. Elgammal, A., Duraiswami, R., Harwood, D., et al.: Background and foreground modeling using nonparametric kernel density estimation for visual surveillance. Proceedings of IEEE 90(7), 1151–1163 (2002)

7. Yan-fen, M.A.O., Peng-fei, S.H.I.: Kernel Density Estimation for Dynamic Scene Modeling. Journal of Data Acquisition & Processing 19(4), 391–394 (2004)
8. Chen, G., Zhou, H.-Q., Yan, J.-F.: Real-time Back-ground Subtraction and Moving Shadow Detection Based on UKF Modeling. Journal of Image and Graphics 14(5), 931–937 (2009)
9. Wei, C.-A., Jiang, S.-D., Sun, C.: Small Target Tracking in Forward Looking Infrared Iimagery Based on Kernel Density Estimation. Journal of Harbin Engineering University 30, 763–767 (2009)
10. Gu, J.-D., Liu, Z., Zhang, Z.-Y.: A Novel Moving Object Segmenta-tion Algorithm Using Kernel Density Estimation and Edge Information. Journal of Computer-Aided Design & Computer Graphics 21(2), 223–228 (2009)
11. Zhu, Y.-S., Guo, C.-M.: Research of Correlation Tracking Algorithm Based on Correlation Coefficient. Journal of Image and Graphics 9(8), 963–967 (2004)

A Belief Based Correlated Topic Model for Semantic Region Analysis in Far-Field Video Surveillance Systems

Jialing Zou, Xiaogang Chen, Pengxu Wei, Zhenjun Han, and Jianbin Jiao*

University of Chinese Academy of Sciences, Beijing, China
jiaojb@ucas.ac.cn

Abstract. In this paper, a belief based correlated topic model (BCTM) is proposed for the semantic region analysis of pedestrian motion patterns in the crowded scenes. The inputs of the BCTM can be holistic trajectories or fragments of trajectories. By integrating the sources, sinks, and a forest of randomly spanning trees of trajectories as priors, the proposed BCTM improves the learning of semantic regions, significantly. In addition, the model can also cluster topics through modeling relations among topics. Experiments on a large scale data set, which are collected from the crowded New York Grand Central Station, show that the BCTM outperforms the state-of-the-art methods on qualitative results of learning semantic regions.

Keywords: topic models, topic clustering, forests of randomly spanning trees, crowded scenes.

1 Introduction

In a far-field video surveillance system, one of the core tasks is to automatically segment the scenes into the semantic regions and then learn their patterns. Those semantic regions represent dominant paths in the scenes, which indicate global behaviors of the grouped people and are a kind of mid-level description for the structures of the scenes. Therefore, it is important to accurately get those mid-level features of the objects to capture structures of the scenes.

Semantic regions in video surveillance system was explored in single and multiple camera scenarios in previous works [1-6]. [1, 2, 5, 6] mainly defined atomic activities, i.e. local motions, and they modeled global motions as distributions over local motions. In the multi-view systems of [3, 4], behaviors of objects were recognized across different views through graph algorithms. The correlated topic model (CTM) in [7] is one of the hierarchical Bayesian models and topic models in natural language processing, which can model the video data in low dimensions and reduce computation cost. Topic models were explored in previous work [5, 6, 8-14] in large scale video surveillance. [11] explored Probability Latent Semantic Analysis (PLSA) model, which hardly handled the situations with multiple categories. [2, 3, 6, 8, 10] pro-

* Corresponding author.

B. Huet et al. (Eds.): PCM 2013, LNCS 8294, pp. 779–790, 2013.

posed different kinds of improved Latent Dirichlet Allocation (LDA) models [15]. Computation of the posteriori in those models is tractable through Markov Chain Monte Carlo approaches like Gibbs sampling in [16]. In [5, 13], they proposed improved hierarchical dirichlet process models, where the number of the topics was unnecessary to be pre-defined. The CTM was explored in [8], which removed the assumption in LDA that each topic is statistically independent from others.

In general, the approaches of learning semantic regions in video surveillance are in two categories: complete trajectories of objects [6, 10, 17, 18] based, and non-complete trajectories objects or local motions [1, 2, 5, 12] based. The applications of topic models in learning semantic regions are also in two categories: hierarchical Bayesian models [5, 6, 8, 9, 12] based, and nonparametric hierarchical Bayesian models [10, 13] based. The number of topics is automatically determined in nonparametric hierarchical Bayesian models and parametric ones are advanced in computation cost.

1.1 Our Approach

A belief based correlated topic model (BCTM) is proposed in far-field video surveillance to learn semantic regions. It can improve the learning of semantic regions significantly, by integrating a forest of randomly spanning trees, sources and sinks as priors. The inputs of the BCTM can be holistic trajectories or tracklets. A tracklet is a part of a complete trajectory, which represents the motions of an object in a short period of time. A tracklet will be terminated when the scene clutters arises and the occlusions among objects happen. In complex surveillance scenes, tracklets are lack of discriminative power and cannot indicate the complete motion of the objects. In [9], the authors put forward the forest of spanning trees to capture temporal and spatial relations between tracklets to link them into different sets. But the forest spanning trees only pursuit local energy minimization, which will usually capture some tracklets belonging to the different categories. If that happens in top level of the trees, then the final path will be invalid. Therefore, firstly we update the definition of the neighbor sets of spanning trees referring to Markov Random Field (MRF) in [19] to avoid the wrong tracklets linking into the trees in some extents. Those improved spanning trees leave out some extremely spatially close tracklets to speed up finding effective paths. Then we also define the sources and sinks as beliefs of objects because almost every object has a clear belief about where it comes from and where it wants to go. Sources and sinks indicate where objects enter and exit in the surveillance system, which are necessary in complex scenes to help discriminate paths, and was explored in [9, 18, 20]. Without sources and sinks, the CTM doesn't work well in complex scenes because it cannot handle multiple-category problem.

Topics in our model indicate the semantic regions in the crowded scenes. The number of topics in the BCTM is a key issue because it will raise the problem of over-segmentation or under-segmentation. But it is always a trade-off between discriminative power and computing complexity. To get accurate paths, we need to mainly avoid under-segmentation and get sufficient topics to completely describe the scenes. Therefore we novelly ignore the over-segmentation and then use relations between topics to cluster over-segmenting topics . By the way, the computing cost has been already effectively reduced because of the priors of the beliefs and the forest of the spanning trees. Because those priors can create a sparse coefficient matrix.

A large scale of data set including 47351 tracklets is extracted from the New York Grand Central Station by the KLT tracker in [21]. Fig.1 (A) shows the histogram of all tracklet lengths. Most of tracklet lengths range from 1 to 100 frames. Fig.1 (B) shows proportions of four kinds of tracklets. Most of tracklets have incomplete beliefs.

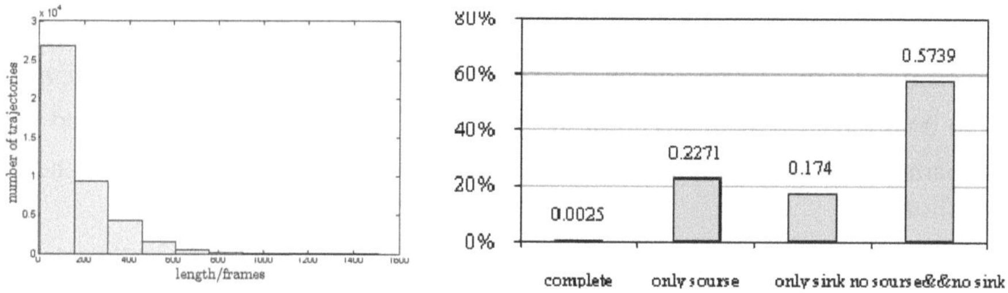

(A) (B)

Fig. 1. (A) The histogram of tracklet lengths. (B) Statistics of sources and sinks of tracklets.

1.2 Related Works

Semantic regions can be used to cluster trajectories, because tracjectories in the same class share the same semantic region. Comparisons of different trajectory clustering methods can be found in [22, 23]. Those non-Bayesian approaches are difficult to incorporate priors such as sources and sinks to improve the learning of the semantic regions. [10] proposed a hierarchical Bayesian model to capture structures of the scenes, but it assumed that trajectories were extracted completely and it ignored the spatial and temporal coherence between tracklets. Therefore, the model did not work well in crowded and complex scene. [5] used hierarchical Bayesian models to learn semantic regions from features of optical flows. It worked well in traffic scenes where activities of objects were observed at different time. However, it is not suitable for the crowed and busy scenes, because different kinds of activities of objects always happen simultaneously.

Some other methods [2, 12, 24] clustered adjacent cells with similar location and motion information for the learning of semantic regions. Those results tended to be in short range and were not accurate for definite scenes. In previous work [24, 25], tracklets were explored to be connected into complete trajectories but not for learning semantic regions. [26] analyzed dynamics of objects in the same crowded scene without topic models, and didn't learn semantic regions as well.

In [8], CTM was explored to capture structures of the scenes. But it did not include priors to improve the learning of semantic regions. In [9, 19], authors used tracklets with topic models to learn semantic regions. But BCTM handles with more topics and finds more dominant paths than [9].

2 The Proposed BCTM Approach

Fig.2 (A) is the graphical representation of the BCTM. The notations of topic modeling in natural language processing are used without loss of generality. Tracklets are treated as documents, and observations on tracklets are treated as words. Words are quantized by tracklets' locations and velocities. Sources and sinks of the scene are known as priors, which are automatically estimated referring to [9, 18, 20]. Fig.2 (B) shows the estimated source and sink regions. An observation of a tracklet has four variables (*x, h, m, z*), where x is a word, h and m are the labels of the sources and sinks of *x, and* z is a latent variable indicating the topic of x. A tracklet's h or m is observed if it starts from a source or exists from a sink. Otherwise, its h or m will need to be inferred. The distribution of document j is specified by η_j. β, δ and κ are hyper-parameters for multinomial distributions of words, sources, and sinks. The joint distribution is

$$
\begin{aligned}
&p(\{(x_{ji}, h_{ji}, m_{ji}, z_{ji})\}, \{\eta_j\} \mid \Sigma, \mu, \beta, \delta, \kappa) \\
&= \prod_j p(\eta_j \mid \Sigma, \mu) \prod_{j,i} p(z_{ji} \mid f(\eta_j)) p(x_{ji} \mid \beta_{z_{ji}})\, p(h_{ji} \mid \delta_{z_{ji}}) p(m_{ji} \mid \kappa_{z_{ji}}),
\end{aligned}
\tag{1}
$$

where j and i are indices of documents and words, η_j is a continuous variable sampled from a multivariate Gaussian distribution $p(\eta \mid \mu, \Sigma)$, and x_{ji}, h_{ji} and m_{ji} are discrete variables sampled from discrete distributions $p(x_{ji} \mid \beta_{z_{ji}})$, $p(h_{ji} \mid \delta_{z_{ji}})$ and $p(m_{ji} \mid \kappa_{z_{ji}})$ respectively. A tracklet k starts at t_k^s and ends at t_k^e. Its starting and ending locations are (x_k^s, y_k^s) and (x_k^e, y_k^e) with velocities $V_k^s = (v_{ik}^s, v_{ik}^s)$ and $V_k^e = (v_{ik}^e, v_{ik}^e)$. $\phi(j)$ is a neighbor set of document j. $k \in \phi(j)$, if it satisfies

$$
\begin{aligned}
&\text{I.} \quad t_j^e \le t_k^s \le t_j^e + T, \\
&\text{II.} \quad S_1 < \sqrt{(x_j^e - x_k^s)^2 + (y_j^e - y_k^s)^2} < S_2, \\
&\text{III.} \quad C_1 < \frac{V_j^e \cdot V_k^s}{\| V_j^e \| \; \| V_k^s \|} < C_2.
\end{aligned}
\tag{2}
$$

Eq. (2) identifies the neighbor sets, which are searched to span the trees and ensures extremely spatially close tracklets are not formed into trees because experiments show those tracklets almost denote different objects' motions. Apparently, those tracklets are lack of coherence in motion, and the accuracy of semantic regions is definitely reduced. Therefore it is reasonable to ignore those tracklets. Spanning a tree is to search in a pre-defined neighbor set and find a minimum weighted path. A corresponding weight function $W(z_j, z_k)$ is defined as

$$W(z_j, z_k) = \begin{cases} \exp\left(\dfrac{V_j^e \cdot V_k^s}{\| V_j^e \| \; \| V_k^s \|} - 1\right), & \text{if I-III, and } z_j = z_k. \\ 0, & \text{otherwise.} \end{cases} \tag{3}$$

Eq. (3) shows only the tracklets which are spatially and temporally close will be connected into the trees.

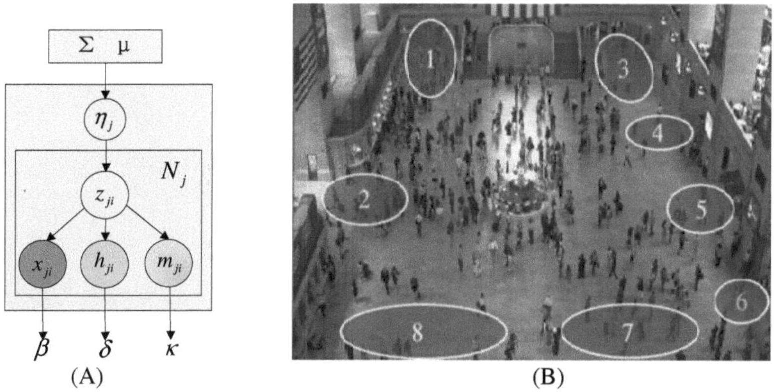

Fig. 2. (A) Graphical representation of the BCTM. *x* is fully observed. *h* and *m* are partly observed. (B) Estimated source and sink regions.

2.1 Topic Clustering

Topics in our model indicate semantic regions in crowded scenes. But the topics are not discriminative enough to represent semantic regions. Therefore, clustering topics is necessary to improve the discriminative ability of the topics through the relations among topics which are modeled by the covariance matrix of topics and the estimated graphs. A graph $\vartheta(\Sigma) = (V, E)$ is formed according to [7]. Vertices V correspond to the mean variables λ_d of topics, $d \in 1, \ldots, K$. Edges E represent the relations and satisfy $(s, t) \in E$ if and only if $S_{st} \neq 0$, where K is the count of the topics, and $S = \Sigma^{-1}$ denotes the inverse covariance matrix, and (s, t) is an element in the matrix S. The approach LASSO in [27] was used to asymptotically estimate the graph $\vartheta(\Sigma)$, which effectively reduced the cost of computation through iteratively optimizing. In [7], covariance matrix was not computed for forming the graphs and the estimated graphs only showed the relevant and irrelevant relations between topics. However, that estimated graphs are not enough for deeply learning relations between topics. In the BCTM, the covariance matrix is incorporated with the estimated graphs to find the positive-relevant, negative-relevant and irrelevant relations between topics, which can indicate topics in the same cluster and over-segmented topics. These relations are corresponding to three definite physical meanings in the crowded scene. Positive-relevant relation indicates that the two topics describe the same path. Negative-relevant relation indicates that the two topics describe different paths, and irrelevant relation indicates that at least one of the two topics describes an over-segmented path.

2.2 Inference

Because there is no advantage of conjugation between low-level and high-level distributions, Markov Chain Monte Carlo methods in [9] is not effective here. [7, 28] used Variable Breaking (VB) algorithm to iteratively optimizes the results to be asymptotically consistent with the real results, therefore we use it to solve the inference problems here. With priors of sources and sinks, the bound function in VB is modified as

$$\begin{aligned} L = & E_q[\log p(\eta \mid \Sigma, \mu)] + E_q(\log p(z \mid \eta)) + E_q[\log p(x \mid z, \beta)] \\ & + E_q[\log p(h \mid z, \delta)] + E_q[\log p(m \mid z, \kappa)] \\ & - E_q[\log q(\eta)] - E_q[\log q(z)]. \end{aligned} \tag{4}$$

The parameters $\hat{\phi}$, $\hat{\kappa}_i$, $\hat{\delta}_i$ are modified as

$$\hat{\phi} \propto \exp(\lambda) * \beta * \delta * \kappa, \tag{5}$$

$$\hat{\kappa}_i \propto \sum_c \phi_{c,i} n_c, \tag{6}$$

$$\hat{\delta}_i \propto \sum_e \phi_{e,i} n_e. \tag{7}$$

Eq. (5) is used to get an updated $K \times V$ topic matrix, where V is the count of the dimensions of word codebook. Eq. (6) and (7) are used to update h and m matrices (with the dimension $K \times L$), where L is set to 8 empirically. If a sink or a source of a tracklet is lost, the topic labels, which are inferred according to the observed words in topic matrices, are explored in h or m matrices to infer the label of the source or sink corresponding to the tracklet.

3 Experiments

Experiments are conducted on a 30 minutes video data collected from the New York's Grand Central Station, which is the same as the data set in [9]. The resolution of the video is 720×480. The scene is divided by 10×10 cells. Directions are quantized into 5 numerical values from 0 to 4, where 0 indicates an object is a static one. Therefore the size of the codebook is $72 \times 48 \times 5$. We define Complete Tracklets as tracklets with both sources and sinks, Semi-Complete Tracklets with either sources or sinks is to be inferred during the tree construction, and Non-SS Tracklets with both sources and sinks are to be captured during the tree construction. Fig.3 shows the statistics of sources and sinks after the tree construction for the three kinds of tracklets. The index of the horizontal axis shows the labels of the sources and sinks. For example, '28' means the label of source is 2 and the label of sink is 8. In total, 23478 trees are created, including 1175 from the Complete Tracklets, 15791 from the Semi-Complete Tracklets, and 6512 from the Non-SS Tracklets.

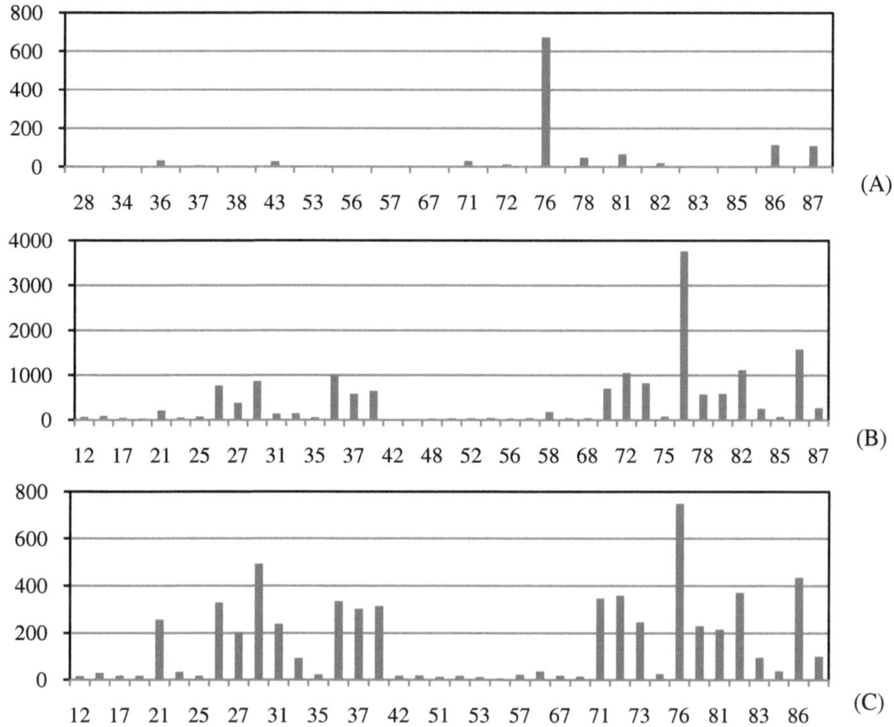

Fig. 3. Statistics of sources and sinks of trees from (A) Complete Tracklets, (B)Semi-Complete Tracklets, (C) Non-SS Tracklets

3.1 Topic Clustering

50 regions are learned in our experiment, 13 of which are randomly selected to show how to perform the topic clustering with the covariance matrix and the estimated graphs, which are showed in Fig.6. Only the relations inside the 13 regions are concerned, leaving out relations with the other 37 regions. Table 1 shows the 13 dimension covariance matrix. Elements along the main diagonal line are non-sensible because only the relations with others are concerned. Three illustrations are given in Table 1, which are in bold and italic font. The element 7.17256 indicates that the relation between topic 3 and topic 5 is positive, which means topics will be merged into one topic. The element -10.874 indicates that topic 3 and topic 12 describe two definitely different paths. The element 0.00413 indicates that at least one of topic 3 and topic 8 is over segmented. Fig.4 shows that the relations of topic 8 with others are smoother than those of topic 3 with others. The numerical values of relation between topic 8 and others approximate to zero, indicating that topic 8 is over segmented, because there is no such structure in the original data set. The estimated graphical relations through LASSO are not completely the same with the relations in the covariance matrix. As a result, we firstly use the estimated graph to describe relevant and

irrelevant relations, and then incorporate with the matrix showed in Table 1 to indicate the positive and negative relations. The estimated graph with modifications according to Table 1 is showed in Fig.5. The dashed lines represent positive relations and solid lines represent negative relations. In addition, some relations, which are not described in the estimated graphs, can also be defined according to Table 1. The topic-clustered results are showed in Fig.7, from which we can clearly see that over-segmented regions and repeated regions are all removed.

Table 1. The Covariance Matrix of the randomly selected 13 topics

12.77929	0.590075	-0.290191	-0.685796	0.143539	0.66297	-0.386788	1.238406	0.569634	-0.473792	4.858148	-3.2946	-0.308319
0.590075	20.398835	-0.141089	-0.708489	0.330114	3.089583	-0.379222	0.896417	1.346482	-0.443425	-0.060972	-3.562985	-0.186173
-0.290191	-0.141089	27.923416	-2.387259	***7.172555***	0.095889	-2.407919	***0.004126***	0.448543	-2.483246	-2.867567	***-10.8741***	7.086671
-0.685796	-0.708489	-2.387259	12.565639	-1.302466	-0.547244	0.243391	-0.11196	-0.232952	-1.414048	-2.374024	4.736406	-1.737755
0.143539	0.330114	7.172555	-1.302466	15.489123	0.461302	-1.169009	0.462392	0.501083	-1.30228	-1.291169	-6.380487	4.559789
0.66297	3.089583	0.095889	-0.547244	0.461302	18.252315	-0.211113	1.064328	1.239994	-0.345946	0.058287	-3.089301	-0.022463
-0.386788	-0.379222	-2.407919	0.243391	-1.169009	-0.211113	18.315791	0.456378	0.057052	-1.186417	-2.164369	2.608233	-1.760639
1.238406	0.896417	0.004126	-0.11196	0.462392	1.064328	0.456378	8.36759	1.037727	0.214755	0.360483	-1.042282	-0.248741
0.569634	1.346482	0.448543	-0.232952	0.501083	1.239994	0.057052	1.037727	9.577117	-0.054914	0.321006	-1.603192	0.223586
-0.473792	-0.443425	-2.483246	-1.414048	-1.30228	-0.345946	-1.186417	0.214755	-0.054914	25.31636	-1.655373	-3.371371	-1.856683
4.858148	-0.060972	-2.867567	-2.374024	-1.291169	0.058287	-2.164369	0.360483	0.321006	-1.655373	31.470841	-9.039292	-2.048835
-3.2946	-3.562985	-10.87409	4.736406	-6.380487	-3.089301	2.608233	-1.042282	-1.603192	-3.371371	-9.039292	32.3616	-7.484945
-0.308319	-0.186173	7.086671	-1.737755	4.559789	-0.022463	-1.760639	-0.248741	0.223586	-1.856683	-2.048835	-7.484945	19.907991

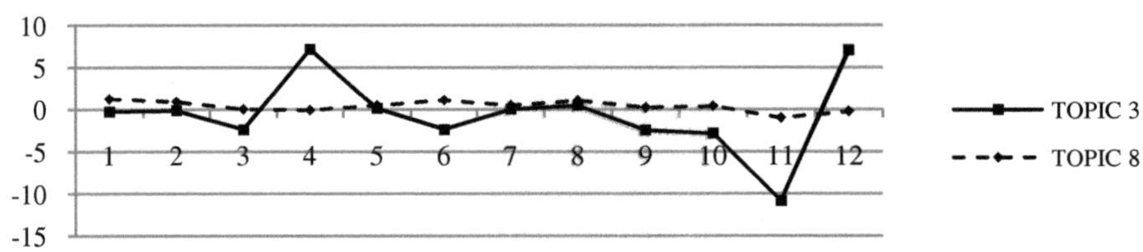

Fig. 4. Statistics of relations of topic 3 and topic 8 with other 12 topics respectively

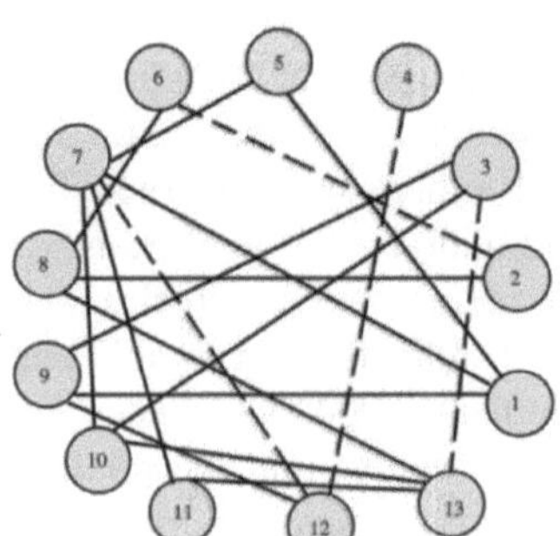

Fig. 5. The estimated graphical representation of 13 randomly selected topics

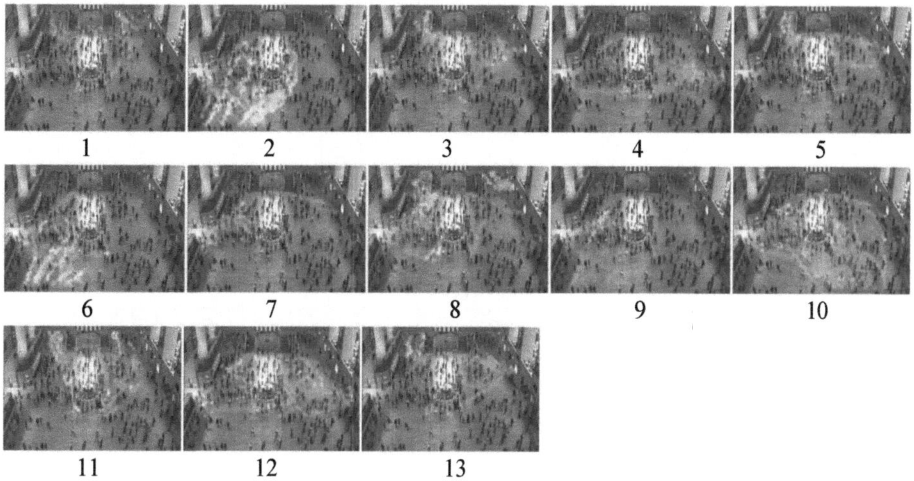

Fig. 6. Learned semantic regions of the randomly selected 13 topics

Fig. 7. The topic-clustered semantic regions

3.2 Learning Semantic Regions

50 regions are learned by BCTM and 24 regions are left after topic clustering. 10 representative regions are shown in Fig.8 (A). It can be clearly seen that almost every topic has one source and one sink. The proposed BCTM finds more dominant paths than [9], e.g. semantic region 1, 5, 6 and 8, which didn't occur in [9], and can't be created through parameter tuning. Because those paths are created by topic-clustering, comparing to the models in [9], which had pre-assumed independences among topics, therefore couldn't do the topic-clustering. Semantic region 5 and 6 in Fig.8 (A) are influenced by noises because of the local optimality of randomly spanning trees. 50 regions are learned by the CTM without beliefs and topic clustering. Fig.8 (B) shows randomly selected 5 semantic regions learned by the CTM. And results of Fig.8 (B) hardly describe dominant paths in the scene. Colors of the paths are randomly shown. Experiments show our BCTM with priors of beliefs and the forest of spanning trees outperform the CTM and the LDA based model in [9].

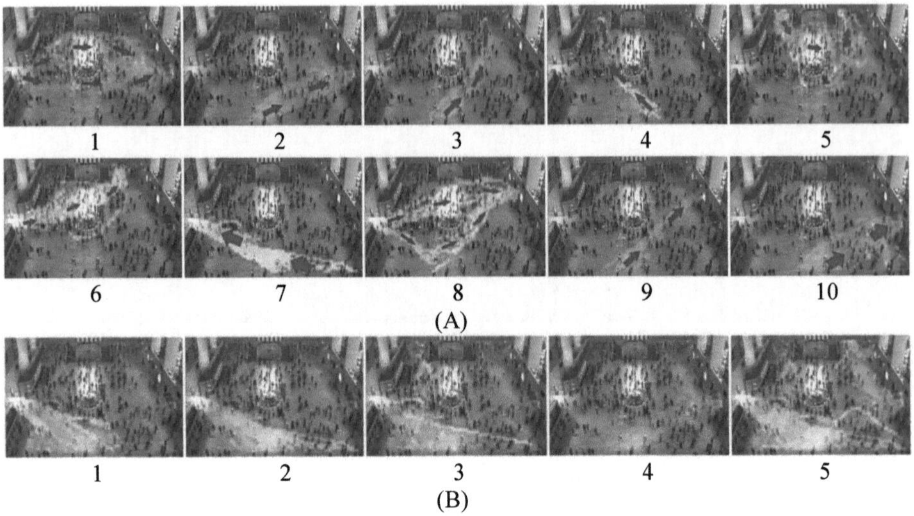

Fig. 8. Representative semantic regions learned by (A) the BCTM, (B) the CTM

4 Discussion and Conclusion

A belief based correlated topic model (BCTM) is proposed to learn the semantic regions. It effectively integrates the beliefs and the forest of randomly spanning trees as priors. Global structures of the scene are well captured by the BCTM. The learned regions obtain good accuracy and have clear semantic interpretation.

Dominant and original paths are found by the BCTM. Some paths are influenced by noises because of the local optimality of randomly spanning trees. Other improved approaches will be explored in the future work.

Acknowledgements. This work is supported in Part by National Basic Research Program of China (973 Program) with Nos. 2011CB706900, 2010CB731800, and National Science Foundation of China with Nos. 61039003, 61271433 and 61202323.

References

1. Li, J., Gong, S., Xiang, T.: Global Behaviour Inference Using Probabilistic Latent Semantic Analysis. In: IEEE Conference on British Machine Vision Conference, pp. 193–202 (2008)
2. Li, J., Gong, S., Xiang, T.: Scene segmentation for behaviour correlation. In: Forsyth, D., Torr, P., Zisserman, A. (eds.) ECCV 2008, Part IV. LNCS, vol. 5305, pp. 383–395. Springer, Heidelberg (2008)
3. Loy, C.C., Xiang, T., Gong, S.: Multi-Camera Activity Correlation Analysis. In: IEEE Conference on Computer Vision and Pattern Recognition, pp. 1988–1995 (2009)
4. Li, J., Gong, S., Xiang, T.: On-the-Fly Global Activity Prediction and Anomaly Detection. In: IEEE International Conference on Computer Vision Workshops, pp. 1330–1337 (2009)

5. Wang, X., Ma, X., Grimson, W.E.L.: Unsupervised Activity Perception in Crowded and Complicated Scenes Using Hierarchical Bayesian Models. IEEE Transactions on Pattern Analysis and Machine Intelligence, 539–555 (2009)
6. Wang, X., Tieu, K., Grimson, W.E.L.: Learning semantic scene models by trajectory analysis. In: Leonardis, A., Bischof, H., Pinz, A. (eds.) ECCV 2006. LNCS, vol. 3953, pp. 110–123. Springer, Heidelberg (2006)
7. Blei, D.M., Lafferty, J.D.: A Correlated Topic Model of Science. Ann. Appl. Stat., 17–35 (2007)
8. Rodriguez, M., Sivic, J., Laptev, I., Audibert, J.Y.: Data-Driven Crowd Analysis in Videos. In: IEEE International Conference on Computer Vision, pp. 1235–1242 (2011)
9. Zhou, B., Wang, X., Tang, X.: Random Field Topic Model for Semantic Region Analysis in Crowded Scenes from Tracklets. In: IEEE Conference on Computer Vision and Pattern Recognition, pp. 3441–3448 (2011)
10. Wang, X., Ma, K.T., Ng, G.W., Grimson, W.E.L.: Trajectory Analysis and Semantic Region Modeling Using Nonparametric Hierarchical Bayesian Models. International Journal of Computer Vision, 287–312 (2011)
11. Deljooi, H., Eftekhari Moghaddam, A.M.: Automatic Image Annotation via the Statistical Semantic Model Based on the Relationship Between the Regions. In: CSI International Symposium on Artificial Intelligence and Signal Processing, pp. 101–106 (2012)
12. Hospedales, T., Gong, S., Xiang, T.: A Markov Clustering Topic Model for Mining Behaviour in Video. In: IEEE International Conference on Computer Vision, pp. 1165–1172 (2009)
13. Hospedales, T., Gong, S., Xiang, T.: Video Behaviour Mining Using a Dynamic Topic Model. International Journal of Computer Vision, 1–21 (2012)
14. Hospedales, T.M., Li, J., Gong, S., Xiang, T.: Identifying Rare and Subtle Behaviors: A Weakly Supervised Joint Topic Model. IEEE Transactions on Pattern Analysis and Machine Intelligence, 2451–2464 (2011)
15. Blei, D.M., Ng, A.Y., Jordan, M.I.: Latent Dirichlet Allocation. Journal of Machine Learning Research, 993–1022 (2003)
16. Gilks, W.R., Richardson, S., Spiegelhalter, D.: Markov Chain Monte Carlo in Practice: Interdisciplinary Statistics. Chapman & Hall/CRC (1995)
17. Hu, W., Xiao, X., Fu, Z., Xie, D., Tan, T., Maybank, S.: A System for Learning Statistical Motion Patterns. IEEE Transactions on Pattern Analysis and Machine Intelligence, 1450–1464 (2006)
18. Makris, D., Ellis, T.: Learning Semantic Scene Models from Observing Activity in Visual Surveillance, pp. 397–408. IEEE Computer Society Press, Los Alamitos (2005)
19. Verbeek, J., Triggs, B.: Region Classification with Markov Field Aspect Models. In: IEEE Conference on Computer Vision and Pattern Recognition, pp. 1–8 (2007)
20. Stauffer, C.: Estimating Tracking Sources and Sinks. In: IEEE Conference on Computer Vision and Pattern Recognition Workshop, pp. 35–42 (2003)
21. Tomasi, C., Kanade, T.: Detection and Tracking of Point Features: School of Computer Science, Carnegie Mellon Univ (1991)
22. Hu, W., Tan, T., Wang, L., Maybank, S.: A Survey on Visual Surveillance of Object Motion and Behaviors. IEEE Transactions on Systems, Man, and Cybernetics, Part C: Applications and Reviews, 334–352 (2004)
23. Morris, B.T., Trivedi, M.M.: A Survey of Vision-Based Trajectory Learning and Analysis for Surveillance. IEEE Transactions on Circuits and Systems for Video Technology, 1114–1127 (2008)

24. Liu, X., Lin, L., Zhu, S.C., Jin, H.: Trajectory Parsing by Cluster Sampling in Spatio-Temporal Graph. IEEE Conference on Computer Vision and Pattern Recognition, 739–746 (2009)
25. Singh, V.K., Wu, B., Nevatia, R.: Pedestrian Tracking by Associating Tracklets Using Detection Residuals. In: IEEE Workshop on Motion and Video Computing, pp. 1–8 (2008)
26. Zhou, B., Wang, X., Tang, X.: Understanding Collective Crowd Behaviors: Learning a Mixture Model of Dynamic Pedestrian-Agents. In: IEEE Conference on Computer Vision and Pattern Recognition, pp. 2871–2878 (2012)
27. Meinshausen, N., Bühlmann, P.: High-Dimensional Graphs and Variable Selection with the Lasso. Annals of Statistics, 1436–1462 (2006)
28. Jordan, M.I., Ghahramani, Z., Jaakkola, T.S., Saul, L.K.: An Introduction to Variational Methods for Graphical Models. Machine Learning, 183–233 (1999)

Weakly Supervised Compressive Tracking with Effective Prediction Model

Yin Liang[1], Xu-Cheng Yin[1,*], Shu Tian[1], and Hong-Wei Hao[2]

[1] School of Computer and Communication Engineering, University of Science and Technology Beijing, Beijing 100083, China
[2] Institute of Automation, Chinese Academy of Sciences, Beijing 100190, China
youxianglyc@sina.com, xuchengyin@ustb.edu.cn, tshu23@gmail.com, hongwei.hao@ia.ac.cn

Abstract. Recently, compressive sensing has been widely used in the field of object tracking. As a typical representative, Compressive Tracking (CT) outperforms the state-of-the-art approaches. But it has a drawback that the scale of object is invariable throughout the whole tracking process. For solving this problem and further improve its performance, in this paper we propose a Weakly Supervised Compressive Tracking (WSCT) approach. Firstly, we introduce an effective prediction model based on optical flow, which could achieve reliable estimation of the position and the scale of current object. Secondly, through the prediction model, the samples around the predicted position of the current frame could be further utilized as weakly supervised information to guide the compressive classifier's updating. In this way, this updating strategy combines the current information and future information to alleviate the phenomenon of object drift. Finally, the updated classifiers are used to locate the final object position. Our WSCT algorithm shows robustness for larger scale changes, especially in film and television videos. Experiments and comparisons on challenging tracking sequences have shown the effectiveness of our approach.

Keywords: Prediction model, object tracking, compressive tracking, weakly supervised tracking.

1 Introduction

Object tracking is a well-studied problem in computer vision due to the proliferation of high-powered computers, the availability of high quality and inexpensive video cameras. Video surveillance, human-computer interaction and video indexing *et al.*[1-4] are all benefit from the solutions of visual tracking. As the tracking problem is always doing with no prior knowledge other than the bounding box defining the object of interest in the first frame, there are two main problems need to be solved to build a reliable tracker, to alleviate drift to the complex variation in scenes caused by illumination change, camera move, occlusion, motion and pose change *et al.*[3,9,19] and to adaptive adjusting the size of the bounding box to fit the object's scale change during the tracking process.

* Corresponding author.

B. Huet et al. (Eds.): PCM 2013, LNCS 8294, pp. 791–802, 2013.

In order to achieve real-time tracking, many solutions were put forward called "tracking by detection" [5-16]. Eigentracking [5] and Frag-Track[6] used static appearance models which were either defined manually or trained using only the first frame, and never update. These methods could not adapt to the significant appearance changes and necessary to know much information about the object before tracking to train the appearance models offline. Consider that, adaptive appearance models with updating during the tracking process, attracted more attention. However, training adaptive appearance models is a difficult task [7]. IVT [8] employed incremental learning to adapt appearance changes while MIL [9] using Multiple Instance Learning to promote for updating adaptive appearance model. In addition, discriminative approaches were proposed which posed the tracking task as a binary classification problem to optimally separate the object from the background. Avidan [10] used support vector machines, while Lepetit [11] applied pixel pairs based randomized trees to discriminate key points. As these methods all training offline, which would not adapt to the complex changes inevitable in video sequences, online boosting methods [12] were proposed to update the classifier during the tracking process. Then, in order to promote the tracker's adaptability and further alleviate drift, Grabner *et al.*[13] proposed semi-supervised boosting method. To further simplified the feature representation, compressive sensing [1-2] has been used for dimensionality reduction. L1-tracker [14] used a sparse linear combination to model target and Li *et al.*[15] further applied orthogonal matching pursuit algorithm to reduce the computational complexity of L1-tracker. Although there are numerous attempts, drift and scale adjust problem are still unsolved.

Compressive Tracking (CT) is a recently proposed simple but efficient algorithm with an appearance model based on features extracted from the compressed domain [3]. It uses a very sparse matrix [17] to select features from the multi-scale image space, then use a naive Bayes classifier to recognize the object and update frame by frame. It runs in real-time and is demonstrated effective to alleviate drift to illumination, occlusion and pose change. However, due to the only use of the previous results, this approach lacks any reliable prediction of the current object scale, and the object may drift since the classifier lacks effective update strategy.

Based on the analysis above, we propose a Weekly Supervised Compressive Tracking approach. We adopt an effective prediction model to reliably estimate the position and the scale of the current object, which help realize scale adjusting. In addition, we put forward an effective update strategy by the guidance of our prediction model, which combines the future information as weekly supervised information and current information to update the tracking classifier.

2 Overview of Weakly Supervised Compressive Tracking

Figure 1 shows the framework of our proposed approach. In each frame, take Frame t as an example, through the prediction model, we can obtain the reliable prediction of the object. Then we combine the samples in Frame t as weakly supervised information and historic samples to update the tracking classifier. Additionally, the prediction model provides the possible location and scale of the tracking object, which could effectively guide sampling in Frame t and realize scale adaptive adjusting. Finally, the tracking classifier determines the object location with the maximal classifier response.

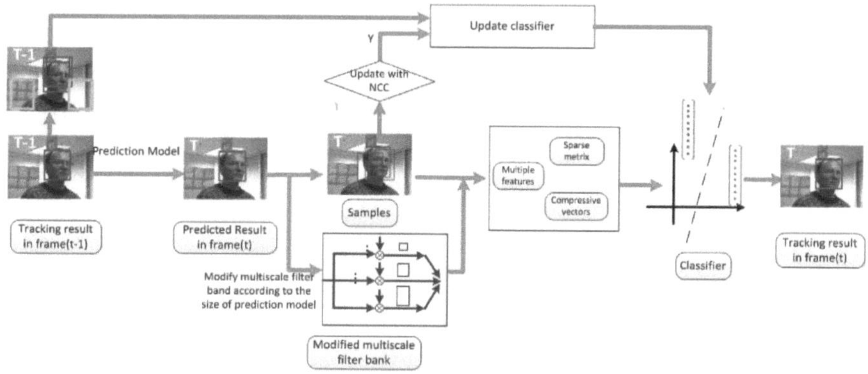

Fig. 1. Overview of the weakly supervised compressive tracking approach

3 Prediction Model Based on Optical Flow

Due to the insufficiency of prior information, sampling procedures in present "tracking-by-detection" approaches are mainly classified into two categories. One is the sliding windows. Though it ensures collecting samples for all possible positions and sizes, the complexity of sampling and detection might be very high, which limits its real-time effects. The other one is recently widely used, which is based on the assumption that object's position is not changed notably in consecutive frames. It samplings around the previous tracking result. It is easy to achieve, but due to the lack of the estimation of current object, the size of the tracking box is invariable. When it comes to the scenes with dramatic changes of object's scale, especially in film and television videos, this method will fail.

In order to solve above problem, achieving scale adaptive adjusting during the tracking process and add more effective prior information to direct sampling, we introduce an effective prediction model to reliably estimate the position and the scale of current object. Additionally, by the guidance of our prediction model, we can further utilize the samples around the predicted position of current frame as weakly supervised information to guide the tracking classifier update, which will help alleviate drift. Figure 2 describes our prediction model.

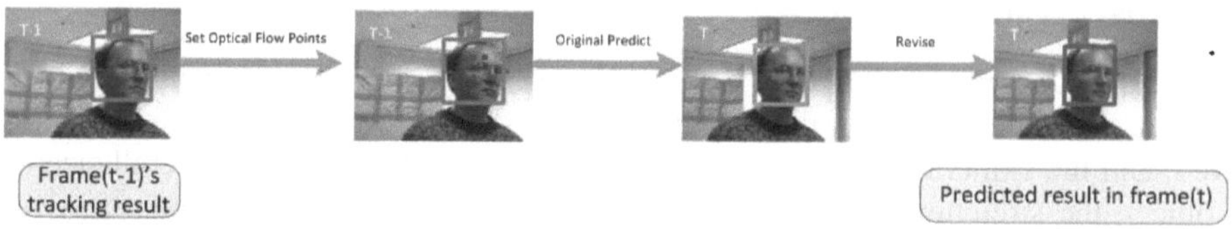

Fig. 2. Prediction Model

3.1 Formalization

We use a rectangular box I to indicate the tracking result. Due to the assumption that object's position is not changed notably in consecutive frames, based on Frame(t-1)'s tracking result $I(t-1)$,the predicted result $I_0(t)$ in Frame t is defined as

$$I_0(t) = f(I(t-1)) \tag{1}$$

Fig. 3. Differences between two consecutive result boxes

As Figure 3 shows the differences between two consecutive result boxes $I_0(t)$ and $I(t-1)$ mainly reflected in scale change and position offset. We use the center point location C and the size S to represent the tracking box I, which

$$I = \begin{bmatrix} C \\ S \end{bmatrix} = \begin{bmatrix} C_x & C_y \\ w & h \end{bmatrix} \tag{2}$$

Then Eq(1) can be further decomposed to the following formula:

$$\begin{aligned} I_0(t) &= \begin{bmatrix} 1 & 0 \\ 0 & \Delta S \end{bmatrix} \times I(t-1) + \begin{bmatrix} \Delta D \\ 0 \end{bmatrix} = \begin{bmatrix} 1 & 0 \\ 0 & \Delta S \end{bmatrix} \times \begin{bmatrix} C(t-1) \\ S(t-1) \end{bmatrix} + \begin{bmatrix} \Delta D \\ 0 \end{bmatrix} \\ &= \begin{bmatrix} C(t-1) + \Delta D \\ \Delta S \times S(t-1) \end{bmatrix} = \begin{bmatrix} C_x(t-1) + \Delta dx & C_y(t-1) + \Delta dy \\ \Delta S \times w & \Delta S \times h \end{bmatrix} \end{aligned} \tag{3}$$

where ΔS describes the scale change between $I_0(t)$ and $I(t-1)$, and $\Delta D = \begin{bmatrix} \Delta dx & \Delta dy \end{bmatrix}$, describes the position offset, which is equal to the offset between $C(t)$ and $C(t-1)$.

3.2 Initial Predicted Result

We use Eq(3) to compute the initial predicted result $I_0(t)$ in current frame. For a video sequence, the scale change of the object is mainly due to the object approximate, removed, and the camera zooming, which will lead to the change of the overall size of the object, the stretching ratio for each local position is the same. As the effectiveness of optical flow in local point prediction, we use Pyramid Lucas-Kanade algorithm [18] with Forward-Backward Error(FB) and Normalized Correlation Coefficient(NCC) scheme, refers to [19], to get reliable corresponding point sets $pt0$ and $pt1$ respectively in Frame(t-1) and Frame t. FB measures the discrepancy between two trajectories as the tracking is performed forward and backward in time, while NCC measuring the template matching degrees, the computational formula for each position is concluded as Eq(4). The detailed descriptions of these two parameters are in [19].

$$ncc(r,c) = \frac{1}{N} \sum_{(u,v)\in T} \frac{t(u,v) - m_t}{\sqrt{{s_t}^2}} \cdot \frac{f(r+u, c+v) - m_f(r,c)}{\sqrt{s_f^2(r,c)}} \tag{4}$$

T is ROI, $N = |T|$, $t(u,v)$ and $f(r+u, c+v)$ respectively represent the gray values in the template and in the shifted ROI of the template at the current position in the image.m_t, ${s_t}^2$,$m_f(r,c)$ and $s_f^2(r,c)$ are the corresponding mean values and variances of the gray values. Then ΔD can be easily obtained by the offset between $pt0$ and $pt1$, while ΔS can be achieved by the mean distance ratio of $pt0$ and $pt1$. Following is the formulization and Figure 4 gives the visual expression. We set $pt0 = \{(x0_i, y0_i)|i = 1, 2, ..., n\}$ and $pt1 = \{(x1_i, y1_i)|i = 1, 2, ..., n\}$.$n$ is the size of $pt0$ and $pt1$.

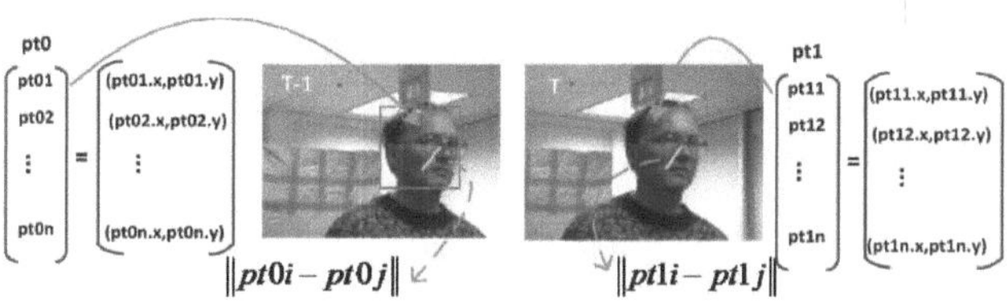

Fig. 4. Visual expression of ΔD and ΔS

$$\Delta D = \left[\Delta dx \ \Delta dy\right], \Delta dx = \frac{1}{n}\sum\nolimits_{i=1}^{n} (x1_i - x0_i), \Delta dy = \frac{1}{n}\sum\nolimits_{i=1}^{n} (y1_i - y0_i) \tag{5}$$

$$\Delta S = \frac{1}{n(n-1)} \sum\nolimits_{i=1}^{n} \sum\nolimits_{j=1|j\neq i}^{n} \frac{\sqrt{(x1_i - x1_j)^2 + (y1_i - y1_j)^2}}{\sqrt{(x0_i - x0_j)^2 + (y0_i - y0_j)^2}} \tag{6}$$

Then we can compute the initial predicted result $I_0(t)$ easily according to Eq(3).

3.3 Correction of the Initial Predicted Result

In order to reduce the error caused by the inaccurate prediction of optical flow, avoid sharp scale changes in consecutive frames, we set two thresholds $T1$, $T2$ ($T1 < T2$) to limit the size changes. We assume S_c is the current predicted scale and S_0 is the size of the defined bounding box of object in the first frame.The correction procedure is: computing R, $R = \frac{S_c}{S_0}$ if $R \notin (T1, T2)$, use the scale of previous tracking result to correct the current predicted scale.

3.4 Correction of the Stability

The above-mentioned prediction and correction processes in current frame are based on the calculated reliable point sets $pt0$ and $pt1$ are not empty, which imply that the current frame is predictable. However, due to the illumination change, object rotation, occlusion, object moving and other complex changes, sometimes, we cannot get the reliable corresponding optical flow point sets. To ensure the smooth track in this case, we add one guarantee condition, $I_0(t) = I(t-1)$, just as the simplified estimate used in most of existing video tracking methods.

When achieving the reliable predicted result $I_0(t)$, we get the object's scale in current frame, then we use this scale to modify the size of bounding box and the feature model, meanwhile, we use $I_0(t)$ to direct sampling to further improve sample quality and update the tracking classifiers.

4 Weakly Supervised Compressive Tracking

4.1 Compressed Dimensionality Reduction

The compressive sensing theories[1-2] demonstrates that the compressible signals, such as natural images or audios, can be almost perfect reconstruction by the projected results of the low-dimensional compressed subspace[1-2], [20]. Thus,we can use a random matrix $R \in R^{m \times n}$ to project data from the high-dimensional image space $x \in R^m$ to a lower-dimensional space $v \in R^n$, where $n \ll m$.

$$v = Rx \tag{7}$$

From compressive sensing theories, when R satisfies Johnson-Lindenstrauss lemma [21], v will preserve almost all the information in x. In order to further achieve real-time property, CT ultimately uses a very sparse matrix, for each row in R, only about c,c $\leq$ 4 nonzero entries. For the limitation of space, the detailed derivation and prove can be find in [3],[17],[21] and[22].

4.2 Low-Dimensional Features

CT uses above described R to generate low-dimensional Haar[23] feature vector v for each sample, $v \in R^n$. Each element v_i in v is a linear combination of spatially distributed multi-scale rectangle features. This obtaining procedure in CT is completed by a feature template, which is computed only once off-line and never change.

4.3 Classification

CT uses a naive Bayes classifier with online update to separate the object from the background. For each sample, its low-dimensional feature vector achieved is $v = (v_1, ..., v_n)^T$. CT assumes that all elements in v are independent distribute, and the prior $P(y=1) = P(y=0), y \in 0, 1$, while $0, 1$ represent the background and the object label respectively. Then the classifier is defined as

$$H(v) = \log(\frac{\prod_{i=1}^{n} p(v_i|y=1)p(y=1)}{\prod_{i=1}^{n} p(v_i|y=0)p(y=0)}) = \sum_{i=1}^{n} \log(\frac{p(v_i|y=1)}{p(v_i|y=0)}) \tag{8}$$

The conditional distributions $P(v_i \mid y=1)$ and $P(v_i \mid y=0)$ in $H(v)$ are assumed to be Gaussian distributed. So

$$P(v_i \mid y=1) \sim N(\mu_i^1, \sigma_i^1), P(v_i \mid y=0) \sim N(\mu_i^0, \sigma_i^0) \tag{9}$$

4.4 Classifier Update

In the classifier updating procedure of Frame $(t-1)$, CT only use the positive and negative samples obtained from current frame to update classifier. In order to alleviate drift further, we put forward an effective update strategy by the guidance of our prediction model. We compute the Normalized Correlation Coefficient $NCC1$ and $NCC2$, respectively represent the template matching result of the predicted result with the previous tracking result and the original defined object in the first frame, when either matching result fits the requirement we set, we combine the samples around the predicted position in Frame t as weakly supervised information and current positive and negative samples to update the tracking classifier. Then, we compute the lower-dimensional feature representation v for each sample z,$z \in R^m$ and update classifier parameters.In this way, we use both current and future information to further direct classifier updating. Algorithm1 is the detailed description of our update strategy.

Algorithm 1. Classifier update

Require: Frame$(t-1)$, Frame t, the tracking result $I(t-1)$ in Frame$(t-1)$ and the groundtruth I in the first frame

1: Set two sample sets P_u and N_u, respectively preserve positive and negative samples. Initially $P_u = \emptyset$ and $N_u = \emptyset$

2: Sampling in Frame$(t-1)$ to get positive samples $D(t-1)^{\alpha} = \{z| \ \|l(z) - I(t-1)\| < \alpha\}$ and negative samples $D(t-1)^{\zeta,\beta} = \{z|\zeta < \|l(z) - I(t-1)\| < \beta\}$, with $\alpha < \zeta < \beta$, then add $D(t-1)^{\alpha}$ to P_u and add $D(t-1)^{\zeta,\beta}$ to N_u

3: Use Prediction Model to compute the predicted result $I_O(t)$ in Frame t

4: Compare $I_O(t)$ with $I(t-1)$ and I, respectively compute $NCC1$ and $NCC2$

5: If $NCC1 > 0.4$ or $NCC2 > 0.4$, go to step 6, else go to step 7

6: Sampling in Frame t to get positive samples $D(t)^{\alpha} = \{z| \ \|l(z) - I_O(t)\| < \alpha\}$ and negative samples $D(t)^{\zeta,\beta} = \{z|\zeta < \|l(z) - I_O(t)\| < \beta\}$, with $\alpha < \zeta < \beta$, then add $D(t)^{\alpha}$ to P_u and add $D(t)^{\zeta,\beta}$ to N_u

7: Extract the features of these two sets of samples P_u and N_u, and update the classifier parameters according to (10)

8: **return** Updated classifier parameters

The parameters$(\mu_i^1, \sigma_i^1, \mu_i^0, \sigma_i^0)$ in Eq(9) are incrementally updated by

$$\mu_i^j \leftarrow \lambda\mu_i^j + (1-\lambda)\mu^j$$

$$\sigma_i^j \leftarrow \sqrt{\lambda(\sigma_i^j)^2 + (1-\lambda)(\sigma^j)^2 + \lambda(1-\lambda)(\mu_i^j - \mu^j)^2} \tag{10}$$

$$\sigma^j = \frac{1}{m^j}\sum\nolimits_{k=0|y=j}^{m^j-1} \sqrt{(v_i(k) - \mu^j)^2}, \mu^j = \frac{1}{m^j}\sum\nolimits_{k=0|y=j}^{m^j-1} v_i(k) \tag{11}$$

where $j = 0$ or 1, $\lambda > 0$ is a learning parameter, m^j is the number of samples according to N_u and P_u. The main steps of our WSCT algorithm are summarized in Algorithm 2.

Algorithm 2. Weakly Supervised Compressive Tracking

Require: Frame t
1: Use Prediction Model to get reliable predicted result $I_0(t)$ in Frame t
2: Sampling around $I_0(t)$, extract features with low dimensionality and update feature model
3: Use classifier H in (8) to each feature vector $v(z)$ and find the tracking position $I(t)$ with the maximal classifier response
4: Update classifier with Algorithm1
5: **return** Tracking position $I(t)$ and classifier parameters

5 Experiments

5.1 Dataset

We evaluate our WSCT approach with original CT and TLD [24] on 22 challenging sequences, which all contain changes in scale and position. The first 13 are publicly [25] and the other 9 are video sequences ours. All video sequences are taken from classic film and television videos, which contain track in and scale abrupt changes. As it often appears that camera switching back and forth between different objects in film and television videos, which is not suit for tracking, all the sequences we captured are fragments without lens switched. Table 1 lists the detailed properties of our interception video sequences. The ground truths of all the sequences are manually annotated by us.

Table 1. Detailed description of our interception video sequences

Name	Frames	Move camera	Background change	Moving	Pose change	Deflection	out-of-plane rotation	Occlusion	Zooming	Scale change
1.Titanic	598	yes	yes	yes	yes	yes	yes	no	no	yes
2.Gossip girl1	247	yes	yes	yes	yes	yes	no	no	yes	yes
3.Gossip girl2	178	no	yes	yes	yes	yes	no	no	yes	yes
4.Friends	613	yes	yes	yes	yes	yes	no	yes	no	yes
5.Harry Potter1	307	yes	no	no	yes	yes	yes	no	no	yes
6.Harry Potter2	491	yes	no	yes	yes	yes	no	no	no	yes
7.Black Swan	487	yes	no	yes	yes	yes	yes	yes	yes	yes
8.King's Speech	645	yes	no	yes	yes	yes	no	yes	yes	yes
9.Gone with the wind	507	no	no	yes	yes	yes	yes	yes	yes	yes

5.2 Experimental Setup and Evaluation Metrics

The tracker is set up only with the manually annotated bounding box in the first frame. As the trackers involving randomness, all reported results are averaged over 10 runs. Parameters in CT and TLD are remained default in source codes provided by the authors. For WSCT parameters, the public parts in both WSCT and CT are equal to CT, while the others are set experimentally. In optical flow prediction, we set 10×10 points in $I(t-1)$, and for thresholds $T1$, $T2$, we found that if the object's size does not dramatically change, $T1$ and $T2$ can be set 0.6 and 1.8. As it is often occurred in film and television videos that the object's scale increases sharply due to object approximate or camera zooms, we just set the lower limit $T1 = 0.6$ for the whole video sequences.

We use 2 quantitative metrics: Precision at a fixed threshold of 0.5 and average center location error ACLE. To precision, $score = \frac{area(ROI_T \cap ROI_G)}{area(ROI_T \cup ROI_G)}$, where ROI_T is the tracking result and ROI_G is the groundtruth. When $score > 0.5$, the tracking result is determined correct. To ACLE, $CLE = \|C_T - C_G\|_2$, while $ACLE = \frac{sum}{\Sigma frames}$. C_T is the center of the tracking result while C_G is the center of the groundtruth. $\|.\|_2$ is Euclidean distance, sum is the sum of CLE for all frames, $\Sigma frames$ is the total number of frames.

5.3 Experimental Results and Analysis

Efficiency of Our Prediction Model. Figure 5 shows the scale matching curves of sequence Gossip girl2. The x axis represent video frames, while the y axis represent the matching degree with groundtruth, which is computed by the expression $\frac{area(ROI_T \cap ROI_G)}{area(ROI_T \cup ROI_G)}$. From it, we can clearly find that our WSCT has a higher scale matching degree with groundtruth than CT which strongly demonstrates the effectiveness of our prediction model in scale adjusting.

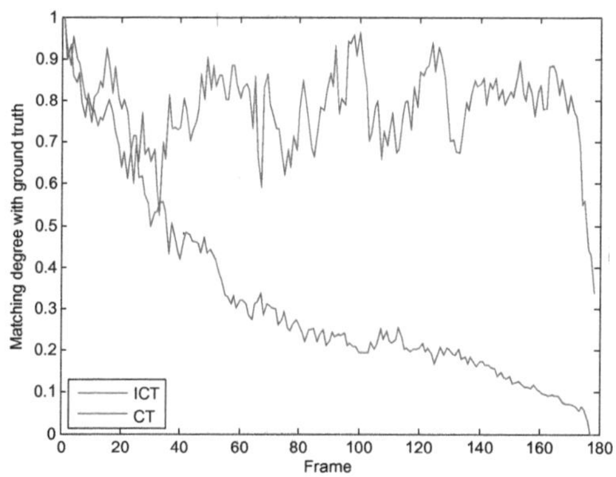

Fig. 5. Scale matching curves for tracking sequence Gossip girl2

Precision and Average Center Location Error (ACLE). Precision and ACLE results are shown in Table 2, the red means the best, while the blue means the second best. As TLD often loses the object, so ACLE of TLD is only a reference for just calculate the frames that TLD could track. Figure 6 shows snapshots for some sample results.

From Table 2, we can see that WSCT achieves the best results on total 22 sequences in terms of both precision and average center location error, and outperforms CT and TLD on 15 sequences. It is clearly that by the guidance of our effective prediction model, the performance of our WSCT has been greatly improved and it is robustness for larger scale changes, especially in film and television videos. As our prediction model mainly based on optical flow, when it comes to be some deviation in optical flow computation, the result might be affected, thus, few results of WSCT lower than CT. For TLD cannot deal with out-of-plane rotation of object, its performances are

Fig. 6. Examples of tracking results

much lower than WSCT and CT on most sequences. However, seq_cubicle, seq_ms and seq_sb are exceptions, in which the objects are not out-of-plane rotation and be occluded completely in many frames by similar obstacles. Thus, WSCT and CT encounter drift but TLD can retrieve the object by redetection after a long time of loss. Through the comparison, Our WSCT shows higher adaptability to scale adjusting and efficiency to further alleviate drift. Our tracker is implemented in MATLAB, on a Intel Dual-Core 2.8GHz CPU with 4GB RAM and runs at 17.55fps, while CT and TLD respectively runs at 19.63fps and at7.75 fps. It can be seen that compare with CT, our approach only sacrifices a little processing time, but obtains a great improvement in effectiveness and adaptability, and it runs much faster than TLD.

Table 2. Precision and ACLE

	Precision (%)			ACLE (pixels)		
Sequence	**WSCT**	**CT**	**TLD**	**WSCT**	**CT**	**TLD**
1.seq_bb	51.8	9.4	41.2	7.7	5.8	40.1
2.seq_cubicle	26.3	23.1	58.8	33.5	22.1	15.4
3.seq_dhb	82.4	80.6	19.6	3.8	5.7	68.5
4.seq_djb	77.1	64.5	21.6	11.6	18.3	21.8
5.seq_dk	89.2	92.9	54.5	4.8	7	10
6.seq_dp	56.1	43.5	45	19.3	22.9	54.2
7.seq_dt	64.4	64	44.3	7.8	7.3	12.5
8.seq_fast	98.1	59.7	1	5.8	22.6	2.5
9.seq_mb	44.6	43	28.1	11.6	11.7	14.7
10.seq_mg	63.5	49.4	48.4	7.1	10.1	39.2
11.seq_ms	47.5	47.2	78.4	17	21.1	16.7
12.seq_sb	16.2	23.6	41.3	20.3	23	16.3
13.seq_villains	58.8	43.4	12.4	10.5	15.9	23.2
14.Titanic	90.9	70.3	76.8	12.4	9.7	13.6
15.Gossip girl1	62.1	46.5	73.7	9.3	9.5	17.6
16.Gossip girl2	89.6	19.3	86.5	8.6	12	35.3
17.Friends	75	65.7	43.7	8.4	11.9	17.7
18.Harry Potter1	88.5	89.9	31.4	8.7	11.2	18.5
19.Harry Potter2	94.3	86.8	89	10.5	9.8	16.9
20.Black Swan	64.6	55.4	63	13.7	31.4	18.7
21.King's Speech	93	88.6	76.7	21.6	28.6	29.9
22.Gone with the wind	94.9	51.5	52.1	7.2	13	11.1
Average	72.3	**59.9**	**56.5**	12.5	**16.1**	**19.6**

6 Conclusion

In this paper, based on the widely used theory of compressive sensing, we propose a Weakly Supervised Compressive Tracking(WSCT) approach. We introduce an effective prediction model to reliably estimate the position and the scale of current object. We use it to effectually guide sampling and realize scale adaptive adjusting. Furthermore, through the prediction model, we combine the current and future information to supervise classifier update. Our approach is robustness for larger scale changes, especially in film and television videos, and could further alleviate drift. Numerous experiments on challenging sequences have demonstrated the effectiveness of our approach.

Acknowledgements. The research was supported by National Natural Science Foundation of China (61105018, 61175020).

References

1. Donoho, D.: Compressed sensing. IEEE Trans. Inform. Theory 52, 1289–1306 (2006)
2. Candes, E., Tao, T.: Near optimal signal recovery from random projections and universal encoding strategies. IEEE Trans. Inform. Theory 52, 5406–5425 (2006)
3. Zhang, K., Zhang, L., Yang, M.-H.: Real-Time Compressive Tracking. In: Fitzgibbon, A., Lazebnik, S., Perona, P., Sato, Y., Schmid, C. (eds.) ECCV 2012, Part III. LNCS, vol. 7574, pp. 864–877. Springer, Heidelberg (2012)
4. Yilmaz, A., Javed, O., Shah, M.: Object Tracking A Survey. ACM Computing Surveys 38(4) (2006)

5. Black, M., Jepson, A.: Eigentracking: Robust Matching and Tracking of Articulated Objects Using a View-Based Representation. IJCV 26(1), 63–84 (1998)
6. Adam, A., Rivlin, E., Shimshoni, I.: Robust Fragments-Based Tracking Using the Integral Histogram. In: CVPR, pp. 198–805 (2006)
7. Jepson, A.D., Fleet, D.J., El-Maraghi, T.F.: Robust Online Appearance Models for Visual Tracking. IEEE T-PAMI 25(10), 1296–1311 (2003)
8. Ross, D., Lim, J., Lin, R., Yang, M.H.: Incremental learning for robust visual tracking. IJCV 77, 125–141 (2008)
9. Babenko, B., Yang, M.H., Belongie, S.: Robust object tracking with online multiple instance learning. IEEE T-PAMI 33(8), 1619–1632 (2011)
10. Avidan, S.: Support vector tracking. IEEE T-PAMI 26, 1064–1072 (2004)
11. Lepetit, V., Lagger, P., Fua, P.: Randomized trees for real-time keypoint recognition. In: CVPR (2007)
12. Grabner, H., Bischof, H.: On-line boosting and vision. In: CVPR, vol. 1, pp. 260–267 (2006)
13. Grabner, H., Leistner, C., Bischof, H.: Semi-supervised on-line boosting for robust tracking. In: Forsyth, D., Torr, P., Zisserman, A. (eds.) ECCV 2008, Part I. LNCS, vol. 5302, pp. 234–247. Springer, Heidelberg (2008)
14. Mei, X., Ling, H.: Robust visual tracking and vehicle classification via sparse representation. PAMI 33, 2259–2272 (2011)
15. Li, H., Shen, C., Shi, Q.: Real-time visual tracking using compressive sensing. In: CVPR, pp. 1305–1312 (2011)
16. Tian, S., Yin, X.C., Xu, X., Hao, H.W.: Transfer Learning Based Compressvie Tracking. In: IJCNN (2013)
17. Candes, E., Tao, T.: Decoding by linear programing. IEEE Trans. Inform. Theory 51, 4203–4215 (2005)
18. Bouguet, J.Y.: Pyramidal implementation of the affine lucas kanade feature tracker description of the algorithm. Intel Corporation (2001)
19. Kalal, Z., Mikolajczyk, K., Matas, J.: Forward-backward error: Automatic detection of tracking failures. In: 20th International Conference on Pattern Recognition, pp. 2756–2759 (2010)
20. Wright, J., Yang, A., Ganesh, A., Sastry, S., Ma, Y.: Robust face recognition via sparse representation. PAMI 31, 210–227 (2009)
21. Achlioptas, D.: Database-friendly random projections: Johnson-Lindenstrauss with binary coins. J. Comput. Syst. Sci. 66, 671–687 (2003)
22. Baraniuk, R., Davenport, M., DeVore, R., Wakin, M.: A simple proof of the restricted isometry property for random matrices. Constr. Approx. 28, 253-263 (2008)
23. Viola, P., Jones, M.: Rapid object detection using a boosted cascade of simple features. In: CVPR, pp. 511–518 (2001)
24. Kalal, Z., Mikolajczyk, K., Matas, J.: Tracking-Learning-Detection. IEEE T-PAMI 34(7), 1409–1422 (2012)
25. Birchfield, S.: Elliptical head tracking using intensity gradients and color histograms. In: CVPR, pp. 232–237 (1998)

Multiple Hypotheses Based Spatial-Temporal Association for Stable Pedestrian Counting

Chengyue Sun[1], Qi Zou[1], Wei Fu[2], and Jinqiao Wang[2]

[1] School of Computer and Information Technology,
Beijing Jiaotong University
{11120475,qzou}@bjtu.edu.cn
[2] National Laboratory of Pattern Recognition,
Institute of Automation, Chinese Academy of Sciences
{wfu,jqwang}@nlpr.ia.ac.cn

Abstract. This paper presents a real-time pedestrian counting approach in dense visual surveillance scenarios with a fixed camera. Firstly, we build a set of generic pedestrian models, which is irrelevant to application scenarios. Feature points are sampled from the edge images of pedestrian, which are further used to determine the potential positions of pedestrian by matching the edges with the head-shoulder template. The template is generated through utilizing the Parzen Windows method on a self-made sample dataset of head-shoulder edge images. Secondly, we refine the frame based counting results with MHT (Multiple Hypothesis Tracking) under a Bayesian framework. And two effective improvements are proposed to resolve the missing tracking problem caused in traditional MHT tracking method. With the data association in the tracking process, false positions in the matching process could be removed and missed ones could be added.

Keywords: pedestrian counting, template matching, Parzen Windows, MCMC, MHT.

1 Introduction

Pedestrian counting has been a long-time researched field, and as one of the most important applications in IVS (Intelligent Video Surveillance), is of great significance in mall security, business data analysis and so on. The occlusion among pedestrians is the biggest challenge in pedestrian counting.

In the frame based counting, the improvements are focused on the detection of the robust features of pedestrians, like the detection of head-shoulder discussed in our paper. But the detection process is severely limited by the quality of the image and the environment, and the meticulous method often needs more processing time, so it is hard to make an effective improvement in the frame based solution. Also frame based feature regression is another way of pedestrian counting. Instead of detecting each single person, it tries to figure out the relationship between the feature of the crowd and pedestrian count.

B. Huet et al. (Eds.): PCM 2013, LNCS 8294, pp. 803–810, 2013.

The temporal continuity of objects is an important reason to achieve stable pedestrian counting. The addition of temporal information can lead to a satisfactory result, and in our case, the MHT is used to realize the association between the hypotheses of different frames.

2 Related Work

In existing pedestrian counting solutions, there are two main approaches: pedestrian detection and feature regression.

For pedestrian detection, Marn et al. [1] detected pedestrians using a HOG descriptor, and the HOG descriptor is trained with visual human. Tao Zhao et al. [2] and Subburaman et al. [3] used a head detector to detect pedestrians. Li et al. [4] improved the accuracy rate of people counting by analysis across multiple cameras. Morerio et al. [5] estimate the count of people relying on accurate camera calibration. However the accuracy of counting depends on the accuracy rate of detection, and when the size of people is small or the environment is more complicated, the counting results degraded greatly. For feature regression, Chan et al. [6] divided the crowd into subdomains with different directions, and then estimated the number of pedestrians in each region with a Gaussian process. Zhao et al. [2] considered the people counting problem as a MAP estimation problem. Multiple artificial models were utilized to represent pedestrians, but the temporal information was not considered.

The temporal information and the space information are complementary for accurate counting. Intuitively, people recognize a pedestrian according to the continued appearance of a feature, so it is reasonable to add the temporal information into the counting model. Xing et al. [7] constructed the global association of tracklets to refine the detected instances. In our case, we use the improved MHT tracker to correct the counting errors after the initial pedestrian number is provided through frame based counting.

3 Our Approach

Before counting and tracking, we perform background subtraction to get the foregrounds of the moving people [8]. Firstly, generic pedestrian models are trained to fits the multiple views of foregrounds, through which potential positions of pedestrians are obtained. Then Markov Chain Monte Carlo (MCMC) is used to search the most reasonable positions and get the pedestrian hypotheses. Finally hypotheses are associated by improved MHT trackers. In the followings, we will detail our approach.

3.1 Generic Pedestrian Models

In order to obtain a set of generic pedestrian models, we sample as much as possible head-shoulder images at first. All the sample images come from the surveillance video, covering different sexes, angles of view and hair styles, and we divide them into four representative categories: front view, side view, long-hair-front view, and unusual-hair-front view, as show in Fig.1. Edges are extracted through running Canny operator over original images, and then the head-shoulder edges are manually refined.

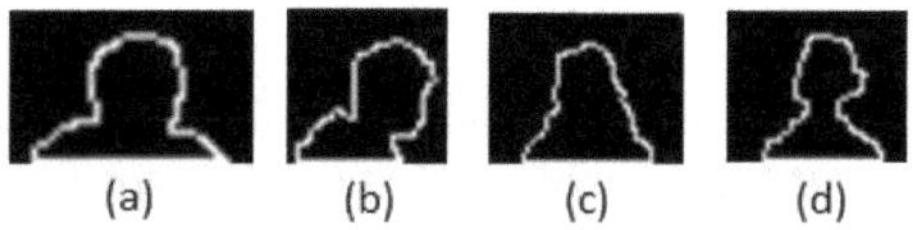

Fig. 1. (a) front view, (b) side view, (c) long-front view, (d) unusual-hair-front view

Different from some previous model, our pedestrian models are not described with a fixed shape, but a collection of the probabilities of the key points' appearances in the image. In our approach, the point set consist of a fixed number of points selected from each sample belonging to the same category uniformly, which can effectively reflect the properties of the head-shoulder shape, as shown in Fig.2. Each category has its own model, and we can also get a different size of model from resized samples to deal with various scenarios. To be specific, the top of pedestrian head is selected as the first key point (i.e., the datum point) for each sample, and it locates on the upper center of the model, Positions of the other key points are benchmarked against the datum point. For example, as shown in Fig.2, there are three samples from the front view, and we select nine key points in each sample. Each point has its corresponding points marked with the same color in the samples, and they reflect the distribution of the roughly same part of the head-shoulder in different samples together. All these key points are mapped into a fixed size of uniform image with the top of head mapped to the upper center as the datum point. Then the Parzen Windows method is used to get the distribution of the key points. Assuming the window function obeys a Gaussian distribution, and the distribution is formulated as follows:

$$P(x) = \frac{1}{N}\sum_{i=1}^{N}\frac{1}{V_N}\varphi\left(\frac{x - x_i}{h_N}\right) \tag{1}$$

Where N represents the number of the sampling points, X is the coordinate of arbitrary point, X_i is the coordinate of sample points, h_N is the variance of the Gaussian distribution, and V_N represents the area of the window function, i.e. the Gaussian distribution in our case.

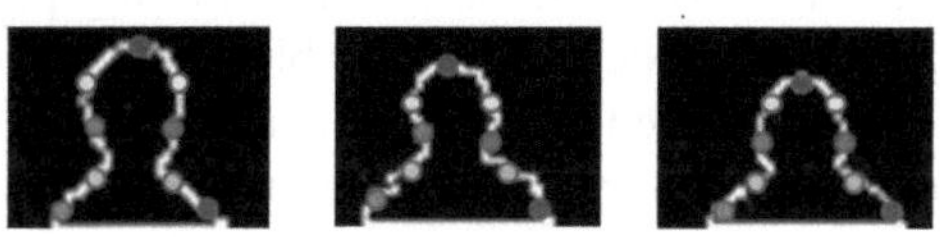

(a) (b) (c)

Fig. 2. The key points are selected as uniform as possible along the edge of the head-shoulder image

3.2 Frame Based Bayesian Counting

To get the hypotheses of pedestrians, we adopt the frame based counting by solving a maximum of a posteriori (MAP) problem [2]. As show in Fig.3, for each frame the potential points detected in the matching progress construct a configuration θ that explains the whole foreground I, and $\theta = \{ n_P, ((x_1, y_1), h_1), \ldots ((x_{n_P}, y_{n_P}), h_{n_P})\}$,where n_P is the number of the potential points, $(x_t, y_t), h_t$ are the position and the height of the t th pedestrian. Then the problem is formulated as follows:

$$\theta^* = \arg\max_{\theta \in \Theta} P(\theta | I) \tag{2}$$

where Θ is the solution space of the configuration, in order to get the optimal configuration that can explain the foreground domain best, the method proposed in [2] is utilized by solving a MCMC problem.

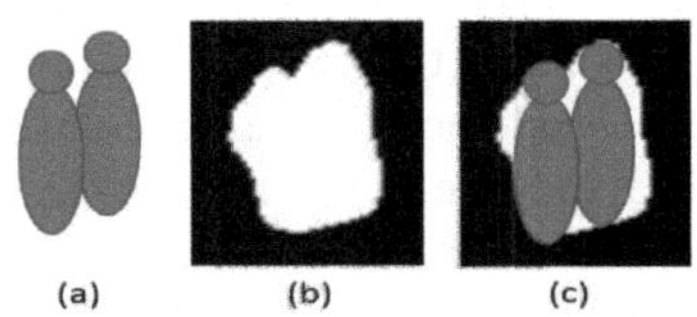

Fig. 3. (a): a configuration that contains two partly overlapped pedestrians. (b): the foreground image of the two pedestrians. (c): the explanation of the configuration to the foreground image.

3.3 Hypothesis Association by Improved MHT

After the MCMC process, the refined potential points where pedestrian candidates are regarded as the input of MHT, and the number of the instances from MHT is the final counting result. Because an instance will not be ensured until it has been tracked for a period of time, i.e., a fixed number of frames, the fault points which are detected sometimes will be removed. On the other hand the stable instances will keep being tracked even it is missed for several frames, so the missing potential points will be filled and the counting result will became more stable and accurate.

In order to deal with the missed hypothesis caused by occlusions in traditional MHT, the image region is divided into two kinds of sub-regions: the continuous regions where no object appears or disappears and the non-continuous regions where some objects may enter into or exit out of the scene.

Hypotheses of instances in the current frame are obtained according to the foregrounds detected and hypotheses in the previous frame. For example, as show in Fig.4, 1H and 2H represent the hypotheses in the previous frame while1F and 2F represent the foregrounds connected regions detected in the current frame. In this case, for each hypothesis in the previous frame, there are three kinds of hypotheses in this current frame. The first kind is denoted as H1-F1, H1-F2, H2-F1, and H2-F2.

For each pair of hypotheses and foreground, the existence possibility of a new hypothesis is measured by their distances. The second ones are the virtual hypotheses denoted as H1-C1 and H2-C2, which are generated according to the speed of a hypothesis if its corresponding foreground is not found in the current frame. The third ones are the end hypotheses represented by H1-E1 and H2-E2, which mean the instances will exit the scene. Besides, the foregrounds themselves in current frame can also generate new hypotheses like H1-N1 and H2-N2, which means new instances have entered into the scenario. Moreover, we set virtual foreground regions represented by C1 and C2 for each virtual hypothesis, E1 and E2 for each end hypothesis, and virtual parent hypotheses represented by R1 and R2 for each new generated instance's hypothesis, so that the preconditions of the assignment problem of the bipartite graph is satisfied, i.e., one vertex in the previous frame corresponds a vertices in current frame and there will be no vertex shared. Regarding the generating probability of the hypothesis as the cost from a hypothesis in previous frame to a foreground region in current frame, the problem can be transformed into an assignment problem, which can be solved through Hungarian algorithm.

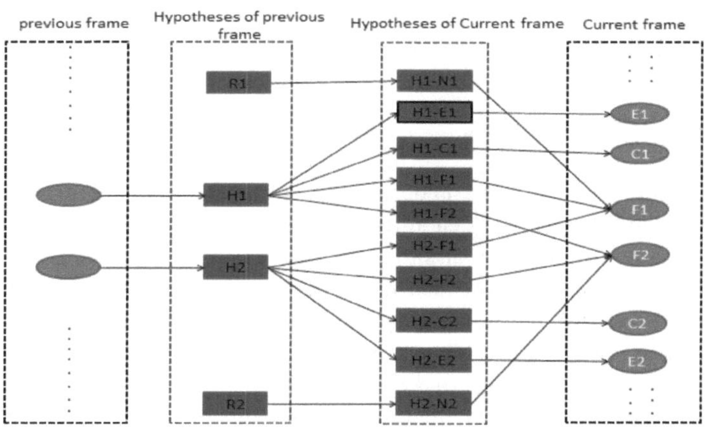

Fig. 4. The illustration to MHT, the red rectangles represent the hypotheses and the blue ellipses represent the foregrounds including the virtual ones

In our case, when the hypotheses of the previous frame are located in the continuing regions, the new generated instances' hypotheses and the end hypotheses are not allowed to be generated so that the hypotheses of the previous frame are forced to continue, even there is no connected foreground region to match it. And when the hypotheses are located in the non-continuous regions, the new generated instances' hypotheses and the end hypotheses are allowed to be generated, which means new instances may be created and the old ones may be destroyed.

In addition, the location estimation in our approach is based on the overlap ratio between the foreground region and the predictive region of the hypothesis. As shown in Fig.5, B1, B2, B3, B4, B5 represent five foreground regions of the instance B in different frames, and A1, A2, A3, A4, A5 represent the corresponding ones to the instance A. We assume B4 is connected with A4. The green one P is the prediction of B3 in the next frame according to its velocity estimation. In order to get the position of the true foreground domain i.e. B4, we move the prediction P along the direction from the center of P to the center of mixture of B4 and A4, until the overlap ratio between P and the mixture of B4 and A4 reaches a maximum value, then the resulted position of P is optimal match of B4.

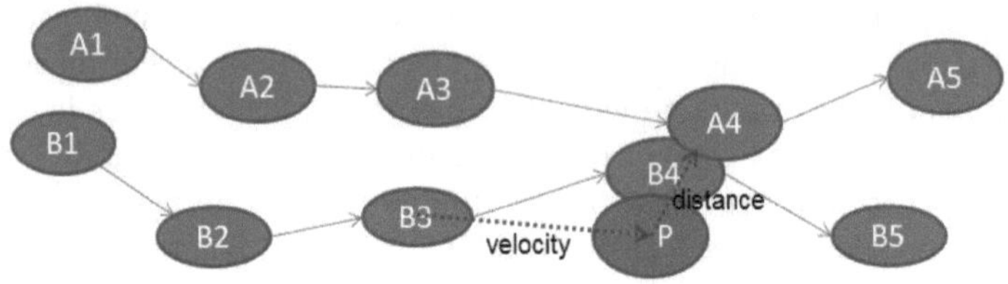

Fig. 5. The velocity is estimated according to the position of the former hypotheses, the more correct velocity lead to more correct prediction P, and the correction to P lead to more correct estimation of the velocity

4 Experiment Results

We test our approach on CASIA Pedestrian Counting Dataset, which is introduced in [9]. The video and images were captured from the stationary digital camcorders, and they reflect many different actual situations such as seasons, weathers, illumination conditions and changes of crowd density. It contains five groups of surveillance videos captured in five different scenes, and each group contains four videos captured in different seasons. We select videos from three groups respectively, which named Scene2_1i, Scene3_1i, Scene5_1 and Scene5_2.

The results are represented in Fig.6. We also compare our approach with [9], and the evaluation criterion is formulated as follows:

$$\text{RMSE}=\sqrt{\frac{1}{n}\sum_{i=1}^{n}(num_g(i)-num_r(i))^2}$$

Where n is the total frames of the tested video, $num_g(i)$ is the ground truth of the ith frame, and $num_r(i)$ is the test result of the ith frame. The comparison with [9] is given in table.1. As shown in Table.1, our apporoach achieve better and stable counting results. For example in Scene3_1 the pedestrian' features are more obvious, especialy the shape of head-shoulder and the pedestrians are occluded for a shor time so that the association of MHT can modify the missed matches, so our apporoach performs better.

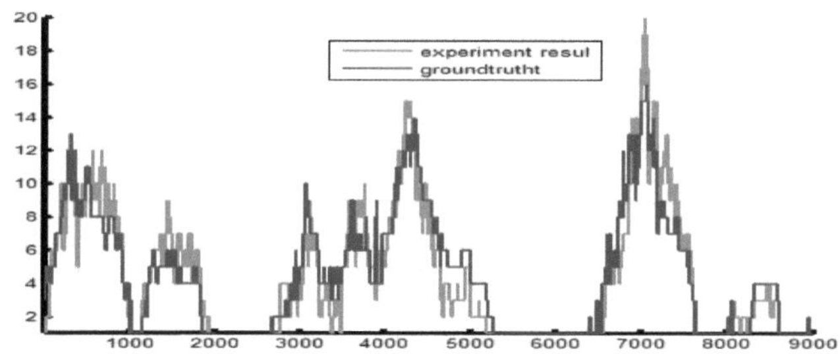

(a) Scene2_1

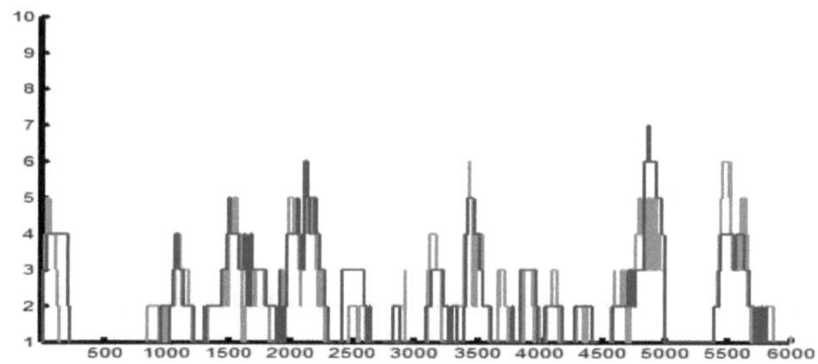

(b) Scene3_1

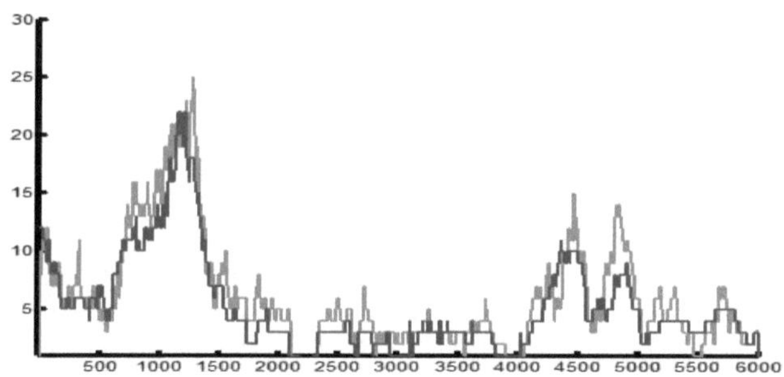

(c) Scene5_1

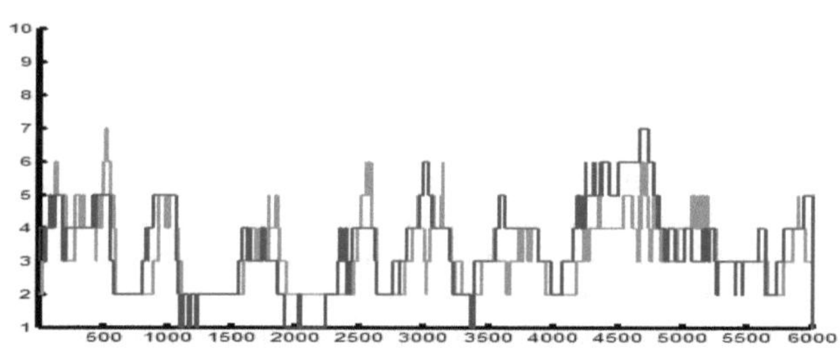

(d) Scene5_2

Fig. 6. Scene5_1 and Scene5_2 were captured in the same location but in different weathers, and the other two were captured in different locations

Table 1. the comparison result of our approach and [9]

Video name	RMSE(our)	RMSE[9]
Scene2_1	1.5475	1.7188
Scene3_1	0.9908	1.2510
Scene5_1	1.9410	2.1684
Scene5_2	1.0408	1.1190

5 Conclusion

We proposed a multiple hypothesis based spatial-temporal association approach for stable pedestrian counting. The experiment results indicate that the association of multiple hypotheses can effectively refine the results from frame based counting, especially for partial occlusions. The improved MHT is efficient to realize the associations of hypotheses in different frames, and performs well in complex environment.

Acknowledgement. This work is supported by Beijing Natural Science Foundation (4112047) and Fundamental Research Funds for the Central Universities (2011JBM026, 2011JBM027),and also supported by 973 Program (2010CB327905) and National Natural Science Foundation of China (61273034, 61070104 and 61272329).

References

1. Marn, J., Vazquez, D., Geronimo, D., Lopez, A.M.: Learning appearance in virtual scenarios for pedestrian detection. In: CVPR (2010)
2. Zhao, T., Nevatia, R.: Bayesian human segmentation in crowded situations. In: CVPR (2003)
3. Subburaman, V.B., Descamps, A., Carincotte, C.: Counting people in the crowd using a generic head detector. In: AVSS (2012)
4. Li, J., Huang, L., Liu, C.: People Counting across Multiple Cameras for Intelligent Video Surveillance. In: AVSS (2012)
5. Morerio, P., Marcenaro, L., Regazzoni, C.S.: People count estimation in small crowds. In: AVSS (2012)
6. Chan, A., Liang, Z.J., Vasconcelos, N.: Privacy preserving crowd monitoring: counting people without people models or tracking. In: CVPR (2008)
7. Xing, J., Ai, H., Lao, S.: Multi-Object Tracking through Occlusions by Local Tracklets Filtering and Global Tracklets Association with Detection Responses. In: CVPR (2009)
8. Barnich, O., Droogenbroeck, M.V.: ViBe: A Universal Background Subtraction Algorithm for Video Sequences. IEEE Transaction on Image Processing (2011)
9. Li, J., Huang, L., Liu, C.: Robust People Counting in Video Surveillance: Dataset and System. In: AVSS (2011)

Block-Based Face Recognition Using WLD

Ting Rui[1], Zijian Yang[1], Fan Liu[2], Suqiang Jiang[3], and Huabing Li[1]

[1] PLA University of Science and Technology, Nanjing, China
[2] Nanjing University of Science and Technology, Nanjing, China
[3] Institute of Computing Technology, Chinese Academy of Sciences, Beijing, China
rtinguu@sohu.com

Abstract. This paper presents a new approach to face recognition which uses a new local descriptor, called Weber Local Descriptor (WLD).To extract local information further, the idea of dividing faces into small regions was adopted. Feature histogram is extracted from every region and concatenated into a single feature vector to efficiently represent the face image. The recognition is performed using a nearest neighbor classifier in the computed feature space with Chi square as a dissimilarity measure. The experiments on ORL, FERET, Yale face database show that the proposed approach is not only better than holistic methods such as PCA, KPCA, 2DPCA but also superior to LBP. Meanwhile, it's robust to pose, noise, facial expressions and lightings.

Keywords: face recognition, Weber local descriptor, histogram, nearest neighbor classifier.

1 Introduction

Automatic face recognition has become a very active topic in computer vision research [1]. A key issue in face recognition is finding efficient descriptors for face appearance. Different holistic methods such as Principal Component Analysis (PCA) [2], Kernel Principal Component Analysis (KPCA)[3] and Two Dimensional Principal Component Analysis (2DPCA) [4] have been studied widely. PCA is a well-known method for feature extraction. By calculating the eigenvectors of the covariance matrix of the original inputs, PCA linearly transforms a high-dimensional input vector into a low-dimensional ones whose components are uncorrelated. Nonlinear PCA has also been developed by using different algorithm [5]. Kernel principal component analysis(KPCA) is one of type of non-linear PCA developed by generalizing the kernel method into PCA [6]. 2DPCA is developed for image feature extraction. As opposed to conventional PCA, 2DPCA is based on 2D matrices rather than 1D vector. That is, the image matrix does not need to be previously transformed into a vector. Instead, an image covariance matrix is constructed directly using the original image matrices and its eigenvectors are derived for image feature extraction [4]. LDA seeks to find a linear transformation by maximizing the between-class variance and minimizing the within-class variance.

Nowadays, local descriptors, which are distinctive and invariant to many kinds of geometric and photometric transformations, have been gaining more and more attention because of their promising performance [7]. And a large variety of methods using

B. Huet et al. (Eds.): PCM 2013, LNCS 8294, pp. 811–822, 2013.

local features have been developed. Existing descriptors can be categorized into two classes: one is a sparse descriptor, which first detects the interest points in a given image, and then samples a local patch and describes its invariant features [8]; the other is a dense descriptor, which extracts local features pixel by pixel over the input image . Scale invariant feature transform (SIFT) [9], and histogram of oriented gradients (HOG) [10] are typical sparse descriptors, while among the most popular dense descriptors are the Gabor wavelet and local binary pattern (LBP)[11]. They are being widely utilized in a large number of applications, such as baseline matching [12], human detection [10], face recognition [12], object recognition [9], and texture classification [11].

In this work, we introduce a new approach for face recognition which considers local information to represent the face image. The face image is first divided into small regions from which the Weber Local Descriptor (WLD) [13] features are extracted and concatenated into a single feature histogram efficiently representing the face image. Here, the Weber Local Descriptor (WLD) is a simple, yet very powerful and robust local descriptor which derived from Weber's Law, a psychological law [14]. WLD is based on the fact that human perception of a pattern depends not only on the change of a stimulus but also on the original intensity of the stimulus. Specifically, WLD consists of two components: differential excitation and orientation. The differential excitation component is a function of the ratio between two terms: one is the relative intensity difference of a current pixel against its neighbors; the other is the intensity of the current pixel. The orientation component is the gradient orientation of the current pixel. For a given image, we use the two components to construct a concatenated WLD histogram. The recognition is performed using a nearest neighbor classifier in the computed feature space with Chi square as a dissimilarity measure. Experiments on ORL, FERET, Yale face database show that WLD show that the proposed approach is not only better than holistic methods such as PCA, KPCA, 2DPCA but also superior to LBP. Meanwhile, it's robust to pose, noise, facial expressions and lightings.

The rest of this paper is organized as follows: in Section 2, we describe the details of the proposed method. In Section 3, we carry out experiments on face image databases to demonstrate the effectiveness and robustness of the method. Section 4 is the discussion and summary. Finally, conclusions are presented in Section 5.

2 Face Recognition with WLD

2.1 Weber's Law

Weber's Law was first described in 1834 by German physiologist E. H. Weber. He observed that the ratio of the increment threshold to the background intensity is a constant [13]. This relationship, which is known as Weber's Law, is shown as the following equation:

$$\frac{\Delta I}{I} = k \tag{1}$$

Where ΔI is the increment threshold, I is initial stimulus intensity and K is proportion and a const number.

Weber's Law, more simply stated, says that the size of a just noticeable difference is a constant proportion of the original stimulus value [13].

From the above formula, we can find that this character is useful in image processing, especially in avoiding the illumination change. Since almost all images are eventually to be observed and interpreted by humans, an ideal digital image processor has to take into account the effects of human psychology and psychophysics, such as the Weber's Law. This is an important new area which is value to be explored by image processing [14].

2.2 Weber Local Descriptor

Weber Local Descriptor is proposed by Jie Chen in the paper [13]. WLD is composed by two components, one is the different excitation(ξ), the other is orientation(Φ)[14].

2.2.1 Differential Excitation

Differential excitation is computed by the original pixel and its neighbor pixels. Specially, a differential excitation $\xi(x_c)$ of a current pixel x_c is computed as illustrated in Fig. 1.

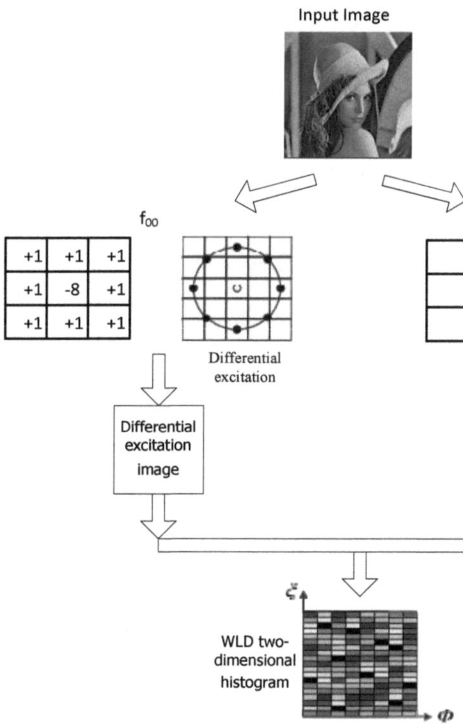

Fig. 1. Construct of WLD two-dimensional histogram

We use the circular filter f_{00} to calculate the differences between its neighbors and the center point. And the formula is shown in the following:

$$v_s^{00} = \sum_{i=0}^{p-1}(\Delta x_i) = \sum_{i=0}^{p-1}(x_i - x_c) \tag{2}$$

Where x_i is the i-th neighbor pixel of the current pixel x_c. The filter operator f_{00} can be extended to use neighborhoods of different sizes. Using circular neighborhoods and bilinearly interpolating the pixel values allow any radius and number of pixels in the neighborhood. For neighborhoods we will use the notation (P,R) which means P sampling points on a circle of radius of R.

Next, we'll compute the ratio of the differences to the intensity of the current point, as is described by the below formula:

$$G_{ratio}(x_c) = \frac{v_s^{00}}{x_c} \tag{3}$$

After that, the differential excitation of the current pixel $\xi(x_c)$ is computed as:

$$\xi(x_c) = \arctan\left[\sum_{i=0}^{p-1}\left(\frac{x_i - x_c}{x_c}\right)\right] \tag{4}$$

Where $\xi(x_c) \in [-\pi/2, \pi/2]$. To calculate the histogram, we divide $\xi(x_c)$ into M intervals $l_m\,(m=0,1,\ldots,M-1)$. Then, we have $l_m = [\eta_{m,l}, \eta_{m,u}]$ for each interval l_m. Here, the lower bound $\eta_{m,l} = (m/M - 1/2)\pi$ and the upper bound $\eta_{m,u} = [(m+1)/M - 1/2]\pi$. For example, $l_0 = [-\pi/2, -\pi/3]$.

2.2.2 Orientation

As shown in Fig. 1, the orientation of WLD is the gradient orientation which is computed as:

$$\theta(x_c) = \arctan\left(\frac{v_s^{11}}{v_s^{10}}\right) \tag{5}$$

Where v_s^{10} and v_s^{11} are the outputs of the filters f_{10} and f_{11}:

$$v_s^{10} = x_5 - x_1, \text{ and } v_s^{11} = x_7 - x_3 \tag{6}$$

For simplicity, θ is further quantized into T dominant orientations. Before the quantization, we perform the mapping $f : \theta \mapsto \theta'$:

$\theta' = \arctan 2(v_s^{11}, v_s^{10}) + \pi$, and

$$\arctan 2\left(v_s^{11}, v_s^{10}\right) = \begin{cases} \theta, & v_s^{11} > 0 \text{ and } v_s^{10} > 0 \\ \pi - \theta, & v_s^{11} > 0 \text{ and } v_s^{10} < 0 \\ \theta - \pi, & v_s^{11} < 0 \text{ and } v_s^{10} < 0 \\ -\theta, & v_s^{11} < 0 \text{ and } v_s^{10} > 0 \end{cases} \tag{7}$$

Where $\theta \in [-\pi/2, \pi/2]$ and $\theta' = [0, 2\pi]$. The quantization function is then as follows:

$$\Phi_t = f_q\left(\theta'\right) = \frac{2t}{T}\pi, \text{ and } t = \bmod\left(\left\lfloor \frac{\theta'}{2\pi/T} + \frac{1}{2} \right\rfloor, T\right) \tag{8}$$

For example, if T=8, these T dominant orientations are $\Phi_t = (t\pi)/4,\ (t = 0,1,\ldots,T-1)$. In other words, those orientations located in the interval $[\Phi_t - \pi/T, \Phi_t + \pi/T]$ are quantized to Φ_t.

2.2.3 Block-Based WLD Histogram

The histogram is one of the variables for many local descriptors [15, 16]. In this part, we describe how to construct a 2D histogram which combines the two components of WLD.

As shown in Fig. 1, we first compute each pixel's differential excitation $\xi(x_c)$ using (4), and orientation Φ_t using (8). Then we get two new image, the differential excitation image and the orientation image. According to the two images, we get the 2D histogram $\{WLD(l_m, \Phi_t)\}$, $(m = 0,1,\ldots,M-1,\ t = 0,1,\ldots,T-1)$ of the original image. Note that the size of this 2D histogram is $T \times M$, where M is the number of intervals of ξ which is described in Section 2.2.1 . In other words, in this 2D histogram, each column corresponds to a dominant orientation Φ_t, and each row corresponds to a differential excitation. Thus, the value of each cell $WLD(l_m, \Phi_t)$ corresponds to the frequency of a certain differential excitation interval l_m and a dominant orientation Φ_t.

To obtain a more discriminative descriptor, the 2D histogram $\{WLD(l_m, \Phi_t)\}$ is further encoded into 1D histogram H. As shown in Fig. 2, we use each row of 2D histogram to form a 1D histogram $H(k),(k=0,1,\ldots,M-1)$, each sub-histogram $H(k)$ corresponding to the sequence of the differential excitation interval l_k .Concatenating the M sub-histograms, we have the 1D histogram: $H = \{H_m\}, m = 0,1,\ldots,M-1$.

To improve the rate of recognition, the differential excitation image and the orientation image is divided into local regions and histograms are calculated from each region independently. For example, the facial image is divided into $N \times N$ regions. For the region $R_{i,j},(i=1,2,\ldots,N, j=1,2,\ldots,N)$, the 2D histogram is $\{WLD(l_m, \Phi_t)\}_{i,j}$, and the 1D histogram is $H_{i,j}$. The histograms $H_{i,j},(i=1,2,\dots,N, j=1,2,\dots,N)$ are then concatenated of form a global description of the face.

2.3 Classifier with Histogram

From the Pattern classification point of view, a usual problem in face recognition is having a plethora of classes and only a few, possibly only one, training sample(s) per class[16]. For this reason, more sophisticated classifiers are not needed but a nearest-neighbor classifier is need. Several possible dissimilarity measures have been proposed for histograms:

Histogram intersection:

$$D(S,M)=\sum_i \min(S_i,M_i) \tag{9}$$

Log-likelihood statistic:

$$L(S,M)=-\sum_i S_i \log M_i \tag{10}$$

Chi square statistic (χ^2):

$$\chi^2(S,M)=\sum_i \frac{(S_i-M_i)^2}{S_i+M_i} \tag{11}$$

All of these measures can be extended to spatially enhanced histogram by summing over i and j.

When the image has been divided into regions, it can be expected that some of the regions contain more useful information than others. For example, eyes seem to be an important cue of in human face recognition.

3 Experimental Results and Analysis

The method of WLD-based face recognition proposed in this paper has been used to the experiments on face databases of ORL, FERET, YALE, respectively, compared with four algorithms of PCA, KPCA, 2DPCA,LBP based on ORL database,and comparison with state-of-the-art methods on FERET database.and the performance of this algorithm has been tested on YALE databases while the expression and illumination change simultaneously. The same processing method has been adopted on the PCA, KPCA, 2DPCA in the experiments (the highest recognition rate has been taken by the nearest neighbor classifier). The polynomial Gaussian kernel has been used for KPCA, of which the index is 0.5. Considering the length of the recognition rate and the histogram, the three experiments on the database have adopted the block sizes of 4 * 4, M = 6, T = 8.

3.1 The Experiment on ORL Databases

ORL face database is consisted of 40 people each with 10 images, in which some images were taken at different times, thus there are different degrees of changes of the facial expressions and facial details of people, besides there is also a considerable degree of variation of face pose, thus the depth rotation and plane rotation can be up to 20°; the scale variation of human faces can be also up to 10%. The resolution of the images has been set unified into112 * 92.

Table 1. Recognition results of different methods on ORL database

P=8,R=2	WLD	PCA	KPCA	2DPCA	LBP
Recognition rate	0.99	0.935	0.925	0.96	0.95

As shown in Table 1, we compare the WLD-based method not only with the holistic methods (PCA, KPCA, 2DPCA) but also with the local methods (LBP). With regard to PCA, KPCA and 2DPCA, we use the top recognition accuracy of each round to compute the average value. From the result we can find that the recognition rate of the WLD-based method is higher than the recognition rate of the other four methods, which to some extent explains the robustness of the WLD-based method on the changes in the expression from minor illumination. Although the pose and sample size are varied, we still get satisfying results.

3.2 Experiments on the FERET Database

The FERET face database is the result of FERET program,which was sponsored by the US Department of Defence through the DARPA program[17].It has become a standard database for testing and evaluating state-of-the-art face recognition algorithms.The proposed method was tested on a subset of the FERET database.This subset includes 1400 images of 200 individuals(each individual has seven images).This subset involves variations in facial expression,illumination,and pose.In our experiment,each image in FERET database was manually cropped and resized to 80*80. In our experiments,we randomly selected four images of each subject for training and the others for testing.All experimental results are average value over ten runs.

3.3 Evaluation without Noise

As shown in Fig. 2, comparative performance is carried out against some classical face recognition schemes, such as PCA, KPCA and 2DPCA, which extract global features. For LBP and WLD,a facial image was divided into 4 * 4 windows and(8,2) neighborhood was selected.The experimental results show that our method outperforms other methods.Although the images in the FERET face database are subject to complex nonlinear changes due to large pose,expression and illumination variations,our proposed method works well in the complex circumstance.

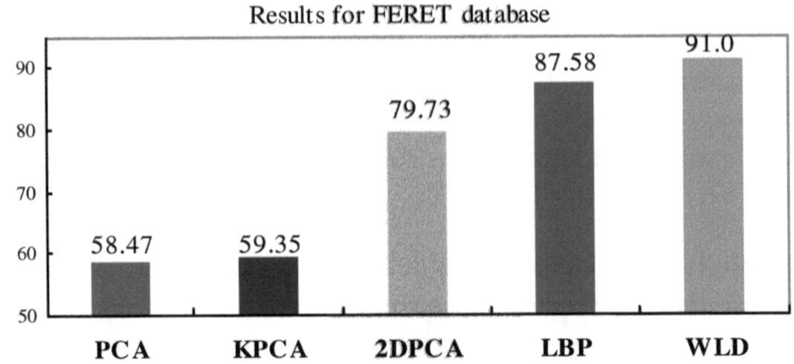

Fig. 2. Results comparison with state-of-the-art methods on FERET database

3.4 Evaluation with Noise

For the comparison of LBP,WLD method,further experiment is conducted on the FERET database with added white Gaussian noise.The results obtained at different noise levels for the three methods are illustrated in Table 2. Here,the column specifies the power of additive noise matrix in decibels relative a watt.The noise matrix is added to the original dataset.In addition, the parameter K, S and σ were fixed at pi/30,5 and 0.6.

Table 2. The accuracy rate (%) of noise sensitivity evaluation

SNR	0	5	10	15	20	25	30	35
LBP	82.1	78.3	73.0	60.6	35.1	9.9	2.0	0.9
WLD	87.6	78.8	69.0	53.7	34.1	18.0	10.8	6.1

It can be noticed that the performance of the two method,degrades as the noise strength increases.When the noise strength reaches a certain level,the advantage of WLD gets greater.The main reasons lie in the following factors:Gaussian smoothing filter and the computation means of differentia excitation using the sum of p-neighbor differences and the ratio to the current pixel.

3.5 The Experiment on Yale Face Database

Yale face database contains 165 images of 15 people and each with 11 face images in different illumination and expression, Figure 3 is one of the 6 images of someone. Before the experiment, each image is manually cropped into the face image of pixel 100x80. In this experiment, the front 5 images of each type are used for training; the later 6 images are used for the test. The obtained experimental results are shown in Table 3.

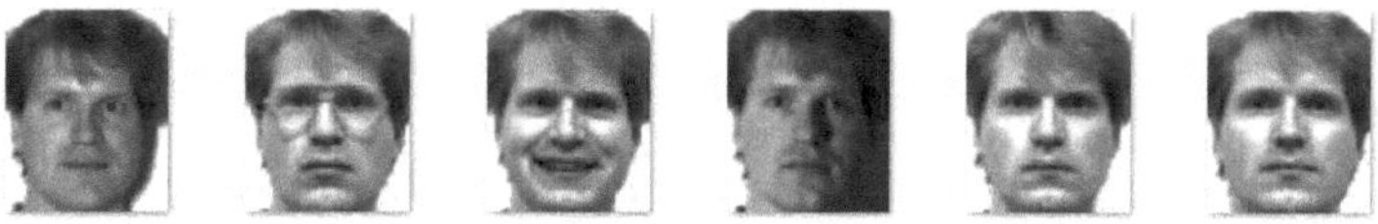

Fig. 3. 6 images of someone of Yale face database

Table 3. Recognition rate comparison of different methods

	WLD	PCA	KPCA	2DPCA	LBP
Recognition rate	0.9861	0.9200	0.9467	0.9600	0.9806

From the above table it is easy to find that the recognition rate of WLD-based face recognition method is superior to the recognition rate of other three methods. This also explains the robustness of WLD face recognition method for illumination and expression variation.

4 Discussion and Summary

The experiment shows that the WLD-based face recognition method mentioned in this paper is superior to the traditional PCA, KPCA, 2DPCA, especially it has strong robustness on expression and illumination changes. In addition, another important advantage of WLD is the strong adaptability because WLD has three adjustable parameters: block number, histograms size (M, T) and Different differential excitation operator (square or round, different radii and sampling points). Then we will discuss the impact of these three parameters on algorithm performance.

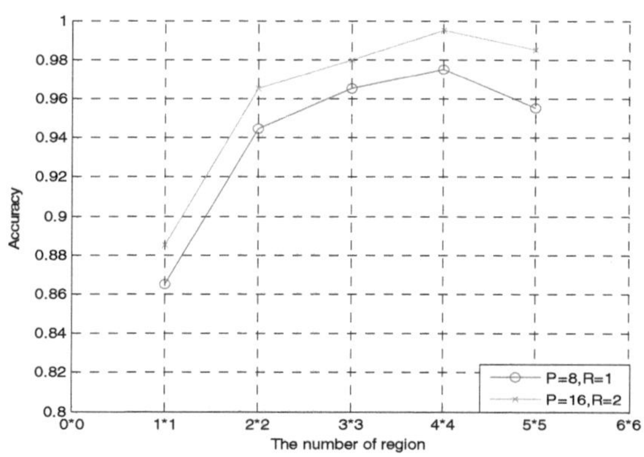

Fig. 4. Fixed two cases of M = 6, T = 8, the sampling points and radius of (8, 1), (16, 2), face blocks vary according to 1×1, 2×2, 3×3, 4×4, 5×5

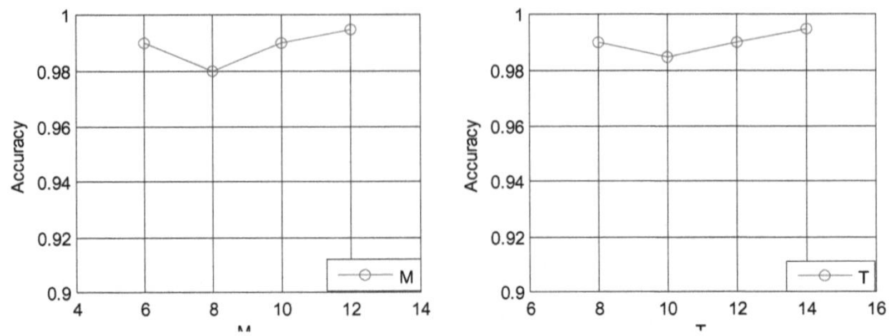

Fig. 5. Fixed block of 4×4 and respectively fixed one of M, T

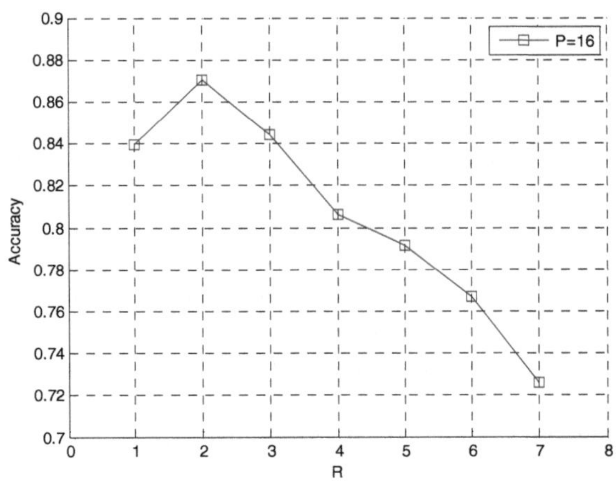

Fig. 6. Fixed block of 4×4, M = 6, T = 8, P = 16 and the radius R is changed

For the sub-block size, the following experiment is designed to test the impact on the recognition rate. Fixed two cases of M = 6, T = 8, the sampling points and radius of (8, 1), (16, 2), face blocks vary according to 1×1, 2×2, 3×3, 4×4, 5×5, the experimental results are shown in Figure 4. For the size of histogram, the fixed block of 4×4, the sampling points and a radius of (8, 2), and respectively fixed one of M, T and change the other, results are shown in Figure 5. ORL database is both used in the two experiments. For the operator variation, due to the good effect of ORL database, in order to more clearly illustrate the impact of operator variation on recognition rate, 1-7 of AR database are selected for training and 14-20 for test, fixed block of 4×4, M = 6, T = 8, P = 16, and the radius R is changed, the results are shown in Figure 6.

Figure 4 shows that: the recognition rate under a human face block circumstance is higher than the case that regardless of the block. However, because of the fixed histogram size (M=6, T=8), each vertical bar of the histogram will become small if there's too much blocks, and it will reduce the statistical reliability of the histogram and affect the recognition rate, and this id the reason why recognition rate begins to decrease when the block is 5*5 in the figure.

Figure 5 shows that: although the recognition rate varies according to the changes of M, T, but the change is small. This indicated that the histogram-based approach has robustness although there will be significant changes in dimension [13]. So in order to reduce the number of dimensions and reduce the amount of computation, it is ideal to make M=6, T=8.

Figure 6 shows the differential excitation operator's (radius and sampling points) influence on the recognition rate. The size of sampling point and radius is decided by the image size and block size since operator is used to extract local information, the local information will be lost if the small image large radius, and the recognition rate will be reduced.

Another advantage of the operator flexibility is multi-scale, different radii are corresponding to different scales. The most direct method to use multi-scale is to adopt the large-scale and the small-scale operators to extract feature information of the image respectively to get more different scales feature information and improve the recognition rate of the image. However, a multi-scale will lead to the increase of the dimension and low operation efficiency inevitably, so the latter studies may take dimensionality reduction into consideration.

5 Conclusions

This article describes the WLD-based face recognition method, and the local information of the image can be effectively extracted according to the use of modular thinking and Weber Local Descriptor (WLD). Experiment shows that: this algorithm is superior to the traditional PCA, KPCA, 2DPCA and LBP, and it has strong robustness on expression and illumination changes. At the same time, the algorithm also has a strong adaptability which can select parameters flexibly according to the image size and the sample number, and this algorithm can not only find the best recognition rate, but also weigh efficiency of algorithm.

Acknowledgment. This work is supported by the National Natural Science foundation under Grants 61070108 and 61070104. It is also funded by 2011BAK07B00.

References

1. O'Toole, A.J., Phillips, P.J., An, X., Dunlop, J.: Demographic effects on estimates of automatic face recognition performance. In: Automatic Face & Gesture Recognition Workshops, vol. 30(3), pp. 169–176 (2011)
2. Bozorgtabar, B., Noorian, F., Rezai Rad, G.A.: Comparison of different PCA based Face Recognition algorithms using Genetic Programming. In: International Symposium on Telecommunications, pp. 801–805 (2010)
3. Wang, Y., Zhang, Y.: The facial expression recognition based on KPCA. In: International Conference on Intelligent Control and Information Processing, pp. 365–368 (2010)

4. Ying, L., Liang, Y.: A human face recognition method by improved Modular 2DPCA. In: IEEE International Symposium on IT in Medicine and Education, pp. 7–11 (2011)
5. Diamantaras, K.I., Kung, S.Y.: Principal Component Neural Networks: Theory and Applications. Wiley, NewYork (1996)
6. Scholkopf, B., Smola, A., Muller, K.R.: Nonlinear component analysis as a Kernel eigenvalue problem. Neural Computing 10, 1299–1319 (1998)
7. Li, J., Allinson, N.M.: A comprehensive review of current local features for computer vision. Neurocomputing 71, 1771–1787 (2008)
8. Mikolajczyk, K., Schmid, C.: A performance evaluation of local descriptors. IEEE Trans. Pattern Anal. Mach. Intell. 27(10), 1615–1630 (2005)
9. David, G.: Lowe, Distinctive image features from scale invariant keypoints. Int. J. Computer Vis. 60(2), 91–110 (2004)
10. Dalal, N., Triggs, B.: Histograms of oriented gradients of human detection. In: CVPR (2005)
11. Ojala, T., Pietikäinen, M., Mäenpää, T.: Multiresolution gray scale and rotation invariant texture analysis with local binary patterns. IEEE Trans. Pattern Anal. Mach. Intell. 24(7), 971–987 (2002)
12. Li, W., Pang, Y., Yuan, Y., Pan, J.: Fully affine invariant SURF for image matching. Neurocomputing 85, 6–10 (2012)
13. Chen, J., Shan, S., et al.: WLD: A robust local image descriptor. IEEE Transactions on Pattern Analysis and Machine Intelligence 32(9), 1705–1720 (2010)
14. Ahilapriyadharshini, R., Arivazhagan, S., Gowthami, M.: Weber Local Descriptor based object recognition. In: IEEE International Conference on Advanced Communication Control and Computing Technologies, pp. 115–119 (2012)
15. Ahonen, T., Hadid, A., Pietikäinen, M.: Face recognition with local binary patterns. In: Pajdla, T., Matas, J(G.) (eds.) ECCV 2004. LNCS, vol. 3021, pp. 469–481. Springer, Heidelberg (2004)
16. DelMarco, S.P., Tom, V., Webb, H.F.: A Theory of Automatic Parameter Selection for Feature Extraction With Application to Feature-Based Multisensor Image Registration. IEEE Trans. on Image Processing 16(11), 2733–2742 (2007)
17. Phillips, P.J., Moon, H., Rizvi, S.A., Rauss, P.J.: The FERET evaluation methodol-ogy for face-recognition algorithms. IEEE Trans. Pattern Anal. Mach. Intell. 22(10), 1090–1104 (2000)

A Wavelet Model of Ganglion Cells Array and Its Application in Image Representation

Dawei Weng[1], Yunhong Wang[1], Hui Wei[2], and Di Huang[1]

[1] IRIP Lab, Beijing Key Laboratory of Digital Media
School of Computer Science and Engineering, Beihang Univ., Beijing, China
[2] Laboratory of Cognitive Model and Algorithm, Department of Computer Science, Fudan Univ., Shanghai, China
daweiweng@cse.buaa.edu.cn, {yhwang,dhuang}@buaa.edu.cn, weihui@fudan.edu.cn

Abstract. In this paper, we explore a new local image descriptor based on the modeling of ganglion cells array at the retina. We first introduce the mathematical model of a single ganglion cell and detailed distribution characteristics of ganglion cells array. From above evidence, we find out that the features of ganglion cells array are similar with wavelets in nature. Hence, a set of novel wavelet basis functions is constructed, which well fit these features. Furthermore, we discover that the modeling wavelets are in possession of tight frame property in its corresponding spatial domain expression, that is, multiple scales, shifts, phases of the spatial wavelet basis functions form a tight frame of the L_2 Hilbert space, which is helpful to extract strongly distinctive and slightly redundant information in the image matching task. Finally, we obtain an efficient descriptor, state-of-the-art ones. Additionally, the proposed descriptor is easier to construct and much faster to compute. Evaluated in the image matching task on the Multi-view Stereo Correspondence Data set, the results demonstrate its effectiveness.

1 Introduction

The feature descriptor is a quantitative description of local image structure and it is expected to sufficiently reflect shape and texture structure characteristics of local image around each interest point. Invariability and completeness are the most important requirements, as they determine the robustness and distinctiveness of the descriptors. In addition, computational efficiency is also pivotal and should not be neglected. Specifically, robustness refers to the capability that descriptors still keep stable when the image is suffering from affine transformation, intensity changes, noise interference and some other adverse conditions. Distinctiveness means that descriptors can catch and reflect the changes and uniqueness of local image structures, i.e., descriptors can effectively distinguish original local image pattern and the its variant. However, robustness and distinctiveness are both conflicting and interdependent to each other. In general, the more information carried by the descriptor, the better distinctiveness, while

B. Huet et al. (Eds.): PCM 2013, LNCS 8294, pp. 823–834, 2013.

the worse robustness, vice versa. Information content that the descriptors hold is determined by two issues, i.e. the interest point detection method and the construction algorithm of descriptors, as blob detection methods always extract these informative points as feature points. Furthermore, the length of descriptors also limits the amount of information carried by the descriptor.

As so far, researchers have explored various local descriptors and their corresponding similarity measurements, which approximately fall into methods based on image gradient, spatial frequency, differentiation and moment invariants, mainly including SIFT [1,2], SURF [3,4], GLOH [5], PCA-SIFT [6], Rotational Image, Shape Contexts [7], Steerable Filters [8], DAISY [9]. Thereinto, these methods based on the statistical histogram of gradient direction have been widely used in a variety of practical applications. Furthermore, among them, the most popular one is SIFT, whose performance is well known and universally acknowledged. Almost all the following descriptors such as SURF, GLOH, PCA-SIFT, Shape Context, DAISY, borrowed the idea from SIFT. The SIFT (or the GLOH) descriptor involves histogram computed over local regions, usually centered on feature points that, however, sometimes also densely sampled for object recognition tasks. Each pixel belonging to the local region contributes non-uniformly to the histogram depending on its location within the local region, and on the orientation and the norm of the image gradient at this position. From the above, the key insight is that SIFT and GLOH should owe much of their strength to the use of gradient orientation histograms, which are relatively robust to distortions, convolving with a kernel simultaneously dampening the noise and giving a measure of invariance to translation. However, they are both computationally demanding. Thus, Speeded Up Robust Features (SURF) [3], much faster than SIFT, had emerged and attained comparable performance. Borrowing the simplification idea of SIFT, SURF simplified the template of Gaussian Second-order Differentiation so as to make the filtering operation of image much simpler and could thus be implemented by straightforward addition and subtraction. In addition, during the construction of descriptor, SURF leveraged the integral image, the intermediate result produced at interest point detection step, to avoid repeated computation. But SURF does away with SIFTs spatial weighting scheme, and all gradients over the local region contribute equally to their respective bins, which introduces damaging artifacts when used for dense matching. For DAISY, it replaces the weighted sums of gradient norms by convolutions of the gradients in specific directions with several Gaussian filters. It efficiently reduces the influence of gradient norms from distant locations, and makes DAISY much faster for dense-matching purposes. Meanwhile it allows the computation of the descriptor in all directions with little overhead. Recently, Matthew Brown proposed a framework to make descriptor by learning technique [10].

Back to the intrinsic problem, how to make the descriptor both robustness and distinctiveness, in other words, how to balance the conflicting requirements is crucial. As we discussed above, they depend on the information contents, carried by the descriptor, and the locations of the key points. In this paper, we mainly focus on the first factor. Our suggestion for the problem is inspired from the findings in neurobiology, electrophysiology, anatomy, and also from our own thinking about valid and compact information representation.

Existing work, involved in research on arrangement manner of ganglion cell array and on corresponding coding way, is reasonably scarce. An interesting and notable electrophysiological finding is that when considering the neural response of single GC, its Receptive Field (RF) falls into CRF and nCRF, but when taking the distribution of all GCs into consideration, we find that GCs are arranged around rings with different radii, and the scale of response function and the interval between neighboring rings both increase exponentially [11].

We therefore introduce a new descriptor that retains robustness of SIFT, GLOH, DAISY and can be computed more quickly while it gives equivalent results. Its shape is closely related to arrangement manner of ganglion cell array. Our paper has two key insights: the first one is that we convolve original image, instead of gradient orientation map so more quickly to assemble the final feature vector by using multi-scale wavelet basis functions obtained from the rigid modeling of ganglion cells spatial arrangement and various multi-scale responses at different locations. The second is that our wavelet convolution kernels are smaller near the interest point and larger away from it, and it is important to note that the corresponding spatial domain form of our multi-scale convolution kernels coincidentally constitute a wavelet tight frame. In our opinion, tight frame may optimally balance the above contradiction of information content carried by the descriptor.

2 Biological Retina and Its Wavelet Modeling

Since Rodiecks [12] DoG model of ganglion cell receptive fields in retina, an enormous amount of experimental and theoretical research has greatly advanced our understanding of retina, especially its ganglion cells. Ganglion cells in the Retina followed into two categories parvocellularly-projecting (P), and magnocellularly-projecting (M) ganglion cells. Based on their differences, together with the anatomically segregated P and M targets in the LGN, and on the basis of neuroanatomical study [13,11] in monkeys, it has even proposed that P and M cells serve distinct functions in the visual system, and the primate visual system may be divided into two pathways.

In the following steps, we focus on the P cells alone. The spatial organization of ganglion cells receptive field and the cells sensitivity to luminance contrast determine the information a ganglion cell sends to the LGN about a visual stimulus. Recent neurophysiological evidence [11] suggests that peak sensitivities of center and surround regions of ganglion cell are inversely proportional to the regions areas, so that integrated contrast sensitivities (contrast gains) are constant across the visual field. Subsequently, we modeled the receptive field organization and sensitivity of the ensemble of ganglion cells as a family of novel wavelets sampling the frequency domain in a log-polar manner. Then, we computed the frame bounds for this family of wavelets in different phase space sampling schemes, and elucidated the conditions under which they form a tight frame. We find that the phase space sampling density provided by the ganglion cells in the retina is sufficient to form an almost tight frame that allows to extract enough information to represent the image.

2.1 Derivation of the Wavelets for Ganglion Cells

In this section, we will derive the following family of 2D wavelets that satisfies the wavelet theory and the neurophysiological constraints for ganglion cells,

$$\psi(x, y, \omega_0, \theta) = \frac{k\sqrt{\pi(k^2+1)}}{\pi\sigma(k^2-1)} \cdot \left[e^{-\frac{x^2+y^2}{2\sigma^2}} \cdot e^{i(\omega_0 \cos\theta \cdot x + \omega_0 \sin\theta \cdot y)} - e^{-\frac{x^2+y^2}{2\sigma^2}} \cdot e^{-\frac{\omega_0^2\sigma^2}{2}} - e^{-\frac{k^2(x^2+y^2)}{2\sigma^2}} \cdot e^{i(\omega_0 \cos\theta x + \omega \sin\theta y)} + e^{-\frac{k^2(x^2+y^2)}{2\sigma^2}} \cdot e^{-\frac{\omega_0^2\sigma^2}{2k^2}} \right] \quad (1)$$

where ω_0 is the radial frequency in radians per unit length and θ is the wavelet orientation in radians, $(1/\sigma)$ represents standard deviation of Gaussian functions, k represents the ratio of two Gaussian standard deviations. The wavelet is centered at $(x = 0, y = 0)$. A discrete ensemble of the modeled wavelet is shown in Fig. 1.

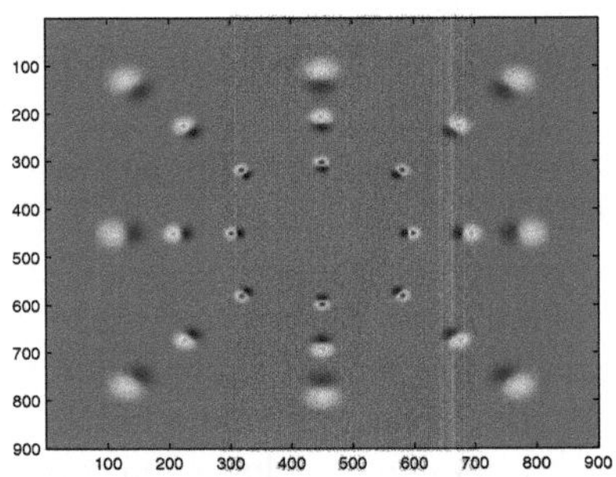

Fig. 1. An ensemble of modeling wavelets

We initiate our derivation with Rodiecks [12] model for ganglion cell,

$$\hat{\psi}(\xi, \nu) = K_c \cdot \exp(-\frac{\xi^2+\nu^2}{2r_c^2}) - K_s \cdot \exp(-\frac{\xi^2+\nu^2}{2r_s^2}) \quad (2)$$

where K_c and K_s equal to peak sensitivities of the center and surround of ganglion cell receptive field respectively, r_c and r_s equal to radii of center and surround respectively expressed as the distances from each midpoint of the region, where sensitivity is greatest, to the point at which sensitivity has fallen to $1/e$ of its maximum. According to L. J. Corners [11] experimental results, the peak sensitivities of centers and surrounds of ganglion cells are both inversely related to the sizes of the regions: larger regions with less sensitivity. Now, the models become the following equation:

$$\hat{\psi}(\xi,\nu) = \frac{1}{2\pi r_c^2}\exp(-\frac{\xi^2+\nu^2}{2r_c^2}) - \frac{1}{2\pi r_s^2}\exp(-\frac{\xi^2+\nu^2}{2r_s^2}) \quad (3)$$

Another important quantitative relation is the value of r_s/r_c. Several evidences from L. J. Corner have suggested that the ratio is constant across the whole retina for both P and M cells.

Next, we executed normalization to make its corresponding spatial domain form satisfy $<\psi,\psi>=1$. According to Parseval equation $<\psi(x,y),\psi(x,y)>= \frac{1}{(2\pi)^2}\cdot <\hat{\psi}(\xi,\nu),\hat{\psi}(\xi,\nu)>$ we arrived at

$$\hat{\psi}(\xi,\nu) = \frac{4\pi\sigma k\sqrt{\pi(k^2+1)}}{k^2-1}\left[\frac{1}{2\pi\sigma^2}e^{-\frac{\xi^2+\nu^2}{2\sigma^2}} - \frac{1}{2\pi k^2\sigma^2}e^{-\frac{\xi^2+\nu^2}{2k^2\sigma^2}}\right] \quad (4)$$

In terms of wavelet theory, admissible wavelets are functions having zero mean, this implies $\parallel \hat{\psi}(\omega=0) \parallel = 0$ is a necessary condition; otherwise the norm will become infinite in the measure of $\frac{d\omega}{\omega}$ as $\omega \to 0$. The d.c. response can be computed with $\xi=0$ and $\nu=0$,

$$\hat{\psi}(\xi=0,\nu=0;\xi_0,\nu_0,\sigma) = \frac{4\pi\sigma k\sqrt{\pi(k^2+1)}}{k^2-1}\left[\frac{1}{2\pi\sigma^2}e^{-\frac{\omega_0^2}{2\sigma^2}} - \frac{1}{2\pi k^2\sigma^2}e^{-\frac{\omega_0^2}{2k^2\sigma^2}}\right] \quad (5)$$

Apparently, the d.c. response is not equal to zero. A family of admissible wavelets is obtained by subtracting this d.c. response. We can further reduce the number of degrees of freedom of the wavelets filter in terms of the spatial frequency bandwidth of the ganglion cell, imposing $\omega_0 = 1/k$, k is fixed to 1.5 in our latter experiments. And now, we can extract the mother wavelet which can generate the whole family of modeling wavelets by rotation and dilation.

2.2 Sampling Scheme and Verification of Frame

Now we have obtained the properly parameterized wavelets equation, we try to discuss the issue of complete representation of image and identify the properties of modeling wavelets. As we all know, if the wavelets are admissible, a function can always be reconstructed from its continuous wavelet transform. Discrete wavelets transformation, as described below, reduces the redundant information for reconstruction, and converts the continuous wavelets to discrete wavelets basis functions,

$$(T^{wav}_{m,n,k,l}f) = a_0^{-m}\int\int dxdyf(x,y)\psi_{\theta_l}(a_0^{-m}x - nb_0, a_0^{-m}y - kb_0) \quad (6)$$

ψ_{θ_l} is a rotated version of the mother wavelet $\psi(x,y)$. $a_0 > 1, \theta_0 > 0$ and $b_0 > 0$ are fixed. The discrete wavelet elementary function is given by

$$\psi_{m,n,k,l}(x,y) = a_0^{-m}\psi_{\theta_l}(a_0^{-m}(x - nb_0a_0^m), a_0^{-m}(y - kb_0a_0^m)) \quad (7)$$

where

$$\psi_{\theta_l}(x,y) = \psi(x\cos(l\theta_0) + y\sin(l\theta_0), -x\sin(l\theta_0) + y\cos(l\theta_0))$$

where ψ is the mother wavelet, θ_0 denotes the step size of each angular rotation, l the index of rotation steps, b_0 the unit spatial interval, and a_0^m the dilation in scale. In the case of orthonormal bases, a function can be reconstructed by a linear superposition of the bases weighted by the wavelet coefficients. In the case of nonorthogonal wavelets, we need to guarantee whether the discrete set of the nonorthogonal wavelets form a frame and how tight the frame is. If there exists a frame, then one can reconstruct f in a numerically stable way from $T^{wav}(f)$ with the dual frame [14].

$$\begin{aligned} f &= \textstyle\sum_{m,n,k,l} < f, \psi_{m,n,k,l} > \tilde{\psi}_{m,n,k,l} \\ &= \textstyle\sum_{m,n,k,l} < f, \tilde{\psi}_{m,n,k,l} > \psi_{m,n,k,l} \end{aligned} \tag{8}$$

Especially, when $0 < A < B < 2$, the wavelets can be treated as if it were an orthonormal basis, and the function can be characterized by the following inversion formula,

$$f \approx \frac{2}{A+B} \sum_{m,n,k,l} < \psi_{m,n,k,l}, f > \psi_{m,n,k,l}. \tag{9}$$

where B/A denotes the tightness of the frame. When $B = A$, the frame becomes a tight frame, the reconstruction using above inversion formula is exact. So, we next identify the sampling scheme of frequency space to mimic primate ganglion cells, and verify whether our wavelets form a tight frame based on above sampling scheme.

First, we insist that the scale of ganglion cell response or the size of receptive field increases exponentially. This insights are roughly based on the following neurophysiological and anatomical findings (considering only P ganglion cell and its corresponding cones receptor). (1) From eccentricity $0^\circ \sim 15^\circ$ of human retina the density of P cells decreases exponentially, correspondingly, density of cones descends exponentially (Majority of P ganglion cells reside in the eccentricity rang $0^\circ \sim 15^\circ$, so our modeling is with in this range). This reflects that the size of receptive field increases exponentially. (2) L. J. Corner [11] investigated the receptive field organization at a wide range of retinal eccentricities and found that R_c and R_s have a steeper increase, especially in the range of $0^\circ \sim 15^\circ$ eccentricities.

In addition, it is known that in the visual cortex, the frequency space is sampled every half or every third of an octave [15,16]. From wavelet theory, suboctave sampling will make the frame tighter [17]. This is due partially to the fact that the phase space lattice, which is a superposition of N lattices, is denser than the lattice whose translation interval is strictly proportional to the scale of the wavelet. The 2D version of the fractionally dilated wavelets is

$$\psi_{\theta_l}^{\eta}(x,y) = 2^{-\frac{2\eta}{N}} \psi_{\theta_l}(2^{-\frac{2\eta}{N}}x, 2^{-\frac{2\eta}{N}}y)$$

with its fourier transform equal to

$$\psi_{\theta_l}^{\eta}(\xi,\nu) = \hat{\psi}_{\theta_l}(2^{\eta/N}\xi, 2^{\eta/N}\nu), \eta = 0, \cdots, N-1. \tag{10}$$

where N is the number of sampling frequency steps per octave and η is the index of the frequency step per octave, $\psi(x,y)$ is the mother wavelet function.

Finally, we have tested many sampling schemes in the experiment. Silvermans [16] neurophysiological finding that sampling interval in spatial frequency is about 1/3 octave in monkeys visual cortex is heuristic for us to set the value of N(scale steps per octave). It is important to note that we adopt the frame conditions of T. S. Lee [18] to verify our modeling wavelets. After the careful scrutiny, we found that under some cases, a discrete family of our wavelets form a approximately tight frame. These experimental results are approximately demonstrated in experiment section.

3 Our Local Descriptor

In this section, we briefly describe the intrinsic difference with methods based on statistical histograms and then introduce our descriptor. Descriptors based on statistical histograms, such as SIFT, GLOH, DAISY, involve different spatial weighting schemes, yet computed histograms can be expected to be very similar. The success of these descriptors can be ascribed to the use of multi-orientations gradients to guarantee the sufficient information content helpful to the distinctiveness of descriptor and to the use of convolution with a kernel to dampen the noise helpful to the robustness of descriptor.

Our descriptor is introduced from the field of signal theory. We utilize a roughly tight frame wavelets to extract information from the original image. As we all know, tight frame, distinct with orthonormal basis not only represent a signal completely but also extract slightly redundant information, which is helpful to the distinctiveness and robustness of the descriptor, respectively. The lattice points are located in many rings with different radii, corresponding wavelet convolution kernel at each grid point is smaller in inner rings and larger in outer rings. This utilization of multi-scale convolution kernels also occurred in DAISY [9] and geometric blur [19], which reflect, from another aspect, its validity. Of course, obviously different from those methods, our work extracts information from the original image without gradient; the convolution kernels are wavelets. Finally, the descriptor consists of wavelet responses at different grid point of sampling lattice. The resulting descriptor is shown in Fig. 2.

3.1 The Proposed Descriptor

Generally, our descriptor resembles the structural distribution and response of ganglion cells. We now give a more formal description of our descriptor. Because we directly extract information from original image, for a given input image, we first determine the number of grid points and compute the filter templates made of the sampling to modeling wavelets. For example, at grid point(ξ_0, ν_0),

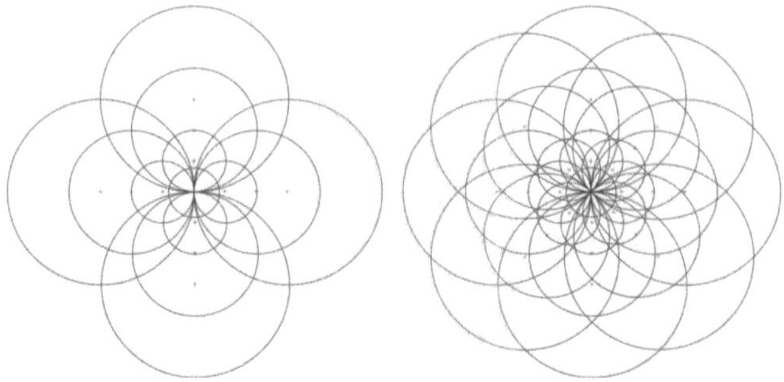

(a) Four orientations (b) Eight orientations

Fig. 2. Our descriptor: Each circle represents a region where the radius is proportional to the standard deviation of the wavelet

we sample the wavelet basic function to form filter template, the size of which depends on the scale of the function. After the acquisition of filter templates, we convolved local region around each grid point with corresponding template, the values of which constitute the final vector,

$$\begin{aligned}\mathcal{D}(u_0, v_0) = \Big[& E_{\Sigma_1}(u_0, v_0), \\ & E_{\Sigma_2} l_1(u_0, v_0, R_1), E_{\Sigma_2} l_2(u_0, v_0, R_1), \cdots E_{\Sigma_2} l_T(u_0, v_0, R_1), \\ & E_{\Sigma_3} l_1(u_0, v_0, R_2), E_{\Sigma_3} l_2(u_0, v_0, R_2), \cdots E_{\Sigma_3} l_T(u_0, v_0, R_2), \\ & \vdots \\ & E_{\Sigma_S} l_1(u_0, v_0, R_{S-1}), E_{\Sigma_S} l_2(u_0, v_0, R_{S-1}), \cdots E_{\Sigma_S} l_T(u_0, v_0, R_{S-1}) \Big]\end{aligned} \tag{11}$$

where $\mathcal{D}(u_0, v_0)$ represents the descriptor at location (u_0, v_0), $l_j(u, v, R)$ represents the location with distance R from (u, v) in the direction j, where the directions are quantized into T values. $E_{\Sigma_i} l_j(u, v, R)$ denotes the value of convolution at location $l_j(u, v, R)$ with scale Σ_i. Finally, there is one point worth to note that when performing convolution at location (u, v) with scale Σ, the convolution value $G_\Sigma(u, v)$ equals to its original value if it is bigger than zero, else equals to zero which is to preserve the polarity of the intensity changes.

3.2 Computational Complexity

The efficiency of our descriptor mainly stems from the fact that computation only depends on the $(S - 1) * T + 1$ times 2D convolutions no gradients computation. Subsequently, we briefly introduce the complexity of both DAISY and our describe that DAISY is much faster than SIFT as shown in [9]. Assuming that the parameters of DAISY are number of layers S' center point computed as one layer, number of grid points at a layer T, number of gradient orientations H, and that the gradients of orientation G_θ can be computed as

$$G_\theta = (\cos\theta\frac{\partial x}{\partial y} + \sin\theta\frac{\partial x}{\partial y})^+ \tag{12}$$

Hence, computing H' gradients needs $2P \times H'$ multiplications and $P \times H$ additions, where p is the size of the image. Constructing the DAISY descriptor requires additional $(S-1) \times H'$ 2D convolutions, or $2(s-1) \times H$ 1D convolutions. In computing our descriptor, assuming that the parameters are the number of Layers (S),center point computed as one layer, number of grid points at a layer (T). Because we don't use gradient, and use different templates at different grid points, totally $(S-1) \times T$ grid points, constructing our descriptor needs $(S-1) \times T + 1$ 2D convolutions.

4 Experiments and Results

In this section, we performed experiments on the examination of tight frame of modeling wavelets family and on the performance verification of the descriptor constructed by the modeling wavelet.

To determine whether the modeling wavelets form a frame and how tight that frame is, we computed their frame bounds using the frame bound condition equations, the mother wavelet equation. Note that, if for the one-octave sampling scheme, the upper bound of the unit spatial interval b_0 is 1 for the single voice wavelets, which can be simply deduced from the Nyquist sampling theorem, else for fractionally dilated wavelets, the upper bound of b_0 could be pushed above 1 because the sampling lattice becomes denser. The frame bounds computed for different sampling densities are shown in Table 1 to Table 5.

Table 1. Frame bounds of our wavelets with κ=1.5 for different numbers of sampling orientations (K), with the number of frequency steps per octave N=1, 3 and the unit spatial sampling interval b_0 =0.8

(a) N=1

K	A	B	B/A
4	6.7483	7.4262	1.1005
8	12.7307	14.0143	1.1008
16	25.4198	28.0684	1.1041
20	31.7603	35.0999	1.1051

(b) N=3

K	A	B	B/A
4	20.6456	21.7831	1.0551
8	39.0503	41.0768	1.0519
16	77.9963	82.2525	1.0545
20	97.5066	102.8043	1.0543

Several general observations can be made from the results. First, the frame becomes tighter with the increase in the number of orientation, frequency steps per octave. Second, from the cues of Table 5 show that only about limit scales are enough to form an almost tight frame. These results suggest that ganglion

Table 2. Frame bounds of our wavelets with κ=1.5 for different numbers of sampling orientations (K), with the number of frequency steps per octave N=3, and the unit spatial sampling interval b_0=1

K	A	B	B/A
4	20.2767	22.152	1.0925
8	38.3139	41.8132	1.0913
16	76.5893	83.6595	1.0923
20	96.7123	104.5896	1.0815

Table 3. Frame bounds of our wavelets with κ=1.5 for different unit spatial sampling intervals b_0, with the number of sampling orientations K=20 and the number of frequency steps per octave N=3

b_0	A	B	B/A
0.25	100.1484	100.1625	1.0001
0.5	101.6703	102.8401	1.0115
0.8	99.3812	105.1292	1.0578
1.0	95.7123	104.5986	1.0928

Table 4. Frame bounds of our wavelets with κ=1.0 for different numbers of sampling orientations (K), with the number of frequency step per octave N=3, and the unit spatial sampling interval b_0=0.8

K	A	B	B/A
4	37.167	39.9718	1.0754
8	64.7577	68.588	1.0591
16	129.4214	137.1644	1.0598
20	161.7742	171.4581	1.05986

Table 5. Frame bounds of our wavelets with κ=1.5 for different ranges of scales, with the number of sampling orientations k=20,and the number of frequency steps per octave N=3,the unit spatial sampling interval b_0=0.8

M	A	B	B/A
[-6 6]	97.2103	103.0219	1.0597
[-7 7]	97.212	103.0794	1.0604
[-8 8]	97.529	102.777	1.0538
[-9 9]	97.5156	102.7944	1.0541
[-10 10]	97.5066	102.8043	1.0543

cells in the retina not only provide a complete representation of the visual world, but also provide an almost tight frame representation through oversampling. In the next section, we will verify the performance of the descriptor constructed on the basis of the modeling ganglion cells array.

In this section, we examine the performance of our descriptor by image patch matching on a large and realistic data set [10]. The data set is from actual 3D correspondences, obtained via a stereo depth map, which allows us to design descriptors for dealing non-planar transformations and illumination changes that result from viewing a true 3D scene. For the results presented in this paper, we used three test sets (Yosemite, Notre Dame, and Liberty) which were obtained by extracting scale and orientation normalized 64*64 patch around DoG interest points. We have compared to Lowe SIFT method.

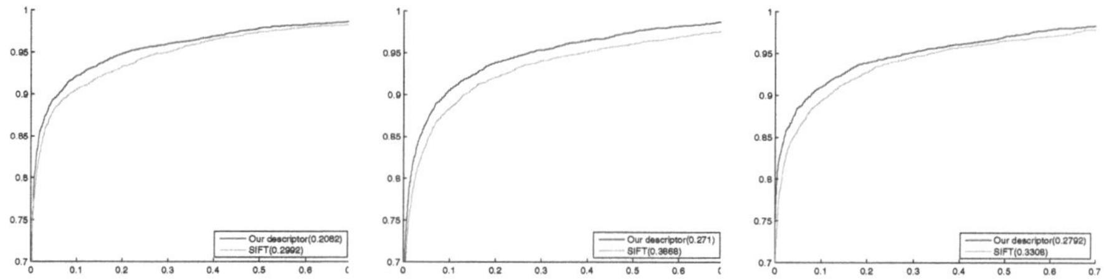

(a) Notre Dame (b) Liberty (c) Yosemite

Fig. 3. ROC curves for SIFT and our descriptor on three different datasets: Notre Dame, Liberty, Yosemite

From the results, we can see that our descriptor has the comparable results with SIFT, but, as described in Section 3.2, has low complexity, and much faster than SIFT and DAISY. Moreover, because the structure of grid points is circular, it brings convenience to solve rotation problem.

5 Conclusions

In this paper, based on the findings of retina in neurophysiology, anatomy and electrophysiology, we modeled the distribution of ganglion cells array and their responses with wavelets. Then, we validated the tight frame property of modeling wavelets by frame theory. As we all know, in visual system of human, ganglion cells play a crucial role in representing visual information. Next, we explore a new descriptor, whose performance achieves state-of-the-art but much faster. In the future, we will further reduce computation cost, and detail the scale and rotation invariant methods.

Acknowledgment. This work is funded by the National Basic Research Program of China (No. 2010CB327902), the National Natural Science Foundation of China (NSFC)(No. 61202237), and the Fundamental Research Funds for the Central Universities.

References

1. Lowe, D.G.: Distinctive image features from scale-invariant keypoints. IJCV 60(2), 91–110 (2004)
2. Lowe, D.G.: Object recognition from local scale-invariant features. In: ICCV, vol. 2, pp. 1150–1157 (1999)
3. Bay, H., Tuytelaars, T., Van Gool, L.: Surf: Speeded up robust features. In: Leonardis, A., Bischof, H., Pinz, A. (eds.) ECCV 2006, Part I. LNCS, vol. 3951, pp. 404–417. Springer, Heidelberg (2006)
4. Bay, H., Ess, A., Tuytelaars, T., Van Gool, L.: Speeded-up robust features (surf). CVIU 110(3), 346–359 (2008)
5. Mikolajczyk, K., Schmid, C.: A performance evaluation of local descriptors. IEEE TPAMI 27(10), 1615–1630 (2005)
6. Ke, Y., Sukthankar, R.: PCA-SIFT: A more distinctive representation for local image descriptors. In: CVPR 2004, vol. 2, pp. II–506 (2004)
7. Belongie, S., Malik, J., Puzicha, J.: Shape matching and object recognition using shape contexts. IEEE TPAMI 24(4), 509–522 (2002)
8. Freeman, W.T., Adelson, E.H.: The design and use of steerable filters. IEEE TPAMI 13(9), 891–906 (1991)
9. Tola, E., Lepetit, V., Fua, P.: Daisy: An efficient dense descriptor applied to wide-baseline stereo. IEEE TPAMI 32(5), 815–830 (2010)
10. Brown, M., Hua, G., Winder, S.: Discriminative learning of local image descriptors. IEEE TPAMI 33(1), 43–57 (2011)
11. Croner, L.J., Kaplan, E.: Receptive fields of p and m ganglion cells across the primate retina. Vision Research 35(1), 7–24 (1995)
12. Rodieck, R.W.: Quantitative analysis of cat retinal ganglion cell response to visual stimuli. Vision Research 5(12), 583–601 (1965)
13. Baizer, J.S., Ungerleider, L.G., Desimone, R.: Organization of visual inputs to the inferior temporal and posterior parietal cortex in macaques. The Journal of Neuroscience 11(1), 168–190 (1991)
14. Daubechies, I.: The wavelet transform, time-frequency localization and signal analysis. IEEE Trans. Infor. Theory 36(5), 961–1005 (1990)
15. Pollen, D., Feldon, S.: Spatial periodicities of periodic complex cells in the visual cortex cluster at one-half octave intervals. Investigative Ophthalmology & Visual Science 18(4), 429–434 (1979)
16. Silverman, M.S., Grosof, D.H., De Valois, R.L., Elfar, S.D.: Spatial-frequency organization in primate striate cortex. Proceedings of the National Academy of Sciences 86(2), 711–715 (1989)
17. Grossmann, A., Kronland-Martinet, R., Morlet, J.: Reading and understanding continuous wavelet transforms. In: Wavelets, pp. 2–20. Springer (1990)
18. Lee, T.S.: Image representation using 2d gabor wavelets. TPAMI 18(10), 959–971 (1996)
19. Berg, A.C., Malik, J.: Geometric blur for template matching. In: CVPR, vol. 1, pp. I–607. IEEE (2001)

Thorough Evaluation of the Discriminabilities of Four Voting Schemes in the Generation of Histogram of Orientation

Ruiguang Hu and Weiming Hu

National Laboratory of Pattern Recognition,
Institute of Automation, Chinese Academy of Science, Beijing, China
{rghu,wmhu}@nlpr.ia.ac.cn

Abstract. In this paper, we thoroughly evaluate four voting schemes for HOG : l_0, l_1, l_2, l_{inf}. Different factors–pyramid level, bin number, angle, similarity metric–were verified separately in our experiments, which were conducted on the challenging PASCAL VOC2011 dataset. The experimental results demonstrate that, best classification performances are not always achieved by l_1 voting scheme. On the contrary, the discrimination abilities of l_0, l_1, l_2 voting schemes are quite comparable at high level; l_0 is better than l_1, and l_2 takes the third place at low levels; l_{inf} is always the last for the reason of information loss. Conclusively, in contrast to those voting schemes utilizing gradient magnitudes, only counting the number of pixels within a bin, as the l_0 voting scheme does, can lead to at least comparable or even better classification performances.

Keywords: HOG, gradient magnitude, voting schemes, discriminability.

1 Introduction

Histogram of Oriented Gradients(HOG) [1], one of the most famous visual features, has achieved great success in many fields of computer vision, such as object recognition and detection, pedestrian detection, intelligent surveillance, etc. The reasons why it has been widely used are mainly that it has quite good discrimination ability and can be gotten easily.

When computed, the conventional process is that, each bin of HOG is voted by corresponding gradient magnitudes of pixels whose gradient angles are within this bin. To the best of our knowledge, there is no publication that clarifies whether this voting scheme is the most suitable or not. Naturally, in this paper, we experimentally evaluate four voting schemes for HOG : l_0, l_1, l_2, l_{inf}.

The evaluation was conducted on the challenging PASCAL VOC2011 dataset [7], which is one of the most important benchmark database in the field of computer vision. 20 categories were all used to evaluate the influences of four voting schemes. Different factors–pyramid level, bin number, angle, similarity metric–were verified separately and thoroughly.

According to the experimental results, we have found that, although l_1 voting scheme can lead to satisfactory performances, it can be replaced by l_0 and l_2

B. Huet et al. (Eds.): PCM 2013, LNCS 8294, pp. 835–845, 2013.

voting schemes. Additionally, without magnitude information, l_0 voting scheme can still result in comparable or even better classification performances. This phenomenon accords completely with the statistical meaning of common histogram generation methods, in which the value of each bin is just the occurring frequency of what is being considered.

The rest of the paper is organized as follows. Section 2 explains clearly the generation of Pyramid of Histograms of Oriented Gradients. Section 3 introduces four voting schemes. Section 4 thoroughly evaluate the four voting schemes under exhaustive setups and give detailed discussion. Section 5 concludes the paper and presents future work.

2 Pyramid of Histograms of Oriented Gradients

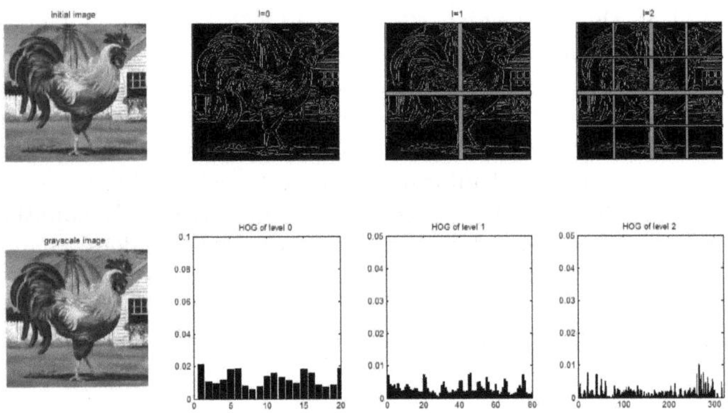

Fig. 1. The generation of Pyramid of Histograms of Oriented Gradients

Histogram of Oriented Gradients is proposed by *Dalal and Triggs*[1] initially for human detection. Their experiments conducted on two challenging datasets have shown that, taking HOG as the shape representation of person can get good detection accuracies. Following their pathbreaking work, *Felzenszwalb et al.* [2] propose the outstanding deformable part-based model to detect objects, which can represent the shape of objects more flexibly and achieve great success in PASCAL VOC competitions.

In HOG-based object detection, the gradient information of all pixels within a detection window are used. Regarding the task of object categorization, this kind of process is not necessary. Alternately, objects' contours are first extracted using the Canny edge detector on the gray-scale version of them, and only gradients of those pixels on the contours are considered when forming the HOG vector. The gradient orientations are equally divided into K bins, and conventionally gradient magnitudes of those pixels whose orientations are within one bin are summed as the value of that bin, which is referred to as the l_1 voting scheme.

The edge HOG of an image cannot capture the information about the spatial layout of the shape. For this purpose, each image is partitioned into increasingly fine sub-regions by repeatedly doubling the number of divisions in both the vertical and horizontal directions. This is a pyramid framework [3] because the number of pixels on the contours in a sub-region at one level is just the sum of the number of those pixels contained in the four sub-regions it is divided into at the next level. An edge HOG is computed for each sub-region at each pyramid level. Consequently, concatenating all edge HOG vectors at all levels sequentially, we get the final PHOG descriptor of an image, and the dimensionality of the final PHOG is $K\sum_{l=0}^{L} 4^l$, where K is the dimensionality of each edge HOG of each sub-region and L is the total level number with the whole image seen as level 0. For instance, if $L = 2$ and $K = 20$, then the dimensionality of the final PHOG is 420, as Fig. 1 shows.

As stated above, actually the numbers of pixels at each level are equivalent but the numbers of edge HOGs are exponential, hence it is necessary to weight edge HOGs of each level appropriately in the final PHOG. According to the discussion on spatial pyramid matching of *Lazebnik et al.*[4], weight for level 0 is $\frac{1}{2^L}$ and weight for level l is $\frac{1}{2^{L-l+1}}$. This means that histograms from finer resolutions are weighted more highly than those at coarser resolutions. Alternatively, the weights may be learned, but the complexity will be increased. The PHOG descriptor is normalized to unity by the L1-norm.

After incorporating spatial information appropriately, PHOG exhibits improved discriminability and has been widely used in object categorization [5].

3 Four Voting Schemes

As we can see in Section 2, the conventional voting scheme for each bin of HOG is the sum rule applying to corresponding gradient magnitudes. Here we expand the operation on gradient information and propose four voting schemes.

Suppose their are n pixels whose gradient angles are within the bin b, m_i is the gradient magnitude of pixel i, vector $M = \{m_1, \cdots, m_i, \cdots, m_n\}$ represents magnitudes of these n pixels, V is the value of bin b, so four voting schemes can be defined:

$$V_{l_0} = \|M\|_0 = n \ . \tag{1}$$

$$V_{l_1} = \|M\|_1 = \sum_{i=1}^{n} m_i \ . \tag{2}$$

$$V_{l_2} = \|M\|_2 = \sqrt{\sum_{i=1}^{n} m_i^2} \ . \tag{3}$$

$$V_{l_{inf}} = \|M\|_\infty = \max_i m_i \ . \tag{4}$$

Concatenating all bin values according to the angle sequence under the pyramid framework, four types of PHOG can be generated, which are referred to as PHOG-l_0,PHOG-l_1,PHOG-l_2,PHOG-l_{inf}. Subsequently, the classification performances leveraged by them were evaluated exhaustively.

4 Experiments

PASCAL VOC2011 dataset was used to evaluate the discrimination ability of four types of PHOG. All objects were first cropped out according to their boundingboxs in annotations to eliminate background influences thoroughly, as shown in Fig. 2. The initial train-set of this dataset were still used for training, and corresponding val-set were used for testing. Each of the 20 categories was verified individually. As to the *person* category, for memory consideration, only 1000 samples were picked out randomly from initial train-set and another 1000 samples were selected randomly from initial val-set, used for training and testing respectively.

Fig. 2. The chopped samples of 20 VOC classes. Each sample stands for one class.

For each category, binary classifiers were trained, and the same number of negative samples as positive samples were picked out from remaining 19 categories averagely, no matter in training or testing. The reason why we adopted this kind of setup is to eliminate sample number bias.

Specifically, when extracting PHOGs, pyramid framework proposed in [3], [4] was utilized. Parameters were as below. Levels: $0, 1, 2$; angles: 360^0, 180^0; bin numbers :$10, 20, 30$.

Four well-known kernels were used as similarity metric-χ^2, RBF, linear, and Histogram Intersection(HI)- for SVM classifiers [6]. Totally, 2560 binary classifiers were implemented. Equal Error Rates(EERs) were reported as classification performance metrics. Note that, in this paper, higher EER value stands for better classification performance. The influences of different factors for the four voting schemes were verified exhaustively, shown as follows.

4.1 Similarity Metric Influence

First, we evaluated the discrimination abilities of four voting schemes under four well-known kernels. Fig. 3 shows the EERs for these kernels when other parameters were fixed. Note that, reminding parameters can also be set to other values, just as what have been done. Additionally, the similarity metric influence can be also found in subsequent figures, from Fig. 4 to Fig. 10.

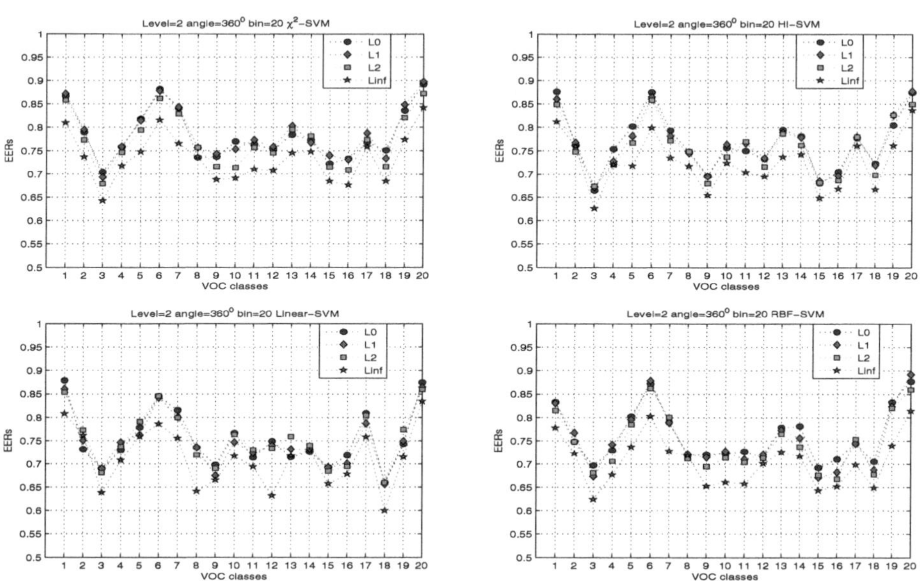

Fig. 3. EERs for different kernels, from up to down: χ^2, HI, linear, RBF. Other parameters: level=2, angle=360^0, bin=20.

4.2 Pyramid Level Influence

Second, we evaluated the discrimination abilities of four voting schemes at three pyramid levels: 2, 1, 0. Totally, Fig. 3, Fig. 6, Fig. 9, Fig. 10 are for L=2; Fig. 4, Fig. 7 are for L=1; and Fig. 5, Fig. 8 are for L=0.

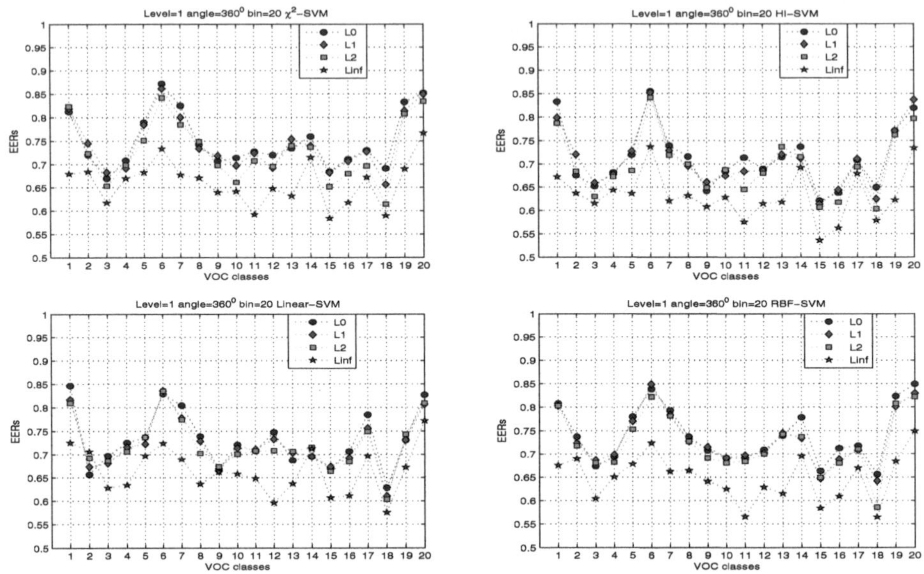

Fig. 4. EERs for level=1. Kernels from up to down: χ^2, HI, linear, RBF. Other parameters: angle=360^0, bin=20.

4.3 Angle Influence

In HOG framework, there are two choices when determining the angle span of each bin: 360^0 and 180^0. When 360^0 is chosen, HOG is said to be signed, and 180^0 chosen means unsigned HOG. We evaluated the discrimination abilities of four voting schemes for both angle choices. Fig. 3, Fig. 4, Fig. 5 Fig. 9, Fig. 10 are for signed HOGs, and Fig. 6, Fig. 7, Fig. 8 are for unsigned HOGs.

4.4 Bin Number Influence

Bin number is another important factor for HOG. If it is too small, votes would be aggregated excessively, then histogram can not represent shape sufficiently; on the contrary, If it is too large, the histogram would be rather sparse and plain, probably reducing its discriminability too. In our experiments, we evaluated three bin number: 10, 20 ,30. Fig. 3, Fig. 4, Fig. 5, Fig. 6, Fig. 7,Fig. 8 are for 20; Fig. 9 is for 10; Fig. 10 is for 30.

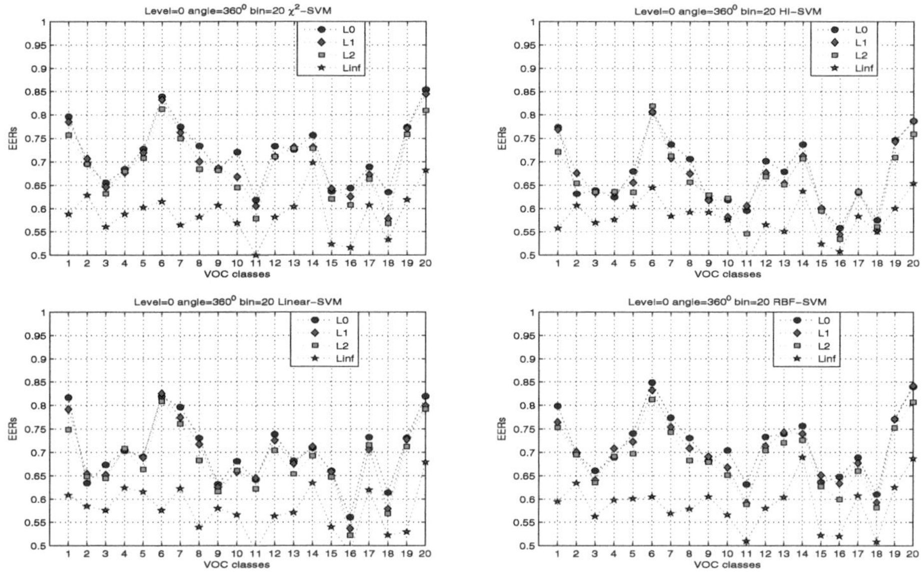

Fig. 5. EERs for level=0. Kernels from up to down: χ^2, HI, linear, RBF. Other parameters: angle=360^0, bin=20.

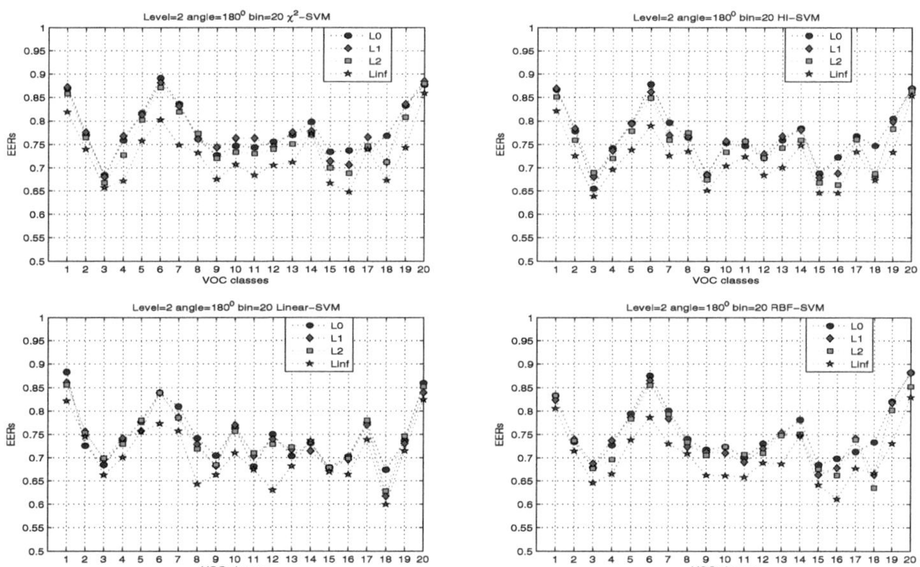

Fig. 6. EERs for angle=180^0. Kernels from up to down: χ^2, HI, linear, RBF. Other parameters: level=2, bin=20.

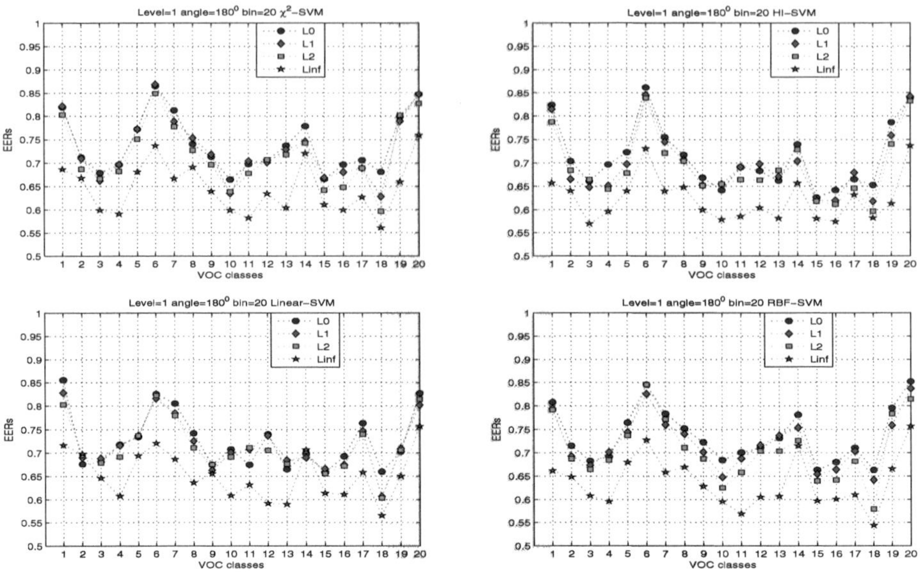

Fig. 7. EERs for angle=180^0. Kernels from up to down: χ^2, HI, linear, RBF. Other parameters: level=1, bin=20.

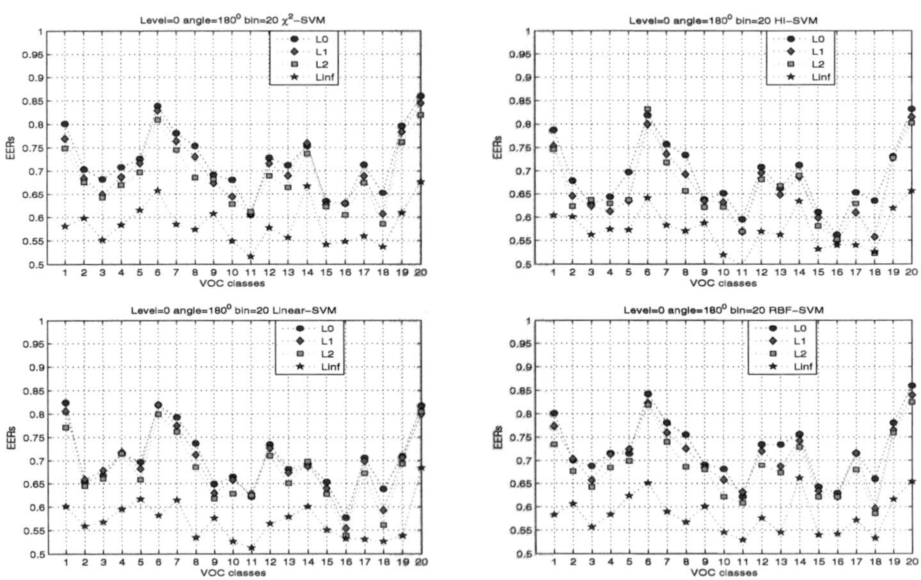

Fig. 8. EERs for angle=180^0. Kernels from up to down: χ^2, HI, linear, RBF. Other parameters: level=0, bin=20.

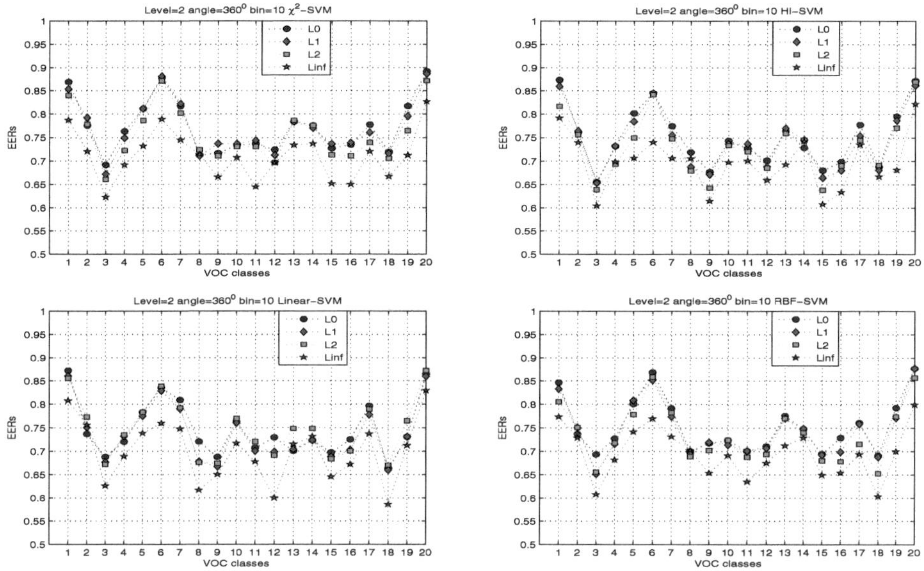

Fig. 9. EERs for bin=10. Kernels from up to down: χ^2, HI, linear, RBF. Other parameters: level=2, angle=360^0.

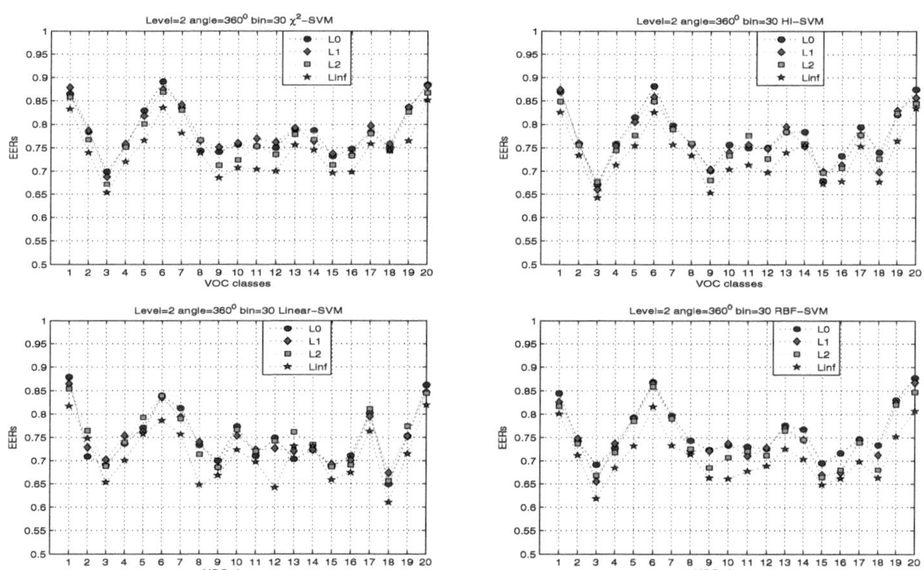

Fig. 10. EERs for bin=30. Kernels from up to down: χ^2, HI, linear, RBF. Other parameters: level=2, angle=360^0.

4.5 Discussion

From the view of similarity metric, all of four kernels result in similar EER curves regarding to 20 categories. Curves of l_0, l_1, l_2 voting schemes are quite close, and crossed by each other in many cases. Curves of l_{inf} voting scheme are always at the bottom, since it discards too much information when generating histograms. Certainly, there are tiny distinction for four different kernels, in the case of χ^2 kernel, l_1 voting scheme results in the best EER values for many object categories at level 2, while in the cases of other three kernels, l_0, l_2 voting schemes result in the best EER values for many object categories asynchronously at level 2, which can be seen clearly in Fig. 3, Fig. 6, Fig. 9, Fig. 10.

From the view of pyramid level, EER values of l_0, l_1, l_2 voting schemes are quite comparable at level 2. At level 1, l_0 voting scheme results in the best EER values, l_1 voting scheme takes the second place, and l_2 voting scheme is the third. This phenomenon is more obvious at level 0 than at level 1.

Angle choices have almost no influence for four voting schemes. In some cases, EER curves are more close for signed HOGs than for unsigned HOGs.

Different bin numbers only lead to changes in the orders of EER values of l_0, l_1, l_2 voting schemes for few categories at level 2, while these EER values are still quite close to each other.

According to comparative analysis from the aspects of similarity metric, pyramid level, angle, and bin number, we can definitely summarize that, each of the l_0, l_1, l_2 voting schemes can result in best performances under certain circumstances, and l_0 voting scheme surpasses the l_1, l_2 voting schemes at low levels and is comparable with them at high levels, no matter which well-known similarity metric or bin number is chosen, no matter the PHOGs are signed or unsigned.

From the perspective of statistics, we can give a clear explanation for this phenomenon. Commonly, when generating a histogram, the value of each bin is just the occurring frequency of what is being considered, and the histogram can be used as probability distribution to promote following process. Concerning the HOG, the number of pixels whose gradient orientation are within an angle bin should be the most appropriate value for that bin, which is referred to as l_0 voting scheme in our paper. Additionally, to a great extent, the l_0 voting scheme for HOG is equivalent to the harding coding scheme for the well-known BOW model, which uses local features, such as SIFT etc., to conduct object categorization.

5 Conclusion and Future Work

We have conducted 2560 classification experiments to thoroughly evaluate the discriminabilities of four voting schemes from the aspects of similarity metric, pyramid level, angle, and bin number. Consequently, we have found that, besides l_1 voting scheme, which is widely used in HOG generation, l_0, l_2 voting schemes can also lead to at least comparable or even better performances.

In the future, four voting schemes will be evaluated in object detection frameworks such as part-based model, and will also be evaluated in local features such as SIFT, which, essentially, is histogram with l_1 voting scheme.

Acknowledgements. This work is partly supported by NSFC (Grant No. 60935002), the National 863 High-Tech R&D Program of China (Grant No. 2012AA012504), the Natural Science Foundation of Beijing (Grant No. 4121003), and The Project Supported by Guangdong Natural Science Foundation (Grant No. S2012020011081).

References

1. Dalal, N., Triggs, B.: Histogram of oriented gradients for human detection. In: CVPR, pp. 886–893 (2005)
2. Felzenszwalb, P.F., Girshick, R.B., McAllester, D.: Object detection with discriminatively trained part-based models. IEEE Transactions on Pattern Analysis and Machine Intelligence 32(9), 1627–1645 (2010)
3. Bosch, A., Zisserman, A., Munoz, X.: Representing shape with a spatial pyramid kernel. In: Proceedings of the 6th International Conference on Image and Video Retrieval, pp. 401–408 (2007)
4. Lazebnik, S., Schmid, C., Ponce, J.: Beyond bags of features: Spatial pyramid matching for recognizing natural scene categories. In: CVPR (2006)
5. Hu, R.G., Xie, N.H., Hu, W.M.: Drug-taking instruments recognition. In: First Asian Conference on Pattern Recognition (2011)
6. Chang, C.C., Lin, C.J.: LIBSVM: a library for support vector machines (2001), http://www.csie.ntu.edu.tw/~cjlin/libsvm/
7. (2011), http://pascallin.ecs.soton.ac.uk/challenges/VOC/voc2011/

Texture Classification via Local Feature Representation of Multi-order Gradients

Hengyang Wei[1], Qingjie Liu[1], Yunhong Wang[1], Chao Zhu[2], and Di Huang[1]

[1] IRIP Lab, Beihang University, Beijing, 100191, China
[2] LIRIS UMR 5205, Ecole Centrale de Lyon, Ecully, 69134, France
patrick934@gmail.com, liuqingjie@cse.buaa.edu.cn,
{yhwang,dhuang}@buaa.edu.cn, chao.zhu@ec-lyon.fr

Abstract. This paper presents a novel method to texture classification using local feature representation of multiple order gradients. Different from the state of the art approaches in literature that make use of the widely-used first order gradient based local descriptors, e.g. LBP, HOG, DAISY, SIFT, etc., we claim that the second order gradient based ones also provide critical contribution to classification performance, and thus propose to use Histogram of Second Order Gradients (HSOG) to describe micro-texton patterns. Both the similarity measurements of first and second order gradients computed by Bag-of-Feature modeling and SVM classifier are combined for decision making. Experimental results achieved on the Outex_TC dataset not only illustrate that the second order gradient based HSOG is effective to classify texture images, but also highlight that multiple order gradient based description by fusing complementary clues of the first and second order gradients is a promising solution to improve the accuracy in texture classification.

Keywords: Local image descriptor, texture classification, multi-order gradients.

1 Introduction

Texture analysis is one of the most popular topics in the field of computer vision due to its scientific challenges and application potential, and it has hence attracted increasing attention in recent years. Texture classification, regarded as one of main problems within texture analysis, refers to assignment of a texture image into one of pre-defined categories. It generally focuses on comprehensive and accurate texture representation, deriving a set of discriminative features from texture images to describe their attributes. If inadequate features are exploited, even the best classifier will fail to achieve a satisfying result. "Good" texture features are expected to own subsequent properties: first, can tolerate the within-class variations, e.g. lighting and rotation changes, while correctly discriminate different classes; second, can be easily extracted from raw data to allow fast processing; finally, lies in a space with moderate dimensionality to avoid high computation cost in classification.

Recent years have witness the progress made by local image descriptors in texture classification. For instance, SIFT is applied to texture classification [6] on the

B. Huet et al. (Eds.): PCM 2013, LNCS 8294, pp. 846–855, 2013.

UIUCTex dataset [16] and the KTH-TIPS dataset [17], using SVM with EMD kernel as the classifier, and achieves an excellent result. Schwartz et al. applied HOG to texture classification [13] with a linear SVM as the classifier on the UMD texture dataset [15], and proposed a strategy of extraction on multiple scales to improve the performance. DAISY is used in texture analysis of terrain classification in [14], while some modifications are made to calculate the descriptor on a grid. With the classifier Random forests, the performance of classification is better than some other texture descriptors at high resolution.

The overwhelming majority of these local image descriptors used in texture analysis are based on the first-order gradient information. However, it still remains a problem whether such information is sufficient to represent variations of local patches. It has been pointed out in the early study [8] that human visual processing should not be explained only by the first order mechanisms that capture the spatio-temporal changes in luminance, and the second order based ones also offer complementary information such as difference of texture and spatial frequency. Similarly some research [9] shows that the second order gradient based information is also important to the performance in computer vision analysis, e.g. object recognition. Unfortunately, second order gradient is rarely investigated in such a topic.

In this paper, in contrast to the state of the art methods on texture classification that make use of the first order gradient based descriptors, we present a novel approach to encode texton patterns through multi-order gradient information, aiming at providing a more comprehensive texture representation and hence achieving improved accuracy. The main contributions of this paper can be summarized as follows.

(1) We propose multi-order gradient based texture representation, which combines the advantages of first and second order gradient based descriptors as well;

(2) To the best of our knowledge, it is the first time to investigate the performance of Histogram of Second Order Gradient (HSOG) in texture classification;

(3) Through the experimental results achieved on the Outex_TC dataset, we prove the effectiveness of the second order gradient based descriptor (i.e. HSOG) as well as its complementation to the first order gradient based ones in texture classification.

The remainder of the paper is organized as follows. Section 2 introduces the widely used first order gradient based descriptors, the second order gradient based HSOG and the Bag of Feature modeling. Section 3 presents and analyzes the experimental results achieved on the Outex_TC dataset. Section 4 concludes the paper with perspectives.

2 Feature Representation

2.1 Local Image Descriptors Based on First-Order Gradient

Several popular local images descriptors are used in the experiment: SIFT, DAISY, HOG and LBP. All of these descriptors are based on first-order gradient of image.

SIFT [2] is one of the most widely used local image descriptors, it is constructed via computing the distribution of gradients at every pixel around the interest point. As one of the state-of-the-art descriptors in object recognition, SIFT has rotation invariance and scale invariance, and the robustness against some luminance variances and small change of view point. Original SIFT is extracted on the interest points detected by Difference-of-Gaussian (DoG) detector, but later other interest point detection strategies have been applied, including dense sampling and so on.

DAISY [5] is a local image descriptor designed for fast computation and dense matching. It uses Gaussian filter to convolve the image, and constructs descriptor with the sum of convolutions, which greatly improves the efficiency of computation. Recently DAISY is applied in object recognition [9], and receives a comparable result with SIFT, while the time of cost is largely reduced.

Dalal and Triggs design the Histogram of Oriented Gradient (HOG) descriptor [4] which was inspired by SIFT and GLOH [3]. HOG also computes the gradients information of pixels around the interest point, which is the same with that of the SIFT. However, the location grids of HOG use both rectangular one and log-polar one.

Proposed by Ojala et al., LBP [1] is a local feature originally designed for texture classification. For every pixel of an image, the LBP code of one pixel is calculated by combining the sign of difference between the pixel and its surroundings. The whole image is then represented by the histogram of the LBP codes, and this histogram is the LBP descriptor of the image. In [10], LBP is applied to face recognition. In which, the regions centered at the interesting points are regarded as patches to do the LBP encoding. The local LBP descriptors extracted from these regions are then combined together to form feature vectors for recognition.

2.2 Histograms of the Second Order Gradients (HSOG)

Histograms of the Second Order Gradients (HSOG) descriptor [8] is a local image descriptor based on second order gradient of the image, which is designed for object recognition. Different from other second order gradient descriptors that using steerable filters, HSOG constructs the descriptor using a map based method. The construction of HSOG can be divided into three steps: (1) construction of the first order Oriented Gradient Maps (OGMs); (2) computation of the second order gradients and (3) spatial pooling.

For the original image regions around the interest points, HSOG constructs the first order OGMs by computing the gradients in different quantized directions. The construction of OGMs provides rotation invariance and robustness to luminance variance. Then the second order gradients are computed on the OGMs by calculating the magnitude and the orientation at every pixel of OGMs. Notice that the orientations of second order gradients are also quantized and keep consistent with the directions of the first order OGMs. After second order gradient is computed, a spatial pooling scheme of DAISY-style is applied to construct the final descriptor. To determine the spatial pooling scheme, there are four parameters: the radius of the image region (R), the number of quantized orientations (N), the number of the rings (CR) and the number of the circles on each ring (C). The region around interest point is divided into $T=CR \times C+1$ circles. The dimensionality of the final descriptor is $(CR \times C+1) \times N^2$.

2.3 Bag-of-Feature (BoF)

Bag-of-Feature [11] is a modeling method that encodes descriptors of an image into an orderless collection based on a visual vocabulary, which is generated usually by applying clustering method on the training data. The BoF method has two main steps: (1) generate visual vocabulary from training data; (2) encode the given image to histogram representation with the generated visual vocabulary. The details of BoF are described as follows.

Firstly, all the descriptors extracted from the training data are input to a clustering algorithm, and a set of cluster centers is computed, where each cluster center is regarded as a visual word in visual vocabulary in the next step. For our experiments, the clustering method used in BoF is k-means, which is often used for various clustering needs.

Then, in the encoding step, the descriptors extracted from an image are assigned to the closest visual words in a certain metric space, and the frequency of assignments of each visual word is counted, with a histogram of which the number of bins is the size of visual vocabulary. So after BoF modeling, each image is represented by a fixed-length vector, and the length equals to the size of visual vocabulary, or to the number of clustering center, in the same meaning.

3 Experimental Results

The texture classification experiments use a standard dataset named Outex_TC_00014 [12]. This dataset contains 68 different textures. For each texture, three images of 746×538 pixels under different illuminants (2856K incandescent CIE A light source (Inca), 2300K horizon sunlight (Horizon), or 4000K fluorescent TL84 (TL84)) are taken. Each image then is divided into 20 non-overlapping images of 128×128 pixels, so the dataset contains 60 images for each texture, or 1360 images for each illuminant, that is to say 4060 images in total. The default training set contains half number of images under Inca illuminant, while the default test set contains half number of images under Horizon and TL84 illuminants.

3.1 Experimental Setup

For each image, a dense sampling grid with a step of 4 pixels is applied to select interest points, and then different descriptors including HSOG are extracted in the regions around the interest points. In the experiment, HSOG is compared with four popular local descriptors: SIFT[1], DAISY[2], HOG[3] and LBP[4]. All these widely used descriptors are based on first order gradient of image. Among these descriptors, HSOG, SIFT, DAISY and HOG can be directly extracted from a region around an

---

[1] Code of SIFT: `http://www.colordescriptors.com/`

[2] Code of DAISY: `http://www.engintola.com/daisy.html`

[3] Code of HOG: `http://www.vlfeat.org/`

[4] Code of LBP: `http://www.cse.oulu.fi/CMV/Downloads/LBPMatlab`

interest point, while the extraction procedure of LBP descriptor is a little different. We apply a mechanisms that allow the extraction of LBP descriptor from a local region around an input interest point, instead of a LBP code at a pixel. Given an interest point, we compute LBP representation of a patch maximum 64×64 pixels, as the LBP descriptor at this interest point. Thus, a set of local descriptors can be extracted from every image of the dataset.

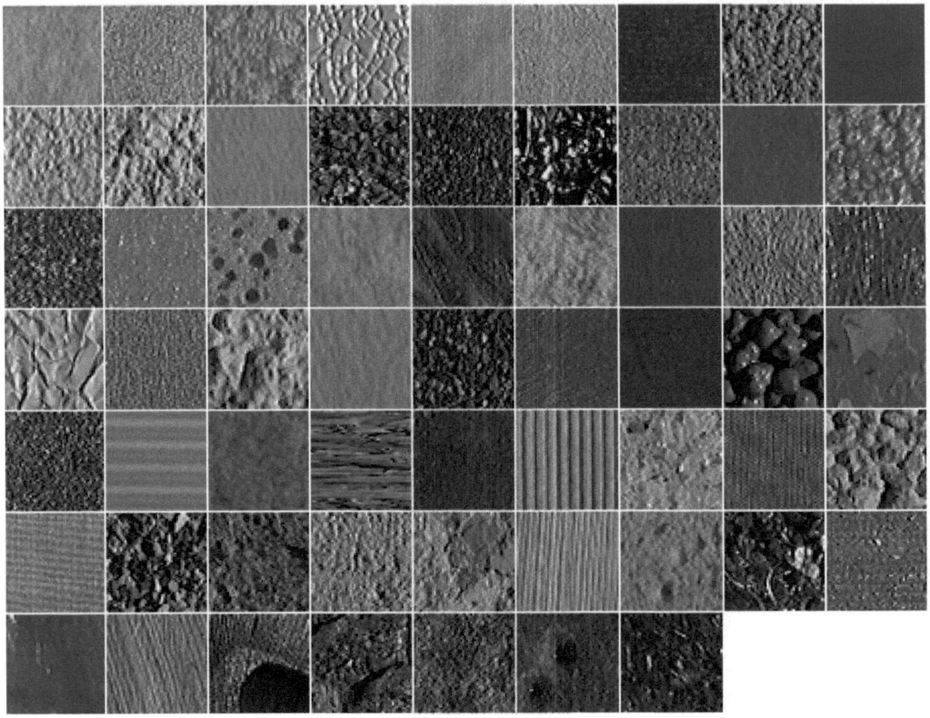

Fig. 1. Textures from dataset Outex_TC_00014

After the extraction of descriptors, an image representation is generated for each image by BoF modeling. In our experiment, a visual vocabulary of 4000 visual words is used for encoding.

With all the images are represented by fixed length feature vectors, a classifier can be applied to perform classification. We use the Support Vector Machine (SVM) algorithm to classify the textures, and use the χ^2 distance (chi-square distance) to measure the similarity between representations of two images. The χ^2 distance is calculated as following:

$$dist_{\chi^2}(F, F') = \sum_{i=1}^{n} \frac{(F_i - F_i')^2}{F_i + F_i'} \tag{1}$$

Where F and F' are two vectors representing images, n is the dimensionality of representation vector. Then the kernel function of χ^2 distance is computed for SVM:

$$K_{\chi^2}(F, F') = e^{-\frac{1}{D} dist_{\chi^2}(F,F')} \tag{2}$$

Where D is the parameter of normalization, and equals to the average of distances between representation vectors of training set.

We use cross-validation to evaluate the performance of the proposed method. The texture images are classified by SVM after tuning SVM parameters by training. The final classification accuracy of the test set is obtained for the evaluation of the descriptor's performance.

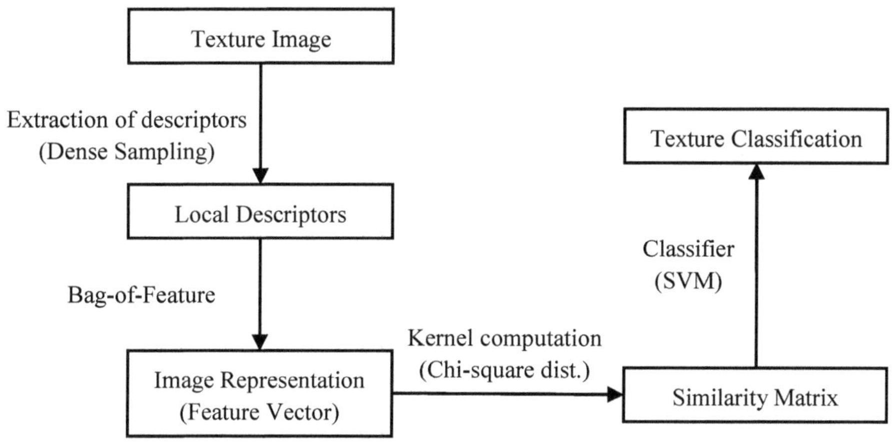

Fig. 2. Flow chart of the experiment

3.2 Selection of the Radius of HSOG

As mentioned in 2.2, HSOG descriptor is determined by four parameters: the radius of the local region (R), the number of quantized orientations (N), the number of the rings (CR), and the number of the circles on each ring (C). Considering HSOG is designed for object recognition, and the features of objects is different from those of textures, for example, the scales of objects is usually larger than that of textures, these parameters should be changed when HSOG is used in texture classification, especially the radius and the number of rings, which are related to the scales of features and the density of the discriminative information. Fig. 3 shows the experimental results.

As we can see, the HSOG descriptor achieves the highest classification accuracy when R=10, CR=4 among all the values of the parameters in experiment. More quantized orientations and more rings may bring a better performance, however it will cause the dimensionality increasing rapidly and leads to more computation cost. Compared with the parameters of HSOG used for object recognition in [8], the parameter R, that is to say the radius of image region, decreases clearly. It is because the

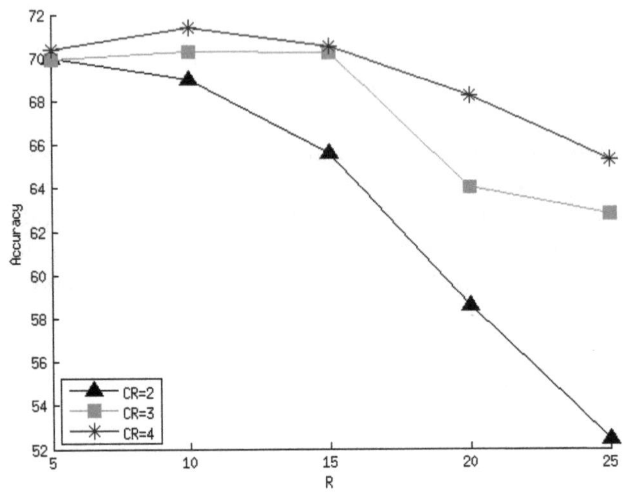

Fig. 3. Classification accuracy of HSOG under different parameters

scales of objects are usually larger than the scales of discriminating textures, and this implies the importance of consistency between parameters of descriptor and the characteristic of image.

3.3 Comparison with First-Order Gradient Descriptors

We apply the same flow path on some other popular first-order gradient descriptors to compare them with HSOG, and Table 1 shows the result. The parameters of HSOG are set to *R*=10, *N*=8, *CR*=4, *C*=4, which have been discussed in 3.2. For DAISY descriptor, the radius of the image region and the number of the rings are chosen the same as HSOG, while other parameters stay the default values of DAISY itself. For LBP, $LBP_{8,3}$ is chosen as it can achieve a relative high performance with a reasonable computation cost. For both SIFT and HOG, as the two descriptors have been used in texture classification [6, 13], default parameters are used as they provide better performance.

As shown in this table, though the performance of HSOG is slightly inferior to that of SIFT and DAISY, the results are on the same level and comparable, showing that the second order gradient is also discriminative and can be used in texture classification. Compared to these four first-order gradient descriptors, whose dimensionalities are from 100 to 300, HSOG descriptor has a dimensionality around 1000, which is relatively high, but still acceptable.

Table 1. Performance comparisons between HSOG descriptor and other first-order gradient descriptors on the Outex_TC_00014 dataset

Descriptor	Performance (%)
SIFT	73.75
DAISY	73.31
HOG	61.32
LBP	69.93
HSOG	71.40

3.4 Fusion of Different Order Gradients Descriptors

To improve the performance of texture classification, we propose a strategy combining different order gradient information of image to provide more discriminative power.

We apply the late fusion to combine HSOG descriptor with other first order gradient descriptor, which has the advantage not increasing the dimensionality of descriptors and also convenient to realize with parallel computation, since the decision values for late fusion of different descriptors can be computed individually. The decision of classification is made by choosing the class with the maximum decision value after fusion. Table 2 shows the final results of fusion between HSOG and different descriptors.

Table 2. Performance of different fusion strategies of multiple order gradient descriptors on the Outex_TC_00014 dataset

Descriptor	Performance (%)
HSOG+DAISY	76.25
HSOG+LBP	72.72
HSOG+SIFT	76.47
HSOG+HOG	75.07

As we can see in Table 2, for each first-order gradient descriptor, the classification accuracy of fusion with HSOG is higher than that of itself. This shows that the HSOG descriptor brings complementary information to those first-order gradient descriptors, and also demonstrates that multiple order gradients provide more discriminative information than single order gradient.

4 Conclusion

In this paper, we evaluate the performance of the second order gradient descriptor HSOG and four widely used first order gradient descriptors, including SIFT, LBP, DAISY and HOG, on texture classification. Experimental results demonstrate the effectiveness of HSOG in texture classification under illumination changes. We also use late fusion between HSOG and first order gradient descriptor to realize the combination of different order gradients, and improve the performance of texture classification, which proves the HSOG provides complementary information to first order gradient descriptors and the effectiveness of using multi-order gradients information in texture classification.

Acknowledgements. This work was supported in part by the National Basic Research Program of China under grant 2010CB327902; the National Natural Science Foundation of China (NSFC) under grant 61202237; and the Fundamental Research Funds for the Central Universities.

References

1. Ojala, T., Pietikainen, M., Maenpaa, T.: Multiresolution gray-scale and rotation invariant texture classification with local binary patterns. IEEE Transactions on Pattern Analysis and Machine Intelligence 24(7), 971–987 (2002)
2. Lowe, D.G.: Distinctive image features from scale-invariant keypoints. International Journal of Computer Vision 60(2), 91–110 (2004)
3. Mikolajczyk, K., Schmid, C.: A performance evaluation of local descriptors. IEEE Transactions on Pattern Analysis and Machine Intelligence 27(10), 1615–1630 (2005)
4. Dalal, N., Triggs, B.: Histograms of oriented gradients for human detection. In: IEEE Computer Society Conference on Computer Vision and Pattern Recognition, CVPR 2005, vol. 1. IEEE (2005)
5. Tola, E., Lepetit, V., Fua, P.: A fast local descriptor for dense matching. In: IEEE Conference on Computer Vision and Pattern Recognition, CVPR 2008. IEEE (2008)
6. Zhang, J., et al.: Local features and kernels for classification of texture and object categories: A comprehensive study. International Journal of Computer Vision 73(2), 213–238 (2007)
7. Smith, A.T., Scott-Samuel, N.E.: First-order and second-order signals combine to improve perceptual accuracy. JOSA A 18(9), 2267–2272 (2001)
8. Huang, D., et al.: HSOG: a novel local descriptor based on histograms of second order gradients for object categorization. In: Proceedings of the 3rd ACM Conference on International Conference on Multimedia Retrieval, ACM (2013)
9. Zhu, C., Bichot, C.-E., Chen, L.: Visual object recognition using daisy descriptor. In: IEEE International Conference on Multimedia and Expo (ICME). IEEE (2011)
10. Ahonen, T., Hadid, A., Pietikäinen, M.: Face recognition with local binary patterns. In: Pajdla, T., Matas, J(G.) (eds.) ECCV 2004. LNCS, vol. 3021, pp. 469–481. Springer, Heidelberg (2004)
11. Csurka, G., et al.: Visual categorization with bags of keypoints. In: Workshop on Statistical Learning in Computer Vision, ECCV, vol. 1 (2004)

12. Ojala, T., et al.: Outex-new framework for empirical evaluation of texture analysis algorithms. In: Proceedings of the 16th International Conference on Pattern Recognition, vol. 1. IEEE (2002)
13. Schwartz, W.R., et al.: A novel feature descriptor based on the shearlet transform. In: 2011 18th IEEE International Conference on Image Processing (ICIP). IEEE (2011)
14. Khan, Y.N., Komma, P., Zell, A.: High resolution visual terrain classification for outdoor robots. In: 2011 IEEE International Conference on Computer Vision Workshops (ICCV Workshops). IEEE (2011)
15. Xu, Y., et al.: Combining powerful local and global statistics for texture description. In: IEEE Conference on Computer Vision and Pattern Recognition, CVPR 2009. IEEE (2009)
16. Lazebnik, S., Schmid, C., Ponce, J.: A sparse texture representation using local affine regions. IEEE Transactions on Pattern Analysis and Machine Intelligence 27(8), 1265–1278 (2005)
17. Hayman, E., Caputo, B., Fritz, M., Eklundh, J.-O.: On the significance of real-world conditions for material classification. In: Pajdla, T., Matas, J(G.) (eds.) ECCV 2004. LNCS, vol. 3024, pp. 253–266. Springer, Heidelberg (2004)

Compact and Robust Image Fingerprints Based on CCA of Local Features*

Yang Jing[1,2] and Fuming Sun[3]

[1] School of Computer and Communication, Hunan University of Technology, Zhuzhou, Hunan, 412008 China
[2] School of Civil Engineering, Hebei University of Engineering, Handan, Hebei, 056038 China
[3] School of Electronic and Information Engineering, Liaoning University of Technology, Jinzhou, 121001, China
yangjing_wp@hebeu.edu.cn, sunwenfriend@gmail.com

Abstract. Image fingerprints are perceptual features or short summaries of a given image. They can be used for identifying image contents just as human fingerprints are used for identification. In this paper, we propose a novel compact image fingerprint extraction method based on canonical correlation analysis (CCA) of local features. After extraction of local SIFT features, we first divide the 128-dimension feature into two segments to form two closely related feature matrices for a given image. Then we apply CCA to the two matrices, and take the quantized canonical coefficient matrix (8×8) as the image fingerprint which is very compact, only 64 bytes for any image of arbitrary size. Experimental results show that the proposed method is discriminative and robust against both normal image processing operations such as compression common geometric transformations including cropping, scaling, and rotation.

Keywords: Image Fingerprinting, Robust Hashing, CCA.

1 Introduction

With the ever-increasing use of multimedia contents, the problems associated with the protection of intellectual property, management of large database and indexation of content are becoming more prominent. Although watermarking has been considered as efficient means to these problems, there are some issues with the use of it, such as the modification of the content and its security. With respect to this, identifying content itself intrinsically based on its own features rather than some embedded identifier can be an alternative solution to these problems. Fingerprints are perceptual features or short summaries of a multimedia object [1–4]. The aim of fingerprinting (also known as multimedia identification, robust hashes, robust signatures) is to provide fast and reliable methods for content identification. It can also be used for image retrieval and

* This work was supported by Science and Technology Program of Hunan Province (2012FJ3035) and National Natural Science Foundation of China (61272214).

B. Huet et al. (Eds.): PCM 2013, LNCS 8294, pp. 856–865, 2013.

other applications [5–7]. It is an emerging research area that is receiving increased attention. A number of applications of multimedia fingerprinting were detailed in [1–4].

Multimedia signal can be represented equivalently in different forms, and undergo various manipulations during distribution that may carry the same perceptual information. This renders bit-sensitive cryptographic hash functions not applicable to multimedia fingerprinting [1–4]. Therefore, fingerprinting must be robust enough to identify two perceptually similar objects as similar and discriminatory enough to identify perceptually different objects as different (pairwise independent).

So far, a number of image fingerprinting methods have been reported [1–3, 8–11, 13–23]. Early fingerprinting methods are not very robust due to the features for producing fingerprints. Most of them are based on block-histogram [8], DCT coefficients [3, 8, 10, 12, 13, 17], etc. Therefore, they only exhibit a limited robustness against normal image processing operations such as low-degree of noise addition, and JPEG compression, while very sensitive to geometric transformations including cropping, scaling, and rotation, etc. For example, although the principal component analysis (PCA) method proposed by [3] is very compact (merely 72 bytes), this method and its extension [21] can not be robust to cropping, scaling, and rotation due to the adoption of DCT coefficients. As is pointed out in [2], the major challenge in content-based authentication is to define a computable feature vector that can capture the major content characteristics from a human perspective. For multimedia fingerprinting, extracting features that allow direct access to the relevant distinguishing information is crucial. For a good fingerprinting system, the features should be both fairly discriminative and robust [1]. Recently, more and more researchers paid attention to more robust features for fingerprint generation, such as visual words [18], shape contexts and local feature points [22], zernike moments and local features [23]. However, compactness of fingerprint are usually neglected.

In this paper, we propose a novel image fingerprint extraction method based on canonical correlation analysis (CCA) of local SIFT features due to its robustness to geometric transformations. At same time, the resultant fingerprint is very compact (only 64 bytes), since we take the quantized canonical coefficient matrix (8×8) as the image fingerprint. Experimental results show that the proposed method is discriminative and robust against common quality-preserving manipulations such as compression, cropping, scaling, and rotation.

The rest of this paper is organized as follows. Section 2 proposes fingerprint generation. Section 3 describes fingerprint matching. Section 4 details experimental results. Finally, conclusion and future work will be given in Section 5.

2 Fingerprint Generation

Proposed by H. Hotelling in [25], canonical correlation analysis (CCA) is a important research subject of multivariate statistical analysis which describes the relationship between two sets of variables (X, Y) by finding linear combinations

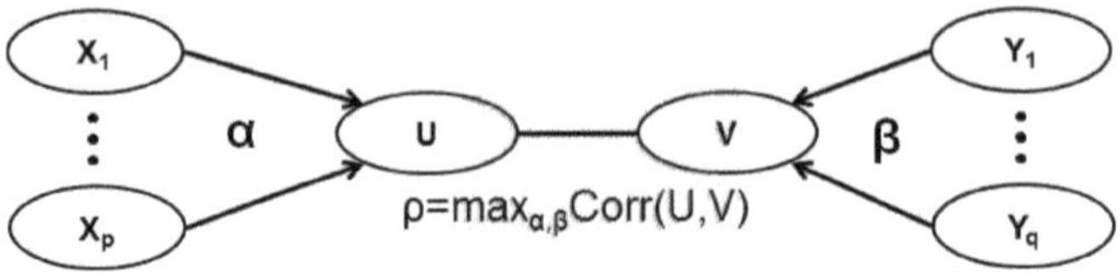

Fig. 1. Canonical Correlation Analysis

of the variables with maximal correlation. As show in Fig.1, CCA tries to seek linear combinations:

$$\begin{aligned} U &= \alpha^T X \\ V &= \beta^T Y \end{aligned} \tag{1}$$

such that the correlation ρ:

$$\rho = max_{\alpha,\beta} corr(U, V) \tag{2}$$

where U, V are the canonical variables, and α, β are called conical loadings, namely canonical coefficient matrix. Since the debut of CCA, a multitude of analyses, adaptations and applications in the field of image analysis and retrieval, have been proposed [26].

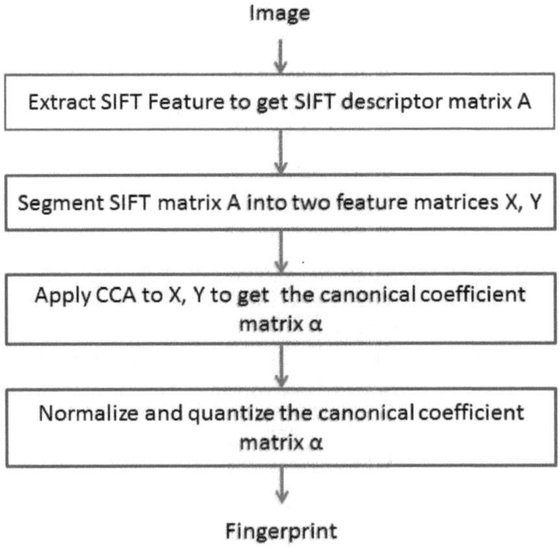

Fig. 2. The procedure of fingerprint generation

In this paper, we propose to use CCA to generate compact image fingerprint. The procedure of our image fingerprint generation is shown in Fig.2.

First, we extract scale-invariant feature transform (SIFT) from images to generate fingerprints since SIFT feature descriptor is invariant to uniform scaling, orientation, and partially invariant to affine distortion and illumination changes [31]. To achieve high efficiency, we use the SiftGPU [32], a GPU implementation of SIFT for fast extraction of SIFT features to obtain a SIFT descriptor matrix A as shown in Eq.(3), where N is the total number of SIFT points for a given image.

$$A = \begin{pmatrix} d_{1,1} & d_{1,2} & \cdots & d_{1,128} \\ d_{2,1} & d_{2,2} & \cdots & d_{2,128} \\ \cdots & \cdots & \cdots & \cdots \\ d_{N,1} & d_{N,2} & \cdots & d_{N,128} \end{pmatrix} \tag{3}$$

After SIFT extraction, we divide the 128-dimension feature into two segments form two closely related feature matrices, i.e., split the SIFT descriptor matrix A into two feature matrices X(consisting of the first n column vectors of A) and Y (consisting of the rest 128-n column vectors of A) as shown in Eq.(4) and Eq.(5) respectively. We set $n = 8$ since the SIFT descriptors are histograms accumulated in groups of 8 directions for each sub-block around the SIFT points.

$$X = \begin{pmatrix} d_{1,1} & d_{1,2} & \cdots & d_{1,n} \\ d_{2,1} & d_{2,2} & \cdots & d_{2,n} \\ \cdots & \cdots & \cdots & \cdots \\ d_{N,1} & d_{N,2} & \cdots & d_{N,n} \end{pmatrix} \tag{4}$$

$$Y = \begin{pmatrix} d_{1,n+1} & d_{1,n+2} & \cdots & d_{1,128} \\ d_{2,n+1} & d_{2,n+2} & \cdots & d_{2,128} \\ \cdots & \cdots & \cdots & \cdots \\ d_{N,n+1} & d_{N,n+2} & \cdots & d_{N,128} \end{pmatrix} \tag{5}$$

Then, we apply CCA to the two feature matrices X, Y with the Eq.(1) and (2). We propose to use the resultant canonical coefficient matrix α to generate fingerprints of images. Since the size of matrix X is $N \times n$ $(n \leq 64)$ and the size of matrix Y is $N \times (128 - n)$, the size of the resultant canonical coefficient matrix α is $n \times n$, which can be denoted as:

$$\alpha = \begin{pmatrix} \alpha_{1,1} & \alpha_{1,2} & \cdots & \alpha_{1,n} \\ \alpha_{2,1} & \alpha_{2,2} & \cdots & \alpha_{2,n} \\ \cdots & \cdots & \cdots & \cdots \\ \alpha_{n,1} & \alpha_{n,2} & \cdots & \alpha_{n,n} \end{pmatrix} \tag{6}$$

Finally, we ℓ_2-normalize the column vectors of α with:

$$\alpha_{i,j} = \frac{\alpha_{i,j}}{\sqrt{\Sigma_i^n \alpha_{i,j}^2}}, \quad j = 1, \ldots, n \tag{7}$$

We propose to use the normalized canonical coefficient matrix α as fingerprint. For more compact fingerprint, compression techniques can be used to reduce the length of fingerprint if necessary. In our experiments, we use the same quantization method as in [3] to quantize each element α of V to one byte, i.e., an integer $\alpha_{i,j} \in [0, 255]$ according to Eq.(8).

$$\alpha_{i,j} = \lfloor 127(1 + \alpha_{i,j}) + 0.5 \rfloor \tag{8}$$

Since we set $n = 8$, the proposed fingerprint is very compact, only 64 bytes after quantization.

3 Fingerprint Matching

According to the fingerprint matching method proposed in [3]: if the similarity S between two fingerprints is above a certain threshold T (which can be determined by experiments), the two fingerprints or images can be considered similar. Therefore, the main procedure of fingerprint matching is to calculate the similarity S between two fingerprints.

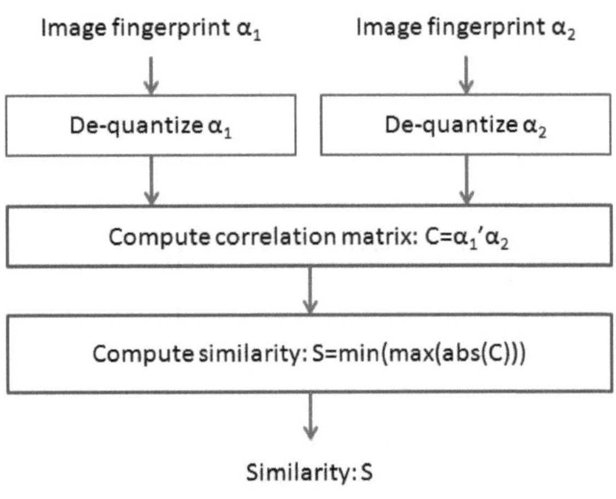

Fig. 3. The procedure of fingerprint matching

The procedure of fingerprint matching is shown in Fig.3. As shown, after getting the two image fingerprints α_1 and α_2 (often generated from images or received from others), we first de-quantize the two fingerprints respectively by using the following method proposed by [3]:

$$\alpha_{i,j} = \frac{\alpha_{i,j}}{127} - 1 \tag{9}$$

where $\alpha_{i,j}$ denotes the element of two image fingerprints α_1 and α_2.

Similar to the main idea of fingerprint matching proposed by [3], if two images are considered similar, the corresponding column vectors of the two canonical coefficient matrices (or fingerprints) should be highly correlative. Therefore, the similarity S can be calculated by computing correlation between each pair of the column vectors.

Let α_1, α_2 denote the eigenvector matrix of original image I_1 and the image I_2 to be tested, calculation of S can be expressed in matrix operations as follows.

First, we multiply the two de-quantized fingerprints to compute the correlation matrix C as (10).

$$C = \alpha_1' \alpha_2 \tag{10}$$

Second, we compute the maximum values of each column of C, then the minimum value of all the maximum values as the similarity S. This can be expressed in Matlab statement:

$$S = min(max(abs(C))) \tag{11}$$

where the function $abs(.)$ returns the absolute values of each element of matrix.

Different from the fingerprint matching method proposed in [3], we do not use only the diagonal elements of the correlation matrix C. This is because after our observation in our experiments, some attacks may change the order of the corresponding pair column vectors of the two canonical coefficient matrices, that is, the diagonal elements do not always reflect the correlation between each pair of the column vectors. Consequently, we use the function $max(.)$ to obtain the maximum values of the column vectors as the correlation coefficients of the pair vectors, and use the minimum value of the correlation coefficients as the similarity S between fingerprints.

4 Experimental Results

In order to evaluate our proposed method, we test it on the well-known "lena" image(512×512) and the $1,000$ images randomly selected from the MIRFlickr-1M dataset [33]. We conduct the two kinds of test: (1) Discriminability test to verify the Discriminability of two fingerprints generated from two visually different images; (2) Robustness test to verify the robustness of fingerprints against compression, cropping, scaling, and rotation.

4.1 Discriminability Test

We randomly select $100,000$ pairs of fingerprints of all the $1,000$ images selected from the MIRFlickr-1M dataset [33], extract fingerprints from them, and calculated the similarities S between them. The histogram of measured S was shown in Fig.4. All the S were in the range between 3.1996×10^{-16} and 0.8292. Only 7 seven out of $100,000$ similarities S are above 0.8. The mean and standard deviation (Std) were 0.5051, 0.0776. As shown in the figure, the histogram closely approaches the ideal random i.i.d. case N(0.5051, 0.0776), which indicates that our fingerprinting method is distinctive, and approximately pairwise-independent.

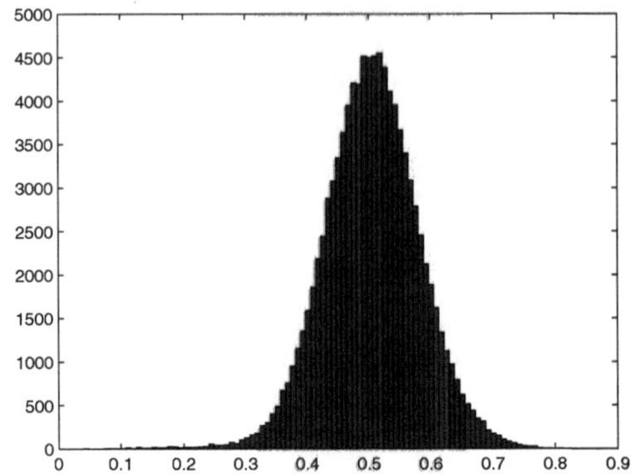

Fig. 4. The result of discriminability test on the 1,000 images randomly selected from the MIRFlickr-1M dataset

Table 1. Mean, standard deviation (Std) of the similarity S for different kinds of JPEG compressions for 1,000 test images

Test	Mean	Std
JPEG(Q=15%)	0.7656	0.1052
JPEG(Q=20%)	0.8421	0.1045
JPEG(Q=25%)	0.9018	0.0965
JPEG(Q=30%)	0.8632	0.0964
JPEG(Q=35%)	0.8645	0.0935
JPEG(Q=40%)	0.8801	0.0998
JPEG(Q=50%)	0.8835	0.0835
JPEG(Q=60%)	0.8659	0.0856
JPEG(Q=70%)	0.8576	0.0892
JPEG(Q=80%)	0.9105	0.0824
JPEG(Q=90%)	0.8743	0.0741
JPEG(Q=100%)	0.8525	0.1093

4.2 Robustness Test

Robustness against Compression: We use the StirMark [24] to compress the image Lena to 12 different JPEG images with different quality levels $Q = [15, 20, 25, 30, 35, 40, 50, 60, 70, 80, 90, 100]$ ranging from 15 to 100. When Q=15, the value of similarity S between the initial image and its JPEG image was 0.7825, and when Q=20, the value was 0.8721. The results are shown in Fig.5.

As for the 1,000 test images, we use the StirMark to generate 12 different JPEG images with same quality levels for each image. The means and standard deviations of the similarity S between the initial images and their corresponding JPEG images are shown in Table 1. We can see from the table that all the means are far above the that (0.5051) of the discriminability test. The above results show that our method is fairly robust against compression.

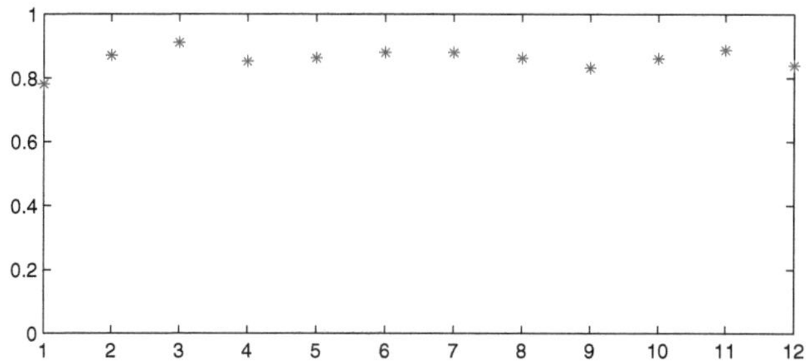

Fig. 5. The result of robustness against JPEG compression test on the image Lena

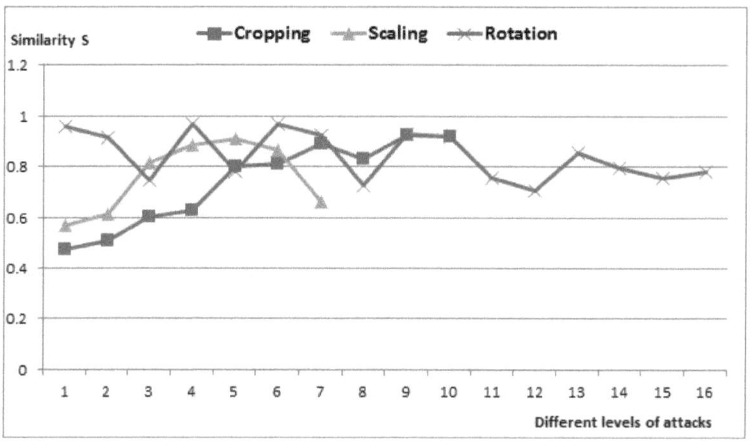

Fig. 6. The result of robustness against cropping, scaling, and rotation tests on the test images

Robustness against Cropping: We use the StirMark [24] to crop the 1,000 test images to 10 different images with different cropping ratios $C_r = [10, 20, 30, 40, 50, 60, 70, 80, 90, 95]$ for each image. When $C_r \leq 40$, the mean values of similarity S between the initial images and their cropping images are all below 0.7, and when $C_r \geq 50$, the value are all above 0.80. The results are shown in Fig.6.

Robustness against Scaling: We use the StirMark [24] to rescale the 1,000 test images to 7 different images with different rescale ratios $S_r = [50, 70, 80, 90, 110, 150, 200]$ for each image. When $S_r \geq 80$ and $S_r \leq 150$, the mean values of similarity S between the initial images and their scaling images are all above 0.8, and when $S_r \geq 200$ or $S_r \leq 70$, the mean values are all below 0.80. The results are shown in Fig.6.

Robustness against Rotation: We use the StirMark [24] to rotate the 1,000 test images to 16 different images with different angle degrees $R_r = [-2, -1, -0.75, -0.5, -0.25, 0.25, 0.5, 0.75, 1, 2, 5, 10, 15, 30, 45, 90]$ for each image. When $S_r \geq 90$ and $S_r \leq 150$, the mean values of similarity S between the initial image and its scaling image are all above 0.7, and when $S_r \geq 200$ or $S_r \leq 75$, the mean values are all below 0.70. The results are shown in Fig.6.

From the above robustness tests, compared with the fingerprinting method proposed in [3] which is robust to JPEG compression while not robust to cropping, scaling, and rotation, we can see that our proposed fingerprinting method can not only generate more compact fingerprints (64 bytes) than the method in [3], but also is robust to cropping (cropping ratio $C_r \geq 50$), scaling (rescale ratio $80 \leq S_r \leq 150$, and rotation due to the CCA of robust SIFT features.

5 Conclusion

In this paper, we propose a novel compact image fingerprint extraction method based on canonical correlation analysis (CCA) of local features. Experimental results show that the proposed method is discriminative and against both normal image processing operations such as compression common geometric transformations including cropping, scaling, and rotation. In the future, we will focus on how to localize of tampered regions based on our proposed fingerprints.

References

1. Seoa, J.S., Haitsmab, J., Kalkerb, T., Yoo, C.D.: A robust image fingerprinting system using the Radon transform. Signal Processing: Image Communication 19(4), 325–339 (2004)
2. Zhu, B.B., Swanson, M.D., Tewfik, A.H.: When seeing isn't believing (multimedia authentication technologies). IEEE Signal Processing Magazine 21(2), 40–49 (2004)
3. Tang, S., Li, J.-t., Zhang, Y.-D.: Compact and Robust Image Hashing. In: Gervasi, O., Gavrilova, M.L., Kumar, V., Laganá, A., Lee, H.P., Mun, Y., Taniar, D., Tan, C.J.K. (eds.) ICCSA 2005. LNCS, vol. 3481, pp. 547–556. Springer, Heidelberg (2005)
4. Tang, S.: Research on multimedia digital signature techniques, PhD Thesis, Institute of Computing Technology, Chinese Academy of Sciences (January 2006)
5. Li, H., Wang, X., Tang, J., Zhao, C.: Combining Global and Local Matching of Multiple Features for Precise Retrieval of Item Images. ACM/Springer Multimedia System Journal 19(1), 37–49 (2013)
6. Tang, S., Zheng, Y.-T., Wang, Y., Chua, T.-S.: Sparse Ensemble Learning for Concept Detection. IEEE Transactions on Multimedia 14(1), 43–54 (2012)
7. Tang, S., Li, J.-T., Li, M., Xie, C., Liu, Y.-Z., Tao, K., Xu, S.-X.: TRECVID 2008 High-Level Feature Extraction By MCG-ICT-CAS. In: Proc. TRECVID 2008 Workshop, Gaithesburg, USA (November 2008)
8. Schneider, M., Chang, S.-F.: A robust content based digital signature for image authentication. In: Proceedings of IEEE ICIP 1996, Lausanne, Switzerland, vol. 3, pp. 227–230 (October 1996)
9. Queluz, M.P.: Authentication of digital images and video: generic models and a new contribution. Signal Processing: Image Communication 16(5), 461–475 (2001)
10. Lin, C.-Y., Chang, S.-F.: A robust image authentication method distinguishing JPEG compression from malicious manipulation. IEEE Trans. Circuits Syst. Video Technol. 11(2), 153–168 (2001)
11. Lin, C.-Y., Chang, S.-F.: Robust digital signature for multimedia authentication. IEEE Circuits and Systems Magazine 3(4), 23–26 (2003)

12. Tang, S., Li, J.-t., Zhang, Y.D.: Compact and Robust Fingerprints Using DCT Coefficients of Key Blocks. In: Marques, J.S., Pérez de la Blanca, N., Pina, P. (eds.) IbPRIA 2005. LNCS, vol. 3523, pp. 521–528. Springer, Heidelberg (2005)
13. Tang, S., Li, J.-T., Zhang, Y.-D.: SSF Fingerprint for Image Authentication: An Incidental Distortion Resistant Scheme. In: ACM Multimedia 2005, Singapore, November 6-11 (2005)
14. Swaminathan, A., Mao, Y., Wu, M.: Robust and secure image hashing. IEEE Trans. Inf. Forensics Security 1(2), 215–230 (2006)
15. Monga, V., Banerjee, A., Evans, B.L.: A clustering based approachto perceptual image hashing. IEEE Trans. Inf. Forensics Security 1(1), 68–79 (2006)
16. Monga, V., Mihcak, M.K.: Robust and secure image hashingvia non-negative matrix factorizations. IEEE Trans. Inf. Forensics Security 2(3), 376–390 (2007)
17. Zhang, Y.-D., Tang, S., Li, J.-T.: Secure and Incidental Distortion Tolerant Digital Signature for Image Authentication. Journal of Computer Science and Technology 22(4), 618–625 (2007)
18. Lu, W., Wu, M.: Multimedia forensic hash based on visual words". In: 17th IEEE International Conference on Image Processing (ICIP), Hong Kong, China, September 26-29 (2010)
19. Ahmed, F., Siyal, M.Y., Abbas, V.U.: A secure and robust hash-based scheme for image authentication. Comput. J. 55(6), 686–701 (2012)
20. Lei, Y., Wang, Y., Huang, J.: Robust image hash in Radon transform domain for authentication. Signal Processing: Image Communication 26(6), 280–288 (2011)
21. Wang, X., Zheng, N., Xue, J., Liu, Z.: A Novel Image Signature Method for Content Authentication. Comput. J. 55(6), 686–701 (2012)
22. Lv, X., Wang, Z.J.: Perceptual image hashing basedon shape contexts and local feature points. IEEE Trans. Inf. Forensics Security 7(3), 1081–1093 (2012)
23. Zhao, Y., Wang, S., Zhang, X., Yao, H.: Robust Hashing for Image Authentication Using Zernike Moments and Local Features. IEEE Trans. Inf.Forensics Security 8(1), 55–63 (2013)
24. Fabien, A.P.: Petitcolas, "Watermarking schemes evaluation". IEEE Signal Processing 17(5), 58–64 (2000)
25. Hotelling, H.: Relations between two sets of variates. Biometrika 28, 312–377 (1936)
26. David, R.: Sparse Canonical Correlation Analysis. In: Technical Report, University College London (August 2007)
27. Bach, F., Jordan, M.: Kernel independent component Analysis. Journal of Machine Leaning Research 3, 1–48 (2002)
28. Hardoon, D.R., Saunders, C., Szedmak, S., Shawe-Taylor, J.: A correlation approach for automatic image annotation. In: Li, X., Zaïane, O.R., Li, Z.-h. (eds.) ADMA 2006. LNCS (LNAI), vol. 4093, pp. 681–692. Springer, Heidelberg (2006)
29. Hardoon, D.R., Shawe-Taylor, J.: KCCA for different levelprecision in content-based image retrieval. In: In Proceedings of Third International Workshop on Content-Based Multimedia Indexing. IRISA, Rennes (2003)
30. Hardoon, D.R., Szedmak, S., Shawe-Taylor, J.: Canonical correlation analysis: an overview with application to learning methods. Neural Computation 16, 2639–2664 (2004)
31. Lowe, D.G.: Distinctive Image Features from Scale-Invariant Keypoints. International Journal of Computer Vision 60(2), 91–110 (2004)
32. Wu, C.: SiftGPU: A GPU Implementation of Scale Invariant Feature Transform (SIFT) (2007), `http://cs.unc.edu/~ccwu/siftgpu`
33. Huiskes, M.J., Thomee, B., Lew, M.S.: New trends and ideas in visual concept detection: the MIR flickr retrieval evaluation initiative. In: MIR 2010, pp. 527–536 (2010), `http://press.liacs.nl/mirflickr`

A Large Scale Audio Fingerprinting System*

Tao Jiang[1], Kang Xiang[2], Jing Lu[2], Rihui Wu[2], Xin Li[3], and Feng Dai[1]

[1] Advanced Computing Research Laboratory, Key Laboratory of Intelligent Information Processing of Chinese Academy of Sciences, Institute of Computing Technology, Chinese Academy of Sciences, Beijing, China
[2] University of Chinese Academy of Sciences, Beijing, 100190, China
{jiangtao01,xiangkang,lvjing,wurihui,fdai}@ict.ac.cn
[3] Department of Computer Science and Technology, Tsinghua University, Beijing, 100084, China
lx_sy@126.com

Abstract. This paper presents a large scale audio fingerprinting system, which can be used to retrieve intended song from more than a million audio tracks. A time offset of several milliseconds could cause a number of audio fingerprints disappear. To decrease this effect, the query clip is converted to several copies with different start times. The audio fingerprints are mapped to hashes. Based on the statistics of the hashes on 25,000 audio tracks, it is found the frequency of occurrence does not follow the uniform distribution. A compact index structure is used to retain the quick access and none memory waste. Our experiments demonstrate the performance of this large scale audio fingerprinting system, and all of the query audio clips are captured from practical environments by various mobile devices. An iOS application is proposed based on this system and its index dataset contains more than 1 million songs.

1 Introduction

Audio fingerprinting [1][2] is a compact content-based representation of audio recording which could be used to efficiently match against other audio files. It is becoming an essential feature for any large scale music information retrieval (MIR) and understanding system. A common use case is query-by-example music recognition. Video-sharing websites want to ensure that the new uploaded videos do not contain unauthorized music. Music indexing tools want to find the metadata of unknown music file. Digital music stores want to remove duplicates in their large scale audio database to save storage.

State-of-the-art audio fingerprinting can be divided into peaks of frequency bands and computer vision feature of spectrogram. In the latter method, the audio file is treated as a series of spectrogram images. Ke et al. [4] apply boosted classifier and local-descriptor based object recognition for music recognition, and these two

* This work is supported by National Nature Science Foundation of China (61102101), National Key Technology Research and Development Program of China (2012BAH06B01, 2012BAH39B02).

B. Huet et al. (Eds.): PCM 2013, LNCS 8294, pp. 866–875, 2013.

methods are common techniques in computer vision. To avoid the learning approach in [4], Baluja et al. [5] directly use the wavelet coefficients of spectrogram, which could efficiently perform image queries on large dataset [10]. The overlapping spectrogram images are transformed into a sparse wavelet representation and the min-hash technique [6] is used to speed up the matching process. In contrast to computer vision method, Wang [1][7] choose a simple way to take audio fingerprints and just focuses on spectrogram peaks. His Shazam is one of the most popular music recognition applications on mobile phone. Ellis and Echo Nest company [8] [11] present Echoprint feature which is similar to Wang's feature. Echoprint relies on the onset in the lowest 8 bands of the MPEG-Audio 32 bands filterbank.

Computer vision features are just be evaluated on a small scale dataset, and get a comparable performance with peaks of frequency bands [3]. The peaks feature has been applied in the music recognition applications on a milliseconds dataset [7] [8]. Nowadays, there are dozen million songs [11] to be tracked, while most of audio fingerprint systems [8] [9][11] do not show the details how to handle large scale music data in an effective way.

This paper presents a large scale audio fingerprinting system based on peaks of frequency bands, which can process dozens of search requests in one second on one million music tracks. The pair of peaks, which are neighboring in both time and frequency domain, are combined to make an audio fingerprint. The audio fingerprint is mapped to a hash in fixed length. Both the query audio clip and the intended song are divided into audio frames, and they may be not synchronized in frames. Their time offset interval is range in 0 and frame shift. Audio fingerprints are sensitive to this interval, and a time offset of several milliseconds could cause a number of them disappear. It is found that the number of robust audio fingerprints and time offset approximate an arcsine distribution. Based on this discovery, two kinds of process are discussed to decrease this effect: 1) get the copies of the query with different start time, and merge the audio fingerprint of them, and 2) get the copies of the target with different start time, and select the robust audio fingerprints from them. If the frequency of occurrence follows the uniform distribution, a matrix is a proper container as the index, which could be quickly accessed. Based on the static of hashes on 25,000 audio tracks, it is found this distribution has a long tail. The matrix would miss amount of hashes with high frequency and wastes memory storage. A compact index structure is created by an extra liner list. The compact index retain the quick access and none memory waste. In our system, the huge index of one million audio songs splits into small ones. This operation does not affect the performance, and brings the benefits of quick update and a shorter processing time.

The rest of the paper is organized as follow. Section 2 shows the overview of the audio fingerprinting system based on peaks of frequency bands. Section 3 describes the sensitiveness of audio fingerprints to the interval time between the query and the target, and two methods to decrease this effect. Section 4 describes the distribution of the audio fingerprints on a large scale dataset, and how to build a compact and quick access index. And then the huge index splits into small ones. Sections 5 demonstrate the performance evaluation on 400,000 songs. Section 6 gives conclusions and future directions.

2 System Overview

Our system is based on Wang's work [1][7] and Ellis's implementation[12]. The audio input is converted to spectrogram, and then the peaks in the frequency banks are taken. The peaks are robust to the kind of spectral modifications and noise encountered in over-the-air recordings. Both the peaks in the intended music and the ones in the ambient noise are retained. It is likely that they are analyzed separately [7]. The peaks in the frequency banks are illustrated in Fig 1, and the audio fingerprint based on pair of peaks is showed in Fig 2.

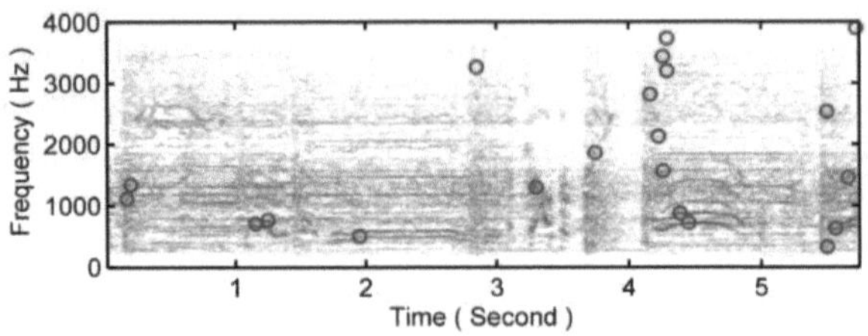

Fig. 1. Peaks in the frequency banks

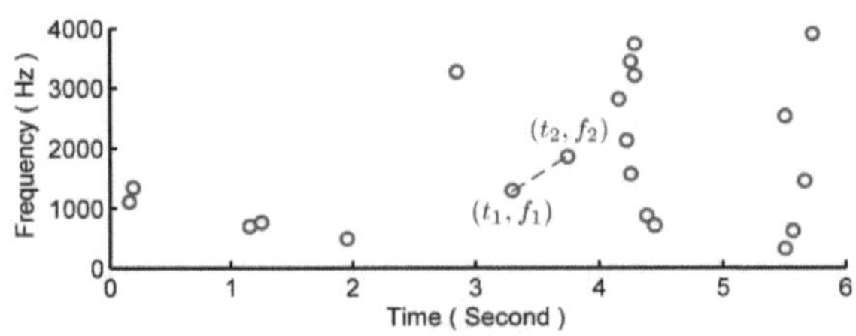

Fig. 2. The audio fingerprint

The hashes of the audio fingerprints are formed from the pairs of peaks, which are close in time-frequency domain. For the pair of peak(t_1, f_1) and k(t_2, f_2), their hash is the triple of f_1, $(f_2 - f_1)$ and $(t_2 - t_1)$. Each of them has been quantized. The combination of this triple is saved as the fingerprint. The hash corresponding value is t_1 and the ID of the intended song.

The sample rate of the audio signal in the query clip and intended music is set to 8 KHz. We use frames that are 256 milliseconds, each of which is taken every 32 milliseconds. The frequency data in spectrogram is divided into 1024 frequency bands. We consider neighboring peaks in adjacent frequency range of 128 units, and timing range of 128 units. The values of the triple f_1 , $(f_2 - f_1)$ and $(t_2 - t_1)$ are represented as 10, 8 and 7 bits respectively to obtain a 25-bit hash. There are most 5 peaks for each frame and 5 pairs for each peak.

Temporal alignment is applied to get rid of false positives, which is proposed by Wang [1]. It is not a complex scheme to obtain high precision for audio retrieval application. The candidate fingerprints are firstly sorted by song ID, and then sorted by time offset to the beginning of the candidate song. By a minimums number of fingerprints matches to convince temporal alignment, most of false positives are rejected.

3 Sensitiveness of Fingerprints to Start Time

The intended song is demonstrated as a series of audio fingerprints, each of which is mapped to a hash. This series is used as the template. All templates are saved in the index. The more robust these fingerprints are, the better performance the MIR system obtains. Our experiments show that audio fingerprints are sensitive to the start time and a time offset of several milliseconds could cause a number of fingerprints disappear. Both the query audio clip and the intended song are divided into audio frames, and their time offset interval is range in 0 and frame shift fs.

Fig 3 shows the relationship between the overlap rate and time offset interval. A 4-minute Chinese folk song is used as the template. Four copies of it are created, and their time offset interval are respectively $0.25fs$, $0.5fs$, $0.75fs$ and fs. The overlap rates demonstrate the audio fingerprints overlap between the template and the copies.

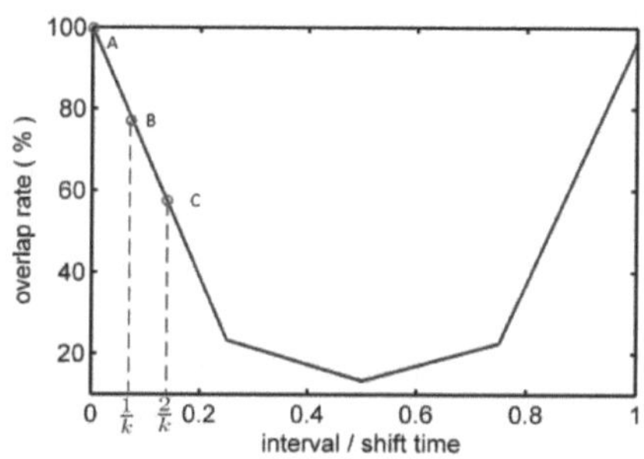

Fig. 3. The relationship between the overlap rate and time offset interval

When the interval = 0 or fs, the overlap rate is 100%; when interval <= 0.5 fs the overlap rate decreases when interval >= 0.5 fs, the overlap rate increases; when interval = 0.5 fs, the overlap rate gets the minimum. The relationship between overlap of audio fingerprints and time offset interval can be seen as a periodic function with period fs, which approximate an arcsine distribution. The start time of the query clip is random. When the frame shift is seen as degree scale, the time offset interval between query clip and the intended song clip is range in 0 and fs. When the interval is close to 0 or fs, these two clips share more fingerprints and the query is easy to be found. When the interval is closer to 0.5 fs, they share less fingerprints and the query is hard to be found.

In order to decrease the effect caused by the interval, two schemes could be carried: 1) get the copies of the query with different start time offset, and merge their fingerprints together [13]; 2) get the copies of the intended song with different start time offset, and select the robust fingerprints which is insensitive to the offset.

The former is an online method, which cost more time to take audio fingerprints. More fingerprints also need an extra search time. This scheme is easy to apply and effective. Kcopies of the query are created, each of which has a time interval fs/K. The distribution in Fig 3 is also divided into K equal segments, and each of copies is associated with one segments. There is at least one copy whose overlap rate is higher than the one in the point B in Fig 3. The fingerprints of K copies are merged together, which could increase the overlap rate.

To obtain the robust fingerprints is another way to decrease the effect caused by the interval, and it could be handled in offline way. K copies of the intended song are created, each of which has a time interval fs/K. The overlap between in point A and C must be lower than that one between point A and B, and this scheme would get lower overlap than the former one. As only a part of fingerprints in intended songs are necessary in the retrieval, the low overlap rate does not directly lead bad performance. One duplicate number is proposed as the threshold to select the robust fingerprints. This parameter should be set by experiments and is associated with the value of K. In order to keep the flexibility and simplification of the system, we select the former method and merge the fingerprints in Kcopies of the query.

4 A Compact Index Structure for a Large Scale Dataset

Query the music by the example is a low level method to manage music data. It is an effective way to find the duplicates in the large scale data, and a natural interface for users to discover their interested and unknown songs. Nowadays, there is a Google-size music on internet, which need to be accessed. As one of the most popular music recognition application, Shazam announced their service had been used to tag five billion songs, TV shows and advertisements. Shazam claimed have more than 250 million users across 200 countries. Echonest builds a bulletin on its Web site [11] to show the numbers of songs and artists, which are known by it and increase minute by minute. Up to now, Echonest has found nearly 35 million songs. Without Shazam and Echonest, there are a number of companies, which also own a large scale music data, such as Apple, EMI, Spotify, Soundcloud, and so on. These companies seldom show the details how to build a large scale MIR system. In this section, we discuss 3 questions about the large scale MIR system: 1) distribution of audio fingerprints on a large scale dataset; 2) build a compact index structure for the large scale fingerprints to achieve the search quickly; 3) split the huge index to small ones, which could keep the quick update of the fresh music, and obtain a shorter processing time.

The audio fingerprints are taken from 25,000 Chinese folk songs, and this dataset is named 25K01-CFS. The average length of these songs is 4 minutes. All the fingerprints are mapped to hashes. The frequency of occurrence of these hashes is showed in Fig 4. The X-axis show the value of the hash, and the Y-axis is the frequency of occurrence. As the length of the hash is 25-bit, its value is in range 0 to $2^{25} - 1$. The max value of occurrence is 17328, which means its corresponding hash 0xC8002 appears 17328 times in these 25,000 songs. There are about 8.77% of the hashes, the occurrence of which are zero, and their corresponding fingerprints do not appear.

The mean and variance of occurrence in 25K01-CFS are 21.68 and 949.80. In practical applications, it is often assumed and wanted that the hash independently occur with uniform probability. From the variance and Fig 4, it could be found that the occurrence of the hash does not follow uniform distribution. If the occurrence is sorted in descending order, a long tail is found obviously in the distribution. When occurrence is sorted in descending order, its max value is much larger than the min value, and its distribution cannot be seen clearly. Fig 5 gives a log version of this distribution.

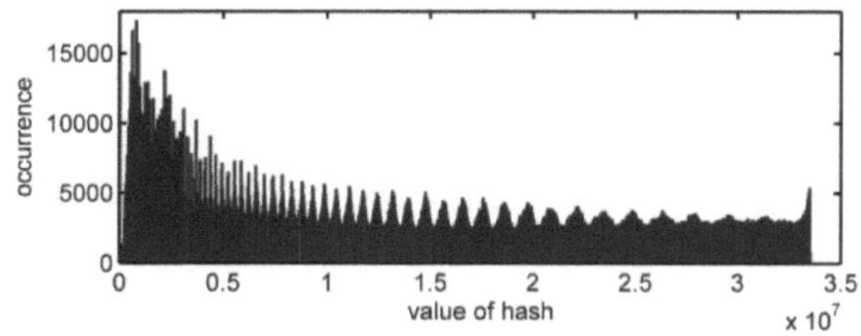

Fig. 4. The occurrence of the hash

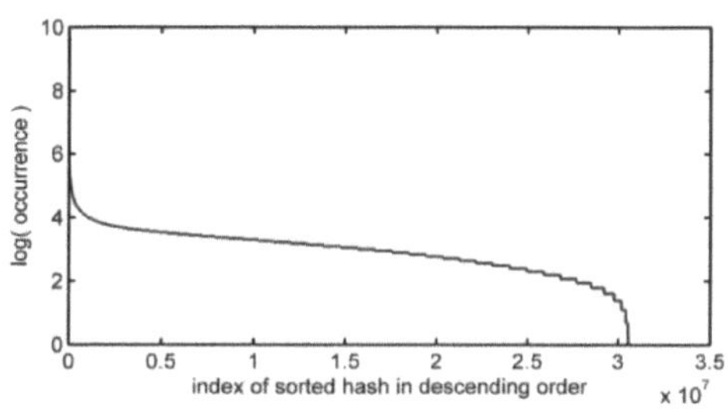

Fig. 5. The log occurrence of the sorted hash in descending order

When the hashes of 25K01-CFS are saved in the matrix M, each of the row numbers is uniquely associated with one hash, and the row number could be used as the index directly. Each of hash in 25K01-CFS can be accessed quickly, and this benefit is much valuable for the retrieval of the large scale dataset. As the floor value of the mean is 21, M is set to keep 21 locations for each hash, while 2.5G bytes memory is occupied by M. Almost on fourth of M is not used to save hash value. Almost one third of fingerprints in 25K01-CFS are discarded, which appears more than 21times. As occurrence of the hashes in 25K01-CFS does not follow uniform distribution, the matrix is not a proper container, and it would lead to the waste of memory storage and the loss of fingerprints.

A compact index structure is created by an extra liner list, which retains the quick access and none waste of memory. This compact index structure is showed in Fig.6. The structure is built in the following three steps:

1. Scan 25K01-CFS and get the occurrence of each hash.
2. Apply memory storage for the list L with fixed-size, and for array A with the sum of all occurrences.

3. Save the occurrence and the storage offset in each row of L, while the row numbers are uniquely associated with one hash code. Based on the storage offset, all the fingerprints of each hash are saved in the corresponding part of A.

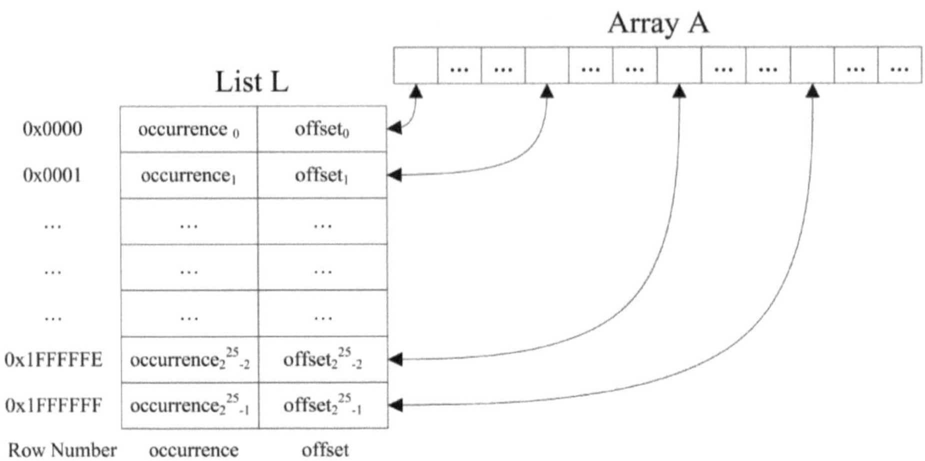

Fig. 6. The compact index structure used in the system

All of the fingerprints can be saved in a huge index, while the memory storage is the only limit. In our system, the fingerprints are not saved in this way and the huge index is spitted to small ones. There are two benefits to do this. Firstly, it is easy to achieve quick update. The fingerprints of the intended song are all saved in one small index, and each of small indexes share none of overlap. There are two benefits to do this spitting. There are new songs coming out at all time and the index should be updated frequently. It takes a long time to modify a huge index, while only a little part of it needs be modified. It takes a short time to modify the small one. And then, the small indexes are easy to process in a parallel way, which could take full advantage of multi-cores of CPU. With these two benefits, the group of small indexes shares the identical performance with the huge index.

5 Performance Evaluation

In this section we present the performance evaluation of our audio fingerprinting system on a large scale dataset. The whole index dataset contains over 1 million songs, and there are a lot of duplicates in it. These duplicates have no effect to users, and both the intended song and its duplicates are the correct result. As these duplicates are unknown, they affect the performance evaluation. We create a subset with 400,000 songs, which are checked to ensure there are no duplicates in them. The songs in subset are randomized to 16 groups, and each group is according to a small index which contains the hashes of audio fingerprints in 25,000 songs. Specs of the index dataset are:

- No. of audio tracks: 400,000
- Resolution: 16 bits
- Average Duration: 4 minutes
- Channel: Two
- Codec: MP3/AAC/unknown
- Container: MP4/MP3
- Sample rate: 44.1 KHz/32 KHz/22 KHz (mixed)
- Type: Chinese folk music/light music (mixed)

A query dataset of 783 Chinese folk songs is used to test the recall of the system. All these songs are over-the-air recordings. Over twenty volunteers joined in the recording task, and both the play sound devices and record mobile devices are their own. The recording is under the complex environments, such as busy streets, crowded lunches, and noisy office rooms. All of the recordings are checked to ensure the music can be heard, although some of them are a little hard to be heard clearly. Some of recordings have been rejected, as they are hard to be heard clearly. This query dataset could show the real practical environments in details. Specs of query dataset are:

- No. of audio tracks: 783
- Resolution: 16 bits
- Average Duration: 4 minutes
- Channel: One/Two (mixed)
- Codec: MP3/AAC/WMA/ADPCM
- Container: MP4/MP3/WMA/WAV
- Sample rete: 44.1 Khz/32 KHz/22 KHz (mixed)
- Type: Chinese folk music
- Play device: over 20 kinds
- Record device: over 20 kinds
- Record environments: busy streets, crowded lunches, noisy office rooms,...

Three segments are separately cut from the beginning, the middle and the end. The duration of the segment is 12 second, and it is further cut to small pieces, which are 4 second and 8 second from the beginning. The top-1 recall of the segments and their pieces on index dataset are showed in table 1.

Table 1. The recall of the system

Segments	Begining	Middle	End	Total
4s	40.1%	45.7%	41.4%	42.4%
8s	63.0%	63.8%	59.3%	62.0%
12s	74.0%	77.2%	75.6%	75.6%

6 Conclusions and Future Work

This paper presents a large-scale audio fingerprinting system based on peaks of frequency bands. Our experiments show that audio fingerprints are sensitive to the start time and a time offset of several milliseconds could cause a number of fingerprints disappear. It is found that the number of robust audio fingerprints and time offset approximate an arcsine distribution. The copies of the query clip with different start times are created to decrease this effect. It is often assumed and wanted that the hash independently occur with uniform probability. Based on the static of the hashes on 25,000 audio tracks, it is found the frequency of occurrence does not follow the uniform distribution and has a long tail when it is sorted in descend order. When these hashes are saved in a matrix, a number of them with high frequency would be discarded and a part of memory storage would be wasted. A compact index structure is created by an extra liner list. The compact index retain the quick access and none memory waste. In our system, the huge index is spitted to a group of small ones, which shares the same performance with the huge index, still keeps the quick update, and obtains a shorter processing time. A query dataset of 783 Chinese folk songs is used to test the recall of the system, and it could show the real practical environments in details. The top-1 recall of the query dataset on a dataset of 400,000 songs is about 75.6%.

We have proposed an iOS application based on this system, and the index dataset is expanded to more than one million songs. This app can be downloaded from `https://itunes.apple.com/cn/app/yin-le-er-duo/id613947449?mt=8`

The evaluation of this paper does not include the test about the false alarm of the system. To a practical retrieval system, which has a large-scale dataset and faces unpredictable environments and input data, it is hard to prepare out-of-set data to evaluate the real ability to refuse the false alarms. When we create this application, the most important purpose is to collect query data which comes from the real users. We have found a number of false alarms in the results. Future work involves the following tasks:

1. Decrease the false alarms.
2. Reduce the duration of the audio query clip. The system now uses query with the fixed-length duration, and its duration should be adaptive based on the input data.

References

1. Wang, A.: An industrial strength audio search algorithm. In: International Conference on Music Information Retrieval (ISMIR), Baltimore, Maryland, USA (2003)
2. Ellis, D.P.W., Whitman, B., Porter, A.: Echoprint: An Open Music Identification Service. In: International Symposium on Music Information Retrieval (ISMIR), Florida, USA (2011)
3. Chandrasekhar, V., Sharifi, M., Ross, D.A.: Survey and Evaluation of Audio Fingerprinting Schemes for Mobile Query-by-Example Applications. In: Proceedings of the 12th International Society for Music Information Retrieval Conference, ISMIR 2011, Miami, Florida, USA, pp. 801–806 (2011)
4. Ke, Y., Hoiem, D., Sukthankar, R.: Computer Vision for Music Identification. In: IEEE Computer Society Conference on Computer Vision and Pattern Recognition, pp. 597–604 (2005)
5. Baluja, S., Covell, M.: Content Fingerprinting Using Wavelets. In: 3rd European Conference on Visual Media Production, CVMP 2006, pp. 198–207 (2006)

6. Seo, J.S., Haitsma, J., Kalker, T.: Linear speed-change resilient audio fingerprinting model. In: Proc. IEEE Workshop on Model based Processing and Coding of Audio, pp. 45–48 (2002)
7. Wang, A.: The Shazam music recognition service. Communications of The ACM 8, 44–48 (2006)
8. http://www.soundhound.com/
9. http://www.shazam.com/
10. Jacobs, C.E., Finkelstein, A., Salesin, D.H.: Fast multiresolution image querying. In: Proceedings of the 22nd Annual Conference on Computer Graphics and Interactive Techniques, pp. 277–286 (1995)
11. http://echonest.com/
12. http://labrosa.ee.columbia.edu/matlab/fingerprint/

Author Index

MIX
Papier aus verantwortungsvollen Quellen
Paper from responsible sources
FSC® C105338

If you have any concerns about our products,
you can contact us on
ProductSafety@springernature.com

In case Publisher is established outside the EU,
the EU authorized representative is:
Springer Nature Customer Service Center GmbH
Europaplatz 3, 69115 Heidelberg, Germany

Printed by Libri Plureos GmbH
in Hamburg, Germany